COMPUTATIONAL SOCIAL SCIENCE

PROCEEDINGS OF THE INTERNATIONAL CONFERENCE ON NEW COMPUTATIONAL SOCIAL SCIENCE (ICNCSS 2020), 25–27 SEPTEMBER 2020, GUANGZHOU, CHINA

Computational Social Science

Editors

Wei Luo
School of Finance and Economics, Nanchang Institute of Technology, China

Maria Ciurea
University of Petrosani, Petrosani, Romania

Santosh Kumar
Department of Education, Kazi Nazrul University, Paschim Bardhaman, West Bengal, India

CRC Press
Taylor & Francis Group
Boca Raton   London   New York   Leiden

CRC Press is an imprint of the
Taylor & Francis Group, an **informa** business

A BALKEMA BOOK

CRC Press/Balkema is an imprint of the Taylor & Francis Group, an informa business

© 2021 Taylor & Francis Group, London, UK

Typeset by MPS Limited, Chennai, India

Library of Congress Cataloging-in-Publication Data

Applied for

Published by: CRC Press/Balkema
Schipholweg 107C, 2316 XC Leiden, The Netherlands
e-mail: Pub.NL@taylorandfrancis.com
www.routledge.com – www.taylorandfrancis.com

ISBN: 978-0-367-70193-2 (Hbk)
ISBN: 978-1-003-14497-7 (eBook)
ISBN: 978-0-367-70195-6 (Pbk)
DOI: 10.1201/9781003144977
https://doi.org/10.1201/9781003144977

Computational Social Science – Luo, Ciurea & Kumar (eds)
© 2021 Taylor & Francis Group, London, ISBN 978-0-367-70193-2

Table of contents

Preface   XVII
Committees   XIX

Session 1. Big data acquisition and analysis

Research on the construction of an intelligent court in the era of big data   3
C. Xu & Z.B. Pei

Architectural design and application of camping big data access platform   11
X.K. He, Z. Wang, H.Z. Zhou & Z.G. Qi

Analysis of the main logistics factors of fresh agricultural products and their
impact on agricultural development   20
Y. Zhang, X.L. Yang, X.H. Chen, B.L. Xiang, H. Yi & C. Liu

An artificial intelligence-based method for judging the quality of spoken
Chinese pronunciation   28
M.J. Guo, Z. Xue & W.X. Gao

Demand prediction for e-commerce multi-warehouse based on machine learning   34
J.H. Du, X. Wang & X.Y. Liu

Design of a depression intervention model: A computational social science approach   45
J. Zhao, Y. Zhao & Y.J. Ruan

A novel high-dimensional abnormal value imputation method for power
grid middle platform data   54
H.Z. Cui, C. Wang, M.S. Xu, Y.Y. Zha & M.M. Zhang

Research on computer forensics based on multiple correlation analysis
technology of multi-source logs   59
X.L. Zhou, H. Tang & J.L. Sun

Research on the mode of network ideological and political education in colleges
and universities driven by big data   66
C. Li, Y. Lv & N. Jiang

Research on university teaching based on artificial intelligence   73
Y.Q. Zhu

Collocational patterns in China English: A corpus-based study   79
L.X. Xia, Y. Xia & C.Y. Liao

Predicting user behavior based on informative lifelogging   86
G.Q. Liu, L.B. Zhang, C.H. Li, Y. Fu & M.U. Rehman

Modern science and technology support evaluation system research for the
construction industry of Jiangsu province   92
Y. Wang

Online teaching in Chinese universities under "COVID-19" epidemic 96
Y.J. Shen

Research on digital library construction 102
B.B. Xu, S.W. Zhu, J.F. Yu, M.J. Zhang, V. Kuzminykh, S.S. Li & J.P. Guo

Study on the value structure analysis and practical case of network user autonomy 108
X. Zhou & L. Tang

Research on comprehensive evaluation of online stores based on user
sentiment analysis 132
H.F. Li & W.J. Chen

Research on fast location algorithm of fault section 136
X.K. He, M.C. Chen, H. Nie, N.H. Zhang & S.Z. Lian

The effect of population aging on the consumption of urban residents – taking
Shanghai as an example 144
Y. Lu & X.Y. Gu

A multiuser virtual service migration scheme based on deep reinforcement
learning in mobile edge computing 154
J.X. Wu, W. Chen, Z.B. Tang & Y.L. Zeng

A study on the design and usability of an intelligent signage system under the
background of human–computer interaction technology 163
H. Wang

Session 2. Integration of qualitative research and quantitative research

An empirical study on US direct investment in PRC and bilateral economic and
trade relationship 179
Z.Y. Zhang & Y.C. Liu

Power grid middle platform data anomaly detection based on two-stream
convolutional neural network 185
H.Z. Cui, C. Wang, M.S. Xu & M.H. Xu

Evaluation of high-tech industry innovation performance using a new
two-stage network DEA model 191
L. Chen, H.H. Hu, D.X. Ling & X.Y. Ye

Empirical analysis of the mechanism of the influence of internal control on
financial performance 197
X.M. Sun, X. Xian, C. Zhang, H. Yu & X. Liu

The attraction of horoscopes: a consensual qualitative research on astrological
personality description 202
Y. Wu & Z.Z. Chen

Construction of automated feedback system of business English correspondence
writing based on discourse analysis 210
D.Q. Liu

Research on the evaluation system of online education service quality:
Evidence from college students 215
J.H. She, Q. Zhang, T. Liu & K. Chen

Can network public opinion predict stock market return? 221
W.F. Feng, S.L. Jia & Y.H. Xiang

Natural language processing features and Critical Thinking assessment a pre-post essay test study in Italian HiEd setting
A. Poce & F. Amenduni
227

Analysis on lending service of paper books of double first-class construction university libraries in COVID-19
H. Tan, X.Y. Cao & Y.H. Zhang
233

The influence of crisis on hotel profitability: A case study of Novel Coronavirus (COVID-19)
X.L. Xian
243

The administrative emergency response research in the COVID-19 outbreaks
T.M. Yan & Z.B. Pei
250

The new trend of strategy in the Indo-Pacific
Y.J. Cao, J.H. Yu & X.L. Zhu
255

Effects of language satisfaction on continued use intention for m-government users in multi-ethnic inhibited multilingual regions: a case study in Xinjiang Uyghur Autonomous Region of China
L.P. Zhu
259

Forecasting of water demand by industry in China in 2020
X.L. Liu
267

An analysis of the factors of university choice for different groups of college students in China
J.B. Xiahou, L.C. Li & X. Cui
271

Research on the experimental teaching reform of "material analysis and test method" based on OBE engineering education mode
L. Xia & J.B. Yu
277

A comparative study of the international students' and the local Chinese students' intercultural sensitivity
B.C. Chen, J.Q. Zhang & X.B. Chen
283

A study of the differences in consumers' green behavior in the selection of tableware
N. Xia & W.H. Chen
293

An integrated FTA-GAHP method for assessment of campus safety: a case study in Puyang, China
H. Wei & X.B. Chen
302

Research on the impact of China's listed companies' financial accounting on innovation: Experience data from listed companies in Shenzhen and Shanghai
Q. Lu, C. Sheng & G. Fu
309

Empirical analysis relationship between NaOH output and economic social development in 1998–2019 based on VAR
J.M. Feng, M.X. Yan & Q.S. Feng
321

Research on the demand for an electric energy substitution policy
W. Tang, P. Wu & Q.K. Tan
329

Influence of entrepreneurial education on the entrepreneurial intentions of university students majoring in sports: A case study from Guangdong Province, China
T. Kang & D.D. Yang
334

Session 3. Cultural studies and humanities

The impact of Internet fiction reading on adolescents' self-concept clarity:
The role of narrative transportation
D.J. Zhang, Z.C. Lei & Z.K. Zhou　341

Characteristic practice of science and engineering in liberal arts colleges
P. Wang　348

The integration of cultural confidence with Bachelor of Translation and Interpreting
education in China
X.M. Qin　353

Study of the communication strategies of multinational internship in the era
of social networks
Y. Liu, W. Tang, L. Li, C. Millman & P.Y. Zhang　356

The present situation and development countermeasures of international
exchanges and cooperation for artificial intelligence postgraduate students in China
M. Yu & X.H. Liao　365

Empirical analysis on the manufacturing regional value chain between China and
"Belt & Road" countries
N.N. Zhu & M.R. Liu　369

Homeland and school geography factors in the social networks of African
college students in Jinan, China
L.S. Wang, H.F. Dong, Y. Fan & T. Wang　376

International experience and cultural intelligence development: A case study
H.J. Xue　382

Theory and case enlightenment of literary canonization in the Occident
B.C. Jiang & J.Y. Jiang　389

The impact of state bureaucracy in the process of democratization
Y.J. Cao　394

The construction of the training model of the trans-cultural cognitive competence
L.Y. Miao & Y. Zhan　401

Social network of female residents in Wulingyuan under the influence
of ecotourism: Scale, connection and structure
F. Zeng & Y.D. Zhong　408

A study of the moderating effect of online shopping experiences on impact
of relationship quality on customer loyalty
X.Y. Meng & X.Q. Wang　419

Research on the "Belt and Road" initiative and the international transmission of
Health Qigong—cultural consensus based on the community of common destiny
T. Kang & D.D. Yang　422

Promoting libraries using social media videos
Y. Zhang　427

"Equilibrium and harmony" and "perfect sincerity"—On the formal aesthetics
in *Zhongyong*
M.Y. Ding　431

Comments on the judgment of the first instance on the compensation case of
Yangzhou QunFa Company from the perspective of critical thinking   436
H.W. Feng & W.M. Ouyang

Marked theme analysis and its implications in academic paper writing   442
Y. Zhan, L.J. Li, L.Y. Miao & Y.L. Li

"Idyllic" shortform video: aesthetic imagination, narrative turn, and modern
anxiety in the new media era   446
J. Liang, S.H. Zhu, J.P. Li & Y.M. Xie

A study of the effectiveness of developing non-English major students' intercultural
communication competence through an English salon—taking an AIB English
Salon as an example   451
F. Wang, X.L. Xu & F. Graciano

Beauty in sadness—research into the concern of life in "Love in a Fallen City"   459
P. Tang

Research on the cultivation of social responsibility of young students from
the perspective of the community of human destiny   463
W.J. Zhou

The progressive way of a community of shared future for mankind   468
J.H. Yu, Y.J. Cao & B.X. Xie

Institutional advantages and spiritual strength in China's fight against COVID-19   473
J. Cao & S.S. Yan

Research into the communication of tea culture from the perspective of international
Chinese education—investigation into the tea culture elements based on
Chinese textbooks   480
Z.Y. Ke & W.H. Zhu

A study on the new humanistic value of "Shaanxi Spirit" in Zhang Zai's "Si Wei"
philosophy and its international promotion strategy   486
Y.G. Wang, C. Wang, W. Guo & J. Zhang

Session 4. Law and education

Research on cross-school sharing practice teaching construction of descriptive
geometry and mechanical drawing courses based on "Internet Plus"   495
X.H. Li, W. Zhao & X.C. Zhou

Research on the integration of entrepreneurship education and professional
education in industrial design specialty   500
X.L. Ma

Study of the practices of action learning for promotion of the innovation
capability of college students   504
Y.X. Liu, P.B. Gao, W.W. Wu & Z. Li

Practice exploration of a three-dimensional tutorial system for cultivating
applied talents in civil engineering under the new engineering
background—a case study from a university in Zhejiang province   508
X.F. Chen, Z.X. Zha & S.S. Wu

Problems and solutions of online learning under an epidemic situation   514
Y. Ma, Q.Z. Yan & J.H. Tian

Exploration of mixed teaching based on MOOC in road and bridge construction
organization technology and management course teaching
W. Li, X.C. Wang & Y.H. Li
520

Functions and practice of instrument & equipment sharing platform for talent
training in higher education
S. Yao, Z.Y. Li, W.B. Liu & Y. Cao
525

Research on the impact of student evaluation of teaching on teacher teaching
performance evaluation in colleges and universities: Illustrated by the example
of Chengdu University of Information Technology
S.Q. Cao & J. Chen
531

Innovation and implementation of multiple teaching strategies in financial
program learning
H.L. Hsu & H.C. Hsu
539

Reform and innovation of the college course teaching mode with an example of the
"Crystal Dislocation" micro-course design
J.J. Gong, J. Sun, P. Zhang, D.L. Wu, J.J. Du, J.C. Gao, J.J. Liu, D.Y. Li & C. Liu
546

Study on the innovation capability of college students based on action learning theory
Y.X. Liu, P.B. Gao, W.W. Wu & Z. Li
551

Development of an interactive learning aid system for cross school sharing of
descriptive geometry and mechanical drawing courses
X.H. Li
556

A study on the cultivation model of information-based teaching ability of
English majors state-funded normal students
C. Dai
561

Discussion of college students' learning burnout from the perspective of three
forces of psychology
X.L. Hu, J.Q. Sha, Y. Wang & Y. Cao
566

Reform and innovation of training mode for master of transportation engineering
under the transformation of new and old kinetic energy
X.D. Tang, X. Lu, Z. Qu & W.H. Wang
571

Research on the impact of the Internet on the informatization teaching reform of
marketing in universities
Y. Zhou, C.H. Jin, J.W. Huang & R.M. Wang
576

Teaching reform and experience of the project-driven teaching and diversified
assessment methods for the mechanical engineering materials course based on CDIO
C.S. Liu, Z.W. Li & J.J. Yang
584

Research into the design and evaluation of information technology in a
learning environment
H.Q. Hu, J. Zhang & C.H. Jin
590

A study of student evaluation of teaching in applied colleges
J.Y. Yuan, P.H. Huang, W.Y. Yan & H.S. Chen
596

The course group construction for the major of information management
and information system
G.M. Wang, J.J. Chai, Y.Q. Li & Y. Hu
603

Construction of the practical teaching system outside campus in local universities
and colleges based on emerging engineering education   609
F.H. Xu, X.H. Zhu & S.H. Wang

Study on countermeasures of the strategy of promoting rural revitalization in
local universities   615
X.Y. Shao

Studies on academic dispute legal features and their settlement by the law
and administration   620
Y.B. Zhou & P.Z. Cao

A study of college students' satisfaction with the online vocational education platform   625
G. Chen & Y.F. Wu

Reform and practice of innovation and entrepreneurship practice education system
for electronic information college students under the perspective of new engineering   630
H.M. Song, Y. Liu & X.H. Sun

A discussion of the definition of quality in higher education   636
Y.X. Ye & Y.J. Chen

Research on student policing management based on political construction of police   641
Z.J. Sheng & Q. Du

Study of the learning mode based on a smart classroom   647
Y.M. Wu

Development of a teaching standard for railway traffic operation management from
the perspective of secondary and higher vocational linkage   651
Q.P. Ye & W.Y. Wu

Research and practice on the construction methods of online and offline
one-stop service systems in colleges and universities   656
S.X. Wang, L.B. Zhu & Y. Xiong

Session 5. Management and economics

Exploration of teaching reform and innovation on the course of advanced
bio-reaction engineering   665
L.Q. Sun & H.J. Gao

Research on the operation mechanism of training quality of economics and
management graduate students based on system dynamics   669
L.C. Li, M.R. Fu, J.J. Xiao, Y.Q. Li & S. Ou

Research on the problems and countermeasures of innovation ability training
mode of economic and management majors   677
B.B. Yu

Exploration on the construction of the gold course of mechanics of materials in
applied undergraduate civil engineering major   682
H.S. Guo, Y.S. Luo, Y.X. Cao, J. Liu & Y. Tu

An analysis of the employment contradiction of tourism management graduates
and some solving countermeasures   688
J.X. Zhang

Exploration and construction of comprehensive training course based on
project oriented
Y. Li & S. Cong
693

Discussion on content design of safety training of public research platform
for biological and medical engineering
L.Q. Sun, M.G. Luo & Y. Cong
697

Predict the adaptability of medical college students returning to school after
COVID-19 using machine learning
Q.T. Xiao, X.R. Liu & J. Jiang
703

Research on tort liability of self-driving cars
Z.B. Pei, S.W. Cai & X.L. Guo
711

Learning modern technology and developing innovative capabilities—teaching
content innovation and teaching practice of modern design course
W. Chen & J. Fan
718

Research on the protection of citizens' rights and interests in the law of
administrative penalties for public security
C. Xu & Z.B. Pei
722

Research on the cognitive needs of mental health in primary and middle
school students and the construction of mental service system
J.M. Fan, X.L. Ni, Y.W. Geng & Y.P. Wang
730

On the approaches to the rural migrant workers' children enjoying equal
access to compulsory education in China
S.S. Yan & J. Cao
734

Research and practice on the reform of electrical basic courses based on the
cultivation of applied technology ability
L.L. Tang, J. Liu & H.Y. Jiang
740

Research on the long-term employment mechanism of college students
D.D. Li
746

On the mode of school enterprise cooperation in the reform of applied talents
training in teaching oriented universities
S.Y. Du & X.F. Peng
752

Exploration on practical teaching reform of logistics engineering based
on new engineering
Y. Fang, Y.S. Luo, W.X. Jiang & C. Gong
758

Institutional isomorphism and professional accounting education development
in Chinese higher education system
G.H. Zhang & X.H. Qu
764

Teaching practice and reform of "process equipment design" course design based
on cultivating team cooperation ability
Z.Y. Duan, H.D. Zhang, H.Y. Zhai, X. Cao, Z. Liu, X.L. Luo & D.Y. Luan
770

Research on the development strategy of Wuhan urban rail transit industry
H.P. Wang, F. Gao & Q.H. Liu
776

An empirical study on factors affecting economic benefit of large-scale pig breeding
W.T. Liu, L. Zhou, G. Fu & C.H. Liu
782

Analysis on the status quo of monitoring terminal subsidies for DSM projects 794
W. Tang, P. Wu & Y. Zhang

How does strategic orientation impact corporate performance: The path
mechanism concerning marketing capability 799
Q.H. Liu, X. Zhao & H.P. Wang

Analysis of Sino-US relations: From the perspective of building a new model
of major-country relations 805
X.T. Xiao

Research on vehicles route problem of rail transit equipment based on improved
ant colony optimization 811
H. Yu, L. Sun, X.Y. Tong & X.J. Zheng

Branding China: What roles can enterprises play in it? 818
Z.J. Yan & J. Wen

Research on China-Singapore trade relation under the strategy of "the
Belt and Road Initiatives" 824
Z.Y. Zhang & Q. Luo

A critical analysis of China's investments, redundant resources and economic
growth in Nigeria 829
S.A. Imanche, Z. Tian & O.T. Tasinda

Application of fuzzy comprehensive evaluation method in university food
safety management evaluation 836
S. Wang, J.Q. Zhang, X.B. Chen & H. Li

Research on the problems and countermeasures of scientific research base
management mode in colleges and universities 841
L. Li

Thoughts on building an enterprise human resource management
information system 846
S.S. Xu & Y. Zhang

Complementary resources and application transformation 850
P. Wang

Building standard domestic service brands in China 856
H.J. Chen, Z.C. Wang, X.W. Wu & W.D. Wang

How sea power serves China's maritime interests 861
H. Zeng

Research and practice on the talent training mode of modern apprenticeship system:
Integration of industry and education, progressive training of ability
and segmental training 867
J. Shan & N.N. Yang

Analysis of high-tech industry capability of the Belt and Road countries - Based
on the revealed comparative advantage index 875
H.Y. Yan, H.N. Qu, X.Y. Ding & L.Y. Zhan

Research on construction of working mechanism of enterprise
knowledge management 881
G.L. Gao, H.Y. Gao, W. Tan, X.H. Pan, X.H. Li, S.C. Jin & Z. Zhang

Research on the positive role of social responsibility in the development of
the e-sports industry   886
X.Y. Zhang

Study on resilience of urban planning from the perspective of urban epidemic
prevention   890
C. Peng & X. Li

Pogge's five proposals to reshape the global order and to eradicate
global poverty   894
J.F. Sun

Research on personalized tourism route based on crowdsourcing model   900
H.F. Li & W.J. Chen

Research on the evaluation system of employee satisfaction based on AHP   905
Y.J. Wu, Y.H. Xiong, X.X. Tang & Y.H. Fu

Research on out-going employment security of out-going migrant workers
under the COVID-19 outbreak   910
J. Chen & R.H. Liu

An integrated research on the benefits, problems, and countermeasures
concerning Wuhan rail transit operation   918
F. Gao, Y. Peng, H.P. Wang & Q.H. Liu

Study on the mechanism of inclusive leadership inspires employees' job
crafting—a moderated double-mediation model   924
K. Du, C.S. Wu & L.G. Zhang

Does job design affects employee knowledge sharing and innovation?—the
moderating role of organizational innovation atmosphere   942
Y.Y. Zhu, C.S. Wu & L.G. Zhang

Analysis on the coupling effects of strategic emerging industry structure and
employment structure in China   956
L. Liu, C.S. Wu, Y.Y. Zhu, B. Ye & N. Yang

Research on poverty reduction effect of health insurance schemes on
multidimensional poverty of agricultural migrants   972
X.J. Lu & Y.N. Wang

Brexit: A challenge or an opportunity for China?   979
H. Zeng

A study inspired by Ukiyo-e   986
Y.T. Xie & H.Y. Dai

Evaluation of a lawyer's professional ability based on fuzzy analytic
hierarchy process   993
Z.B. Wan, A.Q. Li, Z.H. Huang & Z.H. Zhang

Forecasting on China's final consumption in 2020   1001
X.L. Liu

Text quantitative analysis on administrative punishment for the construction
of a case analysis knowledge graph   1005
C.X. Li, J.M. Li & M.J. Du

Exploration and practice of curriculum for ideological and political education:
Taking signals and systems as an example   1013
C. Ji, C.H. Cao, G.Y. Zhang, R. Geng & J. Wang

Author index   1017

XV

Computational Social Science – Luo, Ciurea & Kumar (eds)
© 2021 Taylor & Francis Group, London, ISBN 978-0-367-70193-2

Preface

ICNCSS2020 (The 2020 International Conference on New Computational Social Science) was held on Sept. 25th-27th, 2020 (Virtual Conference) due to the COVID-19 situation and travel restrictions. ICNCSS2020 has been converted into a virtual conference, which was held via the Zoom event platform.

ICNCSS 2020 provided an excellent international platform for all the invited speakers, authors, and participants. The conference enjoys a widespread participation, and we sincerely wish that it would not only serve as an academic forum, but also a good opportunity to establish business cooperation. Any paper and topic around new computational social science was warmly welcomed.

ICNCSS 2020 proceeding tends to collect the most up-to-date, comprehensive, and worldwide state-of-art knowledge on new computational social science. All the accepted papers have been submitted to strict peer-review by 2-4 expert referees, and selected based on originality, significance and clarity for the purpose of the conference. The conference program is extremely rich, profound and featuring high-impact presentations of selected papers and additional late-breaking contributions. We sincerely hope that the conference would not only show the participants a broad overview of the latest research results on related fields, but also provide them with a significant platform for academic connection and exchange.

The Technical Program Committee members have been working very hard to meet the deadline of review. The final conference program consists of 155 papers divided into 5 sessions. The proceedings are published by CRC Press/Balkema (Taylor & Francis Group).

We would like to express our sincere gratitude to all the TPC members and organizers for their hard work, precious time and endeavor preparing for the conference. Our deepest thanks also go to the volunteers and staffs for their long-hours work and generosity they've given to the conference. Last but not least, we would like to thank each and every author, speaker and participant for their great contributions to the success of ICNCSS 2020.

ICNCSS 2020 Organizing Committee

Committees

Honor Chair
Prof. J.J. Luo, *Department of Social Sciences, Wuhan University, China*

Program Chair
Prof. S. Zhao, *Communication University, China*

Editor
Prof. Dr. W. Luo, *School of Finance and Economics, Nanchang Institute of Technology, China*

Co-editor
Prof. Dr. K. Thirugnanam, *Electrical and Computer Engineering Khalifa University of Science and Technology, Abu Dhabi, UAE*
Associate Prof. Dr. M. Ciurea, *University of Petroşani, Romania*
Associate Prof. Dr. S.K. Behera, *Department of Education, Kazi Nazrul University, India*
Assoc. Prof. Y. Xie, *Yibin School of Administration, China*

Technical Program Committee
Dr. Y.M. Zhang, *Dalian Neusoft University of Information, Dalian, China*
Prof. S. Zhao, *Northwestern Polytechnical University, Xi'an, China*
Dr. Z.W. Zhu, *Huaiyin Institute of Technology, Huaiyin, China*
Prof. A.L. Rojas, *Autonomous University of Coahuila, Coahuila, México*
Dr. M.H.B. Haji Razal, *Universiti Sultan Zainal Abidin, Terengganu, Malaysia*
Prof. X. Chen, *University of Electronic Science and Technology of China, Chengdu, China*
Y. Xie, *Party School of CPC Yibin Municipal Committee, China*
Dr. Bonifacio Lamazares, *Valladolid University, Spain*
Dr. Satish Bhagwatrao Aher, *ICAR-Indian Institute of Soil Science, India*
Dr. Mahsa Noori-daryan, *University of Tehran, Iran*
Dr. Salman Nazari-Shirkouhi, *University of Tehran, Iran*
Dr. Ricardo de Jesus Gomes, *Polytechnic of Leiria, Portugal*
Dr. Tse Guan Tan, *Universiti Malaysia Kelantan, Malaysia*
Dr. Biju Theruvil Sayed, *Dhofar University, Sultanate of Oman*
Prof. Ramayah Thurasamy, *Universiti Sains Malaysia, Malaysia*
Prof. H. Zeng, *Brooklyn College of The City University of New York, USA*
Prof. L.D. Milici, *Stefan cel Mare University, Romanian*
Dr. Y.S. Fern, *Multimedia University, Malaysia*
Dr. B. Llamazares, *Universidad de Valladolid, Spain*
Dr. A. Dincer, *Erzincan Binali Yıldırım University, Turkey*
Dr. O.A. Khan, *Computer Technology and Application (CTA), USA*
Prof. D.S. Ma, *Tianjin Polytechnic University, China*
Dr. T.C. Ling, *Graduate School of Business, Universiti Sains Malaysia, Malaysia*
Prof. Z.W. Zhu, *Huaiyin Institute of Technology, China*
Prof. P.P Liu, *Beijing Wuzi University, China*
Assoc. Prof. Dr. T. Hidajat, *STIE Bank BPD Jateng, Indonesia*
Dr. W.X. Su, *Huaqiao University, China*
Dr. M. Dziku, *University of Zielona Gora, Poland*

Dr. H. Yaghoubi, *Iran Maglev Technology, Iran*
Dr. R.R. Jorge, *Technological University of Ciudad Juarez, Mexico*
Dr. X. Lee, Hong Kong *Polytechnic University, Hong Kong*
Dr. M.M. Kim, *Chonbuk National University, Korean*
Dr. X. Ma, *University of Science and Technology of China, China*
Dr. M.V. Raghavendra, *Adama Science & Technology University, Ethiopia*
Dr. P. Velayutham, *Madras Christian College, India*
Dr. M.S. Chen, *Da-Yeh University, Taiwan*
Dr. D. Pathak, *University of Pardubice, Czech Republic*
Dr. I. Minin, *Siberian State Geodesy Academy, Russia*
Dr. M.V. Igor, *SMVD University, India*
Dr. Z. Mahmood, *University of Sydney Technology, Australia*
Dr. N. Yaacob, *University Polytechnic of Bucharest, Romania*
Dr. Q. Chen, *Xihua University, China*
Dr. W. Kavimandan, *Claremont Graduate University, USA*
Dr. K.S. Rajesh, *Defence University College, India*
Dr. M.V. Raghavendra, *Adama Science & Technology University, Ethiopia*
Dr. Z.Y. Jiang, *University of Wollongong, Australian*
Dr. V.K. Jain, *Indian Institute of Technology, India*
Dr. Q.B. Zeng, *Shenzhen University, China*
Dr. F. Keller, *Institute of Electrical Engineering, Slovakia*
Dr. A.V. Brazhnikov, *Siberian Federal University, Russia*
Dr. H. Chung, *National Central University, Taiwan*
Dr. J. Yeh, *Tallinn University of Technology, Estonia*
Dr. O.P. Rishi, *University of Kota, India*
Dr. S. Ehsan, *University of Malaysia, Malaysia*
Prof. Dan Dobrota, *Lucian Blaga University of Sibiu, Romania*
Prof. Syh-Jong Jang, *Asia University, Taiwan*
Prof. Ştefan Ţălu, *The Technical University of Cluj-Napoca, Romania*
Assist. Prof. Dr. Ali Dincer, *Erzincan Binali Yildirim University, Turkey*
Dr. Basil O Saleh, *University of baghdad, Irag*
Dr. L. Amorós-Poveda, *University of Murcia, Spain*
Dr. B. Rojas Hernández, *Univeirisity of Pinar del Río, Cuba*
Dr. Y. Fengxin, *Northwest University of Agriculture & Forest, China*
Prof. Dr. A. Gelbukh, *Natural Language and Text Processing Laboratory, Mexico*
Prof. M. Wang, *College of Chinese Language and Culture, Jinan University, China*
Dr. M. Ebrahimzadeh, *Junior and Senior High Schools, Chekaad Danesh, Iran*
Dr. A. Akanova, *S.Sefullin Kazakh Agrotechnical University, Kazakhstan*
Dr. E. Munirovich Akhmetshin, *Kazan Federal University, Russia*
Dr. J. Honor, *University Of Technology Sydney, Australia*
Dr. S. Bin Badlishah, *Universiti Utara Malaysia, Malaysia*
Dr. M. van Vreden, *North West University, South Africa*
Dr. P. Merello-Gimenez, *Universitat de València, Spain*
Dr. S. Shaw, *Cambridge Assessment, United Kingdom*
Prof. W.C. Hsu, *National Kaoshiung First University of Science and Technology, Taiwan*
Dr. A.R. Masalimova, *Kazan Federal University, Russia*
Dr. C. Nestai Nyoni, *Paray School of Nursing, Lesotho*
Dr. E. Amare Zereffa, *Adama Science and Technology University, Ethiopia*
Prof. Dr. L. Vincenzo Boccia, *Federal University of Bahia, Brazil*
Prof. L.S. Gau, *Asia University, Taiwan*
Dr. Ángel-Antonio San-Blas, *Miguel Hernández University of Elche, Spain*
Dr. C. Cao, *Vrije Universiteit Brussels, China*

Session 1. Big data acquisition and analysis

Computational Social Science – Luo, Ciurea & Kumar (eds)
© *2021 Taylor & Francis Group, London, ISBN 978-0-367-70193-2*

Research on the construction of an intelligent court in the era of big data

C. Xu & Z.B. Pei
Dalian Ocean University, Liaoning, China

ABSTRACT: While working in a court, both author felt the influence of computational science on judicial activities, but also found some problems and shortcomings, such as information security, popularization and implementation, and technical innovation. In order to promote the construction of intelligent courts, this paper first summarizes the impact of big data on judicial work. On the premise of determining the necessity of building an intelligent court, the paper analyzes the causes of the above problems by analyzing relevant data, logical argumentation and summarizing the author's experience and feelings in the court work. Because the author was able to recognize in a timely manner the working characteristics of the court, and have an interaction with the cooperation of the network security company operating personnel exchanges, the countermeasures and suggestions not only consider the key needs of judicial activities, but also consider the law and current situation of technological development, and demonstrate the combination of law and artificial intelligence.

Keywords: Intelligent court, Artificial intelligence, Law.

1 NECESSITY OF BUILDING AN INTELLIGENT COURT

At present, the biggest contradiction in Chinese courts is that judicial supply is far from being able to meet the growing judicial demand of the people. Although China has established a relatively sound mechanism for resolving multiple disputes, the dilemma of increasing cases but fewer people has not been greatly improved. Since the traditional methods are not effective, it is necessary to carry out scientific and technological reform in the judicial field, that is, to build intelligent courts. This kind of court is based on modern artificial intelligence, wisdom about justice for the people, fair justice, adherence to legal rule, system reform and the integration of technological change, in a highly information-based way that supports judicial trials, litigation services, and the administration of justice. It can realize full-service online processing, with the whole process being online with all-round intelligence services. Artificial intelligence, by its very nature, is a simulation of the information process of human thinking. Its subordinate products, such as expert systems, intelligent search, and so on, have a strong auxiliary role for court trial work.

1.1 *The intelligent auxiliary system is conducive to improving the efficiency of the trial*

1.1.1 *Case transfer*
In July 2016, the Supreme People's Court issued the *Guiding Opinions on Comprehensively Promoting the Synchronous Generation and In-depth Application of Electronic Files with Cases of People's Courts*, which set clear requirements for courts across the country to fully realize the synchronous generation and in-depth application of electronic files with cases by the end of 2017. It seems to have achieved remarkable results. As an example, the Higher People's Court of Hainan province has promoted an intelligent assistant case-handling system supported by "information collection + electronic files", which has reduced the time to file a case by 53%, reduced the routine

work by 38%, and increased the trial efficiency by 32%. It is pointed out in some documents that the information center should establish stable and convenient channels for retrieving electronic files in the first instance, support staff of the circuit court to retrieve all electronic files from the district court in real time according to their authority and according to the needs of receiving and handling cases, and support the automatic import of electronic files into the case-handling platform. When conditions permit, circuit courts may, jointly with the higher people's courts of districts, explore the establishment of special channels for retrieving electronic files, so as to realize interconnection and information sharing with the information platform of the courts of districts. The circuit court should urge the district court to cooperate in the production and retrieval of electronic files and other work to ensure that the difficulties in handling cases online are solved at the source.

1.1.2 *Class case retrieval*

People engaged in law have a strong reliance on experience and often need to carry out case retrieval work. By the end of May 2018, the website called "Chinese Judgments Online" had published 4.538 million pieces of judicial documents, with over 15.3 billion visits and an average of more than 330 visits per document. The key to case retrieval is an accurate search. If the judgment documents and laws and regulations are not accurate and complete, the search results will greatly affect the final legal analysis. Generally speaking, legal documents can be as short as 2,000 words or long 100,000 words. According to research conducted by Liang Haiqiang, founder of Jufa Technology, during the retrieval process, users spend 70% of their time determing if the retrieval results meet their own needs."". Therefore the innovative class case retrieval system uses artificial intelligence technology to build a keyword map, in the form of dimension reduction and the written judgment of public data analysis, to effectively improve the trial efficiency.

1.1.3 *Provide evidence quickly*

In a trial, the parties often present a large amount of evidence to support their claims. However, due to various evidences or unreasonable classification and other reasons, the parties may be unable to provide evidence quickly, such as disputes over construction contracts. The payroll standard is the common focus of disputes in construction contract disputes. To be specific, the payroll standard cannot be reached through negotiation between the developer and the builder. In practice, due to the problems stipulated by law, there are many cases in which the developer and the builder do not directly sign the written construction contract, and cases of illegal subcontracting are increasingly common. Although the Supreme People's Court on the trial of construction project construction contract dispute case applicable law questions the interpretation of article 4 of the regulations, illegal subcontracting or the builder having incorrect qualifications and using the name of another qualified construction enterprise results in the contract being declared null and void. However, due to the following provisions in the judicial interpretation, the construction contract is invalid, but if the construction project has passed the acceptance inspection upon completion, the contractor's request to pay the project price according to the contract shall be supported. With the provisions of this law, an illegal subcontracting situation is repeatedly prohibited and ensuing settlement disputes are also more common. Under normal circumstances, the developer will advocate the payroll standard agreed in the bidding agreement, while the builder will advocate the payroll standard agreed in the written contract signed with the subcontractor. In order to support their own litigation claims, both parties will conduct multiple cost appraisals on the project involved in the case, resulting in a large number of evidence materials. Without a pre-trial meeting, much of the court time would have been wasted rummaging and checking through the evidence. However, if the trial activities are intelligent, such as with the installation of a trial voice recognition system and the integration of intelligent applications of electronic files, the corresponding electronic evidence in the electronic files can be accurately retrieved according to the voice instructions of the participants in the proceedings, and displayed synchronously on the display screen in front of all parties, effectively improving the trial efficiency.

1.2 *Big data analysis is conducive to the analysis of regional characteristics*

Due to the influence of history, culture, economy, and other aspects, the regional characteristics of different areas are different, which makes similar judgments have different effects and enforcement problems. Although the Constitution provides that the people's courts exercise judicial power independently in accordance with the law and not subject to interference by public organizations and individuals of administrative organs, social influence cannot be extended to be interpreted as "interference." In terms of jurisprudence, laws originate from society and are determined by the economic foundation of that society. Meanwhile, laws, in turn, maintain the organic unity of the relationship between social relations and social order. Therefore, judges should also consider the social influence within reasonable limits when making judgments and should not be completely divorced from society. Big data technology can combine massive information data to analyze the characteristics of the crowd, regional stream analysis, and so on. According to the reference data provided by the Second Circuit Court of the Supreme People's Court, the Collateral Regulation Disputes account for a large proportion of the cases it accepts, because the responsible area of the Second Circuit Court is in Heilongjiang, Jilin, and Liaoning, namely the three provinces of northeast China. As the most important industrial base in China over time and supported by the strategy of revitalizing the old industrial base in northeast China currently, enterprises in northeast China have frequent economic exchanges, especially loan mortgages, and so disputes caused by material damage are more common. Under the background of this region, the judges of the Second Patrol Court will study the intersections of property law and contract law, and even the theoretical knowledge of the intersections of the criminal law, in disputes over the supervision contract of substance. And through big data analysis they are able to understand the current situation of the industry, to ensure a fair and reasonable judgment.

1.3 *The intelligent physical examination system is conducive to improving the accuracy of judgment documents*

The Internet of judgment documents is an important way to promote judicial fairness and justice. However, due to negligence of proofreading, many open judgment documents have some problems, large and small. Judicial documents are the bridge in communication between judges and parties, the carrier of judicial authority and the embodiment of litigation value. Errors in judgment documents often cause varying levels of impact on the quality handling of cases, and some even lead to the loss of all previous work in the trial. Therefore, many judges and court clerks use the "three times reading" method to proofread correctly. However, document proofreading work is time-consuming. In order to give consideration to the trial efficiency, some courts have adopted an intelligent physical examination system to synchronously correct errors when drafting the judgment documents. With the development of big data technology, the error correction system for judgment documents has become more intelligent and error-free. Taking Jufa Technology's cloud physical examination system for judgment documents as an example, this system can realize intelligent error correction and Internet blocking of judgment documents with one key. It can effectively solve problems in the form, procedure, and content of judgment documents, such as proofreading the name of the judgment document, the identity information of the parties concerned, and correcting errors such as when the date of a trial coincides with a legal holiday.

2 PROBLEMS EXISTING IN THE CONSTRUCTION OF INTELLIGENT COURTS

2.1 *Network information security issues*

Network information security has been widely concerned about the audience. The "wannaRen" incident reflects the current situation of network information security in China and the world. It was a stark warned: network viruses are everywhere, and people are the greatest weakness in the defense of network security. On May 12, 2017, a ransomware virus called "Eternal Blue" hit more than

150 countries and regions, affecting government sectors, medical services, public transportation, postal services, communications, and automobile manufacturing. From the start of 2018 to the middle of September, ransomware attacked more than 2 million terminals, with 17 million attacks. This massive virus attack was not the first of its kind. It reflects not only the breadth of the hacker community, but also the shortcomings of China's early-warning and emergency mechanisms. Some believe that this cyber attack was neither unprecedented nor impossible in the future. No matter how much we try to stop it at its source, at some point in the future, or not-too-distant future, there will be similar or even greater occurrence. Indeed, in the game between the network information security system and the virus, no single party always win, however the reasons can be surmised and the experience learnt from to respond actively to the next test.

2.2 *Promotion and use*

On June 7, 2017, Beijing courts and Jin-Ji pilot courts realized cross-domain filing services. After two years of practical verification, the pilot court introduced the latest cross-domain filing self-service machine, and the whole process from landing the filing platform to submitting the filing application and uploading of evidence materials to examining and filing the case took less than 30 minutes, which greatly improved the efficiency of the litigation service. Therefore, it is necessary to promote online filing and cross-regional filing nationwide. That is to say, in 2019, courts in the Yangtze River Delta region should have realized 100% cross-regional filing within their jurisdiction and across provincial administrative regions. By 2020, all regions in the country should have a 100% online filing rate and 100% coverage of cross-regional filing courts.

Although considerable achievements have been made in the construction of intelligent courts, the uneven and incomplete construction of intelligent courts cannot be ignored at present. According to statistics, 84 percent of courts in China have opened online filing services, and 32 percent of courts in China have provided cross-domain filing services within the province, while most regions fail to provide cross-domain filing services. In addition to the lack of supporting facilities, there are also many practical problems in operation and use. For example, some new guiding robots set up by some courts can answer 5,000 legal questions, but they have been abandoned for various reasons. A complex and simple cases triage system can quickly sort out cases suitable for a non-litigating procedure, quick adjudication procedure, summary procedure, or general procedure, but there has been a failure to make full use of it. The litigant can submit lawsuit materials through the lawsuit service platform, but after examination and verification by the filing staff, the litigant also needs to mail a copy of the paper lawsuit materials.

2.3 *Technical innovation*

Bias, including bias of interest, common bias, and specific bias, is one of the factors affecting judicial impartiality. Due to factors such as interest relationships, racial bias, and the influence of public opinion, judges may make incorrect judgments in determining facts, applying laws, and deductive reasoning, deviating from the essence of the rule of law. Max Weber, a German sociologist, envisioned an ideal judicial system as a "vending machine" with facts on one side of the case and judgments spewed out on the other. In this ideal model, it is indeed possible to eliminate the unfair judgments caused by the personal bias of judges. Therefore, many legal technology companies have invested a lot of human, material, and financial resources to build a machine that strictly observes the legal provisions input by the system and can learn independently, however this has not been realized yet. In order to copy and simulate the judge's judgment through a machine, some companies have used the power of 400–500 people to carry out a commonly used case simulation construction, but only made the 23 simplest cases in the end. The ratio of R&D input to output is so out of balance that many companies are gradually abandoning the development of "intelligent judges."

3 ANALYSIS OF THE CAUSES OF THE PROBLEMS

3.1 *Inadequate network security services*

A shortage of high-end talent is one of the reasons why network security services are not in place. China had 802 million Internet users as of June 2018, according to a statistical report on the Development of the Internet in China. However, in the last three years, China's higher education training of information security professionals was for only more than 30,000 people, that is to say, more than 30,000 people need to ensure the work life of 802 million people online security. Moreover, not every one of these 30,000 people will be of use. In the field of network security services, only high-end talents are scarce, that is, compound talents with mathematical and computer theoretical foundation, programming ability, computer-related technical ability, network security technology, document awareness, and other skills. Based on the background to this industry, major security manufacturers, Internet companies, banks, and so on have a great deal of competition for talent, resulting in industry talent mobility, and so security services cannot be guaranteed. A shortage of high-end technology is also contributing to the lack of network security services. The phenomenon of large companies launching copycat Internet security products to grab the market is neither unique nor new. What's more, as mentioned above, enterprises do not pay sufficient attention to network security issues, and so they do not provide funds to innovate network security products, leading to the innovation being hampered. Therefore with the proliferation of copycat network security products, the cookie-cutter network security mode will be a fatal shortcoming.

3.2 *Economic conditions and ideological problems*

At present, most regions fail to provide trans-provincial and trans-regional registration services, partly due to the economic conditions and partly due to the influence of leadership decisions. Some areas with imperfect economic conditions have problems of insufficient funds, non-renewal, and high maintenance costs, which leaves a big gap compared with the courts in other areas with superior economic conditions. There are also issues with some court leadership that is insufficient to support information construction, and a lack of independent innovation awareness and motivation. In practice, some leading officials in grassroots courts tend to focus only on case-handling and care only about the case-closing rate, and are not good at accumulating and studying data and information. Some leading officials distrust modern tools and still use the traditional manual case-handling mode. The above two reasons have a negative impact on the promotion and implementation of intelligent courts.

3.3 *Unreasonable research and development philosophy*

Many companies are working to build intelligent robots that can carry out a trial in the place of judges. Although a handful of companies around the world have developed models that predict with better than 70 percent accuracy, this is far from the ideal concept of a "robot judge." In fact, the core reason for the delay in development is not that the technology is underdeveloped, but that the concept of a "robot judge," like a perpetual motion machine, is impossible and may not be possible. The operation of the robot strictly follows the system command. Even if an intelligent robot has the ability of autonomous learning, it cannot be separated from the regulation of data and algorithms. An actual trial is a complex project, which not only includes specific parts such as fact finding, law application, and deductive reasoning, but also includes abstract parts such as the judges' empirical intuition, legal interpretation, and discretion, which cannot be converted into algorithms. Take as an example AlphaGo, an intelligent robot that excelled at beating human players on a fixed-format go board. However, if the size of the board was changed, the intelligent robot did not know what to do. Law is an open system, and no two cases are exactly the same case. Even for a series of cases with similar circumstances, different judgment results will be obtained due to the starting point of the statute of limitations, the ownership status of the case involved, delivery method, and

7

other details. Moreover, the work of a judge is not only in choosing applicable law for deductive reasoning based on facts, but also to make a judgment combining legal theory and emotion by integrating the actual situation, social influence, and guiding significance of the parties concerned. Since the essential characteristics of law include uncertainty, and intelligent robots can only work and learn autonomously according to certain data and algorithms, the concept of a "robot judge" is impossible to achieve, and the development direction of "law + artificial intelligence" should be an intelligent auxiliary judicial system.

4 COUNTERMEASURES AND SUGGESTIONS

4.1 *Combination of code regulation and legal regulation*

Laws can regulate human behavior to ensure the development of science and technology in a beneficial direction, therefore it is very necessary to establish an effective legal regulation system. In the process of law continuously standardizing technology and technology continuously influencing law, we find that law can be transformed into code to deal with network disputes directly and efficiently. Therefore, legal technicalization and technical legalization can be regarded as a good way to control risks. Technical legalization is the industry rule of "code is law." The rule that code is law is characterized by the fact that code, as an instruction, can be strictly enforced, eliminating the space for controversy. In this way, the rules on the Internet are guaranteed and people can reasonably predict the consequences of their actions. Therefore, turning laws into codes can regulate illegal behaviors quickly and effectively, and save time and material cost.

Of course, not all laws can be completely translated into code. Besides "rule of code," there is also "rule of law." Although Lawrence Lessig writes in *Code is Law* that code is the cornerstone of the Internet system and has the ability to regulate individual behavior through technical means, the natures of law and code mean that the two cannot be completely transformed. The difference between them lies in the following: in order to be flexible and applicable, the law has some ambiguity, but code, as a kind of instruction, must be accurate and unambiguous; 'aw is mostly a summary of past experience, and so law generally lags behind, but the code must be made in advance; in order to ensure implementation, laws must be made public, but code is valuable in most cases because of its privacy, so we cannot convert the whole laws into code, let alone completely rely on code to regulate risks on the network, because there are also some problems in the governance of code. First, when designing code, technicians should design different forms of code combinations according to local specific laws, but the practical operation is not a small challenge. Second, artificial intelligence relies on input data, and buyers, users, or other subsequent subjects who may provide it with data sources may influence the actual rules of internal operation of the program, so no one can predict the future development trend of intelligent products. Finally, many artificial intelligence algorithms hide discrimination problems, which need to be regulated by laws. To sum up, since the value orientation reflected in the law can guide coders to choose the code, we need to adjust the relevant laws so that the public power of the state can participate in supervision.

4.2 *Policy support*

4.2.1 *Financial support*

It is suggested that relevant departments issue relevant policies to support and adjust the comprehensive promotion and implementation of intelligent court construction. The Second Circuit Court of the Supreme People's Court has the following provisions on the software and hardware facilities of the court: the packing bureau and the information center shall fully guarantee the upgrading and transformation of the platform for handling cases, the equipment for the generation of electronic files, the development of in-depth application functions of electronic files, and the funds needed for the purchase of specialized services. They should provide adequate network bandwidth between the inner and outer net, continuously optimizing the system of indirect performance, enhance the level of information sharing and integration between systems, and focus on the stability of the security information system and ease of use, and through the configuration of high-definition widescreen display, high performance inside and outside the office computer, secret remote wireless network office equipment, etc., to further improve the judge's Internet working conditions, and make full use of the electronic file for the judge by providing software and hardware synchronization support. Local courts can draw lessons from this regulation, rationally allocate office funds, and increase investment in networking systems and related facilities. In addition, it is suggested that relevant departments issue special funds to guarantee the acquisition of relevant hardware and software by local courts.

4.2.2 *Talent support*

The reason why many leading cadres do not actively use intelligent assistive tools is primarily a matter of concept, because they do not trust the safety and accuracy of modern tools. To solve this problem, on the one hand, it is necessary to introduce the information security mechanism of an intelligent auxiliary system and the advantages of big data analysis by lecture training; on the other hand, it is necessary to equip the court with practical talents with relevant professional knowledge to be responsible for the operation and maintenance of the intelligent system. Because of problems in current technology and popularization, the court of law science and technology talent demand is not large, and the technical requirements are not high. However,once China's relevant technology to break through the bottleneck, it will leapfrog into the era of a high degree of integration of law and technology, when the demand for the corresponding talent will occur spurt growth. Therefore it will change the traditional law education, going to a culture method in undergraduate students of science and technology and digital literacy. Most law students in China receive four years of undergraduate law education. If they only learn the law during this period without knowing the knowledge related to artificial intelligence, it will be difficult for them to adapt to the future legal practice led by robots and artificial intelligence after graduation. For this reason, some schools have long foreseen this problem and addressed it. In 2012, Georgetown University Law School began to offer a practical course on technological innovation and legal practice, forming a featured competition program of "Iron Tech Lawyer," aiming to cultivate students' legal development. China can also learn from the beneficial experience of this school and set up elective courses in legal technology in some schools, so as to provide talent for the promotion and implementation of intelligent courts.

4.3 *Developing an intelligent auxiliary judicial system*

As mentioned above, the concept of a "robot judge" is a false proposition, and the most feasible development direction of legal science and technology is an intelligent auxiliary judicial system. The intelligent auxiliary judicial system generally includes two levels. The original intelligent auxiliary judicial system was responsible for the fixed repetitive work in the entire trial activity, such as the trial record, proofreading documents, case filing, mail service, etc., so as to achieve the goal of improving the trial efficiency. The advanced intelligent auxiliary judicial system, based on big data analysis technology, semantic technology, text analysis, and natural language processing, evaluates the value of the established questions by retrieving the case base, and forms professional answers. The results can be used for case prediction, contract analysis, compliance review, and so on. For example, when a judge found that a couple had a joint debt, there was no evidence other than that the borrower and lender gave opposite statements. Now available artificial intelligence technology, through the retrieval of similar cases in the database, the parties time together, their purpose, the types of debt, the law on the basis of a comprehensive cross comparison, can analyze the possibility of a joint debt or not, even estimating the probability of a specific.And these can be the basis on which the judge establishes inner conviction..

At present, most intelligent auxiliary trial systems introduced by Chinese courts are in the early generation, and are used for intelligent guidance, cross-domain filing, complex cases and simple cases triage or linkage preservation, etc., and are not only comprehensive in their popularization scope, but also are not very active in practical uses. In addition to economic conditions and concepts, internal problems such as imperfect system functions and unsmooth links also affect the popularization and use of intelligent judicial assistance systems. Therefore, in order to promote the introduction of intelligent trial assistance systems in local courts, it is necessary for law science and technology companies to shift their research focus from the development of a "robot judge" to the construction of a perfect and secure intelligent trial assistance system, including an initial system and an advanced system.

5 EPILOGUE

Computational science can improve the working efficiency of judicial workers, contribute to judicial openness, actively promote the court to give full play to judicial functions, and promote the modernization of social governance. However, the connotation of an intelligent court is to partly rely on artificial intelligence technology, rather than completely relying on it. Judicial organs are the last line of defense to safeguard social fairness and justice, but judicial justice has relativity and limitations as it is a subjective judgment. When a judge decides a case, he should not only observe the legal provisions, but also consider the overall interests and ethics. The game between these forces cannot be converted into a specific algorithm, and artificial intelligence robots cannot learn human emotions autonomously by analyzing data. Therefore, the construction of intelligent courts is not to build "robot judges," but to develop and perfect a safe intelligent auxiliary trial system, which can share fixed repetitive work for judges, improve trial efficiency, and/or make value assessments on specific matters through big data analysis, and provide reference opinions for judges. Only by following the above research directions, improving the ability to guarantee information security, and strengthening the promotion and implementation of the intelligent trial auxiliary system, can the construction of China's intelligent court achieve its desired effect.

REFERENCES

[1] Liu Wujun. The Rule of Law in China calls for a craftsman carving laws — A Legal Interpretation of Craftsman Spirit [J]. Chinese Justice, 2016(12).
[2] Xiao Na. Research on the Construction of "Intelligent Court" under the Situation of a New Judicial System Reform — A Case Study of The System Court of Reclamation District in H Province [D]. Heilongjiang: Heilongjiang University, 2018.
[3] Shen Zhen. Research on Automatic Generation and Application Method of Electronic Files [J]. Chutian Rule of Law, 2016(12).
[4] Zang Qin Qing, Qiu Chenyi, Research on the Operation Mechanism of "Internet +" Times Intelligent Court — Based on the Empirical Analysis of Hangzhou Internet Court [J]. Youth Times, 2018(34).
[5] Ma Zhizhi and Liu Baolin. Legal Analysis of judicial Application of Artificial Intelligence: Value, Dilemma and Path [J]. Qinghai Social Science, 2018 (05).

Computational Social Science – Luo, Ciurea & Kumar (eds)
© 2021 Taylor & Francis Group, London, ISBN 978-0-367-70193-2

Architectural design and application of camping big data access platform

X.K. He
Duyun Libo Power Supply Bureau of Guizhou Power Grid Co., Ltd., Qiannan, Guizhou Province, China

Z. Wang
School of Electrical Engineering, Sichuan University, Chengdu, Sichuan Province, China

H.Z. Zhou
Duyun Libo Power Supply Bureau of Guizhou Power Grid Co., Ltd., Qiannan, Guizhou Province, China

Z.G. Qi
Hangzhou Harmony Tech. Co., Ltd., Hangzhou, Zhejiang Province, China

ABSTRACT: Aiming at the problems faced in the analysis and processing of the current distribution big data, a large data access platform architecture suitable for distribution networks is proposed. The platform mainly includes the basic layer, the calculate layer and the application layer. The distributed coordination server Zookeeper is the core of basic layer. The distributed file management system HDFS and Ceph, the resource manager Yarn and the observer, which provide consistent services to the entire system, are the most important component. The Spark SQL is the programming interface of the calculation layer. It designs graph calculation GraphX, machine learning ML and MapReduce and other calculation frameworks to complete the calculation tasks of the entire system and provide interface services to the application layer. The application layer takes the HM7000 series as the core, which has the common functions of the power distribution system. We can know from the test results that the annual availability rate of the main station system equipment of the platform reaches 99.9%, the average CPU load rate is 39.6%, the number of accessible workstations is 46, and other key indicators meet the needs of county-level power distribution units.

1 INTRODUCTION

Recent years, the development of modern communication technology has triggered a big data explosion. Big data analysis and processing technologies for data mining, machine learning and knowledge discovery have received widespread attention and become an important means to promote the development of industry technology [1]. Internationally renowned IT companies such as Google, Amazon, Microsoft, IBM, and Facebook have also included big data technology as the key development plan.

With the development of smart grid and the development of network communication technology and sensor technology, the data on the user of the grid has been exploded [2], which play an important role in the distribution network system. Firstly, it will optimize the internal staffing of the enterprise and improve the performance of the enterprise [3–4]. With the application of distribution network big data and deep learning of equipment condition assessment, the power equipment condition assessment in the future will no longer depend on the mode of periodic test, routine maintenance and after-sales repair, but will extract various characteristic parameters of power equipment to carry out real-time evaluation to achieve the right medicine for equipment operation and inspection, improve operation and maintenance efficiency, and reduce equipment operation and maintenance costs [5]. Secondly, it will serve for developing new products and improving the core competitiveness of enterprises. For example, the French Electric Power Research Institute has stored and analyzed some simulation results and used them to load forecasting and gradienting electricity price positioning [6–7]. Lastly, it will serve for social development and predicting the social and economic level of other public utility enterprises [8–10].

Researchers have conducted in-depth discussions and research on the application scope, key technologies, application organizational structure and other aspects of big data, such as the mining of core indicators of county companies based on BDA [11] and the study of policy evaluation system models. But few people design based on big data distribution network data access platform [12].

In order to solve the problem of access and processing of distribution network big data, realize the integration analysis and application of distribution network big data, and improve the level of lean management and intelligence in distribution network, the article proposes a distribution network big data processing and analysis platform architecture. The article analyzes the characteristics of big data in power distribution and the key technologies in the application of big data in power distribution network. We designed a distributed data access and processing platform architecture. The operating mechanism of key components such as distributed coordination processing, distributed file processing, data calculation and processing of the platform is designed. The key indicators of the platform were also tested.

2 DISTRIBUTION NETWORK BIG DATA FEATURES AND KEY TECHNOLOGIES

2.1 *Characteristics of distribution big data*

The big data of electric power possesses the characteristics of "4V" (volume, variety, velocity, and veracity) [13].

(1) Variety

Now, most prefecture-level power companies have multiple distribution management platforms [14], including distribution network automation system, dispatching automation system, power quality monitoring management system and power information management system, etc. This information is the main source of big data for distribution networks which contains a lot of information such as operating mode, voltage quality, equipment operating status, historical records and environmental factors. These data come in a variety of data formats, ranging from structured data to semi-structured and unstructured data.

(2) Velocity

The power network is a real-time network. The transmission and distribution of power energy is completed instantly. The power fault occurs in an instant. The changes in power quality and equipment status are also real-time. According to the different characteristics and importance of the data, the data collection is divided into real-time sampling, semi-real-time sampling and offline sampling. Among them, real-time sampling is the most able to reflect the changes of the switching network, which data changes are also the fastest.

(3) Volume

With the continuous deepening of the distribution network intelligence and the proposal of refined management and precise control of the power grid, the scale of data collection, data dimensions and data sampling frequency have risen rapidly. The amount of data in the distribution network has increased geometrically. According to statistics, a county-level distribution company generates 7.6TB of data per month.

(4) Veracity

Most of the power distribution data collected by the data collection system is normal data, with only a very small amount of abnormal data. These occasional data is an important basis for state maintenance, fault prediction and other intelligent algorithm training sets.

Now, the data processing, the data storage, the data analysis and the data presentation technologies have become the key technologies of modern smart power distribution systems based on big data [15].

2.2 Distribution network big data processing technology

Power distribution big data consists of the real-time data, historical operation records and system data of the equipment itself. These data have a wide range of sources and exist in different databases. The operation management has become a shaft mode [16] (The differences between different units, different departments, and even different professions have led to the emergence of the "data gap"). Complex data type poses huge challenges to the analysis and processing of big data. Therefore, we need to combine the data from different sources, different formats, and different characteristics. Then we need to extract the entities and relationships from them, and use a unified structure to store these data after association and collection to achieve full sharing of big data. Data processing technology includes relational and non-relational data technology, data fusion and integration technology, data extraction technology, filtering technology and cleaning technology. Now, researchers have begun to study the processing technology be suitable for distribution network big data, such as the big data compression technology based on tensor Tucker decomposition studied by Zhao Hongshan, who proposed a compression method for massive heterogeneous data.

In the data access platform, data processing technology is used to process the collected data, including sub-library, partition and sub-table. Sub-database processing is to input some data with high utilization rate into different databases according to certain processing principles, which can improve the utilization rate of some data in the database. Data partition processing is to effectively load data into different files, which can reduce the pressure on large tables, improve data access performance, and make it run better. The platform organically combines structured query language and MapReduce to enhance the performance of data processing in the database and improve the pressure resistance of the data.

2.3 Storage technology of distribution network big data

The relational database can't meet the requirements of big data in processing and storage, which is large capacity, high concurrency and fast response [17]. Common big data storage systems can be divided into three categories. The first one is the file system, such as Google file management system and Lustre file system. The second one is the database system, such as MongDB, HBase, BigTable and DynmoDB. The last one is MapReduce, Druid, Spark and Storm. Different storage modes have different advantages and application. For example, the Google file management system is suitable for distributed data and can be implemented by inexpensive clusters. BigTable can process PB-level non-relational database [18]. However, big data storage model usually combine multiple cloud storage models to process the multiple distribution network data and meet the system requirements. This platform is a combination of distributed file system HDFS and Ceph. The two complement each other to achieve distributed storage of data. Distributed storage transfers the distributed electricity big data to the distributed database through the ETL (extraction transform load) process. Researches on the identification technology, recovery technology, combined index technology and optimized storage structure of non-complete data can improve the access efficiency [19]. ETL contains three parts. Data extraction is to extract the relevant data required by the platform from the existing system. Data transmission, which handles the deviations and errors that occur in the data sources, convert the extracted data into another form to meet the functional requirements of the system. Data loading is to load the processed data into the big data platform.

2.4 Data analysis technology of power distribution big data

In the era of big data, data analysis technology can find modalities and laws hidden in huge amounts of data to provide effective and real reference information for power enterprise decision makers. Distribution network big data analysis methods can be summarized into three types [20]. Type 1: Descriptive analysis method analyzes historical data, asset data and power grid data, which analyzes, interprets and restores the past state of the system or equipment. Type 2: Predictive analysis methods provide prospective analysis. Its users can participate in investment, asset maintenance and grid operation planning. Type 3: The normative analysis method provides users with recommendations on the optimal operation strategy, grid configuration, and route selection under established constraints.

2.5 *Display technology of distribution network big data*

In smart grid big data, the key technologies for displaying data include visualization technology, historical flow, and spatial information flow. Applying these three data presentation technologies to smart grid data processing can allow managers in enterprises to correctly understand the meaning of power data and system operation. The big data access and processing platform uses visualization technology to monitor the operation of the power grid, which can effectively improve the automation level of the power system.

3 DESIGN OF BIG DATA ACCESS AND PROCESSING PLATFORM

A large amount of distribution network operation data and power consumption information are distributed in various systems, which is difficult to obtain. For example, the data of the telemetry and remote signaling of the distribution network terminal, line loss, voltage qualification rate, line (distribution transformer) load rate, and three-phase unbalance rate of the distribution transformer load. They belong to the three remote switch background system, measurement automation system, safety production system and voltage qualification rate module. Processing of these data in the form of tables is often time-consuming and laborious. In particularly, the large volume of data related to the power supply area makes it difficult to be found in time when the indicators are abnormal, which affects the management level of the power grid.

In order to improve the reliability and quality of power supply, improve the service quality of customer service and customers' power consumption experience, distribution units urgently need a lean and visual integrated business platform for distribution. Power distribution units should deeply integrate and mine massive data information from distribution network production systems and marketing and metering automation systems, efficiently integrate distribution network data and marketing data, and build an open, interactive, and efficient distribution cloud platform.

3.1 *The overall design of the big data access platform for distribution*

The storage architecture of the massive data access platform is distributed file management systems HDFS and Ceph, and the data calculation and processing architecture is based on MepReduceV2. The overall architecture of the data access platform is shown in Figure 1, which mainly includes a basic layer, a computing layer and a service layer.

The base layer provides callculative resources and data persistence services, which includes collaborative service Zookeeper, callculative resource management YARN and distributed file management system. The collaborative service Zookeeper provides consistent services for various distributed applications such as HDFS, YARN, and Kafka in the mass data platform cluster of marketing and distribution network production. Callculative Resource Management YARN manages the callculative resources of the entire cluster, and manages and schedules various computing jobs executed by the cluster. The distributed file system provides persistent storage of data for marketing and distribution network production massive data platforms.

The callculative layer provides a unified memory-based iterative callculative architecture, which supports SparkSQL (including DataFrame, DataSet programming interfaces, custom function functions and Hive SQL compatibility), GraphX(graph computing),ML (machine learning) and MapReduce and other commonly used processing frameworks.

The service layer provides services and APIs to users or upper-layer applications. It mainly includes the application of integration of graphics, collection and monitoring, integration of GIS,

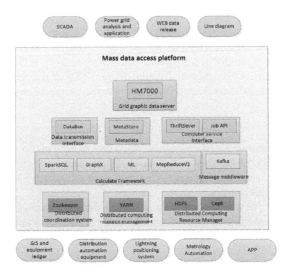

Figure 1.   Overall architecture of massive data access platform.

graphics and real-time data collection, protocol analysis, forwarding, saving, SCADA monitoring applications.

3.2 *Design of basic layer*

In order to access multi-source heterogeneous data with strong real-time sampling, the data access platform adopts a distributed storage and processing scheme. Because programs are prone to deadlocks and inappropriate election competition, collaborative services face huge challenges in distributed application development. The emergence of an open source, distributed service coordinator zookeeper provides new ideas for solving this problem. It has the characteristics of simple use, synchronous replication and order, which can achieve synchronization and configuration maintenance between services.

ZooKeeper sets up an watch window for all read operations. These read operations include: exists (), get Children () and get Data (). The watch window is a one-time trigger. When the state of the watch object changes, the watch event corresponding to this object will be triggered. "Watch" events will be sent to the client asynchronously, and ZooKeeper provides an orderly consistency guarantee for the observation mechanism. In theory, the time the client receives from the "Watch" event is faster than the time it sees the state change of the "watch" object. The watch window is maintained locally by the ZooKeeper server to which the client is connected, so it can be easily set up, managed, and dispatched.

Computer resource manager Yarn is composed of several components: resource manager, node manager, ApplicationMaster, and Container. Container is Yarn's abstraction of computer CPU and memory resources, which all applications will run in. ApplicationMaster is an abstraction of application instances. Its function is to apply for callculative resources from the resource manager and interact with the node manager to perform and monitor specific tasks. Scheduler is a component of the resource manager dedicated to resource management. It is responsible for allocating Container resources on the node manager. The node manager will also continuously send its own resource usage to the resource manager.

The distributed file system provides persistent storage of data for the HM7000. The platform uses HDFS and Ceph as distributed file system components. HDFS (Hadoop Distributed File System) is an implementation of Hadoop abstract file system. HDFS files are distributed on cluster machines, and copies are provided for fault tolerance and reliability. HDFS can store large files and run on commercial software, but it is not suitable for low-latency data and access to a large number of small files, nor does it support multi-party read and write, and arbitrary file modification. In order to make up for the shortcomings of HDFS in storage, Ceph storage system is introduced. Ceph is a unified distributed storage system that can provide better performance, reliability, and scalability. It supports Object (with native APIs, and is also compatible with Swift and S3 APIs), Block (supports thin configuration, snapshots, Clone) and File (Posix interface, support snapshot) three interfaces.

3.3 Calcultative layer

Traditional calculative methods use Map-Reduce-based data mining algorithms, which consumes a lot of resources and the algorithm efficiency is low. The Spark-based memory computing framework overcomes this shortage, allowing Spark to fully utilize the calcultative resources of distributed clusters and process data efficiently.

Spark SQL introduces a new RDD type SchemaRDD, which can be used to define SchemaRDD like a traditional database definition table. It can mix data from different sources in the application, and also has a built-in query optimization framework. After parsing SQL into a logical execution plan, it finally becomes RDD calculation.

Spark GraphX is a distributed graph processing framework. It is based on the Spark platform and provides simple, easy-to-use, and rich interfaces for graph calculation and graph mining, which greatly facilitates the demand for distributed graph processing. All operations on the Graph view will eventually be converted to the RDD operation of its associated Table view to complete. This calculation of a graph is ultimately logically equivalent to a series of RDD conversion processes. Logically, all graph transformations and operations produce a new graph. Physically, GraphX will have a certain degree of optimization of the reuse of constant vertices and edges, which is transparent to users.

The messaging system kafka transfers data from one application to another. Applications only need to focus on the data, no need to pay attention to how the data is transmitted between two or more applications. It is a publish-subscribe model, which is a distributed, partitioned, multi-copy, multi-subscriber, based on the distributed logging system coordinated by zookeeper. Kafka has the advantages of decoupling, redundancy, scalability, high peak processing capacity, and strong recoverability. It is suitable for distributed message queue delivery.

3.4 The layer of service

The service layer adopts the HM7000 platform, which is an application development platform that integrates graphics, collection and monitoring, and integrates GIS, graphics and real-time data collection, protocol analysis, forwarding, saving and SCADA monitoring. The architecture of the HM7000 platform is shown in Figure 2.

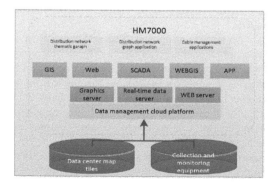

Figure 2.   HM7000 platform architecture.

4 TESTING AND ANALYSIS OF SYSTEM PERFORMANCE

After the design of the data access platform is completed. The platform's web applications, pre-communication management configuration, data collection and communication functions, GIS graphic client, SCDA monitoring part, and data reading and writing functions were tested. The results show that the relevant requirements can be achieved and the platform can run reliably. The performance indexes of the database access platform system are shown in Table 1.

Test indicators show that the system has a low fault rate and a long continuous running time. For power systems with extremely high real-time performance, fault rate and telemetry error are important parameters that characterize the continuity of data collection.

Both system performance and database performance are indispensable. Only when the two work together to form a "bucket" effect can the distribution network data be accessed and processed better. Table 2 shows the index test results of the database.

Table 2 shows that the platform can access 4180 device models, and the number of records of a single device model can reach 1.132 million. Therefore, the platform can fully meet the daily data access and processing needs of grassroots power distribution units.

In order to test the data reading and writing index of the platform, 1000 files with a size of 521MB were used for reading data tests, 1000 files with 512MB were used for writing data tests, 10 mappers and 5 reducers were used to create 1000 files NameNode load test. The test results are shown in Table 3.

Table 1. System performance indicators.

Test Item	Test Results
Annual availability of master station system equipment	99.9%
MTBF(h)	24526h
The average number of automatic hot starts due to accidental faults	1times/3621h
The average load rate of CPU	39.6%
Spare space (root zone)	25.5%
Number of work stations	46
The error rate of telemetry integrated	1.3%
The pass rate of telemetry	98.44%

Table 2. Database indicators.

Test Item	Test Results
Number of equipment models	4180
Number of records for a single device model	113.2
The accuracy rate of system data	99.2%
The integrity rate of data	98.4%

Table 3. Data reading and writing test.

Test Item	Test Results
Average reading speed(MB/s)	55.17
Total reading time(s)	38.702
Average write speed(MB/s)	3.38
Average write(s)	218
Total time(ms)	21176
Maximum map time(ms)	4535.0

From the test results, regardless of the comprehensive indicators of the system, the overall performance of the entire database, and data reading and writing, the data access platform meets the requirements for reliable operation of the current distribution network system.

5 CONCLUSION

This paper analyzes the current status of power distribution big data applications and the 4V characteristics (volume, variety, velocity, and veracity) of big data. The characteristics of big data determine that the data collection technology, data processing and storage technology, data analysis technology and data application technology of power distribution big data have become the key technologies in the application of big data. At present, there are relatively mature technologies in the collection of power distribution big data. Various power distribution management systems, equipment status systems and geographic information systems provide data support for big data, and can not make transformative transformation in a short time. However, there are few studies on big data access platforms based on data processing and storage technology and data analysis technology. The distribution network big data access platform designed in this paper makes up for the "vacuum zone" in distribution network information processing.

This platform architecture integrates massive data of existing production, scheduling and marketing systems. Through the data association analysis and data value mining, the core data integration between the distribution network and the marketing system is initially achieved. Next, the platform will be applied to the management of distribution network fault location and repair, line loss management, and maintenance technical renovation project management, which will help to realize the efficient use of data and promote management improvement and business innovation.

REFERENCES

[1] Wang Xing. Big data analysis: methods and applications. Beijing: Tsinghua University Press, 2013.
[2] Fei Siyuan. A review of the application of big data technology in distribution networks. Chinese Journal of Electrical Engineering, 2018, 38 (01): 85–96 + 345.
[3] Zheng Xiang, Zhao Fengzhan, Yang Rengang, Meng Xiaoli. Evaluation system of reactive power operation of low-voltage distribution network based on big data. Power Grid Technology, 2017, 41 (01): 272–278.
[4] Wu Zaijun, Xu Junjun, Yu Xinghuo, Hu Qinran, Dou Xiaobo, Gu Wei. Review of Active Distribution Network State Estimation Technology. Power System Automation, 2017, 41 (13): 182–191.
[5] Xun Ting, Zhang Keheng, Xue Haoran, Tang Sheng, Tang Chaobo, Han Qiucheng. Comprehensive intelligent analysis and decision-making architecture design of grid regulation data. Power System Protection and Control, 2015, 43 (11): 121–127.
[6] Chen Haiyan, Jin Nong, Ji Cong, Xiong Zheng, Li Kunming. Power user power data analysis technology and typical scenario applications. Power Grid Technology, 2015, 39 (11): 3147–3152.
[7] Marie-Luce PICARD, Pan Xuyang. Looking at the challenges and opportunities brought by big data from the perspective of French public power companies. Power Grid Technology, 2015, 39 (11): 3109–3113.
[8] Shi Changkai, Zhang Bo, Sheng Wanxing, Zhou Meng, Gao Yuan, Li Yuling. Discussion on the functional architecture of distribution network operation and maintenance management and control platform. Power Grid Technology, 2016, 40 (07): 2206–2211.
[9] Cai Jiaming, Xie Ning, Wang Chengmin, Fan Mingtian. Digital Technology and Modeling Method for Distribution Network Planning——A Summary of the Research Results of the 24th International Power Supply Conference. Power Grid Technology, 2019, 43 (06): 2171–2178.
[10] Tian Shiming, Gong Taorong, Huang Xiaoqing, Yu Wenlong. Regional E-GDP value prediction based on big data of electric power. Electric Power Automation Equipment, 2019, 39 (11): 198–204.
[11] Xue Zhenyu, Hu Hanghai, Song Yi, Wu Zhili, Liu Daoxin, Feng Hao, Song Hongfang. Comprehensive evaluation strategy of county companies based on big data analysis. Electric Power Automation Equipment, 2017, 37 (09): 199–204.

[12] Zeng Ming, Wang Yuqing, Yan Tong, Lan Mengxin, Dong Houqi, Zhang Xiaochun, Wang Haolei, Sun Chenjun. Design and model research of renewable energy policy evaluation system based on ubiquitous power Internet of Things platform. Power Grid Technology, 2019, 43 (12): 4263–4273.

[13] Liu Wei, Huang Zhao, Li Peng, Li Mang, Ding Yan. Big data unified support platform system and architecture for intelligent distribution network. Journal of Electrical Engineering and Technology, 2014, 29 (S1): 486–491.

[14] Ding Jianyong, Zhou Kai, Tian Shiming, Pan Mingming. Power supply safety analysis of important users based on big data technology. Power Grid Technology, 2016, 40 (08): 2491–2495.

[15] Wang Shouxiang, Ge Leijiao, Wang Kai. The connotation and key technologies of intelligent power distribution system. Electric Power Automation Equipment, 2016, 36 (06): 1–6.

[16] Zhao Hongshan, Ma Libo. Compression of Big Data in Intelligent Distribution Network Based on Tensor Tucker Decomposition. Chinese Journal of Electrical Engineering, 2019, 39 (16): 4744–4752 + 4976.

[17] Jiang Xiuchen, Sheng Geao. The big data analysis and application of power equipment status. High Voltage Technology, 2018, 44 (04): 1041–1050.

[18] Li Gang, Zhang Bo, Zhao Wenqing, Liu Yunpeng, Gao Shuguo. Data science problems in power equipment condition assessment: challenges and prospects. Power System Automation, 2018, 42 (21): 10–20 + 177.

[19] Ma Zhao, An Ting, Shang Yuwei. Development and development of cutting-edge power distribution technology at home and abroad. Chinese Journal of Electrical Engineering

[20] Chen Haiwen, Wang Shouxiang, Liang Dong, Su Yun. Big data analysis and application of user power saving. Power Grid Technology, 2019, 43 (04): 1345–1354. Journal, 2016, 36 (06): 1552–1567.

Computational Social Science – Luo, Ciurea & Kumar (eds)
© 2021 Taylor & Francis Group, London, ISBN 978-0-367-70193-2

Analysis of the main logistics factors of fresh agricultural products and their impact on agricultural development

Y. Zhang, X.L. Yang, X.H. Chen, B.L. Xiang, H. Yi & C. Liu*
Sichuan Agricultural University, Dujiangyan, China

ABSTRACT: After digesting the relevant domestic and foreign literature and tracking the concerns of fresh agricultural products by business entities, consumers, and relevant experts and scholars in each area of fresh agricultural products, combining fresh agricultural products and agricultural development, this article discusses agricultural economic benefits and logistics quality factors. Three aspects of the industrial market structure are used to construct an evaluation index system for the logistics level of fresh agricultural products. Taking Sichuan Province as the research object, the gray correlation model is used to analyze the main factors affecting the logistics level of fresh agricultural products, and at the same time, a correlation analysis is used to further analyze the correlation of each influencing factor. The article systemically analyzes the main influencing factors, and puts forward development suggestions from multiple aspects to promote the development of agriculture.

Keywords: Fresh agricultural products, Agricultural development, Gray correlation analysis, Correlation analysis, Intelligent logistics.

1 INTRODUCTION

With the acceleration of China's agricultural modernization process and the rise of new retail models of fresh food e-commerce, the demand for fresh agricultural products continues to increase, and consumers' requirements for the quality of fresh agricultural products are increasing daily. At the same time, changes have been made to the fresh agricultural product logistics service industry, which has high demands. The logistics of fresh agricultural products connects the whole process of production, transportation, storage, and trading of fresh agricultural products, and plays an increasingly important role in the development of the agricultural economy. However, the logistics of fresh agricultural products in China started late, and a complete logistics system has not yet been established. The logistics of fresh agricultural products generally exhibits problems such as high cost, high loss, and low efficiency, which restricts the modernization of China's agricultural development.[1] Therefore, it is very important to improve the logistics development level of fresh agricultural products. The identification of the main factors affecting the development of fresh agricultural products logistics and promoting the development of fresh agricultural products are of great significance for accelerating agricultural modernization. As a large agricultural and populous province in China, Sichuan has a large market supply and demand for fresh agricultural products. According to statistics, from 2015 to 2018, the total agricultural product logistics in Sichuan Province increased from 545.47 billion yuan to 656.52 billion yuan,[2] and the demand for logistics continues to increase, making it a major agricultural product logistics province. Therefore, this paper chooses the logistics of fresh agricultural products in Sichuan Province as the research object, and takes the main logistics factors of fresh agricultural products as the starting point to explore

---

*Corresponding author

the development level of fresh agricultural products logistics and its impact on the development of fresh agricultural products, and then analyzes the effect on agricultural production and management. A decision-making reference for promoting agricultural development and solving the "three rural" issues is provided.

2 SELECTION OF THE MAIN LOGISTICS FACTORS OF FRESH AGRICULTURAL PRODUCTS

The outbreak of fresh food e-commerce in recent years has provided an opportunity for the development of a new retail route for fresh agricultural products. The emergence of a variety of new retail models of fresh agricultural products has generated huge demand for logistics and transportation, which requires optimization of the level of fresh agricultural products logistics. After consulting, it was found that most of the current relevant academic papers are based on the logistics efficiency of fresh agricultural products. Huang Fuhua and Jiang Xuelin have set up a fresh agricultural product logistics efficiency evaluation index system from four aspects: logistics scale, logistics loss, logistics cost, and logistics slow-sale[3]; Gong Ruifeng, Wu Hailing, and Peng Xiaohui evaluate the logistics efficiency of regional fresh agricultural products through indicators such as per capita income level, logistics operation level, logistics infrastructure, highway mileage, logistics demand, etc.[4]; Liu Gang evaluates the logistics technology innovation, logistics concept innovation four models and their different combinations that enhance the logistics service innovation of fresh food e-commerce[5]; Wang Cheng, Wang Tao, and Jiang Yuansheng build on the development of fresh agricultural products in the western region from the four aspects of infrastructure, enterprise level, supply and demand, and economic development. Wang Cheng, Wang Tao, Jiang Yuansheng build a fresh agricultural product development evaluation system in the western region from four aspects: infrastructure, enterprise level, supply and demand status, and economic development level[6] Evaluation system.[6] There remain many gaps in the research on the logistics of fresh agricultural products and agricultural development. Therefore, this research combines the development of fresh agricultural products in Sichuan Province and innovatively combines the logistics level of fresh agricultural products with agricultural development. Starting with the three aspects of logistics quality and industrial market structure, the fresh agricultural product logistics factor analysis system of this study is established, including three secondary indicators and 10 tertiary indicators.

2.1 Agricultural economic benefit index

The income of the agricultural economy is an important factor in measuring the level of agricultural development, and the income from the industry directly affects the enthusiasm of suppliers to invest in the industry. This study uses the total value of the agricultural industry in Sichuan Province and the annual per capita income of rural residents in Sichuan Province to reflect the agricultural economic benefits to the province, and explores the development prospects of the fresh agricultural product industry in Sichuan Province.

2.2 Logistics quality indicators

The quality of logistics directly affects the quality of fresh agricultural products and is an important technical factor to ensure the healthy, sustained, and stable development of the fresh agricultural products industry. This study reflects the logistics quality level of Sichuan Province by selecting data on the number of highway operating trucks, total logistics, total logistics costs, postal business outlets, road freight volume, transportation, and warehousing-related personnel.

2.3 Industrial market structure

The balance of the market structure determines the maximum benefit that the industry can achieve. The market composition of fresh agricultural products is analyzed from the aspects of supply and

Table 1. The logistics level evaluation index system of fresh agricultural products.

The first level indicator	The second level indicator	The third level indicator
The logistics level	Agricultural economic benefits (A)	A1: Total value of agricultural industry/100 million yuan A2: Annual per capita income of rural residents/yuan
	Logistics quality factor (B)	B1: Number of trucks in highway operations/10,000 vehicles B2: Total logistics/100 million yuan B3: Total logistics cost/100 million B4: Postal service outlets/offices B5: Road freight volume/10,000 tons B6: Employment related to transportation and storage/10,000 people
	Industrial market structure (C)	C1: Per capita consumption of fresh agricultural products in households/kg C2: Output of fresh agricultural products/10,000 tons

Table 2. Statistics of logistics level indicators in the fresh produce market.

	A1	A2	B1	B2	B3	B4	B5	B6	C1	C2
2014	3068.61	12647	10.29	55000	5327	10767	142132	141.29	220.04	5680.62
2015	3315.51	14561	105.1	57314.5	5522.5	13220	138622	135.91	229.44	5877.47
2016	3701.64	15907	131.06	60800	5752	16532	146046	122.74	232.35	6050.37
2017	4404.2	17264	150.8	63909.5	6318.3	18962	158190	122.8	241.72	6209.21
2018	4153.71	19016	137.8	66024	6440.2	20170	173324	122.37	232.7	6485.71

demand. At the same time, the supply and demand of fresh agricultural products are analyzed at the macrolevel. Therefore, this study chose the per capita household and two indicators of fresh agricultural product consumption and fresh agricultural product output to reflect the market structure of the fresh agricultural product industry.

In summary, the evaluation index system for the main logistics factors of fresh agricultural products is shown in Table 1.

3 EMPIRICAL ANALYSIS

3.1 Data acquisition

In terms of data sources, this study obtains the data from the agricultural industry gross value and annual per capita income of rural residents in Sichuan Province by consulting the Sichuan Statistical Yearbook, so as to understand the agricultural economic benefits; by consulting the National Bureau of Statistics, the Sichuan Logistics Big Data Platform, and the Sichuan Statistical Yearbook, to obtain the number of trucks in highway operations, the total logistics, the total logistics costs, the number of postal business outlets, the volume of road freight, and employment related to transportation and warehousing. The number of people and other data related to logistics quality; the per capita consumption of fresh agricultural products, and the output of fresh agricultural products were obtained from the Sichuan Statistical Yearbook to reflect the market supply and demand of fresh agricultural products. The statistics for various indicators are shown in Table 2.

3.2 Selection of methods and analysis of factors

3.2.1 Selection of method
According to the requirements and purpose of this research, the gray correlation degree[7] was selected as the research method. The gray correlation analysis method is used to measure the

Table 3. Standardization of logistics level indicators for fresh agricultural products.

	A1	A2	B1	B2	B3	B4	B5	B6	C1	C2
2014	0.00	0.00	0.00	0.00	0.00	0.00	0.101	1.000	0.000	0.000
2015	0.185	0.301	0.675	0.210	0.176	0.261	0.000	0.716	0.434	0.245
2016	0.474	0.512	0.860	0.526	0.382	0.613	0.214	0.196	0.568	0.459
2017	1.000	0.725	1.000	0.808	0.890	0.872	0.564	0.227	1.000	0.657
2018	0.812	1.000	0.907	1.000	1.000	1.000	1.000	0.000	0.584	1.000

Table 4. Absolute difference of logistics level indicators for fresh agricultural products.

	A1	A2	B1	B2	B3	B4	B5	B7	C1	C2
2014	1	1	0	1	1	1	0.899	0	1	1
2015	0.815	0.699	1	0.791	0.824	0.791	1	0.284	0.566	0.755
2016	0.526	0.488	0.846	0.474	0.618	0.387	0.786	0.981	0.432	0.541
2017		0.275	0.379	0.192	0.111	0.128	0.436	0.977	0	0.343
2018	0.188	0	0.056	0	0	0	0	1	0.416	0

degree of correlation between factors based on the degree of similarity or difference between the development trends of factors. Its application involves various fields of social science and natural science, and it is an important method of the gray theory system. Due to the small amount of data statistics, the application of gray correlation analysis results in the degree of correlation and trend of change.

3.2.2 Standardization of index values
Since the dimensions of all original data are different, they need to be standardized before analysis to eliminate the data differences between each index. This paper uses the method of min-max-normalization[8] to analyze the data and standardize processing to obtain standardized data for each indicator (see Table 3).

3.2.3 Calculate the absolute difference
The absolute difference can be calculated as the maximum and minimum differences. With the data standardized to 1 as the reference sequence, a difference calculation is performed on each item of data after the standardization, and then the maximum and minimum absolute differences are taken. The treatment conducive to subsequent calculations uses the calculation formulas as follows:

$$\Delta \min = \min_{j} \min_{i} |r_i - r_{ij}| \tag{1}$$

$$\Delta \max = \max_{j} \max_{i} |r_i - r_{ij}| \tag{2}$$

where i represents the year, j represents the evaluation index, rj represents the reference value of the j year, and rij represents the j value of the evaluation index in the i year. According to the calculation formula, the absolute difference of each index can be obtained (see Table 4).

3.2.4 Calculate the gray correlation coefficient
The formula for calculating the gray correlation coefficient is:

$$\varepsilon_i^j = \frac{\min_{j} \min_{i} |r_i - r_{ij}| + p \max_{j} - \max_{i} |r_i| - r_{ij}}{|r_i - r_{ij}| + p \max_{j} - \max_{i} |r_i - r_{ij}|} \tag{3}$$

Table 5. Gray correlation coefficient of fresh agricultural product logistics level indicators.

	A1	A2	B1	B2	B3	B4	B5	B6	C1	C2
2014	0.333	0.333	0.333	0.333	0.333	0.333	0.357	1.000	0.333	0.333
2015	0.380	0.417	0.606	0.389	0.378	0.404	0.333	0.637	0.469	0.398
2016	0.487	0.506	0.781	0.513	0.447	0.564	0.389	0.337	0.536	0.480
2017	1.000	0.645	1.000	0.723	0.820	0.796	0.5534	0.338	1.000	0.593
2018	0.727	1.000	0.844	1.000	1.000	1.000	1.000	0.333	0.546	1.000

Table 6. The correlation degree of fresh agricultural products logistics level indicators.

A1	A2	B1	B2	B3	B4	B5	B6	C1	C2
0.586	0.580	0.713	0.591	0.596	0.619	0.523	0.529	0.577	0.561

Table 7. The correlation between the total benefit of fresh agricultural products and each index.

	A1	A2	B1	B2	B3	B4	B5	B6	C1	C2
Correlation	0.751	0.910	0.864	0.884	0.786	0.901	0.710	−0.936	0.663	0.895

where the ε_i^j is the gray correlation coefficient, which indicates the degree of correlation of the index to the development level of fresh agricultural product logistics, ρ represents the resolution coefficient which reduces the influence of the maximum absolute difference and increases the difference of the correlation coefficient, and the between 0 and 1 This article takes $\rho = 0.5$. According to the above calculation results, it can be seen that $\Delta min = \min_j \min_i |r_i - r_{ij}| = 0$; $\Delta max = \max_j \max_i |r_i - r_{ij}| = 1$; $=1$; the correlation coefficient of each indicator can be calculated by placing it in the gray correlation coefficient formula (see Table 5).

3.2.5 Find the degree of gray correlation
The gray correlation degree represents the degree of correlation between the evaluated series and the reference series. The value ranges from 0 to 1. The larger the value, the greater the degree of correlation. At the e > 0.5 time, it shows that it has a strong correlation between the evaluation index and the development level of fresh agricultural products logistics. The calculation formula of the gray correlation degree is (4):

$$e = \frac{1}{n} \sum_{i=1}^{n} \varepsilon_i^j \qquad (4)$$

The correlation degree of each index is obtained by substituting the data (see Table 6).

3.2.6 Seek correlation
After obtaining the degree of correlation between each index and fresh agricultural products by using the gray correlation degree, in order to further explore the influence of the index and the government, on this basis, a multifactor correlation analysis was carried out. The analysis results are shown in Table 7.

3.3 Analysis of results
From Table 6, the correlation degree of the main influencing factors of fresh agricultural product logistics is greater than 0.5 or close to 0.5, indicating that the evaluation index has a strong correlation with the logistics level of fresh agricultural products, and again proves the scientific nature of

the index selection in this article. According to the results of the influencing factor index correlation degree of fresh agricultural products logistics level, the main factors affecting the logistics level of fresh agricultural products in Sichuan can be found. The article selects the top four factors of the logistics level index gray correlation value as the main factors.

3.3.1 *From the perspective of agricultural economic benefits*

It can be seen from Table 6 that the correlation between the total value of the agricultural industry and the annual income of rural residents is close, 0.586 and 0.581, respectively, with both being greater than 0.5, indicating that the two have a strong correlation with the logistics level of fresh agricultural products degree. At the same time, the correlation degree of the total value of the agricultural industry is slightly greater than the correlation degree of the annual income of rural residents. Therefore, to a certain extent, it can be considered that the total value of the agricultural industry has a greater impact on the logistics level of fresh agricultural products. Therefore more attention should be paid to the development of the total value of the agricultural industry.

3.3.2 *From the perspective of logistics quality factor analysis*

The logistics quality factor is the most direct influencing factor of the logistics level. From the index correlation analysis in Table 6, it can be seen that the gray correlation value of the fresh agricultural product logistics level index ranks in the top four of the number of trucks in highway operations, total logistics, and logistics. The total cost and postal business outlets, among which the number of trucks in highway operations has the largest correlation value, at 0.713, and the following four indicators are 0.591, 0.596, and 0.619, respectively. Therefore, the number of trucks in highway operations, total logistics, total logistics costs, and postal business outlets are the main factors affecting the logistics level of fresh agricultural products. The correlation degree of the other indicators is greater than 0.5, but less than these four indicators, which are secondary influencing factors.

3.3.3 *Analysis from the factor level of the industrial market structure level*

It can be seen from Table 6 that the gray correlation rank of the fresh agricultural product logistics level at the industrial market structure level is the per capita consumption of fresh agricultural products and the output of fresh agricultural products, respectively, at 0.577 and 0.561. Therefore, to a certain extent, the per capita consumption of fresh agricultural products in households is the main influencing factor, and the output of fresh agricultural products is the secondary influencing factor.

4 POLICY SUGGESTION

4.1 *Intensify the construction of infrastructure and accelerate the innovation of logistics technology*

With the rapid development of the Internet, increasing numbers of consumers prefer online shopping, and the logistics industry has developed accordingly. Fresh agricultural products are a special commodity with seasonality and logistics quality factors playing pivotal roles. At the basic level of logistics facilities, the state should earnestly investigate the specific conditions of the facilities and carry out targeted logistics infrastructure construction. The first aim is to continue to improve the connection system of the three logistics channels of "sea, land, and air," with the goal of solving the last mile problem and striving to achieve the best route and the lowest cost efficient logistics. The second aim is to invest in the construction of independent business outlets for fresh agricultural products. The third aim is to accelerate the pace of innovation in logistics technology, improve the level of cold chain logistics, increase investment in technology, reduce redundancy in personnel, and develop smart logistics.

4.2 Increase the output of fresh agricultural products and stimulate the income growth of rural residents

On the one hand, there should be a continuation of increased government investment in the development of fresh agricultural products, improvements in the agricultural production environment and agricultural production equipment, and expansion of the production scale of fresh agricultural products by increasing capital investment. On the other hand, by cultivating new agricultural business entities, the scale of agricultural production and operation should be gradually realized. At the same time, standardized management is implemented to create well-known brands of characteristic agricultural products to improve the output and quality of fresh agricultural products. The state vigorously encourages all provinces to develop fresh agricultural products suitable for local production, not only making full use of land resources, but also increasing farmers' enthusiasm for production. Profit from fresh agricultural products will increase farmers' income, increase the total value of the agricultural industry, and promote agricultural development.

4.3 Promote large-scale operations and adjust the industrial market structure

Although the rural household contract responsibility system has greatly mobilized farmers' enthusiasm for production and increased the output of agricultural products, the increase in the output value of agricultural products brought by it has gradually decreased over time. Moreover, the small-scale decentralized operation of agriculture results in decentralized logistics and large product quality differentiation, which is not conducive to the optimization of logistics resources and the satisfaction of consumer needs. Therefore, it is necessary to promote the large-scale operation of fresh agricultural products to realize the large-scale customization of consumer demand, continuously adjust the industrial market structure, achieve a balance of supply and demand to avoid a state of shortage of supply, stimulate agricultural development, and achieve the optimization of logistics efficiency and logistics resources.

4.4 Ensure the quality of fresh agricultural products and release consumption potential

The quality of fresh agricultural products directly affects consumers' desire to purchase again from the same source. In order to further release the consumption potential and stimulate the fresh agricultural product market, improving the quality of fresh agricultural products is a top priority. First, in order to protect the income of rural residents and maintain the enthusiasm of fresh agricultural product market suppliers, the government can provide relevant fresh agricultural product price subsidies to increase the transfer income of rural residents; the second method is to build a complete agricultural product information network platform, supervising the logistics status of fresh agricultural products through the collection, sorting, and analysis of fresh agricultural products logistics information, and providing farmers and logistics subjects with decision-making and optimized information, which is convenient for further reducing logistics costs and improving logistics quality.

ACKNOWLEDGMENT

This paper is supported by the Research Interest Cultivation Project of Sichuan Agricultural University.

REFERENCES

[1] Huang Fuhua, Jiang Xuelin. Research on the Factors Influencing the Logistics Efficiency of Fresh Agricultural Products and the Improvement Mode. Journal of Beijing Technology and Business University (Social Science Edition), 2017, 32(02): 40–49.

[2] Sichuan Provincial Development and Reform Commission and Sichuan Provincial Logistics Association.

[3] Huang Fuhua, Jiang Xuelin. Research on the Factors Influencing the Logistics Efficiency of Fresh Agricultural Products and the Improvement Mode. Journal of Beijing Technology and Business University (Social Science Edition), 2017, 32(02): 40–49.

[4] Gong Ruifeng, Wu Hailing, Peng Xiaohui. Research on the Factors Influencing the Logistics and Distribution Efficiency of Fresh Agricultural Products in Hunan Province. Journal of Hunan Institute of Humanities, Science and Technology, 2019, 36(02): 86–91.

[5] Liu Gang. Research on the logistics service innovation of fresh agricultural products e-commerce. Business Economics and Management, 2017(03): 12–19.

[6] Wang Cheng, Wang Tao, Jiang Yuansheng. Fresh agricultural product logistics level evaluation and development model selection in the western region. Soft Science, 2014, 28(02): 136–139+144.

[7] Sachin Jambhale, Sudhir Kumar, Sanjeev Kumar. Multi-response optimization of friction stir spot welded joint with gray relational analysis. Materials Today: Proceedings, 2020, 27(Pt 2).

[8] Zhu Jiaqi, Zhao Ruixue, Xu Wei, Ma Jing, Ning Xin, Ma Rong, Meng Fanling. Correlation between reticulum ribosome-binding protein 1 (RRBP1) overexpression and prognosis in cervical squamous cell carcinoma. Bioscience trends, 2020.

Computational Social Science – Luo, Ciurea & Kumar (eds)
© 2021 Taylor & Francis Group, London, ISBN 978-0-367-70193-2

An artificial intelligence-based method for judging the quality of spoken Chinese pronunciation

M.J. Guo & Z. Xue
School of Humanities, Xi'an Shiyou University, China

W.X. Gao
School of Electronic Engineering, Xi'an Shiyou University, China

ABSTRACT: Chinese has increasingly become a global language. Compared with the popular international languages such as English, French, and Spanish, traditional classroom teaching methods are often used in the promotion of Chinese. In particular, oral pronunciation requires face-to-face guidance from teachers, which is not conducive to students' self-study. In response to this problem, this paper proposes to use the most widely used deep convolutional neural network in artificial intelligence technology to judge the standard of the spoken pronunciation of non-Chinese students of native language, and according to the "Mandarin Test Grade Standard" jointly issued by the National Language and Writing Committee and other departments The Chinese phonetic standards in, automatically grade the correctness of students' spoken pronunciation. The activation function of the deep convolutional neural network(DCNN) uses the ReLU function, the loss function uses the cross-entropy cost function, and the training algorithm for the DCNN is also given here. Practical experiments on 100 speech fragments sent by 10 students show that the proposed DCNN is feasible and effective.

1 INSTRUCTIONS

Since Confucius Institute Headquarters (Hanban) started to promote Chinese language and Chinese culture abroad under the framework of global Confucius Institutes, Chinese language learning has increasingly become a popular choice for international students as a second language. In the traditional teaching of Chinese as a foreign language, oral training mostly adopts the classroom teaching model, led by the teacher, using traditional paper teaching materials, and conducting one-to-many or one-to-one training in a limited space and time. In recent years, the situation has changed. The spoken Chinese training system "AhaChinese Spoken Language Training System" jointly developed by the Foreign Research Institute and a commercial organization has appeared, but it still needs to be further improved in some details. To this end, in view of the current Chinese oral training teaching and the existing Chinese oral training system, improve the intelligence of the Chinese oral training system, and the introduction of artificial intelligence technology can improve the accuracy of automatically determining whether the voice is standard, which is beneficial to students. Self-evaluation of oral English learning under the supervision of the teacher.

The characteristics of Chinese speech, especially the pronunciation of students who have acquired Chinese as a second language, are difficult to describe accurately in mathematical language. The rapid development of artificial intelligence technology provides an effective technical means for automatically determining whether the speech is standard. Therefore, this paper proposes to use the most widely used deep convolutional neural networks (DCNN) model in artificial intelligence technology for the recognition and training of spoken Chinese.

Regarding the application of artificial intelligence in recognition, many scholars have conducted related research. Literature[1] proposes a layer-by-layer training method with supervision, which

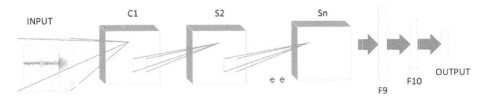

Figure 1. Schematic diagram of DCNN structure.

solves the training problem of neural network by expanding the number of hidden layers. The concept of deep learning was born, and the research on artificial intelligence has also set off an upsurge. Research on neural networks such as convolutional neural networks[2–5] and deep belief networks[6–8] has greatly advanced the understanding of artificial intelligence. Literature[9] proposed an adaptive weighted pooling algorithm for convolutional neural networks, which optimized the feature extraction of the sampling model. Literature[10] improved the DCNN structure based on typical CUDA-CONVNET and improved recognition accuracy. Literature[11] designed an 11-layer DCNN classification model, and gave an adaptive weighted pooling algorithm based on the convolutional neural network, which improved the recognition accuracy and speed of the network.

It can be found from the existing research that few scholars pay attention to the automatic judgment of the spoken speech quality of non-native Chinese students. Based on the existing research, this paper provides a DCNN model and specific parameters suitable for judgment of the spoken Chinese speech quality, Real Experiments show that the proposed model structure is effective.

2 NEURAL NETWORK STRUCTURE DESIGN

2.1 *DCNN structure*

DCNN is a neural network suitable for processing high-dimensional grid structure data, and the waveform of the speech signal has the characteristics of high-dimensional grid structure data. From the structural analysis, DCNN is composed of multiple two-dimensional planes composed of independent neurons. The input data is generally processed by a convolution layer, a sub-sampling layer, and a fully-connection layer to obtain the required recognition result. The convolutional layer and down-sampling layer in DCNN constitute the extraction of speech features. A convolutional layer of DCNN generally includes several feature maps. The neuron weights of the same feature map are shared by the convolution kernel. The sub-sampling forms generally include mean pooling and maximum values. Sampling (max pooling) two forms. The number of convolutional layers in the DCNN model should be determined according to factors such as training data, activation functions, and gradient update algorithms. Chinese speech waveforms are quite different and involve a variety of feature values, so this article sets the convolutional sampling layers to be less than 4 layers. Taking into account that the error will be reduced layer by layer when the error is passed back in the training, too many fully connected layers will affect the effective transmission of errors, so the fully connected layer is set to 2 layers, as shown in Figure 1.

2.2 *Activation function selection and training samples*

The convolutional layer can extract the tiny features of the image, and then combine the local features to form the overall features. The down-sampling layer between the convolutional layers mainly performs down-sampling operations on the convolution results, which preserves useful information while reducing the amount of calculated data and preventing data overfitting. Convolution operation is essentially a linear operation. For linear inseparable data, it needs to be mapped into a high-dimensional space through a nonlinear transformation to achieve accurate data segmentation. The activation function is a means of introducing nonlinearity. There are two commonly used activation

29

Table 1. Comparison table of activation functions.

Function name	analyzing way	Pros and cons
Sigmoid	$f(x) = \frac{1}{1+e^{-x}}$	Advantages: the derivative calculation is simple, and the gradient descent algorithm is easy to implement. Disadvantage: As long as the slope of the function is larger when the independent variable x is near 0, adjusting the parameters according to the gradient will have a better effect.
ReLU	$f(x) = \max(0, x)$	Advantages: The value range is $[0, \infty]$, there is a larger mapping space, and the derivative is relatively simple. When x is greater than 0, the derivative is always 1, completely avoiding the problem of vanishing gradient. Insufficiency: When the value of x is negative, it is equivalent to directly closing the node.

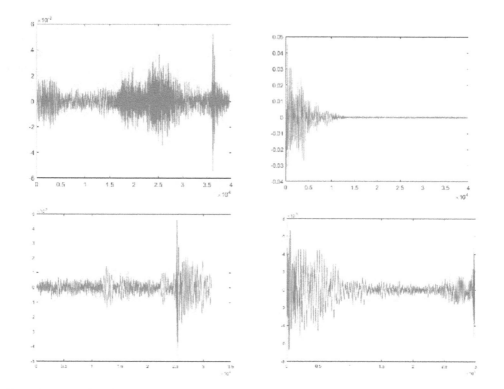

Figure 2. Examples of voice data.

functions, Sigmoid and ReLU. The comparison of Sigmoid and ReLU activation functions is shown in Table 1.

From the comparison in Table 1, it can be found that the derivative of the ReLU function is always 1, and when the input data is complex, it can ensure that the parameters can continue to converge, and the convergence speed is fast. Therefore, ReLU function is more suitable for pronunciation classification in DCNN.

We have established more than 400 hours of Chinese language training phonetic database and calibrated it. Part of the phonetic database data is shown in Figure 2.

Table 2. Comparison of different loss function.

Function name	analyzing way	Features
L1 loss function	$loss = \sum\limits_{i=1}^{N} \lvert y_i - a_i \rvert$	It is not smooth at the zero point, and is seldom used.
L2 loss function	$loss = \sum\limits_{i=1}^{N} (y_i - a_i)^2$	In some cases, explosion gradient can be prevented, but the robustness is weak
Cross entropy cost function	$C = -\dfrac{1}{N} \sum\limits_{i=1}^{N} \left[\begin{array}{l} y_i \ln a_i + \\ (1 - y_i) \ln(1 - a_i) \end{array} \right]$	It is often used in classification problems, and is particularly effective for unbalanced training sets.

2.3 Determine the loss function

The loss function is to calculate the difference between the calculation result of the DCNN and the true value, which determines the changing direction of each weight. The comparison of different loss function is in Table 2.

Since it is not easy to achieve a complete balance of various samples in the constructed speech database, the paper chooses to use the cross-entropy cost function. The neural network training process is as follows:

(1) Initialize network weights with random values;
(2) Input voice samples and calculate the predicted value of output;
(3) Calculate the difference between the calculated value and the true value by using loss function;
(4) Find partial derivative of the loss function for each weight value, and modify the weight value accordingly;
(5) Repeat (2) ∼ (4) until the loss function value is less than the preset value.

3 EXPERIMENT SETUP AND ANALYSIS

The training and prediction of the DCNN model in this paper is carried out under Google's open-source TensorFlow deep learning framework. The specific training scheme is shown in Figure 3.

The regularizer parameter of the convolutional neural network model is 0.0001, and the learning step size α is 0.0001. The optimizer uses the adaptive moment estimation (Adam, adaptive moment estimation) algorithm.

The specific structure of the DCNN for Chinese spoken speech quality recognition is shown in Figure 4. In Figure 4, C1, C3, C5, and C7 are convolutional layers, S2, S4, S6, and S8 are pooling layers, F9 and F10 are fully connected layers. A Dropout layer is added after the fully connected layer, which can improve the generalization and the degree of over-fitting of the network. Dropout layer can reduce the degree of under-fitting. In this paper, the output of Chinese oral training is further refined into three states of "repeat", "good" and "great" according to the standard of Mandarin test.[12] So the number of output neuron is 3, which corresponds to the three states respectively.

Convolutional layer of the DCNN model uses zero padding to prevent the reduction of sound graphics during the convolution operation. Therefore, after convolution, the data length and width remain unchanged, and the depth increases. The pooling layer has no zero padding operation. After pooling, the length and width of the sound graph are reduced, and the depth remains unchanged.

After training, the neural network tested whether the spoken language of 10 international students was standard. Each student tested 10 sentences. The confusion matrix of the experiment is shown in Table 3.

It can be found that the accuracy of the test can meet the needs of automatic assessment of spoken Chinese and guide students to practice.

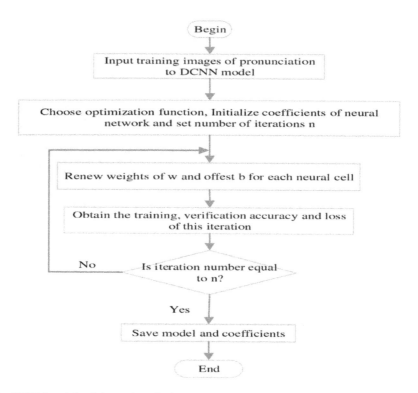

Figure 3. DCNN model training and prediction program process.

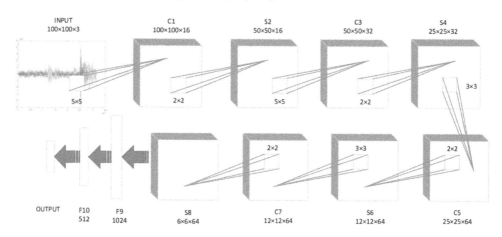

Figure 4. Convolutional neural network structure.

Table 3. Confusion matrix of experiment.

	Test Results: repeat	Test Results: good	Test Results: great	Accuracy
Should be repeat	27	2	1	90%
Should be good	3	36	1	90%
Should be great	0	2	28	93.3%

4 CONCLUSION

(1) DCNN can effectively judge the quality of spoken Chinese speech. When the number of convolution and pooling layers is 3, it can reduce the amount of data input to the fully connected layer and reduce the computational workload while ensuring the recognition accuracy.

(2) The voice image can be used as a basis for identifying the pronunciation quality of spoken language, and the high-accuracy automatic judgment of the voice image can be achieved by training DCNN.

(3) Although the current accuracy rate can guide students' self-learning, it is not enough to replace manual grading completely. In order to improve the recognition accuracy, further research should be conducted on the selection of the size of the convolutional template and the pooling template. Also, the depth of the convolutional layer should be studied further.

ACKNOWLEDGEMENT

This article is for the 2019 Xi'an Shiyou University *Postgraduate Education and Teaching Reform Research Project: Research on the Innovation of Teaching Mode and Quality Evaluation System of the Course "Poetry Recitation and Creation" in Chinese International Communication.*

REFERENCES

[1] Hinton G E, Salakhutdinov R R. Reducing the dimensionality of data with neural networks. Science. 2006. 313(5786): 504–507.

[2] Nguyen C K, Nguyen C T, Masaki N. Tens of thousands of nom character recognition by deep convolution neural networks. 4th International Workshop on Historical Document Imaging and Processing, HIP&#65533; 2017, November 10, 2017–November 11, 2017. Association for Computing Machinery, 2017:37–41.

[3] Shijie J, Ping W, Peiyi J, et al. Research on data augmentation for image classification based on convolution neural networks. 2017 Chinese Automation Congress, CAC 2017, October 20, 2017–October 22, 2017. Institute of Electrical and Electronics Engineers Inc., 2017:4165–4170.

[4] Savchenko A V, Belova N S, Savchenko L V. Fuzzy Analysis and Deep Convolution Neural Networks in Still-to-video Recognition. Optical Memory and Neural Networks (Information Optics). 2018. 27(1): 23–31.

[5] Vu M-T, Beurton-Aimar M, Le V-L. Heritage image classification by convolution neural networks. 1st International Conference on Multimedia Analysis and Pattern Recognition, MAPR 2018, April 5, 2018–April 6, 2018. Institute of Electrical and Electronics Engineers Inc., 2018:1–6.

[6] Chen Y, Zhao X, Jia X. Spectral-Spatial Classification of Hyperspectral Data Based on Deep Belief Network. IEEE Journal of Selected Topics in Applied Earth Observations and Remote Sensing. 2015. 8(6): 2381–2392.

[7] Zhang T, Zhao J, Luo J, et al. Deep belief network for lung nodules diagnosed in CT imaging. International Journal of Performability Engineering. 2017. 13(8): 1358–1370.

[8] Mughees A, Tao L. Multiple deep-belief-network-based spectral-spatial classification of hyperspectral images. Tsinghua Science and Technology. 2019. 24(2): 183–194.

[9] Yao Minghai, Yuan Hui. Surface defect detection based on optimized convolutional neural network. High Technology Letters. 2019. 29(06): 564–569.

[10] Liu Meng-xi, Ju Yong-feng, Gao Wei-xin, et al. Research on X-ray weld defects detection by deep CNN. Transducer and Microsystem Technologies. 2018. 37(05): 37–39+43.

[11] Zhang Z, Wen G, Chen S. Weld image deep learning-based on-line defects detection using convolutional neural networks for Al alloy in robotic arc welding. Elsevier Ltd. 2019. 45

[12] Standard of Mandarin test level (Trial), Ministry of Human Resources and Social Security of the People's Republic of China. May 12, 1999.

Computational Social Science – Luo, Ciurea & Kumar (eds)
© *2021 Taylor & Francis Group, London, ISBN 978-0-367-70193-2*

Demand prediction for e-commerce multi-warehouse based on machine learning

J.H. Du
Management Science and Engineering, Chongqing University, Chongqing, China

X. Wang*
Chongqing University, Chongqing, China

X.Y. Liu
Logistics Engineering, Chongqing University, Chongqing, China

ABSTRACT: Since the prosperity of mobile communication and Internet, E-commerce is booming and it has promoted the explosive growth of express parcel volume. At present, parcel delivery is faced with common problems for E-commerce companies, especially customers have higher requirements for timeliness. Considering that the parcel delivery industry is faced with the characteristics of diversified customer sources and consumer personalized demand, the multi-warehouse inventory can help E-commerce platforms effectively shorten the delivery time and reduce logistics costs by pre-allocating commodity inventory. This paper designs a prediction model of commodity demand based on customer behavior to support the strategy of "commodity first, order later to". Then, it applies data exploration and feature engineering to preprocesses the massive historical order data stored by the e-commerce platform. Finally, the machine learning algorithm of GBDT is applied to analyze the historical order data, to predict and obtain the stock scheme of multi-warehouse for E-commerce platform.

Keywords: Demand prediction, E-commerce, Machine learning, Multi-warehouse inventory.

1 INTRODUCTION

In the era of e-commerce, according to the Report on China Express Development Index 2019 issued by State Post Bureau of The People's Republic of China, the express industry has supported amount of the online retail sales package more than 60 billion (State Post Bureau, 2019) rose from 40 billion in 2017. This has significantly increased commercial vehicle movement in urban areas. Customers have higher requirements on the service quality, effectiveness and timeliness of E-commerce. According to the available data, it can be found that there are fewer and fewer days when the number of signed orders is over 100 million for double 11 shopping carnival. In 2013, it took nine days for more than 100 million packages to be signed. The time was then shortened from six days in 2014 to four days in 2015, 3.6 days in 2016 and 2.8 days in 2018 (Ma, 2018). In order to improve customer satisfaction and loyalty, e-commerce platforms are faced with cost problems by meeting customer requirements and expectations.

Demand forecast (Ren et al., 2019) and sub-warehouse inventory can improve the vehicle loading rate from a central warehouse to sub-warehouse. As a result of multi-warehouse inventory is preconfigured, there is less time from order submission to product acquisition for customers. Demand forecast is the basis of sales plan, replenishment plan and production plan. In the logistics system,

---

*Corresponding author

Figure 1.   Consumer transformation model.

demand forecast can help logistics enterprises rationally allocate resources and make scheduling plans, as well as better control the trend and direction of the future logistics market. Therefore, this research can expand the application of machine learning & prediction and provide optimization solutions for e-commerce platforms.

2   LITERATURE REVIEW

Many scholars have studied the data prediction, and the prediction model mainly adopts the time series analysis method and the causal analysis and a prediction method. ARIMA model is a typical example. It is applied to analyze the time series data for forecasting the spread of the epidemic (Sahai et al., 2020). The principle of ARIMA is that the non-stationary time series is transformed into stationary time series by differential treatment, and then the regression prediction is carried out by model recognition and model order fixing. Ott et al. (2013) designed an evaluation framework for the demand forecasting problem, which consists of statistical indicators and scoring models, allowing analysis of prediction accuracy at different levels. YAO. (1999) reviewed the different combinations of artificial neural networks and evolutionary algorithms, and finally proposed a prediction model combining ANNs and EAs. In the manufacturing industry, Gao et al. (2012) improved GP algorithm to accurately predict the order volume of mechanical products. Thomassey and Happiette (2007) proposed a theory of clustering based on neural networks in the absence of historical data. Zeyu et al. (2020) studied the prediction of customer's re-shopping behavior on the E-commerce platform. In 2015, Alibaba Group ever hold a similar contest based in Tianchi big data, and this paper is also Based on the open source data of Tianchi Big data platform. However, there are few studies on the prediction of regional sales quantity and transform it into distribution volume. Feng et al. (2017) studied the issue of preventive transhipment in the multi-inventory system, proposed horizontal transhipment (LT) and emergency order (EO). The markov decision process was used to describe the issue and found out the timing and quantity of the preventive transhipment decision. Therefore, this research helps enterprises to identify commodities that customers are interested in, help enterprises to identify high-value customers and excavate hidden users. At the same time, it helps E-commerce platform to better carry out the strategy of "commodity first, order later to" and obtain the delivery quantity to allocate resources such as vehicles and human resources in advance.

3   METHODOLOGY

3.1   *Customer demand transformation model*

From relevant characteristics of commodity granularity, many consumer behavior can be extracted, which shows that consumer purchase behavior includes economic and psychological factors. To explain the consumer behavior, this paper builds a consumer funnel mode (shown in Figure 1) based the perspective of sales and existence of the classical theory. Generally speaking, customer behavior on the internet starts from the consumer requirement, then browse, select and purchase action. Therefore, in E-commerce enterprises, the historical data relevant to customer behavior can be applied to predict the production demand quantity, which plays a huge economic value.

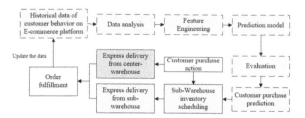

Figure 2.   Diagram of E-commerce demand prediction.

3.2   Flow of prediction model

The flow of Prediction model is designed to describe the prediction actions, as shown in Figure 2. The steps of the model are detailed in Chapter 4. For the predicted high-potential buyers, the products are delivered in advance according to their geographical location, which can improve the timeliness and enhance customer satisfaction. Based on the prediction, these goods are distributed from sub-warehouses to customers, rather than a center-warehouse, when users submit the purchase orders. What's more, the distribution resources can be optimized when a customer takes the action of purchase to reduce delivery cost.

3.3   GBDT algorithm

In the process of training, the model's Boosting will pay more attention to the samples with the wrong classification last time, and give them higher weight to the samples with the wrong classification, to make it easier to identify misclassified samples. For each newly added weak learner in training, a certain weight will be assigned to it, to obtain the final strong classifier. For regression prediction, the algorithm is to fit the residual of the current model, and the algorithm flow is as follows.

Input: Training data set

$$T = \{(x_1, y_1), (x_2, y_2), \ldots, (x_N, y_N)\}, x_i \in \chi \subset R^n, y_i \in \lambda \subset R$$

Output: Boosting decision tree

First, Initialization $f_0(x) = 0$. When $m = 1, 2, \ldots, M, f_0(x) = 0, f_M(x) = \sum_{m=1}^{M} T(x; \Theta_m), f_m(x) = f_{m-1}(x) + T(x; \Theta_m), m = 1, 2, \ldots, M$. Second, at the mth step of the algorithm, Θ_m is computed based on the current model $f_{m-1}(x)$. $\Theta_m = art \min \sum_{i=1}^{N} L(y_i, f_{m-1}(x_i) + T(x_i; \Theta_m))$. Third, parameters for the mth tree is obtained. When the square error is taken, $L(y, f(x)) = (y - f(x))^2$, its losses can be translated into L.

$L(y, f_{m-1}(x) + T(x; \Theta_m)) = [y - f_{m-1}(x) - T(x; \Theta_m)]^2 = [r - T(x; \Theta_m)]^2$.
Finally, Get a regression tree, $f_M(x) = \sum_{m=1}^{M} T(x; \Theta_m)$

4   DEMAND PREDICTION

When customers browse, add a shopping cart and comment on the E-commerce platform, they will leave a record in the background. Therefore, the historical behavior data recorded can be applied to build model features to dig out the purchasing habits of users. In this way, the purchase tendency of users in the next stage can be predicted. At the same time, the order prediction can help enterprises to identify the commodities that customers are more interested in, enhance enterprises to identify high-value customers and dig hidden users.

4.1   Preparing data and customer analysis

The data used in this section is from the Tianchi Competition Data is the background history of the E-commerce platform. In order to protect the information security of customers, the actual sales

Table 1. Relevant characteristics of commodity granularity.

Attribute	Type	Implication	Sample	Attribute	Type	Implication	Sample
date	bigint	Date	20150912	amt_alipay	Double	Trading volume	0
item_id	bigint	SKU ID	132	num_alipay	bigint	Trading orders	0
cate_id	bigint	Leaf ID	18	qty_alipay	bigint	Trading quantity	0
cate_level_id	bigint	Branch ID	12	unum_alipay	bigint	Trading people number	0
brand_id	bigint	Brand ID	203	ztc_pv_ipv	bigint	Times of train guided	0
supplier_id	bigint	Supplier ID	1976	tbk_pv_ipv	bigint	Times of taobao guided	0
pv_ipv	bigint	Browning times	2	ss_pv_ipv	bigint	Times of brown guided	0
pv_uv	bigint	Flow UV	2	jhs_pv_ipv	bigint	Times of JHS guided	0
cart_ipv	bigint	Added purchase times	0	ztc_pv_uv	bigint	Number of train guided	0
cart_uv	bigint	Added purchase count	0	tbk_pv_uv	bigint	Number of taobao guided	0
collect_uv	bigint	Favorites count	0	ss_pv_uv	bigint	Number of brown guided	0
num_gmv	bigint	Purchase orders	0	jhs_pv_uv	bigint	Number of JHS guided	0
amt_gmv	Double	Purchase payment	0	num_alipay_njhs	bigint	Non-JHS orders	0
qty_gmv	bigint	Purchase quantity	0	amt_alipay_njhs	Double	Non-JHS payment	0
unum_gmv	bigint	Purchase UV	0	qty_alipay_njhs	bigint	Non-JHS quantity	0

Table 2. Replenishment cost table.

Attribute	Type	Implication	Sample
item_id	[removed]	Product ID	333442
store_code	String	Sub-warehouse CODE, If it's the national average cost, it is all.	1
a_b	String	The replenishment cost is connected by '_'. i.e. out of stock_ surplus cost	10.44_20.88

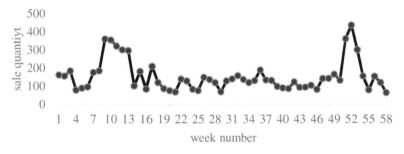

Figure 3. Time series of goods.

volume, cost and page views are generated through certain desensitization processing. The data is described as follows (Table 1):

(1) Item_feature: Commodity granularity characteristics
(2) config: The cost per unit of each item being out of stock and surplus in the national and warehouse area is shown in Table 2.

Based on the user behaviors on the E-commerce platform, the demand prediction model of users is constructed through sample construction and relevant data exploration, to obtain the distribution demand of each region. The original data covers a total of 30 influencing factors and thousands of commodities, including 864,773 pieces of data. The weekly time series diagram of the period from 2014/10/10 to 2015/12/27 is summarized under the condition of commodity date. From Figure 3, it can be found that there is no obvious trend or periodicity in the sales volume of goods, with strong randomness, and the sales volume will increase sharply near double 11 shopping carnival. This

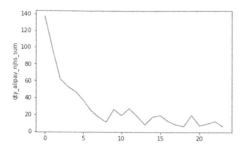

Figure 4. Number of payments.

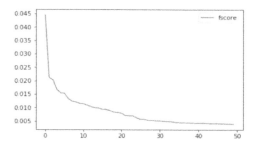

Figure 5. Importance of features.

section mainly studies the distribution demand of goods in the partition. Because of the wide variety of commodities, this research randomly selects a commodity with the number 582 to observe the sales situation at the time series. Because the sales volume during the Double 11 Shopping Carnival will seriously skew the model, and the sales data of the Double 11 Shopping Carnival will be directly eliminated in the process of data cleaning, see Figure 4.

4.2 Characteristic processing

In order to more accurately predict whether consumers will submit orders to buy goods in the following period, data processing should be done for the data based on feature engineering. Amanda Casari pointed out in Feature Engineering that the core of Feature Engineering is Feature processing, including Feature cleaning and Feature conversion. The construction of features not only needs to be combined with business rules, but also needs to consider the preprocessing of data.

This task is a kind of prediction based on time series, and this research needs to pay attention to the historical features, time features and time window features during processing the features. The relevant characteristics of the data are processed as follows:

(1) Store information coding

From the date exploration, the column qty alipay_njhs of data contains a lot of information. It is structurally decomposed and reorganized to adopt brand, supplier and category, and then such characteristics are reorganized and merged again. The combination result, such as cate_id_qty_alipay_njhsbrand_id_qty_alipay_njhssupplier_id_qty_alipay_njhs etc., are obtained.

(2) Commodity classification characteristics

Considering that there is no comparability between commodity categories, such as category ID, warehouse CODE, etc., these numerical data have no meaning of size and need to be redefined. The one_hot alone encoding is applied to do it. For example, The warehouse CODE is 1,2,3,4,5, all, in which region 1 is coded as {1, 0, 0, 0, 0}. Region 2 is coded as {0, 1, 0, 0, 0}.

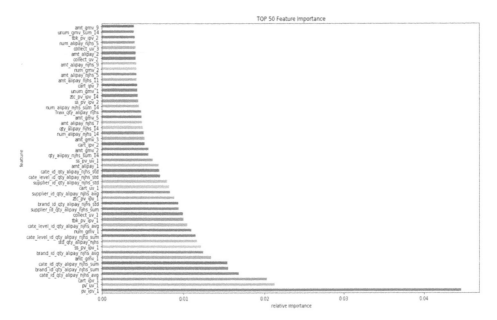

Figure 6.   Sorting of feature importance.

(3) Abnormal value handling

For the original data, there is a lot of noise data. The outlier point is directly deleted. For the missing values, this is filled in with −1.

(4) Window feature structure

Time window sliding feature construction: Based on SKU ID, Leaf ID, Branch ID, Brand ID, Supplier ID, this research extract last X days data relevant to Non-JHS orders, Non-JHS payment and Non-JHS quantity.

(5) Commodity characteristic structure

Eigenvalue construction is to create information that does not exist in the original data, part of which comes from feature reorganization and part from business scenarios. After a series of feature transformation and feature construction, 419 features are finally constructed, but there are only 30 features in the original data. Through the output module of feature importance of the model (Figure 5), the top 50 relatively important features are selected and output. From Sorting chart of feature importance (Figure 6), the feature importance value of pv_ipv is 0.044556, which shows the number of browse is the most important characteristic indicator. The feature importance value of pu_uv is 0.021264, that is to say, pu_uv is the second important characteristic indicator. Feature construction is carried out based on the actual meaning of the scene, where the discount rate (transaction price/auction price) and the average transaction price of commodities are constructed.

4.3   Comparison of prediction models

After the relevant processing of features is completed, the data is divided into a training set and a verification set. The data are respectively brought to various machine learning models, and the appropriate machine learning model is selected through the evaluation criteria of the model.

In this section, three machine learning models, XGboost, Random Forest and GBDT, were used to test respectively according to the task of distribution quantity. By comparing the replenishment

Table 3. XGBoost model parameters.

n estimators	eta	max_depth	subsample	colsample_bytree	seed
500	0.025	7	0.7	0.7	40

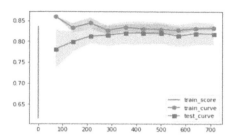

Figure 7.   Learning curve.

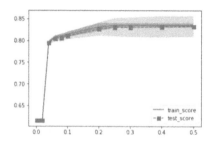

Figure 8.   Validation curve.

cost calculated by corresponding models, GBDT algorithm with the lowest replenishment cost was selected as the final prediction model. The cost is calculated as follows

T_i: National target inventory for product i.
T_{ia}: Target inventory for goods i in sub-warehouse a.
D_i: The real sale quantity of product i in future totally.
D_{ia}: The real sale quantity of product I in sub-warehouse a.
A_i: The replenishment cost of product i, when it is out of stock.
A_{ia}: The replenishment cost of product i, when it is out of stock in sub-warehouse a.
B_i: The inventory cost of product i, when it is left.
B_{ia}: The inventory cost of product i, when it is left in sub-warehouse a.

$$C_R = \sum_{ia} [A_{ia} * MAX(D_{ia} - T_{ia}, 0) + B_{ia} * MAX(T_{ia} - D_{ia}, 0)]$$

First, the XGboost model is used to make predictions. During the prediction process, this study used the grid search CV in Scikit-Learn to adjust the model parameters. Next, try roughing adjustments, and then modify the model slightly. The final model parameters are as follows.

After determining the model parameters, the learning curve and validation curve in Scikit-Learn are used to test the effectiveness of the predictive model. The Learning Curve (Figure 7) can be understood as a reasonable test of model algorithms as the training set increases, in order to help users determine whether the model algorithm is biased or variance problem. The Validation Curve (Figure 8) can be understood as accuracy by examining the different parameters of the model to facilitate our judgment of the overfitting and underfitting of the model.

Table 4. Random Forest model parameters.

n estimators	max_depth	max_features	random_state	min_samples_leaf
500	7	0.8	1024	5

Table 5. GBDT model parameters.

n estimators	max_depth	max_features	subsample	loss	learning_rate
400	6	0.75	0.75	lad	0.01

Table 6. Model replenishment cost and operation time.

Prediction algorithm	Replenishment cost	Run time
XGboost	62830.75	110s
Random Forest	62333.72	79s
GBDT	59276.05	128s

Table 7. Distribution volume in area 1.

item_id	store_code	pred	item_id	store_code	pred
124614	1	38.11	149632	1	26.37
20273	1	32.26	85023	1	26.48
55977	1	32.12	16859	1	34.16
145958	1	23.77	102651	1	22.99
85200	1	17.46	60416	1	14.13
26678	1	28.81	136718	1	29.09
85781	1	22.14	110763	1	28.01
107942	1	30.06	40652	1	18.02
153895	1	37.09	15807	1	6.94
17372	1	13.21	117618	1	28.12
142687	1	26.64	74547	1	19.51
113808	1	10.00	152269	1	22.21
32084	1	12.13	94239	1	30.57
164234	1	12.69	98949	1	23.39
33131	1	21.93	162430	1	38.93
40593	1	21.33	109701	1	21.69
2214	1	22.31	124067	1	21.63
120246	1	29.25	147118	1	4.77
45378	1	24.42	96500	1	10.41
...	...	...	...	...	...

On contrast, Random Forrest and Gradient Lift GBDT are also applies to predict for customer demand forecasting, with the following parameters of the model (Tables 4 and 5).

For the distribution volume results of the sub-warehouse, model comparison is taken here, and the better model is selected by restocking cost and the running time of the model, as shown in Table 6.

In order to pursue the accuracy of prediction, the gradient lift GBDT is selected as the prediction model, and the prediction results include the regional distribution volume. The results of the partial warehouse distribution in Area 1 are presented as a case (Table 7).

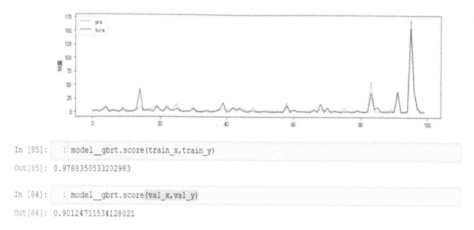

In [85]:   1 model__gbrt.score(train_x,train_y)

Out[85]: 0.9788350533202983

In [84]:   1 model__gbrt.score(val_x,val_y)

Out[84]: 0.90124711534128021

Figure 9.   Model score function.

From the forecast results, we can see the specific sales volume of each item in Region 1. Due to the desensitization of the data, Commodity ID is illustrated here as an example. Specifically, the sale quantity of Commodity ID 124614 in the area 1 is 39 and that of Commodity ID 20273 is 33. The forecast means that the total demand for a single item in the region has been informed in advance. The prediction result can help E-commerce and logistics companies plan the allocation of resources.

4.4 *Model evaluation*

In this paper, the most classical evaluation index Standard Deviation in the regression problem is used to measure the effectiveness of the models. The smaller its value is, the smaller the difference between real value and prediction value, which means the accuracy is good.

From the scoring function of the model (Figure 9), it shows that the model has 97% accuracy in the training set and 90% accuracy on the validation set. The model has not ever had a fitting and underfitting situation. Therefore, the accuracy of the prediction can be considered to be acceptable.

4.5 *Delivery quantity prediction*

After forecasting the volume of open positions, this paper applies the geographical location information recorded by the e-commerce platform to builds the regional distribution volume.

There are some assumptions about the model. sku. (1) The user's geographical location is known; (2) The constructed customers are the user groups; (3) The predicted regional total quantity of individual goods are converted into the delivery quantity of user groups in a certain proportion.

In this paper, the customer groups benchmark used by VRP researchers is used as a reference for the prediction area, according to assumption 1. The 30 user groups were selected as the customers in the prediction area and its location is read by DataMap in Microsoft Office 2016 Excel. Then the customer groups were divided according to the specific geographical location. Although some information is desensitized in the prediction model, such as SKU ID and branch ID, it does not affect the total amount of distribution. The distribution volume of user group is formed by proportional transformation. Taking User group 1 as an example, if the value of proportional transformation is 0.1 and the forecast quantity for SKU ID δ_i is β_i.

(1) User group i needs SKU ID, including $\delta_1, \delta_2, \ldots, \delta_m$; (2) Delivery quantity $\varepsilon_i = \sum_{i=1}^{m} \alpha \beta_i$.

The forecast results can guide the e-commerce platform to deliver goods from the central warehouse to the sub-warehouse in advance, and allocate resources such as vehicles and human

Table 8. Delivery quantity of user group.

User group	x	y	Delivery quantity	User group	x	y	Delivery quantity
1	1300	560	40	16	1150	1250	30
2	1480	600	35	17	1300	1450	45
3	850	790	25	18	1300	1150	40
4	950	520	30	19	1580	1160	50
5	1000	350	40	20	1640	1000	40
6	1410	350	35	21	1920	950	30
7	740	520	40	22	1880	1200	25
8	1000	150	35	23	2200	1390	65
9	400	180	55	24	2150	1120	30
10	500	410	45	25	2250	790	30
11	480	700	60	26	1950	700	25
12	750	1000	50	27	2260	540	30
13	1020	1180	35	28	2580	680	70
14	550	1180	50	29	2360	1000	25
15	690	1480	60	30	2550	1150	40

resources to the sub-warehouse according to the delivery quantity. The results of the delivery quantity conversion of all user groups are shown in the following Table 8.

5 CONCLUSION

In this paper, the customer transformation model promotes the connection and relationship between customer behavior and purchasing goods. The flow of the prediction model shows the process of data processing, which makes the machine learning process cleaner. This research compares three different prediction algorithms and finds the GBDT is most suitable for the demand forecast for the E-commerce platform among them. The original data of e-commerce is converted into feature data that can be recognized and learned by the model. The recombination coding, one-hot coding and sliding coding of time window are adopted. GBDT algorithm is not only able to solve classification, regression and sorting problems effectively, but also has a great advantage that compared with the traditional algorithm. It is able to fit the residual error of the model for many times, and at the same time, it uses the gradient descent method to optimize, which not only improves the accuracy rate but also speeds up the optimization solution of the model. The prediction model of commodity demand based on customer behavior can help E-commerce platform to carry out better the strategy of "commodity first, order later to" and allocate resources such as vehicles and human resources to the sub-warehouse to reduce the cost and face challenges.

ACKNOWLEDGEMENT

This work is supported by the National K&D Program of China (Project No. 2018YFB1403602), and the Fundamental Research Funds for the Central Universities (Project No. 2020CDCGJX019).

REFERENCES

Feng, P., Fung, R. Y. K. & Wu, F. 2017. Preventive transshipment decisions in a multi-location inventory system with dynamic approach. Computers & Industrial Engineering, 104, 1–8.
Gao, L., Xu, S. H. & Ball, M. O. 2012. Managing an Available-to-Promise Assembly System with Dynamic Short-Term Pseudo-Order Forecast. Management Science, 58, 770–790.
Ma, H. 2018. Tmall "double 11" logistics distribution speed up again [Online]. Available: https://news.ycwb.com/2018-11/13/content_30131923.htm [Accessed November 13 2018].

Ott, K., Mensendiek, A. & Gmeinwieser, K. 2013. A framework for economic demand forecast evaluation. OR Insight, 26, 203–218.

Ren, S., Chan, H.-L. & Siqin, T. 2019. Demand forecasting in retail operations for fashionable products: methods, practices, and real case study. Annals of Operations Research, 291, 761–777.

Sahai, A. K., Rath, N., Sood, V. & Singh, M. P. 2020. ARIMA modelling & forecasting of COVID-19 in top five affected countries. Diabetes Metab Syndr, 14, 1419–1427.

State Post Bureau, C. 2019. Report on China express development index 2019 [Online]. Available: http://www.spb.gov.cn/sj/zgkdfzzs/ [Accessed March 27 2020].

Yao., X. 1999. Evolving artificial neural networks. PROCEEDINGS OF THE IEEE, 87, 1423–1447.

Zeyu., L., Jixuan., L., Rujian., C. & DONGMING, C. 2020. Research on prediction of re-shopping behavior of E-commerce customer. Computer Science, 1, 424–428.

Computational Social Science – Luo, Ciurea & Kumar (eds)
© *2021 Taylor & Francis Group, London, ISBN 978-0-367-70193-2*

Design of a depression intervention model: A computational social science approach

J. Zhao, Y. Zhao & Y.J. Ruan
School of Public Policy & Management, Anhui Jianzhu University, Hefei, China

ABSTRACT: Aiming at the serious harm of depression, a depression intervention model from the perspective of computational social science is proposed. The proposed model is based on the social media data of users. The LSTM model was used to analyze the potential depressive tendencies of users, while the psychiatric PHQ-9 depression scale was employed as a tool to diagnose depression and measure depression severity. The LDA model was utilized to identify the topics that users are concerned about, and these topics were considered important sources of their depressive moods. Subsequently, corresponding book prescriptions were provided for the implementation of bibliotherapy to carry out sociological intervention. A quantitative evaluation system for a depression intervention model from the perspective of computational social science was established to an extent in the present study, which promotes the standardization of depression interventions. This system would significantly reduce the consumption of health resources, with real social and economic benefits.

1 INTRODUCTION

The industrialization and urbanization of society in the current economic and social transition period has created in China a basis for depression at the level of "social structure," and the number of pan-depressed people has exceeded 95 million[1]. Depression has reached second among the list of diseases affecting humans, with the first place being held by heart diseases. The self-harm, suicide, and other behaviors of depression patients cause great suffering to themselves and their families, and also bring social harm, which has become a social problem that cannot be ignored. Although the sociological research on depression has been discussed to some extent, the social basis of depression is mainly discusses. There are few studies on the implementation of early interventions for depression from the perspective of sociology. The reason for this is that depression is insidious and easily overlooked and misdiagnosed, and the stigma of patients further aggravates a reduction in the rate of medical treatment. The above reasons have hindered the sociological research into depression. The development of network technology has changed the social form of human existence: People do not only live in traditional three-dimensional real space, but also in virtual cyberspace. People's social behavior in cyberspace has produced massive amounts of data. The formation of big data has brought about the advancement of artificial intelligence technology, giving birth to computing social science, and laying the foundation for sociological intervention in depression. This article explores the sociological intervention model of depression, and analyzes the social information of users on social media by applying deep learning, which is a new generation of artificial intelligence methods, and discovers their hidden depression tendencies. When it is found that the user has a tendency to be depressed, the degree of depression is further diagnosed through the psychiatric PHQ-9 depression scale, and at the same time the topics of concern to the users are explored as a basis for symptomatic bibliotherapy treatment. The results measured by the PHQ-9 scale can also be used as indicators to test the effectiveness of bibliotherapy. The model proposed in this article can not only mine the potential depression tendency of users, conduct online early diagnosis of depression for users, protect patient privacy, and eliminate patients' stigma, but also facilitates large-scale application implementation. At the same time, it presents good usability, normalization, and feasibility.

2 REVIEW OF THE LITERATURE

2.1 *Computational social science*

With the development of technologies such as the Internet and cloud computing, people are increasingly using social media to express their views on social phenomena, share their knowledge and understanding of society, and society has also entered the era of big data. Traditional questionnaire surveys and interviews have been unable to provide guidance for the management and control of the current complex social system. In 2009, 15 scholars including social scientists such as David Lazer and computer scientists such as Alex Pentland published an article entitled "Computational Social Science" in *Science*. At that point, computational social science was officially born[2]. Computational social science is a product of big data. It is a discipline based on the principles of sociology, using natural science and information science tools to reveal the laws of social development, so as to solve social problems[3]. Computational social science is a typical interdisciplinary subject. It has made some breakthroughs in the acquisition and analysis of social factual materials, the research methods, and technical realization methods of complex social systems, and the verification of causality[4]. It provides a new possibility and method for understanding, and even transforming society. Computational social science is a noninvasive research method used for interventional research on depression. It not only helps in early detection, but also helps psychologists to observe them more objectively, rather than relying on the subjective evaluation of patients[5].

2.2 *Bibliotherapy for depression*

Bibliotherapy is based on using literature as a medium, where reading is a means of health care, health preservation, and auxiliary treatment of diseases. It is also a way to enable oneself or guide others to maintain or restore physical and mental health through the study, discussion, and understanding of the content of literature[6].

Bibliotherapy has many advantages over traditional treatments for depression. In 2012, Moss and other scholars published research results showing that bibliotherapy is a cost-effective treatment method[7], which has been widely used in the elderly with symptoms of depression and is also a promising treatment for depression in the elderly. Guarano et al. (2017) published their findings on bibliotherapy data from a 3-month to 3-year follow-up period, illustrating that bibliotherapy can play a unique role in the treatment of severe mental health disorders[8].

Domestic scholars have also carried out research on the effectiveness of bibliotherapy in the treatment of depression. Wei Zhonghua et al. studied 92 depression patients and found that bibliotherapy can effectively alleviate the psychological distress of depression patients and reduce their pain levels[9]. Bibliotherapy, is a good rehabilitation treatment for depression[6]. Yuan Shuai et al. found that bibliotherapy is effective in patients with depression through eight studies with a total of 979 participants[10].

2.3 *Prediction of depression based on social media*

With the rise of the Internet, social media platforms have increasingly become important carriers for people to share ideas and opinions. Therefore, social media provides a way to capture behavioral attributes related to personal thinking, emotions, communication, activities, and social interactions, which could provide a lot of useful information to indicate the onset of depression in individuals[11]. In this context, many researchers have explored how to find clues to users' depression from social media data.

Ricard's research results show that using information provided by community user-generated data in social media can predict depression among social media users[12]. Choudhury and others collected social media data from depression patients using Twitter for large-scale population measurement of depression, and proposed the concept of a social media depression index, proving that the use of social media can be a reliable tool for measuring depression in large-scale populations[13,14]. In order to explore the accuracy of detection and treatment of depression, Vajdi Zakuani established a language annotation corpus project with sentiment analysis. The authors created a large-scale user dataset by collecting the social media data of patients with depression in the Middle East and North Africa, which was subsequently subjected to several analyses in order to understand the psychological and social behaviors of the patients with depression[15].

In related domestic research, Hu Quan believes that it is feasible and effective to use the social media data of active Sina Weibo users to predict their mental health (depression) status[16]. Pengyu Li constructed three classifiers, DBN, BP network, and SVM, based on users' microblog information, and used training samples to identify users with a depression tendency, which verified the effectiveness of the model[17]. Zhenyu Fang used TF-IDF weighted word vectors and Max Pooling to extract lexical features to construct user text semantic vectors, proving that the method of constructing user vectors from word vectors can be used as a new solution for user depression prediction[18]. Meihua Han et al. proposed a new model of bibliotherapy for depression based on "user portraits" in the context of big data[19]. This model was based on the method of machine learning, which comprehensively refines the subjective expression of users' online behavior and depression, calculates the depressive emotion index to obtain the "user portrait," and then pushes the corresponding bibliotherapy resources. Jiahong Qiu used TF-IDF, LDA, and Word2vec to extract the bag-of-words feature, topic feature, and word vector feature, respectively, and build depressed user detection models on text features[20]. The results showed that text features are very feasible in detecting depressed users of social networks.

3   DESIGN OF AN INTERVENTION MODEL FOR DEPRESSION

The present study proposes a bibliotherapy model for depression based on deep learning. The construction of this model involved the following four stages.

(1) Creation of online communities: The online community provides a social media platform which is an infrastructure for analyzing user depression.
(2) Exploring users' potential depression tendency: The LSMT model is used to conduct deep learning on the big data from users' social media, analyzing the language patterns of users' daily expressions, capturing subtle changes in their ways of communicating on social media, and exploring the hidden depression tendencies.
(3) Identification of the sources of users' depressive moods: Extract the user's emotional theme through the LDA model, which is used to identify the source of the user's depression tendency, as the basis for pushing the corresponding book prescriptions, so as to prescribe the right medicine and improve the conversion rate of depression.
(4) Depression diagnosis and bibliotherapy treatment: Use the PHQ-9 Psychiatric Depression Scale as a diagnostic tool to conduct continuous self-evaluation or other evaluations of users who are found to be depressed. According to the evaluation results, the treatment plan and reading prescriptions could be provided to the patients.

4   CONSTRUCTION OF AN DEPRESSION INTERVENTION MODEL FROM THE PERSPECTIVE OF COMPUTATIONAL SOCIAL SCIENCE

4.1   *Creation of online communities*

The online community provides a social media platform on which users can exchange information, share experiences, ask questions, and express opinions. It can also access various social media of

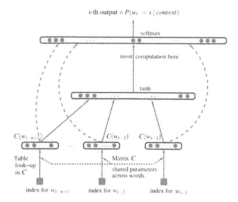

Figure 1. Neural architecture: $f(i, wt-1, \ldots wt-n+1) = g(i, C(wt-1), \ldots C(wt-n+1))$, where g is the neural network and $C(i)$ is the i-th word feather vector.

users, and obtain social media data of online community users under the mode of ethical and moral permission.

4.2 Depressive emotion prediction based on deep learning

(1) Use of word vectors to represent users' social media data

The core of sentiment analysis based on deep learning is distributed word vectors. The word vector is obtained by training a neural network language model (NNLM). NNLM starts with the language model and optimizes the training of the model for the objective function on the neural network. The goal is to predict which word should be next in the word environment, and the parameters obtained by training are the word vectors.

The language model is used to determine the probability $P(w_1, w_2, \cdots, w_t)$ that a given string $P(w_1, w_2, \cdots, w_t)$ is a natural language. According to the Bayesian formula,

$$P(w_1, w_2, \cdots, w_t) = P(w_1)P(w_2 \mid w_1)P(w_3 \mid w_1, w_2)P(w_t \mid w_1, w_2, \cdots, w_{t-1})$$

The meaning of the above probability formula is: after the first word is determined, the probability of the subsequent words appearing in the case of the preceding words is calculated. The objective function is:

$$L(\theta) = \sum_t \log P(w_t \mid w_{t-1}, w_{t-2}, \cdots, w_{t-n+1})$$

The network structure of NNLM is illustrated in Figure 1[21].

Use the word string $w_1, w_2, \cdots, w_t$ to train the model, and continuously adjust the optimization parameters through the method of random gradient promotion, the value of the objective function can be maximized. After the model training is completed, the word vector is obtained. In 2013, Google open sourced the famous tool for obtaining word vectors: Word2Vec. Word2Vec has two commonly used models: CBOW and Skip-Gram. The model CBOW uses the context $w_{t-2}, w_{t-1}, w_{t+1}, w_{t+2}$ of the current word w_t for prediction, while the model Skip-Gram uses the current word w_t to predict its context $w_{t-2}, w_{t-1}, w_{t+1}, w_{t+2}$

We input the result of the previous word segmentation into Word2Vec, to easily get the word vector of the user's social media data.

(2) Prediction of depressive tendencies

In deep neural networks, Hochritt and Schmidhuber proposed the Long Short-Term Memory (LSTM) model[22]. After improvement by Geers et al., it has the characteristics of a "memory

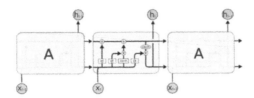

Figure 2. LSTM architecture.

sequence," and can learn the correlation between data context, characterizing or simulating human behaviors and logical development and the cognitive process of neural organization more realistically[23].

Specifically, LSTM introduces a "gate" structure to control the addition and deletion of information, and realizes the function of memory or forgetting. "Gate" is a structure composed of a Sigmoid function and a dot multiplication operation, which can selectively control the passage of information. The value range of the Sigmoid function is [0,1], where the value 0 represents complete discarding of the information, and the value 1 represents the complete passage of the information. Each LSTM unit has three different gate structures, which are called the forget gate, input gate, and output gate. The structure of LSTM is shown in Figure 2 (http://colah.github.io/posts/2015-08-Understanding-LSTMs/). LSTM consists of a series of units connected end to end. In the same layer, each unit will use the output of the previous unit as its input and its output as the input of the next unit. For different unit layers, the output of the previous layer will be used as the input of the next layer.

Based on the current popular deep learning system TensorFlow, after defining the training depth, a depressive mood LSTM model can be easily trained using the word vector obtained above.

4.3 *Identification of the sources of depressive moods*

The user's depression is not groundless, it is a reflection of the real problems he faces, and it comes from the real world. Clues to these issues will inevitably be left in the user's social media emotional expression, including topics that users care about. As the saying goes, "speaking is the voice of the heart," the user's media language is the expression of their thoughts and a direct reflection of their psychological activities. It inevitably reflects the concerns of the speaker, and the inner monologue language is the expression of the user's inner world. Extracting the user's concerns from text involves identifying certain particular user topics or the topics of the text. Data-mining algorithms using the LDA model can be used to help to complete this vision.

LDA is the abbreviation for Latent Dirichlet allocation, which is a method of modeling text with a probability generation model. The LDA model asserts that every word in an article has its formation process. This process starts with "selecting a certain topic with a certain probability," and then "choosing a certain word from this topic with a certain probability." This is the process of word formation. This process can be expressed by the following simplified formula:

$$P(word \mid document) = \sum P(word \mid topic) \times P(topic \mid document)$$

The real problem is the inverse process of this document generation process: We have the bag of words after the document segmentation, we can calculate $P(word \mid document)$ the when it corresponds to any topic for the words in it. Then the theme that the word should correspond to is updated based on these results. With the application of Bayes' law, one could train the information of $P(word \mid topic)$ and $P(topic \mid word)$ on the basis of a large amount of known information $P(word \mid document)$, and subsequently obtain the subject of the document. The topic obtained here is the topic that the user pays attention to. It is closely related to the user's mood and there is a high probability factor that it causes the user's depression. Symptomatic selection of books or related information related to these fields, and the push to users, makes bibliotherapy targeted.

Table 1. The PHQ-9 depression self-measurement table.

Over the last two weeks, how often have the following symptoms appeared in your life?	Not at all	Several days	More than half the days	Nearly every day
Not interested in anything, feel bored	◉	◉	◉	◉
Feeling down, depressed, or hopeless	◉	◉	◉	◉
Always difficult to fall asleep, or excessive sleepiness	◉	◉	◉	◉
Feeling very tired and weak	◉	◉	◉	◉
Poor appetite or overeating	◉	◉	◉	◉
Dissatisfied with yourself, feel you are a failure, or an embarrassment to your family	◉	◉	◉	◉
Inability to concentrate, even when reading newspapers or watching TV, memory loss	◉	◉	◉	◉
Acting or speaking slowly enough to attract people's attention? Or the opposite—being so fidgety or restless that you have been moving around a lot more than usual	◉	◉	◉	◉
Thoughts that it's better to die, or how to hurt yourself	◉	◉	◉	◉

4.4 Depression diagnosis and bibliotherapy treatment

The diagnosis of user depression must also be diagnosed through a self-rating depression scale recognized in psychiatry. Among the most commonly used scales, the PHQ-9 depression self-measurement table has simple entries and easy operation. This test can be completed in less than 2 minutes. The total score for the PHQ-9 depression self-measurement table is an effective tool for assessing the severity of depression[24], with good reliability and validity, and it is worthy of promotion and use in clinics, especially in grassroots community health centers[25–27]. There are nine items in the PHQ-9 depression self-measurement table, which correspond to the nine symptoms required for the DSM-IV major depressive disorder symptom diagnosis. The time window for PHQ-9 is the previous 2 weeks. The time frequencies for the evaluation of each item are "not at all," "several days," "more than half the days," and "nearly every day," with the corresponding scores of 0, 1, 2, and 3, respectively. The PHQ-9 total score indicates the overall severity of depression.

In this model, the PHQ-9 depression self-measurement table is used to diagnose the user's depression. PHQ-9 is shown in Table 1.

In order to test user's depression level, send the PHQ-9 scale to the mobile phone screen of the user with depression tendency, and add the user's scores according to their selections to get the total score. The maximum total score on the scale is 27 points: 0–4 points indicate no depressive symptoms; 5–9 points, depressive symptoms; 10–14 points, obvious depressive symptoms; and 15 points or more can be regarded as severe depressive symptoms.

After the treatment is quantitatively scored, the platform professionals research and select a book or related topics for the paired disorder according to the score and the source of depression to address, or provide offline treatment guidance. Since the causes of depression, genders and ages, education levels, religions and beliefs, and the interests and hobbies of the users are varied, the selection of books for bibliotherapy as well as the offline guidance program should be customized on an individual basis[28].

Diagnosis is according to the scale evaluation to identify the source of depression, the prescription for the symptoms is decided, and then the scale evaluation process continues. This cycle forms the quantitative treatment model of bibliotherapy for depression. This model, under the collaboration of artificial intelligence methods supported by big data, explores users' early potential depression tendency and solves the issue of concealed depression. The Psychiatric Evaluation Scale is used to accurately quantify the evaluation basis, and carry out bibliotherapy for users with depression. To an extent, this research has established a full-process quantitative evaluation system for bibliotherapy to promote the standardization of bibliotherapy for depression.

5 EXPERIMENT ON DEPRESSION PREDICTION

5.1 *Data sets*

The key to this model is the prediction of depressed users. Since there is no publicly published Chinese depression prediction data set, this article uses the public data set on github (https://github.com/peijoy/DetectDepressionInTwitterPosts) as the experimental data set. The data set contains two parts: tweets from users with depression and tweets from users without depression. Among these, the part of the tweets of users with depression is obtained from Twitter using the web crawler TWAIN with the keyword "depression" in English. The crawling results are then manually checked to remove the @ symbol, URL links, and tweets with less than 25 characters to ensure the validity of the data. The final 2345 pieces of data are saved in a csv format file. The tweets of nondepressed users come from the public data set of Kaggle twitter_sentiment, a world-renowned data science competition platform. The data set has a total of 1,578,614 data without any tags.

The experiment selects all 2345 data from users with depression, and 15,000 data from users without depression as experimental data. The selected data are divided into the training set, test set, and validation set at a ratio of 60:20:20. The pretraining vector of the Word2Vec model uses GoogleNews-vectors-negative300.bin pretrained by Google, with the address: https://drive.google.com/uc?id=0B7XkCwpI5KDYNlNUTTlSS21pQmM&export=download

5.2 *Experimental results and analysis*

Experimental environment: Win10 64-bit operating system, 16G memory, Intel i7 2.6 GHz processor. The programming language is Python 3.7.3, and the programming tool is Jupyter 6.0, and the API provided by the Python machine learning library Keras is used to call TensorFlow to build a deep learning neural network model.

This paper builds a four-layer neural network as a classifier to predict user depression. The first layer is an embedding layer, and the second layer is an LSTM layer with 300 memory cells, which returns the sequence. Doing so can prevent the next LSTM layer from receiving randomly scattered data, but rather a sequence. In order to avoid overfitting the model, this study adds a loss layer after each LSTM layer and sets the loss ratio to 20%. The last layer is a fully connected layer, the activation function is Sigmoid, and the output result is 0 or 1, where 0 means nondepressed users, 1 means depressed users. The neural network model was trained for seven rounds, each batch processed 300 pieces of data, and the learning rate was set to 0.0125. The loss function and accuracy rate function of the training set of the model and the loss function and accuracy rate function of the verification set are shown in Figures 3 and 4, respectively.

The experimental results use precision rate, recall rate, F1-score, and confusion matrix as the evaluation criteria for the experiment. The confusion matrix clearly shows how accurate the model is: nine out of 2921 nondepressed users were predicted to be depressed users, and 13 out of 461

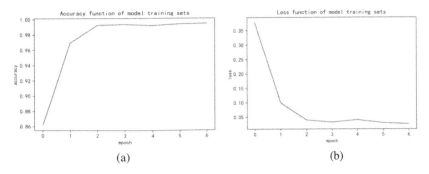

(a)

(b)

Figure 3.   Loss function (a) and accuracy function (b) of the model training sets.

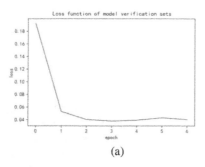
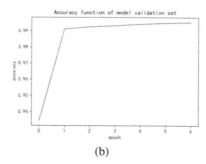

| (a) | (b) |

Figure 4.　Loss function (a) and accuracy function (b) of the model verification sets.

depressed users were predicted to be nondepressed users. The final accuracy of the model was 99.350%. The recall rate and the F1-score are 0.98 and 0.99, respectively. Meanwhile the accuracy of the prediction obtained using the logistic regression method is only 86.37%. The data show that the deep learning method has excellent feasibility for predicting depression through Weibo.

6　CONCLUSIONS

This article proposes a quantitative model of bibliotherapy for depression based on deep learning by automatically obtaining user data from social media, which can greatly save social resources compared with traditional questionnaire surveys. At the same time, it can uncover the potential depression tendency of a large number of users with unknown mental states on social networks, which helps solve the hidden social problem of depression. For users who are predicted to be depressed, the quantitative bibliotherapy based on the PHQ-9 scale is simple and feasible, and it is easy to apply and promote on a large scale. It is a useful discussion on the analysis and solution of social problems in computational social sciences.

REFERENCES

[1] Institute of Depression. *White Paper on Depression in China 2019*. [EB/OL]. https://www. xinli001. com/info/100455855.

[2] D. Lazer, A. Pentland, L. Adamic, et al. (2009). Computational Social Science, Science, Vol.323, No.1, pp.721–723.

[3] Meng Xiaofeng, Zhang Yi. (2019). Computational Social Science Promotes Inter-Disciplinary Researches. *Journal of Social Sciences*, 3–10.

[4] Luo Jun. Computing. (2020). Simulation and Experiment: Three Research Methods of Computational Social Science. *Academic Forum*, 43(01):35–49.

[5] Garcia-Mancilla, J., Ramirez-Marquez, J.E., Lipizzi, C. et al. (2019). Characterizing negative sentiments in at-risk populations via crowd computing: a computational social science approach. *Int J Data Sci Anal* 7, 165–177. https://doi.org/10.1007/s41060-018-0135-9

[6] Wang Bo. 2014.Bibliotherapy, Beijing: Ocean Press:15–16.

[7] Kathryn Moss, Forrest Scogin, Elizabeth Di Napoli, Andrew Presnell. (2012). A self-help behavioral activation treatment for geriatric depressive symptoms. *Aging & Mental Health,* 16(5).

[8] Gualano M R, Bert F, Martorana M, Voglino G, Andriolo V, Thomas R, Gramaglia C, Zeppegno P, Siliquini R. (2017).The long-term effects of bibliotherapy in depression treatment: Systematic review of randomized clinical trials. *Clinical psychology review*, 58.

[9] Wei Zhonghua, Li Hui, Li Yahoing, Yin Huanxin, Zhu Yongxin. (2012). Impact of bibliotherapy on quality of life of patients with depression. *Journal of Nursing Science*, 27 (03):66–68.

[10] Yuan Shuai, Zhou Xinyu, Zhang Yuqing, Zhang Hanpin, Pu Juncai, Yang Lining, Liu Lanxiang, Jiang Xiaofeng, Xie Peng. (2018). Comparative efficacy and acceptability of bibliotherapy for depression

and anxiety disorders in children and adolescents: a meta-analysis of randomized clinical trials. *Neuropsychiatric disease and treatment*, 14.

[11] De Choudhury M, Counts S, Horvitz E. (2013). [ACM Press the 5th Annual ACM Web Science Conference – Paris, France (2013.05.02-2013.05.04)] Proceedings of the 5th Annual ACM Web Science Conference on-WebSci \"13 – *Social media as a measurement tool of depression in populations*. 47–56.

[12] Ricard Benjamin J, Marsch Lisa A, Crosier Benjamin, Hassanpour Saeed. (2018). Exploring the Utility of Community-Generated Social Media Content for Detecting Depression: An Analytical Study on Instagram. *Journal of medical Internet research*, 20(12).

[13] De Choudhury M, Counts S, Horvitz E. (2013). [ACM Press the 5th Annual ACM Web Science Conference – Paris, France (2013.05.02–2013.05.04)] Proceedings of the 5th Annual ACM Web Science Conference on-WebSci \"13 – *Social media as a measurement tool of depression in populations*. 47–56.

[14] Sharath Chandra Guntuku, David B Yaden, Margaret L Kern, Lyle H Ungar, Johannes C Eichstaedt. (2017). Detecting depression and mental illness on social media: an integrative review. *Current Opinion in Behavioral Sciences*, 18.

[15] Wajdi Zaghouani. (2018). A Large-Scale Social Media Corpus for the Detection of Youth Depression (Project Note). *Procedia Computer Science*, 142.

[16] Hu Quan. (2015). *Identification of Internet users mental health based on Sina Micro-blog*. Henan University.

[17] Li Pengyu. (2014). A Detection Model for Identification of Depressed College Students on Weibo Social Network. *Harbin Institute of Technology*.

[18] Fang Zhen-Yu. (2017). Research on depression prediction of micro-blog users based on word embedding method. Hefei University of Technology.

[19] Han Meihua Zhao Jingxiu. (2017). Research on Bibliotherapy Model Based on User Profile—Take Depression as an Example. *Journal of Academic Libraries*, 35(06):105–110.

[20] Qiu Jiahong. (2018). Detecting Depressive Social Network Users Based on Text Mining. *Jiangxi University of Finance and Economics*.

[21] Bengio Y, Ducharme R, Vincent P, Christian Jauvin. (2003). A Neural Probabilistic Language Model. *Journal of Machine Learning Research* 3, 1137–1155.

[22] Hochreiter S, Schmidhuber J. 1997. Long Short-term memory. *Neural Computation*, 9 (8): 1735–1780.

[23] F. A. Gers and J. Schmidhuber, 2000. "Recurrent nets that time and count" in Proc. IJCNN'2000, *Int. Joint Conf. Neural Networks*, Como, Italy.

[24] Du Changjun. (2014). The Study of Major Depressive Disorder Clinical Screening Scales among General Hospital. Tianjin Medical University.

[25] Jin Tao. (2010).*The reliability and validity of patient health questionnaire module (PHQ-9) in Chinese elderly*. Zhejiang University.

[26] Xu Yong, Wu Haisu, Xu Yifeng. 2007. The reliability and validity of patient health questionnaire depression module (PHQ-9) in Chinese elderly. *Shanghai Archives of Psychiatry*: 257–259+276.

[27] Yan Dongmei. (2013). Feasibility analysis of patient health questionnaire depression scale (PHQ-9) in psychosomatic disease clinic of hospital. *Proceedings of Chinese Medical Association, Chinese Society of Psychiatry, The 11th National Psychiatric Conference of the Chinese Medical Association, The 3th Asian Conference on Neuropsychopharmacology. Chinese Medical Association, Chinese Society of Psychiatry*.

[28] Gong Meiling, Cong Zhong. (2011). Typical Cases in Reading Therapy of College Students' Depression and the Compatible Reading Therapy Literatures. *The Library Journal of Shandong*, 33–37.

Computational Social Science – Luo, Ciurea & Kumar (eds)
© 2021 Taylor & Francis Group, London, ISBN 978-0-367-70193-2

A novel high-dimensional abnormal value imputation method for power grid middle platform data

H.Z. Cui
State Grid Jiangsu Electric Power Co., Ltd., Jiangsu Nanjing, China

C. Wang
State Grid Jiangsu Electric Power Co., Ltd. Information & Techcommunication Branch, Jiangsu Nanjing, China

M.S. Xu
Jiangsu Electric Power Information Technology Co., Ltd., Jiangsu Nanjing, China

Y.Y. Zha & M.M. Zhang
State Grid Jiangsu Electric Power Co., Ltd. Information & Techcommunication Branch, Jiangsu Nanjing, China

ABSTRACT: Currently, the building of power grid middle platforms plays an increasingly important role in the development of the smart grid. However, due to the property of real-time sampling, middle platform data suffer from a series of abnormal problems. These contaminated incomplete data greatly hinder grid analysis accuracy and efficiency, and further cause catastrophes for the stable running of the smart grid. To address this challenge, in this paper, a novel abnormal value imputation method for smart power grid middle platform data is designed to deliver more accurate recovered results. Technically, original middle platform data are reconstructed to a high-order tensor form which delivered a more precise representation of time-continuous data. Then, the tensor-based abnormal value imputation model is formulated to a solvable convex function. Furthermore, the alternating direction method of multipliers is developed to optimize the proposed model. Finally, extensive experimental evaluations are conducted in the power grid middle platform dataset to elaborate the superiority of the proposed method by comparison with multiple state-of-the-art abnormal data imputation methods.

1 INTRODUCTION

The construction of the power grid middle platform is a key step in the evolution progress of the traditional power grid into a smart power grid. However, in the process of sampling, due to the wide distribution range of intelligent power equipment, the integrated middle platform data often suffer from abnormalities due to the network delay or signal interference. Figure 1 shows an instance of abnormal middle platform data, the blue line is clean data and the red line is abnormal data. The existence of these abnormal data will affect the accuracy of the data analysis and model formulation, even leading to disastrous consequences. Therefore, how to accurately clean the abnormal data has become a recent focus of research in the field of power big data research.

The traditional methods include regression imputation [1], mean value imputation [2], etc. These methods have high efficiency and low computational complexity. However, when the amount of missing data is too high or the fluctuation is too great, the accuracy of these methods decreases greatly. Methods based on statistical learning [3] and machine learning [4] have greatly improved the accuracy of abnormal data cleaning. However, these methods are mainly aimed at the imputation

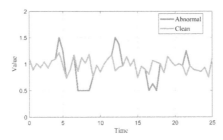

Figure 1. An instance of abnormal time-continuous data.

of traditional data types. And for the power grid middle platform data, its variables have a strong correlation with the sampling time, which is usually directly ignored by the above methods. Based on the application of multivariable regression [5] and deep neural network [6], such a time correlation of data is further utilized. However, for nonlinear problems, multivariable regression usually cannot deliver outstanding results. Meanwhile, the neural network method often requires a large amount of training data and is prone to overfitting for small samples.

Recently, the low-rank tensor representation method has achieved great success in processing high-dimensional self-correlation data [7]. For instance, a third-order tensor is naturally formed by stacking the 2-D middle platform data matrix that has been continuously collected every 15 minutes for 1 year. At the same time as the user's power consumption profiles are usually relatively fixed, the middle platform data tensor shows a strong correlation within its slices, thus revealing the low-rank property of this tensor data. Therefore, using the tensor completion technology, clean data can be recovered from tensor data contaminated by abnormal data, and the accurate recovery of the time-continuous data in the middle platform can be realized.

In this paper, a novel tensor-based abnormal value imputation method for power grid middle platform data is designed to deliver more accurate recovered results for the power grid middle platform. The main contributions of this paper are summarized as:

- The original middle platform data are reconstructed to a high-order tensor which delivers a more precise representation of the time-continuous data. Then, the tensor-based abnormal value imputation model is formulated to a solvable convex function.
- The powerful alternating direction method of multipliers (ADMM)-based convex algorithm is developed to optimize the proposed model.
- Extensive experimental evaluations are conducted in the power grid middle platform dataset thus elaborating the superiority of the proposed method by comparing it with multiple state-of-the-art imputation methods.

2 PROBLEM FORMULATION

In this section, some notations we use in this paper are first introduced. Consider the clean time-continuous middle platform data as a kth-order tensor $\mathcal{X} \in \mathbb{R}^{n_1 \times n_2 \times \cdots \times n_k}$, where $x_{i_1, i_2, \cdots, i_k}$ is the element value. For power grid middle platform data we used in this paper, k is defined as 3. $\mathcal{X}(:; i; j)$, $\mathcal{X}(i; :; j)$, and $\mathcal{X}(i; j; :)$ denote its (i; j)th mode-1, mode-2, and mode-3 fibers, each fiber denotes the data values, the sampling intervals, and the total days of sampling. $\mathcal{X}(:; :; i)$, $\mathcal{X}(:; i; :)$, and $\mathcal{X}(i; :; :)$ denote its i-th frontal, lateral, and horizontal slice, respectively. In the same manner, the observed data which contain abnormal noise values are denoted as $\mathcal{Y} \in \mathbb{R}^{n_1 \times n_2 \times \cdots \times n_k}$, and the noise values are denoted as $\mathcal{S} \in \mathbb{R}^{n_1 \times n_2 \times \cdots \times n_k}$. Therefore, their relationship is formulated as $\mathcal{Y} = \mathcal{X} + \mathcal{S}$. However, this is an ill-conditioned function which is unsolvable due to the imbalance of known variable $\mathcal{Y}$ and unknown variables $\mathcal{X}$ and $\mathcal{S}$. Having formula (1) in mind, the objective problem of abnormal value imputation could be transformed as the following form

$$\min_{\mathcal{X},\mathcal{S}} \operatorname{rank}(\mathcal{X}) + \lambda \|\mathcal{S}\|_0, s.t. \ \mathcal{Y} = \mathcal{X} + \mathcal{S} \tag{1}$$

where $\operatorname{rank}(\mathcal{X})$ is the well-known tensor tubal rank which is used to measure the sparsity of the low-rank tensor, $\|\mathcal{S}\|_0$ is L_0 norm which indicates the abnormal noise value is sparse. However, this function is nonconvex, which means it is still difficult to solve.

3 PROPOSED METHOD

3.1 *Proposed tensor subspace method*

Although the higher-order tensor can well represent the structure feature of multidimensional data, its operation is much more complicated than the common matrix or vector data. In this section, a tensor subspace-based method is proposed to solve this problem. As with the matrix subspace, under the framework of the linear mixture, tensor subspace decomposition can well project the high-dimensional data into a low-dimensional subspace while reducing the computational consumption. This process can be formulated as $\mathcal{X} = \mathcal{A} * \mathcal{E}$, where $\mathcal{A} \in \mathrm{R}^{n_1 \times n_2 \times \cdots \times n_k}$ and $\mathcal{E} \in \mathrm{R}^{n_1 \times n_2 \times \cdots \times n_k}$ are the tensor subspace and orthogonal dictionary, respectively. The operator $*$ is the t-product of two tensors. In this way, the tensor imputation model can be rewritten as

$$\arg\min_{\mathcal{A},\mathcal{S}} \|\mathcal{A}\|_* + \lambda\|\mathcal{S}\|_1, s.t. \ \mathcal{Y} = \mathcal{A} * \mathcal{E} + \mathcal{S}, \mathcal{E} * \mathcal{E}^T = \mathcal{I} \tag{2}$$

In this model, $\|\mathcal{A}\|_*$ is the well-known tensor nuclear norm which is the convex relaxation of tubal rank, and can be deduced by t-SVD of the tensor $\mathcal{A}$. $\|\mathcal{S}\|_1$ is the L_1 norm of noise tensor $\mathcal{S}$, which is the convex relaxation of $\|\mathcal{S}\|_0$. $\mathcal{I}$ is the identity tensor whose first frontal slice is the identity matrix and other frontal slices are all zeros.

3.2 *Model optimization*

With this convex approximation, the original unsolvable nonconvex function is transformed to a convex function which can be easily solved by optimizing the following augmented Lagrange function of the proposed model

$$\mathcal{L}(\mathcal{A},\mathcal{E},\mathcal{S}) = \|\mathcal{A}\|_* + \lambda\|\mathcal{S}\|_1 + \zeta_{\{\mathcal{I}\}}\left(\mathcal{E} * \mathcal{E}^T\right) + \frac{1}{2\mu}\|\mathcal{Y} - \mathcal{A} * \mathcal{E} - \mathcal{S} - \mathcal{G}\|_F^2 \tag{3}$$

Where $\|\cdot\|_F^2$ is the tensor Frobenius norm, $\mathcal{G}$ is the Lagrange multiplier, and $\zeta_{\{\mathcal{I}\}}(\cdot)$ is the indicator function. Employing the ADMM algorithm can efficiently solve the proposed model by

---

Algorithm 1 ADMM for the proposed model.

Require: The data tensor $\mathcal{Y}$ with abnormal values, estimated rank r, parameter λ and μ.
Ensure: The recovered clean data $\mathcal{X}$.
  1: **Initialize:** $\mathcal{A}^{(0)} = 0$, $\mathcal{E}^{(0)} = \left(\mathcal{A}^{(0)}\right)^T * \mathcal{Y}$, $\mathcal{S}^{(0)} = \mathcal{G}^{(0)} = 0$. $t_{max} = \mathrm{T}$.
  2: **for** $t = 1{:}\mathrm{T}$ **do**
  3:       **update** $\mathcal{S}^{(k+1)}, \mathcal{A}^{(k+1)}, \mathcal{E}^{(k+1)}$ by (4–6)
  4:       **update** multiplier $\mathcal{G}^{(k+1)}$ by (7)
  5:       **update** iteration times $t = t + 1$
  6: **end for**
  7: **return** $\mathcal{X} = rec\left(\mathcal{A}^{(k+1)} * \mathcal{E}^{(k+1)}\right)$

---

dividing it into multiple easily solvable subproblems. The overview of the model optimization algorithm is summarized in Algorithm 1. It should be noted that in line 9, the operator $rec(\cdot)$ means to transform the lateral slice into the frontal slice.

Update $\mathcal{S}$: The solution of $\mathcal{S}$ is formulated as

$$\mathcal{S}^{(k+1)} = \arg\min_{\mathcal{S}} \lambda \|\mathcal{S}\|_1 + \frac{1}{2\mu} \left\| \mathcal{Y} - \mathcal{A}^{(k)} * \mathcal{E}^{(k)} - \mathcal{S} - \mathcal{G}^{(k)} \right\|_F^2 \tag{4}$$

which can be quickly solved by the soft-thresholding method.

Update $\mathcal{A}$: The solution of $\mathcal{A}$ is formulated as

$$\mathcal{A}^{(k+1)} = \arg\min_{\mathcal{A}} \|\mathcal{A}\|_* + \frac{1}{2\mu} \left\| \mathcal{Y} - \mathcal{A} * \mathcal{E}^{(k)} - \mathcal{S}^{(k+1)} - \mathcal{G}^{(k)} \right\|_F^2 \tag{5}$$

which can be efficiently solved by the t-SVT algorithm [7].

Update $\mathcal{E}$: The solution of $\mathcal{E}$ is formulated as

$$\mathcal{E}^{(k+1)} = \arg\min_{\mathcal{A}} \zeta_{\{\mathcal{I}\}} \left(\mathcal{E} * \mathcal{E}^T \right) + \frac{1}{2\mu} \left\| \mathcal{Y} - \mathcal{A}^{(k+1)} * \mathcal{E} - \mathcal{S}^{(k+1)} - \mathcal{G}^{(k)} \right\|_F^2 \tag{6}$$

which can be easily solved by the reduced-rank Procrustes rotation algorithm

The update of Lagrange multiplier $\mathcal{G}^{(k+1)}$ is:

$$\mathcal{G}^{(k+1)} = \mathcal{G}^{(k)} + \mathcal{Y} - \mathcal{A}^{(k+1)} * \mathcal{E}^{(k+1)} - \mathcal{S}^{(k+1)} \tag{7}$$

In the process of iterative updates, the noise value is gradually isolated and the original clean value is gradually restored.

4 EXPERIMENTAL EVALUATION

4.1 *Basic configurations*

In our experiment, a middle platform dataset collected by the State Grid Jiangsu Electric Power Company is used for evaluation. It contains power load data of 26 different users, the sampling frequency is four times per hour, and the whole dataset contains data for 180 days. The experiments are run under MATLAB R2017b on a server with dual Xeon 4210 CPU with and 128G RAM. For comparison, the Auto Regressive (AR)- and Long Short-Term Memory (LSTM)-based methods are employed to evaluate our proposed method. The root mean square error (RMSE) and mean absolute percentage error (MAPE) are employed to show the performance of all methods. Before experiments, 10–50% of the original clean dataset is randomly adding noise values.

4.2 *Performance evaluation*

Table 1 shows the results of the RMSE comparison of all methods under the different scene of the sparse contaminated level. All experiments are rerun five times and the mean value is computed to be exhibited. The smaller the RMSE values, the better the data recovery. According to the relative RMSE value of each row in the table, it is clear that compared with the AR and LSTM methods, the proposed method achieved the lowest RMSE value, which indicated that our proposed method achieved the greatest accuracy data recovery.

The same results can be concluded in Table 2, which shows the results of the MAPE comparison of all methods under the different scenes of the sparse contaminated level. Like RMSE, the smaller MAPE gets, the better the performance and the higher the recovery accuracy of the method achieved. From 10–50% of the abnormal rate, the proposed method always delivers the best MAPE value. It is clear that the MAPE value of the proposed approach varies between 0.09 and 0.1, which is a remarkable improvement compared to the comparison methods.

Table 1. The RMSE comparison under the different scenes of the sparse contaminated level.

Abnormal rate	10%	20%	30%	40%	50%
AR	7.9483	8.4271	9.0275	11.5470	13.0716
LSTM	5.5270	5.9022	6.7495	8.2963	10.0029
Proposed method	2.3670	3.3365	5.1663	5.9177	7.7420

Table 2. The MAPE comparison under the different scenes of the sparse contaminated level.

Abnormal rate	10%	20%	30%	40%	50%
AR	0.1309	0.1344	0.1390	0.1408	0.1432
LSTM	0.1211	0.1248	0.1254	0.1289	0.1308
Proposed method	0.0917	0.0931	0.0946	0.0961	0.0990

5 CONCLUSIONS AND FUTURE WORK

In this paper, a novel tensor-based abnormal value imputation method for power grid middle platform data is put forward. First, the original middle platform time-continuous data are reconstructed to a three-order tensor. Then, based on the tensor recovery algorithm, the cleaning problem of middle platform data is formulated as a convex function. The well-known ADMM optimization algorithm is thus employed to solve the proposed convex model. Finally, extensive experimental evaluations are conducted in the power grid middle platform dataset, by comparison with multiple state-of-the-art imputation methods, the superiority of the proposed method is illustrated. For future work, a more sophisticated regularization model will be further devised to improve the efficiency and accuracy of the abnormal data imputation.

ACKNOWLEDGMENTS

This work was supported by the Science and Technology Project of State Grid Co., Ltd. (Research and Application of Service Design and Management Technology based on Data Middle Platform, 5700-202018181A-0-0-00).

REFERENCES

[1] Templ M, Kowarik A, Filzmoser P. Iterative stepwise regression imputation using standard and robust methods. Computational Statistics & Data Analysis, 2011, 55(10): 2793–2806.
[2] Silva-Ramírez E L, Pino-Mejías R, et al. Missing value imputation on missing completely at random data using multilayer perceptrons. Neural Networks, 2011, 24(1): 121–129.
[3] Wang X, Dong X L, Meliou A. Data x-ray: A diagnostic tool for data errors. //SIGMOD/PODS'15. 2015: 1231–1245.
[4] Liu Y, Wen K, Gao Q, et al. SVM based multi-label learning with missing labels for image annotation. Pattern Recognition, 2018, 78: 307–317.
[5] Dilling S, MacVicar B J. Cleaning high-frequency velocity profile data with autoregressive moving average (ARMA) models. Flow Meas Instrum, 2017, 54: 68–81.
[6] Tian Y, Zhang K, Li J, et al. LSTM-based traffic flow prediction with missing data. Neurocomputing, 2018, 318: 297–305.
[7] Lu C, Feng J, Chen Y, et al. Tensor robust principal component analysis with a new tensor nuclear norm. IEEE transactions on pattern analysis and machine intelligence, 2019, 42(4): 925–938.

Computational Social Science – Luo, Ciurea & Kumar (eds)
© 2021 Taylor & Francis Group, London, ISBN 978-0-367-70193-2

Research on computer forensics based on multiple correlation analysis technology of multi-source logs

X.L. Zhou, H. Tang & J.L. Sun
Information Center, Beijing Jiaotong University, Beijing, China

ABSTRACT: Log files are important bases for computer forensics and scenarios reconstruction, on the basis of improving the traditional association algorithm, this paper analyzes the existing and related methods of log forensics, and puts forward a forensics model of multi-source log multi association analysis by using reverse causal association algorithm. The forensics model makes it more suitable for discovering computer crime in time and realizing the reconstruction of computer crime scenarios.

1 INTRODUCTION

Log detection is an important content of daily security detection and an indispensable tool to maintain normal operation of the computer system, with the rapid development of information technology, log analysis and detection are widely used in computer forensics. However, in the process of computer forensics and scenarios reconstruction for network intrusion, most of them only focus on firewall log and intrusion detection system log for intrusion detection and security protection. This kind of computer forensics is one-sided, and a single data source can not constitute a complete and effective evidence chain. Therefore, in the process of computer forensics research, it is the key point to perform correlation analysis on data, make full use of log resources to discover the effective real-time intrusion evidence, reconstruct the intrusion event, and then discover the potential security event and attack intention in time and effectively.

2 THE FORENSICS MODEL OF MULTI-SOURCE LOG ASSOCIATION ANALYSIS

The common logs include host system log, security device log, application log and network log. By collecting and analyzing the log data, we can extract the intrusion clues, so as to reproduce the intrusion scene to corroborate the computer crime. By collecting, filtering, standardizing and analyzing the original log files, we can reconstruct the crime scene and realize the purpose of dynamic forensics (Figure 1).

2.1 *Collection of logs*

First of all, the multi-source log category is selected according to the needs of computer forensics. Then the system log, application log, security device log and network log generated in a specific time node of system operation process can be obtained and stored directly.

2.2 *Multi source log preprocessing*

The multi-source logs are collected from a variety of data sources, which are heterogeneous, with different storage methods and formats. Through the log data preprocessing module, the corresponding attribute characteristics of the logs are extracted according to the needs of forensics. The

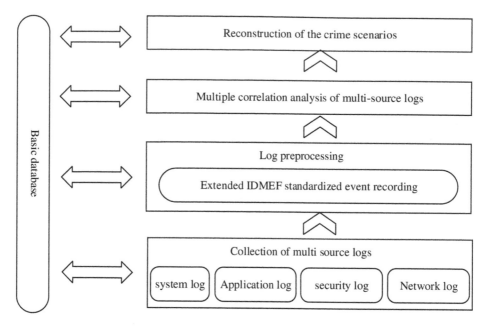

Figure 1.   The forensics model of multi source log association analysis.

multi-source log format is standardized by using the IDMEF data model, and the data is filtered and fused, so as to normalize the format of the collected data, reduce the data redundancy and store the data centrally.

2.3   *Multi source log association analysis*

The attack sequence with internal connection is searched from the massive log records. The information of different types of logs or other logs from different sources is analyzed, and then the problems of repetition and redundancy in log events can be solved by using the association method based on attribute similarity to fuse multiple records generated by the same event. In the whole process of event correlation, no prior knowledge is needed, so it belongs to preliminary correlation analysis.

2.4   *Reconstruction of log association scenario*

On the basis of the preliminary correlation analysis of the log events, the further correlation operation is carried out. According to the causal association algorithm, the deep logical relationship between the logs in each category can be explored. Then we can discover the relationship between the attacks reflected by the security events, reconstruct the computer crime scene, and complete the second correlation analysis of the log events.

3   CORRELATION ANALYSIS BASED ON ATTRIBUTE SIMILARITY

According to the characteristics of multi-source log, such as large quantity, high complexity and relevance of log information, we can judge whether there is correlation between the two log events by comparing the similarity between the attributes and the overall similarity. At present, correlation analysis based on attribute similarity mainly uses the characteristics of a large number of duplicate alarms in IDS output. Thus the similar alarm information can be combined into one, so as to reduce the number of output results and remove redundancy.

3.1 The method of event correlation

Because the log records generated by the same attack usually have similar attributes, the most direct way of log association is to check the similarity between attributes, so as to find the association between logs. The alarm log generated by IDS contains many attributes, from which the common characteristic attributes of log records are selected and their similarity function is defined. The return value of the function is between 0 and 1, and 1 represents an exact match.

In this paper, the IP address including the source IP address and destination IP address, port including source port and destination port, and event type are selected to define the similarity calculation functions.

3.1.1 Definition of similarity calculation function of the IP address

The similarity calculation function of IP address is used to determine whether different logs come from the same attack scenario. Logs with the same source IP address indicate that the attack is from the same device, while logs with the same destination IP address indicate that the target of the attack is the same device. The calculation formula of IP address similarity is as follows:

$$Sim(ip_i, ip_j) = n/32$$

Where 32 represents that the IP address can be converted into a 32-bit binary representation, and n represents the same number of digits in the binary IP address from the highest to the lowest.

3.1.2 Definition of similarity calculation function of port

In the network attack, the intruder will first randomly select the source port to detect whether there is a service that can be used to carry out the attack. However, the destination port is determined and targeted, so we need to carry out correlation analysis for the logs with the same attributes of the source port and the destination port. The port similarity is calculated as follows:

$$Sim(port_i, port_j) = \begin{cases} 1, & port_i = port_j \\ 0, & port_i \neq port_j \end{cases}$$

3.1.3 Definition of similarity calculation function of event type

By judging whether the event type properties of the two event records are the same, their functions can be determined. The event type similarity is calculated as follows:

$$Sim(type_i, type_j) = \begin{cases} 1, & type_i = type_j \\ 0, & type_i \neq type_j \end{cases}$$

3.1.4 Definition of calculation function of similarity fusion

The calculation of the total similarity of log association is the weighted laverage of each attribute similarity, where the weight is the corresponding expectation of using similarity. The calculation formula is as follows:

$$SIM(X, Y) = \frac{\sum_{i=1}^{S} W_i SIM(X_i, Y_i)}{\sum_{i=1}^{S} W_i}$$

$SIM(X, Y)$ is used to calculate the similarity between attributes X_i and Y_i. X and Y are the log events to be associated. W_i is the similarity weight of the ith attribute. The overall similarity of the two alarms is calculated by the formula, and then compared with the defined threshold value of alarm similarity. If it is greater than the predefined threshold value, it means that the two alarms are similar.

Table 1. The attribute weight of network attack.

Log type	Source IP	Destination IP	Source port	Destination port	Time threshold
PortScan	1	1	0	0	0.5
Web attack	1	1	0	1	0.5
R2L	1	1	1	0	0.5
U2R	1	1	0	1	0.5
DoS	1	1	0	0	0.5
DDoS	0	1	0	0	0.5
Other	1	1	1	1	0.5

3.2 *Calculation of attribute weight based on aggregation rules*

Attribute weight is a comprehensive measure of subjective evaluation and objective reflection of the relative importance of attributes. The value of attribute weight affects the result of association directly. The calculation method of subjective weight is too subjective and has large errors. Therefore, the objective weight calculation is used to adjust the weight by combining the aggregation rules, so as to achieve the optimal correlation effect.

For different attacks, the features of concern are also different. By analyzing the various attribute characteristics contained in the log, three attributes including attack source (source IP address), attack target (destination IP address) and event type, are proposed to be selected. According to the aggregation rules and considering the similarity weight in common network attacks, the attribute weight of the attack is determined as shown in Table 1.

Through the correlation analysis of log records, the duplicate log records can be eliminated or reduced. The experimental results show that the alarm data can be reduced by 50%–60%. Although it cannot find out whether there is causal relationship between the alarm information itself, it reduce the input data size of the second association analysis and avoids the interference of non forensic data to the final result.

4 CORRELATION ANALYSIS BASED ON REVERSE CAUSALITY

The preliminary association analysis on forensics data is carried out by using the association analysis algorithm based on attribute similarity. The events with high attribute similarity can be aggregated into the same class by fusing duplicate data and similar data. This association process is simple and easy to realize, but it is unable to deeply mine the essential relationship between events. Therefore, it is necessary to carry out the second association analysis based on causality again, and reconstruct the computer crime scene with the method of multiple association.

4.1 *Analysis of causal association algorithm*

There is causality between the security events generated by each attack step. Peng Ning, who is in University of North Carolina, proposes a causal association algorithm based on premise and consequence. The algorithm defines the premise and consequence of each individual attack firstly, and then matches the consequence of the first attack step with the premise of the next attack step, so as to realize the correlation analysis of security events.

4.1.1 *Super event type*
The super event type uses a triple: $T = (face, prerequisite, sequence)$. It defines the set of several attributes *(Facts)* of each log type, the set of prerequisites for successful attack *(Prerequisite)* and the corresponding set of possible consequences for successful attack *(Consequence)*. The expression

is composed of predicates and logical operators. Any variables that appear in prerequisite and consequence are in the set of *Fact*. For super event e of given event type T, its precondition is expressed as $P(e)$ and result set as $C(e)$.

For example, *SadmindBufferOverflow* is a super event type of buffer overflow attack against sadmind remote control vulnerability, which can be expressed as:

$$SadmindBufferOverflow = (\{VictimIP, VictimPort\}, \{ExistHost(VictimIP) \wedge VulnerableSadmind(VictimIP)\}, \{GainRootAccess(VictimIP)\})$$

So there is: $P(SadmindBufferOverflow) = \{ExistHost(VictimIP), VulnerableSadmind\ (VictimIP)\}$ and $C(SadmindBufferOverflow) = \{GainRootAccess(VictimIP)\}$.

4.1.2 *Super event*

The instantiation of super event type assigns values to each attribute of *Face* according to the information contained in the security event, and describes the prerequisite and consequences of a specific event. There may be multiple *Facts*, corresponding to which there will be multiple *Prerequisites* and multiple *Consequences*. For each *Fact*, the corresponding *Prerequisite* is true, and the corresponding *Consequence* may be true.

For example:

$$hSadmindBOF(\{202.112.154.168, 1235\}, ExistHost(202.112.154.168, 1235) \wedge$$

$$VulnerableSadmind(202.112.154.168, 1235), \{GainRootAccess(202.112.154.168, 1235)\})$$

$$hSadmindBOF = \{(VictimIP = 202.112.154.168, VictimPort = 1235)\}.$$

When the current value of *ExistHost (202.112.154.168)* $\wedge VulnerableSadmind(202.112.154.168)$ is true, the result of *GainRootAccess(202.112.154.168)* may be true. This super event indicates a buffer overflow attack of 202.112.154.168 host, which may cause illegal access to root permissions.

4.1.3 *Correlation of super event*

The super events converted from log records are correlated in this paper. The same super event type can correspond to multiple instantiated super events. We map the super event type T to an instance E_T, and the super event type T' corresponds to an instance $E_{T'}$. Define a set $I = \{p_1, p_2, \cdots, p_n\}$, if *HyperEventTypeCorrelation(T, T')* exists, and $\forall p_i \epsilon I$, which indicates that super event type T and super event type T' can be causal related. This makes $E_T.Consequence.p_i$ and $E_{T'}.Prerequisite.p_i$ have at least one most general oneness replacement θ_i.

For example, two events of computer attack and Trojan upload by the way of executing remote commands are recorded as follows:

$$VulnerableSadmind(202.112.154.168, 1235), \{GainRootAccess(202.112.154.168, 1235)\}),$$

$$Event2 = (\{202.112.154.168, 1235\}, GainRootAccess(202.112.154.168, 1235),$$

$$\{ReadyToLaunchDDOSAttack\}).$$

The descriptions of the two events in the event library are *SadmindBufferOverflow* and *UploadTrojan*. The corresponding causal correlation formula is as follows:

$$C(SadmindBufferOverflow) \cap P(UploadTrojan) = GainRootAccess(IP)$$

Therefore, I = {GainRootAccess(IP)}.

The most general oneness replacement is: *Event1.GainRootAccess.IP= Event2.GainRoot Access.IP.*

Event1.GainRootAccess.IP = 202.112.154.168
Event2.GainRootAccess.IP = 202.112.154.168

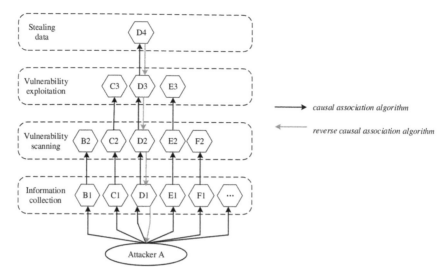

Figure 2. The block diagram for causal association algorithm and reverse causal association algorithm.

Therefore, there is a most general oneness replacement: *202.112.154.168*. Then the computer attack and the upload of Trojan horse can be correlated.

4.2 *Analysis of reverse causal association algorithm*

The traditional causal association algorithm is shown in Fig.2. First, attacker *A* carries out *IPSweep* attack on the target network and scans all hosts in a network segment. Five active hosts *B, C, D, E* and *F* are found in the network segment, then we scan the vulnerability and find out the exploitable vulnerability in host *C, D* and*E*. The buffer overflow attacks are carried out on three hosts with vulnerabilities. Finally, only host *D* is successfully invaded, and the privacy data of host *D* is accessed.

The above process is a complete attack. In the whole attack process, only the successful attack scenario can be used as evidence. However, most of the initial scanning and information collection work of attackers failed to form a complete attack link, which also delayed the association speed and reduced the efficiency of the algorithm.

According to the characteristics of computer forensics, the traditional pure causal association algorithm is improved. We use the reverse causal association algorithm to analyze the victim host, find the last step of the attack, and then find the previous attack depending on the reason for the step. Until the beginning of the attack is found, a complete attack process is formed (Figure 2).

According to the state and influence of the attacked computer, the last event *E* is determined and the prerequisites *P(E)* is obtained. Then, an event *E'* is found in the class where the event is located based on the event information. The result *C(E')* of *E'* is obtained, so that *C(E')* and *P(E)* meet the definition of causal association method. Two events are correlated to form an attack sequence fragment $E' \rightarrow E$. Based on event *E'*, we repeat the above operations until all the log information in this class is associated.

5 CONCLUSION

It is an important method in the field of computer forensics by using log analysis to find effective computer evidence. The computer forensics involves the comprehensive forensics analysis of multiple devices, so it is easier to confuse the event correlation caused by different attacks. Moreover, the log records of attack events are scattered, and adjacent attacks may be mixed with other related records. Therefore, a forensics analysis method based on multiple association is proposed in this paper by the association analysis of multi-source logs. The pure causal association algorithm is improved to eliminate the influence of redundant data and irrelevant logs on association results and efficiency, and a reverse causal association algorithm meeting the identification standards of electronic evidence is proposed. Thus, the interference of invalid attacks on the reconstruction of crime scene is eliminated, the accuracy of scene correlation is improved, and the purpose of reconstruction of computer crime scene is achieved.

REFERENCES

[1] LIU Bixiong. Research and practice on comprehensive analysis technology for multi-source heterogeneous log. Journal of Nanjing University of Information Engineering, 2011, 4: 365–370.
[2] Ya Jing. Towards network threat analysis system based on multi-source security logs. Beijing Jiaotong University, 2014.
[3] Wang Lulu. Research on attack scenarios reconstructing method based on causal correlation and probabilistic correlation. Shanghai Jiao Tong University, 2010.
[4] TANG Channa, FAN Lei. Research on causal correlation of alerts with confidence measurement. Information Security and Communications Privacy, 2009, 6: 83–85.
[5] Pan Anqun. Research on a rule-based approach to network security event correlation. Huazhong University of Science and Technology, 2007.
[6] ZHANG Xiang, HU Changzhen, YIN Wei. Research of network threat analysis technique based on event correlation. Computer Engineering and Applications, 2007, 43(4): 143–145.
[7] Zhang Yi. Research and application of multi-source log analysis technology in computer forensics. Beijing Jiaotong University, 2019.
[8] Liu Xin. Security event extraction methods based on multiple log sources. Harbin Engineering University, 2009.
[9] Tang Xinyu. Design and implementation of network security log visual forensics analysis system. Harbin Engineering University, 2017.

Computational Social Science – Luo, Ciurea & Kumar (eds)
© 2021 Taylor & Francis Group, London, ISBN 978-0-367-70193-2

Research on the mode of network ideological and political education in colleges and universities driven by big data

C. Li
Administration, Harbin, Heilongjiang, China

Y. Lv* & N. Jiang
Machinery, Harbin, Heilongjiang, China

ABSTRACT: At present, a new round of scientific and technological revolution and industrial reform has accelerated its evolution, and we have entered the era of big data, the Internet of Things, the era of networks. Digital change has been promoting national development and social services, in promoting the modernization of China's governance capacity, but also the new era of ideological and political education has put forward greater requirements. The characteristics of "timeliness, richness, precision, and fragmentation" of big data bring opportunities, and its "diversity, mismatch, insecurity, concealment" characteristics also bring many challenges to ideological and political education, and these challenges and opportunities for the era of big data in the ideological and political education initiatives further provide a new perspective and inspiration on how to use the network and new media to do a good job in moral education is particularly important.

1 INTRODUCTION

The informatization of ideological and political work in colleges and universities has become a trend, and according to the IDC's forecast, the global data field will increase from 33 ZB in 2018 to 175 ZB in 2025, therefore it can be said that the field of data use has achieved explosive growth, with China expected to have the world's largest data field by 2025. Correspondingly, this brings opportunities and challenges to ideological and political education. How to free the ideological and political workers in colleges and universities from the heavy transactional work, how to make full use of cutting-edge technology to complete a deep integration of ideological and political work, based on big data to drive the reform of the ideological and political work model of colleges and universities, intelligently and precisely, to meet the growing personalized needs of graduate students, is the core to the transformation and development of the completion of ideological and political work.

2 OPPORTUNITIES FOR EDUCATION IN THE NETWORKS OF IDEOLOGICAL AND POLITICAL WORK IN COLLEGES AND UNIVERSITIES IN THE ERA OF BIG DATA

2.1 *The timeliness of new media accelerates the dissemination of information*

The big data network new media era has truly realized mutual access of data for all people, including micro-blogs, knowledge, circles of friends, qq space, and other network social platforms, for college and university networks of ideological and political education to build a systematic communication stage, to enhance audience and social contacts, through intelligent integration of big data, professional processing, relying on comprehensive decision-making ability, greatly improve the speed of communication to achieve rapid public guidance, but also to broaden the information

---

*Corresponding author

Top 21 Keywords with the Strongest Citation Bursts

Keywords	Year	Strength	Begin	End	2010 - 2019
Ideological Work	2010	1.0518	2010	2013	
Countermeasure	2010	0.9447	2010	2011	
SWOT Analysis	2010	1.5967	2010	2014	
Microblog	2010	1.6669	2011	2014	
Practical Teaching	2010	1.3899	2011	2014	
Innovate	2010	3.1466	2012	2017	
Informatization	2010	0.9788	2013	2016	
University for Nationalities	2010	1.6345	2013	2015	
Colleges and Universities	2010	2.5796	2013	2015	
Ideological and Political Theory Course	2010	2.6197	2014	2016	
WeChat	2010	1.1707	2015	2017	
People Foremost	2010	0.7795	2015	2017	
School	2010	1.1385	2015	2016	
Subjectivity	2010	0.9213	2016	2017	
Ideological and Political Work	2010	1.6408	2017	2019	
College Students from Minority Nationalities	2010	1.2172	2017	2019	
Appetency	2010	0.8174	2017	2019	
Ideological and Political Course Practice Teaching	2010	1.3657	2017	2019	
Chinese Traditional Culture	2010	0.8174	2017	2019	
College Ideological Instruction	2010	3.148	2017	2019	
Ideological and Political Education Work	2010	1.0912	2017	2019	

Figure 1. Time span from 2010 to 2019.

feedback mechanism, we can effectively integrate a comprehensive and accurate understanding of students' growing talents and ideological and psychological dynamics, building up a new media ecosystem linked both upward and downward to enhance its relevance and effectiveness.

2.2 The timeliness of new media accelerates the dissemination of information

The richness of big data broadens the space of ideological and political education of college students, on the basis of the first space of students' campus life and the second space of academic research, the third space of network new media has become the core key point of college students' thought leadership. The content of new media information in the era of big data greatly enriches the content and material of ideological and political education, and the online political participation, network legal morality, network legal consciousness, network technology literacy, and so on have become the new content of network education, which can set up the element of an ideological and political network without restriction, which can provide further value-added services for students' value shaping. Taking CNKI as the retrieval platform, the CSSCI under CAJD was selected, and retrieval fields including "title" and "topic" were used. The time span was from 2010 to 2019 in Figure 1. A total of 489 related articles were published during this period. In the last 10 years, the research themes of ideological and political education in universities have changed from focusing on ideological and political courses and practical teaching initially to focusing on teaching reform, value guidance, and work innovation.

2.3 The fragmentation of information technology breaks the limits of time and space

In today's big data era, due to the rapid development of self-media, everyone can become a reporter, can freely in WeChat, microblog, toutiao, and other types of new media operating platforms to publish information, and expand the form of network media, so that the network media is fragmentary. Students can share information and receive information anytime and anywhere through social platforms, and can break through the space–time limit outside the classroom to receive ideological and political education guidance at any point from the passive acceptance of traditional ideological classroom to online fragmentation active choice search, according to the user's personal characteristics, in turn creating a steady stream of supply information, completely breaking the time and space restrictions, and thus promoting the development and innovation of educational content.

2.4 Big data accuracy provides platform assurance

The precision of big data sets up a dialogue platform for the development of ideological and political education of college students. The students of traditional ideological and political education have a strong passivity, with a general style of education of "flooding with deep water," and students lack the initiative of ideal and belief education. Big data also can drive the network ideological and political change for "some irrigation" with diversified personalized education, through intelligent campus infrastructure, the use of big data platform establishes and improves the ideological and political education "data chain" network, strengthening the linkage mechanism between school departments, for students to establish a set of ideological thinking, including ideological data, that are dynamic, through big data being the future of information. The correlation is analyzed to accurately judge the individual's social behavior and psychological orientation and to make predictions.

3 THE CHALLENGE OF NETWORK THOUGHT AND POLITICS EDUCATION IN THE BIG DATA AGE

3.1 The diversity of information technology affects the value orientation of college students

Different countries, different cultures and different values are constantly blending in the current information age, the directional output of ideology and value orientation has a profound influence on College Students' thinking mode, cultural concept, psychological needs and living habits. College students are a pool of growing talent while they lack independent thinking, and in the face of complex and diverse information conflicts, big data may become the root cause of "group errors," with all expecting big data to usher in a more open society, but without matching rules and sound systems, big data is not able to make society more transparent, but only to make it more isolated. There are many root causes of information asymmetry, such as differences in people's education level, differences in experience, and class differences, which all contribute to information asymmetry. Even with the same objective data and objective and realistic information, people will still come to different conclusions, eventually leading to different behaviors. Some online opinion leaders may promote ideas that don't conform to mainstream values due to capital operation or people's psychological needs for novelty.

3.2 The quality of ideological and political work does not match big data technology

At present, China's college ideological and political educators have made considerable developments, setting up a systematic education work platform, not only for careers to develop, and constantly to improve theoretical literacy, professionalism, and work experience, but also the pursuit of interest development. The ideological and political work team also tend to have a full-time/part-time combination, counselor/class teacher combination, which is increasingly beneficial. However, in the era of network information, the traditional ideological and political education model is constantly under attack, including from micro-blogs, fast hands, and other new media network platforms. The application ability needs to be improved for college students and the network social platform has become an important carrier of college network of ideological and political work.

Ideological and political workers lack the ability to use, integrate, and analyze the big data generated by the smart campus, and they do not have good network information processing and network software application skills, cannot rely on artificial intelligence to visualize students accurately, and are good at using big data, while network technology personnel do not systematically master the ideological and moral quality of education work, and do not meet the corresponding quality requirements.

3.3 *Data tracking causes alienation problems*

In the big data environment, our online life can be tracked, and even our offline life can be tracked. The data-driven world is always open, tracked, monitored, and listened to—because it will always be in a learning state. Unlike general tool technologies, big data has no clear data limits, and data is a very rich and valuable resource, with a variety of data values, versatility, and inexhaustible access. It is precisely because of its diversity of values that the boundaries that define what data collectors can or cannot do becomes very difficult. For example, don't break any laws when it comes to getting your data, it doesn't violate personal privacy, it just predicts and analyzes your preferences based on some of your actions, but that's why it's creating a de facto information blackout. The era of big data has built a virtual society that is different and closely related to the real society, people are active in both real and virtual worlds, and the virtual society's self may be a disguised, modified, "perfect" self, behind the cold data, and the real self's ideological and political problems are hidden due to the lack of face-to-face communication with educators, so that the network field of violations of the law or ethical behavior cannot be effectively restricted. Inevitably, there will some who feel a sense of responsibility to downplay any misbehavior.

3.4 *Concealment weakens moral potential risks*

In the era of big data, college network ideological security is faced with a variety of risks, first, the abuse of data leading to the leakage of personal privacy is typical of the security risks. In the process of dissemination of data information, the data and information released by students may involve ideological problems, and once the data or information are disclosed, there will be a huge risk. The "Prism Gate" incident in the United States is a typical case of big data that analyzes massive amounts of communications data to obtain security intelligence[4].

Big data, recommended by precise algorithms, creates an "information room" for people[5]. To give a simple example, the platform pushes information that is valued by the viewer. For example, if someone was to look at a basketball, then they would receive a variety of basketball messages; this algorithmic selection of recommended information will only make people's minds more and more closed, as if the thoughts are wrapped in a closed "room." Due to the precise individualization alienation, people suffer ideological imprison, so they can't develop freely and comprehensively.

4 THE EDUCATIONAL WORK MODE OF IDEOLOGICAL AND POLITICAL NETWORKS IN COLLEGES AND UNIVERSITIES IN THE ERA OF BIG DATA

4.1 *The orderly flow of technical team and smart campus data*

With the new media network and college students' daily life learning more and more closely linked, the new era of college ideological and political ducation to enhance data, modern intelligent campus environment relies on 5G, new media, and other basic technology platform, while the intelligent campus also becomes the campus life of college students, academic research, learning knowledge, interpersonal communication, leisure and entertainment effective platform, for the scientific, technical, and intelligent ideological and political work to lay a solid foundation for the growth needs of students. Smart campus will individualize the growth needs of college students, and grasp each person's ideological and behavioral state and future trend. Smart campus visualizes

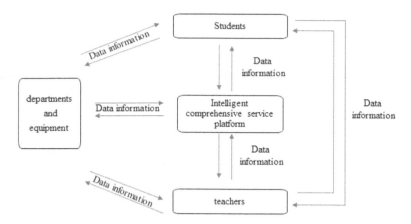

Figure 2.   The flow chart.

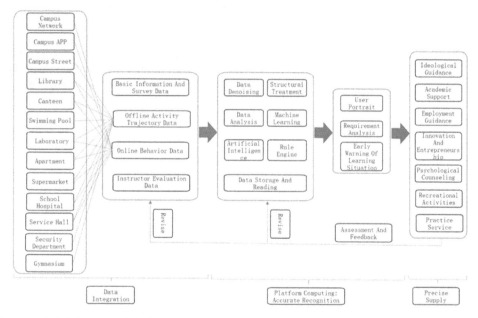

Figure 3.   Block diagram of big data system

the thinking and behavior of college students, and accurately depicts each person's thought and behavior with massive data and real-time feedback technology. Intelligent growth environment will be built to build a smart campus, with the help of artificial intelligence, the Internet of things and other technologies to achieve the precise cost of each student, and through the professional technical team and smart campus platform, all kinds of data of students will flow orderly. The flow chat is shown in Figure 2. The block diagram of big data system is shown in Figure 3.

4.2   *The effective connection between ideal belief and patriotism education*

National Education is an important cornerstone of national rejuvenation and social progress, and a moral and political project that is beneficial to the present and the future. China's education should stand at the strategic height of adhering to and developing socialism with Chinese characteristics in the new era, speeding up the modernization of education, building a strong educational system, and ensuring a good level of education that the people are satisfied with, and answering the fundamental question of who to train and how to train. Therefore, the ideological and political work on the Internet should play a leading role in the ideological and political work, with Xi Jinping's new era of socialist ideology of socialism as a guide, in the education system solidly carrying out patriotic education, through in-depth, lasting, and vivid implementation of the program, unremittingly carrying forward the main theme, spreading positive energy, building an effective carrier, creating a strong atmosphere, stimulating patriotic feelings, rallying national strength, resisting the impact of multicultural and multivalues input, to guide the new era of college students to follow the footsteps of General Secretary Xi., we should adhere to our original intention and mission, live up to our time, take the long march in the new period, and strive to cultivate qualified socialist builders and reliable successors who are fully developed in morality, intelligence, physique, beauty and labor.

4.3 *The organic connection between leading public opinion and the ideological and political aspects of the network*

The construction of an ideological and political network should fully follow the laws of ideological and political work, teaching and educating people, student growth, take students as the center, and building a system of public opinion leadership with precision, effectively improving the affinity and targeted ideological and political work, and striving to enhance organization, leadership, and service. The spirit of college motto should be fully explored through youth learning, youth model advanced deeds report meeting and college students' theoretical propaganda group. And strengthen the love of school education, learn from the excellent typical stories around, learn from the good examples around. Attention should be paid to culture, building a campus cultural promotion brand, strengthening offline activities and online publicity of deep integration, with a variety of activities to attract students to participate, with positive publicity and guidance. Then the attitude and consciousness of college students are gradually changed to learn, helping them overcome the habit of relying on education management in the past, guiding them according to their own interests and abilities, and combined with the needs of society to develop learning goals in line with their personal development needs, complete with the study of valuable knowledge and ability, to help college students to become educated to be active learners, and to help them become their own learning managers.

4.4 *The organization platform and system to build a guarantee*

To promote the quality and efficiency of the work ideological and political network, the spirit of the Fourth Plenary Session of the 19th Central Committee of the Party should be implemented, promoting the modernization of the relevant governance system and governance ability of the ideological and political network, and constantly consolidating the functional orientation of the construction of grass-roots organizations, continuously deepening the construction of positions, content construction, team building, pushing forward reforms, and further, striving to enhance the standardization, scientific, and modern level of network thinking and construction. To build a platform integrating ideological, cultural, interesting, healthy and service with campus website, social software and campus enterprise app. The content of Ideological and political education should be organically and deeply infiltrated into this platform. According to the school's talent training objectives, combined with the school's central work, select materials beneficial to students' outlook on life, values and world, so as to stimulate students' self-consciousness. The aims include to strengthen the construction of the network team, establish the ideological and political education team adapted to the big data ideological and political network, with the leading team of the construction of the network culture of the school, and a good job done in the top-level design of the network construction. Another priority is to cultivate and exercise a management team with good political qualities and high network application development ability and construction and management of network culture, technology research and development and supervision and guidance team. We should cultivate a group of students with good political qualities, high political awareness, a strong sense of responsibility, and related job skills. All the work is carried out to deepen the construction of institutional mechanisms with the spirit of reform and innovation, while strengthening vertical supervision and ensuring responsibility step by step.

5 SUMMARY

Big data has brought opportunities to our network of ideological and political education in terms of the effectiveness of new media information dissemination, the richness of educational space, the fragmentation of space–time restrictions, and the accuracy guarantee of platforms, but it has also brought many challenges to the work of ideological and political education. We should seize this opportunity and challenge to create a new era of an accurate ideological network model: precise positioning of demand, understanding the need for growth; accurate content supply, oriented to top-level innovative personnel training; accurate and effective force, improvement of the level of team work; and accurate design evaluation, while constantly improving the impact of education.

REFERENCES

[1] Bingchen Wang. Study on Innovation of Ideological and Political Education Methods in Colleges and Universities in the Era of Big Data. 2019, 4(3)
[2] Ma Sanxi. Research on the Optimization of Ideological and Political Education in Colleges and Universities under the Background of New Media. 2020, 3(2)
[3] Wang Bo & Li Yu. The Impact of Internet Public Opinion Fragmentation on College Students' Ideological and Political Education and the Countermeasure Analysis [J]. The Guide of Science & Education, 2019(07):84–85.
[4] Leng Wenyong. The Research on Chinese undergraduate Ideological Identification Security in Internet "Micro" Era [D]. The Theory and Method of Moral Education in University,2 018.
[5] SiSi Tong. Development Path of Ideological and Political Education in Colleges and Universities from the Big Data Perspective. 2020, 1533(4)
[6] Jingjing Guo. Study on the Effectiveness of Ideological and Political Education in Colleges and Universities from the Perspective of Marxism Popularization. 2019, 8(5)
[7] Wenwu Zhang. Mechanism and Path of Creating Learning Space of "Micro ideological and political education" in Colleges and Universities in the New Era. 2019, 6(9)

Computational Social Science – Luo, Ciurea & Kumar (eds)
© 2021 Taylor & Francis Group, London, ISBN 978-0-367-70193-2

Research on university teaching based on artificial intelligence

Y.Q. Zhu
School of Electronic Information Engineering, Foshan University, P.R.China

ABSTRACT: With the development of artificial intelligence technology, there are more and more applications of artificial intelligence in the field of education. Using intelligent technology to accelerate the reform of talent training mode and teaching method, and to build a new education system including intelligent learning and interactive learning has a positive significance for cultivating innovative talents adapting to the era of artificial intelligence. Starting from the research of artificial intelligence at home and abroad, discussed the significance and connotation of teaching research in colleges and universities based on artificial intelligence, which provided a reference path for better carrying out teaching research and teaching reform based on artificial intelligence in colleges and universities. Artificial intelligence technology promotes the transformation and reconstruction of education form and education mode, which will inevitably become the development trend of education informatization.

1 INTRODUCTION

Artificial intelligence is a strategic technology to lead the future. The national intelligence education project was stated in the Development Plan of New Generation Artificial Intelligence issued by the State Council, set up artificial intelligence related courses in primary and secondary schools, gradually promote programming education, construct and improve the facilities of Artificial Intelligence Science popularization base, and encourage various forms of artificial intelligence science popularization creation. It can be seen that the arrival of the era of artificial intelligence will inevitably lead to profound changes in the field of education. In his book "artificial intelligence", Li Kaifu pointed out: in the era of artificial intelligence, stylized and repetitive skills that can be mastered only by memory and practice will be the most worthless skills, which can almost certainly be completed by machines; on the contrary, those skills that can best reflect people's comprehensive quality, such as people's comprehensive analysis and decision-making ability for complex systems, and art aesthetic ability and creative thinking of culture, intuition and common sense inspired by life experience and culture, and the ability to interact with others based on one's own emotions (love, hate, enthusiasm, indifference, etc.) These are the most valuable and worth training and learning skills in the era of artificial intelligence [1]. Looking back on the reality of China's higher education, stylized teaching mode and memory knowledge system are common in university teaching. For the era of artificial intelligence education, higher education must reform, introduce teaching mode adapting to artificial intelligence, and strengthen human-computer cooperation, so as to better cultivate talents adapting to the era of artificial intelligence and better adapt to and promote the coordinated development of human-computer society.

2 CURRENT SITUATION OF ARTIFICIAL INTELLIGENCE EDUCATION RESEARCH

In January 2016, the Japanese Cabinet meeting deliberated and passed the fifth basic plan for science and Technology (2016–2020), proposing to realize super intelligent society 5.0. In October 2016, the US government issued the national strategic plan for research and development of artificial intelligence. In December 2016, the British government released the report of artificial intelligence of opportunities and impacts of future decision-making. In July 2017, the Chinese government released the development plan for a new generation of artificial intelligence. These landmark events in the field of artificial intelligence have been widely reported by the media, and have attracted the general attention of the industry and the public, attracting more and more researchers to conduct in-depth exploration of artificial intelligence, and officially announced the rise of the third wave of artificial intelligence, and mankind is entering a new era of artificial intelligence [2].

Under the influence and penetration of artificial intelligence wave, more and more artificial intelligence technologies and applications appear in the field of education. In classroom teaching, Jill Watson, a virtual teaching assistant, was used in more than 300 students' classes to answer students' questions instead of teaching assistants. In the aspect of adaptive learning, Knewton platform can collect the behavior data of learners in the learning process, analyze and predict students' learning interest, knowledge level, learning style and learning progress, so as to provide personalized learning services. In terms of educational robots, Lego's latest generation of programmable intelligent robot product Mindtorms Ev3 has enhanced the interaction with intelligent devices, added Wi Fi module, can connect with IOS and Android devices, control through app, and support human-computer interaction through improved microphone and speaker equipment. In addition, artificial intelligence technology has also been widely used in education scenarios such as examination evaluation and diagnosis, campus management and teaching management, educational decision-making and educational governance. Artificial intelligence technology promotes the transformation and reconstruction of education form and education mode, which has become the inevitable trend of the development of education informatization. The new generation of artificial intelligence development plan issued by the Chinese government in 2017, clearly required the development of intelligent education, the use of intelligent technology to accelerate the reform of talent training mode and teaching methods, build a new education system including intelligent learning and interactive learning, carry out the construction of intelligent campus, and promote the application of artificial intelligence in the whole process of teaching, management and resource construction. Develop three-dimensional comprehensive teaching field and online learning education platform based on big data intelligence. Develop intelligent education assistant, establish intelligent, fast and comprehensive education analysis system, establish learner centered education environment, provide accurate push education service, and realize the customization of daily education and lifelong education [3].

3 TEACHING SIGNIFICANCE BASED ON ARTIFICIAL INTELLIGENCE

3.1 *Artificial intelligence will change the education ecology and create a convenient and efficient service education environment*

Artificial intelligence can provide personalized, diversified and high-quality services for education, and accelerate the construction of a new education system including intelligent learning and interactive learning [4]. In the construction of intelligent campus, it can promote the whole process application of artificial intelligence in teaching, management, evaluation, resource construction and other aspects, develop three-dimensional comprehensive teaching field and online learning education platform based on big data intelligence, develop intelligent education assistant to establish intelligent, fast and comprehensive education analysis system, establish learner centered education environment to provide accurate push realizing the customization of daily education and lifelong education. The efficient service of artificial intelligence will make the campus full of wisdom and warmth.

3.2 *Artificial intelligence will change the way of teaching and learning and build a student-centered personalized learning platform*

Artificial intelligence will subvert the traditional teaching method and realize personalized learning based on complex situations to meet the students' differences [5]. At the same time, students' learning will stimulate teachers' teaching, and teachers and students will truly realize the mutual benefit of teaching and learning. Artificial intelligence education will analyze every student's learning traces, such as the time spent on homework and the difficulty of homework. Teachers can provide targeted guidance according to the different needs of students. Through the data analysis of students' cognitive basis, learning behavior, learning habits, thinking characteristics and other data, provide personalized and customized learning content and methods for each student, design a certain challenging learning task, so as to stimulate students' deep learning desire.

3.3 Teachers' professional focus will shift

Teachers will pay more attention to the cultivation of students' cognitive ability, cooperation ability, innovation ability and professional ability, which support the lifelong development and adapt to the requirements of the times.

3.4 Artificial intelligence will be more people-oriented and construct diversified and integrated teaching resources

The school curriculum system in the era of artificial intelligence should be more diversified, rich and integrated [6]. An era of people-oriented and respect for personality has come. School curriculum should give each student more space to experience. The school should become a learning center, so that students with similar interests can gather together and form a clustering effect.

3.5 Artificial intelligence education will pay more attention to interdisciplinary exploratory learning

Promote the cross integration of artificial intelligence with neuroscience, cognitive science, quantum science, psychology, mathematics, economics, sociology and other related basic disciplines, and cultivate students' key ability and essential character to adapt to the future social development and personal life-long development with rich teaching resources.

3.6 Integration of education and local economic development

The teaching in the era of industrial intelligence will be more combined with the actual situation of local economic and social development, provide wisdom for the development of local economy, and cultivate talents for the development of local economy.

4 CONNOTATION OF COLLEGE TEACHING RESEARCH BASED ON ARTIFICIAL INTELLIGENCE

The goal of teaching research in Colleges and Universities Based on artificial intelligence technology is based on the technology of artificial intelligence, closely combined with the development of local economy, to study the efficient man-machine cooperation ecological teaching system based on the ability improvement of teaching resources system, ubiquitous learning space, multi-level teaching methods, teaching process optimization and automatic learning evaluation system, so as to change teachers' teaching ideas and study The three element teaching community of teachers, students and artificial intelligence, so as to form a teaching mode of informatization and automation which can effectively cultivate innovative talents serving local economic development [7].

4.1 Research on application technology of artificial intelligence education

The research in the field of artificial intelligence education application technology mainly solves the problems of which artificial intelligence technologies can be used in educational application scenarios and how to apply artificial intelligence technologies to educational scenes. For the research in the field of artificial intelligence education application technology, it focuses on how to use the general machine learning algorithm and mature artificial intelligence engine to solve the technical problems in the actual education scene, and pays attention to which education problems, discipline problems and business problems in the actual education application scenarios can be solved by means of artificial intelligence, pay attention to the technical characteristics and Application Strategies of artificial intelligence.

4.2 *Research on teachers' teaching ability in the era of artificial intelligence*

In the era of artificial intelligence, the research of teachers' teaching ability should not only study the professional knowledge, psychological knowledge and communication ability of teachers, but also the ability of how to learn, master, understand, identify and select artificial intelligence technology and scheme. In particular, it should have the ability to summarize and refine the technical requirements in the real education scene, and strengthen the cooperation with artificial intelligence Ability to collaborate.

4.3 *Research on the teaching resource system based on ability improvement*

In the era of artificial intelligence, the research on the construction of teaching resources system based on ability improvement needs to study and analyze which abilities are stylized, repetitive and can be mastered only by memory and practice, and which abilities are based on comprehensive analysis of complex systems and need to be mastered by people. Research and design the teaching resource system based on a certain professional direction, with complex comprehensive analysis ability or can solve the problems of local economic development, and study whether these teaching resource systems are constructed by curriculum or ability module. Study how these teaching resources are expressed, stored and disseminated. Through the construction of teaching resources, it is more convenient to carry out personalized teaching and customized teaching, so as to better carry out the cultivation of personalized and innovative talents.

4.4 *Research on ubiquitous learning space*

The personalized teaching brought by artificial intelligence technology will not only have an impact on the teaching methods and the relationship between teachers and students, but also may have a profound impact on the organizational form of teaching. The class teaching system which has lasted for hundreds of years will be severely challenged, and ubiquitous learning may usher in a real breakthrough. In the era of artificial intelligence, because the teaching content will change from standardization to customization and personalization, students of different ages and with similar learning foundation gather together to carry out a certain teaching activity, which can facilitate teachers to organize teaching. Therefore, mixed age education may become the norm, and the students corresponding to different teaching contents and teaching periods will also change dynamically. The study of ubiquitous learning space has a great significance to better carry out mixed age education. The so-called ubiquitous learning space refers to the space for personalized learning and mixed age learning. It can be physical space, virtual space, or production site. The research of ubiquitous learning space is to study how to construct the space suitable for practical teaching through the dynamic optimization combination of intelligent home, sound, light and electricity in the existing classroom, and how to combine artificial intelligence technology to make students have deep experience in the learning space, study how to use virtual reality technology and computer games to let students enter the environment that is impossible to enter in the real physical space for experience, studies the perceptual space in which the object makes sound in a specific area so that students can communicate with the surrounding environment, and how to better develop practical experimental space (such as laboratory, production workshop), etc. [8].

4.5 *Research on multi level teaching method*

Study the multi-level teaching methods based on principle teaching, technology teaching and operation skill teaching supported by artificial intelligence technology, so that each teaching method can fully mobilize the enthusiasm of teachers, students and artificial intelligence, and jointly and efficiently solve teaching problems.

4.6 *Research on the optimization of teaching process*

Study how teachers cooperate with artificial intelligence in the pre class link to build an efficient personalized teaching scheme, and how to provide guidance for students with weak basic knowledge. Study how to construct personalized learning space, how to monitor the learning process and how to carry out collaborative learning in the link in the class. Study the after-school link, according to the artificial intelligence, how to build personalized learning space, how to monitor the learning process, and how to carry out collaborative learning after the calculation and analysis of the data of the learning platform, to infer the difficulties and puzzles that each student may have, and formulate a personalized counseling program.

4.7 *Research on automatic learning evaluation system*

Study how to use the information system to automatically record, store, calculate and analyze all teaching data before, during and after class, and timely give the existing problems in teaching, and give suggestions and plans for the next stage of teaching. With the teaching process going on, the evaluation of process learning can be given, and the final assessment can also be carried out through the system, so that the teaching evaluation can be carried out with the evaluation scores were accumulated. Study how to use the information system to automatically evaluate the teaching activities such as students' experiment and production practice [9].

4.8 *Research on the ternary teaching community of teachers, students and artificial intelligence*

Study the role orientation of teachers, students and artificial intelligence, study the mechanism of teachers, students and artificial intelligence to jointly promote the improvement of teaching quality and teaching efficiency, and study the system structure of the ternary teaching community of teachers, students and artificial intelligence [10].

5 SUMMARY

In the traditional teaching mode, the teaching space is basically limited in the physical space. In the era of artificial intelligence, in order to more accurately understand some knowledge and grasp the ability, we should not only study and explore in the real physical space, but also give full play to the deep experience and understanding brought by virtual space to students. Because the learning and feeling of virtual space are very different from physical space, how to use virtual space for teaching exploration or effectively integrate physical space and virtual space to solve some teaching problems is a new thing. In order to sum up the mechanism and strategy of effective teaching by using classroom, laboratory, production site, virtual laboratory and game, the ubiquitous learning space should be studied in depth.

In the traditional teaching mode, the main body of teaching was teachers and students, teachers and students are human beings, their thinking mode and values are basically similar, the basis of mutual communication is good, can carry out full communication. However, in the teaching system composed of teachers, students and artificial intelligence, artificial intelligence is another teaching subject. How to effectively communicate among the three, so as to achieve the best teaching quality, is a problem that needs to be deeply studied and practiced.

ACKNOWLEDGEMENT

This paper was supported by "The Higher Education Reform Project of Guangdong Province in 2018 – Research and Practice of Teaching Method Reform Based on Innovation Ability Training" and "2020 Project of Association of Fundamental Computing Education in Chinese Universities – Development and Practice Research of Curriculum of Production Education Collaborative Research (No.: 2020-afcec-016)".

REFERENCES

[1] Kaifu Li, Yonggang Wang, Artificial Intelligence, Cultural Development Press, Beijing, 2017.

[2] Yafei Wang, Bangqi Liu, iResearch Overview of the Application of AI in Education, Modern Educational Technology, 1(2018): 5–11.

[3] Tingting Gao, Jiong Guo, Research Review on the Application of Artificial Intelligence in Education, Modern Educational Technology, 1(2019): 11–17.

[4] Zeyuan Yu, Jinghua Zou, Teaching Reconstruction from the Perspective of Artificial Intelligence, Modern Distance Education Research, 4(2019): 37–46.

[5] Fan Deng, How Do Universities Train Talents in the Era of Artificial Intelligence? Academic Exploration, 9(2019): 143–150.

[6] Xiaojie Quan, Defeng Qiu, On the New Form of Curriculum from the Perspective of Artificial Intelligence, Contemporary Education Sciences, 6(2020): 33–39.

[7] Dejian Liu, A Review of Researches on Reforms of Talents Training in Artificial Intelligence Enabled Colleges and Universities, e-Education Research, 11(2019): 106–113.

[8] Luhai Mei, Construction of Mobile Internet Online Digital Course and Ubiquitous Learning Space Based on Constructivism, Chinese Vocational and Technical Education, 11(2020): 72–76.

[9] Xiaohui Xu, Guyin Li, Xinyin Guo, Design and Construction of Procedural Learning Evaluation System from the Perspective of Effective Learning, Education Teaching Forum, 1(2019): 218–219.

[10] Lei Wang, Ruike Zhang, "Human-Technology" Collusion: Transformation and Orientation of Teachers' Role in the Era of Artificial Intelligence, Journal of Curriculum and Instruction, 02(2020): 56–65.

Computational Social Science – Luo, Ciurea & Kumar (eds)
© 2021 Taylor & Francis Group, London, ISBN 978-0-367-70193-2

Collocational patterns in China English: A corpus-based study

L.X. Xia
Laboratory of Language Engineering and Computing of Guangdong University of Foreign Studies, Guangzhou, China

Y. Xia*
Nanfang College of Sun Yat-Sen University, Guangzhou, China

C.Y. Liao
Center for Lexicographical Studies of Guangdong University of Foreign Studies, Guangzhou, China

ABSTRACT: This study explores the collocational patterns of ten high-frequency adjectives in the Written Corpus of China English (abbreviated to WCCE) with a focus on its collocations with nouns. A corpus-based comparative methodology was adopted to investigate the differences between China English and British English in the use of adjective and noun collocational patterns. The major findings of the study can be stated as follows: (1) Some collocational structures only occur in the Chinese context, such as *economic rebalancing, economic belt, and international airspace.* (2) Some collocational structures, like *economic corridor, economic construction, new energy*, are used more frequently in China English than in British English. (3) The frequency of some collocational structures in BNC is much greater than that in WCCE, in particular, *Old Testament, local election*. The reasons for these different uses of collocational structures with distinct Chinese features might include (but not limited to), (i) the development of China's economy (e.g. *economic belt, economic corridor*); (ii) the differences between China and Britain in politics and society (e.g. *Old Testament, local election*); (iii) the influence of the development of science (e.g. *new energy*); (iv) linguistic transfer (e.g. *old Yao, economic take-off*). The findings show that China English has its distinctive collocational patterns which may suggest that English used in China has undergone a process of nativization.

1 INTRODUCTION

Empirical research in the nativization of English is increasing, among which the difference on lexical meaning is an important study field. (Adegbija 1989; Alm 2003;; Bamiro 1994; Dubey 1991; Lee & Collins 2004; Lee et al. 2009) Most scholars, currently however, are investigating nouns or noun phrases, and less attention was paid to other parts of speech, such as adjectives. In fact, the use of adjectives might exhibit the most typical feature of the nativization of English in Chinese context. When adjectives co-occur with other word classes (such as nouns), the feature of nativization will be more distinctive (Yu & Wen 2010). Therefore, this paper will examine the adjective and noun collocational patterns in WCCE.

2 METHODOLOGY

2.1 *Source of the Data*

The target corpus, WCCE is a collection of samples of written language with a total of 13,962,102 tokens, from four types of genres, i.e. fictions, magzines, newspapers, and academics, designed to

---

*Corresponding author

Table 1. The collocational structures only in the Chinese context.

Collocations	Log-likelihood	Rank	Collocations	Log-likelihood	Rank
Old Tibet	197.01	1	Economic takeoff	50.97	11
Economic work	152.09	2	Many hostess	44.69	12
Other hostess	148.32	3	New CPC	44.50	13
Economic mobilization	105.47	4	Old Wang	44.18	14
Economic belt	85.96	5	Many NGOs	38.32	15
Economic rebalancing	83.24	6	Many netizen	37.39	16
Local govt (debt)	69.97	7	International airspace	34.79	17
First China	69.81	8	New iPhone	32.09	18
New iPad	59.74	9	Other concubine	30.72	19
Old Yao	58.24	10	International metropolis	27.62	20

represent a wide cross-section of China English in modern times. BNC (British English Corpus) is used as a reference for contrast. The 10 most frequently-used adjectives and their hits in the corpus are as follows: other (20440), new (18391), many (14666), economic (14002), first (13805), such (12801), local (10178), old (8885), international (8446), and last (8364).

2.2   Data analysis

The list of statistically significant adjective-noun collocations in the WCCE is based on collocation frequency (≥ 5 hits) and association scores calculated via the loglikelihood test.

2.2.1   Collocational structures that only occur in the Chinese context

Owing to space limit, we will focus on the top 20 collocations in the WCCE (in descending order of Log-likelihood index) (see Table 1).

One very important argument needs to be mentioned here is that, the absence of any collocation in BNC does not imply the phrase in question is problematic or unacceptable, let alone erroneous. On the contrary, it serves as clear evidence that the English used in the Chinese context is, to some extent, undergoing a process of nativization, reflecting the unique Chinese ideology different from the West in economy, politics, culture and other aspects. It will be further discussed in Section 3.

2.2.2   HF Collocations in WCCE

The following table shows the collocational structures that are used more frequently in China English. Due to space limit, we will focus on the top 20 collocations (in descending order of WCCE/BNC ratio) (see Table 2).

In all of the cases, there is no question about the acceptability and comprehensibility of the collocations, as evidenced in their occurrence in the BNC. The more frequent a collocation is, the more likely it is that people from other cultures would encounter it (Liang 2015).

2.2.3   HF collocations in BNC

The following table shows that the frequency of some collocational structures in BNC is much greater than that in WCCE. Due to space limit, we will focus on the top 20 collocations (in descending order of BNC/WCCE ration) (see Table 3).

The overused collocations in BNC will be further discussed in Section 3.

Table 2. HF collocations in WCCE.

HF collocations	Hits per million in WCCE	Hits per million in BNC	WCCE/BNC ratio	Rank
Local media	14.11	0.03	470.32	1
Old China	8.52	0.02	426.15	2
Economic corridor	3.58	0.01	358.11	3
Economic globalization	9.96	0.03	331.85	4
Economic governance	3.15	0.01	315.14	5
Economic data	3.08	0.01	307.98	6
First moon	2.86	0.01	286.49	7
International non-proliferation	2.36	0.01	236.35	8
Economic construction	7.81	0.04	195.17	9
New energy	43.19	0.28	154.24	10
International student	5.87	0.04	146.83	11
Local GDP	1.36	0.01	136.08	12
Local regulation	5.16	0.04	128.92	13
International seabed	1.29	0.01	128.92	14
Economic transition	4.37	0.04	109.22	15
Economic take-off	0.86	0.01	85.95	16
New nexus	0.86	0.01	85.95	17
International board	3.44	0.04	85.95	18
Local tourism	3.37	0.04	84.16	19
International exchange	14.68	0.18	81.57	20

Table 3. HF collocations in BNC

HF collocations	Hits per million in WCCE	Hits per million in BNC	BNC/WCCE ratio	Rank
Local council	5.87	0.07	81.96	1
Local plan	3.9	0.21	18.15	2
New settlement	6.39	0.57	11.15	3
Local paper	2.24	0.21	10.43	4
Local service	2.03	0.21	9.45	5
International rugby	0.61	0.07	8.52	6
Old testament	2.15	0.29	7.50	7
Such matter	4.76	1	4.75	8
Old person	2.52	0.57	4.40	9
Old generation	1.44	0.36	4.02	10
Many thousand	3.25	0.86	3.78	11
International scene	0.79	0.21	3.68	12
New series	2.61	0.72	3.64	13
New scheme	2.02	0.57	3.53	14
Many teacher	1.71	0.5	3.41	15
Such question	2.44	0.72	3.41	16
Local election	3.74	1.15	3.26	17
First sign	2.29	0.86	2.35	18
Such change	2.69	1.15	2.66	19
First edition	2.55	1.15	2.23	20

3 FINDINGS AND DISCUSSION

3.1 *Unique collocations and HF collocations in WCCE*

Both the unique collocations and HF collocations in WCCE fall into five categories, namely, collocations related to: (a) The development of economy (Table 4); (b) Political field (Table 5);

Table 4. CE collocations related to the development of economy

Collocations	Log-likelihood	Collocations	Log-likelihood
Economic belt	85.96	Economic rebalancing	83.24
Economic corridor	118.59	Economic transition	125.63
Economic work	152.09	Economic data	115.44
Economic construction	289.49	International board	72.8
Local GDP	34.58	Economic globalization	413.65
Economic governance	75.9	Local tourism	111.85
Economic mobilization	105.47		

Table 5. CE collocations related to political field.

Collocations	Log-likelihood	Collocations	Log-likelihood
International airspace	34.8	many NGOs	38.32
International seabed	18.4	new CPC	44.5
International non-proliferation	86.16		

Table 6. Collocations related to social factors.

Collocations	Log-likelihood	Collocations	Log-likelihood
International metropolis	26.62	Old China	147.2
International students	122.67	Old Tibet	197
Local media	234.34	Other concubines	30.72
Many hostesses	44.69	Other hostesses	148.32
Many netizen	37.39		

(c) Social factors (Table 6); (d) Culture and technological development (Table 7); (e) Linguistic transfer (Table 8).

The first group concerns collocations related to China's economic development. For example, the pattern, *economic belt*, refers to the Silk Road Economic Belt (SREB), which is the overland interconnecting infrastructure corridors to create a good surrounding political, national defense and environment for China. China has been vigorously reviving the Silk Road Economic Belt in recent years. Hence, the phrase *economic corridor* is also emphasized.

Some collocations reflect the actual economic development status since the Chinese reform and opening-up in 1978. For instance, the basic line of the Communist Party of China focuses on the central task of economic construction. The collocations, such as *economic work*, *economic construction local GDP*, and *economic takeoff*, are stressed accordingly. Then, some emerging economic problems in recent years require new economic policies, concerning *economic governance*, *economic mobilization*, *economic rebalancing*, and *economic transition*

Patterns like *economic data* and *international board* are related to the derivative areas of the economy. From the view of global economic development, the increasing economic integration and interdependence among nations and regions promote the globalization of the economy, thus the expression *economic globalization* highly occurs.

These collocations are strongly associated with the development of the economy in China especially by contrast with its British counterpart. Therefore, it seems obvious why these English terms are needed.

The second category pertains to the political field like the term *international airspace*. This reflects China's foreign policy in territorial integrity. Another term *international seabed* also

Table 7. Collocations related to culture and technological development.

Collocations	Log-likelihood	Collocations	Log-likelihood
New energy	720.78	New Nexus	72
New iPad	59.73	First moon	23.63
New iPhone	32.09		

Table 8. Collocations related to linguistic transfer.

Collocations	Log-likelihood	Collocations	Log-likelihood
Old Wang	44.18	Economic take-off	33.49
Old Yao	58.23	Economic takeoff	50.96

belongs to the same category. In a real context, *international seabed* often occurs in the proper name *International Seabed Authority*, referring to an intergovernmental body established to organize, regulate and administrate all mineral-related activities in the international seabed area beyond the limits of national jurisdiction, an area underlying most of the world's oceans. Similarly, *international non-proliferation* is not often used independently but with other nouns, forming into patterns like *international non-proliferation regime*, *international non-proliferation mechanism*, or *international non-proliferation efforts*.

A few collocations, such as *many NGOs*, *new CPC*, do not appear regularly, but they can still reveal the preference of word choice of China English (CE). For example, *new CPC* in real context does not mean a new Party in China, but a new leading organization of the Party, thus, *new CPC leadership*, *new CPC Central Committee*, are often used.

This group is related to social factors. The past thirty years have witnessed the tremendous changes in China's society in various aspects such as the construction of urbanization (e.g. *international metropolis*), opening to the outside world (e.g. *international student*), the rise of social media (e.g. *local media*, *many netizen*), the discard of the past backwardness (e.g. *old China*, *old Tibet*, *many/other hostesses*, *other concubine*). The development of Chinese society has a great effect on the collocational preference of CE.

This group of collocations is mainly connected with culture and technological development. Advance in science has contributed to the flourishing of the culture. Since the 1940s and 50s, the third revolution of science and technology required more attention to be paid to information technology and the expansion of new energy. Android smartphones and iPhone accordingly came into people's life about a decade ago. Apple Inc subsequently introduced other products such as iPad. Since then, all sorts of intellectual products (e.g. *new iPad*, *new iPhone new Nexus smartphone*) are becoming more popular. In addition, artificial intelligence is boosting the rapid development of satellite astronomy technique. After the first moon probe, *first moonrover*, and *first moon orbiter* in the world, China is striving to break new records in the lunar exploration.

Language is a component of culture. The development of science will, to a certain extent, affect the use of language, in this case, the preference of collocations.

The last group concerns CE user's collocational preferences which have been transferred from the patterns of expressing ideas in Chinese. The linguistic transfer has several branches while transference concerning collocation falls into the semantic field (Yu et al. 2010). The first two examples (*old Wang* and *old Yao*) are terms of address, usually used to call a senior colleague or friend who shares the same level of position in a company or clan. Chinese people have the habit of calling an acquaintance a nickname sometimes in the form of "old/small + family name". In the case of "old Yao", "Yao" is the last name of a person and "old" indicates a relatively elder age. Similar terms such as "small Wang" found in WCCE are used to call a younger person or a junior staff whose family name is "Wang". However, the English address names are in another case due to different conventions. English appellations are relatively simple, apt to call a person commonly by his name. Therefore, such collocations in WCCE cannot be found in British English.

The collocation economic takeoff/take-off (jīngjìténgfēi) in Chinese is a subject-predicate structure. "jīngjì" (economy) is the subject and "téngfēi" (take off) can be actually taken as a verb. When translated into English, téngfēi (take off) is nominalized, in the noun form of the word, i.e. takeoff/take-off. Thus, the phrase economic takeoff/take-off is transferred from the Chinese word-formation.

3.2 Analysis of HF Collocations in BNC

A few examples chosen from Table 2.3 serve to illustrate the HF collocations in BNC such as *local council*, *local election*, *international rugby*, and *Old Testament*.

Various reasons can be enumerated to explain such phenomena, including political factor, eco-social factor, religious factor, and so forth. For example, *local council* is the organ in charge of the local politics and administration in Britain, while such a system is rarely known in China. As for *local election*, it is a way to elect councilors to form the local administrations of the United Kingdom, which is quite different from Chinese electoral system. The phrase *international rugby* demonstrates the cultural difference between Britain and China in that rugby is a popular ball game in Britain while China has ping-pong (table tennis) as its national sport. Moreover, these two countries have different religious faiths. The major religion in the United Kingdom is Christianity. But in China Buddhism and Taoism prevail. Thus, *Old Testament* is less common in Chinese context compared to British English. Britain and China are two distinctive nations in various ways, so there is no doubt that their choice of collocations can be very different.

4 CONCLUSION

This paper illustrates and presents how China English differs from British English in terms of collocational features. Below is a brief summary of the study's major contributions: (1) It elucidated Chinese English speakers' preferences for collocation patterns which contributes substantially to the study of China English. (2) It used authentic and natural data from corpora to investigate the collocational structures in China English, which is a methodological contribution to the study of China English. (3) The findings of the study can be directly applied to the dictionaries which describe different English varieties. (4) The paper explored the structural meanings of the collocational structures in China English which is meaningful to Natural Language Processing with regard to the understanding of natural language.

ACKNOWLEDGMENTS

This work was supported by the Laboratory of Language Engineering and Computing of Guangdong University of Foreign Studies (Grant reference LEC2017ZBKT002: A Study on Grammatical Nativisation of China English), the Humanities and Social Science Planning and Funding Office of Ministry of Education of the P.R.C. under Grant No. 15YJA740048, and the Department of Education of Guangdong Province (grant reference 2017WTSCX32: A Study on the Definitions of Chinese-English Dictionaries for CFL Learners).

REFERENCES

Adegbija, E. 1989. Lexicosemantic variation in Nigerian English. *World Englishes*, 8(2):165–177
Alm, C. O. 2003. English in the ecuadorian commercial context. *World Englishes*, 22(2): 143–158.

Bamiro, E. O.1994. Innovation in Nigerian English. *English Today*, 10(3): 13–15

Dubey, V. S.1991. The lexical style of Indian English newspapers. *World Englishes*, 10(1):19–32

Lee, J. F. K. & Collins, P. 2004. On the Usage of Have, Dare, Need, Ought and Used to in Australian English and Hong Kong English. *World Englishes*, 23(4), 501–513

Lee, N. H., Ping, L. A. & Nomoto, H. 2009. Colloquial Singapore English Got: Functions and Substratal Influences. *World Englishes*, 28(3): 293–318

Liang J. 2015. *Collocational Features of China English: A Corpus-based Contrastive Study*. Doctoral Dissertation of the Hong Kong Institute of Education.

Yu, X. & Wen, Q. 2010. Collocation patterns of evalua-tive adjectives of English in Chinese newspapers. *Foreign Language Teaching and Research*, 254(5): 23–28

Computational Social Science – Luo, Ciurea & Kumar (eds)
© 2021 Taylor & Francis Group, London, ISBN 978-0-367-70193-2

Predicting user behavior based on informative lifelogging

G.Q. Liu
ShenYang Jianzhu University, Liaoning, China

L.B Zhang*, C.H. Li & Y. Fu
General Hospital of North Theater Command, Liaoning, China

M.U. Rehman
Northeastern University, Liaoning, China

ABSTRACT: From 2011 to now, the authors of this paper have published the personal Geo-tagged photo dataset named Informative Lifelogging on the Internet, which has a total of 6000 personal life records. Each dataset includes photos, GPS information, time, description, and address information. At the same time, the authors tagged each dataset to distinguish the behavior classification of these data. In this paper, we introduce the dataset for more researchers, and then the authors mainly discuss how to predict the behavior of individual users based on the dataset collected.

1   INTRODUCTION

Now, in the social computing research area, there are some researchers choosing the continuity of mobile data to understand people's behaviors and movement modes. Researchers usually use two types of datasets.

The first category is the study of behavior analysis, trajectory analysis, identifying events, information retrieval, and so on through personal data collected by wearable devices. This kind of data is characterized by a large amount of data, generally taking GPS and time data as the main recording content. Catha Gurrin published a book entitled *Lifelogging: Personal Big Data*, which details relevant research in this field [1]. There are many studies in the field of LifeLog.

References [2] and [3] study the retrieval of LifeLog data. Reference [4] studies the tagging problem in the process of LifeLog data retrieval. In Reference [5] the team tried to collect other people's Life-Log data through a system called BlogWear. Reference [6] pays attention to people's health through Lifelogging. With the development of science and technology, people record LifeLog data through various devices. With the increase in data volume, Lifelogging has become a research field related to large data and artificial intelligence. In References [7,8,9] Lifelogging is described a problem that needs to be solved by artificial intelligence technology. References [10,11] give a summary of Lifelogging.

The second category is related data obtained through social media software such as Facebook, Weibo, Twitter, and so on. This kind of data is characterized by a small amount of personal data, but a large amount of user data. Malaysian Lifelogging over Twitter is a typical example [12]. The author of this paper collected data on Twitter from 10 million Malaysians, but the data collected did not include GPS information.

However, few researchers have been able to collect LifeLog data and tag all the data for 8 years. The dataset is the main contribution of this thesis.

The overall structure of the study takes the form of four sections, including this Introduction. Section 2 begins by introducing the Lifelog system and the data details the authors have collected.

---

*Corresponding author

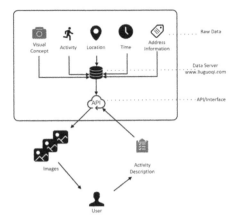

Figure 1.   System structure.

Figure 2.   The interface of the system.

Section 3 analyzes how to predict user behavior by the data, which is the second contribution of this thesis. Finally, the conclusion gives a brief summary and critique of the findings.

2   THE LIFELOG SYSTEM AND THE DATASET

We have developed a system (www.liuguoqi.com) including a background management interface and mobile terminal. Users can submit their own pictures and descriptions by download to the mobile terminal we published on our system. At the same time, the system automatically obtains the user's GPS and time information, and these data are submitted to the data server together. Researchers who want to use the system need to apply for a username and password in advance when submitting data. After obtaining the username and password, anyone can use the system to upload their own data. The structure of the system is similar to that in Reference [3] which is shown in Figure 1.

After submitting data, users can view the data through the web page. Users can also filter data using any keyword or by date. The authors used pagination in the web pages to load pages faster and increase the processing speed of the web page. What we can see now is the data of a user, which are the data used in this experiment. The interface of the system is shown in Figure 2.

Table 1. Names of 14 categories

Behavior type	Detailed description
Working at school	Shows behavior of the user while working at school. For instance, studying or doing any other activity at school.
Teaching	Shows behavior when user is teaching.
Work outside	Shows behavior when user is working. Work can be of any type, we are not concerned about what work the user is doing.
Rest	Shows behavior of user when resting at home or any other place.
Chores	Shows a routine task of the user, especially a household one.
On road	Shows behavior when user is on the road which gives us a clue that user is on the way to somewhere or coming back from somewhere.
Eating at school	Shows behavior of user while eating at school.
Eating at home	Shows behavior of user while eating at home.
Eating outside	Shows behavior of user while eating outside.
Tourism	Shows behavior of user while traveling either for pleasure or business.
Party	Shows behavior while user is having a party with friends whether it be a dinner party, a get-together, or any other type of party.
Entertainment	Shows behavior of user while entertaining himself either by having a meal or drink with friends or by playing or watching any of the user's favorite sports or attending a party which is enjoyable for the user or any other form of entertainment.
Communication	Shows behavior of user while in communication or conversation with someone.
Other	This shows the behavior of the user other than those mentioned above.

At present, the amount of personal data of the author has reached about 6173. The dataset records the things that have been important to the uploader in the past 8 years. Each dataset contains at least one photo, and a description of the photo, which basically describes what the uploader is doing when the photo is uploading. The uploaded photos and information are saved into the database, and we publicized these data through our website system. Compared with the dataset the author previously used in literature [9], we have few data but we have a long period and more rich information. The information in each dataset includes a photo, userid, description, address, behavior, creation time, latitude, and longitude. For the tag of the behavior, we divide all the behavior types into 14 categories by hand. The classification names and details are shown in Table 1.

3 PREDICTING USER BEHAVIOR

The first method we came up with was to use the decision tree algorithm, and therefore we tried to use the ID3 algorithm in the paper. Through the existing personal life data, a decision tree model is established, which can effectively classify and predict the user's behavior according to their personal information. The decision tree model adopts an ID3 algorithm. The core idea of this algorithm is to use the principle of information entropy to select the attributes with the greatest information gain as classification attributes, expand the branches of the decision tree recursively, and complete the construction of the decision tree.

3.1 *Data processing*

Existing user data information includes Position, Time, Description, and Behavior, which are labeled. Position information consists of longitude and latitude, which splices two strings of longitude and latitude into one string as the value of the Position attribute; Time divides them into morning, noon, afternoon, evening, and night by judging the user's time period, such as marking "2016-10-05 11:36:47" as "noon" to further converge data; Description is descriptive information, usually a paragraph. String identifies where the Stanford-Postagger tool is used to determine the part of speech (action verbs, nouns, prepositions) of each value in the string. To assign a value

Table 2. Processed data for experiments

Position	Time	Description	Behavior
124.39168543.175133	Morning	Eat	Eating at home
112.56785837.811481	Morning	Meeting	Working outside
112.56258737.80783	Evening	Taiyuan	Eating outside
123.67241242.232148	Night	Passing	On road

to it, we usually take the action verb in the string, followed by the proper noun, such as getting the part of speech of each word from "Eat at home" and comparing its weight. Finally, we get the word "Eat," whose part of speech is marked as "VB," which belongs to the part of speech with the highest weight in the string; finally, the labeled action classification. The final dataset is as follows in Table 2.

In Table 2, the position is longitude and latitude. Time is the marking results. Description is the key word obtained from the descriptive information. Behavior is the tag we added.

3.2 *Algorithm process*

As the ID3 algorithm is a common algorithm for most researchers, we will not introduce details of it here, we only introduce the algorithm process we designed in the paper.

1. Customized information entropy calculation function for computing information entropy of datasets.
2. Custom data partitioning function for partitioning datasets according to specified values of specified features
3. Step 2's self-dataset is a function input to step 1, which can calculate the information entropy $H(Di)$ of the dataset partitioned by a specified value $(A = ai)$ of a specified feature, and the sample probability of the dataset partitioned by a specified value $(A = ai)$ of a specified feature $|Di/|D.|$
4. The information entropy $H(Di)$ and sample probability $|Di|/|D|$ of the dataset partitioned under each value are calculated, multiplied, and then the empirical conditional entropy $H(D|A)$ of feature A to dataset D is obtained by adding them.
5. Calculate the information gain $g(D, A) = H(D) - H(D|A)$ of feature A to the dataset.
6. By analogy, the information gain of each feature to the dataset is calculated, and the feature with the largest information gain is taken as the best partition feature to get the tree $T1$
7. For each node of $T1$, repeat steps 3–6, select the feature with the greatest information gain, continue to divide the data, and get a new decision tree.
8. The decision tree is completed when the information gain is less than the threshold, or no feature can be partitioned, or all instances under each branch have the same classification. The final structure of the decision tree is shown in Figure 3.

In the decision tree model, 1000 datasets were tested, and 964 datasets were successfully classified. The accuracy of the decision tree model obtained by this method meets the requirements. In the data of classification failure, the data of working at school and teaching, working at school, and working outside are similar. The other data include cases of vague classification and classifications. The experimental results are illustrated with examples below.

3.3 *Analysis of predicted results*

Enter two test data [{longitude: 124.391685, latitude: 43.175133, describe: Eat at home, date: 2016-10-05 11:36:47, behavior: Eating at home}, {longitude: 123.43364, latitude: 41.76892, describe: Meeting in the company. Peaceful discussion, business date: 2017-11-27:29:30, behavior: outside

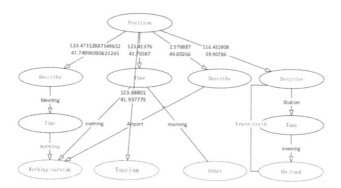

Figure 3. Decision tree structure.

working}]. First, the data are processed according to the process of data processing, and finally the data [{Position: 124.39168543.175133, Time: noon, Describe: Eat, Behavior: Eating at home}, {Position: 123.4336441.76892, Time: afternoon, Describe: Meeting, Behavior: Working outside}] are obtained. The matching process traverses the nodes of the decision tree step by step. First, the position node with the same value is found, then the attribute of the next node or whether it is the result node is judged, and finally the matching result is obtained.

We also found some errors and mistakes when we processed the dataset. There are two types of causes: one is that the value of matching results is ambiguous, such as [123.4734171952262241.66343354247264, morning, null, Working at school] data decision tree matching results are teaching, the boundary of these two kinds of results is very ambiguous, which results in classification errors; the other is that the data are imperfect, such as [123.433197712967]. 1241.661524733374, morning, null, Rest] matching result is working outside, which is caused by the missing description attribute.

4 CONCLUSIONS

This paper is based on the LifeLog dataset collected by our lab itself, which is discontinuous, contains a LifeLog with a richer content, and the time and GPS change of each data can be used as a basis for location changes. We've been very careful about tagging the data, including the user behavior at this location, and the description of the behavior, and at least one photo of each piece of data. The shortcoming of this paper is that we could not propose applications on more than a one-person dataset. Finally, we confirmed the validity of the method using the author's experiences and memory.

The purpose of the current study was to discuss how to predict the behavior with such kinds of Lifelog. We have some conclusions as follows:

1. We published a special Lifelog dataset named informative lifelogging; more research can be done with the dataset.
2. We propose a very effective approach to predicting user behavior in such a kind of dataset.
3. The data we collected and the data people obtain from the SNS website appear to be similar; our work can be used in finding a personal trajectory with both kinds of datasets.
4. How to manage LifeLog data is a big problem currently. In order to distinguish the types of data, the LifeLog data should be classified. Our research work provides a method to manage and retrieve a personal LifeLog.
5. This research has some significance for analyzing user behavior and classifying users, and this kind of research could also be applied to behavioral analysis of criminal groups [13].

ACKNOWLEDGMENT

This work was partially founded by the projects CLB18C050, 20180550252, and 2019-ZD-1039.

REFERENCES

[1] Cathal G., Alan F. S, 2014 LifeLogging: Personal Big Data, http://doras.dcu.ie/19998/1/FnTIR_lifelogging_journal.pdf.

[2] Hyungeun J., Jung-hee R., 2010, Placegram – A Diagrammatic Map for Personal Geotagged Data Browsing. IEEE Transactions on Visualization and Computer Graphics, Vol. 16, NO. 2, MARCH/APRIL.

[3] Van-Tu N, Tu-Khiem L, Liting Z. 2019, LIFER 2.0: Discovering Personal Lifelog Insights using an Interactive Lifelog Retrieval System. CLEF (Working Notes). http://ceur-ws.org/Vol-2380/paper_157.pdf.

[4] Pil Ho K. 2011,Web-based Research Collaboration Service: Crowd Lifelog Research Case Study. 7th International Conference on Next Generation Web Services Practice.

[5] Iori O, Takashi Y 2012, Proposal and Evaluation of User's Actions Distribution Method using Life Streaming Service on Lifelog System. 26th International Conference on Advanced Information Networking and Applications Workshops.

[6] Adrienne H. A, Kevin E, 2013,Andy H. Using Location Lifelogs to Make Meaning of Food and Physical Activity Behaviors. 7th International Conference on Pervasive Computing Technologies for Healthcare and Workshops.

[7] Chelsea D, Madjid M, Paul F, David L-J. 2014,The Big Data Obstacle of Lifelogging. 28th International Conference on Advanced Information Networking and Applications Workshops.

[8] He, K., Zhang, X., Ren, S., Sun, J. 2015, Deep residual learning for image recognition. CoRR abs/1512.03385, http://arxiv.org/abs/1512.03385.

[9] Ren, S., He, K., Girshick, R., Sun, J. 2015 Faster r-cnn: Towards real-time object detection with region proposal networks. In: Proceedings of the 28th International Conference on Neural Information Processing Systems – Volume 1. pp. 91{99. NIPS'15, MIT Press, Cambridge, MA, USA (2015), http://dl.acm.org/citation.cfm?id=2969239.2969250.

[10] Mohammad A.S, Norisma I, Rohana M, 2013, Malaysian Lifelogging over Twitter. First International Conference on Artificial Intelligence, Modelling & Simulation.

[11] Grgić-Hlača, Nina, Redmiles, Elissa M. 2018, Human Perceptions of Fairness in Algorithmic Decision Making: A Case Study of Criminal Risk Prediction.www2018: International World Wide Web Conference.

Computational Social Science – Luo, Ciurea & Kumar (eds)
© 2021 Taylor & Francis Group, London, ISBN 978-0-367-70193-2

Modern science and technology support evaluation system research for the construction industry of Jiangsu province

Y. Wang
Nanjing Institute of Engineering, Jiangsu Province, China

ABSTRACT: Based on the research of modern building industry developments both at home and abroad, combining relevant policy documents, the paper set up a complete set of modern science and technology support evaluation system for the building industry of Jiangsu province. The evaluation system takes "building technology" as the core, builds an ecology and technology environment integrated with "technical innovation", "technology application" and "policy guidance". As a result, this research gives a scientific and accurate assessment for the scientific support ability of modern building industry in Jiangsu province and provide technical support for the development strategies of modernization of building industry in our country.

1 INTRODUCTION

In May 2016, Beijing held the National Science and Technology Innovation Conference, which emphasized the need to deepen reform and innovation and form a vibrant technology management and operation mechanism. Innovation is a systematic project. The innovation chain, industrial chain, capital chain, and policy chain are intertwined and support each other. They must be fully deployed and thoroughly promoted, as innovation leads the overall development and transformation of the construction industry. In order to meet the innovative development requirements put forward in the "Thirteenth Five-Year Plan," from government decision-making and supervision to enterprise management, the application of new technologies, and industry demonstration systems, all aspects of innovation and development should be comprehensive.

2 BUILDING A MODERN TECHNOLOGY SUPPORT SYSTEM FOR THE CONSTRUCTION INDUSTRY

The modern science and technology support system of the construction industry refers to the government as a guide, the enterprise as the main body, the market as a guide, universities and scientific research institutions as participants, and "science and technology" as the driving core to form "technology innovation,", "technology application," and "policy guidance" supports for each other, while supporting a system of synergy between their technologies, talents, and funds. It promotes the construction industry's technological innovations, technological management innovations, integration of new technology applications, and guides the dynamic evaluation system to promote the development of the modernization of the construction industry. The trinity technology ecological environment created by the system is stable, forward-looking, open, and sustainable.

Among them, "scientific and technological innovation" is divided into four categories according to different innovation subjects: property owners, talents, enterprises, and governments. The contents of innovation are intellectual property rights, talent construction, technological innovation, and management mode innovation. This is in line with the transformation of the development factor-led development proposed in the national science and technology innovation strategy for innovation-element-led development; the transformation of the innovation subject from the area of scientific and technological personnel to the public innovation and entrepreneurial interaction area. "Technology application" also includes four aspects, namely standard system, technology system, information technology, and green technology. By studying the applications of these four in the entire life cycle of a building, the objective of deep integration and coordinated management in the promotion and application is achieved. A set of technical and management standards are formulated to promote the modernization of the construction industry of the enterprise from within. "Policy guidance" links "scientific and technological innovation" and "technical application" and serves as a link between implementation and timely feedback. Policy guidance mainly includes the establishment of effective legal protection and policy incentive mechanisms to encourage the development of scientific and technological innovation, provide effective protection for new intellectual property rights and patent mechanisms, and also provide protection for technology applications, thereby establishing an external support system.

3 ESTABLISH A MODERN SCIENCE AND TECHNOLOGY SUPPORT EVALUATION SYSTEM FOR THE CONSTRUCTION INDUSTRY

Based on the overall research into the modernization technology support system of the construction industry, through further analysis of questionnaires (government articles, college articles, enterprise articles, demonstration project articles), it is proposed to construct a modernization technology assessment system for the construction industry. This evaluation system includes evaluation models and evaluation series of standards (demonstration cities, demonstration bases, demonstration projects).

3.1 *Evaluation model*

3.1.1 *Clear evaluation target*
The evaluation object is based on the prototypes of "Science and Technology Innovation," "Technology Application," and "Policy Guidance," and derives three key factors: sociality, technicality, and practicality, which are used as first-level evaluation indicators; further expansion includes sociality including organizational leadership, policy formulation, etc.; technicality including system construction innovation (technical support, supervision mode, platform construction), talent construction, etc.; and practicality is mainly aimed at demonstration bases and demonstration projects.

3.1.2 *Determining the evaluation method*
The evaluation model uses four evaluation methods: the expert scoring method and the questionnaire interview method in the early stage; and the multi-index weight allocation hierarchy method and the mathematical statistics method in the later stage. This is mainly based on the following considerations. Due to the numerous evaluation indicators supported by the modernization technology of the construction industry, the scope is extensive, and the unit magnitude is different, therefore in this part of the quantitative evaluation, each evaluation indicator can be divided into score segments, according to the evaluation criteria, by experts according to the requirements of each system index. The corresponding score is given and then the weight assigned according to multiple indicators to obtain the overall quantitative evaluation result. This evaluation method is simple in operation, high in evaluation, and enhances the adaptability of the evaluation system. Questionnaire survey and interview method, as one of the most important methods of data collection, are applicable to the investigation of user behavior patterns. After obtaining the first-hand information, the data are sorted and analyzed through mathematical statistics to obtain valuable survey results. Therefore, the combination of these four evaluation methods can be used to comprehensively evaluate the modern technology support for the construction industry.

3.1.3 *Determining the building idea*

The establishment of the evaluation model is based on two aspects, quantitative evaluation and qualitative evaluation, where quantitative evaluation mainly targets social and technical aspects through expert scoring to determine the weight of subitem indicators at all levels and qualitative evaluation mainly uses questionnaire surveys and demonstration bases for demonstration projects. The interview method is used for research and interviews. Synthesizing the two parts of the evaluation content gives a total score for the evaluation of the support system for the modernization of the construction industry and providing an evaluation report. The evaluation report includes the status quo of industrial modernization and suggestions for improvement.

3.2 *Evaluation series standard*

The evaluation series of standards is aimed at demonstration cities, demonstration bases, and demonstration projects. The content includes evaluation projects, evaluation contents, scores, and additional explanations. The above evaluation models are used to conduct empirical analysis and research to establish a complete set for Jiangsu Province. Provincial construction industry modernization "technology support" evaluation series of standards (for trial implementation) is used to evaluate its technology demonstration and technology support capabilities, so as to give a scientific and accurate assessment of the science and technology support direction of our province's construction industry modernization, and for the national construction industry modernization. The development strategy provides technical support.

The evaluation results of the provisional science and technology support ability are divided into excellent and good grades, and meet the following requirements:

1. When the total score of the evaluation is 70–80 points, the evaluation is that the technology support ability is good;
2. A total score of the evaluation of 81 points or above is evaluated as excellent technology support capability; when the total score for a demonstration project is 86 points or above it is evaluated as excellent technology support capability.

4 CONCLUSION

Based on the theoretical research into domestic and foreign construction industry modernization science and technology innovation, this article innovatively proposes a three-in-one construction industry modernization science and technology innovation system with "science and technology" as the core of "technology innovation," "technology application," and "policy guidance". In addition, the evaluation modeling and standard are built up through choosing evaluation subjective¡¢deciding evaluation methods and deciding evaluation indexes systems. As a result, the transparency and operability of the evaluation of scientific and technological support capabilities are realized so as to provide technical support for the development strategies of modernization of building industry in our country.

ACKNOWLEDGMENTS

This research was funded by Nanjing Institute of Technology School-level Scientific Research Fund, Project of Introducing Talents to Start Scientific Research Fund (Grant No. YKJ201836), Research on Simulation and Quantification Method of Carbon Emission of Precast Assembly Concrete Building (PC) Based on LCA.

REFERENCES

[1] The Central Committee of the Communist Party of China issued the "Outline of National Innovation-driven Development Strategy". The State Council of the People's Republic of China, 2016, 15: 5–14

[2] Zou Bo, Guo Feng, Wang Xiaohong, Zhang Wei. The mechanism and path of triple helix collaborative innovation. Dialectics of Nature, 2013, 07: 49–54.

[3] Fan Jie, Liu Hanchu. The Impact and Adaptation of the Drive of Scientific and Technological Innovation during the "Thirteenth Five-Year Plan" Period on the Changes of my country's Regional Development Pattern. Economic Geography, 2016, 01:1–9.

Computational Social Science – Luo, Ciurea & Kumar (eds)
© 2021 Taylor & Francis Group, London, ISBN 978-0-367-70193-2

Online teaching in Chinese universities under "COVID-19" epidemic

Y.J. Shen
Shandong Technology and Business University, Yantai City, Shandong Province, China

ABSTRACT: During the period of "COVID-19" epidemic in China, the national large-scale online teaching activities have brought unprecedented teaching methods, communication media, emotional exchange and other issues. Teachers should choose teaching resources reasonably, use technical means concisely, interact with students effectively according to the survey and actively deal with the future challenges of online teaching.

1   GENERAL INSTRUCTIONS

In the face of the predicament of unforeseen epidemic prevention and control, China's Ministry of Education issued a notice to postpone the opening of the spring semester on January 27, 2020. Since then, all levels of education departments and online education institutions have actively responded to and launched the "no suspension plan" (Xie youru et al. 2020).

2   TEACHERS BECOME ANCHORS, FROM CONFUSION TO PROFICIENCY

Although my university has done some MOOC courses in the past few years, they are all completed by individual teacher teams in the form of independent SPOC projects such as golden course and shared course, with a small number and scattered in different online teaching platforms. Facing the "COVID-19" epidemic, the school has clearly stipulated that it should make full preparation for online teaching in advance to ensure that online teaching is carried out according to the original opening date. Although the school's educational administration department provides technical support groups for all kinds of online teaching platforms, most teachers are confused and anxious. Many teachers have joined several platform technical support groups because they do not know which platform is suitable for their courses. The sounding air classroom is a challenge for many teachers.

I have completed the independent SPOC online course of accounting last year and have been running it for a semester. This year, I only need to build my own flipped classroom to use it.

When teachers are preparing for online teaching, they usually face problems in the selection of MOOC platform and communication form with students. When choosing online courses, teachers must browse the courses in advance. Teachers must pay attention to whether the selected network courses conform to the requirements of the school's syllabus, whether they conform to the teaching plan, match with the textbooks currently used, and conform to the learning level of the students in their own classes. If the source course only meets part of the requirements, you need to consider how to make up for the insufficient part (Gong xiaojun et al. 2017). Teachers should be familiar with their own online teaching platform as much as possible, master the authority given by the platform, and use management functions (such as student click rate, video viewing time, online time, accuracy rate of homework completion, etc.) and teaching functions (such as organizing discussions, tests, etc.) to help students learn better.

As China's university MOOC platform has defects in real-time voice and video communication, I chose a simple and commonly used QQ to lecture, and the interaction between teachers and students is also very good. Teaching is not a display activity. Online teaching should start with designing learning activities. Teachers should be designers of learning activities and promoters of ideas, so as to enable students to learn knowledge effectively and acquire knowledge.

3 STUDENTS GROW UP IN ADAPTATION, FROM "GAME FANS" TO "ONLINE COURSE MASTERS"

Nowadays, almost all college students grow up with QQ games. Therefore, it is most appropriate to use QQ as the basic communication tool for online courses during the epidemic period. At this time, the main purpose is not to let students learn the use of a new software, but to let students learn from the software they are familiar with as much as possible. Many applications in QQ (such as announcement, document, assignment, demonstration whiteboard, etc.) can meet the basic needs of teaching.But online teaching is different from online games after all. Although online teaching uses familiar QQ as a basic communication tool, due to the different communication methods of online courses and classroom teaching, students' experience and effects of online classes will also be different from the traditional classroom listening experience.

College students are already adults, have a more mature outlook on life, world outlook and values, and have different understandings and feelings about the new way of teaching. The survey is mainly conducted on the advantages and disadvantages of online courses, students' feelings and adaptability to online courses, and the impact of online teaching methods on normal learning.

3.1 *Investigation of the online classroom learning situation and effect of college students*

This survey takes Shandong Technology and Business University as an example, and uses questionnaire surveys and statistical analysis methods to study the situation and effects of college students' online classroom learning under the epidemic situation.

3.1.1 *Respondents*
Shandong Technology and Business University students who conduct online courses through online teaching methods, involving 673 people, 21 classes, 5 courses.

3.1.2 *Research methods*
The survey determined the research framework and related issues on the basis of literature research and collected data, and formulated a questionnaire with targeted items. By sorting, classifying, and summarizing the data obtained from the questionnaire, and making statistics and comparisons, we further analyze the advantages and disadvantages of online courses, students' feelings and adaptability, and the impact of online teaching methods.

3.1.3 *Investigation results*
The data in this questionnaire are all from 673 college students who received online courses in Shandong Technology and Business University, which is representative. The questionnaire consists of 15 objective questions and 1 subjective question. The objective questions are divided into single-choice questions and multiple-choice questions. The topic involves the use of tools by college students in online courses, the factors that affect learning efficiency in online teaching, the advantages and disadvantages of online course teaching methods, and the differences between online courses and traditional teaching methods. The subject of the subjective survey is to hope that students will make suggestions about online teaching.The specific contents of the survey results are as follows:

In terms of the degree of recognition of online teaching, most (59.63%) college students have an acceptable attitude towards the way of online teaching under the epidemic. 33.94% of people do not like online teaching and only 6.42% like it; The practicality of online courses, about half of people (50.46%) think that online courses are more useful, and some people (35.78%) think it is very practical, and a small number of people (13.76%) think it is not practical; It is said that 65.14% of people think the help is average, 22.02% think it is very helpful, and only 12.84% think it is not very helpful.

In terms of the impact of objective factors on online teaching, there are more college students (433 people) who only use mobile phones for online courses, and 229 people use PCs (desktop computers or tablets and laptops), while those who use both mobile phones and computers There are 105 people, and a small number of people use other tools for online classes (11 people). Regarding the online environment of online teaching, most people (71.56%) use wireless networks (Wi-Fi) for classes, and a small number use mobile data for classes. (27.52%), very few people use both wireless network and mobile phone traffic for class (0.92%). Under the influence of the epidemic, 550 people feel that the delayed start policy has a greater impact on their study and life, and 123 people feel that the delayed start policy has no impact on their study and life; everyone's family environment will also have an impact on online learning. Certainly, 438 people think that the family environment affects their learning, while 235 people think that the family environment does not affect their learning.

In terms of the influence of personal subjective factors, the useful online and offline teaching methods are different, and the students' feelings are also different. In terms of homework, 309 (45.91%) expressed satisfaction with the homework in the online course, and 364 (54.09%) were dissatisfied with the homework; regarding the learning efficiency of the online course, 500(74.3%) considered the learning efficiency to be average, 117 People (17.38%) believe that the efficiency is low, only 56 people are higher (8.32%); in online learning, 148 people said they can listen carefully, 494 people said that their attention will be occasionally distracted, and 31 people said It is impossible to concentrate the attention of learning at all; because the learning process and operation methods of online courses are different from classroom teaching, most students (62.25%) believe that online learning has more complicated tasks and operations than classroom learning.

3.2 Discussion and analysis

Through the investigation of college students, we found that online teaching during the new crown virus plays a certain positive role in student learning, but in comparison, students are still more adapted to the traditional teaching model at this stage, and the following main problems occur: students learn independently Insufficient ability, no learning atmosphere at home, many external disturbance factors, less teacher-student interaction, and less timely question answering. Online teaching is not a direct copy of the traditional classroom, it needs to redesign and reorganize the curriculum according to the syllabus, and repackage the knowledge points. Considering that some students may have weak self-consciousness and self-control ability, teachers should improve online teaching methods and communication methods, and improve online teaching quality.

3.2.1 Clear online teaching goals

Teachers must first announce the requirements of curriculum learning and assessment, so that students can clearly understand the learning objectives, assessment methods of learning achievements, video and courseware learning progress and the time and form of chapter examination, and the time limit for completion (Cui yaqiong et al. 2019).

3.2.2 Design learning activities and interaction

Teachers should try to design learning activities to improve the interaction between teachers and students. Online teaching is different from classroom teaching. Teachers cannot understand each student's listening condition at the same time. Perhaps after a period of online lectures, students will no longer feel that the novelty of home learning will relax their learning. In order to maximize students' attention in class, teachers can increase their collective interaction with students in class (reduce or avoid one to one interaction). The simple and feasible way is to choose small and precise problems closely related to the content of the lecture, so that students are not distracted in the classroom.

3.2.3 *Reasonable arrangements*

Teachers should make reasonable arrangements for online learning activities and slow down their pace appropriately. Taking into account the speed of network transmission, teachers should not speak too fast, so that students can hear clearly. Teachers should not tell the whole story. Students only listen and see the screen, easily distracted and exhausted. Teachers should make full use of teaching platform data, understand learning situation in time and intervene appropriately. Teachers can timely and appropriately disclose the learning process of students on the platform, such as participation, discussion, communication and homework completion data, so as to encourage students' good learning behavior. Taking my online course as an example, I usually log in class QQ group 10 minutes in advance, waiting for students on line, and open asynchronous SPOC courses on MOOC platform of China university. Due to the network jam, I will extend the check-in time appropriately. In online class, I use the functions of limited time discussion, practice in class, questionnaire and online voting, limited time rush to answer, brainstorming and other functions to interact with students, instantly share views and answer classroom questions. Students have the pleasure of learning in games and gradually grow up in adaptation.

4 OTHER IMPORTANT PROBLEMS AND EFFECTS OF ONLINE TEACHING UNDER THE INFLUENCE OF "COVID-19"

4.1 *The problem of "technical jam" in online classroom*

China's online technology has experienced an unprecedented test due to the nationwide large-scale online teaching (simultaneous use of universities, middle schools and primary schools) caused by the new crown epidemic. Previous online courses have no current technical problems (whether platform or network). When online teaching is carried out at the same time in primary and secondary schools across the country, Wechat, QQ, MOOC and other learning platforms and software are also facing unprecedented huge traffic pressure. In the process of online teaching, there are problems such as stuck, flashback, APP crash, network disconnection and so on, which are temporarily interrupted. The technical personnel of the platform shall not have taken measures such as current limiting and peak staggering. Even the powerful QQ has experienced transient anomalies. In my class, some students who come back to their hometown or are in the countryside, because there is no broadband support or good equipment to watch online courses in the local area, cannot log in and technical jamming often occurs.

4.2 *Online mode of teaching and learning*

The teaching place under the influence of the epidemic has changed from the original classroom to the home online learning. If we still use the idea and method of classroom teaching in the past to deal with the network teaching, the effect may be different from the expectation. The teaching organization and learning style changed from centralized to decentralized, from face-to-face to face computer screen, and learning partners changed from multiple to individual. The advantage of this kind of change is that it can cultivate students' ability of independent thinking and arrange their learning methods according to their own needs. The disadvantage is that in the online learning mode, the learning partner changes from multiple people to one person, which is very different from the real normal class scene. In the conventional classroom scene, due to the collective teaching, the students are in a micro environment with each other (Ghazali 2016). For the problems encountered in the learning, there are both questioning at the same table and sharing with the peers, thus forming a kind of visible learning community model. In the classroom, due to the teacher's supervision and other disciplines, even passive learning may have good results, but now at home online learning, the teacher and classroom in the sense of the physical scene do not exist, the peer table does not exist, and learning has completely changed It became autonomous learning without supervision by external forces. If the teacher only uses video or PPT to present the teaching content, and explains it like a broadcaster, there are few other links to inspire, guide and stimulate thinking, and the students only listen passively, then online teaching and learning may be in two independent rhythms, just like walking on two parallel roads that never meet.

4.3 *Emotional communication barrier caused by facing computer screen*

The Internet can transmit sound, but it can't transmit mental interaction and eye contact. Teaching is not a one-way teaching, nor just the transfer of knowledge, but also the exchange of emotion and soul. Teachers and students in the classroom face to face, through language, eyes and other factors to communicate and easy to carry out. Teachers and students face each other in the classroom. They can communicate easily through language, eyes and other factors. Through the computer network teaching, we can only use language and interactive software to communicate with students. It is totally different from facing a group of students, which is inconvenient for emotional communication. Although the teacher will try his best to teach, if the students' learning is not in the state and there is no effective interaction, it is difficult to guide the students' thinking into the cognitive object without good communication with the students. The students only have the ears to listen and the eyes to listen, which is easier to be distracted and tired than the classroom teaching, and may make the effect worse. There is a temperature for people to people and face-to-face communication, and MOOC and Internet can't replace it at present (Bansal & Singh 2016). According to the questionnaire I did in the online classroom, 87% of the students miss the traditional classroom, especially the teaching atmosphere of face-to-face communication between students and teachers, and think that the classroom teaching is better than online teaching.

5 REFLECTIONS AND SUGGESTIONS

5.1 *Understanding of "congestion" of network and platform*

Under the current technical conditions, the large-scale online teaching in China is not as smooth as expected, and the bottleneck is not the network but the platform. The online teaching system is essentially a set of video service system. The video is transmitted to the remote server in real time for students to download and watch. The current 4G network can perfectly support video transmission (let alone local area network). As long as there are 4G placements, online teaching can be carried out. However, a large number of concurrent online video access will challenge the teaching service platform. Due to the limited service resources, when the number of concurrent users reaches a certain number, the video service will be stuck, or even the system will be paralyzed. The solution is that the central server adopts a large-scale cloud computing platform. However, small and medium-sized platforms are unable to do anything in terms of technology and financial resources. Therefore, it is suggested that teachers should choose a platform with massive cloud computing capabilities as much as possible; ordinary universities usually do not need to build their own teaching platform, and should actively use the existing "cloud services". The platform is the hardware, which is becoming the public infrastructure of the information society. Because of the different social service functions, colleges and universities do not need to have their own hardware, but more need to have their own high-quality teaching content.

5.2 *Thinking about "teaching and learning"*

Teachers should make full use of appropriate platforms, select appropriate MOOC resources, make practical teaching plans, seriously think about what is valuable online teaching, how to organize students to effectively learn online, and achieve knowledge goals and ability goals. The simpler the online technology that connects teaching and learning, the better. The network flow is a valuable resource, so in the process of online teaching, traffic should be reduced as much as possible to reduce the problems and waste caused by the lack of network bandwidth. Video images should be as simple and clean as possible. The color and gray level of PPT should not be too high. Illustrations, animation and demonstration parts should be made into vector images as much as possible. Teachers should teach in a quiet environment to avoid unnecessary noise, so as to improve the audio quality and reduce the amount of data.

5.3 *The main body of teaching activities is students*

Under the epidemic situation, online teaching only provides a new learning mode option. Instead of leaving the teaching task to the machine, the technical means are always auxiliary. Teachers should be the instructors, collaborators and practitioners of teaching activities. Online teaching, students do not have the usual class partners, a person in front of the computer screen will have a strong sense of discomfort. At this time, we need more effective interaction between teachers and students, try to use language, pictures, etc. to convey emotion. Teachers should guide students to learn protection knowledge, pay attention to the progress of the epidemic, and independently arrange their own life and study.

6 CONCLUSION

Whether classroom teaching or network teaching, content is the key factor, classroom or network is just a means. High quality teaching content is reflected in knowledge renewal, teaching design and teaching effect. In the future, when evaluating the teaching quality, the education management department can directly extract the relevant teaching content and historical data from the cloud platform. Blended MOOCs have arisen as an alternative model to integrating conventional and online strategies for better teaching and learning in higher education (Yousef et al. 2015). Through the platform data, relevant institutions will know what the school taught in time, and employers will know what the students learned in time, that is, the network makes our education more transparent. This may be a predictable trend.Although the current online teaching divides teachers and students in physical space, the network allows us to integrate into a larger classroom. We are looking forward to the end of the epidemic as soon as possible, but the value of large-scale online teaching will not be reduced. We believe that the future on-site teaching and online teaching will be in parallel for a long time, symbiosis, co-promotion and integrated development.

REFERENCES

Bansal, S., & Singh, P. 2016. Blending active learning in a modified SPOC based classroom. Proceedings of the 2015 IEEE 3rd International Conference on MOOCs, Innovation and Technology in Education, MITE 2015.

Cui yaqiong, Kang shugui, Guo jianmin, Guo caixia. 2019. Teaching practice research on flipped classroom of real variable function. Education Theory and Practice (12): 53–55

Ghazali, N. 2016. The Perception of University Lecturers of Teaching and Learning in Massive Open Online Courses (MOOCS). Journal of Personalized Learning 2(1): 52–57.

Gong xiaojun, Tang yixiang, Tu liming. 2017. Exploration of online and offline mixed teaching methods based on the MOOC platform. Fujian Computer (2): 66–77.

Xie youru, Qiu yi. 2020. Characteristics, problems and innovation of online teaching mode of "no suspension of classes" during epidemic prevention and control. Research on Audio Visual Education (3): 20–28.

Yousef, A. M. F., Chatti, M. A., Schroeder, U., & Wosnitza, M. 2015. A usability evaluation of a blended MOOC environment: An experimental case study. The International Review of Research in Open and Distributed Learning 16 (2): 23–27.

Computational Social Science – Luo, Ciurea & Kumar (eds)
© 2021 Taylor & Francis Group, London, ISBN 978-0-367-70193-2

Research on digital library construction

B.B. Xu, S.W. Zhu, J.F. Yu & M.J. Zhang
Qilu University of Technology (Shandong Academy of Sciences), Jinan, China
Information Research Institute of Shandong Academy of Sciences, Jinan, China
National Technical University of Ukraine "Igor Sikorsky Kyiv Polytechnic Institute", Kyiv, Ukraine

V. Kuzminykh
National Technical University of Ukraine "Igor Sikorsky Kyiv Polytechnic Institute", Kyiv, Ukraine

S.S. Li & J.P. Guo
Qilu University of Technology (Shandong Academy of Sciences), Jinan, China
Information Research Institute of Shandong Academy of Sciences, Jinan, China

ABSTRACT: This paper analyzes the development status of domestic and foreign digital libraries, discusses the problems faced by the development of Chinese digital libraries in light of China's national conditions, and points out the deficiencies in resource construction, platform technology and user participation, and in the diversity of information under the background, considering the technical support of the future digital library, the application of technologies such as semantic web, cloud computing and big data in the digital library is analyzed. Finally, the service development model of the future digital library is discussed, hoping for the construction of the digital library provide reference for development.

1 INTRODUCTION

The origin of the digital library dates back to 1993: the library of congress of the United States was connected to the internet for the first time, pushing the library into the digital era. Since then, the library has entered a new stage of development. As a carrier of knowledge information, libraries carry a large amount of data and diversified information. With the ubiquitous evolution of the knowledge environment today, the emergence of digital libraries has undoubtedly changed the source of information and the way of dissemination of knowledge. As an important part of the information superhighway, it has gradually developed into a standard for measuring a country's knowledge informatization level. Traditional paper-based collection libraries still have places that digital libraries cannot replace because of their special social functions and nearly a thousand years of collection history, but digital libraries provide fast, efficient, and real-time digital resources that make traditional libraries unmatched. Based on the current development status of digital libraries at home and abroad, this article studies the problems faced by the development of digital libraries in China, discusses the technologies on which the future development of digital libraries depends, and looks forward to the future service development model.

2 THE DEVELOPMENT OF DIGITAL LIBRARIES

2.1 *Development status of foreign digital libraries*

In the 1990s, the development of computer technology and network technology brought new opportunities for library digitization. As early as 1945, the famous American science and technology management scientist Bush (V. Bush) proposed to store library documents. Linked to retrieval and the development of computer technology, after a period of time, people in the industry continued to study this field. In 1994, the United States launched the "Digital Library Initiative" project[1], which set off the digital library research. The upsurge, the United States has always been in a leading position in the research of digital libraries, which largely depends on the importance that the United States attaches to the construction of digital libraries. In 1992, the United States included the development of digital libraries as a "national challenge" project[2]. Two years later, the United States announced the "Digital Library Initiative", which was jointly issued by the National Science Foundation, the National Aeronautics and Space Administration, and the Department of Defense Advanced Research Agency. In 1994, the Library of Congress Digitization of collections. "American Memory" project converted all multimedia information related to American history into electronic format[3] in 1995. The knowledge sharing project led by the Library of Congress of the United States-the construction of the World Digital Library has been completed and officially opened in Paris in 2009. It has gathered the cooperation support of many university libraries and research institutions, and 7 languages including Arabic, Simplified Chinese, English, French, etc. are provided for query. In addition, various universities and enterprises in the United States are also actively carrying out the construction of digital libraries. For example, Carnegie Mellon University has established the use of intelligent retrieval technology[4], voice image recognition technology, etc. A comprehensive information digital video library system was developed. Stanford University focused on user interface, information sharing and search, etc. to study the interoperability of heterogeneous network information, and built an integrated digital library system. Google released the Google print service in 2004, which can complete scanning, optical character recognition and storage[5]. By 2013, more than 15 million books have been scanned. This service directly promotes the digitization of information objects while ignoring its copyright issues. In 2013, The U.S. Digital Public Library went online in 2013, connecting thousands of public and private libraries, archives, museums, etc., using multi-dimensional display methods such as time, location, and themes to provide online information to the public and provide developers provide API interfaces to integrate teaching, culture and innovation content. In addition to the United States, Japan, the United Kingdom, Canada, Australia and other countries are constantly building and developing digital libraries. The European Union Digital Library was officially launched in Brussels in 2008. Looking at the development of foreign digital libraries nowadays, there has been a transformation in services, from the initial focus on collection information to user-oriented customization, and added some new functions, such as online classroom sharing services, subject service platforms, courses Management, etc. In the construction of digital library service platform, with the continuous improvement of network technology and computer technology, the construction of digital library service platform has also continuously introduced new technical methods. In addition to Web technology, database technology and other foundations, virtual the continuous introduction of real technology, cloud computing technology, big data, semantic analysis, etc., has played a supporting role in the continuous development and progress of digital libraries. At the same time, the development of mobile terminal equipment has also made library resources more mobile and friendly.

2.2 *Development status of china digital library*

The construction of China's digital library began in the mid-1990s, marked by the implementation of the "China National Experimental Digital Library Project". After the vigorous support of national policies in recent years and the efforts of all relevant units, China's digital library construction has also made great progress. In 1998, the National Project for the implementation of the "Chinese Digital Library Project", the project's goal is to build a very large-scale high-quality Chinese information resource library group, and included in the national "Tenth Five-Year Plan"[1]. In 2000, the "National Science and Technology Books and Documents Center" was established, consisting of a number of libraries including the Library of the Chinese Academy of Sciences and the Library of the Chinese Academy of Agricultural Sciences, and established a collection of scientific and technological documents to promote the sharing of scientific and technological documents. In 2003, the National Library ALEPH500 computer integrated management system was put into operation. In recent years, the development of the National Digital Library of China has focused on mobile terminal equipment services. With the advancement of mobile Internet technology, Chinese universities and regional libraries have begun to launch mobile service functions. However, compared with foreign service forms with diversified personalities, China's service forms are still relatively single, without mature standards, and regional differences are obvious. For innovative services, such as online course sharing, subject service platforms and other Chinese digital libraries have been involved, but not yet mature enough, for example, in June 2014, Shanghai Jiao Tong University and the "Mooc" platform-"Future learn" in London sign a contract and reach a partnership to learn the British Library courses across half the globe. In terms of digital library platform construction, Chinese digital library research and development personnel are also constantly applying new theories and new technologies to the construction of digital libraries, such as Wu Gexin[6] has conducted research on Linked Data in the aggregation of digital library resources, Ou Shiyan[7] designed and implemented a framework for linking data-oriented semantic digital library resource description and organization in order to solve the problem of heterogeneity of library digital resources, Yuan Yuan and Ling Hui[8] discussed the virtual environment of digital library driven by cloud computing technology. Yu Xin and Wang Jingyi[9] researched the architecture of the digital library cloud service platform based on cloud computing technology, constructed the digital library cloud service platform, and studied the service process and operating mechanism of the digital library cloud service platform.

3 TECHNICAL SUPPORT FOR FUTURE DIGITAL LIBRARIES

To solve the problem of digital library construction, technology is the key. The emergence of the new generation of World Wide Web-Semantic Web, the popularization and application of cloud computing platforms, and the advent of the era of big data have provided new opportunities for the development of digital libraries. The future digital libraries will also support these technologies and provide broader and more advanced services.

3.1 *Semantic web technology provides automatic query and retrieval services for digital libraries*

The idea of semantic web is adopt a new way to define and link web data[10], so that it can effectively realize data discovery, automatic processing, integration and reuse in different application scenarios. Semantic web has been widely used in foreign digital libraries. It can provide new information organization and processing capabilities for digital libraries. With the continuous deepening of research on semantic web in China, it will eventually bring China's future digital libraries with to benefit. Firstly, to explain the structural framework of the semantic web, which consists of seven layers, the structure is shown in Figure 1.

- The first layer: Unicode and URI, responsible for processing resource encoding and resource identification.
- The second layer: XML + NS + xmlschema, which grammatically represents the content and structure of the data, and separates the network information representation form, data structure and content.
- The third layer: RDF + rdfschema, to achieve the description of web resources.
- The fourth layer: Ontology vocabulary, defining concepts and relationships between concepts.
- The fifth to seventh layers: Logic, Proof, Trust, where Logic provides axioms and inference rules to prove validity, Proof exchange and digital signature to establish Trust relationship.

The application of semantic technology to the construction of digital libraries can improve the browsing and searching capabilities of resources and the effectiveness of information search. For example, first, use information retrieval tools to identify the semantic parts of scientific research works, provide users with more detailed and customized resource descriptions of electronic publications, and provide automatic analysis of electronic resources. Second, use semantic technology to

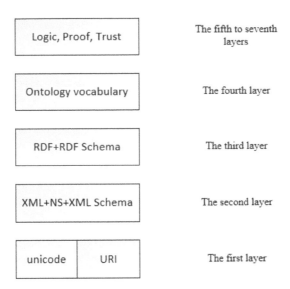

	The fifth to seventh layers
Logic, Proof, Trust	
Ontology vocabulary	The fourth layer
RDF+RDF Schema	The third layer
XML+NS+XML Schema	The second layer
unicode | URI	The first layer

Figure 1.   Semantic web structure framework.

perform semantic analysis on user queries to significantly improve query efficiency and accuracy. Third, share knowledge by reconstructing collaborative methods in semantic technology. Fourth, use the semantic analysis technology of linked data to solve the problem of resource heterogeneity, etc.

3.2   *Cloud computing technology provides a new direction for the co-construction and sharing of digital libraries*

Currently, one of the main problems facing digital libraries in China is the co-construction and sharing of resources, and cloud computing is an architecture model of shared technology. It integrates distributed computing and parallel computing resource provision modes, which can provide data resources. Massive cloud storage space, and provide safe, reliable, and personalized on-demand services. It uses cross-platform technology to harvest and manage metadata for heterogeneous databases. In this mode, the digital resources of different platforms can be managed in a unified manner, provide users with cloud storage, and provide personalized customized services according to their needs, so as to realize the co-construction and sharing of digital libraries. There are roughly three types of cloud computing according to service types[11]: IaaS with infrastructure as a service, PaaS with platform as a service and SaaS with software as a service. IaaS can build storage and data centers, and PaaS can build database storage for digital libraries With the construction of application platforms, SaaS can construct and manage digital library search services and unified integrated management software.

3.3   *Big data allows digital libraries to enter an era of massive data sharing, application and management*

Big data is a general term for massive data, including structured data, semi-structured data, and unstructured data. Big data technology is a general term for related technologies that analyze and apply massive data. The potential value of the analysis and mining of massive information by big data technology will become an important force to promote the development and transformation of digital libraries. It is not only manifested in the analysis and management of massive digital document information, but also in the analysis and mining of user behavior data. Through the analysis of structured and semi-structured data and information of user behavior to form potential service needs, service expansion and service innovation are carried out based on this, and personalized customized services are implemented to promote the transformation of digital libraries to user-centric, for example, the Shanghai Library conducted in-depth mining and analysis of the data on the loan and return of books in 2014, and found that the circulation rate was 91.6%. It uses big data to analyze the books in circulation and uncirculated books, and finds the popular book types and bibliography. By analyzing the age and gender of readers, guide the types of future books based on the characteristics of the reader group and the types of popular titles. The use of big data analysis is conducive to the development and construction of digital libraries from the perspective of readers. This will be an important measure to solve the problem of lack of user participation.

4 THE SERVICE DEVELOPMENT MODEL OF THE FUTURE DIGITAL LIBRARY

The library carries a large amount of knowledge and information and is the third space of the city. As a learning and communication center, the library is developing from reading to learning and communication. In recent years, the addition of LibGuides, Sakai, Moocs and other subject sharing related services adds new elements to the library platform, and the future service model will also change significantly. The future service model will also have major changes.

4.1 *The future digital library will face users*

User research has gradually become a research hotspot in the field of digital libraries. International conferences on digital libraries such as JCDL, ECDL and ICDL all take user research as the theme of the conference, mainly involving user interface research and user behavior research. Different habits have different needs of users, and guiding the construction of digital libraries based on the needs of users can effectively play the leading role of users. For user interface research, the development of visualization technology has promoted user interface research to a new level. The idealized interface presented by the mindset and emotional characteristics requires in-depth analysis of user needs. For the study of user behavior, it reflects a process of cognition and reaction of users to the things they touch. Using data mining technology to analyze user concerns, experience and preferences is the main method of user behavior research. Use effective research on user behavior to predict user behavior and guide the construction of digital library platform functions and interfaces.

4.2 *Combining the future digital library with education*

Libraries are places for learning and communication. As the most important builders of libraries, colleges and universities still focus on education. Therefore, the combination of digital libraries and education in the future will become a major development trend of digital libraries. Online classrooms and subject sharing services have been research hotspots in recent years. The tools and platforms that provide these services include LibGuides, Sakai and Moocs. LibGuides is a subject navigation tool launched by SpringShare. It has LibGuides widgets and application program interfaces. It is currently widely used in digital library service platforms of universities at home and abroad. It can integrate multiple resources and service forms to provide users with a station -Style subject service experience, the main functions are browsing, subject tags and classification, RSS customization, video embedding, user reviews, information evaluation, online communication, etc. The introduction and promotion of LibGuides will play an important role in promoting the subject service. Sakai is an open source online collaboration and learning environment. It provides courses and teaching management services. It integrates nearly 30 tools. It provides tools such as personal workspace, personal website establishment, database connection, grade book, test and quiz. Documents are not detailed enough, etc. Currently, it is mainly used in foreign regions and few domestic applications, but the rich functions provided by it still expand the ideas for the development of this field. Moocs is a new curriculum model that has emerged in recent years. It has gathered many colleges to cooperate and share and provide free open classrooms. It has created a new model of distance education learning and led the direction of future learning models.

It has many advantages over traditional classroom teaching. In the future, it will develop into an indispensable part of digital libraries in colleges and universities.

4.3 *The future digital library will pay more attention to the construction of intelligent space*

With the development of the Internet, artificial intelligence, smart cities, etc., smart life will no longer be extravagant and imaginative. People pay more attention to user experience. For digital libraries, users' reading environment, reading fun and reading comfort will also be more and more attention. Future digital libraries may have workplaces, 3D printing equipment, digital brain maps, etc.[12]. In addition, wearable devices, smart robot services, etc. can also be applied to digital libraries to bring innovative experiences to users.

5 CONCLUSION

On the whole, the development of China's digital library has achieved certain results, and it will go through a long period in the future. This article makes a comparative study on the development status of China and foreign countries, and summarizes the problems faced by the development of digital libraries in China. This article's elaboration of future digital library technology support is based on the hot research topics in recent years and the problems facing the development of digital libraries in China. The technologies supporting the construction of digital library platforms are not limited to this. During the construction process, we will do in-depth research in this field. I believe that the future digital library will gradually develop towards intelligence and sharing, and provide readers with more comprehensive, innovative, convenient and comfortable services.

ACKNOWLEDGMENT

This work is supported financially by the 2019 International Science and Technology Cooperation Project of Shandong Academy of Sciences (Grant NO. 2019GHZD13, 2019GHPY11), Shandong Province Key R&D Program Soft Science Project (2019RZC01007).

REFERENCES

[1] Li Pei. Digital Library Principles and Applications Higher Education Press, 2004.07.
[2] Jia Xilan, Wang Qiong, Wu Yingmei, "New Thoughts on the Construction of University Digital Library," Journal of University Libraries, 2012.02, pp. 59–63.
[3] American Memory Program http://memory.loc.gov/.
[4] Krassen Stefanov:Digital Libraries as a Social Media Digital. Presentation and Preservation of Cultural and Scientific Heritage, 2014(5): 26–32.
[5] Wikipedia Google Books http://en.wikipedia.org/wiki/Google%E5%9C%96%E6%9B%B8.
[6] Wu Gexin, Research on Resource Aggregation and Service of Digital Library Based on Linked Data, Central China Normal University 2013.10.
[7] Ou Shiyan, "Design and Implementation of Linked Data-Oriented Semantic Digital Library Resource Description and Organization Framework," Chinese Journal of Library Science, 2012.11, p. 58–71.
[8] Yuan Yuan, Ling Hui. "Exploration of Constructing the Virtualized Environment of Digital Library Driven by the Cloud Computing Technology" Information Theory and Practice, 2010, 12, p. 119–123.
[9] Yu Xin, Wang Jingyi. "Research on the Cloud Services Platform Architecture of Digital Library Based on Cloud Computing Technology", Information Science, 2011, 7: 1049–1053.
[10] Hou Jiti, Cheng Huirong, "Review of Researches on Digital Library based on Semantic Web Abroad", Library Information Work, 2011, 03.
[11] Han Pu,Shen Si,Lu Gaofei. "The Application and Progress of Cloud Computing in Digital Library at Home and Abroad". Modern Information, 2012.5: 67–70.
[12] http://news.sciencenet.cn/sbhtmlnews/2015/1/296294.shtm.
[13] Terence k.huwe:Building Digital Libraries.Computers in Libraries, 2014(7):23–26.

Computational Social Science – Luo, Ciurea & Kumar (eds)
© 2021 Taylor & Francis Group, London, ISBN 978-0-367-70193-2

Study on the value structure analysis and practical case of network user autonomy

X. Zhou & L. Tang
School of Journalism, Fudan University, Shanghai, China

ABSTRACT: Based on the combination of qualitative and quantitative research methods, a comparative analysis of Chinese and American social platforms was conducted. This article generalizes an analytical framework of Internet user autonomy in 18 elements. Based on the case study of Bilibili, it also tries to explore deeper into the status quo and limitations of user autonomy during the process of value co-creation between platforms and users. In addition, it discusses how to improve and return users' autonomy step by step in order to create a win-win developing generation of the Internet.

1   GENERAL INSTRUCTIONS

At the end of 2019, the father of the World Wide Web, Tim Berners-Lee launched the "Contract for the Web," which listed nine basic principles and 72 clauses. The core lies in the comprehensiveness of user rights protection. The contract provides a new common vision and action plan for the Internet. Therefore, its launch was supported by more than 150 organizations, including many nonprofit organizations such as Microsoft, Google, Facebook, and the Electronic Frontier Foundation. As the first mutual credit behavior contract of the three major actors of government, enterprise, and individuals in this human society, it aims to better protect the user data which have become an important source of personal fortune and social wealth, to better welcome the arrival of the intelligent era.

2   THE DEFINITION AND CONNOTATIONS OF NETWORK USER AUTONOMY

As the main creators and providers of a huge data wealth, network users have not yet realized the actual ownership and autonomous use of personal data. As a result, personal private property with natural attributes is forcibly transformed into enterprise property and state-owned property by technology, capital, and public power at zero cost, resulting in the absence of major owners of huge social wealth, as well as the belated abuse of individual human data resources. Now that personal data is in a de facto state of none ownership, making it become an objective free public wealth, it falls into what the economist Harding (1968) called the "tragedy of the public grassland"[1]. That is, using public resources for private interests, which made it impossible to achieve the optimal allocation and Pareto optimization of the entire social data resources. This has undoubtedly restricted the development of the entire Internet economy to a great extent, especially the development of a personalized customization economy which is based on users having a high degree of autonomy.

At the current stage, as the most powerful support of personalized customization economy, major intelligent network platforms are constantly digging data wealth from the continuous innovations of technology and function. This paper probes into the concepts and connotations of user autonomy, which is different from the concept of user autonomy in the field of communication[2]. Instead, it focuses on autonomy in the process of value co-creation[3] between users and major intelligent platforms. The author attempts to integrate the relevant concepts of informatics, management, and economics with communication, pushing the general process of information dissemination to the front end of information production and the dissemination value chain, in order to observe and analyze the follow-up information transmission based on the information value creation process.

A typical example is the app store platform created by Apple, whose Micro crowd value creation and public value sharing of the Internet economy provides an unprecedented value carrier and distribution system which is highly comprehensive, for public goods and commodities. And it was this system, on the one hand, that has changed the process and mode of value creation and production, that is, to iterate the traditional value creation process and mode of commercialization, corporatization, or personalization through the co-creation of users and platforms. On the other hand, it also synchronously changes the identity of recipients or consumers, so we have to change the autonomy of users in the definition and theoretical analysis framework of right accordingly. This paper explores and gradually constructs a new analysis idea and framework from the perspective of value co-creators. Although there are many omissions, new attempts are always dynamically improved during the process. This paper explores one of the possible new directions.

The theory of value co-creation was put forward by CK Prahalad and V Ramaswamy in 2000. It was pointed out that in the future, competition among enterprises will rely on a new method of value creation, that is, the theory that consumers and enterprises jointly create value with individuals at the center. Value co-creation has witnessed great development in the Internet economy, which has changed the role of users from passive value buyers to active value-creating participants. Value co-creation means that all social individuals and their social groups can actively participate in value co-creation through sharing network, thus forming an open transaction-matching process including value consensus, value symbiosis, value sharing, and value win-win. Therefore, the effective interaction of various stakeholders based on the user's position has become the core power of value co-creation and Internet economic development. However, the structural framework and value boundary of this core power are not clear currently. This paper attempts to analyze the general theoretical connotation and framework structure of user autonomy, and at the same time, deeply analyze the typical case of Bilibili, which is the largest original community of secondary cultural video in China, and explore the similarities and rules, with a focus of empowering users with an innovative theory and practical mechanism.

2.1 *Academic research and concept definition of user autonomy*

From June 6, 2017, to November 19, 2019, before the paper was completed, a cross-database subject search was conducted on Wanfang.com and CNKI with "User Autonomy" (the scope covers thesis topics, keywords, and content abstracts), whose search results for fuzzy matching remained at zero all the time.

After further analyzing relevant data, the author found that, on the one hand, domestic scholars' research on "user autonomy" accounted for only a small proportion, and up to 67% of the relevant research papers were from the perspective of technology applications and information service business, and only about 21% were from the perspective of communication governance, and only 8.3% clearly defines and classifies user rights and interests. Topics with high relevance include private power, social network users, power game, social network, etc. Zhang Xiaoqiang (2018)[3] has a more representative description from the perspective of network governance. He discussed the private power of the network intermediary represented by platform media. Liu Xiaoping and Deng Wenxiang (2019)[4] used grounded theory to conduct qualitative research on the performance of users during the process of social responsibility co-creation for virtual enterprises and found that the interaction related to value construction can effectively promote cooperation and a win-win situation on the network platform. These two scholars believe that to establish the understanding of consumers through interaction and dialogue to achieve the level of interaction expected by users is a prerequisite for effectively promoting the co-creation of unique values with the cooperation of users, platforms, and enterprises, while the co-creation of social responsibility of virtual enterprises related to the co-creation of values has a positive impact on users' interactive behavior, including enhancing user participation and level of recognition. Previous studies by scholars Fan Shuai, Tian

Zhilong, and Hu Xiaoqing (2017)[5] showed that in the process of social responsibility co-creation for virtual enterprises, the dynamic mechanism of positive effect of user participation stems from the psychological ownership generated in the process of co-creation, and the relationship between user participation behavior and psychological ownership is positively affected by the richness of media ecology, which indirectly affects the user's identification of corporate social responsibility.

On the whole, domestic scholars' cognition of user autonomy in the context of value co-creation in platform economy needs to be enlightened. At present, most research on the rights and interests of users is mainly carried out from the perspectives of management, communication, law, and other social sciences, but less so from the perspectives of technology, engineering, and psychology. Therefore, our interdisciplinary research on technology empowerment, that is, the promotion effect of new technology on user participation in value creation and dissemination, is relatively lagging behind, forming an interesting contrast with the related research of foreign scholars.

In fact, although the research into user autonomy by foreign scholars has just started, there have been few achievements, although it has already involved many disciplines.

Up to November 19, 2019, in the global academic resources database web of science, the author carried out a multidatabase search for "user autonomy" with a time span of 1996 to date, and obtained a total of 4390 papers. Among them, 2144 papers were from the perspective of computer science, which was nearly 50%; and this was followed by engineering, involving 1557 papers, accounting for 35.47%, respectively from the perspectives of technology and engineering. This paper discusses the relationship between user autonomy and machine autonomy, for example, how to improve the identity authentication needed by the user owner and autonomy in the future environment of all IP networks, how to enhance the user autonomy in information search, etc. Healthcare science closely followed with 767 articles, accounting for 17.47%. The main topic was the application of new intelligent technology and new equipment. The results of communication, sociology, and psychology with high relevance to this paper were 332, 295, and 757, accounting for 7.56%, 6.72%, and 17.24%, respectively. Communication science ranked no. 17 and social science was far behind. The 56 papers highly related to this topic were computer science, software science, psychology, communication science, sociology, and management science. Among them, some achievements organically combine the theories and methods of computer science, psychology, and sociology, showing a clear trend of multidisciplinary integration. This is different from the differences found in domestic research.

The two series of research papers published by Hua Ye (2018)[6] are representative. He acutely takes the online platform of Internet enterprises supporting users' innovative services as the research object, and further expands the theory of "value co-creation" by CK Prahalad, CK and V Ramaswamy (2000)[7], reflecting the interdisciplinary perspectives of computer science and management science. It is found that the mechanism of the mobile phone platform is an innovation behavior which significantly affects users with technology (toolkit) and policy (rule), and user dominance, as an important attribute of user innovators, significantly driving users' innovation behaviors. Kim and Da Jung (2016)[8] studied deeper into the principles that should be adopted to design and realize user autonomy in the system-driven personalization process of social media, because Kim believes that personalization greatly affects the way people communicate with each other through the system. At this time more and more social media providers automatically personalize their services according to user data, and the data processing adopts the usual black box mode, with users' autonomy in using the system being hindered. In addition, Klang and Mathias (2006)[9] discussed the relationship between network censorship and user autonomy. Matthew Ball and Vic Callaghan (2012)[10] pointed out that the user's views and concerns on autonomy are highly personalized and disparate, making it necessary to provide adjustable autonomy between intelligent autonomous agents and user autonomy. In the intelligent environment, users should be allowed to increase or reduce the autonomy of agents, so as to find a comfortable balance between giving up and maintaining control, and gaining and losing convenience.

After full reference to the above research results, this paper leaves the perspective and position of specific disciplines and does not integrate the interests of users into the ideas and positions centered on technology subjects, enterprise subjects, or government subjects separately and partially, but breaks away from the multiple interest subjects, and observes and analyzes user autonomy completely based on the user's position. Thus, it is possible to construct a new value analysis idea and system.

3   THE ELEMENT STRUCTURE OF NETWORK USER AUTONOMY

When we analyze the characteristics of a new media network from the perspective of humanism and based on user autonomy, we can not only effectively cover the technical variables, but it will also be more conducive for us to carry out an in-depth and continuous research on their evolution path and value mechanism.

To sum up, based on the logic of users' independent behaviors in the process of co-creation of information value, and according to the 5W elements of creation and dissemination of information and value, namely who, when, what, how, and why, the author initially constructs the analysis framework of the basic connotation of user autonomy, and through the high degree of value co-creation between teachers and students in colleges and universities. We should investigate the most important group of pioneers to carry out further research.

1) Users have the right to choose who to communicate with

It should be emphasized that the potential objects selected independently should cover all users on the same network platform. Otherwise, if we choose within the limited scope after being dominated by the market will or technological hegemony, we will lose the meaning and value of discussing the connotation of user autonomy in the general sense. For example, the communication structure of WeChat has many preset restrictions, which basically belong to the social mode of small circle acquaintances. In other words, on the WeChat platform, now users don't have channels and possibilities for information dissemination and communication for all WeChat users.

2) Users have the right to choose when and whether to communicate simultaneously, as well as the autonomy of interactive communication

In fact, users can independently choose whether to carry out time-sharing or synchronic communication and interaction, which is not only a simple technological innovation and progress, but also a great liberation for the mandatory constraints of time on the interactive participants, which is conducive to achieving a mutually beneficial pattern of multiagent cooperation and a win-win situation.

3) Users have the right to decide the form of information dissemination and interactive communication

This mainly means that users can choose various forms like text, image, sound, video, or multimedia to achieve specific information dissemination and interactive communication. In this sense, the language form of fusion media is a new compound language form itself, which has more advantages in communication efficiency.

4) The autonomy of users in deciding what kind of information to disseminate and communicate

This mainly means that users have the autonomy to spread any information within the scope of law and ethics. Any form of unreasonable or illegal content review will damage the user's autonomy badly.

5) The autonomy of users in choosing the purpose and motivation of information dissemination and communication

This mainly means that users have the right to decide whether or under what conditions the information they disseminate is commercialized. At present, generally speaking, most network users do not have this kind of autonomy. For example, the user's personal location information is transformed into various valuable or free online map service products, but they have no right to interfere or even know anything about it.

To sum up, within the scope of this paper, the ideal state of complete user autonomy means that users have ownership of their personal data on a specific platform, and can independently search, delete, change, integrate, price, and trade their own personal data at any time, independently choose the content, form, object, and time of information dissemination and communication within the whole platform, and at the same time, independently select the content, form, object, and time of information dissemination and communication within the scope of the platform, and independently construct and manage personalized and characteristic communities.

There are two systematic factors influencing the user autonomy under the above definition: first, time constraints; and second, scope constraints.

Specifically, it refers to the extent to which users' autonomy is constrained by time conditions. Take WeChat as an example, on the WeChat platform, the entire platform is not the maximum scope for users to realize peer-to-peer interactive social interaction, but only within the limited range of 500 people. In addition, although WeChat users have the right to delete the "personal photo album," they do not have the right to change the "personal photo album," they have to delete them and then make another change. WeChat users do have the right to delete real-time interactive information, although it is only valid for 2 minutes.

After clarifying the connotation and structural framework of the concept, the following in-depth analysis of user autonomy was gradually carried out.

3.1 *Analysis of the structure of user autonomy*

In this paper, in terms of user size and user activity, we adopt the social network platforms which is in the forefront of the new media industry and has relatively large user autonomy to establish the core coordinates for observing and analyzing the evolution of Internet value form. The first generation of social network platforms is represented by Facebook and WeChat. Compared with the traditional World Wide Web, user autonomy has made great progress, but there remain some limitations. The largest limitation of autonomy lies in the lack of effective participation rights of users on the structure and organizational boundaries of online communities as well as the lack of necessary self-protection rights.

In contrast, the social network platforms based on video content creators, represented by Instagram, YouTube, and Bilibili, has not only a high starting point, but has also been effectively and continuously improved because core users are content creators and value providers.

Taking Bilibili as the object, this paper deeply explores the network evolution path and its value form, which is based on content innovators, so as to put forward the internal logic and development trend of the network evolution plan, as well as the valuation ideas of new value forms. The time and scope of these two system variables are included in the structural analysis of autonomy, and the following coordinates are set up to observe and analyze the existence of user autonomy in state. Independent boundary setting means that users can search and choose the objects of information dissemination and communication independently, and have the power to set up the boundary of interactive communication scope

This means that users can choose real-time or delayed two-way interaction independently.

In Figure 1, the first quadrant is characterized by time-sharing communication and user self-determination. Network users in this field enjoy relatively high autonomy because time-sharing communication removes the time limit, and all users have the right to construct communities with different themes toward all users of the platform and communicate with each other based on independent personality, and equal and free power. In other words, the community in this field is open to all people and can form various public life fields through information communication. Each participant can freely express his opinions and defend his legitimate rights and interests. For example, interest stations in Bilibili, WeChat Mini Programs, and an official account.

Network users in the second quadrant of the above coordinate system also communicate with each other on the basis of independent personality, as well as equal and free rights, but users have no right to construct communities or set community boundaries for all users of the same platform. Power is often limited by technical barriers or commercial barriers to a large extent. However, users can usually set up and construct theme communities within the limits of technical and commercial barriers, so as to establish corresponding public opinion areas. For example: WeChat social group, online community "Hedgehog Commune," and MediaBistro,[11] which is a resource website for international media professionals.

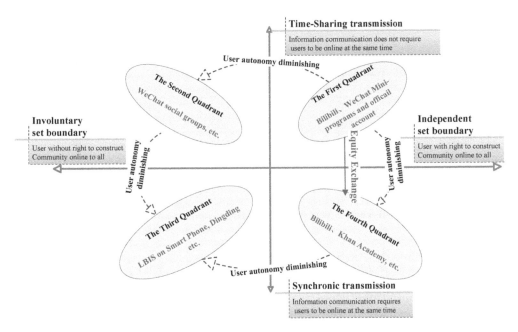

Figure 1. Analysis of user autonomy in digital network.

The fourth quadrant is mainly characterized by synchronic propagation and autonomous demarcation. Compared with the first quadrant, the main reason why users are willing to "sacrifice" the freedom of time-sharing communication is that they have realized the personalized "value transaction" of equivalent or better value to some extent, for example, value recognition, cultural innovation, knowledge and skills learning, and so on. The realization of this kind of "value trading" is often not highly dependent on high technology, but on content innovators and opinion leaders.

Bilibili and Khan Academy are typical representatives of high user autonomy. The former is the largest secondary original video-sharing community in China, while the latter is the largest online university in the world. Their common feature is that the creation and operation of online communities are dominated by content creators rather than platform operators, thus forming a large number of active communities with a self-organizing theme. Later, we analyze the growth and internal law of this kind of self-organizing network community in detail.

The user autonomy in the third quadrant is relatively minimal because users can neither set boundaries nor choose synchronic propagation independently. The typical representative is the public interactive digital media based on user location information, such as subway interactive media, or a commercial social platform for enterprise users such as Dingding.

From the perspective of the overall coordinated structure, the autonomy of network users decreases gradually from the second and fourth quadrants to the third quadrant. The ideal state for users to have a high degree of autonomy should be that users can face all users on a specific platform, rather than a limited part. And users can choose the objects of information dissemination and communication independently, set personalized community structure and boundaries, and choose time-sharing or synchronous communication independently.

To further study the academic definition and specific connotation of user autonomy, the author interviewed 166 Chinese and foreign Internet users in three stages from June 18, 2018, to September 22, 2019, to understand the specific connotation and main variables of network user's independent choice rights based on the user's position, and rights and interests. The structure of interviewees and research objects in the whole research process is as shown in Table 1.

In the first stage of the study, a total of 18 clearly defined and well-defined related elements are extracted. At the same time, it is also found that there are some differences between Chinese

Table 1. Structure of focus group interviewees.

USA: Boston teachers		Teacher (T)	Student (S)	China T/S	USA T/S	Canada T/S	EU T/S	Africa T/S	Asia [Non Chinese] T/S	AU T/S
The first		33	45	6/8	7/9	4/5	6/7	3/6	4/7	3/3
stage 2018/6-9	%	42	58	7/10	9/11	5/6	7/9	3.8/7	5/9	3.8/3.8
The second	Number	15	25	2/4	3/5	2/2	3/4	2/4	2/4	1/2
stage	of people									
2018/10-12	%	37	63	5/10	7.5/12.5	5/5	7.5/10	5/10	5/10	2.5/5

China		Teacher (T)	Student (S)	Beijing T/S	Shanghai T/S	Guangzhou T/S	Anhui T/S	Hunan T/S	Asia [Non-Chinese] T/S	AU T/S
The third	Num	43	45	6/8	7/9	4/5	6/7	3/6	4/7	3/3
stage 2019/2-9	%	48.9	51.1	6.8/9	7.9/10	4.5/5.9	6.8/7.9	3.4/6.8	4.5/7.9	3.4/3.4
Summary	Num	76	90							
	%	45.8	54.2							

Source: Created by the author.

and foreign users on the specific connotation of autonomy, which is mainly reflected in the right to delete or change personal data and the right to price the transaction of personal data. In the second and third stages of the study, the author further analyzed variance on the related differences of the survey data.

In the second stage, based on the interview data of the first stage, the author selected 40 non-Chinese people from 78 interviewees in the first stage (see Table 1 for the nationality distribution), so as to form a comparison group with the research data of Chinese users in the third stage. The basic requirement for the interviewees is that each person has at least one account on Facebook, WeChat, Twitter, WhatsApp, Instagram, and line platforms, who have been logged in at least once a week, and used for more than 1 year. Among the 40 electronic questionnaires, 39 were valid. Based on the data from the questionnaire survey, focus group interview, and expert interview in the first stage, based on the dynamic process of information dissemination and two-way communication on the digital network platform, the author lists the following key elements in the process of user's independent behavior in choosing information form, information dissemination, information processing, and information transaction, and adopts the Likert scale to rank the importance of key elements. In this way, the author gains an in-depth understanding of their preferences and dependence on various social platforms that have achieved the super scale, as well as their cognition and attitude toward the large amount of personal data derived from them. In addition, according to the data from "open subjective questions" in the second stage of the questionnaire survey, we found that "platform scale" and "platform function" are two very important observation dimensions (Tables 2 and 3).

4 THE FRAMEWORK OF NETWORK USER AUTONOMY

In the third stage, focus group interviews and questionnaire surveys were conducted for Chinese local network users. In Fudan University, Tongji University, Peking University, Tsinghua University, Guangzhou University, University of Science and Technology of China, and Hunan University, the teacher focus groups aged 35–65 and the student focus groups aged 18–35 were selected, with 43 and 45 students in each group, respectively, with at least 5 years of network experience and personal email and social accounts, at least one account on WeChat and Facebook, and more than 1 year of experience, logging in at least once a week. The purpose of this paper is to deeply understand

Table 2. Analysis of the structure of related elements of network users' autonomy.

Classification of autonomy	Analysis of elements of autonomy			
The election of information form	Text mainly For example: Twitter	Sound mainly For example: WeChat voice Himalayan FM	Image mainly For example: Instagram	Video mainly For example: YouTube, Bilibili, TikTok
The election of information sharing	For specific individuals For example: WeChat @someone	For specific communities For example: WeChat group announcement	For specific platforms For example: WeChat official account. Front page community of Bilibili	For global networks Waiting to be discovered
The election of information processing	Restricted upload For example: Twitter, WeChat voice message	Unlimited upload restricted For example: Bilibili, YouTube	Limited delete or change For example: WeChat messages are deleted within a limited time	Unlimited delete or change For example: video of Bilibili, WeChat photo album
	Integrating and ordering For example: WeChat payment bill	Nonintegrating and ordering For example: WeChat social personal data collection and map	Restricted searching For example: WeChat search is limited by incomplete historical information; Bilibili is limited by video information analysis technology	Unrestricted searching Waiting to be discovered
	Limited storage Free services	Unlimited storage For example: Cloud personal photo albums	Limited transfer For example: Cross platform data transfer barriers	Unlimited transfer Waiting to be discovered
The election of information transaction	Copyright autonomy For example: WeChat official account, TikTok, Bilibili	Pricing autonomy For example: TikTok, Bilibili	Transaction autonomy For example: TikTok, Bilibili	

Source: Created by the author.

their preferences and dependence on the most frequently used social platforms, as well as their cognition and attitude toward a large amount of personal data derived from them, and to make a comparative analysis with the second stage data.

The Cronbach alpha coefficients of the two parts of the questionnaire were 0.871 and 0.814, respectively, which was much higher than the high reliability standard of 0.7.

In this regard, the author extracted the four groups of comparative data one by one, totaling four groups, including the non-Chinese group and the Chinese group under the same subject stably declining, that is, the P14 in the non-Chinese sequence was paired with the P14 of the Chinese group, so as to carry out the analysis of variance, that is, the mean square quotient (F value) within the group was removed by the intergroup mean square. The values were 6.784 and 6.126, respectively, which were far greater than 1, indicating that there was a statistically significant difference between the two groups, that is, the difference between Chinese and non-Chinese users. However, for the ranking differences of "interactive communication for global network users" and "self-construction of communities within the platform," F values were 1.135 and 1.658, respectively, which are relatively close to 1. Therefore, it can be concluded that the actual differences between Chinese and

Table 3.   Classification list of user autonomy-related elements of digital network platform.

Serial number	Elements of autonomy Select category	Serial number	Elements of autonomy Select category
P_1	Text as the main information carrier (form selection)	P_{10}	Storage capacity of interactive information (processing selection)
P_2	Voice as the main information carrier (form selection)	P_{11}	Independent transfer of personal information (processing selection)
P_3	Image as the main information carrier (form selection)	P_{12}	Self-searching of personal information (processing selection)
P_4	Video as the main information carrier (form selection)	P_{13}	Independent integration of personal information (processing selection)
P_5	Interactive communication for specific individuals (sharing selection)	P_{14}	Personal information deletion or change (processing selection)
P_6	Interactive communication for specific communities (sharing choice)	P_{15}	Self-construction of communities within the platform (processing selection)
P_7	Interactive communication for specific platform (sharing choice)	P_{16}	Ownership of personal information (transaction selection)
P_8	Interactive communication for global network (sharing choice)	P_{17}	Having pricing power of personal information (transaction selection)
P_9	Upload amount of interactive information (processing options)	P_{18}	Trading right of personal information (transaction selection)

Source: Created by the author.

Table 4.   The sorted list of the key elements of network user autonomy.

Sort	Elements of autonomy Select category	Sort	Elements of autonomy Select category
1	Self-construction of communities within the platform (processing selection)	10	Independent integration of personal information (processing selection)
2	Personal information deletion or change (processing selection)	11	Interactive communication for global network (sharing choice)
3	Ownership of personal information (transaction selection)	12	Having pricing power of personal information (transaction selection)
4	Independent integration of personal information (processing selection)	13	Video as the main information carrier (form selection)
5	Independent transfer of personal information (processing selection)	14	Image as the main information carrier (form selection)
6	Interactive communication for specific individuals (sharing selection)	15	Voice as the main information carrier (form selection)
7	Interactive communication for specific communities (sharing choice)	16	Text as the main information carrier (form selection)
8	Self-searching of personal information (processing selection)	17	Storage capacity of interactive information (processing selection)
9	Interactive communication for specific platform (sharing choice)	18	Upload amount of interactive information (processing options)

Source: Created by the author.

non-Chinese users in these two items are not significant enough and have no statistical significance or value. At the same time, the author finds that this result is consistent with the difference between Chinese and non-Chinese users in the first stage of qualitative data.

To sum up, we can fit the research data of the second and third stages, and get the following overall results about the importance ranking of the key elements of user autonomy (Table 4).

Table 5. KMO and Bartlett sphericity test of the structure of digital network user autonomy elements.

Kaiser-Meyer-Olkin measure	0.796
Bartlett Sphericity test approximate chi-square	806.124
df	106
Significance	0.000

Source: Created by the author.

Table 6. The total variance of the validity of principal component interpretation of demand factors.

Serial number (element)		Initial eigenvalues			Extract the principal component characteristic value and load		
		Total	Variance %	Cumulative variance %	Total	Variance %	Cumulative variance %
1	P_{15}	3.686	16.246	16.246	3.686	16.246	16.246
2	P_{14}	2.945	13.496	29.742	2.945	13.496	29.742
3	P_{16}	1.974	11.477	41.219	1.974	11.477	41.219
4	P_{18}	1.764	10.033	51.252	1.764	10.033	51.252
5	P_{11}	1.621	9.987	61.239	1.621	9.987	61.239
6	P_5	1.453	7.169	68.408	1.453	7.169	68.408
7	P_6	1.381	5.223	73.631	1.381	5.223	73.631
8	P_{12}	1.372	4.939	78.570	1.372	4.939	78.570
9	P_7	1.216	3.074	81.644	1.216	3.074	81644
10	P_{13}	1.209	3.042	84.686	1.209	3.042	84.686
11	P_8	1.147	2.887	87.573	1.147	2.887	87.573
12	P_{17}	1.143	2.467	90.040	1.143	2.467	90.040
13	P_4	1.019	2.116	92.156	1.009	2.116	92.156
14	P_3	1.012	2.325	94.481	1.007	2.325	94.481
15	P_2	1.009	2.122	96.603	1.003	2.122	96.603
16	P_1	1.001	2.018	98.621	1.001	2.018	98.621
17	P_{10}	0.247	0.811	99.432			
18	P_9	0.229	0.568	100.000			

Source: Created by the author.

In order to further understand the relationship and essence of the above elements, the author uses SPSS 19.0 software and the factor analysis method to analyze the above 18 elements. Element variables must meet the requirements of KMO measurement and Bartlett test. The author checked the above P_1 to P_{18}, and the results are shown in Table 5.

The results of KMO and Bartlett sphericity test showed that the KMO measure value was 0.796, significance value of the Bartlett sphericity test was 0.000, which is less than the required 0.05, indicating that the data provided by the questionnaire are suitable for factor analysis. Because the structural elements of user autonomy have reasonable explanatory validity, it depends on the corresponding eigenvalue value. Therefore, according to the correlation coefficient matrix of the factor variables, the principal component analysis method is used to extract the factor factors, and the factor whose feature root is greater than 1 is selected as the key factor (Table 6).

The above analysis data show that except for P_{10} and P_9, the other user autonomy variables enter the principal component variables.

From the order of importance and the quantitative analysis of its power subitems, user autonomy can be divided into three levels as described next.

The initial characteristic value of the first level is more than 2, which includes the two most valued rights: self-construction of communities within the platform (P_{15}) and personal information deletion or change (P_{14}). The former involves the independent construction of communication structure and interest relationship among users, while the latter involves the ultimate disposal right of users' personal information. At present, these two important powers are almost not in the hands of users but are controlled by many new media enterprises. Users can neither see the data, nor know how the data are used, nor can they own or manage the data.

In the focus group interview, the author found that in the view of more than 45% of the respondents, the general ignorance of how personal data are generated, specific content and how it is used, as well as the resulting worries, largely restrict their own willingness to create, disseminate, and consume as users of Internet content products.

Therefore, Bilibili as the research object of this paper has unique and typical research value and practical significance. As a video-sharing website, through technical and strategic active decentralization, it not only successfully attracts and effectively inspires users with high professional standards to produce, provide, or share audio-visual content and bullet screen text, but also, in a noncontractual way, has reached a cooperative relationship with these professional users with the same rights and interests with the platform, so as to jointly construct the sustainable development of Bilibili subject interest community.

Due to the double positive incentive of benefit sharing and internalization of cultural values, the empowered users have the ability to continuously brew, cultivate, and maintain cultural collective consciousness and multiple values, and on this basis achieve content filtering consistent with the independent will. It is worth emphasizing that content filtering is different from content review. Although the appearance and results tend to be the same, the internal action logic of the former is bottom-up, in line with the ecological laws and value norms of digital networks, while the latter is top-down, in line with the ecological laws and value norms of public power institutions and commercial institutions. For example, the interest community on the platform of Bilibili is constructed by users independently, not by the platform side from top to bottom. Users have a high degree of autonomy over the videos they upload, including the very critical right of "deleting or changing independently."

The initial eigenvalues of the second level are more than 1 and less than 2, including 14 items in total. Generally speaking, the ranking logic is that information trading autonomy comes first, followed by information sharing autonomy and information dissemination autonomy. Among them, the importance of the "independent pricing" right in the transaction autonomy has been put behind, which indicates that users pay more attention to the independent ownership of copyright and the completion of transactions. To a certain extent, users are not very concerned about transferring the pricing power of personal data to the market or other types of intermediaries.

This is a very interesting data analysis result. To a degree, it can well explain why a large number of users with original content are willing to share personal data including personal views, personal photos, personal behavior tracks, etc. for free, and are even willing to share personal data with market-oriented copyright value, including personal academic achievements, for free. Of course, the fact that users are willing to transfer the pricing power has some deeper external reasons besides subjective will. Through further in-depth interviews, the author found that, to a considerable extent, there is a reverse forcing mechanism, that is, because the information giants collect, use, display, and trade user data without the consent of users, resulting in large-scale established facts, and these facts are bound together with the user's personal interests in various forms, thus forming some extra-legal compulsion and inducement occupies the time and space for users to think and discuss a reasonable price negotiation for their personal data.

A typical example is that it is difficult for users to own or integrate their own network behavior data across platforms. In fact, this not only leads to the imbalance in the distribution of information and interests, but also leads to the public losing the basic right to know and have judgment on the authenticity of big data related to public interests due to the unreasonable and opaque high monopoly of data, which means the potential reality of "data corruption." In other words, public data owners such as Alibaba or Tencent will tamper with or conceal data that are unfavorable to them in case of data monopoly, thus damaging the public interest. The public totally lacks the necessary channels to understand and supervise this activity. Assuming that each user has the ownership and use right of personal data, only users can reach mutual trust and cooperation under certain conditions and circumstances, which makes cross-platform collection and integration of personal data a reality, so as to provide the possibility to verify the authenticity of information released by data monopolists.

The initial eigenvalue of the third level is less than 1, including two items in total, namely, the amount of information storage related to terminal hardware (P_{10}), and the amount of information uploaded related to network speed and software design (P_9).

The above theoretical analysis framework of network user autonomy provides us with a panoramic analysis of user autonomy and a strong theoretical basis and practical tool for the implementation of a "Contract for the Web."

5 EVOLUTION PATH OF NEW MEDIA BASED ON USER AUTONOMY

The author, based on the analysis of the main variables of user autonomy, and taking the three major social platforms of WeChat, Bilibili, and TikTok as examples of the top three users in China, and the user scale of the corresponding variables, roughly drew the distribution of China's network user autonomy. It can be seen intuitively from Figure 2 that, on the whole, the position of the autonomy curve of WeChat users is relatively high, that is, WeChat provides users with relatively more use value.

The hollow breakpoints on the three different colored lines in Figure 2 indicate a loss of functionality or structure. If it is a structural "breakpoint," then it indicates that there is a congenital defect in its communication competitiveness, which may lead users to divert to companies or products without such breakpoints at any time. If it is only a functional "breakpoint," it means that the competitiveness can be effectively enhanced through the gradual improvement of the functional structure. In fact, if the curve shape and height of two similar products are similar, the growth potential of the products and companies with more functional breakpoints is theoretically larger, because it reaches the same user size with more functional defects. Similarly, the concave part in the continuous curve also means that users may be diverted to the same type of companies or products without such concave shape.

In Figure 2, the number of vertices at the top is more, and the overall competitive advantage is more obvious. Bilibili[13] in the figure has an obvious comparative advantage; WeChat[14] with a total number of 1.1 billion users mainly uses the focus as the main communication and communication mode, and "self-construction of communities within the platform," "personal information deletion or change,"[15] and "ownership of personal information." The three factors of user autonomy are significantly less than those of Bilibili and TikTok. There is obvious competitiveness, because these three represent the strategic competitiveness of the new generation of Internet platforms with user autonomy as the core value.

According to the data in Figure 2, we can use two different valuation ideas for comparative analysis.

First, we calculate the user value difference V1 (X) of different social products according to the unified mean value, that is, in the digital network with average tendency, the value of each point is equal regardless of the potential. Therefore, we give each intersection real point a score of 10. Secondly, according to the actual value of the coordinates, the user value difference V2 (X) of different social products is calculated as:

$$V2 = \frac{\sum_{n=r}^{n=1} I_n}{\sum_{n=r}^{n=1} (U_n) \times 18}$$

Among them, I represents the total amount of interactions, U represents the total number of users, and r represents the total number of nonzero real value points.

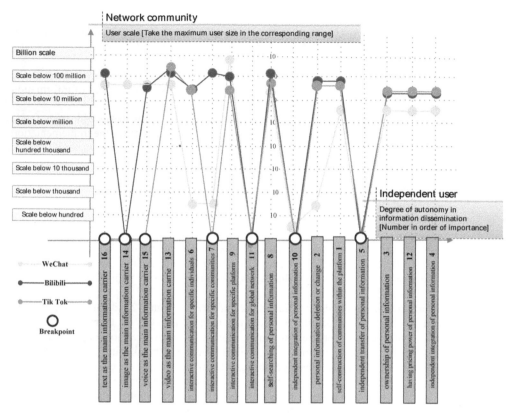

Figure 2. Distribution of user autonomy of three social giants in China.
Source: Created by the author according to public data of Tencent, Bilibili, and TikTok.

V1 (Bilibili) = 10 * 12 = 120; V2 (Bilibili) = {12 real value points accumulation} / 18 / total users

V1 (WeChat) = 10 * 14 = 140; V2 (WeChat) = {14 real value points accumulation} / 18 / total users

V1(TikTok) = 10 * 9 = 90; V2 (TikTok) = {9 real value points accumulation} / 18 / total users

It should be emphasized here that the scope of "user autonomy" in this paper ideally covers all users of a specific network platform, and the specific value is the maximum number of users within a specific range. For the user autonomy that only exists in the minority group, it is measured and valued according to the maximum number of users in the minority group. For example, in the calculation of "interactive communication for specific individuals" of WeChat, due to the equality of two-way communication rights required by interactive communication, the maximum WeChat user group meeting the definition requirements is the personal WeChat group, with the upper limit of 500 users, so the value is 500. Another example is that in calculating WeChat's "text as the main form of information dissemination," because the definition of "user autonomy" is defined as the autonomous dissemination of text information, it is the largest user number of a single WeChat official account. The same is true that the corresponding value of TikTok is the maximum limit of two-way interactive communication on the shaking platform, that is, the maximum user volume of a single TikTok. Because there is no direct interactive communication interface for all users on the platform, it is necessary to enter the specific regionalized community first. The difference between Bilibili and the above two is: first, Bilibili has a relatively stable home page content recommendation position for all users of the platform; second, there is no upper limit for the total number of users in a specific community of Bilibili, which can theoretically realize the interactive communication of all users of the platform.

From this, we can roughly infer that although the user scale of Bilibili is only about one-third of that of WeChat, the strategic valuation of market value is very close. If different elements are given different weights according to the above ranking of the importance of user autonomy elements, the market strategic valuation of Bilibili will exceed that of WeChat, because WeChat ranks behind Bilibili in the top three elements of autonomy that users are most concerned about. In addition, what I want to emphasize about the above figure is that the breakpoint of the connecting line with a value of zero can be regarded as a "key pressure area." On one hand, it represents the short board of competitiveness, on the other hand, it also represents the strategic trend of innovation. In other words, under the general trend of full-featured platforms, if the "breakpoint" cannot be improved in time, it is likely that the whole curve will be dragged down, resulting in the short board effect of structural downward movement. Therefore, innovation must be strategically inclined to the breakpoint. This effectively explains why Bilibili is sought after by the capital of Alibaba and Tencent at the same time, because the level and form of user autonomy are becoming the competitive focus and value core of new media enterprises. Bilibili has an important and unique strategic value: it has already represented a new model of network user autonomy, that is, the methods of power organization and commercial profit.

6 NEW GENERATION NETWORK WITH CONTENT CREATORS AS CORE VALUE

As a video-sharing website, Bilibili was initially based on secondary animation videos, and later developed into a creative video-sharing community with original videos as the main body and secondary meta culture as its characteristics. It has gradually constructed an "emerging" network community dominated by users and realizing self-growth from bottom to top. Every "emerging" community is a potential knowledge or emotion comprehensive value carrier, with dominant and easy to observe group wisdom. Before in-depth analysis and elaboration of the ecological characteristics and group wisdom of Bilibili, we require a comprehensive understanding of Bilibili.

6.1 *The core value of Bilibili is original content users*

The Bilibili bullet screen video network was founded on June 26, 2009, also known as "Bilibili," focusing on the creation, incubation, and sharing of video content works. In this high-frequency interactive community, registered users upload a variety of original videos including animation, music, dance, technology, and other network cultural content, with a bullet screen as the main form of interaction to share and publicly comment on each other. Now, Bilibili is China's largest youth cultural community, because 75% of its users are under the age of 25. Those born in the 1990s and 2000s are the dominant group, and users include a considerable number of primary school students. According to the official data of Bilibili, the average age of new users in the first quarter of 2019 was 21.5 years old. These young people are the driving force of Internet development, so the development potential of Bilibili is very large.

The biggest contribution of Bilibili to China's Internet lies in the construction of a communication mode (i.e., 2.0UGC mode) based on the user's high degree of autonomy and creativity, which takes the content creators and high-quality content enthusiasts as the origin of information dissemination. Every step of its development process is directly derived from the scale expansion of uploader and bullet screen traffic.

By the end of September 2019, the user increment and revenue of Bilibili maintained a high and stable growth rate (Figure 3).

It is worth emphasizing that Bilibili does not specialize in content production at this stage, but creates and provides opportunities for users to "generate and disseminate good content" by reasonably transferring the dominant power of communication to users (mainly referring to the uploaders and sharers of original content, i.e., Uploader), such as UGC[17] and PUGC[18], where the

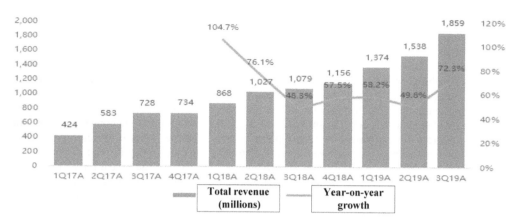

Figure 3. Total revenue and year-on-year growth rate of Bilibili in the third quarter of 2019.
Source: Financial announcement of Bilibili, Tianfeng Securities Research Institute.

former refers to "User Generated Content," which refers to "Professional User Generated Content." In other words, the uploader of Bilibili is responsible for the creation and release of content, and the users of Bilibili decide and disseminate the value of original content by sharing with their responses and thoughts on it. They cooperate to build a community of common interest with the characteristics of "psychological ownership."

There are two factual bases for Bilibili to give way or distribute power to users as described next.

First, Bilibili currently has 33 content partitions, or content communities, which are built on the basis of user wishes and autonomous clusters. The main role of the platform is to observe, analyze, and determine when to upgrade a naturally formed user cluster into a formal interest community of the platform. The specific process is usually derived from the personal preferences of one or several uploaders to create video content or upload and forward their favorite video products, so as to arouse the interest of other users and participate in the interaction of watching movies and bullet screens. When the cluster interaction users are maintained at a certain scale, the platform operators at Bilibili will naturally retain, support, and finally formally set up these theme communities (as shown in Figure 4). Video websites other than Bilibili are basically unified by the platform to construct the content classification system, and users have no right to speak on it.

Second, there are 200 content recommendation sites on the home page of Bilibili computer terminal, only seven are occupied by the platform side of Bilibili, and the remaining 193 locations are all content recommendation places for each interested community.

Third, Bilibili has set a soft threshold for formal members, that is, they must pass a test, so as to reach a certain agreement on the network social behavior, especially the barrage etiquette and community culture, before they are qualified to participate in community interactions. At present, the interactive barrage flow of Bilibili is basically in the state of full overflow. It is necessary to set up a barrage pool and complete the appropriate filtering to ensure that the barrage content does not overflow to affect the normal viewing. According to the author's small-scale sampling, it was found that the direct correlation between the bullet screen of Bilibili and the corresponding video content is more than 65% on average according to the statistical data and text analysis of random sampling from July to September 2019, which is far higher than the 5% of other video-sharing websites. To an extent, this proves the energy difference between the bottom-up user logic and the top-down institutional logic in giving a new media product or platform.

Fourth, Bilibili consciously protects the privacy of Bilibili users, in addition to the normal community interaction, any external stakeholders, including advertisers and academic researchers, cannot contact users, especially influential content creators and opinion leaders, outside the normal community interaction scope, even if both sides have the potential to achieve a win-win situation and mutually beneficial cooperation. Bilibili rather chooses the market-oriented indirect guidance

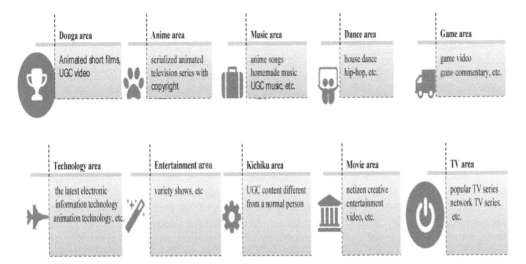

Figure 4.   Core interest communities on Bilibili platform.
Source: Officially provided by Bilibili.
Figure 4 shows the following:

Douga area: Animation adopts a shooting object frame by frame and plays continuously to form a moving image, and uses the painting technique to create the art of life movement, which is a kind of comprehensive art, such as animated movies, where netizens make their own video.

Anime area: This is a Japanese serialized animated television series, which can be viewed as a television series, including both live action and animated television series, and includes copyright video.

Music area: This is an art form and cultural activity that includes the creation of musical works (songs, tunes, symphonies, etc.), performance, evaluation of music, study of the history of music, and music teaching. This includes anime songs and homemade music.

Dance area: Dance is an art form accompanied by music and characterized by rhythmic movements. It includes house dance and hip-hop.

Games area: This community is a gathering place for gamers, a full-content community tailored for tens of millions of gamers. It includes game videos and game commentary.

Technology area: Society is used to science and technology being together, and the task of technology is to apply the results of science to practical problems. This includes the latest electronic information technology and animation technology.

Entertainment area: This contains tragicomedy, various competitions, games, music, dance performances, and appreciation, etc. It includes variety shows.

The Kichiku area: This refers to being out of one's mind and doing something different to a normal person. "The Kichiku" is a relatively common type of original video in China, which uses highly synchronized and rapidly repeating materials to match the rhythm of background music to achieve a funny effect, or combines the rhythm with sound and video (or audio) editing and extremely high-frequency repeating pictures (or sound). This area contains a lot of parodies like this, such as netizen creative entertainment videos.

Movie area: This area focuses on the latest domestic and foreign film information, celebrity film reviews, classic film recommendations, film commentary, and so on, including copyright films.

TV series: This focuses on the latest domestic and foreign TV series information, celebrity TV series reviews, classic TV series recommendations, TV series commentaries, and so on, including popular TV series and network TV series.

tools, such as "soliciting contributions" and "rewards," rather than direct intervention, because they believe that talent cultivation and content creation have their own inherent laws, and it is not appropriate to encourage them to evolve. They also believe that the comprehensive protection of users' privacy and independence is the best incentive strategy.

At present, take the initiative to give the users right to the extent of the above, the similar network video community can hardly find a second one. Based on the above characteristics, the uniqueness of Bilibili is self-evident. The user group with high loyalty and high professionalism has the best return for the idea of empowering or returning rights to users.

As of September, 2019, Bilibili had 128 million active users, with more than 1.1 million monthly active uploading users and an average monthly barrage volume of 2.5 billion. Among them, as of May, 2019, the number of overseas users was about 7 million, and the number of active uploading users was 20,000, making it the largest trending cultural community in China and even Asia. A simple visual inspection of the number of registered users of Bilibili may not be eye-catching, but if you add the registration threshold and then look at such a large number of registered users, it is very important. By the end of September, 2019, 62 million Bilibili users were full members after answering questions. Every member who wants to become a registered member of Bilibili must pass the 100 registration questions of Bilibili.

The initial threshold of Bilibili registered members determines the minimum "professional level" of Bilibili registered members. In other words, the registered members of Bilibili have the knowledge threshold, which is the real reason why the content of the barrage and the related videos always maintain a high degree of correlation. In addition, this has become an important reason why many primary and secondary school students are keen on Bilibili, because it is difficult for their parents to pass the test and become a member of Bilibili to supervise them. In fact, most parents of students have a positive attitude toward Bilibili, because the registered members of Bilibili regard it as their spiritual home. The quality of video content provided by them is very high, and the professional level of the bullet screen content is high. There is a benign upward trend in values within the community.

When the initial threshold, high-quality content, high professional bullet screen social interaction, and strong sense of belonging integrated together, on the one hand, it forms a continuously increasing user attraction, on the other hand, it also forms a continuously increasing comprehensive sink cost, making Bilibili a highly sticky community, and the monthly retention rate of formal members is stable at more than 70%.

For Bilibili, which is a "three-high" network community with high threshold, high viscosity, and high user autonomy, its real value and market competitiveness are no longer limited to the simple daily active user scale or total user scale, because the core of the "three-high" community is not ordinary users, but uploaders and active bullet screen users with the ability of content creation and value shaping. Each uploader and "bullet screen owner" who interacts with the uploader in high-frequency interactions means that the network community builders who can be fully expanded at any time have the potential to produce a complete and unique small world. These small worlds can be either parallel or intersecting. In other words, the competitiveness of Bilibili stems from its unique value concept: dare to reasonably decentralize the power to users, and then gradually return the rights (i.e., take the initiative to remove the interest constraints on users), so as to form a cooperative symbiosis system with dynamic balance of rights and interests between users.

6.2 Structure and ecology of Bilibili based on a high degree of user autonomy

Figure 5 shows the static network relationship between the top 10 uploaders and 50 randomly selected fans in the demon area of Bilibili.

Through the above network analysis, the author initially presents the internal structure of the cooperation and symbiosis system between the Bilibili platform and users, and proves to an extent that the social sharing network represented by Bilibili and based on the core value of content creators, is a parallel network form of "big platform and big community" with platform and community sharing with highly equal rights, that is, there is no right level or right barrier between the platform and each network community. Users can obtain support from various resources on the platform through the community construction, so that the Bilibili platform has an internal self-expansion and optimization mechanism. In other words, with content creators as the core (uploader) on Bilibili, users have the right to construct independent communities on the platform from bottom to top, and there are no barriers between groups. Each community maintains an all-round opening up with the whole platform and other communities, and only takes interest at

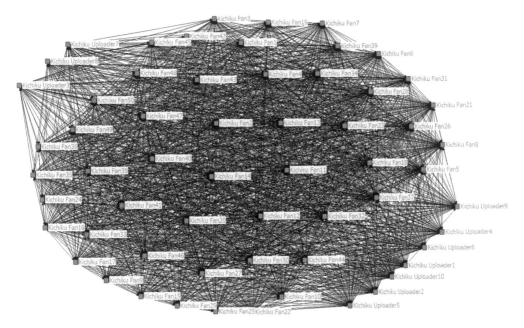

Figure 5.   Network relationship between a top 10 uploader and 50 random fans in the demon area.

the intangible community boundary, forming infinite isomorphic structures with the platform and adjacent communities' parallel network. This is essentially different from the "big platform and small community" network form represented by WeChat. The full-featured large platform holds the advantageous resources, so as to set different rights levels or rights barriers among small communities and distribute resources and rights from top to bottom.

In order to have a better understanding of the above essential differences from sensibility to rationality, we may as well continue to observe the social network relationship between the top ten uploaders and the 50 fans randomly selected from the demon area in the top 10 interest communities of Bilibili. By comparing the above-mentioned social network diagram between uploaders and fans in the demon areas, we can clearly understand that the vitality of Bilibili mainly comes from the high-frequency interaction within each community, rather than from the interregional interaction between different communities, at least for now.

Figure 6 shows that the relationship between the fan users in a specific community on the platform of Bilibili and the main users of the core up on the platform is a loose and nonclose relationship, which is in sharp contrast to the high-density interaction within the demon area in Figure 5. At the same time, it also confirms the relatively independent form between the major communities of Bilibili. It is worth emphasizing that this kind of relative independence is formed naturally based on the user's behavior preference, rather than constructed by platform operators through technology and human design. In fact, except for Bilibili in mainland China, almost all other video-sharing platforms are set up and constructed by platform operators, and user autonomy is very limited.

The reason why the author chooses the demon area is that it has a large scale of users. Compared with other interested communities, users are more diversified and are more likely to pay attention to other communities on the same platform. Therefore, although it is a small sample and the data are relatively static, the above comparison directly shows that the interdistrict interaction between different communities on the Bilibili platform is very intensive, and the high-density interaction of users in Bilibili is mainly concentrated in the interior of each major interest community.

Bilibili, which actively empowers users, also sets an access threshold for empowerment, so that users with high interest and high professional level have priority to participate in interactions.

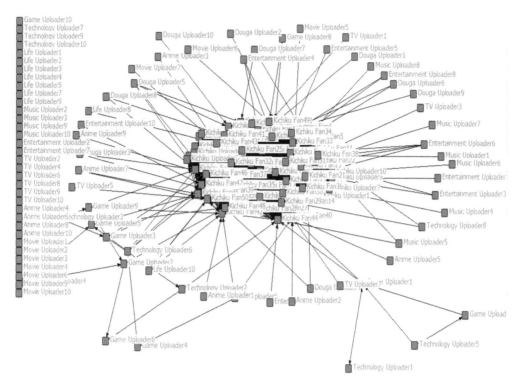

Figure 6. The social network relationship between the top 10 uploaders in Bilibili and the 50 random fans in the demon area.

This figure contains a variety of uploaders: game uploaders, science and technology uploaders, livelihood uploaders, music uploaders, entertainment uploaders, TV series uploaders, fan soap opera uploaders, animation uploaders, film uploaders, and the demon uploader and demon fans.

Bilibili classifies the authority between the uploader, the bullet screen user, and the "tourist type" user. The tourist type user whose main purpose is to consume content rather than interact cannot enjoy interactions with members and cannot send bullet screens to participate in interactions. This measure effectively promotes the inward aggregation and effective interaction of original content users and ensures the enthusiasm of active interactive users while realizing the overall user growth of the platform. The author calls it a "double high type network community," namely "high user autonomy + high proportion of original content." Later, the author elaborates on the unique value form and valuation model of the double high network community.

To sum up, the parallel network under the guidance of equal power is the core feature of the network configuration construction of Bilibili. This kind of parallel network structure also exists between uploading masters in different communities. The following is the social network relationship between the top 10 uploader in the top 10 interest communities on the Bilibili platform.

Here, we first continue to analyze the social network relationship of Bilibili.

The author introduces the time variable in an intuitive and simple way to improve the statistics of the above social network analysis. Therefore, during the 2-year period from June 2017 to September 2019, the author conducted regular continuous sampling observations on the mainstream communities within Bilibili platforms such as games, science and technology, life, animation, music, drama and film, focusing on the theme communities such as demon, national music, star dance, technical house, handicrafts, and freeze frame animation. The author's main research method was to take regular random sampling (including community sampling and video sampling) every 7 days, and collect data by video and screen capture, and conduct in-depth data and text analysis. In the specific

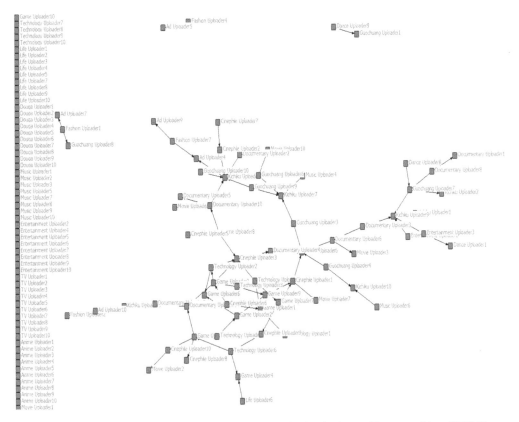

Figure 7. Social network relationship among the top 10 uploaders in the top 10 communities of Bilibili. The Figure 7 contains a variety of uploaders: game uploaders, science and technology uploaders, livelihood uploaders, music uploaders, animation uploaders, dance uploaders, documentary uploaders, amitron uploaders, film uploaders, film and television uploaders, fashion uploaders, advertising uploaders, and the demon uploader.

research process, according to the hot video or music that users watch and send with subtitles, the author spent a continuous 1 hour selecting video to accumulate 5 hours of real-time video recording. At the same time, screenshots were taken every 30 seconds. In each recording material, 30 bullet screen users were randomly selected and marked with the corresponding time code and interactive information summary. At the same time, combined with the public information data on the Bilibili platform, the mutual attention and mutual interaction between them and the relevant video uploader were analyzed one at a time, so as to dynamically record and draw the social network relationship after adding time parameters. As a result, the author found that there are a large number of network forms similar to a "time capsule" in each Bilibili community, that is, with the extension of time, the static social network develops a continuous and in-depth two-way interaction around specific topics among active Bilibili users, and maintains a high interaction intensity for a long time, thus forming an interactive communication density net, until hot topics gradually fade away under normal circumstances. Most of the dynamic "time capsule" disappears with the dynamic updating of Bilibili.

Among them, on average, there are 2.34 "time capsules" per video, while 54% of "time capsules" have a "time tail" with different sizes. The larger the "time tail" is and the more intensive the interaction is, the longer the tail will be. At the same time, the curvature of the tail will be smaller, and the tail will be smoother. Most of the "time tail" will decline suddenly at a certain time node, which indicates that the interaction of a specific topic will suddenly lose the attention of users, rather than gradually weakening, but the reason for this has not been further studied empirically. However, 33.6% of the long tail "time capsule" will regenerate another topic "time capsule" on the decaying long tail, and its regular mechanism is worth further study. The current data show that the probabilities are highly correlated with the user size and the diversity of user preferences for specific topics.

It is the existence of a large number of "time capsules" that maintains the positive activity and stable growth of Bilibili network interest community users, on the other hand, it is significantly different from other video-sharing platforms in terms of communication form and platform value. However, although the analysis and drawing of "time capsules" are public data, it is difficult to complete the security desensitization processing of relevant personal information when it is related to specific topics and specific time codes, and it is easy to ensure the user information is in the right place. Therefore, the author does not disclose any charts or related data. It is hoped that after more appropriate processing of the data, a further in-depth study on the morphology and change mechanism of "time capsules" can be carried out.

It should be noted here that the acquisition and processing of Bilibili video and its users' real-time big data are relatively difficult. First, it is difficult to process video data in a batch and intelligently; second, the barrage data of Bilibili increases and decreases with time. Not only is it different watching the same video screen at different times, but also a considerable part of Bilibili's videos will suffer from excessive barrage volume as part of the barrage never showed up. Therefore, the objective uniqueness of data analysis is limited and challenged. In fact, the internal data-processing department of Bilibili is also restricted by specific constraints such as user data protection, limited technical means, and human resources, and basically focuses on functional development data analysis, while academic data analysis and research are mostly carried out in the form of foreign cooperation.

To some extent, these factors are the data analysis for this paper to be improved, but the author believes that this is not sufficient to affect the overall analysis of the relevant data and the final analysis of the scientific and reasonable conclusions.

As the Bilibili platform actively empowers users, it dilutes the voice and management rights of the organization to an extent, so that users have greater autonomy, and the platform has a more openness. There is no artificial barrier between interest communities, which effectively promotes the rapid diffusion of "time capsules." When the "time capsule" expands to a certain scale, the content subject carried by intensive interaction often spills over the Bilibili platform though not Bilibili personal social channels of user groups of Bilibili, and penetrates into other media networks through the fusion media platform, becoming the seed and cultural power mechanism for Bilibili to create and lead popular culture. for example, the Bilibili 10th anniversary series of activities in *The three-body problem* animation boom. On June 26, 2019, at the time of the 10th anniversary of Bilibili, Liu Cixin, the author of the novel *The three-body problem*, announced that he had officially launched the animation production of *The three-body problem*, and released the concept version of a music video (MV), with Bilibili being the producer of the cartoon. As a result, this has become one of the most important cultural representative topics of the 10th anniversary of Bilibili in public opinion, resulting in a large-scale spillover effect. Taking Weibo topics as an example, according to statistics, the total number of reads in the main venue of the 10th anniversary of Bilibili is 300 million, while the subtopic #*The three-body problem* animation of Bilibili# reached 170 million, becoming a mainstream topic[20] and forming a large "time capsule." Another example is that although the most widely spread "six learning" videos on social networks originate from Bilibili, most of the content is from Weibo and Zhihu. The users of Bilibili are good at content integration, so as to carry out creative adaptation, and finally put the "six learning" into the track of popular culture[21], forming a "time capsule" with a conventional volume but a longer time axis. As the extension research of this paper, the author plans to do more in-depth and detailed research into the video and bullet screen interaction data related to six studies on the platform of the whole of Bilibili, so as to understand the specific formation mechanism of a "time capsule."

Since users are the creators and owners of the "time capsule" on the content platform (the ownership here is mainly reflected in the user's independent upload and deletion or change of video content), there is no artificial barrier between the major communities on the Bilibili platform. The existing autonomy of Bilibili users is equal to that of the Bilibili platform, which means that

Bilibili users are not only focused on the participants of a particular community and are actually introduced as one of the managers of the entire platform. In the near future, when the creativity and autonomy inspired by user autonomy are further developed and mature, Bilibili users are expected to be upgraded to one of the owners of the community they are involved in. Then, the Internet, which is being changed by "Bilibili" and "the Bilibilis," is expected to form a new value-added mechanism and multiagent collaborative governance mode with user autonomy as the core and based on user autonomy.

In fact, this multiagent collaborative governance model has opened the most difficult prelude: as mentioned at the beginning of this article: Sir Tim Berners Lee, the father of the World Wide Web (web), released the *Contract for the Web*[22] on the completion of this paper, calling on governments, companies, and individuals around the world to make specific commitments to protecting the Internet from abuse and to ensure its benefit to mankind. According to the structure of user autonomy and the process of its practical development, the *Contract for the Web* will be very helpful in enhancing the autonomy of Internet users in an all-round way, so as to create a highly integrated network platform of "content production and communication channels," aiming to form a better network era, that is, to use and protect the number of users more intelligently on the basis of a larger scale of users. On the one hand, it attracts and encourages more users to create high-quality content products on their own, and on the other, it provides more personalized and higher value-added information services. In this better network era, the peer-to-peer communication structure based on ordinary users is in urgent need of and facing the pressure of evolution: how to make the content creators upgrade to the core value of the network as soon as possible, so as to promote the formation of a new network value form and its economic model.

7 EPILOGUE

Although there are many obstacles, the global trend to protect the rights and interests of network users is forming. The EU issued *The General Data Protection Regulation* on May 25, 2018. More than a year later, on July 8, 2019, British Airways was fined 183.39 million pounds (about 1.58 billion yuan) for violating *The General Data Protection Regulation*. Facebook has just settled with the government regulatory authorities for its improper use of user data for US $5 billion, and then fell into a series of similar lawsuits again, until the user's autonomy is fully protected and fully respected.

The theoretical analysis framework of network user autonomy proposed in this paper, as well as the evolutionary path of new media based on user autonomy, can make the public understand the connotation and importance of user a

ACKNOWLEDGMENTS

Funding for this project was received form: National Social Science Fund Project "Research on The All-platform Flexible Intervention Governance Model of Network Self-made Audio-visual Programs" (15BWX084), First-class discipline project of School of Journalism, Fudan University.

NOTES

[1] In 1968, the American economist E. Harding published his paper *The tragedy of the public Grassland* in the magazine *Science*. He thought that the public grassland belongs to the collective, the income after grazing sheep and eating grass belongs to the individual, and the use of public resources for personal benefit leads to the "tragedy of public grassland"!

[2] Carpenter, S. (2018). Ten steps in scale development and reporting: A guide for researchers. Communication Methods and Measures, 12(1), 25–44.

[3] Prahalad, CK; Ramaswamy, V, Co-opting customer competence, Harvard Business Review, 78(1). 2000. P79–87.

[4] Virtual CSR Co-creation, Consumer Interaction and Performance Co-creation: A Single Case Study Based on Grounded Theory. Liu Xiaoping, Deng Wenxiang. Management Case Studies And Reviews. 2019(05), P509–520.

[5] Research on Consumer Participation Behaviors in Virtual CSR Co-Creation from the Perspective of Psychological Ownership. Fan Shuai, Tian Zhilong, Hu Xiaoqing, Journal of Management 2017(03). P414–424.

[6] User Service Innovation On Mobile Phone Platforms: Investigating Impacts Of Lead Userness, Toolkit Support, And Design Autonomy. Ye, Hua (Jonathan); Kankanhalli, Atreyi MIS Quarterly, 2018(1). p165–167.

[7] Encouraging Innovations of Quality from User Innovators: An Empirical Study of Mobile Data Services. By Ye, Hua (Jonathan) Service Science. 2018(4), p423–441.

[8] Kim, Da-jung, Design for User Autonomy in the System-Driven Personalization of Social Media, 19th ACM Conference on Computer-Supported Cooperative Work and Social Computing (CSCW), FEB 27-MAR 02, 2016, San Francisco, CA. p155–158.

[9] Virtual censorship: Controlling the public sphere, Klang, Mathias, 7th International Conference on Human Choice and Computers (HCC7): Maribor, Slovenia, Sep 21–23, 2006, Social Informatics: An Information Society for All?: In Remembrance of Rob King, Collection: International Federation for Information Processing, Roll(223), p185–194.

[10] Managing Control, Convenience and Autonomy A Study of Agent Autonomy in Intelligent Environments, Ball, Matthew; Callaghan, Vic, Ambient Intelligence and Smart Environments, Roll(12), 2012, p159–196.

[11] Mediabistro is a website that publishes lists of blogs and journalists. It was founded by Laurel Touby in 1993. Initially, it was a gathering place for professionals in New York's journalism, publishing and other media related industries. Now it has developed into an international media professional resource website.

[12] Research on New Media Enterprise Valuation Model Innovation Based on User Autonomy —Take Bilibili Video Website as an Example. Zhou Xiao, Media economy and management. 2020(01), P180–188

[13] According to the public financial report data of Bilibili in the first three quarters of 2019, details can be seen at http://ir.bilibili.com/

[14] For the data, please refer to the "Research Report on the market prospect and investment opportunities of China's instant messaging industry in 2019–2024" released by the China Business Industry Research Institute, and some data can be found on its official website: http://www.askci.com/news/chanye/20190516/1346051146282.shtml

[15] Hopewell index: https://baijiahao.baidu.com/s?id=1637955651908484784&wfr=spider&for=pc

[16] TikTok vice president of the ByteDance Zhang Yu, launched publicly the data released in the DOU art program. As of July 2019, the number of active users was over 320 million, and the number of short video related TikTok art categories had reached 109 million, with a total playback volume exceeding 608 billion 100 million.

[17] The concept of UGC originated from the Internet field, that is, users display their original content through the Internet platform or provide it to other users, also known as UCC (user created content), which represents a new way for users to use the Internet, that is, from the original focus on downloading to paying equal attention to both downloading and uploading.

[18] PUGC (Professional User Generated Content) refers to "professional user produced content" or "expert produced content", which is an Internet term, which refers to the content production mode combining UGC and PGC in the mobile audio and video industry.

[19] The time node for data collection is April 16, 2019.

[20] Ten years of Bilibili return to zero: the more popular Bilibili is, the less culture of Bilibili is?, Chuang Shi Ji, micro blog article, Internet pointing North. 2019-07-02 16:36:53, https://tech.sina.com.cn/csj/2019-07-03/doc-ihytcitk9191557.shtml

[21] What exactly is the "six learning" that netizens play bad? Official account of curve wrecker: https://www.sohu.com/a/286166968_231759

[22] On November 25, 2019, the *Contract for the Web* proposed and advocated by Tim Berners, the founder of the World Wide Web, has become the first socialized international agreement based on mutual trust and credit among individuals, countries, and enterprises, voluntarily participating in and jointly signing, aiming to obtain the largest scale of mutual benefit and win–win situation in human society. The contract sets out three core principles for government, companies, and individuals to abide by, with a total of nine items, mainly involving the provision of affordable and reliable Internet access services, and the responsibility of respecting citizens' discourse and dignity.

REFERENCES

[1] User Service Innovation On Mobile Phone Platforms: Investigating Impacts Of Lead Userness, Toolkit Support, And Design Autonomy. Ye, Hua (Jonathan); Kankanhalli, Atreyi MIS Quarterly, 2018(1). P165–176.

[2] Encouraging Innovations of Quality from User Innovators: An Empirical Study of Mobile Data Services. By Ye, Hua (Jonathan) Service Science. 2018(4), p423–441.

[3] Kim, Da-jung, Design for User Autonomy in the System-Driven Personalization of Social Media, 19th ACM Conference on Computer-Supported Cooperative Work and Social Computing, Feb 27-Mar 02, 2016, San Francisco, CA. p155–158.

[4] Virtual censorship: Controlling the public sphere, Klang, Mathias, 7th International Conference on Human Choice and Computers (HCC7): Maribor, Slovenia, Sep 21–23, 2006, Social Informatics: An Information Society for All?: In Remembrance Of Rob King, Collection: International Federation for Information Processing, Roll (223), p185–194.

[5] Managing Control, Convenience and Autonomy A Study of Agent Autonomy in Intelligent Environments, Ball, Matthew; Callaghan, Vic, Ambient Intelligence and Smart Environments, Roll(12), 2012, p159–196.

[6] User Acceptance of Computer Technology: A Comparison of Two Theoretical Models. Fred D. Davis, Richard P. Bagozzi, Paul R. Warshaw. Management Science. 1989 (8).

[7] A Scholia-based Document Model for Commons-based Peer Production. Joseph Corneli, Aaron Krowne. Proceedings of the Symposium on Free Culture and the Digital. 2005

[8] Understanding the motivations, participation, and performance of opensource software developers: a longitudinal study of the apache projects. Roberts, Il-Horn Hann, Slaughter. Management Science. 2006

[9] The Ethics of Smart Pills and Self-Acting Devices: Autonomy, Truth-Telling, and Trust at the Dawn of Digital Medicine. Klugman, Craig M.; Dunn, Laura B.; Schwartz, Jack. American Journal of Bioethics. Roll (18), 2018(9). p 38–47.

[10] Benefits and risks of smart home technologies. Wilson, Charlie; Hargreaves, Tom; Hauxwell-Baldwin, Richard. Energy Policy. 2017(103). p72–83.

[11] Minimizing User Cost for Shared Autonomy. Javdani, Shervin; Bagnell, J. Andrew; Srinivasa, Siddhartha S. Eleventh ACM/IEEE International Conference on Human Robot Interaction (HRI'16). p: 621–622. 2016

[12] Balancing autonomy and user control in context-aware systems – a survey. Fourth Annual IEEE International Conference on Pervasive Computing and Communications Workshops, Proceedings. p51–55. 2006.

[13] Prahalad, CK; Ramaswamy, V, Co-opting customer competence, Harvard Business Review, 78(1). 2000. P79–87.

[14] Networked Internet Governance: Contractualization of User Rights and Dependence on Private Power of Internet Intermediaries, Zhang Xiaoqiang. Journalism and Communication Research. 2018(07). P87–108.

[15] Virtual CSR Co-creation, Consumer Interaction and Performance Co-creation: A Single Case Study Based on Grounded Theory. LIU Xiao-ping, Deng Wen-xiang. Journal of Management Case Studies.2019(05), P509–520.

[16] Research on Consumer Participation Behaviors in Virtual CSR Co-Creation from the Perspective of Psychological Ownership. Fan Shuai, Tian Zhilong, Hu Xiaoqing, Chinese Journal of Management, 2017(03). P414–424.

Computational Social Science – Luo, Ciurea & Kumar (eds)
© 2021 Taylor & Francis Group, London, ISBN 978-0-367-70193-2

Research on comprehensive evaluation of online stores based on user sentiment analysis

H.F. Li & W.J. Chen
Dalian Jiaotong University, Dalian, Liaoning, China

ABSTRACT: With the rapid development of the Internet, the e-commerce industry has become more and more popular, resulting in a large number of commodity review data, which provides users with reference value. However, the amount of comment data on the network is often huge, and it is difficult for consumers to extract useful information from these mixed opinions. Therefore, how to effectively analyze the product comment data and extract the user's emotion is the key issue at this stage. This paper proposes a comprehensive evaluation of online stores based on user sentiment analysis, which can accurately identify the evaluation of users on online shopping goods, point out the emotional trend of users, provide reference information for other users, and provide feedback information for businesses. As a reference index for e-commerce website to use recommendation algorithm for personalized recommendation.

Keywords: Hierarchical clustering, Sentiment analysis, Word vector, E-commerce.

1 RESEARCH BACKGROUND AND PURPOSE

With the development of Internet e-commerce websites such as jd.com, taobao.com and amazon.com, online shopping has become a trend. Most users will know other users' evaluation of the goods before they buy the goods, but there are a large number of goods, each product has a large number of user comment data, so that users can not accurately determine the quality of the goods, affecting the purchase of users.

In the field of e-commerce, the main concern is the user's emotional analysis. As the basis of the use of the website recommendation system, through the emotional analysis of data, mining the user's purchase tendency, to determine whether the product recommended by the website conforms to the user's preference or deviates from the user's preference. Through data mining and text mining, real-time user sentiment analysis is provided to reduce time cost, human cost and other costs. It is of great significance to the development of e-commerce industry.

2 PRINCIPLE AND ALGORITHM OF EMOTION ANALYSIS

In the process of text processing, we often encounter the problem of duplicate text. In some cases, these duplicate text data will affect the final text processing results, so we need to de duplicate the data. At present, the most popular de recalculation methods are edit distance algorithm, cosine similarity algorithm and simhash algorithm.

Edit distance is the simplest way to convert string a to string B, that is, the shortest path method. If the conversion method of the two strings is more complex, the difference will be larger, the editing distance will be smaller, and the difference will be smaller. If the two strings are exactly the same and do not need conversion, the editing distance will be 0.

Suppose the edit distance of string A.B is editdis (A.B), then the distance between a and B can be expressed by the following formula:

$$Similarity(A, B) = 1-EditDis(A, B)/max(length(A), length(B))$$

The main steps of cosine similarity algorithm to calculate the similarity between two documents are as follows:

(1) Find out the keyword corresponding to the document.
(2) For the keywords in each document, the word frequency of the keywords is counted, and the word frequency information is analyzed to filter it and extract the keyword set.
(3) The set of word frequency vectors corresponding to the document is generated, which is the feature vector corresponding to the document.
(4) The similarity between feature vectors is calculated, and the higher the value is, the greater the similarity will be.

Compared with the exact match of hash algorithm, the simhash algorithm published by Google can compare the similarity between two texts. Suppose we input a feature set of a document, each feature has a certain weight value. The main principle of the algorithm is as follows:

(1) The vector V of an f-dimension is initialized to 0, and the binary number s of the F-bit is initialized to 0;
(2) Each one-dimensional feature of the document generates the F-bit signature B. by judging whether the value of each bit of the signature B is 0 or 1, the weight value of the corresponding position of the vector V is increased or decreased;
(3) If the i-th element of V is greater than 0, the i-th position of S is 1, otherwise it is 0;
(4) Outputs as signature;

According to the simhash algorithm, the corresponding F-bit of the document is calculated, and the signature is done, and the similarity between the documents is calculated.

3   JUDGMENT AND ACTUAL CALCULATION OF TEXT EMOTION ANALYSIS

There are many repetitions in the comment data, they actually express the same meaning, but the repetitions will affect the model.

In this paper, edit distance is used for subtraction. First, edit distance is used to calculate the similarity matrix of all comments in the annotation data, and then the threshold value of 0.9 is set. If it is higher than the threshold, it will be regarded as duplicate data. Because the editing distance itself is relatively complex and the amount of data is relatively large, this paper uses the parallel computing framework Hadoop.

Data cleaning. Data includes not only repeated information, but also a lot of spam. In order to clean up data, the following rules are mainly used:

Rule 1: comment length is less than 10;
Rule 2: annotations contain specific Adwords;
Rule 3: automatic reply based on specific network reply template;
Rule 4: the comment contains a deleted URL.

In this paper, we use the jeeba wordbook based on Python to segment the product review data in Chinese. Jieba wordbook can be used to segment Chinese words accurately. At the same time, the hidden Markov model based on Viterbi algorithm is used to recognize the unrecorded words. Word segmentation accuracy, then disable word filtering, which includes 9876 Chinese and English stop words.

Delete words, including some useless words, such as date, time, English name and English number combination need further processing. If TF-IDF is higher than the threshold value, select the word as a feature of the document, or the word will be discarded. HowNet describes the concepts represented by English and Chinese words, and reveals the attribute relationship between concepts. There are two very important concepts in HowNet vocabulary, meaning item and sememe.

The positive evaluation words and positive emotion words in HowNet are modified, and then the commendatory words dictionary is added. In the same way, revise the negative emotion words and negative evaluation words, and then add the derogatory words dictionary. The emotional words in commendatory and derogatory dictionaries do not distinguish the intensity of emotional tendency. The emotional tendency value of derogatory words is -1, and the emotional tendency value of commendatory words is 1.

The emotional tendency of the text mainly depends on the emotional words in the text, but it is not enough to judge the emotional tendency only by the emotional words, because the collocation of the emotional words with other words sometimes affects the emotional tendency of the words.

We can find the meaning of negation in HowNet and observe the negation words that often appear in Internet comments. After manual sorting, we can get the dictionary of negation words. We temporarily collect 28 negation words: No, not enough, forbidden, forbidden, abstained, prevented, invisible, etc.

Modify characteristic words with emotional words, and add the characteristics of words to the discourse to explain. The emotional words with different tendency to modify different characteristic words are called collocations. Therefore, we should pay attention to the characteristics of words, modify the characteristics with emotional words, and consider the combination of emotional tendencies, so as to accurately calculate the value of the whole text.

Degree adverbs are usually adverbs that modify emotional words. Emotional words will have an impact on the degree of certain restrictions on the emotional tendency of the text. You can sort by severity.

4 EXPANSION OF EMOTION DICTIONARY BASED ON WORD2VEC

At present, there are two methods to calculate the polarity of emotion: corpus based and emotion dictionary based. This paper mainly studies the method based on emotion dictionary. The validity of emotion polarity calculation method based on emotion dictionary depends on two points, whether the commendatory and derogatory words in emotion dictionary can be accurately separated by word segmentation. Can we find a domain dictionary suitable for the current research field.

By calculating the cosine value between words to measure the correlation between two words. According to the similarity ranking, the top words are added into the e-commerce commodity emotion dictionary. Compared with the traditional method, the method based on word2vec embodies two advantages. On the one hand, the method based on word vector contains rich semantic information, which can automatically learn the potential semantic relationship between words.

5 TEXT PROCESSING

The treatment of double negative sentences. "Double negative sentence" means that there are two negative words in a sentence, usually positive sentences, to strengthen the function of mood. In this kind of negative sentence, the calculation of affective tendency should be based on affective dictionary. If the frequency of negative words is 2, the sentence is considered as double negative structure, i.e. positive. The derogatory distribution is consistent with the positive sentence. There are k negative words before or after the emotional words in the sentence. There is also a form of double negative sentence, that is, a rhetorical question ("do", "how to", etc.) plus the composition of "no", constitutes a rhetorical question, and it is divided into two classes of double negative sentences, the tone is stronger than the voice.

The treatment of degree adverbs. In commodity reviews, we often find that there are many comments with similar meanings, which are different in degree. By combining the tendency value of emotional word W, we can get the tendency value of the whole sentence.

The comprehensive treatment of double negation and degree adverbs. If there are many negative words and degree adverbs in the commodity review to modify the emotional words, such as "I have to say that this facial cleanser is not clean", there are two negative words and a degree adverb in this review, meaning the same as "I have to say that this facial cleanser is not clean"

6 SOLUTIONS TO EXISTING PROBLEMS

Deviation of special text. To provide corpus for emotional analysis research is to solve this problem first, a large number of commodity reviews are full of shopping websites, which must be collected after the web page, then extract the next step of relevant information, and then preprocess the opinion content in this paper, preprocessing includes Chinese word segmentation and vocabulary deactivation. In the process of Chinese classification, although there are many effective and mature methods, there is still no powerful classification technology. In this paper, we introduce several selection methods, construct space vector model, assign proper weight to feature words, and improve the accuracy of emotion calculation as much as possible.

At present, in order to improve their evaluation, many online shopping shops will use some unusual means to find a large number of workrooms to brush the list and then praise them, which leads to the failure of getting a proper evaluation.

In this paper, we want to combine the text emotion analysis technology with the computer IP identification system and the mobile phone identification code system, use the IP and the identification code to record the online shopping information and comment information of "the buyer" registered in the network for a period of time, define the blacklist users through the statistical setting of standards, and pull the studio IP or identification code that swipes the list into the blacklist All information provided by the user shall be removed during statistics.

ACKNOWLEDGEMENT

Natural Science Fund of Education Department of Liaoning Province JDL2019027
Liaoning Natural Science Foundation, 201800177
The subject of educational science planning in Liaoning Province, JG16DB054

REFERENCES

[1] Kai Huang. September 8, 2014, Uncover big data player eBay: guess your purchase desire. 21st century economic report.
[2] Liu Zhizhi, Zhang Quanling. 2014, Review of big data technology research. Journal of Zhejiang University (Engineering Edition), 6(6). 15–17.
[3] Lu Xudong. 2015, Research on Influencing Factors of cross-border e-commerce development in the era of big data. Economic and trade practice, (06). 185–204.
[4] Pak A, Paroubek P. Twitter as a Corpus for Sentiment Analysis and Opinion Minmg.
[5] Zhang Jing, Jin Hao. 2010, Research on automatic judgment of emotional tendency of Chinese words. Computer enginerrring. 36 (23): 194–196.

Computational Social Science – Luo, Ciurea & Kumar (eds)
© 2021 Taylor & Francis Group, London, ISBN 978-0-367-70193-2

Research on fast location algorithm of fault section

X.K. He, M.C. Chen & H. Nie
Duyun Libo Power Supply Bureau of Guizhou Power Grid Co., Ltd, Qiannan, Guizhou Province, China

N.H. Zhang
Hangzhou Harmony Tech. Co., Ltd, Hangzhou, Zhejiang Province, China

S.Z. Lian
School of Electrical Engineering, Sichuan University, Chengdu, Sichuan Province, China

ABSTRACT: With the growth of the power grid, the distribution network, which is the end of the power grid, faces problems such as difficult fault location and labor shortages. This paper proposes a fault location algorithm based on the existing distribution network automation information system. The data for the fault location algorithm come from the equipment ledger, telemetry signals, fault location information system, and lightning location information system. Fault location algorithms include multi-interval optimal fault location algorithms and a combination of deep optimal search and breadth optimal search. A set of fault location systems is designed, and historical fault data are imported for experimental verification. The results show that the fault location system based on this algorithm has accurate system positioning and a fast response, which greatly reduces the time for fault finding.

1 INTRODUCTION

The distribution network is the final link in power transmission, which directly connects users. The safety and efficiency of the power grid system are directly affected by the health status, reliability, and power quality of the distribution network. The characteristics of the distribution network include a complex structure, with many branches, small power supply radius, mixed power supply of overhead lines and cable lines, etc. These characteristics have led to a complicated troubleshooting process for the distribution network, and the time taken to restore power supply has also increased, which further seriously affects the reliability of the power supply[1].Traditionally, when a fault occurs, the employees of the power supply and distribution station search for the fault point level by level, and the expert's judgment also plays an important role in fault search and accident analysis and prevention[2].

According to the "China Power Industry Annual Development Report 2019" released by the China Electricity Council[3], in 2018, China's entire society's electricity consumption was 6.9002 trillion kWh. This shows an increase of 8.4% over the previous year and is also the highest growth rate since 2012, which increased by 1.8 percentage points over the previous year. The rapid increase in the number of users and electricity consumption has led to rapid expansion of the power supply scale and an increasingly complicated structure of power supply and distribution line grids. The increased load makes fault power failure problems prominent. Serious equipment aging, weak fault management, and weak self-healing ability make this situation more severe[4]. In addition, the grass-roots power distribution unit is faced with a shortage of personnel and heavy work load. It takes a lot of time to locate a fault after it occurs, leading to a serious reduction in power supply reliability and a rapid increase in customer complaints.

1.1 Related work

In order to reduce the fault location time and improve the reliability of the power supply, many researchers have conducted a great deal of research into the methods of fault location in the distribution network, who put forward many fault location theories and methods£¬such as multiterminal fault traveling wave location method[5], Bayesian compressed sensing positioning[6], improved impedance method[7], and parameter identification[8]. The application of these theories and methods will greatly reduce the workload of grass-roots distribution units, improve the efficiency of fault finding, and ensure the reliability of the power supply. With the development of big data technology, some fast fault location theories and methods based on electric power big data have been proposed, such as the real-time assessment method of distribution equipment status based on MI-PSO-BP algorithm proposed by Yang Zhichun[9], who applied the neural network algorithm and big data analysis technology to the status assessment of power distribution equipment. This paper analyzes the network big data deriving from distribution and its key technologies, and proposes a new method for rapid fault location. The limited PMU active distribution network fault location method proposed by Zhang Jianlei et al. is not affected by the type and location of the fault[10]. The layered fault location algorithm of distribution network with special load proposed by Gao Fengyang et al. greatly reduces the fault location time and greatly improves the fault accuracy rate[11]. Liu Xin et al.'s 2-norm fault location algorithm based on current deviation can identify the fault location without determining the fault type[12]. Yin Zhihua et al. proposed a fault location algorithm suitable for various fault conditions based on transient zero sequence pole symmetry mode decomposition technology[13]. Wang Chengbin et al. proposed a fault location method for overhead lines based on microPMU[14]. However, these fault location algorithms have the disadvantages of being highly targeted, requiring replacement of existing equipment, and being unable to be applied in a timely manner at this stage.

1.2 Contribution

This paper proposes a fault location algorithm based on the existing information collection system, including multi-interval optimization algorithm and lightning fault location algorithm. The algorithm will greatly reduce the workload of grass-roots distribution units and reduce the time for the power distribution unit to find faults.

2 DATA SOURCE

At present, most prefectures and cities already have complete distribution management systems, including distribution automation systems, dispatch automation systems, production management systems, power quality management monitoring systems, and electricity information collection systems. These data provide support and guarantees for the application of the distribution network big data.[15]

2.1 Distribution equipment account

The distribution network equipment account mainly includes equipment operation parameters, test records, and fault conditions. The distribution network is mainly composed of lines (overhead lines and cable sections), switches (intelligent boundary circuit breakers on columns, intelligent load switches on columns, contact circuit breakers, intelligent load switches, and outlet circuit breakers in stations), and transformers (on-column transformers, tank transformers, main transformers), monitoring equipment (TTU, fault indicators), and other power distribution devices (running rods, switch boxes, branch boxes, power distribution rooms), and other equipment. The equipment operation parameters can reflect the current network operation mode, power distribution network, and equipment status mutation. The test records and fault records of the distribution network equipment can reflect the health status and historical operation status of the equipment, and are the priority criterion in the distribution network fault discrimination algorithm.

2.2 Telemetering, remote signaling and remote control system, measurement automation system

The telemetry data include real-time voltage, current, and active and reactive data of the telemetry object. The telemetry data include the status of the telemetry object, fault indications, and protection actions. They mapped the collected data of FTU, DTU, user boundary switch, and fault indicator to the power grid equipment according to the mapping relationship of measuring points, and connected the related collected data to the real-time data platform. "Three remote" data reflect the distribution of power flow in the distribution network in real time, the real-time parameters of the distribution equipment, and the real-time operation mode of the network. They can identify sudden changes such as in equipment operation modes and in network power flow in a timely manner. They can also be combined with the operation mode before and after the fault, the power flow distribution, and the network topology given by the equipment ledger to become another criterion for fault discrimination.

2.3 Fault location system information

The fault location system information mainly includes fault indicator ID, fault name, fault address, substation, communication status, fault type, fault phase, update time, update status bit, and fault history record table of the fault location system background analysis.

2.4 Lightning positioning information

Lightning positioning information refers to obtaining real-time lightning information in the supply area through the marketing cloud platform server, specifically referring to lightning strike time, latitude and longitude, intensity, lightning current, and polarity.

3 RESEARCH ON THE LOCATION ALGORITHM OF THE DISTRIBUTION NETWORK FAULT SECTION

3.1 Multi-interval optimization algorithm

On the one hand, the multi-interval optimization algorithm analyzes the status of the distribution network switch, the status of the fault indicator, and the topology of the line equipment. On the other hand, it analyzes the real-time collection information of the distribution network switch and the action history information of the fault indicator. The two analysis results are combined to locate the fault when the line is faulty, the system obtains the fault interval D1 according to the real-time collection information of the distribution network switch, and the fault interval D2 according to the action history information of the fault indicator. Then the system uses the device's topological connection relationship to take the common part D3 and its starting point device and end point device for the two intervals. The system notifies the line patrol personnel of the interval, which will shorten the length of the line patrol line to the greatest extent, which helps to find the faulty equipment as soon as possible, thus shortening the repair time of distribution network faults. The flowchart for the multi-interval positioning algorithm is shown in Figure 1.

3.2 Lightning positioning algorithm

3.2.1 Lightning positioning process
The input condition required by the lightning positioning algorithm is the equipment struck by lightning (mainly overhead lines). As long as the device that is directly struck by lightning is located, the range of lightning impact can be located according to the topology relationship.

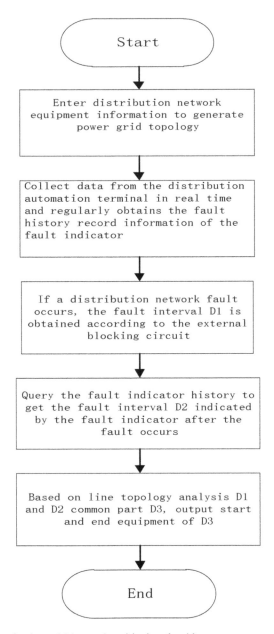

Figure 1. The flowchart for the multi-interval positioning algorithm.

The accidental nature of lightning strikes determines that the equipment subject to lightning strikes also has irregularities. To locate the area affected by the lightning strike, it is necessary to locate whether the overhead line where the equipment struck by the lightning strikes is connected. The reason why the overhead line is used as the criterion is that all the equipment in the model is directly or indirectly connected to the line. If the equipment on the line is struck by lightning, it is likely to affect other equipment on the entire line, and the line is also convenient for searching and positioning.

After sorting the affected lines, the lines with connectivity are sorted into several groups (single lines are grouped separately). If all the lines are individually grouped, the process of sorting the

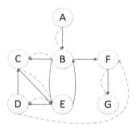

Figure 2.   Depth-first search algorithm.

lines into groups can be omitted. Connectivity here means that the two lines are directly connected through the tower before they are considered to be connected.

The line groups are analyzed and organized with connectivity, querying the model to determine the main road and branch roads, and splitting the entire group of lines into two or more separate line groups if the topology analysis shows that there were branches between the lines.

The lines divided into groups are analyzed separately from the distance to the power point, and the lines of each group are sorted according to the distance. After the lines in the group are sorted, they are grouped and compared, and the lines at the same power point are sorted.

Finally, all devices are sorted as connected directly or indirectly to the line in order. According to the search depth set at the beginning of the algorithm and the stop device, the final device amount is determined, and the device and line data are returned.

3.2.2 *Lightning positioning algorithm*

When calculating the distance between the line and the power point, a topology search based on depth first is used, and a topology search based on breadth first is used when sorting all devices.

(1) Depth-first search

Depth-first search (DFS) is a kind of graph algorithm (Figure 2), which is a traversal algorithm for graphs and trees. It is a classic algorithm in graph theory. The DFS algorithm can be used to generate the corresponding topological sort table of the target graph. Using the topological sort table can easily solve many related graph theory problems, such as the maximum path problem. Heap data structures are generally used to assist in the implementation of the DFS algorithm. The process is to deepen each possible branch path to the point where it can no longer be deepened, and each node can only be visited once.

According to the depth-first search, the access sequence is: A > B > C > E > D > F > G.

(2) Breadth-first search

Breadth-first search (abbreviated as BFS) is a traversal algorithm for connected graphs. This algorithm is also the prototype of many important graph algorithms. Dijkstra's single-source shortest path algorithm and Prim's minimum spanning tree algorithm are similar to BFS. BFS is a blind search method, the purpose is to systematically expand and check all nodes in the graph to find the results. In other words, it does not consider the possible location of the result, and searches the entire picture thoroughly until the result is found. BFS starts at the root node and traverses the nodes of the tree (graph) along the width of the tree (graph). If all nodes are visited, the algorithm is aborted. The queue data structure is generally used to assist in implementing the BFS algorithm. Figure 3 is a schematic diagram of a breadth-first search algorithm.

According to breadth-first search, the access order is: A > B > C > E > F > D > G.

3.2.3 *Line GIS buffer search*

The search process for the line GIS buffer is as follows:

(1) Calculating the coordinate points of each line and defining the buffer size.
(2) Calling the buffer generation algorithm to generate a buffer for each line.

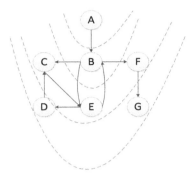

Figure 3.   Breadth-first search algorithm.

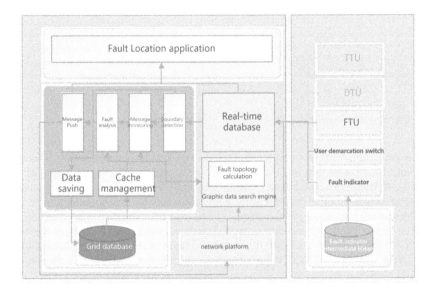

Figure 4.   Technical architecture of a rapid positioning system.

(3) Combining the buffer area of each line as a whole buffer, as the buffer of this line.
(4) Querying the lightning data according to the incoming parameter start and end times of the lightning strike, and determining whether the latitude and longitude values fall within the buffer according to the returned latitude and longitude of the lightning record.

4   TEST AND ANALYSIS

According to the aforementioned research and analysis, a rapid fault location system is designed for a county-level power supply area. The technical architecture of the system is shown in Figure 4.

The functional modules involved in this system are basic information cache management, message monitoring, boundary detection, fault analysis, message push, and data storage. The system adds a fault topology calculation interface to the spatial data engine to support fault analysis topology calculation. After the system design is completed, the multiple failures that occurred in the county company are used as simulated data, including data such as telemetry and telesignal. These

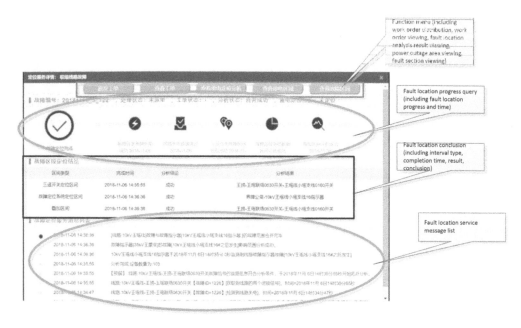

Figure 5. Screenshot of the location of the fault line fault distribution network.

data are imported into the positioning system through the test tool, which simulates the fault section location of the connection line distribution network, the fault section location of the single radiation line distribution network, and the fault section location of the boundary switch. Figure 5 is a screenshot of the fault line location of the connection network distribution network.

From the test results, the location accuracy of the fault location system reaches 99%, and the average fault analysis and response time is about 2 minutes, which greatly reduces the workload of the grass-roots power distribution unit. It is estimated that the successful application of this fault location system will save 73.4% of the troubleshooting time, and reduce the customer complaint rate by about 57.3%, ensuring the reliability and stability of power supply.

5 CONCLUSION

Faced with the problems of difficult fault location and long fault location time in the current distribution network, this paper proposes a fault location algorithm based on existing equipment, including multi-interval location algorithm and lightning location algorithm. Applying this algorithm to the fault location system greatly improves the efficiency of fault location.

The fault location system also has many shortcomings. For example, because the distribution system often changes the operation mode, once the operation mode is changed, it needs to be reset. Therefore, the operation mode identification function will be added to the positioning system in the following research work.

REFERENCES

[1] Ma Zhao, Liang Huishi, Su Jian. Important issues in the planning and operation of active distribution systems. Grid Technology, 2015, 39 (06): 1499–1503.
[2] Jiang Xiuchen, Sheng Geao. The big data analysis and application of power equipment status. High Voltage Technology, 2018, 44 (04): 1041–1050.

[3] China Electricity Council publishes "China Power Industry Annual Development Report 2018". China Electric Power Enterprise Management, 2018 (16): 11.

[4] Ma Zhao, An Ting, Shang Yuwei. Development and development of cutting-edge power distribution technology at home and abroad. Chinese Journal of Electrical Engineering, 2016, 36 (06): 1552–1567

[5] Deng Feng, Li Xinran, Zeng Xiangjun, Li Zewen, Guo Jun, Tang Xin. A Novel Multi-terminal Fault Location Method Based on Traveling Wave Time Difference for Radial Distribution Systems With Distributed Generators. Proceeding of the CSEE, 2018, 38(15): 4399–4409.

[6] Jia Ke, Li Lun, Yang Zhe, Zhao Guankun, Bi Tianshu. Research on fault location of distribution network based on Bayesian compressed sensing theory. Chinese Journal of Electrical Engineering, 2019, 39 (12): 3475–3486.

[7] Dai Zhihui, Wang Xu. Active distribution network fault location algorithm based on improved impedance method. Power Grid Technology, 2017, 41 (06): 2027–2034.

[8] Jia Qingquan, Wang Zhenyu, Wang Ning, Dong Haiyan, Ai Li. Single-phase grounding fault location method for arc suppression coil grounding grid based on parameter identification. Journal of Electrical Engineering, 2016, 31 (23): 77–85

[9] Yang Zhichun, Jing Xiaoping, Le Jian, Shen Yu, Zhang Hao, Yang Fan. Real-time assessment method of distribution equipment status based on MI-PSO-BP algorithm. Electric Power Automation Equipment, 2019, 39 (12): 108–113.

[10] Zhang Jianlei, Gao Zhanjun, Wang Zhiyuan, Sun Xuri, Zhang Song, Li Wensheng, Wei Zhen. Active power distribution network fault location method based on limited μPMU [J/OL]. Power grid technology: 1–10[2020–06–01]. https://doi.org/10.13335/j.1000-3673.pst.2019.2607.

[11] Gao Fengyang, Li Zhaojun, Yuan Cheng, Li Xiaofeng, Qi Xiaodong. Layered fault location method of distribution network with special load [J/OL]. Journal of Southwest Jiaotong University: 1–8[2020–06–01]. http://kns.cnki.net/kcms/detail/51.1277.U.20200116.1252.002.html.

[12] Liu Xin, Teng Huan, Liang Mengke, Teng Deyun. Fault location of active distribution network based on current deviation 2-norm. Journal of Electrical Engineering, 2019, 34(S2): 720–728.

[13] Yin Zhihua, Sun Guoqiang, Ding Jianzhong, Lü Feng, Yuan Haixing, Wei Zhinong. Location technology of single-phase grounding fault section in distribution network based on pole symmetry mode decomposition. Electric Power Automation Equipment, 2019, 39(06): 173–178+191.

[14] Wang Chengbin, Yun Zhihao, Zhang Hengxu, Shi Interview, Ling Ping, Xie Bangpeng. Multi-branch unsupported parameter-free fault location algorithm for distribution network based on micro PMU -3211.

[15] Shao Meiyang, Wu Junyong, Shi Chen, An Ran, Zhu Xiaowen, Huang Xing, Cai Rong. Reactive power optimization of distribution network based on data-driven and deep confidence network. Power System Technology, 2019, 43 (06): 1874.

Computational Social Science – Luo, Ciurea & Kumar (eds)
© 2021 Taylor & Francis Group, London, ISBN 978-0-367-70193-2

The effect of population aging on the consumption of urban residents – taking Shanghai as an example

Y. Lu
School of Economics, Jiangxi University of Finance and Economics

X.Y. Gu
School of Marxism, Jiangxi University of Finance and Economics

ABSTRACT: Shanghai is the earliest area in China that has entered an aging population and is currently threatened by aging. This article will use the relevant time series data from 2002 to 2018, and use the structural vector autoregressive model (SVAR) to analyze the relationship between population aging and consumption expenditure. The results of the study show that the aging population has a significant impact on the consumption expenditure of urban residents in the short term, but not in the long term. In view of the impact of the aging problem on residents, this article will also provide some feasible method.

1   RESEARCH BACKGROUND AND SIGNIFICANCE

1.1   *Research background*

China entered the population aging society at the beginning of the 21st century, and with the spend of time, China's population aging has also shown an increasingly serious upward trend. Until 2018, the national population aging coefficient has reached 11.9%. Population aging is a serious problem that every developed and developing country must face, because it is related to the country's economic growth and stability from the macro perspective. From the micro perspective, population aging is related to the daily expenses of each of our families. The aging coefficient of Shanghai's population has reached 21.8%, so the aging problem is very serious in shanghai. The city has also become one of the most severe places in the country's aging regions.

1.2   *Research significance*

Population size and age structure are a very important factor that affects our country's economic development, so it is very meaningful to study the impact of population aging on the consumption of urban residents. Therefore, this article will take Shanghai as an example to discuss the impact of population aging on the consumption of Shanghai urban residents. Based on the analysis results, some policy recommendations are proposed to alleviate the impact of aging.

2   LITERATURE REVIEW

Tingting Qian (2016) used relevant data from 1978 to 2013 to analyze the relationship between the Shanghai's aging population and the residents' consumption level. It shows that the impact of population aging on consumption levels is negatively correlated. Qian Wan[1] analyzes the influence of changes in the age structure of the Chinese population on the consumption of urban residents. It shows that population aging has a negative impact on the urban residents' consumption.

Qi Hongqian and Yan Hai-chun[2] use the balanced panel data of 31 provinces from 2001 to 2015 to construct a panel smooth transfer regression model, and empirically test the economic development impacted by population aging. It shows that aged tendency of population has a non-linear characteristic on economic growth, and this phenomenon shows a dynamic trend of first decreasing and rising with time. Zhou Han[3] used time series data from 1990 to 2015 to explore the effect of the elderly dependency ratio on urban families' consumption. The results show that aging population has a positive effect on urban families' consumption. Jiang Yu and Quan Mengzhen[4] used inter-provincial data from 2002 to 2015 to study and build a generational alternation model. Studies have shown that as the population aging increases, the level of pension insurance development, and finally promote the urban residents' consumption.

By sorting and summarizing the existing literature, we can find some shortcomings. A large part of the literature is analyzed and studied using national data. The degree of aging varies greatly in various regions of China, so it is flawed to analyze and compare the aging data throughout the country alone. The conclusions of these articles can be classified into three categories. The aging of the population has a positive impact, a negative impact, and no significant impact on the consumption expenditure of urban residents. Therefore, this article selects the data related to aging in Shanghai as an example to analyze the impact of aging population on the consumption expenditure of urban residents. Selecting SVAR model can strongly show the effect of aging population on the future consumption expenditure of urban residents, and it is also very scientific.

3 THE THEORETICAL ANALYSIS OF POPULATION AGING ON RESIDENTS' CONSUMPTION

3.1 Definition of the concept of population aging

Population aging is an irreversible process. The concept of population aging is that, in the case of a decrease in the proportion of newborn babies and a raise in the average life expectancy of the elderly group, it leads to a decrease in the young and middle-aged population and an increase in the elderly group among the whole population. The aging population is mainly manifested in the continuous addition in the number of elderly people over the age of 60, which makes this part unbalanced. In this article, 7% of the total population aged over 65 in the region or country is used as the unified standard for calculation.

3.2 Measurement index of population aging

There are many indicators for calculating the severity of population aging in a region or a country. In order to facilitate observation and analysis, the following mainly uses the elderly population coefficient and the elderly dependency ratio to determine the severity of the aging population.

3.2.1 Coefficient of the elderly population
The coefficient of the elderly group shows the percentage of the population whose age is 65 and above within a certain period of time. This coefficient can succinctly and clearly show the basic situation of population aging in this area, so some scholars directly call it the aging coefficient. Calculated as follows:

$$\text{Coefficient of elderly population} = \frac{\text{Number of elderly people aged 65 and above}}{\text{total population}} \quad (1)$$

3.2.2 Old-age dependency ratio
The old-age dependency ratio shows in a period of time, the percentage of the non-working-age group whose age is 65 and above to the working-age group whose age is 15–64. The old-age

dependency ratio indicates the number of elderly people who need to support in the working-age group. From an economic perspective, this ratio explains the social consequences of the phenomenon of aging population. Calculated as follows:

$$\text{Elderly dependency ratio} = \frac{\text{the number of elderly people aged 65 and above}}{\text{the population of young and middle-aged people aged 15--64}} \quad (2)$$

4 ANALYSIS OF THE STATUS QUO OF POPULATION AGING IN SHANGHAI

4.1 *The causes of population aging*

The essence of an aging population is that, in the age structure, the proportion of the elderly group constantly increasing in the whole population. There are many reasons for the phenomenon of the aging population. China's aging population phenomenon is started by the reform and opening up strategy and socialist modernization strategy after 1978, which have led to rapid economic development, the resolution of people's clothing, food, housing and transportation issues, improved medical care, and a higher standard of living for people. The big improvement makes people's mortality rate lower and relatively improves people's average life expectancy. The birth control policy has significantly reduced the birth rate of babies in China. From the overall data point of view, lower birth rate, mortality rate and higher average life expectancy are the main reasons for China's aging population.

4.2 *Analysis of the status quo of population aging and residents' consumption in Shanghai*

Until 2017 at the end of time, Shanghai Statistical Yearbook data show that Shanghai household population of 1455.13 million, of which 65 over the age of household aged population has reached 317.67 million. The coefficient of the elderly population in Shanghai has reached 21.8%. In comparison, the coefficient of the elderly population in China was only 11.4% in the same year. The severity of the aging population has far exceeded the national standard. According to real data, as early as the end of 1979, the coefficient of the elderly population in Shanghai reached 7.2%.

In Figure 1 shows that the elderly population in Shanghai coefficient in 2010 before the annual growth rate of more gentle, but in 2010 years after the growth of the elderly population coefficient has increased significantly, which means that the aging population in Shanghai severity as time The progress continues to improve. The aging population has a certain impact on the regional economic

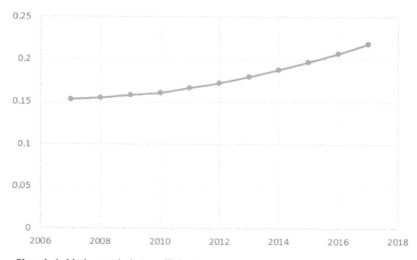

Figure 1.   Shanghai elderly population coefficient map.

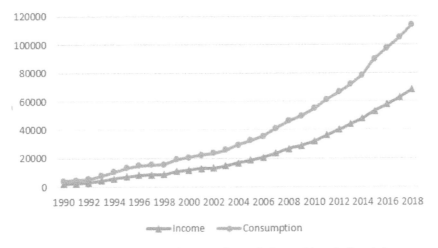

Figure 2. Disposable income and consumption expenditure of urban residents in Shanghai.

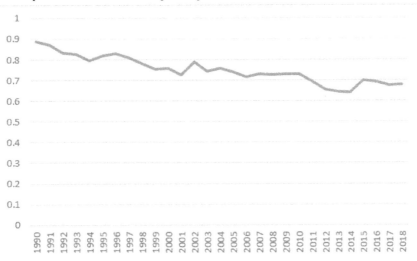

Figure 3. Shanghai urban residents' average consumption trend.

development. Therefore, Shanghai is facing the dual pressure of high population aging rate and high population aging growth rate.

Generally speaking, the consumption of urban residents mainly refers to the final consumption of goods and services of the resident population in the area. The final consumption expenditure is equivalent to the purchaser's price, that is, the price paid by residents to obtain goods or services. Therefore, the data selected in this article are based on the final consumption expenditure.

It can be seen from Figure 2 that from 1990 to 2018, the per capita disposable income and consumption expenditure of Shanghai urban residents showed a linear upward trend, in which the growth rate of per capita disposable income was significantly higher than that of consumer expenditure; the growth rate of consumer expenditure was in 2002 It was relatively slow years ago, after which the growth rate of consumer spending began to accelerate. The main reason is that the income level of the people has increased substantially after that year. Because per capita disposable income has increased significantly and consumer spending has increased slightly, residents' savings are increasing. Therefore, Shanghai urban residents still have a lot of consumption space to dig out.

As can be seen from Figure 3, the average consumption propensity of urban residents in Shanghai shows a downward trend. Although there were large fluctuations in 2002 and 2014, the trend of consumption propensity declined. The average consumption propensity reflects the influence of people's income on consumption expenditure. As a relatively developed area in China, Shanghai has a higher per capita income than most other cities in the country, but urban residents in Shanghai also suffer from insufficient consumption. There are many factors that restrict residents' consumption, such as the widening income gap, consumption habits, and cultural traditions of diligence and frugality. The consumption situation of residents plays an important role in the stability of regional economic growth. In recent years, Shanghai has continuously promoted the reform of its system and industrial structure to guide residents' consumption behavior.

5 EMPIRICAL ANALYSIS OF THE RELATIONSHIP BETWEEN POPULATION AGING AND CONSUMPTION BY SVAR MODEL

5.1 Model setting and variable description

5.1.1 Variable selection and data source description
The main indicators that influencing Shanghai's urban residents' consumption are the following three, the old-age dependency ratio (ODR), gross domestic product (GDP), and urban residents' per capita consumption expenditure (EXPEND), so the data selected in this article are from the Shanghai Municipal Statistics Bureau, they are ODR, GDP, expend, we were taken on a number of variables to get Inodr, Ingdp, Inexpend.

This article will use Eviews10 and stata software for data analysis. The data source is the Shanghai Statistical Yearbook and National Bureau of Statistics data from 2002 to 2018.

5.1.2 Model settings
This paper uses the structure vector autoregressive model, namely SVAR model. The biggest advantage to the SVAR model is that the SVAR model adds a short-term constraint condition, which can produce a certain economic relationship between the variables, so that the model can eliminate the limitations of the random disturbance of the VAR model.

Generally speaking, the constraints of the SVAR model are based on the equations $\frac{n^2-n}{2}$. Because there are 3 variables in this article, the determinant A has 6 equations. According to this three variables, there are three constraints, shown below are the short-term constraints of A and B:

$$A = \begin{pmatrix} a_{11} & a_{12} & a_{13} \\ a_{21} & a_{22} & a_{23} \\ a_{31} & a_{32} & a_{33} \end{pmatrix} \quad B = \begin{pmatrix} b_{11} & b_{12} & b_{13} \\ b_{21} & b_{22} & b_{23} \\ b_{31} & b_{32} & b_{33} \end{pmatrix}$$

In the determinant of A, where a_{13}, a_{21}, a_{23} are all 0. $a_{13} = 0$ means the expenditure of urban residents in the elderly dependency ratio has no effect; $a_{21} = 0$ refers to the gross domestic product in Shanghai had no effect on consumer spending; $a_{23} = 0$ means the Shanghai's gdp has no effect for the elderly dependency ratio; a_{11}, a_{22}, a_{33} are 1. This refers to the influence of variables on the variables themselves. a_{12}, a_{31}, a_{32} are the constraints we add, which is consistent with the theory of t economics, consumer spending will impact gdp. The aging population will have a certain impact on consumer spending and economic development. In the determinant B, b_{11}, b_{22}, b_{33} are constraints, and all others are 0. The constraints imposed on A and B are called "short-term constraints", and the SVAR model is also called short-term SVAR.

5.2 Descriptive analysis

Table 1 shows that the effective sample is 17. The data selected span 2002–2018 years. The average value of Shanghai's gross national product is RMB 17,886.02 billion, the average value of urban residents' per capita consumption expenditure is RMB 24,627.47, and the average value of Shanghai's elderly dependency ratio is 16.34%.

Table 1. Descriptive statistics.

Variable	Observations	Average value	Standard deviation	Minimum value	Maximum
odr	17	16.34	3.47	9.4	21.9
expend	17	24627.47	11304.99	10464	46015
gdp	17	17,886.02	8480.36	5,795.02	32679.87
Ingdp	17	9.67	0.54	8.66	10.39
Dlnexpend	16	0.09	0.04	0.04	0.19
Dlnodr	16	0.01	0.20	−0.38	0.31

Table 2. Unit root test result table.

variable	ADF inspection	1% critical value	5% critical value	10% critical value	P value	Stationarity
Ingdp	−5.641	−3.750	−3.000	−2.630	0.00	Smooth
Inexpend	−0.446	−3.750	−3.000	−2.630	0.9022	Unstable
Inodr	−1.631	−3.750	−3.000	−2.630	0.467	Unstable
Dlnexpend	−4.115	−3.750	−3.000	−2.630	0.001	Smooth
Dlnodr	−4.050	−3.750	−3.000	−2.630	0.001	Smooth

Table 3. Maximum lag order test statistics table.

Lag	LogL	LR	FPE	AIC	HQIC	SBIC
0	24.28	NA	5.8e-06	−3.55	−3.59	−3.43
1	62.08	75.60	5.2e-08	−8.35	−8.53	−7.86
2	71.72	19.26*	7.1e-08	−8.45	−8.77	−7.60
3	NA	NA	−2.7e-25*	NA	NA	NA
4	1107.68	NA	NA	−178.61*	−179.15*	−177.16*

5.3 Stability analysis

The relevant time series data selected in this paper for 2002–2018 has the problem of whether the data is stable, and there is no way to construct SVAR models for those time series data that are not stable. In this paper, all the data obtained are logarithmically processed, because doing this can turn the time series values into a stable sequence. Therefore, we must first conduct stability judgments on Ingdp, Inexpend, and Dlnodr. Unit root test is used to verify the stability of the data. Finally, it is concluded that Ingdp, Inexpend and Dlnodr meet the stationarity test, and the SVAR model can be constructed.

5.4 Selection of optimal lag order

In the analysis of economic issues, the endogenous variables in the past may also have a certain impact on the current endogenous variables, just like the introduction of an economic policy to the results of this policy may have a certain lag Therefore, before establishing the SVAR model, an optimal lag order needs to be determined. Therefore, this paper uses the lag item length test and the Akaike criterion (AIC) to determine the optimal lag order of the SVAR model.

It can be seen from Table 3 that according to a series of test criteria, when the lag order is fourth order, there are 3 variables that meet the standard. Therefore, this paper will adopt the structure vector autoregressive model with lag order of fourth order. Both scientific and reasonable, the next model stability test can be carried out.

Table 4. Results of cointegration test.

Number of cointegration equations	Eigenvalues	Trace statistics	P value
None	0.80	53.50	0.0000
At most 1*	0.66	29.03	0.0003
At most 2*	0.57	12.64	0.0004

Table 5. Granger causality test results.

Null hypothesis	chi2 value	P value
Dlnodr does not Granger Cause Dlnexpend	1.04	0.60
Dlnexpend does not Granger Cause Dlnodr	2.03	0.36
Ingdp does not Granger Cause Dlnodr	6.60	0.04
Dlnodr does not Granger Cause Ingdp	2.23	0.33

5.5   Co-integration test

It can be seen from Table 4 that according to the Johansen cointegration test method, it is found that there is a cointegration relationship among Ingdp, Dlnodr, and Dlnexpense, which shows there is a long-term stable relationship among these variables.

5.6   Granger causality test

Granger causality test is primarily used to test whether there is a causal relation between the variables from Table 5, accepting Dlnodr does not granger cause Dlnexpend, accepting Dlnodr does not granger cause Ingdp, which shows in Shanghai the phenomenon of population aging is not the main reason affecting the expenditure of urban residents, and it also shows that population aging is not the main factor affecting the economic development of Shanghai. Rejecting Ingdp does not granger cause Dlnodr, which shows that a good development in a region will make the local people's material living standards improve, which will gradually become the cause of the aging phenomenon.

5.7   Stability test of SVAR model

It can be seen from FIG. 4 that the black dots in the figure are all within the unit circle with a radius of 1, and the black dots represent the reciprocal of the AR root in the stationarity test. All are within the unit circle, indicating that the VAR model is stable and scientific. The next impulse response analysis can be performed.

5.8   Impulse response analysis

The impulse response graph is to observe the impulse effect produced by each variable in the system impacting itself or other variables. Impulse response can analyze the future impact trend of variables in the system on itself or other variables under a unit change. The future trend of the observed variable can be predicted intuitively through the impulse response graph.

Considering that the strategic layout of China's economic and social development is implemented with a five-year plan, the SVAR model selects an impulse response function with a lag of five periods to illustrate the impact of population aging on urban residents' short-term consumption expenditure and GDP impact.

It can be seen from Figure 5 that the influence of the aging population to the Shanghai's urban residents on expenditure has a more obvious positive impact at the beginning of the first period, and then there has been a gradual downward trend in the second half of the first period It became a

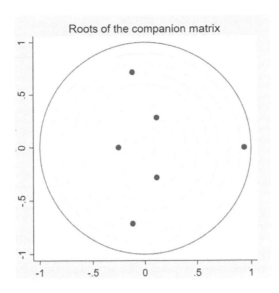

Figure 4.   SVAR model stability test chart.

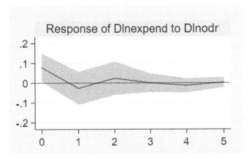

Figure 5.   Short-term impact of population aging on urban residents' consumption expenditure.

negative shock and reached the highest value of negative shock at the end of the first period. There was a rebound at the beginning of the second period, which changed from negative to positive in the second half of the second period. At the end of the third period, it returned to the negative shock, and returned to 0 until the end of the fifth period. Through the observation and analysis of empirical results, we can find that the impact of population aging on the consumption expenditure of urban residents has changed from a positive effect at the beginning to a negative effect, and then from a negative to a positive effect, and finally from a negative effect to a positive influence, creating a cycle. This is in line with the Granger causality test results, indicating that the aging population has a certain impact on the expenditure of Shanghai's urban residents, but this is not the main reason to affect the expenditure of urban residents in Shanghai. When the aged tendency of population appear, which makes household consumption expenditure in the lag 1 within the period with relatively large fluctuations, gradually reduce the lag period after the volatility. This is also in line with real life situations. When the phenomenon occurs, the proportion of the elderly group in this region starts to expand, resulting in a reduction in the expenditure of urban residents. The emergence of aging has also led to the continuous tune-up of the elderly industrial structure in Shanghai. Many products that meet the needs of the elderly population will appear on the market. Manufacturers will continue to expand the supply of products in order to obtain sufficient short-term profits, which lead to increase in residents' spending. However, after this consumption momentum has passed, in one hand, the consumption demand of the elderly population is limited,

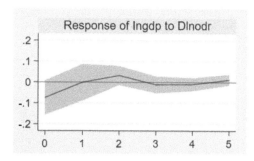

Figure 6.   Short-term impact of population aging on the GDP of Shanghai.

on the other hand, the excess supply in the market has made the market weak, and it has led to a reduction in the consumption expenditure of urban residents.

It can be seen from Figure 6 that the impact of aging population to the Shanghai's GDP had a negative effect and reached the lowest value at the beginning of the first period, but then the impact of negative shocks gradually decreased in the second period. It becomes a positive shock at the beginning and reaches the maximum value of the positive shock at the end of the second period. At the terminal of the third period, it became a negative shock again, and continued until the middle of the fourth period. Then it rose and became a positive shock. Through empirical observation and analysis, the impact of aged tendency of population to the Shanghai's GDP has changed from negative to positive, and then from positive to negative, and then tends to zero. This is also in line with the Granger causality test results. The aging population is not the main reason affecting the Shanghai's economic development. When the coefficient elderly group raised by one percentage point, the GDP of Shanghai area had a significant negative growth before the lagging period, and a positive growth after the lagging period, and then it fluctuated around the trend of 0. This also conforms to the general law in real life. With the gradually enhancing of the elderly group in Shanghai, the short-term labor supply is insufficient, and the shortage of labor makes the productivity of the Shanghai area insufficient. The lack of productivity directly leads to a decline in Shanghai's GDP. However, with the spend of time, the government has launched new economic policies, the industrial structure has been continuously adjusted, and the injection of new labor has made the market adapt to the changes in demand brought about by the increase in the elderly population, resulting in the aging of the population. The impact of economic growth is getting smaller.

The aging population has advantages and disadvantages. It is a contradiction in our society. The impact of the aging population to the economic development at the beginning is more obvious, but the aging of the population will continue to stimulate all aspects of society and make people adapt to the changing environment. Eventually the negative impact of population aging on economic development will become smaller and smaller.

6   CONCLUSION

In the short term, population aging has both positive and negative effects on urban residents. As the population ageing continues to increase, for one thing, society will produce products that meet the value needs of the elderly, such as silver hair industries£¬elderly clubs and nursing homes, and health care products that can cause the elderly to consume. For another, our country's social security system is becoming more and more perfect, and it has gradually formed a triple model of the state's basic pension, personal savings pension and enterprise supplementary pension. The improvement of the pension system has enabled a large part of the elderly population to have more sufficient capacity and income to achieve their material consumption than before. The two macroscopic reasons are the increase in residents' consumption expenditure. conversely, in one side, the consumption demand of the elderly population is limited after all. When there are too many elderly products in the society, the supply of products and services will exceed the declining demand, resulting in overcapacity. In other side, most elderly people leave more or less money for their children, which is the traditional culture of China. The microscopic reasons of these two aspects have reduced the consumption expenditure of residents.

Society is like a market. At the beginning, the aging population will disrupt the rhythm of consumers' lives and affect consumers' lives. However, as time goes by, consumers adapt to the rhythm of an aging society, and they adapt to changing environments and conditions, so that the impact of aging population on the consumption of urban residents becomes less significant. The aging of the population has advantages and disadvantages for society. At the beginning, people were unfamiliar with aging. The negative effects of aging caused a short-term economic regression, but at the same time, people were also recognizing aging and population aging. It will continue to stimulate the adjustment of industrial structure and the adjustment of economic policies and plans. It is this optimization that keeps pace with the times, so that the final aging of the population will have less and less effect on economic growth.

ACKNOWLEDGEMENTS

I would particularly like to thank the thesis supervisor for his care and guidance. The tutor gave me great help and guidance during the writing of my thesis. From the beginning of the topic selection to the finalization of the final draft, he has offered me many valuable suggestions.

REFERENCES

[1] Wan Qian. Research on the impact of China's aging population on urban residents' consumption. Sichuan Normal University, 2019.

[2] Qi Hongqian, Yan Hai-chun. Population aging suppress China's economic growth yet? Economic Review, 2018 (06): 28–40.

[3] Zhou Han. Research on the effect of population aging on urban household consumption based on SVAR model. Jiangxi University of Finance and Economics, 2019.

[4] Jiang Yu, Quan Mengzhen. China's population structure, pension insurance and residents' consumption. Economic Survey, 2018, 35(01): 131–137.

[5] Qian Tingting. A Study on the Impact of Shanghai's Population Aging on Consumption Level. Northwest Population, 2016, 37(06): 78–85.

[6] Leng Jianfei, Huang Shi. A study on the impact of changes in China's population age structure on the consumption structure of urban residents. Consumer Economics, 2016, 32(06): 16–21.

[7] Shang Haiyan. The Impact of China's Demographic Changes on Residents' Consumption – An Empirical Study Based on Data from 1990 to 2017. Business Economics Research, 2019(05): 39–42.

[8] Liu Hongmei, Zhou Xiao, Wu river, Wang Keqiang. An Empirical Study of Shanghai population aging on household savings. East China Economic Management, 2018, 32 (04): 19–25.

[9] Wang Yong, Zhou Han. Research on the effect of population aging on urban household consumption level. Shanghai Economic Research, 2019(05): 84–91.

[10] Dai Jinhui, Ma Shucai. Empirical test of the effect of population aging on residents' consumption behavior. Statistics and Decision, 2017(21): 82–86.

[11] Tu Qi. Study on the threshold effect of population aging on the consumption of urban residents]. Commercial Economic Research, 2018(18): 36–39.

[12] Wu Jing. Research on the impact of China's aging population on the consumption structure of urban residents. Modern Trade Industry, 2017(25):113–115.

[13] Tour soldiers, Cai Yuanfei. Dynamic analysis of the impact of population aging on economic growth – Based on the Panel VAR analysis of the empirical model. Economics and Management, 2017, 31 (01): 22–29.

[14] Lu Yang. An empirical study on the influence of regional industrial structure adjustment on the transmission mechanism of monetary policy – based on panel VAR model. Exploration of Economic Problems, 2016(10):10–17.

Computational Social Science – Luo, Ciurea & Kumar (eds)
© 2021 Taylor & Francis Group, London, ISBN 978-0-367-70193-2

A multiuser virtual service migration scheme based on deep reinforcement learning in mobile edge computing

J.X. Wu, W. Chen, Z.B. Tang & Y.L. Zeng
Donghua University in Shanghai, China

ABSTRACT: The advantage of mobile edge computing (MEC) is that the computing and storage resources can be distributed in all parts of the network, and the deployment node of MEC also meets the requirements of an application for low latency. However, user mobility may make it far from the edge server which undertakes the application task, resulting in inevitable service interruption. In this paper, a new virtual machine (VM) service migration scheme supporting mobility is proposed. Our scheme is realized in three aspects: (1) Some VMs in the related edge server can host the multiuser application tasks. The VM migration strategy can properly migrate the user's tasks, reduce the user-perceived delay, and ameliorate the quality of service (QoS); (2) The system dynamically allocates resources to users, including bandwidth resources and computing resources, which affects the perceived delay of users; (3) We further propose a multiuser service migration scheme based on deep reinforcement learning (DRL), which can reduce the large state space and realize fast decision-making. We conduct extensive experiments, which show that using the DRL algorithm outperforms the classical RL algorithm and some other baseline algorithm.

Keywords: Mobile edge computing, Service migration, Virtual machine, Multiuser mobility, Deep reinforcement learning, Aware delay, Resource allocation.

1 INTRODUCTION

Compared with the traditional centralized cloud computing system, mobile edge computing (MEC) [1] can significantly lessen the service access delay, so there are many delays to sensitive and data-intensive mobile applications, such as connected vehicles [2] and virtual/augmented reality [3]. However, one of the important challenges of MEC is to solve the mobility problem by service migration. This paper focuses on multiuser service migration and dynamic resource allocation to ensure the continuity of user services. When users move in various geographical areas, the system enables their services to migrate to maintain the benefits of MEC. One question to consider is when and where to migrate services [4]. In the actual scenario, considering the different number of tasks that users need to process, the system resources allocated to users should also be allocated on demand. MEC virtual machines (VMs) have deployed the user terminals nearby for data application. If the user is transferred geographically, the VM on MEC needs to be migrated along with it. What needs to be considered is that even if the VM migration only takes a few seconds, the real-time applications can only be processed at the terminal or other remote VMs. We need to create an algorithm to determine whether the VM needs to be migrated. In addition, we need to find a suitable MEC node to migrate to meet the needs of the user application's quality of service (QoS) and to reduce consumption of network resources. We noticed that the user's movement has no memory, and it is a sequential decision-making process [5]. Therefore the DRL algorithm is suitable to solve large-scale service migration problems.

With the rapid development of machine learning, because it does not require prior knowledge and low complexity computing, scholars have applied it to the field of service migration in recent years.

Shan Cao et al. [6] proposed an efficient service migration algorithm by RL called MIG-RL. It used a proxy to learn the optimal policy and determine when and where services should be migrated. Cheng Zhang et al. [7] proposed a deep Q-learning neural network algorithm to plan task migration strategy without knowing the user's moving process. Gao Zhipeng et al. [8] used reinforcement learning to build a service migration model. And VM plays a decision-making role, decides whether to migrate or not, and finds the optimal path in the migration process. In the field of resource allocation, a resource allocation method [9] was proposed based on DRL to reduce the system cost. In addition, a resource allocation scheme based on DRL was proposed [10] to reduce the average user awareness delay and balance system resources. However, they considered single-user mobility scenarios, which cannot meet the actual application scenarios. For a multiuser MEC system, the author [11] considered centralized multiple users computational resource allocation to solve the end-to-end (E2E) service reliability and studied distributed communication and computational resource management scheme to optimize system cost [12]. However, the multiuser mobility issue was overlooked.

Therefore, our work discusses the multiuser mobility scenarios based on DRL and system resource allocation study. In summary, the main contributions of this paper are:

- In order to reduce the average cost of long-term service migration as much as possible, this paper proposes a multiuser service migration scheme in the MEC system, in which each user randomly moves and learns dynamic service migration strategy. In addition, we also consider dynamic resource allocation in different slots, which can affect the long-term average cost.
- Using the Deep Q Network (DQN) algorithm, a DRL algorithm for dynamic service migration is designed. Compared with the traditional reinforcement learning (RL) algorithm, Q Learning, it can quickly make migration decisions in a large state space and does not require the prior knowledge of the environment.
- Numerical experiments are carried out to compare DQN with Q Learning and baseline algorithms, without migration and Greedy Migration algorithms, based on different user numbers. Then, we prove that multiusers service migration based on DQN can obtain better results with different system parameters.

The remainder of this paper is organized as follows. Section 2 describes the system model. Section 3 describes multiusers service migration algorithm based on DRL. Then, Section 4 illustrates the numerical results, and finally, Section 5 concludes the paper.

2 SYSTEM MODEL

This paper considers the multiusers dynamic virtual service migration problem based on VMs in an MEC environment. An MEC environment consists of edge nodes (ENs) and mobile users as illustrated in Figure 1. For simplify the model, this paper assumes that the user moves in this environment remain associated with the same VM (user-to-VM mapping) throughout the entire movement process. Each EN includes a base station, an edge server, and a monitoring unit. The base station is used to communicate with users, the edge server provides computing resources for users, and the monitoring unit tracks GPS signals sent by users, respectively. In addition, the software defined network (SDN) controller [10] formulates an appropriate migration strategy for maintaining stable services in this system architecture. Our work mainly focuses on the dynamic VM migration strategy when the user moves out of the current EN service area, and the corresponding service can be dynamically migrated to an appropriate new EN to optimize the service delay of mobile users.

2.1 *Model view*

In our proposed system, all services should be provided by the VM running on the EN. Some tasks are offloaded to related VMs and processed through a wireless access network. Each VM shares the

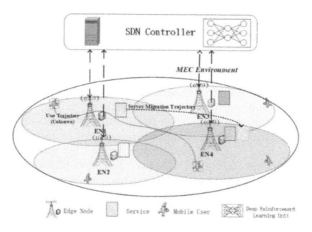

Figure 1. Multiple users mobile system architecture.

same physical resources as other VMs in the same EN. The SDN controller collects state information from the environment to formulate an appropriate VM migration strategy. For example, the user moves away current EN in Figure 1, which may lead to an increase in communication delay and a decrease in the QoS. At this time, the SDN controller captures the user's location $L_{u_i}(k)$ at time slot i and executes corresponding policies to guide VMs to migrate to appropriate EN, so as to improve communication quality and reduce round-trip delay. For generality, we divide the users' tasks into different levels. The higher the user task level value is set, the higher the execution priority is. As a result, the system allocates more resources to high-priority tasks. This paper mainly studies the allocation of base station bandwidth resources and edge server computing resources. The task size of users is denoted as λ_i. We use time slots to represent iteration episodes in learning process and use them interchangeably below, where time slots are denoted by an integer $K \in t = 1, 2, \ldots, T$. and let $u_i \in u = \{u_1, u_2, \ldots, u_n\}$, $e_j \in e = \{e_1, e_2, \ldots, e_m\}$, $v_i \in v = \{v_1, v_2, \ldots, v_n\}$, which are the set of mobile users; the set of ENs with communication and computing capability; the set of VMs hosted in nodes map with users, use the same index. According to the allocation of resources for users that is determined by the level of user tasks, we mark the allocated resources in order from low to high levels, including computing resources $f_i \in \{f_1, f_2, \ldots, f_n\}$ and bandwidth resource $b_i \in \{b_1, b_2, \ldots, b_n\}$. We also denote each edge node e/s computing resource capacity and bandwidth resources as constant F and B, respectively.

2.2 Problem formulation

Different migration decisions lead to different system costs. This also needs to weighing the pros and cons, so VM migration is divided into three quantifiable variables: communication delay, computing delay, and migration delay. According to the above three variables, we propose an algorithm to determine whether the virtual machine needs to be migrated, and find a suitable EN for migration.

Communication Delay. The communication delay comprises of two parts: 1) the uplink communication D^{up} delay between the user and the corresponding node, and 2) the downlink communication delay D^{down} for the result of the processing offloading task returned by EN.

We assume that the uplink transmission of users in the same cell uses orthogonal frequency division multiplexing (OFDM) technology, which can effectively mitigate channel interference [13]. D^{up} is relevant to the free-space path loss model, and can be written as [14, 15]:

$$D_i^{up}(k) = \frac{\lambda_i}{b_{i,j}(k) \log_2 \left(1 + \frac{P_i h_i(k)}{N_0}\right)}, \tag{1}$$

where $b_{i,j}(k)$, P_i, and N_0 are the wireless channel bandwidth allocated to u_i in cell of e_i, the transmit power of the u_i and the channel noise power, respectively. The limitations of the allocated bandwidth resources $b_{i,j}(k)$ are denoted as

$$\sum_{i=1}^{n} b_{i,j}(k) \cdot 1\{L_{v_i}(k) = L_{e_j}(k) \leq B \tag{2}$$

where the Iverson bracket $1\{\cdot\}$ is equivalent to 1 when the condition is satisfied, in which the VM u_i and node e_j are in the same location in time slot k. Otherwise, it is equivalent to 0. Furthermore, $h_i(k)$ is the time-varying communication channel between user i and BS, which can be denoted as

$$h_i(k) = \tau d_i^{-2}(k), \tag{3}$$

where τ expresses the communication channel power at the reference distance that is equal to 1 meter and $d_i^{-2}(k)$ is the space distance within slot k between user u_i and BS. $d_i^{-2}(k)$ can be denoted as

$$d_i(k) = \sqrt{\Delta H^2 + \|q_i(k)\|^2}, \tag{4}$$

where ΔH is the fixed relative altitude and $q_i(k)$ is the relative distance on the horizon plane between BS hosting VM v_i and user u_i. Since the result of the processing offloading task returned by EN is too small, we assume the downlink communication delay can be neglected, $D^{down} \approx 0$.

Computing Delay. Since the computing resources allocated to users are related to user task levels, the computing delay $D_i^{com}(k)$ can be defined as

$$D_i^{com}(k) = \frac{\lambda_i f_0}{f_{i,j}(k)}, \tag{5}$$

where f_0 is the number of CPU cycles per input bit needed for computing, thus λ_i, f_0 is the total number of CPU cycles for offloading task λ_i. Moreover, $f_{i,j}(k)$ is resource allocated to v_i in e_i for slot k and can be indicated as

$$\sum_{i=1}^{n} f_{i,j}(k) \cdot 1\{L_{v_i}(k) = L_{e_j}(k)\} \leq F \tag{6}$$

where F [cycles/s] is the CPU computing frequency for each ES.

Migration Delay. The migration delay mainly depends on whether the position of the VM has changed from time slot $k-1$ to the time slot k. Therefore, the migration delay of v_i is defined as:

$$D_i^{mig}(k) = \frac{s_i}{b_0} \cdot 1\{L_{v_i}(k) \neq L_{v_i}(k-1)\}, \tag{7}$$

where parameters s_i and b_0 are the migrated data size of v_i and wire bandwidth of node e_j, respectively. We assume the amount of data and the wire bandwidth are set to a fixed value to simplify the model.

In short, based on Eqs. (1), (4), and (6), $D_{total}(k)$ is obtained as:

$$D^{total}(k) = \sum_{i=1}^{n} D_i^{up}(k) + D_i^{com}(k) + D_i^{mig}(k). \tag{8}$$

157

Problem Definition. Our goal is to minimize the cost of the delay. Briefly, the total cost is the average sum of the delay $D_{total}(k)$. Therefore, the average cost function of the multiusers service migration problem in MEC environment is defined as follows:

$$\min \ C = \frac{1}{T} \sum_{k=1}^{T} D_{total}(k)$$

$$St : \sum_{i=1}^{n} b_{i,j}(k) \cdot 1\{L_{v_i}(k) = L_{e_j}(k)\} \leq B \tag{9}$$

$$\sum_{i=1}^{n} f_{i,j}(k) \cdot 1\{L_{v_i}(k) = L_{e_j}(k)\} \leq F$$

with the restrictions in Eqs. (2) and (3).

As the VM continues to migrate, resource allocation changes dynamically. This problem is NP hard. Instead of using a traditional optimization method, this paper proposes a novel deep reinforcement learning (DRL)-based multiusers service migration algorithm to find the optimal solution.

3 MULTIUSERS SERVICE MIGRATION ALGORITHM BASED ON DEEP REINFORCEMENT LEARNING

In this section, we propose a DRL based on the DQN [16] framework to minimize the cost of awareness delay for mobile users. The SDN controller selects an action to migrate the relevant VM to the appropriate ENs by collecting the environment parameters. The SDN controller has no prior knowledge of MEC environment, which means that the user's trajectory is unknown. Therefore, the learning process is completely modeless. In an unstable MEC environment, the DQN algorithm combined with deep neural network and QL can make a fast decision.

3.1 *RL settings*

There are three key elements, state, action, and reward, in the RL system which can be defined as:

State. The state vector can be represented as $s_k = \{L_u(k), f_u(k), b_u(k)\}$, where $L_u(k) = [L_{u_1}(k), L_{u_2}(k), \ldots, L_{u_n}(k)]$ is a vector of the location of user u_i in time slot k; $f_u(k) = [f_{u_1}, f_{u_2}, \ldots, f_{u_n}]$ is a vector of allocated computing resources to u_i in time slot k; $b_u(k) = [b_{u_1}(k), b_{u_2}(k), \ldots, b_{u_n}(k)]$ is a vector of allocated bandwidth resources to user u_i in time slot k.

Action. The action vector can be given as $a_k = \{L_v(k)\}$, where $L_v(k) = [L_{v_1}(k), L_{v_2}(k), \ldots, L_{v_n}(k)]$ is a vector of the location of the location of corresponding VM v in time slot k.

Reward. In each episode k, after choosing a feasible action a_k, the RL agent can obtain a certain reward $r_k(s_k, a_k)$. Due to the special nature of reinforcement learning algorithms with maximum rewards, we design the cost function as the negative reward increase after a decision to migration corresponding VM. Thus, we redefine the optimization reward as $r_k = -\frac{1}{T} \sum_{k=1}^{T} \sum_{i=1}^{n} D_i^{up}(k) + D_i^{com}(k) + D_i^{mig}(k)$.

The agent gets an immediate reward r_k after each interaction with the MEC system. Moreover, the goal of this agent is to achieve the maximum long-term discounted reward $R_k = r_k + \gamma R_{k+1}$ where γ is the discounted factor $\gamma \in (0, 1)$.

3.2 *Deep Q Network method*

Based on DQN that can effectively solve the large state space and action space characteristics, our paper utilizes the DQN algorithm to solve the problem of multiuser service migration. According to any selected action, the agent can get the corresponding Q function $Q(s_k, a_k)$. The proposed DRL based DQN is shown in the algorithm in Table 1. To estimate the Q function, the DQN algorithm

Table 1. Deep reinforcement learning-based DQN algorithm.

1: Initialize replay memory D to capacity H, evaluation network with parameters θ and target network with parameters $\theta = \theta$
2: for each episode do $e = 1, 2, \ldots, E_{max}$ do
3:   Initialize state $s_1 = \{L_u(1), f_u(1), b_u(1)\}$
4:   for each step t do
5:     given probability ϵ, either chose action a_k or select $a_k = argmax_a Q(s_k, a; \theta)$
6:     Execute action a_k in emulator which command the location of the VMs migration
7:     Obverse new state s_{k+1} and calculate reward r_k
8:     Store transition (s_k, a_k, r_k, s_{k+1}) in D
9:     Sample randomly mini-batch of transition (s_k, a_k, r_k, s_{k+1}) from D
13:     Set $y_k = \begin{cases} r_k + \gamma \max\limits_{a_{k+1}} Q'(s_{k+1}, a_{k+1}; \theta^-) & episod erminatesat k+1 \\ y_k & O.W. \end{cases}$
15:     Perform gradient decent using TD error function $L(\theta)$ with respect to θ
16:     Reset $\theta^- = \theta$
17:   end for
18: end for

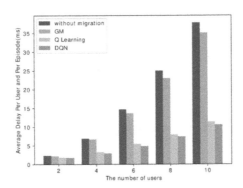

Figure 2. Average delay under different numbers of users.

uses target network, which provides stable $Q(s_k, a_k, \theta^-)$ as

$$y^{target}(k) = r_k + \gamma \max_{a_{k+1}} Q'(s_{k+1}, a_{k+1}; \theta^-) \qquad (10)$$

where θ^- is the network parameters used as the weights of target network which will be updated to the weights of evaluation network θ. The objective of training is to minimize the loss function [17] $L(\theta)$, can be written as

$$L(\theta) = E[(r_k + \gamma \max_{a_{k+1}} Q'(s_{k+1}, a_{k+1}; \theta^-) - Q(s_k, a_k; \theta))^2] \qquad (11)$$

The loss function $L(\theta)$ uses the TD error [18] method which can calculate the error between the estimated value and the target value. At the beginning, there are the same parameters between the target network and the prediction network. After learning, the target network θ^- uses a stationary multistep update and the evaluation network $L(\theta)$ uses a single step update. the algorithm in Table 1 summarizes the multiusers service migration scheme based on DQN. DQN uses the weights θ to predict the Q function. The agent chooses the action to maximizes the Q value at the state s_k.

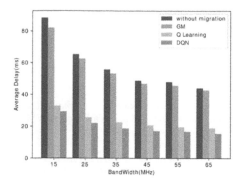

Figure 3. Average delay under different bandwidths.

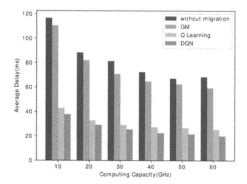

Figure 4. Average delay under different computing capacities.

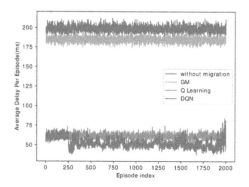

Figure 5. Average delay per episode.

4 PERFORMANCE EVALUATION

In this section, numerical simulation is evaluated for the performance of multiusers service migration based on DQN. Moreover, we compare the migration performance of DQN with classical Q Learning algorithm and baseline algorithm, Greedy Migration algorithm (GM) and Without Migration. In GM, each corresponding VM is migrated to the location closest to the user in each slot k. In addition, in Without Migration, each corresponding VM is always in the initial location, no migration occurs. We use Python to build a simulation environment and import the TensorFlow [19] toolkit for DQN training. We set the users' trajectory to be random. At the beginning of each

160

iteration, the location of users and VMs will be reset. For the generality of the experiment, we set the task size of user λ_i obeys uniform distribution between (100, 800) kilobits, and sort tasks from the smallest to largest. What's more, P_i and N_0 are set at 1 db and 1 w in the communication delay model; $\tau(w)$, $\Delta H(m)$ and f_0 are set at 10^{-5} w, 5 m, and 50 Hz in the computing delay model; and b_0 and γ_c are set at 1 Gb/s in the migration delay model. Also, to implement the DQN algorithm, we use a double fully connected layer to build a neural network and each hidden layer is set to 35 units. And we set the time of episodes to equal 2000.

The objective of this paper is to develop a VM migration scheme that is able to optimize the average users' delay. In Figure 2, we set some parameters which are $B = 15\,MHz$, $f = 20\,GHz$. Then, this experiment calculates the expected value of Eq. (9) and obtains the average delay per user per episode, denoted as C/N. With the number of users increasing in the MEC system, the average service time keeps rising. This is because the number of users increases leading to a decrease in system resources allocated to each user, which affects the average awareness delay of each user. In addition, we can clearly see that using DQN for multiuser service migration has the best performance. The ranking of the performance of the four algorithms is DQN > Q Learning > GM > without migration.

In Figure 3, we set parameters of users' number and bandwidth, $N = 6$, $B = 15\,MHz$. The results show that the average delay decreases with the EN bandwidth resources increase. When the bandwidth resources increase, the relative resources obtained by users are better satisfied, which can effectively reduce the average awareness delay of users. The experimental results of Figure 4 are similar to those in Figure 3, and the average aware delay of users is effectively alleviated with the computing resource increase. Moreover, it can clearly be seen that the multiusers service migration using DQN can get the best performance in Figures 3 and 4.

In Figure 5, we set the number of users to 8 and set the bandwidth and computing resources in each EN to 15 MHz and 20 GHz. The performance of average delay in one iteration is given. We can observe that the performance of DQN is best compared to the other baseline algorithms as the iteration grows. The loss function eventually converges to the stable state.

5   CONCLUSION

In this paper, we studied a multiusers virtual service migration problem in the MEC environment, where a SDN controller makes a migration decision and commands the corresponding VM to migrate to a suitable EN, which ensures continuous and reliable services for users. We also consider different user tasks and classify different levels for them. According to different user levels, each node dynamically allocates system resources for the accessed users. Subsequently, we propose a multiusers virtual service migration scheme based on the emerging DQN algorithm to minimize the average total cost. Based on the ability of a DRL adaptive learning environment, our algorithm can also better adapt to the complex and unpredictable MEC environment. Experimental results show that compared with Q Learning and baseline algorithms, DQN can achieve better performance. In the future, we try to do research in the field of system energy efficiency within the service migration scenario.

REFERENCES

[1] N. Abbas, Y. Zhang, A. Taherkordi, and T. Skeie, "Mobile Edge Computing: A Survey," IEEE Internet Things J., vol. 5, no. 1, pp. 450–465, 2018.
[2] N. Lu, N. Cheng, N. Zhang, X. Shen, and J. W. Mark, "Connected vehicles: Solutions and challenges," IEEE Internet Things J., vol. 1, no. 4, pp. 289–299, 2014.
[3] I. Bisio, A. Delfino, F. Lavagetto, and A. Sciarrone, "Enabling IoT for In-Home Rehabilitation: Accelerometer Signals Classification Methods for Activity and Movement Recognition," IEEE Internet Things J., vol. 4, no. 1, pp. 135–146, 2017.
[4] K. Ha et al., "Adaptive VM Handoff Across Cloudlets," Tech. Report-CMU-CS-15-113, no. June, 2015.

[5] Y. Zhai, Y. Wang, I. You, J. Yuan, Y. Ren, and X. Shan, "A DHT and MDP-based mobility manage-ment scheme for large-scale mobile internet," 2011 IEEE Conf. Comput. Commun. Work. INFOCOM WKSHPS 2011, pp. 379–384, 2011.

[6] S. Cao, Y. Wang, and C. Xu, "Service Migrations in the Cloud for Mobile Accesses: A Reinforcement Learning Approach," 2017 IEEE Int. Conf. Networking, Archit. Storage, NAS 2017 – Proc., 2017.

[7] C. Zhang and Z. Zheng, "Task migration for mobile edge computing using deep reinforcement learning," Futur. Gener. Comput. Syst., vol. 96, pp. 111–118, 2019.

[8] Z. Gao, Q. Jiao, K. Xiao, Q. Wang, Z. Mo, and Y. Yang, "Deep reinforcement learning based service migration strategy for edge computing," Proc. – 13th IEEE Int. Conf. Serv. Syst. Eng. SOSE 2019, 10th Int. Work. Jt. Cloud Comput. JCC 2019 2019 IEEE Int. Work. Cloud Comput. Robot. Syst. CCRS 2019, pp. 116–121, 2019.

[9] N. Din, H. Chen, and D. Khan, "Mobility-aware resource allocation in multi-access edge com-puting using deep reinforcement learning," Proc. – 2019 IEEE Intl Conf Parallel Distrib. Process. with Appl. Big Data Cloud Comput. Sustain. Comput. Commun. Soc. Comput. Networking, ISPA/BDCloud/SustainCom/SocialCom 2019, pp. 202–209, 2019.

[10] J. Wang, L. Zhao, J. Liu, and N. Kato, "Smart Resource Allocation for Mobile Edge Computing: A Deep Reinforcement Learning Approach," IEEE Trans. Emerg. Top. Comput., vol. 6750, no. c, pp. 1–1, 2019.

[11] Y. Sun, S. Zhou, and J. Xu, "EMM: Energy-aware mobility management for mobile edge computing in ultra dense networks," IEEE J. Sel. Areas Commun., vol. 35, no. 11, pp. 2637–2646, 2017.

[12] M. C. Computing, X. Chen, L. Jiao, and W. Li, "Efficient Multi-User Computation Offloading for," IEEE/ACM Trans. Netw., vol. 24, no. 5, pp. 2795–2808, 2016.

[13] S. Deb and P. Monogioudis, "Learning-based uplink interference management in 4G LTE cellular systems," IEEE/ACM Trans. Netw., vol. 23, no. 2, pp. 398–411, 2015.

[14] G. Feng et al., "UAV-assisted wireless relay networks for mobile offloading and trajectory optimization," Peer-to-Peer Netw. Appl., vol. 12, no. 6, pp. 1820–1834, 2019.

[15] Y. Zeng and R. Zhang, "Energy-Efficient UAV Communication with Trajectory Optimization," IEEE Trans. Wirel. Commun., vol. 16, no. 6, pp. 3747–3760, 2017.

[16] V. Mnih et al., "Playing Atari with Deep Reinforcement Learning," pp. 1–9, 2013.

[17] V. Mnih et al., "Human-level control through deep reinforcement learning," Nature, vol. 518, no. 7540, pp. 529–533, 2015.

[18] H. Sasaki, T. Horiuchi, and S. Kato, "A study on vision-based mobile robot learning by deep Q-network," 2017 56th Annu. Conf. Soc. Instrum. Control Eng. Japan, SICE 2017, vol. 2017-November, pp. 799–804, 2017.

[19] M. Abadi et al., "TensorFlow: Large-Scale Machine Learning on Heterogeneous Distributed Systems," 2016.

Computational Social Science – Luo, Ciurea & Kumar (eds)
© 2021 Taylor & Francis Group, London, ISBN 978-0-367-70193-2

A study on the design and usability of an intelligent signage system under the background of human–computer interaction technology

H. Wang
Dalian Ocean University, China

ABSTRACT: As the economy develops in a rapid way in modern society and the global cultures intermingle and collide with each other, the application of traditional visual guidance design in public space is unable to adapt to the requirement of the rapid development of modern society anymore, while the emergence of an intelligent signage system meets society's needs for a signage system and the more complex and advanced services in a public space. Among these, the design of an intelligent signage system plays an important role. The issue of how to enable people to use an intelligent signage system conveniently and willingly and how to meet the needs of different people has become a common concern of large-scale commercial entities all over the world. In this paper, an experiment is established and the conclusion is achieved through a user survey of Dalian Hang Lung Plaza, on the basis of the large-scale commercial complex's intelligent signage system. In this way, consumers can not only experience a more comfortable and convenient shopping environment when shopping, but also consumer satisfaction in regard to the large-scale business complex can be improved through a more efficient shopping experience, thus achieving a higher commercial value. Therefore, the amount of sales of the large-scale commercial complex is further improved, and meanwhile, the overall image of the enterprise is promoted by improving the design of the intelligent signage system in the complex, while the comfort and convenience of consumers in are also enhanced.

Keywords: Human–computer interaction design, Intelligent, Perceptual characteristics of a signage system, Usability.

1 INTRODUCTION

1.1 *Study purpose and significance*

All aspects of city life are changing with the rapid development of China's economy and the acceleration of urbanization. In the meantime, as the population structure of large-scale cities changes, the supporting services of the cities are changing, and the traditional shopping malls are gradually being replaced by more inclusive commercial complexes, and their internal service settings should be more in line with the requirements of the current generation, so as to make people's shopping experience more comfortable and convenient. The purpose of this paper is to establish a new method which combines human–computer interaction technology and intelligent signage system design. Through this intelligent signage system, people can obtain all the service information for a commercial complex by themselves. At the same time, this method avoids expensive labor costs, the pollution caused by construction materials, and other problems caused by traditional signage systems. It has contributed to the development of a commercial complex and the establishment of a marketing strategy and other basic information. The significance of this study is as follows.

First, we learn about the human–computer interaction technology and the principle of intelligent signage system based on the existing literature and reference material, and also about intelligent signage systems and investigate the role of an intelligent signage system in a commercial complex through study.

Second, the essential factors that constitute the characteristics of an intelligent signage system in commercial complexes are obtained by investigating the theories and advanced studies related to human–computer interaction technology.

Third, an empirical test is carried out on the relationship between the availability and value of an intelligent signage system and the satisfaction and action intention.

Finally, the differences with existing studies are proposed through the study results, and the basis for the establishment of a marketing strategy using an intelligent signage system in commercial complexes is provided.

1.2 *Current status of the study*

As stated by the economists Eric Brynjolfsson and Andrew McAfee, the industrial revolution also exacerbates inequality, especially as it may upend the labor market. With automation gradually replacing manpower throughout the economy, the net displacement of workers by machines could widen the gap between capital returns and labor output. The changes to the social structure caused by the fourth industrial revolution are being transformed into allowing users to control themselves with the help of machines. That is, the way to query information independently without employee interference.

For the purpose of creating more free space and a clearer flow of activities in a hall, the signage design team of SchirnKunsthalle Germany designed an intelligent guidance system. This not only saves signage space, but also opens up new space by using new signage methods. The main body of the signage system software is the RGB computer control system, which is responsible for producing all kinds of tones. The development of modern electronic technology not only provides infinite possibilities for the design of a signage system, but also equips signage designers with a new type of signage.

An intelligent signage system for the blind and visually impaired has been designed in Rome, Italy, to help blind people and tourists orientate themselves. A total of 1260 electronic tags were inserted into the walkway and connected to the network. The antenna at the top of a blind man's walking stick activates the RFID chip, which transmits signals over the wireless radio frequency signal to his smartphone, where the phone's database displays the current location. Through the Bluetooth headset on the phone, blind people receive information about their location and surroundings. The road using this RFID chip is 2000 m long.

The world-wide total of interactive kiosks excluding vending machines, ATMs, and tablets reached \$9.22 billion in 2019, up 17.6% from a year earlier, surpassing the previous year's growth rate, based on data from the American *2019 Kiosk Market Survey Report*. Different groups of readers have different views on the most promising kiosk technology, however, users and kiosk hardware and software makers agree that the integration between touchscreen and mobile devices and remote machine management are the three most promising technologies. These three technologies demonstrate that the expansion of e-commerce and mobile ordering is driving the growth of global retail commerce.

Looking around the world, traditional shopping malls are gradually being replaced by large-scale commercial complexes in a business war, where people now want a more diversified and more comprehensive one-stop service, so that food and drink and amusement outlets are concentrated in large-scale commercial complexes, thus attracting more consumers.

2 THE COMBINATION OF HUMAN–COMPUTER INTERACTION TECHNOLOGY AND INTELLIGENT SIGNAGE SYSTEM

2.1 *Human–computer interaction technology*

Alan Dix, a professor at the School of Computing of Staffordshire University, says that human–computer interaction studies look at the way people interact with computers. What we study is the

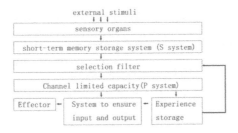

Figure 1.   Human processing of information.

interaction between the user and the system. The system can be the interface of various hardware and software to communicate with the user. The interface can be as small as a smartwatch, as large as the dashboard of a giant freighter or airplane, or the dashboard of an airport control tower. These are visible examples of human–computer interaction interfaces that can allow people to operate and use them. The human–computer interface is designed to enhance a user's understanding of the system (i.e., a mental model) for the sake of system availability and user-friendliness.

The general mode of people's information processing is: external stimuli inputs in parallel through a variety of sensory organs, the perceptual information is saved first through the short-term memory storage system (S system), and then is selected through the selection filter, so that one part reaches the brain, while the other part of the information does not enter the central pivot, so as to avoid the central pivot accepting too much being overburdened. Finally, the outlined in Figure 1 is performed.

Therefore, only when human–computer interaction technology designs interactive products by relying on this kind of process can it solve problems efficiently.

2.2   Intelligent signage system

Kiosks originated in Turkey and Persia, and are perfected in Islamic architecture. In modern times, they mainly consist of outdoor booths such as domes and parks, and small shops, and they are one of the types of technology-based self-service (TBSS). Technology-based self-service refers to all technical means by which consumers can directly collect services through machines without the help of staff, and its includes kiosks, mobile banking, ATM, and so on. The greatest characteristic of the Internet age is interactivity, and the intelligent signage system involves the two-way reception of information by the user or consumer. If the designed product can give the user a result the first time that the user gives the designed product an action, then the interactivity can greatly enhance the user's pleasure and sense of reward. Consequently, from a marketing point of view the mall's friendliness to consumers can be enhanced if this kind of interaction is put into the mall's signage system, so that consumers can independently search for the information they need. In terms of the rapid development of traditional push-button mobile phones to smartphones, the traditional two-dimensional static signage system design cannot meet the city's construction and economic development gradually. There is a large touch screen similar to a mobile phone's screen for the intelligent signage system, and consumers are able to communicate information through the interface, thus not only respecting the user's usage habits but also reducing the cost of learning.

3   PERCEPTION CHARACTERISTICS OF AN INTELLIGENT SIGNAGE SYSTEM USING HUMAN–COMPUTER INTERACTION TECHNOLOGY

In the process of human operation, the machine transmits information to the sensory organs (eyes, ears, etc.) through a monitor. After the information is processed by the central nervous system, it directs the motor system (hands, feet, etc.) to use the controller of the machine and change the state of the machine. In this way, the information passed from the machine returns back to the machine through this "link," thus forming a closed-loop system (Figure 2). Human–machine external environment factors (temperature, illumination, noise, as well as vibration, etc.) also affect and interfere with the efficiency of the system. Therefore, in a broad sense, the man–machine system is also called a man–machine–environment system.

Obviously, to make the above closed-loop system work effectively, many sensory synergies are required in the human body. People's senses include sight, hearing, taste, smell, skin sense, motion sense, balance, and so on. These feelings of the human body can not only accept information from the external environment, but also can sense the position in which they are located.

In addition, the relative number of stimuli can also affect the results of interactive behavior, with this relative number being called sensory threshold, and the external stimulus cannot be lower than the minimum of the threshold, in the case that the user does not feel the stimulus; nor can it exceed the maximum of the threshold, due to damage to the user's sensory organs. Moreover, when two or more stimuli are produced simultaneously, they must differ in intensity to an extent to ensure a sense of difference.

3.1 Visual perception characteristics

About 80–90% of the information in people's cognition of the world is obtained through the visual system, and as a result, the visual system is the most important way to make the human body react through human–computer interactions. Visual perception characteristics are now described. First, the highest value of the visual threshold is $(2.2–5.7) \times 10{-}17J$ and the lowest value is $(2.2–5.7) \times 10{-}8J$. The design principles for all elements of an interactive screen should be within this range to meet the most appropriate requirements for stimuli. Second, the three elements that make up the visual phenomenon are light, objects, and eyes. Light is the main visual production but the famous British psychologist Gregory believed that the visual element should also include human knowledge and experience, for example, an inverted triangle is unstable, an inclined plane object will fall, the weight of a light object is light, and the weight of a dark object is heavy, and so on. In the end, the conditions of the visual field, visual acuity, visual range, and so on should also be considered. In brief, visual perception, a kind of body perception which has the greatest influence on users in the field of human–computer interactions, is an important breakthrough point from ordinary feelings to enjoying interactions. Users can expect more from tactile perception and auditory perception on the basis of excellent visual perception.

3.2 Tactile perception characteristics

Human skin contains many sensors and nerves, and moreover, the maximum degree of freedom a hand can achieve is very high, and therefore, tactile perception can convey more delicate and real

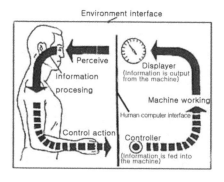

Figure 2. Closed-loop system.

information than visual perception. In 2002, a paper published by Marc O. Ernst in *Nature* stated that human visual deviations rely directly on tactile information correction, which can be perceived simultaneously with visual information and has the ability to sense in all directions.

The tactile information achieved in the process of human–computer interactions can be divided into two kinds of information, that is, surface information and muscle motion perception information. Surface information can be felt through the contact between the human skin and the intelligent terminal, however, muscle motion perception information can be obtained through the position of the limbs and the force generated by movement. Therefore, tactile perception feedback can be widely applied in people with visual impairment.

3.3 *Auditory perception characteristics*

The second most important channel (the first is the sense of vision) of the human body is the sense of hearing. The sense of hearing is characterized by the sense of azimuth, loudness, tone, and so on. Familiar auditory perception can also help users understand a product and adapt to an unfamiliar environment, make different sound effects by means of "echo location", and create a different sense of a scene experience, playing an important role in the user's use of the product.

4 THE DESIGN COMPONENT OF AN INTELLIGENT SIGNAGE SYSTEM UNDER THE BACKGROUND OF HUMAN–COMPUTER INTERACTION TECHNOLOGY

4.1 *Technical design elements*

The touch-screen all-in-one intelligent signage system comes in four sizes: 32-inch, 42-inch, 47-inch, and 55-inch (Figure 3). For the configuration, the infrared, optional capacitors, as well as optical touch mode are generally used, there are two touch points, with an optional six or 10 points. Each touchscreen category has its merits and demerits, which also depend on the customer's multiple needs. In terms of contrast ratio, the ratio is generally 4000:1, its resolution is 1920 × 1080p, Core I5 processor (optional i3, i7) is applied, 4G memory CPU, hard disk 500G, and solid-state hard disks 128G and 64G are optional. A safe, convenient, Internet-based quality service is created with 4 mm physical toughened safety glass and RJ-45 wired network, and IEEE802.11 standard wireless WIFI.

In addition, a good signage method is provided for people with hearing impairment by GPS navigation technology, which can be a great concern in the design of a signage system. The Munich airport signage system is specially designed to provide a tactile signage map for blind people. The signage system has sharp contrast and signal colors, so that tourists with limited vision can also recognize it. Japan's Tokyo Metro has introduced a platform ramp lift in the design of a signage system for the disabled to read.

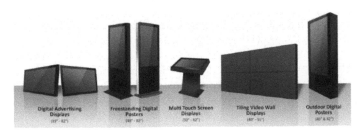

Figure 3.   Touchscreen intelligent signage systems.

4.2 Functional design elements

An intelligent signage system generates an intelligent navigation route in real time in the navigation process, showing the pathway through animation on a large screen, vividly displaying the effects of layer changing and taking lifts, and so on, which can be used to replace the manual navigation service and enhance the service image of a shopping mall. The system's convenient intelligent electronic map navigation function helps customers find their destination quickly, saving their time and enhancing the user's consumption efficiency, and thus enhancing the shopping mall's sales ability.

Whenever the store information in the mall changes, the manager only needs to log into the background program of the intelligent signage system of the mall to make a simple operation to enable the system to automatically update and refresh the path, and in this way, the management time and cost of the mall are reduced, and the duplication of labor and manmade mistakes are avoided. The intelligent signage system of a shopping mall integrates the membership system of the shopping mall so that customers can query their membership information in real time, such as member registration guidance, member point inquiry exchange, as well as member courtesy description.

The user's whole operation process can be realized based on a large-scale touchscreen, providing users with a high-tech fashion experience and greatly improving the mall's brand image (Table 1).

4.3 Emotional design elements

Professor Norman generalizes the influence level of products on human emotion as follows: visceral level, behavior level, and reflective level from the point of view of emotional design (Figure 4), and the three progressive.

The visceral level is the basic sensory perception, as well as the physical emotional arousal. At the visceral level, the user is concerned with the look and feel of the product at first contact. As far as an intelligent signage system is concerned, users pay attention not only to the appearance of the machine, but also to the design of the user interface. A good UI design has the important features of saving time and energy and providing convenience. The appearance structure, color, background color, terrain structure, figure, and so on of the intelligent signage system can be said to be the characteristic elements of the machine design. The excellence of the machine design can produce influences on the image of the machine, thus further boosting the sales volume of the shopping mall. When the intelligent signage system provides a service for the users in the shopping mall, the users' satisfaction at the perceptual level can make them have a more pleasant experience, thus the users will favor the shopping mall more generally.

The behavior level is to judge the product through the sensory organ. The behavior level can be further divided into four parts: purpose → communication → behavior → result, for example, the purpose: the user is looking for a restaurant in the shopping mall. The users seek help through the dining room display function on the intelligent signage system, thus the communication is produced. Behaviors: typing and searching. Result: The users discover a restaurant that they are interested in and get specific information. The users will not only feel visually comfortable at this stage, but will also be sensitive that the machine is operating through them and will experience whether the overall logic of the system is clear.

The reflective level is the highest level, and this is a description of the users' feelings after using the product.

5 USABILITY ANALYSIS OF AN INTELLIGENT SIGNAGE SYSTEM UNDER THE BACKGROUND OF HUMAN–COMPUTER INTERACTION TECHNOLOGY

In order to improve the experience of the intelligent signage system design, this paper makes a more in-depth analysis of the intelligent signage system and user experience on the basis of the

Table 1. Function design of an intelligent signage system.

Type	Category	Function	Realization result
Basic module	Home page	Standby interface	Advertisement picture, activity express, and other picture advertisement alternately scroll
	Brand list and search function	Search by floor	List all the brands on the designated floor
		Search by category	Show brand by category
		Search by brand letter	Click on the alphabetic list to display the search content and list the brands in the first letter of the pre-set brand
	Mall map signage system	Floor map signage	The 2D floor map, showing the shop location on the floor and the brand logo, can be scaled. Including less than 10 sets of 2D map floor plan design, customers provide shop floor CAD map and other materials The 3D floor map, showing the shop location on the floor and the brand logo, can be rotated and scaled. (Optional) Including less than 10 sets of 3D map modeling, customers provide floor CAD drawings and other materials
		Intelligent redrawing	When a shop adjustment is made, the map does not need to be remade, the space can be divided (limited to 2D) according to needs through simple editing, intelligently redrawing the map, and it is easy to split and merge shops
		Public facilities	Display the corresponding public facilities marks according to the floor
		Automatic routing	Based on the designated store triggered by customers, the system automatically generates the traffic routes from the customer's location to the brand store, and escalators and elevators can be chosen
		Cross-floor route finding	Automatically calculate the shortest route, intelligently cue route
	Brand module	Brand classification	Classify based on the brand categories, such as shopping and catering, customers are required to provide classification, and the category name can be customized
		Brand identification	Brand logo
		Shop introduction	Shop number, shop sign, real photo, business hours, and text description
		Preferential information	Brand discount activity display
	Food	Food information	Animation shows the discount menu information, click on the menu to see the details of the discount activities, click on the picture, then route guidance is shown for customers to help them enter the Building directory module
	About the shopping mall	Project information	Mall introduction and picture and text content
		Information on the surroundings	Graphic and text content based on traffic information
Background development	Background function module development		1. Background service editor; 2. Advertising picture release management; 3. Brand store release management; 4 Floor map management; 5. Map editor; 6. Database design; 7. Article management on shopping mall
Value-added function	Parking lot car lookup system		Find the car based on the parking space number, the same as the shop guide Find the car based on the license plate number and show the guiding route (the existing parking system is required to provide an interface)
	Catering queuing system		Enter dining list, choose table size, queue up, input mobile phone number thus there will be a text message reminder for when it is the customer's turn [requires a short message platform collaboration and the cooperation of food and beverage shop client software and hardware (networking with the signage network)]. The quotation does not include terminal hardware
	Multilanguage	English and other languages	Provide a choice of English and other languages at the interface

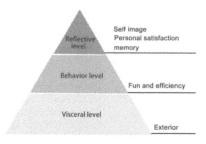

Figure 4.   Level of influence of products on human emotions.

application effect of the intelligent signage system at Dalian Hang Lung Plaza. The reason for choosing Dalian was that it is one of the most densely populated cities in Liaoning Province and an important port, industrial, trade, and tourism city in northeast China. In 2018, Dalian's retail sales of consumer goods for the whole year totaled RMB 388.01 billion, with an increase of 7.8% over the previous year, putting it in a leading position in the three provinces in the northeast of China. If we want to promote the intelligent signage system in the three provinces in the northeast of China, it will be the first test point, and Dalian Hang Lung Plaza is a relatively new commercial complex with urban construction in recent years, and therefore, the author puts the focus here.

5.1   Intelligent test method for the design of an intelligent signage system

5.1.1   The purpose of the experiment
There is a big difference in the purpose of using the intelligent signage system as the users' identities are different and their reasons for coming to the mall are also different.

The consumers with college or higher educational levels and aged from 20 to 40 are more capable and cognizant of using an intelligent signage system based on the analysis of the basic information of the participants, and these consumers are more acceptable of digital products, thus they are the main target group of the design. Four tasks were assigned to the participants in the study: search for brands, search for restaurants, search for service facilities, and search for vehicles. The four tasks were the four most frequently searched options for shoppers who came to the mall and the four most representative services for consumers when they used them; the subsequent usability measurements obtained the information they needed from the four items.

5.1.2   The contents of the experiment
Dalian Hang Lung Plaza over 12 hours was the range of this experiment, and the tests of the four tasks using intelligent signage system was done on consumers of different genders, educational backgrounds, ages, and occupations (Table 2). In the experiment, each experimenter was asked to determine which of the four items they wanted in the commercial complex, and to independently complete the search for the location of the brand store, the location of the restaurant, the location of service facilities such as toilets or elevators, and the location of their vehicles in the parking lot, and finally, the conclusions were made.

5.1.3   Experimental results
The results of this experiment show that the majority of consumers who are looking for designated brand stores and restaurants are young and highly educated consumers with clear consumption patterns, who are the main consumers of FMCG (fast-moving consumer goods) in shopping malls. The proportion of business people among the consumers who search for services has increased, indicating that large-scale commercial complexes are a place for business people to discuss their businesses. Searching for vehicles is one of the best services reflected by consumers in a modern commercial complex compared with traditional shopping malls, which allows consumers to locate their vehicles most quickly when they are lost in a complex underground parking area. From this experiment, we can see that in order to attract the target consumer group's attention from traditional

Table 2. Basic information on the four missions of the participants.

Number of samples	Attribute	Search brand 92 people		Search restaurant 78 people		Search service facility 47 people		Find a vehicle 418 people	
		Number of times	%	Number of times	%	Number of times	%	Number of times	%
Sex	Male	45	48.9	31	39.7	27	58.7	197	47.1
	Female	47	51.1	47	60.3	19	41.3	221	52.9
Level of education	Postgraduate and above	13	14.3	13	16.9	4	9.1	53	12.9
	Junior college and undergraduate	74	81.3	62	80.5	35	79.5	333	81.2
	High school	3	3.3	2	2.6	2	4.5	20	4.9
	Junior middle school and below	1	1.1		0.	3	6.8	4	1.0
Age	10–19	14	15.4	5	6.5	2	4.3	49	11.8
	20–29	62	68.1	54	70.1	32	68.1	276	66.5
	30–39	12	13.2	10	13	6	12.8	60	14.5
	40–49	1	1.1	7	9.1	5	10.6	24	5.8
	50–59	1	1.1		0.	1	2.1	2	0.5
	Above 60	1	1.1	1	1.3	1	2.1	4	1.0
Occupation	Student	66	71.7	42	53.8	15	32.6	6	1.4
	Peasant	2	2.2	2	2.6		0.	2	0.5
	Worker	4	4.3	2	7.7	4	8.5	18	4.3
	Businessperson	6	6.5	3	29.5	18	38.3	259	62.1
	Civil servant	3	3.3	6	2.6	2	4.3	59	14.1
	Faculty and staff	7	7.6	2	2.6	1	2.1	31	7.4
	Freelance occupation	1	1.1	1	1.3	2	4.3	22	5.3
	Other	3	3.3	2	2.6	4	8.5	20	4.8

shopping malls to the new-type commercial complexes, these new-type commercial complexes need to set up more practical and convenient services, and take into account the sensitivity of the target consumer group to the emerging technology and the growing aesthetics demand.

5.2 Usability measurement method for an intelligent signage system design

The design of an intelligent signage system can be divided into two major aspects, namely, usability experience and functional value experience, based on human–computer interaction technology. Usability experience is the effectiveness, efficiency, and satisfaction that users experience when using a product to achieve a specific goal. From the above analysis, we can see that the usability experience is the theoretical basis of an intelligent signage system. Functional value experience is the users' feedback to the product after using the newly launched products. The value of the function can be simply divided into two parts: user value and business value. As an experience-oriented product, it should put user value first, although sometimes there is a trade-off between business value and user value. These two combined results can affect the users' shopping experience in the mall.

5.2.1 The contents of the experiment

1. Validity experiment

In this experiment, the experimenters were asked to look for a designated restaurant through an intelligent signage system and to evaluate the validity based on the task's completion situation, and

Table 3. Comparison of the validity and efficiency values of the four tasks.

Variable	Status	N	Efficiency	Mean value	Standard deviation	Mean difference	t value	Significance (P)
Search brand	Failure	6	●	0.6451	2.3064	0.1936	2.5	0.0192
	Partial completion	17	○					
	Completion	39	◎					
Search restaurant	Failure	7	●	0.6851	2.2664	0.2336	2.5	0.0232
	Partial completion	18	○					
	Completion	29	◎					
Search service facility	Failure	4	●	0.75	2.2015	0.2985	2.5	0.0297
	Partial completion	8	○					
	Completion	20	◎					
Find a vehicle	Failure	18	●	0.8713	2.0802	0.4198	2.5	0.0417
	Partial completion	35	○					
	Completion	223	◎					

these data were achieved by observing the users' actions. Task execution results are divided into failure, partial completion, and completion.

(1) Failure

If the experimenter decided that he or she could not complete the task and gave up the search, or that he or she could not complete the task within the time limit, it was marked as a failure.

(2) Partial completion

The experimenter performed only part of the operation and did not complete the action by entering the designated restaurant. For example, the experimenter entered the classification interface of the restaurant but did not find the designated restaurant in the classification; or he or she found a map of the floor where the designated restaurant was located, but didn't find the exact location of the restaurant. That is partial completion. The reason for having a "partial completion" category is because it is different from 100% completion, but it cannot be defined as a failure.

(3) Completion

This is easy to understand, by entering the interface with all the details of the designated restaurant within a limited period of time.

2. Efficiency experiment

Efficiency can be measured over time, with timing of the operation by the experimenter.

Users can start the timer from the home page, and end it when the experimenter announces that they have completed it, or when the time limit has elapsed.

Significance of completing four tasks. The data obtained were collated and the differences in the four tasks were studied using an independent sample t-test. From Table 2, we can see that the users' demand for the four tasks is significantly different, $P<0.05$ (see also Table 3). Compared with the other three tasks, the task of finding a vehicle is particularly significant. Consequently, it shows that the intelligent signage system is more in line with the users' expectations in terms of the function of finding a vehicle through the whole operating system.

3. Satisfaction experiment

Satisfaction needs to be collected by the experimenters' self-rating scale since it involves the subjective evaluation of the experimenters. Moreover, it is further subdivided into five specific issues: alleviating brainwork, vision matching, fair use, acceptance, and familiar use method. The scores awarded are 1–5 points.

4. User value

User value, referring to the functional value perceived by the users, also needs to be achieved after being evaluated by the experimenters.

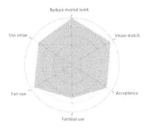

Figure 5.   Statistical analysis of brand search.

Figure 6.   Statistical analysis of restaurant search.

Figure 7.   Statistical analysis of service facility search.

Since what we do is design of the signage system, we learn about the users' value by asking if it's helpful in the process of finding the route. The scores awarded are 1–5 points.

5.2.2   *Experiment methods*
This is a five-point scale, the Likert scale, established by Rensis Likert. The score is divided into five scales (the full score is 5).

Heuristic evaluation: the goal of the Heuristic evaluation, created by Jakob Nielsen and his colleagues, is to identify usability problems in the design.

5.2.3   *Experimental results*
Based on the experimental results, the overall evaluation of the design concept of the intelligent signage system in Hang Lung Plaza can be intuitively understood through the comprehensive statistical analysis of the evaluation data of the subjects under the four tasks (Figures 5–9). A small number of individual evaluation indicators score low, which may be caused by the small sample size, individual cultural level, difference level of understanding, but in general, the data from the investigation in the final mixed market have significance for the future commercial value, when the usability and efficiency of users using the intelligent signage system have significant results ($P<0.05$), the interface of the optimal intelligent signage system can be designed with the references to the data result in Figure 9.

Figure 8.   Statistical analysis of search for a vehicle.

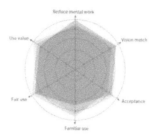

Figure 9.   Comprehensive statistical analysis of hexagonal graphs.

6   CONCLUSION

First, users feel that the intelligent signage system is very useful according to the survey as their time spent in looking for shops, public facilities, and vehicles is reduced and therefore, it can be judged that this is the main factor affecting commercial value. It is necessary to develop, maintain, and manage an intelligent signage system suitable for a large-scale commercial complex, in order to enable consumers to save time when using it. In addition, the business value can also be influenced by the validity, the reason being that with the development of IT technology, the number of consumers who will use intelligent terminal devices increases, and it is not difficult to learn how to use an intelligent signage system. The intelligent visual guide system is used to carry out activities or add additional functions to induce users' interest in shops. Whether the screen design of the interface is good quality or not can greatly shorten the search time and reduce the error in the search process. It is necessary to study, develop, and design more diligently as the intelligent signage system will take the place of traditional guidance systems.

The weakness of the intelligent signage system is that it is not easy for the disabled and the elderly to use it. Therefore, it should simplify the search, with voice prompts or assign staff to participate in the service, in order to make it easier for these vulnerable groups to use the intelligent signage system.

Second, the commercial value is the most important factor for the user using the intelligent signage system in a shopping mall and that influences their satisfaction. The higher the users' value is, the higher the consumer satisfaction with the intelligent signage system will be. Usability has a greater impact in terms of the usability and functional value, and therefore using intelligent signage systems makes people feel comfortable and happy and also has a more positive impact on user satisfaction than functional value elements.

Third, user value has a positive impact on functional value according to the survey. The higher the perceived value of consumers is, the greater the positive impact on the sustainable use, priority, re-access intention, and oral intention of the intelligent signage system will be. Based on this judgment, in order to ensure loyal customers, shopping malls using an intelligent signage system should work out a marketing plan to form an emotional value for consumers. For consumers who have not used it, they should be guided by the use of intelligent signage systems to gain a positive experience. In the meantime, users can ask for help at any time in the use process and staff need to pay attention to this requirement, so as not to let users feel dissatisfied with the system.

On the basis of the motivation for an intelligent signage system applied in a shopping mall, the usability characteristics of the intelligent signage system including validity, efficiency, and satisfaction are different, and the user value and commercial value in the functional value are also different. In conclusion, human–computer interaction technology is expected to enter a new era in the design of intelligent signage systems.

ACKNOWLEDGMENTS

The authors acknowledge the 2019 Liaoning Province Colleges and Universities Overseas (Overseas) Training Project (2019GJWYB019) and the 2019 research project of the Agricultural and Forestry Working Committee of the Chinese Academy of Degree and Graduate Education (2019-NLZX-YB52)

REFERENCES

[1] Ahn, T., Ryu, S. & Han, I. 2007. The impact of web quality and playfulness on user acceptance of online retailing. Information and Management 44: 263–275.
[2] Fang, X. & Tian, Y.H. 2019. A Study on Interactive Digitization of Modern Subway Signage System. Wuhan: Wuhan University of Technology, February.
[3] Hou, W.J. & Wu, C.J. 2015. Gestures interaction research based on the data analysis for smart watch. Packaging Engineering 36(22): 13–21.
[4] Kraft, C. 2014. User experience innovation. Epress.
[5] Wang, Y. 2017. A Study on Urban Visual Signage System Design under Interactive Concept. Chongqing: Chongqing Institute of Humanities and Science and Technology.
[6] Xu, Z.J. & Liu, J. 2018. An Empirical Study on Cross-cultural Cognition of Signage System Based on Eye Movement Data. Beijing: Beijing University of Technology.
[7] Yeon Joo, K. & Jung Kyo, L. 2013. A study on Status and Trends of the Space marketing in Retail shop using Digital-installation. Journal of Korean Society of Design Studies.
[8] Jinhee, K. 2018. A study on consumer acceptance intention of unmanned order payment systems of foodservice companies. Tourism Research Journal.
[9] Yukyung, L. & Jinkyung, P. 2004. Formative Elements of the Directional Sign System for the Effective Information Transmission of the Shopping Mall Complex. Archives of Design Research.
[10] Yonggeum, K. 2018. A Study on the Acceptance Factors of In-Store Kiosk and Mobile Order Payment Service. Seoul Yonsei University.
[11] Juyoung, M. 2009. A Study on the Satisfaction Level of Visual Elements of Buttons in Kiosk Interface (GUI) Design. Ewha Womans University.
[12] Sooyeon, L. 2014. A Study on the Properties and Consumer Response of Digital Signage Shopping Service. Seoul Yonsei University.
[13] Insu, C. 2018. 2018 Korea Trend. Korea Economic Daily.
[14] Sam & Parkers. 2017. Start Line Design Trend of the 4th Industrial Revolution 2017. National Institute of Design Promotion.

Session 2. Integration of qualitative research and quantitative research

Computational Social Science – Luo, Ciurea & Kumar (eds)
© *2021 Taylor & Francis Group, London, ISBN 978-0-367-70193-2*

An empirical study on US direct investment in PRC and bilateral economic and trade relationship

Z.Y. Zhang & Y.C. Liu
College of International Trade and Economics, Jilin University of Finance and Economics, Changchun, PR China

ABSTRACT: To be the largest developing and developed countries within the world, PRC and USA are highly complementary and mutually beneficial in trade and economic cooperation. The impact of American direct investment in China on Sino-U.S. trade is studied in this article. The United States has become China's most important foreign investment country and trade partner, and made a very important contribution to China's economic development. Twenty seven years of relevant annual data from 1992 to 2018 is selected for the study. The study about the proportions of China's relative imports to America, China's relative exports to America, American direct investment in China and empirical testing and analysis of the relationship among the three variables found that there is a long-term equilibrium co-integration relationship. It is hoped that suggestions and countermeasures will be made to work out the problems of Sino-US trade imbalance and increasing Sino-US trade frictions.

Keywords: US direct investment in PRC China-US economic trade relations, Granger causality.

1 GENERAL INSTRUCTION

In the forty years after the founding of diplomatic relations between PRC and USA, the trade and economic relations between the two have developed quickly, and the mutual influence degree and interdependence between them has also developed with the quick growth in trade volumes. With the step of reform and opening up, China's absorption of foreign direct investment scale has been expanded continually. The rapid foreign direct investment growth has promoted China's economic development greatly. Moreover, US-funded enterprises are playing an increasingly important role in China, providing corresponding impetus for China's economic development. However, while the Sino-US trade is developing rapidly, there have been some discordant factors in the economic and trade between them. The number of Sino-US economic and trade frictions is increasing day after day, and the scope is wider and wider. All these have given us a lot of inspiration.

2 DATA SOURCE AND ITS PROCESSING

The sample interval of the relevant variables selected in this article is from 1992 to 2018, and the data comes from UN Comtrade Database, the National Bureau of Statistics, and China Statistical Yearbook. The objects of quantitative analysis are specifically following three time series:

$$M = m_u/m_w \qquad (1)$$

M is relative imports from America, that is, the proportion of China's total imports from America to China's total imports from the world.

$$X = x_u/x_w \qquad (2)$$

Table 1. ADF unit root test.

variable	Inspection form (C,T,k)	ADF	Prob.	variable	Inspection form (C,T,k)	ADF	Prob.	Conclusion
OFDI	$(C,T,0)$	−5.942	0.00	Δ	(C,N,0)	−11.878	0.00	$I(1)$
X	$(C,N,5)$	−2.801	0.075	Δ	(N,N,0)	−10.029	0.00	$I(1)$
M	$(N,N,0)$	−1.161	0.218	Δ	(N,N,0)	−5.318	0.00	$I(1)$

Note: The test form (C, T, k) represents the intercept term, trend term and lag order respectively; Δ represents the first-order difference; Prob. is the p-value of MacKinnon one-tailed test.

X is China's relative exports to America, that is, China's total exports from America account for China's total exports percentage from the world.

$$OFDI = ofdi_c/gdp_c \qquad (3)$$

OFDI stands for American direct investment in China, which is expressed as a percentage of China's GDP that China utilizes U.S. foreign direct investment actually.

3 TIME SERIES STATIONARITY AND COINTEGRATION RELATIONSHIP TEST

3.1 ADF unit root test

Since time series are mostly non-stationary series, they cannot be used in the statistical method of stationary time series. The OLS method requires that the data must be stationary, and there is no significant correlation between the series. Otherwise "false" will appear in the analysis. "Return" phenomenon. Therefore, before the regression analysis, the stationarity of the series must be tested first. Therefore, within this paper, ADF unit root test is used to test the stationarity and single integral order of the above three variables. According to the method of selecting the optimal test form, this paper found that the linearity is better. Table one shows the results of ADF unit root test.

As shown in the table, the original series of X and M are nonstationary, but their 1st-order difference series can pass the significance test. The original series and the first-order difference series of OFDI are both stationary. Therefore, M, X and OFDI are all first-order single-integration sequences, and the long term balanced cointegration relationship among variables can be tested through the co-integration relationship.

3.2 Cointegration relationship test

Johansen co-integration relationship test way is used to investigate if there is a cointegration relationship among M, X and OFDI sequences. The first is to determine the optimal lag interval. According to the method of Pang Deliang, Hong Yu (2009)[1]and Hong Yu (2010)[2], an unconstrained vector autoregressive (VAR) model including M, X and OFDI sequences is established. The aim is to estimate the VAR model and analyze its lag results. According to the Akaike Minimum Information Criterion (AIC) and Schwarz Minimum Information Criterion (SIC), the optimal number of lag periods for VAR is selected p = 5; then, p-1 is determined as the minimum cointegration test. Therefore, the optimal lag interval of the overall sample from 1992 to 2018 is 1–4, and the corresponding optimal lag order for VECM is 4.

The second is to determine the optimal way of co-integration relationship test. Based on the model selection method of Pang Deliang and Hong Yu (2009), corresponding error correction models ((VECM) were established for each of the five model forms that may have co-integration relations. In addition, according to the trace statistics (Trace) and the maximum feature (Max-Eigenvalue) Two statistics, Table 2 reported the summary results of the cointegration relationship test during the

Table 2. Johansen cointegration relationship test results.

Inspection form	Sequence space / Cointegration space:	No additional items / No intercept No a linear trend	No additional items / With intercept No a linear trend	Has a linear trend / With intercept No a linear trend	Has a linear trend / With intercept Has a linear trend	Secondary trend / With intercept Has a linear trend
Lags interval:	Trace	1	3	2	3	3
1 to 4	Max-Eig	1	3	2	3	3
Series: M,X,OFDI	AIC	−29.5757	−29.7096	−29.9815	−30.0680*	−2.99 E+0
	SC	−27.4928	−27.5771	−27.7498	−27.7868*	−27.5454

Explanation: * indicates the best cointegration test form for final recognition.

sample period. Since the optimal additional item form of the Johansen cointegration relationship test directly corresponded to the optimal form of VECM, each additional term form had established a corresponding VECM with a lag order of 4. The optimal model form was selected according to the AIC and SC statistics of different error correction models, and their results showed that the optimal model form is the sequence space has a linear trend, and cointegration Space contains the form of VECM with a linear trend in the intercept.

4 GRANGER CAUSALITY TEST

The existence of a co-integration relationship does not necessarily indicate that there is a Granger causality from the independent variable to the dependent one, and the short term and long term Granger causality may not be the same. It is necessary to make a distinction between short term and long term tests of Granger causality, and study the relationship and degree of independent variables and dependent variables.

4.1 Short-term Granger causality test

The short-term Granger causality test is to impose Wald constraints on the lags of the independent variables of VECM. If the $\chi 2$ statistic with the accompanying probability is less than 0.05, the null hypothesis is rejected. The hypothesis that the corresponding independent variable is the dependent variable short-term Granger reason is accepted. The results of Granger causality test based on the optimal form of VECM are shown in Table 3. From the table we can see:

In the short term, there is a two ways opposite causal relationship between OFDI and M. Firstly, M is the short-term Granger cause of OFDI, which has a short-term positive effect. Secondly, OFDI is the short-term Granger cause of M, which has short-term negative effects.

4.2 Long-term Granger causality test

Table 4 reported the long-term Granger causality test results based on the optimal form error correction model.

On the base of VECM, the significance of the long term relationship is judged first, and the generalized impulse response function is established on the basis of it. The long-term convergence value of the impulse response function is not given for those that fail the significance test. In this study, all impulse response function curves converged to a certain constant around the 100th period after the impact, indicating that the model system of this study is stable. It can be calculated by calculating the convergence value of each impulse response function at the 100th period. The sign

Table 3. Short-term Granger causality test

Dependent / Independent	ΔMt Chi-sq	SE	ΔXt Chi-sq	SE	Δ OFDIt Chi-sq	SE
ΔMt-1 ΔMt-2 ΔMt-3 ΔM t-4			4.084 (0.395)		23.840 (0.0001)	0.050
ΔXt-1 ΔXt-2 ΔXt-3 ΔX t-4	4.287 (0.369)				3.222 (0.521)	
ΔOFDIt-1 Δ OFDIt-2 ΔOFDIt-3 ΔOFDI t-4	17.78 (0.001)	−42.996	3.417 (0.491)			

Note: (1) χ^2-stat represents the χ^2 statistic with 2 degrees of freedom; (2) L = 1, 2, 3, 4 represent the lag period of the variable; (3) () is the corresponding adjoint probability; (4) tL in the subscript of the independent variable represents the corresponding The optimal lag order of the error correction model; (5) If this short-term effect cannot pass the test at the 10% level, the corresponding lag parameter is omitted from the table.

Table 4. Long-term Granger causality test

Dependent / Independent	ΔMt F-stats	LE	ΔXt F-stats	LE	ΔOFDIt F-stats	LE
εt-1	12.855 (0.007)		5.868 (0.042)		0.365 (0.563)	
εt-1, ΔMt-1, ΔMt-2, ΔMt-3, ΔMt-4			1.320 (0.341)		4.600 (0.032)	0.0004
εt-1, ΔXt-1, ΔXt-2, ΔXt-3, ΔXt-4	3.161 (0.078)	−0.0072			0.7546 (0.582)	
εt-1, ΔOFDIt-1, ΔOFDIt-2, ΔOFDIt-3, ΔOFDIt-4	4.489 (0.034)	−0.0002	0.879 (0.517)			

Note: (1) The F statistic given in the table, () is the corresponding adjoint probability; (2) LE represents the long-term effect, which is the convergence value of the corresponding generalized impulse function after 100 periods. Because the absolute value of the convergence value of some impulse response functions is less than 0.001, 4 significant digits are retained after the decimal point. (3) t-L in the subscript of the independent variable represents the optimal lag order of the corresponding error correction model, L = 1, 2, 3, 4.

of the long-term total effect is judged positive or negative. If the long-term effect level is below 0.05, it is considered significant, and the generalized impulse response function curve is shown in Figures 1 and 2:

It can be seen from Table 4 that in the long-run, the causal relationship between OFDI and M is consistent with the short-term, and there is also a two-way opposite causal relationship.

Firstly, M is the long-term Granger cause of OFDI and has a long-term positive effect and this explanation ability is relatively strong. This is because with the fast development of China's economy within recent years, China's import and export volumes have continued to increase. In order to follow up its own development needs, China needs to import technological and advanced equipment from USA, which has attracted a large number of foreign companies to China. Investment also provides huge business opportunities for multinational companies in USA. Therefore, as the relative imports of Sino-US trade increase, US direct investment in China will also increase.

Secondly, OFDI is the long-term Granger cause of M and has long-term negative effects, but this explanatory effect is weak. In the long-term, China's relative import trade to USA promotes US direct investment in China, but US foreign direct investment cannot cause the long-term further expansion of Sino-US relative import trade, leading to a decrease in relative imports. This is because with the increase of U.S. investment in China, it has provided domestic advanced technology,

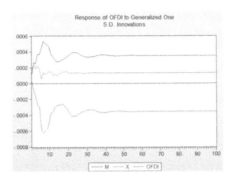

Figure 1.   The response of M to shock.

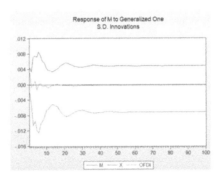

Figure 2.   The response of OFDI to shock.
Note: The horizontal axis represents the number of lag periods of impact action, and the vertical axis represents the degree of impact.

equipment, capital, and management mode. Therefore, products originally imported from abroad can be produced in the domestic market, which has led to more American direct investment in China. It has replaced China's relative imports from America in the long run. Therefore, both parties have a certain degree of substitution.

Finally, X is the long-term Granger cause of M and has long-term negative effects. It stated that as China's relative exports to USA increase, China's relative imports to USA will weaken.

5   CONCLUSIONS AND RECOMMENDATIONS

The first conclusion of this paper is that these three variables of M, X, and OFDI are all first-order single integer sequences, and there is a long term equilibrium co-integration relationships among them. Secondly, the short term and long term Granger causality tests showed that M and OFDI had a two ways opposite Granger causality and X had a one-way Granger causality with M, which may be different from our general understanding. Thirdly, through the above analysis, it can be concluded that the increase in China's imports to USA will cause an increase US direct investment in China, which is beneficial to both parties. Therefore, if the US government continues to expand trade disputes, it will not only damage China's interests, but also cause many US investment companies in China to suffer losses more.

To solve the imbalance in Sino-US trade, it needs to be considered from many aspects. First of all, the United States has always implemented strict export control policies against China. This has always been a major issue that has plagued the development of Sino-US trade relations. Therefore, the United States should expand its exports to China and ease the trade gap. Secondly, we should encourage Chinese companies to go global and allow more of them to invest in USA.

ACKNOWLEDGEMENTS

This paper was supported by the project of Jilin Provincial Social Science Fund (2019N26, 2020J58).

REFERENCES

[1] Pang Deliang Hong Yu. 2009(1). Oil price shock, endogenous technological progress and Japan's economic growth. Modern Japanese Economy
[2] Hong Yu. 2010(1). "Korea's oil imports, real exchange rate and endogenous economic Growth". Northeast Asia Forum
[3] Qi Yao, Hong Yu. 2013. FDI inflow, import and export and China's relative welfare level. International Conference
[4] Fan Haijun. 2012. The impact of US direct investment in China on Sino-US trade imbalance. Jilin University
[5] Guan Ting. 2012. Research on the Impact of US Direct Investment in China on Sino-US Trade. Shandong Normal University
[6] Song Hong. 2019(06). Development and Prospects of Sino-US Economic and Trade Relations. International Economic Review, 74–99 6.

Computational Social Science – Luo, Ciurea & Kumar (eds)
© 2021 Taylor & Francis Group, London, ISBN 978-0-367-70193-2

Power grid middle platform data anomaly detection based on two-stream convolutional neural network

H.Z. Cui
State Grid Jiangsu Electric Power Co., LTD., Jiangsu Nanjing, China

C. Wang
State Grid Jiangsu Electric Power Co., LTD. Information & Techcommunication Branch, Jiangsu Nanjing, China

M.S. Xu
Jiangsu Electric Power Information Technology Co., LTD., Jiangsu Nanjing, China

M.H. Xu
State Grid Jiangsu Electric Power Co., LTD. Information & Techcommunication Branch, China

ABSTRACT: With the rapid development of the Internet, network anomaly detection has increasingly become a problem worthy of attention. To effectively detect anomalies in the network, this paper describes a two-stream convolutional neural network (TS-CNN) anomaly data detection method. First, TS-CNN uses multiple 1-D convolutional layers to extract features from 1-D network data, and generate 1-D feature vectors through pooling; at the same time, TS-CNN preprocesses 1-D network data to generate 2-D grayscale images. Multiple 2-D convolutional layers are used to extract features from grayscale images, and 1-D feature vectors are generated through a pooling layer and a fully connected layer; finally, the two feature vectors are spliced through a fully connected layer to output the detection results. Experimental results show that the detection method based on TS-CNN can effectively detect network abnormal data.

1 INTRODUCTION

With the continuous development of science and technology, the Internet has become a basic communication technology required by all social fields. Network traffic is the carrier of information and data exchange between entities in the network space. No matter what form of network communication is, it will leave a mark on the network traffic. How to dynamically perceive the abnormal attack behavior that may be occurring in the network by analyzing different traffic characteristics is of great significance for maintaining the security of cyberspace. However, network intrusion detection methods based on fixed rules have the problem that they cannot detect new types of attacks. As a reliable means of detecting network attacks, anomaly detection can detect unknown new types of attacks promptly, thereby greatly improving network security. Therefore, network anomaly detection has gradually become a research hotspot in academia in recent years.

Among anomaly detection algorithms, algorithms for network data anomalies have developed rapidly. Liu et al. [1] used the wavelet analysis method. The wavelet analysis method divides a series of values from the sensor in a period into high-frequency and low-frequency components to determine whether it contains abnormal data. However, this algorithm is not suitable for real-time detection of early warning for every value passed into the sensor. Martin et al. [2] proposed the Markov model. In the production process, the value transmitted by the sensor in the machine can be regarded as a random process composed of a series of discrete values in time and state. Due

Figure 1. Data preprocessing.

to the local correlation of the program operation, it can be seen that a series of values passed by the sensor satisfy the Markov chain of, especially aftereffect. Therefore, the Markov model can perform abnormal data detection on the 1-D data passed by the network traffic. The research on outlier detection algorithms in multi-dimensional data mainly includes principal component analysis (PCA) method, hybrid outlier detection algorithm based on angle variance for high-dimensional data (HODA) [3], etc. The use of dimensionality reduction method-principal component analysis method is better for detecting abnormal data in thousands of networks. In network data detection, multi-dimensional data has a certain complexity, and some data cannot be processed by dimensionality reduction methods. The HODA algorithm has a certain reference for anomaly detection and has a better detection effect for outlier abnormal data points.

Neural networks have strong function approximation capabilities, adaptive learning capabilities, and fault tolerance [4]. It can flexibly establish detection standards suitable for the current data based on the data obtained, to make judgments about the possibility of abnormalities. Moreover, in the case of small sample data, higher accuracy can also be obtained through the transfer learning method [5].

2 GENERAL INSTRUCTIONS

2.1 *Data preprocessing*

Data preprocessing refers to the process of input data from the original network data flow to the input image of CNN. The preprocessing of the proposed model is shown in Figure 1. The preprocessing consists of three steps, including data cleaning, session extraction, and final image generation. The following is an introduction to each step.

Data cleaning refers to filtering the original data, leaving only the parts that are of interest to us. In the data cleaning process, data packets are classified, leaving only data packets containing important network protocols such as IP protocols, which reduces the amount of data and facilitates subsequent processing. The session extraction process refers to merging scattered IP message segments into network sessions according to the characteristics of the session layer protocol to separate different sessions. At the same time, there may be different applications using the network, and the session extraction part separates the different connections, thereby reducing the possibility of misjudgment due to network congestion. Finally, compare the generated session with the labels in the data set, and label the sessions with the correct labels.

The final image generation part refers to the conversion of the processed session into a two-dimensional image that can be directly processed by CNN. Cut the data processed in the previous step to a fixed length and arrange them together as the output vertically. If the value of each byte is regarded as the grayscale representation of the corresponding pixel, a fixed size 2-D single-channel grayscale image can be obtained. The grayscale image generated after preprocessing is stored in the form of a 2-D matrix. According to the grayscale value of each pixel, the elements in the matrix are compressed to the interval from zero to one and converted to real number type.

2.2 *TS-CNN model*

The two-stream convolutional neural network (TS-CNN) model proposed in this paper is shown in Figure 2. The layer identified in blue is the convolutional layer, the layer identified in green is the

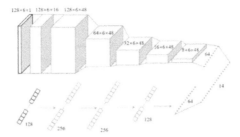

Figure 2. TS-CNN structure diagram.

pooling layer, the first layer is the input layer, the layer identified in purple is the fully connected layer, and the last is the softmax classifier [6].

Suppose X_{nj} is the original matrix input to the convolutional layer. P_i represents the feature matrix of the i_{th} layer ($P_0 = X_{nj}$). The generation process of P_j is:

$$P_j = f \left(b_j + \sum_{i=1}^{n} P_i \times W_{ij} \right) \qquad (1)$$

In (1), j represents a neuron, and n input signals are simultaneously input to the neuron j. The weight value of the input layer P is w_{ij}, the bias value is b_j, and the output is P_j. Where $f(x)$ is the activation function, usually linear correction function (Rectified Linear Unit, ReLU), sigmoid function, tanh (x) function, radial basis function, etc. The tanh (x) function is used here. The main purpose of the pooling layer is to reduce the dimensionality of the feature matrix while ensuring that the downward trend of features remains unchanged [7].

$$O_i = \text{pooling}(P_i - 1) \qquad (2)$$

Through the transfer calculation of the convolutional layer and the pooling layer, the feature matrix is classified by the fully connected layer to obtain the probability distribution Y where g_i is the category of the i_{th} label. As shown in (1), the original matrix (P_0) is calculated through multiple convolutions to obtain a new model of characteristic expression Y:

$$Y(i) = P(S = g_i | P_0 : (W, b)) \qquad (3)$$

The training target calculates the minimum loss function (W, b), the loss function is:

$$\text{NLL}(W, b) = -\sum_{i=1}^{|Y|} \log Y(i) \qquad (4)$$

During the training process, the model may describe the training data too finely, so that the generalization ability of the abnormal data or other new data tested is not strong, which is called the overfitting problem. The L2 norm is added to the loss function to control overfitting, and the parameter d is used to control the intensity of over-fitting:

$$E(W, b) = L(W, b) + \frac{\lambda}{2} W^T W \qquad (5)$$

After the input matrix X_{nj} is forward propagated, there is a residual error between the expected value and the calculated value of the loss function. In the training process, the stochastic gradient descent method is used to back-propagate the residual to update the trainable parameters (W, b)

Table 1. Specific Parameters Used in TS-CNN model.

Number of layers	Type	Number of convolution kernels (size)/number of neurons	Number of layers	Type	Number of convolution kernels (size)/number of neurons
1	conv + tanh	16(3 × 3)	1	conv + tanh	256(3 × 1)
2	conv + tanh	48(3 × 3)	2	conv + tanh	256(3 × 1)
3	pooling	2 × 1	3	pooling	2 × 1
4	conv + tanh	32(3 × 3)	4	fully connected	64
5	conv + tanh	16(3 × 3)			
6	pooling	1 × 2			
7	fully connected	64			
8	fully connected			128	
9	softmax classifier			14	

layer by layer during the propagation process. Therefore, the learning intensity rate parameter μ is used to control the intensity of backpropagation:

$$W_i = W_i - \mu \frac{\partial E(W,b)}{\partial W_i}, b_i = b_i - \mu \frac{\partial E(W,b)}{\partial b_i} \tag{6}$$

The specific parameters of TS-CNN are shown in Table 1.

3 EXPERIMENTAL EVALUATION

All experiments in this article are performed on a desktop with Intel Core I7-4790K, GeForce 1080Ti GPU, and 24GRAM. The experimental environment is the Python3.5 and TensorFlow1.14 platforms. The experimental data is KDD99. It contains nearly 4.9 million connection vectors, and each vector contains 41 features. There are about 300,000 attack instances in the test set of the KDD99 data set, and we select 14 representative attacks.

3.1 *Experimental steps*

First, the original 1-D data is preprocessed into a 2-D gray image. Second, 20% of the original 1-D data and 2-D images are used as the training set, and the remaining 80% are used as the test set. Third, input the training set into the network for training. Finally, the test set is tested by the trained model to obtain the detection result.

3.2 *Results and analysis*

Generally, the pros and cons of the intrusion detection system can be evaluated by three indicators: detection rate (DR, also known as recall rate, 1-false positive rate), false alarm rate (FAR, also known as the false positive rate), and accuracy (ACC). These evaluation indicators can be obtained from the confusion matrix of the model. These evaluation indicators can be obtained from the confusion matrix of the model. Take the binary classification model as an example, its confusion matrix is shown in Table 2.

The calculation formulas for DR, ACC, FAR are as follows.

$$DR = \frac{TP}{TP + FN}, ACC = \frac{TP + TN}{TP + FP + FN + TN}, FAR = \frac{FP}{FP + TN} \tag{7}$$

Table 2. Confusion Matrix.

Ground truth	Detection result	
	True	False
true	TP	FN
false	FP	TN

Table 3. The Comparison of ACC, DR and FAR of Different Algorithms on the Dataset KDD99.

Model	ACC	DR	FAR
ALL-AGL	94.55%	92.23%	0.52%
MHCVF	98.46%	64.22%	0.06%
HAST-II	98.79%	95.73%	0.03%
TS-CNN	99.32%	99.11%	0.01%

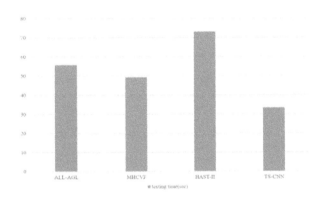

Figure 3. Comparison of testing time of different algorithms.

In this paper, ALL-AGL, MHCVF, and HAST-II are selected as comparative experiments for comparison. To ensure the fairness of comparison, we use the same operating environment and training sample sampling rate for all algorithms. The test results of different algorithms on the KDD99 data set are shown in Table 3.

Although the ACC and DR of ALL-AGL both reached more than 90%, the false alarm rate reached 0.5%. The FAR of MHCVF reaches a satisfactory level, but the detection rate is only 68.2%, which is difficult to meet the daily abnormal detection. HAST-II is currently an excellent algorithm for anomaly detection, and all indicators have reached satisfactory results. The TS-CNN mentioned in this paper uses the advantages of deep learning to fully learn the hidden features in the network traffic data and reaches the highest in each indicator. At the same time, we also compared the time complexity of the four algorithms. As shown in Table 4, our time complexity is smaller than the other three comparison algorithms.

In summary, the proposed TS-CNN model is better than the current mainstream anomaly detection algorithms. It can realize real-time monitoring in a gigabit network, and it can also be applied to a 10-Gigabit network with low network traffic.

4 CONCLUSION

This paper introduces a dual-stream CNN based on deep learning, which detects network traffic anomaly by learning features of 1-D network traffic data and transformed 2-D gray-scale image data. Through experimental comparison, our algorithm is superior to the state-of-art algorithms in all indicators. However, the attack methods against CNNs may be applied to attack this model. The output of the CNN has been proved to be discontinuous, so a minimal disturbance to the input is likely to have a greater impact on the output classification results. How to combine the characteristics of network data packets to effectively defend against neural network attacks is one of the research focuses in the field of network anomaly detection.

ACKNOWLEDGEMENTS

This work was supported by the Science and Technology Project of State Grid Co., LTD (Research and Application of Service Design and Management Technology based on Data Middle Platform, 5700-202018181A-0-0-00).

REFERENCES

[1] Liu C, Gryllias K. A semi-supervised Support Vector Data Description-based fault detection method for rolling element bearings based on cyclic spectral analysis. Mechanical Systems and Signal Processing, online, 2020.

[2] Boldt M, Borg A, Ickin S, et al. Anomaly detection of event sequences using multiple temporal resolutions and Markov chains. KNOWL INF SYST, 2020, 62(2): 669–686.

[3] Zhang W, Dong X, Li H, et al. Unsupervised Detection of Abnormal Electricity Consumption Behavior Based on Feature Engineering. IEEE Access, 2020, 8: 55483–55500.

[4] Naseer S, Saleem Y, Khalid S, et al. Enhanced network anomaly detection based on deep neural networks. IEEE Access, 2018, 6: 48231–48246.

[5] Tian Z, Luo C, Qiu J, et al. A distributed deep learning system for web attack detection on edge devices. IEEE Transactions on Industrial Informatics, 2019, 16(3): 1963–1971.

[6] Vinayakumar R, Alazab M, Soman K P, et al. Deep learning approach for intelligent intrusion detection system. IEEE Access, 2019, 7: 41525–41550.

[7] Andresini G, Appice A, Di Mauro N, et al. Multi-Channel Deep Feature Learning for Intrusion Detection. IEEE Access, 2020, 8: 53346–53359.

Computational Social Science – Luo, Ciurea & Kumar (eds)
© *2021 Taylor & Francis Group, London, ISBN 978-0-367-70193-2*

Evaluation of high-tech industry innovation performance using a new two-stage network DEA model

L. Chen, H.H. Hu & D.X. Ling
School of Economics and Management, Southeast University,
Nanjing, Jiangsu, China

X.Y. Ye
Department of Landscape Architecture and Urban Planning,
Texas A&M University, Texas, USA

ABSTRACT: This article describes a new two-stage network DEA model. This model considers shared input, additional input, and undesirable output on a concept of cooperative game theory. Based on this model, we evaluate the overall and substage innovation performance of China's high-tech industry from 2006 to 2015. The results display that: (1) the spatial distribution of innovation efficiency of high-tech industry is extremely uneven in China. The efficiency mean value in the south of China for the decade is 0.791, which is the best, and central China performed worst with 0.328 average overall efficiency. (2) Eleven provinces, making up 44% of all provinces, had overall efficiency mean values between 0.3 to 0.6 over the decade. There are six districts with an overall efficiency value above 0.8, including five provinces in the southwest, meanwhile the values in northwestern and central areas of China are below 0.3. (3) the availability of the high-tech industry still depends on the efficiency in the production stage in China, and most Chinese provinces have good substage coordination with low overall efficiency.

1 INTRODUCTION

The high-technology industry is an industry that produces new, innovative, and technologically advanced products featuring technological change, high input of scientific research, development expenditure, and employment. High-quality innovation and high value-added activity are the crucial characteristics of the high-tech industry, along with the long development time and high risks. Even if innovation research is successful, renewal of the production process is essential. Hence, the two stages of high-tech industry activities are coherent. Therefore, the overall efficiency of high-tech industry innovation activities has been widely studied and discussed. For the sake of discussion, our definition of the innovation process includes the research and manufacturing stages in the high-tech industry.

Data envelopment analysis (DEA) is excellent and popular for measuring input–output efficiency. The basic idea of DEA is to identify the best practices of peer decision-making units (DMUs). The two-stage network DEA model is an improved DEA model with only two stages, in which the output of the first stage becomes the intermediate variable of the system as the input of the second stage (Kao and Hwang, 2008). As for how to deal with the connection between the two phases, Cook et al. (2010) chose additive efficiency decomposition while Liang et al. (2008) used the multiplicative method.

Inputs in the first stage are usually shared by the two stages. In some cases, the shared inputs are too special to be split up and allocated to the two-stage operations. In the high-tech industry, some researchers and equipment are shared in both the innovation and production stages. Yu and Lin

(2008) initially introduced the shared input into the network DEA evaluation model and established the multiactivity network data envelopment analysis model. Based on this model, Chen et al. (2010) considered the limitations of sharing input allocation proportion and developed an additive two-stage input DEA model with shared resources. Chen et al. (2018) first considered the shared resources in a two-stage DEA model to estimate the innovation and manufacturing efficiencies for the Chinese high-tech industry. Considering prior researches (Cook and Hababou, 2001), although shared resources have been widely considered in the DEA model, very few scholars pay attention to the issue for the high-tech industry. Research by Kao and Hwang (2008) and Chen et al. (2018) is closest to the research topic of our research, but we have two crucial improvements compared with their papers, one is to consider the undesirable output, and the other is to use the cooperative game theory to connect two substages.

2 METHODOLOGY

Considering the characteristics of the high-tech industry, we introduced undesirable output and additional input into Liang et al. (2008)'s multiplicative DEA model with shared input, and finally, put forward a revised multiplicative two-stage network DEA model.

The structure of our network DEA is shown in Figure 1. There are some inputs shared by two stages and several additional inputs used in stage two. At the same time, we considered several undesired outputs in the first stage. Each $DMU_j(j=1,\ldots,n)$ has m input denoted by $x_{ij}(i=1,\ldots,m)$, some proportion of the m input is the only input to the first stage, denoted as $x_{i_1j}(i_1 \in I_1)$. Meanwhile another proportion of input, denoted by $x_{i_ij}(i_2 \in I_2)$, is shared by two phases, where $I_1 \cup I_2 = \{1, 2, \ldots, m\}$. We assume that x_{ij} are divided into $\alpha_{i,j}x_{ij}$ and $(1 - \alpha_{i,j})x_{ij}$ $(0 < \alpha_{i,j} < 1)$, corresponding to the portions of shared input used by the first and second stages, respectively. Similar to the constraints in Cook and Hababou (2001), $\alpha_{i,j}$ is confined to a certain range, namely $L_{i,j} \leq \alpha_{i,j} \leq U_{i,j}$.

$\overline{b}_{d_1j}(d_1 = 1, 2, \ldots t_1)$ is modeled as the undesired output of subsystem 1. According to the method of Seiford and Zhu (2002), we transformed the undesirable output by multiplying "-1" first, then we construct an accepted translation vector M to let the negative undesirable output value become positive. That is, $\overline{b}_{d_1j} = -b_{d_1j} + M$, M should be a properly large number. At the same time, each $DMU_j(j=1,\ldots,n)$ at the first stage has output denoted by $b_{d_2j}(d_2 = 1, 2, \ldots, t_2)$, which then become the inputs to substage 2. $b_{d_2j}(d_2 = 1, 2, \ldots, t_2)$ is considered as an intermediate measurement. In subsystem 2, there are some additional inputs denoted by $q_{lj}(l = 1, 2, \ldots, z)$, the desired outputs of the whole system are indicated by $y_{rj}(r = 1, 2, \ldots, s)$. The solution model of this paper is shown in the model (1).

$$\theta_o^* = \max \theta_o^1 * \left(\sum_{r=1}^{s} u_r y_{ro} + U^2 \right)$$

s.t.

$$\sum_{i_1=1}^{m_1} v_{i_1} x_{i_1o} - \sum_{i_1=1}^{m_1} \beta_{i_1o} x_{i_1o} + \sum_{d_2=1}^{t_2} w_{d_2} b_{d_2o} + \sum_{l=1}^{z} p_l q_{lo} = 1$$

$$\sum_{d_1=1}^{t_1} w_{d_1} \overline{b}_{d_1j} + \sum_{d_2=1}^{t_2} w_{d_2} b_{d_2j} + U^1 - \left(\sum_{i_1=1}^{m_1} \beta_{i_1j} x_{i_1j} + \sum_{i_2=1}^{m_2} v_{i_2} x_{i_2j} \right) \leq 0 \qquad (1)$$

$$\sum_{r=1}^{s} u_r y_{rj} + U^2 - \left(\sum_{i_1=1}^{m_1} v_{i_1} x_{i_1j} - \sum_{i_1=1}^{m_1} \beta_{i_1j} x_{i_1j} + \sum_{d_2=1}^{t_2} w_{d_2} b_{d_2j} + \sum_{l=1}^{z} p_l q_{lj} \right) \leq 0$$

$$\sum_{d_1=1}^{t_1} w_{d_1} \overline{b}_{d_1o} + \sum_{d_2=1}^{t_2} w_{d_2}^1 b_{d_2o} + U^1 - \theta_o^1 \left(\sum_{i_1=1}^{m_1} \beta_{i_1o} x_{i_1o} + \sum_{i_2=1}^{m_2} v_{i_2} x_{i_2o} \right) = 0$$

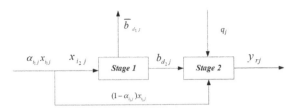

Figure 1. The structure of the DEA model with shared input, additional input, and undesired output.

$$\theta_o^1 \in [\theta_o^{1min}, \theta_o^{1max}]$$
$$L_{i_1 j} v_{i_1} \le \beta_{i_1 j} \le U_{i_1 j} v_{i_1}, \forall j$$
$$w_{d_1}, w_{d_2}, v_{i_1}, v_{i_2}, u_r \ge 0, U^1, U^2 free$$
$$d_1 = 1, 2, \ldots, t_1; d_2 = 1, 2, \ldots, t_2; i_1 = 1, 2, \ldots, m_1;$$
$$i_2 = 1, 2, \ldots, m_2, r = 1, 2, \ldots, s; l = 1, 2, \ldots, z; j = 1, 2, \ldots n.$$

According to Kao and Hwang (2008)'s assumption and Liang et al. (2008)'s centralized model, we presume that $w_{d_2}^1 = w_{d_2}^2 = w_{d_2}$ for all $d_2 = 1, 2, \cdots, t_2$. Also $v_{i_1}^1 = v_{i_1}^2 = v_{i_1}$ for all $i_1 \in I_1$ is assumed because these are the same types of input. In the cooperative model, we use $\theta_o* = \theta_o^1 * \theta_o^2$ to calculate the final efficiency value. The multiplicative model is nonlinear programming, which could not be transferred into a linear model by Charnes-Cooper (C-C) transformation. We treat the first-stage efficiency θ_o^1 as a variable, with a range of $\theta_o^1 \in [\theta_o^{1\,min}, \theta_o^{1\,max}]$, indicating that the global optimal solution θ_o^* is the function of θ_o^1, an independent variable. A one-dimensional search method is a way to solve the function. The upper bound $\theta_o^{2\,max}$ of θ_o^2 is the constraint for $\theta_o^{1\,min}$, so, $\theta_o^{1\,min}$ can be solved. Then, we can determine the range of θ_o^1 and get the nonlinear program, denoting $\alpha_{i_1 j} v_{i_1} = \beta_{i_1 j}$. By applying the C-C transformation we get the final model (1).

In model (1), we let $\theta_o^{1k} = \theta_o^{1\,max} - k \times \varepsilon, \{k = 0, 1, 2, \ldots [k^{max} + 1]\}$, ε (for example $\varepsilon = 10^{-6}$) is the step length from $\theta_o^{1\,min}$ to $\theta_o^{1\,max}$ and $[k^{max}]$ is the largest integer not greater than $[(\theta_o^{1\,max} - \theta_o^{1\,min})/\varepsilon]$. We can get the optimal value θ_o^* by walking along with all values of k. We rendered the matching optimal objective value for the model (1) by formula (2):

$$\theta_o^* = \max_k \theta_o^{1k} \times \left(\sum_{r=1}^{s} u_r y_{ro} + U^2 \right) = \theta_o^{1k*} \times \left(\sum_{r=1}^{s} u_r y_{ro} + U^2 \right) \qquad (2)$$

Formula (2) is the optimal overall efficiency. Based on the result, we can get θ_0^{1k*} for the optimal efficiency of substage 1, then we can use $\theta_0^{2k*} = \theta_o^* / \theta_o^{1k*}$ to calculate the optimal efficiency for substage 2.

3 EMPIRICAL STUDY

The first stage is the R&D stage, which is the innovation research stage of the high-tech industry. During this period, material, human, and capital inputs are the main factors involved. The material input means fixed-asset investments such as workshops and equipment. We select a new fixed asset to represent this indicator. The human input refers to the scientific research personnel, which is represented by R&D labor. The capital input is the funds involved in the R&D stage. We use the indicator R&D cost which is the sum of innovation internal expenditure, technology introduction expenditure, and technology transformation expenditure. We believe that R&D labor and fixed investment are also crucial in the second stage. Therefore, we treat them as shared resources distributed in two stages in some proportion. The main purpose of this period is to obtain scientific and technological innovations, such as technology patents. The number of patent applications is on behalf of the enterprise's efforts in the R&D stage. Some patent applications will be authorized and can be used as an important input in the next commercial transformation phase, while some

Table 1. Input, output, and intermediate output indicators.

Types	Variables	Indicators
Input	Fixed asset investment	Fixed asset investments (shared) in R&D activities
	R&D labor	Human input (shared) in R&D activities, the full-time equivalent of R&D personnel
	R&D cost	R&D internal expenditure + technology introduction expenditure + technology transformation expenditure
Additional input	Manufacturing labor	Manpower input in production activities
Intermediate output	Number of new authorized patents	The results of corporate R&D activities
Final output	New product sales	R&D activity conversion results
	New product export sales	The degree of recognition of the entire technological production activity
Undesirable output	Number of unauthorized patents	Number of invention patent applications minus the number of regional invention patents

will not be approved and are considered as undesirable output. The second stage is the commercial stage, which is to transform the innovations of the first stage into sales. The manufacturing labor is the additional input that is only used at this stage. We choose new product sales and new product export sales as the final output of the overall system. The variables and indicators definitions are shown in Table 1.

This paper measures the high-tech industry by the panel data of 25 Chinese provinces from 2006 to 2015. The data sources are the annual *China High-tech Statistical Yearbook*, *China Science and Technology Statistical Yearbook*, *China Industrial Statistical Yearbook*, and *China Statistical Yearbook*. We estimated the results using MATLAB12.0 software.

From the province perspective in Table 2, the following findings can be concluded. First, there are 11 provinces, covering 44% of all provinces in China, with an overall efficiency mean value between 0.3–0.6 over the decade. There are six districts with overall efficiencies of more than 0.8, including Beijing, Tianjin, Shanxi, Guangdong, Yunnan, and Gansu. Among these, Beijing, Tianjin, and Guangdong have political and economic advantages which bring them rich resources regarding the high-tech industry as a core economic growth point. They have maintained high-efficiency thanks to industrial agglomeration, a talented labor pool, and policy support. However, Gansu, Yunnan, and Shanxi are those provinces that lack deep pockets, which means that they get maximum output with less input in the high-tech industry. On the contrary, there are five provinces, Hebei, Liaoning, Heilongjiang, Hubei, and Shaanxi, showing low overall efficiency in the high-tech industry ($\theta \leq 0.3$). These provinces are from the southwest, northwest, and central areas in China, sharing common characteristics of economic recession, resource depletion, and brain drain. Second, the average conversion efficiency is greater than the innovation stage efficiency. Among the 25 provinces we studied, only five (accounting for 20% of the total) are more efficient in the innovation phase than in the production transformation phase. This conclusion is completely contrary to the conclusion of Chen et al. (2018). We believe that our conclusion is more convincing and more realistic. The production and transformation capability of China's high-tech industries is undoubtedly very efficient around the world, which is related to China's low labor costs and scale of production.

By comparing the efficiency trends of the two substages, we can see that the trend of the production and R&D stages of the high-tech industry is consistent with the overall efficiency, which increases first and then decreases. However, the efficiency in the innovation stage witnessed a 10-year growth, but was always lower than that in the production transformation stage, indicating that the efficiency of the high-tech industry still depends on the efficiencies in the production stage in China. This could be related to the low labor cost and scale effect in the production stage.

We constructed a regional efficiency coordination index to analyze the relationship between overall efficiency and substage efficiency, where L = innovation efficiency/conversion efficiency −1.

Table 2. Mean innovation efficiency value of different provinces in China (2006–2015).

Province	Overall efficiency	Innovation stage efficiency	Conversion stage efficiency	Province	Overall efficiency	Innovation stage efficiency	Conversion stage efficiency
Beijing	0.951	1	0.951	Shandong	0.443	0.655	0.648
Tianjin	0.792	0.792	1	Henan	0.314	0.705	0.491
Hebei	0.265	0.378	0.715	Hubei	0.231	0.352	0.633
Shanxi	0.801	0.921	0.863	Hunan	0.384	0.711	0.539
Liaoning	0.256	0.495	0.480	Guangdong	1	1	1
Jilin	0.383	0.579	0.652	Guangxi	0.596	0.753	0.767
Heilongjiang	0.403	0.443	0.936	Chongqing	0.557	0.680	0.841
Shanghai	0.709	0.795	0.891	Sichuan	0.282	0.474	0.602
Jiangsu	0.766	0.824	0.882	Guizhou	0.562	0.647	0.783
Zhejiang	0.400	0.650	0.566	Yunnan	0.971	0.974	0.997
Anhui	0.354	0.744	0.512	Shaanxi	0.361	0.440	0.871
Fujian	0.435	0.448	0.980	Gansu	1	1	1
Jiangxi	0.197	0.279	0.691	**Mean**	**0.537**	**0.670**	**0.772**

When $L \geq 0$, it indicates that the conversion efficiency is higher than the innovation efficiency in this region and vice versa; the closer L is to 0, the better the match between the conversion efficiency and the innovation efficiency is in this area. Taking the overall efficiency of each region as the abscissa and the efficiency coordination degree of each substage as the ordinate, we draw the efficiency coordination degree matrix of each province in China. We take the overall innovation efficiency as 0.6 and the coordination degree as 0 and 1 as the dividing lines, creating four categories of these 25 provinces in China. Type I (overall efficiency ≥ 0.6 and $0 \leq L < 1$) contains those nine provinces including Beijing, Tianjin, Shanxi, Yunnan, Guizhou, Guangdong, Jiangsu, and Shanghai, that maintain high overall efficiency while achieving two-process coordination. Type II (overall efficiency <0.6 and $L > 1$) contains Heilongjiang, Shaanxi, and Fujian, which perform poorly in both efficiency and coordination. Types III and IV are provinces that show good coordination with low overall efficiency. The difference between the two types is reflected in the comparison between innovation efficiency and conversion efficiency. Type III has a higher conversion efficiency while type IV has a higher innovation efficiency. Thirteen provinces fall into these two categories—more than half of the provinces in our study.

4 CONCLUSIONS

This paper constructs a new cooperative game theory two-stage network data envelopment analysis model with considerations of shared input, additional input, and undesirable output, following the peculiarities of the high-tech industry innovation and commercial process. The contributions of this paper are mainly proven in the following three points. First, this paper first applied cooperative game theory two-stage network data envelopment analysis with considerations of shared input, additional input, and undesirable output. Second, we have fully considered the characteristics of the high-tech industry and measured the performance efficiency of the high-tech industry in the last decade in China. The results are in line with the actual situation and have a reasonable economic explanation. Certainly, the factors of the high-tech industry are affected by policies, macroeconomic fluctuations, and other aspects which are worth exploring in depth.

ACKNOWLEDGEMENTS

The research is supported by National Natural Science Funds of China (Nos. 71473037).

REFERENCES

Chen Y, Du J, David Sherman H, Zhu J: DEA model with shared resources and efficiency decomposition. Eur J Oper Res 2010, 207:339–349.

Chen X, Liu Z, Zhu Q: Performance evaluation of China's high-tech innovation process: Analysis based on the innovation value chain. Technovation 2018, 74–75:42–53.

Cook WD, Hababou M: Sales performance measurement in bank branches. Omega 2001, 29:299–307.

Cook WD, Zhu J, Bi G, Yang F: Network DEA: Additive efficiency decomposition. Eur J Oper Res 2010, 207:1122–1129.

Kao C, Hwang S: Efficiency decomposition in two-stage data envelopment analysis: An application to non-life insurance companies in Taiwan. Eur J Oper Res 2008, 185:418–429.

Liang L, Cook WD, Zhu J: DEA models for two-stage processes: Game approach and efficiency decomposition. Naval Research Logistics (NRL) 2008, 55:643–653.

Seiford LM, Zhu J: Modeling undesirable factors in efficiency evaluation. Eur J Oper Res 2002, 142:16–20.

Yu M, Lin ETJ: Efficiency and effectiveness in railway performance using a multi-activity network DEA model. Omega 2008, 36:1005–1017.

Computational Social Science – Luo, Ciurea & Kumar (eds)
© 2021 Taylor & Francis Group, London, ISBN 978-0-367-70193-2

Empirical analysis of the mechanism of the influence of internal control on financial performance

X.M. Sun*, X. Xian, C. Zhang, H. Yu & X. Liu
Canvard College, Beijing Technology and Business University, Beijing, China

ABSTRACT: As a product of the network era, the electronic technology industry has become an important driving force in China's national economy. Based on the sample data of several listed companies in the electronic technology industry, this paper obtains the research results on the impact of internal control on financial performance through empirical analysis.

Keywords: Electronic technology industry, Financial performance, Internal control.

1 QUESTIONS RAISED

Financial scandals occur among the world-famous large enterprises one after another, bringing heavy blows to the capital market and enterprises. The SOX Act came into existence in July 2002 through the COSO Management Committee of the United States, marking the fundamental change in the legal thought of American Securities: from disclosure to substantive management.

1.1 *Empirical research*

At the end of 2018, Kangmei Pharmaceutical Co., Ltd., a listed company, committed income fraud by means of false bank documents and business vouchers. At the theoretical level, this paper selects 55 electronic technology enterprises as the research objects, studies the influencing mechanism of internal control on financial performance, and provides a reference for related research into electronic technology enterprises.

2 THEORETICAL ANALYSIS AND RESEARCH HYPOTHESIS

2.1 *Theoretical analysis*

2.1.1 *Connotation and evaluation of internal control*
The DIB internal control index includes five indicators: risk supervision, risk assessment, internal environment, information, and communication and control activities, so it can comprehensively measure the quality of internal control. Since its publication, the index has been widely recognized by the industry and wider society. Therefore, this paper uses the internal control index of the DIB database as the standard to quantify the internal control quality of enterprises. The value range of the index is 0–1000. The larger the value, the better the internal control level of the enterprise.

2.1.2 *Financial performance indicators*
There are two kinds of standards for enterprises to measure their financial performance, one is accounting indicators, and the other is market indicators. In this paper, it is based on the ratio of the company's profitability to conduct further analysis, so as to judge the profitability of the enterprise.

---

*Corresponding author

2.1.3 Asset liability ratio

The asset liability ratio is the percentage of the total liabilities of the enterprise and the total assets of the enterprise. It indicates how much of the company's assets are in debt due to the investment of creditors, and reflects the ability of the enterprise to repay its debts. The higher the ability is, the more stable the risk level of the enterprise's loans can be reflected to an extent. According to the above point of view, we have the following hypothesis: the degree of internal control is positively proportional to financial performance.

2.1.4 Internal supervision

When enterprises strengthen the level of internal control, through continuous inspection of the implementation and improvement of the internal control mechanism, corresponding supervision is carried out on the designated aspects in the operation and management activities, and the corresponding supervision and inspection report can be obtained, which can have clear evaluation indicators for their own operation and management, and realize the most effective evaluation. This method not only improves the cohesion between enterprise departments, but also has a positive impact on the internal control mechanism and the cost is controlled effectively, to maximize the economic benefits of enterprises, and then improve the financial performance of enterprises.

2.2 Research design

2.2.1 Variable design

Explained variable: due to the uncertainty of China's capital market, compared with the market index and accounting index, the financial performance of enterprises can be more accurately reflected. Therefore, the rate of return on total assets is adopted to quantify the financial performance of enterprises, so as to avoid the impact of market changes. Explanatory variable: we plan to refer to the parameters of the DIB company and use its internal control index (IC) to judge the specific situation of its enterprise's internal control level. Control variables: in order to analyze the accuracy and authenticity of the results, this paper selects the enterprise size (size = the natural logarithm of the total assets at the end of the period) and the asset liability ratio (Lev = total liabilities at the end of the period/total assets at the end of the period) as the control variables.

2.2.2 Model design

Based on the above assumptions and variable design, the model is established as follows:

$$ROA = \beta 0 + \beta 1\ IC + \beta 2\ Size + \beta 3\ Lev + \varepsilon$$

where β i is a constant and ε is a residual.

3 EMPIRICAL ANALYSIS ON THE MECHANISM OF THE INFLUENCE OF INTERNAL CONTROL ON FINANCIAL PERFORMANCE

3.1 Descriptive statistics

It can be seen from Table 1 that in the data samples of 55 sample companies of the electronic technology industry from 2014 to 2017, the minimum value of return on capital (ROA) is −1.324100, the maximum value is 0.247900, the average value is 0.02754000, and the standard deviation is 0.10459094, which indicates that there are differences in the financial performance of China's electronic technology industry enterprises. However, the internal control index (IC) of these enterprises during 2014–2017 is 0, the maximum is 846.0500, the average is 641.187364, and the standard deviation is 121.9052105. So we can find that the internal control level of the electronic technology industry is good, however, the internal control level within the industry is quite different. The minimum size of the company is 18.478785, the maximum is 26.268868, and the average value

Table 1. Descriptive statistics of data.

Descriptive statistics

	Number of cases	Minimum value	Maximum value	Average value	Standard deviation
ROA (%)	220	−1.324100	0.247900	0.02754000	0.104590948
IC	220	0.000000	846.050000	641.18736364	121.905210546
Size	220	18.478785	26.268868	22.34642811	1.402432085
Lev (%)	220	0.094400	0.983900	0.41297091	0.189929386

Data sources: Ruisi database and Guotai Junan database.

Table 2. Variable correlation analysis.

Relevance

	ROA (%)	IC	Size	Lev (%)
ROA (%)	1			
IC	0	1		
Size	0.112	0.272**	1	
Lev (%)	−0.298**	0.013	0.469**	1

Data sources: Ruisi database and Guotai Junan database.

Table 3. Variable regression analysis.

R	R-squared	Adjusted R-squared	Standard estimate error	F	Significance
0.573[a]	0.329	0.319	0.086296091	35.233	0.000[b]

	B	Standard error	β	T	P	tolerance	VIF
IC	0.000	0.000	0.417	7.131	0.000	0.909	1.100
Size	0.014	0.005	0.181	2.739	0.007	0.709	1.410
Lev (%)	−0.214	0.035	−0.389	−6.101	0.000	0.765	1.306

Data sources: Ruisi database and Guotai Junan database.

is 22.34642811, which means that the size difference among the selected sample data individuals is not large; the minimum value of the asset liability ratio (Lev) is 0.094400, the maximum is 0.983900, the average is 0.41297091, and the standard deviation is 0.189929386, which indicates that the financial risk of listed enterprises in electronic technology industry is large.

3.2 Correlation analysis

According to the correlation analysis results of the indicators reported in Table 2, the correlation coefficient between IC and ROA is 0.461, which can temporarily confirm the above hypothesis: there is a positive correlation between internal control and corporate financial performance.

3.3 Regression analysis

According to the results of the regression analysis in Table 3, we can see that the correlation coefficient of internal control index is 0.417, which means that if the internal control of enterprises in the electronic technology industry is increased by 1 unit, ROA will be increased by 0.417. The T value of internal control is 7.131, and the corresponding P value is 0.000, which is lower than 0.01, indicating that there is a positive correlation between internal control and ROA at the significance level of 1%. This shows that the more effective the internal control of the electronic technology industry, the better its financial performance; the correlation coefficient of company size is 0.181, which means that when the control variable of company size increases by 1%, the financial performance can be further improved by 1.239986%. However, the correlation coefficient of asset liability ratio is −0.389. As a control variable, the asset liability ratio has not passed the significance test, which shows that there is no significant correlation between the asset liability ratio and the financial performance of listed companies in the electronic technology innovation industry.

4 ANALYSIS OF THE MECHANISM OF THE INFLUENCE OF INTERNAL CONTROL ON FINANCIAL PERFORMANCE

4.1 Qualitative analysis of data results

4.1.1 Qualitative analysis of descriptive analysis results

Through the descriptive analysis of sample data, we can draw the following conclusions: first, the scale difference among sample enterprises is not very obvious. Second, the financial risk of sample enterprises in the electronic, science, and technology innovation industry is relatively large. Scientific and technological innovation enterprises are faced with higher risks, and it is difficult for companies to obtain credit in securities institutions. Therefore, it is difficult to obtain external capital investment. Borrowing funds through other channels can bring about a financial leverage effect and higher interest payments.

4.1.2 Qualitative analysis of correlation analysis and regression analysis results

The results of correlation analysis and regression analysis show that the higher the internal control index is, the higher the rate of return on assets will be, and the stronger the ability of enterprises to obtain income. When the internal control system is more effective, the management level is relatively high and the utilization degree of enterprise resources is higher. In terms of material procurement, the higher level of internal control means that the cost of materials can be better controlled. In terms of funds, a better level of internal control makes the utilization rate of funds higher. In terms of corporate reputation, stronger internal control can support and guarantee the good image of the enterprise in the public mind, and the brand value of the enterprise can also be maintained and benefit from this.

4.2 Influence mechanism

4.2.1 Internal control activities

Through the establishment of a reasonable organizational structure, development strategy, and human resources policies, it is helpful to enhance the efficiency of business activities and competitive strength, and then enhance the value of enterprises and improve financial performance.

4.2.2 Risk assessment

By establishing an information system combined with its own characteristics, strengthening risk assessment and control, and identifying its potential internal and external risks, the corresponding risk-coping strategies are determined. Enterprises can prevent internal control failure, strengthen internal control, and improve enterprise financial performance. This is particularly important for the electronic technology industry which is subject to high financial risk and high uncertainty.

4.2.3 *Control activities*

The implementation of authorization approval control, budget control, property protection control, and other measures can strengthen the enterprise control activities, strengthen its control and protection of its own production and operation activities, and improve its financial performance.

5 CONCLUSIONS AND SUGGESTIONS

Through the data descriptive statistics, correlation analysis, and regression analysis of 55 listed electronic technology companies from 2014 to 2017, we can draw the following conclusions: the relationship between DIB internal control index and financial performance of listed companies in the electronic technology industry has a positive correlation, the more efficient internal control, the better financial performance of enterprises. Therefore, the following suggestions are put forward under this premise. From the perspective of the enterprise itself, the focus of strengthening the level of internal control is to attach importance to the degree of control for each employee; moreover, the standard of rewards and punishments should be established to check the implementation of internal controls of each employee.

Externally, first, the national finance, taxation, and audit and other government regulatory departments can implement their own responsibilities and strengthen the supervision of enterprise management; second, the external supervision departments should strengthen the punishment level of the third party of audit and strictly pursue the responsibility.

ACKNOWLEDGMENT

This paper is supported by 2019 Beijing Higher Education Undergraduate Teaching Reform and innovation project: Research and practice of international talents training of finance major in the new era (No.: 201913630001).

REFERENCES

[1] Mulan Xia, "Current research status and comments on financial restatement based on literature analysis of domestic journals" Hebei Enterprises 2018, PP.21–22.

[2] Min Yuan, "Financial Statement Restatement and Internal Control Evaluation of Financial Statements: Based on Dell Case Analysis", Accounting Research, 2012 (04), PP. 30–37+96.

[3] Miklós Pakurár, Hossam Haddad, János Nagy, József Popp, Judit Oláh. The Impact of Supply Chain Integration and Internal Control on Financial Performance in the Jordanian Banking Sector. Sustainability, 2019,11 (5).

Computational Social Science – Luo, Ciurea & Kumar (eds)
© 2021 Taylor & Francis Group, London, ISBN 978-0-367-70193-2

The attraction of horoscopes: A consensual qualitative research on astrological personality description

Y. Wu
School of Marxism, Huazhong University of Science and Technology

Z.Z. Chen
Mental Health Center, Central China Normal University

ABSTRACT: Astrological personality description is a main attraction of astrology and while sharing several common purposes with psychological personality profiling, it gained greater popularity
among youth and laypersons. We conducted a consensual qualitative research on the astrological
personality descriptions to explore whether there is a special style of personality description that
may have contributed to its continued popularity. We collected texts from seven influential Chinese
astrology websites or astrology sections from internet portals. Our analysis formulated a framework
of statement style of astrological personality descriptions including seven domains with seventeen
categories. Several domains unique to horoscope could be the difference between Astrological
personality description and personality psychology. The internal structure, domain implication and
their relationship with horoscope attraction are discussed.

1 INTRODUCTION

Despite the rigorousness and scientific underpinnings of modern personality psychology, astrology
in its various forms remained a popular leftover from earlier times, a striking cultural phenomenon
that has garnered increasing research attention in China (Dong & Kong, 2003; Su & Zheng, 2005;
Wu, 2013) and the West (Sax, 1999) alike.

   Given the general assumption that astrology has no scientific validity, its continued popularity
has sparked spirited debate concerning the roots of its attraction (Dong & Kong, 2003; Wu, 2013).
Our purposes here are to conduct a systematic analysis of the astrological personality description
texts and to build on this analysis to offer a reconsideration of the relationship between psychologist'
and astrologist' personality description, one that addresses their fundamental differences, identifies
the aspects that gave astrology glamor and attraction, and helps the psychologists to refine their
ways of presenting personality descriptive texts.

1.1 *Personality description in personality psychology*

Putting a premium on accuracy and rigorousness, mainstream personality psychologists usually
incline toward a strict correspondence with the body of empirical evidence accrued concerning
certain personality dispositions and are generally careful so as not to "over-interpret" the test
score. In the end, to keep the accuracy of the description, simple, decontextualized, and atomistic
sentences that erased unique individuality and personal significance are especially favored, which,
unfortunately, often cause ambiguity to different people (Mayer, 1977). Gergen, Hepburn and
Fisher (1986) found through experiments that people with high linguistic sophistication could
easily interpret the content of a same personality statement in personality test as describing any
personality trait; even statements opposite in meaning can be construed as instantiation of the same
trait. Reading personality description of this kind, a sophisticated and self-reflective layperson
would find it superficial and wanting in profundity, and more so if the targets of description

are the readers themselves, because people often consider their own personality as richer, more multifaceted and less predictable than that of others (Beer & Watson, 2008; Mirels, Stevens, Greblo, & Yurek, 1998; Sande, Goethals, & Radloff, 1988).

1.2 *Personality description in astrology*

Unlike mainstream personality psychology, the symbolism of astrology is often invoked as a metaphor in showing meaning and preserving individual wholeness in the process of personality description (Kozlova, 2012; Mayer, 1977; Sax, 1999). Moreover, the postulate that one's personality stemmed from mystified horoscopic influence suggest a reified and concrete personality "entity" from which an individual's behavioral characteristics emanate, a very intuitive ontological notion that coincided with the human propensity to reify constructs that are essentially metaphors (Barclay, 1997). All these factors could lead to a holistic, metaphor-laden, and coherent (albeit inaccurate) narrative of a person's personality, and even by a cursory inspection one could notice its differences with psychologists' texts. The wording, pacing, style, and the terms are all different.

Therefore, it is likely that personality description in astrology may possess certain linguistic and paralinguistic features that confer credibility and prestige, making the description attractive to the layperson. However, the focus of early research has been the accuracy and reliability of astrological personality description, as well as the possible mechanisms. Psychologists systematically compared the personality description given by astrology and the personality test, and found that despite a paucity of positive results that are not replicable, the astrological personality description does not correspond with one's real personality (Carlson, 1985; Clarke, Gabriels, & Barnes, 1996; Fichten & Sunerton, 1983; Hentschel & Kiessling, 1985; Jackson & Fiebert, 1980; Su & Zheng, 2005; Wyman & Vyse, 2008). Another branch of studies considered the possible mechanisms underlying the accuracy of astrological personality description, such as the weather (Tyson, 1977; Veno & Pamment, 1979), the exposure to astrological vocabularies (Dong, 2004; M. Hamilton, 2001; M. M. Hamilton, 1995) and cognitive bias (e.g. positive test strategy, Davies, 2003). Most of these attempts culminated in failure, yet Hamilton (1995) did find that people familiar with astrology deemed astrological personality description more accurate than did those less familiar with astrology. In sum, the studies from early research indicated that the accuracy of personality description in astrology cannot be validated, rendering the issue of astrology's perceived credibility and attractiveness among layperson even more relevant.

1.3 *Contributors to the belief in astrological personality description*

Both intra-individual and contextual factors may influence people's belief in astrology. The bulk of the research on intra-individual factors include, for instance, studies on gender (Sosis, Strickland, & Haley, 1980), self-attribution or Barnum effect (Glick, Gottesman, & Jolton, 1989; Meehl, 1956), locus of control (Fichten & Sunerton,1983; Sosis et al., 1980), and self-uncertainty (Lillqvist & Lindeman, 1998; Lindeman, 1998). The contextual factors also include the textual, or linguistic, features characterizing the astrological products. Building on Adorno and colleagues' content analysis of astrological products and their hypotheses, Svensen and White (1995) collected astrological fortune-telling texts published by the astrologer Carroll Righter in Los Angelis Times over 3 months, and tested these hypotheses using keywords methods. The results confirmed the hypotheses (features such as a tone of authority and anxiety-arousal) and replicated previous research on astrology columns (Weimann, 1982).

However, this line of research has several limitations. Most empirical studies on astrology column only used simple content-analysis methods and did not consider the ways in which astrologically derived information is presented (Svenson & White, 1995; Weimann, 1982). Besides, past qualitative studies on astrology tend to consider the astrology column that are mostly about fortune telling, to our knowledge, no study has qualitatively analyzed the astrological personality description.

1.4 *The present study*

It is our purpose in this study to explore the textual features in astrological personality description, and the ways in which personality traits or characteristics are articulated and verbalized. We are especially interested in mythical reference and contradictory statements, as well as any other potential features.

So far as we know, this is the first qualitative study on astrological text. Consensual qualitative research (CQR; Hill et al., 2005; Hill, Thompson, & Williams, 1997) was employed for the current study. The aim of this method is to generate conceptualizations, such as structure, domain and categories, of the materials through systematic analysis of participants' experience or observations. As an experiential method, CQR only considers the information explicitly presented in material, and did not actively infer any "hidden" meaning or assumptions as some other methods did (Rennie, 2012). In addition, CQR highlights the participation of multiple researchers. The method expressly requires researchers reaching a consensus before next steps should begin. In certain time points, external auditors who have never participated in discussion will independently review all the progress, lest group dynamics should divert the direction of study (Hill et al., 1997).

2 METHOD

2.1 *Text selection*

Consistent with the aim of this study, we select texts devoted to describing personality of people with different sun sign, in main internet portals of China. The criteria are that the website should contain texts of all twelve signs and the personality description should be about the people with a particular sun sign, but not discussion about any subtype (e.g. text entitled "Aries with type O blood" or "Aries born in the year of Ox", the latter referring to an intersection of Western and Chinese zodiac). To avoid the unnecessary noise in language style and theme, we only select websites or columns where description texts of all twelve signs are included. In the end, the selected texts are from astrology columns of internet portals or astrology websites, which are Sina Constellation Channel (http://astro.sina.com.cn/), Constellation House (http://www.xzw.com), Tencent Constellation Channel (http://astro.fashion.qq.com/), Sohu Constellation Channel (http://astro.sohu.com), No. 1 Constellation Network (http://www.d1xz.net), NetEase Constellation Channel (http://lady.163.com/astro/) and Ifeng Constellation Channel (http://astro.ifeng.com).

2.2 *Text preparation*

To keep consistency, and to ensure that the contents of the texts are related with personality description, we deleted irrelevant contents from the texts. Our guiding principle during text preparation is that the remained texts should be, in any form, descriptions of the personality of people born with different signs. There are, however, several ambiguous cases, in which the contents are potentially or theoretically related with one's personality. For example, one's potential "partner" is theoretically contingent on one's personality. In these cases, so long as they did not actually depict one's dispositions, we delete these sentences (e.g. "You should find a strong person as partner").

2.3 *Procedure of analysis*

CQR data analysis relied heavily on reaching consensus by multiple researchers, in order to reduce implicit bias of researchers and to better understand the complex issues at hand (Hill et al., 1997). To obtain varied perspective, it requires a research team and independent auditors. In the current study, the researcher team consisted of three members and two external auditors. The research team received trainings on general qualitative research methodology and learned CQR method by carefully reading the Hill et al. (1997) article, in conjunction with several research studies using CQR methods. The three members varied in their familiarity with astrology, with one member reading about astrology regularly and another two generally less familiar with anything astrological. This combination could obviate the bias stemming either from too much or too little prior knowledge about astrology.

The three members meet on a regular basis and discuss their understanding about the texts with mutual respect. Inconsistent understandings of the texts were thoroughly discussed until reaching a consensus. The texts were analyzed by the aid of Microsoft Excel and Nvivo 8.

Hill et al. (1997) described the key procedures of CQR, including: (1) develop domains for the data, (2) extract core ideas, and (3) cross-analysis of the domains across cases. However, the original procedure was largely for conducting CQR in counseling settings, the main sources being interviews with participants. In current study, the data sources are online texts from different websites. Considering that the texts from the same website have similar layout and linguistic style, we treated each website as a case, equivalent to the concept of participant in original CQR procedure. Moreover, since our aim is to analyze the style with which astrological texts describe one's personality, we should focus more on the linguistic style of personality statements than on the actual content of the personality traits being described. To ease the analysis, we decided after a discussion that we set the level of analysis at sentence level, discarding any additional information at the level of sentence group or paragraph level. Thus, each sentence should belong to a domain. Only in the rare occasions where a sentence is too short to have independent meanings, should two or more sentences be seen as one unit. If a sentence should belong to two domains, we first redefine the domain so as to make sure that one sentence only belongs to one domain; if that fails, we put the sentence into a temporary domain named "double membership". At the end of domain development, we reevaluated these sentences. The rare cases whose domain membership are still in dispute will be put into the category of "other".

After domain development, we consulted Hill (Hill, personal communication) and decided not to extract core ideas of the sentences in each domain, because the single sentence is concise and simple enough for analysis. The research team followed suggested procedure of CQR, and conduct cross-analysis and stability check. After the completion of these steps, two external auditors reviewed our domains and categories to ensure that we categorized the sentences correctly and the naming of domains and categories is appropriate.

3 RESULTS

We split the texts into 2237 sentence units, and our data analysis yielded seven domains or main styles of describing personality. After reevaluation of controversial sentence unit, 25 sentence units remained at "double membership" and another 12 remained uncategorized due to its conciseness or vagueness of meaning. Consistent with Hill et al. (1997, 2005), 16 categories were classified as belonging to "general" findings (abbreviated as "G"; true for texts from all seven websites); five categories were classified as belonging to "typical" findings (abbreviated as "T"; as is the case for texts from four to six websites); and one category classified as belonging to "variant" findings (abbreviated "V"; as is the case in texts from two to three websites).

3.1 Independent definition

This domain refers to the sentences that define one's personality traits by abstract and summarizing words or concepts and may give examples or emphasize the extent to which a person fits the trait description. Four categories are further categorization of the sentences by whether they emphasized the degree of trait expression, or whether they gave examples. Directly describing or summarizing one's personality characteristics, this type of statements is the most fundamental type in personality description.

3.2 Transcendental analogy

This domain refers to the sentences that liken one's personality to certain transcendental symbolic system (e.g. the planetary orbit, constellation), and may assert that this system determined one's personality. The domain consisted of two categories, the simple and the standard. The former merely describe the structure and mechanisms of the symbolic system and its changes, whereas the latter, aside from similar contents in the simple transcendental analogy sentence, made further assertions as to the influence of factors within the system on one's personality. Considering that personality concepts in psychology are generally nominal in nature, the explicit reference to transcendental influence of this domain might be a characterizing feature of astrological personality description.

3.3 Explanatory statement

Sentences in this domain analyze personality phenomenon present in persons with certain sign, explaining the source of their personality (trait or dynamics). The three categories differed in the types of reason given. Sentences belonging to independent explanation sought to explain personality phenomenon by clearly giving a source, whereas indirect explanation use indirect statements in its explanation. In cases where both statements are used, we treat them as instances of multiple explanations.

3.4 Conditional statement

Sentences in this domain state the prerequisite of a personality phenomenon, such as "If your family life is very happy and satisfying, then you would be faithful to your love". They touch upon the situational dependency of the personality traits they are describing. This type of statements is relatively simple and we did not categorize its sentences in cross-analysis.

3.5 Comparative statement

Comparative statements describe people's personality by comparing them with other signs or other subtypes. The standard of comparison includes but are not limited to other signs or different blood types within the same sign. The numbers of standards constitute the two categories, with a category referring to those having only one standard of comparison and another referring to those having multiple standards. The multiple standards in the latter type could be explicitly stated, but it could also include cases where it is not, e.g. "people with _____ (name of a sign) is the most _____ (name of traits) in all twelve signs".

3.6 Dialectical statement

This domain consists of four categories, which are transition, contradiction, situational transition and proportion limitation. The similarity of them lies in that they all emphasize that a person's personality has multiple aspects, whereas the difference among the categories is the relationship between these aspects.

Transition refers to the personality statements that transit from one trait to another, and the two traits are not intrinsically related. The two parts of the statement are usually linked by using a conjunction word, such as "but". Contradictory statements, on the other hand, highlight that the individual embodied the two opposite of a same personality trait, and sometimes further mention that the two opposites can be reconciled or interconvertible in due situation. Another manifestation of contradictory statements is to point out the contrast between an inner layer of a person's disposition and the outside appearance or manifestation, such as "you appear cold and chilly but inwardly you are warm-hearted and kind".

Similar to the transition, statements that falls within the category of situational transition describe a situational restraint on the manifestation of a personality trait. At last, proportion limitation is a little different. Statements of this category first state that an individual has embodied one personality trait, and in the other half, it changes the extent or proportion of the manifestation of the trait. Usually the trait first described would be put to an extreme, often with an opposite valence.

3.7 Indirect statement

The structure of indirect statements is relatively loose. Statement sentences in this domain do not directly point out the personality trait one embodies, but only describe various personality phenomenon, and it is left to readers to judge which personality trait is at play. We further categorized the indirect statements into six categories by the type of phenomenon they described, which are need, schema, attitude, behavior, emotional experience and potential.

The category of need includes sentences describing the specific need of an individual or giving suggestions of how to cater to these needs. Statements in the category of schema invoke a personality schema (e.g. "you possess a leader's boldness") to suggest the individual's personality. Similarly, statements in the categories of emotional experience and potential describe one's emotional characteristics or experiences, and one's behavior potential. It should be noted that the potential is different from ability, with the latter more aligned with problem solving than the potential of a motivational dynamics.

4 DISCUSSION

4.1 Stylistic features of astrological personality description

In the present qualitative study, we conduct a consensual qualitative research on astrological personality descriptive texts, and through discussions by the research team, we developed a classification consisting of 7 domains and 21 categories. Auditors in our study checked the classification during analysis process, and the stability check in the final stage confirmed the representativeness of our model.

Aside from the simple and direct description of a personality trait (independent definition), astrological texts may depict one's personality by comparing with other zodiac signs (comparative statements), and for complex and nuanced personality phenomenon, may set conditions for its happening (conditional statements). Another type of device may indirectly describe one's personality by describing one's need, attitude, behavior, emotional experience, as well as potentials (indirect statements). Moreover, the dialectical statements and transcendental analogy could both make the texts appear subtler and nuanced in contents and the ways in which they describe a personality profile. The impact of various styles of language on social cognition is well validated in social psychology, and a full discussion is far beyond the scope of the current study. Here we just focus on the several most pronounced aspects of astrological texts, as compared with psychologists' personality profiling.

Characteristic of astrological texts, the transcendental analogy could clearly reflect its astrological lineage. This type of statements would explicitly ascribe one's personality phenomena and traits to one's membership of a specific sign or to various horoscopic and meteorological factors that may appear mysterious and often metaphysical. For example, a typical sentence would say, "Hot summer gives Leos courage". This way of presentation may suggest that the theoretical postulates of astrological personality description did not treat the name of each sun sign and its corresponding personality manifestation as merely symbolic; quite the contrary, it suggest that the astrological texts assumed a transcendental, out-of-experience essence (e.g. planetary movements, angles or positions) of each personality type that determined, at least partly, the personal disposition. This mystification might be one of the attractions of astrological personality description.

In addition, sentences using transcendental analogy attribute the manifestation of personality to external forces that are uncontrollable and mysterious, which dovetail nicely with the layperson's general impression that personality is complex, mysterious and inscrutable (Sande et al., 1988). Lacking knowledge in personality and social psychology, the public may lose sense of control when encountering personality phenomenon that is influenced by innate dispositions. On the other hand, the situational influence on people's seemingly idiosyncratic behavior is often wrongly ascribed to one's stable personality (such as in fundamental attribution bias; Ross, 1977), probably leading one to perceive personality as a constantly changing enigma. To make the situation even worse, for people less familiar with psychological theories, the unconscious motivations and implicit cognitive processes (Payne & Gawronski, 2010) that have been psychologists' focus since Freud's time would also appear especially baffling, as these factors are generally not accessible through introspection. Indeed, study showed that astrologers have exploited the layperson's propensity to mystify one's personality. Svenson and White (1995) found that in astrological fortune-telling, astrologers tried to appear authoritative in their language use, and tended to suggest that uncontrollable external forces would determine the reader's fate.

Another source of astrological texts' attraction might be the metaphysical nature of astrology, obviating the necessity to keep the concordance between the personality descriptive texts and reality. Needless to consider whether its description fits one's personality, astrological personality description may make sophisticated personality statements that are theoretically consistent and complete (Kozlova, 2011; Sax, 1999), and thus appearing "smooth" and more attractive.

4.2 Implications for psychological personality description

It is critical for psychologists to learn certain ways of presenting personality in natural languages, to answer the expectation of layperson. Currently, personality psychologists, especially in China, place a higher premium on summarizing personality phenomenon through mathematical methods, and are relatively less concerned about the interpretation of personality test score. It is usually left to readers to infer as well as to construct a fuller picture of their personality by reading the introduction of different dimensions and their test scores on that dimension, together with a short description of behavioral characteristics of people scoring extremely high or low in each continuum. These short descriptions are often prosaic cluster of behavioral characteristics, and little or no effort is devoted to integrating a trait and its corresponding behavioral pattern in description texts.

5 CONCLUSION

Seven categories of stylistic features were identified in astrological personality description texts.
Stylistic features of the astrological personality description texts may have contributed to the surface attractiveness of astrology.

ACKNOWLEDGEMENT

This research was supported by the Fundamental Research Funds for the Central Universities (grant No. 2019WKYXQN18). We wish to thank T. J. Xu and Y. X. Duan for their contribution to the present study.

REFERENCES

Barclay, M. W. (1997). The metaphoric foundation of literal language: Towards a theory of the reification and meaning of psychological constructs. Theory & Psychology, 7(3), 355–372.

Beer, A., & Watson, D. (2008). Asymmetry in judgements of personality: Others are less differentiated than the self. Journal of Personality, 76(3), 535–559. doi: 10.1111/j.1467-6494.2008.00495.x

Carlson, S. (1985). A double-blind test of astrology. Nature, 318(5), 419–425. doi: 10.1038/318419a0

Clarke, D., Gabriels, T., & Barnes, J. (1996). Astrological signs as determinants of extroversion and emotionality: An empirical study. The Journal of Psychology, 130(2), 131–140.

Davies, M. F. (2003). Confirmatory bias in the evaluation of personality descriptions: Positive test strategies and output interference. Journal of Personality and Social Psychology, 85(4), 736–744. doi: 10.1037/0022-3514.85.4.736.

Dong, X. (2004). The impact of astrology on Chinese students: A study of personality psychology. (Unpublished master's thesis). East China Normal University, Shanghai.

Dong, X. & Kong, K. (2003). Xingxiangxue de renge xinlixue yanjiu [Studies on astrology in personality psychology]. Psychological Science, 26(4), 701–703.

Fichten, C. S., & Sunerton, B. (1983). Popular horoscopes and the "Barnum Effect". The Journal of Psychology, 114(1), 123–134. doi: 10.1080/00223980.1983.9915405.

Gergen, K. J., Hepburn, A., & Fisher, D. C. (1986). Hermeneutics of personality description. Journal of Personality and Social Psychology, 50(6), 1261–1270. doi: 10.1037//0022-3514.50.6.1261.

Glick, P., Gottesman, D., & Jolton, J. (1989). The fault is not in the stars: Susceptibility of skeptics and believers in astrology to the Barnum Effect. Personality and Social Psychology Bulletin, 15(4), 572–583. doi: 10.1177/0146167289154010.

Hamilton, M. (2001). Who believes in astrology? Effect of favorableness of astrologically derived personality descriptions on acceptance of astrology. Personality and Individual Differences, 31(6), 895–902. doi: 10.1016/S0191-8869(00)00191-4.

Hamilton, M. M. (1995). Incorporation of astrology-based personality information into long-term self-concept. Journal of Social Behavior & Personality, 10(3), 707–718.

Hentschel, U., & Kiessling, M. (1985). Season of birth and personality: Another instance of noncorrespondence. The Journal of Social Psychology, 125(5), 577–585. doi: 10.1080/00224545.1985.9712031.

Hill, C. E., Knox, S., Thompson, B. J., Williams, E. N., Hess, S. A., & Ladany, N. (2005). Consensual qualitative research: An update. Journal of Counseling Psychology, 52(2), 196–205. doi: 10.1037/0022-0167.52.2.196.

Hill, C. E., Thompson, B. J., & Williams, E. N. (1997). A guide to conducting consensual qualitative research. The Counseling Psychologist, 25(4), 517–572. doi: 10.1177/0011000097254001.

Jackson, M., & Fiebert, M. S. (1980). Introversion-extroversion and astrology. The Journal of Psychology, 105(2), 155–156. doi: 10.1080/00223980.1980.9915145.

Kozlova, E. (2011). A phenomenological inquiry into the process and effects of finding meaning with astrological symbolism. (Unpublished Doctoral Dissertation), Institute of Transpersonal Psychology, Palo Alto, California.

Lillqvist, O., & Lindeman, M. (1998). Belief in astrology as a strategy for self-verification and coping with negative life-events. European Psychologist, 3(3), 202–208. doi: 10.1027/1016-9040.3.3.202

Lindeman, M. (1998). Motivation, cognition and pseudoscience. Scandinavian Journal of Psychology, 39(4), 257–265. doi: 10.1111/1467-9450.00085.

Mayer, M. H. (1977). A holistic perspective on meaning and identity: Astrological metaphor as a language of personality in psychotherapy. (Doctoral Dissertation), Saybrook University, San Francisco.

Meehl, P. E. (1956). Wanted-A good cookbook. American Psychologist, 11(6), 263–272. doi: 10.1037/h0044164.

Mirels, H. L., Stevens, F., Greblo, P., & Yurek, D. L. (1998). Differentiation in personality description of the self and the others. Personality and Individual Differences, 25(4), 663–681. doi: 10.1016/s0191-8869(98)00084-1.

Payne, B. K., & Gawronski, B. (2010). A history of implicit social cognition: Where is it coming from? Where is it now? Where is it going? In B. Gawronski & B. K. Payne (Eds.), Handbook of implicit social cognition. New York: Guilford Publication, Inc.

Rennie, D. L. (2012). Qualitative research as methodical hermeneutics. Psychological Methods, 17(3), 385–398. doi: 10.1037/a0029250.

Ross, L. (1977). The intuitive psychologist and his shortcomings: Distortions in the attribution process. In L. Berkowitz (Ed.), Advances in experimental social psychology (Vol. 10, pp. 173–220). New York: Academic Press, Inc.

Sande, G. N., Goethals, G. R., & Radloff, C. E. (1988). Perceiving one's own traits and others': The multifaceted self. Journal of Personality and Social Psychology, 54(1), 13–20. doi: 10.1037//0022-3514.54.1.13

Sax, D. (1999). An astrological theory of personality. (Unpublished Doctotal Dissertation), California Institute of Integral Studies, San Francisco, California.

Sosis, R. H., Strickland, B. R., & Haley, W. E. (1980). Perceived locus of control and beliefs about astrology. The Journal of Social Psychology, 110(1), 65–71. doi: 10.1080/00224545.1980.9924223

Su, D. & Zheng, Y. (2005). Zhiyi xingzuo jueding renge tezhi de shizhengyanjiu [Astrological signs as determinants of personality traits: An empirical study]. Psychological Science, 28(1), 220–221.

Svensen, S., & White, K. (1995). A content analysis of horoscope. Genetic, Social, and General Psychology Monographs, 121(1), 5–38.

Tyson, G. A. (1977). Astrology or season of birth: A "split-sphere" test. The Journal of Psychology, 95(2), 285–287. doi: 10.1080/00223980.1977.9915891.

Veno, A., & Pamment, P. (1979). Astrological factors and personality: A southern hemisphere replication. The Journal of Psychology, 101(1), 73–77. doi: 10.1080/00223980.1979.9915054.

Weimann, G. (1982). The prophecy that never fails: On the uses and gratifications of horoscope reading. Sociological Inquiry, 52(4), 274–290.

Wu, Y. (2013). Social psychological mechanisms of horoscope-related behaviors: Effects of attitude and personality consistency (Unpublished master's thesis). Central China Normal University, Wuhan.

Wunder, E. (2003). Self-attribution, sun-sign traits, and the alleged role of favourableness as a moderator variable: Long-term effect or artefact? Personality and Individual Differences, 35(8), 1783–1789. doi: 10.1016/s0191-8869(03)00002-3.

Wyman, A. J., & Vyse, S. (2008). Science versus the stars: A double-blind test of the validity of the NEO five-factor inventory and computer-generated astrological natal charts. The Journal of General Psychology, 135(3), 287–300. doi: 10.3200/GENP.135.3.287-300.

Computational Social Science – Luo, Ciurea & Kumar (eds)
© 2021 Taylor & Francis Group, London, ISBN 978-0-367-70193-2

Construction of automated feedback system of business English correspondence writing based on discourse analysis

D.Q. Liu
School of English Language and Literature, Xi'an Fanyi University, Shaanxi, China

ABSTRACT: Some shortcomings of traditional writing feedback boost the development of automated writing feedback system. There are AWE system in America and www.pigai.org in China. However, there is lack of an automated writing feedback system designed especially for business English correspondence, which is one of the main courses of business English majors. The study is based upon several theories of linguistics and psychology, including reliability and validity, interlanguage, error analysis, etc. Moreover, several analytical methods are employed, including text analysis, latent semantic analysis, genre analysis, and qualitative and quantitative analysis, to carry out orderly quantization analysis of local contents of foreign trade terms, syntax characteristics, overall contents of generic structure, and top-level features of foreign business correspondence. This attempt provides the objective and feasible concept of and access to the construction of the business English correspondence intelligent evaluation system. It will undoubtedly play a positive role in improving students' writing ability and optimizing feedback modes of business English correspondence writing.

1 INTRODUCTION

As a specific branch of ESP (English for Specific Purposes), business English meets the communication requirements of workplace, involving all aspects of business activities. Business English correspondence is a functional and indispensable tool of delivering business information to trade parties in a written format like letters, e-mails, telegrams, telexes, faxes, etc. in the process of establishing trade relation, inquiry, offer, counter-offer, placing an order, payment, packing, shipment, insurance, complaint, claim, etc.

In the context of "the Belt and Road Initiative", China's external exchanges, especially its interactions with foreign corporations, have been on the increase day by day, which calls for more internationalized business English talents, and provides more unprecedented opportunities for the development of business English in China. According to 2019 statistics, more than 300 Chinese universities or colleges offer bachelor's degree courses for business English majors, one of which is business English correspondence. But how to improve the writing competence of business English correspondence?

Writing feedback is an essential link in the process of getting inspiration, modifying problems, and finally improving students' English writing level. Traditional feedback methods include teacher feedback, peer feedback, and self feedback. Their (especially teacher feedback) weaknesses lie in time consuming, slow speed, low efficiency, intense subjectivity, etc. Therefore, traditional writing feedback should give some place to a constant and automated writing feedback mode – automated writing evaluation (AWE).

2 PRIOR RESEARCH ON AUTOMATED ENGLISH WRITING FEEDBACK

Different from Automated Essay Scoring (AES), which is widely used for scoring essays in high-stakes testing like TOEFL, AWE systems are commonly used for lower-stakes writing instructions. It consists of two most critical components, a scoring engine and a formative feedback engine based on a branch of artificial intelligence called natural language processing. According to Vantage Learning (2007), AWE has been employed in American schools for more than 20 years, and there is significant improvement in writing performance and language skills. Grime & Warschauer (2010) presents that AWE simplifies classroom management. Teachers appear more relaxed. Students are more autonomous, they are more motivated to write and revise, and their writing portfolios are conveniently organized.

Many Chinese scholars and English teachers also published related primary research articles and secondary research articles. They affirm the positive role of AWE for teachers and students in English writing. But they find that there are vast differences in grammar and diction between Chinese students and native English students. That is to say, the AWE systems are not fully applicable in Chinese schools. Then in 2010, a website *www.pigai.org* was launched and quickly got good reviews. Based on the comparison between students' compositions and the corpus, it provides scores and comments by a certain algorithm. In this system, specific feedback and suggestions outweigh the ratings, because the former enables students to know how to make progress.

However, a problem with *www.pigai.org* is its limited corpus, which only covers standard English not English for Specific Purposes. Therefore, it is not a good choice for getting effective feedback for business English correspondence. To some extent, a system to offer intelligent and thorough feedback to business English correspondence is still a blank field.

This paper serves to develop an automated system for business English correspondence feedback, which will have significance in standardizing and improving students' business English correspondence writing, in promoting teachers' business English correspondence teaching, and in evaluating large-scale business English writing tests like BEC, ETIC, LCCI business English test.

3 RELATIVE THEORIES OF AUTOMATED ENGLISH WRITING FEEDBACK

As a matter of fact, the scientificity and feasibility of automated writing feedback can find support in many second language acquisition (SLA) theories and psychological theories.

3.1 *Reliability and validity*

Scoring validity is the critical point of all language tests, which plays a vital role in the reliability of scoring results and the feasibility of scoring procedures.

3.2 *Interlanguage theory*

Interlanguage theory was proposed by American linguist Larry Selinker in 1969. Interlanguage refers to the transitional language of the second language learners between their mother tongue and the target language. It is under continuous development and change, and gradually approaching the target language.

3.3 *Error analysis*

Based on the interlanguage hypothesis, error analysis was proposed by S. P. Corder. Through error analysis, it is possible to foresee the difficulties in the target language learning, to find out the reasons for its formation, to put forward solutions, and to improve the accuracy of learners' English expression.

Table 1. Research methods of business English correspondence automated feedback system.

	Research Contents	Analysis Methods
Local contents	Terminology	Text analysis
	Syntactic properties	Latent semantic analysis
Global contents	Genre structure	Genre analysis
	Upper level features	Qualitative and quantitative analysis

3.4 Monitor model

The monitoring model refers to five SLA hypothesis theories put forward by Stephen D. Krashen since the mid-1980s. He put forward the concept of "comprehensible input", often expressed by the formula "i + 1", which means that the amount of effective information input should be larger than the current language level of learners. Intelligent writing modification suggestions can provide more sufficient and effective feedback than traditional teacher feedback. It not only meets the requirements of "comprehensible input" but also "comfortable input".

3.5 Affective filter hypothesis

The purpose of the affective filter hypothesis is to reduce anxiety in the process of second language acquisition. Sometimes, inappropriate teacher feedback may cause anxiety and fear to academically poor students. Automated English writing feedback may well reduce students' pre-writing and post-writing psychological burden.

4 THE METHODOLOGY FOR THE RESEARCH

As a unique functional discourse in business activities, business English correspondence has its characteristics in terms of expression, structure, and length. Based on the study of its terminology, syntactic properties, genre structure, and upper-level features, this paper explores the ways to achieve the purpose of automated feedback system construction of business English correspondence.

4.1 Terminology

The use of terminology (like complaint, non-firm offer, draft, etc.) and abbreviations (like CIF, HQ, etc.) can save writing time and printing space, making the writing of business English correspondence more natural and professional.

4.2 Syntactic properties

The syntactic properties of business English correspondence are mainly reflected in the relationship between cohesion and coherence. Cohesion can be thought of as all the grammatical and lexical links that link one part of a text to another. Coherence can be considered as how meanings and sequences of ideas relate to each other.

4.3 Genre structure

Genre is a more macroscopic concept than vocabulary and syntax. Its main feature is to differentiate styles of works. Since the 1980s, genre analysis has attracted more and more attention from applied linguists. As a representative of genre analysis, John Swales (1990) believes that the standardization of genre structure is the most important performance of a specific text and the key to distinguish

Table 2. The structural features of business English correspondence.

Move	Step	Sentence Pattern
M1 Introduction	S1 Indicating the intention of writing	E.g.1 I feel awfully sorry to tell you that… E.g.2 I am writing to make a complaint regarding. . .
	S2 Responding to past business dealings	E.g.1 Thank you for your letter of December 3th, 2018. E.g.2 Thank you very much for the offer and samples of your perfume you've kindly sent to our head office.
	S3 Explaining information sources	E.g.1 Your company has been recommended to me by… E.g.2 Your letter dated... has been passed to me by…
M2 Details	S1 Handling specific business	E.g.1 Since the big season is coming soon, we suggest delivering the first 50% before November 30th and the rest before December 10th. E.g.2 So we have no other choice but to file a compensation of 5% of the total amount of the contract.
M3 Response	S1 Asking for confirmation	E.g.1 We feel you will be able to provide us with a satisfactory explanation. E.g.2 If the above arrangements are convenient for you, please feel free to inform us.
	S2 Welcoming to each other	E.g.1 We are looking forward to your joining and wish you a success at the fair.
M4 Conclusion	S1 Blessing each other	E.g.1 Hope your business thrives.
	S2 Showing your expectation	E.g.1 We sincerely hope the volume of trade between us will be even greater in the future. E.g.2 We would appreciate your more attention to future shipments.

it from other genres. Based on the analysis of business English corpus, this paper employs the "move approach" to analyse the genre characteristics of business English correspondence, so as to determine the criteria of genre structure in the automated feedback system.

4.4 *Top-level features*

Apart from above-mentioned characteristics, business English correspondence is also characterized by some upper-level features like social context, voice, macro standard, etc. For example, voice may reflect personal attitude and temperament in writing the text. Zhao C. G. (2013) and (2017) stated that voice was a significant predictor of TOEFL essay scores, explaining about 25% of the score variances.

5 CONCLUSION

Although it is impossible for the writing automated feedback system to completely replace the manual marking in a short period of time in China, the computer technology plus artificial intelligence will unquestionably play an increasingly important role in the field of language testing.

This paper is bound to contain errors, omissions, and misrepresentations given the author's limited knowledge, and further studies should focus on the establishment of a multidimensional statistical model.

REFERENCES

Burstein, J., Kukich, K., Wolff, S. Lu, C., & Chodorow, M. (2001), Enriching Automated Essay Scoring Using Discourse Marking. Eric, 8.

Grime, D., & Warschauer, M. (2010), Utility in a fallible tool: A multi-site case of automated writing evaluation. The Journal of Technology, Learning and Assessment, 6.

Rudner, L., & Gagne, P. (2001), An Overview of Three Approaches to Scoring Written Essays by Computer. Eric, 7 (26): 8.

Swales, J. (1990), Genre Analysis: English in Academic and Research Settings. Cambridge: Cambridge University Press.

Valenti, S., Neri, F., & Cucchiarelli, A. (2003), An Overview of Current Research on Automated Essay Grading. Journal of Information Technology Education, 2: 323.

Vantage Learning. (2007), MY Access! Efficacy Report [EB/OL]. http://www.vantagelearning.com/school/research /myaccess. htmal.

Zhao, C. G. (2013), Measuring authorial voice strength in L2 argumentative writing: The development and validation of an analytic rubric. Language Testing, 30(2): 201–230.

Zhao, C. G. (2017), Voice in timed L2 argumentative essay writing. Assessing Writing, (31): 73–83.

Computational Social Science – Luo, Ciurea & Kumar (eds)
© *2021 Taylor & Francis Group, London, ISBN 978-0-367-70193-2*

Research on the evaluation system of online education service quality: Evidence from college students

J.H. She, Q. Zhang & T. Liu
Capital University of Economics & Business, Beijing, China

K. Chen
The Open University of China, Beijing, China

ABSTRACT: Based on the traditional evaluation method of service quality and network service quality, this paper develops an evaluation system of online education service quality, which is verified by factor analysis and regression analysis. It is concluded that college students pay more attention to tangible, empathic, and resource-effective services when using online education. This is helpful for improvement of the quality of online education service and more efficient future teaching.

1 INTRODUCTION

China's online education began at the end of the 20th century with the prosperity of China's Internet technology. With the progress of network bandwidth technology and the emergence of new teaching forms, such as live broadcasts and short videos, online education is booming. The impact of this epidemic has also accelerated the vigorous network industry. However, the quality of the online education service has started to become uneven, with frequent problems such as random teaching, lack of supervision, and difficulty in refunding fees. There is an urgent need to evaluate the service quality of online education, feedback user experiences, and improve the service quality.

As the main user of online education, college students have diverse learning needs, with varied knowledge and experience, and they can evaluate the quality of online education services more accurately. Therefore, this study chose college students as the research subjects to construct and verify the evaluation indicators of the quality of online education services, to provide strong theoretical guidance for improving the quality of online education services, to promote the high-quality development of online education services, and to improve the education system.

2 THEORETICAL REVIEW

Levitt [1] first proposed the concept of service quality, and then Gronroos [2], a Finnish scholar, put forward the concept of customer-perceived service quality, which was favored by researchers. In respect of e-learning, an e-learning service quality evaluation model [3] and learning management system evaluation model (Sevgi Ozkan et al. 2008) [4] are proposed from six dimensions of learners and teachers. Conole (2016) [5] put forward the 7cs framework for evaluation around the interactive indicators of online education services. The research into an evaluation index of online education service quality also began to rise in China (Zhang Yi 2003) [6]. Subsequently, four primary indicators, namely system architecture, education resources, interaction mode, and market environment, were used to evaluate the quality of online education services (Huang Wei et al. 2016) [7]. With the deepening of perspectives, the scope of research topics is also expanding

Table 1. Validity analysis of evaluation variables.

KMO and Bartlett test		
Kaiser-Meyer-Olkin test		0.891
Bartlett test	Approximate chi-square	2625.933
	df	153
	Sig.	0.00

(Zhang Qingtang, Cao Wei 2016; Wei Shunping, Cheng Gang 2017) [8, 9]. With the deepening of research and technological progress, evaluation tools and perspectives are becoming more and more novel, and service resources, processes, innovation, and other indicators can be obtained (DU Jing et al. 2019) [10]. However, there are insufficient or measurement errors in the research because of the nondifferential evaluation methods.

3 MODEL DEVELOPING

There are three commonly used evaluation methods for online education service quality: SERVQUAL (service quality) evaluation method [7], also known as the "Expectation-Perception" model ;Mik Wisniewski 1996; Simon Nyeck et al. 1996)[6]; the SERVPERF (service performance) evaluation method [5], and the nondifference evaluation method (Brown et al. 1993) [16]. On this basis, we introduce the WEBQUAL evaluation method that is a network service quality evaluation method [4] and develop the evaluation system of online education service quality including tangibility, reliability, interactivity, security, assurance, empathy, and resource effectiveness. These evaluation systems can not only avoid the measurement error of indexes, but also reduce the data measurement error of nondifference evaluation methods, and effectively ensure the measurement accuracy.

4 EMPIRICAL RESEARCH

4.1 *Data sources*

The items of the questionnaire were modified on the basis of the SERVPERF scale. There were 18 questions in the questionnaire, all of which were scored by a Likert 5-point scale. All the questionnaires were distributed online and 261 valid questionnaires were collected. The overall reliability coefficient α of the questionnaire is 0.912, and the overall reliability is high. The α value of seven potential variables is 0.839, and the α value of each variable is between 0.675–0.891. Not only is the overall evaluation ability of the seven indexes good, but also each variable has greater availability.

The scale of the online education evaluation system is a combination of SERVPERF and WEBQUAL, with high content validity. KMO and Bartlett tests of structural validity are also within the required range. The results are shown in Table 1.

4.2 *Descriptive statistics*

From the results of all the data, males accounted for 24.90% and females accounted for 75.10%; among them, the possibility of using online education service was higher in senior grade, accounting for 52.11%. Among all kinds of online education services, the use of online higher education services is as high as 79.69%, which is also consistent with the actual use of online education services. In addition, according to the results of the previous survey, the choice motivation for online higher education services, entrance, their own hobbies and interests, research ranked in the

Table 2. Rotation component matrix A of evaluation variables (The shaded part in this table is the result corresponding to the seven common factors and their respective observation variables according to theoretical hypothesis).

	Ingredients							Observation variables	Latent variable
	1	2	3	4	5	6	7		
Q1	0.109	0.839	0.101	−0.155	0.153	0.045	0.113	Ease of use	Tangibles
Q2	0.185	0.883	0.002	0.141	0.164	0.102	0.016	Superior performance	
Q3	0.177	0.871	0.091	0.142	0.149	0.106	0.113	Page design	
Q4	0.301	0.377	0.123	−0.001	0.638	0.127	0.203	Reliable service commitment	Reliability
Q5	0.248	0.314	0.069	0.221	0.771	0.135	0.045	Reliable learning plan	
Q6	0.077	0.037	0.480	0.167	0.562	0.296	0.191	Service response	
Q7	0.167	0.094	0.835	0.263	0.086	0.142	0.112	Interaction after class	Interactivity
Q8	0.136	0.087	0.858	0.197	0.117	0.090	0.160	Interaction in class	
Q9	0.318	0.210	0.175	0.020	0.328	0.664	0.161	Transaction security	Security
Q10	0.134	0.088	0.140	0.142	0.092	0.878	0.208	Privacy security	
Q11	0.137	0.040	0.234	0.258	0.242	0.176	0.764	After sales support	Assurance
Q12	0.350	0.261	0.152	−0.006	0.041	0.281	0.692	Teacher security	
Q13	0.242	0.099	0.201	0.834	0.181	0.093	0.115	Personalized service	Empathy
Q14	0.176	−0.013	0.302	0.835	0.061	0.080	.089	Caring	
Q15	0.701	0.207	0.067	0.084	0.351	0.068	0.290	Course content	Resource effectiveness
Q16	0.812	0.041	0.107	0.132	0.119	0.253	0.014	Access	
Q17	0.782	0.223	0.154	0.110	0.147	0.100	0.083	Feedback mechanism	
Q18	0.676	0.190	0.109	0.313	0.055	0.038	0.236	Abundant resources	

Extraction method: Principal component
Rotation method: orthogonal rotation method with Kaiser standardization.
a. Rotation converges after seven iterations

top three. This shows that college students are more independent, which is why they were chosen as the research subjects.

4.3 *Factor analysis*

Factor analysis is the key to index construction research. Factor analysis is divided into two steps: exploratory factor analysis and confirmatory factor analysis. The first is exploratory factor analysis. According to the needs of the study, seven common factors were extracted by PCA, and the cumulative variance interpretation rate of the seven common factors was 79.48%. However, this does not correspond each factor to each observed variable. In order to better explain the practical significance of common factors, Table 2 is obtained after the factors are rotated.

The results in Table 2 clearly show that the latent variable and the observed variable correspond to the theoretical hypothesis. In addition, the factor load of each measurement index is higher than 0.5, and other dimensions are significantly lower than 0.5. There is no case in which the two factor load is high, indicating that the internal structure of the questionnaire is clear and the overall structure validity is high.

Secondly, confirmatory factor analysis is used to verify whether the results of the exploratory factor analysis are accurate. According to the requirements of Mplus variable name setting, the English word lowercase initial of the latent variable is taken as variable name, and the resource validity recorded as re; the observation variables under each latent variable are represented by a lowercase initial plus number, with the numbers starting from 1. The operation result of Mplus is shown in Figure 1.

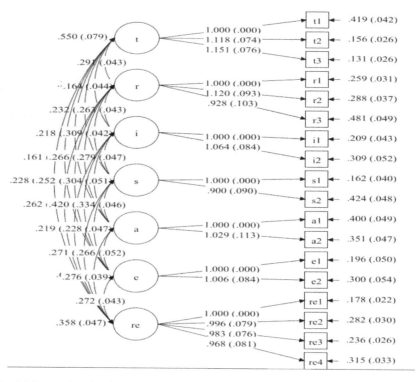

Figure 1.   Mplus results of exploratory factor analysis.
*The number outside the brackets indicates the path coefficient, and the number inside the brackets is the standard error.

Table 3.   Model goodness-of-fit index.

Model	X^2	Df	X^2/df	P-value of X^2	RMSEA	CFI	TLI
1	273.926	114	2.403	0.0000	0.073	0.937	0.916

The path map clearly shows the corresponding relationship between the seven indicators and their respective dimensions, including specific coefficients. However, to more clearly explain whether the model assumptions between the two are accurate, we need to use relevant indicators.

Generally speaking, the Chi-square value is large, the P test is significant; RMSEA is below 0.8; CFI and TLI are above 0.9, and it is considered that the goodness-of-fit of the model is good. The results in Table 3 meet the requirements, which indicates that the corresponding relationship between the online education service quality indicators and their respective observation variables is correct.

4.4   Regression analysis

In order to further verify the effectiveness of the evaluation indicators, we take the quality of the online education service as the dependent variable and seven indicators as the independent variables to conduct regression quantitative processing on the relationship between them. It has been proved that they are significantly correlated at the 0.01 level. From the perspective of the overall model, F-test is significant and the adjusted R^2 is 0.734, and the explanatory power of independent variable

Table 4. Model result report.

Model	R	R^2	Adjusted R^2	Standard deviation	Statistics changed					
					R^2	F	df_1	df_2	Sig.F
1	0.861^a	0.741	0.734	0.280	0.741	103.501	7	253	0.000

a. Predictive variables: (constant), assurance, safety, reliability, empathy, interactivity, tangibility, resource effectiveness.

Table 5. Regression coefficient result report.

Model	Nonstandardized coefficient		Standard coefficient			Confidence interval (95.0%)		Collinearity statistics
	β	Standard error	Adjusted	t	Sig.	Lower limit	Upper limit	VIF
1 (Constant)	3.831	0.017		221.144	0.000	3.797	3.866	
Resource effectiveness	0.186	0.017	0.343	10.735	0.000	0.152	0.221	1.000
Tangibles	0.250	0.017	0.460	14.383	0.000	0.215	0.284	1.000
Interactivity	0.100	0.017	0.185	5.784	0.000	0.066	0.135	1.000
Empathy	0.204	0.017	0.375	11.728	0.000	0.169	0.238	1.000
Reliability	0.094	0.017	0.173	5.408	0.000	0.060	0.128	1.000
Security	0.175	0.017	0.322	10.070	0.000	0.141	0.209	1.000
Assurance	0.174	0.017	0.321	10.035	0.000	0.140	0.208	1.000

Dependent variable: quality of service.

to the dependent variable is more than 60%. Therefore, it is considered that the regression model has a good degree of fitting. See Table 4 for the model results.

In addition, the results of the regression coefficient show that VIF is 1, which is in the standard range, and there is no multicollinearity (Table 5).

From the regression model, we can not only get the regression equation through the specific coefficient value: the quality of online education service = 3.831 + 0.186 * resource effectiveness + 0.25 * tangibility + 0.1 * interactivity + 0.204 * empathy + 0.094 * reliability + 0.175 * Security + 0.174 * assurance; we can also arrange the seven indicators into tangibility and migration according to the impact intensity of the coefficient on the quality of online education service from large to small. Sentiment, resource effectiveness, safety, security, interactivity, and reliability are also conducive to the improvement of online education services in the future.

To sum up, both theory and empirical test have proved that the seven evaluation indicators constructed in this study are effective, with high reliability and validity, and the impact of the seven indicators on online service quality evaluation is different, which lays a solid theoretical foundation for the future evaluation of online education service quality.

5 CONCLUSIONS AND PROSPECTS

From the current research results, the improvement in the quality of the online education service in China should focus on the tangible, empathy, and resource effectiveness, which may be related to diversified needs, high quality, and ability of college students, with more emphasis on emotional experience. At the same time, we should also pay attention to the security of online education services, enhance service commitment, and ensure reliability. The future development of e-education services is expected. Future research can be evaluated with other user groups or managers to ensure the completeness of results. The improvement of online education service quality is still the focus of the online education industry. The follow-up research on the effectiveness of indicators remains to be further explored.

ACKNOWLEDGEMENTS

The work described in this paper was supported by 2019-2020 Capital University of Economics and Business Course Construction Project, 2020 Key Projects of Education Reform of Capital University of Economics and Business, and 2019 Major Project of the Open University of China:"Research on the large-scale online ideological and political education in the Open University of China" (No: D19J0001).

REFERENCES

[1] Levitt T. Production-line approach to service. Harvard Business Review, 1972, 50:42–52.
[2] Gronroos, Christian. A Service Quality Model and its Marketing Implications. European Journal of Marketing, 1984, 18(4):36–44.
[3] Sun P.C., Tsai R.J., Finger G, et al. What drives a successful e-Learning? An empirical investigation of the critical factors influencing learner satisfaction. Computers & Education, 2008, 50(4):0-1202.
[4] Barnes S J , Vidgen R T . An Integrative Approach to the Assessment of E-Commerce Quality. Journal of Electronic Commerce Research, 2002, 3:114–127.
[5] Cronin J.J., Taylor S.A. Measuring Service Quality: A Reexamination and Extension. Journal of Marketing, 1992, 56(3):55–68.
[6] Donnelly M , Wisniewski M , Dalrymple J F , et al. Measuring service quality in local government: the SERVQUAL approach. International Journal of Public Sector Management, 1995, 8(7):15–20.
[7] Parasuraman A, Zeithaml Valarie A , Berry Leonard L. SERVQUAL: A Multiple-Item Scale for Measuring Consumer Perceptions of Service Quality. Journal of Retailing, 1988, 64(1):12–40.
[8] Wisniewski, Mik. Measuring service quality in the public sector: The potential for SERVQUAL. Total Quality Management, 1996, 7(4):357–366.

Computational Social Science – Luo, Ciurea & Kumar (eds)
© 2021 Taylor & Francis Group, London, ISBN 978-0-367-70193-2

Can network public opinion predict stock market return?

W.F. Feng & S.L. Jia
School of Economics and Management, Lanzhou University of Technology, LanZhou, China

Y.H. Xiang
School of Life Science and Engineering, Lanzhou University of Technology, LanZhou, China

ABSTRACT: Stock return prediction is an important financial topic that has gained a great deal of attention from both researchers and investors. This paper investigated the relationship between network public opinion and the return prediction of the Shanghai Stock Exchange (SSE) 180 index by support vector regression (SVR). The findings show that predicted stock return direction was successful for 19 out of 22 samples, and the accuracy of prediction was 86.36 percent. These findings suggest that network public opinion has a very strong relationship to stock performances, and the stock market is sensitive to network public opinion.

1 INTRODUCTION

In recent years, networking has become an important method for producing and spreading public opinion Textmining techniques and machine learning algorithms were applied successfully to quantify the impact of network public opinion on stock market returns and to predict changing trends Kong et al. [1] applied textmining techniques to quantify network public opinion as an impact factor, and then used the VAR model to analyze the influence on the Chinese stock market. Kao et al. [2] introduced SVM to forecast stock prices and the experimental results showed that their proposed approach was suitable and outperformed other prediction models and machine learning methods.

This paper investigated the relationship between network public opinion and the stock return prediction using SVR The research objects are the text contents of forum posts on the Oriental Fortune Network in China; we transformed and quantified unstructured text contents of forum posts into a structured characteristic matrix and created a network public opinion indicator with textmining technology We then built the SVR model to forecast the direction for the stock return of the Shanghai Stock Exchange (SSE) 180 index by creating the network public opinion indicator.

The remainder of the paper is organized as follows Section 2 introduces the proposed methodology, including the models of SVM and SVR and data collection used in this study. Section 3 describes the textmining process and the creation of a network public opinion indicator. In Section 4, the experimental results are reported and discussed. Finally, Section 5 contains the concluding remarks.

2 METHODOLOGY AND DATA

2.1 *Methodology*

Support vector machines (SVM) and support vector regression (SVR) are new machine learning algorithms based on Vapnik's [3] structural risk minimization (SRM) principle along with a very useful regression model SVR uses the elementary idea of the kernel function mapping and the

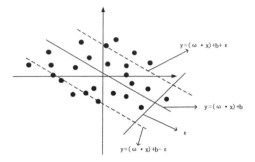

Figure 1.   Linear SVR using an ε-insensitive zone

principle of SRM to transform given examples of data (the training data) into points of higher dimensional feature space and to find the linear optimal separate feature vectors that can maximize the margin The research data of SVM were divided into two categories: training data and testing data (predicting data) The main principle of the linear optimal separate SVR is illustrated in Figure 1.

The ε-insensitive zone is between the dotted lines on the diagram, and denotes the precise degree at which the boundaries on generalization ability apply. The points on the dotted line were used to confirm the regression surface. The training data within the ε-insensitive zone were deemed correct, and are not important to the regression function and contributed to the error loss function. Those outside the ε-insensitive zone were regarded as incorrect and called support vectors. The regression surface was determined only by support vectors. Given a training data set,

$$\{(x_i, y_i)|x_i \in X = R^d, y_i \in R, i = 1, \cdots, n\} \tag{1}$$

where x_i denotes a data point of the space of the input patterns X; y_i denotes output outcome. Linear regression function $f(x)$ can be expressed as:

$$f(x) = \omega^T \varphi(x_i) + b \quad \omega \in X, \quad b \in R \tag{2}$$

where ω denotes a weight vector; $\varphi(x)$ denotes the mapping function; and b is a bias. However, not all the training data are linearly separable. Formula (3) can be reformulated as: minimize $\frac{1}{2}\|\omega\|^2$

$$\text{subject to} \begin{cases} y_i - \omega^T \varphi(x_i) - b \leq \varepsilon \\ \omega^T \varphi(x_i) + b - y_i \leq \varepsilon \end{cases} \tag{3}$$

However, not all the training data are linearly separable To address this problem, slack variables ξ_i and ξ_i^* were introduced Formula (4) can be reformulated as:

minimize $\frac{1}{2}\|\omega\|^2 + C\sum_{i=1}^{n} \left(\xi_i + \xi_i^*\right)$

$$\text{subject to} \begin{cases} y_i - \omega^T \phi(x_i) - b \leq \varepsilon + \xi_i \\ \omega^T \phi(x_i) + b - y_i \leq \varepsilon + \xi_i^*, \ i = 1, \ldots, n \\ \xi_i, \xi_i^* \geq 0 \end{cases} \tag{4}$$

where parameter C controls the trade-off between structural risk and the training error; ξ_i and ξ_i^* are slack variables, which denote the lower and upper training errors. By introducing Lagrange multipliers, formula (5) can be quadratically optimized. The optimization problem is given by

$$\text{minimize } \frac{1}{2}\sum_{i,j}^{n}(\alpha_i^* - \alpha_i)(\alpha_j^* - \alpha_j)\phi(x_i)^T\phi(x_j) + \varepsilon\sum_{i=1}^{n}(\alpha_i^* + \alpha_i) - \sum_{i=1}^{n}y_i(\alpha_i^* - \alpha_i)$$

$$\text{subject to }\begin{cases} \sum_{i=1}^{n}(\alpha_i - \alpha_i^*) = 0 \\ \alpha_i, \alpha_i^* \in [0.C] \end{cases} \tag{5}$$

By changing the equation

$$\omega = \sum_{i=1}^{n}(\alpha_i - \alpha_i^*)\phi(x_i)$$

$$f(x) = \sum_{i=1}^{n}(\alpha_i - \alpha_i^*)\phi(x_i)^T\phi(x_j) + b \tag{6}$$

Consequently, introducing Lagrange theory and kernel function, SVR function can be shown by:

$$f(x) = \sum_{i=1}^{n}(\alpha_i - \alpha_i^*)K(x_i, x) + b \tag{7}$$

where $K(x_i, x)$ denotes the kernel function that the value equals the internal product of x_i and x_j in the feature space $\phi(x_i)$ and $\phi(x_j)$, $K(x_i, x_j) = \phi(x_i)^T\phi(x_j)$.

2.2 Data

The data set was based on the text contents of daily forum posts on the Oriental Fortune Network in China, obtained by web crawler technology. The data set included a total of 1.7 million daily observations. The entire data covers the period from January 1, 2014 to December 31, 2014. In order to get the most meaningful information from observations that contained complex and unreliable information, this study preprocessed data and removed the noise. Approximately 3.1 million text data were able to be used. The second data set shows daily observations taken from SSE 180 index.

3 NETWORK PUBLIC OPINION INDICATOR CONSTRUCTION

Most forum post texts are unstructured data We converted unstructured text data into structural feature vectors, and then constructed a network public opinion indicator to make the prediction.

3.1 Chinese word segmentation

It is very easy for a computer to read English words but it is different for Chinese because there are no spaces between the words. Chinese word segmentation separates the text applied to the corresponding word by a segmentation algorithm in order for the computer to understand and manage the information. Therefore to make the computer capable of dealing with Chinese sentences it needs to separate words in sentences

Table 1. The generated indicator data of the matrix.

Date	Limit	Rise	Positive	Negative	To buy	To sell	Price	...
1/7	45.5221	42.5703	94.5845	81.5464	105.5626	70.9695	38.4997	...
2/7	32.8123	28.3927	107.7002	63.1331	106.7878	55.3924	32.1918	...
3/7	77.6324	37.1806	100.8870	65.3468	132.5849	46.4266	44.2536	...
...	...	...	...	...	...	...	...	...
30/12	1196.005	376.9591	1481.651	1466.642	1530.831	704.3989	626.1385	...

Note: #/* date; # day; * month, respectively

3.2 *Stop-words removal, synonyms, and approximate synonym combinations*

In order to reduce the noise, synonyms and approximate synonyms are replaced with corresponding words, and multiple synonyms and approximate synonyms are combined as one word. Meanwhile stop-words were removed from the segmentation results because they were common words without deeper meaning. There are about 4500 stop-words in Chinese, the words did not affect the accuracy of the prediction, but they did reduce the vector dimension primarily by stop-words removal, synonyms, and approximate synonym combinations.

3.3 *Text vectorization*

Unstructured text information was converted into a vector of numeric representation, called text vectorization, which is the most critical step in text mining. The most immediate way is that the dimensions of the vector are associated with all the text entries, so each text is represented as a vector in an *n*-dimension vector space. To improve the accuracy of categorization and the quality of feature selection, this paper chose the term frequency-inverse document frequency (TF-IDF) formula to measure the importance of a word throughout the document. The principle of TF-IDF is that a word or phrase in one article that has a high frequency (TF) and rarely appears in other articles (IDF), means a stronger classification ability of a word or phrase.

3.4 *Dimensionality reduction*

After the text is expressed as vector representations, and a text dependency matrix is built for all the text terms, the results are generally high-dimensional. The reduction in vector dimensions can make quantitative results more reasonable, therefore improving the accuracy of the regression model. The methods of synonyms, approximate synonym combinations, and feature word extraction have been widely applied to reduce the high dimensions of the eigenvectors. The former appeared above in 3.2. Therefore, the feature words with the highest frequency were chosen for this paper as being characteristic words that should have an influence on the stock market.

3.5 *Network public opinion indicator construction*

After a series of above-mentioned processing, vast amounts of unstructured texts were converted into a numeric matrix and a network public opinion indicator was also constructed. The generated indicator data of the matrix are shown in Table 1.

4 EXPERIMENTAL RESULTS AND DISCUSSION

Network public opinion was considered as the input variable of the SVR, and the corresponding stock return as the output variable in order to train the SVR model. The results of the stock return prediction are presented in Figure 2.

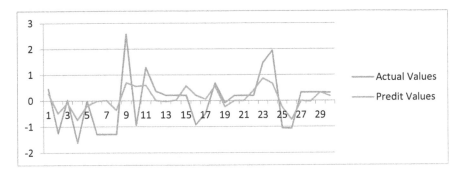

Figure 2. Actual values and predicted valuesNote: The horizontal axis represents data (day).

Table 2. The direction of actual values and predicted values.

Date	Actual values	Predicted values	Correct?	Date	Actual values	Predicted values	Correct?
1/12	Rise	Rise	✓	16/12	Rise	Fall	✓
2/12	Rise	Rise	✓	17/12	Fall	Rise	✓
3/12	Rise	Rise	×	18/12	Rise	Rise	✓
4/12	Rise	Rise	✓	19/12	Rise	Rise	✓
5/12	Rise	Rise	✓	22/12	Fall	Fall	✓
8/12	fall	fall	✓	23/12	Fall	Fall	✓
9/12	Rise	Rise	✓	24/12	Rise	Rise	✓
10/12	fall	Rise	✓	25/12	Rise	Rise	✓
11/12	Rise	Rise	✓	26/12	Rise	Fall	✓
12/12	Rise	Rise	×	29/12	Rise	Rise	×
15/12	Rise	Rise	✓	30/12	Rise	Rise	✓

Note: #/* date; # day; * month, respectively. ✓ shows concordance of the predicted values with the actual ones while × denotes inconsistent.

For a further discussion Table 2 gives a visual representation of the direction of actual values and predicted values of stock return.

As shown in Table 2 only four return values of the predictions and actual values are opposite This shows that the stock return prediction of rising and falling was wrong. The direction of the rest of the predicted values and actual values corresponded. The accuracy of the predictions was 86.36 percent (19/22).

The results suggest a strong correlation and relatively consistent trend between network public opinion and the stock market return of the SSE 180 index. The reasons are as follows: first, network public opinion is regarded as a channel to vent investor sentiment. Investor sentiment as a subjective expectation gives rise to a change in mood under the influence of investors' own characteristics from external information, which is reflected in investment decisions. Second, since 2009, the reforms of offline inquiry systems have broken Chinese government regulations in stock pricing to create conditions that provide an advantage to the network. Network public opinion has become an essential way of looking for investment information. Finally, the network has already become one of the important channels of capital market message dissemination in the information era. Herd behavior is very apparent in the Chinese stock market. In retail, the investors leading structure lacks the ability to perceive network public opinion, which has been regarded as the major reason for investment errors. As a result, network public opinion has already become a significant factor in impacting the direction of stock returns.

5 CONCLUSIONS

The findings from this study may provide some insights—first, because the effects of network public opinion increase continuously in the security market, combining quantitative analysis of other stock indicators, which can solve the investment information asymmetry problem. Second, the application of text mining in securities investments will effectively reduce missing soft information about listed companies and promote the success rate of investments. Finally, due to the complexity of stock market information, apart from network public opinion, other major information should be considered. Additional research will further test the relationship between network public opinion and stock performances, by excavating more information of different types and from other sources.

ACKNOWLEDGMENTS

This work was supported by the National Natural Science Foundation of China [No. 71473036] and the Social Science Foundation of Gansu Province [No. 19YB141].

REFERENCES

[1] Kong, X. Y., BI, X. C., Zhang, S. G., 2016. Financial news and prediction for stock market: an empirical analysis based on data mining techniques. Journal of Applied Statistics and Management 35(2):215–224.
[2] Kao, L. J., Chiu, C. C., Lu, C. J., et al., 2013. A hybrid approach by integrating wavelet-based feature extraction with MARS and SVR for stock index forecasting. Decision Support Systems 54(3): 1228–1244.
[3] Bystritsky A, Craske M, Maidenberg E, Vapnik T, Shapiro D. Ambulatory monitoring of panic patients during regular activity: a preliminary report.[J]. Biological psychiatry, 1995, 38(10).
[4] Vapnik, V. N., Vapnik, V., 1998. Statistical learning theory [M]. New York: Wiley.
[5] Zhao, L. L., Zhao, Q. Q., Yang, J., 2012. Quantitative analysis of the impact of news on stock market. *Journal of Shandong University* 47(7): 70–75.
[6] Hsu, S. H., Hsieh, J. J. P. A., Chih, T. C., et al., 2009. A two-stage architecture for stock price forecasting by integrating self-organizing map and support vector regression. Expert Systems with Applications 36(4): 7947–7951.

Computational Social Science – Luo, Ciurea & Kumar (eds)
© *2021 Taylor & Francis Group, London, ISBN 978-0-367-70193-2*

Natural language processing features and Critical Thinking assessment a pre-post essay test study in Italian HiEd setting

A. Poce & F. Amenduni
Department of Educational Sciences, Roma Tre University, Rome, Italy

ABSTRACT: Critical Thinking (CT) is considered a desirable learning outcome at all levels of education, included Higher Education (HE). Experts agree with the idea that CT assessment should include both Multiple Choice and Constructed Response Task (CRT) kinds of items (Hyytinen et al. 2015; Ku 2009; Liu et al. 2014) in order to retrieve the complexity of the CT construct. Nevertheless, CRT tasks are poorly used because of the costs of scoring and reliability issues. According to Liu, Frankel and Roohr (2014) automatic assessment of open-ended answers could be a viable solution to these concerns. In recent years, many attempts have been carried out in order to develop and validate tools for the automatic assessment of CT related-skills. Most of these studies applied Natural Language Processing techniques (NLP) to English written texts and there are a few attempts to generalize these techniques to other languages.

Therefore, this research was aimed at understanding which NLP features are more associated with six CT sub-dimensions (Poce 2017) as assessed by human raters in essays written in Italian language. The study used a corpus of pre-post essays written in Italian language by 200 students who attended a Master Degree University course in "Experimental Education and School Assessment". The essays were assessed both by human raters and by a computerized tool which automatically calculates different kinds of NLP features. Pearson correlations were calculated to explore the strength of the association between NLP features and CT sub-skills assed by human raters. Non-parametric Test U Mann Whitney was used to explore average difference scores between pre-post test scores in CT sub-skills (as assessed by human raters) and NLP features. We found out a strong correlation between the NLP feature "elaboration" and the CT sub-skills relevance ($r = 0,89$) and importance ($r = 0,75$). Elaboration was calculated as an inverted measure of verbatim copying. As expected, syntax complexity positively relates with Language skills as assessed by human rater whilst repetition negatively correlates with the same skill. It was also possible to retrieve similar trends in pre-posttest between scores provided by human raters and NLP features. Limitation and future developments will be discussed.

1 INTRODUCTION

Critical Thinking (CT) is considered a desirable learning outcome at all the levels of education, included Higher Education (HE), according to both economic (OECD 2012) cultural (UNESCO 2015) and educational research-oriented organizations (IEA 2018). In response to the Bologna Declaration of 1999 aimed at developing a comparable degree system among European countries, the Tuning Projects identified different general and subject specific skills to be developed in HE students, by including CT (Gilpin & Wagenaar 2008). The AHELO project (Assessment of Higher Education Learning Outcomes) carried out by OECD (2012) also included CT as one of the general skills that should be assessed at an international level. Thus, reflecting upon CT assessment choices is necessary at least for two reasons: firstly, CT is considered a desirable learning outcome for European HE students and should be recognized in a comparable way, according the Bologna Strategy; secondly, research is necessary to understand which teaching strategy can foster CT skills in HE. As asserted by Rear in a recent review (2019), the assessment of CT has become a significant challenge with a number of standardized test available. Assessment tests could be classified in different ways. Hyytinen and colleagues (2015) differentiated self-report from performance-based measurements. Moreover, the performance-based measurements can be classified into multiple-choice (MC) tests/questionnaires and constructed response tasks (CRT). Another way to classify CT assessment is to distinguish between assessment tools focused on CT as a process or as an outcome (Garrison et al. 2001).

Authors agree with the idea that CT assessment should include both MC and CRT kinds of items (Hyytinen et al. 2015; Ku 2009; Liu et al. 2014) in order to validly retrieve the complexity of the different CT dimensions.

Nevertheless, CRT are poorly used because of the costs of scoring and reliability issues. According to Liu, Frankel and Roohr (2014) automatic assessment of open-ended answers could be a viable solution to these concerns. In recent years, many attempts have been carried out in order to develop and validate tools for the automatic assessment of CT related-skills, such as argumentation (Zhu et al. 2020), reflective writing (Ullman 2019), discourse coherence quality (Burstein et al. 2013) and use of evidence (Rahimi et al. 2017). Most of these studies applied Natural Language Processing techniques (NLP) to English written texts and there are a few attempts to generalize these techniques to other languages.

According to the above mentioned premises, the main goals of this research are:

1. understanding which NLP features are best associated with six CT sub-dimensions, as assessed by human raters in essays written in Italian;
2. seeing how those NLP features change within the pre-post tests designs.

2 METHODS

2.1 *Learning activities aimed at stimulating Critical Thinking Skills*

The experimentation was carried out within a Master Degree University course in "Experimental Education and School Assessment" at the Department of Educational Sciences (Roma Tre University). The University course lasted 9 months and students were involved in different kinds of activities designed to foster students CT throughout two semesters.

In the first semester, students were required to individually search for and assess Open Educational Resources (OERs) on the topics related to 21st skills and Museum Education for primary school children. Students used a rubric for the OERs assessment developed in the context of the European Erasmus Plus Project "OpenVirtual Mobility" (Poce et al. 2019). This activity was aimed both at stimulating *evaluative* skills and preparing students for the second semester not-mandatory assignments where students were given the possibility to design collaboratively their own OERS, following the design principles of the Project-Based Learning methodology (Sasson et al. 2018).

Out of 200 students, 40 students voluntarily participated in the OER design activity by working in 8 groups. OERs, produced by the students, were assessed by supervisors through the same rubric students used in the first semester to assess the OERs retrieved from the web.

2.2 *Data collection*

The study used a corpus of pre-post essays written in Italian language by 200 students who attended the Master Degree University course in "Experimental Education and School Assessment" described in the previous paragraph. Students were asked to read an extract of the "Dialogue concerning two chief world systems" (Galilei 1632) entitled "Origin of the nerves according to Aristoteles and according to the doctors" (p. 107–8). Students were asked to write an essay by including in their arguments the answers to the following six questions:

1. What are the two opposite positions regarding the origin of the nerves described in the text?
2. What are the differences between the methods supported by Simplicio and Sagredo?
3. What does the "principle of authority" consist of? When is it explicitly referred to in the text and when is it implicit?

Table 1. Associations between NLP indicators and CT sub-skills assessed by human raters calculated through r Pearson Correlation. *indicates sign. <0,05 **indicates sign. <0,001.

CT subskills		NLP indicators						
		N.	Hapax	Lexical	TfxIDF	Syntax	Repetition	Elaboration
assessed		words		Extens.		complex		
by human	Grammar	0,226*						0,463**
experts	Language	0,224*				0,225*	0,240*	0,654**
	UoL		0,229*	0,217*	0,283**			
	Argument.	0,310**				0,279**	−0,224*	0,754**
	Relevance	0,300**				0,199*		0,933**
	Importance	0,230*				0,212*		0,890**
	Critical Evaluation	0,271**				0,223*		0,750**
	Novelty	0,357**				0,271**		0,647**

4. Why do you think the episode was settled in the Republic of Venice?
5. In your opinion, has the principle of authority affected scientific discoveries throughout history? If so, how?
6. Choose one or more elements in the passage that, in your opinion, have played a role in the development of scientific knowledge in the modern and contemporary world. Explain the reasons for your choice.

2.3 Data analysis

In this preliminary analysis, 50 (Female = 45; Male = 5) pre-post tests were included: thus, the corpus is composed by 100 essays. All the essays were assessed both by human raters and through a tool which automatically calculates different kinds of NLP features.

Human raters assessed all the essays based on a rubric composed by six macro-indicators on a scale from 1 to 5: use of the language, argumentation, relevance, importance, critical evaluation and novelty (based on Poce 2017). At the same time, different NLP features were automatically measured: corpus length, mean sentence length, hapax (Poce 2012), lexical extension (Crossley et al. 2011), readability (Vacca 1972), syntax complexity (Yang et al. 2015), repetition, elaboration (Chang & Ku 2015) and TD-IDF (Salton, & McGill 1983). All these features were previously studied in association to the six CT sub-skills theorized by Poce (2017). Pearson correlations were calculated to explore the strength of the association between NLP features and CT sub-skills assed by human raters. Non-parametric Test U Mann Whitney was used to explore average difference scores between pre-post test scores in CT sub-skills (as assessed by human raters) and NLP features.

3 RESULTS

Table 1 shows which NLP indicators are associated with CT sub-skills assessed by human experts. In coherence with our expectations, syntax complexity positively relates with Language skills ($r = 0,225$; sign. <0,05) as assessed by human rater whilst repetition negatively correlates with the same skill ($r = −0,240$; sign. <0,05). Hapax ($r = 0,229$; sign. <0,05), lexical extension ($r = 0,217$ sign. <0,05) and TfxIDF ($r = 0,283$; sign. <0,01) positively correlates with the general assessment of the Use of Language (UOL).

Argumentation (as assessed by human experts) strongly correlates with elaboration NLP ($r = 0,754$; sign <0,01); a moderate positive correlation was registered also with number of words ($r = 0,310$ sign. <0,01) and syntax complexity ($r = 0,279$; sign. <0,01), and a moderate negative correlation was registered with repetition ($r = −0,224$ sign. <0,05). Elaboration (NLP) strongly correlates with Relevance ($r = 0,933$, sign <0,01) Importance ($r = 0,890$, sign. <0,01)

Table 2. Level of internal coherence of NLP indicators calculated through r Pearson Correlation. ∗indicates sign. <0,05 ∗∗indicates sign. <0,01.

	Hapax	Lexical Extens.	TfxIDF	Syntax complex	Elaboration
Hapax	1	0,977**	0,360**	−0,456**	
Lexical Extens	0,977**	1	0,364**	−0,499**	
TfxIDF	0,360**	0,364**	1	−0,484**	
Syntax complex	−0,456**	−0,499**	−0,484**	1	0,222*
Elaboration				0,222*	1

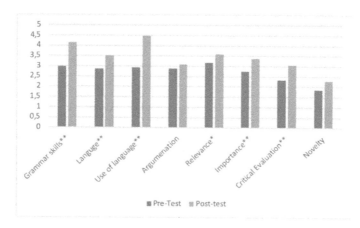

Figure 1. Differences in Pre and Post-tests in CT sub-skills average scores. ∗indicates sign. <0,05 ∗∗indicates sign. <0,01 that it was calculated with the non-parametric test U Mann Whitney.

Critical Evaluation (r = 0,750; sign. <0,01) and Novelty (r = 0,647 sign. <0,01) as assessed by human raters. Syntax complexity (NLP) moderately correlates with Relevance (r = 0,199, sign <0,05) Importance (r = 0,212, sign. <0,05) Critical Evaluation (r = 0,223; sign. <0,05) and Novelty (r = 0,271 sign. <0,01) as assessed by human experts. Finally, number of words moderately correlates with Relevance (r = 0,300, sign <0,01) Importance (r = 0,230, sign. <0,05) Critical Evaluation (r = 0,271; sign. <0,01) and Novelty (r = 0,357 sign. <0,01) as assessed by human experts.

The Table 2 shows the level of internal coherence of NLP indicators. Hapax strongly correlates with Lexical Extension (r = 0,977 sign <0,01) and TfxIDF (r = 0,360 sign <0,01) and negatively correlates with syntax complexity (r = −0,456 sign <0,01).

Similarly, Lexical Extension negatively correlates with syntax complexity (r = −0,499 sign <0,01) and positively correlate with TfxIDF (r = 0,364 sign. <0,01).

Syntax complexity positively correlates with Elaboration (r = 0,222 sign. <0,05) and negatively with TfxIDF (r = 0,484; sign 0,01). None of these indicators correlate with exam grades. The only indicator that correlates with the exam grade is "syntax complexity" (r = 0,298 sign <0,01).

Figure 1 shows the differences in Pre and Post-tests in CT sub-skills average scores. We can see that participants received in average higher scores in the post-tests compared to the pre-test, although differences are not statistically significant for argumentation and novelty average scores.

Figure 2 shows the differences in Pre and Post-tests in NLP indicators average scores. It is possible to detect a growth in the Hapax index and Lexical Extension, which is consistent with the increasement of the Use of the Language average score obtained through human evaluators' assessment. Elaboration also improves, which is interconnected with most of the CT subskills, especially with Relevance and Importance.

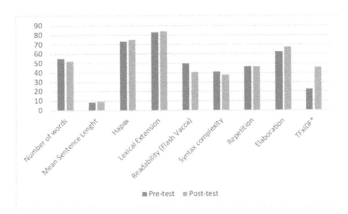

Figure 2. Differences in Pre and Post-tests in NLP indexes. *indicates sign. <0,05 **indicates sign. <0,01 that it was calculated with the non-parametric test U Mann Whitney.

4 DISCUSSION AND FINAL REMARKS

Although in the last years different research works have been carried out in order to automatically assess CT related skills in open-ended answers through NLP techniques, most of these analyses were applied to English written texts. This exploratory work was aimed at verifying if NLP features commonly used to automatically assess CT related skills in English texts could be also applied to assess the same skills in Italian written-essays.

We obtained some encouraging results: firstly, we found out a strong correlation between the NLP feature Elaboration and the CT sub-skills Relevance ($r = 0,89$) and Importance ($r = 0,75$). Elaboration was calculated as an inverted measure of *verbatim copying,* which means that students copy word for word from the texts provided. Previous studies have found that *verbatim copying* is not as beneficial as reorganized note taking, and it is less beneficial for comprehension (Chang & Ku 2015). We have also found different correlations among NLP indicators and the Use of Language sub-skill as assessed by human raters. As expected, syntax complexity positively relates with Language skills as assessed by human raters whilst repetition negatively correlates with the same skill. Even if NLP tendentially does not correlate with exam grades, a significant correlation was found between the exam grade and "syntax complexity" ($r = 0,298$ sign <0,01), suggesting room for improvements in terms of predictive validity. It was also possible to retrieve similar trends in pre-posttest between scores provided by human raters and NLP features. Indeed, we found a growth in the Hapax index and Lexical Extension, which is consistent with the increasement of the Use of the Language average score obtained through human evaluators' assessment. Elaboration also improves, which is interconnected with most of the CT subskills, especially with Relevance and Importance.

This study was exploratory in nature and we acknowledge its limitations: in future studies, we would need to expand the sample size, by including the remaining 300 essays collected. We have also explored a limited number of NLP indicators because of the difficulties to find out Open Tools for Italian Language to be incorporated in our automatic system.

In future studies, we are going to use larger *corpora* and to test new NLP features. These improvements will allow us to carry out more sophisticated statistical analysis such as structural equation modelling and Latent Factor Analysis (MacArthur et al. 2019).

ACKNOWLEDGEMENT

A. Poce coordinated the research presented in this paper. Research group is composed by the authors of the contribution that was edited in the following order: A. Poce par. 1; 2.1; 2.2; 4. F. Amenduni par. 2.3, 3.

REFERENCES

Burstein, J., Tetreault, J., Chodorow, M., Blanchard, D., & Andreyev, S. (2013). 16 Automated Evaluation of Discourse Coherence Quality in Essay Writing. Handbook of automated essay evaluation: Current applications and new directions, 267.

Chang, W. C., & Ku, Y. M. (2015). The effects of note-taking skills instruction on elementary students' reading. The Journal of Educational Research, 108(4), 278–291.

Crossley, S. A., Weston, J. L., McLain Sullivan, S. T., & McNamara, D. S. (2011). The development of writing proficiency as a function of grade level: A linguistic analysis. Written Communication, 28(3), 282–311.

Garrison, D. R., Anderson, T., & Archer, W. (2001). Critical thinking and computer conferencing: A model and tool to assess cognitive presence. Retrieved from: https://auspace.athabascau.ca/bitstream/handle/2149/740/critical_thinking_and_comput?sequence=1.

Gilpin, A., & Wagenaar, R. (2008). Approaches to Teaching, Learning and Assessment in Competence Based Degree Programmes. Tuning Educational Structures in Europe. Universities' Contribution to the Bologna Process. An Introduction, 91–118.

Hyytinen, H., Nissinen, K., Ursin, J., Toom, A., & Lindblom-Ylänne, S. (2015). Problematising the equivalence of the test results of performance-based critical thinking tests for undergraduate students. Studies in Educational Evaluation, 44, 1–8.

IEA (2018) IEA International Computer and Information Literacy Study 2018 assessment framework. Springer. Switerland.

Ku, K. Y. (2009). Assessing students' critical thinking performance: Urging for measurements using multi-response format. Thinking skills and creativity, 4(1), 70–76.

Liu, O. L., Frankel, L., & Roohr, K. C. (2014). Assessing critical thinking in higher education: Current state and directions for next – generation assessment. ETS Research Report Series, 2014(1), 1–23.

MacArthur, C. A., Jennings, A., & Philippakos, Z. A. (2019). Which linguistic features predict quality of argumentative writing for college basic writers, and how do those features change with instruction?. Reading and Writing, 32(6), 1553–1574.

OECD (2012). Assessment of Higher Education Learning Outcomes – Feasibility Study Report. Volume 1 – Design and Implementation. Retrieved from: http://www.oecd.org/education/skills-beyond-school/AHELOFSReportVolume1.pdf.

Poce, A. (Ed.). (2012). Contributi per la definizione di una tecnologia critica: un'esperienza di valutazione. F. Angeli.

Poce, A., Amenduni, F., Re, M. R., & De Medio, C. (2019). Establishing a MOOC Quality Assurance Framework–A Case Study. Open Praxis, 11(4), 451–460.

Rahimi, Z., Litman, D., Correnti, R., Wang, E., & Matsumura, L. C. (2017). Assessing students' use of evidence and organization in response-to-text writing: Using natural language processing for rubric-based automated scoring. International Journal of Artificial Intelligence in Education, 27(4), 694–728.

Rear, D. (2019). One size fits all? The limitations of standardised assessment in critical thinking. Assessment & Evaluation in Higher Education, 44(5), 664–675.

Salton, G. and McGill, M.J. (1983) Introduction to Modern Information Retrieval. McGraw-Hill Book Co., New York.

Sasson, I., Yehuda, I., & Malkinson, N. (2018). Fostering the skills of critical thinking and question-posing in a project-based learning environment. Thinking Skills and Creativity, 29, 203–212.

Ullmann, T. D. (2019). Automated analysis of reflection in writing: Validating machine learning approaches. International Journal of Artificial Intelligence in Education, 29(2), 217–257.

UNESCO (2015). The future of learning 2: what kind of learning for the 21st century skills? Retrieved from https://unesdoc.unesco.org/ark:/48223/pf0000242996.

Vacca, F. (1972). Per una critica quantitativa: romanzi a chilometri, Il Messaggero.

Yang, W., Lu, X., & Weigle, S. C. (2015). Different topics, different discourse: Relationships among writing topic, measures of syntactic complexity, and judgments of writing quality. Journal of Second Language Writing, 28, 53–67.

Zhu, M., Liu, O. L., & Lee, H. S. (2020). The effect of automated feedback on revision behavior and learning gains in formative assessment of scientific argument writing. Computers & Education, 143, 103668.

Computational Social Science – Luo, Ciurea & Kumar (eds)
© *2021 Taylor & Francis Group, London, ISBN 978-0-367-70193-2*

Analysis on lending service of paper books of double first-class construction university libraries in COVID-19

H. Tan, X.Y. Cao & Y.H. Zhang
University of Electronic Science and Technology of China, Chengdu, China

ABSTRACT: [Purpose/significance] How to carry out lending services of paper books during the COVID-19 pandemic to meet the needs of teachers and students for teaching and scientific research is an important issue that deserves attention. [Method/process] With the method of bibliometric analysis, the lending service of paper books of double first-class construction university libraries in China was investigated after the outbreak. It has been sorted into key elements from the investigation, and shows three characteristic features which are process specification of "online" appointment, "offline" book delivery service, and "pandemic prevention" requirements in place. [Result/conclusion] Three aspects are enlightened, including taking collection "electronification" into consideration, innovating lending service as "online appointment + offline delivery" model, and promoting the public health safety as the focus of library management. It provides a reference for domestic and foreign university libraries to carry out lending services of paper books during the pandemic.

1 INTRODUCTION

In December 2019, there was a COVID-19 outbreak in Wuhan, Hubei, and then spread across the country. All provinces quickly launched a first-level response to major public health emergencies, and people across the country unanimously joined the fight against the pandemic. In order to implement the Party Central Committee and the State Council's decision-making on pandemic prevention and control, and in accordance with the requirements of the Ministry of Education for the prevention and control of new coronavirus infection pneumonia, university libraries issued notices of temporary closure or postponement of opening after the outbreak. The universities were originally scheduled to open in February. Due to the severe situation of the pandemic, all universities fulfilled their commitments to "suspend classes without stopping courses" and launched online teaching. University libraries, as an important support in teaching and scientific research, have no "function" missing in the pandemic situation. According to the unified requirements of "suspending the opening but not suspending the service," university libraries during the closure innovated the form of patrons' service in order to ensure uninterrupted service for scientific research and teaching under pandemic prevention and control.

The lending service of paper books has always been the most important traditional patron service of university libraries. In recent years, with the development of the Internet and information resources, e-books have had a greater impact than paper books. There has always been attention given to the differences and comparison of reading between paper books and e-books, from comparison of studies on visual fatigue and perception[1] to recently published meta-analysis[2]. Paper books remain an important information carrier tool in the digital information era[3], and most university libraries have also proposed improvement measures and plans to increase the borrowing rate of paper books[4]. Relevant research from college groups shows that paper book reading remains the mainstream in universities[5]. A survey by Direct Textbook shows that 70% of college students prefer paper textbooks[6]. On the other hand, the US Consumer News and Business Channel (CNBC) gave a report at the end of 2019 entitled "Why do physical books still outperform e-books," pointing out that today's global market for paper books remains strong[7]. It can be seen therefore that the demand for paper book lending cannot be ignored. Under the pandemic situation, it remains a very necessary patron service for university libraries. How do university libraries break through the limitations under the current situation of temporary closure or postponement of opening, to meet patrons' borrowing needs for paper books and support teachers and students in teaching and scientific research? This paper investigates and analyzes the current status of the lending service for paper books of double first-class construction university libraries in China during the pandemic, and then sorts and analyzes the current status of the lending service work.

2 THE INVESTIGATION AND ANALYSIS ON THE LENDING SERVICE OF PAPER BOOKS OF DOUBLE FIRST-CLASS CONSTRUCTION OF UNIVERSITY LIBRARIES UNDER THE PANDEMIC RESTRICTIONS

2.1 An online survey about the lending service of paper books of double first-class construction of university libraries

An online survey was used as the method from select 23 January 2020 to 24 March 2020 during the period of the pandemic, to investigate whether libraries of all 42 double first-class construction universities in China carried out a lending service for books.

In order to ensure teachers and students' teaching and scientific research, 42 double first-class construction university libraries in China adjusted and innovated their lending services. Among them, 19 libraries were launched successfully during the pandemic. Further analysis of the lending service work of paper books in 19 libraries was carried out. The key points are summarized in Tables 1 and 2. (The ranking of university libraries is sorted according to the order of their service opening time.)

2.2 Sorting out the key elements of the investigation

Based on the above research, the authors summarize the key elements of the above 19 libraries carrying out their lending services of paper books, as follows.

2.2.1 Service opening time

University libraries during the pandemic successfully launched their lending service of paper books after the closure. Peking University Library[8] first took the lead in launching the "book delivery to the building" service to establish a lending service of paper books on February 3, 2020. According to the order of the opening service time, during the period from January 23, 2020 to March 29, 2020, nine university libraries successfully opened their paper books lending service; during March 1, 2020 to March 24, 2020, ten university libraries also conducted their lending services. Judging by the length of the service running time, 18 university libraries provided a lending service of paper books since opening. The entrusted borrowing service of paper books carried out by the Library of University of Science and Technology of China (USTC) was effective from February 17, 2020 to March 22, 2020. It is currently the only one that has ended the lending service during the period of survey. According to a notice published on its website, the University of USTC[9] gradually resumed normal operation on March 23, 2020. The entrusted borrowing service of paper books has been shut down since. Patrons are advised to use e-copies of collections to meet demand.

2.2.2 Service name

All the above 19 university libraries have lending services of paper books. Each service name is different. In terms of creativity of service name, some are catchy and appealing to patrons. For example, the "book delivery to the building" service from Peking University Library and the "book delivery" service from the Library of University of Electronic Science and Technology of China

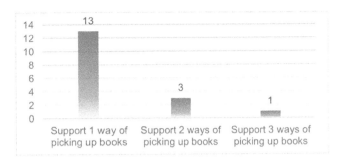

Figure 1.　Comparison of ways in which double first-class construction university libraries supported the paper books lending service.

(UESTC), "delivery" has the meaning of "takeaway". As for the content of the service name, 18 university libraries are aimed at reserving and lending services for paper books. Only Shandong University Library has a consignment service for the "electronic collection of paper books." It does not need to notify patrons to pick up or distribute books at designated locations thanks to this kind of service. The entire service process can be completed online, which is a major advantage of the lending service of paper books under pandemic conditions.

2.2.3　*Service reservation method*
The loan-service reservation methods of 19 university libraries mainly include book-to-door service system submission applications, email applications, QQ reservations, WeChat public account platform reservations, "personal library" account delegation applications, other "online" reservation methods, and telephone reservations. The email application method is a more traditional online reservation method, and the WeChat public account platform is a relatively new "online" reservation channel, which is popular and convenient for people. At present, there are three university libraries, which are Tianjin University Library, Library of UESTC, and Hunan University Library, supporting WeChat public account platforms for appointments. The surveyed university libraries adopted the "online" reservation method as the main method and the "offline" reservation method as the supplementary one, which solved the problem that patrons could not go to the library to apply for lending services due to geographical factors during the pandemic prevention and control.

2.2.4　*Pick-up methods of service*
The pick-up methods mainly include the following three types: patrons pick up books at the library, books delivered by librarians at designated locations, and express delivery. As Shandong University Library developed an entrustment service of electronic books for picking up online, and Yunnan University did not clearly specified the method of book collection, the author focuses on the analysis of the other 17 double first-class construction university libraries involved in the pick-up method of paper books.

According to the statistics, 13 libraries support one type, three libraries support two types, and one library supports three types, as shown in Figure 1. The Chongqing University Library was at that time the only one in the survey which supported three methods of pick-up. For faculties and students who do not have convenient access to their university and have urgent need for paper books, it provides an express delivery service[10]. Thirteen libraries supported pick-up at the library, eight libraries supported distribution of books by librarians, and only one library supported express delivery. The statistics about the three methods for pick-up are shown in Figure 2.

2.2.5　*Service object*
Most libraries during the pandemic mainly serve faculties and students. The service targets of some libraries were expanded to the local population. For example, Yunnan University Library provided

Table 1. The survey of paper books lending services of double first-class construction of university libraries.

No	Univ. Library Name	Service name	Service opening time	Service reservation method	Service notification method	Pick-up methods of Service	Service object	Others
1	Peking Univ. Library	"Book delivery to the building" service	From 2020-02-03	Submits a loan request online	Unspecified	Books delivered by librarians at designated locations	Faculties and students on campus	Packed in plastic bags
2	Dalian Univ. of Technology Library	Borrow online in deed urgently	From 2020-02-08	Online appointment, telephone appointment		Contactless delivery	Faculties and students	
3	Xiamen Univ. Library	Urgent need for loan service of paper books	From 2020-02-11	Online appointment	Phone or email notification	Pick up books at designated locations on campus	Faculties and students on campus	Do not accept cross-campus entrustment
4	Univ. of Science and Technology of China Library	Paper books entrusted loan service	From 2020-02-17 to 2020-03-22	Apply at "Personal Library" account	Phone notification campus	Picking up at the library	Faculties and students on	
5	Tianjin University Library	Book lending service	From 2020-02-18	Mail application (WeChat and website)	Phone notification	Picking up at the library	Faculties and students stay at campus	Provide relevant identification when picking up
6	Shandong University Library	Entrusted service of "Electronic Collection of Paper Books"	From 2020-02-21 campus card	Online appointment to 4 volumes	Email notification	Send to patrons by email	Faculties and students holding an official	Scope of application: Jinan and Qingdao collections; apply up
7	Library of Shanghai Jiaotong University	Book Borrowing Reservation Service	From 2020-02-24	Mail application	Phone notification	Pick up books at library or deliver at designated locations	Prioritize the needs of online teaching for teachers	Only in the main library of Minhang Campus
8	Univ. of Electronic Science and Technology of China Library	"Book Delivery" Service	From 2020-02-25	Telephone appointment, online appointment (WeChat)	Phone notification	Deliver at designated locations on campus	Faculties and students on campus	
9	Tongji University German Library	Online book lending service	From 2020-02-28	Online appointment	Phone or email notification	Pick up at library's security	Faculties and students working days	Pick-up does not exceed 3
10	Yunnan University Library	Paper books loan service	From 2020-03-03	Contact staff by phone, QQ, email	Unspecified	Unspecified	Readers in Kunming	Only loan, not for return

Table 2. The survey of paper books lending service of double first-class construction of university libraries.

No	Univ. Library Name	Service name	Service opening time	Service reservation method	Service notification method	Pick-up methods of Service	Service object	Others
11	Xi'an Jiaotong University Library	Book loan service	From 2020-03-03	Mail application	Phone notification	Pick up books at library or deliver at designated locations	Faculties and students on campus	
12	Northwestern Polytechnical University Library	Paper books and literature loan service	From 2020-03-04	Mail application	Phone notification	Pick up books at library or deliver at designated locations	Faculties and students on campus	Do not accept cross-campus entrustment
13	Nanjing University Xianlin Campus Library	Paper Books reserve and loan Service	From 2020-03-04	Online appointment (website and QR code)	Phone notification	Pick up books at library	Students living in Xianlin campus, teachers in need	Apply up to 5 volumes
14	Chongqing University Library	Paper Books Contactless reservation and lending Service	From 2020-03-09	Online appointment(email)	Phone or email notification	Pick up books at library or deliver by librarians or Express delivery	Faculties and students	Appointment will be cancelled automatically after the notice is issued one week; apply up to 5 volumes
15	Hunan University Library	Book reservation service	From 2020-03-09	Reserve at WeChat		Pick up books at library	Patrons in need	Pick up books at library within 7 days after notification
16	Central South University Library	Paper books loan service of "online reservation and pick-up by patrons"	From 2020-03-11	Online appointment	Phone or email notification	Signed and picked up by the doorman of corresponding campus' library	Local faculties in Changsha	Apply up to 5 volumes
17	Beijing Normal University Library	The appointment of book loan service	From 2020-03-16	Online appointment	email notification	Pick up books at library	Faculties and students	Specified borrowing time
18	Southeast University Library	"Local entrusted loan" Service	From 2020-03-18	Apply at "Personal Library" account	Phone or email notification	Pick up at self-service machine	Faculties who resume work	The entrusted book will only be kept for 7 days
19	Zhejiang University Library	Open for book reservation	From 2020-03-23	Apply by email		Pick up books at a designated place	Faculties and students on campus	Pick up books within two working days, cross-campus service is not provided.

Figure 2.　Chart of ways of picking up paper books in double-first-class construction university libraries.

services for patrons who were already in Kunming; the service targets of Central South University Library referred specifically to the faculties and staff in Changsha local campus. In addition, some libraries focused on "in demand" and "faculties" as a clear service object. For example, Shanghai Jiaotong University Library gave priority to meeting teachers' needs for online teaching; the service targets of the Nanjing University Xianlin Campus Library referred to students living in the Xianlin Campus and teachers in need; Hunan University Library emphasized that service objects are aimed at those in need and Southeast University also referred to faculties.

2.2.6　*Others*

First, the lending service of paper books did not provide cross-campus lending services during the pandemic, for example Xiamen University Library, Northwestern Polytechnical University Library, etc. The second is to avoid the waste of book resources. Although patrons failed to pick up books for personal reasons after making an appointment or commissioning, some libraries clearly stipulated a specific book retention period. Tongji University German Library stipulated that the pick-up time was only 3 working days; the Southeast University Library stipulated that the books were only reserved for 7 days. The third is to rationally allocate book resources. Some libraries stipulated the maximum number of applications for paper books. For example, the library of Shandong University stipulated a maximum of four volumes, and the Nanjing University Library, Chongqing University Library, and Central South University Library all stipulated a maximum of five volumes.

3　ANALYSIS OF CHARACTERISTICS OF THE LENDING SERVICE OF PAPER BOOKS FOR DOUBLE FIRST-CLASS CONSTRUCTION UNIVERSITY LIBRARIES UNDER PANDEMIC RESTRICTIONS

3.1　*Process specification of "online" appointments*

In the investigation, 19 libraries under pandemic restrictions had launched "online" appointments to solve the problem that patrons were unable to apply to the library due to geographical factors, so that patrons were not subject to time and space limitations, and for a convenience service provided by the library[11]. Firstly, in order to facilitate patrons to apply, to solve the problem that patrons were not clear about the process of reservation, and effectively increase reservations and borrowing volume, some libraries gave detailed explanations of the process. An example was the flow chart for the borrowing service online provided by Tongji University German Library. This was more standardized than others. It could check the collection catalog to confirm that the required book status was "available," then record the information about the book, confirm whether the patron met the loan conditions, and then submit the application and wait for the librarian to notify the client. The entire flowchart is comprehensive, clear, and easy to operate, as shown in Figure 3.

　　Second, from the perspective of application forms provided by librarians, there are four libraries, including Beijing Normal University Library, Dalian University of Technology Library, Tongji

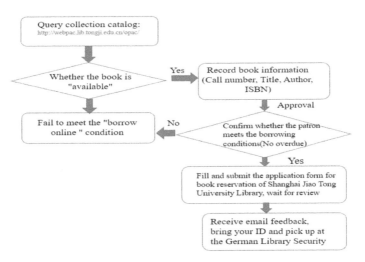

Figure 3.　Flowchart for the online borrowing service for books of Tongji University German Library[12].

Table 3.　The application form for book reservation of Shanghai Jiao Tong University Library.

Serial number	Title	Call number	Collection place	Student/faculty Name	Telephone number	Book delivery number	location	Remarks

University German Library, and Shanghai Jiaotong University Library, which provided special document application forms or Excel application forms, accounting for 21%. Judging from the content of the application forms, they contain two major items: "reader information" and "book-related information." "Reader information" contains personal information such as student ID, name, cell phone number, email address, department, etc. "Book-related information" includes book title, author, call number, collection place, etc. The application forms of the first two libraries contain approximately the same content. Compared with the application form of the online borrowing service of Tongji University German Library, the "commitment" requirement is added and this item needs to be signed by the patron. An autographed application form must also be filed in the German Library. The entire process is very formal and standardized. In addition to the simple "reader information" and "book information," the application form of the Shanghai Jiaotong University Library adds the location information for pick-up, the necessary information is all arranged in a column table, which is clear and simple, as shown in Table 3.

3.2 "Offline" book delivery service

According to the statistics from the above survey, pick-up in the library as a traditional borrowing mode was most used. However, eight libraries provided book delivery services, including Peking University Library, Dalian University of Technology Library, Shanghai Jiaotong University Library, Xiamen University Library, Chongqing University Library, University of UESTC, Xi'an Jiaotong University Library, and Northwestern Polytechnical University Library. Due to closure of libraries and the regulations for pandemic prevention and control, the entrances to universities during the pandemic were strictly controlled, and patrons and express delivery were unable to enter the campus. Therefore "offline" book delivery was an effective way to solve the problem that patrons could not access paper books. It is a major innovation and feature in the lending service of paper books. University librarians have played the role of "takeaway" deliveryman for delivering books on campus. For example, the library of Northwestern Polytechnical University provided a door-to-door book delivery service on campus, and the library of Xi'an Jiaotong University provided a book delivery service within three campuses.

3.3 *"Pandemic prevention" requirements in place*

While carrying out the lending service of paper books, some libraries had put pandemic prevention and control measures in place. Southeast University Library[13] stipulated that patrons must wear a mask and cooperate with the pandemic prevention and control management when entering the library for pick-up; the library of University of Science and Technology of China gradually resumed normal operations on March 23, stipulating that all admission staff must receive temperature measurements and wear masks, and strictly implement swiping cards and registration for admission, otherwise entry was not permitted[9]. As for books delivered by librarians, some librarians took safety precautions during delivery and disinfected the books so that patrons felt reassured. For example, Peking University Library stipulated that needed books should be packed in plastic bags; Library of UESTC[14] also stipulated that, to ensure safety, all books were sterilized for delivery and packed in plastic bags; Chongqing University Library[15] stipulated that, in addition to wearing masks and gloves and sterilizing books, librarians also used special pockets to pack the books in order to ensure pandemic prevention and control.

4 ENLIGHTENMENT AND REFLECTIONS ON THE LENDING SERVICE OF PAPER BOOKS OF UNIVERSITY LIBRARIES IN CHINA

4.1 *Take collection "electronification" into consideration*

The transformation of paper books into "electronic ones" carried out by some double first-class construction university libraries is worthy of reference. Among them are Tsinghua University Library, Library of Beijing University of Aeronautics and Astronautics, Sichuan University Library, Lanzhou University Library, which are relatively well placed for "electronic teaching reference" services. For example, Tsinghua University Library[16] had already established a special "Teaching Staff Service Platform" to provide an electronic teaching reference service for faculties and students to meet borrowing needs. In addition, the Library of Beijing University of Aeronautics and Astronautics launched a full range of electronic resource services. "BcLibrary"[17] pushed and remotely accessed resources from four modules. Faculties and students of that university can access e-books remotely at any time and any place.

These libraries have carried out "electronic teaching reference" services and a comprehensive collection of "electronic resource plus paper books" to make up for the lack of paper books available, which are very suitable for an innovation of rapid access to library resources and effectively ensure the reader's research and learning under pandemic. Therefore, it is a good lesson for university libraries to make collection construction "electronic."

4.2 *Innovate lending service as "online appointment + offline delivery" model*

The lending service of paper books carried out by 19 double first-class university libraries is a service that breaks through the barriers of the pandemic, meeting the needs of paper books lending, and supporting faculties and students in teaching and scientific research. The "online appointment + offline delivery" mode of paper books has become an innovative form of service in emergency situations, which has solved the difficulty of university libraries not being able to carry out normal lending services of paper books due to their temporary closure under the pandemic. As we all know, the "takeaway delivery" service is an effective popular service model that meets people's shopping needs in e-commerce services, especially in the context of advocating the antipandemic message to "stay at home." The "offline" book delivery service of university libraries is an effective adoption of the "takeaway delivery" service from other industries. This "offline" book delivery service form is different to the traditional model of self-service lending for patrons, and is an innovative form for university librarians who walk out of the library and actively serve patrons. Therefore, the "online appointment + offline delivery" service form is an effective way for the lending service of paper books to continue under pandemic restrictions.

4.3 *Promote public health safety to the focus of library management*

University libraries should take the pandemic as an opportunity to include public health safety in the management category, to improve risk prevention and control, to strengthen the library response to pandemic measures, and to increase library emergency management capabilities.

One method is to take body temperature measurements before patrons enter the library, to increase the infrared nonsensing temperature entry detection, to strengthen the body temperature monitoring, and report to the university information platform for patrons who exceed the temperature measurement at the time. Patrons with a body temperature exceeding 37.3° C will be forbidden entry. The second method is to normalize the mass media propaganda work for the prevention and control of COVID-19. Librarians and patrons are required to wear masks when working and studying in libraries. The library can also warmly remind patrons of eye-catching and lively slogans by posting various public health safety signs and other forms, such as publicizing the "seven-step method" to wash hands, paying attention to coughing and sneezing etiquette—when sneezing or coughing, cover your nose and mouth with a tissue, or use the elbows to cover it. The third method is to equip the hall with sanitation and disinfection materials. Libraries should be equipped with professional book disinfection machines to enable the disinfection of returned books, to improve the disinfection and sanitation of public areas, and to increase the frequency of cleaning and disinfection of high-use facilities, such as elevators, self-service machines, and query and retrieval computers. In addition, it is necessary to ensure the reasonable provision of alcohol-containing disinfectant wipes, masks, disinfection cotton balls, and other materials, besides the library bookshelves or toilets.

5   CONCLUSIONS

University libraries have been at the forefront of pandemic prevention and control, innovating the lending service of books for patrons, contributed to serving a good role in scientific research and teaching, and strived to help the victory in the fight against the pandemic.

It is an enlightenment of the above three aspects for university libraries, which are taking collection "electronification" into consideration, innovating lending services as "online appointment + offline delivery" models, and promoting public health safety to the focus of library management. In this paper, it is expected that the research and analysis can provide reference for the implementation of lending services of paper books for domestic and foreign university libraries under COVID-19 restrictions.

REFERENCES

[1] Hanho Jeong. A comparison of the influence of electronic books and paper books on reading comprehension, eye fatigue, and perception. The Electronic Library, 2012, 30(3):390-408.
[2] Yiren Kong, Young Sik Seo, Ling Zhai. Comparison of reading performance on screen and on paper: A meta-analysis.Computers in Education, 2018, 123:138–149.
[3] Li Wei .The Value and Service Countermeasures of Print Resources in the University Library in the Digital Information Ages Library and Information Service, 2017(3): 79–85.
[4] Li Li Juan. Analysis of Strategies to Improve the Borrowing Rate of Paper Documents in University Libraries. Office Operations.2018, (5):175.

[5] Zhi Keke, Liu Hua. Comparative Study on the Borrowing of Same Title Comparisons E-books and Print Books—Taking the Class H and Class I of Shanghai University Library as an Example. Library Development, 2017(6): 46–52.

[6] Gong Chen. Survey shows: 72% of college students prefer paper books to e-books. Papermaking information, 2016(4):74.

[7] Physical books still outsell e-books-and here's why[EB/OL] .[2019-09-19].https://www.cnbc.com/2019/09/19/physical-books-still-outsell-e-books-and-heres-why.html

[8] Library's notice on adjusting service methods before the start of school[EB/OL].[2020-01-31]. https://www.lib.pku.edu.cn/portal/cn/news/0000002112

[9] Notice of normal operation of the library during pandemic prevention and control[EB/OL].[2020-03-23].http://lib.ustc.edu.cn/cat_news/.

[10] Announcement on the opening of the "contactless book lending for paper books" service [EB/OL].[2020-03-09].http://lib.cqu.edu.cn/news/newsdetail/2473

[11] Lu Hong Ying, Tong Yun Juan. Discussion on Personalized Service Mode of Book Reservation Library Work and Study, 2009,(3):104–106.

[12] Tongji University German Library Online Book Lending Service Launched[EB/OL].[2020-02-28].https://www.lib.tongji.edu.cn/index.php?classid=11979&newsid=30759&t=show

[13] How to borrow books by faculties when the library has been closed during the pandemic [EB/OL].[2020-03-18]. https://mp.weixin.qq.com/s/CPnWeQLW0-UJ9qNJcBLdMA

[14] The library opens the "book delivery" service to fight against pandemic [EB/OL].[2020-02-25]. http://www.lib.uestc.edu.cn/news?id=2171

[15] Library "Zero Distance" Service (2)-"Contactless Appointment Lending of Paper Books" [EB/OL].[2020-03-15]. http://lib.cqu.edu.cn/news/newsdetail/2478

[16] Guide to the Library's Electronic Teaching Services during Pandemic Prevention and Control [EB/OL].[2020-02-13].https://www.lib.tsinghua.edu.cn/dra/news/annoucement/7876

[17] Building" BcLibrary" to escort the teaching and scientific research of teachers and students [EB/OL].[2020-03-03].https://lib.buaa.edu.cn/newsinfo?cid=1&id=20307&pid=46

Computational Social Science – Luo, Ciurea & Kumar (eds)
© *2021 Taylor & Francis Group, London, ISBN 978-0-367-70193-2*

The influence of crisis on hotel profitability: A case study of Novel Coronavirus (COVID-19)

X.L. Xian

Department of Management, Shenzhen Institute of Information Technology, China

ABSTRACT: This paper is to investigate the factors of hotel profitability by introducing a new construct: Crisis Severity into a Partial Least Model and examine to what extent different factors affect hotel profitability under the severe pandemic of COVID-19. Data were collected from different hotel managers across China. From the results of PLS-SEM, Hotel Locations, Tourist Destination, Crisis Severity significantly influence Hotel Performance. This paper contributes to the defining the determinants toward hotel profitability and integrated Crisis Severity as a new variable. It provides a deeper insight for hotel business or tourism industries to better quantify hotel profitability.

1 INTRODUCTION

COVID-19, a respiratory illness with pneumonia-like symptoms, was declared an international public Health Emergency in January 2020. Because of its rapid speed and scary scale of transmission, more than 20,311,350 cases have now been reported to WHO, from over 100 countries and territories. Not only threatened the loss of life, COVID-19 caused tremendous economic losses (Foster K.A. & Matthew. K.A., 2020). The epidemic risk of the COVID-19 impacted travel volumes and global hotel revenue is expected to decrease as tourists adopt a pessimistic cautious approach. The vulnerability in tourism stems from possible political, social, nature-driven disasters and infectious diseases that deter tourists from traveling (Sonmez, Apostolopoulos, and Tarlow, 1999).

With the outbreak of COVID-19, the occupancy rate of hotels in mainland China dropped dramatically from its highest 70% on January 14th to 17% on January 26th. Thus, it is vital to understand the impacts of crisis and consequent influence. This research is to investigate the factors of hotel profitability by introducing a new construct: crisis severity into a Partial Least Modeling and explain how hotel profitability be determined by different factors under the severe pandemic of COVID-19. This paper unifies former research approaches into one single model. And this research will help provide insights regarding the determinants of hotel performance and serve to enlighten those in hotel industry, specifically those working in the field of crisis management.

2 LITERATURE REVIEW

2.1 *Prior research on hotel profitability*

Hotel Profitability determinants have been studied by various approaches. Prior studies examined that there are external determinants and internal determinants in the hotel sector.

External factors such as economic growth, national policies, political situation, economic crisis and other factors affect performance and profitability of hotels. Hotel business is extremely sensitive and easily influenced by economic crisis (Megoutas & Sfakianakis, 2013). The global financial crisis has weakened the revenue of hotel industry (KapiKi, 2012).

Internal factors such as hotel size, age and leverage influence on tourism companies' profitability (Agiomirgianakis, Magoutas, Sfakianakis, 2012; Ben, Aissa & Goaied, 2014). Moreover, various financial structures can affect availability of investment capital (Chan, Cheung & Law, 2012). According to Javanovic (1982) and Wernerfelt (1984), the key determinants of a company's success accounts to its resources and capabilities. These factors constitutes of intellectual capital and innovation, capitalization, size and financial resources (Agiomirgianakis et al., 2012). In order to achieve profitability, room occupancy must also be considered (Cheng, 2013; Kim, Cho, & Brymer, 2013).

2.2 *Crisis and profitability*

A crisis is considered as an accident that is crucial to a firm's reputation or its growth, profit or survival (Stafford, Yu & Armoo, 2002). If not well prepared, a crisis will result in loss of revenue. Chen et al. (2005) confirmed Taiwanese hotel stock performance was influenced by unexpected events, such as the SARS outbreak. Soemodinoto et al. (2001) estimate a decrease of 6,680 tourist arrivals and more than 320,000 USD loss tourism caused by large-scale riots in Indonesia. Blake and Sinclair (2003) used a computable equilibrium model to investigate the impact of the 9.11 terrorist attacks.

3   METHODOLOGY

3.1 *Sampling and method*

The research applied an adapted version of the instruments by Ruben and Milagros (2018) to evaluate the determinants. And the measurements for the Crisis Severity is first developed in this study. Questionnaires were sent out to different hotel personnel across China. Online survey was posted for two weeks on a popular questionnaire platform in China (www.wjx.com) and customized only be filled out by employees from the hotel industry. By conducting online surveys in April 2020, 200 valid data were collected over 324 sets of data and the valid response rate was 61.7%. Partial Least Squares regression is adopted to examine the data. PLS can fit multiple response variables within one model. It does not assure the predictors are fixed, making it more robust to measure uncertainty. Therefore, PLS (SmartPLS 3.2) was applied to test the proposed research model in this study.

3.2 *Hypotheses*

Prior research verified that there are four factors in hotel profitability and they are the hotel characteristics, location, competition and destination factor (Ruben & Milagros, 2018).

In this research, Crisis Severity (CS), a new construct, will be investigated into the primary determinants identified in previous studies and examined whether it may have positive effects on hotel performance. Figure 1 illustrates the research model.

3.2.1 *Hotel characteristics*
Dominant factors of hotel performance includes hotel characteristics and capital structure. Since size can generate economies of scale (Demsetz, 1973, 1974), it is the most frequently used variable. Another factor is the capital structure. Greater liquidity reduces the default risk and increases hotel profit (Kim et al., 2007). Indebtedness affects profitability adversely (Liu and Hung, 2006). Therefore, the formative indicators of Hotel Characteristics are: Rooms, Liquidity, Debt.
H1: Hotel Characteristics (HC) has a direct effect on Hotel Performance (HP).

3.2.2 *Hotel location*
Studies confirmed that proximity to nodes of attraction positively affects hotel price or profitability (Sainaghi, 2011; Lee & Jang, 2012). Honma and Hu (2012) verified that the proximity to airports

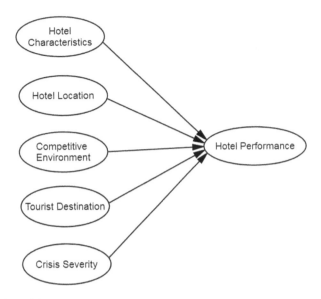

Figure 1. Research model.

influences hotel efficiency. And agglomeration and urbanization (Chun & Kalnins, 2001; Chnina et al., 2005) also affect hotel profitability. Thus, the formative indicators of Hotel Locations are: Density, Airport and CBD.

H2: Hotel Location (HL) has a direct effect on Hotel Performance (HP)

3.2.3 *Competitive environment*
Lado-Sestayo et al. (2016) verified that market structure is important in hotel profitability. The formative indicators of Competitive Environment are: Market Share and Reputation.

H3: Competitive Environment (CE) has a direct effect on Hotel Performance (HP)

3.2.4 *Tourist destination*
Tourist destination could account for nearly ninety percent of hotel survival (Lado-Sestayo et al., 2016). And the seasonality was verified as a factor that could decrease hotel performance (Butler, 2001; Ganguin & Bilardello, 2005). The formative indicators of Tourist Destination are: Occupancy Rate and Seasonality.

H4: Tourist Destination (TD) has a direct effect on Hotel Performance (HP)

3.2.5 *Crisis severity*
Three items are used to test the Crisis Severity of a particular city. They are: the number of confirmed cases in that particular city, its ranking in the total number of confirmed cases national wide and mortality rate of COVID-19 in that city.

H5: Crisis Severity (CS) has a direct effect on Hotel Performance (HP).

4 DATA ANALYSIS

4.1 *Characteristics of samples*

From Table 1, 40.5% of hotel respondents are males, 59.5% are females. Among their various positions, captains were most (28.5%). And respondents aged 26–40 accounted for 77.5%. Only slight difference was witnessed between 3 Stars, 4 Stars and 5 Stars, respectively 28%, 29.5% and 28%.

Table 1. Demographic information (n = 200).

		Frequency	Percentage
Gender	Male	81	40.5
	Female	119	59.5
Occupation	Staff	35	17.5
	Captain	57	28.5
	Manager	40	20.0
	Vice Head of Department	30	15.0
	Head of Department	27	13.5
	Director	5	2.5
	Deputy General Manager	3	1.5
	General Manager	3	1.5
Age	18–25	32	16.0
	26–30	68	34.0
	31–40	87	44.0
	41–55	12	6.0
District (Top 10)	Shanghai	29	14.5
	Zhe Jiang Province	19	9.5
	Beijing	17	8.5
	Shan Dong Province	16	8.0
	Jiang Shu Province	13	6.5
	Hu Bei Province	12	6.0
	He Bei Province	12	6.0
	Guang Dong Province	10	5.0
	He Nan Province	10	5.0
	Shan Xi Province	9	4.5
Hotel Stars	None	11	5.5
	1 Star	2	1.0
	2 Stars	16	8.0
	3 Stars	56	28.0
	4 Stars	59	29.5
	5 Stars	56	28.0

4.2 Reliability and construct validity

Table 2 presents the standard deviations, means, excess kurtosis and skewness of the measurement items. The data reached the recommended level as all Outer loadings of constructs exceed 0.7.

Table 3 indicates a high degree of reliability and validity as the values of Cronbachs' alpha, CR are higher than 0.7, and AVE values exceed 0.5.

Table 4 indicates that the data are relatively independent of one another.

4.3 PLS results

Figure 2 demonstrates the PLS Results in the research model. Bootstrapping was performed using 200 responses to 5000.

5 CONCLUSION

From the PLS-SEM results, the R2 value is 0.510, meaning the research model explaining up to 51.0% in Hotel Profitability. The P values of HL, TD and CS are less than 0.05, meaning these three variables significantly influence HP. Thus, the Hypothesis 2, 4 and 5 are supported, but H1 and H3 are unsupported. The Beta Value of Tourist Destination weighed most (0.586), meaning

Table 2. Mean, Standard Deviation, Excess Kurtosis and Skewness.

	Mean	Std. Dev	Excess Kurtosis	Skewness	PLS Loadings
HC1	3.465	1.679	−0.667	0.499	**0.717**
HC2	4.635	1.234	−0.040	−0.306	**0.942**
HC3	4.530	1.330	−0.757	−0.039	**0.713**
HL1	4.995	1.302	−0.668	−0.278	**0.786**
HL2	4.560	1.535	−0.356	−0.603	**0.754**
HL3	2.135	1.605	0.864	1.371	**0.865**
CE1	4.635	1.171	−0.276	−0.047	**0.963**
CE2	5.145	1.369	−0.427	−0.441	**0.919**
TD1	3.125	1.543	−0.178	0.703	**0.996**
TD2	4.575	1.501	−0.632	−0.230	**0.867**
CS1	3.355	1.637	−0.050	0.760	**0.844**
CS2	3.110	1.693	−0.018	0.785	**0.751**
CS3	2.870	1.727	0.055	0.959	**0.734**
HP1	2.525	1.367	−0.030	0.740	**0.924**
HP2	2.650	1.441	0.222	0.883	**0.912**
HP3	2.650	1.314	0.053	0.682	**0.892**

Table 3. Cronbachs' alpha, CR, and AVE.

	Cronbachs' Alpha	CR	AVE
Competitive Environment (CE)	0.875	0.939	0.886
Crisis Severity (CS)	0.873	0.745	0.634
Hotel Characteristics (HC)	0.757	0.791	0.768
Hotel Location (HL)	0.770	0.782	0.646
Hotel Performance (HP)	0.895	0.935	0.827
Tourist Destination (TD)	0.715	0.752	0.768

Table 4. Square Roots of AVEs.

	CE	CS	HC	HL	HP	TD
CE	0.941					
CS	−0.104	0.659				
HC	0.616	−0.107	0.684			
HL	0.338	0.006	0.319	0.632		
HP	0.247	−0.174	0.254	0.419	0.909	
TD	0.234	−0.148	0.228	0.339	0.679	0.707

it is a key factor to explain hotel profitability. The Beta Value of CS (−0.084) was a negative figure, demonstrating that the higher the crisis severity in a district, the more it decreased the hotel profitability in that area.

From the sample characteristics, it's found that the occupancy rate of hotels in mainland China dropped dramatically after the COVID-19 outbreak. Approximately 81% of the hotels fell below the occupancy rate of 50%, in which nearly half couldn't achieve 20% (Table 6).

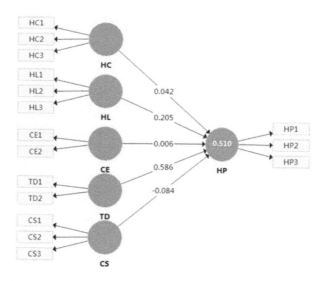

Figure 2. SmartPLS results.

Table 5. Results of PLS-SEM Analysis in research model.

Factor → Hotel Performance (HP)	Beta Value	P-Value	Conclusion
H1: Hotel Characteristics (HC) → HP	0.042	0.573	Reject
H2: Hotel Location (HL) → HP	0.205	0.001	**Accept**
H3: Competitive Environment (CE) → HP	0.006	0.921	Reject
H4: Tourist Destination (TD) → HP	0.586	0.000	**Accept**
H5: Crisis Severity(CS) → HP	−0.084	0.038	**Accept**

Table 6. Hotel occupancy rate in China.

		Frequency	Percentage
Occupancy Rate	0–10%	26	13.0
	11–20%	50	25.0
	21–35%	63	31.5
	36–50%	23	11.5
	51–65%	15	7.5
	66–80%	18	9.0
	81–100%	5	2.5

6 LIMITATIONS & FUTURE RESEARCH

There are some limitations in this study. Firstly, besides Hotel Characteristics, Tourist Destination and Crisis Severity, there might be other potential constructs such as government policies, economic growth that this research did not evaluate. Secondly, multi-group analysis (MGA) can considered to further evaluate dataset by dividing hotels into several groups including hotel stars, district etc. Comparison can be more easily found and model results may differ across certain variables.

ACKNOWLEDGEMENTS

Fund Project: Empirical Analysis of Influencing Factors of Information Education Acceptance in the New Media Era: PLS-SEM Model, 2019GXJK246.

REFERENCES

Alan, K.M., Bala, R. and Matthew. C.H. 2005. The Effects of SARS on the Hong Kong Tourism Industry: An Empirical Evaluation. *Asia Pacific Tourism Association*, 10(1): 85–95.

Aissa S.B. and Goaied, M.2016. Determinants of Tunisian Hotel Profitability: The Role of Managerial Efficiency. *Tourism Management*, 52(2): 478–487.

Hung, W.T., Shang, J.K. and Wang, F.C. 2010, Pricing Determinants in the Hotel Industry: Quantile Regression Analysis. *International Journal of Hospitality Management*, 29(3): 378–384.

Ivanov, S. and Zhechev, V. 2012, Hotel Revenue Management – A Critical Literature review, *Turizam: znanstveno-Stru cni Casopis*, 60(2): 175–197.

Lado-Sestayo, R., Otero-González, L., Vivel-Búa, M. and Martorell-Cunill, O. 2016, Impact of Location on Profitability in the Spanish Hotel Sector, *Tourism Management*, 52(2): 405–415.

Ming, H.C. 2011, The Response of Hotel Performance to International Tourism Development and Crisis Events, *International Journal of Hospitality Management*, 30(2): 200–212.

Ruben, L.S & Milagros, V.B.2018. Profitability in the Hotel Sector: A PLS approach. *Journal of Hospitality and Tourism Technology*, 9(3): 455–470.

Samuel, S.K. Hejin, C. and Heesung, L. 2005, The Effects of SARS on the Korean Hotel Industry and Measures to Overcome the Crisis: A Case Stud of Six Korean Five-Star Hotels, *Asia Pacific Journal of Tourism Research*, 10(4): 369–377.

Yang, Z. and Cai, J. 2016, Do Regional Factors Matter? Determinants of Hotel Industry Performance in China, *Tourism Management*, 52(5): 242–253.

Computational Social Science – Luo, Ciurea & Kumar (eds)
© *2021 Taylor & Francis Group, London, ISBN 978-0-367-70193-2*

The administrative emergency response research in the COVID-19 outbreaks

T.M. Yan & Z.B. Pei
Dalian Ocean University, Dalian, China

ABSTRACT: The COVID-19 outbreak has brought a great test to the people of Wuhan and even the whole China. The importance of administrative emergency response is highlighted at this moment. Administrative emergency response requires the government to take appropriate emergency measures in case of emergency to ensure public order and the safety of public life. Through the response to the COVID-19 epidemic, it can be seen that China still has deficiencies in the administrative emergency response system. There are a series of problems, such as blocked information exchange and insufficient emergency response. Based on the analysis of the administrative emergency response measures in this epidemic, combined with the actual situation, and based on the legislative purpose of the Emergency Response Act, this paper proposes corresponding solutions for different problems, in order to achieve the goal of solving the practical problems.

1 LEGALITY OF ADMINISTRATIVE EMERGENCY RIGHT

Since the outbreak of COVID-19, emergency plans have been activated in all provinces across the country. In such an abnormal period, in order to prevent the sudden outbreak of the epidemic from causing greater damage to the public interests, the government initiated the administrative emergency response system and exercised emergency powers. The purpose of protecting the greater interest is achieved by damaging the lesser interest. Administrative emergency system refers to a system in which administrative organs take a series of emergency measures to control and eliminate the negative effects of emergencies. Administrative emergency right is an abnormal administrative right to respond to emergencies in the emergency period. Under the system of administrative emergency response, exercising the right of administrative emergency response will inevitably damage part of the interests of citizens.

The rule of law government requires that all acts or measures of the government should be carried out in accordance with the law, and the principle of rule of law should be adhered to in the process of performing their duties. The administrative emergency response system should also have its legal basis. The *Constitution* stipulates that the state may restrict citizens' rights in order to meet the needs of public interests, which is the constitutional source of administrative emergency response system and the legal basis of administrative emergency response system. At the legal level, the *Emergency Response Law*, as the first law in the field of emergency management in China, plays a vital role in the establishment of the administrative emergency system, making the administrative emergency actions of the government legally actionable. At the level of administrative regulations, there are *Regulations on Emergency Response to Public Health Emergencies* and other administrative regulations formulated in response to public health emergencies. These relevant laws and regulations constitute the legal norm system of China's administrative emergency system.

2 PROBLEMS OF CURRENT ADMINISTRATIVE EMERGENCY RESPONSE SYSTEM

Although China's administrative emergency response system has its own legal norm system, its legal system is still incomplete. In the specific implementation process, the disadvantages brought by this incomplete legal norm system appear. During the response to the COVID-19 epidemic, many defects that could only be found in the actual implementation were exposed. These shortcomings remind us that China's administrative emergency response system needs to be further improved, and we hope to reduce losses and improve efficiency in the future response to emergencies.

2.1 Lack of experience in emergency response

The government exercises the right of administrative emergency response in accordance with the law, which is the guidance provided by the legal system, but the specific emergency measures should be taken, still need to rely on the past experience to make a judgment. As administrative emergency usually occurs in emergencies, leaders are required to decide on emergency measures and formulate emergency plans according to a series of characteristics such as specific time, place and specific reasons when an emergency occurs. Emergencies are accidental and unpredictable. In addition to specific types of emergencies that often occur in certain regions due to geographical characteristics, emergencies are infrequent and the accumulation of experience can only come from previous emergency measures. For example, the current COVID-19 outbreak is similar to the SARS outbreak in 2003. Therefore, the emergency plan for COVID-19 response can appropriately draw lessons from the SARS outbreak of that year. However, this is a special case. In general, common ground cannot be found between emergencies. Different problems require different countermeasures. Due to the contingency and uncertainty of emergencies, it is difficult to accumulate relevant experience of emergency response measures. Experience in dealing with emergencies is valuable, but there is very little to draw on when past experience is needed. It is precisely because of the accumulated difficulties of experience that making emergency decisions or judgments requires such experience in dealing with similar incidents in the past that we are often unable to make appropriate decisions when we need to make emergency measures.

2.2 The flow of information is not smooth

When there is a special or unusual situation, timely communication and communication is the key to prevent the situation from worsening, find a solution in time and make a solution plan. However, due to the traditional mode or habit of administrative management, the information exchange within the management system is still dominated by the system of layer upon layer reporting. This way of information transmission has one shortcoming, that is, it takes a long time. In the event of an emergency, time is very valuable, and wasting time means delaying the formulation of a solution. On the one hand, due to the fear of taking responsibility and consequences, and out of the psychology of not being willing or afraid, the administrative organs may delay or even conceal the behaviors of reporting the true situation. On the other hand, the disclosure of government information is particularly important. In 2019, the Regulations on The Disclosure of Government Information was amended, and article 27 was adopted to relax the requirements for information disclosure and lower the threshold for citizens to apply for information disclosure. Its purpose is to promote the government information disclosure work, hoping to build an open and transparent management environment. The unimpeded information exchange is reflected in two aspects: first, the information exchange within the administrative management system is not smooth, leading to the failure to make an emergency decision in a timely manner; Second, the management system internal and external information exchange is not smooth, information disclosure is not timely, leading to the timely protection of citizens, thus causing serious consequences.

2.3 Inadequate protection of emergency legal interests

Administrative emergency measures are taken to maintain the public order in emergency situations and to protect the interests of the majority of the society. In the process of taking emergency measures, some compulsory measures may damage the legitimate rights and interests of a small number of people or individuals. In this extraordinary period, the government holds the public power and is in a strong position, while the citizens whose rights are restricted are in a weak position. As a result, the legitimate rights and interests of citizens are easily not guaranteed. Although the government exercises the right of administrative emergency response according to law, it still needs to avoid excessive damage to the legitimate rights and interests of citizens when taking administrative emergency measures. The implementation of administrative ACTS should abide by the principle of proportionality. As one of the basic principles that administrative ACTS must abide by, the principle of proportionality is of self-evident importance. The purpose of proportionality principle is to limit public power and protect citizens' legitimate rights and interests. Administrative ACTS that violate the basic principles are flawed. Illegal and unreasonable administrative ACTS will damage the credibility of the government and hinder the process of building a law-based government.

3 PROPOSE SOLUTIONS

Aiming at the above problems, the author starts from the legislative purpose of *Emergency Response Law* and puts forward solutions. It is expected that while perfecting the administrative emergency system, it can also provide some Suggestions for improving the legal system of administrative emergency, improve the level and efficiency of all aspects of administrative emergency management, which is conducive to saving the national resources and social resources, and speeding up the process of China's socialist modernization.

3.1 From response to prevention

In the actual formulation of administrative emergency measures, appropriate measures are often not taken because of lack of experience, and some measures cannot have ideal effect or may even have adverse effects. Or, due to lack of experience, the emergency plan is not timely enough to take the right emergency measures in the first place of an emergency. According to the previous emergencies, it can be concluded that emergencies are generally divided into four stages: symptom period, onset period, peak period and recession period. However, due to the lack of relevant experience of the leaders, they did not pay enough attention to the outbreak during the warning period and did not take corresponding emergency measures, which led to the outbreak of COVID-19. Thus it can be seen that at present, attention is paid to the handling of emergencies only after the severity of the situation, and emergency measures are also focused on how to deal with the situation, while neglecting to take preventive measures at the warning stage.

The *Emergency Response Law* is a law focusing on the prevention of emergencies. Whole law article seventy, the first is referred to for the purpose of the law in order to prevent and reduce incidents occur, and the second chapter "prevention and emergency preparedness" is the largest number of a chapter in law, which can infer is the purpose of the lawmakers on prevention before the occurrence of an emergency and contingency measures of the preparatory work, rather than wait for after the incident happened to take a series of emergency measures. Based on this legislative concept, the author believes that a big data model of emergency measures can be established. When similar or related situations occur, the system will respond to the data and give appropriate Suggestions for emergency measures. This can solve the problem of misjudgment due to lack of experience and improper administrative emergency measures. The emergency response should not only focus on the post-event response, but also strengthen the prevent prevention. Big data model can play an important role in the early prevention and control process of emergencies. In the past, emergency treatment measures were taken only after serious consequences were generated. In fact, it can be seen from the legislative ideas reflected in the Emergency Response Law that prevention beforehand is more important than response afterwards. At the same time, the cost of prevention work is far less than the cost of response work after the event. Therefore, we need to change our concept and focus on cultivating the concept of prevention of emergencies, so as to better solve the problem of emergency response.

3.2 *Improve the current information response mechanism*

In this response to the COVID-19 outbreak, the shortcomings of the information exchange mechanism within the administrative management system have also been exposed. In the initial stage of coVID-19, the government can only obtain information through a single channel. To determine whether a patient has been diagnosed with COVID-19, it must analyze the patient's pathological report before reaching a conclusion. After the diagnosis must be reported level by level, and then make the response measures. The mechanism of reflecting information is relatively single, and the current electronic case system is not developed, which cannot unify effective information and make accurate judgment in a short time. The author thinks that we should perfect the present information reflection mechanism and establish a more convenient, more perfect and more rapid information exchange mechanism that can reflect the problem. Whether in an emergency or a normal social order, information reflection and communication are very important for social governance. Obtaining information quickly and accurately is a powerful guarantee for formulating a reasonable emergency plan in time.

In order to perfect the current information reflection mechanism, the author thinks that the following two points should be done. The first step is to break the traditional mode of information transmission step by step. Timeliness of information is very important when responding to an emergency. In the early stages of a COVID-19 outbreak, a series of tests are required after the emergence of a suspected case, the condition can be verified only after the test results are obtained, and the source of infection can be isolated. The time required for this series of work is not short, and if the time is delayed due to the process of information reflection, more serious infections will occur. For public health emergencies, disease early warning system is particularly important in the process of prevention. Multiple source disease warning systems are more timely and accurate than single-source disease warning systems in determining the presence of a disease that could trigger a public health event. At present, the single source of early warning information can only take the diagnostic information from hospitals as the basis for the occurrence of diseases, which makes the efficiency of the early warning system very limited. If we can increase the sources and channels of information and expand the amount of information, we can improve the work efficiency of the disease warning system, so as to make more accurate and rapid judgments.

3.3 *Strengthen oversight over the right to administrative emergency response*

In order to make the government exercise the power of administrative emergency in an appropriate way and abide by the principle of proportion, the author thinks that the supervision of the power of administrative emergency should be strengthened. First, strengthen the supervision of the Constitution. *The constitution* is the fundamental law of our country, and the legal source of the administrative emergency power also starts from the Constitution. The government should exercise the administrative emergency power under the framework of the Constitution. Therefore, the exercise of administrative emergency power cannot go beyond the scope of the Constitution, and the Constitution can play a supervisory role in the exercise of administrative emergency power. A series of relevant laws and regulations formulated in accordance with the concept of the Constitution constitute the legal norm system of China's administrative emergency system, which makes the power source of administrative emergency more systematic. Secondly, strengthen the supervision of the people's Congress to the government at the same level. Article 16 of the Law on Response to Emergencies stipulates that when the government decides on an administrative emergency plan, it shall report it to the standing Committee of the People's Congress at the corresponding level for the record. This is a way of prior supervision. However, the effect of this mode of supervision in practical application is not ideal, so it is necessary to strengthen this mode of supervision. In order to make the government attach importance to this kind of filing procedure, the punishment for neglecting the filing procedure can be increased, each link and process of filing can be clearly defined, and the responsibility of each step can be implemented to the individual. In this way, we can strengthen the supervision effect of the people's Congress on the government at the same level, and avoid the situation that the record filing system is not in existence. Finally, we should strengthen supervision over the government. According to the provisions of the *Supervision Law*, if an administrative organ commits an improper or illegal act, it may be held accountable by the supervisory organ. The main function of supervisory organs is to supervise the exercise of public powers, and the administrative emergency power exercised by the government is one of the public powers, and whether there is violation of the principle of proportion in the exercise of emergency rights is also the content scope of supervision

4 CONCLUSION

The COVID-19 outbreak has made public health emergencies difficult to respond to, and the country and its people have made tremendous efforts and sacrifices to contain the outbreak and prevent its more serious consequences. If we can improve the level of prevention and control of administrative emergency and related supporting facilities, we can reduce losses and save resources. Taking administrative emergency measures is the first measure to deal with emergencies, and perfecting the short boards in the current administrative emergency work is a great help to the future emergency response work.

REFERENCES

[1] Jiang Fei. Administrative Emergency Right – "Double-edged Sword" in dealing with Emergencies. Legal System and Society, 2010(13):58.
[2] See Article 13 of the Constitution of the People's Republic of China for details.
[3] Wang jianxin. Research on the basic concept of emergency policing. Public security science, 2014, 8(00):276–284.
[4] Xiao jiye. Reform and shaping of administrative emergency power mode in the era of big data. Journal of hubei university (philosophy and social science edition), 2019, 46(05):135–142.
[5] Guo Xianghong. Legal Interpretation of "Wartime control". Shanxi Science and Technology News, 2020-03-26(A07).
[6] Chang Jian. Human rights protection in epidemic prevention and control. Academia, 2020(02):35–49.
[7] Chen Chen. Reflections on the Legal Issues of Guaranteeing Civil Rights by Administrative Emergency Requisition in Emergencies. Regional Governance, 2019(52):108–110.
[8] Wang siting. Research on the legal regulation and improvement of administrative emergency right in China. Modern business and trade industry, 2020, 41(16):137–138.

Computational Social Science – Luo, Ciurea & Kumar (eds)
© 2021 Taylor & Francis Group, London, ISBN 978-0-367-70193-2

The new trend of strategy in the Indo-Pacific

Y.J. Cao, J.H. Yu & X.L. Zhu
Research Center for Indian Ocean Island Countries, South China University of Technology, Guangzhou, China

ABSTRACT: The concept of "Indo-Pacific" is the most popular diplomatic term in the field of international relations in the past 20 years. It has changed from a vocabulary in Japanese politicians' speeches to a regional and geographic concept adopted by many countries, integrating geopolitics, maritime security, development economics, ethnology and many other subjects. The conceptualization of the "Indo-Pacific" is a complicated historical process. Geopolitical scholars are arguing whether the world today is the "Pacific Era", the "Chinese Century", or the "Indo-Pacific Era". Therefore, this research will mainly focus on the role of the India, Japan, China, or even European countries which cannot underestimate in the Indo-Pacific, which explores the new trend of strategy in Indo-Pacific in the future.

1 INTRODUCTION

In the past 20 years, one of the major changes in global shipping has been to balance traffic back and forth through the eastbound route of the Pacific Ocean and the westbound route of the Indian Ocean-Suez Canal-Panama Canal. Due to the expansion of the Panama Canal in 2016, the rejuvenated eastbound route has had an important impact on the strategic significance and competitive advantage of the relationship between the United States and its Asian allies, as well as China's BRI The "Indo-Pacific" is the second partial concept of "order" in the geopolitics of rim. Therefore, whether the international community can adapt and adapt to the complexity of geopolitics and development will depend to a certain extent on their continuous the ability to react quickly to changing reality. It originated from the rise of the ASEAN multilateral trading mechanism in the late 1990s. At that time, there are three structural characteristics in the Asia-Pacific region: (1) U.S.-led, U.S., Japan and U.S., Australia, South Korea, New Zealand and the alliance system of U.S.; (2) ASEAN-led regional cooperation system; (3) East Asia and Pacific Rim regional cooperation system where China exerts great influence. This regional governance model emerged from the Asian financial crisis in 1997. For example, "10+3" and "10+6", Tokyo politics prefers the latter mode of cooperation, adding three democratic countries with common values to balance China's influence in Southeast Asia (Terada 2010). It is estimated that by 2050, the Indo-Pacific countries will account for most of the global GDP and economic growth. The Indo-Pacific region will become the fulcrum for the transfer of power between the old and the new, dominate the process of regional historical development, and become the centre of global power structure and order.

2 THE ROLE OF CHINA IN INDO-PACIFIC

Experts and scholars have formed a series of views on the Indo-Pacific region's cognition and views. John Hemmings (2019) thinks that the Indo-Pacific region is a key part of the global economy. As its importance grows, and geopolitical and maritime rivalries emerge, geopolitical factors will be the ultimate determinant of China's maritime expansion. The Chinese government will use commercial trade and state-owned shipping companies to increase its influence in local governments and ports. This has aroused concerns among countries in the region about China's use of debt diplomacy to promote the Belt and Road Initiative. The maritime power and influence directly determine the strategic position of this country. Nearly, 67% of the earth's surface area is the ocean. Since the 16th century, whoever has the ability to control the ocean means taking the lead in global commerce and trade competition, thereby providing preconditions for the primitive accumulation of capital and achieving global economic growth. Take the initiative in competition. Powerful basic entities, such as land area, natural resources, population, etc., these key indicators to measure the comprehensive strength of a country are the core elements for maintaining hegemony of a great power. Compared with the above-mentioned hard power, culture is the most important soft power indicator of a big country, and its influence on a country's historical process is the longest and most vital. Since the 1970s, containerization has been the most important driving force of globalization, its impact is greater than the sum of all trade liberalization agreements. In these decades, the nighty percent of non-bulk goods worldwide have been transported in containers. China has at least become a top port trader-the largest exporter of containerized goods. In 2014, it has mastered 26% of the world's port trade volume and 36 million containers, making it the world's second largest port trading country. In the container trade from 1999 to 2003, among the 25 large ports in the world, 15 countries such as China, Japan and New Zealand accounted for 15, and China accounted for 7 of these 15 ports, which are mainly concentrated in the Yangtze River Delta, Bohai Bay and Pearl River Delta regions (Levinson 2008).

Secondly, China will achieve the goal of controlling local ports through maritime trade competition. This is another major national strategy since China's Deng Xiaoping reform and opening up. It will create a foreseeable future. In the foreseeable future, China will provide transportation routes to core markets by investing in commercial projects such as the Nicaragua Canal and the Kra Isthmus Canal in Thailand to dispel the doubts of other big countries about China's Belt and Road Initiative, such as SLOC in the Indo-Pacific region. The key to China's maritime trade control is the international competitiveness of its coastal ports. Globalization supports Deng Xiaoping's reform and opening policy in the post-Mao era. The growth areas of its development and the east-west cooperation route are determined by the policies of the port and shipping industry. The government's strategy of assisting domestic companies to identify and utilize global chains of value to promote port development is correct. It is consistent with the free trade system, facilitates maritime trade and investment by Asian partners, and encourages companies to restructure their Asian production networks. In order to enhance the competitiveness of port trade, the government has issued a series of incentive measures, which focus on the integration and cooperation of seaports, the establishment of large land-sea transportation channels, and the two-way opening strategy of coastal ports (Notteboom & Zhongzhen 2017)

3   THE BELT AND ROAD INITIATIVE FOR THE STRATGIES OF INDO-PACIFIC IN INDIA AND JAPAN

China's Belt and Road Initiative will collide with India's international and regional strategies. For example, China will control the west coast of the Indian Ocean through Gwadar Port, and the Indian government should pay attention to it to ensure energy and trade supply. Although India has strategic doubts about China's Belt and Road Initiative, it has also adopted some projects to improve relations with countries in the Indian Ocean region and West African countries. Japan is also taking similar measures in the Indian Ocean and West African regions. The Monsoon Plan and the Asian-African Growth Corridor provide space for cooperation. Shinzo Abe (2007) stated, "Japan has rediscovered India as a partner with common interests and values. I hope that we will work together to develop into friends, and the rich, prosperous and free Indo-Pacific Ocean will be the Open to all." He also discussed the importance of the right to freedom of maritime navigation, strategic interests, maritime channel security, and foreign direct investment to the delicate relationship between India and Japan. The possibility of the Tokyo-Osaka Economic Corridor and the Delhi-Mumbai Economic Corridor is a huge breakthrough. It can be seen from Abe's speech that strategic interests, maritime channel security, and infrastructure investment have become the main elements of Indo-Pacific regional cooperation. It predicts the future development trend of the Indo-Pacific region. Japan wants to play a stronger role in the construction of the Indo-Pacific geopolitical concept and regional cooperation. In 2017, the Asian Development Bank (ADB) pointed out in the report of "Meeting Asia's Infrastructure Needs" that only twice the amount of funds in this region can be used for infrastructure construction and poverty alleviation in order to make the index show a growth trend. The regional infrastructure expenditure that year was approximately 567 billion pounds, and the region needed 1.29 trillion pounds in infrastructure expenditure. It is very certain that India-Japan cooperation focuses on development and foreign direct investment. Shinzo Abe proposed the "Free and Prosperous Indo-Pacific Strategy" at the Sixth African Development Conference (FOIP) in 2016. (Prime Minister of Japan and His Cabinet 2016). He emphasized the importance of rules for the regional strategy. Whatever, from the Indian Ocean to the South China Sea, the Indo-Pacific region is essential to Japan's economic prosperity and traditional security. Japan's trade relies on the SLOC, and 80% of Japan's oil imports need to go through this route, which can provide broader space and opportunities for economic growth and social development in the Middle East, Africa, South Asia and Southeast Asia. The essence of Japan's Indo-Pacific strategy can be summarized as "the formation of free security diamonds between the United States, southern China, ASEAN, Australia and India" (Koga 2018). Meanwhile, the U.S.-Japan alliance as the axis of Japan's security also establishes rules as the core regional international order. The competition between the Belt and Road Initiative and the Indo-Pacific strategy is comprehensive, covering political, economic, cultural, technological, and military aspects. Therefore, the rise of the Indo-Pacific region is as important to the entire world as the rise of China. Based on the orderly interests and concerns of geopolitics, countries in the region will benefit from the healthy competition between the Belt and Road Initiative and the Indo-Pacific strategy.

However, the most important thing is that the role of France cannot be ignored in the Indian Ocean, South Pacific, and Far East regions, France's lack of discussion on Indo-Pacific strategy. Since 2013, France has paid more and more attention to the Indo-Pacific region. By 2016, the attitude toward the region had changed from "returning" to "emphasis", reinterpreting France's interest areas, and defining the Indo-Pacific region as a region of practical significance in French geopolitics and geo-economics. The Indo-Pacific strategies of India, the United States, Australia and other countries are strategically coordinated to promote regional stability, maritime security and regional conflicts with these major countries. Therefore, traditional security is often the most necessary. France is a good example. They are involved in resolving regional crises and ensuring the safety of major air routes, and have achieved certain results in combating terrorism, extremism and organized crime (Hemmings 2019). To strengthen France's presence in the Indo-Pacific region through military means to demonstrate and give security commitments to allies. Consolidate and develop a comprehensive and strategic partnership with the region, attach importance to bilateral and small multilateral cooperation, and unite NATO partner countries and major countries in the Indo-Pacific region to cope with common risks and challenges. Strengthen participation in regional organizations, promote the development of multilateralism, maintain the balance of regional power, and prevent any hegemony.

4   CONCLUSION

However, the most important thing is that the role of France cannot be ignored in the Indian Ocean, South Pacific, and Far East regions, France's lack of discussion on Indo-Pacific strategy. Since 2013, France has paid more and more attention to the Indo-Pacific region. By 2016, the attitude toward the region had changed from "returning" to "emphasis", reinterpreting France's interest areas, and defining the Indo-Pacific as a region of practical significance in French geopolitics and geo-economics. The Indo-Pacific strategies of the United States, India, Japan, Australia and other countries are strategically coordinated to promote regional stability, maritime security, and the regional conflict with the four countries. Security interests tend to be substantive, which they are deeply involved in resolving regional crises, ensuring the safety of major air routes, fighting terrorism, extremism and organized crime, and emphasizing the importance of security and prosperity in the Indo-Pacific region to France. Thus, by strengthening its military power, France was able to increase its presence in the Indo-Pacific and to demonstrate to its Allies and to give security assurances. We will consolidate and develop comprehensive strategic partnerships with the region, attach importance to bilateral and multilateral cooperation in small areas, and unite NATO partners and major countries in the region of Indo-Pacific to address common risks and challenges. We should strengthen the participation of regional organizations, promote multilateralism, maintain regional balance of power and prevent hegemonism.

ACKNOWLEDGEMENTS

Funding: This work is supported by "the Fundamental Research Funds for the Central Universities" in South China University of Technology, project NO: XYMS202008.

REFERENCES

Address by Prime Minister Shinzo Abe at the Opening Session of the Sixth Tokyo International Conference on African Development (TICAD VI). 2016. Prime Minister of Japan and His Cabinet, retrieved 28 February 2019, https://japan.kantei.go.jp/97_abe/statement/201608/1218850_11013.html.

'Confluence of the Two Seas: Speech by HE Mr Shinzo Abe, Prime Minister at the Parliament of the Republic of India', Ministry of Foreign Affairs of Japan website, 2007, retrieved 17 October 2018, http://www.mofa.go.jp/region/asia-paci/pmv0708/speech-2.html.

Hemmings, J. 2019. Infrastructure, Ideas, and Strategy in the Indo-Pacific, London: The Henry Jackson Society, p.7–p.10.

Koga, K. 2018. 'Honing Japan's regional strategy: Tokyo's Indo-Pacific vision could do with sharper teeth', Policy Forum, retrieved 28 February 2019, https://www.academia.edu/37999137/Honing_Japans_regional_strategy_Tokyos_IndoPacific_vision_could_do_with_sharper_teeth.

Levinson, M. 2008. The Box: How the Shipping Container Made the World Smaller and the World Economy Bigger (Princeton University Press, 2008).

Notteboom, T. and Zhongzhen, Y. 2017. 'Port Governance in China since 2004: Institutional Layering and the Growing Impact of Broader Policies', Research in Transportation Business and Management 22 (2017), pp. 184–200.

Computational Social Science – Luo, Ciurea & Kumar (eds)
© 2021 Taylor & Francis Group, London, ISBN 978-0-367-70193-2

Effects of language satisfaction on continued use intention for m-government users in multi-ethnic inhibited multilingual regions: A case study in Xinjiang Uyghur Autonomous Region of China

L.P. Zhu
Minzu University of China, Beijing, China

ABSTRACT: Users' continued use of mobile government (m-government) applications is essential for deriving the benefits of m-government services. For multi-ethnic inhibited multilingual regions, the kinds of the language on an m-government application are vital for its acceptance. To investigate whether and how users' satisfaction with the kinds of the language on m-government applications˜a special emotion, referred to as language satisfaction (LS)˜would influence the continued use intention, I developed a research model on the sequential relationship between LS and m-government continuance intention. Empirical results based on the data collected from Xinjiang Uyghur Autonomous Region of China (abbreviated Xinjiang) demonstrate the proposed research model. The results reveal that LS indirectly positively related to continued use intention of m-government users through users' cognition of the language-related attributes of m-government applications, implicating the importance of the language used on an m-government application to the adoption of the m-government in a multilingual region. Based on the findings, language policy suggestions are proposed to enhance the ethnic minority citizens' continuing adoption of m-government applications in Xinjiang.

1 INTRODUCTION

Mobile government (m-government), which is defined as an extension or evolution of e-government through utilization of mobile technologies for public service delivery (ITU, OECD 2011), makes it possible to deliver public service to rural areas and disadvantaged people via mobile telephone, Personal Digital Assistant (PDA), laptop and wireless internet infrastructure, thus protecting information equity and narrowing digital gap between economically developed and developing regions in the world. The value of an m-government application lies in its continued use. So it is important to understand factors that influence continued use intention of m-government users.

M-government adoption has been one of the hot issues in the m-government research fields since the concept of m-government was proposed (Wang, 2016). Researchers found that cross-cultural differences impact consumers' perception of m-government adoption behavior. For instance, perceived empathy positively drives Bangladeshi consumers to adopt m-government, which is not the driving force for Canadian and German consumers (Shareef et al., 2016). Therefore, it is important to take demographics and cultural norms into consideration in the research of m-government adoption.

China is a multi-nationality country. Cultural differences exist among different regions. In the past decades, scholars in China have made several important empirical findings on the research of m-government adoption generally (Guo & Zhu, 2015; Guo & Zhu, 2016; Zhu & Guo, 2016; Zheng & Zhao, 2016; Li et al., 2018). However, there was a lack of knowledge about the m-government adoption in the multi-ethnic inhabited multilingual regions.

In this paper, I focus on the continued use intention towards m-government applications in Xinjiang Uyghur Autonomous Region (abbreviated Xinjiang), one of the minority nationality autonomous regions in northwest China. There are 56 ethnic groups in Xinjiang. Among the major ethnic groups, *Uyghur*, *Han*, and *Kazakh* rank top three in terms of population, all have their own languages and characters. The m-government applications in Xinjiang provide three types of languages, national common language (Chinese), Uyghur language and Kazakh language. Languages are a kind of representational symbols, representing native speakers' culture. People usually have a natural sense of closeness to their own national language or languages that they have learnt and often used since childhood. However, it is not clear how much effect the language used on m-government applications has on the continuance of the applications. I wondered whether users' emotion at the kinds of the language on m-government applications might influence their continued use intension and if it might how? The goal of this study is to investigate the influence path of users' satisfaction emotion at the m-government language on the intension to use the m-government continuously.

The remainder of the paper is organized as follows. Section 2 presents a literature review, including an introduction of m-government development in China, and discussion of factors influencing the m-government adoption. Theoretical perspective, proposed research model and hypotheses of this study are described in Section 3. The research methodology is presented in Section 4, followed by empirical results and analysis. Finally, Section 6 presents the conclusion, discussion of findings and implications, and limitations and future research directions.

2 LITERATURE REVIEW

Most of the m-government adoption studies are based on existing basic theories and classic models on information technology (IT) adoption, for example, Theory of Rationed Action (TRA) (Fishbein and Ajzen, 1975), Theory of Planned Behavior (TPB) (Ajzen, 1991), Diffusion of Innovations (DOI)/Innovation Diffusion Theory (IDT) (Rogers, 1995), and Technology Acceptance Model (TAM) (Davis, 1989), etc. More complex models extended the classic models by integrating new constructs on social, human and cultural factors, and adding more technological factors. Venkatesh et al. (2003) examined eight prominent models of IT adoption, formulating the unified theory of acceptance and use of technology (UTAUT). Dwivedi et al. (2017) examined nine theoretical models of IT adoption, developing a unified model of e-government adoption (UMEGA).

In the study of factors influencing the intention to use the m-government, many studies concerned with the influence of individuals' pre-adoption beliefs and attitudes on the intention to use the m-government. Most of the studies adapted the above IT adoption models to specific m-government contexts in their studies (Hung et al., 2013; Ohme, 2014; Liu et al., 2014; Abaza & Saif, 2015; Ahmad & Khalid, 2017). However, factors influencing the potential adopters' intention to use m-government services and the users' continuance intention are different, as there exists a distinction between individuals' pre-adoption and post-adoption beliefs and attitudes (Karahanna et al., 1999). The investigation results of Zheng et al. (2016) show that respondents' demand for the APPs and the acceptance of new technologies are the determinants for the intention to use m-government APPs, while after use feedback - satisfaction and PU, impacted respondents' continuance intention. Several empirical studies have revealed that gratification or satisfaction was the key determinant for m-government users' continuance intention (Guo & Zhu, 2016; Zhu & Guo, 2016; Li et al., 2018). These studies analyzed the antecedent or the determinants of satisfaction. Nevertheless, there has thus far been relatively little research on how satisfaction, or more generally emotion, affects the m-government users' continuance intention.

3 RESEARCH MODEL AND HYPOTHESES

3.1 *Research model*

People's emotion at the m-government language is likely to have carry-over effects on the judgment or cognition of the language-related attributes of the m-government applications during the use

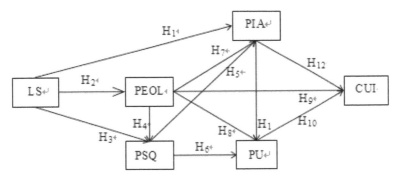

Figure 1. Research model.

period. Satisfaction, a primary emotion (Kemper, 1987), has positive effect on the cognition of the language-related attributes, in turn the cognition would be directly linked to the continuance intention. Figure 1 presents the research model I developed. Language satisfaction (LS) is indirectly related to continued use intention (CUI) via perceived ease of learning (PEOL), perceived search & query convenience (PSQC), perceived information acquisition (PIA) and perceived usefulness (PU). PEOL, PSQC, PIA and PU interrelate with each other with sequential order.

3.2 Hypotheses

Language is the main carrier of m-government information. To acquire information on an m-government application, people need knowledge and skills on both language and mobile Internet applications. Among these knowledge and skills, language knowledge and skills have the attributes of "meta-knowledge" and "meta-skill", the basis for individuals to acquire most of other knowledge and skills (Wang, 2018) such as mobile Internet knowledge and application skills. LS implies that people have control over the language on m-government applications. It has a say on the language-related cognitions, and contributes to the positive cognition about the language-related attributes of m-government applications ‾ PEOL, PSQC, and PIA. Hence, the following path hypotheses are proposed:

H_1: LS positively affects the PIA of m-government.
H_2: LS positively affects the PEOL of m-government.
H_3: LS positively affects the PSQC of m-government.

For m-government users, search & query are rational behavior with the purpose to acquire information. People often search for the information of interested through eye scanning of the messages on m-government applications, or use keywords queries to find the information they need. In other words, information acquisition is mainly realized by search & query, with the latter antedating the former. As a result, cognition of the search & query attribute of m-government applications would positively influence the cognition of the information acquisition results. If people can obtain the information they need, based on which they are more likely to make the benefit of information in their lives, they would form the cognition that the m-government application is useful to them. Hence, cognition of the information acquisition results would positively influence cognition of the PU of m-government. As the search & query attribute is the key determinant of information acquisition, cognition of the search & query attribute of m-government applications would also positively influence the cognition of the PU of m-government. To summarize, the following hypotheses are proposed:

H_5: PSQC positively affects the PIA of m-government.
H_6: PSQC positively affects the PU of m-government.
H_{11}: PIA positively affects the PU of m-government.

Furthermore, if the result of an act is pleasant or beneficial, people tend to do so more often, in other words, they decide to repeat the act in the future (Elster, 2007). Perception of information acquisition can be viewed as a reward for m-government users who intend to acquire information, reinforcing their CUI for m-government applications. Hence, the following path hypothesis is proposed:

H_{12}: PIA positively affects the CUI for m-government.

Previous studies suggest that ease of use and ease of learning are strongly related. Whiteside, et al. (1985) concludes that ease of learning and ease of use are congruent. Ease of learning is also regarded as one substratum of the ease of use construct by Davis (1989). As a consequence, the relationship among perceived ease of use (PEOU), PU and the continuance usage also holds for that of PEOL, PU and the CUI. Specifically, both PEOU and PU positively affect the CUI (Davis, 1989; Belanche et al., 2012; Liu et al., 2014), and PU mediates between PEOU and CUI (Zhu & Guo, 2016; Guo & Zhu, 2016). Therefore, the following hypotheses are proposed:

H_8: PEOL positively affects the PU of m-government.
H_9: PEOL positively affects the CUI for m-government.
H_{10}: PU of m-government positively affects the CUI for m-government.

In addition, PEOL is people's subjective appraisal of effort in learning to use the m-government applications involving learning to use the functions (such as the search & query function) of the m-government applications to acquire information. For m-government applications, learning comes first, and then comes the usage. Since former comings constitute a reference frame for those latter comings, cognition with learning period may influence cognition with using period. The easier it is to learn the m-government applications, the less effort needed to use the applications to search, inquire and finally find the information of interest, thus contributing to more positive appraisal of the search & query function and its outcome ‾ information acquisition. Hence, the following hypotheses are proposed:

H_4: PEOL positively affects the PSQC of m-government.
H_7: PEOL positively affects the PIA of m-government.

4 METHODOLOGY

4.1 Data acquisition

To test the research model, data were collected through questionnaires on m-government applications targeted to the residents in Xinjiang. Both electronic form and paper form of questionnaires are available. The Electronic questionnaires were sent through WeChat. Both QR code and website link are provided to access the electronic questionnaire. Paper questionnaires are issued by students from Xinjiang when they returned back to their hometown on vacation. The questionnaire has been subjected to pretesting through a pilot survey of 13 ethnic minority college student respondents and been modified according to their feedback. After that, the questionnaire was administered to 346 citizens in Xinjiang.

4.2 Sample

Of the 346 questionnaires, 343 were valid. Of the 343 valid questionnaires, 164 were from those who had previously used m-government, and 179 were from those who had not used. The 164 valid questionnaires formed the sample which was used for this study.

Of the 164 valid questionnaires, male users composed 40.9% of the sample and female users 59.1%. Users are mainly young and middle-aged (82.9%). The user group whose ages lie between 18 and 25 is the dominant, accounting for 59.1% of the sample. In terms of ethnic structure,

Table 1. Operationalization of variables.

Variable	Operational Definition	How measured
LS	Users' affect with the kinds of the language (national, minority or bilingual) on the m-government application.	Adapted from Spreng et al.'s (1996) information satisfaction scale.
PEOL	Users' perception of the difficulty in learning to use the m-government application.	Adapted from one item from Davis et al.'s (1989) perceived ease of use scale.
PSQC	Users' perception of the convenience of m-government application in searching and querying for information.	New scale developed.
PIA	Users' perception that they can obtain the required information by using the m-government application.	New scale developed.
PU	Users' perception of the usefulness of m-government application for their study/work/life.	Adapted from one item from Davis et al.'s (1989) PU scale.
CUI	Users' intention to continue using the m-government application.	Adapted from one item from Liu et al.'s (2014) intention scale.

Uyghurs account for 58.5% of the users, followed by Han and Kazakh, accounting for 25% and 7.3%, respectively. In terms of educational level, undergraduate degree ranks first, accounting for 62.2% of the sample, following by graduate degree and above, accounting for 17.7%. Among the most frequently used m-government applications, WeChat is the most popular application, which accounts for 63.4% of the sample. The study is responsible for the sample.

4.3 *Variable measurement*

Operations with single statements of feelings or beliefs were employed to measure each variable of the model. The single-item scale for variables PSQC and PIA were self-designed. The single-item scale for other variables are selected from prevalidated measures in IS or m-government use, and reworded to relate specifically to this study. Operational definitions and sources for the variables are presented in Table 1.

5 EMPIRICAL RESULTS

The research model in Figure 1 was tested using AMOS 24.0 structural equation model (SEM) software. BC bootstrapped 95% confidence intervals (CI) of the total indirect effects, and specific indirect effects were derived from 1,000 bootstrap resamples. Model fit indicators presented good values (ξ^2/df= 2.44, P-Value = .866 (> 0.05); NFI = .999; IFI = 1.004; TLI = 1.020; CFI = 1.000; RMR = .008; RMSEA = .000; 90% CI of RMSEA (.000, .068), PCLOSE = .921).

Path analysis results are presented in Figure 2. All the direct effects proposed in the model were significant confirming the 12 hypotheses. The results reveal high explained variance of PIA ($R^2 = 0.622$), PU ($R^2 = 0.549$) and CUI ($R^2 = 0.545$). PIA was explained by PSQC ($\beta = .397$), convenience LS ($\beta = .329$), and PEOL ($\beta = .188$). PU was explained by PIA ($\beta = .235$), PEOL ($\beta = .255$), and PSQC ($\beta = .351$). CUI was predicted by PIA ($\beta = .443$), PEOL ($\beta = .240$), and PU ($\beta = .157$). PSQC ($R^2 = 0.516$), which was explained by LS ($\beta = .394$) and PEOL ($\beta = .429$), also has considerable level of explained variance ($R^2 = 0.516$). Finally, LS alone explained 27.2% of variance of the PEOL, thus was a significant predictor of the PEOL.

Results of the mediator effects are presented in Table 2. Apart from PEOL that does not have indirect effect, the mediated effects between LS and the other variables are all significant as the

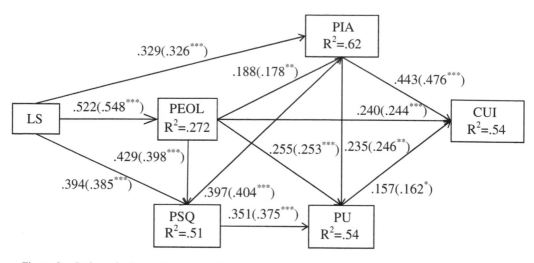

Figure 2. Path analysis results. Notes: Values outside parentheses: standardized coefficient; values in parentheses: unstandardized results. *** p<0.001, ** p<0.01, * p<0.05.

Table 2. Standardized indirect, direct and total effects and the corresponding BC 95% CI.

	Indirect Effects		Direct Effects		Total Effects	
	Values	CI	Values	CI	Values	CI
LS ◇PSQC	.224	[.118, .372]	.394	[.229, .549]	.618	[.469, .760]
LS ◇PIA	.344	[.226, .514]	.329	[.149, .475]	.672	[.539, .771]
LS ◇PU	.508	[.369, .653]	–	–	.508	[.369, .653]
LS ◇CUI	.503	[.380, .628]	–	–	.503	[.380, .628]

CIs corresponding to the indirect effects do not include zero. LS, via PSQC, PEOL, PU and PIA, explains 25% of the variance in CUI. LS, via PSQC, PEOL, and PIA, explain 26% of the variance in PU. The LS-CUI relationship and LS-PU relationship are fully mediated. LS has direct as well as indirect effects on PSQC and on PIA. The direct effect of LS on PSQC is much larger than the indirect effect, accounting for 64% of the total effects. The indirect effect of LS on PIA accounts for 51% of the total effects, slightly larger than the direct effect.

6 CONCLUSION AND DISCUSSION

6.1 *Conclusion*

This study investigated how m-government users' emotion at the kinds of the language on m-government applications influences their continued use intention. Empirical results in Xinjiang reveal that m-government users' emotion of language satisfaction on m-government applications has significant indirect effect on their CUI through cognition of the language relevant attributes of m-government applications, namely, PEOL, PSQC, PIA, and PU.

Results of this study reveal the importance of the m-government language on users' continued use intention in a multilingual region. From the study results, it can be inferred that if users experience positive emotions such as satisfaction, content and happiness in the process of using m-government applications, they are more inclined to use these applications continuously.

6.2 Implications

The practical implications for m-government services policy makers are summarized as follows.

In order to enhance citizens' acceptance and continuing adoption of m-government services, it is essential that policy makers suit the languages of m-government systems to local conditions, and keep the language use strategy accord with the national language development strategy at the same time. Currently, Department of Language and Information Management, Ministry of Education of the People's Republic of China is promoting the effective implementation of the "National Common Language Popularization Project", aiming to realize the goal of "basically popularizing the national common language and writing throughout the country by 2020". Nevertheless, the popularity of national common language is relatively low in Xinjiang, especially in the countryside, suburbs and towns. Even in the cities, citizens of the same minority nationality may have different level of Chinese proficiency. Some people have learned national common language since childhood and often use them in their daily life, while others began to learn the national common language after adulthood feel it difficult to fully understand and use the national common language freely. Under these backgrounds, I suggest m-government applications provide national common language and ethnic minority language bilingual edition in recent several years. In this way, ethnic minority people have not only got better understanding of the government information through their own language, but also got more contact with national common language environment. This may facilitate their learning national common language, help create a close affinity with national common language emotionally to overcome the language barrier psychologically and eventually be able to use m-government applications in whichever language. Thus enhanced m-government adoption will be expected.

ACKNOWLEDGEMENT

The study is partly supported by General projects of the 13th five year plan of the National Committee of Language, Ministry of Education of the People's Republic of China (YB135-59) and the National Planning Office of Philosophy and Social Science in China (17BGL199).

REFERENCES

Abaza, M., & Saif, F. 2015. The adoption of mobile government services in developing countries. *International Journal of Computer Science Issues*, 12(1), 137–145.

Ahmad, S., & Khalid, K. 2017. The adoption of M-government services from the user's perspectives: empirical evidence from the United Arab Emirates. *International Journal of Information Management*, 37, 367–379.

Ajzen, I. 1991. The theory of planned behavior. *Organizational Behavior and Human Decision Processes*, 50, 179–211.

Belanche, D., Casaló, L., & Flavián, C. 2012. Integrating trust and personal values into the technology acceptance model: the case of e-government services adoption. *Cuadernos de Economía y Dirección de la Empresa*, 15, 192–204.

Davis, F. 1989. Perceived usefulness, perceived ease of use and user acceptance of information technology. *MIS Quarterly*, 13(3): 319–339.

Dwivedi, Y., Rana, N., Janssen, M., Lal, B., Williams, M., & Clement, M. 2017. An empirical validation of a unified model of electronic government adoption (UMEGA). *Government Information Quarterly*, 34, 211–230.

Elster J. 2007. *Explaining social behavior: more nuts and bolts for the social sciences*. Translated by: Liu, J., He, S. J., Xiong, C. 2019. Chongqing: Chongqing University Press.

Fishbein M., & Ajzen, I. 1975. *Belief, attitude, intention and behaviour: an introduction to theory and research.* Reading: Addison-Wesley.

Guo J., & Zhu D. 2015. User adoption model and empirical analysis of mobile government services based on trust. *Ruan Ke Xue*, 29(12), 108–110.

Guo J., & Zhu D. 2016. The research on service quality and users' intention to reuse mobile government services. *Research on Library Science*, 2, 64–70.

Hung,S., Chang, C., & Kuo, S. 2013. User acceptance of mobile e-government services: an empirical study. *Government Information Quarterly*, 30(1): 33–44.

Karahanna, E., Straub, D., & Chervany, N. 1999. Information technology adoption across time: a cross-sectional comparison of pre-adoption and post-adoption beliefs. *MIS Quarterly*, 23(2), 183–213.

Kemper, T. 1987. How many emotions are there? Wedding the social and the autonomic components. *The American Journal of Sociology*, 93(2), 263–289.

Li, Y., Yang, S., Chen, Y.,& Yao, J. 2018. Effects of perceived online–offline integration and internet censorship on mobile government microblogging service continuance: a gratification perspective. *Government Information Quarterly*, 35, 588–598.

Liu, Y., Li, H., Kostakos, V., Goncalves, J., Hosio, S., & Hu, F. 2014. An empirical investigation of mobile government adoption in rural China - a case study in Zhejiang province. *Government Information Quarterly*, 31(3), 432–442.

OECD/International Telecommunication Union. 2011. *M-Government: Mobile Technologies for Responsive Governments and Connected Societies*, OECD Publishing. http://dx.doi.org/10.1787/9789264118706-en.

Ohme, J. 2014. The acceptance of mobile government from a citizens' perspective: Identifying perceived risks and perceived benefits. *Mobile Media & Communication*, 2(3), 298–317.

Rogers, E. M. 1995. *Diffusion of innovations* (4th ed.). New York: Free Press.

Shareef, M., Dwivedi, Y., Laumer, S., & Archer, N. 2016. Citizens' adoption behavior of mobile government (mGov): a cross-cultural study. *Information Systems Management*, 33(3), 268–283.

Spreng,R., MacKenzie,S., & Olshavsky,R. 1996. A reexamination of the determinants of consumer satisfaction. *Journal of Marketing*, 60, 15–32.

Venkatesh, V., Morris, M., Davis, G., & Davis, F. 2003. User acceptance of information technology: towards a unified view. *MIS Quarterly*, 27(3), 425–478.

Wang Changlin. 2016. A literature review of the research development and hotspot of mobile government at home and aboard. *Journal of Management*, 29(2), 50–56.

Wang, H. 2018. The role of language as human capital on economic growth. *Chinese Journal of Language Policy and Planning*, 2, 89–96.

Whiteside, J., Jones, S., Levy, P.S., & Wixon, D. 1985. User performance with command, menu, and iconic interfaces. *ACM SIGCHI Bulletin*, 16(4), 185–191.

Zheng Y., & Zhao J. 2016. Citizens' use of Gov mobile Apps and their determinants: a study to first-tier cities in China. *Journal of Public Administration*, 6, 23–43,197.

Zhu D., & Guo J. 2016. Research on users' satisfaction of mobile government based on TAM model. *Qing Bao Ke Xue*, 34(7), 141–146.

Computational Social Science – Luo, Ciurea & Kumar (eds)
© 2021 Taylor & Francis Group, London, ISBN 978-0-367-70193-2

Forecasting of water demand by industry in China in 2020

X.L. Liu
Academy of Mathematics and Systems Science, Chinese Academy of Sciences, Beijing, China
Center for Forecasting Science, Chinese Academy of Sciences, Beijing, China
University of Chinese Academy of Sciences, Beijing, China

ABSTRACT: The forecasting of demand for water in advance is critical for water supply planning. This paper analyzes water consumption by industry in China. With multiple regression analysis, time series analysis, and expert experience methods, the water demand from industry in China in 2020 was forecast. The results show that, without the COVID-19 out broke the total demand for water in China in 2020 will be about 599.01 billion m^3, a slight decrease of 1.44 billion m^3 compared with 2019. Among these, agricultural water demand will account for about 60.2%, and industrial water demand will account for about 20.6%. For domestic and environmental water demand, the proportions will be 15.2% and 4.0%, respectively.

1 INTRODUCTION

China is severely lack in water resources. It has been pointed out that water problems, which are mainly water shortages, will result in the economic growth rate decreasing by 1–2% in China, which is greater than the impact of energy prices increase and a decline in foreign investment, and water shortage has become a main factor restricting China's economic and social development. The analysis of water consumption by industry and forecasting of water demand will provide a scientific reference to achieve the targets of regulating total water consumption and water use intensity in China, which is of great significance to the macro-control of the discrepancy between supply and demand of water resources and the realization of the coordinated development of economic and social elements and water resources. Currently, research has been carried out on water demand forecasting using a fuzzy model (Sen & Altunkaynak, 2009), a CCNN model (Firat et al., 2010), a SVM model (Herrera et al., 2010), a genetic algorithm model (Nasseri et al., 2011), a neural network model (Ajbar et al., 2013), etc. However, these models generally do not have a high prediction accuracy, and errors are usually higher than 5%, which is not conducive to analyzing how the factors affect water demand. Research inside China has been mainly applied to the multivariate prediction model (Liu et al., 2012; Wang & Sun, 2012) and models considering the inherent law of water use and time trend (Li & Chen, 2010; Zai et al., 2009). Liu et al. (2012) and Wang et al. (2012) concluded that the combined forecasting model had better forecasting accuracy than a single model; and its forecasting error was generally less than 3%. Meanwhile the forecasting error of models which only consider the inherent law of water use and time trends was generally higher than 5%. In this paper, multiple regression analysis, time series analysis, and expert experience methods (Liu, 2018) were applied to forecast water demand by industry in China in 2020.

2 WATER CONSUMPTION ANALYSIS

In 2018, the total water consumption was 601.55 billion m^3, which is 2.79 billion m^3 less than in 2017 (Figure 1). Compared with 2017, agricultural water consumption decreased by 7.33 billion m^3, industrial water consumption decreased by 1.54 billion m^3, domestic water consumption increased

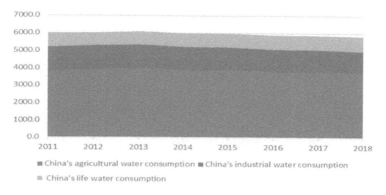

Figure 1. China's water consumption by industry.

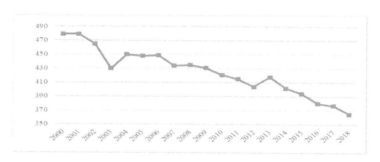

Figure 2. The average water consumption per mu of farmland irrigation in China.

by 2.18 billion m³, and ecological water supply increased by 3.09 billion m³. In 2018, per capita water consumption in China dropped to 432 m³, with 10 provinces (or municipalities) including Tianjin, Beijing, Shanxi, and Shandong having a per capita water consumption of less than 300 m³. A total of 75 billion tons of waste water was discharged in China in 2018, which was 0.8 percent less than in 2017.

2.1 China's agricultural water consumption

From 2000 to 2018, China's agricultural water consumption increased in the first few years and then decreased, reaching a peak of 392.15 billion m³ in 2013, and then decreased year by year. Compared with 2017, it further decreased to 369.31 billion m³ in 2018, and its proportion of total water consumption also decreased from 68.8% in 2000 to 61.4%, indicating that China's agricultural water consumption efficiency had significantly improved. The average water consumption per mu of farmland irrigation kept declining, from 479 m³ in 2000 to 365 m³ in 2018 (Figure 2). In 2018, the effective utilization coefficient of farmland irrigation water in China was 0.554, which was 0.06 less than in 2017, and 0.23 lower than the average level in developed countries. In 2020, the completion of China's high-standard farmland construction task, the development of high-efficiency water-saving irrigation, and the further strengthening of farmland water conservancy construction will promote the further improvement of China's agricultural water efficiency.

2.2 China's industrial water consumption

From 2011 to 2018, the share of industrial water consumption in China's total water consumption declined year on year, accounting for 21.0% in 2018, which was 0.1 percentage point lower than

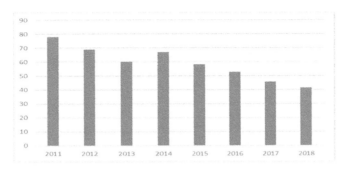

Figure 3. China's water consumption per 10,000 yuan of industrial added value.

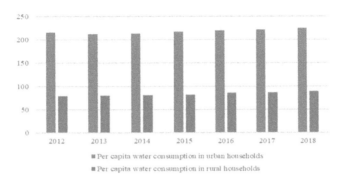

Per capita water consumption in urban households
Per capita water consumption in rural households

Figure 4. China's per capita water consumption in urban and rural households (unit: liter).

in 2015. The water consumed by the thermal power (including dc cooling power generation), steel, textile, papermaking, petrochemical and chemical industries, food and fermentation, and other high-water-consumption industries accounted for about 50% of the industrial water used. In 2018, China's water consumption per 10,000 yuan of industrial added value was 41.3 m^3, which was 36.7 m^3 lower compared with that in 2011 (Figure 3).

From 2011 to 2017, China's water consumption by the iron and steel industry had declined from 2.62 billion m^3 to 2.56 billion m^3, the new water consumption per ton of steel declined from 4.07 m^3 to 3.23 m^3, decreasing by about 20.6%. China's new water consumption in the paper industry declined from 4.56 billion m^3 in 2011 to 2.898 billion m^3 in 2015, and water consumption per 10,000 RMB of output value had declined from 67.4 m^3 to 40.6 m^3, decreasing by 39.76%. With the improvements in water use efficiency in major water-use industries and the upgrading and optimization of the internal structure of industrial sectors, China's industrial water consumption is expected to decrease slightly in 2020.

2.3 *China's life water consumption*

The growing population and urbanization rate are the main reasons why the domestic water consumption has increased. China's total population increased from 1236.26 million to 1395.38 million during 2000 to 2018, the urban population increased from 394.49 million to 831.37 million, and the urbanization rate rose from 31.9% to 59.6%. Per capita water consumption in urban households was 225 liters per day and in rural households was 89 liters per day (Figure 4). With the acceleration of China's urbanization process and the growth of its population, it is expected that China's domestic water consumption will continue to increase in 2020.

2.4 China's water quality

In 2018, the overall surface water quality of the entire country remained stable. Among the 262,000 km length of rivers, the length of rivers of class I–III accounted for 81.6%, increasing 1.0 percentage points from 2017, and 5.5% of class V, decreasing 1.3 percentage points. By the end of 2018, China's municipal sewage treatment capacity reached 167 million m^3 per day, with a cumulative sewage treatment capacity of 51.9 billion m^3, and the application of renewable water conservancy increased to 7.58 billion m^3. As an important part of the ecological civilization construction, ecological water supplement will be paid more attention to in the 13th Five-Year period. The water supplement for ecological environment will increase steadily in 2020.

3 CHINA'S WATER DEMAND FORECASTING

Considering China's economic growth, the adjustment of industrial structure, urbanization process, and the fluctuation of water efficiency in different industries, China's total water demand and four components of it were forecast with regression analysis, time series analysis, and expert experience methods.

The main results are as follows: without the COVID-19 out broke in 2020, the total water demand in China will be about 599.01 billion m^3, slightly less than in 2019. As for the four kinds of water demand, agricultural water demand will be about 360.6 billion m^3 in 2020, accounting for 60.2% of the total water demand. Industrial water demand will be about 123.20 billion m^3, accounting for about 20.6 percent of the total. Domestic water demand will be about 91.03 billion m^3, accounting for 15.2% of the total, and ecological water supplement will be about 24.18 billion m^3, accounting for 4.0% of the total.

ACKNOWLEDGMENTS

The author would like to acknowledge support from the National Natural Science Foundation of China (No.71874184).

REFERENCES

Ajbar, A. H. & Ali, E. M. 2013. Prediction of municipal water production in touristic Mecca City in Saudi Arabia using neural networks. *Journal of King Saud University-Engineering Sciences.*

Firat, M. Turan, M. E. & Yurdusev, M. A. 2010. Comparative analysis of neural network techniques for predicting water consumption time series. *Journal of Hydrology* 384: 46–51.

Herrera, M. Torgo, L. Izquierdo, J. & Perez-Garcia, R. 2010. Predictive models for forecasting hourly urban water demand. *Journal of Hydrology* 387: 141–150.

Li, K.B. & Chen, S.F. 2010. Robust control of the investment in urban water supply and demand system. *Journal of System Science and Mathematical Science* (1): 22–32. (In Chinese).

Liu, X.L. 2018. Forecasting on China"s total water demand in 2018 (2018), *Advances in Intelligent Systems Research* (151): 254–257.

Liu, Z.X. Zhang, X. & Wang, Y.H. 2012. Forecast of urban residential domestic water consumption in Baotou, *Journal of Water Resources & Water Engineering* 23:67–70. (In Chinese).

Nasseri, M. Moeini, A. & Tabesh, M. 2011. Forecasting monthly urban water demand using Extended Kalman Filter and Genetic Programming. *Expert Systems with Applications* 38: 7387–7395.

Sen, Z. & Altunkaynak, A. 2009. Fuzzy system modelling of drinking water consumption prediction. *Expert Systems with Applications* 36: 11745–11752.

Wang, S. & Sun, Y.F. 2012. Urban water consumption prediction based on a partial least-square model coupled with stepwise regression. *Journal of Safety and Environment* 12:170–173. (In Chinese).

Zai, S.M. Guo, D.D. & Wen, J. 2009. A prediction of irrigation water in People"s Victory Canal Irrigation District based on the least squares support vector machine model. *China Rural Water and Hydropower* (12): 49–51. (In Chinese).

Computational Social Science – Luo, Ciurea & Kumar (eds)
© 2021 Taylor & Francis Group, London, ISBN 978-0-367-70193-2

An analysis of the factors of university choice for different groups of college students in China

J.B. Xiahou, L.C. Li & X. Cui
Xiamen University, Xiamen, China

ABSTRACT: Teachers, popularity, disciplines, and professional reputation play decisive roles in college students' choice of learning. Schools with strong teachers can recruit many excellent talents; the distance between schools and families is also an indispensable judgment factor when students choose schools; different family social backgrounds and different family cultural backgrounds have a subtle influence on students' school choice. The process of students' choice of colleges and universities is an individual subjective rational choice, resulting the most suitable choice for their own interests made by individuals out of their own circumstances, different resources, and varied needs and preferences. It has sufficient rationality and practicality.

1 INTRODUCTION

The main method of this study was a questionnaire. The "questionnaire on admission willingness of freshmen" was distributed to seven colleges and universities and independent colleges in mainland China, including one in each of 985 and 211 colleges and universities, one general public undergraduate university, one public higher vocational school, one independent college, one private undergraduate university, and one higher vocational college, which take freshmen as the research object. A total of 1403 samples were obtained. In this study, 16 school choice reasons were designed, each of which was arranged according to four levels of "completely unimportant," "not very important," "important," and "very important," so that the investigators could see their importance.

In this study, China's ordinary colleges and universities are divided into "985" colleges and universities, "211" colleges and universities, general public undergraduate colleges, independent colleges, public higher vocational colleges, and private higher vocational colleges. School selection mainly refers to the individual's choice of the above six types of universities when competing in their college entrance examination. As for the variables of parents' education level, this study divides them into three levels: the first level being undergraduate, graduate and above, the second level being vocational college, secondary vocational school, high school, and the third level being junior high school, primary school, and below. With regard to the variables of family social stratum, this study divides the family social stratum into three levels, namely, the dominant class, the middle class, and the lower class.

In this study, social science statistical software SPSS was used to analyze the survey data. The importance of college students' reasons for choosing a school was sorted and explained in groups. This study analyzes the reasons why college students from different cultural backgrounds choose schools, and finally the distribution of higher education resources among college students at different social levels is analyzed.

2 ANALYSIS OF THE IMPORTANCE OF COLLEGE STUDENTS' REASONS FOR CHOOSING A SCHOOL

First, 16 factors of college students' school choice were simply weighted, and the importance index obtained. According to the magnitude of the importance index, the importance of these 16 factors was ranked. Then, with the help of SPSS software, the main components of these 16 factors were analyzed. Through the analysis, we obtained three groups of school selection factors, which are also explained.

2.1 *Importance index method*

In the questionnaire for freshmen, there was the following question for students to answer: "How important are the following reasons when you choose your current university?" We listed 16 possible reasons and asked students to evaluate their importance. The importance is divided into completely unimportant, less important, important, and very important. In order to compare the importance of 16 possible causes more intuitively, we used the statistical method of important indicators to calculate the importance index of each cause, and on this basis, we ranked the importance of the 16 possible causes.

There are two things to do before calculating the importance index of each cause.

First, the importance degree should be graded, specifically: 1 = completely unimportant, 2 = less important, 3 = important, and 4 = very important. Second, using SPSS13.0 statistical analysis software to gain statistics from the answers of the urban and rural freshmen to the importance of the 16 possible reasons.

The formula of the importance index for each item was calculated as follows: importance index $= \sum ai \times xi / 3$, where ai is the weighting coefficient of the i-th level among the four importance levels from completely unimportant to very important, when i equals 1, 2, 3, 4, and ai equals 0, 1, 2, 3, respectively. xi represents the percentage of the number of people whose importance is the i-th grade to answer an item in the four importance grades from completely unimportant to very important in the total number of effective answers of the item. $\times 1$ = percentage of "completely unimportant" answers, $\times 2$ = the percentage of "less important" answers, $\times 3$ = the percentage of "important" answers, $\times 4$ = the percentage of "very important" answers.

Table 1 shows the statistics of the importance index and ranking of the freshmen to 16 possible reasons for school selection. The results show that the five items of high importance for school choice by freshmen are employment prospect, student enrollment score, academic and professional reputation, university teachers, and global popularity; the less important items include the influence or suggestion of friends, classmates, senior students (elder sisters), special items, closer to home, farther away from home, having relatives and friends in the university.

2.2 *Principal component method*

PASS software was used to process the survey data of freshmen in all kinds of universities. The 16 reasons for school selection were taken as 16 variables for principal component analysis.

As shown in Table 2, after KMO test and Bartley sphere test, the KMO test coefficient is 0.915>0.5, and the output P value is 0 (less than 0.05), indicating that the survey data are independent and factor analysis can be carried out.

As shown in Table 3, from the total table of explanatory variation, we can see that the eigenvalues of three main elements are greater than 1, so we extract the first three main components.

Combined with the output of rotating element matrix in Table 4, the data with a correlation coefficient less than 0.3 will not be considered generally. It can be seen that the interpretation degree of the main element 1 to 10 variables is greater than 30%. Here, we take the three reasons for school choice with the highest degree of explanation from the three main elements: (1) university teachers, school popularity, school discipline, and professional reputation. There are five variables in which the interpretation degree is more than 30% in the main element (2). We take the three

Table 1. Statistics of importance index and ranking of 16 possible reasons for school choice.

Number	Reasons for school choice	%	order
1	Influence or advice from parents, family members, or other relatives	61.03457	9
2	The influence or suggestion of middle school teachers and nonrelative elders	52.81008	11
3	Influence or suggestion of friends, classmates, and seniors	50.66764	12
4	Reputation of the school	62.13333	8
5	Far from home	43.29922	15
6	Closer to home	46.79767	14
7	Have family and friends in the university	38.21316	16
8	Campus environment, facilities, and equipment	64.60756	6
9	Special projects (such as laboratories, special classes or enrollment by major categories, etc.)	48.5756	13
10	University faculty	65.28384	4
11	School popularity	65.23636	5
12	Discipline and professional reputation	66.30382	3
13	Tuition and cost of living factors	59.10528	10
14	Employment prospects	69.4948	1
15	School enrollment score section	67.56103	2
16	The degree of economic development of the university location	64.05039	7

Table 2. KMO and Bartley test.

Type	Value
Kaiser-Meyer-Olkin measurement sampling suitability	0.915
Bartlett's spherical test	8757.832
df	120
Significance (P value)	0.00

reasons for school choice with the highest degree of explanation from the three main elements (2) far away from home, close to home, and having relatives and friends in the college/university. For main element (3), similarly, we choose three reasons for school selection: suggestions from parents, family members or other relatives, suggestions from nonrelative elders, suggestions from friends and classmates.

The first group is the most important factor for school choice. Whether a university has a first-class teaching staff determines its characteristics, the quality of personnel training, and the level of discipline. The school is based on talents and teachers. Teachers are the basis of education, the foundation of a developing education, the source of enhanced education, and the comprehensive embodiment of the school's core competitiveness. The key to the success of colleges and universities lies in teachers who are full of passion and professionalism. Second, the university has a high reputation, which not only attracts better quality teachers, but also attracts more excellent candidates. The academic and professional reputation of the school is also very important. The authors participated in the enrollment work and enrollment publicity work many times. The most important thing of many examinees is whether they can apply for the most popular major of a school. Students go to university to have a good career in the future. To be able to go to a university's best major of school is a good start to their career.

The second group of reasons for school choice also accounts for a certain proportion. Many students may feel homesick. Some parents are afraid that their children will travel too far, so they are reluctant to let their children choose a school far from home. Even if they can go to 985 colleges

Table 3. Total explanatory variation.

Element	Starting eigenvalue			Cyclic sum of squares loading		
	Total	Variation %	Accumulation %	Total	Variation %	Accumulation %
1	6.222	38.89	38.89	4.674	29.212	29.212
2	1.936	12.099	50.989	2.313	14.459	43.671
3	1.084	6.777	57.766	2.255	14.096	57.766
4	0.861	5.381	63.147			
5	0.815	5.096	68.244			
6	0.673	4.208	72.452			
7	0.605	3.778	76.23			
8	0.573	3.581	79.812			
9	0.522	3.26	83.072			
10	0.494	3.089	86.161			
11	0.461	2.884	89.045			
12	0.446	2.786	91.832			
13	0.373	2.33	94.162			
14	0.35	2.186	96.348			
15	0.324	2.024	98.372			
16	0.26	1.628	100			

Table 4. Matrix of rotating element.

Reasons for school choice	Element		
	1	2	3
Influence or advice from parents, family members, or other relatives	0.209	0.141	0.781
The influence or suggestion of middle school teachers and nonrelative elders	0.206	0.288	0.761
Influence or suggestion of friends, classmates, and seniors	0.219	0.307	0.716
Reputation of the school	0.593	0.063	0.422
Far from home	0.058	0.714	0.233
Closer to home	0.083	0.676	0.245
Have family and friends in the university	0.091	0.790	0.136
Campus environment, facilities, and equipment	0.698	0.083	0.098
Special projects (such as laboratories, special classes, or enrollment by major categories, etc.)	0.436	0.573	0.009
University faculty	0.803	0.155	0.096
School popularity	0.805	0.183	0.125
Discipline and professional reputation	0.812	0.100	0.121
Tuition and cost of living factors	0.573	0.312	0.204
Employment prospects	0.753	0.034	0.239
School enrollment score section	0.646	−0.025	0.299
The degree of economic development of the university location	0.479	0.139	0.077

and universities far away from home, they may choose 211 colleges and universities in the province and often go home to visit. This is the case with a college classmate of one of the authors, who is worried about his daughter's study in the province. However, some students, especially boys, will choose a school far from home, so that they can go out and experience the wider world, and exercise their freedom. That is what I thought at that time. In addition, Chinese families are so closely connected that they want familiar friends or relatives to go to the same school.

The last group of reasons for school selection is also understandable. Now many parents will help their children fill in their wishes, give them advice, plan their life path, and students will follow their parents' advice. In addition, before applying for the examination, they will ask the seniors for advice and ask about filling in the application. Comparing the importance of school choice factors obtained by the two methods, the top six factors of importance are employment prospects, school enrollment score, discipline and professional reputation, university teachers, school popularity, and campus environment, facilities and equipment.

3 ANALYSIS OF THE REASONS FOR COLLEGE STUDENTS' CHOOSING SCHOOLS WITH DIFFERENT CULTURAL BACKGROUNDS

The cultural background of the family has a great influence on children's access to different types of colleges and universities. Based on the statistics of the distribution of freshmen from different cultural backgrounds in different types of colleges and universities, it can be seen that with an improvement in their parents' education level, the proportion of their children entering colleges and universities of "211" level and above has increased significantly as a whole. Among the parents, the proportion of children of families with higher education who attend universities of "211" level or above is significantly higher than that of other families, especially the freshmen whose family cultural background is graduate students or above. With the improvement in the parents' education level, the proportion of freshmen in general public universities and public higher vocational colleges is obviously decreasing. From the statistics of the importance index and ranking of the 16 possible reasons for school choice of freshmen from different cultural backgrounds (as shown in Table 5), it can be seen that the reasons for school choice of freshmen with different cultural backgrounds are different.

The employment prospects are the most important factor for the freshmen of all cultural backgrounds when choosing universities.

The college enrollment score is a difficult condition for college entrance, so it is listed as the second most important factor by students from all cultural backgrounds. If the score does not reach the minimum score line for a university, there is no way to enter that university.

In general, families with low educational backgrounds also have low income, so they pay more attention to the factors of tuition and living expenses. Although the importance of this factor ranked lower in the eyes of freshmen from all cultural backgrounds (7, 10, and 10, respectively), freshmen from families with only junior high school, primary school, and lower education background advanced the importance of this factor by three places compared with the other two groups. For freshmen whose parents have a bachelor's degree or higher level of education, they are more obedient to their parents' opinions in choosing a school. The importance of their parents' opinions is two to three point higher than the other two groups. This is probably due to the fact that parents are highly educated and have a high level of insight, discretion, and judgment.

The importance of the economic development degree of the university location in families with different cultural backgrounds varies greatly. Children with good family culture background put this factor for school choice in third place. It may be that these families think that places with good economic development will have more opportunities in the future, which is beneficial to their children's future career development. However, families with poor family cultural background put the importance of this factor in eighth place. This may be because the cost of living in areas with good economic development is also high, and families with low economic income worry about this.

4 CONCLUSION AND DISCUSSION

According to this analysis, the six most important factors for high school graduates in choosing higher education institutions are employment prospects, school enrollment score, discipline and

Table 5. Comparison of college students' reasons for choosing schools from different cultural backgrounds.

Reasons for school choice	Junior high school, primary school, and below		Vocational college, secondary vocational school, high school		Undergraduate, graduate, and above	
	Proportion %	Order	Proportion %	Order	Proportion %	Order
Influence or advice from parents, family members, or other relatives	59.5	10	58.3	9	65.4	7
The influence or suggestion of middle school teachers and nonrelative elders	52.2	11	51.9	11	51.4	11
Influence or suggestion of friends, classmates, and seniors	51.2	12	48.5	12	48.5	13
Reputation of the school	60	9	60.6	8	64.7	8
Far from home	41.2	15	43.8	15	41.8	15
Closer to home	46.7	13	46.2	14	43.8	14
Have family and friends in the university	36.5	16	38.1	16	38.8	16
Campus environment, facilities, and equipment	63.4	5	62.3	7	66.1	4
Special projects (such as laboratories, special classes, or enrollment by major categories, etc.)	45.7	14	48.0	13	50.8	12
University faculty	63.4	6	64.0	4	65.9	6
School popularity	64.8	3	62.8	6	64.7	9
Discipline and professional reputation	64.7	4	65.4	3	66.1	5
Tuition and cost of living factors	60.9	7	57.0	10	51.8	10
Employment prospects	68.1	1	67.7	1	70.7	1
School enrollment score section	65.7	2	66.9	2	68.1	2
The degree of economic development of the university location	60.8	8	63.1	5	68.0	3

professional reputation, university teachers, school popularity, and campus environment, facilities and equipment.

The cultural background of the family has a great influence on children's access to different types of colleges and universities. With the improvement of parents' education level, the proportion of their children entering universities above the "211" level is increasingly obvious. Students from different cultural backgrounds choose different schools for different reasons.

REFERENCES

[1] Kaiser, M. (1974). Kaiser-Meyer-Olkin measure for identity correlation matrix. Journal of the Royal Statistical Society, 52, 296–298.
[2] Le, T.D.; Robinson, L.J.; Dobele, A.R. Understanding high school students use of choice factors and word-of-mouth information sources in university selection. Stud. Higher Educ. 2020, 45, 808–818.
[3] Simões, Cláudia, and Ana Maria Soares. 2010. "Applying to Higher Education: Information Sources and Choice Factors." Studies in Higher Education 35 (4): 371–389.
[4] Tobias, S., & Carlson, J. E. (1969). Brief report: Bartlett's test of sphericity and chance findings in factor analysis. Multivariate Behavioral Research, 4, 375–377.
[5] Verma, J.P. Data Analysis in Management with SPSS Software; chap. 10. Springer; 2013.

Computational Social Science – Luo, Ciurea & Kumar (eds)
© 2021 Taylor & Francis Group, London, ISBN 978-0-367-70193-2

Research on the experimental teaching reform of "material analysis and test method" based on OBE engineering education mode

L. Xia & J.B. Yu*
School of Materials Science and Engineering, Harbin Institute of Technology (Weihai), Weihai, China

ABSTRACT: According to the concept of OBE engineering education, a "flipped classroom" has been established for the experimental teaching of "material analysis and testing methods" under the CDIO mode. An experimental teaching design of the teacher–student interaction is pioneered. Multidimensional experimental comprehensive performance assessment systems basing on student-centered and experimental teaching quality supervision mechanism are formulated. These systems rely on reformed experimental teaching content, opening laboratories and online experimental teaching platforms, and focusing on cultivating students to master the operation of large precision instruments, such as scanning electron microscopes, X-ray diffraction and transmission electron microscopes, and the method of analyzing the microstructure of materials. The comprehensive abilities of students, including hands-on practical ability, innovation, analysis and testing are greatly improved, laying a solid foundation for the graduation design of the senior year.

Keywords: OBE, Material analysis and testing methods, Experimental teaching, CDIO.

1 INTRODUCTION

OBE is the abbreviation for outcome-based education [1], which first appeared in the basic education reform of the United States and Australia. From the 1980s to the early 1990s, OBE was a very popular term in American education. Currently, it is an internationally recognized educational mode and also provides valuable ideas for the reform of higher educational modes in China [2–4]. Its core concepts are "student-centric," "result-oriented," and "continuous improvement of teaching methods."

The "Materials Analysis and Testing Method" experimental course is a professional basic experimental course for undergraduates majoring in materials, a very practical experimental course, containing scanning electron microscope (SEM), X-ray diffraction (XRD), and transmission electron microscope (TEM) experiments, and other materials microstructure analysis experiment courses. The experiments take 10 hours. Through the study of this experimental course, students can strengthen their understanding and grasp of theoretical knowledge, and master the structure and working principles of large-scale instruments such as scanning electron microscopes, X-ray diffractometers, and transmission electron microscopes. At the same time, they can also preliminarily master the operation of large-scale instruments and equipments, so that they can comprehend the analysis and detection methods of the material's microstructure, analyze the formation mechanism of the material's microstructure, improve the process, and research new materials [5]. Therefore, it is necessary to reform the traditional experimental teaching mode, which is very important in improving students' grasp of modern analytical testing methods and beneficial to the improvement of their practical ability and innovative consciousness. This also improves their ability to comprehensively apply knowledge and solve problems [6].

---

*Corresponding author

2 PROBLEMS WITH TRADITIONAL EXPERIMENTAL TEACHING

2.1 *Single laboratory equipments that cannot meet the requirements of the experiments*

The main pieces of experimental equipment (SEM, XRD, TEM) are all large-scale precision equipments, that are complicated and difficult to operate. Therefore, the relevant experimental content is mainly demonstration, and it is impossible for each student to operate the equipment independently. In the previous experimental teaching mode, students previewed relevant experimental content before the experiment, and the experimental teacher explained the experimental content and demonstrated the operation of the experimental equipment. After class, students submitted their experimental report. However, there are a large number of students and the laboratory space is limited, so students need to be divided into several groups during the experimental process, which increases their workload. At the same time, due to the limitation of the experimental time, the demonstration accounted for a large proportion of the class, and the students cannot acquire adequate experimental content in the time allowed. Consequently, students are in a passive state of learning, which causes students to participate in the experimental teaching activities of "Material Analysis and Testing Methods" in a hurried manner, which is not conducive to improvement of students' thinking and hands-on ability, affecting the effective combination of theoretical teaching and experimental teaching, and not meeting the requirements of experimental teaching.

2.2 *Lack of experimental hours and neglect of students' learning attitudes*

In the process of formulating a new undergraduate training plan, the theoretical teaching hours for "Material Analysis and Testing Methods" have been further reduced from the original 64 hours to 32 hours at present. The experimental teaching hours have been reduced to 10 hours or less, while students need to master a relatively large number of analytical testing methods. This presents difficulties for the organization and development of related experimental teaching, including how to teach the testing techniques and methods that students need to master during the limited experimental teaching process. In addition, in the experimental teaching process, the experimental teaching means are single, which is mainly manifested as follows: the teaching experiments are mainly demonstration experiments. The experimental teacher spoke in front of the experimental equipment, and the students listened and took notes. There is little or no teacher–student interaction, neglecting students' initiative [7,8]. As a consequence, students are not impressed with the experiment. The analysis of experimental organization in the experimental report is always the same, and students' learning is very poor.

2.3 *The simple examination means of experiments*

At present, the experiment score partially accounts for the usual performance, which is mainly evaluated according to the attendance rate of students in experimental class. It is attributed to the single teaching method in the experimental teaching process, where students cannot do the experiments, leading to a difficulty in the evaluation of students' hands-on practical ability. Therefore, when the results are the same or similar, the real performance of students is not reflected. More significantly, the experimental results are mainly reflected by the scores of experimental reports, which are primarily composed of experimental results, experimental principles, experimental procedures, material microstructure analysis, and experimental experience. Among these, experimental purposes, principles, and procedures can be copied mechanically from the experiment instruction. However, material microstructure, analysis, and experience, as the most critical parts, are similar or even copied from each other. This is because students cannot do the experiments themselves. Therefore, the experiment reports fail to arouse students' initiative and enthusiasm, as well as failing to reflect their practical and innovation ability. Hence this traditional method of experimental assessment can only partly reflect the students' mastery of analytical test methods.

3 EXPERIMENTAL TEACHING REFORM OF THE "MATERIAL ANALYSIS METHOD" BASED ON THE OBE ENGINEERING TEACHING MODE

Outcomes-based education (OBE) refers to the goal of teaching design and teaching, which represents the final learning outcomes of students through the education process. The implementation principles of OBE are as follows:

(1) Clear focus

The experimental teaching of the material analysis test method should focus on the final learning outcomes that students can achieve after completing the course, and guide students to focus on these learning outcomes. Experimental teachers must explain clearly and devote themselves to helping students develop knowledge, ability, and level, so that they can achieve the expected results.

(2) Expansion of opportunity

The individual differences of each student should be fully considered in the experimental teaching process, and every student should have the opportunity to achieve learning outcomes in terms of time and laboratory resources. Teachers should not provide all students with the same learning opportunities at the same time in the same way, but should be more flexible to meet the personalized requirements of students, so that students have the opportunity to prove what they have learned and show their learning results. If students have the right learning opportunities, they will achieve the expected learning results.

(3) Improvement of expectations

The experimental teacher should improve the expectations of students for the class, and develop challenging executive standards to encourage deep learning and promote a more efficient learning process. There are three aspects to improving expectations. First, improve the executive standard and promote the students to reach a higher level after completing the course; second, eliminate the additional conditions for success and encourage students to achieve peak performance; Third, set up high-level teaching content to guide students to strive for high standards.

(4) Reverse design

The final goal refers to the final learning achievement or peak achievement. Therefore, the experimental teaching curriculum and activities can be designed and carried out in reverse. Experimental teaching is designed from the final learning achievement backward to make sure the appropriateness of all teaching toward the peak achievement. Therefore, the starting point of experimental teaching of this course is to achieve peak results, rather than what teachers want to teach [9–13].

3.1 Reform of experiment teaching content

The goal of experimental teaching is to make sure the focus is transformed from the traditional way of mastering the working principle and operation of large-scale instruments such as scanning electron microscopes, X-ray diffractometers, and transmission electron microscopes. The student's ability is cultivated during the analysis process of material microstructure. The specific reform methods are as follows, the latest cutting-edge technologies such as two-dimensional graphene materials, advanced ceramics, nanomaterials, etc. are added to the experiment teaching content [14], and 5–10 groups of comprehensive material analysis experiments are designed based on the research projects by the teachers. The experimental items include scanning electron microscope, X-ray diffraction experiment, projection electron microscope, and other analysis methods. Before the experiment, the students are divided into 6–7 groups, with the group members free to mix according to their interests and hobbies, therefore they are free to choose comprehensive experimental items and complete the comprehensive experimental report later.

3.2 Reform of the experiment teaching mode

According to the experiences in experimental teaching, the authors found that students have individual differences in relatively obscure professional knowledge, instrument structure, and operation process. This is mainly reflected in that some students are quick to understand and have strong practical ability, while others have a solid theoretical foundation. In order to ensure that every student can master the operation of large-scale analytical instruments and the analysis of microstructure, a network experimental teaching platform of "material analysis method" was established. This platform includes the experimental instruction module, large-scale equipment operation video module, 3D animation module of "material analysis method," typical cases of analysis module, and mutual communication module. Through this platform, students not only can change the traditional experimental teaching mode, but also can change the traditional laboratory teaching location into dormitories, canteens, parks, beds, and other places to facilitate students to watch and learn the operation of various large-scale instruments and equipment at any time. As a result, students can change the traditional fixed experimental teaching time to students' self-arranged time for efficient learning and improvement of their learning efficiency.

3.3 Innovative experimental teaching based on the CDIO engineering education concept

The CDIO (conceive/design/implement/operate) engineering education concept [15,16] is a new achievement of international engineering education reform, which was jointly researched by four universities including MIT. The CDIO concept emphasizes "learning by doing," which involves the use of comprehensive design projects as close as possible to the actual engineering to teach and cultivate students' enthusiasm for learning. In the process of experimental teaching of material analysis method, the engineering practice ability of students is cultivated from the process of conception, design, implementation, and operation of scientific research projects of theoretical teachers. For example, the research and development project for modified graphene can be divided into four stages in the experimental teaching process: the opening stage, the middle stage, the conclusion stage, and the result display stage. In the opening stage, the students mainly consult the relevant materials and demonstrate the experimental scheme. They design and realize the experimental scheme using free team cooperation. Finally, they select the optimal experimental scheme for analysis and testing, which includes the preparation of graphene samples. Therefore, in innovative experimental teaching, the main emphasis is on the cultivation of comprehensive quality.

3.4 Comprehensive experimental performance assessment

As shown in Figure 1, the problem that the traditional experimental teaching assessment is singular, and students' comprehensive ability to apply analysis and testing cannot be assessed accurately. To avoid this scenario a new comprehensive experimental performance assessment system was developed. In this new comprehensive experimental score evaluation system, the traditional attendance score and experimental report score are weakened, at 5 points and 15 points, respectively, the experimental operation score is increased to 15 points, the preclass question score is 15 points, and the defense score is 50 points. The experiment attendance mainly investigates the students' attendance in the experiment, and the experiment operation mainly investigates the students watching the experiment operation video and answering questions related to the operation of the experiment equipment on the network experiment teaching platform. The results of the experiment report mainly investigate the students' experimental data recording and whether the written part is completed correctly.

4 CONCLUSION

According to the OBE engineering education concept, the experimental teaching of "materials analysis and testing methods" is student-centered, and trains students to master the operation of large

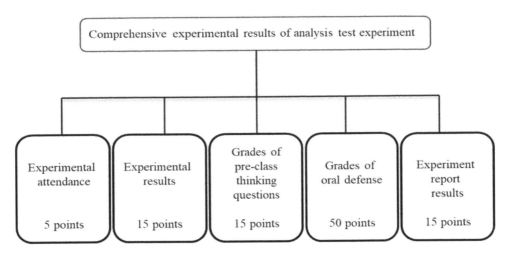

Figure 1. Comprehensive evaluation of analytical test experiments.

precision instruments such as scanning electron microscopes, X-ray diffraction, and transmission electron microscopes. A "flipped classroom" under the CDIO model has been established, which is based on reform of the experimental teaching content, opening laboratory and online experimental teaching platform. The experimental teaching design with teacher–student interaction is developed. The comprehensive experimental performance evaluation system and experimental teaching quality supervision mechanism have improved students' understanding, including hands-on practical ability, innovation, analysis, and testing techniques, which lay a firm foundation for the graduation design and future scientific research.

REFERENCES

[1] Y. Xu, 2017, Research on undergraduate teaching curriculum reform based on OBE teaching model, Course teaching research. (17) 208.
[2] H.M. Yang, F.Z. Duan, C. Yang, et al. 2019, "Material analysis and testing techniques" based on OBE engineering education model, Preliminary study on curriculum construction and reform. Course teaching research. (18) 49–52.
[3] C.Y. Wang, Z.J. Yin, Z.Q. Li, et al. 2018, Reform and practice of training mode for material science and engineering professionals based on OBE concept, Journal of Heilongjiang university of engineering. (1) 73–76.
[4] Y.N. Cui, 2018, Discussion on curriculum reform based on OBE concept, Automobile education. 54–55.
[5] J. Wu, J. Chi, Q. Zhang, et al. 2018, Experimental teaching reform of "Material analysis and testing techniques", Horizon of science and technology. (25) 60–61.
[6] Z.G. Zhang, 2019, Curriculum reform and practice of modern analytical methods for materials, Practice-exploration. (6) 52–54.
[7] B. Wang, M.L. Wang, W. Liu, et al. 2019, Practice teaching reform of electronic technology based on OBE concept, Laboratory research and exploration. (10) 52–54.
[8] C.Y. Zhou, Y. Liu, T. Hong, et al. 2016, The research of step-by-step practical teaching based on output-oriented OBE, Laboratory research and exploration. (35) 206–208.
[9] K. Zhang, B.G. Zheng, J.H. Cui, et al. 2019, Exploration on the construction of environment engineering virtual simulation experiment center based on the OBE model, Experimental technology and management. (36) 270–273.
[10] S.X. Zhou, Z. Han, Q. Huang, 2018, Research and exploration on practical teaching of material mechanics based on CDIO-OBE engineering education mode, Laboratory research and exploration. (37) 176–179.
[11] S.D. Yao, Z.G. Fang, L. Chen, et al. 2019, The application of virtual simulation in OBE practice teaching and innovation entrepreneurship, Experimental technology and management. (36) 229–233.

[12] Z.M. Liu, 45 (2018), Reform of basic teaching of material science based on the OBE concept, Guangzhou chemical industry. 197–198.

[13] W. Zhao, M.Y. Liu, X.B. Liu, et al. 2018, The combination of virtual and real teaching system practice based on the OBE concept, Experimental technology and management. (35) 185–189.

[14] H.H. Liu, X.X. Zhang, J. Song, et al. 2017, The role of materials science and engineering specialty integrated design experiments in the cultivation for undergraduates' innovation ability, Education teaching BBS. (24) 275–276.

[15] Z.X. Guo, G. Xian, J. Xiong, 2019, Experimental teaching method of composites based on CDIO mode, Laboratory research and exploration. (38) 202–205.

[16] X.X. Yang, 2016, The application of CDIO mode to innovative practice project guidance in aerospace discipline, Experimental technology and management. (33) 18–20.

Computational Social Science – Luo, Ciurea & Kumar (eds)
© *2021 Taylor & Francis Group, London, ISBN 978-0-367-70193-2*

A comparative study of the international students' and the local Chinese students' intercultural sensitivity

B.C. Chen
School of International Studies, University of Science and Technology Liaoning, Anshan, Liaoning, China

J.Q. Zhang & X.B. Chen*
School of Electronics and Information Engineering, University of Science and Technology Liaoning, Anshan, Liaoning, China

ABSTRACT: The current research employed quantitative and qualitative approaches to investigate international students' and local Chinese students' intercultural sensitivity. The quantitative results show that the international students' intercultural sensitivity was much higher than that of the local Chinese students. A structured written interview was conducted to shed light on the main elements that affect each dimension of intercultural sensitivity. The findings reveal that the Chinese students focused more on foreign language competence and scope of knowledge, while the international students concentrated on the affective aspects in intercultural communication. The two empirical results of this study have significant implications for EFL teaching reform in China.

Keywords: Intercultural sensitivity, Local Chinese students, International students, Chinese context.

1 INTRODUCTION

Intercultural communication (IC) has been a hot topic worldwide for a number of decades and intercultural communication competence (ICC) is of growing importance with economic globalization and widespread migration because interactions with culturally different people for personal or professional purposes have become much more common than ever before (Czerwionka et al. 2015). Since Hymes (1972) put forward the concept of communicative competence, numerous research have been done on IC or ICC in the Western context (Hymes 1972, Ruben 1976, Bennett 1986, 1993, Gudykunst 1993, Chen & Starosta 1997, 1998a, b, 2000, Fritz et al. 2002, 2005, Deardorff 2004, 2006, 2012, Rissanen et al. 2016). Despite its late start, Chinese scholars have made considerable efforts to do research on IC or ICC since the 1980s (Lin 1999). According to Shi & Zhu (2015), more than 13,000 research papers on IC or ICC studies were published in China's academic journals between 2005–2014. However, empirical research is still not the mainstream intercultural communication research method among Chinese scholars (Hu 2005, Peng 2008, 2010, 2011, Hu 2011, Shi & Shan 2016). Most scholars focus on the importance or definitions of ICC, but few of them discuss its assessment using scientific research methods (Peng et al. 2015). Research on intercultural sensitivity (IS), a main dimension of ICC (Fritz et al. 2002), is also woefully insufficient in this sense. With China fully opening up, growing numbers of international students have been attracted to study in Chinese universities, but little research on international students' IS in a Chinese context has been done. We attempt to investigate the international students' and local Chinese students' IS in a Chinese context by both quantitative and qualitative approaches so that our research outcomes can be shared by international scholars.

---

*Corresponding author

2   LITERATURE REVIEW

The *Silent Language* by Edward T. Hall published in 1959 is commonly regarded as the corner-stone of intercultural communication as a discipline. We adopt the term "intercultural," while other researchers might use "cross-cultural" or "multicultural" instead. As to the definitions or constructs of ICC, opinions vary somewhat among scholars. Ruben's (1976) communication theory related to ICC involves seven components among which displays of respect, empathy, and tolerance for ambiguity belong to the affective realm. Gudykunst (1993) believes knowledge, skills, and moti-vation are fundamental to defining intercultural competence. Chen and Starosta (2000) think that cognitive, affective, and behavioral ability are the building blocks of ICC. Fantini (2000, 2006) considers knowledge, attitudes, skills, and awareness as the main dimensions of ICC. Deardorff (2006) points out that ICC is the ability to communicate both effectively and appropriately in an intercultural context based on the individual's intercultural knowledge, skills, and attitudes. We understand that Gudykunst's "motivation," Fantini's "attitudes and awareness," and Deardorff's "attitudes" all refer to the affective aspect of ICC. Though scholars utilize different terms, they generally agree that the affective aspect is an indispensable part of ICC.

2.1   *ICC components: a focus on intercultural sensitivity*

Chen & Starosta (1998) believe that people with higher intercultural sensitivity are highly motivated to understand, appreciate, and accept cultural differences. Hammer, Bennett, & Wiseman (2003) view it as "the ability to both discriminate and experience cultural differences." Fritz et al. (2002) refer to intercultural sensitivity as a main dimension of ICC which has earned increasing attention in different academic fields. Byram (1997) maintains that attitudes are the essential component of ICC since noticing and understanding cultural difference is fundamental to the development of ICC. Hammer (2011) argues that intercultural sensitivity is the very foundation of intercultural competence, and even synonymous with it. We understand intercultural sensitivity as a prerequisite for successful intercultural communication because, if people are sensitive to cultural differences, such as different world views, social customs, values, religious beliefs, and taboos, they will respect or at least tolerate these differences, which makes intercultural communication possible. In a word, intercultural sensitivity has tremendous significance in ICC and intercultural sensitivity research results definitely reveal ICC to a large extent.

2.2   *Intercultural sensitivity research in the Chinese context*

Peng (2006) studied the influences of nationality and profession on intercultural sensitivity and found that Chinese subjects' intercultural sensitivity is much higher than that of Thai subjects. Peng (2007) found that students who major in English have higher intercultural sensitivity than those who major in other subjects. It seems that the foreign language learning experience is very important in building intercultural sensitivity. Shao and Chen (2011) investigated high school English teachers' intercultural sensitivity. They found that although the majority of English teachers themselves have a relatively high intercultural sensitivity, they fail to do qualified culture teaching because of a lack of training and guidance, and the heavy burden of language teaching. Hu (2011) investigated 175 undergraduates' intercultural sensitivity. Her results show that the mean of respect for cultural difference is relatively high, while the mean of interaction engagement is the lowest among the five dimensions. Qu and Dou (2014) reveal that ICC and IS are positively related. Compared to the huge population and China's urgent need for intercultural talents as the second largest economic power in the world, the research efforts have not been great and the results do not demonstrate the whole picture of the status quo within the Chinese context.

2.3   *Current goal and research questions*

The current study attempts to investigate the IS of Chinese students and international students in the same university in China. Although Bennett's (1993) developmental model of IS has been found to be interculturally valid (Rissanen et al. 2016), the intercultural sensitivity scale (ISS) developed by Chen and Starosta (2000) is preferred in this study since the ISS is comprised of five dimensions which are more interrelated. After the quantitative research data were collected, a structured written interview was conducted with the same two types of students about the very five dimensions of ISS. By making use of the two empirical research methods, we intended to reach the following goals:

(1) Make a comparison between the intercultural sensitivity levels of the two types of students in the Chinese context;
(2) Shed light on the elements that affect each building block of intercultural sensitivity, respectively;
(3) Discuss practical considerations for training students to become interculturally competent based on the first and the second empirical results.

3 METHODS

3.1 *Questionnaire and participants*

The questionnaire was adopted from Chen and Starosta's ISS. The first part of the questionnaire was designed to collect personal information such as age, gender, and personal intercultural experience, if any, and nationality for the international students. The second part involved the five dimensions of ISS, that is, interaction engagement (7 items), respect for cultural differences (6 items), interaction confidence (5 items), interaction enjoyment (3 items), and interaction attentiveness (3 items). Each item was scored according to the five-point Likert scale. For questions 1, 2, 3, 4, 6, 7, 10, 11, 14, 16, 17, 18, 22, 23, and 24, a "strongly agree," an "agree," an "uncertain," a "disagree," and a "strongly disagree" answer received 5, 4, 3, 2, and 1 point(s), respectively. For questions 5, 8, 9, 12, 13, 15, 19, 20, and 21, a "strongly agree," an "agree," an "uncertain," a "disagree," and a "strongly disagree" answer received 1, 2, 3, 4, and 5 point(s), respectively, because these nine questions achieved negative effects. 1 point represented the very low level, 2 points represented the low level, 3 points represented the average level, 4 points represented the high level, and 5 points represented the very high level of intercultural sensitivity in each item.

Ninety-eight university students, whose ages ranged from 19 to 23, participated in this study, among them there were 62 Chinese students (44 males, 18 females) and 36 international students (28 males, 8 females). The 62 Chinese students majored in mining engineering. All participants were juniors from a university located in Liaoning province, China. None of the Chinese students had had any overseas experience, but they had completed a 32-hour intercultural communication course in addition to a college English course. They had never learnt any other foreign languages except English. Among the 36 international students, there were 17 from Zimbabwe, 2 from Croatia, 10 from Nigeria, 1 from South Africa, 1 from Ghana, 1 from Zambia, 1 from Rwanda, 1 from Congo, 1 from Morocco, and 1 from Kenya. They majored in either natural sciences or social sciences in which lectures were given both in English and Chinese. Besides English and Chinese, all the international students could speak a third language. All the participants were well informed that the participation was voluntary and the collected data were intended only for research purposes.

3.2 *Quantitative results*

SPSS 20.0 was employed to calculate the collected data. The mean score of all 24 items for the international students was 4.00 (between 4 and 5), which indicates that the international students' IS was high. The mean score of all 24 items for the Chinese students was 3.0934 (between 3 and 4), which indicates that the Chinese students' IS was average. To demonstrate the specific gap between the two types of students' intercultural sensitivity, their performances in each dimension were calculated and listed in Table 1.

Table 1. Descriptive analysis of each dimension.

Dimension	n		Mean		SD	
	International	Chinese	International	Chinese	International	Chinese
Interaction engagement	36	26	4.03	3.52	.377	.504
Respect for cultural differences	36	26	4.44	2.76	.607	.619
Interaction confidence	36	62	3.78	2.95	.638	.556
Interaction enjoyment	36	62	4.39	2.44	.766	.617
Interaction attentiveness	36	62	4.00	3.73	.633	.705
Column width	36	62	4.03	3.52	.377	.504

As is shown in Table 1, although the two types of students did not differ statistically in "Interaction engagement" and "Interaction attentiveness," the international students performed better than the Chinese students in all five dimensions, which accounts for their higher total score; that is, the international students' intercultural sensitivity was higher than that of the local Chinese students. The two types of students constituted a sharp contrast in "Respect for cultural difference" and "Interaction enjoyment." The international students held a notable lead in these two factors, which contributes a great deal to their much higher intercultural sensitivity. The Chinese students did not have real intercultural communication experience, which might make them fall behind the international students in these two dimensions, and they were not as confident as their international counterparts when it came to intercultural communication. What indeed affects each dimension of their intercultural sensitivity? We conducted a structured written interview with the same 62 Chinese students and 36 international students.

3.3 Listing and numbering

Since the purpose of the interview was to reveal the elements that affect each dimension of their intercultural sensitivity respectively, the interview questions must perfectly align with the five building blocks of the intercultural sensitivity scale. Considering the time-consuming nature of interviewing 98 students, a structured written interview was a workable solution. Since the use of a foreign language provides students with an authentic opportunity for foreign language use (Czerwionka 2015), the written interview was conducted in English. The students were interviewed on the following five questions:

(1) What affects your interaction engagement most?
(2) What affects your respect for cultural differences most?
(3) What affects your interaction confidence most?
(4) What affects your interaction enjoyment most?
(5) What affects your interaction attentiveness most?

The first author was present when the two types of students answered the interview questions to ensure all the interview questions in English were well understood. Each student was prepared with a copy of the ISS in which the specific items comprising each dimension were clarified. All the interviewees were clear what the five interview questions meant. They were asked to write down five English words or phrases as an answer to each question. Each word or phrase stood for an element that affects each dimension of their intercultural sensitivity. They might have plenty of ideas, but they were told to describe the top five elements only. They were informed that they could repeat their answers to different questions if they thought it necessary. They were allowed to refer to their cell phones for words they were not sure how to spell, but they could not discuss answers among the interviewees. The written interviews were also anonymous, and the students were encouraged to take their time and think carefully. Ninety-eight written interviews were distributed and returned.

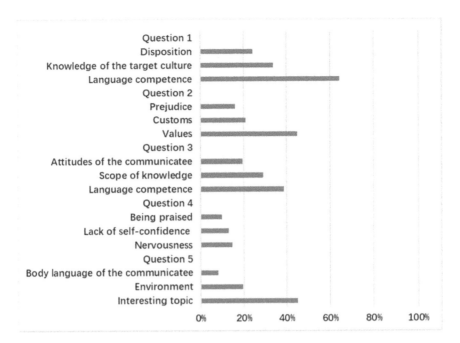

Figure 1. Percentage of the top three themes mentioned by the Chinese students to each question.

3.4 *Approach to analysis*

The first author grouped what the students had written by themes following a grounded approach into a qualitative, thematic analysis (Corbin & Strauss 2008; Glaser & Strauss 1967). The answers of the Chinese students and the international students were grouped separately in order to draw a comparison. For each given question, the total possible student answers was 62 or 36, respectively, for the Chinese and international students, indicating that all the students mentioned the particular theme. Since the research purpose was to obtain an understanding of the elements that affect the students' intercultural sensitivity, a strong emphasis was laid on the content rather than the word choice or part of speech because the students' English language proficiency varied. For example, language ability, language proficiency, foreign language competence, English fluency, oral English, and vocabulary written by different students were grouped under the theme "language competence"; character, conservative, shy, introverted, outgoing, and extroverted, disposition fell into the category "disposition." Since the Chinese students had never learnt any other foreign languages except English, they took it for granted that communication with people from different cultures must use a foreign language. It was understood that the "language competence, etc." mentioned by Chinese students meant foreign language competence or English language competence. The authors selected one of the terms written by a student or some students from the category as the theme name when doing the final calculation. The corresponding author reviewed the themes identified by the first author to enhance the validity. Since the students were not given any hints about what they should write, the answers to the same question were very varied. According to the above-mentioned grouping method, the top three themes that the two types of students shared, respectively, are presented in Figures 1 and 2.

As presented in Figure 1, language competence and scope of knowledge of the target culture seemed the top priority for the Chinese students to consider when it came to interaction engagement. Disposition, such as extroverted or introverted personality, also played a decisive role in this dimension for the Chinese students, which indicates that some of them might fail to notice differences in interaction engagements with people from their native culture and from foreign cultures. For interaction confidence, they also focused on language competence and knowledge. As

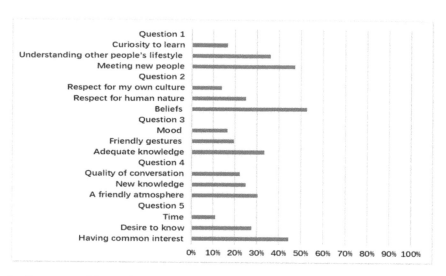

Figure 2. Percentage of the top three themes mentioned by the international students to each question.

for respect for cultural differences, though many students did mention values and customs, they might just consider the two-items part of knowledge of the target culture. Chinese students' inter-cultural sensitivity was much lower than the intercultural students, perhaps because they thought foreign language competence and scope of knowledge of the target culture were the main elements that decided successful intercultural communication, and it seemed that they were not as highly motivated to communicate with culturally and linguistically different people if they didn't think they were sufficiently linguistically competent. They were not fully aware of attitudes, which both Byram (1997) and Deardorff (2004) consider to be the fundamental component of ICC.

As Figure 2 illustrates, the strong desire to meet new people and understand other people's life style enabled the international students to be actively involved in interaction engagement. As to respect for cultural differences, the international students also behaved quite differently because they mentioned beliefs a lot compared with the nonreligious Chinese students. They demanded respect for human nature and their own culture, which demonstrated a strong awareness of commonalities and differences as human beings. Both the international students and the Chinese students thought the scope of knowledge helped to build up interaction confidence. The Chinese students mainly mentioned the negative aspects that affected their interaction enjoyment, while the international students mainly mentioned the positive ones. Just as Chen and Starosta (2000) put it, the international students were "moving to develop empathic ability to accept and adapt" differences among cultures (p. 5). As to interaction attentiveness, the Chinese students stressed interesting topics, that is, the communication content should be interesting enough to maintain their attention, while the international students focused on having common interests, which is more communication oriented.

4   DISCUSSION

The international students' intercultural sensitivity was much higher than that of the Chinese students. Just as Deardorff (2012) described, the international students developed a greater degree of intercultural sensitivity because they try different cultures and "situate their identity in a broader context." Bloom & Miranda (2015) revealed that "the study abroad group reported greater global mindedness, intercultural communication and openness to diversity." Hauerwas et al. (2017) maintained that studying abroad has many benefits for students. People not only develop their communication ability, but also give a thought to the appropriateness and effectiveness of their communication in a foreign context (Deardorff 2012). The results of this research further support the ideas of the above scholars. Besides being exposed to foreign culture, most of the international students in this study spoke at least three languages. According to Peng (2006), foreign language learning experience is equally important to IS development. They were less prejudiced and more open to different social customs and cultural values because they have experienced the enjoyment of being able to communicate with culturally and linguistically different people. In a word, double exposure to Chinese language and culture in both academic environments and social settings (Czerwionka et al. 2015) helps a great deal to increase the international students' intercultural sensitivity. They were much more interculturally sensitive than their Chinese counterparts by displaying more open-mindedness and interaction involvement (Chen and Starosta 2000).

The local Chinese students' intercultural sensitivity was lower than that of the international students. The absence of real intercultural experience is definitely an important contributing factor. Chinese teaching style might also contribute to these results. It is not unusual for the teacher to dominate the classroom in China, since the teacher is considered responsible for the transmission of knowledge. Consequently, students are expected to sit quietly, listen to the teacher attentively, and take notes carefully. Interrupting the teacher and asking questions are not normal practice in China because the teacher is regarded as the absolute authority in the classroom (Resende & Ji 2012). As a result, Chinese students are not as well trained in interacting with people which the intercultural sensitivity scale emphasizes. Whenever they think of "intercultural communication," they first think of their foreign language proficiency.

Furthermore, since the national college English proficiency test focuses on language points for the sake of reliability, test-oriented EFL teaching cannot be totally avoided in some provincial universities. Some teachers are just used to dwelling on grammatical rules and, in consequence, culture instruction is synonymous with foreign background information introduction in many EFL classrooms. With increasingly easy access to online learning resources, according to Wang (2014), students prefer to be taught intercultural knowledge and intercultural communication skills rather than factual knowledge which they can learn by themselves.

The Chinese students' relatively low intercultural sensitivity suggests that the EFL teaching strategy in China needs changing urgently. The importance of intercultural competence has been generally recognized in the Western world (Busse & Krause 2016) and the development of intercultural competence is regarded as the primary goal in foreign language teaching (Gobel & Helmke 2010). In China, although Hu (2011) claims that university students have various ways of exposing themselves to foreign cultures, many students mainly learn a foreign language and its culture from a foreign language classroom (Peng et al. 2015). If educators and teachers fail to understand "all language learning is culture learning" (Heath,1983, p. 5) and culture is a personal issue as well as a professional concern (Deardorff 2012), students are denied the full experience as the teaching goal or teaching methodology falls behind the current requirements. To cultivate global-minded talents with high intercultural competence (Sun & Bennett 2017), foreign language teaching reform in China nowadays is more pressing than ever before. First, teachers should be trained or reeducated to make them qualify for this new challenging task. An acknowledged aim of teacher education is to develop their intercultural competence with intercultural sensitivity as its core (Rissanen et al. 2016) because teachers' intercultural competence has an impact on their intercultural teaching quality (Gobel & Helmke 2010). Given this age of globalization and the EFL classroom as the most important channel for students to gain intercultural competence, it is essential for Chinese EFL teachers to take the development of intercultural competence as the main teaching goal and design specific teaching strategies for that purpose.

5 LIMITATIONS AND FUTURE RESEARCH

The current study is limited mainly in three aspects. First, the numbers of both Chinese and international students are rather small, which makes this study impossible to fully disclose the status quo of both the Chinese students' and the international students' intercultural sensitivity in a Chinese context. In addition, the sample of international students was not representative enough, with the absence of subjects from other parts of the world, such as Western Europe or North America. Second, both the quantitative and the qualitative research methods relied on students' self-report data. Third, English language proficiency of the two types of students was not measured and the correlation between their English language proficiency and their intercultural sensitivity was not interpreted. Follow-up relevant research can be done by examining the correlation between the international students' Chinese teachers' intercultural sensitivity and their teaching quality. The correlation between the international students' Chinese language proficiency and their intercultural sensitivity is also a topic worth investigating.

6 CONCLUSIONS

The implications of this research can be summarized as follows. First, the results suggest that being exposed to foreign context both culturally and linguistically is the most efficient way to develop intercultural sensitivity, which is consistent with previous research (Peng 2006, Deardorff 2012, Czerwionka et al. 2015, Hauerwas et al. 2017). Second, it lays bare that the overemphasis on language learning in EFL classrooms in China makes Chinese students extremely sensitive to foreign language competence and knowledge when it comes to intercultural communication. They were still not conscious of the importance of other components of ICC such as attitudes. Third, given what the Chinese students and the international students focused on in the written interview based on the five building blocks of ISS, the international students were more interculturally minded because they noticed the affective aspect along with knowledge when they thought of interactions with culturally different people. Finally, the results of the two empirical efforts in this study have wide significance in promoting Chinese EFL teachers and educators to reflect on current teaching activities and providing them with a new perspective when they devise an education goal or teaching plan in the future.

ACKNOWLEDGMENTS

The research reported herein was supported by the NSFC of China under Grants Nos. 71571091, 71771112, and 71371092; University of Science and Technology Liaoning Talent Project Grants No.:601011507-03.

REFERENCES

Bennett, M.J. 1986. A developmental approach to training for intercultural sensitivity. International Journal of Intercultural Relations 2: 179–196.
Bennett, M.J. 1993. Education for the Intercultural Experience. Yarmouth: Intercultural Press.
Bloom, M. & Miranda, A. 2015. Intercultural sensitivity through short-term study abroad. Language and Intercultural Communication 15(4): 567–580.
Busse, V. & Krause, U-M. 2016. Instructional methods and languages in class: A comparison of two teaching approaches and two teaching languages in the field of intercultural learning. Learning and Instruction 42: 83–94.
Byram, M. 1997. Teaching and assessing intercultural communicative competence. Clevedon: Multilingual Matters.
Chen, G.M. & Starosta, W.J. 1997. A review of the concept of intercultural sensitivity. Human Communication 1:1–16.
Chen, G.M. & Starosta, W.J. 1998a. Foundations of intercultural communication. Boston: Allyn & Bacon.
Chen, G. M., & Starosta, W. J. 1998b. A review of the concept of intercultural awareness. Human Communication 9:27–54.

Chen, G.M. & Starosta, W.J. 2000. The development and validation of the intercultural sensitivity scale. Human Communication 3: 1–22.

Corbin, J. & Strauss, A. 2008. Basics of qualitative research: Techniques and procedures for developing grounded theory. Thousand Oaks, CA: Sage.

Czerwionka, L. Artamonova, T. & Barbosa, M. 2015. Intercultural knowledge development: evidence from student interviews during short-term study abroad. International Journal of Intercultural Relations 49, 80–99.

Deardorff, D. 2004. The identification and assessment of intercultural competence as a student outcome of international education at institutions of higher education in the United States. Raleigh: North Carolina State University. (Unpublished dissertation).

Deardorff, D. 2006. Identification and assessment of intercultural competence as a student outcome of internationalization. Journal of Studies in International Education 10(3): 241–266.

Deardorff, D. 2012. Intercultural competence in the 21st century: perspectives, issues, application. In B. Breninger, & T. Kaltenbacher (Eds.), Creating cultural synergies: Multidisciplinary perspectives on interculturality and interreligiosity. Newcastle upon Tyne: Cambridge Scholars Publishing.

Fantini, A.E. 2000. A central concern: Developing intercultural competence. SIT occasional papers series 1: 25–42.

Fantini, A.E. 2006. Exploring and assessing intercultural competence. CSD Research Report.

Fritz, W. Graf, A. Hentze, J. Mollenberg, A. & Chen, G.M. 2005. An examination of Chen and Starosta's model of intercultural sensitivity in Germany and the United States. Intercultural Communication Studies 1: 53–65.

Fritz, W. Mollenberg, A. & Chen, G.M. 2002. Measuring intercultural sensitivity in different cultural context. Intercultural Communication Studies 11: 165–176.

Glaser, B. & Strauss, A. 1967. The discovery of grounded theory: Strategies for qualitative research. New York: Aldine.

Gobel, K. & Helmke, A. 2010. Intercultural learning in English as a foreign language instruction: The importance of teachers' intercultural experience and the usefulness of precise instructional directives. Teaching and Teacher Education 26: 1571–1582.

Gudykunst, W.B. 1993. Toward a theory of effective interpersonal and intergroup communication: An anxiety/uncertainty management (AUM) perspective. Wiseman Richard L. Koester Jolene. intercultural Communication Competence. international & Intercultural Communication Annual 33–71.

Hall, E.T. 1959. The Silent Language. New York: Doubleday & Company, Inc.

Hammer, M.R. 2011. Additional cross-cultural validity testing of the intercultural development inventory. International Journal of Intercultural relations 35: 474–487.

Hammer, M.R. Bennett, M.J. & Wiseman, R. 2003. Measuring intercultural sensitivity: The intercultural development inventory. International Journal of Intercultural Relations 27: 421–443.

Hauerwas, L.B., Skawinski, S.F., & Ryan, L.B. 2017. The longitudinal impact of teaching abroad: An analysis of intercultural development. Teaching and Teacher Education 67: 202–213.

Heath, S.B. 1983. Ways with words: Language, life and work in communities and classrooms. Cambridge: Cambridge University Press.

Hu, W.Z. 2005. On empirical research in intercultural communication. Foreign language Teaching and Research 5: 323–327.

Hu, Y. 2011. An investigation into undergraduates' intercultural sensitivity. Foreign Language World 3: 68–73.

Hymes, D. 1972. On communicative competence, Sociolinguistics 269293: 269–293.

Lin, D.J. 1999. The historical development of American intercultural communication research and its enlightenment. Journal of Fujian Normal University (Social Science Edition) 2: 82–88.

Peng, R.Z. Wu, W.P. & Fan, W.W. 2015. A comprehensive evaluation of Chinese College students' intercultural competence. International Journal of Intercultural Relations 47: 143–157.

Peng, S.Y. 2006. Influence of nationality and profession on intercultural sensitivity. Journal of Zhejiang University (Humanities & Social Sciences) 1: 74–80.

Peng, S.Y. 2007. Intercultural sensitivity: a comparison of English majors and non-English majors. Journal of Ningxia University (Humanities & Social Sciences Edition) 1:171–176.

Peng, S.Y. 2008. International mainstream of intercultural communication research and empirical methods. Foreign Language of China 5(5): 96–103.

Peng, S.Y. 2010. Status quo and Dilemma of intercultural research in China. Shantou University Journal (Humanities & Social Sciences) 4: 14–18.

Peng, S.Y. 2011. Differences between intercultural communication research in China and International mainstream research. Journal of Guangzhou University (Social Science Edition) 2: 65–69.

Qu, N.N. & Dou, Q. 2014. A study of foreign language learners' cross-cultural communication competence and cross-cultural intercultural sensitivity. Foreign Language and Literature 6: 167–170.

Resende, M. & Ji, Y.H. 2012. A practical Course in intercultural communication. Shanghai: Shanghai Foreign Language Education Press.

Rissanen, I. Kuusisto, E. & Kuusisto, A. 2016. Developing teachers' intercultural sensitivity: case study on a pilot course in Finnish teacher education. Teaching and Teacher Education 59: 446–456.

Ruben, B.D. 1976. Assessing communication competence for intercultural adaptation. Group & Organization Management 1(3): 334–354.

Shao, S.Y. & Chen J.L. 2011. An investigation into senior high school teachers' intercultural sensitivity. Foreign Language Re-search 3: 144–147.

Shi, X.S. & Shan, X.H. 2016. An exploratory study on research methods of intercultural communication studies published in SSCI journals in the last five years. Foreign language Teaching and Research 4: 594–605.

Shi, X.S. & Zhu, X.B. 2015. Retrospect and prospect of intercultural communication studies in China in last ten years. Foreign Language of China 12(6): 58–64.

Sun, Y.Z. & Bennett, M.J. 2017. Toward intercultural education: a dialog between professor Sun Youzhong and Dr. Janet Bennett. Foreign Languages and Their Teaching 2: 1–8.

Wang, T. 2014. The application of general education and intercultural communication in a 'news-listening' class. Language and Intercultural Communication 14(1): 125–131.

Computational Social Science – Luo, Ciurea & Kumar (eds)
© 2021 Taylor & Francis Group, London, ISBN 978-0-367-70193-2

A study of the differences in consumers' green behavior in the selection of tableware

N. Xia & W.H. Chen
Management School, Wuhan University of Technology, Wuhan, China

ABSTRACT: Recently, consumer satisfaction improvements and the rise of take-out platforms provide people with new ways of catering and consumption. While bringing convenience to people, it produces a large amount of catering waste, and causes tremendous pressure on the ecological environment. This serious pollution has encouraged people to increase their awareness of environmental protection and explore a green, healthy, and sustainable way to consume. Based on motivation theory and research into differences in consumer behavior, this paper explores the motivations that influence consumers' green choices of tableware from the perspective of catering enterprises. The results show that values and green tableware supply by catering enterprises can stimulate the willingness to use green tableware, and functional applicability and a green consumption attitude can improve the willingness to buy green tableware. Meanwhile, personal environmental inclinations and functional applicability are significantly higher than marketing factors and social symbolic functions.

1 INTRODUCTION

The emergence of contemporary environmentalism can be traced back to the late 1960s and early 1970s. Since the first use of disposable tableware in China's railway in 1986, it has been widely used in the catering industry with its advantages of convenience, low price, simplicity, and wide availability. While bringing convenience to people's lives, disposable tableware also poses a huge threat to urban safety, tourist attractions, and the ecological environment. The 17th Congress clearly put forward that we should attach importance to the ecological environment, strengthen the construction of ecological civilization, change the consumption pattern, and that the industrial mode of economic development was at a cost to the environment in the past, gradually forming a resource-saving and environment-friendly industrial structure, and increase customers' awareness of green consumption. In March 2019, the EU overwhelmingly issued an important bill, which stipulates that in 2021 the use of disposable plastic products will be totally banned, including disposable tableware, plastic meal boxes, plastic bags, straws, cups, etc. (Huang et al. 2020). According to the data, more than 8 million tons of plastic waste products flow into the oceans every year, causing serious damage to the marine ecological environment and posing an enormous threat to the survival of all kinds of marine creatures (Fieschi & Pretato 2018).

Green tableware refers to a kind of tableware in which the materials used are harmless to the human body, the production process is pollution-free to the environment, and the quality of the product fully meets the requirements of the national food hygiene standards. After the use, it has easy recovery, disposal, manageability, and degradability. As a typical project to solve the tableware pollution in the 21st century, the rational exploitation and extensive utilization of green tableware can actively respond to the policy, meeting the requirements of the sustainable development strategy of resources, and curbing the "white pollution" brought about by disposable tableware. The analysis of the motive sources for green consumption of tableware is conducive to promoting the process of banning disposable foamed plastic tableware, reducing the resource costs of eating out, and controlling the pollution and waste caused by disposable tableware. Motivation is the internal force that drives behavior. Psychologists believe that motivation is the internal need or impetus to stimulate, regulate, maintain, and guide people's behavior. Currently, research on green consumption focuses on the formation mechanism of green consumption behavior from the perspective of psychological factors or consumption psychological activities based on cognitive psychology (Vlosky et al. 1999; Cordano & Frieze 2000). The research into the influence of consumers' personal characteristics, learning ability, and cognitive processing mode on green consumption behavior does not touch on the formation mechanism of green consumption of tableware in depth. The difference in internal motivation leads to differences in consumers' choices of tableware. Based on the perspective of catering enterprises, the study into the influences of consumer behavior differences on the choice of tableware has enriched the relevant theories of influencing factors of consumer behavior, green consumption behavior, and greenness of tableware, promoted the development of the green market for tableware, and helped catering enterprises to improve their environmental responsibilities.

2 LITERATURE REVIEW

2.1 Classification of consumer behavior

Consumer behavior is guided by intrinsic motivation. Through summarizing the existing literature, the types of consumer behavior can be divided into psychological appeal, value appeal, cultural appeal, and emotional appeal. Green consumption belongs to the field of environmental responsibility. Psychological appeal is reflected in consumers' willingness and attitude toward green consumption. Strong awareness of environmental protection and green willingness can trigger consumers' green consumption behavior (Abdul-Muhmin 2007). Green consumerism will influence the purchasing decision-making process of consumers. Individuals who advocate green consumption will give priority to the environmental friendliness of products when making purchasing decisions and are inclined to pay extra for green products (Peattie 1999). When consumers believe that a green consumption pattern is a responsible behavior for the ecological environment, they are more inclined to make green consumption decisions. Individual values are stable and lasting beliefs or opinions formed in long-term social practice. Environmental values are the inevitable value orientation for the coordinated development of man and nature, which are mainly divided into altruistic values, social values, and ecological values. There is a significant correlation between individual environmental values and green traffic (McMillan et al. 2004), green product purchase intention (Gallagher & Muehlegger 2008), and ecological consumption behavior (Sundqvist 2005). Environmental education can improve citizens' awareness of environmental protection and their ability to fulfill environmental obligations. In addition, the advancement of ecological values education can encourage consumers to pay greater attention to pollution, so as to make green consumption choices (Olson 2013). Culture refers to beliefs, values, and customs at a specific time that are commonly acquired and observed by the majority of people in a specific society (Hoyer & MacInnis 2008). Culture is the carrier of history, which is rooted in people's thoughts and expressed through daily behaviors, speeches, interests, and ways of thinking, including at the spiritual, social, and material levels (Wang & Lin 2009). Culture has layers, and psychological culture is the integration of values, spiritual beliefs, national emotions, and thinking modes, while spirit–material integration culture mainly refers to the spiritual products created by people (Mooij & Hofstede 2011). A green and healthy lifestyle and the cultural concept of harmonious coexistence between man and nature can stimulate consumers' green consumption behaviors. As a kind of inner need of people, emotions influence people's behaviors (Borges et al. 2016). There are positive and negative emotions. When consumers have a strong preference for green products or green consumption behavior, or green consumption has a certain emotional significance for consumers, they will have a distinct tendency in their choice of tableware.

2.2 Causes of differences in consumer behavior

Green consumption behavior is an important part of environmental behavior. Stern believed that the influencing factors of environmental behavior can be divided into four categories, including:

(1) attitude variables: attitude, knowledge, consumption motivation, values, etc.; (2) situational factors: advertising, subjective norms, environmental atmosphere, policies and regulations, behavioral costs, legal systems, incentives and costs, etc. (3) individual ability variables: economic status, social image and status, individual knowledge and skills, etc.; and (4) customs and conventions (Stern 2000). Narasimhan (2003) believes that the factors influencing consumers' green consumption behaviors include values, attitudes, knowledge, subjective norms, perceived behavior control, situational factors, social and demographic characteristics, and other factors. Balderjahn (1988) divided the factors affecting consumers' sustainable consumption behavior into values, environmental attitudes, consumer psychology, propaganda and education, situational factors, and social demographic variables. Uddin & Khan (2018) proposed that the factors of green consumption behavior can be divided into five categories, including: (1) individual psychological variables: environmental attitude, environmental knowledge, green consumption cognition, environmental concern, environmental awareness, environmental belief, perceived behavioral effectiveness and environmental responsibility; (2) external situational variables: role influence (such as persuasion and demonstration), laws and regulations, incentives, information publicity, subjective norms, behavioral costs, and perceived behavioral control; (3) social and cultural factors: values, consumption concepts, and consumption customs; (4) demographic variables: gender, age, education, marital status, and income level; and (5) other factors: price sensitivity, consumer innovation (Gupta & Ogden 2009), involvement, media orientation (Coad et al. 2009), group pressure, and health awareness. On the basis of summarizing previous studies, this paper divides the factors that influence consumers' green consumption behavior into four categories: individual environmental inclination, marketing factors, functional applicability, and social symbolic function.

2.2.1 Individual environmental inclination

The research into individual environmental inclination is mainly based on the perspective of consumers, considered from four aspects including values, individual responsibility, self-efficacy, and green consumption attitude. Values are systematic views of beliefs, inclinations, propositions, and attitudes, which can be seen as a method to evaluate principles, criteria, and measures (Wang et al. 2018). The values of green consumerism first developed in Western developed countries, where economic growth provides an opportunity for consumers to change their consumption patterns. Consumers in developed countries comprehensively promote the value of green consumption and advocate a "truly green" lifestyle (Young et al. 2010). Individual responsibility is a psychological process. Moral-driven factors can influence consumers' green consumption decisions. The environment was rated as the most important factor in decision-making. When consumers realize that it is their responsibility to prevent pollution and fulfill environmental obligations, they are more inclined to contribute to the necessary actions. On the contrary, when consumers have a weak sense of individual responsibility, they will maintain a negative attitude toward the environment. Self-efficacy is an environmental behavior variable that measures green consumption behavior. Individuals have a clear cognition of their own behaviors toward the environment. When consumers believe that their environmental behaviors are obviously effective in improving the ecological environment, they are more inclined to make green consumption decisions (Straughan & Roberts 1999). Attitude is a stable psychological tendency toward a specific object. Green consumption awareness and environmental protection values of consumers can change their purchase intentions, make them focus on green products, and then choose environment-friendly products when making purchasing decisions.

2.2.2 Marketing factors

Based on the perspective of catering enterprises, marketing factors involve the influence of supply, catering environment, propaganda, and education on green consumption behaviors regarding tableware. Advocating green consumption and providing green tableware are the manifestations of social responsibility undertaken by catering enterprises. The green tableware provided by catering enterprises can stimulate consumers' demand for it and increase consumers' desire for green consumption. When catering enterprises do not provide disposable tableware, only a small number of consumers who are very concerned about environmental issues and have high quality-of-life requirements will take the initiative of asking the catering enterprises to provide green tableware and even change their purchasing behavior. The decoration style and atmosphere of catering enterprises are a reflection of the enterprise culture and values. An empirical study proved that the factors influencing the choice of dining places mainly include food safety, catering environment, convenience, food taste, and service (Hamilton 2003). Music mode and light color affect people's mood. When the decoration style and environmental atmosphere of catering enterprises conforms to the concept of green health, it will stimulate consumers' psychology and desire for green consumption of tableware (Shin et al. 2015). Green consumption behavior is influenced and interfered by external factors, which are mainly divided into information stimulus, policy stimulus, and economic stimulus. Environmental protection information can stimulate consumers at a psychological level to enhance their awareness of environmental protection, which can then be translated into demands and expressed through behaviors (Winett et al. 1985).

2.2.3 *Functional applicability*

The functional attributes of a product refer to the performance and characteristics that can meet the needs of consumers, which are the most basic source of consumers' purchasing motivation (Lee 2009). The commodity is the organic unity of its value and use value, and the functional attribute of the product is the centralized reflection of the use value of products. The function attribute of green tableware lies in its sensory and use functions. Product value can be divided into functional, experiential, and perceptual values. Functional value is mainly dependent on the physical attributes of the product to meet consumers' low-level needs, such as security, physiology, and social contact. The functional attributes of the product mainly include four dimensions: the style and design of the product, quality, performance, durability, applicability, and the main components and characteristics of the product. The functional applicability mainly refers to the core valued of the product and its ability to meet the basic practical needs of consumers. The differentiation of functional attributes is one of the ways to achieve product and target market differentiation, and also has an impact on consumers' decision-making behavior (Kim et al. 2001). Green tableware is used as a substitute for common tableware. Its functional attributes are reflected in its usability, and the quality and effectiveness of tableware are improved through green production technology and environment-protective materials.

2.2.4 *Social symbolic function*

With the development of the consumer goods market and the diversification of consumer demand, products are no longer limited to meeting consumers' needs for use, but are gradually endowed with special symbolic meaning, especially luxury goods, cultural products, electronic products, and cosmetics. Symbolic consumption behavior, also known as "conspicuous consumption" or "face consumption," is a special purchase behavior of consumers for some symbolic products. Symbolic purchasing behavior has a certain symbolic meaning. Consumers can manifest their own values, economic status, personality, and beliefs by purchasing specific products. Symbolic buying behavior is a behavior with high-value perception of products, that is, consumers satisfy their inner demands for certain emotions by consuming some symbolic products (Zhang & Jolibert 2000). In terms of luxury consumption, China's traditional "face view" influences the decision to buy luxury goods (Escalas & Bettman 2005). Consumers with certain economic strength will take "reputation" and a "sense of identity matching" into consideration when making consumption decisions. There is abundant research on the factors affecting conspicuous consumption, including internal and external factors. Internal factors refer to reference groups, cultural differences, and the symbolic significance of goods and services. External factors include the consciousness of face, material enjoyment, and status consumption (Qattan & Khasawneh 2020). According to the research into college students' conspicuous consumption motivation, it was found that personality display, group identification, brand preference, individuality seeking, face preservation, interpersonal communication, subjective social class (Neave et al. 2020), and self-consistency

Table 1. Statistical table of the gender of respondents.

Gender	Sample size	Percentage
Male	68	40.48%
Female	100	59.52%

Table 2. Statistical table of the age of respondents.

Age	Sample size	Percentage
Under 25	68	40.47%
25–35	47	27.98%
36–45	35	20.83%
46–55	10	5.95%
Over 55	8	4.77%

influence their conspicuous consumption behavior (Amatulli et al. 2020). Cheron took young Polish consumers as the research object and divided conspicuous consumption motivation into five dimensions: material enjoyment, social status, personal relationship adjustment, ostentation, and intragroup communication. Therefore, on the basis of existing literature, this study divides the symbolic factors that influence consumers' green consumption behavior of tableware into social image status, individual self-expression, and group affiliation.

3 RESEARCH DESIGN AND VARIABLE SELECTION

In this paper, we explore the differences in green consumption behavior in the choice of tableware, and divide the factors into four categories, including individual environmental inclination, marketing factors, functional applicability, and social symbolic function. Then, we explore the relationship between the factors and green tableware use intention as well as purchase intention. In order to expand the sample range and monitor sample quality, this study was conducted by issuing questionnaires on the Internet, of which 168 valid ones were retrieved, including 68 male respondents and 100 female respondents. The consumption intention of green tableware was set up with three items. Thirteen items were set for individual environmental inclination, which respectively reflected the values ($X11$), self-efficacy ($X12$), individual responsibility ($X13$), and green consumption attitude ($X14$). Thirteen items were set for marketing factors, which respectively reflected the supply ($X21$), catering environment ($X22$), and propaganda and education ($X23$). There were five items for functional applicability, which respectively reflected the use function ($X31$) and the green function ($X32$). There were eleven items for the social symbolic function, which respectively reflected group affiliation ($X41$), social image status ($X42$), and individual self-expression ($X43$).

4 RESULTS ANALYSIS

4.1 *Descriptive statistical analysis*

As shown in Tables 1 and 2, male respondents account for the total number of 40.48%. The majority of respondents are under 35 years old, accounting for 68.45% of the total.

Table 3. Questionnaire reliability test table.

Variable	Factors	Scale projects	Number of items	α
The independent variables	Personal environmental inclination	X11	3	0.659
		X12	3	0.738
		X13	4	0.874
		X14	3	0.689
	Marketing factors	X21	5	0.779
		X22	2	0.917
		X23	6	0.798
	Functional applicability	X31	3	0.790
		X32	2	0.809
	Social symbolic function	X41	4	0.946
		X42	4	0.933
		X43	3	0.854
The dependent variable	Green choice of tableware		3	0.850

4.2 The reliability analysis

In this research, the Cronbach α coefficient was used to measure the reliability of variables. When $\alpha < 0.35$, it is of low reliability, indicating that the scale is not acceptable; $0.35 < \alpha < 0.7$ indicates that the scale is acceptable; and $\alpha > 0.7$ indicates high reliability of the scale. The coefficients of variables are shown in Table 3.

As shown in Table 3, the coefficient of values, individual responsibility, self-efficacy, and green consumption attitude were 0.659, 0.859, 0.738, and 0.689, respectively. The coefficients of supply, catering environment, and propaganda and education were 0.779, 0.917, and 0.798, respectively. The coefficients of use function and green function were 0.790 and 0.809, respectively. The coefficients of individual self-expression, group affiliation, and social status and image were 0.854, 0.946, and 0.933, respectively. The coefficients in this table are greater than 0.35, and the reliability value of most variables is greater than 0.7, which is a high reliability.

4.3 Correlation analysis

Correlation analysis mainly measures the relationship between values, individual responsibility, self-efficacy, green consumption attitude, supply, catering environment, propaganda and education, functional applicability and social symbolic function, and consumers' choice of green tableware. Pearson correlation coefficient was selected when measuring the correlation between variables. When the correlation coefficient $r>0$, it is considered that there is a positive correlation between the two variables. When $r<0$, the two variables are considered to be negatively correlated. When the absolute value of r is greater than 0.8, there is a strong correlation between the two variables; when the absolute value of r is less than 0.3, the correlation between the two variables is weak.

As shown in Table 4, the correlation coefficients between the use function, supply and consumers' values, and the willingness to use green tableware were as high as 0.703, 0.664, and 0.534, showing a high correlation. There was a positive correlation between self-efficacy, green consumption attitude, propaganda and education of catering enterprises, green functions, and willingness to use green tableware, where the correlation coefficients were 0.447, 0.495, 0.476, and 0.448, respectively. There was no correlation between individual self-expression, catering environment, and consumers' willingness to use, and the correlation coefficients were 0.298 and 0.276, respectively.

As shown in Table 5, the correlation coefficient between the use function and the purchase intention of green tableware was as high as 0.624, showing a high correlation. There was a positive correlation between consumers' values, green consumption attitudes, green function and buying intentions of green tableware, with correlation coefficients of 0.427, 0.436, and 0.425, respectively.

Table 4. Correlation coefficient between variables and willingness to use green tableware.

Factors	Scale projects	r
Personal environmental inclination	X11	0.534
	X12	0.447
	X13	0.351
	X14	0.495
Marketing factors	X21	0.664
	X22	0.298
	X23	0.476
Functional applicability	X31	0.703
	X32	0.448
Social symbolic function	X41	0.314
	X42	0.305
	X43	0.276

Table 5. The correlation coefficient between variables and purchase intention.

Factors	Scale projects	r
Personal environmental inclination	X11	0.427
	X12	0.376
	X13	0.327
	X14	0.436
Marketing factors	X21	0.379
	X22	0.225
	X23	0.217
Functional applicability	X31	0.624
	X32	0.425
Social symbolic function	X41	0.169
	X42	0.178
	X43	0.319

There was no correlation between catering environment, propaganda and education, social status and image, group affiliation, and consumers' purchase intention, and the correlation coefficients were 0.225, 0.217, 0.178, and 0.169, respectively.

5   CONCLUSION

There are several conclusions obtained from the above analysis, as outlined here.

First, the correlation analysis shows that the use function, supply, and values can positively influence consumers' willingness to use green tableware. However, there is no obvious relationship between individual self-expression, catering environment, and consumers' willingness to use green tableware. The use function of tableware, values, green consumption attitude, green function, and the purchase intention of green tableware are positively correlated, especially the use function of tableware, while there was no correlation between the propaganda and education, catering environment, social status and image, group affiliation, and the purchase intention of consumers.

Second, through lateral correlation, it can be found that the personal environmental inclination will have a greater impact on consumers' purchase intentions of green tableware than marketing factors. The influence of functional applicability on consumers' willingness to use and purchase green tableware is greater than the social symbolic function. Meanwhile, there is an attitude–behavior gap in the choice of green tableware through longitudinal comparison. That is, although consumers realize the importance of green consumption and environmental protection, this awareness is difficult to translate into buying decisions.

Finally, based on the above analysis, it can be seen that promoting the greening of tableware mainly relies on policy incentives and catering enterprises to take the initiative to assume social responsibilities, and provide green tableware. Meanwhile, the government should take incentive measures to encourage catering enterprises to assume social responsibility.

REFERENCES

Abdul-Muhmin, A.G. 2007. Explaining Consumers' Willing to be Environmentally Friendly. International Journal of Consumer Studies 31: 237–247.

Amatulli, C., Angelis, M.D. & Donato, C. 2020. An investigation on the effectiveness of hedonic versus utilitarian message appeals in luxury product communication. Psychology & Marketing 37(4).

Balderjahn, I. 1988. Personality variables and environmental attitudes as predictors of ecologically responsible consumption patterns. Elsevier 17(1).

Borges, J.A.R., Tauer, L.W. & Lansink, A.G.J.M.O. 2016. Using the theory of planned behavior to identify key beliefs underlying Brazilian cattle farmers' intention to use improved natural grassland: A MIMIC modelling approach. Land Use Policy 55.

Coad, A., Haan, P.D. & Woersdorfer, J.S. 2009. Consumer support for environmental policies: An application to purchases of green cars. Ecological Economics 68(7): 2078–2086.

Cordano, M. & Frieze, I.H. 2000. Pollution reduction preferences of US environmental managers: Applying Ajzen's theory of planned behavior. Academy of Management journal 43(4): 627–641.

Escalas, J.E. & Bettman, J.R. 2005. Self-construal, reference groups, and brand meaning. Journal of consumer research 32(3): 378–389.

Fieschi, M. & Pretato, U. 2018. Role of compostable tableware in food service and waste management. A life cycle assessment study. Waste Management 73:14–25.

Gallagher, K.S. & Muehlegger, E. 2008. Giving green to get green? Incentives and consumer adoption of hybrid vehicle technology. Journal of Environmental Economics and Management 61(1).

Gupta, S. & Ogden, D.T. 2009. To buy or not to buy? A social dilemma perspective on green buying. Journal of Consumer Marketing 26(6).

Hamilton, R.W. 2003. Why Do People Suggest What They Do Not Want? Using Context Effects to Influence Others' Choices. Journal of Consumer Research 29(4).

Hoyer, W.D. & MacInnis, D.J. 2008. Consumer Behavior. 5th ed.

Huang, Q., Chen, G.W., Wang, Y.F., Chen, S.Q., Xu, L.X., Wang, R. 2020. Modelling the global impact of China's ban on plastic waste imports. Resources, Conservation & Recycling: 154.

Kim, C.K., Han, D. & Park, S.B. 2001. The effect of brand personality and brand identification on brand loyalty: Applying the theory of social identification. Japanese Psychological Research 43(4).

Lee, K. 2009. Gender differences in Hong Kong adolescent consumers' green purchasing behavior. Journal of Consumer Marketing 26(2).

McMillan, E.E, Wright, T. & Beazley K. 2004. Impact of a university-level environmental studies class on students' values. The Journal of Environmental Education 35(3): 19–27.

Mooij, M.D. & Hofstede, G. 2011. Cross-Cultural Consumer Behavior: A Review of Research Findings. Journal of International Consumer Marketing 23:3–4, 181–192.

Narasimhan, Y. 2003. The Link between Green Purchasing Decisions and Measures of Environmental Consciousness. Ohio University.

Neave, L., Tzemou, E. & Fastoso, F. 2020. Seeking attention versus seeking approval: How conspicuous consumption differs between grandiose and vulnerable narcissists. Psychology & Marketing 37(3).

Olson, E.L. 2013. It's not easy being green: the effects of attribute tradeoffs on green product preference and choice. Journal of the Academy of Marketing Science 41(2): 171–184.

Peattie, K. 1999. Trappings versus substance in the greening of marketing planning. Journal of Strategic Marketing 7 (2): 131–148.

Qattan, J. & Khasawneh, M.A. 2020. The Psychological Motivations of Online Conspicuous Consumption: A Qualitative Study. International Journal of E-Business Research (IJEBR) 16(2).

Shin, Y.B., Woo, S.H., Kim, D.H., Kim, J., Kim, J.J. & Park, J.Y. 2015. The effect on emotions and brain activity by the direct/indirect lighting in the residential environment. Neuroscience Letters 584.

Stern, P.C. 2000. Toward a coherent theory of environmentally significant behavior. Journal of Social issues 56(3): 407–424.

Straughan, R.D. & Roberts, J.A. 1999. Environmental Segmentation alternatives: a look at green consumer behavior in the new millennium. Journal of Consumer Marketing. 16(6): 558–575.

Sundqvist, A.T.S. 2005. Subjective norms, attitudes and intentions of Finnish consumers in buying organic food. British Food Journal : 107(11).

Uddin, S.M.F. & Khan, M.N. 2018. Young Consumer's Green Purchasing Behavior: Opportunities for Green Marketing. Journal of Global Marketing 31(4).

Vlosky, R.P., Ozanne, L.K. & Fontenot, R.J. 1999. A conceptual model of US consumer willingness-to-pay for environmentally certified wood products. Journal of Consumer Marketing 16(2): 122–136.

Wang, C.L. & Lin, X.H. 2009. Migration of Chinese Consumption Values: Traditions, Modernization, and Cultural Renaissance. Journal of Business Ethics 88(3).

Wang, S.Y., Wang, J., Wang, Y., Yan, J. & Li, J. 2018. Environmental knowledge and consumers' intentions to visit green hotels: the mediating role of consumption values. Journal of Travel & Tourism Marketing 35(9).

Winett, R.A., Leckliter I.N., Chinn, D.E., Stahl, B. & Love S.Q. 1985. Effects of Television Modeling on Residential Energy Conservation. Journal of Applied Behavior Analysis 18(1): 33–44.

Young, W., Hwang, K., McDonald, S., & Oates, C.J. 2010. Sustainable consumption: green consumer behaviour when purchasing products. Sustainable development 18(1): 20–31.

Zhang, M.X. & Jolibert, A. 2000. Culture chinoise traditionnelle et comportements de consommation. Décisions marketing: 85–92.

Computational Social Science – Luo, Ciurea & Kumar (eds)
© *2021 Taylor & Francis Group, London, ISBN 978-0-367-70193-2*

An integrated FTA-GAHP method for assessment of campus safety: A case study in Puyang, China

H. Wei & X.B. Chen*
School of Electronics and Information Engineering, University of Science and Technology Liaoning, Anshan, Liaoning, China

ABSTRACT: The paper presents the results of safety assessment of crowd stampedes at the Experimental Primary School in Puyang, China. A methodology is developed for assessment of campus safety based on the FTA-GAHP method, which combines the advantages of the traditional fault tree analysis technique and a grey analytic hierarchy process approach for accurate estimation of weight coefficients. Our results indicate that the primary cause is insufficient hardware facilities; students' faulty behaviors are the direct cause; the secondary cause is the lack of safety awareness. Moreover, an imperfect safety management system was the root cause of this accident. These findings are consistent with the actual investigation results of the Puyang County Government, which further proves that the FTA-GAHP method is simple, yet reasonable and effective to assess safety level of crowd stampedes on primary and middle schools. Some precautionary measures have also been identified. Thus, this work affords a decision support for the preparative scheme of campus safety management in China, as well as elsewhere in the world where there are large-scale gatherings.

Keywords: Campus safety, Crowd stampedes, Fault tree analysis, Grey analytic hierarchy process, Safety assessment.

1   INTRODUCTION

Based on current population growth trends, Lutz (2010) predicted that the global population will increase from its current 7 billion to 8–10 billion by 2050. Population growth will affect, to some extent, the health risks associated with crowds or large-scale gatherings. Such gatherings can result in higher rates of diseases, pollution, and particularly crowd stampede disasters (Haghani & Sarvi, 2019). According to incomplete statistics, over 40 crowd stampedes have occurred since 2007, causing 2683 deaths and more than 10,000 injuries (Rutty, 2004). For instance, in 2014, a stampede at Shanghai New Year's Eve celebrations in China, caused 36 deaths and more than 47 injuries (BBC News, 2015); in 2015, at least 2177 people were killed and 934 injured at the annual pilgrimage in Saudi Arabia (The Express Tribune, 2015). Crowd stampedes have been determined as a major hazard which could occur during religious mass gatherings in India, as well as elsewhere in the world where crowds swarm (Illiyas, Mani, Pradeepkumar, & Mohan, 2013). Therefore, research on crowd stampede safety assessment has become a dominant issue.

Crowd stampedes show representative characteristics compared with most accidents. A combination of multiple factors, such as deficient facilities or buildings design, lack of crowd management, and a real or perceived threat by others, may result in a stampede. In order to reflect the behavior of pedestrian flow, many models are proposed to describe the nonlinear dynamics relationship between pedestrians. For example, Helbing et al. (2000) proposed the social force model intended to explain the interaction within a panicked crowd, which has been improved by many researchers

---

*Corresponding author

to analyze pedestrian dynamics (Cao, Liu, & Chraibi, Zhang, & Song, 2019). After this, the cellular automaton model reflected simpler interaction rules and calculation efficiency (Liu, Li, & Su, 2019). Similarly, simulations and experimental studies on crowd movement also have been conducted to improve the physical models (Wang, Chen, & Jin, Li, & Wang, 2019). These original models focus only on certain parts of crowd management. Foreign scholars have paid less attention to stampede accidents so far due to fewer occurrences or no public reporting abroad. However, it is different in China, where stampede accidents occur frequently, especially in primary and middle schools. In order to fill this gap, this paper presents the results of a safety assessment of a stampede accident at the Experimental Primary School in Puyang, China.

A methodology is developed for safety assessment of stampede accidents based on the FTA-GAHP method, which combines the advantages of a traditional fault tree analysis technique and the grey analytic hierarchy process approach. The combined FTA-GAHP method can not only eliminate subjective errors, but also ensures comprehensive and systematic evaluation factors, while enabling the new model to make predictions for the decision-making process. This work is expected to present a good application of the combined FTA-GAHP method for safety assessment of stampede accidents at primary and middle schools, as well as proving that the method is relatively simple, yet reasonable and effective.

This article is organized as follows. Section 2 proposes the combined FTA-GAHP methodology, and its related concepts are introduced. In addition, a numeric application of the proposed method is described in Section 3. Finally, Section 4 concludes our work in this paper.

2 METHODOLOGY

2.1 The fault tree analysis model

In nature, the FTA model, a special tree-like logic diagram, is provided to graphically illustrate system failures, combining different events (i.e., the main, intermediate, and basic events), logic gates, and symbols to illustrate various causalities. The typical fault tree construction consists of the following steps.

Step I. Investigate the accident. Step II. Determine the main event. Step III. Identify all direct and indirect factors related to the main event, including equipment failure, management errors, and environmental factors, etc. Step IV. Clarify the logical, causal relationships among the main event (represented by a special rectangle), intermediate events (represented by common rectangles), and basic events (represented by circles). Step V. Construct the fault tree, that is, connect the above events with AND gates and OR gates in order to demonstrate the logical deductive relationship between them.

AND gate: An output event only occurs if and only if all input events occur at the same time. OR gate: An output event can occur when there is at least one of all input events occurs.

2.2 Grey analytic hierarchy process evaluation model

In GAHP, a typical 1–9-scale method is proposed to construct pairwise comparison matrices (Table 1), and the following steps are applied to calculate the weight vector and consistency ratio (CR). Step I. Construct comparison matrices based on experts' evaluation score. Step II. Verify the rationality of the comparison matrix based on the consistency ratio (CR).

$$CR = \frac{CI}{RCI} \tag{1}$$

$$CI = \frac{\lambda_{\max} - n}{n - 1} \tag{2}$$

where the random consistency index (RCI) can be obtained from Table 2, $\lambda_{\max}$ is the maximum eigenvalue of the comparison matrix, and n is the corresponding dimension.

Table 1. The scale and definition of the comparison matrix.

Scale	Definition
1	Extremely unimportant
3	Unimportant
5	General
7	Important
9	Extremely important
2, 4, 6, 8	The importance level lies between the above two adjacent scales

Table 2. The random consistency index (RCI) for various n.

n	1	2	3	4	5	6	7	8	9
RCI	0.00	0.00	0.58	0.90	1.12	1.24	1.32	1.41	1.45

Step III. Calculate the relative weight column vector w_i.

$$w_i = \sum_{j=1}^{l} (p_h \cdot w_{ij}) \tag{3}$$

where, $w_{ij} = M_{ij} / \sum_{k=1}^{n} M_{kj}$, $M_{ij} = \sqrt[n]{a_{i1}, a_{i2}, \ldots, a_{in}}$, and $\sum_{j=1}^{l} p_j = 1$.

Step IV. Determine the evaluation grey cluster, that is, identify the grades, grey number, and whitening weight function of the grey cluster. Based on the FTA approach, the grey cluster has been classified into five levels: (1) "Very good" ($e = 1$) corresponding to the grey number $\otimes_1 \in [0, 9, \infty)$, (2) "Good" ($e = 2$) corresponding to the grey number $\otimes_2 \in [0, 7, 14)$, (3) "General" ($e = 3$) corresponding to the grey number $\otimes_3 \in [0, 5, 10)$, (4) "Poor" ($e = 4$) corresponding to the grey number $\otimes_4 \in [0, 3, 6)$, and (5) "Very poor" ($e = 5$) corresponding to the grey number $\otimes_5 \in [0, 1, 2)$. The corresponding whitening weight functions are shown in formulas (7)–(11), respectively. Thus, the comment set of evaluation index can be defined as $U = [9\ 7\ 5\ 3\ 1]^T$.

$$f_{\otimes_1}(x) = \begin{cases} x/9 & 0 < x < 9 \\ 1, & x \geq 9 \\ 0, & others \end{cases} \tag{4}$$

$$f_{\otimes_2}(x) = \begin{cases} 1, & 0 < x \leq 7 \\ \frac{14-x}{7}, & 7 < x \leq 14 \\ 0, & others \end{cases} \tag{5}$$

$$f_{\otimes_3}(x) = \begin{cases} 1, & 0 < x \leq 5 \\ \frac{10-x}{5}, & 5 < x \leq 10 \\ 0, & others \end{cases} \tag{6}$$

$$f_{\otimes_4}(x) = \begin{cases} 1, & 0 < x \leq 3 \\ \frac{6-x}{3}, & 3 < x \leq 6 \\ 0, & others \end{cases} \tag{7}$$

$$f_{\otimes_5}(x) = \begin{cases} 1, & 0 < x \leq 1 \\ 2 - x, & 1 < x \leq 2 \\ 0, & others \end{cases} \tag{8}$$

Step V. Calculate the grey evaluation coefficient. The grey cluster evaluation coefficient is defined as:

$$X_{ie} = \sum_{j=1}^{l} f_{\otimes_e}(d_{ij}) \tag{9}$$

where d_{ij} represents the evaluation score of a certain index V_i by the j-th expert, and the total grey cluster evaluation coefficient should be defined as:

$$X_i = \sum_{e=1}^{5} X_{ie} \tag{10}$$

The grey evaluation weight vector of e-th evaluation grey cluster is defined as:

$$r_{ie} = X_{ie}/X_i \tag{11}$$

Thus, the grey evaluation weight vector for the m−th subindicator of i−th indicator is $r_{ime} = (r_{im1}, r_{im2}, r_{im3}, r_{im4}, r_{im5})$, and the grey evaluation weight matrix could be defined as $R_i = [r_{i1}, r_{i2}, \ldots, r_{im}]$.

Step VI. Calculate the comprehensive grey evaluation values.

$$C_i = w_i^T \cdot R_i = (C_{i1}, C_{i2}, C_{i3}, C_{i4}, C_{i5}) \tag{12}$$

$$C = w^T \cdot R = (C_1, C_2, C_3, C_4, C_5) \tag{13}$$

According to the maximum principle, the comprehensive grey evaluation values can be determined by formula (13. Finally, based on the grey cluster level, comprehensive evaluation values S is used to comprehensively evaluate the system of the evaluated object.

$$S = C \cdot U \tag{14}$$

3 CASE STUDY

The proposed model of analyzing a stampede accident was put into practice in a primary school in Puyang, China. At 9:00 am on March 22, 2017, a stampede accident occurred in the primary school, causing 22 students to be injured. One was killed on the way to hospital and five were seriously injured (China Central Television, 2017). What caused this painful stampede accident? For the application, an expert team was formed from two principals of the primary and middle school, a director of investment promotion, a librarian, and two managers from different departments. These experts were organized to participate in a deep interview, and their opinions were derived from the interview results. The indicator and subindicator to be used in the model were determined by analyzing the practical cases of stampede accidents in Chinese primary and middle schools from 2002 to 2017. Pairwise comparison matrices used to calculate weight vectors corresponding with each indicator and subindicator were also determined by the same team. The application was carried out based on the steps proposed in Section 2.

Following the above investigation, based on the logical deduction method, human factors, buildings factors, management factors, and emergency situations were identified as the intermediate events. The fault tree of the stampede accident at the primary school in Puyang is shown in Figure 1. All basic events of this stampede accident are listed in Table 3.

The hierarchical structure model, based on the fault tree of the stampede accident at the primary school in Puyang was established and classified into three layers: a target layer (V), first-layer indicator $(V_i, i = 1, 2, \ldots, 5)$, and second-layer indicator $(V_{ik}, k \in \{1, 2, \ldots, 6\})$, as shown in Table 3. In order to obtain the indicator and subindicator weight vectors, the pairwise comparison matrix

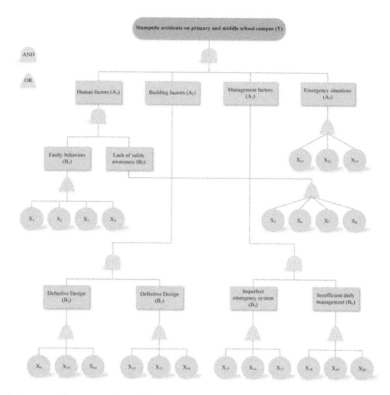

Figure 1.   Fault tree of the stampede accident at the Experimental Primary School in Puyang.

for the first-layer indicator is formed by the j−th expert based on the aforementioned scale (Table 1). According to formulas (1)–(3), the weight vector w_j can be calculated as follows:

$$w_j = [0.2059, 0.1471, 0.2059, 0.2353, 0.2059]^T, \qquad (15)$$

where, the consistency ratio (CR) is $0 < 0.1$, which demonstrates that the pairwise comparison matrix is reasonable and has satisfactory consistency. However, the above results only represent the weight vector given by an expert's opinion. By combining the weight values of each expert, refer to formula (3), the weight vector of each indicator listed in Table 3 can be obtained as follows:

$$w = [0.2212 \ 0.2051 \ 0.1927 \ 0.1870 \ 0.1940]^T$$

$$w_1 = [0.2714 \ 0.2438 \ 0.2972 \ 0.1876]^T$$
$$w_2 = [0.2909 \ 0.1876 \ 0.2906 \ 0.2309]^T$$
$$w_3 = [0.1772 \ 0.1764 \ 0.1602 \ 0.1432 \ 0.1741 \ 0.1689]^T$$
$$w_4 = [0.1656 \ 0.1431 \ 0.1772 \ 0.1660 \ 0.1413 \ 0.2068]^T$$
$$w_5 = [0.3569 \ 0.3357 \ 0.3074]^T$$

In this section, six experts were required to evaluate each subindicator based on the aforementioned five grey clusters, respectively. Thereby, the grey evaluation weight matrix can be obtained by using formulas (4)–(11).

Then, the comprehensive evaluation results corresponding to the first-layer indicator ($V_i, i = 1, 2, \ldots, 5$) are calculated using formulas (12) and (13).

$$C_1 = w_1 \cdot R_1 = [0.2480 \ 0.3773 \ 0.2751 \ 0.0796 \ 0.0201]$$

306

Table 3. The AHP model of the stampede accident at the Experimental Primary School in Puyang.

First-layer indicator	Weight	Second-layer indicator	Weight
Faulty behaviors (V_1)	0.2212	• Small child prank (V_{11})	0.2714
		• Reverse motion (V_{12})	0.2438
		• Making a fuss (V_{13})	0.2972
		• Bending over to tie laces (V_{14})	0.1876
Lack of safety awareness (V_2)	0.2051	• Stairway blocked (V_{21})	0.2909
		• Failure to identify risk factors (V_{22})	0.1876
		• Panic psychology (V_{23})	0.2906
		• Lack of responsibility of teachers (V_{24})	0.2309
Insufficient hardware facilities (V_3)	0.1927	• Defective design of the corridor (V_{31})	0.1772
		• No emergency channels (V_{32})	0.1764
		• The number of toilets is not up to standard (V_{33})	0.1602
		• Unreasonable distribution of classrooms (V_{34})	0.1432
		• The ground is not slip resistant (V_{35})	0.1741
		• Lack of emergency equipment (V_{36})	0.1689
Imperfect safety management system (V_4)	0.1870	• Insufficient training in safety education (V_{41})	0.1656
		• No emergency plan (V_{42})	0.1431
		• No emergency drill (V_{43})	0.1772
		• Excessive class size (V_{44})	0.1660
		• Violation of the exam arrangement (V_{45})	0.1413
		• No safety inspection system (V_{46})	0.2068
Emergency situations (V_5)	0.1940	• Fire disaster (V_{51})	0.3569
		• Earthquake (V_{52})	0.3357
		• Power cut (V_{53})	0.3074

Similarly,

$$C_2 = W_2 \cdot R_2 = [0.1924\ 0.3434\ 0.3077\ 0.1468\ 0.0096]$$

$$C_3 = W_3 \cdot R_3 = [0.1834\ 0.3411\ 0.2973\ 0.1394\ 0.0389]$$

$$C_4 = W_4 \cdot R_4 = [0.2275\ 0.3677\ 0.3034\ 0.1016\ 0]$$

$$C_5 = W_5 \cdot R_5 = [0.2864\ 0.3878\ 0.2604\ 0.0653\ 0]$$

and

$$C = W \cdot R = [0.2278\ 0.3636\ 0.2885\ 0.1062\ 0.0139]$$

Finally, according to maximum principle, the comprehensive grey evaluation values are determined by formula (14).

$$S = C \cdot U = 6.3707$$

The grey evaluation values corresponding to the first-layer indicator ($V_i, i = 1, 2, \ldots, 5$) are illustrated in Figure 2, which shows that the primary cause is the insufficient hardware facilities (V_3). This is also consistent with the actual investigation results of the Puyang County Government, that is, the cause of the stampede accident was the short break time and insufficient toilets.

4 CONCLUSION

The FTA-GAHP-based methodology is developed for risk assessment of stampede accidents at primary and middle schools. This research has provided a good application of the combined FTA-GAHP method for assessing the stampede accident at the Experimental Primary School in Puyang, China. First, the FTA technique was utilized to explore how each basic event influences the main event. Second, considering the inaccuracy of the quantitative analysis of the FTA, the GAHP approach was introduced to transform the fault tree model into a grey analytic hierarchy model and

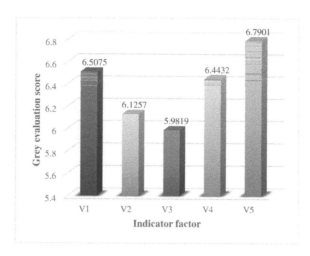

Figure 2. The grey evaluation values corresponding to the first-layer indicators.

conduct a qualitative analysis. Thereby, a more comprehensive result of risks of stampede accidents at primary and middle schools was obtained. Finally, this work further proves that the combined FTA-GAHP method is relatively simple, yet reasonable and effective for assessing campus security management in China.

ACKNOWLEDGMENTS

The research reported herein was supported by the NSFC of China under Grant Nos. 71571091 and 71771112.

REFERENCES

BBC News. (2015). Shanghai new year crush kills 36. Retrieved from https://www.bbc.com/news/world-asia-china-30646918

Cao, S., Liu, X., Chraibi, M., Zhang, P., & Song, W. (2019). Characteristics of pedestrian's evacuation in a room under invisible conditions. International Journal of Disaster Risk Reduction, 41, 101295.

Haghani, M., & Sarvi, M. (2019). Heterogeneity of decision strategy in collective escape of human crowds: On identifying the optimum composition. International Journal of Disaster Risk Reduction, 35, 101064.

Helbing, D., Farkas, I., & Vicsek, T. (2000). Simulating dynamical features of escape panic. Nature, 407, 87–90.

Illiyas, F.T., Mani, S.K., Pradeepkumar, A.P., & Mohan, K. (2013). Human stampedes during religious festivals: a comparative review of mass gathering emergencies in India. International Journal of Disaster Risk Reduction, 5, 10–18.

Liu Y., Li F., & Su Y. (2019). Critical Factors Influencing the Evolution of Companies' Environmental Behavior: An Agent-Based Computational Economic Approach. SAGE Open, 9(1), 2158244019832687.

Lutz, W., & KC,c S. (2010). Dimensions of global population projections: what do we know about future population trends and structures? Philosophical Transactions of the Royal Society B: Biological Sciences, 365(1554), 2779—2791.

Rutty, G.N. (2004). Essentials of Autopsy Practice. London, LON: Springer.

The Express Tribune. (2015). Hajj stampede death toll rises to 2,177. Retrieved from https://tribune.com.pk/story/976079/hajj-stampede-death-toll-rises-to-2177

Wang, J., Chen, M., Jin, B., Li, J., & Wang, Z. (2019). Propagation characteristics of the pedestrian shockwave in dense Crowd: Experiment and simulation. International Journal of Disaster Risk Reduction, 40, 101287.

Computational Social Science – Luo, Ciurea & Kumar (eds)
© 2021 Taylor & Francis Group, London, ISBN 978-0-367-70193-2

Research on the impact of China's listed companies' financial accounting on innovation: Experience data from listed companies in Shenzhen and Shanghai

Q. Lu & C. Sheng
College of Management of Sichuan Agricultural University, Wenjiang, Sichuan, China

G. Fu*
Business School of Sichuan Agricultural University, Dujiangyan, Sichuan, China

ABSTRACT: In the process of changing from being factor-driven to innovation-driven in China, there is a complex impact of debt as an important index in the financial accounting of enterprises. We analyze the innovation level of the A-share-listed companies in Shanghai and Shenzhen with relevant data. We found that: (1) When other conditions remain unchanged, the asset liability ratio of the enterprises and the innovation level are inverted U-type relations. (2) The inverted U-type relationship will weaken with the enhancement of enterprise growth and is highlighted by the gradual expansion of the enterprise size. (3) Whether the company is state-owned or not has no obvious influence on this inverted U-type relationship. We suggest enterprises regulate the level of debt and improve their innovation ability by: (1) Paying attention to the level of debt and adjust their financing decision-making; (2) Improving the management system and optimizing the enterprise management; and (3) Building supporting facilities and enhancing innovation capabilities.

1 INTRODUCTION

In recent years, the process of China's economy turning to high-quality development innovation has increasingly become the focus of people's attention. The country also has issued innovation subsidies such as the national high-technology research and development program (863 program) and the torch plan, in order to directly promote the pace of national innovation from a policy perspective, at the same time, ZTE, Huawei, and a series of positive and negative cases also show us that no innovations in the development will be restricted or preventing companies from taking the next step. At present, there are a lot of researches on the factors affecting enterprise innovation, for example, research has been carried out on the internal human resources, equity, capital structure and external official turnover and government subsidies. However, as a main form of financing, there is still no clear and unified conclusion on how to use debt to play a role in the innovation level of enterprises. Most researchers hold the view that debt has a single positive or negative effect on the level of innovation, and lack sustained attention to changes in corporate debt levels, however this article considers debt ratio changes over a long time in the process of research, combining previous analysis experience, studying, respectively, the periods of different debt levels of sample enterprises, finally drawing a comprehensive conclusion. At the same time, most of the existing research only includes innovation input or innovation output, ignoring the effects of enterprise innovation, or ignoring the enterprise investment intention and efforts. However, this article will take both into consideration, adopting patent quantity of innovation output indicators and the measure of corporate innovation input, and also including the structure of the enterprise assets related to the innovation. Therefore this article comprehensively judges the impact of corporate debt on innovation in all aspects.

---

*Corresponding author

2 RESEARCH REVIEW

2.1 *Innovation and enterprise innovation*

Innovation, as its name implies, is the entire process of creating something new. For enterprises, innovation not only refers to all-round innovation in technology and production skills, but also includes summarizing and learning from previous experience in the production process to eliminate drawbacks in current work. There have been many definitions of enterprise innovation given by domestic and foreign scholars. Xiao Hailan et al. (2014), when studying the relationship between debt and an enterprise innovation model, divided enterprise innovation into exploratory innovation and conventional innovation, which respectively represent creative R&D and daily R&D carried out by enterprises in order to enhance their competitiveness, and represented by innovation R&D and cutting-edge R&D. Li Houjian and Liu Peisen (2018) summarized exploratory innovation and developmental innovation as dual innovation to explore the role of human capital in enterprise innovation. Yang Mingjing et al. (2019) divided innovation into strategic innovation and substantive innovation. Substantive innovation refers to the innovation carried out in order to enable enterprises to make technological progress and gain competitive advantage. Strategic innovation refers to innovation that only pursues speed and quantity for other purposes. In addition to the above-mentioned R&D numerical to measure the level of enterprise innovation, scholars also use innovation input, innovation output, and innovation risk to define the level of enterprise innovation. Innovation input includes the investment of capital and manpower in the whole innovation process. Innovation output is often represented by the number of patent applications, total factor productivity, and other indicators. However, innovation risk is quite special. Some scholars believe that innovation risk refers to the possibility that innovation behavior will bring losses to the enterprise (Yang et al., 2011), and the judgment is based on whether the enterprise's innovation investment can bring profits to the enterprise (Wang Yuze et. al., 2019).

2.2 *Influencing factors of enterprise innovation*

2.2.1 *Internal influencing factors of enterprise innovation*

Due to the high risk and high investment characteristics of enterprise innovation, the innovation decisions made by management should be closely aligned with the internal situation of the company to ensure the successful completion of innovation activities and create value for the enterprise. The internal factors influencing the level of enterprise innovation are as follows. (1) The influence of enterprise employee compensation as a whole, on the one hand, the increase in employees' salary may occupy resources that enterprises use to invest in innovation, leading to a decrease in innovation investment, however, on the other hand, it should also be noted that an increase in employees' salary will make employees strive to improve their technical level and carry out innovation activities independently to retain their jobs.

At the same time, because of the rising cost of labor, enterprises are more willing to carry out innovation investment in technology production to improve the existing low capacity and other problems, so as to achieve technological transformation and upgrading and reduce labor costs (Tang Manping et al., 2019). From the macro point of view, a rise in the minimum wage standard can increase the consumer demand and business volume of enterprises from the source, and also stimulate enterprises to carry out technological innovation (Wang Xiaoxia et al., 2016). For senior managers of enterprises, increasing their salary not only gives them more room to tolerate mistakes and enables them to dare to innovate, but also alleviates the agency problem to a certain extent and inhibits the adverse impact of major shareholders on enterprise innovation (Gao Wenliang, 2018). (2) Human capital structure: the staff is the key to company operations and creation, from the perspective of the whole enterprise there is a process of learning and communication among employees within the enterprise, and skilled workers can not only make innovations in technology and methods by themselves, but also highlight problems and inspiration for specialized R&D personnel, so the diverse skills of production workers can improve the level of exploratory innovation.

Meanwhile, some studies have shown that women are more likely to facilitate the communication process between people, and the gender diversity of nonproductive workers effectively promotes the bidirectional innovation of enterprises (Li Houjian and Peisen, 2016). Some scholars also explored the CEO, a special group within an enterprise, and found that the higher the academic experience of the CEO, the higher the innovation level of the enterprise. In particular, a CEO with experience in natural science research experience can enhance the innovation awareness of the enterprise from the inside and promote the innovation development of the enterprise through the industry–university–research combination technology (Zhang Xiaoliang et al., 2019). (3) Ownership concentration: there is no consensus on the influence of ownership concentration on the level of enterprise innovation. Some believe that an increase in the shareholding ratio of major shareholders inhibits the level of enterprise innovation. Due to the complexity, uncertainty, and high risk of innovation activities, the owners of enterprises are more inclined to carry out projects with stable investment returns to avoid shrinkage of their assets. It remains difficult to reverse the decision of shareholders with a large proportion of shares, even if the company's senior executives have innovation beliefs. This situation exists in companies with multiple major shareholders, which shows the phenomenon of "excessive supervision" by multiple major shareholders and multiple managers (Zhu Bing et al., 2008). Some believe that ownership concentration has a positive influence on the innovation level of enterprises (Sun Zhaobin et al., 2006). They proposed that the aggregation of decision-making power brought about by equity concentration could eliminate the influence of risk-averse shareholders and enable enterprises to increase innovation input. At the same time, however, some scholars have proposed that the ownership concentration has a U-shaped effect on enterprise innovation, in other words, in the initial process of equity concentration, shareholders and executives will hinder enterprise innovation for the purposes of equity value and interests, however, with further intensification of the ownership concentration and corporate owners sense of belonging, they are willing to invest in the innovation and progress of the enterprise. At this point, the innovation level of the enterprise is enhanced (Shu Qian et al., 2013). (4) Pledge of share right: the pledge of share right refers to the act of shareholders of a company pledging their own corporate equity for the purpose of obtaining funds. In order to ensure the value and appreciation of equity is not diluted, it will inevitably inhibit the innovation behavior of the enterprise. Li Changqing et al. (2018) found that when the equity pledge rate is high and the distance to the open line is close, the increase in the equity pledge of controlling shareholders will reduce the innovation input of enterprises. At the same time, Yang Mingjing et al. (2019) also found that, under the unstable monetary policy, the stock pledge of controlling shareholders has a more obvious restraining effect on the innovation level of enterprises, which seriously affects the technological progress of enterprises.

2.2.2 *External factors influencing enterprise innovation*

The external environment of an enterprise is an important factor that determines its innovation and creation. (1) Political factors: the political environment is the most important prerequisite for the macro environment of an enterprise operation. The attitude of a government toward innovation investment determines the intensity of enterprise innovation investment, and the degree of protection of a government for innovation achievements determines the enthusiasm of the public for research and development. A change of government officials is one of the most influential factors for enterprise innovation. On the one hand, the change of government officials helps to break the power rent-seeking relationship formed by the long-term tenure of officials and reduces the occurrence of corruption. On the other hand, bringing new opportunities for local enterprises can promote enterprises' R&D investment. In the process of political environment discussion, corruption is inseparable from this factor. Enterprises, in order to be able to have smooth operation in a local area, may be subject to bribery that occurs when a company receives preferential policies and assistance from the government, which can improve its operation to an extent, gradually increasing investment in innovation, however, as the level of corruption worsens, the resources for innovation investment of enterprises are depleted and become insufficient, which will eventually reduce the intensity of innovation. This draws the conclusion that there is an inverted U-shaped relationship between the level of corruption and enterprise R&D investment (Li Houjian and Zongyi, 2014).

Research shows that since the 18th National Congress of the Communist Party of China (CPC), China's increasing anticorruption drive has made it much more expensive for companies to seek political influence, thus using capital for corruption and innovation. (2) Government subsidies: at present, China is in a critical period of innovation-driven construction. Whether the government's innovation subsidies for enterprises can play its due role is a key step for China to further build an innovation-driven country. Existing studies show that the impact of government subsidies on enterprise innovation is difficult to determine. Those who believe that government subsidies can promote enterprise innovation believe that the direct financial help provided by government can not only increase the amount of capital enterprises investment in innovation, but also the government's help is a way for companies to brand their innovation projects, attract other investors to provide funds, lead the whole industry's promotion of innovation projects, and contribute to the formation of the whole society innovation (Bronzini et al., 2014; Guo Yue, 2018). However, those who hold the opposite view believe that the government's support for enterprise innovation will send a signal to the market, which will increase the cost of enterprise innovation. Meanwhile, the government's funding will also crowd out the capital that enterprises originally planned to invest in innovation, which is not conducive to the development of enterprise innovation (Boeing, 2016). (3) Bank–enterprise relationship: enterprise innovation needs substantial and ongoing financial support, and bank loans are one of the most important sources of enterprise innovation funds, and a harmonious relationship with banks determines whether an enterprise has sufficient funds to support innovation and R&D. Bank credit is one of the components of bank debt. Different from an ordinary bank loan, this enables enterprises to borrow money from banks according to the pre-agreed financing costs in case of financial difficulties, so as to support the innovative research and development of enterprises. If the senior managers are related to the bank, they can not only increase the loan amount and extend the loan term, but also break the original information barrier between the enterprise and the fund provider, which is conducive to the improvement of the innovation level of the enterprise. What needs to be pointed out in particular is that some companies bribe and take bribes from bank officials in order to raise funds. This kind of corruption will not only occupy a large amount of corporate funds, but also the funds raised are mostly for small and short-term loans, which cannot support the innovation input of enterprises (Zhang Xuan et al., 2017).

3 RESEARCH HYPOTHESIS

Since the innovation process of enterprises is characterized by long time, high cost, and unstable income, the innovation process of an enterprise needs more and longer financial support than usual. Debt financing, one of the two major channels of enterprise financing, is the focus of our attention. What impact does the debt level of an enterprise have on enterprise innovation? The forms of corporate debt include bank loan, commercial credit, bond financing, and finance lease, among commercial credit and bank loan are the main sources of corporate debt. Commercial credit refers to obtaining short-term financial support from upstream and downstream operators of an enterprise by relying on its own brand effect and long-term accumulated reputation. This financing method lacks effective constraints and is difficult to support the long-term capital demand for innovation. Compared with bank loans, although constrained by contracts and fixed financing costs, bank loans can raise more long-term funds for enterprise innovation and increase enterprise innovation input (Xiao Hailan et al., 2014). On the whole, the influence of debt on the innovation level of enterprises is not singular. Under the circumstances of less debt levels in an enterprise, the enterprise has not made full use of the leverage effect, increasing the debt can create more funds for innovative research and development, so that the enterprise can achieve fruitful innovation results, however, after a certain amount of debt accumulation, the debt ratio of the enterprise reaches the level that causes the enterprise to have a debt crisis, and therefore companies will need to consider debt servicing, projects such as innovation, which are difficult to determine the revenue period and revenue amount, will slow down investment or even abandon investment to some extent, which will depress the innovation drive of enterprises.

Based on this, the hypothesis of this paper is put forward that there is an inverted U-shaped relationship between the corporate debt ratio and the innovation level.

4 RESEARCH DESIGN

4.1 *Data sources*

This paper used CSMAR, RESSET database, and China Research Data Service (CNRDS) to collect the financial and nonfinancial data of A-shares-listed companies in Shanghai and Shenzhen from 2012 to 2017, among which the patent data are from CNRDS (CNRDS data came from students of SWUFE). In this paper, special ST and *ST enterprises, financial enterprises, enterprises with data for less than 3 years in the sample period, listed enterprises after 2012, and enterprises with significant abnormalities in financial status were removed, including: (1) enterprises with total assets less than 0; (2) enterprises with total assets less than total liabilities; (3) enterprises whose total assets were smaller than their fixed assets, intangible assets, and current assets; and (4) enterprises with total liabilities less than 0, less than long-term liabilities and short-term liabilities. The analysis tries to prove that there is a significant correlation between the level of corporate debt and innovation, so as to provide a theoretical basis for the future process of corporate debt adjustment to improve the degree of innovation.

4.2 *Variable selection and research model*

Explained variable: There are many ways to reflect the level of enterprise innovation. On the one hand, the innovation input is directly represented by the amount of R&D investment and the calculated expenditure of manpower investment, equipment and so on in the enterprise investment process, On the other hand, the number of patent applications, the increment of intangible assets, the turnover of technology market and other data can be used to reflect the innovation achievements of enterprises (Yu Yihua et al., 2018). At the same time, the possibility that innovative activities can enhance business performance by the enterprise can be included as a measure of investment risk (Wang Yuze et al., 2019). However, it is not enough to look at the willingness and enthusiasm of innovation if we want to see how substantial the benefits of innovation are to the enterprise, and more attention should be paid to the display of enterprise innovation results. Therefore, in this paper, the number of patent applications in the current year is selected to represent the innovation level. However, in order to ensure the reliability of the research, the ratio of innovation investment and intangible assets is still introduced to replace the number of patent applications as the explained variable in the robustness examination later in this paper.

Explain variable: liabilities as an important source of funds for innovation and creation are the focus of this paper. There have been many measures of debt in previous studies, Xiao Hailan et al. (2014), selected indicators of bank loan/total assets and commercial credit/total assets to measure the impact of two different financing channels on R&D investment of enterprise innovation, and Wang Yuze et al. (2019) added the overall leverage ratio, such as asset–liability ratio, long-term liabilities/total assets and short-term liabilities/total assets, and analyzed the impact of these indicators on the innovation level of the enterprise to draw conclusions. Combined with previous research results, this paper determines the corporate debt level as a whole, and selects asset–liability ratio equal to total liabilities/total assets as an important explain variable.

Control variables: the cash flow, profitability, operating capacity, and asset ratio of an enterprise reflect the internal capital supply and circulation capacity of the enterprise, and are closely related to the capital source of enterprise innovation. Growth, enterprise age, and shareholding ratio of senior executives reflect the innovation will of the enterprise from the perspective of its basic situation. Government subsidies can change the innovation level of an enterprise from external social factors. Therefore, in order to eliminate the influence of other factors on the level of enterprise innovation and ensure the accuracy of the research, the influencing factors of enterprise innovation analyzed

Table 1. Main independent variable definition.

Code symbol	Variable name	Operational definition
Cf	Cash flow	Net operating cash flows = revenue – out-of-pocket cost – income tax
Pro	Profitability	Net profit growth rate = added value of net profit/last year net profit
Turn	Operating capacity	Current asset turnover = sales revenue/average balance of current assets
Gro	Growth	Total assets growth rate = added value of total assets/last year total assets
Fix	Asset ratio	Fixed assets ratio = fixed assets/total assets
The age	Number of years of establishment	Enterprise age
The state	Property type	Whether is the state holding
Exe	Executives at stake	CEO share-holding rate = the number of shares held by executives/total shares
Gov	Government subsidy	Total amount of government subsidy received throughout the year

above are combined. Nine indicators are selected for control from the perspectives of cash flow, profitability, operating capacity, growth, asset ratio, enterprise age, type of property right, executive shareholding, and government subsidy, as shown in Table 1.

Based on the above assumptions, the following research model is constructed in this paper:

$$\text{patent}_{it} = \beta_0 + \beta_1 \times \text{lev}_{it} + \beta_2 \times \text{lev}_{it}^2 + X_j \times \mu + \alpha_m + I_n + \varepsilon_{it} \qquad (1)$$

where $patent_{it}$ = the patent achievements of i enterprise in year t; and lev_{it} = the asset–liability ratio of i company in year t. And X_j = the set of multiple control variables selected at the enterprise level; αm = industry fixed effect; in = shear stiffness number; and εit = random error.

At the same time, based on the theoretical analysis and summary of previous experience, it is concluded that the level of enterprise innovation is likely to be associated with the square of the enterprise asset–liability ratio, therefore lev_{it}^2 is introduced into the model for analysis as the term is introduced into the model for analysis.

4.3 Descriptive statistical analysis

Table 2 shows the results of a descriptive statistical analysis of the sample size. According to the data in the table, the average annual number of patent applications of enterprises is 7.68, and the maximum number is 340. Correspondingly, some enterprises have no patent output for some years. The huge gap in this index shows not only the difference in the realization degree of innovation achievements of various enterprises, but also the strong difference in the attention degree of different enterprises to innovation, which is the premise to promote the degree of innovation of enterprises from the perspective of enterprises themselves. The logarithmic average of the listed companies in the sample after adding 1 is 17.49. The increase in this value reflects that the company invests heavily in research and development, and its analysis is one of the links in the follow-up robustness test in this paper. The ratio of intangible assets refers to the proportion of intangible assets in the total assets of an enterprise. Intangible assets include patent rights and trademark rights owned by the enterprise, which can reflect the achievement of technological assets obtained by the enterprise through innovation to a certain extent. In the sample data, the top five companies in terms of intangible assets ratio are all greater than 0.70 and in the division of the industry are handling and transportation industries and road transportation industries, which can reflect the degree of dependence on technology in the daily production and operation of these two types of companies.

Table 2. Descriptive statistical analysis table.

Variable	Obs	Mean	Std. dev.	Min	Max
Patent	4959	7.680	18.685	0.000	340.000
RD	4958	17.486	1.496	8.006	28.084
Innovation	4959	0.051	0.056	0.000	0.787
Lev	4959	7.680	18.685	0.000	340.000
Cf	4959	0.415	0.191	0.022	0.995
Pro	4959	18.932	1.431	14.868	22.218
Turn	4876	0.219	2.313	5.371	6.688
Gro	4959	1.370	0.943	0.243	5.509
Fix	4959	0.202	0.374	0.190	2.426
The age	4959	0.245	0.148	0.011	0.664
The state	4959	2.633	0.405	0.693	3.497
Exe	4959	0.417	0.493	0.000	1.000
Gov	4959	8.268	8.135	0.000	20.741

5 ANALYSIS OF EMPIRICAL RESULTS

5.1 Analysis of regression results

In the process of regression, fixed-effect regression was used to estimate model (1). Table 3 shows the result of the baseline regression, and the second column is the result of adding the relevant control variables of the enterprise to the first column. The results show that positive estimated coefficient of asset-liability ratio and negative estimated coefficient of the squared asset-liability ratio both can guarantee the significance level of 1%. This indicates that there is an inverted U-shaped correlation between the asset–liability ratio and the number of patent filings, which is consistent with the previous assumption.

In terms of control variables, at the level of 10%, the profitability, operation ability, enterprise age, ownership, and government subsidies have no significant impact on the number of patents. The cash flow at 1% has a significant positive impact on the number of patents, while the growth, asset ratio, and shareholding ratio of executives at 1% have a significant negative impact on the number of patents. These indicate that the increase in corporate cash flow will promote corporate innovation, and the stronger the enterprise growth, the larger the proportion of fixed assets and the larger the shareholding ratio of senior executives, the lower the innovation level of the enterprise.

5.2 Robustness test

Because the number of patent applications can only measure the innovation level of an enterprise from the perspective of innovation output, the originally explained variables are now replaced by the proportion of enterprise intangible assets (innovation) and research and development investment of enterprise (RD) for two times, respectively, from two aspects of enterprise innovation and innovation investment to judge, variable substitution results are shown in Table 4. Inspection results in the aspect of control variables are different, in addition to the consistent impact of enterprise growth and fixed assets ratio on the two, the net profit growth rate and current asset turnover ratio have a positive impact on the proportion of intangible assets at the significance level of 1%, enterprise ownership under the 5% level has a negative effect on intangible assets. For enterprise innovation investment, cash flow, enterprise age, executives' shareholding proportion, and the government innovation subsidies all have positive effects at the significance level of 1%. Although the test results of the control variables are slightly different, we can see that both the proportion of intangible assets and R&D investment show an inverted U-shaped relationship with the enterprise's assets and liabilities, which is consistent with the previous hypothesis.

Table 3. Results of baseline regression.

Variables	Patent	Patent
Lev	40.490 * * *	27.414 * * *
	(4.823)	(6.173)
Lev^2	34.175 * * *	31.527 * * *
	(5.523)	(6.949)
Cf		1.986 * * *
		(0.221)
Pro		0.098
		(0.110)
Turn		0.062
		(0.413)
Gro		2.787 * * *
		(0.854)
Fix		19.170 * * *
		(2.485)
The age		0.116
		(0.804)
The state		0.671
		(0.733)
Exe		0.138 * * *
		(0.041)
Gov		3.638 * * *
		(0.210)
Industry fixed effect	Yes	Yes
Regional fixed effect	Yes	Yes
Observations	6402	4837
R-squared	0.154	0.268

Note: ***, **, and *, respectively, represent the significance levels of 1%, 5%, and 10%.

Table 4. Robustness test of substitution-dependent variables.

Variables	Innovation	RD
Lev	0.060 * * *	1.636 * * *
	(0.016)	(0.386)
Lev^2	0.054 * * *	1.880 * * *
	(0.018)	(0.434)
Control variables	Yes	Yes
Industry fixed effect	Yes	Yes
Regional fixed effect	Yes	Yes
Observations	4837	4836
R-squared	0.426	0.546

Note: ***, **, and *, respectively, represent the significance levels of 1%, 5%, and 10%.

5.3 Heterogeneity analysis

5.3.1 Divide the research into groups according to the growth of enterprises

Taking the average growth rate of total assets as the limit, the enterprise is divided into two types of enterprises with strong growth and weak growth. Regression analysis was carried out retrospectively to judge whether the debt of different growth enterprises had different influences on the innovation

Table 5. Empirical results grouped by enterprise growth.

Variables	The overall sample	Growth enterprise	Non-growth companies
Lev	27.414 * * *	37.058 * *	27.405 * * *
	(6.173)	(17.156)	(7.037)
Lev^2	31.527 * * *	46.709 * *	31.579 * * *
	(6.949)	(19.129)	(7.895)
Control variables	Yes	Yes	Yes
Industry fixed effect	Yes	Yes	Yes
Regional fixed effect	Yes	Yes	Yes
Observations	4837	1178	3595
R-squared	0.268	0.286	0.290

Note: ***, **, and *, respectively, represent the significance levels of 1%, 5%, and 10%.

level of enterprises. The empirical results in Table 5 shows that despite both enterprises with strong growth and enterprises with weak growth showing an inverted U-shaped structure for the asset–liability ratio, it is obvious that the correlation is at the significance level of 5% for companies with strong growth, and at the significance level of 1% for companies with weak growth. In other words, this inverted U-shaped structure is more obvious in nongrowth enterprises. This is because, when corporate debt levels are low, growing companies rely on existing technology as a result of their own growth status being good and invest more funds in existing production and operation. However, when the debt level exceeds the critical value of enterprises, nongrowth enterprises are prone to fall into financial crisis and bankruptcy risk, thus greatly reducing their investment in innovation. In the same situation, growth companies can rely on rapid growth in total assets to sustain a slow decline in innovation spending.

5.3.2 Group research according to different enterprise ownership

It is a unique feature of China's economy to divide enterprises into state-owned enterprises and private enterprises. Due to the high level of state support and the reflection of national policies, state-owned enterprises are often of interest to researchers. Table 6 shows the empirical results of grouping according to the structure of enterprise property rights. The difference between state-owned and nonstate-owned enterprises cannot affect the inverted U-shaped relationship between asset–liability ratio and innovation level of enterprises, which disputes the opinion of some scholars that state-owned enterprises are weaker in innovation than nonstate-owned enterprises. The state-owned nature of enterprises will make managers more dependent on government support, at the same time ignoring corporate earnings and long-term performance and turning to social performance and short-term on-the-job performance, but in the current process of innovation-driven national construction, on the one hand, the government will provide direct financial or policy support for enterprises' innovation projects, on the other hand, enterprises' innovation and creation in order to cater to the national policy is also a manifestation of fulfilling social responsibility. These effects are enough to reverse the disadvantages of state-owned enterprises in innovation, and achieve the same effect as ordinary private enterprises (Table 6).

5.3.3 Grouping research according to enterprise scale

The enterprises are grouped into regressions according to their total assets, and the results are shown in Table 7. Compared with small enterprises, the inverted U-shaped relationship between the number of patent applications of large enterprises and the asset–liability ratio exists at the significance level of 1%, which is more significant. First, large enterprises are more likely to obtain external financing support for innovative projects due to their high level of assets, relatively

Table 6. Empirical results of grouping according to the structure of enterprise property rights.

Variables	The overall sample	State-owned enterprises	Nonstate-owned enterprise
Lev	27.414 * * *	49.841 * * *	24.491 * * *
	(6.173)	(12.226)	(8.031)
Lev^2	31.527 * * *	46.686 * * *	33.625 * * *
	(6.949)	(12.472)	(10.001)
Control variables	Yes	Yes	Yes
Industry fixed effect	Yes	Yes	Yes
Regional fixed effect	Yes	Yes	Yes
Observations	4837	2033	2797
R-squared	0.268	0.399	0.299

Note: ***, **, and *, respectively, represent the significance levels of 1%, 5%, and 10%.

Table 7. Empirical results grouped by enterprise scale.

Variables	The overall sample	Large companies	Small businesses
Lev	27.414 * * *	41.781 * * *	9.437 *
	(6.173)	(15.640)	(4.846)
Lev^2	31.527 * * *	45.469 * * *	11.917 *
	(6.949)	(15.580)	(6.192)
Control variables	Yes	Yes	Yes
Industry fixed effect	Yes	Yes	Yes
Regional fixed effect	Yes	Yes	Yes
Observations	4837	2371	2437
R-squared	0.268	0.316	0.255

Note: ***, **, and *, respectively, represent the significance levels of 1%, 5%, and 10%.

perfect management system, and better reputation. Second, the self-protection ability of large enterprises can also prevent the innovation achievements being stolen by others and driving the market demand to ensure that the innovation achievements make profits for themselves. Finally, as large enterprises have larger production scale and business volume, they can achieve a scale effect for enterprises, which makes the asset–liability ratio have a stronger influence on the innovation of large enterprises.

6 RESEARCH CONCLUSIONS AND SUGGESTIONS

Based on the data of share-listed companies in Shanghai and Shenzhen from 2012 to 2017, this paper examines the impact of corporate debt on the innovation level. The research finds that: (1) When other conditions remain unchanged, the asset–liability ratio has an inverted U-shaped relationship with the innovation level, that is, within a certain range, the increase in the asset–liability ratio will increase the innovation level of the enterprise. However, beyond this range, the increase in the asset–liability ratio will actually lead to a weakening of the innovation level. (2) The growth of enterprises will weaken this inverted U-shaped relationship, while a gradual expansion of enterprise size will make this inverted U-shaped relationship more significant. (3) Whether the enterprise is state-owned or not has no obvious influence on this inverted U-shaped relationship.

Based on the above conclusions, the following suggestions are put forward for enterprises to reasonably control the debt level in the future and continuously improve their innovation ability.

(1) Pay attention to the debt level and adjust financing decisions. The critical value of liabilities, the source of liabilities, and the corresponding structure ratio of each industry and enterprise are the key contents of our liability monitoring process. Low-debt enterprises should give full play to the leverage effect to raise capital. Enterprises with high debts should take active and passive measures to deleverage, implement the requirements of supply-side structural reform, reduce the possibility of financial crisis, transform debt financing into equity financing, expand the financial market, and improve market management, so as to provide economic support for enterprise innovation. (2) Improve the management system and optimize enterprise management. For Chinese enterprises, most modern and advanced enterprise systems exist in large enterprises, and the complete construction of middle and small-sized enterprises' systems, especially accounting systems, is of great urgency. The excellent financial accounting system of enterprises can provide continuous capital supply for enterprise innovation and avoid the harm brought about by financial risks to enterprises. At the same time, an improvement of the enterprise management system also enables enterprises to better cater to the corresponding national innovation and entrepreneurship policies and fulfill their corporate social responsibilities while developing new technological products. (3) To build supporting facilities and enhance innovation capacity. On the one hand, an enterprise should provide a stable internal environment for its innovation process, including full support from the enterprise owner, management, and ordinary employees, as well as the supply of profits for its existing business, so as to avoid blind investment in the existing business and to support innovation projects. On the other hand, continuous external support should also be provided for innovation activities. It is not only necessary to make a good plan for enterprise innovation projects, making good use of government subsidy policies and tax incentives, but also to properly cooperate with external institutions of enterprises, so as to prepare for innovation activities in law, technology, and other aspects together.

REFERENCES

Boeing, Philipp, 2016. The allocation and subsidies: evidence from listed firms. Research policy, 45(9): 1774–1789.

Bronzini, Raffaello, and Eleonora Iachini, 2014. Are there any incentives for R&D effective? Evidence from a regression discontinuity approach. American economic journal: economic policy, 6(4): 100–134.

Gao Wenliang, 2018. Major shareholder, CEO compensation incentive and enterprise innovation. Science and technology management research, 38(18):100–106.

Guo Yue, 2018. Signal transmission mechanism of government innovation subsidy and enterprise innovation. China industrial economy, (09):98–116.

Li Changqing, Li Yukun, Li Maoliang, 2018. Equity pledge of controlling shareholders and innovation input of enterprises. Financial research, (07):143–157.

Li Houjian, Liu Peisen, 2016. Research on the influence of diversity of human capital structure on enterprise innovation. Research in science of science,36(09):1694–1707.

Li Houjian, Zhang Zongyi, 2014. Tenure of local officials, corruption and investment in enterprise research and development. Science research, 32(05):744–757.

Shu Qian, Chen Zhiya, 2013. Affect the governance structure of Chinese manufacturing enterprise R&D input factors. Science and management science and technology, (9): 97–106.

Sun Zhaobin, 2006. Equity concentration, equity checks and balances and technical efficiency of listed companies. Management world, (07):115–124.

Tang Manping, Li Houjian, 2019. Enterprise scale, minimum wage and R&D investment. Research and development management, 31(01):44–55.

Wang Xiaoxia, Jiang Dianchun, Li Lei, 2016. Will the rise of minimum wage force manufacturing enterprises to transform and upgrade? Empirical analysis based on patent application data. Financial and economic research, 44(12):126–137.

Wang Yuze, Luo Nengsheng, Liu Wenbin, 2019. What leverage ratio is conducive to enterprise innovation. China industrial economy (03):138–155.

Xiao Hailan, Tang Qingquan, Zhou Meihua, 2014. Influence of debt on innovation investment model of enterprise: An empirical study based on R&D heterogeneity. Scientific research management, 35(10):77–85.

Yang Mingjing, Cheng Xiaoke, Zhong kai, 2019. Research on the influence of equity pledge on enterprise innovation: Based on the analysis of the regulatory effect of monetary policy uncertainty. Financial and economic research, 45(02):139–152.

Yang, P. C., H. M. Wee et al, 2011. Mitigating the Hi-tech products risks due to rapid technological innovation. Omega, 33 (4) 6:456–463.

Yu Yihua, Zhao Qifeng, Ju Xiaosheng, 2018. Executive of inventors and enterprise innovation. China industrial economy, (03):136–154.

Zhang Xiaoliang, Yang Hailong, Tang Xiaofei, 2019. CEO academic experience and enterprise innovation. Scientific research management, 40(02):154–163.

Zhang Xuan, Liu Beibei, Wang Ting, Li Chuntao, 2017. Credit rent-seeking, financing constraints and enterprise innovation. Economic research, 52(05):161–174.

Zhu Bing, Zhang Xiaoliang, Zheng Xiaojia, 2008. Multiple major shareholders and enterprise innovation. Management world, 34(07):151–165.

Computational Social Science – Luo, Ciurea & Kumar (eds)
© 2021 Taylor & Francis Group, London, ISBN 978-0-367-70193-2

Empirical analysis relationship between NaOH output and economic social development in 1998–2019 based on VAR

J.M. Feng & M.X. Yan
SINOPEC Jianghan Salt Chemical Hubei Co., Ltd, Qianjiang, Hubei, China

Q.S. Feng*
Industrial and Commercial Bank of China, Minhang Branch, Shanghai, China

ABSTRACT: Based on the yearly data for the caustic soda output during 1998–2019 in China, as well as the industrial added-value index, urbanization rates, and exchange rates (USD/CNY), based on the VAR model, Johansen co-integration tests, Granger causality tests, impulse response, and variance decomposition, this quantitative thesis studies the relationship between caustic soda output and economic social development. The results show that, among the four variables there exists a stable co-integration relationship over the long term. When LnIGDPI, LnCZHL, and LnEX increased by 1%, respectively, in period t – 1, LnNaOH increased –0.1695%, 0.7237%, and –0.1628% in period t. LnNaOH, LnCZHL, and LnEX all Granger caused LnIGDPI. The caustic soda industry and economic social development show mutual promotion. The influence of the economic social development on LnNaOH has a lag. Based on the economic social development plan, the caustic soda industry plan should be made in advance, to better promote the economic social development.

1 INTRODUCTION

Caustic soda is used widely in industries such as petrochemical, papermaking, textile, printing and dyeing, medicine, alumina, water treatment, etc.

Driven by the steady promotion of new urbanization, the market demand for caustic soda downstream products has maintained rapid growth. The Chinese urbanization rate was 60.6% in 2019, with reference to developed countries in the world, for example, the US urbanization rate was 82.058% in 2017 (CEIC, 2020). With the reform of the household registration system in China, the Chinese urbanization rate will be increased further in the future.

The fluctuation of RMB exchange rate will affect the competitiveness of China's commodities in the international market, and then affect alumina, textiles, and other export-oriented industries, so affecting the demand for caustic soda.

This thesis studies the relationship between caustic soda output and the economic social development during 1998–2019 in China, in order to forecast the caustic soda output prospects, and to put forward some suggestions for the development of the caustic soda industry.

2 DATA SOURCE AND PROCESSING

The caustic soda (100%) output (NaOH) is 10,000 tons. The economic development is measured by the industrial added-value index (1978 = 100) (IGDPI) which is closely related to caustic soda use. Social development is measured by the urbanization rate (%) (CZHL), with the annual average datum of the exchange rates (USD/CNY) as the variable of EX.

---

*Corresponding author

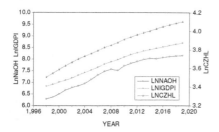

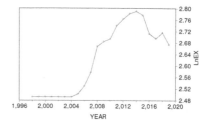

Figure 1.   LnNaOH, LnIGDPI, and LnCZHL during 1998–2019.

Figure 2.   LnEX in 1998–2019.

Table 1.   ADF test results (P value) of LnNaOH, LnIGDPI, LnCZHL, and LnEX.

	NaOH	IGDPI	CZHL	EX
Ln	0.9935	0.5574	0.8177	0.5632
D(Ln)	0.0127	0.2835	0.0299	0.3502
DD(Ln)	0.0215	0.0007	0.0018	0.0027

Data were obtained from the "China Statistical Yearbook."

Taking the natural logarithm for the data of the four variables, the corresponding symbols are LnNaOH, LnIGDPI, LnCZHL, and LnEX.

Eviews 7.2 software was used.

3   STATUS OF NaOH PRODUCTION AND ECONOMIC SOCIAL DEVELOPMENT

During 1998–2019, in China, caustic soda output increased by 6.42 times, with an average annual growth rate of 9.26%. The industrial added-value index (1978 = 100) increased 6.45 times, and the average annual growth rate was 9.28% (Figure 1). This showed that the caustic soda output is closely related to the industrial added-value index (1978 = 100), and the correlation coefficient was 0.9937.

The Chinese urbanization rate was 33.4% in 1998, increasing to 60.6% in 2019, and the correlation coefficient was 0.9922 between the caustic soda output and the urbanization rates.

After the exchange rate reform in 2005, the exchange rates (USD/CNY) increased to 2014, and decreased at a general trend in 2015–2019 (Figure 2)

4   EMPIRICAL ANALYSIS

4.1   *ADF test*

First, an ADF test (Tie-mei Gao, 2009) was performed. The time trend and intercept were selected. The lag period was determined by SIC, and the results are shown in Table 1.

In Table 1, the P values of DDLnNaOH, DDLnIGDPI, DDLnCZHL, and DDLnEX were all less than 0.05. At the 5% significance level, were the same order stable time series. Co-integration tests were carried out.

4.2   *VAR model*

Based on the VAR model, endogenous variables were LnNaOH, LnIGDPI, LnCZHL, and LnEX, and the exogenous variable was C. Determining the optimal lag period of VAR was done according to the majority principle. The results are shown in Table 2. The optimal lag period of VAR was 1.

Table 2. Lag period of VAR model.

Lag	LogL	LR	FPE	AIC	SC	HQ
0	123.7466	NA	7.41e-11	−11.97466	−11.77552	−11.93579
1	247.8862	186.2094*	1.56e-15*	−22.78862*	−21.79289*	−22.59424*
2	261.0323	14.46076	2.62e-15	−22.50323	−20.71092	−22.15336

Table 3. Johansen co-integration test results (5%).

	No. of CE(s)	Eigenvalue	Statistic	Critical value	Prob.**
Trace	None *	0.925075	101.6839	47.85613	0.0000
	At most 1 *	0.779541	47.26716	29.79707	0.0002
	At most 2 *	0.495313	15.51428	15.49471	0.0497
Maximum Eigenvalue	None *	0.925075	54.41675	27.58434	0.0000
	At most 1 *	0.779541	31.75287	21.13162	0.0011
	At most 2 *	0.495313	14.36017	14.26460	0.0483

The inverse roots of the AR characteristic polynomial were all less than 1, which shows that the VAR (1) model was stable. The VAR (1) model is shown in Formula (1):

$$LnNAOH = 0.9272 \times LnNaOH(-1) - 0.1695 \times LnIGDPI(-1) \tag{1}$$

$$+0.7237 \times LnCZHL(-1) - 0.1628 \times LnEX(-1) - 0.3956$$

For this VAR (1) model, $R^2 = 0.9915$. According to the result of the VAR, the three variables in China, which are industrial added-value index, urbanization rates, and exchange rates (USD/CNY), all influence the caustic soda output.

The impact of the industrial added-value index (1978 = 100) (LnIGDPI) in period $t - 1$ was negative. It showed the higher the industrial added value of the previous period, the more the caustic soda output in the respective period would decrease.

The impact of the urbanization rate (LnCZHL) in period $t - 1$ was positive. It showed urbanization stimulating the demand for caustic soda.

The impact of exchange rates (USD/CNY) (LnEX) in period $t - 1$ was negative. RMB appreciation is not conducive to export of foreign trade enterprises, impact on alumina, textile, and other downstream industries of caustic soda, thus, the NaOH output in period t was affected.

The above three variables in period $t - 1$, when the rest was unchanged, LnIGDPI, LnCZHL, and LnEX increased by 1%, respectively, LnNaOH increased −0.1695%, 0.7237%, and −0.1628%, respectively, in period t.

4.3 Co-integration tests

Based on the VAR (1) model being stable, Johansen co-integration tests were performed for the four research variables. The exogenous variable was C, and the lag period was 0. Intercept and no trend were selected. At the 5% significance level, the results are shown in Table 3.

There existed three co-integration relationships in the long term among the four variables. Formula (2) was one of the co-integration equations:

$$LnNaOH = -4.1341LnIGDPI + 21.2066LnCZHL + 0.5051LnEX \tag{2}$$

The above co-integration equation demonstrated the relationship between the caustic soda output and the economic social development in the same period.

Table 4. Granger causality tests results (5%).

Null hypothesis:	Prob.	Null hypothesis:	Prob.
LNIGDPI does not Granger Cause LNNAOH	0.9876	LNCZHL does not Granger Cause LNIGDPI	0.0243
LNNAOH does not Granger Cause LNIGDPI	0.0193	LNIGDPI does not Granger Cause LNCZHL	0.3904
LNCZHL does not Granger Cause LNNAOH	0.5138	LNEX does not Granger Cause LNIGDPI	0.0749
LNNAOH does not Granger Cause LNCZHL	0.7747	LNIGDPI does not Granger Cause LNEX	0.6341
LNEX does not Granger Cause LNNAOH	0.2830	LNEX does not Granger Cause LNCZHL	0.2550
LNNAOH does not Granger Cause LNEX	0.2189	LNCZHL does not Granger Cause LNEX	0.7013

The three variables in period t, when the rest was unchanged, LnIGDPI, LnCZHL, and LnEX increased by 1%, respectively, LnNaOH increased −4.1341%, 21.2066%, and 0.5051%, respectively, in the same period t. Among them, the urbanization rate had the greatest impact. This showed that stimulating domestic demand would stimulate economic growth. The increase in caustic soda output was an example.

4.4 Granger causality tests

Based on the VAR (1) model, Granger causality tests were performed. The lag period was 1 as for VAR, and the results are shown in Table 4.

During the study period, there was no causal two-way relationship between the caustic soda output, urbanization rates, and exchange rates (USD/CNY).

The caustic soda output, urbanization rates, and exchange rates (USD/CNY) all Granger Caused the industrial added-value index.

4.5 Impulse response

Figure 3 shows the responses of the industrial added-value index, urbanization rates, and exchange rates (USD/CNY) to the caustic soda output. Responses to disturbance with one standard deviation of LnNaOH were as follows:

LnIGDPI, 0.0109 in period 1, then increased to 0.0172 in period 3 which was the highest, and then decreased to 0.0046 in period 10, all were positive responses.

LnCZHL, 0.0007 in period 1, 0.0006 in period 2, then gradually increased to 0.0025 in period 8, and maintained stable to period 10, all were positive responses.

LnEX, 0.0094 in period 1, then increased to 0.0378 in period 4 which was the highest, and then decreased gradually to 0.0006 in period 10, all were positive responses.

When the caustic soda output increased, the above responses showed that the industrial added-value and urbanization rates would increase, and RMB would appreciate.

Figure 4 shows the responses of the caustic soda output to the industrial added-value index, urbanization rates, and exchange rates (USD/CNY).

In period 1, due to the disturbance with one standard deviation of LnIGDPI, LnCZHL, and LnEX, the responses of LnNaOH were all 0. This showed that the influence of the industrial added-value index, urbanization rates, and exchange rates on the caustic soda output had a lag. There is no immediate response when caustic soda enterprises have a disturbance, because there is a time cycle from the start of design and construction to the operation.

Responses to disturbance with one standard deviation of LnIGDPI, no response in period 1, then decreased gradually to −0.0016 in period 3 which was the lowest, and then increased gradually, and all were positive responses from period 7 to period 10. The responses to LnEX disturbance were the same trend as the responses to LnIGDPI disturbance. This showed when the industrial added value increased, and RMB appreciated, and the caustic soda output would decrease in the short term and then increase at a later stage.

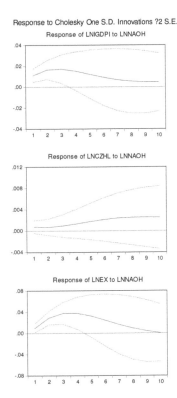

Response to Cholesky One S.D. Innovations ?2 S.E.

Figure 3.   Responses to NaOH of IGDPI, CZHL, EX.

Responses were positive to the disturbance with one standard deviation of LnCZHL. There was no response in period 1, then increased gradually to 0.0071 in period 5, which was the highest, and then decreased gradually to 0.0044 in period 10.

This showed that the urbanization in China had increased the caustic soda output.

4.6   *Variance decomposition*

Variance decomposition of LnNaOH was performed first, and the results are shown in Table 5.

In period 1, the change of LnNaOH was explained by itself. The self-explanation ratio decreased gradually in subsequent periods, and was 96.69% in period 10.

The explanation ratio of the change of LnNaOH by LnIGDPI, LnCZHL, and LnEX increased gradually in subsequent periods. The impact of LnCZHL on LnNaOH was relatively large, at 2.86% in period 10. The impact of LnIGDPI on LnNaOH was relatively small. This was consistent with the VAR model. This showed that urbanization in China had a great influence on the development of the caustic soda industry.

The variance decomposition of LnIGDPI and LnCZHL then was performed, and the results are shown in Table 6.

In period 1, the change of LnIGDPI and LnCZHL was not explained by LnNaOH. The LnNaOH explanation ratio increased gradually in subsequent periods, at 11.81% and 11.16%, respectively, in period 10. The LnNaOH explanation ratio was relatively large.

The self-explanation ratio of LnIGDPI was the largest, at 100% in period 1, and decreased gradually in subsequent periods, to 76.35% in period 10. The LnCZHL explanation ratio of LnIGDPI increased rapidly in subsequent periods. The LnEX explanation ratio of LnIGDPI was relatively small.

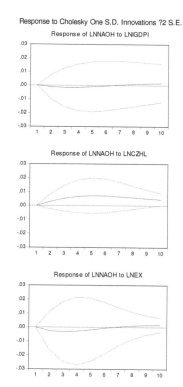

Response to Cholesky One S.D. Innovations ?2 S.E.

Response of LNNAOH to LNIGDPI

Response of LNNAOH to LNCZHL

Response of LNNAOH to LNEX

Figure 4.　Responses to IGDPI, CZHL, EX of NaOH.

Table 5.　Variance decomposition of LnNaOH.

Period	S.E.	LNNAOH	LNIGDPI	LNCZHL	LNEX
1	0.060923	100.0000	0.000000	0.000000	0.000000
2	0.081284	99.75287	0.018162	0.132100	0.096864
3	0.092057	99.33292	0.042770	0.424506	0.199806
4	0.097497	98.84160	0.061045	0.836938	0.260421
5	0.100057	98.34770	0.068444	1.305245	0.278612
6	0.101223	97.89552	0.068347	1.760203	0.275928
7	0.101795	97.50540	0.068316	2.152004	0.274277
8	0.102135	97.17960	0.074953	2.461464	0.283980
9	0.102391	96.91181	0.090848	2.693630	0.303714
10	0.102624	96.69387	0.114689	2.864569	0.326872

In period 1, the change of LnCZHL was not explained by LnIGDPI. The LnIGDPI explanation ratio of LnCZHL increased from 0.01% in period 2 to 12.57% in period 10. The LnIGDPI explanation ratio of LnCZHL was relatively large. The self-explanation ratio of LnCZHL decreased gradually in subsequent periods, from 100.00% in period 1 to 72.12% in period 10. The LnEX explanation ratio of LnCZHL was relatively small, from 0.00% in period 1 to 4.14% in period 10.

This showed that the development of the caustic soda industry steadily promoted the growth of industrial added value and the process of urbanization.

Table 6. Variance decomposition of LnIGDPI and LnCZHL.

Period	LNNAOH	LNIGDPI	LNCZHL	LNEX	Period	LNNAOH	LNIGDPI	LNCZHL	LNEX
1	0.000000	100.0000	0.000000	0.000000	1	0.000000	0.000000	100.0000	0.000000
2	5.626976	93.46191	0.459541	0.451576	2	0.112092	0.010719	99.55916	0.318029
3	9.942654	87.79525	1.380644	0.881451	3	0.078804	0.389436	98.54946	0.982304
4	12.12968	84.06728	2.656566	1.146468	4	0.378453	1.630222	96.18331	1.808012
5	12.89314	81.67494	4.177395	1.254522	5	1.457211	3.703356	92.25106	2.588375
6	12.91407	80.03973	5.791200	1.254999	6	3.276617	6.128378	87.39889	3.196114
7	12.64900	78.80550	7.337488	1.208015	7	5.465464	8.401457	82.52286	3.610218
8	12.33071	77.80986	8.698502	1.160929	8	7.641839	10.24909	78.23729	3.871780
9	12.04313	76.99954	9.821477	1.135862	9	9.570835	11.61852	74.77609	4.034552
10	11.81064	76.35087	10.70564	1.132847	10	11.16353	12.57488	72.12040	4.141185

5 CONCLUSIONS AND SUGGESTION

5.1 *Research conclusions*

This quantitative thesis studied the long-term equilibrium and dynamic interaction relationship between caustic soda output and economic social development. The research variables were caustic soda output (NaOH), the industrial added-value index (IGDPI), the urbanization rates (CZHL), and the exchange rates (USD/CNY) (EX) during 1998–2019 in China, using unit root tests, VAR model, Johansen co-integration tests, Granger causality tests, impulse response, and variance decomposition. The conclusions are as follows:

(1) There is a long-term stable relationship between the caustic soda output, the industrial added-value index, the urbanization rates, and the exchange rates (USD/CNY).

(2) According to the results of the VAR, the three variables in period $t-1$, when the rest was unchanged, LnIGDPI, LnCZHL, and LnEX increased by 1%, respectively, LnNaOH increased –0.1695%, 0.7237%, and –0.1628%, respectively, in period t. This showed that urbanization stimulated the demand for caustic soda. It also showed that RMB appreciation affected caustic soda output by impacting on alumina, textile, and other downstream foreign trade enterprises.

(3) There was no two-way causal relationship between the caustic soda output, urbanization rates, and exchange rates The caustic soda output, urbanization rates, and exchange rates (USD/CNY) all Granger caused the industrial added-value index.

(4) The impulse response showed that, when the NaOH output increased, the industrial added value and urbanization rates would increase, and RMB would appreciate. The influence of the latter three variables on the caustic soda output had a lag. When the industrial added value increased, and RMB appreciated, the caustic soda output decreased in the short term and increased at a later stage. The urbanization in China increased the caustic soda output.

(5) The variance decomposition showed that urbanization in China had a great influence on the development of the caustic soda industry. At the same time, the development of the caustic soda industry steadily promoted the growth of industrial added value and the process of urbanization.

5.2 *Policy suggestion*

The caustic soda industry and economic social development show mutual promotion. Between the two there exists a stable quantitative relationship over the long term. The influence of urbanization on the caustic soda output had a lag. Therefore, the caustic soda industry plan should be made in advance based on the economic social development plan, to better promote economic social development.

REFERENCES

[1] CEIC. 2020. US: Population and Urbanization Statistics. CEIC: Macro & Micro Economic Data [EB/OL]. https://www.ceicdata.com/zh-hans/united-states/population-and-urbanization-statistics?page=9.

[2] Tie-mei Gao. 2009. Econometric Analysis and Modeling: Application and Examples of EViews, second ed., Tsinghua University Press, Beijing, 2009. (In Chinese).

Computational Social Science – Luo, Ciurea & Kumar (eds)
© 2021 Taylor & Francis Group, London, ISBN 978-0-367-70193-2

Research on the demand for an electric energy substitution policy

W. Tang*, P. Wu* & Q.K. Tan
State Grid Energy Research Institute Co., Ltd., Beijing, China

ABSTRACT: This paper introduces the significance of the implementation of electric energy substitution, describes the status quo of the implementation of electric energy substitution in recent years, analyzes the factors restricting the development of electric energy substitution, and puts forward eight policy requirements for the development of electric energy substitution, including the coordination of the work of various ministries and commissions, the establishment of a collaborative mechanism, and the improvement of the market mechanism and technical standards.

Keywords: Power substitution, Policy, Demand.

1 INTRODUCTION

The implementation of electric energy substitution can improve the level of electrification of the whole society, improve the living standard of residents and social production efficiency, optimize the ecological environment, and improve air quality, at the same time, it can also reduce the dependence on foreign energy and ensure China's energy security. The development of China's electric energy substitution is good, and there has been a lot of academic research in this area. For example, References 1–10 in this chapter have studied the technology, economy, project implementation, and application effect of power substitution, but there are not many literatures systematically studying the policy demands of electric energy substitution in the future.

2 IMPLEMENTATION OF ELECTRIC ENERGY SUBSTITUTION

In May 2016, the National Energy Administration and other eight ministries and commissions jointly issued the guiding opinions on promoting electric energy substitution. In November of the same year, the National Energy Administration issued the 13th five year plan for electric power development (2016–2010), which clearly proposed that by 2020, the new electricity consumption of electric energy substitution would be 450 billion kw · h. During the "13th five year plan" period, the State Grid Corporation closely focused on the overall goals of national energy green development and air pollution control, and took power substitution as the strategic focus of the company's market development. From 2016 to 2019, the State Grid Business Zone completed 533.5 billion kw · h of replacement electricity. It is estimated that, by 2020, the amount of electric energy substitution will reach 190 billion kw · h, which will exceed the national planning task. In recent years, the growth scale of power substitution is shown in Figure 1.

As an effective means to improve the ecological environment, electric energy substitution has an obvious effect in controlling air pollution and improving air quality. During the 13th Five Year Plan period, the effective implementation of power substitution work reduced carbon dioxide emissions by 531.89 million tons and nitrogen oxide emissions by 16.8 million tons, promoted more than 169,000 projects, and increased terminal energy consumption by 2.5 percentage points, making

---

*Corresponding author

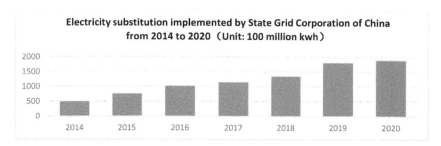

Figure 1. Electricity substitution implemented by State Grid Corporation of China from 2014 to 2020.

positive contributions to promoting social energy conservation and emission reduction, winning the blue sky defense war, and optimizing the terminal energy consumption structure. According to the 2019 annual report on China's policies and actions to cope with climate change issued by the Ministry of Ecology and Environment, China's carbon dioxide emissions per unit of GDP in 2018 decreased by 4.0% compared with the previous year and were 45.8% lower than in 2005, realizing the commitment of reducing carbon intensity by 40–45% in 2020 compared with 2005. Some experts estimate that the contribution of electricity substitution to the decline of carbon intensity in China is greater than 30%.

3 ANALYSIS OF THE RESTRICTIVE FACTORS OF ALTERNATIVE DEVELOPMENT OF ELECTRIC ENERGY

3.1 *The coverage and strength of power substitution support policies need to be optimized and strengthened*

In recent years, there have been more than 1000 environmental protection policies issued by governments at all levels, but there are few policies to support environmental protection. Most of the existing subsidy policies are partial and special subsidies, which have not yet formed a systematic whole-process subsidy for investment, construction, and operation of power substitution projects, and some of them are "one-size-fits-all" modes, which need to be improved and optimized. Affected by the current COVID epidemic and international trade disputes, the macroeconomic operation is not optimistic. It is more difficult for the government to issue and implement subsidy policies, and subsidies in some fields have declined.

3.2 *The problem of insufficient economy still exists in some fields of electric energy substitution*

With the fierce competition in the energy market, the international and domestic oil and gas prices continue to drop, and gas and other energy use methods are developing rapidly. Due to the lack of environmental cost in the development and use of energy in China, the advantages of clean, safe, and easy control of electric energy have been weakened, and the price competitiveness is biased. General industrial and commercial electricity prices continue to decline, which improves the economy of electric energy substitution to an extent. However, the problem of large initial investment in power substitution equipment remains prominent. Compared with other energy sources, the operating cost advantage of some power substitution projects is not prominent. The development and application cycle of new technology and new equipment is long, and the breakthrough and penetration of new technology still needs a long time. Under the background of economic growth decline and the Sino–US trade war, the enthusiasm of industrial and commercial enterprises to expand production is not high, and the willingness to implement electric energy substitution is declining.

3.3 *Great pressure on investment recovery*

The infrastructure of electric energy substitution is weak, and the State Grid Corporation of China has invested a lot of money in constructing the supporting power grid. However, the overall economy of the substitution project is weak, the load rate of the substation area is not high, and the power consumption of users is low. The relevant distribution equipment is not fully utilized, and the investment recovery pressure is great.

4 DEMAND ANALYSIS OF THE POWER SUBSTITUTION POLICY

(1) All ministries and commissions make overall plans to enlarge the policy effect. Power substitution involves a wide range of areas and many fields. Therefore, we should establish an overall planning mechanism and integrate various professional forces to promote the development of electric energy substitution. The first is to establish an overall working mechanism for the development of electric energy substitution, so as to promote the work of electric energy substitution as a whole. It is suggested that the Energy Administration, Ministry of Finance, Ministry of Housing and Urban Rural Development, Ministry of Environmental Protection, Ministry of Communications, Ministry of Agriculture, Ministry of Education, and Ministry of Industry and Information Technology should make overall planning for alternative fields such as electric heating, electric vehicles, shore power, industrial manufacturing, agricultural electrification, etc. We will strengthen planning and guidance, jointly carry out power substitution and pollution control, industrial development, and new energy development, and break down barriers between departments and energy sources. Second, it is suggested that all ministries and commissions should put forward clear targets on power substitution in the "14th five year plan."

(2) Establish a collaborative mechanism between governments and enterprises at all levels to jointly assist in the work of substitution. First, governments at all levels should adjust measures to local conditions, find out the potential according to the characteristics of regional energy consumption, and promote regional power substitution work as a whole. Second, all departments should perform their respective duties and work together to boost the development.

(3) Perfect market transaction and encourage market cultivation. First, we will improve the trading mechanism of the electricity market, promote the development of the spot market linking power substitution and clean energy, and release the price elasticity of demand. The second is to innovate the operation mode of electric energy substitution, implement the modes of energy contract management, third-party operation and maintenance, "package transaction" and PPP, and encourage the units of power substitution projects to actively apply for corporate bonds and low-interest loans, and adopt the PPP mode to solve financing problems. Third, encourage the innovation of the auxiliary service mechanism, and allow the power and heat production enterprises and users who provide peak sharing services to invest in the construction of regenerative electric boilers to obtain reasonable compensation income. Fourth, encourage power substitution enterprises to carry out bilateral negotiation or centralized bidding direct transaction with wind power and other clean energy power generation enterprises, so as to reduce the cost of alternative power projects. The fifth is to encourage the grid enterprises as the leading force to open up upstream and downstream of the electric energy substitution industry chain, relying on the ubiquitous power of the Internet of Things to build a multifunctional ecosystem of government, manufacturers, power grid, and users, and then rely on the market mechanism to promote the sustainable development of power substitution and achieve win–win results for all parties.

(4) We should strengthen the support of subsidy funds and support the implementation of alternative projects with a weak economy. First, we will introduce more stringent restrictive and prohibitive environmental protection standards for decentralized coal-fired and fuel oil facilities, intensify implementation, clarify the time limit for renewal and elimination, and strictly implement environmental protection and energy efficiency standards. Second, we will improve the green certificate trading mechanism, carbon tax, and other policies to reduce the use of traditional fossil energy. The third is to establish laws and regulations and a monitoring and evaluation system, and make full use of big data and other technical means to gradually carry out the construction of a monitoring and evaluation system. The fourth is to optimize the price mechanism of electric energy substitution, establish the price mechanism of linkage

between the generation side and power consumption side, promote the rationalization of the price comparison relationship between energy sources, and promote the cost advantage of power substitution projects with no economy but broad prospects. The fifth is to increase the scope of green home appliances.

(5) We should increase support and promote the upgrading of equipment, technology, and standards of electric energy substitution. First, increase investment in scientific research and promote the quality of equipment and facilities. Second, the advanced technology and equipment manufacturing enterprises will be given preferential tax and fee reductions, so as to improve the scientific and technological research and development ability and manufacturing level of equipment manufacturing enterprises, and reduce equipment manufacturing costs and operating costs. Third, financial subsidies should be given to the power substitution projects with large investment in the early stage, high operating costs, and significant energy saving and emission reduction effects. Fourth, for the construction field, transportation and other key application fields of electric energy substitution, organize enterprises and research institutions to carry out the standardization work of electric heating, shore power, and other equipment professional fields. Fifth, the government should establish a strict safety responsibility tracking mechanism for electrical energy substitute equipment to improve its quality and safety.

(6) We should increase support for the construction of alternative infrastructure for electric energy. The first is to increase support for the construction of the supporting power grid. It is suggested that electric energy should be included in the central financial fund instead of supporting investment, and a special central financial fund should be set up to share the cost of the whole society. Second, it is suggested that the investment in shore power engineering facilities should be included in the state-owned capital operation budget of central enterprises.

(7) Develop new forms of power substitution and tap the potential of substitution. First, in combination with the strategy of new smart city construction, encourage in-depth development of energy consumption big data mining. It is encouraged to identify and mine new alternative areas for potential customers based on big data technology and artificial intelligence model. Second, enterprises in key industries are required to cooperate with power grid companies to carry out the application of compensation mechanisms and promotion strategies such as "power substitution + demand response," "power substitution + energy storage," and "power substitution + energy service" based on energy Internet and intelligent control, so as to realize the interactive response between electric energy replacement load and the power grid, improve the flexibility and stability of power system, and reduce the energy cost to customers. The third is to encourage pilot units to carry out carbon emission reduction verification of power substitution projects and participate in national carbon emission reduction transactions.

(8) Adjust measures to local conditions to achieve differentiation. First, there are great differences in the surrounding environments of power substitution in different regions, so the policy of power substitution should be implemented according to the specific situation. The second is to focus on rural power substitution. In rural areas that are high coal areas, we should consider such factors as "going to the city to go upstairs," farmers' economic capacity, and other factors, combined with the rural natural resources conditions, relying on the beautiful rural construction, optimize the alternative route of electric energy, such as guiding the centralized area to adopt the centralized electric heating technology route, so as to avoid investment waste. Third, according to the demand for diversified energy, various regions can adopt a method of combining electricity and multiple energy supply to carry out alternative work.

REFERENCES

[1] Design of school power alternative scheme based on load decomposition. Li Yong, Zhang Xu. Zhejiang electric power. 2018 (10).
[2] Research on comprehensive evaluation model of power substitution based on cost utility analysis. Zhou Qian, he Qing, Du Dongmei. Power and energy. 2018 (04).

[3] Economic analysis of electric energy substitution for electric cookers based on multi-agent. Fu chengran, Chen Rongjun, he Yongxiu. Northeast electric power technology. 2018 (04).

[4] Power substitution analysis and Strategy Research of new residential quarters in Beijing. Gao Xin, Du Aixia, Hu caie, Wang Jian, Ren jiancong. Power supply and consumption. 2017 (08).

[5] Research on post evaluation of power substitution projects. Ren Xueliang, Mou Qiang, pan Guangxu, Liu Weidong, Yang Weijin, Li Yizhen, sun Xinghua. Energy saving and environmental protection. 2017 (04).

[6] Research on electric energy substitution space of cooking utensils in Gansu Province. Jiao Jie, he Yongxiu. Power demand side management. 2017 (02).

[7] Empirical Study on power substitution based on cost utility analysis. Zeng Qingxin. Enterprise technology development. 2016 (26).

[8] Empirical Study on power substitution based on cost utility analysis. Yan Qingyou, Zhu Mingming, Tang Xinfa. Operation research and management. 2015 (06).

[9] Market promotion strategy and implementation of power substitution. Zhang Jing, Zhang Jianzhong. Power demand side management. 2015 (01).

[10] Qing Zhongfa, he Xianyu. Analysis and Research on the economy and environmental protection of all electric kitchen. Power demand side management, 2020, 22 (03): 33–37.

Computational Social Science – Luo, Ciurea & Kumar (eds)
© 2021 Taylor & Francis Group, London, ISBN 978-0-367-70193-2

Influence of entrepreneurial education on the entrepreneurial intentions of university students majoring in sports: A case study from Guangdong Province, China

T. Kang
Sun Yat-sen University, Guangzhou, Guangdong, China

D.D. Yang
Shinawatra University, Bangkok, Thailand

ABSTRACT: This article aims to improve the entrepreneurial intentions of sports college students and mobilize the initiative of sports college students' self-employment, and to settle the factors of entrepreneurial education influencing the entrepreneurial intentions of university students majoring in sports in Guangdong Province. This research used a questionnaire to collect data from sports college students at universities in Guangdong Province. The research employed multiple regressions to test hypothesis. The main research conclusions were as follows: (1) Student attitude could effectively promote the entrepreneurial intentions of sports college students. Entrepreneurial education participation had a significant positive impact on sports college students' entrepreneurial intentions. (2) Education method can effectively promote the entrepreneurial intentions of sports college students. Entrepreneurial planning contests and entrepreneurial lectures have a significant positive impact on sports college students' entrepreneurial intentions. (3) Investments in educational resources can effectively promote the entrepreneurial intentions of sports college students.

1 INTRODUCTION

Innovation and entrepreneurial activities are the foundation of social and economic development (Shumpeter, 1936). China has begun to bring innovation and entrepreneurial actions to the peak of its national strategy. In the process of cultivating the national innovation system, entrepreneurial education must start the process. Only through the development of entrepreneurial education can talents be cultivated with innovative awareness and entrepreneurial spirit, essentially to promote innovation and entrepreneurial activities throughout the country (Zhang & Nie, 2006). Under this background, entrepreneurial education has received unprecedented attention. Therefore, how to carry out entrepreneurial education and cultivate entrepreneurial talents has become a hot topic.

Guangdong is an economic province which has abundant entrepreneurial resources. Driven by China's entrepreneurial background, it encourages university students to actively start their own businesses. There is a difference between sports college students and other college students: sports cultivate a spiritual quality of hard work. Not only do they possess solid professional skills, but they also have rich theoretical knowledge and comprehensive qualities. It integrates "body" and "education" ingeniously. In order to adapt to the requirements of society, it is necessary to combine the professional skills of sports college students with reality. It has become a priority condition for their business, and it is very important to uncover the entrepreneurial intentions of sports college students. Entrepreneurial intentions can be cultivated through entrepreneurial education. Therefore, it is necessary to improve the entrepreneurial education system by continuously deepening the reforms of entrepreneurial education in sports colleges.

2 RESEARCH DESIGN

2.1 Research object

This research selected four universities in Guangdong Province, China: Guangzhou Sport University, South China Normal University, Shao guan College, and Ling nan Teachers College. The university students majoring in sports of four universities as the research object were the subjects. Then, the Yamane formula was used to calculate the sample size of 396 students.

2.2 Research questions

(1) What is the relationship of entrepreneurial education and entrepreneurial intentions? (2) What are the main factors that affect the entrepreneurial intentions of university students majoring in sports? (3) How can the entrepreneurial intentions of university students majoring in sports be improved?

2.3 Variables

(1) Independent variable. In the theory of planned behavior, there are three elements: behavior attitude, subjective norm, and perception behavior control. Basically, the independent variables are intended to be student attitude, education method, and investment in educational resources.

Student attitude included entrepreneurial education participation and entrepreneurial education satisfaction. The education method included entrepreneurial course teaching, entrepreneurial planning contests, and entrepreneurial lectures. Investment in educational resources included teacher quality, leadership attention, financial support, and hardware facilities.

(2) Dependent variable. Many scholars constructed a multidimensional theoretical model of entrepreneurial intention to analyze the factors that affect intention; this research has not been scientifically verified and has not been widely applied. Therefore, this research still studies entrepreneurial intention as a single variable.

2.4 Research hypothesis

H1: Student attitude has a significant impact on entrepreneurial intentions.

SubH1a: Entrepreneurial education participation has a significant impact on entrepreneurial intentions. SubH1b: Entrepreneurial education satisfaction has a significant impact on entrepreneurial intentions.

H2: Education method has a significant impact on entrepreneurial intention.

SubH2a: Entrepreneurial course teaching has a significant impact on entrepreneurial intentions. SubH2b: Entrepreneurial planning contests have a significant impact on entrepreneurial intentions. SubH2c: Entrepreneurial lectures have a significant impact on entrepreneurial intentions.

H3: Investment in educational resources has a significant impact on entrepreneurial intention.

SubH3a: The quality of teachers in entrepreneurial education has a significant impact on entrepreneurial intentions. SubH3b: Leadership attention has a significant impact on entrepreneurial intentions. SubH3c: Financial support has a significant impact on entrepreneurial intentions. SubH3d: The hardware facilities of entrepreneurial education have a significant impact on entrepreneurial intentions.

2.5 Research method

This research used a questionnaire survey to collect data from 396 sports college students who currently were studying in four different universities. Regression analysis was used to explore the relationship between entrepreneurial education and entrepreneurial intentions, then the hypotheses were tested.

Table 1. Regression analysis results of student attitude on entrepreneurial intentions.

Variable	Model 1a		Model 1b		Model 1c	
	Beta	t	Beta	t	Beta	t
Gender	.079	.1.521	.080	.178	.084*	.1.993
Profession	.130*	2.478	.138**	3.026	.146**	3.396
Educational background	.228***	4.343	.251***	5.480	.266***	6.172
Entrepreneurial education received Received	.214***	4.105	.022	.437	.074	.1.551
Student attitude			.535***	11.133		
Entrepreneurial education participation					.588***	10.157
Entrepreneurial Education satisfaction					.620	1.109
F	7.648***		32.882***		39.078***	
Adjusted R^2	.065		.293		.372	
ΔR^2	.074		.228		.308	

2.6 Data analysis

Data obtained from questionnaires was coded and decoded in the SPSS.25 software program for computation. The researcher employed regression analysis to test data variables and the hypotheses.

3 RESULTS AND ANALYSIS

3.1 Regression analysis of the influence of students' attitude on entrepreneurial intentions

It can be seen from Table 1 that the regression coefficient of the overall student attitude affecting entrepreneurial intentions is 0.535 (P<0.001), indicating that the overall attitude of students significantly affects their entrepreneurial intentions positively, and verified hypothesis H1. The regression coefficient of entrepreneurial education participation affecting entrepreneurial intention is 0.588 (P<0.001), indicating that entrepreneurial education participation significantly affects entrepreneurial intentions, verifying hypothesis H1a. The regression coefficient of entrepreneurial education satisfaction affecting entrepreneurial intentions is 0.620 (P=0.268), indicating that entrepreneurial education satisfaction has no significant effect on entrepreneurial intention. Hypothesis H1b has not been verified.

3.2 Regression analysis of the influence of education method on entrepreneurial intentions

It can be seen from Table 2 that the regression coefficient of the education method overall impact on entrepreneurial intentions is 0.465 (P<0.001), indicating that the education method overall significantly and positively affects entrepreneurial intentions, and verified hypothesis H2. The regression coefficient of entrepreneurial course teaching affecting entrepreneurial intentions is 0.90 (P=0.274), indicating that entrepreneurial course teaching has no significant effect on entrepreneurial intentions. Therefore, hypothesis H2a has not been verified. The regression coefficient of entrepreneurial planning contests affecting entrepreneurial intentions is 0.172 (P<0.01) and the regression coefficient of entrepreneurial lectures affecting entrepreneurial intention is 0.233 (P<0.01), indicating that entrepreneurial planning contests and entrepreneurial lectures significantly influence entrepreneurial intentions. Therefore, hypotheses H2b and H2c are verified.

3.3 Regression analysis of the influence of investment in educational resources on entrepreneurial intentions

It can be seen from Table 3 that the regression coefficient of overall investment in educational resources affecting entrepreneurial intentions is 0.402 (P<0.001), indicating that the overall investment in educational resources significantly affects the entrepreneurial intentions positively, and

Table 2. Regression analysis results of education method on entrepreneurial intentions.

Variable	Model 1a		Model 1b		Model 1c	
	Beta	t	Beta	t	Beta	t
Gender	.079	.1.521	.044	.942	.042	.890
Profession	.130*	2.478	.132	1.806	.135*	2.861
Education background gBackground	.228***	4.343	.160**	3.351	.160**	3.222
Entrepreneurial education Education Received	.214***	4.105	.030	.588	.034	.665
Education method			.465***	9.687		
Entrepreneurial course teaching					.090	1.095
Entrepreneurialplan contests Contests					.172**	2.827
Entrepreneurial lectures Lectures					.233**	2.689
F	7.648***		26.378***		18.908***	
Adjusted R²	.065		.248		.246	
ΔR²	.074		.074		.185	

Table 3. Regression analysis results of investment in educational resources on entrepreneurial intentions.

Variable	Model 1a		Model 1b		Model 1c	
	Beta	t	Beta	t	Beta	t
Gender	.079	.1.521	.041	.868	.043	.904
Profession	.130*	2.478	.081	1.683	.100	1.968
Education background	.228***	4.343	.137**	2.775	.150**	2.913
Entrepreneurial education received	.214***	4.105	.095	1.893	.115*	2.238
Investment in educational resources Resource resources			.402***	8.393		
Teacher quality					.128*	2.520
Leadership attention					.055	.672
Financial support					.179**	2.985
Hardware facilities					.178	1.836
F	7.648***		21.322***		13.906***	
Adjusted R²	.065		.209		.211	
Δ R²	.074		.145		.154	

verifies hypothesis H3. The regression coefficient of leadership attention affecting entrepreneurial intentions is 0.055 (P=0.502), and the regression coefficient of hardware facilities affecting entrepreneurial intentions is 0.178 (P=0.067), indicating that leadership attention and hardware facilities have no significant effect on entrepreneurial intentions. So, hypothesis H3b and H3d has not been verified. The regression coefficient of teacher quality affecting entrepreneurial intentions is 0.128 (P<0.05), and the regression coefficient of financial support affecting entrepreneurial intentions is 0.179 (P<0.01), and so hypotheses H3a and H3c are verified.

4 CONCLUSION

Based on the relevant literatures with entrepreneurial education and entrepreneurial intentions both home and abroad, this research has built a theoretical model of entrepreneurial education on entrepreneurial intentions. We discussed the influence of entrepreneurial education on entrepreneurial intentions of sports college students. The main research conclusions were as follows: (1) Student attitude could effectively promote the entrepreneurial intentions of sports college students. (2) Education method could effectively promote the entrepreneurial intentions of sports college students. Sports college students possess different professional skills, and by participating in entrepreneurial planning contests and entrepreneurial activities, they use entrepreneurial practice platforms to demonstrate their own entrepreneurial skills.

(3) Investment in educational resources could effectively promote the entrepreneurial intentions of sports college students. The higher the quality of teachers, the deeper the entrepreneurial knowledge and understanding of the students. They combine the actual entrepreneurial situation of and its disadvantages while teaching.

REFERENCES

[1] Ajzen, I. 1991. The theory of planned behavior. Organizational behavior and human decision processes, 50(2), 179–211.
[2] Bird, B. 1988. Implementing entrepreneurial ideas: The case for intention. Academy of management review, 13(3), 442–453.
[3] Bandura, A. 1989. Human agency in social cognitive theory. American psychologist, 44(9), 1175.
[4] Chen Jun. 2007. Research progress in social cognitive theory. Journal of Social Psychology, Vol.22 (1–2):59–62.
[5] Crant, J.M.1996. The proactive personality scale as a predictor of entrepreneurial intentions. Journal of small business management, 34, 42–49.
[6] Feng Xiaotian. 2009. Sociological research methods. Beijing: Renmin University of China Press.
[7] Fishbein, M. 1963. An investigation of the relationships between beliefs about an object and the attitude toward that object. Human relations, 16(3), 233–239.
[8] Kirzner, I. M. 1997. Entrepreneurial discovery and the competitive market process: An Austrian approach. Journal of economic Literature, 35(1), 60–85.
[9] Pittaway, L. and Cope, J. 2007. Entrepreneurial education: A systematic review of the evidence. International small business journal, 25(5), 479–510.

Session 3. Cultural studies and humanities

Computational Social Science – Luo, Ciurea & Kumar (eds)
© *2021 Taylor & Francis Group, London, ISBN 978-0-367-70193-2*

The impact of Internet fiction reading on adolescents' self-concept clarity: The role of narrative transportation

D.J. Zhang*
School of Marxism, Huazhong University of Science and Technology, Wuhan, China
Key Laboratory of Adolescent Cyberpsychology and Behavior (CCNU), Ministry of Education, Wuhan, China

Z.C. Lei
School of Marxism, Huazhong University of Science and Technology, Wuhan, China

Z.K. Zhou
Key Laboratory of Adolescent Cyberpsychology and Behavior (CCNU), Ministry of Education, Wuhan, China

ABSTRACT: This study was aimed at investigating whether Internet fiction reading had a negative impact on adolescents' self-concept clarity, and tested the mediating role of narrative transportation. A sample of 745 adolescent students (366 females, 379 males; mean age = 14.47, SD age = 1.56 years, range: 10–17 years) were recruited. They completed questionnaires including Internet fiction reading scale (frequency and engagement), self-concept clarity scale, and narrative transportation scale. The results showed that Internet fiction reading could negatively influence self-concept clarity significantly, and totally through the mediating effects of narrative transportation. The present findings suggested the importance of Internet fiction reading and narrative transportation for adolescents' self-concept development in China.

Keywords: Internet fiction reading, Self-concept clarity, Narrative transportation.

1 INTRODUCTION

With the rapid development of technology and society, Internet usage has become increasingly rich and deep. Internet fiction, as a new type of entertainment activity online, has become one of the main activities for adolescents in China. According to the 45rd Chinese statistic report on Internet development (2020), approximately 4.52 billion people were reading Internet fiction, and this number is continuing to increase with the rapid growth of mobile device usage. In particular, adolescents are quickly becoming the main consumers of Internet fiction, with 38.3% of them reading fiction on the Internet (;NNIC, 2020; Zhang et al. 2017). At the same time, studies about narrative persuasion demonstrated story-telling could change individual attitude, values, and perceived self-concept, as Van Laer et al. (2019) used a meta-analysis to demonstrate the persuasive effect of digital storytelling. What's more, adolescence is an important period for individual self-concept development, especially a large number of studies claiming that teenagers' Internet usage affects their self-concept development (Omar et al. 2014; Zhou et al. 2017).

According to the self-concept fragmentation hypothesis, the Internet makes it easier for an individual to shape a variety of possible selves, with more possibilities for the individual to envisage various ideas, values, and thoughts, which are likely to change their personality, and increase the risk of self-concept confusion (Valkenburg & Peter 2008, 2011). Internet fiction refers to the novels displayed on the Internet or a digital device, that have not been subject to professional editing, and that are posted on the web by author (Zhang et al. 2017). On the one hand, the

---

*Corresponding author

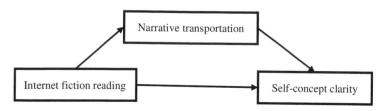

Figure 1.   The hypothesis model.

Internet fiction content, as a kind of leisure reading, is full of daydream-like, fantastical, and unreal characters (with super powers, legendary experiences, and so on), which allows adolescents to lose themselves in the described world easily. On the other hand, similar to Internet games, Internet fiction provides young people with an "idealized world" where they are able to experience other people's lives by reading, which is for adolescents can be an attractive illusion. These self-images can cause issues in real life, causing self-concept confusion. Thus, Internet fiction reading can have a negative influence on adolescents' self-concept. The first question of this study is to discover whether Internet fiction reading influences adolescents' self-concept clarity, and therefore whether the fragmentation hypothesis is supported or not?

Transportation theory (Green & Brock 2000) refers to a process entailing emotional engagement, generation of mental imagery, and attention while reading stories. Narrative transportation is the main impact mechanism researchers focused. . Several empirical evidences in the fields of consumer psychology highlight narrative transportation as a mediating mechanism of persuasion (Hamby et al. 2016; Thompson et al. 2018; Zhang et al. 2017). Transportation into a narrative world is similar to having a real traveling experience, becoming lost in the story and unable to realize what is happening in the real world (Escalas 2004). One study into Internet fiction addiction demonstrated that narrative transportation acted as a mediating role between a neurotic personality and Internet fiction addiction of Chinese college students (Zhang et al. 2017). Thus, the second hypothesis of this study was that narrative transportation played a mediating role between the relationship of Internet fiction reading and self-concept clarity of Chinese teenage students.

In conclusion, the present study theoretically provides a new perspective for studying the association between Internet fiction reading and adolescents' self-concept clarity, which was based on the self-concept fragmentation hypothesis. This study investigated the relationship between adolescents' Internet fiction reading and students' self-concept clarity, and the underlying mediating mechanism. First, we examined whether Internet fiction reading impacted self-concept clarity positively or negatively. Second, we examined whether the relationship between Internet fiction reading and self-concept clarity would be mediated by narrative transportation. The hypotheses model is shown in Figure 1.

2   METHOD

2.1   Participants and procedure

We recruited participants from two middle schools during the spring semester of 2018. Four classes were selected in each grade by convenience sampling. A total of 745 students (366 females and 379 males) participated in the study, with 491 having had an experience of reading Internet fiction. Students' age ranged from 10 to 17 years (mean age = 14.47 years, SD = 1.56 years) in this study.

2.2   Measures

2.2.1   Internet fiction reading
Based on the social networking and entertainment involvement scale (Ellison et al. 2007; Lin et al. 2016), the Internet fiction reading scale consists of two parts: reading frequency and involvement.

Reading frequencies include fiction reading years and numbers. The involvement scale includes five items, such as "I invested a lot of time and energy on reading Internet fiction." We asked participants to indicate their agreement with each question on a five-point scale (1 = strongly disagree, 5 = strongly agree). Scores on these items were centralized, and then summed to a total score, with higher scores indicating higher reading experience. Cronbach's α value for this measure was 0.88.

2.2.2 Self-concept clarity

Self-concept clarity was assessed using Campbell's scale (Campbell 1996). Participants answered 12 items on a five-point scale (1 = never, 5 = always). A higher average score indicated lower self-concept clarity. This measure had good reliability and validity in previous relevant studies in Chinese adolescents (Liu et al. 2017; Niu et al. 2016). Cronbach's α value for this scale was 0.83 in this study.

2.2.3 Narrative transportation

Narrative transportation was assessed by Green's scale (Green & Brock 2000). This scale contained three dimensions: affect, cognitive, and imaginative. Participants answered 10 items on a five-point scale (1 = never, 5 = always). This measure had good reliability and validity for adolescents in previous relevant studies (Bourgeon et al. 2019; Osanai, & Kusumi 2016). Cronbach's α value for this study was 0.87.

2.3 Statistical analysis

First, we conducted the analyses of descriptive statistics, the independent samples T test, and correlation. Next, a simple mediation model analysis (PROCESS SPSS macro, model 4; Hayes, 2017) was conducted to test the proposed model (provided visually in Figure 1). In addition, the SPSS macro PROCESS (Hayes 2017) has been widely used and was appropriated to test the simple mediation model in various studies (Fadoir et al. 2019; Niu et al. 2016). We also controlled the adolescents' age and gender before the regression analysis with supervisors.

3 RESULTS

3.1 Descriptive statistics and correlation analyses

Descriptive statistics analyses found that 491 students (65.9%) reported they had experienced reading Internet fiction, and the average reading time was 1.71 ± 1.91 years and the average number of books read was 21.57 ± 11.65. The independent samples T test showed that there was a significant difference between students who had read Internet fiction and students who had never read Internet fiction ($t = -2.85$, P < 0.05), as students who had reading experience reported lower self-concept clarity in this study. The correlation analyses found that Internet fiction reading was negative correlated with self-concept clarity ($r = -0.19$, P < 0.000), and positively correlated with narrative transportation ($r = 0.48$, P < 0.000). The narrative transportation was negative correlated with self-concept clarity ($r = -0.24$, P < 0.000). The descriptive statistics and correlations of all measures are presented in Table 1.

3.2 The mediating model analysis

As Internet fiction reading, narrative transportation, and self-concept clarity were significantly correlated with each other, we used the mediating model to test whether the narrative transportation mediated the relationship between the Internet fiction reading and self-concept clarity. The SPSS macro PROCESS was used in this study. First, after controlling gender and age as conditions, the regression analysis found that Internet fiction reading positively predicted the narrative

Table 1. Means, standard deviations, and correlations.

	M	SD	1	2	3	4	5
Age	14.47	1.56	1				
Gender	–	–	–0.02	1			
Internet fiction reading	–0.15	0.70	0.33***	0.10*	1		
Narrative transportation	2.78	0.78	0.15	0.14**	0.48***	1	
Self-concept clarity	2.02	0.73	–0.23***	–0.10**	–0.19***	–0.24***	1

Note.*P < 0.05. **P < 0.01, ***P < 0.001.

Table 2. The regression analysis.

Model			Fitting index			Regression coefficient	
	Dependent	Predictor	R	R2	F	β	t
Model 1	Gender	SCC	0.30	0.09	21.44***	–0.17	–3.21**
	Age					–0.06	–5.25***
	IFR					–0.11	–2.77*
Model 2	Gender	NT	0.49	0.24	51.77***	0.17	2.80**
	Age					0.01	0.23
	IFR					0.56	11.91***
Model 3	Gender	SCC	0.34	0.11	15.75***	–0.11	–1.91***
	Age					–0.10	–5.06*
	IFR					–0.3	–0.74
	NT					–0.18	–4.36***

Note. N = 495. Each column is a regression model that predicts the criterion at the top of the column. Gender was dummy coded such that 0 = male and 1 = female. IFR = Internet fiction reading; NR = narrative transportation; SCC = Self-concept clarity; *P < 0.05. **P < 0.01, *** P < 0.001.

transportation (β = 0.56, P < 0.000) and negatively predict the adolescents' self-concept clarity (β = –0.11, P < 0.05) in models 1 and 2. Then, for model 3, the narrative transportation and Internet fiction reading were added as independent variables together, and the regression analysis found that Internet fiction reading did not significantly predict the adolescents' self-concept clarity (β= –0.04, P > 0.05), but the narrative transportation predicted the adolescents' self-concept clarity (β= –0.18, P < 0.000). Therefore, Internet fiction reading through narrative transportation influenced the self-concept clarity. The regression analysis results are presented in Table 2.

4 DISCUSSION

This study extended the research into adolescents' Internet fiction reading behavior. Although the relationship between adolescents' Internet usage and self-concept clarity is well established in the aspects of social network and games (e.g., Appel et al. 2018; Niu et al. 2016; Omar et al. 2014; Quinones, & Kakabadse 2015; Zhou et al. 2017), the link between Internet fiction reading behavior and self-concept clarity remains unclear, and the mechanism underlying this association is also unclear. What's more, to our knowledge, this is the first study to investigate the role of narrative transportation in the association between adolescent' Internet fiction reading and self-concept clarity in China.

As expected, the independent samples T test results showed there was a significant difference between students who had read Internet fiction and students who had not (t =–2.85, P < 0.05) in their self-concept clarity. That means that adolescents with high fiction reading frequency showed lower levels of self-concept clarity, supporting the self-concept differentiation hypothesis. In addition, Internet fiction reading and narrative transportation were both negatively correlated with self-concept clarity. These results indicate that adolescents who like reading fiction on the Internet have a greater risk of self-concept confusion through novels. An earlier study also demonstrated the narrative transportation effect in the digital era (Van Laer et al. 2019), and another that people became lost in the narrative when reading (Golke et al. 2019; Shedlosky-shoemaker et al. 2014). Consistent with Valkenburg and Peter (2008, 2011) there is a proposal that Internet fiction can make individuals more easily shape a variety of possible selves, and increase their risk of self-concept confusion. Taking a further step beyond previous studies, we found that adolescents who were highly involved in Internet fiction reading were less likely to clearly and confidently describe who they were.

Consistent with our expectations, we found the effect of Internet fiction reading on self-concept clarity was totally through narrative transportation. According to transportation theory (Green 2000), when teenagers read Internet fiction they are attracted to the characters, fantastical world, or superpowers of these stories, and become lost in the unreal world, becoming unconscious of their real lives, and avoid the difficulties or negative experiences in their daily lives. At the same time, previous studies have confirmed narrative transportation in the digital commercial and market aspects. The mediating role of narrative transportation in Internet fiction reading is a new way to explain how fiction influences individual self-concept. Also, this result has reminded the government and education-related authorities of the high importance of Internet literature's influence on teenagers.

This study also had some limitations that future research could address. First, this study used a cross-sectional design to investigate the relationship between Internet fiction reading and adolescent' self-concept clarity. In future research, longitudinal and experimental designs would help researchers disentangle causal links between adolescents' Internet fiction reading and self-concept clarity to better investigate the direction of effects. Second, although this study showed significant relationships among Internet fiction reading, self-concept clarity, and narrative transportation in adolescence, we cannot exclude other related variables in the psychological selfconstruction process with online reading behavior. Based on the previous limitations, future research could examine the role of other potential variables (narrative and reader features), in terms of both structural and personality variables, to better understand the complicated reading mechanism. Third, self-report measures were used in this study, and therefore multiple informant sources should be used in future studies.

In conclusion, this study focused on the Internet activity of adolescents' fiction reading behavior. The results found that reading behavior can make teenagers lose their sense of reality, as they become transported into the stories. This study has contributed to highlighting the link between Internet fiction reading and self-concept clarity, suggesting an important direction for future study. The psychological mechanism of how Internet fiction affected individual self-construct will help researchers and those in education (including parents and teachers) pay greater attention to teenager's online reading behavior.

5 CONCLUSION

In conclusion, the present study theoretically provides a new perspective for studying the association between Internet fiction reading and adolescents' self-concept clarity based on the self-concept fragmentation hypothesis. The main results were as follow:

(1) Internet fiction reading was negatively associated with adolescent' self-concept clarity significantly.
(2) Narrative transportation played the mediating role between the relationship of Internet fiction reading and self-concept clarity in Chinese adolescents.

ACKNOWLEDGMENT

This article is one of the final achievements of a project supported by Basic Research Fund for Universities in China (Project ID: 2019WKYXQN025). The president of this project is Dong-jing Zhang.

REFERENCES

Appel, M., Schreiner, C., Weber, S., Mara, M., & Gnambs, T. (2018). Intensity of Facebook use is associated with lower self-concept clarity: cross-sectional and longitudinal evidence. Journal of Media Psychology Theories Methods & Applications.

Bourgeon, D., Maud, D., Jarrier, E., & Christine, P. (2019). Narrative transportation and transmedia consumption experience in the cultural field. International Journal of Arts Management, 21(2), 21–42.

CNNIC. (2020a). The 45rd Chinese statistic report on Internet development. 2020-4-28, from https://www.cnnic.com.cn/hlwfzyj/hlwxzbg/hlwtjbg/202004/P020200428596599037028.pdf.

CNNIC. (2020b). The 2019 national juvenile Internet usage report. 2020-5-13, from https://www.cnnic.com.cn/hlwfzyj/hlwxzbg/qsnbg/202005/P020200513370410784435.pdf

Ellison, N. B., Steinfield, C., & Lampe, C. (2007). The benefits of Facebook "friends": Social capital and college students' use of online social network sites. Journal of Computer-Mediated Communication, 12 (4), 1143–1168.

Escalas, J. E. (2004). Narrative processing: building consumer connections to brands. Journal of Consumer Psychology, 14(1), 168–180.

Fadoir, N. A., Lutz-Zoi, G. J., & Goodnight, J. A. (2019). Psychopathy and suicide: The mediating effects of emotional and behavioral dysregulation. Personality and Individual Differences, 142(1–6).

Golke, S. , Hagen, R. , & Wittwer, Jörg. (2019). Lost in narrative? the effect of informative narratives on text comprehension and metacomprehension accuracy. Learning and Instruction, 60, 1–19.

Green, M. C., & Brock, T. C. (2000). The role of transportation in the persuasiveness of public narratives. Journal of Personality & Social Psychology,79(5), 701.

Hamby, A., Brinberg, D., & Daniloski, K. (2016). Reflecting on the journey: mechanisms in narrative persuasion. Journal of Consumer Psychology, S105774081630050X

Hayes, A. F. (2017). Introduction to mediation, moderation, and conditional process analysis: A regression based approach (2nd ed.). New York, NY: Guilford Press.

Lin, J. S., Sung, Y., & Chen, K. J. (2016). Social television: examining the antecedents and consequences of connected TV viewing. Computers in Human Behavior, 58, 171–178.

Liu, Q. Q., Niu, G. F., Fan, C. Y., & Zhou, Z. K. (2017). Passive use of social network site and its relationships with self-esteem and self-concept clarity: A moderated mediation analysis. Acta Psychologica Sinica, 49(1), 60–71.

Niu, G. N., Sun, X. J., Zhou, Z. K., Tian, Y., Liu, Q. Q., & Lian, S. L. (2016). The Effect of Adolescents' Social Networking Site Use on Self-concept Clarity: The Mediating Role of Social Comparison. Journal of Psychological Science, 39(1), 97–102

Omar, S. Z., Daud, A., Hassan, M. S., Bolong, J., & Teimmouri, M. (2014). Children internet usage: opportunities for self development ☆. Procedia - Social and Behavioral Sciences, 155(155), 75–80.

Osanai, H., & Kusumi, T. (2016). Reliability and validity of the narrative transportation scale in Japanese. Japanese Journal of Personality, 25(1), 50–61.

Quinones, C., & Kakabadse, N. K. (2015). Self-concept clarity, social support, and compulsive internet use: a study of the US and the UAE. Computers in Human Behavior, 44(44), 347–356.

Shedlosky-shoemaker, R., Costabile, K. A., & Arkin, R. M. (2014). Self-expansion through fictional characters. Self & Identity,13(5), 556–578.

Thompson, J. M., Teasdale, B., Duncan, S., Evert, V. E. B., Budelmann, F., & Maguire, L., et al. (2018). Individual differences in transportation into narrative drama. Review of General Psychology, 22(2), 210–219.

Van Laer, T., Feiereisen, S., & Visconti, L. (2019). Storytelling in the digital era: A meta-analysis of relevant moderators of the narrative transportation effect. Journal of Business Research, 96, 135–146.

Valkenburg, P. M., & Peter, J. (2008). Adolescents' identity experiments on the internet: consequences for social competence and self-concept unity. Communication Research, 35(2), 208–231.

Valkenburg, P. M., & Peter, J. (2011). Online communication among adolescents: an integrated model of its attraction, opportunities, and risks. Journal of Adolescent Health, 48(2), 121–127.

Zhang. D. J., Zhou, Z. K., Lei. Y. J., Niu. G. F., Zhu. X. W., & Xie, X. C. (2017). The relationship between neuroticism and internet fiction addiction of college students: the mediating effects of narrative transportation and flow experience. Journal of Psychological Science. 040(005), 1154–1160.

Zhou, Z. K., Niu, G. F., Liu, Q. Q., & Chen, W. (2017). Chapter 5 – internet use and self-development in Chinese culture. Boundaries of Self & Reality Online, 75–96.

Computational Social Science – Luo, Ciurea & Kumar (eds)
© 2021 Taylor & Francis Group, London, ISBN 978-0-367-70193-2

Characteristic practice of science and engineering in liberal arts colleges

P. Wang
Fuzhou University of International Studies and Trade, Fuzhou, China

ABSTRACT: This paper narrates the origins of the environment of development and issues about engineering courses in our university. It analyzes issues about how to improve personal training and correct faultiness in the current situation. According to specific engineering areas, using TCP/IP protocol principle and network programming for example, this paper discusses the methods about accelerating the process of electronic information engineering and states the reason why it is necessary and significant to improve our teaching methods. Finally, it concludes with the processes and achievements of this decade.

1   THE SUBJECT DEVELOPMENT

National prosperity depends on talents, talents prosperity depends on education, education prosperity depends on the teacher, and the teacher prosperity depends on normal education [1]. "Normal university" only literally should be understood to foster "just one teacher and model of the talent school," so the domestic various normal university usually has a long history, and the teaching standard in society has high credibility.

Fuzhou Institute of foreign languages and foreign trade was established with the approval of the Ministry of Education and has an independent issuing country. A private full-time regular undergraduate college with recognized academic qualifications, it is a foreign language business school featuring foreign languages, business management, commerce, electronic information technology, animation, and other major categories. A liberal arts-based institution at the beginning of the school, relying on the teachers of Fujian Normal University, has a high starting point and standardized running.

The school is positioned as an application-oriented university and participates in the personnel training of higher vocational colleges. In the process of transformation, how to run the applied engineering specialty well is our research the question.

2   CHARACTERISTICS CHANGE

Arts and engineering belong to different subject categories, as do heavy science theory and method, and heavy engineering technology and process. Engineering is the application of the principle of basic science, combined with the production practice of accumulated experience and technology development of subjects, the main cultivation of practical application ability of technical personnel.

The science teacher position will be responsible for the teaching and the task of knowledge, a process of training students on the theoretical basis. Development of instruments, writing, expression, and communication are requirements [2]. The engineer position will be responsible for project implementation and technology research and development tasks. Training process must pay attention to practical ability.

Only from the electronic information (EI) engineering specialty name, society will, because of the school's one hundred years of history, trust its quality of teaching. Engineering was born from the ability to solve practical problems.

3 ENGINEER TRAINING

In science, the focus of engineering majors is how to highlight the engineering characteristics, develop distinctive directions, and improve competitiveness? Our professional teachers should be diligent in thinking through problems and long-term effort direction.

In the early construction of electronic information engineering specialty, we focus on the laboratory construction, to the professional laboratory building area of 3,134 square meters, instruments, and equipment of RMB 7.8 million yuan. At the same time, constantly adjusting is different from the normal kind of curriculum structure [3], increasing the proportion of practice class, practice credits is greater than 30% of the total credit, in order to improve the employment rate in the professional direction and increase various modules for students to take as elective courses. The whole training process uses the series means to strengthen the students' beginning ability:

(1) Classes teacher: From freshman year began to shift to appoint professional teachers and graduate students for vice class teachers, mainly to help solve the problems of learning, science, and technology.
(2) Students' 10th science day: The 10th science day once a year is to decorate project tasks, display comparison of students' science and technology works, the commendation of the advanced individual. The purpose of this paper is to keep the fine tradition of institution.
(3) Innovation laboratory: To provide students with good science and technology activities, with independent projects of scientific research and participation in all kinds of high-level professional competitions as the means of cultivating engineering quality and innovation spirit.
(4) To promote teachers' scientific research: Encourage teachers' active declaration research (vertical and horizontal), only the teacher who first masters the new technology can teach good students, and at the same time, attract outstanding students to participate in scientific research projects and lead students in science and technology activities.
(5) Pay attention to the graduation design: Design work is divided into hardware design class, software design class, engineering design class, and teachers by the school and enterprise of two parts.

4 PROFESSIONAL CHARACTERISTICS

Electronic and information engineering is a broad caliber engineering major, the coherent main subject for electronic science and technology, information and communication engineering, computer science and technology. The professionals obtain employment in good conditions, in posts involved in telecommunications, energy, transportation, finance, economics, trade, military, science, education, social security, all levels of government agencies, and other fields. In the surrounding university are set up the professional situations, of how to improve the competitiveness? Professional characteristics are very important!

A feature is a thing or significant things different from other things in style or form, which is the things of the emergence and development of certain specific environmental factors: it is their unique things. The photoelectric and information engineering college, by some related discipline system, research centers, with the provincial, the Ministry of Education key laboratory, the related interdisciplinary rich experience in teaching faculty and outstanding scientific research platform, is the professional development that has laid a good foundation.

We keep the direction of the popularization of the basis of running with their history and resource advantages put forward: keep "computer technology of electronic and information engineering application and research and development," "management information technology based software engineering application and research and development," and focuses on the traditional "application of the embedded system and research and development" as the main direction of the characteristics. Popular explanation is based on embedded system in partial hardware training direction, said of the embedded system platform that contains a single-chip microcomputer technology, ARM technology, FPGA technology, DSP technology, virtual instrument technology on five hardware platforms.

5 TEACHING REFORM

Einstein once said: "interest is the best teacher." Electronic information or related professional knowledge updates quickly. If in students there is not this professional strong study interest, it is hard to imagine the students after graduation can meet the demand of talents society needs [4]. In the teaching process, it is the most important still to improve the students' interest in professional knowledge.

In today's rapid development of the information age, the electronic information technology at home and abroad and the development of science and technology play a very important role in the [5], the electronic information courses also has its own particular characteristics and nature of [6]. The Internet originated in the United States, is the biggest influence in the world, and the most closely related to people's lives. It changed people's daily communication and the communication mode of communication networks; the network is the TCP/IP protocol. Now, there are all kinds of computer networks, protocol types of many courses, although they have different emphasis but also including the TCP/IP protocol principle and network programming for the core principle about agreement, more important, in our school, this course is electronic and information engineering, network engineering major required courses, and many related professional elective courses.

The problem is that the translation versions of the Chinese teaching materials mostly exist in statements obscure, and are wrong in a lot of problems. And the core content of this course is composed of 0, 1 see touched agreement, teaching in some key points and difficulties, with the traditional teaching method of interpretation and is time-consuming, the effect is not good, should the study be supporting the corresponding teaching means, and the concrete measures are as follows:

(1) According to relevant course content repetition, revised curriculum system, to make the curriculum standard, the standard experimental program process, expand the course, each class hour will be cut down the network technology, the OSI protocol, integrated wiring technology part of the course content into the course teaching, which rises to digest the good effect.
(2) According to the teaching material is not ideal, it is only through the teachers more than their reading, their diligent practice. Teaching theory will be part of the focus on the OSI protocol and TCP/IP layer, application layer in the theory of interpretation teaching will focus on practical programming link. Generally speaking, a good course, at least after three rounds of the teaching experience, is a clear understanding to the teaching order and emphasis.
(3) In theory, how to contact the practice problems, due to the course is very suitable for the project drive method to improve the students' understanding and interest, the whole teaching processes are in various applications for projects, three people in a group, each network programming point of view is different, but in the end can be consolidated into a perfect application.
(4) According to the agreement which is difficult to understand, as shown in Figure 1, explaining in routing principle stage, the network simulation software to understand routing principle. As shown in Figure 2, the interpretation agreement stage uses caught analysis software confirmed agreement, so that the original relative tasteless 0, 1 agreement theory becomes lively and interesting.

Practice has proved that through the reform of teaching methods and means, students of the course have a strong interest in mastering digital communication system related agreement knowledge and theory, are familiar with the network programming methods and means, have strong analysis and

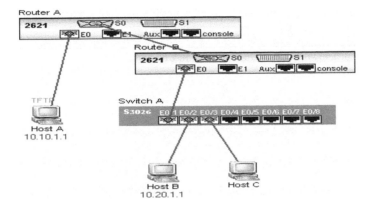

Figure 1. Network simulation software.

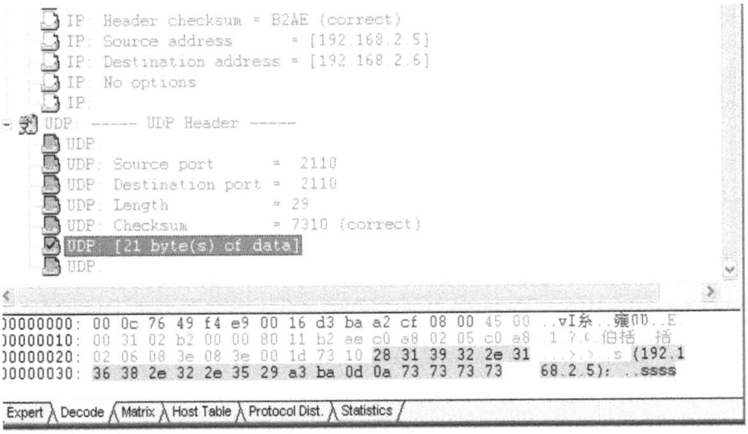

IP Header checksum = B2AE (correct)
IP Source address   = [192.168.2.5]
IP Destination address = [192.168.2.6]
IP No options
IP
UDP ----- UDP Header -----
UDP
UDP Source port   = 2110
UDP Destination port = 2110
UDP Length   = 29
UDP Checksum   = 7310 (correct)
UDP [21 byte(s) of data]
UDP

)0000000: 00 0c 76 49 f4 e9 00 16 d3 ba a2 cf 08 00 45 00   ..vI系..雍ΠΠ..E.
)0000010: 00 31 02 b2 00 00 80 11 b2 ae c0 a8 02 05 c0 a8   .1.?.(.伯括 括
)0000020: 02 06 08 3e 08 3e 00 1d 73 10 28 31 39 32 2e 31   . . > . > s (192.1
)0000030: 36 38 2e 32 2e 35 29 a3 ba 0d 0a 73 73 73 73 73   68.2.5): ..ssss

Expert \ Decode \ Matrix \ Host Table \ Protocol Dist. \ Statistics \

Figure 2. Caught analysis software.

the ability to solve practical problems, and cultivate a good professional quality and innovation consciousness.

6 CONSTRUCTION EFFECT

Through the joint efforts of all teachers, electronic information engineering construction has obtained certain results: in 2005, EI was upgraded to the school subjects (signal and information processing); also in 2005, communication and information systems, computer application technology, radio physics, and nuclear physics were approved for four master degree programs. In 2007, EI was assessed at a provincial electrical and electronic experimental teaching demonstration center; in 2008, it was assessed at provincial professional characteristics.

Of the more than 20 full-time teachers, those forty years old accounted for 48%, with the professional formal education teachers accounting for 85%, those with a masters degree or above accounted for 48% of teachers, those teachers with doctoral degrees accounted for 22%, and high title teachers accounted for 35%. In recent years the high-level academic papers published hundreds of articles, such as by EI and ISTP, and included international papers, patented items, and writing materials. Dozens of hosts were asked to participate in research projects for more items and the provincial science and technology progress prize.

In 10 years, electronic information engineering professional trained graduate students were 899 people, turnover rate was about 20%, an employment rate greater than 90%. The professional students of nearly 80 people won the national (province) "challenge cup," the national electronic design contest (province), winning the first, second, and third places.

7 CONCLUSION

Specialty construction is a long-term process. In the present size environment there may be many difficulties, which need all the teachers' hard work and selfless dedication, to only follow the law of engineering education, closely related social needs, and concise unique professional direction to highlight the characteristics of a school, realizing training subject basis of down-to-earth, practice ability of high-quality engineering senior talents training target [7].

ACKNOWLEDGMENT

Fund Project: Fujian Electronic Information Engineering Specialty Construction Point (L200802008); TCP/IP principles and network programming teaching methods and reform of teaching methods (I201007020).

REFERENCES

[1] Qu Tiehua, FanTao. Higher normal education system in the late qing dynasty free characteristics analysis. Journal of Hebei normal university, 2009, (3): 23–27.
[2] ZhangYanFen. Analyses the development of normal colleges recessive course. Journal of education theory and practice, 2010, 21 (3): 13–16.
[3] Soloing. In one hundred normal course cultural tradition and the characteristic. Journal of Jiangsu education research, 2010, (4:21–23).
[4] XiongJieFeng, electronic class undergraduate course teaching in teaching students according to their aptitude thinking. China science and technology information, 2008, (12): 252–253.
[5] QinLi, ZhangWenDong, XiongJiJun, Liu Jun, RenYongfeng, LiuWenyi. Electronic professional students' innovative and practical ability training mode analysis. Journal of bei university, 2008, (6): 27–29.
[6] Jiang Shuhua, Li Mingqiu, Zhang Chen clean. Electronics course examination system reform of the exploration and practice of science and technology information. J, 2009: 252–253.
[7] Wu Ling tin, and Zhan Jie, Zhou Renlong. Electronic information science and technology major curriculum system characteristics. Journal of contemporary education theory and practice, 2010, 2 (1): 105–107.

Computational Social Science – Luo, Ciurea & Kumar (eds)
© *2021 Taylor & Francis Group, London, ISBN 978-0-367-70193-2*

The integration of cultural confidence with Bachelor of Translation and Interpreting education in China

X.M. Qin
Xi'an Fanyi University, Xi'an City, China

ABSTRACT: This paper focuses on the fact that Chinese President Xi Jingping, on July 1, 2016, proposed that China must remain confident in the path, theories, system, and culture of Chinese socialism, of which cultural confidence is the foundation of the 95th anniversary conference of the founding of the CPC. Therefore, the establishment of cultural self-confidence is one of the main tasks of talent cultivation in colleges and universities in the new era, especially for students majoring in Translation and Interpreting. This paper presents the current situation of the cultural education in BTI (Bachelor of Translation and Interpreting) from the cultural confidence viewpoint. It also attempts to find the causes and gives suggestions from a talent program, curriculum, and teachers in order to cultivate students' correct cultural values and sense, which can make a contribution to their inheritance and the spreading of Chinese culture around the world.

1 INTRODUCTION

Chinese President Xi Jinping proposed "to enhance the confidence in Chinese cultural and values" on July 1, 2016. He also stated at the 95th anniversary conference of the founding of the CPC that we must remain confident in the path, theories, system, and culture of Chinese socialism, of which cultural self-confidence is the foundation. Therefore, the establishment of cultural self-confidence is one of the main tasks of talent cultivation in colleges and universities in the new era, especially for students majoring in Translation and Interpreting [1]. Translation, as a cross-cultural activity, plays an important role in cultural exchange and orientation, therefore, we needs to use the great foundation of humanistic knowledge and qualities, and first learn from China's excellent cultural traditions, to portray the consciousness of Chinese culture with self-esteem and self-confidence. Only in this way can we show the Chinese culture, values, and ideals to the world using foreign languages to get this message across.

2 BTI EDUCATION AND CULTURAL CONFIDENCE

The "Teaching Requirements for BTI Education" (hereinafter referred to as "Requirements") issued in 2011 set forth clear training objectives for the majors of English and Translation and Interpreting. They give a clear description of the quality of translation talents and an interpretation of the cultural needs of the training process of translation talents. Translation is the bridge between any two languages. It is not only the cognition and transformation between language symbols, but also the interpretation of the culture behind the language. Therefore, the main teaching task should not be only on the language, but also on the culture reflected by the language. A student of translation and interpreting should be familiar with the cultures in the English-speaking countries and also China. At the same time, he/she should be able to communicate with people in the two different cultures, which is called cross-cultural communication.

"Cross-cultural communication" is a topic that has always been a part of BTI education. The "Requirements" also clearly stipulate that teaching in BTI education should focus on cultivating students' quality of cross-cultural communication. This cannot be done without the input of Chinese language and culture. There can be a false idea that we need only the input of Chinese culture as we are always in the Chinese environment. Also, if only English culture is input, it can be described as one-way communication but not cross-culture communication. The Chinese input is also important in two-way communication, which must be contain the input from both English-speaking and Chinese cultures.

3   THE CURRENT SITUATION OF CHINESE CULTURAL INHERITANCE IN BTI EDUCATION

In the 1980s and 1990s, Mr. Fei Xiaotong put forward the concept of "cultural consciousness" with the background of a complex global cultural impact, that is, "people, living in a certain culture have 'self-knowledge' of their culture, can understand its origin, formation process, characteristics and development trend, without any meaning of cultural return " [2]. In 2000, the "aphasia of Chinese culture," a defect in English teaching in China, was first raised in Guangming Daily by Congcong, pointing out that with China's reforming and opening up, under the influence of the new upsurge of learning English, Chinese culture is in a "marginal state" in foreign language education. Some have discussed foreign language education policy issues as well as the Chinese cultural heritage and the cultural awareness in China [3,4]. At present, it is an undeniable fact that the Chinese language and cultural foundations are becoming increasingly weakened because of the "English first" and "Chinese zero situation" in teaching and curriculum structure.

4   INNOVATIVE APPROACHES FROM THE PERSPECTIVE OF CULTURAL CONFIDENCE

4.1   *Training program*

BTI education has made great achievements since its establishment, but it has also caused some problems. Zhong Weihe pointed out in "China Translation" in 2014 that there were seven major problems in China's BTI and MT, one of which was that it is inappropriate in the training program for BTI education. The point is that it is difficult to train the professional talents if there is not a clear training objective. It takes a long time to identify the differences between an English major and a major in Translation and Interpreting. Language, which is like a tool, can help us to communicate with others from different countries. Translation and interpreting ,act like a bridge, and can create a link to other countries, with culture playing a very important role. In the new era, cultural confidence is one of the main educational tasks of colleges and universities in China. Under this premise, we should re-examine the training program for BTI and clarify the goals from the perspective of cultural confidence. This will create for BTI a bright and unique future.

4.2   *Curriculum*

For a long time, "English first" was emphasized in translation teaching. It is a pity that there are few curricula about the Chinese language, Chinese culture, and Chinese studies. The students educated with this curriculum will only be able to carry out English translation as a limited tool, without the cultural transmission, which inevitably leads to a "deficit" [5]. BTI education should focus on the balance between Chinese and English. Efforts should be made to solve the problem of unbalanced language structure between translation teaching and learning. We can learn English well but "if it does not serve the native language, it cannot be converted into real Chinese information, and cannot achieve a balance with Chinese" [6], and the real value of English teaching and learning will be greatly diminished. In addition, students majoring in English or Translation and Interpreting, will have more opportunities to understand information from English-speaking countries that have different life styles and values. They would be affected by these different life style and values easily. Therefore, it is necessary to increase the establishment of related courses such as Chinese language, Chinese culture, and basic knowledge of Chinese studies. In the teaching process, the comparison of and induction to two languages and cultures should be continuously strengthened to gradually improve students' intercultural communication ability.

4.3 Teaching staff

Teachers play an important role in integrating cultural confidence with BTI education, and this cross-cultural ability should be strengthened. Teachers in BTI education not only need to teach the general language to students, but also to spread positively the culture and help students to have balanced and accurate cultural views. They, as a guide for students in learning, should not only pay attention to the basic skills of foreign language teaching, but also strengthen students' Chinese cultural self-knowledge and exert a positive influence on the establishment of students' accurate cultural views. For example, there is a lesson of "The Way to Rainy Mountain" taken from Advanced English, a core course of BIT education. It is an English essay with great difficulty in understanding cultural aspects. In order to enable students to have a better understanding of the complex emotions expressed in this essay, teachers can encourage students to have a discussion about the Chinese and Indian cultures. Students should then be allowed to independently rewrite the text as a script, shoot a micro-movie for about 10 minutes, upload it to the learning platform, and discuss and learn from the shooting effect and content. Through this activity, students have a good understanding of the author's pride, sense of loss, and sense of belonging in the text. At the same time, they would have a deep learning of Chinese and Indian cultures with certain aesthetic, reflective, and critical abilities.

5 CONCLUSION

Translation is regarded as one of the most important ways for the circulation and accumulation of cultural capital, with the important task of cultural exchange among all ethnic groups [7]. In the new era, China is showing its image to the world in the recent years. This requires us to tell Chinese stories well, use Chinese voices well, and show the world the real China. This is the social responsibility and historical mission of BTI education. In the process of training translation students, Chinese culture cannot be ignored in the study of learning and mastering English. The correct training target for BTI should be set up, with a reasonable curriculum, and teachers' active guidance, so that students' cultural confidence can be continuously enhanced during the cross-cultural learning in order to make them become truly qualified cross-cultural communicators.

REFERENCES

[1] Cai Yongliang. Linguistic Study and Translation Study in China's Drastic Turn. Foreign Language World. 172 (2016) 6–11.
[2] Fei Xiaotong. Culture and Culture Awareness, Qunyan Press, Beijing, 2007.
[3] Sun Juan. Native Chinese Culture and College English Teaching. Journal of Huzhou Vocational and Technological College. 10 (2007) 50–52.
[4] Zhu Zhengwu. On Cultural Confidence in Translation Efforts. Foreign Language Education. 37 (2016) 83–85.
[5] Tong Minqiang. The Translation Teaching and Chinese Cultural Literacy. Journal of Xi'an International Studies University.23 (2015) 112–115.
[6] Xu Hong, Li Wenge. The Influence of Discourse on Translation from Different Perspectives. Foreign Language Research. 181 (2014) 99–102.
[7] Zeng Wenxiong. The Manipulation of Translational Cultural Capital and Contextual Interference. Foreign Language Education. 33 (2012) 83–88.

Computational Social Science – Luo, Ciurea & Kumar (eds)
© 2021 Taylor & Francis Group, London, ISBN 978-0-367-70193-2

Study of the communication strategies of multinational internship in the era of social networks

Y. Liu
Wuhan University of Science and Technology, Wuhan, Hubei
Birmingham City University, Birmingham, UK

W. Tang & L. Li
Wuhan University of Science and Technology, Wuhan, Hubei

C. Millman
Birmingham City University, Birmingham, UK

P.Y. Zhang
Wuhan University of Science and Technology, Wuhan, Hubei

ABSTRACT: This paper discusses the communication and satisfaction issues during internship at three universities, Wuhan University of Science & Technology (WUST), Washburn University of U.S. (WU), and PXL Technology University of Belgium. It summarizes the communication efficiency of different communication methods, as well as the communication problems during the whole process. Through questionnaire analysis, including correlation analysis and regression analysis, it summarizes the main factors which influence internship satisfaction; these are process communication and guidance, company's support, and cultural and language differences. It provides some useful management strategies for better cross-cultural cooperation.

1 INTRODUCTION

Globalization and internationalization in the sphere of education are caused by growing demand for international economic activity, intercultural understanding caused by the global nature of global processes, and a means to exchange knowledge, technologies, and information, and form an objective dynamically developing process (Tetiana Obolenska, Olena Tsyrkun, 2016). This means that international cooperation of universities is crucial for higher education and social development. Currently China's international cooperation between universities includes teacher and student exchanges, transcript transferring, collaborative schools, cooperative research, international meetings, and so on. The international internship between WUST, WU, PXL, and other enterprises is a creative, challenging complex cooperation model, which has faced a lot of communication problems. This paper took three countries' international internship as the research case, discussing different stages of communication tools and their effects, and the factors' impacts on student satisfaction. In order to improve the satisfaction and better management strategies of international exchange and cooperation in the future, the value of international exchanges and cooperation, especially in the social media era, should be increased. Unlike earlier communication models, people in different countries can communicate with each other more conveniently and at lower cost thanks to the mobile communication technology development. At present, based on the mobile communication tools, cross-country experiences of students can cover many aspects, such as academic learning, personal life, and culture (Shenghong Liu, 2018). College students from different countries prefer to use different social communication software. Most European and American students like to use WhatsApp, but Chinese students prefer to use WeChat and QQ. WhatsApp is a direct and simple way to talk, whereas WeChat focuses on the personal relationship construction (Jin Xiaoling, Feng Huihui, Zhou Zhongyun, 2017).

Many authors have highlighted various aspects of the importance of internships performed during university studies. International education can not only improve undergraduates' comprehensive capability, but also provide educational benefits of increased international knowledge and intercultural skills for staff and overseas students (Knight, 2006). It also can provide both intellectual and cultural bene?ts to students which can help in their future employment [3]. The short-term international internship often has the potential to broaden, enrich, or augment student learning and personal development (Pence & Macgillivray, 2008). In the long term, this can lead to the development of human capital and increase the competitiveness fundamentals of a dynamic economy based on knowledge (Alina Simona, 2016).

In formal cross-culture communication, Mitra Madanchian and Hamed Taherdoost (2016) describe email as playing a crucial role in establishing and maintaining business relationships, both within a company and with external contacts. Web-meetings undoubtedly are one of quick ways to improve students' foreign language ability. Shi Weilin (2015) thought "cross-culture online communication" can solve the problems of Chinese lower-level students' foreign language speaking skills.

Hawes & Kealey (1981) considered Western educational approaches may not recognize the complexity of students' cultural backgrounds. Similarly, an interpersonal orientation that embodies respect and curiosity toward others is an essential characteristic of people ready to listen to others, become acquainted with them, and seek to understand their viewpoint. Perceptiveness and receptiveness, as well as the capacity for action derived from understanding, are essential elements in successfully acquiring cultural empathy (Cui & Van Den Berg, 1991; Batey & Lupi, 2012). Most Chinese experts found it very important to give cross-culture knowledge to students, which can reduce misunderstandings among the students from different countries (Tan Yu, 2015; Wang Qihe, Jiang Dehong, Guo Wenxia, 2016).

Julie Matthews and Meredith Lawley's (2011) findings showed that students were satisfied with their international education experience and that the internship/work-integrated learning experience enhanced their satisfaction, which is good way to increase culture understanding and blending, and to enrich the global educational system (Lv Yiwen, Liu Mingzhen, 2017).

International higher education leaders must recognize that student learning and developmental outcomes are accomplished through high-level engagement of students in both classroom and beyond the classroom experiences, as well as by infusing international perspectives into students' experience both at home and across borders (Roberts & Komives, 2016)

2 BACKGROUND TO INTERNATIONAL INTERNSHIP

2.1 *Communication models*

Communication was mainly among three universities professors and students, consisting of external communication and internal communication. Internal communication was between one country's professor and its students, and external communication was that between students or professors from different countries, and communication with companies. The whole process of internship is manifested in the Figure 1 as follows:

2.2 *Four stages of international internship*

(1) The first stage of internship: Grouping

Since 2007, WUST has begun cooperating with WU, and the first international internship between these two universities has started. In 2014, PXL University also entered into this project. At the start of the year professors from the three university start a web-meeting to discuss each

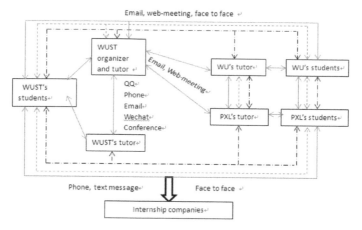

Figure 1. International internship communication model.

year's plan, and then begin to recruit students for the internship. When all the students are decided on, the three parties' professors exchange the students' information, name, gender, major, and so on through emails. According to the students' interests and majors, they are divided into different groups; each group had a foreign tutor, and WUST's tutor.

(2) The second stage: Defining research directions

After successful grouping, students and professors hold the first web-meeting together. Through this meeting, students get to know each other better (besides names and majors), they can see the other students' faces, and can talk to each other.

Each group gets to start their own project after the web-meeting, with WUST students being told more details of the visiting company, and then the WU and PXL students find similar companies in their countries, which can help them to make some comparisons in later research. In the meantime they exchange their enterprise investigation ideas, and discuss some problems they may face during the internship. They keep in touch with each other frequently; each group member invests a lot to time and effort in this internship.

At the same time, students talk to their tutors to find some research directions in order to match the company's requirements based on theory. Also in this stage, email and web-meetings are the mainly communication methods among students and tutors.

(3) The third stage: Field survey

WU and PXL professors and students go to WUST in the middle of May each year. The three universities' students finally can communicate with each other face to face. Each group goes to the target company with their tutors for about 5–7 days. They interview different members of staff in the company, talk with managers to acquire more information of the company, and prepare for the report.

(4) The fourth stage: Presentation

The final presentation is primarily a formal speech in front of WUST's Vice President, Dean of School of Business, the other department leaders of WUST, and students. Here they can show their half-year outcome, as well as it being a good opportunity to practice their oral skills. At the end of the presentation, all students are awarded certification and a gift from the other two universities.

3 QUESTIONNAIRE AND DATA ANALYSIS

3.1 Questionnaire design

According to a classified summary of the literature, combined with discussions with teachers, and open interviews with some of the students, to learn about the effectiveness of communication

Table 1. Communication tools effect's ranking in different stages.

Stages Ranking Tool	The first stage	The second stage		The third stage	The fourth stage
		Among professors	Among students		
Email	①	③	②	④	③
Web-meeting	②	——	①	——	——
QQ	⑤	②	——	⑤	⑤
WeChat/WhatsApp	⑦	⑤	——	⑥	——
Formal conference	③	④	——	②	①
Informal conference	④	①	——	①	②
Phone/text message	⑥	——	——	③	④

scheduling, we designed a questionnaire survey of 25 problems between enterprises and related personnel, teachers, and team members combined with practice. The problems are divided into several areas: the influence of cultural differences on the student satisfaction, communication and guidance for student satisfaction, and the effect of enterprise coordination on student satisfaction.

Enterprise support consists of two aspects: one is the leader's support, which means whether the leaders welcome students is heartfelt or just done as expected; the subordinate cooperation, which means whether the staff were happy to cooperate with the students, such as answering students' questions, helping students to figure the companies' situations, and so on.

The communication and guidance represents the tutors' help for students when students have problems during the internship.

Three hypotheses are proposed in this paper:

H1: cultural differences are negatively related to students' satisfaction, and the influence is significant.

H2: process communication guidance is positively correlated with student satisfaction, and has a significant impact.

H3: enterprise support is positively correlated with student satisfaction, and the impact is significant.

A total of 150 students in the joint practice were investigated, and after the questionnaire review, 138 questionnaires finally qualified. The reliability and validity of all the questionnaires were tested, and the reliability index value was 0.748 and greater than 0.7, which indicated that the reliability of the research data was very good. The KMO value was 0.81, the Bart's sphericity was 2316.286, and the validity was also good, making it suitable for further analysis.

3.2 Data analysis

Over the whole internship process, professors and students communicated with each other mainly through email, web-meetings, formal and informal conferences, phone, and text messages. Each communication method has different effects at different stages according to the analysis. In the past, people enjoyed the use of formal emails to exchange their ideas in the business context. Now people, especially the young, prefer to use quicker and more convenient ways to communicate in the new social media era, such as with WhatsApp, WeChat, and QQ. According to the Table 1, it shows the importance of different communication tools in each stage.

(1) First stage

Email is the best communication method in the first stage of the internship because the three parties needed the others' information precisely and completely, such as their name both in English and Chinese, and a brief introduction to the investigation companies. The other communication

Table 2. Correlation analysis.

	Students' satisfaction
Enterprise support	0.682**
Culture and language difference	−0.306**
Communication and guidance	0.405**

* P<0.05 ** P<0.01.

methods did not reach that goal. In fact, email plays a crucial role in establishing and maintaining business relationships, both within a company and with external contacts.

And then a formal web-meeting can help students get to know each other, including the personalities of individuals. Students can become a little excited in web-meetings, and it is an appropriate way to communicate for young people. Online informal communication of each group can improve Chinese students' oral English ability.

Offline formal and informal conferences are between professors and students in one university at this stage. Students will get to understand internship objects and missions, matters which need attention. In WUST, QQ is the key tool to give out the information for collective activities, and WeChat, telephone took the assistant role.

(2) The second stage

During the second stage, each party has all the information about the other two parties. However, foreign professors and students are still in their own countries, and there has been no face-to-face communication at this stage. Therefore students prefer web-meetings in the external communication circle, and it is also more convenient, intuitionalist, and has a quick response. Another aspect is that it is more flexible in the internal communication circle. Students and professors like to discuss research problems through offline free conversations. Second, talking with each other using QQ is free and without time limitations, therefore people can find some creative ideas through chatting on QQ. However, if the research direction is decided through discussion, it needs to be formalized by email, or by uploading QQ document or an offline formal meeting.

(3) The last two stages

In the third stage, foreign students arrived in Wuhan, and the three universities' professors and students have more time to communicate with each other through offline, informal, and formal meetings and talking. In the meantime, companies prefer to confirm interview times through telephone messages, email, and occasionally QQ.

In the final stage, internship rehearsal and final presentation are reported both informally and formally through public methods. The Dean of School and other departments of WUST can be informed through email, telephone, and/or invitation letter.

4 FACTORS INFLUENCING STUDENTS' SATISFACTION

4.1 Correlation analysis

Through factor analysis, four factors are acquired, based on principal component extraction for each question, and these are: enterprise support, communication and guidance, cultural and linguistic differences, and students' satisfaction.

This paper performed correlation analysis of these four factors, with the conclusions as shown in Table 2.

From Table 2 it is seen that the relation between the students' satisfaction and cultural and linguistic differences is −0.306, and presents a significance level of 0.01, thus indicating a significant negative correlation between students' satisfaction and cultural and linguistic differences. The correlation coefficient between students' satisfaction and communication and guidance was 0.405,

Table 3. Regression analysis.

	N-Std		Std.							
	Regression coefficient	Std	Beta	t	P	VIF	R^2	Adjusted R^2	F
Constant	2.419	0.677	–	3.574	0.001**	–	0.519	0.503	31.682**
Enterprise support	0.446	0.134	0.525	3.338	0.001**	4.523			
Culture and linguistic difference	–0.284	0.286	–0.177	–0.993	0.323	5.837			
Communication and guidance	0.413	0.134	0.409	3.085	0.003**	3.209			
Dependent variable: Students' Satisfaction									

* p<0.05 ** p<0.01

and showed a significance level of 0.01, which indicated that there was a significant positive correlation between satisfaction and communication and guidance. The correlation coefficient between students' satisfaction and enterprise support was 0.682, and showed a significance level of 0.01, which indicated that it was a significant positive correlation between the students' satisfaction and the enterprise support, and verified the first part of the hypothesis.

4.2 Regression analysis

From Table 3, the regression analysis took the communication and guidance, enterprise support, cultural and language differences as independent variables, and students' satisfaction as the dependent variable. From Table 3, the model of R^2 of 0.519 means that the communication and guidance, enterprise support, cultural and language differences can explain 51.9% of the reasons for the change in students' satisfaction. It shows the communication and guidance, enterprise support, cultural and language differences had a relationship with students' satisfaction. The model formula is students' satisfaction = 2.419+ 0.446* (enterprise support) + 0.413*(communication and guidance) –0.284* language and cultural differences. In addition, it is found that the VIF values of the model are all less than 5, which means that there is no collinearity problem and the model is better.

The final analysis shows that the regression coefficient value of communication and guidance is 0.318, and the P value is 0.009, which is less than 0.01 and means that communication and guidance will have a significant positive impact on student satisfaction. The regression coefficient value of enterprise support is 0.446, and the P value is 0.001, which is less than 0.01, and means that the enterprise support has a significant positive impact on student satisfaction. The regression value of cultural and linguistic differences is –0.284, and the P value is 0.323, which is greater than 0.05, and means that cultural and linguistic differences do not have an impact on student satisfaction.

It was found that communication and guidance, and enterprise support have a significant positive impact on joint practice satisfaction. However, cultural and linguistic differences do not have a significant impact on student satisfaction.

5 DISCUSSION

(1) Through the questionnaire analysis, we interviewed the students before the arrival of foreign students. Students like to use email to exchange ideas, whereas after the foreign students arrived in China, they tended to use informal face-to-face communication. Even in the Internet era, people can communication online, but if they want to discuss a serious topic, or if they want to get a quick response, they still prefer face-to-face communication.

(2) Even network video conferences can give students the opportunity to see each other, however they complained about the interruptions of this process, because the campus network was not good enough. In addition, our communication software and computer systems are not compatible with those in foreign countries, for example, Chinese students use the QQ video conference, and foreign students use Skype software.

(3) Although the research results support first half of the hypothesis, namely the cultural and linguistic differences and student satisfaction being negatively correlated in terms of correlation analysis, this was not significant according to the regression analysis. It seemed to conflict with our common knowledge, as we always thought cultural and language differences may be the biggest barriers in the process of cross-cultural communication. There are three assumed reasons for this: first, the students are very young, and they easily understand and accept to alien cultures; second, the internship mainly involved business interviews, and finding some good solutions for the enterprises, in the business world, there some common rules, and compared to the other two factors, cultural and linguistic difference is not that important. Finally, because the students who completed the questionnaire are mainly Chinese students, they are not confident in their English ability, so they wanted to improve this through the internship. Therefore their expectations on English communication can be a little low, which reduces the language difference effect on the students' satisfaction.

(4) Enterprise support and communication and guidance have a positive impact on students' satisfaction, and are very significant. This means that in the future the management of these two aspects should be strengthened.

6 COMMUNICATION STRATEGIES

6.1 *Performing the whole internship process management and communication*

The internship process, objectives, tasks, and research direction should be determined before the formal internship each year. Therefore some formal papers and documents should be prepared and updated to increase students' rational thinking. In the meantime interns' emotional knowledge will be improved through past pictures and videos. It is especially beneficial to invite the previous year's interns to share their experiences.

Tutors should guide interns in every period of internship to help students complete the internship efficiently. According to the statistics, some tutors did not give students full instruction in the whole process; this leads to a lower evaluation of the internship by some students. Tutors should invest more time and energy into the internship, such as answering interns' questions, visit companies with students and helping interns to write their internship reports.

6.2 *Choose the appropriate companies for the internship*

From the correlation and regression analysis, we all know that enterprise support is most important to student satisfaction, which means we should take some measures to promote the international internship.

(1) Moderate-sized companies

The size of the internship company should be moderate, not too large and not too small. If the company is too large intern students will not understand the whole company deeply and thoroughly in a short time and will not have the ability to solve the problems. If the company is too small, students may not have sufficient interest, or the problems of the company may be not universal.

(2) Similar companies can be found in foreign countries

It is better if similar companies in other countries can be found. The foreign students can understand the situation of the target internship company more easily, which can reduce the communication problems with Chinese students. And students can establish a comparison model and provide some good management strategies to the internship company.

(3) Support from the company's leader is very important.

Generally speaking, with more support from the leaders, staff will be more cooperative and more patient in answering questions. Tutors and students can get to know more about the company, which means they can offer more practical suggestions.

6.3 *Improve students' English communication ability, and strengthen the cultural compatibility*

The current college English training is still focused on the examination, and daily communication remains somewhat insufficient, however, in the future, the college English education process should place greater emphasis on practical training. In addition, cross-cultural knowledge training and interpretation can be carried out specifically. The introduction to culture and the information materials of cultural differences and the training of some cross-cultural communication skills can help students integrate into the practice project better.

7 CONTRIBUTIONS

This paper discussed the communication and satisfaction issues during an international internship among three universities, including Wuhan University of Science & Technology (WUST), Washburn University of U.S. (WU), and PXL Technology University of Belgium. Few papers have discussed international internships among three universities from three continents.

The article has summarized the communication efficiency of different communication methods, as well as the communication problems throughout the whole process. Through questionnaire analysis, including correlation analysis and regression analysis, it has summarized the main factors influencing internship satisfaction, including process communication and guidance, company support, and culture and language differences.

In addition, this paper has also provided some good management strategies to improve international internship communication efficiency and student satisfaction. Furthermore, these communication strategies may be helpful in other international cooperation projects.

Last but not least, it reminded us that culture difference remains one of the most important factors when people from different countries communicate with each other.

ACKNOWLEDGMENT

This paper was supported by a project of Teaching & Research of Wuhan Science & Technology (2019x041), and a project of Humanities and Social Sciences in Hubei Province(17CYYO4). Professors Yunchuan Zhang and Michael Stoica are thanked for their support.

REFERENCES

Alina Simona (2016). The role of international internships conducted during academic studies in development of entrepreneurial skills. Bulletin of the Transilvania University of Braşov Series V, 9 (58): 169–174.

Batey, Jacqueline J., & Lupi, Marsha H. (2012). Reflections on Student Interns' Cultural Awareness Developed through a Short-Term International Internship. Teacher Education Quarterly, 39:25–44.

Clegg J (2015). Internationalisation in online distance learning postgraduate education: a case study on student views on learning alongside students from other countries. Innovations in Education & Teaching International, 52(2):137–147.

Cui, G.,& Van Den, Berg S. Testing the construct validity of intercultural effectiveness. International Journal of Intercultural Relations, 1991, 15:227–241.

Hawes F., & Kealey, D.J. (1981) An empirical study of Canadian technical assistance: Adaptation and effectiveness on overseas assignment. International Journal of Intercultural Relations, 4: 239–258.

Knight J (2006). Higher Education Crossing Borders. A Guide to the Implications of the General Agreement on Trade in Services (GATS) for Cross-Border Education. Commonwealth of Learning, 27(2):383–94.

Julie Matthews, Meredith Lawley. (2011) Student satisfaction, teacher internships, and the case for a critical approach to international education. Discourse Studies in the Cultural Politics of Education, 32(5):687–698.

Jin Xiaoling, Feng Huihui, Zhou Zhongyun.(2017)An Empirical Study on Healthcare Information Diffusion Behavior in WeChat Moments. Journal of Management Science, 30(1):73–82.

Lv Yiwen, Liu Mingzhen (2017) Educational exchange and scientific research cooperation to promote innovation and development in Colleges and Universities——Interview with Joseph, President of Tel Aviv University, Israel. Journal of World Education, (2):3–5.

Liu, Yong, Xu, Li.(2015) Study on the Management Strategy of Chinese overseas students based on the cooperation of WUST and WU. Proceedings of the International Conference on Social Science and Technology Education book series: Advances in Social Education and Humanities Research, (04). pp. 999–1001.

Madanchian M, Taherdoost H. (2016) Perusing of Organizational Culture Effects on E-Mail Communication. Procedia Technology, 22:1076–1083.

Pan, Kailing, Liu, Yong, Yin, Xiaolong (2013) Study on the Students' Satisfaction of IBA Class under bilingual teaching model. Education and Vocation, (01):179–181.

Pence, H. M., & Macgillivray, I.K (2008) The impact of an international field experience on preservice teachers. Teaching and Teacher Education: An International Journal of Research and Studies, 24(1):14–25.

Roberts D C, Komives S R (2016) Internationalizing Student Learning and Development. New Directions for Higher Education. (175):9–21.

Shi Weilin, Wang Yi (2015) Research on integration and optimization of international exchange and cooperation projects from the perspective of information technology. China Audiovisual Education, (8):128–132.

Shenghong Liu (2017) The Research of the Mobile-phone's social function to Meet the customer needs. Media, (16):46–48.

Tetiana Obolenska, Olena Tsyrkun (2016) Conceptual Approaches to International Cooperation between Higher Education Institutions, International Economic Policy, 2(25):40–58.

Tan Yu (2015) Research on the concept and strategy of College Students' intercultural competence training – from the perspective of cultural awareness and intercultural communication. Contemporary Education and Culture, (6):63–68.

Wang Qihe, Jiang Dehong, Guo Wenxia (2016) Obstacles and solutions of cross cultural communication in school enterprise cooperation. Education in Henan: Higher Education, (6):41–43.

Computational Social Science – Luo, Ciurea & Kumar (eds)
© *2021 Taylor & Francis Group, London, ISBN 978-0-367-70193-2*

The present situation and development countermeasures of international exchanges and cooperation for artificial intelligence postgraduate students in China

M. Yu & X.H. Liao
School of Public Administration, Sichuan University, Chengdu, Sichuan, China

ABSTRACT: International exchange is an important measure to promote the cultivation of artificial intelligence graduate students. This article aims to analyze and discuss the current situation, and to put forward suggestions for the development countermeasures of international exchanges and cooperation of artificial intelligence postgraduate students in China. The results show that there are problems such as the low participation in various international academic exchanges, insufficient cross-border collaborations, and inadequate attractiveness to talents in China. The present situation and international trends illustrate the compelling need to strengthen the international exchanges and cooperation of postgraduate students in the field of artificial intelligence. Finally, this article proposes that the government, enterprises, and universities should cooperate closely to jointly promote the international exchange and cooperation of artificial intelligence graduate students.

1 INTRODUCTION

Artificial Intelligence (AI) is one of the core technologies in the fourth industrial revolution. The development of AI has profound effects on industrial transformation and economic development in China, which relies on professionals with innovation ability and good knowledge in the field of AI. However, there is a shortage of AI professionals worldwide. Thus, strengthening the training of AI talents has been seen as one of the important development strategies of many countries. To achieve the expected goal of this strategy, it is indispensable to learning from the overseas vanguard technology.

In 2020, the Chinese Ministry of Education issued Several Documents on "Double First-Class" Constructing Universities to Promote Discipline Integration and Acceleration of Training Postgraduates in Artificial Intelligence, where attention was called to focusing on the international forefront research and the domestic shortcomings in AI development, to increase the support for international co-cultivation of AI doctoral students, and to encourage international exchanges between high-level AI talents and to add depth to their cooperation.

Briefly, to improve the development of Chinese AI technology, it is important to analyze the current situation of international exchanges and cooperation for AI postgraduate students and explore how to promote it.

2 INTERNATIONAL AND DOMESTIC SITUATION

From 2012 to 2015, there were 7851 international students studying computer science and other related courses in the United States. In 2015, more than one-third of international students at New York University were studying AI-related courses. In 2018, 25 European countries signed the Declaration of Cooperation on Artificial Intelligence, aiming at ensuring the competitiveness of Europe in AI research and development through international exchanges of high-level talents and

Table 1. The quantity of papers co-authored by researchers of China and other countries in the field of machine learning (data source: https://www.aminer.cn/research_report /).

Cooperating countries	Number of papers	Total cited	Average cited	Number of authors
Sino–US	511	26,694	52	819
Sino–UK	44	1398	32	73
Sino–Singapore	36	1189	33	56
Sino–Australia	31	744	24	42
Sino–India	22	1123	51	19

cooperation in the AI industry. In addition, in Preparing for the Future of Artificial Intelligence and National Artificial Intelligence Strategic Planning released by the American government, how to attract talents and promote international exchanges was a key point. As stated above, AI technology has been extensively valued by many countries. International exchanges and cooperation of AI-related postgraduates have become an international trend.

In China, there are currently four main ways for AI postgraduate students to conduct international exchanges. First, visiting scholars are exchanged between universities of different countries to promote the flow of university talents. Second, the construction of international joint research centers, the conduct of international seminars, and the initiation of international joint programs, including training, scientific research, and internship provide cooperation opportunities. Among them, Xidian University and Zhejiang University are relatively successful in this practice. Third, universities have carried out technical cooperation with well-known multinational companies such as Microsoft, Siemens, Mitsubishi, etc., thus providing students with a good communication environment and practice opportunities. Fourth, students are encouraged to participate in international AI competitions.

According to statistics from the Artificial Intelligence Research Center of Tsinghua University, the number of postgraduate students participating in SCIGIR and ICML international forums in 2019 was 133, which was a small size compared with the total number of AI postgraduate students.

Table 1 illustrates the top 5 quantity of papers co-authored by researchers from China and other countries in the field of machine learning. As a way to achieve AI, machine learning is one of the main AI research fields in China. According to these statistics, 1009 authors were from China, but postgraduate students accounted for less than 10%, which reflects the lack of international exchanges and cooperation for AI postgraduate students, or that their international exchanges and cooperation were limited in their effectiveness.

In addition, among current internationally influential AI scholars, 11% obtained their degrees in Chinese universities, which is second only to the 44% for American universities. However, the total number of AI researchers with strong capabilities in China is ranked sixth in the world. This indicates that the brain-drain problem is serious.

3 DEVELOPMENT COUNTERMEASURES

The favorable international exchange environment requires the cooperation of universities, enterprises, and governments to formulate policies and regulations, offer financial support, develop strong industrial support, and form positive interactions.

First, from the perspective of universities, it is necessary to strengthen foreign language teaching that can lay a foundation for international exchanges and cooperation. In addition, universities could offer access to advanced AI-related online courses from renowned foreign universities for non-mobile domestic postgraduate students. More postgraduate students should be encouraged to participate in international high-level academic forums to learn at the academic frontier. Also, more international professors and teachers could be invited to exchange new ideas and research trends. At the same time, universities could send excellent teaching staff abroad to study as academic visitors and bring technologies and experience back for Chinese students. Finally, while excellent students are chosen to go to overseas universities for academic exchange or further studies, universities should pay attention to the brain-drain problem. To alleviate this problem, universities, on the one hand, could consider the national awareness and patriotic spirit of chosen students, and on the other hand, they could provide more financial support for postgraduate students by signing an Employment Agreement for Returned Students to study abroad.

Second, enterprises are supposed to provide funding support for the scientific research base and related industries, as measure to precede talent transportation. On the one hand, they should cooperate with universities to establish platforms for turning theoretical outcomes to practice as it is beneficial to attract foreign scholars to China for further study. On the other hand, internship opportunities should be offered for postgraduate students to work in international joint research and development project teams. This builds a bridge between postgraduates and the best international teams.

Third, the government has the responsibility to issue policies, supply funds, and build bridges to ensure effective international exchanges for postgraduate students. For example, they should formulate programs and special fund plans for international exchanges, and organize their implementation, supervision, and administration. It is suggested to increase the funding investment of AI-related international exchanges, and to finance these through multiple channels. It is also important to provide attractive employment opportunities with postgraduate students graduating and returning from other countries, and launch the government's funds to support entrepreneurial investments. In addition, more support should be put into the construction of research centers, science and technology parks, and advanced AI laboratories so that more international top talents are attracted to and stay in China.

4 SUMMARY

Compared with some developed countries, postgraduate education in the field of AI in China got off to a late start. The talent training system is not sound enough, and ways to enhance international exchange are still being explored. To cultivate competitive internationalized talents, the international exchange of postgraduates is obviously important. Through analyzing current statistics and related documents, it can be seen that the international exchanges and cooperation of postgraduate students in the field of AI is a general international trend, among which America, the UK, and other countries with advanced AI technology have the most frequent international exchanges. The international exchange and cooperation of AI postgraduate students in China is insufficient, especially in international joint training.

The government should actively improve the policies and regulations for overseas exchanges and lead national exchange and cooperation projects; enterprises should actively build platforms and strengthen financial support; universities need to integrate resources and encourage graduate students to obtain doctoral degrees abroad. Through a collaborative cooperation between universities, enterprises, and the government, the international exchange of AI graduate students will be normalized and the quality of the exchanges will be improved.

REFERENCES

[1] W. Wilkinson, I. Podhorska, A. Siekelova. Does the Growth of Artificial Intelligence and Automation Shape Talent Attraction and Retention? J. Addleton Academic Publishers. 1(2019)7.
[2] Element AI. Global AI Talent Report 2019 on https://jfgagne.ai/talent-2019/
[3] CB Insights AI 100: The Artificial Intelligence Startups Redefining Industries on https://www.cbinsights.com/research/artificial-intelligence-top-startups
[4] Student library 2019. On https: //gct.aminer.cn/ eb/gallery/ detail/eb/ 5d4527 b3530 c7 0 eb4e2a79b9

[5] O. M. Sarah. Will China lead the world in AI by 2030? J. Nature, 7770 (2019)572.

[6] M. Grove, T Croft. Learning to Be a Postgraduate Tutor in a Mathematics Support Centre, J. International Journal of Research in Undergraduate Mathematics Education. 2(2019)5.

[7] M. Fu. Can China's Brain Drain to the United States be Reversed in the Trump Era?: Trends in the Movements of American-Trained Chinese STEM Talent and its Implications, J. The Global Studies Journal, 4(2018)10.

Computational Social Science – Luo, Ciurea & Kumar (eds)
© *2021 Taylor & Francis Group, London, ISBN 978-0-367-70193-2*

Empirical analysis on the manufacturing regional value chain between China and "Belt & Road" countries

N.N. Zhu & M.R. Liu
School of Economics and Management, Beijing University of Technology, China

ABSTRACT: This article conducts research on manufacturing regional value chain (RVC) between China and "Belt and Road" ("B&R") countries and its relevance to the upgrading of China's manufacturing industry. Based on the calculation of Kaplinsky upgrade index and RVC–Position index, this paper makes an empirical analysis with the above indicator results and a random effect model. The empirical results show that when KUP increases by 1%, RVC_Position will positively increase by 0.24%, which means that the domestic industrial chain upgrading of China's manufacturing industry has a positive impact on embedding in manufacturing RVC and rising up to the global value chain.

1 INTRODUCTION

It has been seven years since the "B&R" Initiative was put forward. With the deepening of the bilateral trade integration between China and the countries along the route, the foundation for building RVC is already in place. The "B&R" Manufacturing RVC led by China has already promoted the development of manufacturing trade within the region and thereby stabilized and enhanced China's position in RVC.

In recent years, more and more researchers have studied RVC, especially the "B&R" RVC. The concept of RVC is to take industrial upgrading and mid-to-high-end development as the goal, unite with neighboring emerging countries or regions that are highly complementary in industries, and a kind of cross-regional network organization that connects the regional aspects of production, sales, recycling, and other processes in order to realize the value of goods or services (Wei & Wang, 2016). For the analysis of how to carry out the layout, it was considered that mutual benefit and a win–win situation should be the foundation, strengthening cooperation with countries along the route, and that China's manufacturing industry has the ability to lead the joint construction of RVC (Xu & Na 2018). By using the KPWW method and the value chain position indicator, Chinese manufacturing was considered as capable of leading and co-constructing the RVC (Xie 2018). Based on the total trade accounting framework of WWZ, calculating VSS & GVC status index and other indicators to analyze the realistic basis of China's leading construction of the equipment manufacturing RVC (Li et al. 2020). On the basis of combing and researching the related literature, this paper focuses on the status quo of the construction of the "B&R" manufacturing RVC and its relevance to the upgrading of China's domestic manufacturing value chain.

The paper's structure is as follows. It starts with a statement on the theoretical framework, which was illustrated through the relationship between the Kaplinsky upgrade index and RVC–Position index, then we use a random effect model and panel data to make an empirical analysis, and finally get the results: China's manufacturing industry upgrade has a positive effect on embedding into RVC. It is also helpful to China's climbing up of the manufacturing RVC, and China has the ability to become the chain leader.

2 THEORETICAL FRAMEWORK

Based on the research method provided (Qiuchen 2019), this paper proposes a theoretical framework: Upgrading of the manufacturing industry within the domestic value chain has a positive effect on its construction and embedding in the RVC and this section explains the Kaplinsky upgrade index, RVC_Position index and the relationship between the two important indicators.

2.1 *The capabilities for upgrading in China's manufacturing industries in the RVC—based on Kaplinsky upgrade index analysis*

The measurement of the Kaplinsky index was proposed as it needs to consider two product factors, namely relative price and market share. The trend of these two determines whether an industry and product are upgraded or downgraded (Kaplinsky & Readman 2005). In order to analyze the upgrading capabilities of China's manufacturing industry and carry out the empirical analysis, the paper calculates the product upgrade index in this section.

China's manufacturing industry market share (RRX):

$$RRX^{ni}_{t,t-1} = \frac{RX^{ni}_t}{RX^{ni}_{t-1}} = \frac{X^{ni}_t / X^{ri}_t}{X^{ni}_{t-1} / X^{ri}_{t-1}} \tag{1}$$

First, calculate the market share RRX of China's various subindustries relative to the "B&R" area according to Eq. 1, as shown in Table 1.

China's manufacturing industry relative price (RRP):

$$RRP^{ni}_{t,t-1} = \frac{RP^{ni}_t}{RP^{ni}_{t-1}} = \frac{\frac{P^{ni}_t}{P^{ri}_t}}{\frac{P^{ni}_{t-1}}{P^{ri}_{t-1}}} = \frac{\frac{X^{ni}_t / Q^{ni}_t}{X^{ri}_t / Q^{ri}_t}}{\frac{X^{ni}_{t-1} / Q^{ni}_{t-1}}{X^{ri}_{t-1} / Q^{ri}_{t-1}}} \tag{2}$$

Then according to Eq. 2, calculate the relative price RRP of various industries in China relative to the "B&R" region, as shown in Table 2.

Table 1. China's manufacturing subindustries' RRX.

Indicator/year	2013	2014	2015	2016	2017
Chemicals	1.07	1.10	1.13	0.95	1.51
Manufactured goods classified by material	1.01	1.13	1.17	0.86	0.86
Machinery and transport equipment	1.07	1.00	1.13	0.83	0.96
Miscellaneous manufactured articles	1.11	0.96	1.25	0.66	0.82

Table 2. China's manufacturing subindustries' RRP.

Indicator/year	2013	2014	2015	2016	2017
Chemicals	0.96	0.98	1.04	0.96	1.15
Manufactured goods classified by material	1.12	0.96	0.99	0.98	2.43
Machinery and transport equipment	0.74	0.83	1.32	1.71	0.26
Miscellaneous manufactured articles	1.26	1.17	0.96	0.87	1.01

Table 3. China's manufacturing subindustries' upgrade index KUP.

Indicator/year	2013	2014	2015	2016	2017
Chemicals	0.82	0.84	0.89	0.95	1.00
Manufactured goods classified by material	0.86	0.95	0.98	0.92	1.00
Machinery and transport equipment	0.91	0.91	0.97	1.00	0.61
Miscellaneous manufactured articles	1.00	1.87	0.98	0.77	0.91

China's manufacturing industry upgrade index KUP:

$$Kaplisnky\_Upgrade\_Index_{t,t-1}^{n} = \frac{\left\{ \sum_{i=1}^{k} X_t^{ni} | RRX_{t,t-1}^{ni} > 1 \cap RRP_{t,t-1}^{ni} > 1 \right\}}{\sum_{i=1}^{k} X_t^{ni}} \qquad (3)$$

Note that the range of this index would be [0,1], and the closer the result is to 1, the more products that can be upgraded. When the index is 1, it means that all export products of China are in the process of upgrading during the relevant period; conversely, the closer to 0, the weaker the upgrading ability. When the index is 0, it means China has no product upgrades during that period. Depending on the two previously calculated indicators, the result can be worked out using Eq. 3, as shown in Table 3.

Based on the obtained results, a subindustry analysis of China's manufacturing industry can be carried out. In recent years, chemical products have been in the process of continuous upgrading. By 2017, the KUP had reached the boundary value of 1, and all products in the industry had been upgraded. The manufactured goods are classified chiefly by material products upgrade situation, similar to that of chemicals, and the index is constantly upgraded to 1. The difference is that the overall value of this industry is higher than that of chemicals, and the upgrading situation is more optimistic. The KUP of machinery and transport equipment rose to the highest value of 1 in 2016, and then fell sharply to 0.61. It can be seen that some of the products in this industry have begun to degrade after all upgrades, and the upgrade capability has been suppressed. As early as 2013, all miscellaneous manufactured products were upgraded, and there was a decline afterwards. In 2017, it rebounded to 0.91. Although the upgrade ability declined in the middle period, it is still strong overall, and it is expected to realize upgrades for all products again.

2.2 Analysis of the degree of Chinese manufacturing embeddedness in the RVC

A global value chain division of labor status indicator (GVC_Position) was proposed to measure the position and competitiveness of a country's industry in the global value chain from the perspective of division of labor status and added value (Koopman et al. 2010).

This paper focuses on the RVC of the "B&R," so this indicator is selected to reduce the caliber from the original global value chain to the RVC, so as to judge the embedding of Chinese manufacturing in the "B&R" region. Circumstances was regarded as a reference to analyze the feasibility of joint construction.

$$RVC\_Position_{ir} = \ln\left(1 + \frac{IV_{ir}}{E_{ir}}\right) - \ln\left(1 + \frac{FV_{ir}}{E_{ir}}\right) \qquad (4)$$

Note that "i" indicates manufacturing and "r" indicates China. Among these, IV represents the indirect value-added export value of manufacturing in China; FV represents the foreign added value included in the export value of China's manufacturing final products; E represents the added value of China's manufacturing exports. If RVC_Position>0, that is, IV>FV, more of a country's intermediate goods are exported than imported, indicating that China's manufacturing industry is

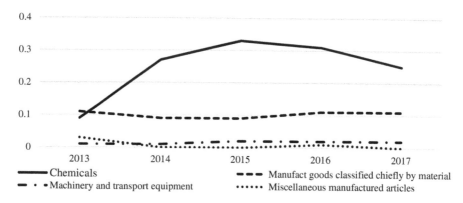

Figure 1.　China's manufacturing subindustries' RVC_Position change.

in the upstream link of the RVC and can create more added value; if RVC_Position<0, that is IV< FV, a country imports more intermediate goods rather than exports, indicating that China's manufacturing industry is in the downstream link of the RVC and its ability to create added value is relatively weak. Therefore, the larger the value of the RVC_Position, the better the embedding ability, and vice versa. The calculation results of Eq. 4 are shown in Table 4.

To facilitate the analysis, a line chart is for Table 4 has been created to visually show the degree of embedding of China's manufacturing industry as a whole and for the four subindustries in the value chain of the "B&R" region, as shown in Figure 1.

It can be seen from Figure 1 that: first, the RVC_Position of chemicals has the most obvious and significant change. It has experienced great ups and downs in the past few years, from the lowest value of 0.09 in 2013 to the highest value of 0.33 in 2015, which has made a huge contribution to the consolidation of China's position in the "B&R" RVC; the value then dropped to 0.25 in 2017. Although it is declining, it is still much higher than the initial value. China's chemicals have a relatively strong and embedded capacity in the RVC. When the product of manufactured goods classified chiefly by material is higher than the product of machinery and transport equipment, the change trend of the two industries is basically the same, and the values of both remain greater than 0, which shows that the two have enhanced their embedding in the RVC to varying degrees. The image of miscellaneous manufactured articles shows that they are less than 0 in the process of change, which shows that this kind of product in China has poor embedding capabilities and is still in the lower reaches of the value chain. They are more often imported, and there is still a lot of room for development. Through the specific measurement of indicators, combined with some realistic factors such as control variables, the empirical analysis of this theoretical framework is carried out in the following section.

Table 4.　China's manufacturing subindustries' RVC_Position.

Indicator/year	2013	2014	2015	2016	2017
Chemicals	0.09	0.27	0.33	0.31	0.25
Manufactured goods classified by material	0.11	0.09	0.09	0.11	0.11
Machinery and transport equipment	0.01	0.01	0.02	0.02	0.02
Miscellaneous manufactured articles	0.03	0.00	0.00	0.01	0.00

Table 5. Selection and description of variables.

Variable type	Indicator	Expression	Expectation
Explained variable	RVC embedded degree	RVC	+
Explanatory variable	Kaplinsky upgrade index	KUP	+
	R&D expenses	RD	+
	Institutional conditions	INS	+
Control variable	Professional human capital	HR	+
	Open to the outside world	EX	+
	Overall labor productivity	YL	+

Table 6. Hausman test result.

Variable	Coefficients		(b-B) Difference	Sqrt(diag (V_b-V_B)) S.E.
	(b) fe	(B) re		
KUP	0.0499314	0.2427745	−0.1928432	0.1018064
HR	−0.2083329	−0.4691477	0.2608148	0.1502979
RD	0.2829658	0.409231	−0.1262653	0.0810142
EX	0.1186507	−1.754169	0.2940676	0.1999768
INS	−0.18458	−1.031498	0.8469177	0.482921
YL	−0.0021414	−0.001675	−0.0004664	0.0001959

Chi2(6) = (b-B)'[(V_b - V_B)ˆ(-1)](b-B) = 4.28
Prob > chi2 = 0.6393.

3 EMPIRICAL ANALISIS AND RESULTS

3.1 Variables selection and model setting

For the indicators measured in this paper, the variables required by the model are selected, and the specific explanations are shown in Table 5.

Based on the selected variables and the panel data obtained, in order to analyze the influence of the Kaplinsky upgrade index on the degree of embedding of Chinese manufacturing in the "B&R" initiative, we established the following model:

$$RVC_{it} = \beta_0 + \beta_1 KUP_{it} + \beta_2 HR_{it} + \beta_3 RD_{it} + \beta_4 EX_{it} + \beta_5 INS_{it} + \beta_6 YL_{it} + \varepsilon_{it} \qquad (5)$$

*Note that: β_0 as constant term, ε_{it} as random disturbance. The subscript i [1,4] represents the subindustries of China's manufacturing; and t [1,5] represents the year.

3.2 Hausman test

The Hausman test is to solve the problem of whether the random effect model or the fixed effect model is used in the panel data, and it is used to eliminate the possibility of using random effects (Hausman 1978). The test process is performed by Stata, and the result shows that the P value is relatively large at 0.6393, which is greater than 0.05. Therefore, the random effects model would be better (Table 6).

3.3 Regression analysis

The random effects model regression was performed on the model through Stata, and the results are shown in Table 7.

Table 7. Random effects model regression results.

Variable	Coefficient	Std. error	Z-value	Prob.
KUP	0.2427745	0.1188887	2.04	0.041
HR	0.4691477	0.1557261	−3.01	0.003
RD	0.409231	0.124157	3.30	0.001
EX	0.1754169	0.1105413	−1.59	0.113
INS	0.031498	0.1290751	−7.99	0.000
YL	0.001675	0.0006608	−2.54	0.011
_cons	0.2261035	0.1118487	2.02	0.043
R-squared	0.9022			
Z-statistic	119.95			
Prob.	0.0000			

The Hausman is used test to judge whether there is an individual effect in the model, and to judge whether to choose the fixed effect or random effect to regress the model. The result shows that it does not exist. Therefore, the random effect model is adopted. The coefficients of the regression result are all positive and consistent with expectations. Also, the explanatory variable KUP has a larger value of 0.2427745, and the P-value obtained from the regression is 0.0000. It can be seen that each variable has a significant influence on the explained variable. In summary, in order to adapt to this model, choosing random effects regression has the best and most significant effect.

4 SUMMARY

4.1 *The degree of embeddedness of China's manufacturing industry in the "Belt and Road" Initiative Region is obvious and will develop well*

Through calculation of the Kaplinsky upgrade index and the RVC_Position, the results show that chemical products are embedded in the RVC to the greatest extent, the upgrade is also better than before, and it is in the process of continuous upgrading. In addition, the upgrading ability of manufactured goods is classified chiefly by material product and is also continuously strengthening, however it is slightly inferior to chemical products. The upgrading index for the machinery and transport equipment fluctuates. The latest results show a significant downward trend, and the upgrading index of miscellaneous manufactured products also fluctuates but is smaller than the former. Overall, China's manufacturing industries have strong upgrading capabilities.

4.2 *It is possible for China to became the chain leader of "B&R" regional value chain*

Through regression analysis of the econometric model, the results show that the explanatory variable has a positive and significant influence on the explained variable, that is, when KUP increases by 1%, RVC_Position will positively increase by 0.24%. It can be seen that China's own manufacturing strength is strong, supplemented by good product upgrades, and will positively help its embedding in and climbing up the value chain. These positive factors also make it possible for China to become the chain owner of the RVC.

REFERENCES

Hausman J. A. 1978. Specification tests in econometrics. *Econometrics* 46(6): 1251–1271.
Kaplinsky, R. & Readman, J. 2005. Globalization and upgrading: What can (and cannot) be learnt from international trade statistics in the wood furniture sector? *Industrial & Corporate Change* 14: No. 4.

Koopman, R., Powers, W., Wang, Z., & Wei, S. J. 2010. Give credit where credit is due: Tracing value added in global production chains. *NBER Working Paper*: No. 16426.

Li, Yan, Gao, Yaxue & Huang Qingbo. 2020. The collaborative construction of regional value chains between China and the "Belt and Road" countries: evidence from the equipment manufacturing industry. *International Trade* 2020(01): 4–14.

Pan, Qiuchen. 2019. Research on the impact of global value chain embedding on the transformation and upgrading of China's equipment manufacturing industry. *World Economic Research* 2019 (09): 78–96&135–136.

Xu, Poling & Na, Zhenfang. 2018. The industrial chain layout of China's manufacturing industry in the "Belt and Road" – Complementary Competitive Advantages and Intermediate Trade Perspectives. *Northeast Asia Forum* 03(137): 88–109.

Xie, Yuping. 2018. Research on the regional value chain of China's leading manufacturing industry under the "One Belt and One Road" initiative. *Jiangxi: Jiangxi University of Finance and Economics*.

Wei, Long & Wang, Lei. 2016. From embedding in the global value chain to leading the regional value chain-the economic feasibility analysis of the "Belt and Road" strategy. *International Trade Issues* 2016(05): 104–115.

Computational Social Science – Luo, Ciurea & Kumar (eds)
© *2021 Taylor & Francis Group, London, ISBN 978-0-367-70193-2*

Homeland and school geography factors in the social networks of African college students in Jinan, China

L.S. Wang
Urban and Rural Planning Department of the School of Civil Engineering and Architecture, University of Jinan, and Urban and Rural Space Development and Planning Research Center of Jinan University, Jinan, Shandong, China

H.F. Dong
Urban and Rural Planning Department, School of Civil Engineering and Architecture, University of Jinan, Jinan, Shandong, China

Y. Fan*
School of Civil Engineering and Architecture, Jinan University and Urban and Rural Space Development and Planning Research Center, University of Jinan, Jinan, Shandong, China

T. Wang
Institute of Geographical Sciences and Natural Resources Research, Chinese Academy of Sciences, Beijing, China

ABSTRACT: Based on the data obtained from a social communication survey of African college students in Jinan, this paper uses Ucinet to construct the social network based on the strength of ties to map to their homeland and school. The results show that: (1) The geographical proximity of the homeland is not the main factor of the social network of African college students in China, but plays a more important role in high-frequency communication. (2) Increasing urban expansion has hindered the interpersonal relationship of African college students in China. Jinan University Town has not played its due role in promoting mutual sharing and communication among different universities. (3) Most African college students regard the current geographical proximity factors as an important factor in social communication.

1 INTRODUCTION

China has become the third largest host of international students in the world and the second largest destination for African students. Like international students from China, African college students in China also face the problem of gradual adaptation to the language, culture, and values of the host nation. In this process of adaptation, international students often form a group internalized social network because of their similarities to each other and their differences in the country of study. For this type of social network, it is only in the last two years that academics have carried out analysis and research based on survey data such as questionnaires and microphones, and have paid attention to the influences involved.

Of all the influencing factors of social networks, geography is always an important one. The concept of the social network was spatially defined from its inception, and its creator, Radcliffe-Brown, made it a focus on how culture dictates the behavior of members within the spatial boundaries of tribe, village, etc. At present, domestic scholars also place the study of social networks of special groups in the background of geographical disciplines. In fact, social networks in the traditional

---

*Corresponding author

sense are inseparable from the influence of geopolitical factors. For the group of African college students coming to China, the geopolitical factors involved in the composition of the social network have two main levels: one is the "home" factor, which is shown by the nationality of the foreign student before coming to China. Considering the geopolitical role in the social networks, there are assumptions that the closer the geographic location between countries, the closer the culture between countries, and the more international students between countries have interpersonal contacts in China. The second factor is the "school" factor, that is, the location of the school where the international students come to in China. Considering the role of geography in interpersonal networks, there are assumptions that the closer the geographic distance between schools, the more convenient the transportation connection between schools, and the more interpersonal contacts there are between schools for international students. This article starts under this hypothetical background, taking the social network of African college students in Jinan as the object, observing and analyzing the geographical mapping of the country to which the international students belong and the school, in order to outline this typical and special interpersonal network and the geographical factors involved in its formation.

2 MATERIALS AND RESEARCH METHODS

2.1 *Study area selection*

This paper selected Jinan as an example. Jinan is the political, cultural, and educational center of Shandong Province, with 48 universities. As of the end of May 2016, there were 2937 permanent foreign residents in Jinan, with foreign students accounting for 55.7% of the total number. In the first half of 2016, the number of international students from the South Africa was 132, ranking it second in the ranking of the number of international students in universities in Jinan.

2.2 *Data sources*

The data in this paper come from the statistics obtained from a questionnaire. The questionnaire was designed to ask international students about their own nationality, school, and time in China, the nationality and school of their top three friends with whom they have the most contact, and the language they use to communicate with their friends and the main factors that they think affect the frequency of their contacts. The questionnaire was designed in English on the questionnaire star platform, and was distributed through the WeChat group of African college students in Jinan from May to June 2018, with the respondents responding online. A total of 105 questionnaires were returned, all of which were valid questionnaires, involving 40 African countries and 17 campuses (including branch campuses) of universities.

2.3 *Research methods*

The focus of this paper is mapping the social networks of international students coming to China to the geographic information of their countries of origin and schools. First, information on the country and school of the top three friends with whom the respondent had the most contact was counted, and duplicate data were overlaid; then, the number of mutual connections between the country and the school was counted by establishing a matrix and converted into the strength of the connection; next, the Ucinet software was used to build a network of inter-State and interschool connectivity, which is also the social network with the geographical location of the state and the school; finally, the nodes and links of the two linkage networks were adjusted and implemented in the geographic drawing. The observation of this drawing leads to an analysis of the correlation between strong (thick line) and weak (thin line) connections and geographic distance.

When constructing the matrix, the principle of a gravity model to build a mathematical model of the national and school attention indexm (Formula 1) was used. In the case of the data for African countries, for example, country A is the respondent's country and has the number of friends corresponding to the respondent's country B. These two data jointly characterize the closeness between the two, and a set of correlation values between various African countries is obtained.

$$R_a = \frac{A}{\sum_{i=1}^{n} i/n} \qquad \text{(Formula 1)}$$

In this formula: Ra is the proportion of "country a" mentioned by respondents in "a country" to the average number of mentions in "country a"; "A" is the number of countries searched by "a country"; "n" is the number of other countries. To avoid the impact of too much zero data in the questionnaire results on this study, the number of countries with zero data were removed from "n." In addition, among the respondents, there were more Cape Verde students and more countries involved in friends, resulting in the country's nodes in the social network being too strong. Therefore, the contact countries with lower rankings were deleted, and the number of remaining contact countries was the same as the number of the respondents ranked second. The survey data show that on average each country is associated with 3.95 countries.

3   RESULTS AND ANALYSIS

3.1   *The geographical proximity is not strong in the national connection network based on social interaction*

Using the above research methods, a mapping of contact networks on the map of Africa was formed (Figure 1). According to four levels of high-intensity linkages, sub-high-intensity linkages, low-intensity linkages, very low-intensity linkages, and high-intensity linkages occur only between Cape Verde and Congo-Brazzaville, which are separated by several countries. A total of 17 groups of sub-high-intensity contact countries, with the immediate neighboring countries being Zambia and Zimbabwe, South Africa and Zimbabwe, Zambia and Tanzania, and four groups of neighboring countries one country apart, including Tanzania and Zimbabwe; a total of 39 groups of low-intensity contact countries, and six groups of immediate neighboring countries, including Rwanda and Tanzania, and eight groups of neighboring countries one country apart, including Burundi and Kenya; a total of 50 groups of low-intensity contact countries, with the immediate neighboring countries being Sao Tome and Principe and Nigeria in nine groups, and seven groups of neighboring countries one country apart, including Ghana and Liberia. When a strait is encountered, only the two countries that are perpendicular and nearest to the strait are considered in the proximate and neighbor statistics above.

In summary, proximate and neighbor groups account for about 35% of the total number of groups in each strong connection; about 41% of the groups with sub-high-intensity linkages are in proximate and neighbor countries, which is the highest in each classification class. This suggests that the geographical proximity of the country is not an absolute factor in the interpersonal network of African international students coming to China. However, the geographical proximity of the homeland plays a more prominent role in the more connected interpersonal interactions.

3.2   *The school connection network based on interpersonal interaction matches the development of urban space*

Using the above research method, the mapping of the linkage network on the map of the central city of Jinan was formed (Figure 2). There are three parts in the central city of Jinan: the central part is the main part of the city, to the west is the western part of Changqing University City, and to the east is the eastern part of Tangye New City (Eastern New City). According to the three types of high-intensity, medium-intensity, and low-intensity, the school contact network based on interpersonal communication is divided. There are two groups of high-intensity contacts, both of which appear in the main urban area. These are Shandong University's Baotu Spring Campus

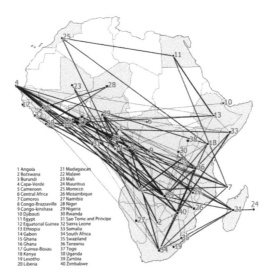

Figure 1. Network diagram of African countries based on the communication frequency of African students coming to China.

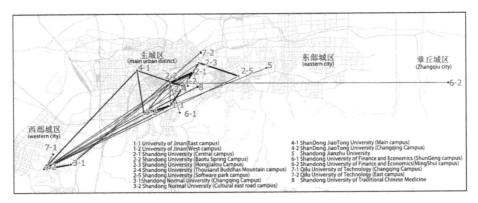

Figure 2. Network diagram of universities based on the communication frequency of African students coming to China.

and Central Campus, Jinan University's East Campus and West Campus. There are a total of nine groups of medium-intensity contacts, of which there are six groups within the main city district, namely, the University of Jinan (West Campus) and ShanDong JiaoTong University (Main Campus), the ShanDong JiaoTong University (Main Campus) and Shandong University (Thousand Buddhas Mountain Campus), the University of Jinan (East Campus) and Shandong University (Hongjialou Campus), the Shandong University (Hongjialou Campus) and Shandong University (Software Park Campus), the University of Jinan (West Campus) and Shandong University (Baotu Spring Campus), the University of Jinan (West Campus) and Shandong University of Traditional Chinese Medicine. There is a group in the western city, namely, the Shandong Normal University (Changqing Campus) and ShanDong JiaoTong University (Changqing Campus). There are two groups linking the main urban district with the western city, namely, the ShanDong JiaoTong University (Changqing Campus) and ShanDong JiaoTong University (Main Campus), the ShanDong JiaoTong University (Changqing Campus) and University of Jinan (East Campus). There are 18 groups of low-intensity contacts, of which eight groups are found within the main urban district,

379

Note: The joined coils are the lower-level subgroups of neighboring countries, and the dashed circles are the lower-level subgroups separated by one country.

Figure 3. Distribution of condensed subgroups in African countries based on the communication frequency of African students coming to China.

one group is found in the western city, eight groups are found between the main urban district and the western city, and one group is found between the main urban district and Zhangqiu city.

On the whole, the urban form obviously affects the connection network between schools. This is because all of the high-intensity and about 67% of the medium-intensity linkages occur within the main urban district. The connections between the main urban district and the peripheral urban areas (the eastern urban and the western urban) mainly appear in low-intensity linkages, accounting for 50% of all low-intensity linkages. This suggests that the sprawling spatial expansion of cities has blocked the interpersonal connections of incoming African students to some extent. In addition, it should be noted that, first, high-intensity contacts occur between different campuses of the same university, indicating that the connections between African students in the same university are not constrained by the geographical proximity of the branch campus. Second, there is no extensive internal contact among the universities in University Town in the western urban. In this regard alone, the University City has not played its due role in promoting mutual sharing and complementarity among universities.

3.3 *International students' perceived social network influences verify the above results*

The questionnaire also surveyed the main factors (multiple choice questions) that African students thought affected their social network after they came to China. Sorted by the number of times selected, these factors are "study in the same campus" (64 times), "having the same language" (51 times), "adjacent to Jinan's residence" (39 times), "the location of the motherland is close" (25 times), and "same religion" (20 times). Such questionnaire statistics validate the reasonableness of the above-mentioned network generation between countries and universities based on the social network of international students. Most African students in China do not see "the location of the motherland is close" as an important factor in facilitating interpersonal communication. On the contrary, the current geographical proximity factor, such as "study in the same campus," is more important after coming to China.

4 DISCUSSION

Geographic proximity features, based on countries neighboring or spacing out a country, are not evident in interpersonal social networks of national connections. However, a subgroup analysis of the network via the "Cliques" command in Ucinet found that many countries with the same geographic location or close proximity (including spaced countries) were in close subgroups (Figure 3). Moreover, the phenomenon of two-by-two, gradual upward cascading is very obvious. Among the 12 lowest-level subgroups of the two countries, South Africa and Zimbabwe, Malawi and Zambia, Sao Tome and Principe and Gabon, Central Africa and Djibouti, Congo-Kinshasa and Namibia neighbor each other. In addition, Angola and Congo-Brazzaville, Egypt and Niger, Botswana and Mauritius, Mali and Togo, are countries that are neighboring to or spaced out in the same parent cohesive subgroup. This shows that countries with the same geographical location do not have a direct relationship with each other. However, geographical proximity plays a role in enabling African countries to play a role of sameness and similarity in social networks based on the frequency of international student contacts.

Note: The joined coils are the lower-level subgroups of neighboring countries, and the dashed circles are the lower-level subgroups separated by one country.

5 CONCLUSION

In this paper, we use Ucinet to construct the social network based on the strength of connections and observe the geographic mapping of international students in their home countries to their schools. We found that: (1) The geographical proximity of the country is not an absolute factor in the interpersonal network of international students coming to China from Africa. However, the geographical proximity of the homeland plays a more prominent role in the more connected interpersonal interactions. An analysis of cohesive subgroups found that, while many countries of the same geographical location do not have direct interrelationships, particularly high-intensity connections, a large proportion of countries of the same geographical location are in the same subgroup. (2) To a certain extent, the sprawling spatial expansion of cities has blocked the interpersonal connections of incoming African students. The connection between African students in the same university has not been constrained by the geographical proximity of the branch campus. The university city in Jinan has not played its role in promoting mutual sharing and complementarity among universities. (3) The ranking of interpersonal influencing factors in the questionnaire indicates that most African students in China do not see "the location of the motherland is close" as an important factor in facilitating interpersonal communication. On the contrary, the current geographical proximity factor, such as "study in the same campus," is more important after coming to China. For international students living in foreign countries, although the social communication method of information and communication is becoming increasingly popular, it is difficult to override the traditional geographical proximity.

REFERENCES

Cewenjiabu. 2018. Survey investigation on interpersonal relations of Mongolian students in China. Shandong University.

Feng Jian, Zhao Nian. The social space and network of homosexual groups in context of postmodern geography: A case study of Beijing. Acta Geographica Sinica. 71(10):1815–1832.

Hu Jiongmei. Analysis of factors which influence the cross-cultural interpersonal communication of the Central Asia students in China. Journal of Xinjiang Vocational University. 22(02):65–67+77.

Hu Yve. 2018. A Study on Interpersonal Relationship of Overseas Students in China from the Perspective of Recessive Curriculum. Anhui University.

LV JING. 2013. Acculturation strategies of Chinese international students in the US from the perspective of social networks. Xi'an International Studies University.

Zhang En. 2014. Social network spatial structure analysis and research. Jiangxi University of Science and Technology.

Computational Social Science – Luo, Ciurea & Kumar (eds)
© *2021 Taylor & Francis Group, London, ISBN 978-0-367-70193-2*

International experience and cultural intelligence development: A case study

H.J. Xue
Jiangxi Police Institute, Jiangxi, China

ABSTRACT: Based on a semi-structured in-depth individual interview, this paper explores the correlation between international experience and cultural intelligence. It shows that *"habitus"* and cultural capital at home lay a foundation for Victor's international career and illustrates that international education and experience have a positive impact on the interviewee's career development in multicultural backgrounds.

1 INTRODUCTION

Accompanying the globalization of the economy, internationalization of higher education and international education have become an essential part of the educational system in every country. International education and experience bring the recipients many benefits, including improved English language proficiency and research abilities, wider social networks, richer cultural and world knowledge, and improved cultural intelligence (CQ). This paper examines how Victor developed his CQ through international education and experience and ultimately successfully enhanced his career as an expatriate in multicultural environments.

2 AN OVERVIEW OF INTERNATIONAL EDUCATION AND CULTURAL INTELLIGENCE

International education and experience play an increasingly important role against the background of globalization in the new century. A past study has shown that international education underpins *Young*'s career significantly, thus reaping lifelong benefits [1]. Likewise, *Dong Xiang* has accumulated rich cultural and symbolic capital from his higher education and subsequent work experience in the UK, and enjoys ever-lasting benefits through capital transformation. *Dong* proves that his UK education and experience have a positive and durable impact on his career enhancement in China, thus realizing upward social mobility [2]. Because of the high-quality and well-accepted foreign degrees and certificates, international education is highly valued by Chinese society, and its obvious advantages lie in the recipients' good English proficiency, scientific research and problem-solving abilities, widened social network, intercultural communication competence and social and work experience in multicultural backgrounds [3].

Cultural intelligence is referred to as "a system of interacting knowledge and skills, linked by cultural metacognition that allows people to adapt to, select and shape the cultural aspects of their environment" [4] and it includes three elements—cultural knowledge, cross-cultural skills, and cultural metacognition. CQ is defined as an individual's capability of functioning effectively in culturally diverse environments and consists of cognitive, metacognitive, motivational, and behavioral dimensions [5]. The construct of CQ was first introduced to encompass an individual's ability to effectively function in culturally diverse settings, answering the question why some effortlessly adapt and adjust in cross-cultural situations while others struggle [6]. CQ has been demonstrated to be essential to expatriates' adjustment and adaptation in cross-cultural migration and communication. It is the core element in one's successful adaptation to a foreign society and culture in receiving international education. It is generally believed that studying abroad is instrumental in advancing and developing expatriates' acculturation and adjustment skills. Through a questionnaire with international students in diverse study abroad programs from 33 countries, Ott examines the impact of the level of immersion on developing expatriate acculturation and adjustment skills including self-efficacy, CQ, and willingness to communicate. It proves that it is important and beneficial to study abroad but suggests that students should be encouraged to enroll in programs offering a deeper level of immersion for better self-efficacy, cultural knowledge, and skills [7].

CQ is the key to survival, adaptation, and development in multinational enterprises in a multicultural context. CQ enhances the likelihood of social exchange between culturally diverse individuals and therefore has a positive effect on the relationship between. Imai and Gelfand argue that CQ is one of the highly relevant predictors of an affective performance outcome in a culturally diverse environment. Individual creativity is high in the knowledge condition when individual CQ is high [8]. A moderate level of individual CQ can reduce task conflict and enhance individual creativity, thus influencing the task conflict–creativity relationship in culturally diverse environments and revealing a moderating role. Bogilovic's study illustrates that the relationship between task conflict and creativity in multicultural collaboration is more positive when employees have a moderate level of CQ [9]. Using the above-reviewed literature as a theoretical base, this paper studies how international education and experience impact Victor's CQ and career development in culturally diverse environments.

3 RESEARCH DESIGN

As one paper from "Exploring the Impact of Study-abroad Experience on Career Development," a "13th Five-Year Plan" Education Science Key Project in Jiangxi Province in 2019, this study principally adopts a qualitative method of a semi-structured in-depth individual interview. The interview consists of three parts: (A) demographics (basic data), (B) 23 open-ended questions, and (C) 10 closed questions. Part A aims to understand the interviewee's basic information such as where they are from, where and how long they have been in international education, their current job, and so on. Part B comprises questions in four areas, covering personal background, education, career and experience, and evaluation. Part C is based on a Likert interval measurement scale, aiming at soliciting quantitative date to complement and supplement qualitative data for a comprehensive understanding of Victor's international experience and CQ, thus improving the research validity and reliability.

Victor, the subject in this case, took an MBA program in Johns-Hopkins University (JHU), USA, between 2009 and 2011. By the time he was interviewed for this study, he had completed his international education for about 8 years. He took the first role of the Chief Financial Officer in a famous American Consultation Organization for 6 years after being awarded an MBA degree and returning to China. Currently Victor is working as a Financial Officer in a United Nations (UN) Office in Africa. Due to the fact that he far from China, the interview with Victor was through a WeChat video call and lasted 75 minutes. Shortly after the call, the author transcribed the data for this study. The latter part of this paper analyzes the subject's international experience and CQ development, with the "real voice" heard all through, which reveals the uniqueness of this empirical study.

4 INTERNATIONAL EXPERIENCE AND CULTURAL INTELLIGENCE DEVELOPMENT: A CASE STUDY

4.1 *Socioeconomic background and education in China*

Victor was born to an intellectual family, rich in cultural and social capital at home, his parents being university graduates in the 1980s. His father later took a doctoral research program in education management and was awarded a PhD degree. He has been the Director of a provincial educational research institute for 17 years. Victor's grandfather was an expert in aeronautics and astronautics, a senior engineer, and his grandmother was a vice-director of a hospital. Born as the only son in such a family, Victor cultivated a habit of learning and acquired much cultural knowledge since he was young. Such *"habitus"* (one of Pierre Bourdieu's key theoretical concepts, referring to "systems of durable, transposable dispositions" [10]; it structures an agent's subsequent perceptions, thoughts, attitudes, responses, judgments, evaluation, decisions, actions, and behaviors) shaped Victor's character and nurtured his disposition and temperature, working jointly with great capital at home, the so-called sources of objective power or "trump cards," and it put him in a favorable position from the start [11, 12].

Victor studied in key schools from primary to senior-middle school, and went on to a foreign language school for secondary education. *"I feel that being admitted to the foreign language school was a turning-point in my education, it was very competitive."* Because of the parental educational background, there was no hired tutor at home, Victor finished learning "Junior English" and picked up "New Concept English" in middle-school under the guidance and tutorship of his father. When asked whether this way of learning English helped him to lay a good foundation or form the habit of learning the language, Victor remarked: *"It's a psychological hint. Because when I first started learning English, I felt it difficult and was resistant ... When I passed the first stage, I felt it was a more familiar thing, and with this psychological hint, I feel English is what I am good at. I've never fallen behind others ever since."* "Parental educational background" in human capital, "the frequency of contact" in social capital, and "good parenting," "mutual communication," and "learning at home" in family–school collaboration are shown as important variables that have significant positive effects on children's growth [13]. Consistent with such results, although Victor was unwilling to learn English at the beginning, parental engagement, instruction, and guidance played a vital role in his later liking and learning English well, and finally using English as the work language in multicultural environments overseas. In 2003, Victor passed the entrance examination successfully and went to a famous "985" university, majoring in international accounting. Since he had always dreamed of studying abroad as a university student, he took GMAT and IELTS, with IELTS 7.5 points.

4.2   *International education in the USA and work experience*

Victor went to JHU for an MBA program in Health Care Management in 2009. After 2 years study, he graduated with a degree and returned to work in China. Although he had worked for an American company in Shanghai for 2 years after graduation from the university, Victor had little savings because of the short period of work and average salary. Fortunately, he succeeded in winning a 50% university scholarship and his family covered the rest, including tuition fees and living costs. As mentioned above, Victor comes from an intellectual family with considerable social and cultural capital at home. In this sense, he had two trump cards of *"habitus"* and "capital" from the start [11]. Families lacking financial resources or from low socio-economic backgrounds cannot afford to pay the expensive US tuition fees and living expenses, in addition to supporting him in not working for two years.

While in America, Victor chose home stay, namely staying in the local home with other international students rather than Chinese in this case. On one hand, he made great endeavors to complete his postgraduate program. On the other hand, he tried to integrate into the local culture and make friends with local and international students. After class, he had a part-time job in Internet platform maintenance on the campus, further strengthening communication with foreign students and teachers and promoting his understanding of diverse cultures. When it comes to the greatest benefit he gained from studying in the US, Victor commented: *"it expanded the sphere of my culture and knowledge, broadened my vision, strengthened my confidence, which gave me a sense of satisfaction."* Before returning home, Victor was interviewed for several jobs in America and was eventually employed by a prominent American consultation company. Upon being awarded his MBA degree, he returned to work in the Beijing office of that American company. For 6 years, from 2012 to

2018, he acted as the Chief Financial Officer in China, with the first year contract signed alone, and subsequently he was rewarded with a permanent contract. Prior to being admitted to the UN as an international civil servant in 2019, after leaving the consulting company, Victor went to Beijing Creative Company to be the Chief Financial Officer for some time, and then worked in Lenovo for a period, both adding up to almost a year. During that period, he had been preparing to take tests in public service organized by the UN in his spare time. After successfully passing the written tests and interview, he went to work as a UN Financial Officer in Africa in March 2019. Up till then, he had worked for about 8 years. Concerning the closed-question of "the importance of international education to career development," Victor gave the highest score on the Likert five-point scale, indicating his full affirmation. Likewise, he thinks himself "financially better off as a result of studying abroad," giving "5" points.

4.3 International experience and cultural intelligence development

Because he had planned to take an MBA program in the USA, Victor took a job as a management trainee in an American industrial company located in Shanghai for 2 years after graduation with a BSc, the first job in his career. Management trainee is a method. To be more exact, the company use this method to recruit promising young graduates as management trainees, equivalent to having future corporate management talent reserve, and it normally enrolls only several every year. At the interview, Victor said he had comparatively more contacts with intermediate- and senior-level management as a trainee, and he was also given some rotating opportunities to handle businesses from different angles at that time.

His role as Chief Financial Officer in the China Office of the US consulting corporation after completing his US study enabled Victor to accumulate a wealth of international experience. This American company has over 80 years of experience and global influence, which is rooted in three core areas of strategic consultation, leadership enhancement, and global analysis. The China Office receives clients at home and abroad, engaging in domestic and foreign market investigation and analysis, prediction and related consulting services, and providing managements and enterprises with in-depth studies and professional insight. Victor worked at the company for 6 years, moving from Beijing to Shenzhen and Bangkok, Thailand, thus enriching his intercultural experience greatly. He used the company's international standards to manage the team and had collaborations with enterprises and schools and other corporations in China and other countries. When collaborating with Tsinghua University, for example, Victor, as a department manager, represented the company to practice mock interviews with graduates and provide other career advice. At work in Bangkok, a foreign country, Victor needed to enrich his world and cultural knowledge even more, otherwise he would have been unable deal with the local people.

Victor took over managing a team after 6 months in the American consulting company. When asked about his greatest benefit for working there, he believed it was his all-around experiences as an intermediate-level manager. Victor remarked: "*I just completed my MBA, and then I immediately got the practical experience of management and the stage of management, therefore I accumulated much practical experience in the field. This was the greatest benefit.*" Compared with the size of the consulting company in China, Lenovo and the UN are both mega-corporations. He thinks that even though he is more experienced and capable than before, he is only a little cog in a big machine in the UN. Although his past experience, due to different corporate natures, impacts his UN work indirectly, Victor has inherited the mentality at work, the way of learning and communication from international companies. In his view, going to work in the UN is a turning-point in his career enhancement.

5 DISCUSSION

By scrutinizing the individual case of Victor's CQ development, three points worthy of discussion are described.

1 Family factors played a vital role in Victor's developing CQ in this case. He was born into an intellectual family, and his father cultivated his interest in learning English and encouraged him to enrich his cultural knowledge from an early age. As a result of such enlightened education, Victor likes listening to English songs and music very much. When his parents bought a CD player and headphones, Victor would use red packets and pocket money to buy original English musical CDs and listen to them, to such a degree that he would completely forget food and sleep. Obviously, his father's earlier enlightenment in English enabled Victor to develop and extend his interest in the language further by listening to English songs and music, thus improving his English proficiency and getting to know the world. Victor thinks such an environment at home gave him a psychological advantage that he had the ability to learn English well and communicate with people in English in a foreign cultural background, and that it was a very interesting thing to do. This study shows that Victor changed from his initial resistance to learning English to liking the language, and ultimately he tried his best to learn English well and gained admission to a foreign language school. Entering the provincial key middle-school was the biggest turning-point in his education. His mother's easy-going and communicative personality also had a positive impact on him. Rich social resources and cultural capital at home prepare middle-class children for the transition to higher education, leading to their easy academic progress, a successful path through higher education, and into work [14]. In this light, Victor's "*habitus*" and the joint work of home and school led him to the subsequent academic and career success from the beginning. In other words, "*habitus*" and cultural capital were the two trump cards determining his winning at the game [1516].

2 Victor chose an MBA program at JHU in the USA for his international education. In his words: "*MBA is a professional learning environment. To learn in such a setting is to learn how to communicate with others ... In a different cultural environment, in the English environment, to communicate both at and off work.*" Throughout the whole interview, Victor repeatedly stressed the importance of communication in diverse cultural contexts. Chinese students at UK higher education like to stay together, the so-called "togetherness," a phenomenon viewed as disadvantageous in international education. Although being with fellow compatriots helps to settle in at an early stage and assists in a quicker transition to foreign education and cultures, it hinders students' learning progress and integration into the local society and culture, thus causing communication problems [3]. By striking contrast, Victor preferred to mix with American and other international students, thus greatly improving his intercultural communicative competence. Even for accommodation, he chose to live in a local home and kept in close touch with the host family, rejecting the Chinese "togetherness." Furthermore, Victor tried to communicate with foreign students and teachers by having a part-time job on campus to absorb foreign cultures.

Without a doubt, engaging in the local culture and mixing with local people is the best way to acquire host cultural knowledge and improve intercultural communication skills. When asked about the impact of international education on his career development, Victor listed two areas of support: first, because of JHU's good reputation, it is globally acknowledged; and second, an MBA degree improved his communication competence and skills in the English cultural environment. Victor commented on his personal international education: "*if I had not been to JHU for an MBA degree after working in that American company in Shanghai, I would not have had the opportunity to work in the American consulting company, nor in the UN.*" As can be seen, study in the USA opened up Victor's international professional career.

3 To accumulate international experiences was the best way for Victor to develop his cultural intelligence. First, after graduating from the university, Victor went to work in an American company. His American bosses had profound cultural knowledge, rich international and managerial experience, handled tasks with efficiency, and got along with others in a friendly manner. Through his 2-years' work as a management trainee, he acquired the corporate culture, formed a good working habit, and came to like the multicultural work environment, which was a good beginning to his career. Second, Victor enriched his social experience and cultural knowledge while taking US education. When interviewed, he fully acknowledged becoming "more knowledgeable about the

world as a result of studying abroad," awarding "5" on the Likert scale for this statement. Following this was the 6-years work experience at the American consultation company, particularly the stay in Bangkok, Thailand. As an intermediate-level manager in a multinational company, Victor inherited the corporate culture and management style and had contact with people with culturally diverse backgrounds in Southeast Asia and other foreign countries, which boosted his CQ to an even greater degree.

Finally, the accumulation of cross-cultural knowledge and the ability to work in a multicultural environment have empowered him to secure a permanent contract in the UN after a 2-year application process, written tests, and an interview for the selection of international civil servants worldwide. He stood out as the only Asian in the UN office in that African country. Victor believes that, as an internationalized Chinese, he can find his best value where the Chinese and Western cultures meet. At his interview, he affirmed that each stage of his experience impacted his CQ development in a positive way. He remarked: *"My intercultural communication competence and skills are principally attributed to my experiences. From my own point of view, I maintain more curiosity about foreign cultures."*

6 CONCLUSIONS

Based on a semi-structured in-depth individual interview, this paper explores how his international experience impacted Victor's CQ, and especially his international career development. While studying in the US, Victor was interested in determining the focus of life abroad, and culture constituted a big challenge, thus he suffered from disorientation in how to arrange his academic studies and cultural learning, English learning, social networking, and the like. However, he quickly overrode the "cultural shock" by communicating with his parents and friends, thus enjoying the legacy of rich social resources and cultural capital at home, which reveals the significant impact of *"habitus"* and cultural capital at home. The two trump cards of *"habitus"* and cultural capital created an unshakable foundation for Victor's future international career [1, 17].

His international experiences included 2 years as a management trainee in an American company after his Chinese higher education and 6 years of work experience in a US consulting company after his MBA study in America, which accustomed Victor to enjoy experiencing diverse cultural backgrounds. He acquired a multinational corporate culture, management model, work style, and especially improved his CQ and cross-cultural communication competence and skills. At a personal level, the interviewee maintains his curiosity about foreign cultures and endeavors to understand, integrate, and enrich his experiences. His absorption of foreign culture and international experiences made him encounter the "reverse culture shock" even more keenly than the "culture shock" upon his arrival in the US. To conclude, *"habitus,"* cultural capital at home, and international experience impacted Victor's CQ positively, with CQ underpinning his international career development.

REFERENCES

[1] Xue, H. International Education and Lifetime Benefits: a Case Study. Advances in Social Science, Education and Humanities Research[C], 2019 (347): 253–256.

[2] International Education and Career Enhancement: A Case Study. Education Research Monthly, China, 2020 (9): 86–92 (In Chinese).

[3] Xue, H. Cultural Adaptation and Personal Capital Formation: The Experiences of Chinese Students in UK Higher Education[M]. Shanghai: Shanghai Jiao Tong University Press, China, 2011.

[4] Thomas, D. C., Stahl, G. K., et al. Cultural Intelligence: Domain and Assessment. International Journal of Cross Cultural Management, 2008, 8 (2): 123–143: 126.

[5] Ang, S. & Van Dyne, L. Handbook of Cultural Intelligence: Theory, Measurement and Application. Armonk, NY: ME. Sharpe, 2008.

[6] Earley, P. & Ang, S. Cultural Intelligence: Individual Interactions across Cultures[M]. Stanford, CA: Standard University Press, 2003.

[7] Ott, D. Study Abroad and the Development of Expatriate Acculturation and Adjustment Skills[A]. Between Cultures and Paradigms: Intercultural Competence & Managerial Intelligence[C]. Warwick: The University of Warwick, 2014.

[8] Imai, L. & Gelfand, M. J. The Culturally Intelligent Negotiator: The Impact of Cultural Intelligence on Negotiation Sequences and Outcomes. Organizational Behavior and Human Decision Processes, 2010, 112 (2): 83–98.

[9] Bogilovic, S. From Heaven to Hell: Task Conflict, Creativity and Cultural Intelligence[A]. Between Cultures and Paradigms: Intercultural Competence & Managerial Intelligence. Warwick: The University of Warwick, 2014.

[10] Bourdieu, P. Outline of a Theory of Practice. Translated by R. Nice, Cambridge: CUP, 1977: 72.

[11] Bourdieu, P. Distinction: A Social Critique of the Judgment of Taste. Translated by R. Nice, London: Routledge, 1984.

[12] Bourdieu, P. Sociology in Question. Translated by R. Nice, London: Sage, 1993.

[13] Wu, C., Zhang, J. & Wang, M. Power of Parental Involvement: Family Capital, School-family Partnership and Child Development Education Research Monthly, China, 2014 (3): 15–27(In Chinese).

[14] Power, S., Edwards, T., Whitty, G. & Wigfall, V. Education and the Middle Class. Buckingham: Open University Press, 2003.

[15] Bourdieu, P. 'The Forms of Capital'. In Halsey, A. H., Lauder, H., Brown, P. & Wells, A. S. eds., Education: Culture, Economy, Society. Oxford: Oxford University Press, 1997: 46–58.

[16] Bourdieu, P. 'Cultural Reproduction and Social Reproduction'. In Karabel, J. & Halsey, A. H. eds., Power and Ideology in Education, Oxford University Press, 1977: 487–511.

Computational Social Science – Luo, Ciurea & Kumar (eds)
© 2021 Taylor & Francis Group, London, ISBN 978-0-367-70193-2

Theory and case enlightenment of literary canonization in the Occident

B.C. Jiang & J.Y. Jiang
School of Humanities, Huazhong University of Science and Technology, Wuhan, Hubei, China

ABSTRACT: Since the 1960s, Western academic circles have discussed "literary canon" and "literary canonization." This paper first describes the meaning and standard of literary canon, and then discusses the canonization methods and characteristics in the Occident by taking the canonization process of Shakespeare and Thoreau as examples and explores the enlightenment of Western literary canonization theory for Chinese literature canonization studies. It is possible to break the Western centralism concept in the process of canon composition and realize the canon reconstruction of cross-cultural boundaries by drawing on Bloom's revisionist theory.

1 INTRODUCTION

Literary canon refers to exemplary, authoritative, and enduring masterpieces, which are the symbols of literary achievements and the essence of culture in every era. How a literary work becomes a canon is not a new topic in Western literary theory and cultural research. Since the 1960s, the issue of "literary canon" and "literary canonization" has been heatedly discussed. In the early 1990s, the Dutch scholar Douwe Fukkema came to China to lecture and promote the awakening of the classical consciousness of Chinese literary theory. In the late 1990s, the Chinese literary research circles began to study literary canonization. Initially, various literary canon series were published, such as the eight-volume *Hundred Chinese Literature Classics* by Xie Mian and Qian Liqun. Then the discussion of Chinese contemporary literature canon and even the ancient literature canon was developed. Thereafter, the study of Chinese literature canonization has been fruitful into the 21st century.

However, although there is a consensus on what a canon is, more research about the canonization of contemporary literature has yet to be done. Through the Western theoretical discussion on literary canon and the canonization process examination of Shakespeare and Thoreau, this paper summarizes the basic methods and characteristics of literary canonization in the Occident, providing some useful enlightenment and references for the study of Chinese literature canonization.

2 THE CONCEPT AND EVOLUTION PROCESS OF CLASSIC AND CANON

Like most of the classics, the formation of two words—classic and canon—also experienced a long evolution process. Classic, derived from Latin *classicus*, is a term used by ancient Roman tax officials to distinguish tax grades. It was not until the Renaissance that people began to use it to evaluate writers. It was then extended to model and standard. Later, people linked classic with ancient and came up with the term "classic antiquity." As a result, the Ancient Greek and Roman writers became classical authors. Then the word "Classicism" was evolved and further spread and penetrated into other cultural and artistic fields.

The term Canon in Western literary theory originally derived from a Greek word *Kanôn*, which means the length of reeds or sticks. Then, after the Reformation and the Renaissance, canon referred to the Bible or other orthodox texts recording the holy truth. It was not until the end of the 19th century that the word canon first appeared in the 1885 edition of the *Encyclopedia Britannica*.

After that, the meaning of canon went beyond the scope of biblical canon and extended to various fields of Western culture. Some pioneering works in the English world were also called canons of English classics. For example, Charles Dickens' novels and Tocqueville's *De la démocratie en Amérique* translated into English.

In Chinese academia, there has been no precise definition of canon since ancient times. Canon is written as "Jing" and "Dian," and often used interchangeably in ancient books. The chapter of *Zong Jing* in *Literary Mind and the Carving of Dragons* said, "The so-called Jing is the eternal and absolute truth that cannot be changed." Another book *Er Ya* defined Dian as laws. In this way, before Jing and Dian were used together, they had already gained the meaning of truth and laws. Subsequently, Jing and Dian were combined to indicate those classic Confucian works in *Han Shu*. In modern times, canon usually means normative exemplary laws and paradigms. It not only refers to the authoritative texts handed down throughout history and tested by time in the form of words or other symbols, but also contains the cultural norms in such texts which restrict people's thinking, feelings, and behaviors.

It can be seen that both classic and canon have gone through thousands of years of complex evolution, and it was very late for them to acquire their modern meanings. The long history of the formation of classics has prepared sufficient conditions for future generations to carefully examine, read, appreciate, and determine the ancient classical writers in the modern sense. In other words, it is very difficult to test the real classics only in the contemporary era or in limited times.

3 EVALUATION CRITERIA ON CANONIZATION

So, what are the characteristics of literary classics? What are their evaluation criteria? Combined with the domestic and foreign studies of classicization in recent years, the standards of literary classics are summarized as follows. First, literary classics should be original, that is, classics should have substantial creativity. Harold Bloom once said in *The Western Canon*, "to ask what makes the author and the works canonical. The answer, more often than not, has turned out to be strangeness, a mode of originality that either cannot be assimilated, or that so assimilates us that we cease to see it as strange." Second, literary classics should have profound cultural heritage and aesthetics. According to Huang Manjun, the literary classics reveal a kind of far-reaching cultural connotation and human nature. "From the aesthetic point of view, they should have poetic connotation; from the perspective of national characteristics, they also have epic value in the history of national literature." In addition, literary classics should be a model of an era or national language. Just like the importance of Shakespeare in English, Pushkin in Russian, Schiller and Goethe in German; their literary classics not only reflect the cultural characteristics and spirit of the nation, but also become a model of the national language.

Canonization means the process of being transformed into a canon. Through the discussion of how a canon is formed, there are two points of view, one is the canonization theory of essentialism and the other is the theory of constructivism. The former infers that the formation of canon is mainly determined by internal factors, while the latter believes this is due to external factors.

The theory of essentialism classicization holds that the reason why a literary work becomes a canon is entirely determined by the excellent character of the work itself, also based on the aesthetic elements that is contains. The tragedy of ancient Greece marked the peak of the development of ancient Greek art because of the artistic spirit and aesthetic style of the work itself. Marx once commented on the ancient Greek tragedy as a highly inaccessible template and norms. The greatness of Shakespeare is that his works are engraved with the highest achievements of aesthetic and moral quality. In Harold Bloom's view, the core reasons why William Shakespeare became a classic were his keen insight, profound thought, and extraordinary power in language. He also criticized the supporters of constructivism, because they "cannot illuminate him with a new doctrine"; Shakespeare is always ahead of them, "conceptually and imagistically."

Constructivists believe that a canon is mainly invented or produced by external factors, not aesthetics inside the works. Some even refuse to recognize the aesthetic principle that literary classics have universal validity. They think that the reason why classics are regarded as classics is that the values advocated by them are in line with the mainstream ideology, and are forced to be carried out due to the need for political power, thus being legalized and naturalized by the mainstream cultural circle and intellectual elites of the times. As a representative of the canonization theory of constructivism, Bourdieu claimed that he abandoned the essentialism dilemma of the internal and external dichotomy of literary classics and put forward the theory of cultural production field. Although he acknowledged the aesthetic essence of literature itself, he still advocated constructivism theory. He believed that the formation of classics reflected a sort of negotiation relationship between various location elements in the cultural production field, and also reflected the interactive relationship between the cultural production field and other fields.

4   SPECIFIC CASES IN THE OCCIDENT OF CANONIZATION

Both the internal and external factors influencing the formation of canon vary greatly due to different times and sociocultural contexts. Therefore, the formation of representative classic literary texts should be analyzed in detail. There is no universal classic, but Shakespeare as a classic seems to be indisputable. Shakespeare did not reach the status of being so highly respected in his life, but he experienced a relatively smooth process of canonization.

4.1   *What makes Shakespeare classical*

Judging from the aesthetic value of his works, Shakespeare surpassed all other Western authors in cognitive acuity, linguistic vitality, and creative talent, putting him at the center of the classics. These three endowments combine to form an ontological passion—the capacity for joy. Since the success of *Love's Labour's Lost* in 1594, Shakespeare showed great creativity and deep thought in the richness and diversity of his language and in his characterization. His Hamlet, Lear, Falstaff, Rosalind, Othello, Macbeth, and other classic images have been successfully transplanted into the world literature and attracted attention beyond the limitations of time, place, and translation. The major figures of Spain's golden age—Cervantes, Vega, Molina, and Gongora—all brought to Spanish literature a strong Baroque style of Shakespearean and romantic art. Milton's first poem, published in his early twenties, is one of Shakespeare's second folio to the book. Shakespeare had died 16 years previously, but his reputation remained, and he had undergone the canonization process that began in the eighteenth century with Dryden, Pope, and Dr. Johnson, and in the early stages of romanticism, which was also a movement to deify Shakespeare. Goethe's affirmation of Shakespeare strengthened his position in Germany. Hegel was a great admirer of Shakespeare, and he created characters through his wit and imagination that fully aroused our interest in sinners, the most vulgar of lowlife, and fools. Late French romanticism brought a strong Shakespearean influence to French literature, which was particularly strong in Stendhal and Hugo. Manzoni, the Italian novelist, was a Shakespearean writer, and while Tolstoy had angry arguments regarding Shakespeare, his own art borrowed from Shakespeare's sense of character. Pushkin and Turgenev were important Shakespearean critics in the 19th century. Ibsen was brilliant, but he could not avoid Shakespeare. In the American Renaissance, Shakespeare's influence is most evident in the works of Melville, Emerson, and Hawthorne. Shakespeare was not abandoned by the wave of modernism and new criticism of the 20th century, and his work has long been the subject of study.

Of course, the canonization of Shakespeare also shows the complex interests of the literary field. While paying attention to the cultural value of Shakespeare's works, the economic value in the process of canonization cannot be avoided. Printing technology plays a role in the process of canonization. At the start of the printing age, Shakespeare's works were reprinted in large numbers. Although this did not mean that his works had eternal literary value, it was closely related to his active integration into the mainstream culture, because he had been highly praised before his death, especially by the aristocracy and intellectuals. Great economic interests promoted the mass printing of Shakespeare's plays, making them widely circulated and widely read, criticized, and accepted by the English intellectual class, which resulted in him gaining a great reputation. The publication of Shakespeare's first folio in 1623 was an important symbol of the canonization of the printing industry. By the 18th and 19th centuries, an academic commentary on a series of Shakespeare's works had been published, further enhancing his reputation and spreading it globally. Thus, Shakespeare went through the process of re-canonization.

4.2 *How Walden became a canon*

Compared with Shakespeare's works, *Walden*, regarded as a classic of American literature, has undergone a relatively tortuous and slow process of canonization. Compared with the genius of his contemporaries Hawthorne, Melville, and Emerson, his prose seems more in keeping with the fashion of the 20th century, when fiction and travel notes were in vogue. At that time, many American readers did not fully understand the idea and style of this book. They thought of it as a prose poem, which sometimes felt new and original, and sometimes felt strange. Influenced by Emerson's transcendentalism, *Walden* reflects the connection between nature and human spirit, emphasizing the creative side of the spirit of the natural universe, which can give people the enjoyment of beauty and edification of morality. Yet this expression of transcendentalist philosophy, and the writer's way of life, was once seen as bizarre and unrealistic. It was not until the early 20th century that the study of American literature in the 19th century began to emerge in the United States. Later, due to the Great Depression, Thoreau's philosophy of a simple life was respected, and *Walden* also became known and accepted by more people. Francis Otto Matthiessen thinks Thoreau's text understand up, concise and accurate, no Victorian distractions, abstract and obscure, and emotional appeal, paying attention to use of the local language, especially puns, etc., making the product a "satirical humor." This established the literary status of *Walden* in the American Renaissance. With the rise of global ecological trends, *Walden* is considered to have far-reaching ecological significance, and Thoreau is regarded as the pioneer of the world's environmental protection movement and ecologism. In addition, after the Second World War, Mahatma Gandhi and Martin Luther King's "Civil Rights Movement" ideas were widely spread, Thoreau's "Civil Disobedience" thought was re-explored and had importance attached to it, which had an important role in promoting the canonization of his works.

From these two cases, we can see that the canonization of Shakespeare's works was at the right time. Besides the artistic value of his works, it is inseparable from the promotion of printing technology, economic interests, and the praise of intellectual elites. Of course, there was a phenomenon of devaluation in the process. However, due to the artistic value of his works, it has been valued and studied by literary circles all over the world. As for the canonization of Thoreau's *Walden*, on the one hand, its inherent artistic spirit and aesthetic style are unique. On the other hand, the historical opportunity and the needs of social development in later generations contributed to its success. We call this the combination of the implied potential value of the work and the practical needs. However, this kind of external condition is inevitable, which Thoreau himself cannot predict. Therefore, no matter what role external factors play in the process of canonization, the artistic value and ideological and cultural connotation of literary works are the basic conditions for the formation of classics. If a good work wants to be a classic, its artistic value comes first, and its status in history fluctuates, mainly because it is closely related to the artistic interest demand of later generations. As for the ideological and cultural value contained in the work, it is closely related to the political, economic, and cultural needs of later generations. If both can be released at the same time in an era, it will inevitably become a classic and may constitute the cultural spirit and ideology of the era. At this time, state, national, political, and cultural power as well as economic interests are infiltrated into it.

5 CONCLUSION

It can be seen that although literary canonization is composed of internal and external factors, external factors are uncontrollable and cannot be met. Therefore, when referring to the "canonization study" of Chinese contemporary literature, we should see whether our literary research is from the aspects of literary originality, cultural heritage, aesthetics, and national language representativeness. This includes whether the works of well-known writers who reached a conclusion in the 20th century have originality, cultural heritage, and aesthetic value in the history of Chinese literature and world literature development? Do they represent the highest level of contemporary literary language and become a model of modern Chinese? If not, what is the direction of our future efforts? Perhaps after we have made clear these questions, Chinese literary circles will emerge with more classic works that can represent the essence of national literature and enter the treasure of world literature.

In the present era of cross-culture and cross-literature interaction, we should view the position of Chinese literature in the world from a broader perspective of human civilization. Also, Chinese scholars should draw lessons from Bloom's revisionist theory, inherit the criticism of cultural modernity and literary modernism, combining the skepticism of postmodernism with the attempt of "decanonization" of cultural studies, so as to devote themselves to the study of Chinese literature and realize the reconstruction of classics across cultural boundaries.

REFERENCES

[1] Bloom, H. 1994. *The Western canon*. New York: Harcourt Brace & Company.

[2] Calvo, I. 1986. Why read the classics. *The New York review of books* 33(15).

[3] Fokkema, D. 2005. All classics are equal, but some are more equal than others. *Comparative literature in China* 61(4): 51–60.

[4] Huang, M.J. 2004. The canonization of the Chinese new literature in 20th century. *Literary review* 04(4): 108–114.

[5] Kolbas, E.D. 2001. *Critical theory and the literary canon*. Boulder: Westview Press.

[6] Matthiessen F.O. 1941. *American Renaissance: Art and expression in the age of Emerson and Whitman*. New York: Oxford University Press.

[7] Tao, D.F. 2004. Literary classics and cultural power. *Comparative literature in China* 56(3): 58–74.

Computational Social Science – Luo, Ciurea & Kumar (eds)
© 2021 Taylor & Francis Group, London, ISBN 978-0-367-70193-2

The impact of state bureaucracy in the process of democratization

Y.J. Cao
South China University of Technology, Guangzhou, China

ABSTRACT: It is worth emphasizing that, after the Second World War, the process of democratization is either taking place or has occurred in most states. At the same time, the concept of modern countries is pure and simple, but they are more bureaucratic than other older governmental organizations (Waters, Tony 2015). Therefore, on the basis of understanding the proposition that "state bureaucracy is necessary and dangerous," this essay discusses states bureaucracy and the consolidation of democratization. Taking Japan as an example, it can be seen that bureaucracy is deeply rooted in Japan's constitutional system. There is a common concept of bureaucracy in Japan that bureaucrats have too much prestige and influence in policy making. However, it is necessary to consider and discuss whether there is a hidden danger behind the bureaucracy or whether it plays an indispensable role in a country.

1 INTRODUCTION

Modern state has developed permanent bureaucracies as the primary instrument for their consolidation. In most academic definitions and general descriptions, the notion of bureaucracy and democracy is generally considered as an opposing approach which is provided for the governance of a state (Etzioni, 1983). On the one hand, bureaucracy is usually defined as an effective management and governance system of national public projects as a mandatory condition. While it is legitimate, it is generally largely indifferent to the individual citizen's wills and demands. At the same time, bureaucracies are often hierarchical, even autocratic forms of management. Nevertheless, bureaucracy is not as depressing as it is described. In fact, the bureaucratic form of logical institutionalized management can ensure the equal treatment of citizens and give citizen's comprehensive records and satisfactory reasons, which can promote their correct decisions in the public sector (Kenneth and Deth, 2005). On the other hand, the governing institutions of democracies are supposed to respond to the wishes of the public and to try to turn these preferences into positive results (Kenneth and Deth, 2005). However, Rose (1974) noted that in traditional representative democracy, the link between voting and policy choice is not as clear as most democrats believe it to be. In addition, the purpose of the public vote is uniform. There are also unrealistic expectations in inconsistent situations. Therefore, it requires leaders to be returned and bureaucrats to make policy decisions (Caplan, 2007). Therefore, it is vital that bureaucrats carry out policies, make rules and regulations, and manage the people and other government departments. At the same time, it may be overlooked that bureaucracy can also be also dangerous. Although the division of labor is very operative, it can also have some harmful effects on a democratic country. For example, scandals and corruption have arisen. Japan was the first country in Asia to join to group of industrialized democracies, providing a prominent example of a highly democratic state bureaucracy outside the Western states. Thus, as Pempel (1992) emphasizes "Rather than America's democracy or other industrialized democracies, Japanese bureaucracy is the prominent success" (p.19). In particular, Japanese bureaucracy plays a central role in the deploying and implementation of many changes and reforms in Japan. It is an appropriate example to illustrate the essence of bureaucracy as the deployer or promoter of these changes. Secondly, the close ties between Japan's bureaucracy and its political leadership are a notable contrast with some other countries. In these countries, officials and politicians are reckoned to play quite independent and uncoordinated roles. Third, Japan's conscious restrictions on the scope of its government bureaucracy are considered inevitable. Instead, Japan provides the most prominent example of a systematic reduction of bureaucracy. Finally, Japanese bureaucracy has undergone several conscious reconstructions in its history, including direct imitation of the Prussian political model and an organized reconfiguration of American bureaucracy, which is a first-rate case study of the interaction among traditional concepts in bureaucracy.

2 HISTORICAL BACKGROUND

2.1 *The Meiji era (1868–1912)*

According to the records of the Meiji era, the early Meiji government implemented a series of measures to lay a foundation for the development of truly progressive state bureaucracy. During the Meiji era, profound changes took place in all aspects of society. To achieve the goals of a wealthy country and a strong military, Japan's leaders intended to modernize the country. In 1868, the Meiji government took the Oath of the Charter to communicate and implement these changes throughout Japan. This brief document outlined the government's intentions and policies and laid the groundwork for all reforms to be carried out in the coming decades. At the end of the Meiji era, Japanese reformers succeeded in curbing bottom-up resistance, while building a modern state and a world power.

2.2 *Taisho democracy (1912–1926)*

The Taisho era was also known as the Taisho democracy because of an unexpected atmosphere of radical liberalism in Japan after decades of Meiji dictatorship. A famous political figure coined the term "Da Zheng democracy." Doctor Yoshino was a professor in law and politics. He made extensive observations and traveled to the West. Then, he returned to Japan and began to write. A series of articles were written by him to promote free democracy in Japan (Japanese Taisho democracy: 1912–1926, n.d.). Yoshino (2005, pp. 838) wrote that:

> *"The fundamental prerequisite for perfecting constitutional government, especially in politically backwards nations, is the cultivation of knowledge and virtue among the general population. This is not a task that can be accomplished in a day. Think of the situation in our own country."*

Taisho Democracy in Japan: 1912–1926 (n.d.) describes Japan's rush to form a constitutional government before its civic and economic foundations were ready. The end result was that they experienced various failures. However, to return to the old path of authoritarianism was not desired. They could only take the path of reform and progress cheerfully and surely. In conclusion, the most important method is to rely not only on politicians but also on the collaborative efforts of leaders, thinkers, and educators in all fields of society. As these ideas spread, a mass movement for political change emerged in Japan. The Labor Union began massive strikes involving protests against political injustice, the unequal status of workers, raising objections to treaty negotiations, and a discussion of whether Japan should participate in the First World War took place. The number of strikes increased from 108 in 1914 to 417 in 1918 (Japanese Taisho democracy: 1912–1926, n.d.). Regardless of the fact that democracy began to emerge gradually, the impact of the Great Depression of 1929 forced talk of democracy to temporarily be withdrawn from Japan's political arena at the time. From the 1930s to the end of the war in 1945, Japan experienced a period of military rule. During this period, a new type of bureaucracy known as Kakushin Kanryo appeared.

2.3 *Post-war bureaucracy*

Japan's bureaucracy began to emerge as actively after the war. During this period, regardless of the fact that the economic bureaucracy established a strong system, the economy as the central figure was controlled and governed by empowering each ministry. Because personnel management can organize and promote the healthy competition of resources and power between departments, the entire bureaucratic system can be perfected using competition. Therefore, the early experience with democracy is one of the reasons why Japan was able to make a transition to democracy relatively smoothly in the post-Second World War period. According to Pempel (1992): "Japan's contemporary bureaucracy is structurally similar to many of its counterparts in Western Europe, although it is typically smaller than any other" (p. 23). This is very meaningful because it is usually equipped with national elites. The bureaucracy instilled a sense of mission which enhanced the political power. It has close relations with the ruling party and the main opposition parties, but the individual members of the bureaucracy rarely participate directly in party or electoral politics during their time in office. One thing that differs from the state bureaucracy in the United States is that many high-level positions change hands in the United States as the presidential position changes. In Japan, the renewal of politicians is very fast but the bureaucracy is stable for a long time. Japanese bureaucracy is also a particularly small and constantly re-examined body of government. Japanese politics is a kind of pluralistic politics and Japanese bureaucracy is part of this political system. It is worth mentioning that Japanese pluralism is different from American pluralism, because Japanese bureaucracy plays a more important role in Japanese politics than that in America.

3 STATE BUREAUCRACY IS ESSENTIAL IN JAPAN

Bureaucracy is a way to achieve democracy in Japan. It is for this reason that the occupying authorities have shown unparalleled shrewdness in making effective use of the bureaucratic machinery that has allowed a form of democracy to survive in Japan. There are indications that the origins of bureaucracy go back to the Tokugawa period in Japanese history (1603–1868). Meanwhile, Japan's political system, policies, and decision-making processes are all embodied in Japanese bureaucracy (ISP, 2014). At that time, the lords of the provinces appointed selected warriors as their attendants, who gradually became bureaucrats in the 17th century. They were disarmed, deprived of land ownership, forced to move to a castle town, and handled the running of systems in times of peace, such as the prevention of crime, taxation, flood control, infrastructure development, management of political relations in the Tokugawa bakufu (ISP, 2014) Therefore, according to the development and evolution of Japanese bureaucracy, it is worth noting that Japanese bureaucracy has a meritorious pattern. Therefore, elites have the highest social status among all the four existing classes of the Japanese society. Nonetheless, it makes them feel all powerful as members of the official elite of society, resulting in some abusing their positions and causing some personal scandals. On the other hand, the consolidation of bureaucracy is a fairly important factor. Japan's bureaucracy was finally consolidated in the latest period of the Meiji era, and also played an important role in mobilizing resources to promote economic growth and political participation after the Second World War. Meanwhile, the relationship between Japan's contemporary bureaucracy and the ruling Liberal Democratic Party (LDP) and bureaucracy in history is well known. The LDP, which ruled from 1955 to 1993, is characterized by a large number of former bureaucrats among its members. The LDP tightly controls promotions within the bureaucracy, making it almost impossible for party bureaucrats to rise to unacceptably high positions. This makes the interdependence between bureaucracy and government even stronger. In addition, bureaucracy plays an important part in the decision-making process, preparing bills for the party, which are amended for submission to Congress.

In terms of the economy, Seymour Martin Lipset (1959) focused on the importance of economic development as a necessary precondition for democracy for a long time. For example, Kosai (2008) found that Japan began economic and political reconstruction under the American-led Allied military occupation until 1952. A lasting impact of post-war reconstruction was the beginning of Japan's remarkable economic performance. According to Kosai (2008), "It is estimated that the Japanese economy grew at the tremendous rate of almost 10 percent per year between 1945 and 1973" (pp. 494–537). It can be illustrated the democratization in Japan progressed along with economic growth in the years during and after the Allied occupation, with a series of constitutional guarantees in the constitution of 1947 (Kosai, 2008). At the same time, after the Second World War, Japan's democratic transformation and consolidation took shape. Japan's democratic consolidation was based on improved civil services. A good civil service can guarantee the basic civil rights of its citizens. The autonomous government system operating was safeguarded under the rule of law and a democratic and fair electoral system. There are also government agencies that had extensive experience in the early stages. These basic standards are reviewed as essential. At the same time, economic performance and social equality have also been adopted. Therefore, Japan's democratization over this period may be accounted for by the Japanese economic success of the post-war years. However, it cannot deny the importance of bureaucracy in Japan. For a long time, Japan's bureaucracy has been better at keeping an eye on the needs of the Japanese people than their own elected representatives. This clearly demonstrates the importance of bureaucracy in Japanese political life. Second, because Japanese bureaucrats experience a fairly orthodox career model, the Seniority is mainly based on the personnel of equivalent record of formal schooling to evaluate. Seniority seems to prevail within the merit system, and there is little lateral mobility. Another aspect of bureaucratic homogeneity is that civil servants remain in their original departments. This state of affairs has led to loyalty to the various ministries and factions. This is not very different from the typical school faction. As a consequence, some small groups are clearly inclined to cultivate factionalism among Japan's top bureaucrats. However, such factionalism does not breed a great risk of corruption. However, given all of this, it is difficult to dismiss the importance of groups within Japan's bureaucracy. The homogeneity of social background, education, and career model often leads to the independence of the whole bureaucracy and weakens the unseen danger of collusion. However, the emphasis on school cliques also demonstrates the importance of groupings within the bureaucracy. This reliance on groups may reveal the habitual weaknesses of individualism and bureaucracy in Japanese society. However, Japan's current situation draws our attention to the fact that this group dependency is the result of the "vestiges of feudalism" and is clearly the result of a relatively recent imposition on Japan as a democracy. In any case, we can doubt the importance of individuals or groups in Japanese bureaucracy as there is no concept of individualism in Japanese culture. At the same time the group phenomenon is also defined as a result of an imposed "feudal residue." Therefore, Japanese bureaucracy is relatively necessary for professional Japanese government.

4   STATE BUREAUCRACY IS DANGEROUS IN JAPAN

Although bureaucratic impressions in many states have not been good for a long time, if public bureaucracies have good order then this will be close to Weber's ideal type. Therefore, there is a more feasible way to implement it. Collaboration between bureaucracies and informal organizations can be used to increase bureaucratic efficiency and customer responsiveness. For example, informal organizations have opened up so-called "channels" for officials who accept bribes. This makes it easy to fall into the practice of violating democratic principles. As a result, some new democracies today are threatened by nepotism and bureaucracy. National civil servants also belong to bureaucracies. The national civil servants are also in the category of bureaucracy. It is a good and loyal position for the authorities and state. The appointment of civil servants should never be a way for bureaucrats to make a profit. However, according to the theory of nepotism, some bureaucracies are used by politicians to achieve their own political goals. This is inconsistent with nonpersonal, politically neutral, dominant, and universal Weber's theory. Nepotism refers to a bad relationship between politicians and the aides created by the use of public office to obtain greater power or money in the bureaucracy using family relationships. Under the influence of political nepotism, the government has also become "the only beneficiary." The government acts as a customer and distributes benefits to society in the form of public work, money, contracts, and pensions in exchange for public support for the public. In other words, nepotism is a form of sponsorship for bureaucracies, as summed up in an old proverb: "When a winner is spoiled, it can easily become a form of corruption." According to Shendrikova (2014), in May 2014, Japanese Prime Minister Shinzo Abe decided to establish the Cabinet Personnel Bureau, a new institution managed by Katsunobu Kato, one of Abe's closest assistants. Undoubtedly, this strategy is to strengthen his overwhelming control over the Japanese bureaucracy for many years.

Recently, scandals involving Japanese civil servants have exposed a series of bureaucratic mistakes and failures (Mulgan, 2018), for example, forging, concealing, and denying the facts of official documents, editing and tampering with key parts of administrative records privately, giving false testimony or refusing to testify before Congress, "discovering" missing documents, generating misleading information, and sexual harassment allegations. In most of the cases listed above, ministers colluded or were aware of these incidents. The impact of these abuses has greatly increased the dangerous nature of bureaucracy. Prime Minister Shinzo Abe, for example, has lost his reputation and credibility, leading to his cabinet's lowest approval ratings since his second government came to power. The public impression, however, is that many leaders who should have gone have not, from Taro Aso, Finance Minister, to the Prime minister himself, for example. The unease has even spread to Japan's ruling Liberal Democratic Party, where the public has expressed sympathy with bureaucrats with deep doubts about whether the scandal has been fully exposed. Kaneko (1999), an academic economist, said the developments revealed a deeper "structural problem" in Mr Abe's government, which he blamed on a new system of senior appointments led by the Cabinet Bureau of Personnel Affairs. It not only allows the Prime Minister and Chief Cabinet Secretary to be closely involved in the personnel affairs of the country's civil servants, but also fosters bureaucrats who "speculate" about what Kantei (the prime minister's office) wants. As a result, the neutrality, fairness, moral standing, and competence of the Japanese bureaucracy have been eroded. Hideaki Tanaka (2009) agrees, saying that "the fundamental problem of Japan's government bureaucracy is that public servants' expertise is being belittled and politicised." Another reason is that the Prime Minister's administrative status has risen to the pinnacle of government power. Today, not only is the Prime Minister himself exercising unprecedented power, but so are many of the key supporters of the scandal that have engulfed the bureaucracy and the broader Abe government. Kaneko (1999) believes that these scandals are a symptom of the growing phenomenon of "crony capitalism" in Japan, which is thriving under the "Prime Minister-dominated politics." This has had a negative impact on Japan's economy, as the government supports uncompetitive industries led by people close to the Prime Minister. Crony capitalism refers to an economic system in which the usual democratic checks and balances do not work, with the result that benefits are distributed to people close to those in power. In today's Japan, it involves offering favors to the Prime Minister's friends and suppressing opposition or criticism. In a similar vein, Koichi Nakano, a Japanese political scientist, writes in his latest book that "Prime Minister Abe appropriated state funds for his own benefit and for the benefit of his network." However, in the case of Japan, there are some dangerous factors in the democratic process, such as corruption and scandals, in the Japanese bureaucracy. However, it is true that the speed at which the bureaucracy is expanding at both the national and local levels in Japan has been astonishing, and certainly demonstrates the role that bureaucracy plays in Japanese life and government. For example, the size of Japan's bureaucracy almost tripled in the 20 years from 1931 to 1951. Statistics show that in 1965 about one in 12 workers in Japan was a member of the bureaucracy, the most recent statistics available in 1982 put the total number of national and local bureaucracies in Japan at more than 5 million. Japan's national bureaucracy has remained relatively stable over the past 17 years. Overall, Japan's bureaucracy is still growing fast. Under normal circumstances, this huge growth and expansion of Japan's post-war bureaucracy to some extent shows that the nature of Japanese bureaucracy as part of Japanese society and culture is generally accepted today. Therefore, this mainly shows that there are some dangers related to the state bureaucracy. Corruption, bureaucratic scandals, and other negative factors will increase the risks to the country's bureaucracy. Of course, these dangers could actually be avoided through highly professional and detailed decentralization and legalization. According to the above, even Japan has certain shortcomings and risks to its bureaucratic system, however, according to what was discussed earlier in this article, Japan is also considered to be a well-functioning bureaucratic state. It is therefore necessary to recognize that state bureaucracy is necessary and also that it can be dangerous.

5 CONCLUSIONS

In conclusion, the definitions of "state" and "bureaucracy" are absolutely necessary, because the state is defined in part by the characteristics of the bureaucratic organization, nonindividualism, and sovereign claims. Bureaucratic institutions, in their ideal form, are nonpersonal and transparent. Therefore, countries with less bureaucratic functions necessarily have much lower national capabilities. However, the common perception that bureaucracies are inefficient is partly a consequence of their working well which will often still lead to frustration. However, Weber's bureaucratic theory (1905) emphasizes that it not only has a comparative advantage in terms of expertise and feasibility, but also categorizes it as the dominant source of the decline of the class system, including feudalism and other forms of unfair social relations within individual countries. In bureaucratic organizations, universal rules and procedures dominate, making personal status or relationships insignificant (Mulder, 2017). In this form, bureaucracy is the epitome of generalized standards. Despite the hidden dangers in bureaucracy, the government system is based on law and requires bureaucracies to perform their roles. Therefore, the bureaucratization of a country is not entirely positive, however it provides the foundation of democratization, because it removed the feudal, plutocratic, and hereditary administration. Japan's Meiji restoration, driven by powerful modernization and ambition, promoted centralization, weakened aristocratic rule, and established a powerful bureaucracy. Japan is now a wealthy, non-English-speaking democracy, due to its long history of bureaucracy. Therefore, bureaucracy is necessary, but a function of bureaucracy, administrative formalism of bureaucracy, should be reduced, the number of rules should be reduced, and there should be less discretionary power and performance accountability system in the organization should be increased. However, this essay only describes the situation in Japan, and so has certain limitations.

REFERENCES

Caplan, B. (2007). The Myth of the Rational Voter: Why Democracies Choose Bad Policies. Princeton: Princeton University Press.

Etzioni, H. E. (1983). Bureaucracy and Democracy: A Political Dilemma London: Routledge and Kegan Paul.

Facing History and Ourselves (n.d.). Meiji Period in Japan. [Online]. Available at: https://www.facinghistory.org/nanjing-atrocities/nation-building/meiji-period-japan [Accessed 12 December 2019].

Facing History and Ourselves (n.d.). Taisho Democracy in Japan: 1912–1926, [Online]. Available at: https://www.facinghistory.org/nanjing-atrocities/nation-building/taisho-democracy-japan-1912-1926 [Accessed 12 December 2019].

Hideaki, T. H. (2009). The Civil Service System and Governance, The Tokyo Foundation for Policy Research. [Online]. Available at: http://www.tokyofoundation.org/en/articles/2009/the-civil-service-system-and-governance [Accessed 12 December 2019].

Kaneko, M. (1999). The Political Economy of the Safety Net. Seefutii Netto no Seiji Keizagaku, Tokyo: Chikuma Shinsho.

Kenneth, N. and Deth, J. W. V. (2005). Foundations of comparative politics: Democracies of the modern world. 2nd ed. New York: Cambridge University Press.

Kosai, Y. (2008). The Postwar Japanese Economy, 1945–1973. 6. The Twentieth Century, online ed., ed. Peter Duus, trans. Andrew Goble, pp. 494–537. New York: Cambridge University Press.

Lipset, S. (1959). Some Social Requisites of Democracy: Economic Development and Political Legitimacy. The American Political Science Review, 53(1), 69–105. [Online]. Available at: https://www.jstor.org/stable/1951731 [Accessed 12 December 2019].

Mulgan, A. G. (2018). What's wrong with Japan's bureaucrats? East Asia Forum. [Online]. Available at: http://www.eastasiaforum.org/2018/05/03/whats-wrong-with-japans-bureaucrats/#more-124403 [Accessed 12 December 2019].

Pempel, T. (1992). Bureaucracy in Japan. PS: Political Science and Politics, 25(1), 19–24. [Online]. Available at: https://www.jstor.org/stable/419570 [Accessed 12 December 2019].

Rose, R. (1974). The Problem of Party Government. London: Macmillan.

Shendrikova, D. (2014). Japanese Bureaucracy: Descent from Heaven, ISP.

Yoshino. (2005). Sources of Japanese Tradition. In Wm. Theodore de Bary, C. Gluck, and A. Tiedemann (Eds.). 2nd ed. New York: Columbia University Press, pp. 838.

Computational Social Science – Luo, Ciurea & Kumar (eds)
© 2021 Taylor & Francis Group, London, ISBN 978-0-367-70193-2

The construction of the training model of the trans-cultural cognitive competence

L.Y. Miao & Y. Zhan
Foreign Language Teaching and Research Section of the Rocket Force University of Engineering, Xi'an, P.R. China

ABSTRACT: The development of trans-cultural cognitive competence is the basis for the cultivation of communicators' trans-cultural competence. With reference to professor Ren Yuhai's study on trans-cultural cognitive competence, this paper constructs the training model of trans-cultural cognitive competence. Firstly, through the comparison and contrast among multi-culture, cross-culture and trans-culture, the paper illustrates definitions of trans-culture, trans-cultural competence and trans-cultural cognitive competence, respectively. Then, it designs the training model of trans-cultural cognitive competence in foreign language teaching, which consists of knowledge instruction, classroom practice, critical thinking and innovative practice, aiming to train the new international talents who will be able to transcend the cultural boundaries and have successful interactions with people from other cultures.

1 INTRODUCTION

"Homogeneous society", "ethnic group" and "cultural boundary" are three characteristics of traditional cultural concepts, forming the basis of multiculturalism and cross-culture identity [1]. However, both the multi-culture focusing on the static comparison of cultural differences and the cross-culture concerning the interculturality deny the interactions between different cultures in the deep structure. The fast development of computer technology, population migration, the inter-infiltration of different lifestyles and cultures in the world require human beings in the age of globalization to break the traditional homogenization and localization, transcend the framework of native culture, give play to individual creativeness, integrate different cultures and establish a positive interaction between cultures. Trans-cultural competence is the basic ability for individuals to adapt to the global development, carry out successful communications and obtain long-term development. Based on professor Ren Yuhai's multi-construct model of trans-cultural competence, this paper tries to build a training model of trans-cultural cognitive competence in foreign language teaching.

2 TRANS-CULTURE, TRANS-CULTURAL COMPETENCE AND TRANS-CULTURAL COGNITIVE COMPETENCE

2.1 *Trans-culture*

Trans-culture refers to the open system that provides symbol selection for existing cultures and their established symbol system.[2] It does not mean to abandon individuals' native culture, nor to deny other cultures simply, but to eliminate the culture boundaries, free individuals from the symbols, mental tendency and prejudice they get accustomed to, seek and save the most meaningful and valuable parts in every cultures during the interaction, integrate different cultures and symbolic systems, and ultimately create an ideal communication field.

Trans-culture is different from multi-culture and cross-culture. The following figures show multi-culture (Figure 1), cross-culture (Figure 2) and trans-culture (Figure 3), respectively.

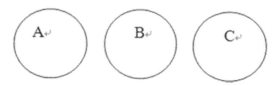

Figure 1.

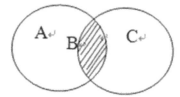

Figure 2.

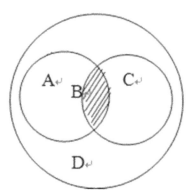

Figure 3.

Multiculturalism advocates culture equality and emphasizes loyalty and adherence to the native culture. It respects the differences between different cultures, recognizes the independence of each culture, and opposes judging other cultures according to one cultural standard. Cross-culture refers to the interaction between groups or individuals with two or more different backgrounds. In the communication, communicators have a full and correct understanding of cultural phenomena, customs and habits that are quite different from that of their own culture, and accept and adapt to other cultures in a spirit of openness and inclusiveness. Figure 1 reflects culture equality and culture independence. Figure 2 cites Wai-Lam Yip's chart on "cultural fusion and divergence" approved by professor Cai Zongqi. The two circles A and B are equal in size. The shaded area C representing the cultural similarities is also equally distributed. Professor Cai analyzed that Wai-Lam Yip might want to use this diagram to highlight that any true "cross-culture" communication must be carried out on the basis of mutual equality and we should not promote our own culture by denying or degrading other cultures.[3] Figure 3 is drawn by Professor Cai Zongqi to show trans-cultural perspective. Professor Cai believes that we should jump out areas A and B which only reflect culture differences and area C which only reflects culture similarities, then establish a more inclusive vision D. [4] The presentation of trans-cultural perspective will help communicators transcend the cultural boundaries, integrate many cultural traditions and symbol systems, and finally obtain a brand-new and positive way of living.

2.2 *Trans-cultural competence*

Based on the above analysis, it can be concluded that trans-culture does not mean to abandon the native culture and choose another one, nor to deliberately stick to the native culture and reject other cultures, but to transcend the boundaries of the two cultures and build a more inclusive and shared "third culture". Individual's ability of overcoming all the cultural and racial prejudices, accepting cultural similarities and differences, integrating many cultural elements and building a mutually beneficial interactive environment is known as trans-cultural competence.

Professor Ren Yuhai pointed out that the trans-cultural competence is a comprehensive system with different dimensions. It involves trans-cultural cognitive competence, trans-cultural communicative competence, trans-cultural ethical competence, trans-cultural creative competence, etc. Thus, the cultivation of individual's trans-cultural competence requires the comprehensive development of all the above abilities. Meanwhile it is imperative to construct the training model of trans-cultural competence in all dimensions.

2.3 *Trans-cultural cognitive competence*

Cognition is an information processing activity for individuals to learn the objective world, including sensation, perception, memory, imagination, etc. Trans-cultural cognition includes two stages: trans-cultural perception and trans-cultural awareness.

People's cognition of objective world starts from their own perception to the objects they want to learn. There will be no cognition without self-perception activities. For example, in our mind operation process, when a new object is presented to us, we will get the integrated information of this object through our sensing system. This information will be stored in our memory so that we can use it when we need to apply language to describe and interpret this object. The reserved information in our mind is just our cognition to this object. The more complete the information gained through perception, the more comprehensive the cognition of the new object is. As Linell Davis says:"Entering strange culture means entering strange world of perception." [5] When cross-cultural communicators enter a new culture, various differences in clothing, food, housing and other aspects bring them different perceptual experiences. Differences in traditional concepts, evaluation standards and living habits will cause individual perceptual barriers. When perceptual barriers appear, subsequent communications will fail to proceed smoothly.

The process of perception usually includes two stages: receiving information and investing meanings to the information obtained at the previous stage. There is little difference in individual perception at the first stage. The second stage is realized in a specific cultural environment. The specific material culture and ecological environment influence the individual perception orientation and perception mode to some extent. Although the influence of culture on perception is mainly reflected at the second stage, human beings have been able to adapt to environmental changes since the ancient times and they will consciously adjust their way of processing information they obtained in order to survive. Therefore, when individuals perceive new cultural information, they can try to adjust their original cognitive modes like information classification and meaning evaluation. Human being's subjective initiative and self-regulation provide the condition for the development of trans-cultural perception. Trans-cultural perception refers to "individual ability of transcending the limitations of one or more cultural perceptions, carrying out perceptual regulations properly, enriching perceptual experience through expanding perceptual field, and realizing perceptual interaction with people from other cultures". [1]

Trans-cultural awareness is a conscious and active way of cognition for individuals. With an open and inclusive attitude, individuals will be sensitive to the new cultural environment, consciously explore the different aspects between the new and original culture, then regulate the cognitive process. To cultivate the trans-cultural awareness, people need to increase the breadth and depth of contact with different cultures, experience the conceptual difference brought about by cultural exchanges, enhance the awareness of self-reflection and self-monitoring, and make timely adjustment to one's cognition and thinking mode. [1] Conversely, one is less likely to have trans-culture awareness if he is overly dependent on the original culture and rejects the new. Through perception and awareness, individuals can give full play to their subjective initiative and cultural consciousness, experience critical reflection, and realize trans-cultural cognition. Trans-cultural cognitive competence is the basis for trans-cultural communication and trans-cultural creation.

3 CONSTRUCTION OF THE TRAINING MODEL OF THE TRANS-CULTURAL COGNITIVE COMPETENCE

Based on Ren Yuhai's study on trans-cultural cognitive competence and Zhang Delu's exploration on the training mode of trans-cultural communicative competence, this paper attempts to construct the training model of the trans-cultural cognitive competence in foreign language teaching. The cultivation of this ability will lay a foundation for the development of communicators' trans-cultural competence in the age of globalization, which consists of knowledge instruction, classroom practice, critical thinking and innovative practice.

3.1 Knowledge instruction

Through knowledge instruction, communicators' perceptual openness, perceptual insight and perceptual receptivity will be improved. Perceptual openness refers to that communicators can "consciously control and overcome the perceptual barriers caused by cultural prejudice and stereotypes". Perceptual insight means that communicators can "sense and perceive the similarities and differences of various cultural information and perception modes". Perceptual receptivity is the ability to "accept and tolerate the differences in perceived objects and ways of different cultures". [1] To promote the above three kinds of competence, it is necessary to introduce communication rules, customs, values, philosophy, religion, emotional patterns and other culture-related knowledge to language learners so that they will not make biased judgment or evaluation on other cultural information in the future communications. Meanwhile, acquiring the culture-related knowledge can help to extend the breadth of communicators' cultural perception, make them learn the difference of various cultural perceptions and guide them to avoid imposing their own perception on other cultures.

The knowledge imparted to students can be generally divided into four categories: (1) knowledge of the native culture; (2) knowledge of the target culture; (3) universal knowledge of the world; (4) the open and inclusive mind, tolerability to new cultures and universally recognized morality. One of the most important factors in this teaching process lies in the instructors themselves. First, teachers should have reasonable perceptual expectation and strong perceptual tolerability without prejudice to other different cultures. Secondly, teachers should be quite familiar with the native culture and the target culture, knowing well about the origins, formation process, and characteristics and developing trend of both cultures. Thirdly, teachers should constantly update their knowledge reserves and improve their trans-cultural cognitive competence because culture is not a fixed and static entity but can be constantly generated and transformed into mew modes.

3.2 Classroom practice

Based on the knowledge acquired at the previous stage, students' perceptual integration ability is to be cultivated in classroom practice. They will learn to transcend their original perceptual mode and reintegrate the perceptual information from different cultures. At the same time, teachers need to help students cultivate the ability of perceptual interpretation, leading students to transcend the limitations of the native cultural perception, discard the absolute judgment of either-or, reevaluate and reinterpret the meaning of perceptual information from a "third party" perspective by comparing and contrasting the perceptive modes between native and target culture. Additionally, students' perceptual interaction skill should be enhanced through introducing other culture-related information to them as well as designing and organizing activities that could develop student's empathy. [1] The cultivation of the above ability can be realized through the real classroom teaching and the second classroom activities.

The real classroom teaching for cultivating trans-cultural cognitive competence includes the following steps: (1) Teachers assign learning task before class. They need to divide students into different groups and ask students to learn more than two cultures assigned by the teacher to the group. Students are required to have a comprehensive self-study, systematically analyze and integrate the cultural information according to their own understanding, compose the speech draft delivered in the class time with group members and prepare related auxiliary PPT. (2) Each group makes a presentation in class time. The whole class discusses the root cause for the culture differences. Then teachers assign related topics for further discussion. In this process, "students tend to look at strange cultures from their own cultural perspectives and ask questions that surprise or confuse them in new cultures." [6] (3) After class, students are asked to complete a composition independently to express their own opinions on the topics discussed in the class. They are required to propose solutions to the issues discussed and interpret their own views from the perspective of "third party" through integrating the cultural information obtained from the classroom speech and topic discussions and adjusting their views, positions as well as attitudes from a trans-cultural perspective.

The second classroom mainly takes the Model United Nations as the main activity form. Students are divided into groups representing different countries, studying the politics, economy, geography, history and military of the country they represent as well as their attitude towards the topic of the meeting. Even if the country they represent has no obvious attitude to the issue discussed, the representatives can study countries with similar political and economical conditions and form alliance with those countries to fight for the rights and interests of the country they represent. If their representative country has no voice in the international arena, the representatives can also make their countries shine on the stage of the Model United Nations through their own efforts. In this process, students will take an appropriate psychological transformation in perceptional interpretation and evaluation. The collision of ideas and interweaving of multiple thoughts let students have a perceptional communication with new cultures. Students transcend the limits of the original perceptional modes, enrich their perceptional experience, extend their trans-cultural cognitive field and improve their trans-cultural cognitive competence.

3.3 Critical thinking

At the critical thinking stage, students reflect, criticize, evaluate what they have learned and the learning activities they took part in and constantly develop their own knowledge, during which their trans-cultural awareness is improved. Students will realize that they will perceive, classify and interpret the target culture in a different way from people coming from other cultures, then they will adopt a broader perspective to adjust their recognition to the target culture so that to gain a deeper understanding of the new culture.

The teaching mode of this stage is as follows: (1) Conduct the online discussion. Based on Internet technology, teacher-student communication and student-student communication are carried out by means of WeChat group, microblog, email, etc. Discussions and debates are conducted on such issues as whether the cultural information collected by individuals or groups is complete, credible and fair, whether the various viewpoints are correct, and whether the evaluation criteria are scientific and reasonable. The whole process is not limited by time and space. All participants can constantly search and collect new cultural information, adjust and modify existing cognitive schemata, transcend the cultural boundaries and form a deeper understanding to the different cultures. After the discussion, teachers can guide students to evaluate the discussion result and the process and make a summary to the whole online discussion activity. (2) Submit the written reports. In the report, students are required to summarize what they have learned, make their personal conclusions as well as analyze and evaluate their own views.

3.4 *Innovative practice*

No matter in the classroom practice or in real life, students will come across a different cultural context form the one they have learned and analyzed in previous classroom teaching. At this moment they need to apply knowledge they acquired to the new trans-cultural field, integrate the cultural symbols from native culture and the new culture, analyze and solve problems in a more inclusive and shared way.

The innovative practice mainly includes the following steps: (1) Analyze new contextual information. Students need to be clear about three kinds of contextual changes: change of the communicators, change of the communication field and change of the communication approach. (2) Construct an inclusive and innovative cognitive model. Students learn to collect the information of the target culture, flexibly apply the knowledge, methods and skills acquired to integrate the new cultural information. Meanwhile, they need to break the binary opposition between the original culture and the new culture to construct the "third culture" that transcends the existing cognitive framework. (3) Make analysis and reflection. In the form of group discussion, students analyze, reflect and evaluate whether individuals or groups provide inclusive and flexible classification and judgment to the new cultural information, and whether the solutions to the conflicts between cultures are feasible and innovative. (4) Conduct a summative evaluation. Students have a summative evaluation to the theories and methods adopted in the discussion of the new context information and related cultural topics, laying a good foundation for the future trans-cultural communication in multiple contexts.

4   CONCLUSION

Based on Professor Ren Yuhai's study on the trans-cultural cognitive competence and professor Zhang Delu's research on the training model of trans-cultural communicative competence, this paper constructs a training model of trans-cultural cognitive competence. Through learning and practice, trans-cultural cognitive subject can break through the cultural boundaries, reintegrate the perceptual information and construct an inclusive and shared "third culture", which will help communicators lay a solid foundation for the success of subsequent trans-cultural communication. The construction of the training model of the trans-cultural cognitive competence is the initial stage of the author's theoretical exploration, which will be continuously verified and improved through continuous teaching practice in the future.

REFERENCES

[1] Ren Yuhai. Globalization, Identity and Trans-cultural Competence. Nanjing: Nanjing University Press, 2015.
[2] Zhang Delu &Hao Xinggang. Exploring a Training Model of Trans-cultural Communicative Competence. *Foreign Language World*, 2019(5):57.
[3] Cai Zongqi. Inner-cultural, Cross-cultural and Trans-cultural perspectives in Comparative Study of Western and Chinese Culture. *Literary Theory Research*, 2009, (4): 23.

[4] Davis, Linell·Doing Culture: Cross-cultural Communication in Action. Beijing: Foreign Language Teaching and Research Press, 2001.

[5] Gao Yihong. The Cultivation of Cross-culture Awareness and Self-reflection Competence — "Language and Culture" & "Cross-culture Communication" Curriculum Teaching Philosophy and Practice. *Foreign Language Teaching in China*, 2008, (2):59–68.

Computational Social Science – Luo, Ciurea & Kumar (eds)
© 2021 Taylor & Francis Group, London, ISBN 978-0-367-70193-2

Social network of female residents in Wulingyuan under the influence of ecotourism: Scale, connection and structure

F. Zeng & Y.D. Zhong

School of Tourism, Central South University of Forestry and Technology, Changsha, Hunan, China

ABSTRACT: In this paper, the social network of female residents in Wulingyuan World Natural Heritage Site was studied through questionnaire survey and semi-structured interview. The quantitative data and qualitative analysis of female residents in Yuanjiajie Village, Longweiba Village and Maershan Village in Wulingyuan District were combined with sampling questionnaire and snowball interview. By analyzing the data of network scale, network centrality and network density, we found that the social network of female residents in different locations of Wulingyuan varies in scale, connection and structure. Results show that the ecotourism brings about gradually expanded social interaction scope and diversified social relations for female residents in Wulingyuan. The increasingly-enhanced heterogeneity of social networks makes the largest social network scale, both strong and weak ties, and the most stable network structure of female residents in Yuanjiajie Village, followed by those in Longweiba Village and Maershan Village.

1 INTRODUCTION

Social network, representing all kinds of social relations, refers to a relatively stable system of relations between individual members of a society as a result of their interaction. Through these social relations, various people or organizations from casual acquaintances to intimate family relationships are connected (Qin 2013). A social network is a collection of the nodes (social actors) and their connections (relationships between the actors)among various nodes (Lin 2009). In 2008, Bhat et al. analyzed the centrality, embeddedness and density of the marketing network of New Zealand's tourist destination by way of social network analysis, and the results showed that the effects on marketing varied with different positions of actors in the network and with differences existing in social network structure (Maestre-Andrés et al. 2016); Scott et al. analyzed the network structure and network cohesion of four different types of tourist destinations such as Australia and Victoria by using social network analysis, and pointed out that the network scale and industrialization were positively correlated with community cohesion (Scott et al. 2008); Palvovich made a literature survey and in-depth interview on the changes of strong ties and weak ties in New Zealand's tourist destination Waitomo from 1887 to 2000 based on the theory of strong and weak ties in social network analysis. It is found that strong ties can effectively enhance the cohesion of tourist destinations and promote the harmonious and rapid development of tourist destinations, while weak ties can tap various potential social resources and increase new social ties. Only with the balanced development of strong and weak ties can the tourist destination maintain a good competitive advantage (Pavlovich 2003). Yang Lijun's study of the specific impact of social network structure changes on the cultural landscape variation of ancient towns was the first one to apply the social network analysis in the field of tourism (Yang & Yang 2005). Later, some discussions arose successively on the combination of social network analysis and tourism research. At present, little research has looked at the application of social network analysis techniques to female residents in tourist destinations (Quezada-Sarmiento et al. 2018). Only a few scholars take the social capital theory as the basis to study the possession and use of social capital of females in rural areas, and

Figure 1.   Map of the study site (The green area is the core area of the Wulingyuan Natural Heritage Site).

then study the scale, density and structure of its social network (Chen & Huang 2006; Jing 2011; Saufi et al. 2017; Zhou & Zhou 2013).

This study takes the social network of female residents in the Wulingyuan World Natural Heritage Site as the research object to explore the impact of ecotourism on its scale and structure. Over the past three decades, Wulingyuan World Natural Heritage Site has achieved a tourist reception scale increasing from 580,000 visitors in 1989 to 30.2889 million visitors in 2018, with the comprehensive tourism income growing from 25 million yuan in 1989 to 26.252 billion yuan in 2018. By the end of 2018, there were 29 star-level hotels in the region, including 3 five-star hotels, 6 four-star hotels and 22 three-star hotels. There were 115 social hotels (more than 80 beds each), 161 family hotels (less than 80 beds each) and 206 characteristic inns. There were more than 42,465 reception beds in the region, of which 7,834 were star-rated, which greatly improved the tourist reception capacity. The research samples include Yuanjiajie Village, Longweiba Village and Maershan Village in Wulingyuan District, Zhangjiajie City, which are situated inside the core area, edge and outside of the World Natural Heritage Site respectively. Among them, Yuanjiajie Village is located in the core scenic area of Wulingyuan World Natural Heritage Site, Longweiba Village is adjacent to Zhangjiajie National Forest Park, and Maershan Village is about 40km away from Zhangjiajie National Forest Park.

2   RESEARCH METHOD

In this study, three villages in Wulingyuan World Natural Heritage Site were selected as case sites. Three in-depth investigations, including questionnaire survey and semi-structured interview, have been conducted on the female residents of communities in the case sites from the perspective of ego network (Figure 1).

2.1   Questionnaire survey

The researchers paid three in-depth survey visits of 27 days in total to three villages of Yuanjiajie Village, Longweiba Village and Maershan Village in Wulingyuan World Natural Heritage Site in September, December 2016 and March 2017 respectively. 350 questionnaires were randomly distributed, and 276 valid ones were collected, of which the variable was the place of residence.

2.2   Semi-structured interview

During the field research, 72 community female residents were interviewed and recorded starting from an innkeeper by using a snowball sampling method, and the data was analyzed and summarized by the qualitative research method. Most of the interviewees are engaged in tourism-related industries.

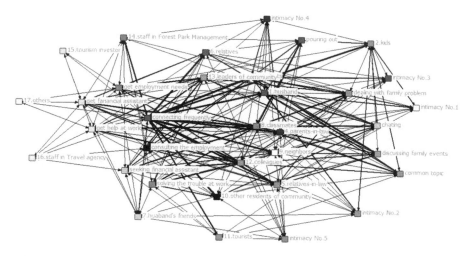

Figure 2.   Social network on Female residents of Yuanjiajie Village.

2.3   *Data analysis*

In this paper, the methods of network scale analysis, network centrality analysis and network density analysis in social network analysis are adopted. Network scale refers to the actors interacting with each other in various ways at ordinary times, which is a variable that reflects the unit of network capital. Degree centrality is the most direct measure index to describe node centrality in network analysis. The greater the degree of a node, the higher the degree centrality of the node, and the more important the node is in the network. Network density is an important feature of the social network structure. Female's ego social network density reflects the average degree of interaction between community female residents and other actors in the network (Aynaud et al. 2013; Knoke & Yang 2012).

3   ANALYSIS OF SOCIAL NETWORK OF FEMALE RESIDENTS

3.1   *Network scale analysis*

This study measures the ego network of female residents in the community using the "position generation method". Based on the measurement of the ego network scale of female residents in different places of residence, it is found that the ego network of female residents in Yuanjiajie Village (Figure 2) contains 17 network nodes. In addition to family members, relatives-in-law and neighbors, their communication objects also include leaders of the community, tourists, staff in Forest Park Management, staff in the travel agency and tourism investor; the ego network of Longweiba Village (Figure 3) contains 15 network nodes, with communication objects including family members, relatives-in-law, neighbors, leaders of the community, tourists and staff in Forest Park Management; the ego network of Maershan Village (Figure 4) contains 14 network nodes, and most of the communication objects are relatives-in-law and neighbors.

The ego network diagram of female residents of Yuanjiajie Village shows a path of connection spreading from the center to the edge, with stable network structure; that of Longweiba Village shows a path of connection spreading from right to left, with relatively stable network structure; that of Maershan Village shows a path of connection completely inclined to one side, with relatively unstable network structure.

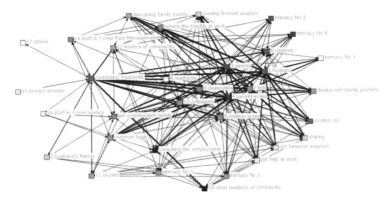

Figure 3.   Social network on Female residents of Longweiba Village.

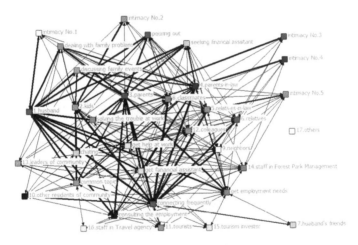

Figure 4.   Social network on Female residents of Maershan Village.

3.2   *Network centrality analysis*

Upon the statistics of the questionnaire, UCINET is adopted to analyze the three indexes, i.e. the degree centrality, the closeness centrality and the betweenness centrality involved in the social relations individual network of female residents in Wulingyuan World Natural Heritage Site.

3.2.1   *Degree centrality analysis*

Degree centrality is the most direct measure index to describe node centrality in network analysis. The greater the degree of a node, the higher the degree centrality of the node, and the more important the node is in the network (Fu 2012).

According to the statistical results of the degree centrality of female residents' ego network in three villages, the network concentration of Yuanjiajie Village, Longweiba Village and Maershan Village is 25.62%, 22.97% and 21.78%, respectively (Table 1). The ego network of female residents in Yuanjiajie Village has the highest degree centrality, followed by Longweiba Village and Maershan Village, indicating that the female residents in Yuanjiajie Village are in an important position in the social network and play a crucial part in the network connection.

Table 1. Descriptive statistics.

		Yuanjiajie Village		Longweiba Village		Maershan Village	
		Degree	NrmDegree	Degree	NrmDegree	Degree	NrmDegree
1	Mean	192.000	6.190	196.118	7.337	200.176	6.740
2	Std Dev	173.700	5.600	206.411	7.722	210.965	7.103
3	Sum	6528.000	210.445	6668.000	249.458	6806.000	229.158
4	Variance	30171.529	31.355	42605.633	59.631	44506.027	50.455
5	SSQ	2279208.00	2368.647	2756304.00	3857.707	2875606.00	3259.992
6	MCSSQ	1025832.00	1066.087	1448591.50	2027.440	1513205.00	1715.477
7	Euc Norm	1509.705	48.669	1660.212	62.110	1695.761	57.096
8	Minimum	10.000	0.322	0.000	0.000	2.000	0.067
9	Maximum	940.000	30.303	774.000	28.956	809.000	27.239
	Network Centralization	25.62%		22.97%		21.78%	

Table 2. Descriptive statistics.

		Yuanjiajie Village		Longweiba Village		Maershan Village	
		Between-ness	nBetween-ness	Between-ness	nBetween-ness	Between-ness	nBetween-ness
1	Mean	12.412	2.351	12.706	2.406	13.471	2.551
2	Std Dev	8.214	1.556	10.935	2.071	15.258	2.236
3	Sum	422.000	79.924	432.000	81.818	458.000	86.742
4	Variance	67.470	2.420	119.576	4.289	232.820	8.351
5	SSQ	7531.758	270.165	9554.529	342.722	14085.402	505.244
6	MCSSQ	2293.994	82.286	4065.588	145.833	7915.873	283.943
7	Euc Norm	86.786	16.437	97.747	18.513	118.682	22.478
8	Minimum	0.460	0.087	0.000	0.000	0.000	0.000
9	Maximum	33.619	6.367	39.331	7.449	89.046	16.865
	Network Centralization	4.14%		5.20%		14.75%	

3.2.2 Betweenness centrality analysis

The betweenness centrality measures the degree of "middleman" of the node, that is, the degree of media. The betweenness centrality represents the betweenness of the whole network. If an actor is on the path between many other two nodes in a network, the actor can be considered to occupy an important position, because he/she has some control over the interactions between the other two actors (Niu et al. 2013).

According to the statistical results of the betweenness centrality of female residents' ego network in three villages, the betweenness centrality of the ego network of female residents in Yuanjiajie Village, Longweiba Village and Maershan Village is 4.14%, 5.2% and 14.75%, respectively (Table 2). Comparing the three villages, the ties between the female residents of Yuanjiajie Village and other network members lie in the most central position, and they play the most important role in the network, followed by Longweiba Village, while the ties between female residents of Maershan Village and other network members are outside the central position, with an unobvious role in the network.

3.2.3 Closeness centrality analysis

Closeness centrality is a measure of not being controlled by others that is, to put it informally, indicating how close a node is to all other nodes in the network. The closeness centrality represents

Table 3. Closeness statistics.

		Yuanjiajie Village		Longweiba Village		Maershan Village	
		Farness	nCloseness	Farness	nCloseness	Farness	nCloseness
1	Mean	57.824	57.885	92.182	36.063	59.941	56.096
2	Std Dev	7.131	6.661	8.081	3.029	8.530	7.421
3	Sum	1966.000	1968.082	3042.00	1190.093	2038.000	1907.250
4	Variance	50.851	44.369	65.300	9.172	72.761	55.070
5	SSQ	115410.000	115430.531	282572.00	43221.520	124634.000	108860.719
6	MCSSQ	1728.941	1508.537	2154.909	302.678	2473.882	1872.383
7	Euc Norm	339.720	339.751	531.575	207.898	353.035	329.940
8	Minimum	49.000	44.000	81.000	29.730	49.000	40.741
9	Maximum	75.000	67.347	111.000	40.741	81.000	67.347
	Network Centralization	19.80%		21.32%		23.55%	

Table 4. The density comparison of the ego network.

Place of residence	Density	Standard deviation
Yuanjiajie Village	0.2941	0.3987
Longweiba Village	0.1536	0.5878
Maershan Village	0.0775	0.3103

the closeness centralization trend of the overall network. In a social network, the closeness centrality considers the average length of the shortest paths from each node to all other nodes. In other words, the closer the node is to other nodes, the lower its centrality will be (Niu et al.).

According to the statistical results of the closeness centrality of female residents' ego network in three villages, the closeness centrality of the ego network of female residents in Yuanjiajie Village, Longweiba Village and Maershan Village is 19.80%, 21.32% and 23.55%, respectively (Table 3). Comparing the three villages, the ego network of female residents in Yuanjiajie Village has the lowest closeness centrality, indicating that the female residents of Yuanjiajie Village have the shortest distance from other nodes in the network, and are in the most central position in the network. Longweiba Village comes second. The ego network of female residents in Maershan Village has the highest closeness centrality, indicating that the female residents of Maershan Village have the farthest distance from other nodes in the network, and are not at the central position of the network.

3.3 Network density analysis

A social network with high network density features closer relations between network nodes, and more ways to obtain information and share resources. In general, the more frequent the connections between the social network nodes, the greater the overall density of the network, and the greater the influence of the network on the attitude and behavior of actors therein. Moreover, actors can obtain more social resources and effective information as well, and the reciprocity of network resources will be relatively better (Knoke & Yang 2012).

According to the results of measuring research objects by place of residences as shown in Table 4, the ego network density of female residents of Yuanjiajie Village is 0.2941, which means relatively high overall network density and strong social network influence; that of Longweiba Village is 0.1536, with social network influence relatively weakened; that of Maershan Village is only 0.0775, with the weakest social network influence. The increase of network density indicates closer

413

relations between social network nodes, more frequent interaction degree, and better circulation of information and resources.

4 RESULTS AND DISCUSSION

4.1 *Social network varies in scale*

According to the results of the ego network scale of female residents in the Wulingyuan World Natural Heritage Site, the social network structure of female residents in Yuanjiajie Village contains the most network nodes, followed by Longweiba Village and Marshan Village. This suggests that Yuanjiajie Village, featuring the highest ecotourism development degree, has the largest social network scale of female residents, while Maershan Village, which is the farthest from the core scenic area and has a relatively low degree of ecotourism development, has the smallest social network scale of female residents. The 17 nodes in the ego network of female residents in Yuanjiajie Village include family members and relatives based on kinship, neighbors and leaders of community based on the geographical relationship, as well as tourists, staff in Forest Park Management, staff in the travel agencies and tourism investors based on industry and tourism relationships. The 15 nodes in that of Longweiba Village include family members and relatives based on kinship, neighbors and leaders of community based on the geographical relationship, as well as tourists, staff in Forest Park Management and staff in the travel agencies based on industry and tourism relationships. The 14 nodes in that of Maershan Village mainly include family members and relatives based on kinship, neighbors and leaders of community based geographical relationships, as well as tourists and staff in Forest Park Management based on industry and tourism relationships, with less interaction with staff in the travel agencies.

Relying on the core scenic areas and central towns of Zhangjiajie National Forest Park, Wulingyuan World Natural Heritage Site has developed several agricultural sightseeing tours, ecological tours, outdoor experience tours, created several characteristic farming and handicraft products, and promoted agricultural production to serve tourism, agricultural products to turn into tourism commodities, and farmers to become tourism practitioners, so as to constantly lengthen the industrial chain. Residents who go out also gradually return home to participate in tourism development. Community residents can involve in the construction and development of scenic areas directly by way of being employed in scenic areas with normal wage income. or indirectly by renting out houses, selling petty commodities, opening restaurants or providing accommodation, etc. Poor households can even buy shares with houses, or work in inns and share dividends on a pro-rata basis. In the process of community participation in tourism, local tourism enterprises, travel agencies, hotels and restaurants can attract and accept a large number of the rural female labor force. Upon the realization of non-agricultural transfer, women are no longer confined to farming, or even family work. They have the opportunity to get in touch with the broad world outside, interact with more people, and constantly expand their social network.

The survey results show that 80% of the families in Yuanjiajie Village participate in ecotourism mainly through family hotels and restaurants, and almost every participating family involves women; about half of the families in Longweiba Village participate in ecotourism mainly through family hotels or selling tourist souvenirs, and almost all families involve women; only about 30% of the families in Maershan Village participate in ecotourism mainly through selling tourist souvenirs or being tour guides, and almost all families also involve women. This suggests that the development of ecotourism has a direct impact on the social network scale of female residents in the community. The higher the degree of participation in eco-tourism, the richer the social relations, and the larger the social interaction scope and the social network scale will be.

4.2 *Social network varies in connection*

There also exist differences in the centrality of the ego network of female residents in the Wulingyuan World Natural Heritage Site. Yuanjiajie Village, situated at the core scenic area and with the highest ecotourism development degree, has the highest degree centrality of ego network of female residents but the lowest closeness and betweenness centrality; Maershan Village, which is farthest from the core scenic area and has the lowest ecotourism development degree, has the lowest degree centrality of ego network of female residents but the highest closeness and betweenness centrality. This suggests that the female residents of Yuanjiajie Village are in the most central and important position in the social network, with the shortest distance from other nodes of the network, thus playing an important role in the network connection. Longweiba Village comes second. The female residents of Maershan Village have the farthest distance from other nodes of the network, with an unobvious role in the network connection. In comparison, the female residents of Yuanjiajie Village have the widest social interaction scope, the highest position in the network and the highest social capital stock, thus playing the most obvious role in the network.

According to the results of the network centrality analysis of female residents in Wulingyuan World Natural Heritage Site, neighbors, husband and colleagues are the core nodes in the ego network of female residents in Yuanjiajie Village, with the degree centrality of 0.087, 0.060, 0.059 and 0.062 respectively. Among them, "connecting frequently" is the spreading node, and "neighbors", "husband" and "colleagues" are the gathering nodes. In the ego network of female residents in Longweiba Village, husband, neighbors and parents-in-law are the core nodes of the network, with the degree centrality of 0.118, 0.106 and 0.075 respectively. Among them, "connecting frequently" is the spreading node, and "husband", "neighbors" and "parents-in-law" are the gathering nodes. In the ego network of female residents in Maershan Village, husband, parents and parents-in-law are the core nodes of the network, with the degree centrality of 0.116, 0.113 and 0.079 respectively. Among them, "connecting frequently" is the spreading node, and "husband", "parents" and "parents-in-law" are the gathering nodes (Table 5).

American sociologist Granovetter believed that human social network can be divided into strong ties and weak ties . The strong tie is a stable, deep and intimate social relation connected by emotional factors; by contrast, the weak tie is a flexible and extensive social relation, which is a brief social contact between two actors, without strong emotional support (Granovetter 1983). The result of the network centrality analysis demonstrates that there exist differences in the strong ties and weak ties in the social network of female residents in the Wulingyuan community. The social network of female residents in Maershan Village mainly includes strong ties based on kinship, with unobvious weak ties. The social network of female residents in Longweiba Village mainly includes strong ties based on kinship and geography, with certain weak ties. The social network of female residents in Yuanjiajie Village has both strong ties based on kinship and geography and weak ties based on industry relationships. This suggests that the ecotourism development has a direct impact on the social network connection of female residents. After the non-agricultural transfer of female residents through tourism management, in addition to the strong ties based on consanguinity, kinship and geography, the weak ties based on industry and tourism relationships have also been expanded. Tourists and tourism participants are the subjects of the weak ties of female residents in the Wulingyuan World Natural Heritage Site, therefore, female residents in Yuanjiajie Village,which features the highest ecotourism development degree, has the most frequent weak ties, followed by Longweiba Village. The weak ties of female residents in Maershan Village are relatively not obvious.

4.3 *Social network varies in structure*

Network density can measure the relaxation degree of the network structure and the interconnection degree between members. There exists a significant difference in the ego network density of female residents in the Wulingyuan World Natural Heritage Site. Yuanjiajie Village, situated at the core scenic area, has the highest ego network density, followed by Longweiba Village and Maershan Village. This indicates that the higher the ecotourism development degree, the more social connections of female residents in the community; the higher the connection degree with other actors in the social network, the greater the social network density, and the more stable the network structure will be. Moreover, the ego network scale diagram also shows the most uniform

Table 5. Degree centrality.

	Yuanjiajie				Longweiba				Maershan		
	1 Degree	2 Nrm Degree	3 Share		1 Degree	2 Nrm Degree	3 Share		1 Degree	2 Nrm Degree	3 Share
18	940.000	30.303	0.144	18	809.000	27.239	0.119	1	774.000	28.956	0.116
9	569.000	18.343	0.087	1	804.000	27.071	0.118	3	752.000	28.133	0.113
1	390.000	12.573	0.060	9	719.000	24.209	0.106	18	728.000	27.235	0.109
12	388.000	12.508	0.059	4	510.000	17.172	0.075	4	530.000	19.828	0.079
13	309.000	9.961	0.047	2	465.000	15.657	0.068	2	460.000	17.209	0.069
3	246.000	7.930	0.038	24	216.000	7.273	0.032	24	214.000	8.006	0.032
2	236.000	7.608	0.036	25	212.000	7.138	0.031	25	210.000	7.856	0.031
24	213.000	6.867	0.033	26	211.000	7.104	0.031	26	209.000	7.819	0.031
8	210.000	6.770	0.032	27	205.000	6.902	0.030	34	203.000	7.594	0.030
28	202.000	6.512	0.031	28	200.000	6.734	0.029	31	203.000	7.594	0.030
25	198.000	6.383	0.030	34	199.000	6.700	0.029	27	203.000	7.594	0.030
27	195.000	6.286	0.030	29	199.000	6.700	0.029	28	202.000	7.557	0.030
4	192.000	6.190	0.029	30	198.000	6.667	0.029	32	201.000	7.52	0.030
26	186.000	5.996	0.028	31	197.000	6.633	0.029	30	200.000	7.482	0.030
5	176.000	5.674	0.027	33	197.000	6.633	0.029	33	198.000	7.407	0.030
29	176.000	5.674	0.027	32	194.000	6.532	0.029	29	195.000	7.295	0.029
32	174.000	5.609	0.027	8	158.000	5.32	0.023	5	171.000	6.397	0.026
10	162.000	5.222	0.025	3	134.000	4.512	0.020	8	145.000	5.425	0.022
33	156.000	5.029	0.024	5	130.000	4.377	0.019	9	102.000	3.816	0.015
30	155.000	4.997	0.024	19	114.000	3.838	0.017	6	98.000	3.666	0.015
6	154.000	4.965	0.024	12	107.000	3.603	0.016	19	79.000	2.955	0.012
31	153.000	4.932	0.023	6	104.000	3.502	0.015	12	79.000	2.955	0.012
34	153.000	4.932	0.023	13	91.000	3.064	0.013	20	77.000	2.881	0.012
19	103.000	3.320	0.016	10	83.000	2.795	0.012	22	75.000	2.806	0.011
20	80.000	2.579	0.012	20	78.000	2.626	0.011	21	75.000	2.806	0.011
21	79.000	2.547	0.012	21	73.000	2.458	0.011	23	62.000	2.319	0.009
14	74.000	2.386	0.011	22	63.000	2.121	0.009	13	59.000	2.207	0.009
11	64.000	2.063	0.010	23	38.000	1.279	0.006	10	56.000	2.095	0.008
22	63.000	2.031	0.010	11	34.000	1.145	0.005	14	34.000	1.272	0.005
7	46.000	1.483	0.007	7	33.000	1.111	0.005	11	33.000	1.235	0.005
23	38.000	1.225	0.006	14	11.000	0.370	0.002	7	30.000	1.122	0.004
16	21.000	0.677	0.003	16	11.000	0.370	0.002	15	6.000	0.224	0.001
15	17.000	0.548	0.003	17	7.000	0.236	0.001	16	5.000	0.187	0.001
17	10.000	0.322	0.002	15	2.000	0.067	0.000	17	0.000	0.000	0.000

Note: 1. Husband; 2. Children; 3. Parents; 4. Parents-in-law; 5. Husband's relatives; 6. Wife's relatives; 7. Husband's classmates and friends; 8. Their Wife's classmates and friends; 9. Neighbors; 10. Other community residents; 11. Tourists; 12. Colleagues; 13. Village committee leaders; 14. Staff in Forest Park Management; 15. Tourism investors; 16. Travel agency staff; 17. Others; 18. Frequent contact; 19. Intimacy (1); 20. Intimacy (2); 21. Intimacy (3); 22. intimacy (4); 24. Chatting; 25, Common topics; 26. Pouring out; 27. Seeking financial assistance; 28. Discussing family events; 29. Dealing with family conflicts; 30. Solving work problems; 31. Employment consultation; 32. Getting financial assistance; 33. Getting help at work.

distribution in the ego network scale diagram of female residents in Yuanjiajie Village, followed by Longweiba Village. The ego network scale diagram of female residents in Maershan Village is the least uniform. This suggests that the Yuanjiajie Village has the most stable female social network structure, while Maershan Village has the most unstable one.

The social network structure is determined by the network density. The higher the density, the richer the heterogeneous information is in the network, and the greater the span of the network resources will be. When the heterogeneity of the social relations network overcomes the repetition of network resources, the ability to control social resources and the possibility of approaching multiple information is also improved continuously, and the social network structure is more stable. The entry of tourists and tourism participants has brought more heterogeneous information and resources to the social network of female residents in Wulingyuan World Natural Heritage Site, which enhances the heterogeneity of their social network, and enriches the social resources absorbed from it. The social network gradually presents the phenomenon of network structure expansion, scale increase, capstone raising, and social capital stock growth.

Since Yuanjiajie Village is located in the core scenic area, female residents have the highest participation in ecotourism and communicate most frequently with tourists and tourism participants everyday. Tourists worldwide with different cultural traditions, educational backgrounds and living environments, as well as various tourism participants have provided abundant heterogeneous information and resources for the social network of female residents in Yuanjiajie Village. Their social network density keeps rising, and the social network structure becomes increasingly stable. In comparison, Maershan Village is located in the remote area far from the core scenic area, so the participation of the female residents in ecotourism is relatively low. Their social interaction is mainly based on traditional kinship and geographical relationships, and there are relatively few heterogeneous information and resources in the social network, with relatively low social network density and unstable social network structure.

5 CONCLUSION

By investigating the social network of female residents in three villages in different locations of the core scenic area of Wulingyuan World Natural Heritage Site, this study finds that the ecotourism development degree has a direct impact on the scale, connection and structure of the social network of the female residents.

(1) Under the influence of ecotourism, the social interaction scope of the female residents in the Wulingyuan World Natural Heritage Site has gradually expanded. The social network of female residents in Yuanjiajie Village, which is situated at the core scenic spot and features the highest ecotourism development degree, has the most effectively connected members and the largest social network scale. Maershan Village, which is far away from the scenic area and least affected by ecotourism, has the least effectively connected members and the smallest social network scale.

(2) Under the influence of ecotourism, the social relations of the female residents in the Wulingyuan World Natural Heritage Site are gradually diversified. The social network of female residents in Yuanjiajie Village has both strong ties based on kinship and geography and weak ties based on industry relationships. That of Longweiba Village mainly includes strong ties based on kinship and geography, with certain weak ties. That of Maershan Village mainly includes strong ties, with unobvious weak ties.

(3) Under the influence of ecotourism, the heterogeneity of the social network structure of the female residents in the Wulingyuan World Natural Heritage Site is increasingly enhanced. Yuanjiajie Village, which is in the core scenic spot, has the highest social network density and the most stable network structure. Longweiba Village comes second. Maershan Village has the lowest social network density and relatively unstable network structure.

The research results of the social network of female residents in Wulingyuan World Natural Heritage Site are conducive to understanding the reasons and trends of the changes of local women's ideas and behavior, with a certain guiding significance for solving the problems of employment, education, rights and interests of rural women in China. However, the research objects of this paper only include three villages. Wulingyuan tourism has been quite mature after undergoing more than 30 years of development, so the impact of local ecotourism development on the social network of female residents is not universal. There may exist a shortcoming of overgeneralization. It is hoped that more tourism destinations with different development degree can be selected in future research, so as to more comprehensively reveal the impact of ecotourism on the social network of female residents.

REFERENCES

Aynaud T, Fleury E, Guillaume J L, et al. 2013. Communities in Evolving Networks: Definitions, Detection, and Analysis Techniques. Dynamics On and Of Complex Networks (Vol. 2).

Chen X.Q., and F.C. Huang. 2006. Research on tourism spatial structure and its optimization: a network analysis. Geography and Geo-Information Science 22(5):75–80.

Fu L.D. 2012. Complex network centrality measurement and community detection algorithm research. Xidian University.

Granovetter M. 1983. The strength of weak ties: a network theory revisited. Sociological Theory 1(6):201–233.

Jing X.F. 2011. Study on social support network of dongxiang women. Journal of Anhui Agricultural Sciences 2011(11): 6806–6810

Knoke D., and S. Yang. 2012. Social network analysis. Truth & Wisdom Press.

Lin J.R. 2009. Social network analysis: theory, method and application. Beijing Normal University Press.

Maestre-Andrés S., L. Calvet-Mir, and J.C.J.M. van den Bergh. 2016. Sociocultural valuation of ecosystem services to improve protected area management: a multi-method approach applied to Catalonia, Spain. Regional Environmental Change 16(3):717–731.

Niu J.W., B. Dai, L.M. Sun, et al. 2013. PQBCF: A P2P query algorithm based on betweenness centrality forwarding in opportunities networks. Acta Electronica Sinica 41(9):1815–1820.

Niu J.W., J.K. Guo, and C. Tong. A data transmission method for opportunistic mobile sensor networks based on closeness centrality: CN.

Pavlovich K. 2003. The evolution and transformation of a tourism destination network: the Waltomo Caves, New Zealand. Tourism Management 24:203–216.

Qin J. 2013. Research of development and application value for frontier theory of social network. Technology and Innovation Management 34(2):128–131.

Quezada-Sarmiento, P. A., Suasnavas-Rodriguez, M. G., Chango-Canaveral, P. M., Gonzaga-Vallejo, C., & Calderon-Cordova, C. A. 2018. Used of social networks and web application to design and promote the ecotourism route in the Southern Amazon of Ecuador. 2018 13th Iberian Conference on Information Systems and Technologies (CISTI).

Saufi, A., Andilolo, I. R., Othman, N., & Lew, A. A. 2017. Community Social Capital in the Ecotourism Development of Gunung Padang Site, Cianjur, West Java. Springer Singapore.

Scott N., C. Cooper, and R. Baggio. 2008. Destination networks: four Australian cases. Annals of Tourism Research 35(1):169–188.

Yang L.J., and B. Yang. 2005. The variations of cultural landscape and the changes of the structure of social network in the old towns of region of rivers nad lakes in the south of lower reaches of yangtze river – a case study of the old-town of jinze and liantang in the suburban of shanghai. Fujian Architecture & Construction 2005(2):30–32.

Zhou Y., and Z.A. Zhou. 2013. Analysis of possession and use of social capital of female hui nationality in rural area. Journal of Chongqing University of Posts and Telecommunications (Social Science Edition) 2013(2): 104–108.

Computational Social Science – Luo, Ciurea & Kumar (eds)
© 2021 Taylor & Francis Group, London, ISBN 978-0-367-70193-2

A study of the moderating effect of online shopping experiences on impact of relationship quality on customer loyalty

X.Y. Meng
Zhuhai College of Jilin University, Zhuhai, China

X.Q. Wang*
Zhuhai City Polytechnic, Zhuhai, China

ABSTRACT: This paper aims to examine the online purchase experience's moderating effect on the relationship between online customer relationship quality and online customer loyalty. Our results demonstrate that a higher level of online purchase experience does not strengthen the impact of online relationship quality on customer loyalty. Another finding showed that there is a mediating impact between customer loyalty and relationship quality.

1 INTRODUCTION

Understanding customers can result in generating on-going income and profit. As a consequence, the online vendors' main purpose is to enhance customers' positive responses, including trust and relationship quality.

This study investigates whether online shopping experiences have a moderating effect on the relationship between online relationship quality and customer loyalty.

1.1 *Online relationship quality*

Most previous studies on relationship marketing use two or three different dimensions including satisfaction commitment and trust, to conceptualize the construct of relationship quality.

Customers' satisfaction is the core of the vendor's target of relationship management in the online marketing environment. Customer e-satisfaction has become increasingly critical in marketing literature in recent times. Since evaluations from customers appear on every transaction experience, e-satisfaction is defined as a pleasant and satisfied order process in this study through several purchase experiences, resulting in a general assessment of online sellers.

Many studies have recognized the importance of trust as an indication of a positive relationship in marketing. The main two aspects of trust in this study include perceived trustworthiness and trusting behaviors.

Relationship commitment has some characteristics of relationship strength, and some researchers believe that commitment is not sufficient to create a strong relationship because it does not have all of the dimensions of relationship strength.

Therefore, we need a more comprehensive construct to describe the customer relationship in an online selling situation. In our research, relationship quality is defined as comprehensive measure composed of customer trust, commitment, and customer satisfaction. These three dimensions reflect a general evaluation of the depth and strength of the customer relationship.

---

*Corresponding author

Table 1. Testing results.

Path 1	Hypothesis	Estimates	T value	Supported
Online relationship quality◇ customer loyalty	H2	0.739	10.781***	Y

*Y = supported; ** N = not supported.

1.2 Online purchase experience

Most research has demonstrated that a previous online shopping experience can affect customers' purchase intentions for online shopping in the future. To be more specific, if someone has bought some products on the Internet, their purchasing experience can influence their willingness to continue to buy products through e-commerce.

As a consequence, online customers who have more online purchasing experience are likely to have more confidence in their using a new service than customers who have less experience or do not sufficient specific evaluation methods. Hence we hypothesize that an online shopping experience will moderate the influence of the online relationship quality on customer loyalty.

2 RESEARCH GAP FOR THIS RESEARCH

This study was designed in an attempt to provide a model as to what kinds of measurement scales of online vendor improve the online relationship quality and customer loyalty. In this framework, online relationship quality proved to have an impact on customer loyalty.

This study is investigates whether the impacts of relationship quality are contingent on online purchase experience. An examination of such moderators enables marketers to understand what kinds of e-commerce store attributes in online relationship quality are expected to be more or less effective.

Hypothesis 1: A higher level of online relationship quality can lead to a higher level of customer loyalty.

Hypothesis 2: A higher level of online purchase experience has a moderating effect on the impact of online customer relationship quality on customer loyalty.

3 MODEL AND HYPHOTHESES TESTING

This study showed that customer satisfaction and commitment have a significant positive effect on customer loyalty. The measures of "satisfaction" and "commitment" are dimensions related to relationship quality. We believe that a high level of online relationship quality may enhance customer voluntary behavior.

Another important finding is that a higher level of online relationship quality can increase customer loyalty. To explain these results, we argue that a stronger relationship between online vendors and their customers broadens the online vendor's social network, which in turn, plays a dominant role in the level of customer loyalty.

As described in Table 1, hypothesis 1 is fully supported by the test result.

As shown in Table 2, online shopping experience is not a moderator between online relationship quality and customer loyalty.

In this study, online shopping experience was used to test whether it was a moderator of the relationship between online customers' loyalty and the online relationship quality.

The results fully reject moderation of the online shopping experience. Consumers who perceive the same online relationship quality have no different effects on loyalty behavior even if their online shopping experience shows a significant difference.

Table 2. Moderated regression analysis for customer loyalty.

Variables	Customer loyalty	
	ß	Sig.
Online relationship quality (X)	0.554***	0.000
Online shopping experience (Z2)	0.335***	0.000
(X)*(Z2)	0.017	0.447

Note: $*p < .05$ $**p < .01$ $***p < .001$.

4 CONCLUSIONS

In this study, online relationship quality is the main driver of customer loyalty, and their relationship is not affected by their online shopping experience. Thus, an online shopping experience was not a moderator of the relationship between online relationship quality and customer loyalty.

One important contribution of this research is the findings of the relationship between online relationship quality and customer loyalty. A higher online relationship quality can lead to a higher level of customer loyalty Therefore, the implication of the research model has potential to assist online vendors in understanding how customers evaluate the online relationship quality and how to influence customer loyalty to different extents.

Last but not the least, this study establishes that an online shopping experience does not have a moderating effect between online relationship quality and customer loyalty.

REFERENCE

[1] Crosby, L.A. Relationship quality in services selling: an interpersonal influence perspective. *Journal of Marketing, 54*(3), 68–81.

Computational Social Science – Luo, Ciurea & Kumar (eds)
© 2021 Taylor & Francis Group, London, ISBN 978-0-367-70193-2

Research on the "Belt and Road" initiative and the international transmission of Health Qigong—cultural consensus based on the community of common destiny

T. Kang
Sun Yat-sen University, Guangzhou, Guangdong, China

D.D. Yang
Shinawatra University, Bangkok, Thailand

ABSTRACT: In order to build diversified communication platforms, to spread excellent national cultures, thus it will be recognized by the world. The "Belt and Road" initiative provided the background to combining historical materialist dialectics and logical analysis to study the international transmission of Health Qigong. According to the research, Health Qigong has similar characteristics to education, spanning time and space, and being people-oriented. Based on the "Belt and Road" Initiative, the transmission strategy of Health Qigong includes strengthening scientific research and guiding transmission practice; broadening channels of transmission and enhancing the radiant power of culture; using the concept of a "community of shared destiny," strengthening international exchanges, promoting in-depth development, and leading the international transmission of Health Qigong culture.

1 INTRODUCTION

The "Belt and Road" Initiative is a great concept that was put forward by General Secretary Xi Jinping. It is a major strategic decision made by the CPC Central Committee and the State Council to coordinate domestic and international affairs. It is of great and far-reaching significance in realizing China's all-round opening up to the outside world, promoting the rejuvenation of the Chinese nation. With the deepening of China's "Belt and Road" initiative, the increasing soft power of China and strengthening of the research on the transmission of Chinese traditional sports can help to enhance the diplomatic charm of Chinese traditional culture. It also promotes friendship and cooperation between China and other countries, by highlighting the dominant position of Chinese traditional culture. Health Qigong is an integral part of Chinese traditional culture, which combines limb movement, breathing, and psychological adjustment, while containing the spirit of Chinese culture. "Harmony between Heaven and Man" and "Internal and External Harmony" reflect the traditional Chinese way of maintaining a healthy life. As a representative of the excellent traditional health-preserving culture of China, it is greatly favored by practitioners both at home and abroad. Therefore, led by the "Belt and Road" initiative, it is especially important and urgent to strengthen research into the transmission of Health Qigong culture.

2 THE IMPORTANT INFLUENCE OF THE "BELT AND ROAD" INITIATIVE ON THE INTERNATIONAL TRANSMISSION OF HEALTH QIGONG

2.1 *Building the common vision of the "Belt and Road" lays a good political foundation for the international transmission of Health Qigong*

In order to promote the development and prosperity of the global economy and culture, China has put forward the "Belt and Road" initiative. Under the new landscape of global politics and economy, General Secretary Xi Jinping, based on reality and with great foresight, put forward the idea of the "community of shared destiny" to lead the new world political and economic order in the future. It is advocated that the international situation should be moderated through economic interdependence, and the international system and mechanism should be adopted to maintain and regulate the interdependent relationship so as to safeguard the common interests. Among these, the cultural community is an important goal of this concept, as culture is the basis of promoting economic prosperity and political mutual trust. Therefore, as a representative of the outstanding traditional culture of the Chinese nation, Qigong, with its internal spirit, external simplicity, and easy use, can promote international cultural exchange and political mutual trust better than can sports.

2.2 *The "Belt and Road" initiative raises the international influence of Health Qigong culture*

The flourishing of culture is a distinctive symbol and important supporting quality of a strong country. As General Secretary Xi Jinping stressed in his report to the Nineteenth National Congress of the Communist Party of China (CPC), "Culture is the soul of a country and a nation. Without a high degree of cultural self-confidence and prosperity, there will be no great rejuvenation of the Chinese nation." Health Qigong international transmission is not only a way for the international community to understand China's various sports and health preservation methods, such as Yi Jin Jing (changing tendon exercise), Wu Qin Xi (five mimic-animal exercise), Liu ZiJue (the six-character formula), and Ba DuanJin (eight-sectioned exercise), but also a way for the international community to truly understand the traditional cultural spirit that China has created over thousands of years. Under the "Belt and Road" initiative, the gradual strengthening of cultural exchange and cooperation along the route has promoted the cultural charm and international influence of Health Qigong itself, allowing it to play a positive role in carrying forward the outstanding Chinese traditional culture, strengthening Chinese and foreign sports exchanges, and boosting local social and economic development.

2.3 *The macro strategy of the "Belt and Road" initiative provides a great opportunity for the international transmission of Health Qigong*

The "Belt and Road" initiative has accelerated the pace of opening up between China and other countries, resulting in a multi-field, multi-level, and all-round opening up, with all-round cooperation and integration in the fields of economy, trade, transportation, culture, education, information, tourism, science and technology, and resources. It has brought about a good opportunity for the international spread of Qigong. The international transmission of Health Qigong is a huge cross-cultural transmission activity. Rational cross-cultural transmission must act in the same way as a dialogue and is a dynamic method of transmission featuring mutual respect and understanding. Therefore, it is necessary to actively promote the international academic exchange of Health Qigong in order to create the opportunity for mutual exchange and understanding for the promotion of Qigong in the countries along the "Belt and Road." Through a series of academic research activities, we can communicate, learn, and integrate with each other, providing support for the dissemination of Health Qigong along the route at an academic level.

3 THE FEATURES OF HEALTH QIGONG INTERNATIONAL TRANSMISSION

3.1 *Education: symbol dissemination and entity dissemination*

The construction and dissemination of culture are based on symbols. Symbols are the carriers of information and are used to refer to specific things which are different from the forms of expression of the carriers. The entity transmission is the main mode of transmission of Qigong culture. It is true that code transmission has been the major focus since transmission has been taking place, while entity transmission has not been taken seriously. With the development of film and television technology, movements can be propagated by means of symbols. Videos, pictures, books, and so on are symbols of movements. However, the real physical action transmission cannot be replaced by code transmission, and so the main form of Qigong overseas transmission is field teaching by expatriate teachers, because face-to-face teaching cannot be replaced by video and/or books. In addition, with the profound cultural connotations of the Health Qigong movement, such as power, vigor, spirit, and so on, sometimes they cannot be properly expressed using words or demonstrations, and they cannot be substituted by code dissemination. Therefore, in the process of Qigong culture dissemination, if there is no entity and its dissemination, code dissemination will not be generated.

3.2 Spreading space: crossing the boundary of space–time

"Cultural space" is a place with specific cultural significance, which is especially set up or reserved for a certain cultural activity, during which traditional cultural activities or events will be held periodically or irregularly according to the occurrence of that particular event. Culture depends on the demonstration, storage, and solidification of space to obtain recognition and continuation, while cultural space depends on the nourishment of the culture to attain meaning. Cultural space consists of three elements: time, space, and culture. Humans have always accumulated knowledge for health preservation. Health Qigong has a profound cultural heritage, with the effects of curing illness and strengthening the body and health. Besides carrying vast and rich health preservation information, its oral traditions, folklore activities, folk knowledge, and other cultural forms also preserve a large amount of space–time knowledge, concepts, and consciousness. At the same time, due to the general site requirements, coupled with its gentle movements, it has gained popularity among fitness enthusiasts both at home and abroad.

3.3 Harmony between heaven and man: people-oriented transmission

The existence and dissemination of Qigong culture needs human society to provide it with endless "human resources," which means that it inherently has a dependence on human society. First is the adherence to human culture, which can be understood from the content and dissemination process of Qigong culture. Within its content, Health Qigong performs and implicates the outstanding culture of human health preservation. As the dissemination of Qigong culture is people-oriented, the subjectivity of humans is reflected in the fact that they are not only the transmitters of the information, but also the receivers as the encoder of symbols or texts. One of the most important functions of the transmission of Qigong culture is to carry out the socialized education of the individual, shaping them into a product that meets the requirements of Qigong culture, thus becoming a person belonging to the "Qigong culture."

4 THE OVERALL PLAN FOR HEALTH QIGONG INTERNATIONAL TRANSMISSION UNDER THE "BELT AND ROAD" INITIATIVE

4.1 Strengthen scientific research, guide transmission practice

The prerequisite for the international transmission and development of Qigong is to conduct in-depth scientific research. At present, Health Qigong cause is undergoing a new period of development, and so we must have a profound understanding of the importance and urgency of strengthening the underlying scientific research. To this end, we need to give full play to the guidance and impetus of the theory, carry out research into the theory and practice of Qigong in an organized and planned way, explore and grasp the factors behind the spread of Qigong culture in the new period, continuously improve the working ability and theoretical research level of the competent departments and researchers, and push the cause of Qigong into a new high level. It is necessary to give free rein to the guidance of experts who are proficient in professional theories and have practical experience. They can theoretically demonstrate the cultural project in an all-round way, and form an operational and complete theory, providing a theoretical basis and policy advice for promoting the dissemination of Qigong culture, and thus helping relevant departments to formulate applicable policies, regulations, and implementation plans.

4.2 *Broaden transmission channel, promote cultural radiant power*

Strengthening public opinion is of great significance in encouraging the public to pay greater attention to the international transmission of Qigong. In addition to the connectivity of infrastructure and the construction of the Silk Road Fund financing platform, cultural exchanges, as one of the main lines of the "Belt and Road," have also attracted widespread attention. As culture is the soul of "Belt and Road" construction in China, the China Health Qigong Association in the course of its work has always taken the dissemination and exchange of the Qigong culture as its key area, and has accomplished its various missions by organizing international forums, international competitions, sending out coaches to other countries, and holding training courses. In foreign cultural exchanges, Health Qigong culture has been displayed comprehensively through Qigong training teaching, tours, and in other ways; second, all kinds of events and activities should be held as platforms for the Qigong culture to "go global." Events advertising and displaying Qigong cultural transmission have been endowed with a distinctive connotation, in addition, the transmission content and media have been integrated through sports events through this "global language."

4.3 *Enhance international transmission, promote deep development*

Through the international exchange of Health Qigong, professionals in China and other countries along the route can understand each other's sports and medicine culture resources, health-preserving culture characteristics, and individuality, as well as developing the potential of their respective cultures. Different sports and medicine cultures can establish a relationship of mutual understanding, appreciation, learning, and complementarity, which is an indispensable condition to realizing the international spread of Qigong. Since 2017, the International Federation of Health Qigong has designated the second Sunday of August as the annual "World Health Qigong Day." On August 13, 2017, the First World Health Qigong Day was held synchronously in 43 countries and regions of five continents. The slogan of the event was "Let me teach you to practice Qigong." With the support of 87 member associations, the festivals were staged in dozens of countries and regions. Activities such as the "World Health Qigong Day" have set up a platform to showcase the skills of Qigong enthusiasts. At the same time, it has also attracted more young people, white-collar workers, and sports lovers to participate in Qigong projects.

ACKNOWLEDGMENT

This work was supported by National Social Science Foundation Youth Project No: 17CTY006.

REFERENCES

[1] Chen Gang. 2016. A Study on International Transmission Strategies of Sports Culture in China. Sports Culture Guide Journal (07): 8–11+ 22.

[2] Health Qigong Management Center of State General Administration of Sport. 2007. Training Textbook for Social Sports Instructors of Health Qigong. Beijing: People's Sports Publishing House.

[3] Health Qigong Management Center of State General Administration of Sport. 2007. Study on the Effect of Four Kinds of Health Qigong. Beijing: People's Sports Publishing House.

[4] He Huaxing. 2010. Research on the Dissemination of Intangible Cultural Heritage. East China Normal University: 65.

[5] Tong Shimin, Yu Dinghai & Wang Meijuan. 2011. Research on Current Situation of Overseas Extension of Health Qigong. Journal of Shandong Institute of Physical Education 27 (06): 24–28.

[6] Wang Yigui. 2015. The "Belt and Road": Opportunity and Challenge. Beijing: the people's Publishing Press: 12–14.

[7] Xue Wenzhong. 2017. Research on the Basic System of International Transmission of Traditional Chinese Sports under the Strategy of the "Belt and Road". Journal of Nanjing Institute of Physical Education (Social Sciences Edition) 31(02): 36–40.

[8] Zhao Jianguo. 2009. Discussion on Classification of Entity Dissemination and Dissemination. Journalism and Dissemination Research 16 (04): 92–96 +110.

Computational Social Science – Luo, Ciurea & Kumar (eds)
© 2021 Taylor & Francis Group, London, ISBN 978-0-367-70193-2

Promoting libraries using social media videos

Y. Zhang
Library of Beijing University of Civil Engineering and Architecture, Beijing, China

ABSTRACT: Social media videos have emerged as a new mode of public communication in recent years. They are widely used, require only basic equipment, are easy to share, and are interactive. Libraries should try to provide social media video services, such as live stream communication and short video clips, to help people become acquainted with and visit libraries.

1 INTRODUCTION

With increasing access to the Internet and electronic modes of reading in recent years, people can now consume content online at any time and from anywhere. This has led to a reduced reliance of readers on libraries, and has led to a significant decline in the number of visitors to libraries and the duration of their visits. Students in higher education are most receptive to new information and means of dissemination. According to The People's Daily newspaper, on the 20th World Reading day, the number of books borrowed from libraries in higher education reached a new low for the last decade. Exemplary is Peking University Library, where a total of 620,000 books were borrowed in 2014, in contrast to 1.07 million borrowed from in 2006. A survey by journalists from the Wuhan Morning Post of seven higher education institutions found that the numbers of borrowers had dropped by 30% in five years due to the impact of electronic media [1]. To increase the number of visitors and the appeal of libraries, it is important to carry out promotional campaigns in various formats.

2 MAINS FORMS OF LIBRARY PROMOTION TO DATE

Promotional campaigns for libraries refer to the use of physical, virtual, online, and offline forms of services and technologies to advertise libraries and promotes the services offered by them to encourage more teaching staff, students, and readers in the public at large to frequent them. The following forms of library promotion are common at present:

Offline physical promotion. This is a series of publicity campaigns organised by staff members and volunteers that involves printing and distributing pamphlets, inviting academics to give lectures, and organising visits to the library. The advantage is that readers can be directly exposed to the library, but the disadvantage is the limited scale of impact owing to lack of space, funds, and human resources.

Social media publicity using text is a form of promotion via such platforms as Weibo, WeChat, and Zhihu. It is not confined by time and space, and takes full advantage of the popularity of social media. However, the library as a topic of discussion is less appealing than more eye-catching issues, and struggles to trend, which is critical to popularity on social media. For example, the official Weibo of Tsinghua University Library was created in 2010 but has only 50,000 followers, and gained fewer than 400 in 2019. Few of its 6,000-odd posts have received one or two comments. On Zhihu, a trendy question-and-answer website, only 7,000 people follow of library-related topics, whereas a common topic such as Chinese history has 230,000 followers. This is a large gap.

Videos on social media should be explored as a new way of promoting library services.

3 ANALYSIS OF PROMOTION BY LIBRARIES WITH VIDEOS ON SOCIAL MEDIA

Social media videos are more interactive than conventional modes of library promotion. Readers can see different parts of the library in videos and carry out targeted discussions with other interested browsers. This can create a strong sense of community. At present, the following forms of social media videos are being used by libraries in China:

3.1 Online live video streaming

3.1.1 Characteristics

Live video streaming is a form of online content delivery using video streaming technology via webpages or mobile phones. Viewers can send comments in real time that directly appear on the screen as moving text or captions. The most popular webpage streaming originated in Twitch in the United States. In China, Fengyun introduced live streaming in 2012 and many webpage-based video streaming websites were offered in the next two years. The following are advantages of promoting libraries with caption-based Internet live streaming:

A wide audience. The China Internet Network Information Center released its 39th "Statistical Report on the Development of China's Internet Network" in 2017, which reported that there were 344 million Internet live stream users in the country. The "in-depth survey of Internet livestreaming industry, and predictive report on its investment prospects between 2016 and 2020" published by the Industry Research Centre of the Chinese Investment Consultancy mentioned that 49% of people watching livestreaming were between 20 and 29 years of age, 28% were between 30 and 39 years, and 12% were above 40 years old. An audience base consisting of young adults and the middle-aged is ideal for library promotion.

Streaming is easy to operate and is unconfined by time, space, or special equipment. With mobile devices becoming increasingly accessible, most live streaming websites support mobile streaming. Compared with the conventional television or live PC streaming, mobile streaming incurs a small cost of equipment. All that is needed for real-time streaming is a smartphone with a camera.

A ground-breaking mode of promotion needs to be interactive, down to earth, and entertaining. Conventional modes of promotion involve one-way distribution of textual content or simple questions and answers. In live video streaming, texts simultaneously appear on the screen of the host and the audience. This has allowed hosts to interact with audiences and answer queries with no delay. The audience can also post its suggestions and opinions on the content being streamed, and communicate with other members. This creates an atmosphere of socialised discussion, during which time the relation between the hosts and the audience is one of equals, multi-directional, and interactive.

3.1.2 Issues with promotion using live video streaming

It is difficult to obtain a large group of audience on live video streaming. This is commonly encountered in situations where members of library staff stream live by applying for personal accounts. Although staff are familiar with the layout of libraries, specialised knowledge and standard work practices can provide the audience with a professional introduction and answers to their questions, and this mode of streaming lacks specific appeal and a distinctive theme. It tends to become a tedious show-around or a simple video shoot that cannot achieve satisfactory results. Consider as an example the Douyu live streaming offered by the library of Beijing University of Civil Engineering and Architecture. Only a dozen users were found to engage the live streaming room every hour. Most library-related live streaming accounts on Douyu were followed by fewer than 100 people, and most of them had stopped streaming. Jennifer Koerber claimed in the Library Journal that the Meridian Library District of Idaho attempted to stream live on Facebook, and the first stream was watched by only two users. Its total number of viewers were between 200 and 300.

After a period of streaming, the number of real-time viewers increased to 14 and the total number rose to around 500[2].

Funding-related issues hinder the use of celebrity hosts for library promotion. On live streaming platforms, 80% of the browsing data are shared by 20% of the hosts, who enjoy far greater popularity than other hosts. These hosts are referred to as celebrity hosts. When they stream live, thousands or even tens of thousands of people watch in real time, generating hundreds of thousands of captions in a streaming session. Therefore, if a celebrity host is invited to the library for a promotional campaign, a large number of viewers can be attracted. However, as these hosts are driven by profit, it may significant increase cost to make it prohibitively high for many libraries.

3.2 *Short video promotion*

3.2.1 *Characteristics*

With increasing coverage of 4G networks and growing demands for high-speed communication, the ways in which the general public receives information are numerous and fragmented. Readily accessible and updated information, and efficient online communication that is mobile and inter-connected have become the mainstream modes of online social life today. Since 2017, short video applications such as Bilibili and TikTok have witnessed explosive growth in their numbers of users. TikTok has 150 million daily active users and 300 million monthly active users in China. There are two modes of library promotion on these sites.

The first mode consists of specialised and streamlined short videos represented by Bilibili, which uses professional video production software and producers, and releases short video clips of 1–5 minutes featuring a certain part of the library by employing delicate design and special effects. This mode has a distinctive theme, where texts and contents are professionally planned or outsourced to professional video companies or promoters. This mode contains certain performance elements. On the Bilibili website, videos with high ratings are clicked tens of thousands of times and receive a substantial number of comments. For example, a video with the title "Which province of China has the most keen readers? Statistics of library visitors" was clicked 69,000 times, and received 1,343 captions and 584 comments. "How on Earth is Tsinghua University Library? Prerequisite for advanced productivity! You can't help but study in it," attracted 13,000 clicks, 44 captions, and 44 comments.

The second major mode of video promotion is life oriented, random, and decentralised, and is characterised by TikTok. The relevant videos are short and concise, and last from 15 seconds to a minute. They share interesting stories from libraries and do not have a distinctive theme. The stories are about everyday life or surprising events in libraries. A search of the term "library" as keyword yields the 10 most watched videos that have thousands, if not hundreds of thousands, of likes and comments. The top-ranked video, not professionally made and titled "Tianjin Binhai New District Library," was liked by 600,000 people and received over 6,000 comments. "Jiaotong University Library at 10 pm" ordered by Shanghai Jiaotong University was like by 200,000 people and commented on more than 4,000 times. These videos were significantly more popular than the corresponding videos on Bilibili.

3.2.2 *Disadvantages of short video promotion*

Short videos exclusively produced for publicity need to satisfy certain standards in terms of plan-ning, shooting, and special effects. Professional video editing and post-processing are also required for them. Libraries in China struggle to meet these standards with the human resources they have at present. To achieve satisfactory results, a team of professional producers needs to be hired, which requires a significant amount of funding. Most TikTok short videos do not have specific planners or a central ideal to promote libraries, where the library is only a provider of a venue and cannot itself provide any specialised information. The audience cannot obtain adequate knowledge of the library and often watches the promotional clips out of curiosity. In addition, short video clips are large in number. A search on Bilibili of the keywords "library" yields nearly 1,000 results, and TikTok delivers even more results. They are not effectively categorised or ranked. Ways of enabling promotional videos of libraries to stand out from a sea of clips and attract the user's attention is worth considering for parties who commission them.

4 CONCLUSIONS

In an age where the Internet and social media have emerged as the most important means of learning, entertainment, and communication for the general public in China, social media video promotion has become a useful channel of distribution with convenient, quick, and highly interactive content, and large numbers of users and viewers. Libraries should use this opportunity to carry out and produce videos of various forms and contents to promote themselves to users at large. Short videos on Bilibili and TikTok are more compatible with people's needs to socialise and entertain than online live streaming. Therefore, libraries should lean towards content provision through these channels. The production of videos should be made as professional as possible and specialised production teams and content producers should be introduced where possible to attract more followers of content. Moreover, new ways of promotion should be explored and videos should be distributed in more platforms to publicity adapt to the de-centralised online social networks. In this way, multi-directional interactions with readers can be established to promote libraries and induce more members of the public to visit them.

REFERENCES

[1] Cheng Yongjuan. A study on the causes and coping strategies for the drop in library users in higher education. Sci-Tech Information Development & Economy, 2015, 25(18): 68–70.
[2] Jennifer Koerber. Live from the Library. [2017-04-10]. http://lj.libraryjournal.com/2017/04/marketing/live-from-the-library/

Computational Social Science – Luo, Ciurea & Kumar (eds)
© 2021 Taylor & Francis Group, London, ISBN 978-0-367-70193-2

"Equilibrium and harmony" and "perfect sincerity"—On the formal aesthetics in *Zhongyong*

M.Y. Ding

Department of Chinese Language and Literature, Huazhong University of Science and Technology, Wuhan, P.R. China

ABSTRACT: In the framework of three forms and life dimensions, the formal aesthetics of *Zhongyong* can be divided into two kinds. The equilibrium and harmony of exemplary persons provide the ideal model of ordinary people to follow their own nature and control their emotions under the ritual propriety. Also, perfect sincerity of the sage is the ultimate model for everyone to complete the aspects and create the art form of internal beauty based on his metaphysical order. Therefore, the formal aesthetics of *Zhongyong* emphasizes the formal unity contained in the life and metaphysical form.

1 INTRODUCTION

As an important category of philosophy and aesthetics, "form" has multiple types and expressions. There may be two problems in the current domestic studies of form, the predominance of Western frames and values in Chinese studies and the absence of a metaphysical dimension, which may neglect the origin of formal aesthetics and Chinese theoretical traditions, making it difficult for further formal comparison. This raises a series of questions, such as: What is the Chinese formal tradition? What are the first cause and final purpose of form? In what sense is a comparison between Chinese and Western formal aesthetics possible?

Hence, it is indispensable to explore a reasonable, coherent, and necessary framework for text analysis in the different cultures, and therefore this article proposes the "three forms" [1] to generalize the basis statuses of form in Chinese and Western traditions. In brief, the "three forms" are the metaphysical form, natural form, and art form. The first is the metaphysical being, and is the origin of the other forms; the natural form is created by the creator, is the basic element of the material world, and can provide the physical reference for the artist; and art form, for humans, represents the understanding of world and the creation of meaning as the final purpose. In general, this framework can touch with the cause and purpose of forms and demonstrate the differences and commonalities of understanding of formal aesthetics between the different traditions. Perhaps most importantly, in this framework, the life dimension can not only dominate the generation of artifacts containing the initiative creation of the individual, but also can be viewed as the center of three forms. Therefore, this article explores formal aesthetics through the three forms and life dimension.

Therefore, in the following I would choose selected *Zhongyong* (中庸, *The Doctrine of the Mean*) written by Tsze-sze (子思) as a demonstration text, to analyze the concrete idea of formal aesthetics from the perspective of the "three forms" and life dimension. The tasks of this paper is tripartite: first, to generalize the "three forms" and different levels of life in *Zhongyong*; second, discuss the "equilibrium and harmony" as an ideal form; and third, discuss the formal influences of perfect sincerity in natural and art forms.

2 THE FORMAL EXPRESSIONS AND THREE KINDS OF LIFE

This part focuses on the two core concepts, metaphysical form and life, to give general basic formal expressions and life dimension. In a nutshell, in *Zhongyong*, the metaphysical form is Tao (道, the path or way), and the related expressions include Heaven (天, *tian*) and Earth (地, *di*); the natural form is matter (物, *wu*) and the art form contains two kinds: the ordinary life based on "ritual propriety" (禮, *li*), and the types of literature and art based on "wen" (文, the refinement) [2].

The expressions of life in *Zhongyong* can be divided into three levels: the nature (性, *xing*) of ordinary people, sincerity (誠, *cheng*) of exemplary persons (君子, *junzi*), and perfect sincerity (至誠, *zhicheng*) of the sage (聖人, *shengren*), corresponding to the different understandings of form, respectively. Unlike the nature of everyone, of course, sincerity and perfect sincerity can be regard as the same ideal over all. First, the nature, the classical expression of pre-Qin Confucianism, means the natural endowment and temperament. At the start of *Zhongyong*, Tsze-sze claims that the Tao is to cultivate the individual talents obeying their own natures; and the process of realizing the natural endowment is education. "What Heaven has conferred is called THE NATURE; an accordance with this nature is called THE PATH *of duty*; the regulation of this path is called INSTRUCTION." (天命之謂性,率性之謂道, 修道之謂教) [3].

As the metaphysical form, Heaven is the origin of nature. In the process of realizing this nature, exemplary persons should regulate their emotion within nature to meet the requirement of "ritual propriety," because emotion has two opposing moral tendencies: good and evil.

Second, sincerity means the ideal status of life of exemplary persons. "Sincerity is the way of Heaven. The attainment of sincerity is the way of men. He who possesses sincerity, is he who, without an effort, hits what is right, and apprehends, without the exercise of thought—he is the sage who naturally and easily embodies the *right* way" [3]. By this notion of "sincerity" Chu Hsi (朱熹) means "being true and genuine without falseness" [4]. As such, the life statuses of exemplary persons contain sincerity and "equilibrium and harmony," which are discussed in the next section.

Finally, "the perfect sincerity" of the sage melds life and metaphysical form, and even the three forms. As Chu Hsi said, "in the world, perfect sincerity is referred to as the genuineness of the sage's virtue; nothing in the world can add to [it]" [4]. Because of this absolute status, perfect sincerity can affect the other two forms, which is discussed in the third section.

3 THE RELATIONSHIP AMONG THE THREE FORMS: EQUILIBRIUM AND HARMONY

This part focuses on "equilibrium and harmony," as the unity among the three forms. However, what is the originality of form in *Zhongyong*? Generally, it is a truism that the unity among forms is the basic truth of Confucianism. For example, Confucius holds "my way (道, *dao*) is bound together with one continuous strand" (吾道一以貫之) [5]. So I would rethink this question from the perspective of life in *Zhongyong*.

Regarding the relationship among the three forms, *Zhongyong* emphasizes the relationship between internal emotion and external etiquette, which is different to understanding the unity between metaphysical and physical forms. In other words, the author seeks to realize equilibrium and harmony as the ideal status of life and form by fulfilling the personal nature beyond the Tao as the absolute being in Confucius's formal idea of unity.

In the opening chapter, after the account of "Heaven-The Nature-The Path-Instruction," Tsze-sze discusses how emotion operates the work of forms. This section is divided into four parts. The first part focuses on the formless affection of Tao for exemplary persons. As the metaphysical being, Tao also is an invisible and instant being, rather than a definite doctrine, so exemplary persons should constantly retain self-examination, including respect and awe, at all times, especially in when alone. Obviously, there is a concrete and special description of the deterrent force of Tao, which could provide for the relatively complicated relationships between Tao and humans, to further the formal ideas in Confucianism.

In the she second part, Tsze-sze describes equilibrium and harmony. The term "equilibrium and harmony" contains "equilibrium" (中, *zhong*) and "harmony"(和, *he*).This equilibrium mainly means the emotions, including pleasure, anger, sorrow, or joy, and softens the meaning of endowment in nature. As for "equilibrium," Chu Hsi explains that "when they have not yet arisen, this is nature. It does not deviate to one side or the other, so it is called 'equilibrium'" [4] and because of the formless affection of Tao, exemplary persons can control their emotion or nature in their inner life. Also, "harmony" means "the emotions are correct and there is nothing which is unreasonable or perverse" (Chu Hsi) [4]. From equilibrium to harmony, nature depended on self-control and had a harmonious relationship under the combined action of Tao and "instruction." In fact, the formal aesthetics of harmony is the representational idea in Confucianism, while Tsze-sze focuses on the inner peace based on nature or emotion as the ontology rather than the general understanding of harmonious form. Furthermore, as for the term "zhong" (中), Confucius tends to view it as the central way to adjust the binary opposition, but it seems evident that, on the level of the inner life, Tsze-sze understands this term as the original cause of formal generation instead of the term Tao.

In the third part, Tsze-sze demonstrates that equilibrium and harmony, respectively, are the creator and purpose of form, and they represent the ideal formal relationship of Confucianism. First, Tsze-sze argues that the "equilibrium" is "the great foundation of the world" (大本, *daben*) [4], "harmony" is the all-pervading way of the world, because reason and instruction are for emotion, including pleasure, anger, sorrow, and joy (Zheng Xuan) [5]. Tsze-sze emphasizes that "equilibrium and harmony" can maintain "a happy order," and the material form "will be nourished and flourish" [3].

In sum, on the level of nature or emotion, *Zhongyong* regards "equilibrium and harmony" as the one containing life and ontology with three forms. Comparatively speaking, the three forms in *Zhouyi* have a definite boundary, including Tao, image (象, *xiang*), and the images of the trigrams (卦象, *guaxiang*) [6], and the expression in *The Analects of Confucius* is relatively general, while *Zhongyong* demonstrates how exemplary persons can exalt the positions of life to the metaphysical role by controlling his their own emotions, which is the model of "instruction" for ordinary people.

4 THE FORMAL INFLUENCES OF THE PERFECT SINCERITY TO THE OTHER KINDS OF FORMS

On the basis of the above, the aim of this section is to explore the formal influences of perfect sincerity. Both Sincerity and perfect Sincerity have their own formal influences, and exemplary persons and sages are both ideal life statuses, so the two expressions are interlinked at a quite ideal level for ordinary people. To be specific, sincerity could realize "self-completion" (成己, *chengji*) and "completes other men and things" (成物, *chengwu*). Yet perfect sincerity can also have this influence, and so this section focuses on the perfect sincerity, which can not only complete the self and other things, including natural form and art form, but also "assist the transforming and nourishing power of Heaven and Earth" [3]. Especially, as the metaphysical form, perfect sincerity "can give their full development to the nature of men and things" [3], because the sage "is spoken of as knowing with unquestioned enlightenment and resting in unquestioned correctness" (Chu Hsi) [4]. All in all, the perfect sincerity can assist the functions of heaven and earth, according with the goodness of "production and reproduction" (生生, *shengsheng*) in *Zhouyi*. Hence, a more detailed discussion about the affects of perfect sincerity on the natural and art forms is required.

4.1 *Completing things*

The power of completing things (natural form) is interconnected with self-completion, since the union is affected by the external form and internal life, which emphasizes the relationship between nature and things associated with the perfect sincerity.

Specifically, how does perfect sincerity complete things? Tsze-sze describes the whole of the formless process. First, the perfect sincerity, as a metaphysical being, is "unceasing." And then there are a series of metaphysical characters of the perfect sincerity: long-lasting, efficacious, far-reaching, large and substantial, and high and brilliant. Correspondingly, perfect sincerity completes things from its three kinds of characters, respectively: "Being large and substantial is how it contains things. Being high and brilliant is how it spreads over things. Being far-reaching and long-lasting is how it completes things" [4]. This detailed description can enrich the Confucian understanding of creating and completing things through the different levels of ways.

Furthermore, by its very nature of perfectness and goodness, Tsze-sze also describes how perfect sincerity rebuilt and completed partial things (曲, *qu*) with all kinds of imperfect forms. The perfection of the sage can promote the sage to "raise his nature's whole substance and complete it" [4]. Similarly, the goodness of the sage also could create the part's perfection in each case (Chu Hsi) [4]. It is worth noting that "this sincerity becomes apparent," which means that "it has a form." As Chu Hsi explains, "form (形, *xing*) is what accumulates within and is manifest without" [4]. Therefore, this saying reminds us that affection of perfect sincerity to natural things is from an internal life to external form. Finally, there is the exact purpose of completing things. For example, the purpose of "transform" (化, *hua*) is the goodness of nature from the evil in Kong Yingda's understanding [5].

4.2 The establishing order of art form

As we discussed earlier, there are two main kinds of art forms in *Zhongyong*: social ritual propriety, and specific types of literature and art. For the former, perfect sincerity is "the great invariable relations" (大經, *dajing*). In particular, "in the world, only [someone with] perfect cheng is deemed able to bring order to the great fabric of the social structure of the world, to establish the great foundation of the world" [4]. Here the social structure of the world is the form of society. The Jing (經) and Lun (綸) are of different forms in the silk, which could be due to the metaphysical form and "the complete substance of nature" (大本, *daben*). Since perfect sincerity of the sage is beyond falseness, so, for the relationship between people, "each exhausts their proper genuineness and all can be considered as models for later generations of the world" [4]. This "model" (法, *fa*) is the metaphysical form for the social forms, which includes the institutional system and civilized etiquette. In the framework of three forms, this model contains metaphysical virtue, the nature of things, and the ancient tradition, which is another expression of the art form in the historical dimension. For example, "Chung-ni handed down the doctrines of Yao and Shun, as if they had been his ancestors, and elegantly displayed the regulations of Wan and Wu, taking them as his model. Above, he harmonized with the times of heaven, be was conformed to the water and land" [3]. Yao, Shun, Wen and Wu, are the ancient sages of the model; "the times of heaven" is the metaphysical expression; and the water and land are the natural forms.

Furthermore, the art form also contains the types of literature and art. As for the five virtues of "perfect sagacity," Tsze-sze lists the fifth as "to be orderly, principled, careful and discerning." The term "orderly" also means "wen" It is important that he also highlights the principle of being careful and discerning. So we could conclude that the wen (order) should be considered from the perspective of the whole of tradition containing various forms. Therefore, Tsze-sze holds the "internal beauty" (內美, *neimei*) as the standard of form for acquiring the virtue or sincerity: "He is plain but not tiresome, simple but refined, gentle but principled" [4].

In the concrete discussion, Tsze-sze refers to "over her brocade garments she wore a plain, unlined coat" in *Shijing* to explain the "internal beauty," which is related to "the application of color is to the unadorned" [7]. These accounts emphasize the relationship between internal life and external forms rather than the relationship based on Tao as the absolute metaphysical form in *The Analects of Confucius*. However, under sincerity, the formal relationship has changed into the problem of self-cultivation about the emotion.

On the one hand, this nature of "internal beauty" derives from the perfect sincerity. For example, Tsze-sze refers to "the decree of Heaven—Ah! How profound and unending!" in *Shijing*. The term "profound" (穆, *mu*) means ceaseless beauty (Kong Yingda) [5], and is different to the external form. Therefore, the process of generating and changing in beauty is from perfect sincerity to the visual form of beauty.

On the other hand, the external form also could be viewed as a reference to sincerity. In general, since perfect sincerity is formless or imperceptible, it is necessary to hold the balance between the two forms, according to the way of equilibrium (中道, *Zhongdao*). For example, "one's basic disposition and refinement are in appropriate balance" [7]. The disposition (質, *zhi*) is inner life, and the refinement (文, *wen*) is "art form." As a result, in the context of public education, the external form (art form) could show the formless form of perfect sincerity for everyone to realize its goodness and perfection.

5 SUMMARY

In the framework of the "three forms," the formal idea of *Zhongyong* is closely bound up with the life dimension. Concretely speaking, the equilibrium and harmony of exemplary persons are a model for ordinary people to follow their own nature and control their emotions. And then perfect sincerity of the sage is the ultimate model and purpose for the ordinary people and example person, to complete the things and perfect the partial through the different ways, including the art form. Comparatively speaking, both equilibrium and harmony, and perfect sincerity are those containing the life and metaphysical form, so the formal aesthetics of *Zhongyong*, to a great extent, provides a pure and identical understanding for Confucianism and Chinese tradition, which differs from the relatively clear kinds of form in *Zhouyi*, and the relatively general description in *The Analects of Confucius*. In this sense, this article shows the special ideas of form in *Zhongyong* as those of the Confucian and Chinese classics.

With the account based on the framework of this article, scholars may have come to recognize its rationality and necessity. Therefore, in closing, I would suggest that the framework of the three forms and life dimension should be considered and is applicable for formal studies, especially for the comparison of formal aesthetics between the Eastern and Western traditions.

NOTE

It should be added here that the translation of the *Zhongyong* in this article is mainly based on those of James Legge, and Ian Johnston & Wang Ping. In the concrete discussion, I chose and unify the more suitable expressions from these two translation versions. Furthermore, as for some other translations of key words, such as "exemplary persons" and "ritual propriety," I borrow from Roger T. Ames & Henry Rosemont according to the formal studies in this article. I have also added some Chinese characters and phonetic spellings to the key references to help retain the original meanings.

REFERENCES

[1] Wang Qiankun, "The Language of Awe", *Poetry, Calligraphy, Painting*. 2013(4): 105–114. (in Chinese).
[2] Peter K. Bol, *"This Culture of Ours": Intellectual Transitions in T'ang and Sung China*. Stanford University Press, 1992.
[3] James Legge, *The Chinese Classics* (Vol I), *The Doctrine of the Mean (Zhong Yong)*. Taipei: Smc Publishing Inc., 1991.
[4] Ian Johnston, & Wang Ping. *Daxue and Zhongyong: Bilingual Edition*. Hong Kong: The Chinese University of Hong Kong Press, 2012.
[5] Zheng Xuan, & Kong Yingda, *The Correct Meanings of the Records of Rites (Liji Zhengyi*. Shanghai: Shanghai Classics Publishing House, 2008. (in Chinese).
[6] Ding Maoyuan, "Image and Life: on the Formal Concept of *Zhou yi*", *Chuanshan Journal*. 2018(2): 88–97. (in Chinese).
[7] Roger T. Ames, & Henry Rosemont, JR., *The Analects of Confucius: A Philosophical Translation*. New York: Ballantine Books, 1998.

Computational Social Science – Luo, Ciurea & Kumar (eds)
© 2021 Taylor & Francis Group, London, ISBN 978-0-367-70193-2

Comments on the judgment of the first instance on the compensation case of Yangzhou Qunfa Company from the perspective of critical thinking

H.W. Feng & W.M. Ouyang
Shanghai University of Political Science and Law, China

ABSTRACT: On Saturday, May 14, 2016, because Chang Xuehong and Shi Jimei, staff members of the inspection department of Yangzhou Qunfa Heat Exchanger Co., Ltd. were unwilling to work overtime in the afternoon, the company failed to deliver goods on time and compensated the customer, Youmeng company, with liquidated damages of 120,000 yuan. Qunfa sued Chang Xuehong and Shi Jimei for compensation for the company's losses caused by refusing to work overtime. In the first instance, the People's Court of Hanjiang District of Yangzhou City sentenced Chang Xuehong and Shi Jimei to bear 15% of the losses incurred by Qunfa, that is, 18,000 yuan. On April 29, 2020, Yangzhou Hanjiang Court took this case as a typical case to deliver a speech at a press conference, which aroused heated discussion in the wider society and became the focus of public opinion. This paper comments on the judgment of the first instance of the case from the perspective of critical thinking and points out its logic errors.

1 INTRODUCTION

On Saturday, May 14, 2016, because Chang Xuehong and Shi Jimei, staff members of the inspection department of Yangzhou Qunfa Heat Exchanger Co., Ltd., were unwilling to work overtime in the afternoon, the company failed to deliver some goods on time and compensated their customer, Youmeng Company, with liquidated damages of 120,000 yuan. Qunfa then sued Chang Xuehong and Shi Jimei for compensation for the company's losses caused by them refusing to work overtime. In the first instance, the People's Court of Hanjiang District of Yangzhou City sentenced Chang Xuehong and Shi Jimei to bear 15% of the losses incurred by Qunfa, that is, 18,000 yuan. On April 29, 2020, Yangzhou Hanjiang Court took this case as a typical case to deliver a speech at a press conference, which aroused heated discussion in the wider society and became the focus of public opinion.

In the first instance judgment of the case, the judge held that the employer and employee should fully perform their respective obligations in accordance with the labor contract. Regarding the dispute focus of whether Chang Xuehong and Shi Jimei should be held responsible for the loss caused by the delay in delivery of goods, the judge held that Chang Xuehong and Shi Jimei had faults in the above-mentioned losses and should bear corresponding responsibilities, and conducted legal logic reasoning. From the perspective of critical thinking, this paper carefully analyzes the fact identification and legal logic reasoning in the first instance judgment of the case, and points out the mistakes in the legal logic, so as to distinguish right from wrong and highlight social fairness and justice.

2 THE TOPIC AND CONCLUSION

In the case of the labor dispute of Qunfa Company vs. Chang Meihong and Shi Jimei, the appeal of Qunfa Company is that the appellee shall jointly compensate the appellant for economic loss of 120,000 yuan. The reason is that Chang Meihong and Shi Jimei should be responsible for the loss caused by the delay in delivery of goods by Qunfa Company.

The topic of the first instance judgment of the case is: should Chang Meihong and Shi Jimei be responsible for the losses caused by Qunfa Company's delay in the delivery of goods? The conclusion is: Chang Meihong and Shi Jimei should be responsible for the losses caused by Qunfa Company's delay in delivery of goods. That is to say, the first instance judgment partially supports the appeal of Qunfa Company.

3 THE REASONS

According to the judgment of the first instance, the judge held that Chang Xuehong and Shi Jimei had fault in the above-mentioned losses and should bear the corresponding responsibility for three reasons, as described below.

First, combined with the overtime stipulation agreed in the labor contract between Chang Xuehong and Shi Shi Jimei, as well as the record of vacation transfer in the attendance record, we can confirm that Chang Xuehong and Shi Jimei have agreed that Qunfa Company can arrange overtime work according to the production task (Chang Xuehong and Shi Jimei have the right to ask for compensatory leave or overtime pay by Qunfa Company).

Second, in the case that taking time off after working overtime has been confirmed as an integral part of the labor contract through the actual behavior of both parties, even though Qunfa Company has not provided evidence to prove that they had reached an agreement with the trade union on overtime work, considering Chang Xuehong and Shi Jimei had agreed to work overtime, Qunfa Company had reason to expect that when the production task was urgent, Xuehong and Shimei would agree with the overtime requirements of Qunfa Company, while Chang Xuehong and Shi Jimei still chose to refuse to work overtime although they knew that the production task was urgent. Even if they did not do this intentionally, they were at least guilty of gross negligence.

Third, the loss caused by the delay in delivery of the company in this case generally belongs to the business risk of the enterprise and should be borne by Qunfa Company itself. However, considering that Chang Xuehong and Shi Jimei were the inspection workers who were necessary to fulfill the delivery obligations on time, in the case that Qunfa Company had urgent production tasks and could protect the legitimate rights and interests of Chang Xuehong and Shi Jimei by arranging to switch their working day, Chang Xuehong and Shi Jimei still refused to work overtime. They were not aware of the risks that the employers may face. They had no sense of ownership. They have a certain fault for the losses, so they should bear the corresponding fault liability.

According to the above three reasons, the judge decided that Chang Xuehong and Shi Jimei should bear 15% of the compensation liability for the loss of Qunfa, that is, 18,000 yuan, taking into account their income level, the employer's management negligence, and the extent of the damage.

Therefore, the judgment of the first instance is as follows: part of the claims of Qunfa Company have a factual and legal basis, which is supported by the court of first instance. In accordance with the provisions of Article 29 of the Labor Contract Law of the People's Republic of China, it was decided that: 1. Chang Xuehong and Shi Jimei should jointly pay 18,000 yuan of compensation to Yangzhou Qunfa Heat Exchanger Co., Ltd. within 10 days from the effective date of the judgment; 2. Other claims of Yangzhou Qunfa Heat Exchanger Co., Ltd were rejected.

In the remainder of this paper, from the perspective of critical thinking, the three reasons proposed in the judgment of the first instance are analyzed using legal logic.

4 COMMENT ON THE FIRST REASON

The first reason for the judgment of first instance is as follows:

Combined with the overtime stipulation agreed in the labor contract between Chang Xuehong and Shi Jimei, as well as the record of vacation transfer in the attendance record, we can confirm that Chang Xuehong and Shi Jimei have agreed that Qunfa Company can arrange overtime work according to the production task (Chang Xuehong and Shi Jimei have the right to ask for compensatory leave or overtime pay by Qunfa Company).

The reason is actually a subtopic. The conclusion of this subtopic is that Chang Xuehong and Shi Jimei have agreed that Qunfa Company can arrange overtime work according to their production tasks. There are two reasons for this. One is the overtime stipulation agreed in the labor contract between Chang Xuehong and Shi Jimei, that is to say, the labor contract has the stipulation of overtime work, and the second is the record of compensatory leave in the attendance record.

According to the above two reasons, it can be logically concluded that Chang Xuehong and Shi Jimei had previously agreed that Qunfa Company could arrange overtime according to the production tasks. However, this does not mean that Chang Xuehong and Shi Jimei must agree to the present and future overtime requirements. Therefore, it is uncertain whether Chang Xuehong and Shi Jimei definitely agreed to any overtime work arranged by Qunfa Company according to the production tasks.

Here, the judgment of the first instance committed a logical fallacy called an irrelevant conclusion. Therefore, the conclusion of the subtopic is wrong.

5 COMMENT ON THE SECOND REASON

The second reason for the judgment of first instance is as follows:

In the case that taking time off after working overtime has been confirmed as an integral part of the labor contract through the actual behavior of both parties, even though Qunfa Company has not provided evidence to prove that they had reached an agreement with the trade union on overtime work, considering Chang Xuehong and Shi Jimei had agreed to work overtime, Qunfa Company had reason to expect that when the production task was urgent, Xuehong and Shimei would agree with the overtime requirements of Qunfa company, while Chang Xuehong and Shi Jimei still chose to refuse to work overtime even though they knew that the production task was urgent. Even if they did not do this intentionally, they are at least guilty of gross negligence.

This reason is also a subtopic.

The conclusion of this subtopic is that Chang Xuehong and Shi Jimei are at least guilty of gross negligence in dealing with the losses caused by refusing to work overtime, even if they did not intentionally do so. The reasons are as follows: (1) taking compensatory leave after working overtime has been recognized as an integral part of the labor contract through the actual behavior of both parties; (2) even if Qunfa company did not provide evidence to prove that its overtime arrangement had been agreed with the trade union, considering the fact that both Chang Xuehong and Shi Jimei agreed to work overtime before, Qunfa company has reason to expect Chang Xuehong and Shi Jimei to agree with Qunfa's working overtime requirements when the production task is urgent; (3) Chang Xuehong and Shi Jimei still chose to refuse to work overtime even though they knew that the production task was urgent.

According to the provisions of Article 8 of the implementation of the regulations of the State Council on the working hours of employees, that is, if the working hours are extended, the enterprise should pay wages to the employees or arrange compensatory leave in accordance with the provisions of Article 44 of the Labor Law of the People's Republic of China, the enterprise should pay overtime wages to the employees or arrange compensatory leave for them. Therefore, the enterprise has the right to choose between paying wages or arranging compensatory leave. That is to say, enterprises can take compensatory leave without paying wages after overtime work, while employees have no right to choose whether to ask for compensatory leave or pay wages after overtime work. Therefore, the first reason for the above subtopic is tenable.

Article 41 of the Labor Law of the People's Republic of China stipulates as follows:

The employing unit may extend working hours due to the requirements of its production or business after consultation with the trade union and laborers, but the extended working time for a day shall generally not exceed one hour; if such an extension is called for due to special reasons, the extended hours shall not exceed three hours a day under the condition that the health of laborers is guaranteed. However, the total extension in a month shall not exceed 36 hours.

According to Article 41, enterprises need to negotiate with the trade union and workers in advance to require employees to work overtime. That is to say, enterprises should consult with both trade unions and employees to require employees to work overtime. In other words, employees have the right to agree to overtime work, but also have the right not to agree to overtime, that is, employees have the right to consent to overtime. This right to consent to overtime of employees shall not be extinguished by whether they have agreed to work overtime in the past. To put it in greater detail, it is not because working overtime was agreed to in the past, that they must agree to work overtime at that time and in the future, and because they did not agree to work overtime in the past, does not mean they must not agree to work overtime at that time and in the future. Whether they agreed to work overtime in the past or not does not affect exercise of the right to consent to work overtime at that time and in the future.

In this case, we cannot think that Chang Xuehong and Shi Jimei would definitely agree to the overtime work requirements of Qunfa company just because they both agreed to work overtime in the past. Of course, Qunfa Company can expect Chang Xuehong and Shi Jimei to agree to work overtime. However, expectation is only a unilateral wish of Qunfa, and it does not mean that Chang Xuehong and Shi Jimei are obliged to meet such expectations of Qunfa Company. In other words, Qunfa Company can ask its employees Chang Xuehong and Shi Jimei to work overtime, but Chang Xuehong and Shi Jimei may not accept the overtime requirements.

Since Qunfa Company has not provided evidence to prove that its overtime arrangement was reached in an agreement with the trade union, it should bear the adverse consequences of proof failure, that is, it must accept the legal fact that Qunfa Company has not obtained the consent of the trade union for overtime work. Since there is no agreement from the trade union, Qunfa Company cannot force its employees, Chang Xuehong and Shi Jimei, to work overtime. To say the least, even if the trade union agrees with the overtime work requirements of Qunfa Company, it should also negotiate with the employees according to law. If the employees do not agree to work overtime, Qunfa Company cannot force them to work overtime.

Article 41 of the Labor Law of the People's Republic of China on the extension of working hours is not absolute. Article 42 of the Labor Law of the People's Republic of China stipulates the following exceptions:

The extension of working hours shall not be subject to restriction of the provisions of Article 41 of this Law under any of the following circumstances:

(1) Where emergency measures are necessary in the event of a natural disaster, accident, or other reason that threatens the life and health of laborers and the safety of the property;
(2) Where prompt quick repair is needed in the event of a breakdown of production equipment, transportation lines, or public facilities that affects production and public interests; and
(3) Other circumstances as stipulated by laws, administrative rules, and regulations.

At the same time, Article 7 of the State Council "Measures for the Implementation of the Provisions on Working Hours of Staff and workers" also stipulates the following exceptions:

In case of any of the following special circumstances and urgent tasks, the extension of working hours shall not be restricted by the provisions of Article 6 of these measures:

(1) Occurrence of natural disasters, accidents, or other reasons, where the safety and health of the people and the state's assets are in serious danger and need to be treated urgently;

(2) Failure of production equipment, transportation lines, and public equipment, which affects production and public interests and must be repaired promptly;

(3) The period of production suspension on statutory holidays or public holidays must be used to overhaul or maintain equipment;

(4) In order to fulfill the urgent tasks of national defense or other urgent production tasks arranged by the higher authorities outside the state plan, or the urgent tasks of purchasing, transporting, and processing agricultural and sideline products by commercial and supply and marketing enterprises in peak season.

Under what kind of case does Qunfa Company require its employees Chang Xuehong and Shi Jimei to work overtime? Qunfa Company signed two purchase contracts with Shenzhen Youmeng Industrial Co., Ltd. on April 15, 2016 and April 26, 2016, respectively, stipulating that Qunfa Company should deliver the product to the designated dock on May 15, 2016. On the morning of May 14, 2016, most of the product inspection work was completed. Chang Xuehong and Shi Jimei were required to work overtime in the afternoon of May 14, 2016 (Saturday), which was originally a rest time, to complete the remaining product inspection work. This overtime work belongs to a common emergency of the company itself. It is neither an exception to the provisions of Article 42 of the Labor Law of the People's Republic of China, nor a special case or urgent task specified in Article 7 of the measures for the implementation of the provisions on the working hours of employees issued by the State Council. Therefore, Qunfa Company should negotiate with the trade union and workers in accordance with the law, and the workers have the right to refuse to work overtime. Therefore, Chang Xuehong and Shi Jimei have no fault or responsibility for the losses caused by the delay in the delivery.

6 COMMENT ON THE THIRD REASON

The third reason for the judgment of first instance is as follows:

The loss caused by the delay in delivery of the company in this case generally belongs to the business risk of the enterprise and should be borne by Qunfa Company itself. However, considering that Chang Xuehong and Shi Jimei are the inspection workers who are necessary to fulfill the delivery obligations on time, in the case that Qunfa Company having urgent production tasks and being able to protect the legitimate rights and interests of Chang Xuehong and Shi Jimei by switching the working day, Chang Xuehong and Shi Jimei still refused to work overtime. They were not aware of the risks that the employers may face. They had no sense of ownership. They have a certain fault for the losses, so they should bear the corresponding fault liability.

In Qunfa Company, in addition to Chang Xuehong and Shi Jimei, there are other product inspectors. The inspection staff necessary to fulfill the delivery obligations on time were not necessarily Chang Xuehong and Shi Jimei, who could have been replaced by other inspection staff. In other words, the inspection work was indeed necessary to fulfill the delivery obligations on time, but the inspectors available were not only Chang Xuehong and Shi Jimei. Therefore, it is wrong to hold that Chang Xuehong and Shi Jimei, as the inspection staff necessary to fulfill the delivery obligations on time, do not conform to the facts. Here, the judgment of the first instance has committed a logical fallacy called disguised replacement of concept.

It has been pointed out previously that the urgent production task of Qunfa Company was not an exception provided by law, nor was it a special case or urgent task provided by regulations, but was caused by improper arrangement of Qunfa company's work and personnel. This was due to the poor management of Qunfa Company, and the responsibility lies with the company, not with the employees.

In this case, the Qunfa Company should take a sincere attitude to negotiating with its workers, which it did not. It is wrong to say that the legitimate rights and interests of Chang Xuehong and Shi Jimei can be protected by arranging compensatory leave, because according to Article 44 of the Labor Law of the People's Republic of China, Qunfa Company should pay wages higher than the wages of workers during normal working hours, while Qunfa Company are only willing to arrange compensatory leave without providing other compensation methods. Meanwhile, laborers have the right to agree to work overtime in accordance with the law. They can either agree to work overtime or refuse. Therefore, this method, that is, the Qunfa Company taking compensatory leave to compensate laborers for overtime work, is to protect the rights and interests of the company, and damage the legitimate rights and interests of laborers.

As an employing unit, Qunfa Company should predict the business risk of the enterprise, and have the responsibility to take positive measures such as adjusting the production plan and personnel arrangement of the enterprise to resolve the possible risks. As employees of the enterprise, Chang Xuehong and Shi Jimei have the responsibility to obey the work arrangement in normal working hours, but they do not replace the responsibility and obligations of the enterprise to resolve these risks; otherwise, the responsibility of the enterprise's managers and laborers is misplaced.

As for the so-called "sense of ownership," it is suspected that the enterprise has carried out moral kidnapping of employees. In an enterprise that is not willing to pay overtime wages, enterprise managers do not care about the gains and losses of laborers' interests, but require laborers to carry forward the "sense of ownership". It is not only obviously unfair, and not commensurate with the rights and responsibilities, it also illustrates the low standards of an organization, with high standards for individuals, and the judge's double standards highlight that its values are seriously deviated from the justice and rule of law in our country with socialist core values.

Therefore, the third reason for the judge's argument that Chang Xuehong and Shi Jimei have fault is not valid. Chang Xuehong and Shi Jimei have no fault and should not bear the corresponding fault liability.

7 CONCLUSIONS

In conclusion, the judgment of the first instance of the compensation case of Qunfa Company not only made many legal logic mistakes, but also obviously deviated from the two socialist core values of justice and rule of law. The mistakes made in the judgment are not only relatively poor, but also contrary to common sense. If it was not intentionally done by the judge, it was caused by the judge's low legal logic reasoning.

Judging from the behavior of Hanjiang Court of Yangzhou, which took this case as a typical case to discuss at a press conference, the judges of the first instance court are less likely to do this intentionally, and their legal logic quality is more likely to be poor. It is inevitable that such a low-level and disordered judgment of the first instance would not be accepted by society. Fortunately, the judgment of the second instance of Yangzhou Intermediate People's Court of Jiangsu Province made a negative conclusion on the judgment of the first instance, which recovered the negative social impact caused by the judgment of the first instance to an extent.

Thus it can be seen that the logical quality and values of judges are of great importance to the maintenance of social fairness and justice.

REFERENCES

Legal Book Editorial Department of China Labor and Social Security Press, Labor Law of the People's Republic of China, *China Labor and Social Security Press*, February 1, 2019.
Ministry of Labor, PRC, Measures for the Implementation of the Regulations of the State Council on the Working Hours of Employees, *http://blog.chinalawedu.com/falvfagui/fg23051/152935.shtml*, July 12, 2020.
Weimin Ouyanga, Haowen Feng, Critical Analysis On the Apology Letter of Lin Jianhua, In *Proceedings of 7th International Conference on Applied Social Science*, December 19-21, 2018, Mexico City, Mexico.
M. Neil Browne, Stuart M. Keeley, translated by Zhao Yufang, *Asking the Right Questions: A Guide to Critical Thinking (10th Edition)*, China Machine Press, Published in December 2012.

Computational Social Science – Luo, Ciurea & Kumar (eds)
© 2021 Taylor & Francis Group, London, ISBN 978-0-367-70193-2

Marked theme analysis and its implications in academic paper writing

Y. Zhan, L.J. Li, L.Y. Miao & Y.L. Li
Xi'an High-tech Research Institute, Hongqing, Xi'an city, P.R. China

ABSTRACT: Halliday holds that there is a marked theme and an unmarked one from the perspective of functional linguistics. The marked theme is not consistent with the subject of the sentence; instead it is the beginning of the topic. In this study, we have modified Davies framework for a marked theme with contextual functions. Then the collected data from both Chinese magazines and EI index papers were counted sentence by sentence to calculate the proportion of contextual function (CF) marked themes. We compared and contrasted the percentages of CF marked themes, from the abstract to conclusion section, involved in both Chinese papers and English papers. Data analysis shows that the frequency of the marked theme differs from section to section both in Chinese and English papers; and the CF marked theme is more frequently applied in Chinese papers for its linguistic features, which will guide our academic paper writing.

1 INTRODUCTION

It is more likely for Chinese and foreign scholars in various fields to exchange their ideas and research work by published papers and international conferences due to the process of internationalization, both of which are based on paper writing. However, most Chinese writers "translate" their papers into English rather than writing it in English directly, resulting in awkward expressions and ambiguous descriptions because of the language differences. Therefore, more and more linguists realize that the social and cultural environment of language use and the influence of language communities must be taken into account in language or discourse descriptions (Swales, 1990). In this way the language can be used correctly. Nowadays, a large number of researchers have become dedicated to the study of the textual organization function as well as the surface linguistic means of genre. These studies show that the use of different means will be determined by the function of the genre. The marked theme–rheme structure is an important means to form the features of a generic structure. In this paper, we have studied Halliday's theme theory, collected research papers both in Chinese and English, and then applied Davies' categories of generic marked themes to compare and contrast the textual functions of marked themes in different languages. Based on the data analysis, we reveal the differences in thematic patterns between English and Chinese, which will guide English academic paper writing.

2 THEORETICAL BASES

2.1 *Thematic theory*

Halliday defines theme from the perspective of function rather than the distribution of components. He holds that the theme is the starting point of information, with which the clause is concerned, and rheme refers to the remainder of the message, in which the theme is developed. In his view, theme can be divided into nonmarked and marked themes. The nonmarked theme is consistent with the subject in traditional grammar. Examples are as follows:

1) Matlab / SIMULINK 仿真软件是一种比较理想的仿真平台。 (nonmarked theme)
2) MATLAB/SIMULINK is a very ideal simulation platform. (nonmarked theme)

Table 1. The CF marked thematic classifications.

Function	English expressions	Chinese expressions
Location in time	Recently, later, as, before, initially, first	目前，同时，此时，首先介绍，最好讨论，前者
Location in place	Here, there, in this paper, from the above discussion	在该仿真中，在文章中，通过上面的讨论可知
Addition	That is, in other words, furthermore, indeed, plus, besides	例如，即，另外，此外，再者
Contrast/concession	However, instead, by contrast	但是，然而
Cause	Thus, therefore, consequently, in order to, because, since	由于，为了，因为，因此
Means	By using, using, in this way, based on	通过，根据
Condition	If, in this case, then, in practice, assuming, perhaps	如果在这种情况下，假如，可能会得出
Validation	Usually, as is known, in general, evidently, clearly, from the previous study	通常，实际上，众所周知，从上述图标可清楚看出算法的有效性
Viewpoint	Apparently, surprisingly	显然，当然

3) 在此基础上，就可设定结构参数。(marked theme)
4) In engineering practice, we can select a proper mono-discipline. (marked theme)

Examples (3) and (4) show that it is helpful for us to understand and grasp the stylistic features of a genre by recognizing that the marked theme at the beginning of a sentence usually has a special textual function.

2.2 Marked theme

2.2.1 The categories of the marked theme
The marked theme refers to the one that is not the subject of a clause, which can be further divided into two subcategories due to its different functions. When the theme is the complement of a sentence, or it is in the inversion structure, it does not play the role of organizing the text; while when the theme is to show time, place, instrument, transition, etc. it can be used to organize the text.

2.2.2 Davies' classification of CF marked theme
Davies (1989) pointed out that the contextual function (CF) marked theme is mainly used to express the contextual functions of time, place, transition, condition, and affirmation, which can be divided into nine categories. In this paper, we have collected academic papers both in English and Chinese, based on which we modified the CF marked thematic classifications as shown in Table 1.

3 CF MARKED THEME IN ACADEMIC PAPERS

3.1 Research subject and method

In this paper, we focus on Chinese and English academic papers and analyze them from the perspective of marked themes. As for the Chinese academic papers, we collected them from journal articles of Xi'an High-tech Research Institute; for English ones, we collected papers from journal articles of EI retrieval. We then analyzed the CF marked theme involved respectively to compare and contrast the similarities and differences.

Table 2. The frequency of CF marked themes used in each part of the paper.

	Abstract (%)	Introduction (%)	Method (%)	Results and discussion (%)	Conclusion (%)
Chinese	6.6	25.7	34.7	28.1	4.9
English	1.8	11	51.6	32.5	3.3

3.2 *The use of the CF marked theme in academic papers*

The results showed that 46% of the total sentences had a CF marked theme in Chinese papers and 19% in English papers. In Chinese papers, the frequency of a CF marked theme in the abstract, introduction, experiment, result and discussion, and conclusion was 37.8%, 56.9%, 47.2%, 46.7%, and 56%, respectively, while in English, the percentage was 11.1%, 18.5%, 16.4%, 26.5%, and 11.3%.

3.3 *The frequency of the CF marked theme used in each section of an academic paper*

In academic papers, each section plays a different role in the discourse, therefore, the distribution of CF marked themes varies a great deal. Eiler states that whether the CF marked theme is used or not depends on its content. Its real functional meaning is conveyed only when we combine the CF marked theme with textual content. An academic paper usually consists of the following five parts: abstract, introduction, experiment, result and discussion, and conclusion. Table 2 shows the frequency of CF marked themes used in each part of the paper.

4 RESULTS AND DISCUSSION

From the data, we can see that the CF marked theme is used more frequently in Chinese than English texts. This is mainly due to the differences in sentence structures between Chinese and English. For example, in English, nonmarked positions, such as adverbials of time, place, and manner, are usually placed at the end of a sentence. In Chinese, most of these adverbials are located at the beginning or in the middle of a sentence, rather than at the end of a sentence. Therefore, the CF marked theme which is used to express time, space, manner, and means in an Chinese academic paper plays the dominant role in sentences. For example, we often use "now,", "in recent years," "recently," "first," "second," and so on to show time; and use "through" and "according to" to show ways or methods.

We also found that CF marked themes are used in the introduction, experiments, and results and discussion sections. In both English and Chinese papers, the CF marked theme is the most frequently used, accounting for 51.6% and 34.7%, respectively, followed by the results and discussion, and the summary and conclusion sections.

The data show that compared with other sections, the CF marked theme is most frequently used in the experimental section of both Chinese and English academic papers, accounting for 51.6% and 34.7%, respectively. This may be determined by the dominant role it plays in the academic paper. As for the paper, the main purpose is to tell your readers the way you applied the research and to let them understand your experiment and testify whether it is effective or not. the experimental section is used to explain the research method and research procedure, therefore it is relatively complex. Most use some marked thematic structures such as "in an emergency," "in simulation experiment," "to solve," and "therefore."

The abstract, the condensed form of the full text of the paper, is used to introduce the main content of the paper, which consists of four components, namely, research background, research purpose, research methods, results and conclusions, and points out the innovation of the paper. According to frequency statistics, the CF marked thematic structure is least used in both English and Chinese academic papers, accounting for 6.6% and 1.8%, respectively. This is mainly due to the differences in the textual functions of the various sections. The abstract, as a mini-form of the whole paper, is comparatively independent. One of its linguistic features is that it is concise. Therefore, in an English academic paper, the nonmarked theme–rheme structure is frequently used. For example, This paper presents... ; The new algorithm was applied to... ; The results are as follows... ; The conclusion will be given... etc. In contrast, the CF marked theme is more commonly used in Chinese papers. We often use "at present," or "in recent years" to highlight the research background or research significance. In the introductory section, we often use words such as "by/through," "according to," "first," and "second" to explain the methods used in the study and to introduce the experimental steps.

In an academic paper, the conclusion section, located at the end of the paper, summarizes the whole paper, to be specific, it sums up the method used, the major findings, the contributions, as well as its theoretical and practical significance. Due to its relatively simple content, the CF marked theme is least used, especially in English papers.

5 MARKED THEME FEATURES AND ITS IMPLICATIONS FOR ENGLISH ACADEMIC PAPER WRITING

According to systemic-functional grammar, the relationship between genre and language is that of content and expression. This means that language can be used to realize the function of the genre, while the genre determines the use of language. An academic paper is a kind of writing which writers use to express their research work in the field and to exchange their ideas with other researchers. Therefore, the use of language is very important to ensure the effective communication. In this paper, we collected both Chinese and English academic papers, analyzed their use of marked theme, and revealed their similarities and differences. The results show that the marked theme can be used to realize its textual function, while at the same time it is affected by its textual function. Comparatively speaking, the CF marked theme is more frequently used in Chinese academic papers. To master the linguistic features of Chinese and English academic papers helps us eliminate the influence of our mother tongue and avoids word-for-word translation, which can guide us to present ideas precisely, hence, ensuring effective communication in the research field.

REFERENCES

[1] Davies, F. Introducing Reading. Penguin English. 1995.
[2] Davies, F. Designing a Writing Syllabus in English for Academic Purpose: Process and Product. In P. Robinson. Ed. Academic Writing: Process and Product. London: Modern English Publications/ British Council. 1998.
[3] Eiler, M. Thematic Distribution as a Heuristic for Written Discourse Function. In B. Couture, ed. Functional Approaches to Writing: Research Perspectives. Norwood, NJ: Ablex. 1986.
[4] Halliday, M. A. K. "Systemic Background" in W.S. Greaves & J. D. Benson (eds.) Systemic Perspectives in Discourse. 1983.
[5] Halliday, M. A. K. Introduction to Functional Grammar. London: Arnold. 1985/1994/2004.
[6] Robert, A. Day. and Barbara Gastel. How to write and Publish Scientific Papers (Photocopy edition) Peking: Peking University Press, 2007.
[7] Swales, J. Genre Analysis: English in Academic and Research Settings. Cambridge: Cambridge University Press. 1990.
[8] Qilong Chen. Introduction to Systematic Functional Grammar. Shantou: Shantou University Press.1994.
[9] Cong Cong. English for International Academic Communication. Nanjing: Nanjing University Press. 1977.
[10] Guowen Huang. Research on English Language. Guangzhou: Zhongshan University Press. 1999.
[11] Bojiang Zhang, Mei Fang. Functional Studies of Chinese Grammar. Nanchang: Jiangxi Education Press. 1996.
[12] Yongsheng Zhu, Shiqing Yan. Reflections on Systematic Functional Linguistics. Shanghai: Shanghai Education Press. 2001.

Computational Social Science – Luo, Ciurea & Kumar (eds)
© *2021 Taylor & Francis Group, London, ISBN 978-0-367-70193-2*

"Idyllic" shortform video: Aesthetic imagination, narrative turn, and modern anxiety in the new media era

J. Liang, S.H. Zhu, J.P. Li & Y.M. Xie
Network and Educational Technology Center, Jinan University, Guangzhou, China

ABSTRACT: The rapid development of shortform video in the new media era has become a new opportunity for the development of the Internet. At the same time, the homogeneity and vulgarity of shortform video have become serious. "Idyllic" shortform video has gained a huge following on shortform video platforms. The thesis analyzes why "idyllic" shortform video is popular both in the West and East, and provides ideas of "nature" aesthetics, micronarrative strategy, and modern anxieties about those videos.

1 INTRODUCTION

Shortform video is developing very fast in the new media era, becoming one of the most important ways for people to socialize, entertain, and receive information. YouTube and Netflix have become the most popular platforms worldwide. A survey has shown that YouTube and Netflix are the two most popular video-streaming platforms among teenagers. At a marketing event in May 2019, YouTube CEO Susan Wojcicki revealed that YouTube currently has 2 billion monthly active users. Politicians, news organizations, companies, celebrities, and people from every walk of life are using YouTube. The 45th "Statistical Report on Internet Development in China" (in Chinese) claims the scale of shortform video users has reached 773 million, accounting for 85.6% of the total Internet users (2020). At the same time the production field was full of homogeneity, routines, and vulgarity content. Original high-quality content gradually was replaced by core competitiveness, videos illustrating an idyllic life in rural areas have garnered millions of followers; this paper will provides ideas for "nature" aesthetics, a micronarrative strategy, and modern anxieties about those videos.

2 AESTHETIC IMAGINATION

2.1 *Text narrative: "nature" aesthetics in ancient poetry*

Nature is the home to people's souls. In particular, in the creation of literati, nature is a space with an aesthetically completed structure, which is regarded as the beginning and end of beauty. Therefore, the literary person naturally has a sense of friendliness and belonging to nature.

Regardless of being from the East or West, there is a sense of closeness to nature. Traditionally, in China and the East, the lifestyle itself was nature-friendly. Since ancient times, China has been an agricultural society based on profound rural characteristics, with agriculture as the mainstay of self-sufficiency, which has lasted for thousands of years. Land is an important object of worship, and landscape poetry has been a material for narration since ancient times. From the Qing Dynasty to the Tang Dynasty, Tao Yuanming, Xie Lingyun, and other poets have created natural poetry, and developed in the direction of pursuing pure natural poetry, expanding the aesthetic ideals. In the traditional imagery, the landscape contains a love for the natural homeland and the self-pursuit of life and individuality. Normally, when creating poetry, they are often frustrated by the reality, this is not only a lively and quiet description of an idyllic scenery, it is also a portrayal of the inner peace and tranquility of the narrative subject. They sought solace in the landscape, expressing dissatisfaction with reality and yearning for a peaceful life.

In the West, nature is an object that can release love and thought naturally, arouse people's aesthetic consciousness, and provide poetic lyricism. Although the history of natural poetry creation in the West is shorter than that in China, the trend of natural creative poetry that developed in the romantic literary movement has also played an outstanding role in its creation.

Romantic writers in the 19th century praised the vast nature and tried to convey mature emotions through natural aesthetic expressions. Natural poetry in Western literature tried to connect emotional flow and faithful narrative with a natural view of nature.

2.2 Image narrative: "nature" themes in landscape painting

Poets use words to describe mountains, rivers, and foods, and painters directly use natural images to construct a painting. Chinese landscape paintings appeared around the 4th century, with European landscape paintings appearing in the Netherlands in the 17th century. Chinese landscape painting began more than 1000 years before Western landscape paintings.

Chinese landscape paintings are the artist's expression of the idea of "the unity of nature and man" since ancient times, with man as a part of nature, not its ruler. Painters express personal emotions in their works, and emphasize the harmonious coexistence of humans and nature. Li Zehou mentioned that the art of the Han Dynasty carefully portrayed various occasions, characters, objects, and even general parts of daily life, such as barns, stoves, pigsties, and chicken coops (2014), reflecting a positive concern and affirmation of nature and life.

Landscape painting exploded during the Dutch Golden Age. Western painters of the 17th century turned their attention to the daily life of ordinary citizens and colorful natural scenery. Gone were the paintings of religious subjects and instead a whole new market for all kinds of secular subjects emerged. They regard real life as the source of artistic creation; they carefully describe Dutch landscape features, common rooms, villages, characters, and rural life, such as trees, windmills, or a cloud-filled sky, showing the Dutch people's enthusiasm for their daily life, and praising their struggle to conquer nature.

Landscape plays an important role within the discourses of both Western and Chinese art. Both Chinese and Western landscape paintings take "nature" as the narrative theme, depicting human activities in nature, depicting the harmonious relationship between humans and nature, and expressing the emotional communication between humans and nature.

When agricultural societies became more industrialized and urban, the more industrialized and technological they were, the more prominent and important the relationship between humans and nature became. The most precious "nature" and "harmony" lost in people's social life can be compensated for in the tranquil, peaceful, and beautiful landscape art, which evokes the common visual and emotional cognition.

3 NARRATIVE TURN

3.1 Micronarratives: easier to spread in the new media era

Compared with the grand narrative of major themes of "global or comprehensive cultural narrative mode, which organizes and explains knowledge and experience" (Stephens and McCallum 2013), the main theme of the micronarrative is the preferred daily life, which has concise form, short length, emphasized fragments, and is easy to understand.

Postmodernists attempt to replace grand narratives by focusing on local contexts and the diversity of human experience. They believe that there are "multiple theoretical viewpoints" (Peters 2001) rather than grand theories.

Jean François Lyotard proposes that grand narratives should give way to micronarrative, or more moderately "localized" narratives. He argues in his *Post-modern Condition: A Report on*

Knowledge that as citizens of a fragmented society, we have fallen back onto micronarratives, each having a limited context to understand our world better in today's post-modern literature (Mannava and Sciences 2014).

As mentioned earlier, smart phones have become popular, and mobile networks are developing rapidly in the era of new media. Technological innovation has changed the social structure, the nature of social life, our personal consciousness, and our relationship with time and space. The popularity of social media and the endless release, sharing, and reposting of status updates show that our enthusiasm for diverse narratives has not diminished. The pursuit of complete, epic, and heavy grand narrative usually takes a long time for deep learning, while the simplicity, mobility, and fragmentation of shortform video can satisfy the public's scattered reading need, ubiquitous learning, and rapid social sharing. Therefore, the media technology of the new media era eliminates the complete grand narrative to an extent, and provides space for the development of a micronarrative.

People pay more attention to individual life and feelings, and focus on the personal micronarrative of emotions, which dilutes serious thinking. People only need to express their feelings, without serious speculation and rigorous logical order, so micronarrative has become the main narrative paradigm of shortform video. The country lifestyle videos constructed within a short time, and the relaxed and happy atmosphere have become the new trend of micronarratives that the audience loves.

Li Ziqi is a typical micronarrative blogger on YouTube (11.6 million subscribers) and Weibo (26 million followers). Her videos are based on Chinese rural scenes, presenting a kind of country lifestyle. She often wears modified Chinese traditional clothes and shows the production process of Chinese food and handicrafts.

Her videos have similar narrative characteristics. They are very short, each of which is less than 10 minutes, they are nondramatic narratives without strong conflicts, the theme of her videos is the harmonious coexistence of nature and humans, she uses natural resources to make food and handicrafts, and the product materials are all from her garden or nature. She seldom speaks in the videos, background music sounds are generally composed of ambient sounds, light music, and natural sounds, with brief subtitles, as she creates a world of tranquility and peace, and presents a reclusive life with her micronarrative style.

3.2 *Digital expression: establish parasocial relationships on social media*

Social media has led to a new era of digital expression. Millennials tend to feel more comfortable expressing themselves online.

One research pointed out that 55.2% of shortform video users are composed of those born "post-80s" and "post-90s" (Zhang & Luo 2019) This generation has more energy for self-exploration and development, to pursue personal freedom and self-realization, and has a strong spirit of innovation. At the same time, the environment that grows up with the development of information technology also has an impact on cognitive development and survival interest. The empowerment of new media technology enables personal discourse to rebel against the power of traditional media; online digital expression has established an important field for the social expression of young people.

Growing up immersed in a fast-paced, mobile-first landscape has given youth almost unending access to information and communication. The rise of online social platforms like Douyin (TikTok) and the video app Bilibili have given young people new creative outlets for individualistic self-expression via viral shortform video. Audiences establish para-social relationships with the world that bloggers create in their videos. They praise or question, debate and think, positive or negative, opinions are gathered here; the influence of technology has given young people an unprecedented degree of connectivity among themselves and with the rest of the population.

4 MODERN ANXIETY

4.1 *Urban diseases: country-life landscape consumption*

It is undeniable that there is a huge change in life now from the past, with economic and societal developments, and the degree of industrialization and commercialization in modern society continues to increase. Urban diseases refer to the predicament of living in major metropolises, ranging from traffic jams, to air quality, to resource shortages. For young people living an urban life, many are disillusioned with today's ever-changing industrialized consumerist society and tired of the city, and the revival of traditional culture and country life has become a way of lifestyle for the younger generation to escape their reality.

People's evasion of urban pressure has made pastoral shortform video popular in the new media era. Information technology provides a platform for constructing a spiritual homeland. The theme of this type of video is pastoral and food, mostly in the countryside or in the private kitchen, without too many modern implements. They use everything in nature, washing vegetables in river water, cutting meat with a hatchet on a stone, cooking bamboo rice on a fire by the river, etc., and they have constructed an online pastoral landscape of the new media era. This countryside is shaped as an idyllic living place, portrayed as a community with beautiful scenery and more neighbors.

Li Ziqi's videos describe a poetically pastoral world and lifestyle. The nondramatic narrative style and production activities which obey the laws of nature create a world of tranquility, presenting images of bamboo forests, mountains, mist, snow, and other images with movie-like pictures, and uses calligraphy, couplets, lanterns, HanFu (Chinese traditional clothing), paper-cutting, firecrackers, and other symbols to record daily life. She has constructed a peaceful and natural pastoral landscape with fairy-tale-like elegant performances, presenting a dream fairy-tale world that is away from the hustle and bustle. In those videos, the lifestyles of humans can be self-sufficient and they can live in harmony with nature, and the idyllic scenery and aesthetics exhibited enable the busy modern people to get a spiritual escape. Some comments claim that Li Ziqi's videos show the leisurely pastoral life he had always dreamed of.

The French philosopher Baudrillard pointed out that we are living in a new world, where the old industrial order based on labor and material production is gradually being replaced by a social reproduction order based on the replication of information and the dissemination of images. Images and information symbols are becoming the main source of our understanding of reality, we are living in a world that has been copied and recreated, becoming trapped in the midst of a "hyper-reality" (Baudrillard 2000).

A data website indicates that Li Ziqi's average video views had reached 13.7 million by April 11, 2020 (Influencer 2020). The estimated audience age is mostly among those 18–34, living in cities, enjoying the convenience of technology, however, they do not understand the original process of goods and where they come from. Li's videos bring them a supreme visual experience, which was fascinated by a peaceful world without realizing the hard work behind it. People feel satisfaction through watching rural physical work, and they also generate alternative satisfaction because of the nonreplicability of this life. The online landscape video has become an outlet to release emotions. For them, Li's video is a utopian escape from the reality of city life.

4.2 *Slow life: concept in a fast modern society*

The concept of a slow life originated in Italy in the late 1980s. It advocates a healthy and harmonious life, emphasizes the essence of life, and pursues a harmonious coexistence between man and nature. "Slow" is a balanced life attitude and philosophy, hoping that people seek balance with society, environment, thought, and spirit.

The fast pace of the post-industrial society has resulted in rapid development of cities. The mass production of machinery has replaced most of hand-made products, and fast food has replaced most home-cooked food. In cities and virtual environments, anxiety and stress levels have been shown to rise, which is called a "natural deficiency." People desire contact with nature and "slowness," so the videos that create those scenes and images with emotions are very popular.

Audiences can perceive "slowness" in Li Ziqi's pastoral videos. In some videos, the production of food can take a whole day, or even a month. For example, making bamboo beds, swings, and silk quilts, you can see time elapse over several days, and persimmons and bacon takes a month or several months to grow, and you can see the seasons changing from the videos. The long-term persistence of producing high-quality content with slow time shows the natural mentality of the video producer and the real scenes of living in harmony with nature. In addition, animal images also appear in her videos sometimes, such as pet lambs or puppies. These images contribute to creating a quiet, warm, and happy atmosphere.

This brings the audience into a harmonious and balanced situation of nature and people, which relieves anxiety and achieved an effect of healing the soul.

5 CONCLUSION

In the new media era, most shortform video content is difficult to generate new meaning, it easily disappears, and hinders the renewal and reconstruction of culture.

This thesis analyzes why "idyllic" shortform videos are popular both in the West and East, and provides ideas of "nature" aesthetics, micronarrative strategy, and modern anxieties about those videos. From text to image and video, the digital generation prefers to establish parasocial relationships on social media, and micronarratives are easier to spread in the new media era.

This paper reveals that in modern society, people desire contact with nature and "slowness," so the videos that create those scenes and images with emotions are very popular.

Although viewers often question the reality of "idyllic" videos', such as the use of heavy filters, or whether they reflect the reality of rural life, we can see that this fantasy world is based on real-world knowledge and comes from a genuine desire for the pastoral ideal, which can relieve the anxieties of modern society.

Overall, this thesis has argued that "idyllic" shortform videos grasp the narrative characteristics of the times and the emotional needs of modern society and cultivate high-quality content, and they will play a more important role in communicating traditional culture in the new media era in the future.

REFERENCES

Baudrillard, J. (2000), The Consumer Society, Nanjing: Nanjing University Press.
CNNIC (2020), The 45th "Statistical Report on Internet Development in China" (in Chinese), [online] Available at: http://www.cac.gov.cn/2020-04/27/c_1589535470378587.htm
Influencer. (2020, February 22). Liziqi YouTube Stats & Analytics Dashboard. Influencer. https://www.noxinfluencer.com/youtube/channel/UCoC47do520os_4DBMEFGg4A
Li Z.H. (2014), The path of beauty, ShangHai: SDX Joint Publishing Company.
Mannava, S.J.P.-S. and B. Sciences (2014). "Micro-narratives compensating the omissions of grand historical narratives." 158: 320—325.
Peters, M. A. (2001). Poststructuralism, Marxism, and neoliberalism: Between theory and politics, Rowman & Littlefield.
Stephens, J. and R. McCallum (2013). Retelling stories, framing culture: traditional story and metanarratives in children"s literature, Routledge.
Zhang, T.L. & Luo, J. (2019). Report of Short Video User Value Research 2018—2019.MEDIA.5.p.11.

Computational Social Science – Luo, Ciurea & Kumar (eds)
© 2021 Taylor & Francis Group, London, ISBN 978-0-367-70193-2

A study of the effectiveness of developing non-English major students' intercultural communication competence through an English salon—taking an AIB English Salon as an example

F. Wang & X.L. Xu
Guangdong AIB Polytechnic College, Guangzhou, Guangdong, China

F. Graciano
University of Texas at San Antonio, San Antonio, Texas, USA

ABSTRACT: This article discusses a survey of non-English major college students and the effective use of English Salon activities to develop teaching methods for intercultural communication competence (ICC). This study identifies problems students may have developing ICC. The paper also identifies applied teaching methods which can overcome the educational deficiencies associated with non-English major students. By the use of student engagement activities, non-English major students can increase their language learning and develop their intercultural communication competence. The paper also identifies various teaching methods and activities which can be used to cultivate and promote intercultural communication competence.

1 INTRODUCTION

Intercultural communicative competence is the ability to communicate effectively with people from different cultural backgrounds in a multicultural setting (Bai 2016). With the rapid development of globalization, the world economic and cultural integration strongly promotes the communication and development of Chinese and Western cultures. Only through competent intercultural communication can people from different cultures communicate effectively and appropriately in the upcoming global society. "The requirement of college English teaching" issued by the Educational Department in 2007 also points out the importance of intercultural communicative competence (Fan 2019). Intercultural communication teaching became an integral part of students' curriculum for foreign language learning.

English Salon has created a communication environment of linguistic comprehensive application for students which is freer, easier, and closer to life so that students can put their knowledge into practice (Wang et al. 2012). Through constant practice, students can improve English listening and speaking as well as other skills. In 2011, English Salon was established in Guangdong AIB Polytechnic College and since then it has successfully organized 228 English Salons with a total number of attendees exceeding 22,000 people. Taking AIB English Salon as an example, this paper explores the effectiveness of developing non-English majors intercultural communication competence through English Salon.

2 OBJECT AND METHOD

2.1 *The research object*

Empirical research was conducted in a total of 227 nonEnglish majors who took part in the AIB English Salon from 2017 to 2019. A total of 300 questionnaires were distributed, among which 281 were recovered, and 227 valid questionnaires were collected—the effective rate was 80.74%.

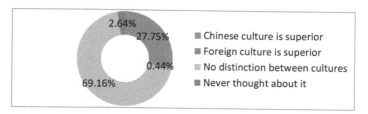

Figure 1.   Do you think our Chinese culture is superior to foreign culture?

Table 1.   Are you interested in learning about the cultural differences between China and other countries in terms of values and lifestyles?

Item	Very interested	Somewhat interested	Not very interested	Not at all interested
%	30.84%	66.08%	2.64%	0.44%

2.2   The research method

Given the research objective, we chose the method of a questionnaire, as well as quantitative and qualitative research methods. Based on the By-ram intercultural communication model and other related theories, the questionnaire is divided into two parts The first part is a survey of the inter-cultural communication competence of non-English majors, including the basic information on students and the test of intercultural communication competence. There are 13 multiple-choice questions, which are investigated from the four dimensions of attitude, knowledge, skill and aware-ness respectively (Byram 1997). The second part is the English Salon cross-cultural communication survey, which consists of a total of five questions. The cross-cultural communication studied in this paper mainly focuses on communication with English-speaking countries, and the culture referred to involves all aspects of the basic daily life needs of English-speaking countries.

3   RESULTS AND ANALYSIS

Figure 1 shows that 69.16% of students believe that there is no distinction between cultures, 27.75% think that Chinese culture is superior, and only 0.44% think that foreign culture is superior; with 2.64% of the students never having thought about it. This shows that most students have a rational understanding of the merits and demerits of cultures.

According to Table 1, 97.36% of the students were interested in learning about the cultural differences between China and other countries in terms of values and lifestyles, 3.08% were not interested. This shows most students preferred to understand the differences between cultures and desired to enhance their cross-cultural awareness.

As Table 2 shows, 41.41% of the students were very happy to have a chance to communicate with people from other cultures, 25.99% were comfortable and willing to communicate, and 32.16% were willing to communicate but a little shy. Only 0.44% were unwilling to communicate and felt uneasy. The survey shows that the majority of students would like to communicate with foreigners even if they lacked the confidence to talk with them.

The college students' reasons for learning English were very diverse. Paying more attention to their own interests, helpful in making choices for future professional development, and taking exams were the main reasons for students learning English. Cross-culture learning ranks fourth and professional learning ranks fifth as Figure 2 shows.

As Table 3 shows, 68.72% of the students were acceptable of the differences between different cultures, 29.52% said it depended on the situation, and 0.88% of the students had never thought

Table 2.　How do you feel when you communicate with people from other cultures?

Item	Very happy to have a chance to communicate with them	Comfortable and willing to communicate	Willing to communicate but a little shy	Unwilling to communicate and feeling uneasy
%	41.41%	25.99%	32.16%	0.44%

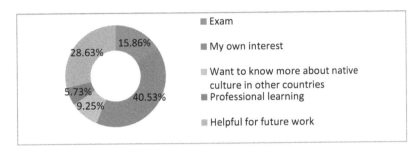

Figure 2.　Reasons for learning English.

Table 3.　Could you tolerate the differences between different cultures?

Item	Acceptable	Totally unacceptable	Depends on the situation	Never thought about it
%	68.72%	0.88%	29.52%	0.88%

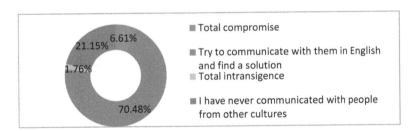

Figure 3.　How do you solve the conflict between your own culture and other cultures?

about it. Only 0.88% found it totally unacceptable. This shows that the majority of students have an open view and an inclusiveness of cultural differences.

It is necessary to cultivate students' intercultural communication abilities on how to deal with a cultural shock. Figure 3 shows that 6.61% of the students chose total compromise, 70.48% try to communicate in English and find a solution, and 21.45% chose total intransigence. This shows that most students can be flexible in dealing with cultural conflicts.

As Figure 4 shows, an overwhelming amount of students responded that they "know a little" (55.51%) about foreign culture. This is an identified issue with non-English-speaking students. Non-English-speaking students lack the ability to study and understand foreign culture. Not sure (19.82%) responses indicate students are not competent on this topic. Responses of know (13.66%) and know a lot (3.52%) indicate that a few students may consider themselves to be knowledgeable about foreign culture. Unknown (7.49%) responses reveal students who did not study or considered culture as part of their curriculum.

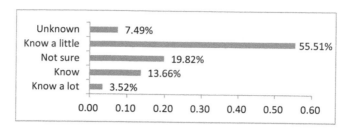

Figure 4. Do you know anything about the culture of English-speaking countries?

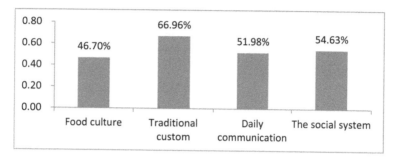

Figure 5. What aspects of foreign culture do you least understand? (multiple choice).

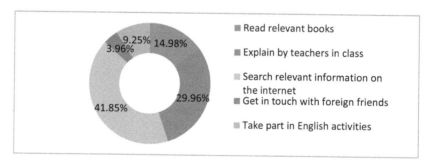

Figure 6. How do you learn about foreign culture?

Figure 5 shows that 66.96% of the students least understand the traditional custom of foreign culture, and then "the social system," "daily communication," and "food culture" follow. They are in a balanced proportion, therefore the students should learn more about foreign cultural knowledge and raise their sensitivity to cultural factors.

In the survey "How do you learn about foreign culture?" (Figure 6), 14.98% of students read relevant books, 29.86% were taught in class, 41.85% searched relevant information on the Internet, 3.96% got in touch with foreign friends, and 9.25% took part in English activities such as English Salon.

As shown in Table 4, 3.96% of the respondents had no problems at all speaking English to people from other cultures, 51.38% had some obstacles, and 35.68% thought it was a big challenge. Only 8.37% thought it was a huge problem. This shows that most students have difficulties in communicating with foreigners.

There are several reasons for student's poor intercultural communication skills. According to Figure 7, 42.73% of students thought little knowledge makes it difficult to speak, 42.49% believed that they were not good at spoken English and afraid of making mistakes, 7.93% had no time to

454

Table 4. Do you feel comfortable speaking English to people from other cultures?

Item	No problems at all	Somewhat	A big challenge	Not at all
%	3.96%	51.98%	35.68%	8.37%

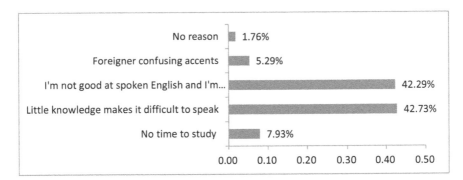

Figure 7. The most important reason for your poor intercultural communication skills.

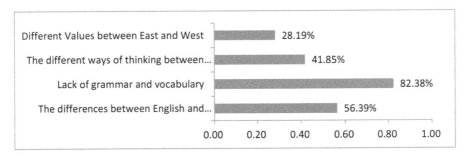

Figure 8. Main reasons for intercultural communication barriers between you and English-speaking people (multiple choices).

study, and 5.29% were confused about foreigner accents. Lack of knowledge and psychological effects can lead to students' poor intercultural communication skills.

As for the main reasons for intercultural communication barriers between students and English-speaking people (Figure 8), 82.38% of the respondents had a lack of grammar and vocabulary, 56.39% believed the differences between English and Chinese languages caused communication barriers, 41.85% considered the different ways of thinking between East and West, and 28.19% regarded the different values between East and West.

According to Figure 9, 66.08% of the students went through practicing more basic listening, speaking, reading, and writing skills in order to improve their intercultural communication skills. A total of 18.06% chose to practice through language training, 13.22% chose to understand certain language and cultural background knowledge, and only 2.64% attended cross-cultural communication training regularly.

Table 5 shows a range of 13.66% to 24.67% with a mean average attendance of 22.03% (range of 6–9 times) for active participation for every English Salon. The standard deviation of each attendance level is 3.68%, which demonstrates a consistent and steady level of participation. The English Salon is usually held 12–14 times each semester.

According to Table 6, a combined total of 92.07% of students thought that English Salon was very helpful and helpful in their improvement of intercultural communication competence. A breakdown

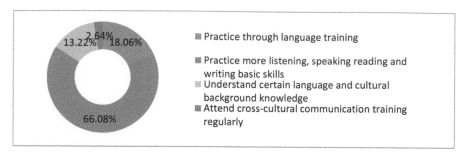

- ■ Practice through language training
- ■ Practice more listening, speaking reading and writing basic skills
- ▨ Understand certain language and cultural background knowledge
- ■ Attend cross-cultural communication training regularly

13.22% 2.64% 18.06%
66.08%

Figure 9.　How do you start to improve your intercultural communication skills?

Table 5.　How many times a semester have you participated in English Salon?

Item	1–3 times	3–6 times	6–9 times	9–12 times	12–14 times
%	24.67%	18.94%	22.03%	13.66%	20.70%

Table 6.　Is English Salon helpful to improve your intercultural communication competence?

Item	Very helpful	Helpful	Not sure	Not helpful
%	30.40%	61.67%	7.49%	0.44%

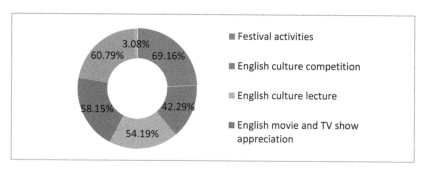

- ■ Festival activities
- ■ English culture competition
- ▨ English culture lecture
- ■ English movie and TV show appreciation

3.08%
60.79%　69.16%
58.15%　42.29%
54.19%

Figure 10.　Your favorite English Salon cultural program (multiple choice).

of very helpful (30.40%) and helpful (61.67%) represents a high majority of students, with 7.49% of students not sure whether English Salon was helpful, and only 0.44% thinking English Salon was not helpful.

English Salon introduces Western cultural knowledge to students through various activities. As shown in Figure 10, the favorite English Salon cultural programs are festival activities and English cultural experience exchange activities which ranked first and second, accounting for 69.16% and 60.79%, respectively. English movie and TV show appreciation, English culture lecture, and English culture competition ranked third, fourth, and fifth, respectively. Only 3.08% loved other activities such as cultural group discussion and cross-cultural communication experience sharing.

Figure 11 shows the wide variety of cultural topics in use by English Salon. The wide variety of topics presents a problem in identifying the best and most popular topics which students can use. With a rating of 10 the top 3 popular topics were "Watching movies, learning English," "Social etiquette," and "English songs contest." These topics are used because they are easy for first-time language learners to use. The next most popular topics were "Christmas party" and

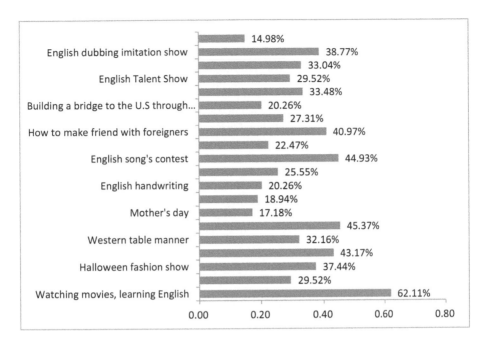

Figure 11.   Your favorite cultural topics of English Salon activities (multiple choice).

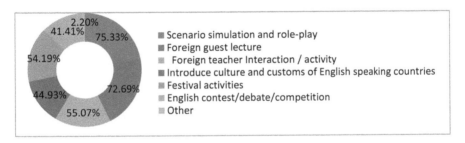

Figure 12.   How do you think English Salon can create a better cross-cultural atmosphere? (multiple choice).

"Halloween fashion show" which ranked fourth and seventh because the students would like to experience tradition Western festival activities to learn Western culture. "How to make friends with foreigners", "English dubbing imitation show," and "Say goodbye to Chinglish" ranked fifth, sixth, and eighth, respectively. Those topics allow students to have active learning interaction to reduce the intercultural communication barriers. "Excellent English film and television recommendation" ranked ninth and allows the students to positively recommend their favorite film and TV shows. The students were also interested in food culture: "Western table manners."

As Figure 12 shows, Scenario simulation and role play (75.33%) was recorded as a popular method to create a cross-cultural atmosphere. Students found Foreign Guest Lectures (72.69%) to be the second most popular method in promoting cross-culture. Foreign Teacher Interaction/Activity (55.07%) was demonstrated to be accepted and useful for learning. Culture and Customs (44.93%) allows students to comfortably explore different cultures in a tolerant atmosphere. Festival Activities (54.19%), English contest/debate/competition (41.41%) and others (2.20%) completed.

4 CONCLUSION

This survey attempts to address problems associated with non-English majors and intercultural communication competence. Non-English major students have relatively little knowledge of intercultural communication competence, which is mainly reflected in the following aspects: most students have a neutral cross-cultural attitude, limited cross-cultural awareness, lack of intercultural communication knowledge, and low intercultural communication skills. The main reasons for these problems are: lack of concentrated cultural teaching methods and inadequate development of an intercultural cultural communication learning environment. This causes students to become unaware of cross-cultural knowledge, and lack initiative and enthusiasm for cultural learning. Individual differences in foreign language students' personality or psychological factors have an influence on intercultural communication competence.

English Salon can be helpful in developing college students' intercultural communication ability in the following aspects: (1) English Salon introduces Western cultural knowledge to students through various activities such as festival activities, English culture lectures, and English movie and TV show appreciation; these learning activities deepen students' cultural knowledge and culture awareness so as to improve their ability to communicate in English correctly and appropriately; (2) English Salon holds English cultural experience exchange activities and English culture competitions, such as English song's contest, foreign guest lectures to help students master the necessary language skills and intercultural communication skills; (3) English Salon creates a better cross-cultural atmosphere and provides opportunities to students in intercultural practice. Students become engaged in active learning by using various cultural scenario simulations, role-play, and group discussions to encourage their intercultural communication. In conclusion, English Salon teaching methods may be used effectively for promoting intercultural communication of foreign language students.

REFERENCES

Bai Wenxia. 2016. *Research on the Status Quo of the Intercultural Communicative Competence of Non-English major University Students.* Tian Jin: TianJin Normal University.

Byram M. 1997. Teaching and Assessing Intercultural Communication Competence. *New York: Multilingual Matters*, 4:23–44.

Fan Hui. 2019. *The Strategy Research on the Intercultural Communicative Competence of Non-English majors in Daqing City.* Daqing: Northeast Petroleum University.

Wang Fen, Graciano Fernando. 2012. Use of English Corner and English Salon as a Language Learning Program at Polytechnic College Based on BPM, *Advances in Education Research*, 11: 150–154.

Beauty in sadness—research into the concern of life in "Love in a Fallen City"

P. Tang
Wuhan Donghu University, Wuhan City, Hubei Province, China

ABSTRACT: "Love In a Fallen City," is a classic masterpiece by Zhang Ailing. Its uniqueness lies in the conscious concern for life and the full manifestation of real human nature in the work. This paper takes a deep observation on the unique "wonderful" life and the real and living human nature in "Love in a Fallen City" from the perspective of Maslow's humanistic psychology.

1 INTRODUCTION

Literature is the study of human beings, because it is written by people, for people to read. This point of view has almost become a common understanding in contemporary Chinese literary circles, reflecting that we have realized that the concern and expression of life care is as important to literature as the concern and expression of ultimate concern to religion. The ultimate concerns of religion are "Where do I come from? Where am I going?" Then the life concern of literature is "Who am I? Why am I alive? How should I live?" It can be said that we judge the quality of a literary work from the perspective of whether it reflects life and human nature, rather than whether it reflects the mainstream of society in the final analysis. If a literary work is divorced from the expression and analysis of the basic problem of life "why people live and how people should live," even if this work can be very popular for a while, it will never be popular for a long time. Therefore, to some extent, we can say that the concern and performance of life care is the common pursuit of the world's literature classics.

"Love In a Fallen City," written by the famous female writer Zhang Ailing, is a masterpiece of concern for life, which is also the difference and brilliance between "Love in a Fallen City" and those literary works of "for politics" and "for government," and it is also the fundamental reason why it has been popular for a long time and remains so to this day.

2 "WONDERFUL" LIFE IN THE SAD TIMES"

Bai Liusu, a 28-year-old divorced woman, had no choice but to live in her mother's home. Bai Liusu had tolerated sarcasm and mockery from her brother and sister-in-law for a long time. Later, her brother and sister-in-law unsympathetically tried to ask her to be widowed to her husband who had been divorced for 7 or 8 years and who had just passed away. This ridiculous and absurd proposal was unbearable to Bai Liusu. However, Bai Liusu was almost driven to a death by her family under the poison of the long-term thought that a woman should be faithful to her husband unto death. One day, under the recommendation of Mrs. Xu, she met a "playboy" Fan Liuyuan. From then on, she tried to keep her destiny in her own hands, so she took great pains to embark on a path of self-salvation. Then she met a rival who was equal to her, Fan Liuyuan, who was also a shrewd guy full of tricks. As a result, he was able to understand the minds of most women. Therefore, the two "extremely smart" guys began a show of playing tricks in the hope of achieving their respective wishes. However, in the end, the result of their competition was that they achieved their own internal needs, but neither of them took the lead. Readers can see a "wonderful" life between two real, undisguised "non-gentlemen" with true temperaments in this part. Although their words and deeds have a certain purpose, many of which are not sincere, it is a good thing that their actions do not involve other people, and even if they are dishonorable, they do no harm to anyone in the end. It can only be regarded as a game of two love masters manipulating their respective love. From the perspective of modern people, this is nothing more than a flirting trick of young men and women in love, which can't be criticized from a moral level. Obviously, Zhang Ailing's intention is not to criticize, but to extend her tentacles to the deep surface of our human nature by this "wonderful" life, so as to reveal the different aspects of human nature. She reveals the basic problems of life to the readers inadvertently, that is why Fan and Bai are alive, and how to live, which is clearly presented to the readers in "Love in a Fallen City" and also reflects her consciousness of the concern for life in her literary works. From the perspective of Maslow's humanistic psychology, the inner needs of humans play an important role in their inner world, which is the internal motivation of human behavior choices. The deliberate choice of life in "Love in a Fallen City" is in line with the internal needs of Bai Liusul and Fan Liuyuan, which reflects their inner desire. This is the merit of the "splendor" of this dishonorable life.

And the final ending of "Love in a Fallen City" also made this dishonorable "wonderful" perfect in a sense. Fan and Bai finally met their internal needs and came together after exhausting their efforts and means. Although it was written in the book that "the fall of Hong Kong has fulfilled her," whose fulfillment came at the cost of the fall of a city, the two of them had a "peak experience" after their self-realization needs met at this moment.

When Fan Liuyuan told Saheyini that Bai Liusu was his wife and asked when they would get married, Bai Liusu's response was: "she didn't say a word with shedding tears." It was undeniable that this was the moment that Bai Liusul had been waiting for for a long time. At this moment, she must have experienced the pleasure and happiness brought by the "peak experience," who was immersed in it. When Fan Liuyuan finally said that he would like to marry Bai Liusu, he may have had the same experience. Because, at this time, he finally put down all kinds of masks and restored his true self in front of his lover. In addition, he mentioned several times that "For life or for death, however separated, To our wives we pledged our word. We held their hands, We are to grow old together with them" from "the book of songs." His words more or less reflected the real needs of his heart, which shows that the prodigal son also had the expectation of eternal and beautiful love.

3   THE PUBLICITY OF HUMAN NATURE IN SAD TIMES

The most successful character in "Love in a Fallen City" is Bai Liusu, the heroine, who was praised as "your specialty is to bow your head" in the novel. However, when she pretended to admit that she was "the most useless person," Fan LiuYuan added "useless woman is the most powerful woman." From this description, we are sure that Bai Liusu is by no means a useless person, who appeared as a young woman divorced for 7 or 8 years in the novel. As soon as she appeared, she broke the readers' stereotype of Chinese women. In China, the traditional concept that "a woman should be faithful to her husband unto death," "the three obedience and the four virtues" confined Chinese women to be men's "vassals." In the old society, there were not many women who dared to choose their own partner for marriage or said no to their husbands. Bai Liusu was undoubtedly a rebel of that time. And then her words and deeds showed that she was not a useless person, but a smart and kind woman who had her own opinions and was good at seizing opportunities.

Later, when she looked for a boyfriend for her younger sister, she understood that Fan Liuyuan was really interested in her. At this time she began to have the idea of planning for herself. Until Mrs. Xu invited her to Hong Kong, Bai Liusu knew that it was all the tricks of the prodigal Fan Liuyuan. But she was very glad to have such an arrangement, because if she could seize the opportunity, she would be able to get rid of the sarcasm and mockery from her mother's family and get a new life. And so she was willing to gamble on her future. With such a gamble to meet his self-realization needs, she did not hesitate to offend her younger sister and went to Hong Kong.

From the perspective of Maslow's humanistic psychology, people's internal needs include security needs, which require people to have security of life. However, Bai Liusu's divorce and the sarcasm and mockery from her mother's family made her life difficult, and the living guarantee was about to be lost. The opportunity to go to Hong Kong became the last "life-saving opportunity" for her to regain a new life. Only by seizing it could her security needs be met. This gambling behavior also shows her unique aspect in human nature. As a result, it turned out that Fan Liuyuan was really fond of her. So the two love masters began a game of playing hard to get. In this scene of tricks, Bai Liusu brought her ingenuity and intelligence into full play. In their repeated contests, Bai Liusu always kept the bottom line of "remarrying and following herself," because she always understood what her internal needs were. She had never been the mistress of a "playboy," but wanted to become the legitimate wife of the prodigal son. Finally, Fan Liuyuan agreed to marry Bai Liusu, and she also achieved her wish. The hierarchy of needs theory of Maslow's humanistic psychology once again proves that people also have the need of belonging and being loved, that is, everyone has a desire to be loved by others. Bai Liusu's desire to become Fan Liuyuan's legitimate wife is powerful proof of meeting this need from the bottom of her heart. And her step-by-step behavior fully reflected her initiative in holding her own life path and choice firmly in her own hands, and also well showed the basic question of "why to live and how to live" for herself.

4 ANALYSIS OF THE CHARACTER IMAGES OF BAI LIUSU AND FAN LIUYUAN

In general, Bai Liusu was not a good woman in our traditional sense, because she betrayed the virtues of traditional Chinese women such as "a woman should be faithful to her husband unto death," "the three obedience and the four virtues," and she was cunning, scheming, and full of tricks. It was written in the novel that "if Bai Liusu is a thoroughly good woman, then Fan Liuyuan will not notice her." But we can't deny her kindness. She took good care of Mrs. Xu's family of four on the ship to Hong Kong, as well as protecting of her servant A. Li and her children during the fall of Hong Kong. All that reflected the goodness of human nature. Therefore she was more real and flesh-and-blood than the characters portrayed in many literary works. It was due to the fact that the author did not depict the formulaic characters, but tried to publicize human nature in his works.

As for the hero Fan Liuyuan, who was mostly a side story in the novel. He was a "playboy" and prodigal son who was well educated in Western culture. He had always been so popular among the women that he formed a sense of superiority for a long time, which made him look down upon all women. Fan Liuyuan hoped that Bai Liusu would become his mistress initially just to meet his physiological needs, but when he repeatedly took actions and tricks to seduce Bai Liusu without success, he finally understood that Bai Liusu was not a woman willing to become a mistress. She was a smart person with the same scheming and means as him, but she was even better than him. Over time, in the process of pitting his wits against with Bai Liusu, the prodigal son, Fan Liuyuan also produced his true feelings. His feelings for Bai Liusu rose from the level of meeting the physiological needs at the beginning, to the level of longing for "holding hands and growing old together" to obtain the need of belonging and love.

It is believed that he had true love for Bai Liusu, but he was the man who hoped others would love him more than he loved others. Therefore he hoped that Bai Liusu would love him wholeheartedly from the very beginning, and he tried a lot of tricks on her, but also brought a lot of suffering to Bai Liusu. He reflected the prodigal aspect of his human nature and it was also in line with his background and identity of being educated in Western culture. But when he realized that Bai Liusu was not a social butterfly, nor a mistress, he began to have a real longing for Bai Liusu and his eternal love, which also reflected the good aspect of his human nature. This character image is different from the heroic image of "tall, big, and complete" in many works. Although he is not perfect, he is vividly shown in front of readers. His behaviors and choices well reflect his thinking on the basic problems of life. Therefore he is more persuasive than those hero images to a certain extent, and the author's conscious concern for life is once again clear.

5 CONCLUSIONS

From what has been discussed above, the author believes that the reason why "Love In a Fallen City" has become a masterpiece of Zhang Ailing lies in the conscious expression and concern of life and human nature in the work, which makes the love story of Bai Liusu and Fan Liuyuan more real and convincing, and makes this short story have enduring charm.

REFERENCES

[1] Zhang Ailing. Selected works of Zhang Ailing. Beijing: Beijing Yanshan Publishing House,

[2] Shuijing. Making up for Zhang Ailing. Jinan: Shandong Pictorial Publishing House, 2004.

[3] Liu Fengjie, Xue Wen, Huang Yurong. Zhang Ailing's world of Images. Yinchuan: Ningxia People's Publishing House, 2006.

[4] Tong Qingbing, Cheng Zhengmin. A Course in Literary Psychology. Beijing: Higher Education Press, 2001.

[5] Xu Yueyue, Ren Qiang. Research on the imperfect female Images in Zhang Ailing's Novels. Journal of Hubei University of Economics (Humanities and Social Sciences), 2020.3, 83–85.

[6] Xie Likai. Research on the Evolution of female consciousness in Zhang Ailing's novels. Journal of Suihua University, 2008 (02), 69–73.

[7] Liu Ting. A brief Analysis of the Modern characteristics of White tassel. Literatures. 2010 (04), 11–12.

[8] Yang Xiaolan. The Power Behind the Weakness—Analysis of the Image of the Heroine Bai Liusu in "Love in a Fallen City". Social Sciences Review, 2004.4, 111–112.

[9] Shi Yufeng. A Lonely Man Seeking True Love—Comment on the image of Fan Liuyuan in "Love in a Fallen City", Masterpieces Review, 2012(17), 33–35.

Computational Social Science – Luo, Ciurea & Kumar (eds)
© 2021 Taylor & Francis Group, London, ISBN 978-0-367-70193-2

Research on the cultivation of social responsibility of young students from the perspective of the community of human destiny

W.J. Zhou

Shandong Vocational and Technical University of International Studies, Rizhao, China

ABSTRACT: As China enters a new era of its socialist construction, the CPC Central Committee with Comrade Xi Jinping at its core, based on the current new situation, has proposed the idea of promoting the construction of a community of human destiny, and seeks happiness and development for the people in the overall situation. In a speech at the conference marking the 100th anniversary of the May 4th Movement, Xi Jinping pointed out, "The times call for responsibility and China's revitalization lies in the youth" [1]. Therefore, it is necessary to cultivate young students' sense of social responsibility in the new era. Currently, in a profound and complex international environment, colleges and universities should educate young students on values, establish their concept of social responsibility, and cultivate young students' sense of social responsibility in a targeted manner, so that they can correctly understand the laws of social development, the future and destiny of our country, nation, and people, as well as their own social responsibilities. Young students are expected to be a responsible generation and contribute to the building of a community of human destiny.

1 INTRODUCTION

The international situation is changing currently. Since the 18th National Congress of the CPC, the CPC Central Committee has put forward a basic strategy for building a community of shared future for mankind by judging the current situation, and expounded on our country's responsibility and accountability from a global development perspective. Young students are an important force to promote this social progress, a solid pillar to realize the nation's rejuvenation and national development, carrying the mission of the times. In his speech at the conference commemorating the 100th anniversary of the *May 4th Movement*, General Secretary Xi Jinping encouraged the Chinese youth to cherish this era, and take on the mission of the era, experience in it, and grow in due diligence. Therefore, taking the role of young students in the development of the country as markers, this paper explores the cultivation of young students' sense of social responsibility from the perspective of the community of human destiny.

2 YOUNG STUDENTS' ROLE IN CONSTRUCTING A COMMUNITY OF HUMAN DESTINY

"A nation will prosper only when its young people thrive." It is the important task of history and the nation to cultivate a younger generation which is responsible, accountable, and ideal. At the historical moment when China faced a severe disaster in modern times, young students stood up and raised the banner of the *May 4th Movement*. After the founding of a new China, the majority of young people devoted themselves to the construction of socialism and building the foundation of national development with youthfulness and vigor. The fate of the Chinese youth has always been closely connected with the rise and fall of the nation as a whole. At present, the global situation is complicated and changeable, and the new normal of the Internet has brought convenience

to our lives while producing many negative aspects, weakening young students' sense of social responsibility. Therefore, while cultivating students' professional abilities, college education should pay attention to the cultivation of students' ideological and moral spirit as well as the humanistic spirit. Through reinforcing patriotic feelings, strengthening the sense of social responsibility, we can lay a foundation of talent for the construction of a community of human destiny. Young students' role in the construction of the community of human destiny is reflected in the following aspects.

2.1 *Young students are new talents in the era of the national rejuvenation*

In recent years, Chinese youth have vigorously prospered in various fields such as science and technology, sports, humanities, and the Internet, having played a key role in the construction of socialism with Chinese characteristics in the new era. With the continuous efforts of the whole society, our country has become a socialist developing country. China is becoming stronger, and its development speed is increasing. This is inseparable from the efforts of our young people [2]. At present, the world is undergoing major development, with great changes and adjustments. Under this situation, China is facing unprecedented opportunities and challenges to realize the Chinese dream and make China a global power. This great dream needs all our compatriots to work together from top to bottom. Therefore, young students need to strengthen their ideals and beliefs both in thinking and action, cultivate their skills, consciously shoulder responsibility, and closely combine the realization of personal ideals with the great rejuvenation of the country, to be the new talents in the course of the national rejuvenation.

2.2 *College students are fully developed successors of the course of socialism*

General Secretary Xi Jinping has pointed out the social role of young students on many occasions. He believes that young students should play an exemplary role in youth groups, and young students should be the key pillar to achieving the great rejuvenation of the nation, realizing the flame-passing of the CPC, and achieving the long-term stability of the country [3]. Cao Yuan, China's youngest top scientist, solved global problems that have plagued the world's physics community for 107 years, and successfully realized the "superconductivity experiment" of graphene. At the start of 2020, the COVID-19 epidemic began spreading. Young students, together with the majority of staff preventing the epidemic, were not afraid of hardships and changes, and proved by their actions that the Chinese youth in the new era can carry out their duties. Colleges and universities provide fertile ground for the growth of young people. In the new era, college education should continuously improve the ideological and cultural qualities of young people, hone their will, and shape their characters in practice, cultivate young students with an international perspective, and grasp the pulse of the times in the process of globalization, so as to build a solid foundation of youthfulness. Young students are the practitioners of the Chinese dream, as well as the builders under the concept of the community of human destiny. The responsibility for the development of human society needs to be undertaken and practiced by young students.

2.3 *Young students are the carriers and advocates of the world civilization*

From a global perspective, there are still contradictions among countries in the world today. The main manifestation of this invisible war is the cultural war between countries, which is reflected in the value orientation and ideology of cultural differences between countries. Based on this status quo, the Communist Party of China put forward the concept of a community of human destiny, in order to build a common sense of value for all people, advocate for countries to build cooperation mechanisms, alleviate the cultural conflicts between countries, and achieve cultural sharing among all countries for mutual benefit. As a university that trains young students, it should also bear a sense of the community of human destiny. In the process of training talent, we should not only cultivate professionals with professional knowledge, but also teach young students to actively undertake the great rejuvenation cause of the Chinese nation in the process of comprehensive development, and participate in the cause of human development [4]. College education, as the main way to cultivate talents, is also an important tool for spreading Chinese and foreign cultures. It realizes the equal exchange and mutual learning of human civilizations in universities, and makes young students the carriers and communicators of world culture. In recent years, China has opened Confucius Institutes around the world as a platform to spread traditional Chinese culture, so that students and people around the world can learn about the profound heritage and cultural self-confidence of traditional Chinese culture, and let young students see China's responsibility and accountability as a global power.

3 THE CULTIVATION OF YOUNG STUDENTS' SOCIAL RESPONSIBILITY IN THE CONSTRUCTION OF THE HUMAN DESTINY COMMUNITY

The Chinese Communist Party's idea of building a community of shared future for mankind in the new period is China's outlook and expectation for the future development of the world. This reflects the Chinese nation's consciousness of closely linking its own destiny with global development, and China's contribution to the common development of the world as a great power. Young people are the most active and creative advanced forces in society, they advance with the times, while having a unique perspective and judgment on the world situation and national development. However, young people are in a critical period of life development, and their thoughts are not sufficiently mature, and their social experience is also insufficient. They are susceptible to following wrong information. Therefore, college education should correctly grasp the characteristics of youth growth, starting from the cultivation of their social responsibility, lead the life of young students, synchronize it with the country in the ideological field, and promote the construction of a community of human destiny in a broad perspective.

3.1 *Cultivate young students' sense of social responsibility based on Marxist theory*

During the Chinese revolution and constructional practice, we always adhere to the Marxist theory as the guiding ideology, and the scientific world outlook and methodology as the guidance for young students to learn and practice education. Marxist theory plays an irreplaceable role in guiding young people's thoughts and actions. If young students want to establish a sense of responsibility in the construction of a community of human destiny, they must first establish scientific and dialectical thinking. Through systematic study of Marxist theory, they should learn to use dialectical thinking to look at problems, analyze and solve problems, as well as understand the relationships between things with connected thinking [5]. From the perspective of the community of human destiny, the world is something that is constantly connected. Countries in the world have close links in various aspects such as the economy, politics, and culture. Each part has a unique development model. The forms of influence and mutual restraint need to be treated by young students with dialectical thinking. The destiny of all countries in the world today is shared, and many issues need to be resolved together. Young students should stand on the relationship between the whole and parts, realize the global forces of the world and their role, realizing that they are part of the world. In order to realize the Chinese Dream, young students should take the initiative and make their contributions.

3.2 *Strengthen the education on national security for young students*

With the advancement of economic globalization, China's connection with countries around the world is getting closer, while the contradictions between countries have emerged accordingly. As a part of the community of human destiny, national security has become an important consideration for safeguarding national interests. Unlike the traditional national security category, with the development of economy and modernization, today's national security education includes not only military and political education, but also economic, cultural, technological, energy, and many other aspects, all of which have come under the category of national security education. The interconnectedness of various countries has promoted economic development and improved people's standards of living, while also aggravating environmental and ecological problems to varying degrees. Global ecological problems have gradually become major issues faced by people in all countries. Therefore, under the reality that the scope of national security education is constantly enriched, it is necessary to realize the national comprehensive security education. In addition, considering that our country is a multiethnic country, corresponding national security education is also needed for different ethnic groups, so that each ethnic group can work for the common interests of the Chinese nation.

3.3 *To establish young students' sense of social responsibility with cultural confidence*

At the 19th National Congress of the Communist Party of China, the CPC proposed following through the path of socialist cultural development with Chinese characteristics, stimulating the creative vitality of the entire nation, and building a socialist cultural power. The excellent Chinese traditional culture embodies the wisdom and vigor of the Chinese nation for thousands of years. To cultivate young students' sense of social responsibility, it is necessary for young students to embrace and be baptized by the harmonious symbiosis advocated in Chinese traditional culture, and then they consciously become the heir and disseminator of Chinese excellent culture. As a generation that inherits the past and ushers in the future, young students have the potential for creativity and vitality. While inheriting the Chinese civilization, they will transform and creatively develop traditional culture and innovation, and enhance the sense of national responsibility. The spread and construction of socialist culture with Chinese characteristics needs to be based on traditional culture, promoting its integration with the cultures of other countries on the basis of inheriting and developing traditional culture, and realizing the inheritance and innovation of Chinese culture in terms of content and form through exchanges and mutual learning. Therefore, we can constantly enhance the international influence of Chinese traditional culture.

Since the introduction of the idea of the community of human destiny, it has gradually become another expression of the traditional cultural thought of the Chinese nation. This is also an innovation under the fusion of Marxist theory and Chinese excellent culture. It is a manifestation of China's national characteristics in the development process, as well as an international expression of the process of international development. It represents the image of China and shows the strength of China. Young students should understand the idea of a community of human destiny based on an international perspective, and use their own strength to guide the Chinese people and the people of all countries in the world to establish a correct view of international order and governance, so as to achieve harmonious development in all the countries of the world. The responsibility of the people is also the mission of the times that Chinese youth cannot shirk from.

In addition to the above-mentioned methods, young students should also be trained in their responsibilities and accountabilities in international ecological environment governance. As ecological environment governance has become an important issue faced by people all over the world, concepts of new ecological development and governance have become the focus during our country's development process. Cultivating the concept of green sustainable development for young students and guiding students to participate in the governance of the ecological environment is also an important aspect for young students participating in the development of the community of human destiny.

4 CONCLUSION

In the current complex and ever-changing international situation, the links between countries in the world are getting closer, and each country is developing with its own characteristics. China has proposed the establishment of a community of human destiny under this globalization background. As an important construction force of the Chinese nation, young students should shoulder their responsibilities and fulfill missions in realizing the great rejuvenation of the Chinese nation. In the process of promoting the construction of a community of human destiny, young students should also actively participate in the construction of the community of human destiny and establish a firm sense of social responsibility, so as to contribute to the development of the country as well as the wider world.

REFERENCES

[1] Xi Jinping, *Speech at the Conference to Commemorate the 100th Anniversary of the May 4th Movement.* Beijing: People's Publishing House, 2019.

[2] Sun Yixue, Luo Zheng. Tagore: *Constructing a Community of Human Destiny Based on the Cultural Community of East and West. Journal of North China University of Technology,* 2020, 32(03): 58–63.

[3] Song Ruisi. *Governance of Great Powers: Community of Human Destiny—Based on Agamben's Theory of Life Politics. Legal System Expo,* 2020(16):239–240.

[4] Wang Yiqi, Yu Wu, Liang Yingli. *Thoughts on "Going out" of Chinese Films Based on the Community of Human Destiny from the Perspective of Communication. Art Review,* 2020(10):178–179.

[5] Liu Wenge, Yang Zhiwen. *The Mechanism and Path of Local Participation in the Construction of the Community of Human Destiny in China—Analysis and Enlightenment Based on the Case of Yiwu. Zhejiang Academic Journal,* 2019(02):12–17.

Computational Social Science – Luo, Ciurea & Kumar (eds)
© *2021 Taylor & Francis Group, London, ISBN 978-0-367-70193-2*

The progressive way of a community of shared future for mankind

J.H. Yu, Y.J. Cao & B.X. Xie
*Research Center for Indian Ocean Island Countries, South China University of Technology, Guangzhou,
China*

ABSTRACT: The idea of building a community of common destiny is the current top-level aim of China's diplomacy, which embodies the grand vision for the progress of the whole world, and by which Xi provides a Chinese blueprint for the solution to most of the problems of the international community presently, while expressing China's best wishes and pursuits for an orderly world structure.

1 INTRODUCTION

The idea of a "community of common destiny" was already applied to illustrate Chinese new perspectives on the trends of the world in a Chinese official document which was introduced in 2011. This idea inherits the preponderant "Under Heaven" (tianxia) feeling that has been cultivated in the Chinese thought for thousands of years, and contains the traditional Chinese political philosophy of "living in harmony with other nations" (xiehe wanbang) and "great harmony under heaven" (tianxia datong). The theory of a Chinese social system acts as a strong ideological guide for the evolution of the vision of the community of common destiny. The concept of diplomacy and development is an innovation of the New China's outstanding diplomatic thoughts.

"It should find a new angle of view of the community of common destiny, being indeed in this together and pursuing mutual benefit, asking for new paths to intensify interactions with various cultures, exploring new connotations in the corporate benefits and values of the whole of humanity, and seeking new paths to cope with the corporate challenges by cooperating all over of the world and pursuing corporate development" [1].

The 18th National Congress report put forward the idea of a community of common destiny, pointing out "we should promote a sense of building a community of common destiny. A nation should balance the reasonable wishes of other countries when chasing its own profits, and it should pursue the corporate development of each country when promoting China's own development" [2].

Since November 8, 2012, President Xi has continuously explained and added new meaning to the concept of a community of common destiny by enriching and developing its theoretical connotations.

General Secretary Xi gave a speech to MSIIR (Moscow State Institute of International Relations) in 2013, first bringing up this diplomacy vision on the Chinese reflection of the prospect of humanity: "Countries are more interconnected and interdependent than ever before in the world, human beings live in a global village, living in the same space of history meets reality. More and more, it has become a community of common destiny containing each other" [3].

On November 28, 2014, President Xi put forward a plan to build a neighborhood area community of common destiny of an external strategic pattern in the central meeting on foreign affairs.

General Secretary Xi gave a speech at BFAAC (Boao Forum for Asia Annual Conference) in March 2015, in which he said:

> There is only one earth for mankind, and nations live in one world. Asia has exerted an important influence on the world; Asia cannot develop in isolation from the world. Facing the complex international and regional environment, we need to understand where the world order is going, keep up with

history. We will foster a development environment that is conducive to both Asia and the world, by going to the regional community in Asia, pursue the aim of community of common destiny [4].

In this way, Xi Jinping clarified the relationship between the regional community in Asia and the wider global community.

General Secretary Xi gave a speech about "creating a new mutually beneficial partnership and community of common destiny for humanity" at the general debate of the UN General Assembly marking the 70th anniversary of the founding of the UN in September 2015. He first linked the building of a community of common destiny for humanity and a new international order, and systematically elaborated on the five pillars of a humanities community.

On November 30, 2015, when addressing the opening ceremony of the Paris Climate Change Conference, President Xi Jinping regarded the Paris Agreement as an inspiration to build a community of common destiny for humanity and called on each country to act together to pursue it.

On January 18, 2017, General Secretary Xi gave a speech about "Jointly building the community of common destiny for humanity" at the UN Headquarters in Geneva, in which he said:

> Let the fire of peace be passed to the next generation. Let the country achieve sustainable development. Realizing national political democratization and social development, this is the common aspiration of every country and people in the world; it is also the historical mission of every politician. And the development model that China offers to the world is: Allow other countries to hitch a ride on China and achieve the shared fruits of human civilization and eradicate poverty and crime [5].

2 THE UNDERSTANDING OF "A COMMUNITY OF SHARED FUTURE FOR MANKIND"

With President Xi's repeated in-depth explanation, from a community of common destiny between countries, to a community of shared future within a region then to a community of common destiny for humanity, the levels and connotations of the concept have been enriched. Interrelated and interdependent, each with a particular focus, covering multiple fields such as politics, security, development, civilization, ecology, and cyberspace. China forges a community of common destiny not only with many neighboring countries and southern states, but also with other northern states such as New Zealand, France, and Germany. In addition to advocating the Asian community of common destiny for humanity, China has also raised such ideas as the Sino-ASEAN Community of Shared Future, the Sino-Arab community of common destiny for humanity, the China-African community of common destiny, and the Sino-Latin American community of common destiny.

The Earth is the common home of all nations and peoples. it is irreplaceable. There are still many contradictions and conflicts in the world today, but country's modernization, people's happiness, and health are world trends. Never before has China been closer to realizing the great rejuvenation of the Chinese nation as it is today, and never before has it been so closely linked to the future of the outside world as it is today. Building a community of common destiny for humanity is a lofty cause that focuses on the progress of all mankind and aims at both China's modernization and the world's harmony and diversity. In short, a community of common destiny for humanity refers to the forming of an interdependent and interlinked state in which every country in the world should take into account the development requirements and goals of other states in the pursuit of its interests, seek common development and cooperation, enhance the common interests of all mankind, and jointly promote the progress of all mankind. Specifically, a community of common destiny for humanity refers to the association of all countries in the world characterized by such features as interdependence, equality, win–win situations, and inclusiveness.

The first feature is interdependence. Globalization is a reality of the time and space changes we face in our lives, it is also the butterfly effect of events in distant places. It affects us more directly and more quickly than ever before. "Through scientific and technological revolution, religious reform, ideological enlightenment, world trade, and colonial expansion, the world has gradually evolved from isolation into a whole, thus truly forming the human community as a whole" [6]. All countries have formed a bond of interests due to interdependence, and they all rise and fall together.

No country can pursue its interests and ignore the interests of other countries and public interests around the world.

The second feature is equality. A community of common destiny for humanity means that all countries treat and respect each other as equals.

Countries differ in land area, strength, and level of economic and social development. However, they all enjoy equal status in the international community and are qualified to participate in various international affairs. Today's international affairs are becoming increasingly global, which cannot be solved by relying on the power of one country alone. Therefore, the construction of a community with a shared future for mankind must rely on the common participation of the people of every country in the world to promote political democratization, equal opportunities for employment and development, and social harmony among all countries [7].

The third feature is a win–win situation. "If a country wants to achieve the happiness of its people, social harmony, and national strength, it must ensure that other countries have stable internal order and their people live and work in peace and contentment, and aspire to and agree with its development model" [8]. In the face of financial crises, terrorism, natural disasters, climate change, and other global issues, it is hardly possible for any country to achieve autonomy. All countries in the world need to pull together. The correct way for this is that major countries support minor countries, and rich countries aid poor countries. All countries can achieve common development by helping each other.

The fourth feature is inclusiveness. The world is rich and colorful. And all countries should fully respect the diversity of different nationalities, religions, and civilizations, and fully respect the social system and development path of one another's independent choices. "The course of civilization development, the mode of economic development, and the choice of national roads in all countries are diversified, and there is no distinction between advanced and backward, scientific, and nonscientific, democratic and nondemocratic. Instead of war and confrontation, there should be dialogue and exchanges between different civilizations" [9]. The construction of the future world order should be with the coexistence of multiple world powers and regional civilization centers, rather than the patent and private plots of a single country. Every country can freely integrate into and participate in this historical development process.

3 SIGNIFICANCE

The concept of a community of common destiny for humanity grows from a strategic judgment made by China based on the modernization of the country and the prosperity of its people and world trends. It serves the fundamental interests and national development needs of the people of China and the rest of the world. It bears distinctive Chinese characteristics and the aspirations and visions of all mankind, and has an important guiding significance on modern-day foreign relations and the national modernization of China.

First, the vision of building a community with a common destiny for humanity pointed out the direction for sustainable and high-quality development of human civilization. At the commemoration of the 95th anniversary of the Communist Party of China, President Xi made it quite clear: "Realizing national modernization is a long-term strategic plan made by successive Chinese governments. It is also the goal of the Chinese Party and people, and China's national policy makers will strive for it. At the same time, keeping away from war and safeguarding world peace is also an international obligation that China has fulfilled. China will hold high the banner of civilization and progress, and contribute its share to a multi-polar world, political democratization, and the building of a harmonious world" [10].

Second, the vision of forging a community of common destiny for humanity provides a driving force for restoring the international standing and national pride that the country once had. On November 29, 2012, President Xi specified the important idea of "China's National Development Goals" when he viewed "The Road toward Renewal" exhibition. "Realizing the great rejuvenation of the country and nation has been the common aspiration of all Chinese people for more than

100 years, and it is also the fundamental driving force for China's national modernization. This embodies the core interests of the people of all ethnic groups in China, and is the historical mission for which the Party and the people have worked hard in the new era" [11]. He emphasized that, "The realization of China's national development goals is closely related to a stable and peaceful development environment at home and abroad. Therefore, Chinese governments at all levels should take a global and international perspective in their development planning and coordinate regional and foreign development" [11].

Third, the concept of building a community of common destiny for humanity lays the theoretical foundation for the preliminarily establishment of the theoretical system of China's foreign policy formulation and diplomatic strategic planning. Since the 18th National Congress of the Communist Party of China (CPC), the fifth generation of central leadership headed by General Secretary Xi has conducted scientific research and judgment on domestic and international situations, has vigorously promoted innovation in diplomatic theory and practice, stressed that China's external work should have distinctive Chinese characteristics and a Chinese style, and expounded comprehensively and thoroughly on the objectives, principles, and pathways of China's diplomacy in the new situation, initially establishing the basic framework of major country diplomacy with Chinese characteristics.

Fourth, the vision of building a community of common destiny for humanity prepares a guide to action for comprehensively enhancing the practice of major country diplomacy with Chinese characteristics. In the face of complex and ever-changing international situations with new opportunities and challenges, China's major country diplomacy with Chinese characteristics aims at the great rejuvenation of the Chinese nation and the construction of a community with a shared future for mankind, forges ahead, and has built a comprehensive, multilevel, and three-dimensional diplomatic layout, creating a new phase in China's diplomatic work.

4 CONCLUSION

The concept of a community of a shared future for mankind is the inheritance and updated version of the concept of peaceful diplomacy that China has always maintained, reflecting the ethical pursuit and sense of responsibility for the future of the world that has been constant in Chinese diplomacy. It is the responsibility of Chinese diplomacy to uphold fairness, defend justice, and promote righteousness. It is the mission of China's diplomacy to maintain the peace and development of the world and to promote the international order in a more fair and reasonable direction. It is the vision of China's diplomacy to take the common interests of mankind as the starting point and reflect on the prospects and future for the world. With the rise of its national strength, China's advocacy and practice of building a community of shared future for mankind will have a wider international influence. The theoretical connotations of building this community will also be enriched and developed, further highlighting its great significance.

ACKNOWLEDGMENTS

Funding: This work is supported by "the Fundamental Research Funds for the Central Universities" in South China University of Technology, project No: XYMS202008.

REFERENCES

[1] Information Office of the State Council of the People's Republic of China, China's Peaceful Development. Beijing: People's Publishing House, 2011, p24.
[2] Hu J T, Firmly march on the path of socialism with Chinese characteristics and strive to complete the building of a moderately prosperous society in all respects: report to the National Congress of the 18th CPC, People's Daily, and 2012-11-18 (1).

[3] Xi J P, The governance of China, Beijing: Foreign Languages Press, 2014, p272.

[4] Xi J P, Towards a community of shared future and a new future for Asia, People's Daily, 2015-03-29 (2).

[5] Xi J P, Work together to build a community of Shared Future for Mankind: Speech Given at the United Nations Headquarters in Geneva, People's Daily, 2017-01-20 (2).

[6] Xi J P, Work together to promote openness inclusiveness and peaceful development: the speech was given at dinner hosted by the lord mayor of the city of London, People's Daily, 2015-10-23 (2).

[7] Xi J P, Towards a community of common destiny and a new future for Asia: keynote speech was given at the Boao Forum for Asia Annual Conference 2015, People's Daily, 2015-03-28(2).

[8] Xi J P, Work together to maintain world peace and security: speech given at the opening ceremony of the World Peace Forum, People's Daily, 2012-07-08 (2).

[9] Xi J P, Speeches at the Series of Summits Marking the 70th Anniversary of the United Nations, Beijing: People's Publishing House, 2015, p18.

[10] Xi J P, Speeches at the Series of Summits Marking the 70th Anniversary of the United Nations, Beijing: People's Publishing House, 2015, p15.

[11] Xi J P, The governance of China, Beijing: Foreign Languages Press, 2014, p36.

Computational Social Science – Luo, Ciurea & Kumar (eds)
© 2021 Taylor & Francis Group, London, ISBN 978-0-367-70193-2

Institutional advantages and spiritual strength in China's fight against COVID-19

J. Cao & S.S. Yan
College of Marxism, Zaozhuang University, Zaozhuang, Shandong, China

ABSTRACT: Responding to COVID-19 has been a "major test" of China's social governance system and governance capacity, and the institutional advantages and spiritual strength in China's fight against COVID-19 are a beneficial instrument for achieving significant strategic results in epidemic control, showing the world Chinese speed, efficiency, and strength. Seen from China's vivid practice therein, the fundamental political advantage of the CPC's centralized and unified leadership, the institutional advantage of concentrating resources on major issues, and the clear-cut advantage of being centered on the people have become effective institutional positions for the final victory over the virus. Moreover, the Chinese patriotic devotion accumulated for millennia and the Chinese solidarity and struggle have provided great endogenous strength for this final victory. A deep understanding of the institutional advantages and spiritual strength in China's fight against COVID-19 is of far-reaching practical significance for strengthening confidence in China's system and culture, and deepening Chinese patriotism.

1 INTRODUCTION

General Secretary Xi Jinping said, "Achieving national rejuvenation will be no walk in the park; it will take more than drum beating and gong clanging to get there. Realizing our great dream demands a great struggle." The sudden COVID-19 epidemic, as a major public health emergency which has spread faster and wider than any other since the founding of New China, has proven to be extremely difficult to contain, and it is a cruel war faced by China in the decisive period of building a moderately prosperous society in all respects and the critical period of realizing the great rejuvenation of the Chinese nation. Faced with this unexpected problem, under the strong leadership of the CPC Central Committee with Xi Jinping at the core, the whole nation has stood together, helped one another, and put up an unprecedented fight against COVID-19, thereby leading a positive struggle in virus prevention and control, achieving significant strategic results in the battle against COVID-19, and answering fully such questions as why the CPC is "capable," why socialism with Chinese characteristics is "good," and why the Chinese people are "competent" in depth. Facts have proven that the clear practice of China's fight against COVID-19 and the positive results achieved by China demonstrate not only the significant advantage of the system of socialism with Chinese characteristics, but also the vast strength formed by the Chinese spirit. The two have complemented each other, making a joint effort and solid defense to defeat the virus, and provided a strong institutional guarantee and spiritual support for winning the battle against COVID-19.

2 THE PRACTICE OF CHINA'S FIGHT AGAINST COVID-19 FULLY DEMONSTRATES THE INSTITUTIONAL ADVANTAGE OF SOCIALISM WITH CHINESE CHARACTERISTICS

The institutional advantage is the greatest advantage of a country, which can only be tested in reality. As a scientific institutional system developed by the CPC and the people through long-term practices and exploration, the system of socialism with Chinese characteristics has tested the "Three Golden Principles," namely, upholding the CPC's centralized and unified leadership, concentrating resources on major issues, and adhering to a people-centered approach in the "fire test" of the fight against COVID-19, showing incomparable vitality and great superiority. It is a beneficial weapon and an institutional key to securing our victory in defeating the "plague" of COVID-19. The COVID-19 outbreak gave us a breathtaking open/practice lesson on institutional confidence. It helped to not only enhance people's conscious recognition of and firm confidence in the system of socialism with Chinese characteristics, but also translated China's institutional advantage into the strong result of epidemic prevention and control in a better manner, and strengthened our determination and confidence to win the battle against COVID-19.

2.1 Upholding the CPC's centralized and unified leadership is the fundamental political advantage of China's fight against COVID-19

The CPC's leadership is the most essential attribute of socialism with Chinese characteristics, and the greatest strength of the system. At the fourth plenary session of the 19th CPC Central Committee, 13 significant advantages of China's system and national governance system were systematically summarized, in which, upholding the CPC's centralized and unified leadership was put in the first place, highlighting its dominant position and critical role. General Secretary Xi Jinping said, "The Party exercises overall leadership over all areas of endeavor in every part of the country." The CPC's centralized and unified leadership was, is, and will continue to be the fundamental political guarantee for us to overcome any difficulties and obstacles. Only by upholding the CPC's centralized and unified leadership can we give full play to the Party's core and leading role in providing overall leadership and coordinating the efforts of all involved, better pooling the wisdom and strength of the whole Party, and ensuring that the whole Party can lead the whole nation to successfully cope with various risks, challenges, and various threats on the basis of unity and solidarity in action.

Since the outbreak of COVID-19, the CPC Central Committee has attached great importance to epidemic prevention and control, made scientific researches and decisive decisions, and strengthened centralized and unified leadership over epidemic prevention and control. As the core of the CPC Central Committee and the whole Party, General Secretary Xi Jinping has been deeply involved with the epidemic situation, taken command and made arrangements personally, convened and presided over the meetings of the Political Bureau of the CPC Central Committee and its standing committee more than once, listened to reports and delivered a series of importance speeches, made thematic research and arrangements for epidemic prevention and control, and given instructions and ordered mobilization. In the meantime, the CPC Central Committee has provided overall leadership, coordinated the efforts of all involved, put forward the general requirements of "Remaining Confident, Coming Together in Solidarity, Adopting a Science-Based Approach, and Taking Targeted Measures," specified the "four early" prevention and control requirements and the "four centralized" treatment requirements, set up a Leading Group for COVID-19 Prevention and Control to make timely researches and arrangements, sent a central steering group to Hubei and other severely affected areas to guide and oversee epidemic control, given play to the overall coordination role of the Joint Prevention and Control Mechanism of the State Council, and strengthened policy coordination and material allocation to effectively promote the efficient and orderly progress of epidemic prevention and control. Moreover, the CPC Central Committee issued the *Notice on Strengthening Party Leadership and Providing Strong Political Guarantees for Winning the Epidemic Prevention and Control Battle*, and put forward specific requirements and responsibility rules for strengthening the Party leadership on the severe conditions of epidemic prevention and control. "The epidemic situation is a command, and prevention and control is responsibility." Resolutely obeying the unified command and arrangement of the CPC Central Committee, Party organizations at all levels and all Party members and cadres have stepped forward, led the way, fought heroically, and acted proactively in the battle against the epidemic, given full play to their roles as fighting bastions and exemplary vanguards, properly fulfilled joint prevention and control and other tasks, becoming the hard core in China's fight against COVID-19, and enabled the bright red Party flag to flutter on the frontline in the fight against the epidemic. From ICUs to urban and rural communities, from workshops to research institutes, there have been CPC members charging forward everywhere. They have fulfilled the responsibility for defending the country, and vividly interpreted the CPC members' original aspirations, mission, and responsibility. Faced with the "major test" of the epidemic, the centralized and unified leadership of the CPC Central Committee with Comrade Xi Jinping at the core has played the roles of "backbone" and "fixing star." It is a stability anchor and source of confidence for us to win the battle against COVID-19, pointing out the direction and providing guidance for epidemic prevention and control.

2.2 Concentrating resources on major issues is the significant institutional advantage of China's fight against COVID-19

China's state system and national governance system have significant advantages in many areas. In one important aspect, China insists on the overall interests of the whole country, stimulates initiative of all aspects, and concentrates resources on major issues. General Secretary Xi Jinping emphasized, "Our biggest advantage is that our socialist system can concentrate resources on major issues. It is a magic key for us to achieve great success." The historical practice over the past 70 years since the founding of New China has proven that we created a miracle of rapid economic development and a miracle of long-term social stability, successfully coping with various risks and challenges including turbulent situations, floods, SARS, and earthquakes, making the remarkable achievement of "The Governance of China," and finally achieving the tremendous transformation of standing up, growing rich, and becoming strong, mainly because of the institutional advantages of the system of socialism with Chinese characteristics, that is, concentrating resources on major, good, practical, and difficult issues.

Insisting on the overall interests of the whole country and concentrating resources on major issues is not only the significant advantage of the system of socialism with Chinese characteristics, but also an important means and method to ensure our complete victory over the virus. As General Secretary Xi Jinping said, "To prevent and control the epidemic, we should insist on the overall interests of the whole country. Party committees and governments at all levels must resolutely obey the unified command, coordination, and dispatch of the CPC Central Committee, and strictly enforce orders and prohibitions." Gathering collective strength can achieve success. Since the outbreak of COVID-19, the CPC Central Committee gave an overall consideration for the epidemic prevention and control, made the strategic decision to give "top priority" to Wuhan and Hubei as they were "decisive battlegrounds," and emphasized that victory in Wuhan would ensure victory in Hubei, and ultimately victory across the country. In the critical stage of epidemic prevention and control, faced with the overloaded operation of medical resources, a severe shortage of protective supplies, and other negative impacts in Wuhan and Hubei, we gave full play to the institutional advantage of concentrating resources on major issues, and mobilized the whole country to support Hubei and Wuhan. Under the unified command and dispatch of the CPC Central Committee, 19 provinces rendered assistance to 16 prefecture-level cities in Hubei other than Wuhan in the form of pairing assistance. Vehicles loaded with medical supplies and daily necessities left for Wuhan from all parts of the country, without halting to rest; 346 medical teams from all over the country, including more than 42,000 medical workers, assembled quickly and rushed to the frontline; more than 40,000 builders speaking different dialects raced against time, day and night, completed the construction of Huoshenshan Hospital and Leishenshan Hospital in less than ten days. All parts of the country took unified action with the affected areas, transforming the institutional advantage of socialism with Chinese characteristics into an antiepidemic consciousness, and showed strong commitment. Like a classic blockbuster, the Chinese "Answer Sheet" amazed the world again. "The high speed and massive scale of China's response is unprecedented, and it showed the advantage of China's system. China's experience is worth learning for other countries," WHO Director General Tedros Adhanom Ghebreyesus remarked. His remark hit the nail on the head, and profoundly revealed the institutional strength supporting China's fight against COVID-19 behind the "Chinese speed" and "Chinese miracle." Through the epidemic test, the unique advantage (concentrating resources on major issues) of the socialist system and the strong and efficient organization and mobilization ability of the Chinese government have been embodied in all respects, and it has allowed us to feel the importance of the system to the concerted fight against COVID-19 in a more intuitive way.

2.3 *Adhering to a people-centered approach is the clear-cut standing advantage of China's fight against COVID-19*

Affinity to the people is the inherent distinctive characteristic of Marxism, and holding the people's ground is the consistent value orientation of a Marxist governing party. As the CPC is a Marxist governing party, ultimately, the CPC's original aspiration and mission, as well as the construction of the system of socialism with Chinese characteristics, take hold of the people's ground as the logical starting point. In particular, since the 18th CPC National Congress, General Secretary Xi Jinping has said more than once that "The people are the creators of history; they are the fundamental force that determines our Party and country's future." "Popular support is the biggest politics." "Respect the principal status of the people, and maintain the flesh-and-blood ties with the masses." He exhorted us to take the people's stand as the fundamental stand for upholding and developing socialism with Chinese characteristics in the new era. In essence, these are the latest interpretation and innovative development of affinity to the people of Marxism, which run through all activities and aspects of Xi Jinping's governance, and reflect the people-centered development philosophy and the contemporary CPC members' spirit of devotion to the people and sense of responsibility.

In the concrete practice of the fight against COVID-19, our Party has always adhered to the people-centered antiepidemic idea, sparing no effort in safeguarding the people's lives and health as the starting point and foothold of the epidemic control, and has truly realized the concept of all for the people, all relying upon the people. The supremacy of the people is foremost, and the fight against the epidemic is for the people. At the outset, General Secretary Xi Jinping specifically instructed, "Top priority should be given to the people's safety and physical health." He required Party committees and governments at all levels, as well as relevant government departments, to make all-out efforts to treat patients and save lives by raising admission and cure rates and lowering infection and fatality rates, with all medical treatment costs borne by the nation, and sparing no efforts in safeguarding the masses' most important, pressing and fundamental rights to life, safety, and health. Thus, we can see that about 1/10th of China's intensive care force was sent to Wuhan, and about 1/4 of the "life-saving" ECMO equipment was gathered in Hubei. Meanwhile, to reduce the impact of the epidemic on people's productivity and livelihoods, and meet the basic living needs of the population in this unprecedented period, it was necessary to actively propel the orderly work resumption of businesses on the premise of resolutely continuing epidemic prevention and control to ensure the regular supply of civilians' necessities including "vegetables," "rice," "fruits," as well as coal, electricity, oil, gas, and water. The second aspect is the power of the people. The fight against the epidemic relies on the people. Historical materialism holds that the masses are the real heroes, and requires us to adhere to the mass viewpoint and the mass line. The Chinese revolution, construction, and reform have proven that time and again our Party is rooted in the people, and its blood lies in the people; the masses have great wisdom and strength; we must unswervingly take the mass line as our Party's lifeline and fundamental work line. In terms of the fight against the epidemic, it is essentially the people's war led by the Party. In other words, the people are the main force in our victory over the epidemic. Therefore, General Secretary Xi Jinping has given instructions time and again, and required Party organizations at all levels and all Party members and cadres to rely firmly on and fully mobilize the masses, to build an impregnable fortress for epidemic prevention and control, and to resolutely win the people's all-out war against the virus. At the call of the CPC Central Committee, the broad masses, with a common aspiration, made joint efforts to fight against the epidemic, and pooled the mighty power of being united as one. In sharp contrast, Western capitalist countries like the United States, under the guise of "freedom," "democracy," and "human rights," actually pursue the cold value of "capital first." They would rather sacrifice the masses' lives and health than let those capital interest groups pay for the costly epidemic prevention. Some even adopted negative antiepidemic strategies like herd immunity in disregard of the fundamental interests of the people, which clearly exposed the hypocrisy and antipeople nature of the Western capitalist system. These are utterly different from China's value standpoint of putting the people's interests and lives first. The distinction is self-evident.

3 CHINA'S FIGHT AGAINST COVID-19 VIVIDLY INTERPRETS THE GREAT SPIRITUAL STRENGTH OF THE CHINESE NATION

"A person without a spirit cannot stand tall; a country without a spirit cannot be strong." The national spirit, as the soul on which a nation survives and develops, is the spiritual power for a nation to stand firm and move forward, and epitomizes the cohesion, centripetal force, and vitality of a nation. For five millennia, the Chinese nation has continued to thrive after undergoing trials and tribulations, stayed firm and indomitable in face of disasters, and passed the national torch from generation to generation. One important reason for this is that China has a great national spirit accumulated for generations. The sudden emergence of COVID-19 has greatly stimulated the spiritual potential of the Chinese nation accumulated for millennia, and enabled the Chinese national spirit to be refined, enriched, and sublimated again in the fight against COVID-19. Moreover, 1.4 billion Chinese have externalized the great national spirit with patriotism at its core into the conscious action of a collective fight against the virus, and released earth-shattering tremendous energy on a practical level.

3.1 *Patriotic dedication is the spiritual core supporting China's fight against COVID-19*

As the deepest historical feeling of the Chinese nation, patriotism is the heart and soul of the Chinese nation, which has been deeply rooted in the blood of the Chinese nation and built the backbone of our nation. The Chinese nation has maintained a glorious tradition of patriotic devotion since ancient times. Such patriotic sayings as "Dash ahead for the safety of the country, regardless of personal safety," "Rise and fall of a nation rests with every one of its citizens," "In line with the conviction that I will do whatever it takes to serve my country even at the cost of my own life, regardless of fortune or misfortune to myself," "The state at stake, he would give his last breath. Would a home-going soul fear to face death?" reflect the Chinese patriotism and value pursuit of serving the country and people and preferring honor to life.

In the great fight against COVID-19, the patriotic devotion spirit has stimulated the Chinese people to defeat the virus in one mind. It is the spiritual pillar supporting China's final victory over the epidemic. Encouraged by the patriotic devotion spirit, the Chinese people have transformed their patriotism and ambition into the devotion of serving the country regardless of danger, difficulty, or sacrifice. In the process, a large number of impressive heroes and touching deeds have emerged, composing a paean to the fight against the virus. Someone asked, "Doctor, what are you going to do?" The doctor replied, "Fighting against the epidemic to save lives!" He asked, "What if you couldn't come back?" The doctor answered, "I'll still go!" This is an epitome of the Chinese patriotic devotion spirit from the perspective of the medical staff. Medical workers rose to the challenge to protect others' lives at the cost of their own, showing their benevolence and selflessness; the People's Liberation Army took on the heavy responsibility bravely, and protected the common people with great love, showing their political character of being loyal to the Party and the people; researchers and grassroots cadres remained at their posts day and night, and enacted their vows with action, showing their responsibility to the Party and the people; volunteers worked uncomplainingly to sow hope with love, without consideration of pay, showing the volunteer spirit of sincere dedication and mutual assistance; overseas Chinese donated money and materials, stood together with us through the storm and stress, and provided warmth with true feelings, showing their attachment and devotion to the motherland... These touching scenes and impressive stories in the process of fighting against the epidemic have demonstrated the profound background of the fine traditional Chinese culture. The Chinese people shared the same fate with the country, closely linked their feelings with the national feelings, and interpreted the true meaning of the patriotic devotion spirit with practical action. History and the people will never forget 84-year-old academician Zhong Nanshan who was entrusted with the mission at the critical moment; academician Zhang Boli who left his "gallbladder" in Wuhan, but still resolutely plunged into the fray; President Zhang Dingyu who suffered from amyotrophic lateral sclerosis (ALS), but still charged ahead to race against the disease; as well as those faces hurt due to wearing masks. We were even more gratified and touched to see that nearly one-third of the medical teams aiding Hubei were "post-1990s" and "post-2000s". Casting off their inexperience, they volunteered to join the antiepidemic frontline, became the new force in the fight against the virus, and made their youth shine in the fight with their sincerity, integrity, and benevolence. No one is born a hero, yet their selflessness made them fearless. "They are not angels in white, but a group of kids dressed in uniforms, saving lives from death like their predecessors." Their choice of giving up their own families for everyone and putting national interests above individual interests not only is the important spiritual symbol and cultural gene of the Chinese nation, but also adds a richer connotation of the times to patriotism.

3.2 *Solidarity and struggle are the source of power supporting China's fight against COVID-19*

In the long course of social and historical development, the Chinese people have gradually cultivated and formed a solidarity and struggle spirit of coming together in solidarity and pursuing continuous self-improvement, and become an important engine propelling the rebirth of the Chinese nation, which always pools the Chinese strength to overcome difficulties whenever the nation is in adversity. From the fight against floods in 1998, the fight against SARS in 2003, the earthquake relief in 2008 to the fight against COVID-19 in 2020, the Chinese nation pulled through time and again in reliance upon the solidarity spirit: "When people are determined, they can overcome anything" and the struggle spirit: "As heaven maintains vigor through movements, a gentleman should constantly strive for self-perfection" rooted in the Chinese people's blood.

General Secretary Xi Jinping said, "The Chinese nation has experienced many ordeals in its history, but has never been overwhelmed. Instead, it has become more and more courageous, growing up and rising up from the hardships." In the spring of 2020, faced with this devastating disease, under the strong leadership of the CPC Central Committee, the Chinese people, irrespective of their gender, age, and occupation, worked together to overcome difficulties, and remained closely united like the seeds of a pomegranate that stick together. Adhering to the struggle spirit of indomitability and unremitting self-improvement, they jointly built a steel bulwark with joint prevention and control and society-wide efforts. Considering the national interest and the general situation, the heroic people in Wuhan overcame numerous difficulties, made a great sacrifice, and firmly contained the epidemic on the first battlefield by means of "quarantine." The general public assumed their responsibilities, and worked in unison against the virus. They canceled visits to friends and relatives and other gatherings, and consciously quarantined themselves at home so as not to "hinder" the epidemic prevention and control. Plunging into the fray without hesitation, medical workers braved the threat of infection to fight against the virus, saved numerous dying patients, and truly played the role of a mainstay at the frontline. Community workers, grassroots officials, public security officers, as well as customs officers worked day and night, and weaved a gigantic net for epidemic prevention and control to avoid giving the virus any hiding place. Media workers, couriers, sanitation workers, take-away deliverymen, and volunteers remained at their posts to devote themselves to the fight against the epidemic, conveying warmth and strength from workers of all trades and professions. The manufacturers of all sorts of supplies returned to work in a timely manner, and put in extra hours, thereby ensuring the adequate supply of epidemic prevention supplies and residents' daily necessities, effectively alleviating the masses' pressing needs... sticking together on a bumpy road and helping one another to overcome the difficulties. Inspired by the solidarity and struggle spirit of the Chinese nation, the primeval power of 1.4 billion people thoroughly emerged, and formed an indestructible and invincible force, which became the source supporting China's fight against COVID-19. Facing the global spread of the virus, China always upholds the vision of a community with shared future for mankind, and calls on the international community to work together to fight against the epidemic. In addition, China has actively participated in global epidemic prevention and control, and offered help to the best of its ability. So far, China has provided emergency assistance for more than 100 countries, as well as international and regional organizations, dispatched medical expert teams to such countries as Iran, Italy, and Pakistan, and donated USD 50 million to the WHO in support of the international cooperation against COVID-19. These moves won high praise from the international community, and further reflected the Chinese solidarity and struggle spirit: "When a disaster occurs in one location, help comes from all sides."

REFERENCES

[1] Xi, Jinping. 2019. Developing fighting spirit and enhancing fighting ability to strive for "Two Centenary Goals" [N]. People's Daily.
[2] Xi, Jinping. 2017. Secure a decisive victory in building a moderately prosperous society in all respects and strive for the great success of socialism with Chinese characteristics for a new era – report at the 19th National Congress of the Communist Party of China [N]. People's Daily.
[3] Xi, Jinping. 2017. *Xi Jinping: The Governance of China, Volume II* [M]. Beijing: Foreign Languages Press, 2017.
[4] Xi, Jinping. 2020. Speech at the meeting of the Standing Committee of the Political Bureau of the CPC Central Committee on response to COVID-19 [J]. Qiushi.
[5] Xi, Jinping. 2016. *Excerpts of Xi Jinping's Remarks on Strict Governance over the Party* [M]. Beijing: Central Party Literature Press.
[6] Xi, Jinping. 2018. Speech at the conference to mark the 200th anniversary of the birth of Karl Marx [N]. People's Daily.
[7] Xi, Jinping. 2016. Speech at a ceremony commemorating the 80th anniversary of the victory of the Long March [N]. People's Daily.
[8] Xi, Jinping. 2020. Speech at a meeting to advance the work on coordinating COVID-19 prevention and control and socio-economic development [N]. People's Daily.

Computational Social Science – Luo, Ciurea & Kumar (eds)
© 2021 Taylor & Francis Group, London, ISBN 978-0-367-70193-2

Research into the communication of tea culture from the perspective of international Chinese education—investigation into the tea culture elements based on Chinese textbooks

Z.Y. Ke & W.H. Zhu
School of Humanities, Xi'an Shiyou University, Xi'an, China

ABSTRACT: In the Internet + era, the spread of tea culture has become increasingly diversified. Tea culture is a combination of tea art and spirit. In the implementation of international Chinese teaching activities, it is particularly important to choose appropriate textbooks. Based on the statistics of tea cultural elements in 100 Chinese textbooks, this paper analyzes the characteristics of its selection and shortcomings. In response to these problems, the author gives some suggestions in order to promote the spread of tea culture and the development of overseas Chinese education.

1 INTRODUCTION

Tea is a labor product, with tea culture originating in China. Tea culture means the cultural characteristics formed in the process of drinking tea activities. Its spiritual connotation is a combination of the habits of making tea, enjoying tea, smelling tea, drinking tea, and tasting tea with Chinese cultural connotations and etiquette. A cultural phenomenon with distinctive Chinese cultural characteristics, or a phenomenon of etiquette, can reflect material civilization and spiritual civilization in a certain period of time. The elements of tea culture mainly include the tea ceremony, tea spirit, tea books, tea sets, tea paintings, tea studies, tea stories, tea art, etc. Taking tea as a vehicle for spreading Chinese culture through this carrier is an organic fusion of tea and culture. As the cradle of world tea culture, Chinese tea has increasingly become one of the most important elements of cultural communication in international Chinese teaching. International Chinese education is essentially language teaching, and the process of language teaching cannot be separated from cultural education. As textbooks are an indispensable communication medium in the teaching process, the transmission of rich tea culture is an important way to promote international Chinese and cultural learning.

2 STATISTICS ON TEA CULTURE IN CURRENT INTERNATIONAL CHINESE TEXTBOOKS

2.1 *Statistics on the appearance of tea culture content in international Chinese textbooks*

In these statistics, several texts in a book are calculated based on the number of occurrences, without distinguishing between main texts and subtexts. The number of times that tea culture elements appear in one form in a text is counted once.

Of the 100 textbooks in this survey, 35 involved tea culture elements, accounting for 35%, and 65 did not involve tea culture elements, accounting for 65%. Among them, there were 69 comprehensive textbooks, 15 reading textbooks, six listening and speaking textbooks, six reading and writing textbooks, and four cultural textbooks. Of these, the proportion of comprehensive textbooks involving tea culture was 35%; reading textbooks 33%; listening and speaking classes 33%; reading and writing classes 17%; and culture classes 74%.

The textbooks related to tea culture mainly included the "Developing Chinese" series of comprehensive and intermediate-to-advanced reading, "Road to Success" series, "New Chinese Textbook" series, "HSK Standard Course 1–4" series, "Chinese Traditional Culture and Modern Life" series, "Boya Chinese," "Common Knowledge of Chinese Culture," "Intensive Course of Chinese Language 2," "Short-term Listening Chinese," and another 35 textbooks. Tea culture elements appeared 73 times: after-class exercises 27 times, text fragments 20 times, whole texts 18 times, words and phrases four times, and readings after class four times, giving a total of five kinds of forms. The proportions were 37%, 27.4%, 24.7%, 5.5%, and 5.5%.

2.2 *Statistics of tea types in international Chinese textbooks*

There are many types of tea in China, divided into green tea, black tea, oolong tea, black tea, white tea, yellow tea, and scented tea, etc. Based on the analysis and statistics of these six categories, the types of tea that appear repeatedly in one book are counted once, of which green tea appears 17 times. This means that of the 35 textbooks that contain tea culture elements in statistics, green tea is mentioned in 17 books.

"Others" mainly refers to elements that are not in the seven major tea types, such as big bowl tea and butter tea, and 29 textbooks involve these elements, accounting for 82.9%, and appearing 218 times in total. Green tea is the most familiar tea. Among the 35 textbooks, 17 mentioned green tea, accounting for 48.6%, a total of 113 times; black tea and scented tea were highly popularized, and 15 textbooks mentioned these two teas, accounting for 42.9%. Black tea appeared 53 times in total, scented tea appeared 68 times; again, oolong tea and dark tea, of which 11 books mentioned green tea, accounting for 31.4%, a total of 46 occurrences, six mentioned black tea, accounting for 17.1%, with a total of 16 occurrences. White tea and yellow tea, as uncommon teas, are not mentioned in the textbooks.

2.3 *Statistics of tea cultural elements in international Chinese textbooks*

As the cradle of world tea culture, Chinese tea culture is rich in content. Elements of tea culture include the tea ceremony, tea spirit, tea books, tea sets, tea studies, tea stories, tea art, and so on. In the textbook survey, the following statistical data are created on the elements of tea culture from these aspects. Others refers to the elements that are not included in the following seven major tea culture elements, but simply use the word "tea" to summarize and relate to tea. Works included "Tea House" by Mr. Lao She and "The Classic of Tea" by Mr. Lu Yu.

There are 27 other elements involved in 35 textbooks, accounting for 77.1%, a total of 91 times; tea art appears in 14 textbooks, accounting for 40%, with a total of 65 occurrences. The second is the tea story element, which is mentioned in 12 textbooks, accounting for 34.3%, a total of 25 times; again, it is tea set, with tea set elements in seven textbooks, accounting for 20%, with a total of 18 times; five textbooks include tea ceremony culture, accounting for 14.3%, a total of 13 times; tea books and tea spirits are mentioned in four textbooks, accounting for 11.4%, of which tea books appear nine times, and tea spirits seven times; finally, there was one textbook where refreshments were mentioned, 2.9% of the total, and appeared once in total.

3 DEFICIENCIES IN THE INTRODUCTION OF TEA CULTURE IN INTERNATIONAL CHINESE TEXTBOOKS

3.1 *Tea culture elements lack scientific and reasonable arrangement in textbooks*

The arrangement of tea culture content did not follow the principle of gradual progress, such as the "Boya Chinese" series. As a staged international Chinese textbook, "Boya Chinese" only mentions tea culture in one text in the volume of "Boya Chinese Intermediate · Sprint" in its four-stage content arrangement, but in other stages the textbooks are not covered, and there is a lack of phased arrangements for tea culture learning, so that learners cannot gradually understand the tea culture content when using this set of textbooks.

Textbooks have a low emphasis on the output of tea culture content. Of the 100 textbooks surveyed in this study, only 35% involved tea culture elements. Among the 35 textbooks, the texts specifically introducing tea culture and after-class extended reading account for 32.2%, and only the fragments in the texts or the words related to tea culture in after-class exercises account for 67.8%. Compared with other traditional cultures involved in the textbook, the proportion is low and the importance of tea culture is ignored.

The proportion of different types of tea introduced in textbooks is unevenly distributed. In the statistics of textbooks, we can see that the overall introduction of tea is the most frequent, followed by green tea, black tea, and scented tea. In daily life, these three types of tea are also the most common and popular. Therefore, within the cultural dissemination, the selection of teaching materials also focuses on these three types. The arrangement of green tea and black tea is slightly attenuated, while white tea and yellow tea are never mentioned.

3.2 The introduction and consolidation design of tea culture in textbooks is not perfect

In many texts, tea culture is not mentioned or discussed very much, but tea culture appears many times in the exercises after class. According to the above analysis, the tea culture that appears in the textbook in the form of the whole text accounted for 24.7%, the tea culture mentioned in the text fragments accounted for 26%, and the tea culture involved in the after-school exercises was as high as 38.4%. Through the study of the text, students cannot learn enough relevant knowledge to answer the after-school exercises. Such an arrangement of learning makes students prone to swallowing dates when encountering the content of tea culture.

The principles of shallow and deep learning of cultural knowledge are not well implemented. For example, in "Developing Chinese—Intermediate Reading Course 2," the eighth unit, article 4: the knowledge of drinking tea, the whole article introduces the related knowledge of tea culture in detail from the choice of water for making tea, to the types of tea, and the difference between tea sets, etc. However, in the arrangement of the exercises after class, it only arranges to judge the correctness of the sentence based on the article. The practice of filling in the blanks did not appear to encourage students to think and explore further.

3.3 The tea culture content in the textbook fails to reflect the conversion between cultural teaching and language teaching

In the course of cultural teaching, the textbook simply inculcates some theoretical knowledge and explains the theoretical knowledge on the surface of tea culture, but the details of the intake of the specific tea ceremony, tea art, and tea spirit are insufficient, so that learners cannot form a specific picture in their mind during the process of learning theoretical knowledge, and then apply this culture in practice.

Language learning cannot be separated from practice and specific communication activities. However, the textbook lacks operational space for simple cultural activities between learners when arranging contents related to tea culture elements. In the textbook of this research, the tea culture elements are simply output in text content, and there is no mention of how to transition to communication through written learning or possible communication activities, so arrangements are made.

3.4 The material selection of tea culture in textbooks is relatively narrow

Among the textbooks related to tea culture in this survey, from the perspective of the types of tea, the textbooks emphasize the introduction of familiar green tea, black tea, scented tea, and oolong tea, ignoring the niche white tea and yellow tea; from the perspective of tea culture elements, the textbook focuses on the introduction of tea art, tea stories, tea utensils, and tea ceremonies, but there is less of an introduction to refreshments, tea books, and tea spirit, and tea painting elements are not mentioned.

The tea culture in the textbook failed to integrate with the development of the times and keep up with the pace of the times. The tea culture in the textbooks describe the customs and meanings passed down by people in ancient times, informing of the humanistic spirit of the past, and failure to keep up with the changes of the times and update the content. Therefore, in the international Chinese textbooks, the selection of tea culture materials is too narrow and limited to adapt to flexible Chinese teaching.

4 SOME THOUGHTS AND SUGGESTIONS ON THE ARRANGEMENT OF TEA CULTURAL ELEMENTS IN INTERNATIONAL CHINESE TEXTBOOKS

4.1 *Arrange the tea culture content in the textbook scientifically and reasonably*

The learning of Chinese is a long-term process, and the arrangement of the content of textbooks should be in line with the gradual and orderly learning process in which the educated take in knowledge. In international Chinese teaching, there are three different stages, namely primary, intermediate, and advanced, corresponding to different contents. According to the stage of the learners, the characters, words, and sentences involved in tea culture teaching are compiled and arranged reasonably.

In the selection of textbook content, more and more attention should be paid to the dissemination of cultural content, especially the knowledge related to tea culture. People not only realize the importance of language education and communication, but also actively integrate cultural education. Chinese tea culture has formed a magnificent value system, and the integration of Confucianism, Buddhism, and Taoism over thousands of years has made it a unique representative of Chinese traditional culture.

4.2 *Adopt diversified forms to introduce tea culture and improve the design of the introduction of tea culture content*

In addition to the existing text and reading output, we should not stay on the single and traditional mode of doing exercises after class. In addition to the existing theoretical forms such as texts, readings, and exercises, the textbook can also add practical forms. In some cultural textbooks, some film and television materials related to tea culture can even be used to show tea cultural knowledge in a more comprehensive and vivid way. The combination of the two can better stimulate learners' interest in the subject.

Students learn knowledge should follow the i + 1 output theory, by the shallow and deep, in a series of teaching materials, such as "the road to success" series, "the development of Chinese" series of teaching materials, for the output of the tea culture, from junior to senior, from the surface to the deep, giving students a new and deeper knowledge than the original culture idea output, so as to allow students to continuously strengthen the cognition of the tea culture and understand cultural connotations.

4.3 *Realize the conversion of cultural knowledge into communication skills, and vigorously promote tea culture to serve language teaching*

Content design of cultural activities should be included in cultural textbooks. The method of learning cultural knowledge should not be limited to the explanation of knowledge in class, students also can learn and comprehend Chinese culture through cultural practice activities. Cultural textbooks should combine cultural knowledge with cultural activities in order to guide students to learn and apply cultural knowledge in practical activities. During the activity, communication between students, as well as between students and teachers, with tea culture as the carrier, can transform cultural knowledge into communicative skills, so as to promote the improvement of learners' Chinese ability and realize the fundamental purpose of Chinese teaching.

After the text, the discussion section is added, which reflects the principle of transforming cultural knowledge into communicative skills. International Chinese teaching itself is language teaching, with an emphasis on communication. In the process of language teaching, culture unconsciously infiltrates it. Similarly, in the process of cultural teaching, cultural knowledge should also be transformed into communicative skills. After learning the text, the content is discussed through teacher–student dialogue and student–student dialogue. In this way, the teacher can not only check the students' understanding of Chinese culture, but also deepen the students' understanding of Chinese culture and increase cross-cultural knowledge by discussing the similarities and differences of different cultures.

4.4 *Enrich the material of tea culture in the textbook to make it more comprehensive and perfect*

The theoretical knowledge of tea culture should advance with the times, as the wheels of time are rolling forward. Therefore, in the selection and arrangement of international Chinese teaching materials, it is necessary to adhere to the idea of keeping pace with the times, seek breakthroughs and innovations in development, and always pay attention to social development in the process of disseminating cultural knowledge. Change, integrate the traditional cultural value with the current era, consider the actual needs of the development of the era and the new value concept it confers on tea culture, constantly update the content, make "new" knowledge, enrich the concept of tea culture, and realize its overall improvement of international Chinese teaching.

Increase the ratio of other types of tea in the textbook, so that learners can learn more about the values of different teas, and thus have a more comprehensive understanding of the cultural and social values behind the tea culture. At present, there are many arrangements for tea stories, tea art, tea set, and other elements in the textbook, but other elements are less involved. For the lack of spirit and value contained in tea culture, it is easy for students to have insufficient understanding of tea culture knowledge and limitations in thinking when they understand international Chinese teaching activities. Therefore, attention should be paid to this point in the compilation of textbooks, and materials should be continuously enriched from various aspects so that students can acquire different knowledge and concepts of tea culture.

5 CONCLUSION

This article analyzes the shortcomings of tea culture content layout in current textbooks from the three aspects of tea culture appearance form, tea types, and tea culture elements in textbooks. Aimed at the shortcomings in the current textbooks puts forward how to lay out the deficiency of the teaching material content, perfecting the introduction of the tea culture, rich tea culture, transform tea culture knowledge into communicative skills, and vigorously promote tea culture to serve language teaching. In the process of international Chinese education within the Internet + background, the spread of tea culture can give people a deeper cultural connection and reflect different teaching effects. In addition, the contents involved in tea culture in teaching Chinese as a foreign language include the inherent characteristics of China's tea culture. In cross-cultural communication, good interactions can be generated when communicating with foreign tea culture, so as to better improve learners' cross-cultural communication ability and improve Chinese learning efficiency. Chinese tea culture will inevitably gain the attention of more Chinese learners by virtue of its unique global and national charm, and will play an increasingly important role in the international Chinese teaching process.

ACKNOWLEDGMENTS

This article was written for the 2019 Xi'an Shiyou University postgraduate education and teaching Reform Research Project: Research on the Innovation of Teaching Mode and Quality Evaluation System of the Course "Poetry Recitation and Creation" in Chinese International Communication.

REFERENCES

[1] Yu Lu. The Classic of Tea. Yunnan: Yunnan People's Publishing House, 2011:71–451.
[2] Mei Jin. On the introduction of tea culture in international Chinese textbooks.Tea in Fujian, 2018, 40(04):312.
[3] Jiao Jin. Chinese Tea Culture and Chinese International Education. Sichuan Normal University, 2015.
[4] Xin Jiang, Yi Jiang. The Classic of Tea. Changsha: Hunan People's Publishing House, 2009:5, 21, 39.
[5] Jiao Wang. On the introduction of tea culture in international Chinese textbooks. Tea in Fujian, 2017, 39(05):265-266.
[6] Xiaobing Zhou, Nan Chen, Jin Guo. Overview of global Chinese textbooks based on the textbook library. Overseas Chinese Education, 2015(02):225–234.

Computational Social Science – Luo, Ciurea & Kumar (eds)
© 2021 Taylor & Francis Group, London, ISBN 978-0-367-70193-2

A study on the new humanistic value of "Shaanxi Spirit" in Zhang Zai's "Si Wei" philosophy and its international promotion strategy

Y.G. Wang, C. Wang, W. Guo & J. Zhang
Tai Yigong Town, Chang'an District, Xi'an, Xi'an Fanyi University, China

ABSTRACT: The excellent Chinese traditional culture has the characters of rich philosophical thoughts, humanistic spirit, educational thoughts and moral concepts and it can inspire people to know and transform the world further, to govern their counties effectively, and reconstruct their morality deeply. In order to explore the new humanistic value of Chinese traditional culture and make it play a greater role in serving the whole world better, the author read and studied a lot of materials carefully on the basis of plentiful practice and finally found that this topic has the new value of "Win-win Cooperation, Harmonious Coexistence, Mutual Respect and Mutual Tolerance". The core issue to be solved in this paper is to explore how to promote this value to the world from Chinese perspective (cultural strategy, communication strategy, talent strategy, academic strategy, interaction strategy) and international perspective (community strategy and collaborative interaction strategy) to finally form an effective consensus of "global governance".

Keywords: Traditional culture, Win-win cooperation, Global governance.

1   GENERAL INSTRUCTIONS

Zhang Zai, a famous scholar in the Northern Song Dynasty of China, left one of his precious ideological legacies£"To Set the Heart for Heaven and Earth, To Set the Life for People, To Continue the Doctrines for the Saints, To Create the Peace for All Ages ". Since each sentence begins with the word "Wei", it is abbreviated as "Si Wei". This implies that Zhang Zai's thoughts of peace have been fully embodied in the initiative of the vision of "a community of shared future for the world". It is the supreme goal to promote the peaceful development of mankind, and it is also a good medicine to promote the permanent peace of all countries and regions in the world and to pursue the "Universal Harmony" in the era of globalization.

2   THE NEW HUMANISTIC VALUE OF "SHAANXI SPIRIT" IN ZHANG ZAI'S "SI WEI" PHILOSOPHY

2.1   *"To Set the Heart for Heaven and Earth"*

2.1.1   *The historical connotation of "To Set the Heart for Heaven and Earth"*
The "Heart" in "To Set the Heart for Heaven and Earth" refers to the moral and spiritual value of human beings. It is to establish moral standards for the human society, and form people's certain values. In a word, "To Set the Heart for Heaven and Earth" means to establish a set of the core value system and cultural order with "charity", "courtesy" and "loyalty" for the world.

2.1.2   *The new humanistic value of "To Set the Heart for Heaven and Earth"*
"To Set the Heart for Heaven and Earth" is to establish a standard of value for the human society. It reveals the significance of exploring the world and paints a blueprint for win-win cooperation in the

context of a community with a Shared future for the world. The connotation of doctrine described by Zhang Zai is the way of win-win cooperation, harmonious coexistence, mutual respect and mutual tolerance based on a community with a Shared future for the world. Therefore, "To Set the Heart for Heaven and Earth" provides a feasible "way" for friendly exchanges between countries and regions around the world.

2.1.3 *"To Set the Heart for Heaven and Earth" reflecting Xi Jinping's global vision of governing a safe country*

Zhang Zai said that everything has its own rules, and people should expand their thinking to understand the world and govern the country with rational thinking according to the development rules of things. In the face of the complex issues of globalization, Xi Jinping knew about the complex world in a scientific method, and put forward the "Upholding multilateralism and democratization in international relations, with an open, cooperation, win-win mind planning development, unswervingly pushing economic globalization towards open, inclusive, balanced and win-win direction, and promoting the building of an open world economy."

2.2 *"To Set the Life for People"*

2.2.1 *The historical connotation of "To Set the Life for People"*

"Living" in "To Set the Life for People" refers to all living sentient beings, and "life" refers to all living things. Life consists of two levels, namely "morality life" and "natural life". Neither of these two levels can be harmed or abandoned. They must be established in a way that allows them to live in a natural state without human intervention.

2.2.2 *The new humanistic value of "To Set the Life for People"*

"To Set the Life for People" means to think about the life and movement of the common people and to let them lead a normal life. Under the background of "a community with a Shared future for mankind" and from the perspective of "To Set the Life for People", we have examined the destiny of the people of all countries and ethnic groups in the world. We should work together in a candid manner to seek common ground while reserving differences from a diversified perspectives, and work together to solve common problems so that more people will benefit from the fruits of cooperation and development.

2.3 *"To Continue the Doctrines for the Saints"*

2.3.1 *Historical connotation of "To Continue the Doctrines for the Saints"*

The "past saints" here refer to the confucians represented by Confucius and Mencius. "the Doctrines" refers to the Doctrines promoted by Confucius and Mencius. Confucius was the master of Confucianism, while the masters of Lian school, Luo school, Guan school and Min school in the Song Dynasty promoted Confucianism to a new stage. "To Continue the Doctrines for the Saints" not only inherits the excellent Confucian academic tradition, but also regards itself as the inheritor and propagator of the Confucian tradition.

2.3.2 *The new humanistic value of "To Continue the Doctrines for the Saints"*

"To Continue the Doctrines for the Saints" opened a new era of harmonious coexistence. It has the significance of pooling strength, reducing conflicts, promoting concerted efforts, harmonious coexistence, mutual tolerance and common development. We should respect each other's history, culture, customs, and development paths chosen by them. Only in this way can all our ethnic groups live in harmony under different civilizations and different development models and achieve common development, so as to usher in a beautiful era of mutual learning and realize the eternal development of mankind.

2.3.3 *"To Continue the Doctrines for the Saints" demonstrating the excellent traditional Chinese culture and cultural confidence*

The excellent traditional Chinese culture is the lifeblood of our spirit, an important source of fostering core socialist values, and a solid foundation for the Chinese nation to stand firm in the global cultural turmoil. With the promotion of the position of China in the international society, we should push different civilization dialogue, continuously study, comparison and reference, promote the Chinese excellent culture to the world, let the other people around the world benefit in the Chinese culture and Chinese wisdom, to realize Xi Jinping's earnest wishes of "Combing the Chinese excellent traditional culture and outstanding culture of all countries in the world to benefit the world."

2.4 *"To Create the Peace for All Ages"*

2.4.1 *The historical connotation of "To Create the Peace for All Ages"*

"To Create" in "To Create the Peace for all ages" means "The Expectation". "The Peace" is the most profound concept in Confucian scholarship. To realize Zhang Zai's cultural ideal of "To Create the Peace for All Ages", the modern people who are confused and ignorant can return to the homelands of human spirit with frankness and sincerity.

2.4.2 *The new humanistic value of "To Create the Peace for All Ages"*

"To Create the Peace for All Ages" means not only the construction of the current social harmony, but also the hope to lay a stable and peaceful foundation for generations to come, full of great political aspirations. "Infiltration in Chinese excellent culture tradition "Universal Harmony" thoughts that the lofty ideals of real freedom, equality and fraternity among the humanity embodies the destination of development of human society. It should be the most valuable things in the core connotation of the globalization, and is worth inheriting and carrying forward forever for us". It reflects China's pursuit of common peace and development for mankind and permanent peace for the country and its people. It is also China's pursuit of "Universal Harmony" in the context of globalization.

2.4.3 *"To Create the Peace for all ages" highlighting Xi Jinping's ideal of "a community with a Shared future for the world"*

Traditional Chinese wisdom is an important cultural resource for building a community with a Shared future for mankind. To promote the building of a community with a Shared future for mankind is the embodiment of "Universal Harmony". The ideal of "Universal Harmony" is embodied in the practice and sublimation of fine traditional Chinese culture by General Secretary Xi Jinping. Zhang Zai said, "All the people are my brothers and sisters; all things and I are the offspring of heaven and earth." This reflects the wisdom and insight of the Chinese ancestors, and provides rich ideological and cultural resources for today's practice of building a community with a Shared future for mankind.

In a word, Zhang Zai's "Si Wei" Philosophy, involving the pursuit of people's life principle, spiritual value, life meaning, academic tradition and political ideal, expresses Zhang Zai's broad mind. Zhang Zai's "Si Wei" Philosophy not only has great influence in history, but also has important value in today's real life, which is the most precious spiritual wealth and cultural resource in the history of Shaan'xi province. Today, promoting and developing the cultural spirit of Zhang Zai's "Si Wei" Philosophy has important value and reference significance for us to correct the social atmosphere, promote the construction of harmonious society and spiritual civilization, and promote world peace and development.

3   THE INTERNATIONAL PROMOTION PATH OF ZHANG ZAI'S "SI WEI" PHILOSOPHY

3.1   Chinese promotion strategy

3.1.1   Cultural strategy

By constructing the cultural framework of "Shaanxi Spirit" of Zhang Zai's "Si Wei" Philosophy, the internationalization of "Shaan'xi Spirit" of Zhang Zai's "Si Wei" Philosophy is promoted. The frame of ideology, materiality and communication constitutes the cultural frame. The conceptual framework here mainly refers to the values, people's livelihood, politics, education and economy embodied in "Shaan'xi Spirit" of Zhang Zai's "Si Wei" Philosophy. The frame of materiality refers to build the materialized image building of "Shaan'xi Spirit" of Zhang Zai's "Si Wei" Philosophy; the frame of communication refers to the communication system composed of publishing institutions, propaganda institutions and network media.

3.1.2   Communication strategy

Through the construction of tourism platform, communication platform and information platform, we promote the "Shaanxi Spirit" of Zhang Zai's "Si Wei" Philosophy to spread outward.

Tourism platform. It includes the network tourism platform and the reality tourism platform. With the help of the Internet information communication platform, the network tourism platform adopts mobile phones, computers, televisions and other terminals to spread tourism culture through images, sound, text and other media. The realistic tourism platform is based on Zhang Zai's cultural relics, such as Zhang Zai Tomb, Zhang Zai Temple, Zhang Zai Museum, Zhang Zai Jingtian relics, etc., to develop tourism resources and attract tourists to actually appreciate Zhang Zai culture.

Communication platform. The exchange platform is to provide a place for academic exchanges for people who care about the spread of "Shaanxi spirit" of Zhang Zai's "Si Wei" Philosophy, such as cultural scholars, industrialists, ordinary people, etc.

Information platform. With the help of modern network information technology, the network information platform is established to present Zhang Zai culture in the network information platform, so as to carry forward the "Shaanxi spirit" of Zhang Zai culture and its value of "creating more benefits for people".

3.1.3   Talent strategy

Talent is the key to the internationalization of Zhang Zai's culture. Talents include three kinds of talents: research type, industrial type and communication type. The responsibility of research-oriented talents is to deeply interpret the cultural connotation of Zhang Zai and explore its practical significance. The duty of industrial talents is to present the theoretical achievements of research talents in the form of material objects. The responsibility of communication type is to introduce Zhang Zai culture to the outside world with the help of modern communication media, such as the Internet, newspapers, television and radio. The aim of talent classification is to establish a database of talents, based on which to explore the strategy of talent cultivation and development.

3.1.4   Academic strategy

The "Shaanxi spirit " embodied in Zhang Zai's "Si Wei" Philosophy is rich in universal values, especially the thoughts of Win-win Cooperation, Benefit for the People, Harmonious Co-existence and Universal Harmony, which undoubtedly has far-reaching historical value and great practical significance for the realization of the "community with a shared future for mankind" initiative. How to promote the Zhang Zai's "Si Wei" Philosophy needs academic research. Academic institutions shall establish a Zhang Zai cultural research community, integrate the research resources of relevant institutions of higher learning and scientific research institutions, and carry out joint research in conjunction with non-governmental institutions.

3.1.5 *Interaction strategy*

The internationalization of Zhang Zai's "Si Wei" Philosophy is a systematic project, which requires the government, the society and the market to perform their respective functions, and interact with each other and work together. Zhang Zai culture internationalization should be in line with the market guide and the principle of mutual benefit, and balancing the interests of all parties to follow the rules of cultural exchange and cultural market operation rule, give full play to the market in the allocation of cultural resources and adjust action, arouse the enthusiasm of domestic related parties to integrate Zhang Zai's culture with diplomatic, economic and trade exchanges on the basis of Zhang Zai's cultural exchange and spread, so as to promote coordinated development situation of the economic development as the goal, and realize mutual benefit and win-win of the concerned parties.

3.2 *International promotion strategy*

3.2.1 *Community strategy*

We should construct the international translation subject community and translation communication community of Zhang Zai's "Si Wei" Philosophy. The translation community of Zhang Zai's "Si Wei" Philosophy consists of domestic researchers of Zhang Zai's thoughts, translators at home and abroad, and foreign sinologists who study Zhang Zai's thought. Domestic and foreign translators should translate the core connotation of Zhang Zai's "Si Wei" Philosophy into the target country according to the national conditions of the target country. The translation and communication community mainly consists of Chinese and foreign cultural communication people, including media, publishing circles, public opinion circles, etc., whose main duty is to make the core connotation of "Shaanxi Spirit" in Zhang Zai's "Si Wei" Philosophy accepted, recognized, settled and rooted in the target country through Chinese and foreign communication organizations.

3.2.2 *Collaborative interaction strategy*

(1) Identity mechanism. Identity is the premise of communication. Only when each party recognizes the culture of the other party can communication occur. This is cultural value identity. We should promote the "community with a shared future for mankind" and build a shared cultural value based on the "Shaanxi Spirit" in Zhang Zai's "Si Wei" Philosophy, namely win-win cooperation, benefit the people, harmonious coexistence, and universal harmony in the world. To make people of all countries and regions in the world recognize this cultural value, we should establish a policy-oriented mechanism, incentive mechanism and safeguard mechanism for realizing the interests of people of all countries and regions in the world, so as to lay the material foundation for the attraction and cohesion of this Shared cultural value.

(2) Interaction mechanism. From the perspective of scope, social interaction includes the interaction between people, individuals and groups, groups and groups, as well as the interaction between nations and countries. The establishment of interaction mechanisms among countries and regions in the world should follow the principles of equal consultation, mutual benefit and complementarity to carry out communication, consultation and exchange activities at multiple levels and through multiple channels. The establishment of an interactive mechanism is conducive to enhance cultural exchanges, mutual trust and connectivity, and economic connectivity, so as to realize cultural and economic complementarity and common prosperity.

4 CONCLUSION

In the report of the 18th National Congress of the COMMUNIST Party of China (CPC), it is proposed to create "a new situation in which the international influence of Chinese culture is constantly increasing". In order to enhance the cultural soft power of Shaanxi province, the Shaanxi provincial government proposed that "Shaanxi culture should go out and expand the influence of Shaanxi province". The "Shaanxi spirit" of Zhang Zai's "Si Wei" Philosophy is a beautiful name card of Shaanxi culture. In order to provide reference for relevant government departments to make decisions, this study puts forward some Suggestions from both domestic and international perspectives on how to build this business card overseas.

ACKNOWLEDGEMENT

I would like to express my gratitude to all those who have helped me during the writing of this paper. I gratefully acknowledge the help of Professors W.Wang and C.Wang. I do appreciate their patience, encouragement, and professional instructions during my paper writing. Also, I would like to thank my parents and my wife, who did a lot of housework and took care of my little daughter to provide the enough time for me during my writing the paper.

The project of this paper is supported by the major theoretical and practical issues in foreign language studies funds from Shaanxi Social Science Association.

Project Name: A Study on the New Humanistic Value of "Shaanxi Spirit" in Zhang Zai's "Si Wei" Philosophy and Its International Promotion Strategy

Serial Number: 20WY—23

REFERENCES

[1] Qu Sheming. Inheritance and translation of the Chinese spirit of "Zhang Zai Si Wei" in the context of "One Belt and One Road". Journal of baoji college of arts and sciences (social science edition), 2017 (3).

[2] Zhao Fujie. On the Basic Spirit of Guan Xue. Journal of Northwest University, 2005 (6).

[3] Zhang Zai. Zhang Zai Ji. Beijing: Zhonghua Book Company, 1978.

[4] Chao Gejin. Attaching importance to our oral inheritance. People's Daily, March 21, 2016 (7).

[5] Deng Zhiwen. On the Mechanism and Orientation of Cultural Identity. Journal of Changsha University of Science and Technology (Social Science edition), 2015 (2): 30-34.

[6] Zhang zai.Zhang Zaiji.Beijing: Zhonghua Book Company, 2006.

[7] Cong Chunxia. Chinese Excellent Traditional Culture and the Cultural Confidence. The Red Flag Manuscripts, 2020.6 http://theory.gmw.cn/2020-06/26/content_33941112.htm

[8] Jin Minqing. "Xi Jinping's Wisdom of Peaceful Development" People's Forum, 2017 http://www.qstheory.cn/zdwz/2017-08/.

Session 4. Law and education

Computational Social Science – Luo, Ciurea & Kumar (eds)
© 2021 Taylor & Francis Group, London, ISBN 978-0-367-70193-2

Research on cross-school sharing practice teaching construction of descriptive geometry and mechanical drawing courses based on "Internet Plus"

X.H. Li*, W. Zhao & X.C. Zhou
School of Mechanical Engineering and Automation, Northeastern University, Shenyang Liaoning, P.R. China

ABSTRACT: The cross school study of descriptive geometry and mechanical drawing based on "Internet plus" is a new teaching mode. It takes advantage of the Internet platform to break the traditional teaching mode. Students can cross the geographical and time constraints, and choose excellent course resources of our university according to the agreement of the higher schools, and earn credits. In the practical teaching of mechanical drawing courses, based on many years of curriculum reform and research, our school strives to guide students to master the method of analyzing engineering expression problems from the course content, and cultivate the ability of engineering students to express mechanical structure and solve practical engineering problems from a mechanical design idea. It has advanced and strong demonstration effects, and has received good feedback from teachers and students who have selected the cross course.

Keywords: Descriptive geometry and mechanical drawing, Cross school sharing, Internet plus.

1 INTRODUCTION

"Internet plus" refers to the use of the Internet platform and information and communication technology to combine the Internet with all walks of life, including traditional industries, so as to create a new ecology in new fields. Generally speaking, "Internet plus" means "Internet and each traditional industry" [1, 2]. This is equivalent to adding a pair of "Internet" wings to a traditional industry, to help the traditional industry in the new era of optimization and upgrading. Cross-school study is a model exploration of "Internet + education". It uses the technological and platform advantages of the Internet to break the traditional teaching model. It relies on a third-party network teaching management platform and provides excellent courses in the early stages of universities [3]. On the basis of this, students can transcend geographical restrictions, and according to the agreement between universities, choose the quality courses of other universities to study and earn credits.

The practice of mechanical drawing is a very important teaching and training link in the teaching of mechanical drawing and a very important node in the transition from basic theoretical knowledge of drawing to the practical ability of drawing and reading [4]. How to rationally design the practical teaching content of mechanical drawing, improve the teaching links, and improve the teaching quality of the practical links of mechanical drawing courses, so that the audience students can really improve their practical ability based on the basic theoretical knowledge of the teaching materials, so as to truly master the expression of surveying and mapping , The analysis method of reading pictures lays a solid foundation for the subsequent mechanical professional courses and future job application needs. It is a very important teaching reform research content in the teaching and teaching reform exploration of graphics course teachers, teaching assistant managers, and experimenters.

---

*Corresponding author

2 CURRENT SITUATION OF CROSS SCHOOL STUDY IN THE PRACTICE OF LINKING DESCRIPTIVE GEOMETRY AND MECHANICAL DRAWING

At present, engineering education in colleges and universities generally focuses on the teaching and reform research of basic theoretical knowledge, while ignoring the teaching, training, and cultivation of practical skills in courses. According to statistics on the current situation of this course in various colleges and universities, the teaching of theoretical knowledge accounts for most of the class hours of the entire curriculum. Generally, the extremely limited practical teaching links and content are only arranged for students to observe or demonstrate in the teaching process. Demonstration, students do not pay much attention to this part of the content study, do not in-depth exploration, resulting in students' practical skills seldom or even no training and promotion, to a large extent deviated from the principles and goals of this part of the content.

Currently, "Descriptive Geometry and Mechanical Drawing" is the first technical basic course that engineering students come into contact with. In terms of the long-term and continuity of cultivating innovative talents, this course stands at the source of the talent training stream. The cultivation of top-notch innovative talents in the national education plan plays an extremely important basic role. Taking mechanical drawing courses in our school as an example, they cover the first and second semesters of the freshman year. According to the tradition going back many years, the first semester includes the basic knowledge of drawing geometry, and the second semester teaches the drawing of actual parts and assemblies and the course design content [5]. The whole teaching and practical operation do not only need to prepare a large number of teaching surveying and mapping models, parts, assembly, measuring tools, and laboratory space, but also need a fixed time, which invests a lot of financial resources, manpower, and time. With the expansion of school enrollment, the existing surveying and mapping hardware environment can no longer meet the needs of teaching activities, and students cannot conduct surveying and mapping operations anytime and anywhere to complete the practice of setting content. Through the teaching activities in recent years, exchanges and discussions with students, and statistics of the graduation design drawing and engineering drawing understanding of senior students in recent years, the teaching team members believe that it is imperative to reform the current practical teaching content and methods. It is of great significance to complete the reform of the training mode of innovative talents by reforming the practical links and teaching content of graphics courses, meet the learning needs of students' theory and practical content, and cultivate students' creative conception ability and spatial thinking ability. Extend research results to other engineering courses.

From the perspective of teaching staff, some teachers simply teach drawing theory knowledge, the content and practical operation methods of the lectures cannot be closely integrated with practical engineering applications, and the teaching methods and methods are single. The learning personality of different levels and different types of students has resulted in low learning enthusiasm, and the practical operation is just a simple imitation.

With the development of "double first-class" university construction in the country in recent years, higher-level requirements have been put forward for the training of people, but the current training method lacks description and in-depth enlightening education of relevant content from multiple directions and multiple angles. As a result, the students' mechanical drawing knowledge theory and practice are seriously out of touch, and they are not deeply compatible with the goals and tasks of the national "double first-class" construction work. [6].

3 CROSS SCHOOL STUDY COUNTERMEASURES FOR THE PRACTICE LINK OF DESCRIPTIVE GEOMETRY AND MECHANICAL DRAWING COURSE BASED ON "INTERNET PLUS"

In view of the problems existing in current cartographic courses, it is necessary to carry out reform research on practical teaching content and operational means for practical links between basic cartography courses in combination with the national guidelines and guidelines on the training requirements of engineering innovative talents. Reform of practical and interactive teaching should be carried out to encourage students to become more engineering conscious, cultivate practical skills, and improve their innovation skills, while perfecting their basic theoretical knowledge and mastering practical operation skills. Through the implementation of the corresponding teaching reform research, attempting to solve the conflict between the general education of students and the national implementation of innovative talent training methods and means, creating the original innovation of engineering top-level innovative talent training, trying experimentation with personalized teaching of basic engineering courses have been tried first. Therefore, it is obvious and urgent that relevant content teaching and reforms to solve the bottleneck problems existing in the current teaching content and methods should be undertaken.

In the construction of a practical teaching link for the mechanical drawing course in our school, due to the limited funds for course construction, the teaching and research center organized front-line teachers to develop the teaching practice platform for the mechanical drawing course, fully considering the virtualization, simulation, and network of resource construction [7]. To some extent, this teaching practice platform breaks through the simple operation demonstration by teachers in the traditional classroom, and students' passive observation mode, and initially establishes the tasks of students' independent practice and simulation operation platform. The practice shows that this reform project not only effectively makes up for the shortage of course funds, completes the task of course reform and construction with high quality, but also provides students with personalized practical content learning, which facilitates and promotes students to master the corresponding knowledge and skills of mechanical drawing faster and better.

In practical teaching, it is suggested that the teachers and the teaching team, on the basis of understanding and assessment of the students, conduct graded and hierarchical teaching management with the students, and set up a variety of practical contents, modes, and multi-level operation schemes according to the practical links, such as course mapping and drawing for students to choose according to their own conditions. It is suggested that students also set up small teams to learn from each other's strengths and make up for each other's weaknesses in the practical operation process. Students are encouraged to compare and analyze the structure expression problems in mechanical drawing through various expression schemes. For example, on the basis of teaching the surveying and mapping assembly problems of the reducer in the drawing practice, guiding students to organize various parts of the data in the theoretical teaching, discussing the objections in the drawing expression, modifying and improving the defects in the practical operation method, and organizing the students to check each other. On the basis of mutual examination and evaluation, the teacher will comment on the common problems one by one, so that students can know what they are and why they are, and realize the rapid improvement of students' practical application ability.It is proposed to use advanced modern educational technology and means to broaden students' scope of knowledge, fully mobilize their enthusiasm and initiative, train their spirit of scientific exploration and innovation in the learning process, and improve their ability to analyze problems independently and solve practical engineering problems.

In order to achieve the goal of cultivating talents that are in line with modern engineering production and application, grasp the country's goal of "double first-class" construction of collaborative innovation personnel training, complete the training of skilled innovation personnel, and add mechanical drawing to the existing undergraduate teaching and training program Practical teaching hours are also very necessary. By increasing the class hours, the training intensity of students' drawing, reading, and other practical contents can be improved, and the professional ability and level of students' creative design and expression in mechanical parts design also can be improved. It is required that teachers of mechanical drawing should adopt flexible and diversified training methods in their teaching activities, teach in accordance with their aptitude, train innovative and applied talents under new engineering conditions and double first-class construction conditions, and inspire and cultivate students' scientific thinking and creative ability by combining skills training.

4 ADOPT AN OFFLINE FLIPPED CLASSROOM TO PROMOTE KNOWLEDGE ABSORPTION AND INTERNALIZATION, AND IMPROVE THE MAIN POSITION OF STUDENTS

Students still need teachers' guidance and help to understand and internalize the knowledge gained in network teaching. Therefore, the flipped classroom is the best supplement and improvement to online teaching. Flipping to break the traditional classroom teaching under the teacher's class to teach students knowledge, complete absorption and internalization of knowledge model, knowledge transfer, online learning is not bound by space and time, the internalization of knowledge is in the offline real classroom, students use class time to finish homework or practice content, teachers discover the problems of students, and counseling is targeted, better helping students to master knowledge and improve their ability [8]. This model can improve students' learning initiative, making students become the subject of learning and teachers become the students' mentors.

In the network classroom, communication is crucial, which is also the advantage of the network. Through communication, the internalization and absorption of knowledge can be improved, learning interest and investment can be enhanced, and effective utilization of educational resources can be improved. Formal synchronous learning classes at a fixed time and location can be established, and informal classes such as discussion boards, grouping areas, and reading groups can be established to meet the different needs of participants. Everyone here can be a transmitter or beneficiary of knowledge. In this method of communication, it can be one-to-one, one-to-many, many-to-one, many-to-many, and so on, making the whole network classroom active and becoming a dynamic and diversified online classroom with free speech, personality displays, interactive learning, and common improvement.

5 SUMMARY

In the situation of "internet plus," in order to improve the quality of talent training, colleges and universities need to re-examine and adjust the positioning of talent training objectives in the process of professional certification, which is bound to put forward new requirements for the teaching model. The case of sharing and construction of an engineering graphics course in the MOOC and SPOC mode analyzed and summarized the specific implementation of the cross-school teaching mode in the course teaching reform, so as to facilitate the smooth development of school transformation and professional certification. As mechanical drawing course teachers, they should also actively further master more up-to-date course content and related expansion materials, and actively explore new teaching methods for the graphics practice link while improving their own professional ability and quality, and explore the suitable implementation of student drawing, manual mapping, drawing, computer drawing, dismantling, and other practical links of the operation process. In the process of teaching, the teachers should actively communicate with students, improve teaching methods, and improve the quality of teaching construction of mechanical drawing course, especially in practice.

ACKNOWLEDGMENT

This research was supported by Education and Teaching Reform Project of Liaoning Province, China (No. [2018] 471 Letter from Liaoning Education Department) and Education and Teaching Reform Project of Northeastern University of China (No. [2018] 39 Teaching Documents of Northeastern University); Teacher Development Special Project of Northeast University of China (No. DDJFZ202005).

REFERENCES

[1] Z.H. Liu, X.B. Tu, 2016, Research and practice of the innovative teaching mode of mechanical drawing in internet plus era, J. Dezhou Univ. (04) 102–106.

[2] Z. Zhang, P.P. Yan, M. Zhou, Y. Li, Z.W. Wang, 2017, Design and practice of interactive teaching research project based on internet platform in double first-rate background, J. Archit. Educ. Inst. High. Learn. (05) 20–23.

[3] J. Chen, S.Y. Jing, 2015, A preliminary exploration on flipped classroom model of "engineering drawing" course in the internet + era, Sci. Educ. Art. Collect. (11) 51–53.

[4] C.F. Yu, 2019, Adhering to and responding to teaching in the era of "Internet plus" - research on the teaching mode of cartography course in the era of mobile Internet, J. High. Educ. (06) 113–116.

[5] X.F. Wang, H.Z. Mu, L.J. Xue, X.H. Niu, 2018, Thought on the teaching reformation of the engineering drawing combined with the internet plus based on the massive open online course, J. Graph. (03) 605–609.

[6] H.T. Cheng, G.X. Wu, 2017, Research and development of virtual experiment system of mechanical drawing in the "internet plus" era, J. Hanshan Norm. Univ. (06) 71–74.

[7] G. Chen, W.Q. Lan, C.Y. Yu, F.D. Yang, 2017, On the reform of the course mode of engineering drawing based on online education, J. High. Educ. (18) 119–121.

[8] Y. Yu, D.S Xia, B. Yuan, C.G. Sun, 2019, Exploration of the teaching mode of "Internet plus" mechanical drawing course based on blackboard platform, Educ. Mod. (6) 199–200.

Computational Social Science – Luo, Ciurea & Kumar (eds)
© 2021 Taylor & Francis Group, London, ISBN 978-0-367-70193-2

Research on the integration of entrepreneurship education and professional education in industrial design specialty

X.L. Ma
College of fine arts and design, Jinan University, Jinan, China

ABSTRACT: The fundamental goal of entrepreneurship education is to improve the quality of talents and enhance the innovative spirit, entrepreneurial consciousness and entrepreneurial ability of college students. Through the establishment of a teaching platform integrating various disciplines, a new curriculum system aiming at cultivating the ability of innovation and entrepreneurship is constructed, double-qualified teachers are introduced, and students' innovation and entrepreneurship are guaranteed institutionally. The talent training of industrial design major has become a bridge connecting engineering and liberal arts, engineering and art, and classroom education and off-class practice, so as to realize interdisciplinary crossing and optimize students' knowledge structure. Increase the success rate of entrepreneurship.

Keywords: Industrial design, Entrepreneurship education, Professional education.

1   RESEARCH STATUS ABROAD

Compared with Foreign Countries, entrepreneurship education in China starts late and develops imperfectly. Many people regard "entrepreneurship" as a kind of practice and neglect the cultivation of students' consciousness, psychological quality and ability.

China's Engineering Education Quality Report, released by the Ministry of Education in 2015, pointed out that engineering graduates are deficient in innovation, analysis and problem-solving skills in engineering, in the communication ability, team cooperation ability and other aspects to be further improved. Integrating entrepreneurship education into the training system of industrial design professionals is not only to train students' product design ability, but also to train students' market thinking, effectively integrate market, customer needs and social and cultural background into product design. This is not only a new model for the development of entrepreneurship education, but also an important means for engineering education to meet the market demand and improve the quality of personnel training.

Entrepreneurship education can be traced back to 1947, when Harvard Business School Myles Mace offered a "start-up Management" course for second-year MBA students, Management of New Enterprises. Over the next 30 years, entrepreneurship education was concentrated in business schools, with management courses as the vehicle, and mainly targeted at senior management talent. Until the 1970s, the unemployment rate in the United States remained high, and entrepreneurship education gradually went out of business schools and became a compulsory or optional part of the university's professional setting. Entrepreneurship education is defined by the United Nations Educational, Scientific and Cultural Organization as, "It is equally important for salaried people to develop pioneering individuals, as institutions or individuals are placing increasing emphasis on initiative, risk-taking, entrepreneurial and independent working skills, as well as technical, social and managerial skills, in addition to requiring success in their careers. ". The basic aim of entrepreneurship education is to improve the quality of talents and enhance the innovative spirit, consciousness and ability of college students.

Western entrepreneurship education has experienced nearly 70 years of development, and universities have become the main body of entrepreneurship education, and formed a set of Entrepreneurship Education Courses, entrepreneurship degree granting, entrepreneurship academic journals and Entrepreneurship Research Center in one perfect education system. In the developed countries of Europe and America, the idea and practice of entrepreneurship education have been incorporated into the process of professional education. The United States is leading the way in the integration of entrepreneurship education and Engineering Education. Under the profound influence of "academic capitalism" and "triple helix theory", to meet the market demand becomes the connection point between entrepreneurship education and Engineering Professional Education.

2 DOMESTIC DEVELOPMENT STATUS AND DEVELOPMENT STRATEGY

China's entrepreneurship education started relatively late, and after it germinated in the late 1990s, it experienced a transformation from entrepreneurship competition to entrepreneurship education, from teacher training to student training, from classroom teaching to various models, there are four stages of development: The transition from knowledge to liberal education. In the past 10 years, more attention has been paid to the research of college students' entrepreneurship education, entrepreneurship ability, entrepreneurship policy, training mode, etc., the research on integrating entrepreneurship education into interdisciplinary professional education and integrating entrepreneurship education with professional education mainly focuses on two aspects:

First, draw on the successful experience of American universities, and put forward the significance and path of integrating entrepreneurship education into professional education, such as Huang Zhaoxin, wang Zhiqiang's "on the integration of Entrepreneurship Education and professional education in colleges and universities" clarifies the relationship among innovation, entrepreneurship, entrepreneurship education and professional education, this paper introduces the three integration models of American entrepreneurship education and professional education: Magnet Model, radiation model and mixed model; Xu Xiaozhou and Zang lingling in the integration of entrepreneurship education and Engineering Education (2014, No. 4), this paper introduces the Educational Model of American Oulin Institute of Technology, which is called "Oulin Triangle", that is, on the basis of engineering education, entrepreneurial education and humanities and social art education are integrated Ruan Junhua puts forward the significance and path of the integration of entrepreneurship education into the training system of engineering talents in the article "the significance and path of entrepreneurship education for Engineering Students" (research on higher engineering education, No. 5, 2016). A practical study on integrating entrepreneurship education into specific professional education.

The second is the practical research on the integration of entrepreneurship education into professional education, which is represented by Jilin University Liu Yan, Yan Guodong and Meng Wei, based on the interdisciplinary resources of biology, pharmacy, medicine, Chemistry and Engineering, the idea of innovative entrepreneurship education is integrated into the undergraduate talent training program, innovative and pioneering education should be integrated into professional courses and professional practice teaching to cultivate innovative and pioneering talents in biopharmaceutical industry.

The industrial design emphasizes the user-centered design method, the product development procedure of multi-discipline and multi-specialty integration, and the entrepreneurial education is organically integrated into the training system of industrial design professionals, more conducive to professional education in an all-round, all-process to enhance the innovative entrepreneurial ability of students.

3 METHODS AND APPROACHES OF INTEGRATING PROFESSIONAL EDUCATION AND ENTREPRENEURSHIP EDUCATION

Integrating Entrepreneurship Education into talent training program and classroom teaching on the basis of industrial design specialty to improve students' ability of innovation and entrepreneurship in professional education. At the same time, we should strengthen the specialized knowledge and method training for students with innovative spirit and entrepreneurial potential, create a platform for them to stand out and improve the success rate of entrepreneurship.

The concrete implementation plan is as follows:

1. Divides the curriculum into the foundation, the specialized, the development 3 platforms. It has 6 modules: Humanities and Social Sciences, natural sciences, professional foundation, professional core, professional development and capacity development. Realizing multiple choices of compulsory and elective courses.
2. Optimize the students' knowledge structure, enlarge the basic courses of art for the students of industrial design of engineering, enhance their art theory accomplishment, set up sketch, constitution principle, design drawing expression technique and so on. To Guide students to acquire the necessary interdisciplinary thinking mode, the perspective of problem-finding and the ability of problem-solving in the process of interdisciplinary knowledge learning.
3. Introducing social forces to promote entrepreneurship education. Entrepreneurship education is practical and comprehensive, especially in entrepreneurship education contains a lot of tacit knowledge, the need for mentors to "teach by example", set out to establish a full-time and multi-disciplinary school professional teachers mainly, part-time team is entrepreneurs, entrepreneurs, entrepreneurial success, venture capitalists and so on, so that students have the ability to find change, the ability to form a team, the ability to integrate resources and other comprehensive quality.
4. To guide students to conduct market-oriented research, through the "Challenge Cup" entrepreneurial competition, robot competition, electronic design competition, engineering design innovation competition and other scientific and Entrepreneurial Competition for college students, in-depth exploration of creative, creative and innovative projects, and targeted support. To be supportive, to be mentoring.
5. To promote technology transfer and transfer of achievements by universities to the owners of entrepreneurial projects; to establish a reasonable credit system for innovation and entrepreneurship, establish a credit conversion system for innovation and entrepreneurship, and convert patents obtained and innovation and entrepreneurship practices into credits, so that independently developed products and other practical achievements can be used as part of graduation design; and to allow university students to leave school and start their own businesses while retaining their academic status, and to return to school whether they succeed or not; To set up scholarships for Innovation and Entrepreneurship Education, and to recognize outstanding innovation and Entrepreneurship talents and create a good environment for entrepreneurship during the implementation of the academic year awards and awards. The "tutorial system" gives undergraduates the opportunity to participate in scientific research activities, carry out scientific and technological innovation experiments, and strengthen the cultivation of undergraduates' scientific consciousness and innovation ability with good discipline conditions.

Through the above methods, we can stimulate more entrepreneurship based on technological progress and invention creation, and promote the transition of entrepreneurship education from business model to product development model. It is helpful to transfer the training goal of industrial design talents to application end and service end, face the market demand more, adapt to the market development and change, face the social concern and need, effectively improve the quality of talent training; It will help to guide more engineering college students with professional and technical ability to participate in entrepreneurship and effectively improve the overall success rate of entrepreneurship.

4 SIGNIFICANCE OF RESEARCH

First, to lead the atmosphere of "two pioneering", to promote the employment of college students and youth groups to improve the quality and success rate of entrepreneurship. Through a variety of activities, to create a communication platform to disseminate entrepreneurship, share entrepreneurial experiences, develop entrepreneurial ideas and find entrepreneurial teams.

Second, cultivate the contingent of teachers for Entrepreneurship Education. A team of entrepreneurship training teachers and a team of entrepreneurship guidance teachers. We should cultivate a group of key researchers and strive to publish a certain number of academic papers on entrepreneurship education in core journals every year, so as to raise the level of entrepreneurship talent development in institutions of higher learning.

Third, the establishment of regional entrepreneurship ecosystem. To give full play to the functions of the college as an incubator for production, study and research, through the integration of various local resources, including government agencies such as the Youth League committee, the People's Social Welfare Department and the Department of Education, financial institutions such as banks and investment companies, scientific research institutes in universities and various enterprises, we provide policy support, resource matching and guidance for the entrepreneurial team, and try our best to make entrepreneurship become the new growth point of regional economic transformation and sustainable development.

Fourth, entrepreneurship leads to employment. Through the relevant curriculum system to improve the overall quality of students and entrepreneurial ability to help with initiative, creativity and innovation. On the other hand, it also pays attention to the improvement of students' entrepreneurial skills and actual combat ability in order to promote their real entrepreneurship and create new jobs and wealth for society.

ACKNOWLEDGEMENT

Fund project: Shandong province teaching reform project "research and practice on the integration of college entrepreneurship education and professional education based on the conversion of old and new driving forces" (No. M2018X205).

REFERENCES

[1] Zou Jianfen. 2011, An analysis on the development and cultivation of college students' entrepreneurial ability. College education management, (5):91–95.
[2] Liu Jieming,Li You, Chen Xu. 2018, Mechanism of mutual promotion between innovation ability and patent ability of college students. Monthly journal of science and technology innovation, (29):33–34, 37.
[3] Yang Zhixiong. 2008, Research on influencing factors of college students' entrepreneurial power – based on the investigation and analysis of students in high schools and universities in zhangzhou region. Chuangxin and chuangye education, (9):80–86.
[4] Peng Zhengxia, Lu Genshu, Kang Hui. 2012, Influence of individual and social environment factors on college students' entrepreneurial intention. Research on higher engineering education, (4):75–82.
[5] Li Cunjin, Yan Yongjing, Yang Qing. 2013, Empirical analysis on factors influencing the formation of college students' innovative thinking ability. Journal of technology and economics, (32):29–35.
[6] Jiang Kaidong, Zhu Jianqiong. 2015, Research on college students' entrepreneurial orientation and university collaboration mechanism. China higher education research, (1): 54–58.

Computational Social Science – Luo, Ciurea & Kumar (eds)
© 2021 Taylor & Francis Group, London, ISBN 978-0-367-70193-2

Study of the practices of action learning for promotion of the innovation capability of college students

Y.X. Liu & P.B. Gao*
School of Economics and Management, Harbin Institute of Technology at Weihai, Weihai, China

W.W. Wu
School of Management, Harbin Institute of Technology, Harbin, China

Z. Li
China Railway Large Maintenance Machinery Co., Ltd., Kunming, China

ABSTRACT: College students are the major human resources for companies, and thus the development of their innovation capability is increasingly important. Based on action learning theory, this paper proposes several suggestions to leverage action learning to promote the innovation capability of college students, including optimizing of the action learning problem, the action learning group, the action learning facilitator, and the action learning process. The conclusions offer a refreshing perspective on how to employ action learning effectively.

1   INTRODUCTION

Nowadays, the changes to technology development and consumer demands are increasingly rapid, which requires firms to look for a competitive advantage for their survival. Innovation has been widely accepted as a critical means for firms to achieve a sustained competitive advantage because it provides firms with several strategic advantages such as decreased costs and increased quality [1]. In order to achieve successful innovation, firms should have high-quality human resources with a high level of innovation capability. College students are the major human resources for companies, and thus the development of their innovation capability is increasingly important [2, 3].

However, the literature to date has paid more attention to how to evaluate instead of how to develop the innovation capability of college students. Higher education plays an important role for college students pursuing professional careers [4]. While the traditional teaching approaches focus largely on letting students absorb programmed knowledge, they do little to prepare them for the innovative reformulation of that knowledge [5]. Letting students know how to fit the challenges they will face by using what they have learned flexibly is important for the improvement of innovation capability, and action learning settings are the most effective in which to learn [6]. Thus, action learning may have a close relationship with the innovation capability of college students. Based on this understanding, this paper proposes several suggestions for how to leverage action learning to improve college students' innovation capability.

After this introduction, the remainder of this paper is structured as follows. The next section reviews the related action learning literature. Section 3 proposes several suggestions from four aspects, including optimizing the action learning problem, the action learning group, the action learning facilitator, and the action learning process. The final section contains the concluding remarks.

---

*Corresponding author

2 LITERATURE REVIEW

From the traditional view, learning has been assumed to be the process of knowledge transmission, and after the transmission, students can put this knowledge into their own intentions [7]. The learning described in the traditional view is an external process, and learning from the teacher's own knowledge can be helpful it may also be insufficient if there are no opportunities for students to conceptualize this knowledge [8]. Under such circumstances, when college students embark upon their professional careers, they may find themselves disappointed because they realize that the acquired knowledge is less relevant or fails to translate into career advancement. In fact, traditional teaching approaches focus more on developing the capacity to recognize what is needed to attain the grade than on the creative thinking. In this way, students adopt surface learning with limited potential for the development of their innovation capability.

Action learning could identify the possibility for students producing new knowledge by themselves rather than just receiving it passively [9]. Action learning is an effective problem-solving teaching approach proposed by Reg Revans. Contrary to the traditional methodology, action learning offer a means of blending programmed knowledge and authentic experience for students to explore the creative space of their own [10]. In short, action learning is learning from specific problem through group discussion, trial and error, and discovery by discussing with other participants. From the definition, it can be seen that action learning is not only an individual work, but provides a flexible and systematic method for students to use their knowledge [11]. Because action learning is closely correlated with students' capacity to use their knowledge, it immediately contributes to the development of their innovation capability.

Zuber-Skerritt (2002) concludes the main characteristics of action learning, such as collaborating, being open, sharing ideas, experiential learning, reflecting on practice, and so on [12]. These characteristics indicate that the action learning involves an active process. When no one knows the solution, or when no one knows the way forward in resolving a complex problem, action learning can be used to obtain a greater advantage. By comparison with traditional teaching approaches, action learners are more likely to acquire more knowledge on the problem and how to resolve it [13].

3 THE PRACTICES OF ACTION LEARNING FOR THE PROMOTION OF INNOVATION CAPABILITY

3.1 *Optimizing of the action learning problem*

This problem is at the core of action learning, which provides students with the target of their learning [14]. Therefore, to start action learning, the teacher should identify the problem first, and further ensure the problem is related to the students' innovation capability. There are three criteria for teachers to find an appropriate question. First, the proposed problem should have no simple determinate solution or single correct answer. Second, the challenge produced by the problem should require students to go beyond their current knowledge domain, but should also not be so large or broad that there is little chance of students accomplishing it.

At the same time, the sense of relevance with innovation capability of college students is also a critical concern in choosing the problem. The proposed problem should require students learn to account for the complexities of real-world issues. Such a question can give the students access to more information, which encourages them to develop a broad understanding of the varied factors that contributes to the forming of systems thinking. The aim of the question should lead students to understand the influencing factors of the problem, the root causes of the problem, and the constraints that will affect how the problem can be resolved.

3.2 *Optimizing of the action learning group*

The action learning group typically consists of the students tasked with the problem, which may include part or all of the class members. To make action learning more effective, ensuring that there are group members with different knowledge backgrounds is important [15]. Therefore, the group members should be incorporated through careful selection. After selecting the appropriate participants, they should be given some basic information, such as the goal of action learning, the backgrounds of other participants, and the task they need to accomplish. If necessary, a instruction manual can be developed to help all participants take part in action learning.

To further support the drawing out of the group's diversity, the teacher should work on creating a culture that is supportive of trust, critical thinking, and accountability [16]. This kind of culture could help students to act freely and respond to the actions of others energetically, enabling students to interact with each other and add their diversity of knowledge to the group's discussion. Without such a kind of generative culture, the advantage that the group's diversity offers as a means of generating and exploring novel insights is greatly diminished.

3.3 *Optimizing of the action learning facilitator*

The action learning facilitator, typically the course teacher, is responsible for maintaining the group's focus on their learning through leading discussions. More importantly, a facilitator must focus more on fostering students' innovative learning through idea generation and appraisal of their action-taking rather than on telling students how to resolve the problem [17]. Although there is no common standard for the selection of the facilitator, there are some basic competence requirements, including possessing experience in the proposed problem, ability to resolve conflict, and the ability to guide students to discover deeper issues.

The other major function of the facilitator is ensuring that students remain focused on the problem rather than distracting from the learning process. This requires teachers to foster effective communication, collaboration, and coordination among group members. In traditional teaching approaches, the teacher's advanced knowledge of the problem usually provides her or his credibility to lead the course. However, in action learning, teachers must curb their enthusiasm to share what they know, and help students to involve group discussion. Therefore, being a facilitator requires that they can depart widely from their traditional higher education roles and responsibilities.

3.4 *Optimizing of the action learning process*

In action learning, the teacher should ensure that the process is in line with the aim of improving innovation capability. Therefore, in the early stages of selecting the problem, the teacher should explicitly discuss the probability of whether the problem can spark the students' critical thinking with relevant stakeholders. After selecting the problem, the teacher should provide students with some antecedent resources, such as textbooks, readings, and a schedule.

In addition, in the action learning process, the teacher should guide students to participate in the discussion of the problem, which is the central theme throughout the action learning. Through intense discussions, students are given the rich experience of the difficulty of integrating different opinions. In this case, students can learn from trying to recognize the possible weaknesses in their own thinking, and by possibly developing alternative solutions to develop their innovation capability.

4 CONCLUSION

Action learning involves the method of learning, and using this method to learn. The action learning process respects and builds upon each participant's independence, and it is less structured because it responds to the variety of college students and problems on which it draws. Therefore, action learning can be used as an effective teaching methodology to help students to conceptualize what they have learned, and it has quickly emerged as one of the most powerful and effective tools employed by colleges worldwide. Following this logic, we propose several suggestions, including optimizing of the action learning problem, the action learning group, the action learning facilitator, and the action learning process, to ensure that action learning fully play its role in the promotion of college students' innovation capability.

Although this paper presents some important points regarding the role of action learning as an important antecedent of the innovation capability of college students, it also suffers some limitations and could be extended in several ways. First, the practices provided by this paper should be tested by empirical research, such as a questionnaire survey or case study, in order to investigate the effect or the mechanism of the impact of action learning practices on the innovation capability of college students. Second, this paper only focuses on action learning itself, and does not take external environmental factors into consideration, however, action learning practices can be affected by the external environment. Future research is needed to advance optimizing of the action learning practices from an external perspective.

ACKNOWLEDGMENT

This research was financially supported by the Research Project of Postgraduate Education Reform in Harbin Institute of Technology, and the Research Project of Postgraduate Education and Teaching Reform in Harbin Institute of Technology (Weihai).

REFERENCES

[1] T. Anning-Dorson, Innovation and competitive advantage creation: The role of organisational leadership in service firms from emerging markets. Int. Market. Rev. 35 (2018) 580–600.

[2] P. Martín, K. Potočnik, A.B. Fras, Determinants of students' innovation in higher education. Stud. High. Educ. 42 (2017) 1229–1243.

[3] L. Fan, M. Mahmood, M.A. Uddin, Supportive Chinese supervisor, innovative international students: A social exchange theory perspective. Asia Pac. Educ. Rev. 20 (2019) 101–115.

[4] Z. Zhong, D. Hu, F. Zheng, et al. Relationship between information-seeking behavior and innovative behavior in Chinese nursing students. Nurse Educ. Today 63 (2018) 1–5.

[5] C. Brazee, D. Lopp, Innovative learning/learning innovation: Using action learning projects to develop students' industry mindset. Int. J. Innov. Sci. 4 (2012) 155–172.

[6] L. Baron, Authentic leadership and mindfulness development through action learning. J. Manage. Psychol. 31 (2016) 296–311.

[7] Y. Cho, T. Marshall Egan, Action learning research: A systematic review and conceptual framework. Hum. Resour. Dev. Rev. 8 (2009) 431–462.

[8] R.K. Yeo, M.J. Marquardt, (Re) Interpreting action, learning, and experience: Integrating action learning and experiential learning for HRD. Hum. Resour. Dev. Q. 26 (2015) 81–107.

[9] Y. Cho, T.M. Egan, The state of the art of action learning research. Adv. Dev. Hum. Resour. 12 (2010) 163–180.

[10] E. De Haan, I. De Ridder, Action learning in practice: How do participants learn?. Consult. Psychol. J.: Pract. Res. 58 (2006) 216–231.

[11] K. Jones, S.A. Sambrook, L. Pittaway, et al. Action learning: How learning transfers from entrepreneurs to small firms. Action Learn: Res. Pract. 11 (2014) 131–166.

[12] O. Zuber-Skerritt, The concept of action learning. Learn Organization 9 (2002) 114–124.

[13] M. Volz-Peacock, B. Carson, M. Marquardt, Action learning and leadership development. Adv. Dev. Hum. Resour. 18 (2016) 318–333.

[14] F. Sofo, R.K. Yeo, J. Villafañe, Optimizing the learning in action learning: Reflective questions, levels of learning, and coaching. Adv. Dev. Hum. Resour. 12 (2010) 205–224.

[15] K.S. Scott, An integrative framework for problem-based learning and action learning: Promoting evidence-based design and evaluation in leadership development. Hum. Resour. Dev. Rev. 16 (2017) 3–34.

[16] H.S. Leonard, M.J. Marquardt, The evidence for the effectiveness of action learning. Action Learn: Res. Pract. 7 (2010) 121–136.

[17] C.M. Leitch, C. McMullan, R.T. Harrison, Leadership development in SMEs: An action learning approach. Action Learn: Res. and Pract. 6 (2009) 243–263.

Computational Social Science – Luo, Ciurea & Kumar (eds)
© 2021 Taylor & Francis Group, London, ISBN 978-0-367-70193-2

Practice exploration of a three-dimensional tutorial system for cultivating applied talents in civil engineering under the new engineering background—a case study from a university in Zhejiang Province

X.F. Chen, Z.X. Zha* & S.S. Wu
Ningbo Institute of Technology, Zhejiang University, Ningbo, Zhejiang

ABSTRACT: In the current implementation of China's higher engineering education "New Engineering" background, the undergraduate tutorial system is conducive to promoting the change in the educational concept and the improvement of education quality, as a new interactive education teaching model and has great significance in promoting the reform of China's higher engineering education. Aiming at the problems existing in the existing tutorial system of undergraduate classes, this paper puts forward the model of a "vertical and horizontal crossing and flow-style three-dimensional" undergraduate tutorial system. This paper constructs the undergraduate tutorial system from multiple perspectives and conducts preliminary practice, by taking the civil engineering major of a university as an example, so as to provide a reference for setting up the undergraduate tutorial system of a civil engineering major in most universities in China.

1 INTRODUCTION

The core of constructing "New Engineering" education is to cultivate innovative talents who can adapt to and even lead future project development, who need to have some new quality and new capacity including innovation consciousness, global view, ecological consciousness, a large interdisciplinary engineering systemic view, lifelong learning ability, critical thinking, imagination and creativity leading the technical progress [1].

"New Engineering" is based on the new requirements of economic development, the international competition in the new situation and puts forward the new requirements on engineering education reform in our country It is different from "traditional engineering" using integration of subjects including the traditional humanities, social science, technology, management, law and economics. "New Engineering" education aims to foster practical, innovative and highquality engineering talentsfacing the future, and facing the world [2].

Civil engineering is characterized by strong applicability, regional characteristics and large social demand for talents. In order to better adapt to the development needs of the civil engineering era, professional talents conforming to the connotation requirements of "New Engineering" have been trained. Generally, the construction of civil engineering major is promoted from the following four aspects: (1) to reform the curriculum system construction based on industrial demand; (2) to promote the renewal of teaching methods with comprehensive ability as the goal; (3) to strengthen the reform of practical teaching with engineering application as the main line; and (4) taking talent training as the center and optimizing the construction of teaching staff. The core of the above construction ideas is inseparable from "teaching and learning" However, the traditional training mode is more limited to the teaching and learning between teachers and students in the first classroom. Therefore, from the student-oriented perspective, there is an urgent need to change

---

*Corresponding author

the mode of teaching and learning and establish a multiangle and multilevel teaching platform. As a new interactive teaching mode conducive to the change in the educational concept and the improvement of educational quality, the undergraduate tutorial system is of great significance in the reform of higher engineering education [3]–[8].

2 ANALYSIS ON THE CURRENT SITUATION OF THE COLLEGE-BASED STUDENT TUTORIAL SYSTEM

In the existing system of undergraduate mentoring, the tutor and counselor have a clear division of responsibilities. The latter is mainly engaged in student daily management and ideological and political education, and tutor guide students to be familiar with college life and learning environment as soon as possible, grasping the learning rule, setting up the correct world outlook, the outlook on life and values, as well as guiding students in mastering knowledge and professional skills. The system has achieved great results in colleges and universities in China, but there remain some problems.

(1) The lack of tutors and limited communication channels between teachers and students.

On the one hand, professional teachers in colleges and universities are facing the dual pressure of teaching and scientific research assessment while acting as tutors [9]. Teachers usually have no time other than for professional teaching and scientific research, which objectively affects the communication between teachers and students, and makes it difficult to give more ideological guidance and growth care to students, so that the interaction between teachers and students remains in a relatively limited "classroom space and time" In addition, the undergraduates in the first and second grades spend most of their time in common or basic courses, rarely having the opportunity to contact and communicate with professional teachers [10], and so the communication channels between teachers and students are limited. On the other hand, it requires a lot of time and energy for the teacher to observe students, and understand students and their needs.

(2) The lack of effective cooperation between counselors and tutors.

The orientation and responsibilities of college tutors and counselors are unclear, and there is a lack of communication between professional teachers and counselors, as well as a lack of unity and cooperation in education, and even the phenomenon of inconsistent educational caliber [11].

(3) The training of civil engineering applied talents lacks effective vertical phased practice guidance.

Under the current tutorial system, the classes managed by each class tutor are parallel natural classes. The connection between students of different grades only occurs in some associations that a few people participate in, and there is a lack of stable and effective vertical communication channels. Good traditions and experiences of senior grades cannot be effectively transmitted to junior students. The training of Civil Engineering applied talents need more longitudinal stage of planning and guidance, such as learning professional knowledge and training learning habits of freshmen or sophomores, mastering the skills, qualification test, and career planning in junior and senior years, engineering practice, subject contests, participation in scientific research, graduation design and graduation practice, innovation and start up business link throughout the different stages.

3 CONSTRUCTION OF THE "VERTICAL AND HORIZONTAL CROSSING, FLOWING WATER THREE-DIMENSIONAL" TUTORIAL SYSTEM

As shown in Figure 1, in view of the civil engineering applied talents training target, a "Vertical and Horizontal Crossing, Flowing Water Three-Dimensional" tutorial system characterized by discipline characteristics and the high management efficiency has been built. This system can

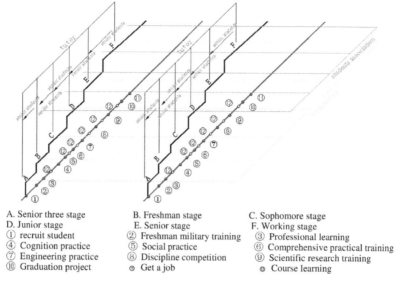

A. Senior three stage
D. Junior stage
① recruit student
④ Cognition practice
⑦ Engineering practice
⑩ Graduation project

B. Freshman stage
E. Senior stage
② Freshman military training
⑤ Social practice
⑧ Discipline competition
⑪ Get a job

C. Sophomore stage
F. Working stage
③ Professional learning
⑥ Comprehensive practical training
⑨ Scientific research training
⑫ Course learning

Figure 1.   Horizontal and horizontal cross-flow system.

deepen professional education and transmit practice experience. The counselors, tutors and out-standing senior students work together and guide students with a flow-type in different stages. The educational goal characterized by whole-staff education, whole-process education and all-round education can truly be achieved.

In this three-dimensional system, there are three type of management teams including vertical management teams consisting of professional tutors, outstanding senior students (class assistants) and other undergraduates, horizontal management teams consisting of counselors and their subor-dinate associations and longitudinal management teams consisting of senior high school students, all students in undergraduate stages and the graduate trainee.

The system has the following characteristics.

(1) By the two-way selection between tutors and students, a class is formed and the students may join or quit the class freely at any time. The peer relationship between undergraduate students of different grades is more conducive to obtaining the potential and subtle influence of knowledge, the transfer, help and guidance of ability and good quality. The flow-type management level is more conducive to the completion of phased learning tasks.

(2) The new tutorial system can improve greatly the work efficiency of the tutors and counselor staff without affecting the education training objective of the students, which frees professional teachers and staff members from the heavy workload of day-to-day management and repetition of professional guidance, thus they can put more effort into the students' innovative education, teaching and research activities.

(3) Based on this tutor working mode, a more open and application-oriented talent training auxiliary information platform consisting of not only college teachers and students but also other teaching assistants, teachers in high school, graduates and other professional persons can be built. This platform can provide more information covering the students' recruiting, training, taking part in the entrance exams for postgraduate schools and obtaining employment, etc.

4   PRACTICAL EXPLORATION OF THE SYSTEM OF VERTICAL AND HORIZONTAL CROSSING, FLOWING WATER THREE-DIMENSIONAL TUTORIAL SYSTEM

On the basis of the system described above, combined with the existing class tutorial system, this paper carries out an active pilot exploration. First, the undergraduate students majoring in civil

Table 1. Pass rates of some courses.

Course name	Pass rate (Grade 2017)	Pass rate (Grade 2018)
C programming fundamentals	86%	87%
Theoretical mechanics (C)	85.5%	88.5%
Calculus (I, II A)	66%	68.5%
Linear algebra (A)	85.5%	88.9%

engineering of class 2018 at our university were selected as the practical subjects to establish the academic tutorial system. For the freshman, the main task of the tutor is to guide the students to adapt to university life and study as soon as possible, and to have a correct understanding of the major. For the sophomore and junior, the main task of the tutor is to focus on the students' knowledge learning and ability training, especially the training in innovation ability. For the senior, the main task of the tutor is to conduct career counseling, academic research and technical development ability training.

Furthermore, practice exploration was carried out from the aspects of the selection of tutor qualifications, the two-way selection between teachers and students, the responsibilities and work requirements of the tutor, the management and assessment of the tutor, and the specific implementation arrangements of the tutorial system, etc. The leading group of the academic tutorial system for undergraduates majoring in civil engineering was established as responsible for the organization and implementation of various works. The working group is composed of the leaders of the civil engineering departments, the key members of the Institute of Geotechnical and Municipal Engineering and the Institute of Structural and Bridge Engineering. The first tutor system consisted of 89 undergraduate students majoring in civil engineering in the 2018 grade, and 24 tutors. The second tutor system consisted of 169 students in the civil engineering undergraduate class of 2019 grade (including engineering management), and 41 tutors.

After nearly a year's implementation of the new tutorial system, it has achieved obvious positive effects in the aspects of discipline competition, the passing rate of main courses and the major transfer.

(1) Discipline Competition: Discipline competition can effectively reflect the "handinhand" peer-type learning effect referred to above. Through the effective combination of different grades and the careful guidance of teachers, the students of the Civil Engineering College have achieved remarkable results in various competitions in 2018 and 2019 after the implementation of the new tutorial system.
In 2018, the Civil Engineering College students were selected as the only finalist of the 11th Zhejiang Province "challenge cup" college student entrepreneurship plan competition and won the silver prize. In 2018, civil engineering students won the first, second and third prizes in the first undergraduate digital construction and engineering management innovation competition in Zhejiang Province. In 2019, the student team won the first prize of BIM5D cost management in the national BIM application skills competition. In 2019, students from the Civil Engineering College won the third prize in the third national college student "Mao Yisheng charity bridge – small bridge project" design competition of the "Huaxi design cup" In 2019, civil engineering students won the third prize in the first national intelligent construction and management innovation competition of college students.

(2) The Passing Rate of Main Courses: A data comparison of the pass rate of major courses of civil engineering major in class 2018 (carry out academic tutorial system) and that of major courses of civil engineering major in class 2017 (not carrying out an academic tutorial system).
As can be seen from Table 1, since the implementation of the tutorial system, the pass rate of civil engineering students of grade 2018 has gradually exceeded that of grade 2017. Among them, the pass rate of a professional basic course called Civil Engineering Drawing was very high, even the proportion of 90 points above increased from 38.3% in 2017 to 66.67% in 2018.

(3) Professional Recognition: since the implementation of the tutorial system, the number of civil engineering students who changed their major has been on the decline, and the degree of professional recognition of freshmen has been greatly improved through the communication with tutors.

(4) Questionnaire Investigation: the questionnaire for the implementation of the academic tutorial system for undergraduate students majoring in civil engineering of grade 2018 was issued and the statistics were completed. There were 100 copies of the questionnaire distributed and 80 were effectively recovered.

According to the statistical results of the effective questionnaires recovered, it is shown that: more than 91% of the respondents had some understanding of the new tutorial system for undergraduates majoring in civil engineering which had been implemented for about one year. At the same time, nearly 77% of the respondents believed that it was necessary to implement the new tutorial system among undergraduates majoring in civil engineering to strengthen the guidance of the whole process of the cultivation of applied talents.

In the survey of who has the greatest impact on their professional learning, nearly 57% of students choose the senior students or class assistants, while less than 20% of students choose the traditional class tutors. Similarly, more respondents believed that under the influence of upperclassmen and classmates, professional recognition is established and gradually strengthened by participating in discipline competitions, scientific research projects, innovation and entrepreneurship competitions and other practical links.

In the survey of the role of tutors in the new tutorial system for undergraduates, 37–58% of students hoped that tutors can provide more help in the four aspects of thought guidance, course learning guidance, career planning and innovation ability cultivation.

In the survey of the shortcomings of the existing new tutorial system, 63% of respondents believed the new tutorial system still needs to be further improved, especially 53% of the respondents believed that the role of the new tutorial senior students needs to be further enhanced, and the learning tasks should be completed by forming a team. At the same time, more respondents wanted greater communication between tutors and students.

5 CONCLUSION

(1) The traditional class tutorial system has obvious deficiencies that affect the construction of civil engineering majors under the new engineering background. In this paper, the vertical and horizontal crossing and flowing-water three-dimensional tutorial system is constructed to focus on the construction of a multi-information linkage platform between teachers and students. It puts forward the "trinity" of counselors, tutors and tutorial senior students providing guidance according to different stages, so as to truly realize whole-staff education, whole-process education and all-round education, and has achieved positive results through the preliminary practice.

(2) The key to the continuous improvement and construction of the new tutorial system is the combination of the "responsibilities, rights and obligation" of counselors, tutors and tutorial senior students" Clearer responsibility and assessment objectives for the tutor should be defined and the corresponding rights and benefits should also be stipulated for tutors to improve their initiative.

(3) The new tutorial system needs to be constantly improved in practice, especially to further enrich the guidance forms of tutorial senior students and meanwhile ensure full communication between the tutors and students simultaneously

(4) The platform construction based on the teacher–student information model is the cornerstone of the future construction of "new engineering" and the construction of a new vertical and horizontal and flowing three-dimensional tutor system is an important part of it.

REFERENCES

[1] Deng-Hua Zhong. 2017, The new connotation of the construction of the engineering and operation. Journal of higher engineering education research, (3): 1–6.

[2] Yang LingMing, Zhang Pinemin, Zhou Ji. 2019, Exploration on the construction of new engineering civil engineering talent cultivation model. Science education research, (5): 96–102.

[3] Hong JunQing, Wu Kun, Yuan XiaoPing, et al. 2019, Construction of civil engineering application-oriented personnel training system under the new engineering background. Economics, management, review, (11): 151–152.

[4] Zhang JiaChao, Jiao LiangBao. 2019, Exploration on the model of "multi-linkage" tutorial system in application-oriented undergraduate colleges. Education modernization, (70): 112–113.

[5] Zhong WeiHui, Shi QingXuan, Huang Ying, et al. 2019, Multi-mode integration and scheme design of undergraduate tutorial system in civil engineering. FORUM, 20–22.

[6] Tang HanQi. Effects, 2019, problems and solutions of implementing the undergraduate tutorial system in Chinese universities. Chongqing higher education research, (4): 98–109.

[7] Li JunBo, Liang LiJuan, He YuXin. 2019, Exploration and research on the mode of undergraduate multi-level whole-course tutorial system. Faculty construction, (22): 58–59.

[8] Tang ShuFeng, Li Jing, He XiaoDong. 2019, Construction of an application-oriented innovative talent cultivation model of undergraduate tutorial system integrating "politics, education, production, learning, research and application". Journal of higher education, (16): 51–53.

[9] Cui Lei, Mao JiangHong, Ge XiaoDan. 2016, Practical research on young professional teachers in colleges and universities as class tutors in talent cultivation.Contemporary education theory and practice, (1).

[10] Gan DaQin. 2015, Research on the work of full-time teachers and class tutors from the perspective of "big ideological and political" – a case study of Fujian Institute of Engineering. Journal of Fujian Institute of Engineering, (10).

[11] Wang ShiYong, Li HongXin. 2016, Construction of collaborative education mechanism between tutors and instructors of specialized classes in colleges and universities. Journal of Anshan Normal University, (6).

Computational Social Science – Luo, Ciurea & Kumar (eds)
© 2021 Taylor & Francis Group, London, ISBN 978-0-367-70193-2

Problems and solutions of online learning under an epidemic situation

Y. Ma, Q.Z. Yan & J.H. Tian
Zhejiang University of Water Resources and Electric Power, Hangzhou, China

ABSTRACT: Since the outbreak of the epidemic at the start of this year, colleges and universities have responded positively to the call of "Suspension of classes and non-stop learning" by the Ministry of Education, such as primary and secondary schools, and have adopted online teaching methods to organize teachers and students to teach and study. This paper mainly introduces an analysis of the main problems in online teaching mode under this background, and proposes the SPOC online course mode of "DingTalk+Chaoxing," that is, using the Chaoxing network teaching platform to customize courses, analyze learning situations and course content, innovate online teaching design, and teaching organization. With this system teachers lead students to learn online and offline actively and then achieve good learning results.

Keywords: Online learning, Live broadcast, Online course, DingTalk, Chaoxing, XueXiTong.

1 INTRODUCTION

The sudden outbreak of the novel coronal pneumonia virus before the Spring Festival this year was ravaging the whole of China. Under the leadership of President Xi Jinping, the Chinese people of all walks of life have made concerted efforts to resolutely win this epidemic prevention and control war. In order to solve the problem of being unable to go to schools, the Ministry of Education issued a notice of "Suspension of classes and non-stop learning" on January 27, calling on schools to conduct online teaching.

After the notice, education departments and universities around the world started to prepare and carry out "online teaching" so that students could have classes at home. Some teachers gave lectures live such as QQ, some used the network platform for online teaching, and others used WeChat to publish homework, and so on. Online teaching and "Suspension of classes and non-stop learning" have become a general trend [1–3] during this epidemic.

Not only did primary and secondary schools adopt the online teaching mode, many colleges and universities also started online teaching afterwards. The greatest impact of the epidemic on students was the shift from studying at school to studying at home. Therefore, how to further enable "effective learning" at home for students is a problem worth pondering. Based on this, this paper first analyzes the factors that affect online teaching of university teachers and home-based learning of college students, puts forward the home-based learning mode of online teaching of SPOC (Small Private Online Course) based on teaching classes, and then shares a practical exploration experience of the multiple existing online courses. It may provide a feasible way for peers to improve the quality of home-based learning of their students.

2 PROBLEMS IN ONLINE LEARNING

2.1 *Poor learning environment and unstable network*

College students live in different places with different networks and learning environments, which leads to different online learning effects. The network configuration in each student's home is different. Students in large cities have better overall network configurations, while some students in rural or remote areas have poorer network configurations, and some can only watch live broadcasts by using mobile phone traffic. Some families can provide a quiet learning environment, while others can only provide a noisy learning environment. Obviously, a quiet learning environment with a good network configuration is more suitable for watching teachers live online.

At present, teachers use many live broadcast platforms, such as DingTalk, Tencent classroom, Mosoteach, Chaoxing, KeTangPai, Cloud classroom of Netease, and so on. However, whether it is the live broadcast platform recommended by universities or chosen by teachers, network congestion occurs at different time stages. The main reason for this is that the corresponding teaching platforms did not expect such a large number of simultaneous online viewers, and their original equipment and technology are insufficient to support the existing application requirements. Although they are adding more and more equipment, there are still occasional network crashes.

Students have access to different learning equipment. Ordinary college students are equipped with personal computers and can learn online. However, there are still a few students who do not have computers. For example, some students thought the winter vacation would be short and left their computers at school. Some students left their computers at home in cities and towns and returned to their hometown in the countryside for the Spring Festival, but were unable to return to the cities and towns because of the epidemic. Some students had computer problems that could not be repaired. Although it is possible to watch a live broadcast on mobile phones, different mobile phone configurations also affect the online viewing experience. For situations requiring the use of computers for exercises, homework, and experiments, the students without computers cannot complete the tasks far from better learning effects.

2.2 Teachers' discrepant capability of online teaching

During the epidemic, the main teaching locations of university teachers were changed from the classroom to the home [2]. The vast majority of front-line teachers have no experience in online teaching creating many psychological worries and anxieties. Also, not all teachers are willing to accept online teaching. At present, teachers can be roughly divided into three categories: (1) backbone teachers with strong online teaching ability; (2) young teachers with relatively strong acceptance but have difficulty participating in online teaching due to multiple unfavorable conditions; and (3) teachers who are in experiencing job burnout and those who are not interested in online teaching and do not accept and adapt to online teaching.

For the first kind of teachers, they are the leaders of online teaching during the epidemic. Some of them may have had relevant teaching experience before. For example, some teachers are the main leaders of national and provincial top-quality courses. They have also taken some online videos and interacted with students frequently and actively online. They have good teaching design and organization, so their teaching effect is the best. The second kind of young teachers find it easy to accept and adapt to new things, and dare to try and innovate. However, due to the constraints of the home network environment and teaching equipment, some teachers also find it difficult to achieve good online teaching results. The third kind of teachers contradicts this online teaching method subjectively and only do perfunctory things, so their teaching effect is the worst.

2.3 Students' poor ability at autonomous learning

Due to the differences in learning ability, attitude, motivation, personality characteristics, and learning habits of each student, not all students are willing to carry out online learning. Especially in the first one or two weeks of live broadcasting, some students had a psychological resistance to this form of class because it was still during the winter vacation. As the epidemic continues and online classes become the norm, this mentality has improved.

It is not enough to simply change the teaching location, such as from classroom teaching to online teaching.. If the teaching design isn't changed fundamentally, some teachers still repeat what the textbook says, and it is difficult to attract the attention of students. With the extension of the epidemic, most courses have adopted online teaching methods. Without paper textbooks and related reference books, the efficiency and effect of doing corresponding exercises, homework, and experiments are also reduced due to the inconvenience of finding data. Students need to stare at the screen during the live broadcasts of various classes and they also need to watch the playback in order to complete tasks such as homework and experiments after class. Watching for a long time can easily make students experience visual and study fatigue. Therefore, many students hold a negative attitude toward this online learning mode.

In addition, the poor ability of self-management and self-supervision leads to the students' weak ability of autonomous learning. Teachers can urge students to adjust their learning attitude and maintain a good learning state in classroom teaching. During online teaching, although teachers remind students, the effect across the screen is poor. Therefore, if there is no scientific planning and effective supervision, it is difficult to achieve the ideal effect of autonomous learning.

3   MOOC AND SPOC COURSES

The network teaching platform is a basic technical platform for online learning, and is a necessary condition for carrying out network teaching or network-assisted teaching. The rise of MOOCs (Massive Open Online Courses) cannot be separated from the support of online teaching platforms. The three major MOOC operating institutions in the United States are Coursera, Udacity, and edX. They did not use the current typical online teaching platform, but developed their platforms independently. After more than 10 years of in-depth research and development, the function and performance of a typical online teaching platform is far superior to that of the MOOC platform, which has just started. In particular, it provides rich tools for teachers to independently build and control online courses, and can meet the management needs of different teaching institutions [1]. MOOC's disadvantages cannot be ignored. The outstanding one is the high dropout rate. There are many reasons for this phenomenon, but the main reason is that there are no restrictions on the enrollment conditions, the number of people, the cost, and so on. In addition, it's difficult to manage MOOC courses due to a large number of students. Even with the Internet, it is difficult to achieve better results of the tutoring for a large number of students due to time and space constraints.

SPOC is a new concept put forward by Harvard University after MOOC, which is called "Post MOOC" (). SPOC=MOOC+Classroom is a mixed teaching method commonly used in elective classes. Unlike MOOC, SPOC has restrictions on the number and conditions of admission, but it is still open and free [3]. It has the characteristics of a small number of people, school registration (charging), more effective and convenient teaching management, more accurate guidance for students, and so on.

Compared with large MOOC courses, in the case of home teaching, SPOC online courses based on classes are an effective online teaching mode under the current epidemic situation, which is convenient for effective management and improvement of the teaching effect. Therefore, teachers can use the network social software to create a group based on the teaching class, use the online teaching resources of the network teaching platform, or directly use the network platform to create courses belonging to their own teaching class, and set and regulate the progress, rhythm, and scoring system of courses according to their own preferences and the needs of their students.

4   SPOC ONLINE COURSE BASED ON "DINGTALK+CHAOXING"

During the epidemic, most online teaching courses have adopted the way of "live broadcast + a network teaching platform." As mentioned above, there are many kinds of live broadcast software. After testing with my colleagues, based on Ali's strong technical strength and equipment, the DingTalk live broadcast platform is sufficient to support a certain number of online live broadcast viewers. And the biggest advantage is support playback, which is convenient for students to view the contents they do not understand at any time after class. Chaoxing (Fanya) is a well-known and large online teaching platform in China, which cooperates with most universities in China. After docking with the educational administration management of colleges and universities, teachers can use this platform to create personalized SPOC online courses.

The SPOC online course mode of "DingTalk + Chaoxing" can be used to solve the three main problems of online learning analyzed in previously.

4.1 Solutions to online teaching problems

4.1.1 Using the platform to customize courses

The equipment of the platform needs to be increased and the technical strength needs to be strengthened as soon as possible. From the perspective of subject teaching and receptor learning, relevant suggestions and requirements are put forward for the platform institution. On the basis of existing functions, platforms should provide more online related resources, evaluation functions, social tools, display platforms, learning management and monitoring, recording teaching and learning processes, big data analysis, and other technical support.

Since most teachers are not familiar with the operation and application of the platform, especially the relevant technical personnel are required to provide teachers with multidimensional online training and guidance, so as to help them better build online courses and better use online interactive tools during live broadcasts.

In addition, it is necessary to advance a proposal and develop a platform more in line with the users' habits from the perspective of teaching and learning, from the teaching rules, teaching tools, learning habits, learning rules, learning ability, and other aspects.

4.1.2 Innovating online teaching design and teaching organization

After the analysis of each type of students' learning ability, knowledge base, cognitive style, and other characteristics, teachers should set teaching objectives, create problem situations, design teaching tasks, integrate teaching resources according to the nature of the curriculum, curriculum content, and curriculum requirements.

According to different course contents, teachers should choose appropriate teaching methods, organize students to have discussions, supervise students' activities, interact with students, answer questions, and solve doubts, and finally combine the platform to summarize, reflect, evaluate, and exchange experience.

Teachers need to get students' personalized reports from big data and provide personalized tutoring, so as to dynamically improve the teaching mode and learning planning. The way of such feedback, tutoring, and improvement can effectively prompt the teaching effect.

4.1.3 Teacher-led online learning model for students

Teachers should explain the status and basic contents of the course and define the learning objectives in the first class. Before each class, teachers set up questions or tasks according to the content and ask students to search for relevant materials. The teacher and students mainly discuss learning problems, and students learn cooperatively and display communication in class. After class, students should make a reflection summary or exchange experiences to form a phased and comprehensive evaluation combining self-evaluation, mutual evaluation (if it is through cooperative learning), and teacher evaluation.

Besides the guidance, assistance, and supervision of teachers, the more students set goals, monitor and manage learning online and offline autonomously, the greater the effect of home-based learning that can be achieved.

4.2 Practice of SPOC online course

The author applies the SPOC online learning mode of "DingTalk + Chaoxing (XueXiTong)" in the course of "SCM principle and interface technology" and "Computer configuration and maintenance."

Before the first class of the courses, the teacher informs the learning situation of the class and the students, and designs classified teaching resources according to the learning habits of different learners. The teacher also records a short video that can be watched repeatedly and mainly explains knowledge points. The teacher also puts the courseware, homework, designed questionnaire and short test, and other interactive resources on the teaching platform.

Then, the teacher builds courses, uploads various learning materials, assignments, and tests on the Chaoxing teaching platform. The teacher broadcasts live, interacts with students in class, and answers questions online using DingTalk. After class, the students watch the live broadcast through DingTalk, and download the Chaoxing's XueXiTong APP through the Chaoxing web platform or mobile phone to submit questionnaires, homework, and short tests. The teacher obtains a statistical analysis of the learning situation through Chaoxing (XueXiTong) as the main basis of the usual process evaluation. In this way, "teachers' online lectures + teachers' and students' online interaction" + "students' online autonomous learning after class" consolidate the corresponding content in each chapter to achieve a virtuous cycle of teaching and learning.

5 CONCLUSIONS

The learning characteristics of students are diversified. Whether the above online teaching mode design is effective depends on the understanding and analysis of students' learning situation. How to properly and accurately locate students' learning and accepting online learning methods is of vital importance. Therefore, the analysis and classification of students' characteristics is one of the focuses and difficult points of online learning. In the process of practical teaching, the project teaching method is adopted, and the students are divided into groups. Thus the learning effect is evaluated according to the division of labor and the quality of its completion.

In addition, various incentive mechanisms to encourage learning are adopted in the teaching process. Theoretically speaking, the more teaching resources are classified, the better each student learns, and the more diverse the evaluation methods are. However, in practice, it is very difficult to achieve this classification. Especially when the number of students is large, it will increase the workload of teachers, and it will be difficult to ensure the objectivity of the evaluation with the increasing workload. How to ensure the objectivity, fairness, and rationality of the evaluation of the learning effect is a difficult problem, which is also the focus of online learning research. Therefore, in the specific implementation process, the first thing to do is a questionnaire survey on students' learning types. After analyzing and adjusting slightly the obtained results, teaching resources should be reasonably classified to the greatest extent possible and evaluation methods should be set reasonably according to the types of students.

At present, some provinces have controlled the epidemic situation well, without increasing cases for several consecutive days, and are assessed as low-risk areas. After meeting the relevant conditions, schools can open. However, most of the provinces that do not meet the requirements still use online teaching. The SPOC online learning mode based on "DingTalk + Chaoxing" shared by the author in the epidemic situation, although there are still some problems such as some students' lack of enthusiasm for learning, occasional problems on the Internet, being not as good as face-to-face teaching, helping students solve problems, and other deficiencies. But it is still a reference online learning mode in the current epidemic situation, and the empirical teaching effect is recognized by most students. Through such practical experience, the author hopes it will be helpful to other teachers who are also teaching online currently.

ACKNOWLEDGMENT

This research was financially supported by Zhejiang University of Water Resources and Electric Power Foundation (2019JG16).

REFERENCES

[1] Zhu Zhiting, Peng Hongchao. 2020, Omnimedia Learning Ecology: A Practical Solution to Cope with Schooling Difficulties during a Large-scale Epidemic. China Educational Technology, 3:1-6.

[2] Song Lingqing, Xu Lin, Li Yaxuan. 2020, Precision Online Teaching + Home Study Model: a Feasible Way to Improve the Quality of Study for Students during Epidemic. China Educational Technology, 3:114–122.

[3] https://baijiahao.baidu.com/s?id=1661049774621798935&wfr=spider&for=pc

[4] Zhu Zhiting, Liu Mingzhuo. 2014. New Trends of Online Learning in the "Post-MOOC Era". Open Education Research, 20(3):36–43.

Computational Social Science – Luo, Ciurea & Kumar (eds)
© 2021 Taylor & Francis Group, London, ISBN 978-0-367-70193-2

Exploration of mixed teaching based on MOOC in road and bridge construction organization technology and management course teaching

W. Li*, X.C. Wang & Y.H. Li
Shenyang University, Liaoning, Shenyang, China

ABSTRACT: The mixed teaching mode and the emergence of massive open online course have triggered a new round of theoretical and practical exploration of classroom teaching reform world-wide. In view of the problems existing in the teaching of Road and Bridge Construction Organization Technology and Management, this paper combines the advantages of classroom teaching and massive open online course platform learning, and puts forward a series of reform measures in teaching contents, teaching methods and practical teaching. The research results have certain guiding significance for further improving the quality of professional teaching.

Keywords: Construction Organization and Management, Mixed Teaching, MOOC, On-line and Off-line Integration.

1 INTRODUCTION

The course of road and bridge construction organization technology and management is a professional course with strong technology. The teaching quality of the course directly affects the quality of personnel training. Through several years of teaching practice, the traditional teaching mode has not adapted to the needs of modern enterprise talents, and a new mixed teaching mode must be established to meet the needs of enterprise talents.

2 MIXED TEACHING MODE BASED ON MOOC

Blended teaching is a comprehensive teaching form that combines traditional classroom, mixed learning and flip classroom whose purpose is to cultivate students' cooperative learning and autonomous learning ability through teaching activities. Students interact anytime and anywhere through learning environment, teachers and resources. Students can independently increase learning time so that to improve teaching quality. Teachers can teach in large classes in traditional classes to improve teaching efficiency. Teaching activities are carried out under the premise of combining classroom and virtual teaching environment to ensure students' autonomous learning and provide students with cooperative learning.

The essence of the mixed teaching method based on MOOC is to rely on MOOC service platform to realize the integration of online and offline. In short, it is online learning and offline interaction. This kind of teaching method can fully realize students' personalized learning, break through the limitation of time and space, and is more conducive to the organization and management of the teaching process. It will change the teacher-centered teaching method into the student-centered teaching method, and the students will change from passive acceptance to active inquiry. In order to make full use of the advantages of network MOOC teaching, the mixed learning combining "network open courses + traditional face-to-face teaching" has become a hot topic.

---

*Corresponding author

The promotion of mixed teaching is of great significance to promote the personalized teaching reform centered on learners, accelerate the implementation of the new talent training mode of "trinity" of value shaping, ability training and knowledge imparting, which is of great significance to cultivate students' active learning ability.

3 ANALYSIS OF THE TEACHING STATUS AND ORIENTATION OF THE COURSE OF ROAD AND BRIDGE CONSTRUCTION ORGANIZATION TECHNOLOGY AND MANAGEMENT

In the curriculum system of this major, the course of road and bridge construction organization technology and management is a practical course, and the course teaching should be close to the actual engineering and professional development trend. Its positioning is: On the basis of theoretical courses and practical teaching of Road and Bridge Engineering Construction Technology and Professional Practice, after the students have mastered the Principles and Methods of Concrete Structure Design, Roadbed and Pavement Engineering, Bridge Engineering and other courses and the theoretical knowledge of construction technology. Through learning the principles and methods of construction organization design compilation, master scientific management methods such as network planning and running water construction, and through the training of curriculum design and graduation design, the goal of being able to engage in the compilation of common engineering construction organization design is achieved.

At present, in the course teaching, the teaching content and teaching form are lagging behind the development of the times, and the teaching material content lags behind the standard, which cannot meet the requirements of the development of highway construction in our country. The introduction of engineering cases about unit engineering construction organization design in the existing teaching materials is relatively scarce. Due to various reasons such as the lagging development of disciplines, the practice teaching links in many colleges and universities are relatively weak, some colleges and universities lack fixed practice bases and practice cooperation units, and the construction of training rooms is relatively backward. As a result, professional practice teaching has always attached more importance to theory than practice. Students have less contact with engineering cases in Road and Bridge Engineering and cannot establish perceptual knowledge.

A considerable proportion of the graduates majoring in Road and Bridge in application-oriented undergraduate universities are engaged in the construction management of the project site. Solid construction theory foundation and project management ability are the basic requirements for students to be competent for work. Therefore, the focus of this curriculum reform should be to combine the teaching process closely with the actual engineering project cases. Imitate the actual working process, strengthen the training of transforming the theoretical knowledge of construction into practical application ability, so that students can be good at compiling practical construction plans according to the actual engineering situation, and carry out the formation of school-enterprise teaching teams and the reform of teaching methods and teaching means according to this idea.

4 IMPLEMENTATION OF MIXED TEACHING PROCESS BASED ON MOOC

4.1 *Reform of teaching content*

The teaching of "Road and Bridge Construction Organization Technology and Management" course should focus on the development requirements of training applied talents and pay attention to the practicability of teaching contents. The selection of teaching materials is very important. It is necessary to basically cover all teaching contents, to select some national excellent teaching materials, and to keep pace with the times and update or replace them in time. In the course reform, attention should be paid to strengthening the application of computer software, introducing BIM and other computer software technologies in due course to assist in the compilation of construction schemes. In the teaching, students are required to skillfully use AutoCAD to draw drawings, use schedule software Project to draw construction schedule, use PKPM construction module and Pin Ming construction software to establish building models, and assist in site plan design and safety calculation of formwork and scaffold. BIM technology has great value in visualization, simulation, coordination, optimization, drawing, cost reduction, quality improvement and management assistance. The application in engineering management has become increasingly common. In teaching, students can assist in the auxiliary preparation of the plan through software In the teaching. It can not only improve their learning enthusiasm, but also produce clear and understandable design results with illustrations and pictures, thus improving students' application level of software technology.

In the process of teaching the course of road and bridge construction organization technology and management, the contents closely related to the engineering practice should be taught according to the characteristics and development trend of the specialty and the course. The theoretical teaching part can make full use of MOOC teaching. A large number of online courses have a thorough analysis of the theory, while the practical teaching part focuses on the case teaching of engineering projects. It is necessary to explain the design theory and ideas in combination with actual engineering. Classroom teaching can be used to demonstrate the design process in detail in order to achieve ideal teaching results. Blended teaching can advance the knowledge imparting process completed in the traditional classroom to the front of the class and be completed independently by students on the Internet, while classroom teaching is mainly a personalized teaching mode that internalizes knowledge, improves ability and integrates the process. It opens the door of personalized education, is also conducive to students' independent thinking, and is conducive to the cultivation of students' innovative spirit and innovative ability.

4.2 *Reform of teaching methods*

The course of road and bridge construction organization technology and management is a very practical course, which adopts mixed teaching mode and heuristic teaching method in the teaching process to guide students to think actively. In the process of learning, students can be arranged to visit and study on site if condition is permitted, so as to improve students' interest in learning and increase their understanding and mastery of relevant professional knowledge. In the aspect of autonomous learning, emphasis is placed on cultivating autonomous learning ability, while teachers focus on guidance. By reserving some problems closely related to actual road and bridge projects, students can actively consult, learn and acquire relevant professional knowledge. Students are selected to give reports in class randomly. By changing "passive learning" into "active learning", students' ability of learning can be improved.

On the basis of online courses, after students understand the basic theories and concepts, the time can increased for answering questions in class. Teachers can teach interactively with students, and fully mobilize the enthusiasm of students to achieve good classroom results. Like other courses, the teaching of this course should also pay attention to the design of teaching effect evaluation methods. Teaching effect evaluation should run through the whole teaching process, including classroom teaching and online courses. After each chapter is completed, a certain amount of homework can be arranged in the operation of MOOC area to understand the students' mastery of knowledge in time.

The open and shared teaching system constructed by MOOC curriculum can promote the integration, exchange and sharing of various excellent teaching resources. MOOC has an interactive teacher-student interaction platform, which can provide teachers and students with multi-functional interactive means, including answering questions, correcting homework, stage evaluation, mid-term examination and final examination and so on, to realize "flip" teaching. In the actual mixed teaching practice, MOOC platform can also be used to implement and promote MOOC courses in the same professional scope of different universities through mutual recognition of credits. Students are encouraged to use network resources to find relevant excellent national excellent network course resources for autonomous learning, and students can also learn the teaching concept of "everyone" from these famous teachers.

4.3 *Reform of practical teaching*

Many colleges and universities have set up corresponding curriculum design courses for Road and Bridge Construction Organization Technology and Management courses. However, students rarely get in touch with actual engineering cases in the learning process of theoretical courses and the time for curriculum design is very short. In the process of theoretical courses, students can make full use of the teaching resources of MOOC's engineering cases, which can come into contact with as many engineering cases as possible in advance before curriculum design, so as to know everything you want to know. In this way, the efficiency of curriculum design can be greatly improved, time can be saved, and the works of curriculum design can be more closer to the actual engineering needs. In addition, the problems encountered by students in the process of learning and practice can be fed back to teachers and schools in time, thus further promoting the reform and practice of education. Colleges and universities can also fully understand the employment needs of the society and drive curriculum reform and professional construction.

Self-built school-enterprise teaching team makes the classroom more smooth in organizing discussions on the selection of construction schemes. The joining of enterprise engineers makes students have more reality substitute sense in the guidance of the comparison and selection of construction schemes and training session. How to become more flexible according to the actual situation on the construction site so as to make the project meet the specification requirements and ensure that the discussion on construction duration and quality is more grounded and practical. By this way, students can understand the skills of knowledge points in application transformation during the discussion. Besides, the establishment of the school-enterprise teaching team can ensure the authenticity and richness of actual engineering cases. By this case, the instructors of both schools and enterprises can also achieve a good interaction of learning from each other and promoting each other in cooperation.

Using off-campus production practice sites to carry out on-site training teaching of "Engineering Construction Organization and Management", students accumulated certain perceptual knowledge of construction site schedule management, site layout, site safety and CI image management by participating in site management and auditing regular work meetings during the internship. Through participating in technical clarification and auditing scheme demonstration, they had a real feeling of the importance of construction scheme. In the training week of curriculum design, Using practical engineering cases, provided necessary basic project materials, undertaking different work tasks in the form of group cooperation and simulating actual work situations, students can learn much more by doing and strengthen their practical ability and engineering awareness.

In the process of guiding the course design of Road and Bridge Construction Organization Technology and Management, teachers should effectively guide and answer the problems encountered by students, leading students to play their main role, learn to use MOOC course to consult literature and learn to solve problems independently. The shared resources of MOOC courses can provide students with a large number of engineering case resources and practical software programs, and have online after-class evaluation and feedback modules.

5 SUMMARY

This paper reconstructs and integrates the traditional teaching and online teaching, combines MOOC learning with classroom learning through effective teaching design, designs a more reasonable mixed teaching mode, cultivates students' autonomous learning ability, mobilizes learning enthusiasm fully, and makes the teaching effect of the course produce a qualitative leap.

ACKNOWLEDGMENT

Author Introduction: Wei Li (1974–), Female, born in Harbin, Heilongjiang Province, Professor, Director of master, Research interests: Road and Railway Engineering, Innovation and Entrepreneurship Education.

Fund Project: Research Project of Undergraduate Teaching Reform in Liaoning Province in 2018: Exploration and Practice of "One Lesson, Two Blends" System of Transformation and Development of Road, Bridge and River Crossing Engineering Specialty Education in Colleges and Universities.

REFERENCES

[1] Li Ruihong. 2018.3, Research on the Application of Mixed Teaching Mode Based on MOOC Platform in Bridge Engineering Construction Technology Course. Journal of Shandong Institude of Commerce and Technology.
[2] Lei L H. 2018.7, Design and Practice of Mixed Teaching Based on MOOC. University Education.
[3] Huang Y C. 2018.9, Research on Construction Organization and Management Teaching of Construction Engineering Based on Practical Course. Modern Vocational Education.
[4] Jiang T P. 2016.11, Some Problems and Improvement Methods in Road and Bridge Construction Organization and Management. Architectural Engineering Technology and Design.
[5] Yuan J. 2019.3, Practice of Flip Class in Mechanized Construction Organization and Management Teaching.Henan Agriculture.

Computational Social Science – Luo, Ciurea & Kumar (eds)
© 2021 Taylor & Francis Group, London, ISBN 978-0-367-70193-2

Functions and practice of instrument & equipment sharing platform for talent training in higher education

S. Yao, Z.Y. Li & W.B. Liu
School of Chemical Engineering, Sichuan University, Chengdu, P. R. China

Y. Cao
College of Life Science & Biotechnology, Mianyang Normal University, Mianyang, P. R. China

ABSTRACT: In colleges and universities, the sharing system of instruments and equipment is beneficial to implement convenient service and management for all users. The administrators and teachers or students achieve unified information sharing through the platform no matter in on-line/off-line teaching, scientific research or entrepreneurship requirements. It also can contribute to realize the connection between network and database, the management end and mutual visit of database, and the communication between users and management end; then the dual purpose of "management and sharing" can be realized at last. Currently, large-scale equipment sharing platform is conducive to stimulate students' enthusiasm for research and entrepreneurship activities by using advanced instruments, and then cultivate students' practical ability, analysis and problem-solving ability. At the same time, the extensive use of similar systems among students is also an important way to create a university education environment and improve the quality of innovative talents training.

Keywords: Sharing platform, Instruments and equipment, Higher education, Talent training.

1 INTRODUCTION

In recent years, the quantity and quality of large-scale instruments and equipment in Chinese colleges and universities have been improved rapidly, which has created favorable conditions for the rapid development of scientific research and further helped the implementation of the national strategic goal of "double top-class" project [1,2]. Under the background of this great construction project, the connotation of talent cultivation in higher education has undergone profound changes in China. The goals of high-level talent education are to cultivate high-quality talents with solid theoretical basis, international vision and strong innovation ability. At present, scientific research or entrepreneurship activities based on postgraduates together with undergraduates are the necessary contents to cultivate "mass entrepreneurship and innovation" (MEI) talents, and also the most direct way to improve individual innovation ability. One of the necessary conditions for the effective implementation of MEI activities is the hardware assurance, which can be understood as common experimental instruments in a narrow sense, as well as all the equipment and related platforms used for MEI activities in a broad sense. In order to meet the requirements of social development and personnel training, the State Council put forward the opinions on the opening up of national major scientific research infrastructure and large-scale scientific research instruments to the society in 2016, which was promoted and implemented by the Ministry of Science and Technology and other relevant departments. After years of continuous construction, many colleges and universities have also built a large-scale instrument sharing platform, which can be easily found on the common public search engines. All of the above are very conducive to the research below.

2 CURRENT STATUS OF INSTRUMENT & EQUIPMENT SHARING PLATFORMS

(1) General trends in their construction and application

With the rapid development of information technology in Chinese colleges and universities, development mode of "Internet plus (+)" provides a new reform way of university informationization. For example, the existed campus platform supported by two-dimensional code of mobile phones together with social networking Apps (e.g. WeChat, Whats, etc) can construct a micro service system for convenient instrument and equipment sharing. The platform is powerful and rich in functions, and can be compatible with many kinds of mobile terminal operations such as query, publicity, appointment, settlement and so on. It has the advantages of simplification of information release, two-way communication and interaction, convenient operation of sharing, and real-time publication and promotion. Its informationalized service & management mode, informationalized resource sharing together with informationalized evaluation methods are still in constant innovation [3].

In recent years, laboratory safety accidents occur frequently in colleges and universities, causing casualties and property losses together with wide public concern. As an important part of higher education system and campus management system, systematic instrument and equipment platform can not only make full use of platform information technology innovation management means, but also effectively promote the standardization of laboratory safety construction, improve laboratory management efficiency and level, and strengthen the construction of laboratory safety system. It is beneficial to promote the unity of safety awareness and risk prevention behaviors, and ensures the normal development of entrepreneurship and scientific research activities. A new mode of laboratory construction and safety management driven by informationization can be finally founded [4].

A large number of virtual and simulation devices have been put into operation. We can make full use of new technologies such as multimedia, artificial and virtual-reality (VR) ways to establish a software and hardware operating environment on the computers to replace the traditional learning or experimental operation, so that students can complete a predetermined experimental project in the virtual environment, and the learning or training effect is even better than that in the real environment. Students can also use virtual experimental equipment to preview the process or predict the results before the experiment, and establish intuitive perceptual knowledge of the experiment. It is useful for them to effectively overcome the blind operation in the actual experiment, reduce the error probability of the experiment, and improve the efficiency of the experiment.

In the era of big data, as a part of the overall construction of "Smart Campus" and "Digital Campus", the administrators of colleges and universities can analyze the data resources about the instruments and equipment in their the whole life cycle through cloud computing, which include their use, management, maintenance, deployment, supply & demand information, etc [5]. Big data technology can be used by us to excavate the laws behind the data, improve the utilization and management of whole equipment resources, promote the highly efficient data sharing, and further save cost and improve output. Even better, many science & technology companies are assisting domestic universities to carry out such construction projects.

(2) Instrument & equipment sharing platforms in Sichuan University (SCU)

According to relevant documents issued by the China Ministry of Education and the needs of the reform & development for Chinese universities, Sichuan University has established a long-term management and operation mechanism to maximize the comprehensive benefits of experimental instruments and equipment, so as to better create more practical and innovative resource conditions for undergraduates and postgraduates to enter scientific laboratories, entrepreneurship teams, and research groups as soon as possible. Based on the Laboratory and Equipment Management office, an open sharing system for experimental instruments and equipment (http://vemc.scu.edu.cn/sfw/e?Page = shareequ. Shareequ & CID = 6) has been established before several years. The equipment mainly covers all the scientific research and teaching experimental equipment (including software), mainly composed of those with the original value higher than RMB 200,000 yuan. The right and opportunity to use the equipment are open to users outside the unit, providing paid services not for profit; the school and secondary units are responsible for the unified management of the equipment at different levels. The instruments and users (teachers, students and off-campus users) are managed by classification, and the incentive and restraint measures are equally important. Especially, an expert group on open sharing of experimental equipment was organized to provide key advice on relevant policies, use of funds and purchasing large-scale equipment. On the internet, a "virtual equipment management center" (VEMC) provides comprehensive information and service support; the user can make an appointment on it very easily, and the person in charge will confirm the applications online rapidly. According to actual situations, the university not only provides necessary policy, personnel, special funds and other conditions for VEMC, but also carries out performance appraisal on the sharing work of secondary units at the end of year.

Besides that, a "Sharing Fund of Experimental Equipment" (SFEE) is established to provide the subsidy for the maintenance of shared equipment, the subsidy on service expense of some users and the reward of excellent units and individuals in sharing work. This system provides the most opportunities for students to use various expensive, large and precise instruments. Based on the needs of students, in principle, no less than 150 machine working hours have been guaranteed for those shared equipment with the value higher than RMB 400,000 yuan in priority every year. Moreover, the "one-card-pass" access control system provides the greatest convenience for students to use the shared instruments, which facilitates the intelligent independent experimental platform for students' MEI activities in SCU (under construction). Furthermore, the project of "Open Sharing Management Platform for Large-scale Instruments Based on Visualization" is also under construction. In addition to these level platforms of the university, some secondary schools/colleges in SCU have also established their own network service platform for sharing instruments and equipment, such as the school of chemical engineering. After years of efforts, a public instrument platform developed by the engineering experiment center of the school has achieved great progress (http://pce.scu.edu.cn/db). There are 49 teachers on the platform who undertake various management and service works, including 6 professors, 27 associate professors (associate researchers, senior engineers) and 13 full-time experimental technicians. Before the successful applications through this network platform, students need to pass the online examination about laboratory safety and instrument operation. After passing the examination (above 90% accuracy), they can make an appointment and then use relevant instruments or labs without paying any fees. This is very conducive to all kinds of students' entrepreneurship and innovation activities together with scientific researches. In detail, Table 1 shows the statistical results of instrument sharing in a certain year.

3 PRACTICE OF CURRENT PLATFORMS FOR TALENT TRAINING

In view of the current situation of postgraduate education in domestic colleges and universities together with the needs of talent cultivation, we creatively introduce large-scale shared instruments and equipment platforms into daily teaching process. Combined with the existing equipment and theoretical study of our school, a series of practical courses based on shared instruments have been set up for graduate students of related major, whether they are academic or professional masters. Here the courses of "Instrumental Analysis" and "Advanced Spectroscopy" are two good examples (as shown in Figure 1). Compared with undergraduate teaching, the requirements for postgraduate education are originally higher and more comprehensive, meanwhile the number of large-scale precision instruments is limited, so it is difficult to achieve large-scale and multiple batch basic teaching. In contrast, postgraduate students have certain introductory knowledge and basic experimental literacy, who are not easy to cause equipment damage, and their total number is small, so it is easy to use related platforms to carry out teaching reform.

To conduct all-round learning and training in principle, we train the students to master standard operation, spectral elucidation, data analysis, daily maintenance and even laboratory management comprehensively, and the concept of "learning by doing" is highlighted in the whole process. In order to fully tap human and material resources and realize complementary advantages of resources

Table 1. Year-end report form for some equipment use on the public instrument platform of the school.

Instruments	Cumulative data from Jan to Dec	
	Use times	Use duration (h)
Gas chromatography	22	2115.44
Calculation sever	13	1092
Microcomputer differential thermal balance	43	698
Fourier Transform Infrared spectroscopy	176	465.01
Atomic absorption spectrophotometer	88	323.41
Ultraviolet visible spectroscopy	6	150
Laser particle size analyzer	22	83.58
Nano-laser particle detector	4	76.5
Automatic specific surface area analyzer	10	40.5
Electrochemical workstation	6	31.5
Atomic fluorescence photometer	1	4
Total accumulated time of instrument use (h)		5284
Total cumulative times of instrument use		399
Average usage duration per month* (h)		528.4
Average usage times per month		39.9

*Note: Considering normal rest in winter and summer vacations of SCU together with various public holidays, the annual working hours are calculated as nearly ten months per year.

belonged to various units, the exploration and practice of large-scale instrument resources across disciplines and platforms is continuously carried out by us for graduate students; the overall project is divided into three levels which include the "foundation, improvement and design" (FID). The practice beginning is based on the basic operation of single equipment, and then the scope and difficulty are gradually leveled up in the following improvement stage. At the highest level, the design topic is assigned by the teacher; the graduate students independently search for information, design scheme and select instruments to complete the tasks within the required time. In the teaching and studying, computer-aided instruction (CAI) is applied and a flexible teaching mechanism for different individuals together with dynamic management are implemented. At the same time, the teachers compile teaching materials, build network resources, condense teaching ideas and improve evaluation system synchronously. The implementation of this project can fully promote the initiative and enthusiasm of students, improve their comprehensive analysis ability and experimental practice ability under the premise of basic experiments, and then improve their ability to solve practical problems independently as well as innovation & scientific research ability. The ultimate goal is aimed to cultivate compound talents with solid foundation, wide knowledge scope, systematic theoretical knowledge, good experimental skills and comprehensive abilities of development & research and process design.

4 SUPPORT FROM SCU

The university encourages students to participate in various competitions at the provincial and national levels based on the sharing platform in various forms, and sets up "student experiment method and equipment function innovation and scientific exploration project" for students together with academic student associations. Encouraged by this, undergraduates usually win different prizes in the national and provincial "Challenge Cup" competitions for innovation practice every year, while many excellent research paper indexed by SCI/EI have been published through their effort. Moreover, according to the relevant incentive ordinances for innovation education of SCU, the corresponding credit reward will be given to the undergraduates who complete the practical activities relying on the shared experimental equipment. Among them, the excellent students can

Figure 1. Teaching and studying scene on the sharing platform.

obtain the qualification certificate of the operation skills of the large precision equipment, the excellent certificate of the operation skills of the large precision equipment and the excellent certificate of the innovation and entrepreneurship practice; the secondary units and management units are responsible for carrying out the operation training of the shared large equipment to students, strengthening the construction of the video training courseware, and issuing the operation skills certificate (qualified and excellent) according to the assessment results for users). Furthermore, the university also provides the "assistant teaching post in laboratory" for students and approves the qualified students to participate in the laboratory teaching together with management. As a reward, they will receive different amounts of subsidies according to their work. For the approved innovation and entrepreneurship experimental projects responsible by the undergraduate individuals or academic student associations, the platform service fee shall be exempted according to the procedures.

5 SUMMARY

Based on our working experience in recent years, this paper takes the construction and management of sharing platform of Sichuan University as an actual example, and its latest developing status, functions and related practice of teaching reform are studied and discussed. In current construction and use of the instrument sharing platform, the requirements in application are treated as the traction and the top-level design is taken as the key point; the platform construction is implemented step by step with the efforts of all participants in China. With the rapid development of informationization in colleges and universities, more and more instrument sharing platforms are put into use for various clients; as the result, more and more teachers, students and off-campus users are benefiting from them. Obviously, their roles in higher education are also growing, which are becoming an indispensable part of modern education.

ACKNOWLEDGEMENTS

This research was financially supported by Personnel Training Quality & Teaching Reform Project of Higher Education in Sichuan Province (2018–2020) and 8th "Reform Project of Higher Education in New Century" of Sichuan University (2018).

REFERENCES

[1] Information on http://www.chinamae.com/shownews_101742_2.html.
[2] Information on http://yqgx.tsinghua.edu.cn/webSite/websiteAction.do?ms=goToIndex.
[3] M.X. Chen, J.L. Xie, J. Zhou, 2018, Development and application of integrated management information system for open instrument laboratory in colleges and universities, Chem. 33, 46–50.
[4] R.Y. Zhang, J.L. Yuan, 2018, Security analysis and improvement countermeasures for terminal devices of the instrument sharing platform, Lab. Sci. 21, 194–195, 199.
[5] S.M. Shi, 2019, Intelligent campus planning and design based on cloud computing and internet of things, Electron. Test. 20, 40–42.

Computational Social Science – Luo, Ciurea & Kumar (eds)
© 2021 Taylor & Francis Group, London, ISBN 978-0-367-70193-2

Research on the impact of student evaluation of teaching on teacher teaching performance evaluation in colleges and universities: Illustrated by the example of Chengdu University of Information Technology

S.Q. Cao & J. Chen*
Chengdu University of Information Technology, Chengdu, China

ABSTRACT: This research is conducted to improve the efficiency of college student evaluation of teaching, so as to prompt teachers' teaching. Through questionnaire and interview methods, this research investigated how students treated student evaluation, and how teachers were affected by student evaluation. The results show that some students have adverse evaluation attitudes, and student evaluation of teaching has influence on teachers' psychology status, teaching methods and compensation. Combined with the specific situation of Chengdu University of Information Technology, this research put forward the adverse impacts and improvement suggestions for the reference of colleges and universities.

Keywords: Student evaluation of teaching, Teacher teaching performance evaluation, Principle of development, Student evaluation scale.

1 INTRODUCTION

With the advancement of modernization and globalization, higher education has experienced the development process from original elite higher education to current mass higher education [1]. Along with this, the management system of higher education also changed from elite management to mass democratic management. For what kind of talents can a university produce largely depends on the quality of university teachers, the management of teachers has always been an important link. Therefore, it is very necessary to make the management of university teachers scientific and service-oriented. As an important task of human resource management in universities, teacher performance evaluation can provide feedback on teachers' own work and have an important impact on teachers' later work performance. Nowadays, universities have gradually established the mutually reinforcing connection between student evaluation of teaching and teacher teaching performance evaluation. As an important part of university management, student evaluation of teaching is conductive to improving teachers' teaching and promoting students' participation in university management.

In order to make the research more targeted, Chengdu University of Information Technology was taken as the research object in this study. From the perspective of college students and college teachers, this study investigated how students treat student evaluation of teaching, and how teachers are affected by student evaluation of teaching. Based on the survey results, the cause for the adverse impact and improvement suggestions were proposed, which can also provide a reference for other colleagues.

---

*Corresponding author

2 THE THEORETICAL FRAMEWORK

2.1 *The performance system and teaching performance evaluation of teachers in universities*

China adopts university organization establishment system. The performance evaluation system of colleague teachers is composed of scientific research and teaching performance [2]. Scientific research takes a large proportion in teacher performance evaluation index. There are two reasons, the one is the progress and update of scientific research achievements play an important role in the development of national science and technology, the other is scientific research evaluation is quite easier to be quantified. In comparison, teaching performance does not have such advantages. There are two main ways to evaluate teachers' teaching, including measuring the quantity of teachers' teaching and measuring the quality of teachers' teaching. However, teacher's teaching quality performance, such as teaching attitude and teaching effect, is not easy to be evaluated. In fact, teachers' teaching quality is always the most difficult part to be evaluated in teacher performance evaluation.

2.2 *The connotation and development of student evaluation of teaching*

Student evaluation of teaching is a kind of comprehensive evaluation from the perspective of learners according to evaluators own standards. Student evaluation of teaching came into being since the emergence of teaching activities. However, it was limited to oral teaching evaluation and impression teaching evaluation in its early stage [3]. In the early 20th century, American colleges were the first to carry out student evaluation of teaching as a formal activity. In the mid-1980s, China has officially incorporated student evaluation into university management system. In 1984, student evaluation of teaching carried out by Beijing Normal University was regarded as the earliest official teaching evaluation activity in China. From then on, student evaluation of teaching became an important indicator to test colleague teachers' teaching performance in China [4]. At the beginning, college teaching evaluation system in China was borrowed from America to some extent. But in the later period, it has been developing and showing some new characteristics.

2.3 *The relationship between student evaluation of teaching and teacher teaching performance evaluation*

Teachers and students are interactive objects in the process of teaching. Student evaluation of teaching and teacher teaching performance evaluation are also interactive links in the process of college management. Student evaluation, as the most widely used evaluation method in universities, has played an important role in the management of college teachers. For the participation of students in teachers' teaching is deep and direct, their evaluation is often very persuasive. Student evaluation of teaching not only plays an important role in testing teachers teaching quality, but also promotes the democratic participation of students in the process of education management.

3 RESEARCH METHODS

3.1 *The conduction of the literature analysis*

Literature is the basis of research. Through searching, reading, sorting and analyzing the relevant literature in the research field, it is conducive to absorbing different arguments and research methods of other researchers. The literature method was adopted at the early stage of this study. It helped obtain some beneficial implications. And in the middle stage of the research, it tends to combine the specific situation of the university to conduct in-depth research in the latter period.

Table 1. Students' attitudes towards the importance of student evaluation of teaching.

Attitude	Number	Proportion
Some necessary	111	52.86%
Necessary	64	30.47%
Unnecessary	35	16.67%

3.2 The formulation of the questionnaires

Questionnaire survey is the most basic way to collect data in investigation, which can always collect data abundantly and efficiently. In this study, two questionnaires were developed, one was based on students' attitudes and the other was based on teachers'. The procedures were as follows. First of all, the general questionnaire frameworks were formulated. Secondly, interviews with the teachers and students were conducted. Thirdly, the contents of the questionnaires were modified according to the results of the interviews. Finally, a preliminary survey was conducted to ensure the rationality of the questionnaires.

3.3 The issuing and recycling of the questionnaires

In this research, two questionnaires were issued. One was 'The influencing factors of student evaluation of teaching in Chengdu University of Information Technology', which was used to investigate students' attitudes. The other questionnaire was 'The impact of student evaluation of teaching on teacher teaching performance evaluation in Chengdu University of Information Technology', which was used to investigate how teachers' attitudes. There were 234 students and 32 teachers who took part in the survey. Then based on the results of the questionnaire surveys, teachers and students were interviewed on certain issues.

3.4 The analysis of the questionnaire data

In this research, SPSS was used to measure the validity and reliability of the results of the questionnaire. Through the measurement, the conclusion about the accuracy and stability of the questionnaire data was drawn. It helped judge the degree to which the questionnaire results reflected the real situation.

4 ANALYSIS OF THE SURVEY RESULTS

Through the collation and analysis of the survey data, the following conclusions are drawn:

Some students do not attach much importance to student evaluation of teaching

The results of the survey show that 30.48% of the students hold it is necessary to for students to evaluate teachers' teaching, 52.86% of the students think it is some necessary, and 16.67% of the students think it is unnecessary. The details are shown in the following Table 1.

4.1 Some students' teaching evaluation methods are not reasonable and objective

According to the results, 46.66% of the students give different evaluation scores to different teachers according to teachers' own performance. 20.48% of the students tend to give higher or lower scores to a few teachers, 29.52% of the students give the same highest scores to all teachers, and 3.33% of students even give scores randomly in the teaching evaluation process. The details are shown in the following Table 2.

Table 2. Different teaching evaluation methods of students.

Method	Number	Proportion
Teachers are given different scores according to their performance	98	46.66%
All teachers are given highest scores	62	29.52%
A few teachers are given high or low scores	43	20.48%
Give scores randomly	7	3.34%

Table 3. The influence of factors affecting student evaluation of teaching.

Serial number	Factor	Average value (0–10)
1	Teacher's teaching attitude	8.75
2	Teacher's knowledge level	7.87
3	Teacher's respect for students	7.74
4	The intimacy between teachers and students	7.7
5	Teacher's teaching method	7.39
6	Student's course learning effect	7.27
7	Student's course interest	7.1
8	The strictness of teacher's management	6.5
9	Student's course grade	6.4
10	Student's course grade	6.03
11	The difficulty level of the course	5.15
12	Teacher's professional title	3.44
13	The length of the evaluation period	3.03

4.2 Teacher's teaching attitude, teacher's knowledge level, teacher's respect for students, the intimacy between teachers and students and teacher's teaching method are important factors affecting student evaluation of teaching

In this research, 13 factors affecting student evaluation of teaching were analyzed. The results of the survey show that the most influential factor for student evaluation of teaching is teacher's teaching attitude, the second influential factor is teacher's knowledge level, and the third influential factor is teacher's respect for students. Besides, the intimacy between teachers and student and teacher's teaching method also have great influences on student evaluation of teaching. The influence force of each factor is shown in the following Table 3.

4.3 Personal characteristic of student and course characteristic are important factors affecting student evaluation of teaching

Based on the Table 3, 13 factors were sorted into the following five types. Among them, teacher's teaching performance, with a weight of 26.25%, is the most influential aspect. In addition, Personal characteristic of student and course characteristic are also influential aspects, and they all reduced the objectivity of student evaluation of teaching. The details are shown in the following Table 4.

4.4 Some teachers hold that there are some deficiencies in student evaluation of teaching

According to the results of the survey, 56.25% of the teachers think that student evaluation of teaching is not accurate enough and needed to be improved. There are 37.5% of the teachers think that student evaluation of teaching is relatively accurate, and only 6.25% of the teachers think that student evaluation of teaching is very accurate. The results also show that no teacher agrees that

Table 4. The different on the impact of different factors of student evaluation of teaching.

Type	Average value (0–10)	Weight
Teacher's teaching performance	7.45	26.25%
Personal characteristic of student	5.17	24.7%
Course characteristic	4.57	22.64%
Personal characteristic of teacher	3.91	18.53%
Arrangement of teaching evaluation activity	3.89	7.9%

Table 5. Teachers' attitudes towards the accuracy of student evaluation.

Attitude	Proportion	Sum
Not accurate enough	56.25%	100%
Relatively accurate	37.5%	
Very accurate	6.25%	
Completely inaccurate	0	

Table 6. The impact of student evaluation of teaching on teachers.

Degree	Proportion	Sum
Certain impact	68.75%	100%
Slight impact	18.75%	
Great impact	12.5%	
No effect	0%	

student evaluation of teaching is completely inaccurate. The details are shown in the following Table 5.

4.5 *Most teachers hold that student evaluation of teaching has a certain impact on university teachers*

68.75% of the teachers think that student evaluation of teaching has a certain impact on teachers; 18.75% of the teachers think it has only a slight influence on university teachers; 12.5% of the teachers think it has a great influence on university teachers. The results also show that no one considers himself to be immune to student evaluation of teaching. In general, almost all teachers think that student evaluation of teaching has an influence on university teachers, but they have different views on the degree of the influence. The details are shown in the following Table 6.

4.6 *Student evaluation of teaching causes changes in teacher's psychological state, teaching method and compensation*

Student evaluation of teaching has impacts on teachers' psychological state, teaching method, compensation, position and professional title. The greatest impact that student evaluation of teaching brings to teachers is the change of teacher's psychological state, such as psychological stress, tension, and invisible supervision. Besides, it also has influences on teacher's compensation and the adjustment of teacher's teaching method. Actually, according to teacher teaching performance systems adopted by different universities, the impact of student evaluation of teaching to teachers will also be different. The details are shown in the following Table 7.

Table 7. The impact of student evaluation of teaching on teachers.

Influenced factor	Average score (0–10)
Psychological state	8.37
The adjustment of teaching method	7.26
Compensation	6.45
position	4. 91
Professional title	4.8

5 ANALYSIS OF THE ADVERSE CAUSES

5.1 *The casual attitudes of some students in evaluating teachers' teaching*

Students have different opinions on the importance of student evaluation of teaching, and only a few students realize the importance of it. Besides, students evaluate teachers' teaching in different ways, and there are only a few students' methods are reasonable. Finally, the inaccurate evaluation results not only affect the objectivity of student evaluation of teaching, but also reduce the enthusiasm of university teachers.

5.2 *The lack of uniformity in the standards of student evaluation*

There are 13 factors affecting student evaluation of teaching. Actually, these factors are composed of five aspects: teacher, curriculum, evaluation items, teaching and student. To evaluate teacher's teaching is the purpose of conducting student evaluation of teaching. However, in addition to teacher's teaching, student's personal characteristic and course characteristic are also important factors influencing students' evaluation.

5.3 *Insufficient colleague propaganda about the importance of student evaluation of teaching*

Student evaluation of teaching is the most important criteria to evaluate teachers' teaching performance, and it always has influences on teachers. However, in interviews with some students, few students know the influence of student evaluation of teaching on teachers. So, it is very necessary to propagandize the importance of student evaluation of teaching to students and change their minds.

5.4 *The lack of unplanned rewards for teachers with high teaching scores*

The basic performance standards for teachers in Chengdu University of Information Technology are divided into half for teaching and half for research. In fact, such standards are too reasonable to make a change. However, the teaching evaluation system lacks unplanned material or spiritual rewards for teachers who are given high scores in student evaluation of teaching. This is a way that needs to be improved.

5.5 *The neglect of the development effect of student evaluation of teaching*

In each semester, Chengdu University of Information Technology conducts student evaluation of teaching. However, student evaluation of teaching is only conducted at the end of each semester, which ignores its development effect. The result of such evaluation method is that teachers can neither get the feedback from students in time nor adjust their teaching methods through their teaching process. So student evaluation of teaching cannot play its developmental role what it could have played.

5.6 The defects of student evaluation scale

On the one hand, there are some important factors missing in the scale, including teacher's respect for students, teacher's knowledge level and student's course interest, which are important indexes for students to evaluate teachers. On the other hand, the statements of some questions in the scale have the characters of polysemy and complexity, such as 'The teaching is simple and in depth, the key and difficult points are prominent, the explanation is vivid and attractive, and it can stimulate students' interest in learning'. The elements it contained are excessive and its expression is also complicated. It is difficult for most students to make objective judgments.

6 SUGGESTIONS FOR IMPROVEMENT

6.1 Strengthen the propaganda on the importance of student evaluation of teaching

Activities to promote the importance of student evaluation of teaching should be reflected in a series of activities in the university. In class-level activities, counselors should publicize the importance and impact of student evaluation of teaching to improve students' understanding of student evaluation. In school-level activities, the colleague should make the propaganda of student evaluation of teaching as an activity of the student assembly to reinforce students' correct beliefs of student evaluation.

6.2 Adjust teacher performance system

The proportion of teaching performance in teacher performance evaluation should be moderate, so that to make teachers evaluated comprehensively. Besides teachers' teaching performance, it should also focus on teachers' performance in scientific research, entrepreneurship and other aspects. As for teachers who are given high scores in student evaluation of teaching, universities should give them spiritual or material rewards, which can also have indirect influence on other colleague teachers.

6.3 Improve the student evaluation scale

As for the content of the score, questions that contain important elements should be added, including 'whether teacher respect students', 'whether teacher's knowledge is universal' and 'whether student is interested in the course'. As for the impression of the score, it should adopt the simplification way to make the improvement. Therefore, the elements of each question should be simplified to improve the effectiveness of the scale.

6.4 Give play to the developmental role of student evaluation of teaching

Nowadays, China pays more and more attention to the developmental principle of education, and the beneficiaries of the principle should include not only students but also teachers. In this occasion, half-term evaluation of teaching should be conducted, which has the function of development. It will be not only beneficial for students to give teachers feedback in time, but also beneficial for teachers to adjust their teaching conditions, so as to promote the positive interaction between teaching and learning.

6.5 Update and upgrade the teaching evaluation system

On the one hand, with the change of the influence of different factors that students care about during evaluating teachers' teaching, universities should update the structure and content of the student evaluation scale according to the development of education and the psychological changes of students. On the other hand, colleagues should keep abreast of the changes of technology.

Colleagues should constantly upgrade the student evaluation system to improve the convenience and secrecy of student evaluation of teaching.

6.6 *Promote the diversification of the subjects participating in teaching evaluation*

Student evaluation of teaching has advantages but it also has disadvantages, such as its subjectivity. In this condition, it is necessary to promote more subjects to participate in teaching evaluation, such as scholars and teachers. Scholars with professional teaching knowledge can always evaluate teachers' teaching objectively and professionally. And teachers' self-evaluation and mutual-evaluation are also conductive to strengthening teachers' reflection and pushing teachers' mutual learning [5].

6.7 *Erect correct conceptions of teaching evaluation to students themselves*

Students should evaluate teachers' teaching objectively on the basis of teachers' teaching performance. The results of student evaluation of are not only feedback of teachers' hard work in teaching, but also related to the fairness of the competition among different teachers. Therefore, students should erect correct conceptions of student evaluation of teaching, and avoid bringing personal emotions into the activity, so as to achieve the goal of maximizing the effectiveness of student evaluation of teaching.

ACKNOWLEDGEMENT

This paper is one of the research results of "The project of 2018–2020 personnel training and teaching reform of general higher education in school level of Chengdu University of Information Technology" (JY2018032).

REFERENCES

[1] B. L. Zhong, X. F. Wang, 2019, Opportunities, challenges and prospects: Universal access of higher education in China, J, China Higher Education Research, 08, 7–13.
[2] M. R. Hua, 2018, Authorized faculty size of universities: historical changes in its management and future reforms, J, Journal of Yangzhou University, 01, 17–22.
[3] J. W. Shi, 2016, Research on student evaluation of teaching and its effectiveness to improve teachers' teaching: Taking N university in Lanzhou for example, D, Chong Qing: Southwest University.
[4] N. Li, 2008, Research on the reliability of student evaluation of teaching related to evaluation of university teachers, D, Beijing: Capital University of Economics and Business.
[5] E. E. Bastista, 2014, The place of colleague evaluation in the approval of colleague teaching: A review of the literature, J, Research in Higher Education 04, 257–271.

Computational Social Science – Luo, Ciurea & Kumar (eds)
© *2021 Taylor & Francis Group, London, ISBN 978-0-367-70193-2*

Innovation and implementation of multiple teaching strategies in financial program learning

H.L. Hsu & H.C. Hsu
School of Business Administration, Baise University, Baise, China

ABSTRACT: This research adopts multi-disciplinary strategies in the financial course of college in Taiwan. Through the process of action research, it explores the learning outcomes of students and the reflection of teachers, so as to improve the teaching practice for teachers. In this study, multi-teaching strategies include problem-based learning, situational case study and interactive game teaching. The research samples are students of college, all of whom are the first-time in the personal financial program. The methodology is questionnaire scale to conduct satisfaction analysis and differences analysis of before and after teaching. Through the satisfaction survey, this study found that most students feel more interesting for learning after problem-based learning with situational case materials, than traditional teaching methods. Students have a high recognition for the learning effects from interactive games. Based on the pre-test and posterior differential analysis, this study found that students who first came into contact with personal financial program have an improvement significantly in their learning outcomes through the implementation of multiple teaching strategies, and the increase in financial vocabulary also reached statistically significant levels. Finally the study evidences that the most students are satisfied with the implementation of the overall teaching strategy.

1 INTRODUCTION

This study uses the financial education curriculum of college in Taiwan as an example to implement a variety of teaching strategies, which are examined and reflected through action research procedures. This study hopes to stimulate students' learning fun through different teaching strategies.

Personal financial literacy is the knowledge and skills to handle personal financial issues, and it is the basic literacy that modern citizens should possess. This literacy includes attention to economic or financial issues, interpretation, attitudes to money, and making the best financial decisions. In 2005, the Organization for Economic Co-operation and Development (OECD) research reported on financial education recommended that financial education should start from schools and advise people to get in touch with financial-related matters as soon as possible [1]. Financial knowledge or management skills are the ability in life, not a difficult professional knowledge. The trend of education reform is to combine education with life. Financial education has an irreplaceable role in this aspect. Through financial education, make people become literate citizens in a civilized society. Therefore, university education should integrate financial management courses in general education, so that young people can get in touch with financial management skills early, and cultivate self-learning and speculative skills focusing on financial issues. Remmele & Seeber (2012) stated that the ability of modern citizens to contemplate should be included in the issues of financial management and economics, and their thinking ability will be more convincing and effective [2].

This study introduces multiple teaching strategies into financial courses, conducts classroom activities in a relatively easy-to-understand manner, enhances students' participation in learning, and gradually guides students to internalize financial management knowledge into their own life skills. Multiple teaching strategies include interactive game learning, problem-based learning, and situational case studies. Explore the results of multiple teaching strategies from teaching implementation and practice.

The teaching practice process uses the action research to conduct a systematic inquiry. That is, the teacher defines the focus study area and collects data during the action, then analyzes and explains the data, and reflects on the course of the action plan. One of the most basic motivations of action research is to improve teaching dilemma and learning quality in the teaching context, support teachers to effectively cope with the difficulties in teaching practice, and to innovate and improve teaching in a reflective process. Action research can help teachers discover the advantages, disadvantages and difficulties of teaching, propose feasible solutions, and modify the feedback found in teaching through the cycle of planning, action, observation, and reflection in the process of teaching design and implementation. At the same time, teachers can observe the students' learning behaviors through the action research process, and establish learning evaluation indicators to understand the learning results and serve as a reference for learning improvement.

Based on the foregoing research motivations, the purpose of this paper is to apply the multi-teaching strategy to the assessment of students 'learning effectiveness and teachers' reflection in general courses about personal finance.

2   LITERATURE REVIEW

This research took the planning, action, observation and reflection of action research as the research process. Action research is a process with procedural steps proposed by Kurt Lewin [3]. It uses systematic research methods to collect relevant data during the teaching process to analyze and interpret information, thereby improving teaching skills and learning effectiveness. Action research is an activity of self-criticism and reflection. The research process in which practical workers solve practical work encountered in the context of their actual teaching work. Research procedures include diagnosing problems, identifying problems, planning solutions, evaluating, and reflecting [4]. The goal of action research is to improve the interactive relationship between teaching and learning.

A variety of teaching strategies can be used in the promotion course of financial education to enhance the interaction effect between teachers and students. Proper application of teaching strategies can effectively support differentiated learning for individual student. This finding can enhance the intention of teachers to increase the professional skills of teaching [5]. The use of multiple teaching strategies can improve the overall learning effectiveness of students. Fatima, Naz, Zafar, Fatima, & Khan (2020) pointed out that different teaching strategies are interesting and useful for most students [6]. The research results show that most students are positively receiving different teaching methods. Furthermore, student feedback is also useful in defining the content of the lessons that need to be improved. Therefore, proper measurement and adjustment can improve the overall quality of the course. Simulation strategies can also be used in multiple strategies to improve the effectiveness of learning. Steven & Mark (2018) implemented games and simulated teaching strategies in classroom and found that the knowledge and learning interest of students have improved [7]. This study also has confirmed a positive correlation between the number of simulations and student performance.

Multi-strategy teaching methods can further stimulate learning fun. For example, the teaching method of PBL allows teachers to drive more in-depth discussions on issues. Through the case discussion or the learning tasks, it can guide the mechanism of teamwork learning for students and enhance the interactive dialogue between students and teachers in teaching and learning. PBL is a self-directed active learning method. When students understand the problem, they know what to learn [8]. The learning process of PBL is centered on students. Teachers design discussion cases, which are real problems that occur in the real world. Students discuss and propose solutions to problems. That's the process of creating knowledge and transforming it into skills by exploring and solving problems. Taking PBL as the original idea, the teaching methods and classroom guidance skills for teachers are more important than their professional knowledge [9].

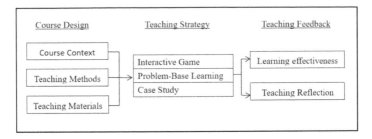

Figure 1. Teaching practice strategy diagram.

Based on past literature, the use of multi-strategy teaching such as PBL case study or simulation games can improve effectiveness on learning and feedback on teaching.

3 RESEARCH METHODOLOGY

3.1 Teaching design and strategy

This study develops a number of teaching strategies. In order to enhance learning initiative and learning fun for students. Different teaching methods are adopted for different learning topics. The teaching practice strategy is shown in Figure 1.

3.2 Research sample

The research object is full-time students of the college in Taiwan. Twenty-five students are enrolled in the course, and four students have dropped out for some reason during the semester. The participants who actually completed the course were 21. There were 7 female students (33%) and 14 males (67%). In terms of the study field that attended by participants, 3 were in the engineering field (14%), 2 were in the design department (10%), 14 were in business management (66%), and 2 were in hospitality (10%).

3.3 Research instrument and methodology

This research adopts action research as teaching practice and quantification as research method. The application of research tools is described below. In order to evaluate the effectiveness of teaching, which is from the perspective of students, this research conducted a teaching satisfaction questionnaire. The questionnaire design refers to relevant literature and other university curriculum questionnaires. The questionnaire used Likert scale. The first part of the questionnaire is a satisfaction survey on the teaching method. There are 8 questions in total, which include indicators for students to reach the core competence of the general course. The second part is the survey on the satisfaction of teaching materials, including the aspects of diversity, practicality, and interest. The questionnaire was reviewed by expert peers to confirm the content and ensure the validity of the questionnaire. The questionnaire used Cronbach's α as a measure of reliability to confirm the stability and internal consistency. The alpha coefficient of the first part of the questionnaire for teaching effectiveness is 0.78, the second part is 0.85, and the total alpha coefficient of the questionnaire is 0.80. The questionnaire is internal consistency and stability.

In addition, in order to enhance the learning interest for first-time financial management beginners, this study introduced the interactive game teaching method and conducted a post-teaching satisfaction survey. There are five items in this questionnaire. The Likert five-point scale is used for analysis.

Moreover, this study refers to literature for questionnaire development to survey the professional learning effectiveness. The content of the questionnaire covers three aspects, including financial

Table 1.   Results of satisfaction survey on PBL teaching methods.

Performance of PBL	Very satisfied	Satisfied	Ordinary	Unsatisfied	Very unsatisfied
1. More fun than traditional lectures	61%	39%	0%	0%	0%
2. Increase learning interest	56%	39%	6%	0%	0%
3. Improve problem solving skills	39%	50%	11%	0%	0%
4. Promote Active Learning	44%	33%	22%	0%	0%
5. Improve expression	44%	33%	22%	0%	0%
6. Promote teamwork	44%	39%	17%	0%	0%
7. Improve relationships	33%	33%	33%	0%	0%
8. Improve classroom participation	44%	33%	17%	0%	0%

management knowledge, financial management goals and investment awareness (Hui-Lin Hsu, 2014). According to the literature, the questionnaire has good construction validity under a large sample. The questionnaire was reviewed by financial experts and scholars to confirm the logic and rationality, and to confirm the validity. The Cronbach's α is used as a measure of reliability. The correlation coefficients of all facets are between 0.60-0.85, and α coefficients are all above 0.80. Learning outcomes were pre-tested and post-tested with self-assessment scales. The score is checked by t test to exam if the difference between before and after learning is significant.

This research was also designed to measure the effectiveness of learning by increasing students' vocabulary on financial management. At the beginning, students were asked to write the vocabulary related to financial management, and at the end of the semester, students were asked to record the vocabulary again that they learned through their studies. Then calculate the average number of vocabulary words separately, and carry out a difference test to confirm the learning effectiveness of financial management.

4   RESEARCH OUTCOMES

4.1   *Teaching reflection*

This research constructs teaching reflection through three satisfaction analyses, including teaching methods, teaching materials and aids, and interactive game teaching.

4.2   *Satisfaction analysis of teaching methods*

In this study, a questionnaire survey was conducted to evaluate the effectiveness of PBL teaching. The survey results are shown in Table 1. All students agreed that teaching with PBL is more interesting than traditional teaching methods. More than 95% of students believe that learning interest has increased, and more than 75% agree that PBL teaching methods help them improve their problem-solving ability, active learning attitude, and improve interpersonal relationships.

4.3   *Satisfaction analysis of case study material*

The results of the percentage analysis on the teaching materials of the case study are shown in Table 2. Most students are sure that the teaching materials are clear and easy to understand, the cases are interesting and close to the real life and they can practice solving problems. The results of this study show that the teaching materials designed for students with different professional backgrounds or first-time financial management courses are easy for students to accept. It proves that well-designed teaching materials are good medium for learning.

Table 2. Results of satisfaction survey on teaching materials.

Performance of teaching materials	Very satisfied	Satisfied	Ordinary	Unsatisfied	Very unsatisfied
1. Clear and easy to understand	50%	33%	17%	0%	0%
2. Case is very interesting	50%	50%	0%	0%	0%
3. Content is diverse	44%	44%	11%	0%	0%
4. Content is practical	50%	33%	17%	0%	0%
5. Close to the real situation	67%	22%	11%	0%	0%
6. Can be operated	44%	39%	17%	0%	0%
7. Content step by step	50%	39%	11%	0%	0%
8. Improve financial ability	28%	44%	28%	0%	0%

Table 3. Results of survey on interactive game learning.

Performance of interactive game learning	Very satisfied	Satisfied	Ordinary	Unsatisfied	Very unsatisfied
1. Understand the significance and importance of investment portfolios	47%	35%	18%	0%	0%
2. Actively adjust portfolio strategy to pursue higher returns	47%	24%	29%	0%	0%
3. Adjusting portfolios through dynamic strategies will generate different returns	65%	18%	18%	0%	0%
4. Inspiring the concept of asset allocation	65%	18%	18%	0%	0%
5. Enhance the fun of financial studies	59%	41%	0%	0%	0%

4.4 Effectiveness of interactive game teaching

The percentage analysis results of the interactive learning effectiveness are shown in Table 3. The results of the questionnaire survey show that more than 80% of the students agreed that the learning goal could be achieved through interactive game learning process, and at the same time, the learning process was fun. However, during the learning process, students were asked if they were willing to make strategic adjustments in order to pursue higher return. Student responses were generally more conservative. This reaction shows that although the teaching process has been simulated, students may still care too much about winning or losing at the end of the game, which may cause them to be afraid to take active actions. Therefore, teachers using interactive simulation as a teaching method need to give students more encouragement to increase their enthusiasm for participation and obtain better simulation experience from their attempts.

4.5 Learning effectiveness

Questionnaires were used as the survey instrument for learning effectiveness. The content of the questionnaire is divided into three parts, the first part is the awareness of financial management, the second part is the personal financial management goals, and the third part is investment management. A total of 21 questionnaires were collected. The results of the survey show that students have made significant progress in three aspects, as shown in Table 4. The post-test averages are greater than the pre-test averages, and after a differential two-tailed t-test, the differences between the post-test and the pre-test averages reached a significant level ($p < .001$). Therefore, the results of this study show that through the implementation of multi-strategy teaching activities, students have made significant progress in learning of financial management.

Table 4. Results of survey on professional learning.

Facet of measurement	Before and after implementation	Average	Standard deviation	tvalue	Significance (P-value)
Cognition of financial management concepts	Pretest	3.36	0.93	-12.49	<.001***
	Posttest	4.60	0.54		
Understanding of financial goals	Pretest	2.81	1.05	-14.59	<.001***
	Posttest	4.41	0.67		
Understanding of investment concepts	Pretest	2.76	1.17	-7.29	<.001***
	Posttest	4.24	0.80		

***$P < .001$.

Table 5. Results of survey on increasing of professional vocabulary.

Pre-Test		Post-Test		Difference Test			
Mean	Standard deviation	Mean	Standard deviation	Mean difference	Percentage increase	t-value	p-value
4.88	2.32	10.29	1.90	5.41	53%	-8.58	.000***

***$p < .001$.

4.6 Analysis of increased professional vocabulary

The students participating in this study were the beginner learners in personal financial management. In order to investigate the progress of students under the multi-teaching strategy, the research design also included a survey of the increase in professional vocabulary. Students were required to write the financial vocabulary based on what they know before the course begins, and then conducted a post-test before the end of the course to observe the actual increase in the number of vocabulary. After data analysis, the study found that the average difference in vocabulary that students could write before and after learning reached 5.41, and the standard deviation decreased from 2.32 to 1.90, showing an improvement of 53%. The test for the difference between the pre-test and post-test averages shows a significant level. Studies have shown that after learning, students' awareness of professional vocabulary has increased significantly, as shown in Table 5.

5 CONCLUSION AND SUGGESTION

This study introduces multiple teaching strategies into the financial education curriculum, using the method of action research as a process, and innovating teaching practice through the process of reflective thinking. The multi-teaching strategy used in this research includes interactive game teaching, PBL teaching, and situational cases. Through the development and evaluation of questionnaires, learning effectiveness for students and reflective for teachers can be explored to obtain feedback and improve teaching practice.

In terms of teaching reflection, the research findings found that most students deem that problem-based teaching is more interesting than traditional teaching methods and can increase motivation for classroom participation. This study introduces situational cases into PBL teaching. The case materials are the dilemmas and problems of real people's physical and financial affairs. The study found that most students believe that learning through PBL cases can improve their problem-solving ability and active learning attitude. This article verifies that PBL is a good teaching method, as in previous literature. In addition, the effectiveness of interactive game teaching shows that the process of interaction and simulation has impressed students and achieved good results in learning. When the lesson was shared at the end of the course, most students could still remember the teaching content of interactive games. It can be seen that the interactive game teaching makes students have a deep impression on learning.

In terms of student learning feedback, the results of questionnaire surveys before and after learning found that students who learn through multiple strategies have made significant progress in learning. Since the students selected for this course are all exposed to the education of finance courses for the first time, this study investigated the degree to which students increased their professional vocabulary. The research results show that the average vocabulary increased by 5.41 after study, showing a 53% increase. After diversified teaching methods, students have made significant progress in the study of personal financial management, not only assimilated into their own knowledge, this knowledge will be applied in life and transformed into practical skills and behaviors.

In order to awaken students' participation and interest in learning, it is important that teachers adopt a variety of innovative teaching techniques. Many literatures in the past have shown that PBL has been applied to various fields of teaching and has performed well. In addition to using PBL teaching methods and situational cases as teaching materials, this study also incorporates interactive game teaching methods and has achieved good results, which can be used as a reference for other researchers in the future. Multiple teaching strategies based on action research also allow teachers to achieve the purpose of reflection and feedback through the implementation process. It can also be of great help in improving teaching practice.

REFERENCES

[1] OECD. (2005). Improving financial literacy: Analysis of issues and policies. Organization for Economic Co-operation and Development.

[2] Remmele, B. Seeber, G. (2012). Integrative economic education to combine citizenship education and financial literacy. Citizenship, Social and Economic Education, 11(3), 189–201.

[3] Lewin K. (1948). Resolving social conflicts. New York: Haper.

[4] McNiff, J. (2017). Action Research: All You Need to Know. London: Sage Pubns.

[5] Hyry-Beihammer, E. K., & Hascher, T. (2015). Multi-grade teaching practices in Austrian and Finnish primary schools. International Journal of Educational Research, 74, 104–113. https://doi.org/10.1016/j.ijer.2015.07.002

[6] Fatima, U., Naz, M., Zafar, H., Fatima, A., & Khan, R. R. (2020). Student's Perception about Modular Teaching and Various Instructional Strategies in the Subject of Obstetrics and Gynecology. Professional Medical Journal, 27(1), 40–45. https://doi.org/10.29309/TPMJ/2020.27.1.3162

[7] Steven D., Mark K. (2018). The impact of simulation activity on student performance. Journal of economics and economic education research, 19(3), 1–10.

[8] Geoffrey E. Mills (2007). Action Research: A Guide for the Teacher Researcher. Pearson Education.

[9] Barrows, S. and Robyn, M. Tamblyn (1980). Problem-based learning: an approach to medical education. New York: Springer.

Computational Social Science – Luo, Ciurea & Kumar (eds)
© 2021 Taylor & Francis Group, London, ISBN 978-0-367-70193-2

Reform and innovation of the college course teaching mode with an example of the "Crystal Dislocation" micro-course design

J.J. Gong*, J. Sun, P. Zhang, D.L. Wu, J.J. Du, J.C. Gao, J.J. Liu, D.Y. Li & C. Liu
School of Mechanical Engineering, Yangzhou University, Yangzhou, P.R. China

ABSTRACT: The rapidly developing world encourages high education researchers to explore new ways to help us successfully socialize and remain competitive. It will become an emerging trend of the synergy of diversified studying and modern information technologies in educational activities. As one of the most promising modern teaching modes, the micro-course is favored by teachers and students because it enables better retention rates, requiring shorter attention spans and resulting in more reinforced quality learning. Also, it is conducive and effective to knowledge creativity, positivism, and time arrangement. Taking the design and application of the micro-course of "Crystal Dislocation," a basic theoretical chapter in the "Materials Science Foundation" (MSF), as an example, the role of the micro-course in the reform and innovation of university teaching model is discussed in this paper. First, the basic information about the micro-course including its concept, development process, and main characteristics is briefly introduced. Second, the technical features and the teaching requirements of "Crystal Dislocation" in MSF are discussed. Finally, based on the content of "Crystal Dislocation," the main process and the application of micro-course production are described. The research results in this paper are valuable for enhancing the strong micro-education growth in the learning system, along with the constant reform and innovation of the university curriculum teaching mode.

1 OVERVIEW OF THE MICRO-COURSE

1.1 *What is a micro-course?*

A micro-course is a video-recorded teaching process. It is conducted by the teacher according to the teaching standards and teaching practice requirements, focusing on a certain knowledge point (generally, it is the key points, difficulties, and doubts in a certain course) or teaching process. It can be seen as an organic combination of various teaching resources [1, 2]. With the overall development trend of information diversification and time modularization in today's society, the micro-course has gradually become a key project in the informatization construction of basic education, and one of the hot research topics in the field of education and teaching reform [3, 4, 5]. The core content of the micro-course is classroom teaching video, supplemented by teaching design, teaching materials and courseware, teaching reflection, testing, feedback, teacher comments, and so on. They are organized and presented in a logical way, creating a small, semistructured, theme-based resource application. Therefore, the micro-course is a new teaching resource developed on the basis of traditional teaching methods.

1.2 *Development of a micro-course*

In 2008, David Penrose created the first one-minute course at San Juan College in new Mexico, calling the "mini-course" [6]. Micro-courses were introduced in China in 2011. Initially, they were

---

*Corresponding author

mainly used in primary and secondary schools. Because the course is interesting, they received very good results and feedbacks. A year later, colleges and universities across the country held a micro-course teaching competition. Since then micro-courses have attracted increasing attention from the education communities and have been gradually promoted in different courses across the country. At present, most colleges in China have set up micro-courses, which complement and integrate with the existing teaching mode, greatly improving the learning efficiency.

1.3 *Characteristics of micro-courses*

By definition, the purpose of a micro-course is to enable learners to learn independently and achieve the best learning effect. Generally speaking, a micro-course only teaches one or two knowledge points, "fragmentation" is its most distinguishing feature. However, a micro-course is aimed at specific target groups and delivers specific knowledge content. Micro-courses themselves need to be systematic, and the knowledge expressed by a set of micro-courses needs to be comprehensive. In general, a micro-course has the following characteristics [7]:

(1) Teachers do not have to appear in the camera. When the teacher is giving lectures, he/she may appear in class or in the form of a voice.

(2) It can be played on network media. The media type of micro-courses is usually video or animation, which can be circulated and played based on the network media.

(3) The course is short. Micro-course videos are usually 5–10 minutes, the shortest ones are only 1–2 minutes, and the longest ones are usually no more than 20 minutes.

(4) Rigorous teaching design. For each knowledge point or skill point, the content design is complete and detailed so that students can understand the content without directly communicating with teachers.

(5) Classic cases are demonstrated. Teaching and learning scenarios combine real and typical cases.

(6) It is supported by relevant resources. A micro-course generally coordinates with related homework exercises, resources, etc.

2 ANALYSIS OF "CRYSTAL DISLOCATION" KNOWLEDGE IN "MATERIALS SCIENCE FOUNDATION"

"Material Science Foundation" (MSF) is a required basic course in a major with strong systematic properties, requirements on rigorous logical thinking, and a practical connection between theory and practice. It is also an important major course for postgraduate entrance examination [8]. Nowadays, universities pay increasing attention to "quality education," and many updated training programs have shortened their teaching hours, including the MSF course. However, the development of students themselves and the needs of society require students to have a solid foundation of professional knowledge, which substantially increases the requirements of course parts. This requirement requires teachers to pay attention to key points and difficulties, and guides students to thinking and solving problems. Teachers need to impart a great deal of knowledge in limited class time. Only by appropriate reform and innovation on the basis of the traditional teaching model, can the teaching of this course adapt to the current social development trend and the teaching needs of MSF.

Dislocation formation is a common phenomenon in crystalline materials such as metals. In engineering applications, dislocation plays a critical role in the mechanical behavior of materials such as plastic deformation, strength, fracture, etc., and also has a great impact on the diffusion and phase transformation process of materials. A crystal dislocation is a boundary line of the local slip region in the crystal, that is, a line defect in the crystal. There are two kinds of typical dislocations: edge dislocations and screw dislocations. The basic concepts of these two dislocations are difficult to distinguish and require strong geometric spatial imagination to learn. On the basis of a clear description of the basic concepts, a more difficult task is to distinguish their geometric relations. There are three important directions in both crystal dislocations: the direction of the dislocation line,

Figure 1.   Typical shots with defects in a micro-course of "Crystal Dislocations".

the direction of the crystal slip, and the direction of the dislocation line movement. Finally, what effect does crystal dislocation have on the growth and application of crystals? This section will be easier for students to understand if it is accompanied by images, animations, and actual examples. It can be seen that the course content of "crystal dislocation" is suitable for the combination of various teaching methods to fully mobilize students' vision, hearing, and attention. Combined with the characteristics of a micro-course mentioned above, this part of the content will be presented to students in the form of a micro-course, which is expected to achieve a good teaching effect.

3   DESIGN AND APPLICATION OF A MICRO-COURSE IN THE KNOWLEDGE OF "CRYSTAL DISLOCATION"

Learning the knowledge of "Crystal Dislocation" is challenging, but may be relatively easy to be understood by improving the teaching method. In this work, the ideology of a micro-course is adapted. Using this teaching method, an attractive opening design is helpful, like showing beautiful dynamic shots of defects, as shown in Figure 1. The lecture starts with the words that "everything has defects, but sometimes defects can also become a special kind of beauty." It also provides soothing background music to create a pleasant learning atmosphere. Then the title of the micro-course and the chapter are displayed.

After the opening, a special defect—dislocation—is introduced, and four basic problems relating to the dislocation are put forward: what a dislocation is, the classification of dislocations, the geometric relation of dislocations, and the application of dislocations. The above part is the importing part. Then we move to the body of the micro-course. First, the concept of crystal dislocation is explained. The premise is to clarify what a crystal is. To realize this, atoms fly in from all directions and form a crystal in a regular arrangement. Then defects appear dynamically in the crystal, arranging as a dislocation. Further, the basic types of dislocations can be divided into edge dislocations and screw dislocations. The formation process and the main differences in the two types of dislocations are illustrated by their dynamic stress process and the structure characteristics, respectively. The geometric relationship between the two dislocations is a difficult point, which needs to be emphasized. During the presentation, a rectangular coordinate system is built. First, the direction of atom movement in the process of dislocation formation is explained through the dynamic process of crystal atom movement. Then the concept of dislocation line is explained and highlighted. As the stress is continuously applied, the dislocation line moves, and an arrow appears to point out its direction. At the same time, the axes in the corresponding direction of the rectangular coordinate system are highlighted. After clarification of the above directions, their pairwise relationships become clear. Thus, the characteristics of the two dislocations can be summarized. Then the similarities and differences of the two dislocations are compared in a table.

The final part is the summary and extension, showing the crystal dislocations in practical applications, which echoes the beginning—that a defect is sometimes a kind of beauty. The knowledge points of this part are summarized, which leads to three reflection questions: 1. What will happen if a edge dislocation and a screw dislocation appear together? 2. What other advantages of dislocations can you think of? 3. What disadvantages may dislocations have in practical applications? The conclusion indicates that we will return to the above three questions next time to stimulate students' interest in following up.

Considering the characteristics of strong systematic properties, high requirement for logical thinking and special imagination, and a close linkage between the theory and practice of the "crystal dislocation" chapter in the MSF course, the teaching process should use a variety of teaching methods in order to provide students with good visual and auditory effects to offer students an imagination space. In the process of teaching, it is suitable to apply a question–answer pattern and multipractice mode to arouse students' enthusiasm and concentrate their attention, adhering to the idea of "students as the main body, teachers as leading." In addition, practical examples should be exhibited to guide students to use the knowledge to analyze questions and develop a solution-seeking thought pattern in students. With integration of the above lesson-teaching requirements and skills, with the support of vivid and understandable images, animations and simulation diagrams, the micro-course enables students to master the concept of crystal dislocations and learn the basic principles, steps, and methods for judging two kinds of dislocations, distinguishing their basic features, and helps students master the basic theory of crystal dislocations. Therefore, this micro-course is very beneficial whether as self-study materials, preview materials, or after-class review materials.

4   SUMMARY

The micro-course plays a great role in the reform and innovation of the teaching mode of college courses. The appropriate design and application of micro-courses is key to the teaching methodology of activating students in learning knowledge. Because of the characteristics of the "Crystal Dislocations" chapter in the MSF course, the teaching process should include a variety of teaching methods. Therefore, the proper design of the micro-course for this chapter is very beneficial to students, whether as self-study materials, preview materials, or after-class review materials. This is unsurprising, as the micro-course reconciles with traditional teaching mode and can help students to enhance their learning enthusiasm and solidify their professional knowledge.

ACKNOWLEDGMENT

This research was supported by grants from the Research Project on Teaching Reform of Yangzhou University (YZUJX2017–7A and YZUJX2019–19B), and Qinglan Engineering Project of Yangzhou University.

REFERENCES

[1] X.H. Liu, LS. Wang. The Analysis on Systematic Development of College Microlecture, Higher Education Studies, 3(2013) 65–70.

[2] S.Z. Yuan, X.F. Mao, Y.L. Y. Wang, Y. Luo. The Application of Micro-lesson in Optics Teaching, 14th Conference on Education and Training in Optics and Photonics.

[3] Z.P. Wang, Y. Luo, Y.L. Qu. Application of Micro-lecture For Engineering Mechanics Experimental Teaching. International Journal of Innovation and Research in Educational Sciences, 4 (2017): 2349–5219.

[4] F.Q. Ying. Research on Blended Learning Mode Based on the Micro-Lecture in Database Application, Review of Computer Engineering Studies, 3(2016): 62–66.

[5] C. Zeng. Application of Micro-lecture in the Teaching Method Reform of Fashion Chromatology, 2016 2nd International Conference on Education Science and Human Development.

[6] Y. Peng. Application of Micro-lecture in Computer Teaching, Earth and Environmental Science 234(2019): 1–6.

[7] J.C. Gao, J.J. Liu, D.Y. Li. Teaching research of basic course of materials subject under the background of engineering education certification–taking Fundamentals of Materials Science as an example, Contemporary Educational Practice and Teaching Research, 2019: 64–65.

Computational Social Science – Luo, Ciurea & Kumar (eds)
© 2021 Taylor & Francis Group, London, ISBN 978-0-367-70193-2

Study on the innovation capability of college students based on action learning theory

Y.X. Liu & P.B. Gao*
School of Economics and Management, Harbin Institute of Technology at Weihai, Weihai, China

W.W. Wu
School of Management, Harbin Institute of Technology, Harbin, China

Z. Li
China Railway Large Maintenance Machinery Co., Ltd. Kunming, Kunming, China

ABSTRACT: College students are the major human resources for societies and organizations, and thus the development of their innovation capability is increasingly important. Based on action learning theory, this paper examines the effect of action learning on the innovation capability of college students, and the results show that action learning exerts a positive impact on their innovation capability. The research conclusion offers a refreshing perspective for research into how to stimulate college students' innovation capability, and also provides beneficial enlightenment for colleges to improve the innovation capability of their students.

1 INTRODUCTION

The rapid changes in technology development and consumer demands have prompted firms to look for a competitive advantage for their survival. Today, innovation has become an important tool for firms to achieve competitive advantages such as decreasing costs and increasing quality. The realization of firms' innovation depends on individual innovation capability. College students are the major human resources for societies and organizations, and thus the development of their innovation capability has become increasingly important [1,2].

However, the literature to date has paid greater attention on how to evaluate instead of how to improve the innovation capability of college students. Improvement of innovation capability is a learning process, and action learning can provide preferable settings in which to learn innovation [3]. Thus, action learning may have a close relationship with the innovation capability of college students. Nevertheless, the role of action learning in the promotion of the innovation capability of college students has not received much attention. Therefore, the main purpose of this paper it to test that relationship.

Following this introduction, the remainder of this paper is structured as follows. The next section offers an overview of the previous research related to action learning, and proposes the hypothesis of this paper. Section 3 provides the research design adopted in this paper. Section 4 presents the basic statistics and regression results. The final section contains discussions and concluding remarks.

2 LITERATURE REVIEW AND HYPOTHESIS DEVELOPMENT

2.1 *Action learning*

From the traditional view, learning has been assumed to be the process of knowledge transmission, and after this transmission, learners are able to apply the knowledge for their own purposes [4].

---

*Corresponding author

The learning described in the traditional view is an external process. Meanwhile action learning identifies the possibility for learners to produce new knowledge by themselves rather than just receiving it passively [5]. Action learning is an effective problem-solving methodology proposed by Reg Revans, which has been used extensively. In short, action learning is learning from a specific problem through group discussion, trial and error, and discovery by discussing the problem. Zuber-Skerritt (2002) concludes the main characteristics of action learning, such as collaborating, being open, sharing ideas, experiential learning, reflecting on practice, and so on [6]. These characteristics indicate that action learning involves an active process. By comparison with traditional teaching methods, action learners are more likely to develop as experts in the problem and how to resolve it [7].

2.2 The relationship between action learning and innovation capability of college students

Action learning is a method that requires the participants to be involved in the solving of some real and complex problems actively to improve their capability [8]. This method is contrary to traditional teaching approaches, which have been criticized as too far from college students' innovation capability. In fact, it is argued that most traditional teaching approaches codify past knowledge. Although learning from the past can be an effective method, it can also be unhelpful if there are no opportunities to apply the required knowledge in reality [9].

Action learning is different to traditional teaching approaches in that it is a process to handle real and complex problems. In the process, the knowledge, experience, and skills owned by college students are used together. The solution can bring in new knowledge, and college students can use the solutions directly [10]. Because action learning aims to generate creative solutions to figure out specific problems, which are closely related with college students' actual knowledge, it can immediately improve college students' innovation capability.

In addition, action learning can also help college students to form systems thinking rather than just looking at the problem from a single perspective [11]. Action learning is a continuous learning process, supported by other participants, with the intention to solve the problem from multiple perspectives. Diverse team working enables participants to generate insightful ideas, and reflecting upon other responses contributes to a more holistic solution [12]. It is critical for college students to have systems thinking, which contributes to the promotion of college students' innovation capability. Following the argument, we hypothesize:

H1: Ceteris paribus, action learning has a positive effect on the innovation capability of college students.

3 RESEARCH DESIGN

3.1 Sample and data collection

This paper tested the proposed hypothesis by using a questionnaire survey, which was conducted in Chinese colleges. We designed a questionnaire that required respondents to complete based on the literature of action learning and innovation capability of college students. In order to discover the potential problems of the questionnaire, a pretest was conducted. After the pretest, a questionnaire was revised according to the suggestions. Then, we sent the questionnaire to 160 college students. Finally, 120 college students provided valid information. This sample size was enough to carry out statistical analysis at the individual level.

3.2 Variables and measures

Referring to previous studies [13,14], this paper developed an instrument to measure action learning. The scale of action learning includes three dimensions, namely orientation, support, and collaboration. The scale of innovation capability of college students was developed based on the

Table 1. The factor analysis result of action learning.

	Factor 1	Factor 2	Factor 3
Collaboration	0.919	0.162	0.119
	0.911	0.173	0.122
	0.850	0.186	0.067
	0.803	0.0237	−0.038
	0.712	0.246	0.066
	0.702	0.343	−0.022
Support	0.227	0.934	0.090
	0.197	0.921	0100
	0.398	0.717	0.102
	0.506	0.642	0.021
Orientation	0.080	0.060	0.927
	0.112	0.112	0.861
	−0.035	0.071	0.729
	0.051	0.004	0.685

research of Chen et al. (2013) and Martín et al. (2017) [15,16]. The scale of innovation capability also includes three dimensions, namely creative thinking capability, creative learning capability, and creative practical ability. Each of the items in the scale used a 5-point Likert scale, with 1 being disagree and 5 agree.

3.3 *Reliability and validity*

Cronbach's α has been commonly used to evaluate the reliability of multi-item scales. The Cronbach's α values for action learning and innovation capability were 0.895 and 0.815, respectively. All values were larger than 0.7, which indicated that the measurement has good reliability.

Factor analysis was employed to test whether the items were loaded on the designed dimension. The results of factor analysis are shown in Tables 1 and 2. From the results, it can be seen that all the items were loaded at the designed dimension. The results verified the construct validity.

4 RESULTS

Table 3 lists the descriptive statistics including means, standard deviation, and correlations. It clearly shows that action learning has significantly positive correlations with the innovation capability of college students.

This paper uses regression analysis to test the proposed hypothesis. Table 4 shows the regression analysis results. From the regression model, we know that action learning has a positive relationship with the innovation capability of college students (β=0.986). Therefore, the hypothesis is supported.

5 DISCUSSION AND CONCLUSION

Action learning provides the logic of learning about learning, and using this to learn. The action learning process respects and builds upon each participant's independence, and it is less structured because it responds to the variety of problems that college students encounter. Therefore, action learning provides a flexible circumstance for college students to improve their innovation capability, and it has quickly emerged as one of the most effective teaching methods used by colleges worldwide.

Table 2. The factor analysis result of innovation capability.

	Factor 1	Factor 2	Factor 3
Creative learning capability	0.909	−0.047	0.094
	0.856	−0.098	−0.114
	0.850	−0.039	−0.064
	0.840	0.087	0.178
	0.838	0.175	0.232
	0.786	0.059	0.142
Creative practical ability	0.012	0.853	0.089
	−0.058	0.820	−0.083
	−0.066	0.783	0.237
	0.012	0.709	0.085
	0.012	0.669	0.138
	0.177	0.646	0.022
Creative thinking capability	0.005	0.147	0.752
	−0.009	0.288	0.746
	0.010	0.347	0.719
	0.199	0.069	0.711
	0.032	−0.011	0.640
	0.154	−0.218	0.635

Table 3. Means, standard deviations and correlations.

Variables	Means	Standard deviation	1	2
Action learning	3.35	0.96	1.00	
Innovation capability of college students	3.49	0.71	0.986**	1.00

Note: **$P<0.01$

Table 4. Regression result.

Model	Unstandardized coefficient	Standard error	Standardized coefficient	t	Sig.
Constant	−1.323	0.075		−17.659	0.000
Action learning	1.339	0.021	0.986	63.648	0.000

Our findings offer an important theoretical contribution. This paper provides theoretical arguments and empirical evidence for illustrating the impact of action learning on the innovation capability of college students. Previous literature has indicated but never tested the effect of action learning on the innovation capability of college students. This paper verifies that action learning can exert positive impacts on the innovation capability of college students. This contributes to the broader literature on innovation capability of college students by identifying the important antecedent, and provides a promising potential direction for colleges to improve the innovation capability of college students. Our findings also contribute to practices. Our findings highlight the value of action learning, such that colleges employing action learning are more likely to improve their students' innovation capability. Hence, colleges should take measures to promote action learning.

Although this paper presents important points in relation to the effect of action learning on the innovation capability of college students, it also has some limitations that could be addressed in future research. First, this paper only describes the direct impact of action learning on the innovation capability of college students, but a limitation is that it does not explain the mediating mechanisms of this relationship. Therefore, in order to get more benefits from action learning with aims to improve the innovation capability of college students, future research is encouraged to investigate the mediating mechanisms.

Second, this paper is only focused on the internal factors, and does not consider external environmental factors. However, the external environment can affect all action learning. Future research is needed to explore the relationships between the action learning and innovation capability of college students with external factors, which would offer a more integrated explanation of how action learning affects the innovation capability of college students.

ACKNOWLEDGMENT

This research was financially supported by the Research Project of Postgraduate Education Reform in Harbin Institute of Technology, and the Research Project of Postgraduate Education and Teaching Reform in Harbin Institute of Technology (Weihai).

REFERENCES

[1] P. Martín, K. Potoènik, A.B. Fras, Determinants of students' innovation in higher education. Stud. High. Educ. 42 (2017) 1229–1243.

[2] L. Fan, M. Mahmood, M.A. Uddin, Supportive Chinese supervisor, innovative international students: A social exchange theory perspective. Asia Pac. Educ. Rev. 20 (2019) 101–115.

[3] L. Baron, Authentic leadership and mindfulness development through action learning. J. Manage. Psychol. 31 (2016) 296–311.

[4] Y. Cho, T. Marshall Egan, Action learning research: A systematic review and conceptual framework. Hum. Resour. Dev. Rev. 8 (2009) 431–462.

[5] Y. Cho, T.M. Egan, The state of the art of action learning research. Adv. in Dev. Hum. Resour. 12 (2010) 163-180.

[6] O. Zuber-Skerritt, The concept of action learning. Learn Organization 9 (2002) 114-124.

[7] M. Volz-Peacock, B. Carson, M. Marquardt, Action learning and leadership development. Adv. in Dev. Hum. Resour. 18 (2016) 318–333.

[8] K.S. Scott, An integrative framework for problem-based learning and action learning: Promoting evidence-based design and evaluation in leadership development. Hum. Resour. Dev. Rev. 16 (2017) 3-34.

[9] H.S. Leonard, M.J. Marquardt, The evidence for the effectiveness of action learning. Action Learn: Res. and Pract. 7 (2010) 121–136.

[10] F. Sofo, R.K. Yeo, J. Villafañe, Optimizing the learning in action learning: Reflective questions, levels of learning, and coaching. Adv. in Dev. Hum. Resour. 12 (2010) 205–224.

[11] C.M. Leitch, C. McMullan, R.T. Harrison, Leadership development in SMEs: An action learning approach. Action Learn: Res. and Pract. 6 (2009) 243–263.

[12] K. Jones, S.A. Sambrook, L. Pittaway, et al. Action learning: How learning transfers from entrepreneurs to small firms. Action Learn: Res. and Pract. 11 (2014) 131–166.

[13] E. De Haan, I. De Ridder, Action learning in practice: How do participants learn?. Consult. Psychol. J.: Pract. and Res. 58 (2006) 216–231.

[14] A. Carmeli, J. Schaubroeck, Organisational crisis-preparedness: The importance of learning from failures. Long Range Plan. 41 (2008) 177–196.

[15] A. Chen, L. Li, X. Li, et al. Study on innovation capability of college students based on extenics and theory of creativity. Procedia Computer Sci. 17 (2013) 1194–1201.

[16] P. Martín, K. Potoènik, A.B. Fras, Determinants of students' innovation in Higher Education. Stud. High. Educ.42 (2017) 1229–1243.

Computational Social Science – Luo, Ciurea & Kumar (eds)
© 2021 Taylor & Francis Group, London, ISBN 978-0-367-70193-2

Development of an interactive learning aid system for cross school sharing of descriptive geometry and mechanical drawing courses

X.H. Li

School of Mechanical Engineering and Automation, Northeastern University, Shenyang Liaoning, P.R. China

ABSTRACT: Descriptive geometry and mechanical drawing is a theoretical and practical course, which aims to cultivate students' ability for spatial imagination and analysis. It needs more practice to be learnt well. In order to optimize the curriculum resources, it is necessary to build a high-quality education resource-sharing system to realize the resource sharing and complementary advantages of colleges and universities in the region. Therefore, the teaching team of the course has built a cross school sharing system for descriptive geometry and mechanical drawing courses, to improve the level of experimental teaching methods and technical information, and to solve the problems of poor resource sharing and insufficient utilization of equipment. It is necessary to improve the level of information and intelligence of student aid, question answering and counseling, and promote the construction and sharing of high-quality experimental teaching resources. At the same time, the courses based on cross school sharing construction and the corresponding mutual learning mode also represent the teaching development direction of descriptive geometry and mechanical drawing and the course construction level of colleges and universities.

Keywords: Descriptive geometry and mechanical drawing, Cross school sharing, Learning aid system.

1 INTRODUCTION

Teaching resources sharing across schools and cross school reading are new teaching modes of "Internet plus education." It helps us break the traditional teaching mode by relying on the advantages of the Internet and the advantages of the platform. Relying on the third-party network teaching management platform, students can cross the geographical restrictions, choose other universities' excellent courses to study and receive credits, and provide students with a wider range of learning. The wide learning space meets the individual learning needs of students in order to promote the sharing of high-quality educational resources, promote educational equity, and promote the reform and innovation of teaching methods and management systems. With the deepening of teaching reform in colleges and universities, great changes have taken place in the teaching concept, from the past indoctrination education to quality education. Under the current situation of national new engineering training, if we want to do well and learn the course of descriptive geometry and mechanical drawing well, we must reform and improve the teaching methods and teaching methods accordingly. The integration of network and multimedia technology provides a broad development space for the teaching reform of the course. Descriptive geometry and mechanical drawing is a very practical course, which aims to cultivate students' ability of spatial imagination and analysis. It needs more practice to learn well.

2 SIGNIFICANCE OF SHARING THE COURSE OF DESCRIPTIVE GEOMETRY AND MECHANICAL DRAWING BETWEEN DIFFERENT SCHOOLS

Modern teaching advocates quality education, which requires that students' initiative and creativity should be stimulated in the process of teaching, and that students' learning motivation should be stimulated. In teaching, a good environment should be created for students' active learning. The resident position of students in the learning process should be highlighted, and students' ability to analyze and solve problems cultivated. The main feature of the descriptive geometry and mechanical drawing course is to correctly express the body structure and relevant national standards of various parts and components by reasonable use of views and expression methods. As an example, it can effectively achieve better teaching results. Teachers often draw views of various models in the classroom, which will take up classroom time. In this way, there is often not much knowledge to be taught in teaching, and at the same time teachers also put in more physical effort.'At the same time, due to the limitations of blackboards, teachers should finish the teaching content of this lesson and update blackboard writing constantly, leading to teachers' inability to reproduce the content of this lesson in the class summary. In order to make the students understand and digest the knowledge more deeply, teachers often have to assign a lot of homework after class. When students do exercises, due to the lack of space imagination ability, some students have difficulty answering questions. If they can't get practical guidance and answer questions, they spend a lot of time, but do not get a better learning effect, and the learning efficiency is low. In view of the current teaching situation of the course, the course team has developed a cross school sharing interactive learning system and applied it in teaching. The purpose of this is to help students in the process of learning to solve doubts, self-test, cultivate students to better establish the ability of spatial thinking and the ability of mutual transformation between the three-dimensional space and the plan.'In the practical teaching of mechanical drawing courses, we try to guide students to master the methods of analyzing engineering expression problems from the content of the courses, and cultivate the ability of students to express the mechanical structure and solve practical engineering problems from the perspective of mechanical design.

3 KNOWLEDGE COVERAGE AND DEVELOPMENT MEASURES OF THE INTERACTIVE LEARNING SYSTEM

The interactive learning system of descriptive geometry and mechanical drawing developed by the course teaching team is mainly based on the course materials and matching problem sets. It includes: putting forward the problem; analyzing the process of solving the problem; demonstrating the steps to solving the problem; displaying the 3D solid modeling in an all-round way; and giving the correct answer to the problem. In the traditional teaching method, chalk and drawing tools were added to the blackboard. Teachers need to prepare a large number of teaching aids, models, etc. in class, and then draw with a ruler on the blackboard. The intensity of teacher work is great, but the amount of information students get is very little. Limited by time, the knowledge points of each class are limited, and cannot meet the requirements of the current teaching development. The use of computer-aided teaching can effectively solve the above-mentioned contradictions. Based on the interactive learning system, students can switch and jump at any time, repeatedly demonstrating and explaining any unfamiliar knowledge points, difficulties, and key points, which is an effective measure to highlight the important and difficult points at this stage. During teaching, the teacher will draw out the plane figure, three-dimensional figure, and solving steps for the contents, with examples and homework exercises of the explanation prepared in advance. When the students can't conceive the three-dimensional space of a figure, they can use the three-dimensional figure to help students imagine and explain the plane figure against the three-dimensional figure.

Multimedia CAI technology is a breakthrough of education modernization, which has the advantages of integration, interaction, real-time, and so on. Especially for the teaching of technical basic courses in colleges and universities, due to the high requirements of courses, tight class hours, and boring content, some students become tired of learning, or even abandon their course. The multimedia teaching with various forms, lively, highly interactive, and friendly interface has made profound changes to the traditional education mode, and also aroused the enthusiasm of students to learn professional courses. The practice shows that by properly introducing multimedia and combining with traditional teaching methods, we can fully mobilize the enthusiasm of teaching and learning, which is conducive to the multidimensional way of thinking and improves the efficiency of learning, and is welcomed by students. Multimedia technology is used to assist teaching, and computer graphics software is used to develop teaching materials. We can choose all kinds of electronic drawings at any time according to the requirements of the teaching situation, and explain them through the network, and use all kinds of electronic materials at any time in the summary stage after class, so as to achieve an effect that traditional teaching cannot achieve. This can create complete freedom from the traditional blackboard drawing, save a lot of time, and can easily organize teaching, increasing the amount of information transfer in the classroom.

The research object of the descriptive geometry and mechanical drawing course is specifically three-dimensional, including various basic assemblies, parts, etc. In traditional teaching, teachers usually use various kinds of teaching physical models to carry out on-the-spot demonstrations. This kind of model has its limitations that can't be eliminated. Its type and quantity ensure good manufacturer design, but the teaching materials change greatly. All the physical models are often out of touch with the teaching materials, so the teachers can't choose flexibly according to the actual situation. Also, the physical model is small and the physical model is small in size and simple in shape, and only some students can clearly observe when there is a demonstration in the classroom. It is also very limited to explain the modeling of the optional physical model, which is not conducive to students' comprehensive understanding of the three-dimensional structure of the spatial shape. For large parts and assemblies, it is also impossible for teachers to demonstrate in class. Even in the workshop, it is impossible to reverse, disassemble, and cut at will, so that students can clearly observe the internal structure and assembly areas. The contradiction between the physical model teaching aids and classroom teaching can be solved by using the electronic three-dimensional model for classroom demonstration.

4   OPERATION MODE OF THE SHARING COURSE ASSISTANT SYSTEM

The interactive system is innovative and practical. The content of the descriptive geometry and mechanical drawing course is relatively abstract, and is difficult for teachers to teach and students to learn. It is an interactive learning system. The content includes: projection characteristics of points, lines and surfaces, drawing methods of three views, intersecting lines, reading combination views, drawing methods and reading of sectional views, drawing methods of threads, drawing methods of gears, etc. In the process of material selection and creativity, it does not rely on copying textbooks, but is a kind of auxiliary teaching method. The main goal is to achieve practicability and serialization as far as possible and to solve a specific problem. In the process of production, first, the material is collected, then the image, sound, and animation are produced and edited. The courseware can be used for lectures, and also for students to review and watch after class. The courseware is intuitive and vivid, the topic is typical, the key point is highlighted, the drawing process is clear, enlightening, and the content selection is flexible, and the operation is convenient. In the process of using it, teachers and students praise it.

In the teaching process of descriptive geometry and mechanical drawing, students are required to complete a large number of homework questions after class, so as to check the quality of classroom teaching. Through the student aid system, the courseware made by the teacher is presented to the students one at a time by the computer. The computer immediately judges whether the students' answers are correct or not, and displays the feedback information or help information. The way of answering questions should also be interactive and diversified. During the operation, the teaching content is divided into several knowledge hopes. After each unit is completed, the students' mastery is checked through questions and exercises. The computer monitors the learning process, or according to the learning situation, decides whether to let them enter the learning of new content or return to relearn. Under the multimedia teaching mode, the teaching content appears to be well illustrated, full of sound and color, and the interactive form is very lively.

5 APPLICATION EFFECT OF A CROSS SCHOOL SHARING COURSE SYSTEM OF DESCRIPTIVE GEOMETRY AND MECHANICAL DRAWING IN OUR UNIVERSITY

The system has been used in more than 70 classes in our school, and the feedback effect is very good. According to the actual needs, teachers can call up all kinds of parts drawings, assembly drawings, and three-dimensional parts drawings at any time, which is no longer limited by time and space as in classroom teaching. For example, when talking about assembly drawing, if teachers need to share the knowledge about the part drawing, they can call all kinds of materials of the part drawing at any time, so that the classroom organization is more flexible, and students will have a thorough understanding. This not only improves the teaching efficiency, but also makes the teaching process more vivid. Using multimedia to teach descriptive geometry and mechanical drawing can effectively realize the interaction between teachers and students, and stimulate students' interest in learning the course. Students view the virtual three-dimensional entity by computer, and put forward their own ideas according to their own imagination. Teachers use software to change students' imagination into virtual reality to verify the results of the students' thought processes. In this way, it can stimulate a strong desire for knowledge in the students' mind, and generate interest in the study of mechanical drawing knowledge, so that the internal and external factors of learning can be effectively combined to better encourage the cognitive initiative of students, and promote the improvement of learning efficiency.

6 SUMMARY

The construction and practice of a highly interactive classroom do not only need an implementing flipped classroom, but are also an important part of classroom practice in improving the credit system of cross school study. It is of great significance to improve the level of undergraduate teaching and the quality of personnel training by gradually changing the concept, gradually exploring and practicing from the education administrative department to the school, and then to the teachers and students. The interactive learning aid system for the descriptive geometry and mechanical drawing course, which is built by the teaching team, puts forward higher requirements for teachers' teaching ability. However, it is very simple to use the system in the classroom, which leaves more time for students and improves teaching efficiency. The use of this system shows that it has achieved a better teaching effect, which is conducive to the implementation of quality education as advocated at present. At the same time, the courses based on cross school sharing construction and the corresponding mutual learning mode also represent the teaching development direction of descriptive geometry and mechanical drawing and the curriculum construction level of colleges and universities.

ACKNOWLEDGMENT

This research was supported by Education and Teaching Reform Project of Liaoning Province, China (No. [2018] 471 Letter from Liaoning Education Department) and Education and Teaching Reform Project of Northeastern University of China (No. [2018] 39 Teaching Documents of Northeastern University); Teacher Development Special Project of Northeast University of China (No. DDJFZ202005).

REFERENCES

[1] H.N. Wu, P. Geng, Q. Wang, J.L. Cui, X.H. Chen, 2015, Practice and study of elective credit cross universities of university physics course, Phys. Eng. (02) 84–86.
[2] T.M. Yang, T.S. Yang, 2015, The adjustment and improvement of internet + based cross-registration credit, Mod. Educ. Manage. (11) 55–58.

[3] B. He, J.C. Li, F.M. Zheng, 2018, Teaching method and art involved in cross-school under credit teaching of machinery design, J. Mach. Design. (35) 224–226.

[4] S.J. Shan, D.Y. Fu, Y.F. Shen, 2017, Internationalization of higher education curriculum based on "Internet plus", West. China. Qual. Educ. (03) 14–15.

[5] Q.Q. Zhang, H.B. Zhang, Z.H. Zhang, 2019, Hybrid teaching mode for engineering mechanics under internet + background, China. Educ. Tech. Equip. (16) 35–37.

[6] S.Y. Zhen, 2017, Thoughts on the construction of cross university humanities course under the background of resource sharing in Colleges and Universities, J. Bohai Univ. (Philos. Soc. Sci. Edit.), (01) 116–118.

Computational Social Science – Luo, Ciurea & Kumar (eds)
© *2021 Taylor & Francis Group, London, ISBN 978-0-367-70193-2*

A study on the cultivation model of information-based teaching ability of English majors state-funded normal students

C. Dai
Sichuan Minzu College, Sichuan, China
Khams Research Institute of Translation on Foreign Educational Literature

ABSTRACT: In-depth development of information technology education background for the teaching ability of teachers in information technology has become one of the skills teachers must adapt to the current needs of the community. For English major's state-funded normal students, as the main reserve force of English teachers, the cultivation of information-based teaching ability of English majors is not satisfactory. Therefore, this paper analyzes the training goal, the course structure, the teaching idea, the teaching method, the teaching practice link, and so on, and puts forward the method of training. It is expected to be helpful in the cultivation of information-based teaching ability of English majors.

1 INTRODUCTION

In the fall of 2007, the State Council decided to implement a policy of publicly funded education for normal students in the normal school, which is directly under the Ministry of Education. The main objective of the policy is to develop education fairly and to train outstanding teachers for rural, remote mountain villages, ethnic minority areas, and other areas, and to strengthen the teaching staff and improve the quality of education so as to promote the all-round development of education. In October 2013, The Ministry of Education promulgated the *opinions of the Ministry of Education on the implementation of the national project to enhance the capacity of primary and secondary school teachers in the application of information technology* (Teacher [2013] No. 13), it is pointed out that the application ability of information technology is the necessary professional ability of teachers in the Information Society. It is necessary to improve teachers' ability to use information technology [1]. The Ministry of Education in June 2016 promulgated the *13th five-year Plan* of education informatization pointed out that, by 2020, China will basically complete the education information system which is suitable to the educational modernization development goal of China, the training of information-based teaching ability of normal students is brought into the curriculum training system [2], the information-based teaching ability of normal students has become one of the important indicators to evaluate the effectiveness of its training. With the globalization of English and the rapid development of information technology, education informatization is the most remarkable educational reform that is going on all over the world at present. Under such circumstances, the process of training English state-funded normal students to seek professional development under the information-based environment is just what the education needs at the present time. Therefore, it is of great importance to study the training model of information-based teaching ability of English majors in state-funded normal schools.

2 AN ANALYSIS OF THE CURRENT SITUATION OF THE CULTIVATION OF THE INFORMATION-BASED TEACHING ABILITY OF ENGLISH MAJORS IN STATE-FUNDED NORMAL STUDENTS

After searching, summarizing, and analyzing the literature, it was found that many local colleges and universities have not achieved good results in developing the information-based teaching ability of English majors. The outstanding problems are that the goal of training the information-based teaching ability of state-funded normal students needs to be improved urgently, the information-based courses of English majors in state-funded normal student are single, the idea and means of information-based teaching are backward, and the information-based teaching practice for English majors in state-funded normal students is weak. These contradictions fully illustrate that there are still many problems in the training model of information-based teaching ability of students in state-funded normal schools.

2.1 *The goal of training the information-based teaching ability of state-funded normal students' needs to be improved urgently*

The aim of cultivation is not only the head of the whole cultivation process, but also the key link to the whole cultivation process. The goal of teaching ability training for English majors in state-funded normal students must be clear, otherwise it will be difficult to achieve the expected results. At present, in many local colleges and universities this training goal is the biggest shortcoming as it is too large and general. As to how to cultivate the core professional teaching ability of English teachers in the information-based environment, there is no specific and clear training objective. In the interview, many students in state-funded normal schools also say they do not know what their teaching objectives are; all these factors fully illustrate that the current training objectives of teaching ability in normal colleges and universities need to be improved.

2.2 *The information-based courses of English majors in state-funded normal student is single*

The information-based teaching skills course is an important part of the educational curriculum system for normal students, The course of information-based teaching ability and skill builds a bridge between the course of discipline, the course of educational theory, and the practice of education, as these courses are an important way for normal students to develop their individual teaching ability. Under the information environment, many colleges and universities, English major state-funded normal students, the course setting of teaching ability and skill is unreasonable. For example the curriculum is to emphasize knowledge over skills, the professional development lacks the support of systematic curriculum, and the consciousness and ability of professional development are not strong. In terms of curriculum development, many normal colleges and universities have set up such courses as pedagogy, psychology, subject teaching method, spoken language training, writing skill training, and modern education technology. Courses such as spoken language training, writing skills training, and modern educational technology are directly related to the training of teaching skills of normal students, and other courses are more about the theory of education [3]. Some English majors only offer modern educational technology courses, but not the integrated courses of information technology and teaching. The obvious manifestation of this tendency is more professional class hours and higher credit, however, the courses of information-based teaching for normal school students have few hours and light credits. The proportion of courses offered in many colleges and universities is seriously out of balance, with only 10% of courses in educational knowledge and information technology. In some colleges, education knowledge courses and information technology courses account for only 5% of the total class hours [4]. In addition, the awareness of English majors state-funded normal students to actively use information technology to promote their professional development is relatively weak. The vast majority of students do not fully grasp the technical means and methods required for teachers' professional development, and only make use of some ready-made network resources for lesson preparation and courseware production. Thus, it reflects a prominent phenomenon, that is, under the information environment, English majors' teaching skill courses and information courses are seriously inadequate. There is an urgent need to construct the curriculum of teaching skills. It is urgent for educational implementers to construct and implement these skills courses and information courses so as to realize the integration of education and teaching skills and information courses.

2.3 *The idea and means of information-based teaching are backward*

Information-based teaching is the process of constructing an information-based teaching environment, using information-based teaching resources to design and carry out information-based teaching activities under the guidance of modern educational thoughts and ideas [5]. Let's take the training of English majors in publicly funded normal students as an example: In the actual training of students, as a result of cognitive thinking, educational implementers are relatively old-fashioned in teaching ideas and backward in teaching methods, as a result, the state-funded normal students in the simulation class use multimedia instead of blackboards, and PPT instead of traditional blackboards to explain the so-called information-based teaching. At the same time, some normal colleges and universities mainly use the multimedia classroom to increase the normal student's informationization teaching ability, that is, the traditional "projector + screen" teaching environment. This kind of backward teaching method is far from the real information-based teaching environment, which leads to the difficulty for normal students to be competent at information-based teaching in primary and secondary schools.

2.4 *The information-based teaching practice for English majors in state-funded normal students is weak*

The aim of information-based teaching ability is to cultivate the application ability of information-based teaching for normal students, the practice link is an important way to cultivate the information-based teaching ability of normal students. At present, for many English majors in normal students, most of their information-based teaching practice links are from the microteaching training. However, microteaching time is usually not started until the second semester of the junior year, which greatly shortens the time for normal students to carry out teaching skills training, so that there is a serious lag in the understanding of the ability of information-based teaching. The second is off-campus educational internships, internships with fewer sessions, and short time internships. Through interviews, we know that most normal colleges and universities put the education internship in the second semester of the junior year or the end of the senior year, each class only going to primary and secondary schools once, with two classes at a time. Moreover, during the 12-week educational practice when students leave school in the last semester of university, some teachers in primary and secondary schools give insufficient guidance to normal students, and the guidance process is superficial, with the instructor not giving any guidance to the students on how to combine the information technology with the teaching. This is out of step with the normal students' teaching ability training under the information environment and the teaching practice of the role transformation from normal students to new teachers.

3 APPROACHES TO THE CULTIVATION OF INFORMATION-BASED TEACHING ABILITY OF ENGLISH MAJORS IN STATE-FUNDED NORMAL STUDENTS

The combination of English teacher training skills and information technology can improve the teaching quality and efficiency, and also can improve students' learning interest. In order to perfect and improve the ability of information-based education of English major's state-funded normal students, we can train the students in the following ways.

3.1 *Perfect the objectives and contents of personnel training*

The goal of talent training cultivates and answers the question of "what kind of person to cultivate," on the objectives and contents of the information technology ability of English majors state-funded normal students, we should adhere to *the standards for information-based teaching ability of normal students* and the *"13th five-year plan" for information-based education*, and refer to *National Primary and Secondary School Teachers Educational Technology Ability Test Outline* to train the English majors state-funded normal students. We should strengthen the ability of English majors in state-funded normal students to apply information technology, and improve their level of application to information technology; to promote educational modernization with educational informationization, actively promote the integration and innovative development of information technology and education technology[6]. As the reserve force of English teachers in the future, the information-based teaching ability of English majors state-funded normal students will directly affect the quality of English education. In order to adapt to the new requirements of the information age for contemporary English teachers, we should cultivate them with good teaching skills and information technology.

3.2 *Optimizing the course setting of information-based teaching and perfecting the system of ability training*

For normal students, the course study is an important place for them to acquire knowledge and improve their ability in the school, the training system of normal university students and the setting of specific courses play a very important role in the training and promotion of its information-based teaching ability. In order to improve the information-based teaching ability of English majors' state-funded normal students almost every normal university offers courses such as "computer foundation and modern education technology," but such information technology courses far from satisfy the needs of students. With the rapid development of education informatization, in order to cultivate new-type teachers who can satisfy the needs of future education, we should offer the required courses such as "web page design and production," "multimedia courseware production," and "micro-course design and production" to the majority of normal students. This would cultivate the ability of normal students to integrate information technology into education and teaching, in order to adapt to the new situation of English majors' normal students' information-based teaching ability training.

3.3 *Promoting the teaching consciousness of information technology and strengthening the idea of integration of subject and technology*

The awareness and attitude of normal students to use information technology are fundamental drives to improving their ability [7]. Strengthening the awareness of information technology of normal school students and active use of advanced science technology in learning and teaching, help to realize the deep integration of subject curriculum and technology. Normal school students themselves should pay more attention to the requirements of the new era for normal school students, think positively about the use of information technology, and constantly enhance their awareness and attitude to the use of information technology. They should explore microteaching, MOOCs, flip classrooms, and other teaching models that incorporate information technology, to explore the combination point suitable for the use of information technology in English subject, and explore the convenience of using information technology for learning and teaching.

3.4 *Strengthening the practice of teaching ability of information technology of English majors in state-funded normal schools*

The information-based teaching practice of normal students is the key to improving their ability with information-based teaching, among them, teaching novitiate, internship, and other links are particularly important, and students can experience the cognitive aspects of information-based teaching and the practical operation process. Schools can use high-definition recording and broadcasting classrooms, online viewing, and other information technology equipment to carry out long-distance novitiate. Students can use a remote observation lesson of primary and secondary school English teachers to take lessons, thus strengthening the normal student informationization teaching ability. Schools can also be targeted each semester or each school year to carry out, for example, Word, PPT, Micro-Course Make, video editing, and other training; hold more teaching simulation contest, lectures, class evaluations, and other activities, to guide English majors state-funded normal students to participate actively so as to improve their technical literacy for information-based teaching.

4 SUMMARY

A study on the cultivation model of information-based teaching ability of English majors state-funded normal students was carried out under the special social policy condition, this is of far-reaching educational impact and deep social significance. In the information age, the traditional cultivation mode of English state-funded teachers must conform to the trend of the times, respond to the reforms of the times, and improve the quality of English majors state-funded normal students. In the process of enhancing their skills, we can make full use of the advantages of the network to provide them with different learning methods and learning content, and develop different learning styles to meet the styles to meet the personalized needs of English majors state-funded normal students.

ACKNOWLEDGMENT

This research was financially supported by Sichuan Minzu College—"A study on the training model of teaching ability of English majors in public-funded normal students under the information environment". Project No.: No. 9[2019] of Sichuan Minzu College.

REFERENCES

[1] Opinions of the Ministry of Education on the implementation of the national project to enhance the information technology application ability of primary and secondary school teachers. [EB/OL]. http://www.moe.edu.cn/publicfiles/business/htmlfiles/moe/s7034/201311/159042.html, 2013-10-25.

[2] Circular of the Ministry of Education of the People's Republic of China, on the publication and distribution of the "thirteenth five-year plan for the informatization of education [EB/OL]. http://www.moe.gov.cn/srcsite/A16/s3342/201606/t20160622 269367.html.

[3] Zhong Ye. The training status of students' teaching skills in normal university under the local government. Sichuan Normal University Master's thesis, (2014) P22–22.

[4] Xianjie Sha, Decai Li. Reflection and reconstruction of teacher education curriculum under the concept of new curriculum. Heilongjiang Researches on Higher Education (2005) P108–110.

[5] Hong Yang, Xiaohua Qian. Research on the cultivate model of information-based teaching ability of normal students under the background of Wisdom Education. Journal of Chengdu Normal University (2017) P28–30.

[6] Youqun Ren, Hanbing Yan. Interpretation of the standards of information-based teaching ability for normal students. Theoretical discussion. (2018) P5–14.

[7] Yunjie Chen. Analysis and promotion of the current situation of normal students 'information-based teaching ability of countermeasure study. Shaanxi Normal University. (2018) P70–71.

Computational Social Science – Luo, Ciurea & Kumar (eds)
© *2021 Taylor & Francis Group, London, ISBN 978-0-367-70193-2*

Discussion of college students' learning burnout from the perspective of three forces of psychology

X.L. Hu & J.Q. Sha*
Shandong Medical College, China

Y. Wang
Taishan Vocational College of Nursing, China

Y. Cao
Shandong Medical College, China

ABSTRACT: Learning burnout is a kind of psychological state of physical and mental exhaustion when students are not interested in learning or become lacking in motivation but have to continue. Learning burnout is common among college students, and has a negative impact on their physical and mental development. From the perspective of psychoanalysis, behaviorism, and humanistic psychology, this paper attempts to analyze the causes of college students' learning burnout, and puts forward some improvement measures, with an expectation to provide some reference for colleges and universities to effectively prevent and intervene in cases of learning burnout in educational teaching and management practice for the promotion of college students' mental health development.

1 INTRODUCTION

The earliest exploration of learning burnout, in the field of psychology, began with the study of occupational burnout. Pines (1980) and Meier (1985) defined learning burnout as a phenomenon in which students are exhausted due to the long-term academic pressure and workload; their enthusiasm for academic work and activities gradually disappears; their attitude toward fellow students is indifferent and alienated; and they have a negative attitude toward learning. Lian Rong, a scholar in China, defined learning burnout as follows: when students have no interest in learning or lack motivation but they have to do it, they will feel bored, resulting in a psychological state of physical and mental fatigue, and further they treat their learning activities passively, causing a series of inappropriate behaviors of avoiding learning. It can be seen that the negative psychological states of learning burnout, such as the fear of learning, the lack of learning motivation, depression, anxiety, and low self-esteem, and the behaviors of not listening to lectures, sleeping in classes, using cell phones, skipping classes, and pursuing material pleasures, will seriously affect the college students' academic success and the development of their physical and mental health.

2 AN ANALYSIS OF THE CAUSES OF LEARNING BURNOUT FROM THE PERSPECTIVE OF THE THREE FORCES OF PSYCHOLOGY

Over more than 100 years since the birth of scientific psychology, there have appeared many influential schools in Western psychology. Among them, psychoanalysis, behaviorism, and humanistic psychology have had the greatest impact, and are called the three major forces of psychology. The following attempts to analyze the causes of college students' learning burnout from the perspective

---

*Corresponding author

of these three forces, aiming to provide theoretical support to improve the teaching and management of college students.

2.1 *Psychoanalysis*

Psychoanalytic psychology was founded by the Austrian psychiatrist and physiologist Sigmund Freud in the late 19th and early 20th centuries. It is a psychological theory gradually formed in the clinical practice of psychotherapy. Freud put forward the personality theory of id, ego, and superego. The id was defined in terms of the most primitive unconsciousness, which looks for a way to pursue gratification according to what Freud called the pleasure principle. The ego represents rationality and correct judgment, controlling and adjusting the id in accordance with what Freud called the reality principle. The superego is related to morality, and follows the principle of the best way to dominate and limit the id. The balance of the three promotes the normal development of an individual personality. If the three are out of balance, individual mental health will be frustrated. From the perspective of psychoanalysis, college students' learning burnout comes from the constant limitations of the ego and the superego on the id impulse of the instinctive pursuit of pleasure. Just as Freud compared the id and the ego, respectively, to a horse and its knight, if the knight wants to control and restrain the unruly horse, he is bound to be strongly resisted, and thus there inevitably will be contradictions and conflicts. The richness of the material well-being life at the present age, not only improves people's living standards, but also brings various temptations. Compared with the boring process of knowledge learning, material enjoyment is obviously more attractive. Many college students begin to indulge in playing with cell phones and electronic games, overeating, and other kinds of entertainment. At the same time, they are clearly aware that their academic tasks must be completed on time, so the gratification of the id is resisted or even hated by the ego and the superego. Thus the individual struggles in the confrontation between the two forces, ability falling short of his wishes, eventually resulting in fatigue and burnout, which will necessarily have a serious negative impact on the study and life of college students.

2.2 *Behaviorism*

In the early 20th century, the American psychologist John B. Watson founded the theory of behaviorism. This school of psychology advocates the objective and experimental study of individual behaviors. It holds that environment determines a person's behavior pattern. Both normal and pathological behaviors are acquired through learning, so the acquired environment can predict and control the individual behaviors. From the view of behaviorism theory, the behavioral expression of learning burnout, such as the lack of learning interest, spiritual decadence, learning inefficiency, truancy, smart phone addiction, delaying learning tasks, and so on, are actually learners' responses to the environmental stimulation, which are a combination of various physical responses made by an individual to adapt to the stimulation. Once learning is replaced by pleasure, and it does not bring immediate adverse feelings, but on the contrary the individual feels happy because there is no need to work hard, he then will have the desire to repeat the behavior. Consequently, during the process of strengthening the stimulation repeatedly, a habit is formed. College students in the growing stage still have immature values and psychological development and lack self-control. Once they get into the habit of avoiding learning, they will be afraid of learning. As soon as they learn, they will show negative emotions such as mental fatigue, frustration, depression, complaints, anxiety, and so on, which will seriously affect their study and life.

2.3 *Humanism*

Humanistic psychology, represented by Maslow and Rogers, originated in the United States in the 1950s and 1960s, and is one of the major schools of contemporary psychology and the most famous theory about it is Maslow's theory of the hierarchy of needs. Maslow studied and classified the basic needs of human beings. He believed that human needs are divided into five levels, like a pyramid, from the bottom to the top. The lowest needs are physiological needs, including breathing, diet, sleep, shelter, etc., which are the needs that people feel should be satisfied first. The second level is about safety and security. The needs of these two levels are the basic ones, that are felt when lacking. The third level is the need for love and a sense of belonging, including people's pursuit of friendship, love, and good relationships with family. The fourth level, which Maslow called esteem, is related to self-esteem, status, freedom, etc. It emphasizes people's desire for recognition and respect from society and others. At the top of the pyramid is self-actualization, which refers to the pursuit of one's personal ideals, truth, goodness, and beauty. At this level, one will try to maximize his personal capacities and make himself the person he expects to be and therefore he will feel the greatest joy and satisfaction. Maslow believed that from low to high, people must first meet the needs of the lower level before they can reach the higher one. Some studies have suggested that the self-esteem level and social support level of college students are negatively correlated with learning burnout. When the need for esteem is not satisfied, one will lack the affirmation of self-worth due to not being recognized, and self abasement will appear. In essence, this is a kind of psychological gap and deficiency. He thinks that he is in an inferior position and lacking ideals, just like a ship without a rudder, living in ignorance, not to mention failing to achieve academic success and self-worth.

3 INTERVENTION MEASURES OF COLLEGE STUDENTS' LEARNING BURNOUT IN THE SCHOOL ENVIRONMENT

The college years are the key period for college students to lay the foundation for their future career, improve their accomplishments and ability, and form systematic values, and an outlook on life and the world. Campus life is the main theme of college time, and the school environment has a decisive influence on college students' learning and life. Therefore, in order to ensure the quality of learning and the physical and mental health of college students, it is of far-reaching practical significance to study, in the school environment, how to take measures to prevent learning burnout.

3.1 Administrative management

College students, who are still immature, tend to be lacking in self-discipline and self-control. They are easily tempted by material pleasures and put aside the main business of learning. According to behaviorism theory, the acquired environment can intervene and control individual behavior. Therefore, strengthening the construction of the learning atmosphere should be an important part of the management of college students, including establishing and improving the school rules and regulations, strictly controlling students' absence, severely punishing the negative behaviors such as delaying homework, skipping classes, playing with cell phones in class, and so on. In addition, it is necessary to set up strict exam discipline and punish cheating in exams.

The construction of a learning atmosphere can also be realized by carrying out various campus cultural activities. Healthy and positive campus activities will help to enrich students' campus life, cultivate their ability, and broaden their vision. All of these will be a great help to the establishment of positive and healthy values of students. For example, all kinds of competitions on professional knowledge, Chinese or English speech contests, or various academic events, can provide wonderful platforms for students to show themselves, by virtue of which, students gain much recognition and encouragement from schools and fellow students, as well as the academic theoretical and practical knowledge. The sense of achievement gained in learning will definitely become a motivation for learning, and it is also a defense and resistance mechanism, for the id, against the pursuit of ease and the escape from hard work. As a result, students will gradually get rid of learning burnout and be more willing to learn. Once they form the habit of loving knowledge, they'll accordingly achieve better academic performances, and this is undoubtedly a favorable and virtuous circle.

3.2 Instructional management

Behaviorism regards environment as a stimulation. In order to adapt to the environment, one will react accordingly, and this will become a habit over time. The application of behaviorism theory in classroom teaching requires the teachers to create a positive and favorable classroom setting for students, shape students' good learning behaviors, and correct any negative ones. A relaxed and natural classroom atmosphere is helpful to arouse students' enthusiasm for learning and achieving good teaching effects. First, the teacher should be amiable, with affection for the students; the teaching language should be fluent and clear, and the teaching methods should be diverse, vivid, and interesting. Furthermore, the teacher should pay close attention to their communication and interactions with the students, and always keep a close eye on the students' responses. For example, the teacher could step into the middle of students, giving timely guidance to students' answers; for the different answers given by the students from different angles, the teacher could give a lot of affirmation and encouragement. Such an attractive class atmosphere will enable students to get a sense of achievement and pleasure from knowledge, thoroughly enjoying the class and knowledge-learning and actively stay away from external material temptations.

In addition, the ideological and political content of classroom teaching is also an effective means to help students establish the correct outlook on life and values. With the help of teaching materials and contents, the importance of learning and correct attitude toward life can be permeated into the teaching process to enhance students' sense of community responsibility and mission. In particular, the ideological and political theory course is an important part of the ideological education for college students. What is most important is to improve students' enthusiasm for learning, and help them to mature into a productive person.

3.3 Career planning

Helping students to carry out scientific and orderly career planning is an effective help for students to realize their self-worth. Once college students have set up their lofty ideals and life goals, they can rid themselves of learning burnout, and form a good habit of independent learning. Thus they will gradually achieve academic success, and finally realize their life goals. In the career planning project, the university administrators should help the college students to associate the importance of learning with their own future, so as to stimulate their motivation for learning. First, students should be guided to look at themselves objectively, treat their own shortcomings and potentials correctly, especially finding their own strengths, so that they will obtain love and respect from others and establish development goals suitable for themselves. And during the pursuit of these objectives, they will gradually realize their self-worth. Furthermore, universities could also set up a scientific training program and professional curriculum framework. It is desirable to invite influential workers in the respective industry or outstanding professionals who have graduated from the same school to return to give speeches to the students, so that they can fully understand the characteristics of their major, and cultivate their interest and enthusiasm for professional learning. In addition, great importance should be put on developing and cultivating the professional commitment of the students; learning a certain major is their own life choice, so they should take the responsibility for it. When professional achievement and social responsibility are linked, and they realize that their profession will become an effective means to benefit the community and serve people, the students will feel a sense of honor and a sense of mission in their heart, which will become a motivation for them to avoid learning burnout, continue to pursue their studies, and further realize their self-worth.

4 SUMMARY

College students are the future builders of the country. It's fair to say that their academic achievements are directly related to the development of the country. Learning must be their main business. Learning burnout, as a negative emotional experience in learning, has a negative impact on the physical and mental health of college students, which must be highly valued by colleges and universities. The authors believe that the problem of learning burnout can be effectively controlled through effective prevention and intervention from the three aspects of administrative management, instructional management, and career planning.

REFERENCES

[1] Pines A, Kafry D. Tedium in College. Paper Presented at the western Psychological Association Meeting, Honolulu, Hawaii, ERIC Document Reproduction Service No.ED192210, 1980.

[2] Meier S F, Schmeck R R. The Burned-out College Student: A Descriptive Profile. Journal of College Student Personal, 1985, 1: 63–69.

[3] Lian Rong, Yang Lixian, Wu Lanhua; Relationship between Professional Commitment and Learning Burnout of Undergraduates and Scales Developing; Acta Psychologica Sinica; 2005, 37(5) (In Chinese).

[4] Catherine Meyer; Translated by Guo Qinglan & Tang Zhian; Criticism of Freud; Shandong People's Publishing House, January 2008 (In Chinese).

[5] John B. Watson; Translated by Guo Benyu; Behaviorism; Commercial Press, July 2019.

[6] A.H. Maslow; Translated by Tang Yi; Maslow's Humanistic Philosophy; Jilin Publishing Group, June 2013, First Edition (In Chinese).

[7] Wang Mingchun, Zhang Lixia; Performance of College Students' Learning Burnout and Intervention Measures; Educational Exploration; 2011, (6) (In Chinese).

[8] Wang Chen; Analysis of College Students' Cell Phone Addiction from the Perspective of Three Forces of Psychology; Legal System and Society; 2019(8) (In Chinese).

[9] Wen Ying; A Study of College Students' Learning Boredom and Corresponding Countermeasures; Journal of LuoYang Institute of Science and Technology (Social Science Edition); 2015, 30(6) (In Chinese).

Computational Social Science – Luo, Ciurea & Kumar (eds)
© 2021 Taylor & Francis Group, London, ISBN 978-0-367-70193-2

Reform and innovation of training mode for master of transportation engineering under the transformation of new and old kinetic energy

X.D. Tang*, X. Lu, Z. Qu & W.H. Wang
Division of Disciplines and Graduate Students, Shandong Jiaotong University, Jinan, China

ABSTRACT: The transportation industry is one of the three major supports for the transformation of new and old kinetic energy. It is of great significance to actively serve the needs of the transportation industry and cultivate high-level applied and interdisciplinary talents, so as to improve the overall quality of talents and consolidate the support for the transformation of new and old kinetic energy. To meet the needs of the transportation industry under the transformation of new and old kinetic energy and to aim at the deficiencies in knowledge ability, engineering practice ability, and application characteristics, a curriculum system for serving transportation industry, a practical system for improving engineering capacity and a quality assurance system for training applied talents need to be established by the reform of the existing curriculum system, engineering practice link and the quality assurance system.

1 INTRODUCTION

High-level applied and interdisciplinary talents with solid foundation, comprehensive quality, strong engineering ability and certain innovation ability are training objectives in the field of transportation engineering. With the rapid development of professional degree education in China, the number of transportation engineering master is increasing, but it is still not suitable for the requirements of the transformation of new and old kinetic energy and transportation power. The master of transportation engineering possesses a distinctive comprehensive and applied character, but the insufficient knowledge and ability, engineering practice ability, and application-oriented characteristics seriously affect and restrict the training quality. To meet the requirements of the transformation of new and old kinetic energy and strengthen the supporting role of transportation, the training mode for master of transportation engineering must reform and innovate.

Following the characteristics of comprehensive and application of transportation engineering and the cross integration with natural science and social science, Shuai et al. put forward that the cultivation of engineering masters needs to establish a multidisciplinary and multi domain knowledge system [1]. Huang et al. thinks that engineering practice is an important segment in the cultivation for master of transportation engineering. The cultivation of engineering ability should strengthen the engineering practice ability, and innovation ability and leadership ability [2]. Zhang believes that the training quality is the core of the engineering master, and the whole process quality assurance system from student source, teaching, management to tracking should be established [3]. Combining with the construction of engineering education certification, training system, and quality assurance system, Zhang [4], Du [5], and Liu [6] analyzed on improving the training quality of master of transportation engineering. Based on the analysis of the social and economic development of Yunnan Province, Chen et al. put forward the optimized strategies for improving the training quality, and pointed out the influence of curriculum design, interdisciplinary education, quality assessment, practical assessment, incentive system and government support on the training quality [7].

---

*Corresponding author

2 QUESTIONS ON THE TRAINING MODE FOR MASTER OF TRANSPORTATION ENGINEERING

Focusing on the requirements of the transformation of new and old kinetic energy and serving the development of the transportation industry and regions is the fundamental principle for the personnel training in the field of transportation. Guided by the construction of a comprehensive transportation system, the construction and upgrading of channels, intelligent guidance, industrial integration, green improvement, and service improvement, and building a high-quality transportation infrastructure network, high-end intelligent transportation network, and efficient transportation service network should be carried out. Talents are the first resource, and transportation engineering masters should grasp basic theory of transportation engineering, the actual work requirements, application and practice, high-level training in transportation engineering expertise, and engage in obvious professional background, rather than academic research.

In the view of the training path and practical performance of master of transportation engineering, the problems of insufficient knowledge and ability, insufficient engineering practice ability, and insufficient application-oriented characteristics are more prominent and should be solved. The deficiency of knowledge ability is composed of lack of the ability to acquire knowledge and the ability to apply knowledge, the former includes the deficiency of the ability to acquire books and documents, to master the progress of engineering application [8]; the deficiency of the ability to acquire the data of experiment, test, and simulation, data analysis processing, comprehensive use of basic and professional knowledge, the advanced technology and methods in the field of transportation engineering, building models by qualitative and quantitative analysis, solving practical problems, and analyzing, writing, expressing, and organizing ability are not enough. The lack of engineering practice ability lies in that the ability to find and analyze problems in the research and technology development practice, the ability to propose effective plans in the research and development process, the ability to systematically analyze and creatively solve complex problems are not enough in the face of huge and complex transportation system engineering. The deficiency of application features lies in that the adaptability to the transportation industry is not enough, the necessary endurance and toughness are lacking, the integration with the industry is insufficient, and the training quality needs to be further improved.

All of above-mentioned questions concentrated in the following aspects: narrow knowledge with unreasonable structure, difficult to meet the knowledge demand of comprehensive transportation system for high-level talents, lack of opportunities for actual project participation, lack of time and content of engineering practice, poor industry adaptability, difficult to play certain role, etc.

3 HOW TO REFORM AND INNOVATE THE TRAINING MODE FOR MASTER OF TRANSPORTATION ENGINEERING

3.1 *Curriculum system reform*

The training curriculum system should be designed based on the professional ability in the field of transportation industry, and the curriculum system should be established according to the idea of industry requirement-training objectives-professional ability-curriculum system-professional ability-goal realization. The curriculum design can be enriched and improved by industry experience and thinking through the employers participating the curriculum. For a long time in the future, China's transportation industry should comply with comprehensive transportation, safe transportation, intelligent transportation, and green transportation. Therefore, there must be relevant specific application courses in the master's training program, and the courses mainly focusing on theory should be adjusted to the courses mainly focusing on application. For example, traffic planning principle, traffic safety engineering, traffic design theory, intelligent transportation system, traffic simulation theory and traffic network balance analysis and model theory should be adjusted to comprehensive traffic planning technology method, traffic safety guarantee technology, traffic design technology, intelligent transportation technology, and traffic network balance analysis technical methods, respectively. Meanwhile, the course of sustainable development strategy should be offered.

3.2 University and enterprise joint training

Industry-university-research cooperation is an effective way to reform the postgraduate training system. Universities have the advantages of knowledge and talents, which can provide enterprises with intellectual support and talents, and promote technological progress. The industries with advantages in capital and practice, can provide services for talent training and achievement transformation in universities, which are mutually beneficial. Through cooperation with the society, the formation of industry-university-research cooperation training mode, not only to meet the needs of society, alleviate the shortage of college teachers, but also improve the quality of graduate education, improve the scientific and technological innovation ability and social service ability of colleges and Universities [9].

For universities, through making use of the advantages of market demand information, practical technical information and other resources, adjusting the training objectives and curriculum settings in time, training talents can meet the needs of the society. The professional practice guidance of enterprise technical talents for graduate students can make up for the deficiency of school education practice. For enterprises, they can use the rich intellectual resources of universities to promote the research and development of enterprises and create benefits for enterprises [10]. Through the joint training of school and enterprise, we can make full use of the superior teaching environment and sufficient technical equipment resources of the school, realize the complementary resources, and combine theory with practice. The establishment of school enterprise joint training platform solves the problem of technical personnel training, promotes the technological progress of relevant enterprises, gives full play to the university function of serving local economic and social development, improves the level of scientific research, and cultivate high-level applied talents for the transportation industry.

3.3 Establishment of effective quality assurance system

Quality is the key to talent training. The quality assurance system involves many segments such as enrollment, training, and degree. Comprehensive, all process, and all factor quality assurance system is anticipated to be established.

Improving the quality of students. It is necessary expand the scope of enrollment, expand the source of students, pay attention to the work experience and performance of examinees, focus on the knowledge structure of students, the ability to analyze and solve practical engineering problems, increase the proportion of interviews, and further increase the autonomy of tutors.

Adjusting the distribution of benefits and improving the training conditions. It is necessary to reduce the proportion of administrative expenses, increase the expenses for teaching and thesis guidance, and actively create and improve training conditions.

Improving the quality of tutors. To ensure the quality of teaching is an important part of improving the quality of training. Those who understand the latest development of engineering technology and communicate well with students should be selected as the first-line teachers, and the full-time professor system for master of transportation engineering should be implemented.

Optimizing the double tutor system. It is necessay to make clear the leading role of tutors, take their advantages in theoretical knowledge, give students strong guidance in topic selection, topic opening, implementation plan formulation, thesis modification, etc. It should be taken the advantages of tutors outside in engineering technology and engineering management practice experience, and solve technical problems. The internal and external tutors should strengthen contact and jointly complete the guidance.

Carrying out the topic selection in combination with the actual project. According to the actual situation, the needs of the enterprises and the scientific research foundation of the tutors, the topic selection is determined, and the systematic correlation between the course learning and the dissertation is strengthened. The topic selection of the thesis should come from the key technical problems that need to be solved urgently by enterprises, so as to improve the practicability and application value of the thesis.

Strictly implemetning the medium-term screening policy. The medium-term assessment and screening mechanism in the training of master of transport engineering, resolutely drop out students who have been identified as unqualified in the mid-term screening should be carried out and maintain the seriousness of the training mechanism.

Implementing expert supervision system. The education quality of master of transportation engineering, including enrollment, training process, curriculum construction, tutor team construction, should be monitored in an all-process way, and the inspection and evaluation mechanism shall be established.

Establish the system of dissertation spot check. It is necessary to improve the quality supervision system, follow the procedure of pre-evaluation-preliminary evaluation-blind evaluation-pre defense- defense, and strictly control the quality of dissertation. For the unqualified theses, the tutors, the tutor group, the defense committee and even the leaders in charge should to held responsibility.

4 SUMMARY

To meet the needs of the transportation industry under the transformation of new and old kinetic energy, through the reform of the existing curriculum system, engineering practice segments, and the quality assurance system ("Three Reforms"), the establishment of a curriculum system aimed at serving the needs of the transportation industry, a practice system oriented by improving engineering capacity and a quality assurance system focusing on the cultivation of applied talents ("Three Systems"), the problems such as the lack of knowledge and ability of master of transportation engineering, engineering practice ability, and application-oriented characteristics ('Three Deficiencies") can be expected to solved, which will improve the training quality of master of transportation engineering, and provide talent support for the healthy and rapid development of transportation industry.

ACKNOWLEDGEMENT

This research was financially supported by Shandong Provincial Graduate Education Quality Improvement Plan Project (SDYZ18021).

REFERENCES

[1] B. Shuai, P. Zhong. Research on master training system of transportation engineering to meet the development requirements of comprehensive transportation system, National Academic Conference on Degree and Graduate Education Evaluation (2012) 392–399. (In Chinese).
[2] J. Huang, L. Zhang, A. Guo. Investigation and analysis on the training of engineering leadership ability of professional degree graduate students, Industry and Information Education 12 (2016) 11–14. (In Chinese).
[3] Z. Zhang. Establishment and realization of education quality assurance system for master of transportation engineering, National Symposium on Postgraduate Education of Engineering (2008) 178–181. (In Chinese).
[4] T. Zhang, X. Yu. Quality evaluation index system of graduation design of transportation under the background of engineering education certification, Education and Teaching Forum 12 (2016) 195–196. (In Chinese).

[5] Y. Du, X. Ye, C. Yang. Exploration and practice of postgraduate training system of transportation engineering, Education and Teaching Forum 37 (2012) 96–98. (In Chinese).

[6] H. Liu, Y. Shen, C. Yang. Construction and consideration of the quality assurance system of engineering postgraduate education. Degree and Postgraduate Education 10 (2004) 25–28. (In Chinese).

[7] F. Chen, X. Yuan, M. Cui, F. Guo, Research on the training mode of transportation engineering graduate students driven by regional development needs, Journal of Kunming University of Science and Technology (Social Science Edition) 15 (2015) 80–85. (In Chinese).

[8] Z. Zhang, Z. Chen, F. Zhou. Innovation and practice of talent training mode guided by social demand: taking environmental science of Chongqing Jiaotong University as an example, Journal of Chongqing Jiaotong University (Social Science Edition) 16 (2016) 126–130. (In Chinese).

[9] H. Shan. Exploring the concept, mode and mechanism of the combination of industry-university-research, Chinese Higher Education 9 (2008) 30–32. (In Chinese).

[10] W. Liu, J. Dong. Exploration and practice of the cooperative training mode of "industry-university-research" for graduate students. Journal of Shanghai University of Science and Technology (Social Science Edition) 38 (2016) 276–280. (In Chinese).

Computational Social Science – Luo, Ciurea & Kumar (eds)
© 2021 Taylor & Francis Group, London, ISBN 978-0-367-70193-2

Research on the impact of the Internet on the informatization teaching reform of marketing in universities

Y. Zhou, C.H. Jin, J.W. Huang* & R.M. Wang
Beijing Information Science & Technology University, Beijing, China

ABSTRACT: In reaction to the impact of the Internet and "big data" on marketing, combining the teaching experience of the front-line, from the four levels: knowledge systematicness, theoretical and academic nature, scientificity of methods and professional practically, a unique four-wheel drive mode of teaching reform: Learning-Research-Method-Application model should be formed, aiming to rethink and reposition the teaching system design and method improvement for the marketing specialty in applied undergraduate colleges, so as to promote the coordinated development of scientific and technological progress and professional talent cultivation.

1   INTRODUCTION

Chinese President Xi Jinping put forward that "when education prospers, the country prospers; when education is strong, the country is strong; higher education is a significant symbol of the development level and potential in a country" [1]. In the Internet era, advanced science and technology, such as big data, artificial intelligence, 5G, and so on, are widely used to bring convenience to production and life, and meanwhile make a huge impact on teaching methods and ideas among front-line educators. The rise of new concepts and marketing methods, such as Internet marketing, digital marketing, KOL (Key Opinion Leader), and precision marketing, and the emergence of interactive tools, such as microblog, WeChat, community service applications, and TikTok, have shortened the marketing distance, updated marketing channels, and enriched communication methods. Instead of being an unchangeable science, marketing is constantly updated and enriched based on classical theories, which requires teachers to reconsider and reorient the teaching methods as well as reconstruct teaching models so that the coordinated development between technology advancement and professional training will be promoted.

BISTU (Beijing Information Science and Technology University), an institution of higher education, featuring information technology, has made great achievements in university information construction in recent years. At present, with the challenges and opportunities of training professional marketing talents against the background of the Internet, it is more necessary and practical to think thoroughly about the methods and ideas for reforming the informative teaching of the marketing specialty.

2   RESEARCH STATUS

2.1   *Overseas research*

Audrey Gilmore and some other researchers, in a paper reviewing and reflecting on the design and delivery of entrepreneurial marketing (EM) education in universities, put forward four key questions, these were: (1) what should be taught; (2) how it should be taught; (3) where it should be

---

*Corresponding author

taught; and (4) who should teach EM [2]. According to Paul Gibbs, marketing practice depends upon technique, while the goal of education is the practical wisdom beyond immediate time horizons [3]. Sarath A. Nonis and some others, whose first research objective in their relevant study was to highlight key developments in the composition of today's college students with an emphasis on marketing students, stated that several key trends in technology, globalization, learning patterns, and student composition have significantly influenced colleges and universities over the past few years [4]. From a historic perspective, Tom Hayes studied marketing of higher education and particularly pointed to the importance of blending strategic planning with marketing and developing integrated marketing systems [5]. Michael L. Klassen mentioned that, when it comes to marketing, the Internet is the "great equalizer," permitting service-based organizations, no matter how successful or well known, the ability to promote their products and services in a more-or-less equal manner [6].

2.2 *Domestic research*

Xu Chonghuan and others explored the introduction of big data research methods and research tools in the field of information science to teaching research in marketing, and built a four-in-one layered teaching implementation platform based on career planning classification. Their suggestion of on-demand teaching implementation which can be realized through measures such as basic classroom teaching and theoretical research, experiments and practical training, and online education, provides a reference for the design and promotion of marketing simulation applications [7]. Wu Ling explored a constructivist teaching model: the 5E teaching model (Engage, Explore, Explain, Elaborate, Evaluate), and applied it to the teaching design of marketing courses [8]. Qiu Yuanyuan analyzed the intelligent teaching model of marketing curriculum in the context of the "Internet + education" era, and advised that a new model should be rebuilt and its design can be viewed from three aspects: teaching frame, teaching content, and imitative practice in teaching [9].

2.3 *Problems in the existing research*

First, in the existing research, the impact of the Internet on traditional marketing as well as the opportunities and challenges that came along have been analyzed in detail. What it lacks is an in-depth discussion on how marketing professionals deal with the problem of detriment and refinement between emerging technologies and traditional theories in the context of the Internet and "big data."

Second, redesign of the teaching model and teaching process, despite a temporary adjustment to connect with the Internet and "big data," cannot be fundamentally integrated with the teaching of marketing courses in a deep and long-term way. More effective implementation steps and specific programs are needed.

Third, as well as the importance of the marketing courses being elaborated, the analyses remain insufficient. The significance of course teaching and the key and difficult points of knowledge that students need to master are not fully understood, let alone the seamless connection between professional personnel training and market demand.

Finally, marketing has not been analyzed as a science. Beside the application and practical significance, the scientific and technical connotation of marketing are not supposed to be ignored, and to conduct an in-depth analysis as an essential discipline in higher-level research is also necessary.

3 BODY

In front-line education work, the importance of and need for reform of the marketing professional courses in colleges and universities are deeply understood. The development of marketing in China lags behind that in Europe and the United States, however, in recent years, due to the governments' attention and the development of disciplines, marketing has become indispensable in the economics and management of universities and colleges. This teaching reform cannot be accomplished overnight. Instead, it should be explored thoroughly in all directions, such as the history formation, classical theory, teaching books, teaching system, scientific research basis, and teaching experience of the subject. More importantly, the reform model needs to be summarized to meets the needs and training goals of national professionals according to student competence and social needs. With the development of the Internet, the Internet of Things, big data, and other information technologies, to reform teaching in colleges and universities needs a serious consideration of the way to closely link the reform and informatization in the Internet era. Some ideas for marketing information reform against the background of the Internet are discussed in this paper from the following four aspects.

3.1 Promoting teaching with learning to consolidate the knowledge systematicness

The traditional concept of education holds that the main task of teachers is to impart knowledge, hence teachers are primarily supposed to teach scientific knowledge. In order to be mentors in the sea of knowledge, teachers need to study theoretical knowledge intensively and master the essence of theory so as to impart scientific knowledge to their students. Therefore, "promoting teaching with learning" means advice for teachers to first learn professional knowledge and improve professional quality. Only by learning the curriculum thoroughly, will teaching be promoted properly and effectively. Choosing a good teaching book is the first step in learning. Although the Internet has brought massive teaching resources, textbooks should not be abandoned, as the quality of resources are can be mixed.

The selection of teaching materials varies from person to person. Some teachers like original textbooks in foreign languages, while some prefer local teaching materials. Taking the textbook *Marketing* as an example, the foreign textbooks and local ones are brought into comparison for analyzing their respective strengths and weaknesses, extracting their essence and discarding the dross to form a distinctive knowledge system. Among numerous excellent textbooks, two are picked. One is *Marketing* (hereinafter referred to as the textbook (1), edited by Wu Jianan and published by Higher Education Press, as the representative of local textbooks, the other is *Principles of Marketing* (hereinafter referred to as textbook (2) by Philip T. Kotler, as the representative of foreign textbooks. By comparing them in Table 1, there is a discussion presented on the similarities and differences between Chinese and foreign textbooks in terms of knowledge system, content structure, case analysis, and teaching ideas.

3.2 Promoting teaching with scientific research to enhance the theoretical and academic nature of the course

The best way to increase the depth of teaching is provided by scientific research feedback. By participating in projects and writing academic articles, teachers can integrate scientific research into teaching, not only to develop their own academic abilities, but more importantly, to combine the results with teaching content to enhance the theoretical and academic nature of the course, so that the students who have research interests and potential can be explored and cultivated by being offering the opportunities of further development.

Domestic marketing research started relatively late compared to other social sciences. Taking academic journals as an example, there are no top journals especially for marketing. According to the statistics of class A journals based on the fourth national discipline assessment, there are in total 34—15 are domestic and 19 are foreign—of Management Science and Engineering (1201) and Business Management (1202), among which there are three with marketing columns, *Management Science*, *Nankai Management Review*, and *Management Review*. In 2018, among all the papers published in the marketing column, 11 were from *Management Science*, accounting for 4.9%; nine were from *Nankai Management Review*, accounting for 6.2%; and 21 were from *Management Reviews*, accounting for 6.0%. The publication rate of marketing in class A journals is low, which means marketing disciplines still need to redouble their efforts.

In terms of scientific research programs which is shown in Table 2, statistically, the numbers of marketing programs approved in by the National natural Science Foundation of China (NSFC), which is one of the highest-ranking and best-funded foundations in China and famous for its high

Table 1. Comparison of Chinese and foreign teaching materials in marketing.

Comparative objects	Textbook 1	Textbook 2
Historical Accumulation	The historical accumulation is relatively shallow: Time from 1978 to 1983 was the period of enlightenment stage of marketing, when it was reintroduced to mainland China. From 1984 to 1994, marketing spread rapidly in China. After 1995, the deeper and broader research and application of marketing theory began to be extended.	The historical accumulation is deep: Marketing, founded in the United States in the 20th century, was formed around 1900 to 1930, and later spread to Europe, Japan, and other countries. Then marketing was born as an independent business management discipline.
Knowledge System	Advantages: Compiled according to the marketing concept, marketing environment analysis, target marketing strategy, SWOT, 4P, and the structure of new concepts in the new field, it is in line with the learning and thinking habits of Chinese students—learn about the context of a discipline first, then every theory or knowledge point, and finally look to the future development prospect of disciplines. Moreover, in the sixth edition of the book, the system is relatively mature. Disadvantages: The knowledge points are disjointed and fragmented.	Advantages: First, the theoretical viewpoints keep pace with the times. For example, some proposals such as the concepts of sustainable marketing, international marketing, and digital marketing combine the theory with the high developing speed of marketing practice. Second, the rich typesetting content is illustrated by a large number of figures that accord well with the theory, which makes the book more comprehensible. Third, using detailed diagrams to intuitively represent procedural content adds to the comprehensibility again, and also reflects its logicality. Disadvantages: Adhering to the consistent characteristics of foreign language teaching materials, the knowledge features a relatively loose system and dot-shaped distribution.
Case Relevancy	There are 174 cases in the book, among which there are 154 in-class cases and 20 after-class cases. The case relevancy is lower, and the proportion of foreign cases and domestic cases is mixed. They are mostly event descriptions and less in-depth analysis.	This book has higher numbers of relevant cases and more novel ones. There are 388 cases in the book, among which there are 290 cases in class and 98 cases after class. In one paragraph mostly several cases expressed in one simple sentence are included, and these cases show a close relation to the key points. The fact that all the cases are examples in recent years makes the book highly up-to-date and practical.
Informatization Level	The book consists of 18 chapters. Hereinto, the parts related to Internet marketing are as follows: Chapter 11.3, Registering Internet domain names; Chapter 12.3, Internet-based pricing strategies; Chapter 13.2, Internet-based mobile cloud distribution; Chapter 13.3, TV shopping and online stores; Chapter 14.2, advertising media; Chapter 14.3, freedom and subsidies in the Internet era; Chapter 14.4, big data and Internet marketing. It can be seen that the informatization level is relatively low, and the background of Internet is loosely combined.	The informatization level is relatively high. All 20 chapters of the book contain cases of informatization-related technical methods and theories such as digital marketing, which means the informatization coverage reaches 100%.

Table 2. List of the marketing programs approved in NSFC in 2018.

Application code	Subject code	Subject name	General program	Youth program
G0207. Marketing	G0207	Marketing	26	27
G020701. Marketing Model	G020701	Marketing Model	21	1
G020702. Consumer Behavior	G020702	Consumer Behavior	3	15
G020703. Marketing Strategy	G020703	Marketing Strategy	8	8

Table 3. The technologies and methods needed for marketing in the context of the Internet.

Basic Methods	Big Data Technologies	Marketing Methods	"Internet" Methods
Factor Analysis Approach, Principal Component Analysis, Cluster Analysis, Multidimensional Scaling, Latent Structure Analysis, Multiple Regression Analysis, Analysis of Variance, Covariance Analysis, Automatic Interference Detection Analysis, Judgment Analysis, Joint Determination Analysis, Normative Correlation Analysis, etc.	Data collection, data access, infrastructure, data processing, statistical analysis, data mining, model prediction and results presentation.	4P, 7P, 11P, 4R, 4C, STP, SWOT, Porter's Five Forces Marketing research: Observation Method, Interview Method, Investigation Method, Experimental Research Demand forecast: Survey of Buyers, Comprehensive Sales Opinion Method, Expert Opinions, Test Marketing, Time Series Analysis Method, Linear Trend, Statistical Demand Analysis, etc.	WeChat, micro blog, QQ and other instant chat tools Community relations Short videos, KOL, online advertising, digital marketing and other communication tools

academic level, strict evaluation and fierce competition, reduced to 53 by nine compared with 2017 (26 major projects and 27 youth projects). More growth has been seen in the marketing disciplines in youth program, and that is also an advantage that young teachers are competing for.

Whether it is a high-level academic paper or a high-level national project, it is an essential impetus for the development of marketing disciplines. The marketing disciplines have been forced to adapt to the development trend and reform the existing research methods by the integration of information features, Internet, big data, and other cross-disciplines.

3.3 Promoting teaching with technology to increase the scientific accuracy of teaching methods

The Internet brings big data, while big data is neither any partial data nor large volumes of data, but all the data. Big data is regarded as having an important role in promoting marketing informatization, providing the data needed for marketing research and forecasting, and facilitating the development of business intelligence and the realization of precision marketing.

Big data analysis includes procedures such as requirement discussion, data extraction, data integration, missing data processing, feature engineering, and model evaluation, which means in the analysis a good foundation of mathematics is required. However, generally, it is mostly liberal art students whose mathematical foundation is weak when a college enrolls students in marketing. Therefore, a full consideration of exceptionally professional curriculums should be taken, which will help students to grasp the methods of data analysis and processing which is shown in Table 3. Meanwhile, besides paying attention to the consistency between the courses taken previously and those taken subsequently, teaching should, aiming to solve scientific and

Table 4. The practice-based project in an industry–university–research combination (IUR).

Project name	The Market Demand and Forecasting Research on the New Products of the Fashion Ski Suits Brand "TWOC"		
Time	Practice Content	Implement Scheme	IUR Results
Jan. 2019	1. Brand founder's introduction to TWOC 2. Watching the promotional video of TWOC 3. Self-introduction; be familiar with the team	1. Initial team grouping—online group and offline group 2.Free choice an independent design of the A and B version of TWOC online questionnaire	1. Understand the process of brand creation, the importance of brand culture, brand logo and brand design; cultivate brand creativity and insight 2. Consolidate market research methods and prediction knowledge
Feb. 2019	1. Market Research on TWOC products of Chinese style series 2. Design of TWOC products of Chinese style	1. Competitive team grouping. Rank each group according to the research results. Select group members in terms of the ranking. Provide team belts 2. Divide the labor in each group. Conduct the research and complete it as required 3. Summarize the research work of TWOC 4. The presentation of the research on "How TWOC combines products with Chinese style"	1. Be Familiar with team competition and cooperation 2. Participate in costume design 3. Learn to make PPT and presentation 4. Conduct market research on the market actuality and consumer demand
Mar. 2019 Apr. 2019	1. The research and product design on "the birth of a TOWC ski suit" Three themes: market research + product design theme + promotion interaction Ultimate goal: design and make a new ski hoodie Task requirements: the team can complete the three challenges within the stipulated time	1. Stage 1 market research 2. Stage 2 product design: product drawing, product design, design optimization and samples making 3. Stage 3 promotion and interaction: new media accounts focus and interaction, we-media operation	1. Consolidate market research data collection and processing methods 2. Know the role of new Internet media in marketing 3. Master the promotion methods of new media 4. Master the application methods of we-media.
May. 2019 Jun. 2019	1. Brand planning and promotion of TWOC's new products	1. Write an advertorial on public accounts to preheat and promote new products for spring and summer 2. Shoot funny TikTok videos, in which the brand information or culture are embedded in any form. 3. Plan a new product sales activity on Taobao with the consideration of the effect and ease of implementation of the activity.	1. Master the marketing mix strategy 2. Plan an online sales activity 3. Design marketing copy writings, posters, videos, etc. 4. Closer teamwork
Jul. 2019		In progress	

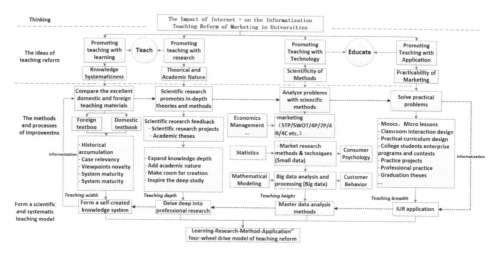

Figure 1. The ideas of marketing informatization teaching reform in colleges and universities against the background of the Internet.

practical problems, cultivate marketing professionals who master information technology with technologies and methods on the theoretical basis.

3.4 *Promoting teaching with applications to enrich the practicability of marketing*

Marketing is considered as an application-oriented discipline. In the teaching reforms, many colleges and universities have combined the Internet to carry out some fruitful activities. Teaching methods such as Moocs and micro lessons are used to break the time and space limitations of traditional classroom teaching. A flipped class model allows students to stand on the stage to participate in teaching. All kinds of college students' entrepreneurship projects and practical courses are also effective methods that provide an impetus to teaching by offering students opportunities to apply their knowledge in practice.

The author advocates the coherent and mature system of teaching projects. As shown in Table 4, taking the open experimental project currently underway this semester as an example, 16 marketing undergraduates are actively involved with the project, which combines the Internet with the marketing positioning, marketing strategy, and competition strategy of the state-owned independent brands, and have rapidly improved their professional knowledge, practical ability, and collaboration ability.

Learning for practice is taken as the fundamental purpose of promoting teaching with application. Only by combining industry, university, and research, and focusing on developing students' ability to use Internet technology and mastering new media marketing methods and strategies, can knowledge be turned into outcomes and the professional talents needed by society be cultivated. Therefore, college students are encouraged to participate in various practical projects and apply their knowledge to solve practical problems in a reasonable and scientific way.

4 CONCLUSION

In the context of the Internet, marketing teaching in application-oriented undergraduate universities is faced with an urgent demand that it be combined organically with informatization. This paper, discussing these four levels, knowledge systematicness, theoretical and academic nature, scientificity of methods, and professional practically, sums up the unique "Learning-Research-Method-Application" four-wheel drive model of teaching reform, aiming to provide ideas for reforming of marketing informatization teaching (Figure 1)

ACKNOWLEDGMENT

This work was supported by the Beijing Information Science & Technology University. (5112010826)

REFERENCES

[1] Jin Pingxi. Speech at a seminar for teachers and students of Peking University, 2018.5.2
[2] Audrey Gilmore, Andrew McAuley, Morgan P. Miles, Hugh Pattinson. Four questions of entrepreneurial marketing education: Perspectives of university educators. Journal of Business Research, 2018. 1–9.
[3] Paul Gibbs. Marketing and education — A clash or a synergy in time? Journal of Business Research, 2007. 1000–1002.
[4] Sarath A. Nonis, Gail I. Hudson, Melodie J. Philhours, Joe K. Teng. Changes in college student composition and implications for marketing education: revisiting predictors of academic success. Journal of Business Research, 2005. 321–329.
[5] Tom Hayes. Delphi study of the future of marketing of higher education. Journal of Business Research, 2007. 927–931.
[6] Michael L. Klassen. Relationship marketing on the Internet: the case of top- and lower-ranked US universities and colleges. Journal of Retailing and Consumer Services, 2002. 81–85.
[7] Xu Chong-huan, Chen Ggui-ting. Research on marketing Teaching Mode Reform and Talent Training under the background of big data. Global Market, 2018, (19), p20-25, DOI£°10.16541/j.cnki.2095-8420.2018.04.009.
[8] Wu Ling. The teaching design of marketing course based on 5E teaching mode. China Market Marketing, 2017, (34), p205-206, DOI:10. 13939/j.cnki.zgsc. 2017. 34. 205.
[9] Qiu Yuan-yuan. Intelligent teaching mode of marketing course in the era of "Internet + Education". Tax Paying, 2017, (031), p123.
[10] Wu Jian-an, Nie Yuan-kun. Marketing. Higher Education Press, Peking, 2017. 11, p16.

Computational Social Science – Luo, Ciurea & Kumar (eds)
© 2021 Taylor & Francis Group, London, ISBN 978-0-367-70193-2

Teaching reform and experience of the project-driven teaching and diversified assessment methods for the mechanical engineering materials course based on CDIO

C.S. Liu, Z.W. Li & J.J. Yang
School of Mechanical & Automotive Engineering, Qingdao University of Technology, Qingdao, China

ABSTRACT: In order to realize the cultivation of various abilities of students, we have experienced the project-driven teaching and diversified assessment methods for the Mechanical Engineering Materials course. The project-driven teaching model is constructed based on the international CDIO higher engineering education, with students as the main body and curriculum projects as the basis, theory and practice are organically combined, and knowledge, ability, and quality are integrated. The project report and self-evaluation, experiment, attendance, comprehensive classroom performance, quiz, and homework are all included in the curriculum evaluation system. Such a diversified assessment method is helpful to improve the students' practical ability and comprehensive quality. The teaching reform and experience shows that the students' active learning and ability training are both achieved.

1 INTRODUCTION

In recent years, China's higher engineering education has made considerable progress. It develops a direct promotion effect to the national economy by training huge numbers of graduates in the engineering field. However, there are many problems in the whole present engineering education, including the following. How to help engineering students to obtain professional knowledge? How to encourage them to develop their innovation and practical ability? Their team cooperation, communication and expression, occupational ethics, and responsibility are also needed. Therefore, the higher engineering educational reform in China is imperative.

Most of the traditional evaluation methods are single classroom tests, which are limited to the examination of understanding and mastering classroom knowledge. These methods are also not conducive to the improvement of students' comprehensive quality, unable to cultivate students' self-study ability, practical ability, innovation ability, critical spirit, and questioning consciousness. However, these abilities and consciousness are more important for students and should be placed in a prominent position. On the one hand, in the postgraduate entrance examination, public basic courses are the main courses. On the other hand, at present, the examination form of professional courses is mainly textbook content. Students form the habit of reviewing lessons in a short time, by which it is easy to ignore the study of professional courses [1,2].

CDIO (Conceive-Design-Implement-Operate), an international engineering education standard, was formulated by international engineering education cooperation organization CDIO committee [3,4]. The organization was initiated by famous universities such as Massachusetts Institute of Technology and Royal Swedish Institute of Technology. The CDIO education model pays attention to the teaching and learning of abstract theoretical knowledge, and emphasizes the importance of doing this in the teaching process It can enhance students' practical ability and cultivate students' innovation consciousness, project design capacity, and team work spirit. The curriculum knowledge structure can be organically linked to the whole process of study and practice [5,6].

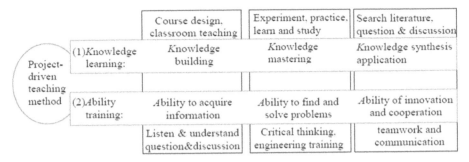

Figure 1. Knowledge learning and ability training based on the projectdriven teaching method.

The "Mechanical Engineering Materials" course is a course that is closely combined with practice. The theoretical knowledge of the course has some difficulties, and the learning initiative of students is not high, and so the rate of failure was high (usually around 20%) for many years. Therefore, it is imperative to carry out teaching reform. In this paper, using the education idea of CDIO as a reference, the project teaching and diversified assessment methods of the "Mechanical Engineering Materials" course were constructed. In the teaching and assessment methods, abilitytraining is the goal, being projectdriven is the basis, and process assessment is emphasized

2 REFORM IDEAS AND PRACTICE

According to CDIO, we should take the engineering materials development project as a background, combining the mechanical engineering discipline knowledge with the CDIO education idea Therefore, we need to refer to the CDIO engineering education program and mechanical engineering professional education for a complete set of reforms. This means planning and reforming the training plan, curriculum outline, teaching methods, practical teaching content, and system elements systematically [7]. Since 2018, the "Mechanical Engineering Materials" course in Qingdao Technology University has begun to actively explore teaching and learning according to CDIO outline standards and requirements, in that the student is the main body, the course projects are the basis, theory and practice are organically combined, and knowledge, ability, and quality are integrated.

2.1 Project-driven teaching method based on CDIO

The teaching mode that is projectdriven can realize two processes at the same time. As shown in Figure 1, one is the knowledge learning process It emphasizes the whole process of building, mastering, and synthesis application of the knowledge. The second one is the training process of the ability, including listening and understanding, engineering training, innovation ability, social ability, and so on Through the practice of the project, the organic integration of the theoretical system, experimental and practical teaching is realized, and the two processes of knowledge learning and ability training are realized simultaneously

According to the teaching content of the course of "Mechanical Engineering Materials," the properties of different steel materials are the main line of the teaching material. Therefore, it is determined that the content of the projects is to design and obtain a type of steel with excellent integrated mechanical properties. According to the project-driven teaching method, the course contents of principles of metallurgy, heat treatment process, metallographic observation, mechanical manufacturing basis, scanning electron microscopy, and other courses are related and integrated. Also, some different alloy materials and their applications in the industry are studied in both the university and factory. Therefore, students get a comprehensive training.

2.2 Diversified assessment method design

In order to pay attention to the learning process, project, experiment, attendance, comprehensive classroom performance, quiz, homework, and other aspects are included in the curriculum evaluation index system. In each assessment, learning attitude, ideology and morality, integrity awareness, team spirit, and many other aspects are considered at the same time. This method can encourage students actively to participate in teaching activities, comprehensively evaluate students' knowledge, ability and quality, and implement the whole process management of students' curriculum learning. Finally, the style of study and examination was improved.

Before the reform, the final results of the course were composed of the final examination experiment, and attendance. After the reform, the final results of the course were composed of the following parts: final examination results, experiment project report and self-evaluation classroom comprehensive performance (discussion, answer questions, etc.) and quiz, attendance, and homework The full score for each part is 100. The initial attendance score is 100 points. The score is reduced when absent. Absenteeism and leave are treated differently. The comprehensive performance of the class is initially evaluated as 60 points to encourage the students to participate in the careful listening, active discussion, answer questions, etc.

2.3 Practice

The practice steps of the project teaching and diversified assessment method are as follows:

(1) *Establishment of the project team group.* The team group is composed of about six students. The students choose a team leader to take charge of the project.
(2) *Project preparation.* According to the teaching objectives of this course, a type of steel with excellent integrated mechanical properties is selected and distributed to each group at the beginning of the course. In the whole teaching process, students analyze the steel material and its properties, and put forward related questions.
(3) *Project practice.* According to the specific tasks of the project, students divide the project into several small tasks, which are completed by each team member. In the process of project implementation, teachers should solve the problems existing in the project implementation, carry out a good job of assistance and supervision to ensure the smooth implementation of the project.
(4) *Analysis and discussion.* In order to achieve the expected objectives of the project, each group regularly analyzes and discusses the implementation results of the project, reaches consensus, and issues the next project plan.
(5) *Report and evaluation.* After completion of the project, each project team reports in public, and the teacher evaluates the completion and quality of the project. At the same time, the evaluation results should also include the self-evaluation scores of the project team members.
(6) *Diversified assessment* Before reforming, the final results of the course were composed of the following parts: final examination results were 80%, attendance was 10%, and experiment was 10%. After the reform, as shown in Figure 2, the final examination results, project report and self-evaluation experiment classroom comprehensive performance and quiz, attendance, and homework were 5%, 15%, 15%, 10%, 5%, and 5%, respectively.

3 RESULTS

After the reform, students' learning enthusiasm and learning effect are significantly improved. The data comparison before and after the reform is shown in Table 1 Thirty students attended the reform experience. There were five team groups and six students in each group.

The final examination score and the final results of the course distribution histogram before and after the reform are shown in Figures 3 and 4 respectively. After the reform, both the final examination and the final course results showed a Gaussian distribution. Only one student's final

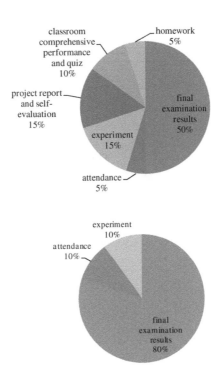

Figure 2. The diversified assessment diagram of final results after (top) and before (bottom) reform.

Table 1. Comparison of results before and after the reform.

| The interval of results | The number and proportion in the interval *before* reform | | The number and proportion in the interval *after* reform | |
	The final examination results	The final results of the course	The final examination results	The final results of the course
0–59	10 (33.3%)	6 (20.0%)	1 (3.3%)	0 (0%)
60–69	8 (26.7%)	8 (26.7%)	1 (3.3%)	1 (3.3%)
70–79	9 (30.0%)	9 (30.0%)	12 (40.0%)	2 (6.7%)
80–89	3 (10.0%)	6 (20.0%)	14 (46.7%)	20 (66.7%)
90–100	0 (0%)	1 (3.3%)	2 (6.7%)	7 (23.3%)

examination score was lower than 59, and all the students reached the standard after the diversified assessment. The "excellent (score = 90–100)" and "good (score = 80–89)" rate reached from 23.3% (3.3%+20.0%) to 90%(23.3%+66.7%).

Through the reform, not only were the students' academic records improved, but also the students' comprehensive ability was exercised. At the end of the term, students give a high evaluation of the course reform.

4 CONCLUSION

In order to achieve the cultivation of practical ability, innovation ability, learning ability, team spirit, and comprehensive evaluation of students, we have reformed the teaching and evaluation methods of

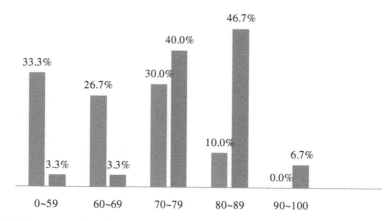

Figure 3.   The histogram of the final examination results distribution before (blue) and after (red) the reform.

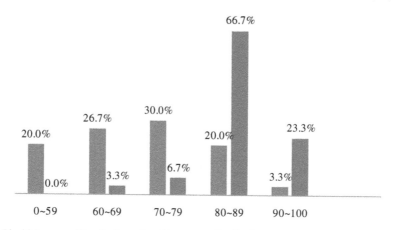

Figure 4.   The histogram of the final results of the course distribution before (blue) and after (red) the reform.

the "Mechanical Engineering Materials" course. Using the idea of CDIO, the project-driven teaching mode was constructed and the diversified assessment method of the whole process was designed. The diversified assessment method is composed of a project report and self-evaluation, experiment, attendance, comprehensive classroom performance, quiz, and homework. The effectiveness of the project-driven teaching and diversified assessment method is illustrated by comparing the results with the feedback from students.

ACKNOWLEDGMENT

The research was sponsored by the Research Project of Teaching Reform in Undergraduate Colleges in Shandong Province (Project No. 2015M087).

REFERENCES

[1] C.H. Li, Y.L. Hou, H.S. Yang, et.al, Research on the Construction and Implementation of Outstanding Engineering Talents Training Pattern, Advances in Social Science Education and Humanities Research, 156 (2017) 55–59.

[2] C.H. Li, Y.L. Hou, G.Y. Liu, Expanding of college students' scientific quality and innovation with the carrier of research program, Adv Mater Research, 171–172 (2011), 90–93.

[3] X.L. Nan, X.F Lu, T.M Guo, et al., Teaching Reform and Experience of Inorganic Nonmetallic Materials Technology Based on the CDIO Concept, Guangzhou Chemical Industry, 5 (2015) 200–202.

[4] W.Y. Chen, The research of the teaching mode based on the concept of CDIO architectural design, Applied Mechanics & Materials, 584–586 (2014) 2753–2756.

[5] J.Y. Wang, X.C. Nie, X.D. Zhang, Study and practice of talent training mode of mechanical specialty based on CDIO project education idea, Equipment Manufacturing Technology, 40 (2012) 218–221 (In Chinese).

[6] Y.Y Li, Teaching method reform based on CDIO concept, New Education, 2017, (33), 144. (In Chinese).

[7] J.B. Liu Z.Y. Yang, Exploration and practice of SE-CDIO educational pattern, I. J. Education and Management Engineering, 3 (2012) 51–56.

Computational Social Science – Luo, Ciurea & Kumar (eds)
© 2021 Taylor & Francis Group, London, ISBN 978-0-367-70193-2

Research into the design and evaluation of information technology in a learning environment

H.Q. Hu*, J. Zhang & C.H. Jin
School of Economics and Management, Beijing Information Science and Technology University, Beijing, China

ABSTRACT: The development of learning environment design and evaluation technology will be increasingly affected by the development of learning concepts, learning space forms, and new information technology breakthroughs. The key developments in learning environment design and evaluation technology will be in learning environment perception, learning space monitoring, and learning space natural interaction technology. The learning environment technology in the Internet environment will more effectively connect learners and learning spaces, learners and other learners, break the constraints of time and space, and introduce new models and paradigms of the Internet, forming a new learning modality.

1 INTRODUCTION

Each revolution from new technologies such as television technology, multimedia technology, the Internet, mobile Internet, and the Internet of Things has brought great changes to the learning environment. The continuous emergence of new technologies has provided a direct means of implementation and a driving force for the development of the learning environment. In recent years, the rapid development of information technology and its continuous application in learning environments have achieved good results.

The learning environment design and evaluation technology is a comprehensive technical embodiment of the new learning environment in the field of Internet education. It is an integrated technology that serves the learning environment elements such as the ideal mode, method, and organizational form of learning activities in the learning space.

With the development of Internet technology, the learning space is constantly expanding, from formal learning spaces such as classrooms, laboratories, multimedia computer rooms, etc. to informal learning spaces such as libraries, study rooms, stadiums, etc., and then to virtual learning spaces. (Including various learning management systems, learning resource platforms, social networking platforms, such as mooc/spoc, etc.) Virtual classroom technology is supporting the continuous integration of physical spaces (formal and informal learning spaces) and virtual spaces to provide learners with a smart learning environment to achieve adaptive and personalized learning. At the same time, the analysis of learning data in the public learning space also provides reference for policy decisions for educational institutions, the education system, and society. It integrates augmented reality technology, rich media technology, and sensor technology. The latest virtual classroom technology, such as learning analysis technology, will represent the development direction of domestic and foreign virtual classroom technology. The design of future learning spaces should fully consider the cognitive load of learners, so that they are controlled within the range that can be carried by students' working memory. Adapt to students' cognitive processes and enhance learning effects. The sense of identity in the learning space will further stimulate their interest in participating in virtual learning activities by improving the sense of identity.

---

*Corresponding author

Under the Internet environment, the traditional closed teaching environment formed since the industrial era has been unable to meet the needs of personalized learning. The teaching and learning methods are facing a change from class-based teaching to information-based teaching. The change provides technical support, and provides a learning space for online and offline integration to achieve interconnected learning and immersive happy learning.

2 LEARNING SPACE ENHANCEMENT TECHNOLOGY

Learning space enhancement technology, specific enhancement technology for typical learning scenarios of three different types of functions, virtual learning environment technology, learning-oriented virtual reality technology, and virtual experiment technology, researching class-oriented learning space and enhanced technology for the construction of learning resource platforms; learning-oriented virtual reality technology, which focuses on solving the research and development and content distribution mechanism of virtual reality learning resources in different disciplines; and for steam education, building a virtual experimental environment, research on virtual experiment construction and interactive technology, and promote students' inquiry learning and discovery learning.

Learning space enhancement method, using SMART model, in the classroom physical learning environment, teaching content presentation (Showing), physical environment management (Manageable), digital resource access (Accessible), real-time interactive(Real-time), teacher and student situational automatic perception (Testing) is enhanced in five dimensions.

How to use interactive and interactive technologies, knowledge models, and learner models to serve virtual classroom technology in the learning space; provide data such as environmental information, learner information, and physical space collected in the learning space for related modeling, interaction, governance, service, etc.

2.1 High-speed two-way video transmission

Combining the results of the construction of existing delivery classrooms, refining user use data to build a new generation of synchronous delivery classrooms based on virtual classroom technology. In the newly constructed synchronized classroom learning space, the experience of teachers and students will be improved and resources will be improved. The delivery and pushing of the cell phone is more efficient and accurate, and teachers and students in many places will be able to conduct teaching and learning more smoothly.

Utilizing high-speed two-way video transmission technology to achieve video cloud services, cloud-based storage, and real-time services for the virtual classroom teacher end; advanced multimedia stream forwarding technology to increase the number of users of virtual video classroom services; increase the real-time interactive presence and A sense of immersion to achieve a high-fidelity classroom environment. Through interactive channels, ensure long-distance interaction between teachers and the system, classrooms, and learners, and improve teaching quality.

2.2 Enhancement of online classrooms

The definition of typical forms and functions of online classrooms, constructing online classroom technology with VR/AR support. Breaking the time and space barriers between classrooms and learners, and realizing the virtualization of the space and time category of classrooms. The openness and distance of courses using mooc and spoc. Features of online, large-scale application, cross-temporal, fragmented, evaluable, and autonomous learning, realizing the large-scale application of virtual classrooms to remote and poor areas; using the characteristics of VR/ER high-immersion technology to break classroom knowledge content and presentation environment. Restrictions on the way of thinking, the realization of the combination of virtual and real knowledge transfer.

2.3 Technology for upgrading multimedia classrooms to smart classrooms

Functions and types of multimedia classrooms and methods for improving their facilities and equipment, integrating intelligent detection, perception, interaction, and other technologies to improve and enhance them into smart classrooms; smart classrooms are based on modern educational principles such as connectivity, learning science, cognitive psychology, and ergonomics, environmental engineering, Internet of Things technology, mobile computing technology, and other multidisciplinary and multitechnology development trends based on the integration of the development of the definition of the function of smart classrooms, and around the needs of educational and teaching activities, research different types of smart classroom types. Form a facility equipment configuration plan and function for different forms and types of smart classrooms.

Physical environment-oriented intelligent environment perception and learner-oriented situational awareness technology. Using sensor IoT technology, real-time collection of a series of environmental parameters such as acoustic, optical, and temperature, to build a multichannel fusion perception model and monitoring environment, through self-learning and intelligence. Analysis functions to achieve intelligent perception of the learning environment.

Smart mobile learning environment technology, relying on the education cloud platform, with mobile devices and smart devices as the main application terminals, fully integrate teaching equipment such as campus networks, electronic whiteboards, projectors, and intelligent central control. Through interactive classrooms, electronic lesson preparation, and lessons Application systems such as post-operation system and entertaining learning system, for k12 education teachers and students, provide intelligent teaching solutions throughout each learning stage.

Utilize the strange changes of multimedia teaching courseware and the fascinating whiteboard special effects function to create a multidirectional interactive space–time platform; provide rich media-based teaching resources and rich classroom teaching modes to achieve real-time information collection and feedback and improve teaching efficiency. Content and note records are saved to the cloud, enabling students to focus and participate in the teaching process.

2.4 Enhanced technology of virtual laboratory

Interactive and remote control method of a virtual laboratory. The virtual laboratory mainly includes two scenarios, one is to build a comprehensive virtual reality simulation experimental environment including virtual experimental content and virtual interactive means through online and local VR/AR. The other is a comprehensive remote experimental environment with remote interactive control functions that can manipulate remote entity experimental equipment.

The virtual reality simulation experimental environment is built on a virtual experimental environment (platform), focusing on the interaction of experimental operations and the simulation of experimental results. VR/AR technology is an important technical component of building a virtual laboratory. The virtual laboratory can assist, partially replace, or even completely replace the relevant software and hardware operating environment of each operation link of the traditional experiment, and the experimenter can complete various experimental projects just like in the real environment. Virtual reality technology can solve the funding of research institutions, the difficulties and pressures generally faced by venues, equipment, and other aspects, the development of virtual experimental teaching can break through the limitations of traditional experiments on "time and space." Whether it is students or teachers, they can enter the virtual laboratory anytime, anywhere without any difficulty.

The comprehensive remote control experiment environment uses remote control technology to operate remote devices through wireless or electrical signals, and the remote control is performed through the network. The experimenter controls the devices and equipment in the laboratory through remote devices and remotely controls the experiment. This can not only liberate the experimenter's space and time, but also avoid the danger caused by some experiments to the experimenter. The advantages of remote control technology in distance education and collaboration can also play a role in the virtual laboratory, which can fully support remote collaboration, remote experiments, collaborative research, and other teaching and research activities.

3 COGNITIVE LAW AND LEARNING EXPERIENCE IN THE NEW LEARNING SPACE

The learning experience in virtual classroom technology, includes: the learner's perception of the design intention and use of the learning space, the way in which the learner quickly and effectively migrates the learning experience and knowledge in the previous learning space, and the learner acquires the expected operation method and operating habits. With a view to approaching the user's way of thinking (using a visual metaphor) and behavior (approaching a natural way of operating), respect the user's differences, and strive to present diversity and construct a learning environment in the design of learning spaces. There is no technical threshold for use, it is easy to form habits, and it can transfer the existing experience of users.

Under the three typical environments of an online learning environment, virtual reality learning environment, and virtual experimental environment, learners' cognitive characteristics and laws, and user experience enhancement methods.

The research focuses on the cognitive load and self-identity rules. It focuses on the three experience levels of perception, feedback, and performance, and on the three typical environments of online learning environment, virtual reality environment, and virtual experimental environment, and studies the characteristics of the learning experience in the learning space.

Cognitive load in the learning space, including the number and layout of things in the learning space, the organization and presentation of multiple media, and the level of prior knowledge of learners, controls the cognitive load within the range that can be carried by students' working memory. Adapt to students' cognitive processes and enhance learning effects. Construct a multi-information learning environment suitable for learners through cognitive load analysis in virtual classroom technology. Simultaneous transmission of tactile information such as sound, images, text, and even force sense through multiple channels The amount of information transmission, rich stimuli and other information are controlled within the scope of the learner's cognitive load, improving learning effectiveness.

The effects of virtual self and identity on learners 'performance in the online learning space and real world, including cognition, attitudes and behaviors, provide a sound online learning environment for the formation of learners' healthy personality. The "virtual" The identity of "identity" affects their learning motivation in the social environment. The lack of identity will result in students being unable to complete learning tasks and interact with others in virtual identities. Therefore, in the online learning space, learners can Characteristics and Ideal Self Construct a virtual self, try various types of self-representation, selectively display the characteristics of the true self, and establish a sound "new" self-image and identity. In the virtual classroom environment, In character design, full consideration is given to help students better adapt to this identity conversion, and to further stimulate their interest in participating in virtual learning activities by improving their identity.

4 LEARNING SPACE EVALUATION TECHNOLOGY

Based on the smart model, a unified evaluation and monitoring system is constructed, and the classroom environment is evaluated. The evaluation content and perspectives include five dimensions of evaluation: teaching content presentation, physical environment management, digital resource acquisition, teacher–student interaction, and automatic scene perception.

4.1 *Design the learner's assessment of learning space*

Based on the smart model, four types of learner feedback evaluation scales for learning spaces were designed. After reliability and validity tests, a tool suitable for the learners' subjective evaluation of learning spaces was formed.

The objective quantitative evaluation method using real-time measurement methods such as the Internet of Things, and the introduction of objective and effective data collection methods for classroom physical environment, atmosphere environment, and user perception, form an objective evaluation model of virtual classroom technology. The research is applicable to large-scale distribution, collection, and analytical evaluation of the method of the online learning environment.

In order to improve the efficiency and convenience of the virtual classroom technology evaluation, an online virtual classroom technology evaluation cloud platform method suitable for distribution, collection, and analysis was researched for large-scale surveys and evaluations.

5 LEARNING SPACE DESIGN OPTIMIZATION TECHNOLOGY BASED ON LEARNER EXPERIENCE

(1) Build a 3D model library in the design of the learning space. Support sketch-based and semantic retrieval of the main environment, facilities, and equipment entities in the learning space. Manage a series of classroom equipment such as desks, whiteboards, chairs, etc., and store the physical model in 3D, its geometric shape, and the combination of semantic associations between it. With its additional attributes such as physics, teaching, and related parameters, the inventory model can be effectively reused.
(2) Automatic evaluation model of the learning space environment. Summarize the corresponding relationship between the structure of learning space elements and requirements, summarize the cost space characteristics of the learning space, and develop various automatic evaluation models of learning space environment.
(3) Learning space design technology. Learning space generation technology can use the 3D model constructed earlier to build learning space VR scenes. These scenes can be used for the purpose of experience, measurement, analysis, and deduction, so that the designer access it there and be more direct. Get the relevant sense and knowledge. Get relevant equipment and environmental space design results. Support the rapid rendering and real-time drawing of virtual classroom design drawings. The constructed learning environment can quickly render actual effects, set environment parameters, and simulate measurements in the simulation space with evaluation to obtain analysis results. Support user interaction and a high-immersion roaming experience in the learning environment. The virtual classroom environment supports user roaming access to get the actual qualitative experience effect by feeling it. Supports automatic evaluation of learning space solutions and automatically feeds back optimization suggestions.

6 SUMMARY

The future development of learning environment design and evaluation technology will be increasingly affected by the development of learning concepts, learning space forms, and new information technology breakthroughs. The key developments in the learning environment design and evaluation technology will be in learning environment perception, learning space monitoring, and learning space natural interaction technology.

The learning environment technology in the Internet environment will more effectively connect learners and learning spaces, learners and other learners, break the constraints of time and space, introduce new models and paradigms of the Internet, and introduce the latest technologies such as virtual reality and augmented realitymultimodal, highly immersive learning environment, forming a new learning modality.

(1) Develop guidelines for the construction of learning spaces. Develop guidelines for the construction of online learning spaces and guidelines for the construction of physics learning spaces.

Compile guidelines for the construction of online learning spaces and guidance for the construction of physics learning spaces, integrate the theoretical results, intelligent technologies, engineering, and development methods that have been obtained, set up test and experimental environments, and evaluate and analyze the application modes, functions, and effectiveness of new technologies in virtual classrooms, and provide a guide to the construction of online learning spaces and physics learning spaces.

(2) Standards and specifications for smart classrooms. Form construction specifications including physical buildings, information networks, power supply and distribution system design, audio system design specifications, video system design, and VR system design; on this basis, the quality of smart classroom environment construction is formed. Policies, goals, responsibilities, and procedures, and through the establishment of relevant systems for process management, quality planning, quality control, quality assurance, and quality improvement, eventually forming a standard system for smart classrooms, providing a standard basis for the automatic assessment of smart classrooms.

ACKNOWLEDGMENT

This research was financially supported by the "Education and Teaching Reform Project of Beijing Information Science and Technology University(2019) 5111910814."

REFERENCES

[1] Demir M, Using online peer assessment in an Instructional Technology and Material Design course through social media, J. Higher Education. 75 (2018):1–16.

[2] Larvin M, E-Learning in surgical education and training, J. Anz Journal of Surgery. 79 (2009)133–137.

[3] Antonenko P, Nichols J. ECLIPSE, Environment for Collaborative Learning Integrating Problem Solving Experiences, J. Instructional Design, 2013.

[4] Baeten M , Dochy F, Struyven K, Students' approaches to learning and assessment preferences in a portfolio-based learning environment, J. Instructional Science. 36 (2008) 359–374.

[5] Wang Q, Huang C, Quek C L, Students' perspectives on the design and implementation of a blended synchronous learning environment, J. Australasian Journal of Educational Technology. 2018, 34.

[6] Brandt C B, Cennamo K , Douglas S, et al, A theoretical framework for the studio as a learning environment, J. International Journal of Technology and Design Education. 23 (2013) 329–348.

Computational Social Science – Luo, Ciurea & Kumar (eds)
© 2021 Taylor & Francis Group, London, ISBN 978-0-367-70193-2

A study of student evaluation of teaching in applied colleges

J.Y. Yuan, P.H. Huang, W.Y. Yan & H.S. Chen
Nanfang College of Sun Yat-sen University, Guangzhou, China

ABSTRACT: As one of the commonly employed methods in teaching quality evaluation, student evaluation of teaching (SET) plays an important role in teaching quality evaluation in colleges and universities. Traditional SET-based research usually takes a small portion of data to measure the influencing factors in SET. However, there are quite a few influencing factors of SET, and the reliability of previous conclusions need verification in a larger volume of data. Based on a bigger value of SET-related data from an applied college, we made an analysis of the influencing factors including students' curriculum performance, teacher information, and curriculum settings. The analysis affirms that students' curriculum performance correlates little with SET, and that teacher information does indeed have some impact on SET, but most importantly, curriculum settings have a heavy impact on SET results.

1 INTRODUCTION

Since the beginning of the 21st century when China's higher education entered the stage of popular education, the quality monitoring and evaluation of higher education also attracted widespread attention. The assessment methods for teaching quality include student evaluation of teaching (SET), teacher's own evaluation, expert evaluation, collection of textual data, and demonstration of teaching effects [1]. Among them, SET has gained much attention from the research community, and has received scholars' positive attitudes in the monitoring of classroom teaching quality. With SET, students are required to provide feedback on the teaching quality of a curriculum according to an evaluation form designed by the teaching unit. As early as 1999, Yukun Chen affirmed the reliability of SET [1]. Shen [2] also believes that SET is more democratic than other evaluation methods, and then further states that not all evaluation can help improve the quality of teaching.

The previous studies on SET were mostly about reliability and effectiveness. Centra believes that the reliability of SET can be measured by the number of students [3]. Chen et al. [4] claimed that the effectiveness of SET is stable based on the time point. Zhou [5] affirmed the effectiveness of SET in that the evaluation results are true and consistent with the teachers' actual state and level, and also that the information reflected in the evaluation results can have practical effects on teaching improvement. Simpson [6] believes that there is a deviation between the results of expert evaluations and student evaluations, which makes the evaluations of students less effective. There are many data and indicators in teaching evaluations that can detect high reliability and validity, but the validity of the data does not mean that all indicators reflected by SET are effective. It can be concluded that it is necessary to consider more influencing factors in the aim of validating the effectiveness of SET.

Traditional analysis methods of SET [7,8] tend to collect an online survey or a small amount of sample data, and thus the general meaning of their analysis results may not be sufficiently convincing. Ma et al. [9] analyzed the effectiveness of SET using big data techniques with data correlation index and Kronbach coefficient. Their research points out that the theoretical research of SET can be used to evaluate whether SET is effective and objective, and can also be used to guide the correct application of SET results. Wang [10] made a simple analysis on the data of SET

Table 1. Interpretation of the collected data.

Data category	Content (range of value)	Notation
SET feedback	Average curriculum SET score (0–100)	g_e
	Curriculum student number	n
Curriculum performance	The average score of all students in a curriculum (0–100)	g_p
Teacher information	Gender (male, female)	
	Academic qualifications (undergraduate, master, and doctoral)	
	Professional title (teaching assistant, lecturer, associate professor, professor)	
	Administrative position	
Curriculum settings	The semester of instruction (1–8)	s
	The instruction department	

in six semesters from an applied college. Long [11] investigated the SET data in one semester of the business school of Shantou University. The data from these two studies come from only one school in a college, and thus may not be rich enough in revealing the characteristics of SET in various subjects.

This paper takes the SET data from an applied college in southern China as the research data set, and investigates the characteristics of SET data in applied colleges with different influencing factors such as majors, students, teachers, and the curriculum factors. To reduce the impact of students' individual factors, this article analyzes the relationship between SET results and the influencing factors on a curriculum basis, where the SET scores from all the students in a curriculum are integrated into one value according to the specified rules in the question forms. The research results can provide new insights for the interpretation of teaching performance evaluation, and provide an empirical basis for decision-making in the cultivation of applied college students.

2 THE COLLECTED DATA SET

With SET, a number of questions from different perspectives are fed back by students to evaluate the teaching performance of a curriculum. In general, a quality indicator is assessed and later the quality indicator is projected into a score. Then, the obtained SET scores are usually averaged on a curriculum basis into one single value for a simple representation of the teaching performance. Although it may not be the most suitable measurement for teaching performance, this averaged value has been widely used by many applied colleges and universities due to its simplicity.

In this paper, data are collected and preprocessed on a curriculum basis into four categories, including SET feedback, curriculum performance, teacher information, and curriculum settings, as illustrated in Table 1. SET feedback includes per curriculum averaged SET scores and the number of students in a curriculum. The curriculum performance is the average score (percentage system) of all students in a curriculum. The teacher information includes gender, academic qualifications, professional title, and administrative position. The curriculum settings include the department of instruction and the semester (1–8). In all, more than 3000 curriculum-based SET records with influencing factor data for six semesters from 2015 to 2018 were collected.

3 DATA ANALYSIS

Based on the above-collected data set, we obtain the SET feedback statistics and analyze their relationship with the influencing factors, including curriculum performance, curriculum settings, and teacher information (see Table 1).

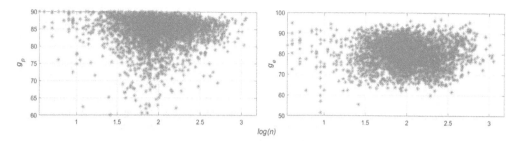

Figure 1.   The impact of student number on g_e and g_p.

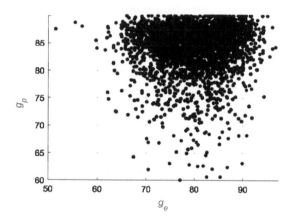

Figure 2.   The correlation between g_e and g_p.

3.1   *Statistics of SET feedback and curriculum performance*

The SET feedback consists of the average curriculum SET score g_e which is derived from all the students' feedback to the curriculum and the student number n in the curriculum. According to previous research, the value of g_e has a negative correlation with n. We confirm this relationship by providing a scattermap as shown in Figure 1(a), where the number of students is used in an exponential form, and each red asterisk denotes a pair of g_e and $log(n)$. It can be observed that g_e shows a decreasing trend with the increment of $log(n)$.

The average performance score g_p of all students in a curriculum is used to represent the students' curriculum performance. The statistics of g_p are first obtained and illustrated in the scattermap in Figure 1(b). In this figure, g_p varies little as the increment of $log(n)$, indicating that no obvious relationship between g_p and student number n. Such a result shows that teachers tend to assess students' performance in a relatively fair manner. Compared with student's SET feedback (see Figure 1(a)), it can be concluded that student's performance in a curriculum is affected little by the size of class, but the students' SET feedback is obviously affected.

3.2   *Correlation between SET feedback and curriculum performance*

The scattermap drawn using g_e and g_p is illustrated in Figure 2. Note that no obvious clue is found. Such an observation indicates that there is no direct relationship between how much teachers assign curriculum performance scores and how much is the student's SET feedback score. The reason may be that, compared to major knowledge, students care more about how a teacher communicates with them, including personal charm, attitude, kindness, patience, etc.

Table 2. The impact of gender.

Gender	ASS	ACP
Female	**84.96**	79.93
Male	83.56	79.97

Table 3. The impact of academic qualification.

	Undergraduate	Master	Doctoral
ASS	84.38	**84.66**	83.73
ACP	**81.44**	79.91	79.93

Table 4. The impact of professional title.

	TA	Lecturer	AP	Prof.
ASS	**84.78**	**84.55**	83.64	83.47
ACP	**80.56**	80.03	79.56	78.08

3.3 The impact of teacher information

The collected teacher information includes gender, academic qualifications, professional title, and administrative position. In this subsection, we take the influencing factor to illustrate their impact on students' SET feedback and curriculum performance, respectively. For simplicity of notation, the average SET score and average curriculum performance across all curriculums corresponding to a specified influencing factor are termed ASS and ACP, respectively.

The impact of teacher gender is illustrated in Table 2. Note that the numbers of curriculums and ACP for different genders are nearly the same, while for ASS, the score is 84.56 for female teachers but 83.56 for male teachers, the difference being 1.40 points. Such a result of gender impact on SET feedbacks indicates that female teachers tend to have better SET performance than male teachers.

Tables 3 and 4 presents the influences from academic qualifications and professional titles, respectively. The teacher's academic qualifications include undergraduate, master, and doctoral, while the professional titles consist of teaching assistant (TA), lecturer, associate professor (AP), and professor. In Table 3, it can be seen that both bachelors and doctors make similar assessments of students' performance, but teachers with a bachelor degree receive the best ASS score. In Table 4, both ASS and ACP decrease with the increment of professional title, the reason for this may be that teachers with higher professional titles are assigned to lecture more theoretical and complex courses. In general, the following observations can be drawn from these two tables: for SET score, the higher the teacher's academic qualification or professional title, the lower the students' SET feedback.

Since it has been stated in earlier research that teachers with administrative roles may receive higher ASS scores, we listed the ASS and ACP scores for teachers with and without administrative rules in Table 5, which tends to confirm the previous statements, although the sample for teachers with administration experiences is not as significant. In addition, we can also note that teachers with administrative experiences assign higher ACP scores by an average of 2.13 points. The reason for this may be that the lecture content and skills of these teachers are more suitable for the cultivation of applied talents. The observations in Table 5 indicate that: (1) teachers may improve the SET feedback by taking some administrative responsibilities to gain a broader educational vision and

Table 5. The impact of administrative experience.

Administrative	ASS	ACP
With	**85.26**	82.11
Without	84.42	79.98

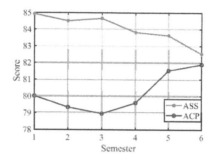

Figure 3. The impact of semester.

some insights in educational skills, and (2) colleges and universities may consider strengthening the training of administrative experience and broader educational visions to the teacher team.

3.4 The impact of curriculum settings

In evaluating the impact of curriculum settings on ASS and ACP, we consider the semester and department of instruction as the influencing factors. Figure 3 shows the trends of ASS and ACP along the increment of semester of instruction. Note that ASS generally presents a descending trend as the increment of instruction semester, while ACP decreases first and then climbs up with the third semester as the valley point.

Figure 3 illustrates an interesting result and we make more analysis here. The SET feedback for the first semester is the highest of all the semesters. The reason is probably that in the first semester, students are curious about fresh new university life and still have enthusiasm in learning inherited from high schools. In the second semester, students start getting used to university life and studies, and get their enthusiasm is not as high as their ACP and ASS in both decades. When the third semester comes, university students starts learning more major-related disciplines and thus they may feel more interesting, explaining why ACP still drops but ASS makes a short rise. Starting from the fourth semester, as students' capabilities and experience become more mature, their ACP has a rising trend. However, during this stage when graduation slowly approaches, students may have expectations on courses related to future jobs, but the traditional methods of lectures makes them bored and derives a lower SET score.

The indication from Figure 3 is that students offer different SET feedbacks during the university lifecyle due to various personal expectations. Such an indication poses a great challenge to universities and colleges and puts forward a strong demand for personalized education, which is crucial for applied colleges. In addition to the question of how students behave in different semesters, we also try to understand whether the department instruction impacts differently on ASS and ACP. Figure 4 shows the scattermap of g_e and g_p in different instruction departments, where the black dots and red dots represent all the curriculums in the data set and the subset curriculums from a single instruction department. Note that there exist some departments with too few curriculums and thus their data are not shown. From this figure, although no clear relationship between g_e and g_p is found, we can see different instruction department composition styles for g_e and g_p.

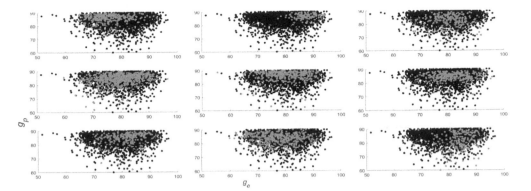

Figure 4. The impact of instruction department, the black dots represent all the curriculums in the data set, and red dots are the subset curriculums from a single instruction department.

Table 6. The impact of instruction department.

Department	No. 2	No. 3	No. 5	No. 7	No. 8	No. 10	No. 12	No. 13	No. 14
ASS	85.75	86.28	82.58	83.98	85.62	84.31	84.55	83.54	81.27
ACP	75.92	87.08	80.62	79.61	79.82	81.39	81.24	77.61	83.81

To further clarify the composition styles, Table 6 is used to illustrate the compositions of ASS and ACP for instruction departments. In Table 6, red, blue, and green are used to represent low, moderate, and high scores, respectively. Specifically, ASS is colored as red, blue and green for values in [0, 83), [83, 85), and [85, 100], respectively, while ACP is segmented by [0, 79), [79, 82), and [82, 100]. Note that most departments have compositions of moderate ASS and ACP, but department No. 2 receives high ASS at the cost of low ACP, while No. 14 obtained a low ASS but high ACP. Such an observation indicates that departments may have various cultural environments, which may be a crucial factor in SET. As a result, the use of SET results should consider the department impact to make a fair SET evaluation.

4 SUMMARY

The student evaluation of teaching has many influencing factors in real-world applications. After making a study of the influencing factors, we affirm that student number, teacher information including gender, academic qualifications, and professional title do have a certain impact. In addition, we also find that administrative experience and curriculum settings have a greater impact. From the observations, three suggestions are provided. First, educational departments cannot make a fair explanation of SET results until the influencing factors are carefully considered. Secondly, cross department evaluation of SET results should consider the impact of instruction department differences. The final and most important suggestion is that personalized education would greatly improve SET feedback.

ACKNOWLEDGMENT

This work was financially supported by the Guangzhou Science and Technology Plan Project of China under grants 201904010276 and 201804010292.

REFERENCES

[1] Y. Chen, Educational Evaluation, Beijing: Peoples' Education Press, 1999, pp. 139–143 (in Chinese).

[2] Y. Shen, Modern education evaluation, Shanghai: East China Normal University Press, 2002, pp. 120–121 (in Chinese).

[3] D. Centola, The Spread of Behavior in an Online Social Network Experiment, Science (New York, N.Y.), 2010, 329, pp. 1194-7.10.1126/science.1185231.

[4] J. Chen & X. Jiang, A Summary of Researches on Students' Evaluation of Teaching at Home and Abroad, Journal of Technical Supervision Education, 2006, 2: 113–116 (in Chinese).

[5] M. Zhou, The existing problems in College students' evaluation of teaching and the reconstruction strategy, Suzhou University, 2011, pp. 9–11 (in Chinese).

[6] R. D. Simpson, Uses and misuses of student evaluations of teaching effectiveness, Innovative Higher Education, 1995, 20(1), pp. 3–5.

[7] J. Xie, W. Li, & C. Zhang, Research on Non-teaching Influencing Factors and Improvement Measures of Deviation in College Students' Evaluation of Teaching, Journal of Heilongjiang Institute of Education, 2019, 38(05), pp. 46–48 (in Chinese).

[8] J. Zhao, H. Tian, & L. Lv, Empirical analysis of influencing factors of students' teaching evaluation in application-oriented university, Journal of Nanchang Institute of Technology, 6, pp. 69–72 (in Chinese).

[9] X. Ma, Analyzing the Effectiveness of Student Evaluation of Teaching from the Perspective of Big Data Mining, Learning Resource and Technology, 2014, 10, pp. 78–83 (in Chinese).

[10] Z. Wang, Data analysis of influencing factors of teaching evaluation management in applied colleges, Journal of Higher Education, 2018, 84(12), pp. 156–158.

[11] Y. Long, An Empirical Analysis of the Factors Affecting the Score of Teachers and Students in Colleges and Universities, Business Accounting, 2019, 6, pp. 116–118 (in Chinese).

Computational Social Science – Luo, Ciurea & Kumar (eds)
© 2021 Taylor & Francis Group, London, ISBN 978-0-367-70193-2

The course group construction for the major of information management and information system

G.M. Wang*, J.J. Chai, Y.Q. Li & Y. Hu
School of Economics and Management, China University of Geosciences, Wuhan, China

ABSTRACT: With the rapid development of various information technologies based on the network, mobile and cloud, the society's needs for the major of information management and information system have also changed. Therefore, the major of information management and information system needs to adjust its construction direction to adapt to the development of the society. This paper first analyzes deficiencies in the construction of information management and information system specialty in School of Economics and Management at China University of Geosciences, and then puts forward the construction of course group, establishes the course group teaching team to improve students' practical ability, and has achieved certain results by the continuous promotion of course group construction. Finally, the paper summarizes and puts forward the direction of further efforts.

1   INTRODUCTION

The major of information management and information system was established in China in 1998 [1]. According to their own advantageous disciplines or professional background, many domestic colleges and universities set up this major. As an interdisciplinary subject integrating information technology and management science, it has strong practicality and innovation [2]. With the rapid development of various information technologies based on network, mobile and cloud, information management plays an increasingly important role in all walks of life [3]. The ever-changing social needs also require this major to adjust its construction direction to adapt to the development of society [4]. However, as an emerging specialty, the construction of information management and information system is still in the stage of exploration. Many colleges and universities only treat it as a simple combination of computer technology and other professional knowledge, which leads to some problems, such as education orientation, curriculum plan, practice links and so on [5]. Therefore, various colleges and universities having this major actively carry out teaching reform, optimizing existing teaching resources and strengthening professional construction, so as to make graduates more in line with the needs of current social development [6,7].

The major of information management and information system has been supported by excellent discipline development and scientific research activities since it was established at China University of Geosciences (CUG) in 2000. However, the development of information technology makes the enterprise transform to the application of agile management aiming at "micro-technology, data-driven and business transformation". Meanwhile, this change requires more strict professional skills and comprehensive quality of information management professionals. Therefore, it is urgent to adjust the orientation of personnel training in the information management and information system specialty [8,9]. For this reason, through the investigation of the previous graduates and the training programs of domestic and foreign universities, we found some problems in the current professional training system, such as lack of coordination between courses, duplicate content from different courses, and weak practicality of the courses. To promote the specialty construction and

---

*Corresponding author

solve the above problems, we focus on the construction of course group and take the construction of teacher team as the key to deepen the reform of undergraduate professional teaching. Besides, we try to promote the complementarity between theory and practice as well as make full use of our strong scientific research capability. Through the integration of scientific research and teaching, the combination of enterprise practice and teaching likewise [10,11], we want to achieve the goal of improving students' practical application ability, social service ability and self-improvement ability.

2 INVESTIGATION

To thoroughly understand the current situation of the development of information management and information system specialty in our university and clarify the direction of specialty improvement, we take graduates of this major in our school and domestic and foreign universities as the research objects, and carry out the following related investigation work. On the one hand, for the specialty graduates, the work feedback survey is processed by organizing talks, questionnaire survey and other ways to understand the graduates' major employment industries, positions, ability needs and suggestions on curriculum, especially to investigate the professional skills or knowledge that graduates need to strengthen at school found in their actual work; On the other hand, we gathered domestic and foreign universities' (e.g. Georgia State University, Florida State University, Fudan University, Dalian University of Technology, University of International Business and Economics, Huazhong University of Science and Technology, Harbin Institute of Technology) related information of information management and information system specialty, and investigated the teaching situation and reform plan of typical universities. Especially we have insight into the training programs, core curriculums and specialty construction ideas under the new situation of these schools.

Through the systematic investigation of the above two aspects, we find that the existing specialty construction plan urgently needs to be adjusted to enhance the correlation between the core specialized courses and the practical operation skills of students from the following three points:

First of all, the survey on feedback of graduates shows that the course design is just around a single course in the core specialized courses of the specialty training plan. It ignores the connection between various courses, and results in the repetition and lack of knowledge points. This not only wastes energy and time on repetitive knowledge points, but also may neglect important knowledge points, which is not conducive to the construction of the overall knowledge architecture. Thus, we should strengthen the relationship between the courses and make overall arrangements in course design.

Secondly, previous graduates generally reflect that the students' abilities to deal with data and implement information system need to be strengthened. The survey shows that 59% of the graduates think that data manipulation abilities needs to be strengthened, and 48% of the graduates think that their system implementation ability needs to be improved. Accordingly, more practical skills training should be provided in the process of expertise learning to improve the students' practical ability.

Thirdly, the specialty aim of information management and information system has gone through four stages, namely "Basic infrastructure", "Internet Business Application", "Mobile Architecture" and "Data-driven" in the past 15 years. The latter stage is the advanced version of the previous stage. On the basis of the research about specialized training plans of information management and information system in many universities at home and abroad, the specialized training plan of our university is relatively lagging behind, and it is still in the stage of "Basic Infrastructure" and "Internet Business Application", while most of the domestic first-class universities have developed to the stage of "Mobile Architecture" and "Data-Driven".

3 SPECIFIC MEASURES

Based on the survey on feedback of graduates and the comparison of the specialized training programs on information management and information system at home and abroad, we realize that deficiencies in the existing training programs and curriculum settings. To adapt to the new technological progress and social needs, and in combination with the disciplinary advantages of management science and engineering in our university, the construction of the curriculum group of information management and information system needs to closely focus on the optimization of the curriculum system and teaching content. Therefore, the teachers of School of economics and management in CUG focus on the construction of curriculum group, starting from the integration of relevant courses and taking the construction of teacher team as the key to deepen the reform of undergraduate professional teaching.

3.1 *Course group construction*

In response to the trend of specialty orientation of major of information management and information system from "Basic Infrastructure" and "Internet Business Application" to "Mobile Architecture" and "Data-Driven" [12], core specialty courses such as "R language and data analysis", "Python data analysis and machine learning" have been added successively to the specialty training plan. These courses can cultivate students' abilities of data acquisition, data analysis and decision making under the data-driven situation. Based on the existing specialty courses, teachers of relevant courses combine with the development of new Internet application and increase the system technology architecture knowledge and business application content in the context of mobile Internet. It makes the course knowledge closely adhered to social demand and specialty programs of top universities.

To optimize the teaching content and straighten the connection between courses, we design and integrate the core courses, and further divide them into three course groups that support each other: "information system design and implementation", "data analysis and optimization decision-making" and "e-commerce operation". The core course group of "information system design and implementation" is located in the cultivation of basic knowledge such as the theory and method of information system design. It requires the students to use computer programming technology to design micro and small systems. This course group includes "Management Information System", "Information System Analysis and Design", "Information System Project Management", "ERP Principle and Application" and "Course Practice". "E-commerce operation" and "data analysis and decision-making" are respectively oriented to the cultivation of students' practical ability in business management application and data-driven management decision-making. The former emphasizes the business management and operation in the emerging Internet, especially in the mobile and social environment, while the latter provides more specific decision-making tools and data processing methods for new business applications. The course group of "e-business operation" includes "E-commerce", "Information System Strategy and Management", "Enterprise E-commerce Management" and "Supply Chain and Logistics Management". These courses are designed to improve business application and management knowledge in social and mobile scenarios, and train students to master business operation and management skills in emerging Internet application scenarios. The course group of "data analysis and decision making" includes "Decision Support System", "Data Mining", "Business Intelligence", "R language and Data Analysis" and "Python Data Analysis and Machine Learning" and "Data Analysis Training". These courses train students' ability of the big data acquisition, business analysis and decision-making by strengthening practical teaching python, data mining, machine learning and other skills.

In each course group, teachers will formulate corresponding syllabus according to the emphasis of each core course. The courses in each course group are from basic to advanced, from easy to difficult, and closely linked. Each course has its own specific objectives, and reaches the objectives of the course group after the completion of all courses. In this way, we can not only avoid wasting teaching time due to the repetitive knowledge points among the courses in each course group, but also enhance the relationship between the course structure and content, as well as help students establish a specialty knowledge structure [13]. Meanwhile, it also provides more teaching hours for the practical courses in each course group and teachers can provide more practical skills training for students, improving their practical ability, especially strengthening their data operation ability and system realization ability [14–15]. All practice contents in each course group will be organized around a project that meets the teaching schedule, so as to realize an organic whole of teaching and practice [16].

3.2 Teaching team construction

Corresponding to the three course groups of the major of information management and information system, we divide teachers into three teaching teams by their professional background and teaching experience. Each teaching team is responsible for the construction of one course group. Each course group is composed of 1–2 well-experienced and high-quality teachers and 2–4 young teachers. The well-experienced teachers lead the course group with responsibility for the construction of all courses and the assignment of teaching tasks. The teaching team of course group adopts the pattern of collective preparation and take turns in class to improve the teaching quality. The teachers in the course group have similar knowledge structures, and they can prepare lessons together, conduct course discussion, study the content of the course and revise the course outline. Teachers should be arranged in each semester according to the situation of teachers to ensure the continuity and development of the teaching system and teachers, mobilizing the enthusiasm of the teachers.

At the same time, our school encourages and supports teachers to participate in professional training and teaching seminars for advanced teaching concepts and methods from domestic excellent educators from this specialty and discuss the development of specialty with them. In this way, teachers can learn from others' strong points and reduce the gap, and improve their own teaching level and knowledge level constantly. In addition, teachers who participating in the training will share the skills they have learned in the training with others and apply them to the teaching of relevant courses.

3.3 Practical ability improvement

In addition to promoting the construction of course group and teachers' team, we also improve the practical ability of students from the following aspects:

(1) Teachers actively sign relevant internship agreements with professional counterparts outside the campus to enable students to enter the enterprise. In enterprises, students can use their knowledge to deal with problems in practical work, enhance their practical ability and improve professional skills. (2) Teachers involve undergraduates in scientific research projects through "academic navigation" activities. At the same time of learning professional knowledge, it can promote students' scientific research ability and improve the students' practical ability of the combination of theory and scientific research. (3) We arrange some teachers with outstanding achievements in relevant fields to compile or translate some textbooks. This measure strengthens the textbook construction of the course system and ensures the cutting edge of teachers' teaching. Meanwhile, teachers will communicate with the students who have completed the internship, ask about their feelings in the internship work, and consult their suggestions for the course offered and arranged. Based on their feedback, we will timely adjust the relevant courses. For example, we will remove the outdated courses and add latest courses and knowledge by arranging special teachers to study and teach as soon as possible. This will enable students to understand the trend of professional development more quickly and grasp the frontier knowledge, ensuring that they always integrate with the needs of the society.

4 ACHIEVEMENTS

As a multi-disciplinary integration of specialty, the major of information management and information system not only needs to teach students professional knowledge and necessary basic computer knowledge in related fields, but also needs to emphasize practice, so that students can use the knowledge they have to find, analyze and solve problems in reality. Promoting the construction of the course group can not only make the specialty orientation clear, reduce the repeated contents among the courses, and strengthen the close connection between the courses, but also integrate the existing teaching resources effectively, strengthen the practice of various professional directions, and achieve the purpose of learning and application. This makes the students more in line with the needs of the current social development.

Some achievements have been obtained after the construction of the course group carried out on the major of information management and information system in CUG as follows:

Students' cognition of the major has improved greatly. Through "entrance education", "internal conversation of each specialty monthly" and "class-adviser system", teachers can solve doubts for students from all aspects constantly, promote students' cognition of the major, and help students to define their own development goals. Gradually, students' understanding of the major has been greatly improved, while students' evaluation of the course has been significantly improved, and the overall performance excellence rate of the students has made greater progress.

Students' practical ability has been improved greatly. In the graduation project of students majoring in information management and information system, students have made significant progress in the completeness of the information system implementation and the ability of data analysis. Simultaneously, students actively apply for various types of scientific research plans and social practice activities of college students, and participate in various kinds of scientific and technological competitions such as national and provincial mathematical modeling competitions and challenge cups. Up to now, students in this major have already won nearly 20 national and provincial awards. For example, students have made great progress in participating in and winning awards in mathematical model competitions and data analysis competitions. More than 10 students won provincial and ministerial awards, including three international awards in 2019 compared with that only three students won provincial-level incentives in 2018.

The employment rate and quality of students has been improved greatly. Through team tasks in the course and enterprise internship, students have improved the cooperation and communication among teammates to a certain extent, and gained some experience in large-scale projects. Therefore, both the quality and quantity of students' employment have been improved greatly.

The teaching team has been enhanced greatly. Our school pays great attention to the construction of the course group of information management and information system specificity, and organizes teachers to attend the observation meeting of undergraduate excellent teaching demonstration course. Teachers have shared their experience in case teaching, classroom interaction design and bilingual teaching through on-site demonstration and explanation. It improves the overall level and quality of teaching teachers and promotes the reform and development of teaching field. More than ten teachers won awards in teaching competitions of our university, two teachers ranked first in the college teaching ranking, and one teacher won ZhuXun Young Teacher Education Award Fund.

This supports for the training program revision work of information management and information systems. Combined with the research results of the project, the training program of the information management and information system specialty has been revised. The latest one is more in line with the actual needs of the society for the students. It can help the students become the talents of the information management specialty in line with the development requirements of the times.

5 CONCLUSIONS

This paper builds the course group for the development of information management and information system specialty in our university after investigation and analysis. We have achieved certain results including establishing curriculum group with curriculum group teachers' team, and improving students' practical ability. Moreover, we will improve the construction of course group as follows:

(1) In the further course group construction, we should not only strengthen the communication between the professional teachers in the department, but also with teachers of other majors,

especially the computer professional teachers, which helps us further improve the construction of course group.

(2) In the process of implementing the course group, we will make more teachers participate in the planning and design of the course group, and adopt the way of communication and discuss to achieve the objectives of the course group. These measures can improve the implementation effect of the course group constantly.

(3) We will actively introduce the needed professional talents urgently. We will also adopt the method of "going out and inviting in". It will not only increase the opportunities for teachers to communicate with the outside world, but also invite relevant well-known experts to give lectures and guide, so as to speed up the training of existing teachers.

(4) We will continue to coordinate with academic workers, class adviser, academic guidance and teachers of the major to make them cooperate with each other. That can achieve full coverage management in a hierarchical way, and better realize the construction of course group.

(5) We will promote cooperation and exchange in teaching and scientific research about using big data laboratory with other colleges, so as to realize the cross integration of majors and disciplines.

REFERENCES

[1] Q.S. Kong, B.Z. Liu and R. Dong, Research and exploration on curriculum system of information management and information system, China Higher Medical Education. 10 (2008) 87–89.

[2] K.Q. Yang, Discussion on awareness of information management and information system specialty, Information Security and Technology. 3 (2012) 24.

[3] C.M. Hao, B. Wu, R.X. Zhang, Conception of professional construction on information management major in financial undergraduate universities, Education Exploration. 4 (2013) 59–60.

[4] S.Y. Jing, Construction and research of database course group of information management and information system, Fujian computer. 30 (2014) 80–81.

[5] R.X. Wang, Exploratory research on characteristic course group of TCM information management and information system, China Medical Journal. 14 (2017) 125–128.

[6] Z.Y. Hu, X. Lei, Research of information management and information system speciality laboratory based on virtual computer, Future Control and Automation, Springer, Berlin, Heidelberg, 2012, pp. 453–460.

[7] S.H. Tian, X.M. Zhang, Research on education system for information management and information system based on three-dimensional model of practice, International Asia Conference on Industrial Engineering and Management Innovation (IEMI2012) Proceedings, Springer, Berlin, Heidelberg, 2013, pp. 1709–1716.

[8] L. Mi, Z.G. Wang, Y. He, Construction and practice of education system for information management and information system major based on advantages and characteristics of Chinese medicine subject, China Medical Herald. 13 (2018) 17–19.

[9] D.J. Wang, Y.L. Zheng, Discussion on the construction of business intelligence curriculum system of information management, Henan science and technology. 23 (2014) 265–266.

[10] S.S. Zhang, Q. Li, Construction of core curriculum system of information management and information system, Modern Economic Information. 19 (2016) 404.

[11] P. Jia, X.P. Hu, Exploration on teaching reform of core course group of information management specialty-Based on the background of combination of engineering and learning, Higher Science Education. 1 (2014) 106–110.

[12] Y.D. Wu, Y.W. Wang, A study of smart construction and information management models of AEC projects in China, International Journal of Simulation: Systems, Science and Technology. 2016 (17) 21–28.

[13] Z.L. Gu, Information management and information system implementation of bilingual teaching professional course exploration and practice, Journal of Zhouyi Research. 4(2014) 36–37.

[14] H. Jiang, X.F. Zhao, Study on the lab teaching system construction of information management and information system profession, 2010 International Conference on E-Health Networking Digital Ecosystems and Technologies (EDT), IEEE. 2(2010) 364–366.

[15] P. Christine, C. Turner, Laboratory information management systems, CRC press, Boca Raton, 2018.

[16] C.L. Yao, X.H. Li, X.J. Tang, Practical teaching reform of computer specialty based on Curriculum Group, Computer Education. 22 (2012) 33–37.

Computational Social Science – Luo, Ciurea & Kumar (eds)
© 2021 Taylor & Francis Group, London, ISBN 978-0-367-70193-2

Construction of the practical teaching system outside campus in local universities and colleges based on emerging engineering education

F.H. Xu
Hubei University of Automotive Technology, Shiyan, Hubei, China

X.H. Zhu
Hubei Sanhuan Automobile Co., Ltd, Shiyan, Hubei, China

S.H. Wang*
Hubei University of Automotive Technology, Shiyan, Hubei, China

ABSTRACT: For the development and transformation needs in local universities and colleges based on emerging engineering education, the practice education system off campus in local universities and colleges is discussed, including the establishment of an effective off-campus practice teaching quality assurance mechanism, and establishing off-campus practice education bases which are suitable for emerging engineering education talents training mode, training the off-campus practice instructors, the construction of practical teaching resources, and so on. These measures are introduced to protect the quality of practical teaching and provide a reference for the development of emerging engineering education in local universities and colleges.

1 GENERAL INSTRUCTIONS

On June 21, 2018, the Ministry of Education held a new-era National Undergraduate Education Conference in Chengdu, Sichuan Province, which emphasized the need to strengthen the construction of new engineering and raise the construction of new engineering to the level of a national education strategy (Ministry of Education 2018). Therefore, as a local university, how should we interpret the new engineering course and carry out the new engineering education reform?

As early as February, April, and June 2017, the Ministry of Education held seminars in Fudan University, Tianjin University, and Beijing, respectively, and formed three program documents to guide the construction of new engineering such as "Fudan consensus," "Tianda action," and "Beijing guidelines" (Fudan consensus 2017; Action route 2017; Guide 2017).

In particular, in May 2017, the National New Engineering Development Summit Forum of local colleges and universities was held in Hunan Institute of Engineering. More than 200 higher education experts and scholars from over 100 colleges and universities conducted in-depth discussions and exchanges around the development path of new engineering in local colleges and universities, including the demands and challenges of the new economy on engineering education in local colleges and universities. At the end of the conference, seven initiatives (drafts) were formed, namely "Hunan Zhejiang initiative" (Department 2017). At the same time, the domestic education sector also carried out exploration into the construction of local universities in the context of new engineering, including the discipline evolution logic (Jianguo and Jun 2017), positioning and paradigm change (Gen et al. 2018), four basic problems (Zhihong 2018), practice education community (Guangzhi and Yanxia 2018) and so on, which points to the direction of new engineering construction in local colleges and universities from the macro-level. This paper focuses on the theme that

---

*Corresponding author

local colleges and universities should play a supporting role in regional economic development and industrial transformation and upgrading. Taking Hubei Automotive Industry College as an example, it discusses the construction of the off-campus practice education system in local universities and colleges.

2 THE GUIDING IDEOLOGY FOR THE OFF-CAMPUS PRACTICE EDUCATION SYSTEM

Based on the background of Dongfeng Motor Corporation, Hubei Institute of Automobile Industry has a strong background in the automobile industry. The school has always adhered to the combination of production, teaching, research, and application, and constantly carried out the exploration and innovation of talent training mode, and initially explored the method of running the school in line with its own characteristics (Shenghuai et al. 2016; Chao 2011), which laid a solid foundation for the implementation of new engineering. Under the current background of new engineering construction, the school has further strengthened the construction of the off-campus practice education system.

Based on the implementation of the spirit of relevant documents of the Ministry of Education and our school, the school further strengthens the practice teaching work and improves the quality of practice teaching. First, starting with the basic principles of the formulation of the undergraduate talent training program, the principles should fully reflect the following aspects: always oriented by serving the regional economic development and the transformation and upgrading of the automobile industry, deepening the practical teaching reform, improving the quality of practical teaching, requiring teachers to strengthen the practical teaching reform of the integration of production and teaching, building the practical teaching system with ability training as the main line, and we should improve the credit of practical teaching, promote the reform and innovation of practical teaching content and teaching mode, explore the integration of theory and practice, encourage the establishment of teaching reform courses based on projects, cases and practical engineering problems, and explore the new practical teaching mode.

3 ESTABLISH AN EFFECTIVE GUARANTEE MECHANISM FOR THE QUALITY OF OFF-CAMPUS PRACTICAL TEACHING

3.1 *Strengthen the organization and leadership of practice work*

Compared with teaching work inside the school, the quality of practice outside the school is difficult to guarantee. For this reason, the secondary colleges of our university set up a Teaching Committee, which is responsible for the organization, management, coordination, inspection, and other aspects of the implementation and monitoring of the practice teaching. Under the teaching working committee of the secondary colleges, a teaching monitoring group of colleges and departments shall be established to check and monitor the practice of the unit, including the preparation of practice teaching, mobilization of practice, implementation of practice plan, guidance of practice, management of practice process, construction of practice base, discipline and safety guarantee measures of practice, and effect of practice, etc.

3.2 *Attach importance to the management of the practice teaching process*

The process of practice teaching includes the establishment of teaching syllabus and practice plan, practice mobilization and safety education, practice teaching guidance, and practice assessment. Only by paying attention to the whole process management of practice teaching and ensuring the effective time of practice, can we ensure the effect of practice, which can be implemented in the following aspects:

(1) Make a scientific syllabus and practice plan. According to the objectives and requirements of the professional training program, teachers should scientifically formulate the practice syllabus and practice instructions, determine the content of the practice, and prepare the practice plan or practice teaching organization plan according to the semester.

(2) Do well in practice mobilization and safety education. Before the practice, teachers must do a good job of the mobilization and safety education of students, arrange the practice tasks, clarify the practice requirements, and explain the purpose and significance of the practice, the content, arrangement, requirements, evaluation, discipline, and safety requirements of the practice to students in detail.

(3) Strengthen the guidance of practice teaching. The instructor is required to do a good job of discipline management in the process of practice, and check and urge students to complete various internship tasks; the instructor is required to strengthen the contact with the internship unit, coordinate the relevant matters of the internship, and ask the internship unit to fill in the "Feedback form of practice teaching quality" after each internship to evaluate the students' internship. For students participating in scattered practice and graduation practice, teachers should strengthen the safety education and management of interns to prevent accidents during practice.

(4) Ensure the effective time and effect of practice. The actual working hours of cognitive practice, professional practice, production practice, and graduation practice shall not be less than 70% of the planned practice days, of which at least three internship units shall be arranged. Each major can break through the limitations of the current curriculum plan, and appropriately extend the students' practice time according to the actual situation, so as to meet the needs of students to improve their practical ability.

(5) Do a good job in assessment. The school should establish a variety of assessment methods for students' practice, and focus on the performance of practice, practice diary, student defense, and other aspects into the comprehensive assessment; a scientific, standardized, unified assessment or scoring standards should be established for each practice, and the practice instructor should evaluate students' performance in strict accordance with the standards, and the performance assessment should have a detailed subrecord; the practice instructor should summarize the practice work earnestly, and improve the content and method of practice teaching continuously to improve the quality of practice work.

4 ESTABLISHMENT OF OFF-CAMPUS PRACTICAL TEACHING AND EDUCATION BASE

4.1 Basic situation of off-campus practical education base

The off-campus practical education base is the carrier to ensure the realization of the quality of practical teaching. According to the characteristics of different specialties and disciplines, each specialty carries out the nearby layout of the professional base construction according to the "531" mode, that is, each specialty has five practice bases, which can undertake the cognition practice, professional practice, graduation practice, etc., three of the five bases should be stable and high-quality, and be able to meet the requirements of professional practice. One is a comprehensive practice base with a certain level of combination of production, learning, and research, each base will be assigned a teacher who is familiar with the base and has a strong sense of responsibility to be the head of the practice base. Each major selects one base as the key and long-term construction core base of the college in the future and determines one demonstration base or provincial practice and training base to carry out the construction.

4.2 Basic situation of the off-campus practical education base

During the construction of Dongfeng commercial vehicle engine plant "practice training base of Hubei Provincial University," our university and the plant jointly set up the "practice training base working committee," "practice training expert committee," "practice training base office," and other institutions, which are responsible for the construction of practice training base, teaching evaluation, teaching reform, and other specific contents include the establishment and improvement of the management and assessment system of the base, the formulation and improvement of various internship and training management systems, such as the measures for the management of part-time teachers and the measures for the monitoring and evaluation of the practice quality of enterprises, all of which will promote the construction and management of the base to be more standardized, institutionalized, and scientific.

Due to the limited investment of enterprises in the construction of the practice base, it is difficult to support its effective operation. Our university seized this opportunity to apply for the approval of the special fund for the construction of Hubei Provincial College Students' internship and training base, which was directly used for the construction of the internship base, so that the quality of the internship of the factory was significantly improved, the reception capacity of the internship was greatly enhanced, and considerable economic benefits were created for the enterprise, thus, a win–win situation has been achieved. Since the provincial practice and training base was approved to be jointly built, the management system and operation mechanism have been further optimized, and greater comprehensive benefits have been achieved. Each year, it receives more than 20,000 college students from more than 80 universities including our university, and new universities have joined the base in succession.

5   THE TRAINING OF THE OFF-CAMPUS PRACTICE INSTRUCTORS

To ensure the quality of practice teaching, our university stipulates that each college should select teachers who are familiar with the business, have rich experience, a strong sense of responsibility, and have certain organizational ability to serve as practice instructors. Professors and associate professors are required to take at least one day of practice every semester.

The team of off-campus practice instructors is built in accordance with the principle of "school enterprise communication, old with new, layered training, two-way improvement." In order to make the off-campus practice instructors become a double-teacher team that can meet the teaching requirements, measures such as carrying out regular training, using the engineering practice education resources of the base, and employing the middle and senior engineering technical personnel of the enterprise to participate in the practice teaching guidance will be used to continuously improve the engineering technology literacy and engineering practice ability of the practice instructors.

To strengthen the cultivation of young teachers' engineering ability, we should carry out the team building of practice guidance teachers, and gradually form a stable team of practice guidance teachers. Each practice should be led by a teacher with rich teaching experience, give full play to the role of "mentoring," gradually improve the teaching level of young practice guidance teachers, and form a team-based, professional, echelon-based, long-term, and institutionalized team of practice guidance teachers. We will open channels for the appointment of part-time teachers, and encourage the college to hire middle and senior managers and engineering technicians who are on-the-job or retired from enterprises to serve as part-time teachers, so that they can participate in the work of guiding students to practice, offering special lectures, guiding graduation design, guiding young teachers to practice, etc.

Young teachers are selected by stages and in batches to carry out engineering practice rotation training in enterprises. We will continue to strengthen the contact with enterprises and R&D units to expand the cooperative relationship between production, learning, and research. We will continue to strengthen scientific and technological cooperation with enterprises and organize discipline leaders and academic leaders to participate in the research on enterprise development strategies and the application of advanced manufacturing technologies. Through jointly undertaking the scientific and technological development, and technological transformation of the enterprise, the ability of young teachers to discover and solve practical problems in engineering is improved. In recent years, the school has undertaken more than 50 million yuan of horizontal technology development projects entrusted by local enterprises, which strongly supports the development of the local economy.

6 SOLUTIONS TO THE PROBLEMS EXISTING IN OFF-CAMPUS PRACTICE

In the process of practical teaching, we often encounter many problems. The common problems are as follows: the planned internship arrangement cannot be carried out normally due to the temporary shutdown of the enterprise due to the influence of the market; the students do not have sufficient time to observe the site due to the compact production tasks of the enterprise; the production line of the enterprise is highly automated; and some advanced production equipment is sealed closed type, so students can only see the shell of the equipment in the workshop, not the actual processing of the equipment; part of the internship base can only provide the teaching mode of cognition internship, but not the production internship, professional practice, and graduation internship. The emergence of these situations has seriously affected the quality of off-campus practical teaching. At present, our school mainly solves the problems from the following aspects.

6.1 Construction of practice and training base in school

To better deal with the uncertainty of practice teaching outside the school, the large-scale practice training base in the school is planned and constructed by the principles of school enterprise cooperation and resource sharing. According to the requirements of specialty construction, we should integrate various school and enterprise resources, introduce typical equipment of industry and enterprise, build several in-school training bases focusing on the cultivation of students' key abilities by means of new construction, optimized combination, and transformation, and build a new out-of-school practical teaching mode combining in- and out-of-school and coexisting in various forms.

6.2 Construction of online and virtual simulation practical teaching resources

The online course resources of off-campus practice teaching include an enterprise introduction video, online practice teaching courses, construction of practice thinking questions, and the opening and sharing of high-quality and excellent teaching resources. Before practice, students can preview the practice through these resources, clarify the purpose, content, and requirements of the practice, timely understand and grasp the doubts and difficulties in the learning process, and quickly integrate into the field practice to improve the practice effect.

The virtual simulation teaching resources of off-campus practice are constructed. We should rely on virtual reality, multimedia, human–computer interaction, database, network communication, and other technologies to carry out the construction of virtual simulation teaching resources, so that students can simulate on-site practice in a highly simulated virtual environment, and enrich the content of practice teaching, reduce costs and risks, and improve the effect of practice teaching.

6.3 Exploration of the new mode of practice teaching outside campus

On the premise of meeting the requirements of professional training, colleges are encouraged to determine the location and time of off-campus practice flexibly according to their own reality, so as to ensure the smooth development and good results of students' practice activities.

Students should be encouraged to combine practical training with graduation project, curriculum substitution, and credit conversion. We should guide the students to actively carry out the technical work concerned by the enterprise at the time of practice, so as to cultivate the talents needed by the enterprise.

We can develop the training workshop together with the off-campus internship enterprises and help the enterprises to transform the idle old workshop area and workshop into the production internship training workshop, which can be used for the enterprise to train the young employees and receive the internship of college students.

7 CONCLUSION

Based on the advantages of the industry, our school has always adhered to the innovation of the school enterprise cooperation mode. Based on cooperation with Dongfeng company and local enterprises, our university has built two national engineering practice education centers, one national off-campus practice education base, 81 production, learning, and research cooperation bases, and more than 160 off-campus practice training bases, among which five bases have won the title of provincial practice training base. These off-campus practice bases cover all specialties of our school and meet the training needs of the talents of our school. The engineering consciousness and practical ability of the students have been exercised in the off-campus practical teaching activities, and the comprehensive quality and innovation consciousness of the students have also been improved. Therefore, the graduates are very popular in the larger society, and the employment rate of the graduates has been ranked as in the forefront of universities affiliated to Hubei Province for four consecutive years.

ACKNOWLEDGMENTS

This project is funded by Hubei Provincial Teaching and research project of colleges and Universities (No. 2014329); The 12th Five Year Plan of Education Science in Hubei Province (No. 2014B188); Research project of Hubei Provincial Department of Education (No. Q20171805; No. D20141802).

REFERENCES

Action route of "new engineering" construction ("Tianda action"). 2017. Research in Higher Education of Engineering 2: 24–25.

Chao, W. 2011. Taking discipline construction as the leading factor to cultivate the comparative advantages of schools. China Higher Education1: 24–26.

Department of education of Hunan Province: The National New Engineering Development Summit Forum of local universities was held in Hunan Institute of engineering [EB/OL]. 2017. http://gov.hnedu.cn/c/2017-05-13/871856.shtml.

Fudan consensus on the construction of "new engineering". 2017. Research in Higher Education of Engineering 1: 10–11.

Gen, C., Qinfang, Z., Huajun, Z. 2018. Orientation and Paradigm Changes in Emerging Engineering Education Construction in Local University. Heilongjiang Researches on Higher Education 7:69–72.

Guangzhi, D., Yanxia, X. 2018. Research on Community Construction of Practical Education in Colleges and Universities. Heilongjiang Researches on Higher Education 12:133–135.

Guide to the construction of new engineering ("Beijing guidelines"). 2017. Research in Higher Education of Engineering 4: 20–21.

Jianguo, X., Jun, Z. 2017. On the Reform and Development of Engineering Education in Local Universities and Colleges Based on Education. Research in Higher Education of Engineering 1 (3):15–19.

Ministry of Education of the People's Republic of China: "The National Conference on undergraduate education in the new era". [EB/OL].2018. http://www.moe.gov.cn/s78/A08/moe_745/201806/t20180621_340586.html.

Shenghuai, W., Fenghua, X., Tukun, L., Yurong, C. 2016. Exploration of Characteristic Practice Teaching Platform Development for Regional Colleges and Universities. Research and Exploration in Laboratory 35(5): 201–204.

Zhihong. L. 2018.The Four Fundamental Issues of New Engineering Discipline Construction for Local Universities. Heilongjiang Researches on Higher Education 12: 40–43.

Computational Social Science – Luo, Ciurea & Kumar (eds)
© *2021 Taylor & Francis Group, London, ISBN 978-0-367-70193-2*

Study on countermeasures of the strategy of promoting rural revitalization in local universities

X.Y. Shao
International Education Institute, Dalian Jiaotong University, Dalian, China

ABSTRACT: The strategy of rural revitalization is a scientific plan of the party and the state for rural development in the new era. Colleges and universities should change their ideas, innovate the training mode of talents, and strengthen their service ability to the rural revitalization. Colleges and universities should also establish a guarantee mechanism to serve the rural revitalization strategy and promote its smooth implementation.

1 INTRODUCTION

The strategy of rural revitalization is a major policy decision at the 19th national congress of the CPC, which indicates the direction for the development of rural areas in the new era. Local colleges and universities, with relatively concentrated knowledge, technology and human resources, should, as the main body of the rural revitalization strategy, take the initiative to shoulder their own social responsibilities, actively serve the rural revitalization, and ensure that the rural revitalization strategy is promoted better and faster.

2 SIGNIFICANCE OF THE STRATEGY OF SERVING RURAL REVITALIZATION BY LOCAL UNIVERSITIES

2.1 *The successful implementation of the rural-revitalization strategy*

In the opinions of the state council of the CPC central committee on the implementation of the strategy of rural revitalization, it is proposed that all sectors should combine their own characteristics, give play to their respective advantages and actively participate in supporting the strategy of rural revitalization. As a national strategy, rural revitalization is a major task in China's economic development. As the training base of agricultural talents and the source of agricultural science and technology research and development, colleges and universities have natural advantages in agricultural science and technology personnel training, agricultural high-tech innovation and promotion, farmers' skill training, and rural governance system improvement. By actively participating in the strategy of rural revitalization, it will definitely promote the development of agricultural economy, increase farmers' production and income, and improve the overall appearance of rural areas.

2.2 *The sustainable development of colleges and universities*

Colleges and universities are responsible for meeting the needs of science and technology support, personnel training and social services. For a long time, influenced by various factors, colleges and universities in China lay stress on talent training and scientific research, while neglecting the important function of serving the society. The direct result of this phenomenon is that the cultivated talents and obtained scientific research results are out of line with the actual needs of society, which to some extent limits their own development and is not conducive to the healthy development of colleges and universities. Rural revitalization can be an effective breakthrough for colleges and universities to get rid of this limitation. By actively participating in the service of rural revitalization and integrating talent cultivation, scientific research and social services into the practice of rural revitalization, it will greatly promote colleges and universities to integrate superior resources and give play to their own characteristics. At the same time, it can also gain direct benefits in obtaining government financial support and expanding employment channels for graduates.

3 THE SHORTAGE OF LOCAL COLLEGES AND UNIVERSITIES SERVING RURAL REVITALIZATION STRATEGY

3.1 *The personnel training shortage*

Generally speaking, in addition to professional education, colleges and universities attach more importance to the education of socialist core values, ideals and beliefs, service and dedication to social education, etc. Although the low proportion of agricultural universities in colleges and universities is one of the reasons, the main reason is that most colleges and universities focus on cultivating new professional farmers as the main body of practical talent education is too little, which leads to the colleges and universities to train agricultural talents with higher education and professional quality for rural areas is far from meeting the needs of the rural revitalization strategy.

3.2 *The lack of initiative of serving the rural revitalization strategy*

Mainly, training and other activities to serve the strategy of rural revitalization have been carried out less and in a single form. At present, the cultural knowledge level of farmers in rural areas is still relatively low, so it is urgent for colleges and universities to carry out more scientific knowledge popularization and practical training of various production and marketing in rural areas. The training of a few colleges or universities or other relevant activity often is a mere formality, not only the content is simple, and fail to be aimed at actual remedy, have very big blindness, because this has little effect.

3.3 *The inperfection of teaching system of colleges and universities*

Firstly, the distribution of college enrollment indicators is not skewed to the source of students from rural areas and western regions, resulting in uneven access to high-quality higher education resources between urban and rural areas and between eastern and western regions. Secondly, universities fail to coordinate with the development of emerging industries in rural economy in terms of regional layout structure, discipline structure, hierarchy structure and specialty setting, which makes it difficult for universities to give full play to their unique advantages in serving the rural revitalization strategy.

4 COUNTERMEASURES OF THE STRATEGY OF SERVING RURAL REVITALIZATION IN LOCAL UNIVERSITIES

4.1 *Changeing idea and training high quality talented person*

Colleges and universities should change their concepts, establish the consciousness of serving the rural revitalization, improve the corresponding service institutions, and give play to the important role of local colleges and universities in serving the rural development. Local colleges and universities should take serving the rural revitalization and promoting rural construction and progress as an important way to serve the society effectively. In the context of rural revitalization strategy, local universities should fully realize their important role in rural revitalization strategy, further expand their survival and development space in the process of assisting rural development, and adjust their functions of serving the society in accordance with the needs of rural revitalization strategy. At the same time, colleges and universities should put the service of rural development in the important position of school development, take the help of rural development as an important part of their development planning, improve the relevant management institutions, guide teachers and students to pay attention to rural revitalization, encourage students to work in rural areas, stimulate teachers to carry out research, and provide intellectual support for rural development. Providing high quality talents is an important responsibility of local universities in the strategy of rural revitalization. To serve the development of rural areas, local colleges and universities should carry out comprehensive reform. In accordance with the needs of rural revitalization and development, they should reform the personnel training mode, specialty setting, training objectives, teaching contents and other aspects, so as to cultivate high-quality talents serving the development of rural areas. First of all, according to the actual needs of rural development, we should further clarify the goal and concept of talent training, and train college students to meet the needs of rural development; Secondly, we should pay close attention to the demand of rural development for various kinds of talents and their changes. On this basis, we should make professional adjustments to improve the effectiveness of talent training. Finally, it is necessary to carry out effective reform of education and teaching methods as well as personnel training mode according to the characteristics and training objectives of various talents required by rural construction. Through the comprehensive reform of local colleges and universities, we will train high-quality talents for the rural revitalization and lay a solid foundation for the rural revitalization strategy.

4.2 *Innovativing talent training mode*

Firstly, focus on training cross-disciplinary talents for rural revitalization. Innovate the mode of personnel training and construct a new system of personnel training. We should speed up the training and training of professional and innovative personnel who are in line with the requirements of The Times and play a leading and driving role, so as to meet the requirements for the implementation of the strategy of rural revitalization. Agriculture-related colleges and universities should give full play to their own advantages and initiative, explore the talent demand and training ways under the background of rural revitalization. Secondly, we should strengthen the professional construction of rural revitalization, encourage agriculture-related colleges and universities to set up innovative classes for rural revitalization, and actively explore the mode of cultivating compound professionals for rural revitalization and cultivate professionals for rural revitalization. Third, we will improve the mechanism of collaborative education for rural revitalization, train multilevel talents for rural revitalization in various forms, and actively build collaborative innovation and synergy. We will encourage agriculture-related colleges and universities to set up institutes, research institutes or research centers for rural revitalization to provide a solid foundation for promoting rural revitalization. Fourth, we will establish a multi-level training system for rural revitalization, train versatile and innovative agricultural talents with applied technologies in modern agriculture, and give full play to the support of new agricultural talents in the process of modern agriculture.

Secondly, we need to accelerate the training of talents for planning and implementation of rural revitalization. Agriculture-related colleges and universities should fulfill their mission of providing personnel support for the implementation of the rural revitalization strategy, optimize the construction of teaching staff, and encourage researchers to participate in the rural revitalization. Government departments should support agriculture-related colleges and universities in carrying out personnel training for rural revitalization planning. In accordance with general secretary xi's requirement that "villages should be like villages" and the knowledge and skills required for rural revitalization planning, a contingent of teachers for personnel training for rural revitalization planning should be established to carry out personnel training for rural revitalization planning in a planned way. We will support agriculture-related colleges and universities in carrying out personnel training for the implementation of the plan for rural revitalization, and set up a contingent of teachers to train the personnel for the implementation of the plan in accordance with the requirements for the implementation of the plan and the knowledge and skills they should possess. We will carry out personnel training for the implementation of the plan for rural revitalization in a planned way. We will explore different ways to encourage talents, and improve the access of researchers from institutions of higher learning and research institutes to rural areas And the system of temporary and part-time employment and innovation and entrepreneurship of enterprises, and the establishment of agricultural industry innovation platform.

4.3 *Strengthening the service ability of colleges and universities to rural revitalization*

Colleges and universities shall conduct extensive and in-depth investigation and investigation on the situation of local rural construction, focusing on the existing problems and main advantages of local rural construction. On this basis, starting from the actual situation of local colleges and universities, they shall determine their own development direction and enhance their ability to serve the rural revitalization strategy. At the same time, it is necessary to make overall arrangements and reasonable planning. On the basis of constantly strengthening communication and interaction with the villages in this region, we should adjust our own specialty setting, discipline setting and talent training program design according to the needs of rural construction and development. In terms of specialty setting, local colleges and universities should embody the characteristics of flexibility and adaptability, take the service of rural construction and development as the basis of specialty setting of local colleges and universities, and build characteristic specialties that help the rural revitalization strategy. College students are the main force in the future social construction, especially in the rural revitalization and national rejuvenation of the great cause still has a long way to go. Therefore, colleges and universities should enhance the ability of college students to serve the rural revitalization strategy in education and teaching. This is a key way for universities to serve the rural revitalization strategy in a sustainable and long-term way. Colleges and universities should give professional guidance to college students according to the construction and development needs of local villages and ensure that they have professional knowledge and ability to devote themselves to the future rural construction. Colleges and universities should be ready for college students ideological and political education work, improve college students' concept of national and social sense of responsibility, help them to clarify the relationship between rural revitalization and national rejuvenation, let they can from a higher perspective sees its future development and the development of national, ethnic, and its associated with country and nation, prepared to body and mind and thoughts, in rural construction.

5 ESTABLISHING A SUPPORT MECHANISM TO SERVE THE RURAL REVITALIZATION STRATEGY

5.1 *Establishing an organizational guarantee mechanism to serve the strategy of rural revitalization*

On the one hand, local colleges and universities should set up the concept of seeking support through services and helping rural revitalization better, set up leading groups to effectively serve rural construction and development in combination with their own reality, and establish corresponding organizations and staff to strengthen the leadership of the strategic work of serving rural revitalization. On the other hand, local colleges and universities can also organize teachers to set up service teams to help rural construction and development, and organize students to set up corresponding social practice teams to effectively serve rural construction and development. Through the establishment of organizational guarantee mechanism, local colleges and universities make positive contributions to the construction and development of rural areas in the new era.

5.2 *Establishing incentive and guidance mechanism to serve the rural revitalization strategy*

Local colleges and universities should set up the incentive and guidance mechanism for serving the rural revitalization strategy from the following three aspects: first, local colleges and universities should strengthen publicity to create a good atmosphere for teachers and students to serve the rural revitalization. We must make it clear in the practical work, the new era of local university service significance of rural construction and development, clear the general teachers and students in the development of the regional rural construction and the important role in strengthening the service strategy of rejuvenating rural teachers and students sense of responsibility and sense of mission, arouse teachers and students to focus on rural construction and development, motivate teachers and students to help rural construction and development, for the local service rural revitalization of college teachers and students create a good atmosphere. Second.

Colleges and universities need to set up a guiding mechanism to serve the rural revitalization strategy from the perspective of guidance and motivation. This is directly related to the effectiveness of the service rural revitalization strategy. Teachers should enrich the setting of human resources incentive for serving the rural revitalization, link subsidies and professional title evaluation with their efforts and contributions to serve the rural revitalization strategy, so as to make teachers actively serve the rural revitalization strategy and further drive the enthusiasm of students. On the part of students, students can link their contribution to the rural revitalization strategy with their credits and comprehensive evaluation, so that students can also assume their social and national responsibilities and make contributions to the rural revitalization strategy with their knowledge after study. Thirdly, local colleges and universities should publicize and report the typical deeds and successful experiences of advanced teams and individuals emerging in the service of rural construction and development through the media such as Internet and newspapers in a timely manner, so as to create a favorable public opinion environment for local colleges and universities to help rural construction and development in the new era.

6 CONCLUSION

Colleges and universities should reflect their own service characteristics and actively explore and try to serve the way of rural revitalization. Proceeding from local conditions and combining their own advantages, we will provide intellectual support and talent guarantee for rural construction and development in the new era, promote the smooth implementation of the strategy of rural revitalization, and make positive contributions to securing a decisive victory in completing the building of a moderately prosperous society in all respects and realizing the Chinese dream of great national rejuvenation.

REFERENCES

[1] Meng tiexin. Research on the path and countermeasures of the strategy of serving rural revitalization by colleges and universities. Science and technology and industry, 2019 (7): 123–126.

[2] Zhang sujie. Analysis on the focus of the strategy of promoting rural revitalization by local universities in the new era. Journal of Beijing agricultural vocational college, 2019 (3): 69–73.

[3] Jiang li, zhou bijing. Research on university service strategy under the background of rural revitalization strategy. Rural science and technology, 2019 (8): 48–49.

[4] Luo pansheng. Research on the strategic countermeasures of promoting the implementation of rural revitalization in colleges and universities. Daqing social sciences, 2019 (8): 137–139.

[5] Zheng baodong et al. Reflections on rural revitalization strategy of agriculture-related universities serving rural areas. Science and technology in Chinese universities, 2018 (12): 7–9.

Computational Social Science – Luo, Ciurea & Kumar (eds)
© 2021 Taylor & Francis Group, London, ISBN 978-0-367-70193-2

Studies on academic dispute legal features and their settlement by the law and administration

Y.B. Zhou & P.Z. Cao*
Department of Law, School of the Public and Administration, Shandong Agricultural University, Taian, China

ABSTRACT: With the globalization of the economy and the development of the Chinese economy as the most second largest in the world, China's academic disputes, which are separate from general civil disputes, have been increasing recently, and from the lessons and resolutions of these cases, China has a long way to go in building a legal system to resolve these cases, and these cases shall affect the resolution issues of future academic disputes. To resolve the cases and confirm the features of academic disputes, we shall legislate special constitutional laws and other laws, and legislate special proceeding rules to resolve this kind of dispute within the university organization. Only the special legislation and organization can hear cases established in the university and wider society.

1 INTRODUCTION

Recently, there has been a case in the university related to a postgraduate student writing a paper that did not meet with the academic criteria and resulted in the student failing to complete their course, and ultimately the student's suicide. The parents of the student make a law suit against the university and the court in the second instance rejected the appeal for the case, with the reasons focusing on the following conditions: first, it is a type of academic dispute which is not a civil dispute; second, there is no fault for the university if the university bears the legal responsibility; third the action of suicide was by an adult with full legal ability. From this case, there were some discussions of how to resolve the academic dispute and the legal features of the academic dispute, from the point of view of law and administration, and also how to avoid such a case and protect the university and also student's legal rights and life. Thereby, in this paper, the author shall discusses these issues, including the academic dispute and its legal features and how to resolve it by law and other means.

2 THE ACADEMIC DISPUTE AND ITS FEATURES

Before discussing the academic dispute, as the above definition of the law, we define a civil dispute as the following: by a common expert's definition, a civil dispute is a kind of social dispute which is related to conflict in people's life, contribution, consumer region, and so on.

---

*Corresponding author

2.1 Disputes and academic disputes

A civil dispute is a dispute with equal parties who disputes the facts that are the focus of civil legal rights and obligations, including property and personality rights. A civil dispute can have the following features: equality of the parties, the contents of the dispute are related to the civil rights of property and personality rights, and free dealing with the dispute by the parties. Therefore, an academic dispute is defined by the following: an academic dispute is about academic issues, such as social science researching and science projects and others with the following features: the contents of disputes are related to academic issues; The Party is a research-oriented party, and its standards are related to academic studies.. From the classification of disputes, there are some criteria such as the party, contents, objective, and so on. Therefore, in general, the classification is about the party criteria and the type of academic dispute is about the university and researching party. The dispute above the case connects with the style of university and researching party.

2.2 Academic dispute features

From the above discussion of the definition, we find that an academic dispute is a civil dispute and that there are special features, and special resolution mechanisms. First, the academic disputes' special features include that the academic dispute is within the compound features of academics and administration, such as the disputes may involve academic issues and others; therefore, the method of resolution is different from general civil disputes; second, the academic history law is very long, with the original academic disputes having been in the UK; third, the resolution of academic disputes shall not utilize only judicial methods and administrative method. However, in China, academic disputes are at the beginning of their researching and practice. Finally, academic disputes are related to the cultures and level of democracy in the university. Of course, the academic controversy is not such a simple description, and in this paper, they are no necessary to discuss it because of the paper topic.

3 THE ACADEMIC DISPUTE PRESENTING METHODS OF RESOLUTION AND TRENDS

With the development of high education of China, China is entering a "large mass step education period" for the student. As a kind of social dispute, an academic dispute are very common, therefore, in universities and researching institutions, there are often cases related to academic disputes.

3.1 The presenting methods for the resolution of disputes

To discuss an academic dispute case, in accordance with the international resolution of disputes, there are four types of resolution as follows: one is self-remedy resolution, using mediation and decisions by the parties themselves. For example, the party signs a contract of civil interest and the other party breaks the contract, and bears responsibility, the party shall sign a compensation contract for the losses willingly, and this style of resolution is called a self-remedy. The second method of dispute resolution is called social remedy, which means that the disputes arise in the social area and the method is dealt with by social groups, such as the civil groups and arbitration, and this method is also called ADR (altered disputes resolution). At present, ADRs are flourishing in the international disputes in commercial society and for these reasons analysis is related to the following issues: the conflict of laws and different courts and law mechanisms, and social culture. For example, if a foreign court hears a sentence for the enforcement and application for execution in another court, the first proceeding is that the foreign court sentence is confirmed by the other court with their sovereignty, and there is a barrier called the judicial barrier by which the resolution

Table 1. The academic disputes and their contents in a university.

Number	Disputes	Contents	Parties	Resolution
1	Paper writing	Copy others	Students	Academic method
2	Test data	Copy others	Professor and students	Academic and civil methods
3	Textbook writing	Copy others and copy others' ideas	Students and professor, including staff writer	Civil methods (civil methods of compensation by IP law)
4	Software designing	Creator and others copy software, ideas, and design	Students and professor, including staff writer	Civil methods (civil methods of compensation by IP law)
5	Science projects and resolution		Researching stall and company	Civil methods (civil methods of compensation by IP law)

(Data origin: public data in the university from 2010 to 2018.)

is very complicated and takes a long time. Therefore, international arbitration using ADR easily resolves this difficulty. The third method is called public remedy, in which case it is resolved by the courts and national organizations, from the aspects that the case resolution is combined with national enforcement and is called the compulsion method. This method is within the features of national enforcement and has compulsion, such as, if the sentence fails to be executed the party can apply for mandatory execution.

3.2 The trends of academic disputes development

From Tables 1, we can see the presenting issues in academic disputes.

From Table Chart 1, the trends of the academic disputes are as follows: the increasing cases and academic contents of are complicated, including civil disputes of compensation, intellectual property, and others, such as trademarks and designing of works writing. By the writer's data of class and researching, the data of the academic disputes trends from 2010 to 2018, and the level of increase are very complicated and the features include: the contents of cases are from the intellectual property and civil contents of personality rights and property; the disputes are related to copy dispute and copying professional ideas, design, and so on.

4 HOW TO RESOLVE THE ACADEMIC DISPUTES BY LAWS AND ADMINISTRATION

In regard to how to resolve the academic dispute in the university and society, the common view is that there are different methods and different legislation.

4.1 The common presenting legislation of resolution of academic disputes and new legislation in time for the regulation of academic disputes

China's legislation is increasing at present, including the Constitution Law and Civil Law of General Rules, which are related to the protection of political and civil rights. However, the academic dispute results in differential cases related to both the political and civil rights, as the above statement, the academic disputes are connected with science issues, such as the creation of science, and others, so there is no suitable method and law to resolve the dispute. In the USA, and also other developed countries, there is special legislation to resolve this kind of case, and China shall legislate to regulate this kind of academic relations and legislate the law to hear cases of this kind, otherwise, there is no legislation and the cases are not heard by the courts and by the law, such as the above case in this paper.

4.2 In accordance with the special legislation, the administration proceeding rules are needed in the university to confirm the academic disputes

At present, there are a many cases in universities, such as Peking University, and the reasons for these cases are absent from the administration of the science and the rule of science. For example, if a student copies others without creation and his own intellectual ideas, this is not permitted by law, and it is within the scope of academic disputes, therefore, the university shall first apply for the administration of the academic disputes and if the case takes place, there are stipulations of disputes and the university shall separate the academic cases from general cases, such as civil cases and tort cases without academic disputes. In fact, in some developed countries, such as Singapore, there is special legislation to confirm an academic dispute, so that the resolution is easy to resolve after the cases take place. This legislation includes the proceeding legislation and substantial legislation to resolve disputes. For example, in Singapore, in universities, there is a special academic commission to deal with this kind of case. However, in China, there is also this type of commission, and, in fact, the function of this commission is not fully utilized.

4.3 To establish a special organization to support this kind of academic dispute

Although there are some organizations to service university students and institutions, special organizations to deal with academic disputes are unavailable in some universities, for example, the academic arbitration in universities. Thus it is very necessary to establish a special organization to deal with academic disputes, including the parties of the case that need support from professional staff and guiding of professional staff, and the financial base supporting this. In this kind of organization, there are rules to support the hearing of cases and the features, criterions and others of the cases, this is first step for the party if the academic disputes to apply for the help from the staff. In other words, establishment of an organization in the university as soon as possible, as in Singapore, is needed to establish a special commission of arbitration for this type of dispute.

5 CONCLUSION

With the globalization of the economy and the development of the Chinese economy as the most second largest economy in the world, China's academic disputes have increased recently. From the lessons and resolutions of these cases, China has a long distance to go in building a legal system to resolve this type of case, above cases will effect the resolution of academic disputes. To resolve these cases and to confirm the features of academic disputes, we shall legislate special laws using constitutional and other laws, and legislate for special proceeding rules to resolve this kind of dispute within the organization of the university. Only by establishing the special legislation and organization to hear case it will win the honor in the university and society.

ACKNOWLEDGMENTS

This paper is a research result of the National Educational Plan of the 13th-Five-Year National Plan of the Studies on the Universities' Academic Disputes on Line arbitration in the "Internet+" Times (No. BIA190196). The authors are thankful for the economic support provided by the National Educational Plan. They also thank other persons who helped in the creation of this paper.

REFERENCES

[1] Jiang Wei. Civil Procedure Law, High Education Press, 2016, Beijing, 2–4.
[2] UCCUSL. Uniform Mediation Act, 14 Word Trade and Arbitration Materials 110–160(No.4, 2002)

[3] Michael E. Schneider & Christopher Kunee, Disputes Resolution in International Electronic Commerce, 14 Journal of Int. Arbitration 6–9 (1997).

[4] Herry H. Perrit, Dispute Resolution in Cyberspace: Demand for New Froms of ADR, http://www.dispute.net/cyberspace2000/ohiostate/perrottl.htm. (Oct.15, 2000)

[5] Barry M. Leiner, A Brief history of the Internet, http://www.isco.org/internet/history/brief.shtm/ (July 30, 2009).

[6] M. Scott Donahey, Current Development in online Dispute Resolution, Journal of Int. Arbitration. P115 (1999).

Computational Social Science – Luo, Ciurea & Kumar (eds)
© *2021 Taylor & Francis Group, London, ISBN 978-0-367-70193-2*

A study of college students' satisfaction with the online vocational education platform

G. Chen & Y.F. Wu
Evergrande School of Management, Wuhan University of Science and Technology, Wuhan, China

ABSTRACT: This paper investigates consumers' satisfaction of online vocational education in China, and gives some suggestions to the online education platforms. Combining the characteristics of online vocational education with consumer satisfaction theory, the authors created a model analyzing satisfaction on online vocational education platforms. Through a survey in Chinese colleges, they collected data from college students who are the main consumers of online vocational education in China. Then, they used SPSS software to conduct descriptive analysis and regression analysis on the data. The results show that all the variables in the model are significantly correlated with the results, and the platform interface and operation, diversified courses, and teaching methods of the platform are the most significant elements with effects on satisfaction. Based on the results of this analysis, three suggestions are proposed for the future development of online vocational education platforms.

1   INTRODUCTION

The Chinese government work report 2019 clearly calls for accelerating the development of a modern vocational education system and using multiple means to stabilize and expand employment. In the report, Prime Minister Li expressed that "the accelerated development of online vocational education has not only reduced China's current employment pressure, but also solved the shortage of professional and technical personnel [1]." It can be seen that online vocational education has become one of the main concerns of the country.

With the continuous expansion of online vocational education in China, young users have gradually become the main consumer group in the field of online vocational education. Among them, college students are a large proportion. Therefore, understanding the factors that influence the satisfaction of college students on online vocational education platforms will help enterprises understand the needs of college students for courses on the online vocational education platform and improve their satisfaction.

2   RELATED WORKS

Dwayne D. Gremler (2002) developed a student satisfaction assessment model to test students' satisfaction with the teaching results of school teachers [2]. Felix T. Mavondo (2004) developed a model of satisfaction based on students' satisfaction on teaching contents, learning methods, technological updates, information resources, student services, and students' attitudes [3]. According to Debnath's research (2005), students hope to find their favorite job with a good salary after finishing their studies. This factor greatly affects students' recognition and satisfaction with the quality of school education and teaching [4]. Li-Wei Mai (2005) studied the factors influencing students' satisfaction with school education, and found that the biggest factor was students' overall impression of the school and its teaching quality [5].

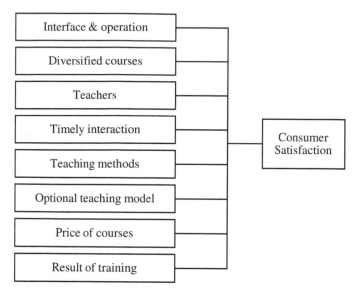

Figure 1. Consumer satisfaction model of the online vocational education platform.

Wu Liu and Xue Yang (2007) established the customer satisfaction index model of Chinese higher education to measure Chinese students' satisfaction with Chinese higher education [6]. Guo-Dong Zhao (2010) established and developed a learning satisfaction model based on teachers, students, system function, and course. Through empirical analysis, it was found that online learning results, timely interaction, quality of teaching content, and easy-operated process are the main factors influencing online customer satisfaction [7]. With the perspective of online users, Ya-Zheng Li (2016) explored the influencing factors of online users' willingness to take online courses, and found that better course quality and rich course content could stimulate users' willingness for payment [8].

3 MODEL BUILDING

According to the characteristics of the online vocational education platform and the preferences of college students, combined with the customer satisfaction model, we built a model of college students' satisfaction with the online vocational education platform shown in Figure 1. Eight influencing factors are included in the model.

4 DATA COLLECTION AND ANALYSIS

Based on the model above, we designed questionnaires and distributed them on the Internet through SO JUMP, and 200 effective answers were obtained.

4.1 *Basic information analysis*

From the data in Table 1, we can see that the age of the sample was less than 30 for all and expenses per month were not high. All the data meet the characteristics of college students in China.

Table 1. Basic information of the sample.

Index	Classes	Number	Percentage (%)
Gender	Male	88	44
	Female	112	56
Age	Under 18	1	0.5
	18–22	190	95
	23–30	9	4.5
	Over 30	0	0
Identity	In school	168	84
	Fresh graduate	30	15
	Past graduates	2	1
	Others		
Expenses per month	Under 800	3	1.5
	800–1000	53	26.5
	1000–1500	90	45
	1500–2000	34	17
	Over 2000	20	10

Table 2. Correlation analysis of variables.

	Model	Nonstandardized coefficients		Standardized coefficients		
		B	Std. error	Beta	t	Sig.
1	(Constant)	0.048	0.328		0.147	0.884
	Interface and operation	0.894	0.089	0.582	10.059	0.00
2	(Constant)	0.039	0.322		0.122	0.903
	Diversified courses	0.896	0.087	0.590	10.285	0.00
3	(Constant)	1.171	0.303		3.862	0.00
	Teachers	0.624	0.087	0.453	7.147	0.00
4	(Constant)	0.928	0.322		2.886	0.004
	Timely interaction	0.666	0.089	0.470	7.485	0.00
5	(Constant)	0.569	0.297		1.915	0.057
	Teaching methods	0.698	0.075	0.554	9.363	0.00
6	(Constant)	0.645	0.300		2.151	0.033
	Optional teaching model	0.679	0.075	0.540	9.020	0.00
7	(Constant)	0.541	0.375		1.443	0.151
	Price of courses	0.695	0.094	−0.466	7.417	0.00
8	(Constant)	0.794	0.357		2.226	0.027
	Result of training	0.644	0.091	0.450	7.0886	0.00

4.2 Linear regression analysis

A linear regression method was used to study the relationship between the influencing factors and satisfaction. We conducted a regression analysis on the relationship between variables with the help of SPSS analysis software to better explain the influence of various factors on the satisfaction of the online education platform. The specific analysis results are shown in Table 2.

The results of Table 2 shows that the standardized coefficient between the interface and operation and satisfaction was 0.582, with a significance of 0.000, which is less than 0.001, indicating that the influence of the interface and operation of the online education platform on the customer satisfaction was significantly positive at the 1% level. In other words, the easier the platform interface and operation were, the higher the consumer satisfaction with the online education platform.

Table 3. The regression coefficient of each factor on satisfaction.

Coefficients

Model	Nonstandardized coefficients		Standardized coefficients		
	B	Std. error	Beta	t	Sig.
1 (Constant)	−0.546	0.346		−1.579	0.116
Interface and operation	0.336	0.149	0.218	2.245	0.026
Diversified courses	0.595	0.156	0.392	3.817	0.000
Teachers	0.159	0.139	0.115	1.146	0.253
Timely interaction	0.009	0.148	0.007	0.063	0.950
Teaching methods	0.332	0.157	0.264	2.109	0.036
Optional teaching model	0.102	0.157	0.081	0.654	0.514
Price of courses	−0.292	0.161	−0.196	−1.811	0.072
Result of training	−0.165	0.142	−0.115	−1.165	0.246

a. Dependent variable: Consumer satisfaction

Using the same method we can conclude that diversified courses, good teachers, timely interaction, flexible teaching methods, optional teaching model, and good results of training are positively correlated with the satisfaction of consumers. Only the price of courses is negatively related with satisfaction of consumers.

4.3 Multivariate regression analysis

We can also use eight influencing factors as independent variables and satisfaction as a dependent variable for multiple linear regression analysis.

According to the data in Table 3, we can see the results of independent variables significance test (using the single-sample t-test), and the last column are the significances of the t-test. The significances of interface and operation, diversified courses, and teaching methods of the model are less than 0.05. This indicates that interface and operation, diversified courses, and teaching methods have the greatest significant influence on the consumers' satisfaction.

5 CONCLUSIONS AND SUGGESTIONS

From the analysis above, we can conclude that:

1. Interface and operation of the online vocational education platform, diversified courses, teachers, timely interaction, teaching methods, optional teaching model, and results of training are positively correlated with the satisfaction of the online vocational education platform by college students.
2. The price of online vocational education platform courses has a negative correlation with the satisfaction of college students using the online vocational education platform.
3. Among the eight independent variables, interface and operation, diversified courses, and teaching methods have the most significant influence on the dependent variable, satisfaction of college students.

We have some suggestions from the concluding of the study.

1. In addition to providing consumers with the basic functions, the online vocational education platform should also update the system in a timely fashion to make the operation easier and

faster, so that users can find the courses and services they need quickly and efficiently. At the same time, consumers' problems and suggestions should be fed back promptly. The platform should prevent losing consumers due to its cumbersome operation.

2. The diversity of platform courses is crucial. Online vocational education platforms must pay attention to the changes in national policies, and offer courses and services that meet the needs of the current labor market

3. A good teaching method can arouse students' concentration and improve their learning efficiency and results. Therefore, the online vocational education platform should provide a variety of teaching methods according to the needs of students. For example, guided teaching, heuristic teaching, case teaching, and combining theory with practice can help students gain a deeper understanding of their future career.

REFERENCES

[1] Information on http://www.gov.cn/premier/2019-03/16/content_5374314.htm

[2] Dwayne D. Gremler. Student Satisfaction Guarantees: An Empirical Examination of Attitudes, Antecedents, and Consequences. Journal of Marketing Education,2002,24(2).

[3] Felix T. Mavondo, Yelena Tsarenko, Mark Gabbott. International and Local Student Satisfaction: Resources and Capabilities Perspective. Journal of Marketing for Higher Education,2004,14(1).

[4] Debnath, Students' Satisfaction in Management Education: Study and Insights. Journal of Marketing for Higher Education, 2005, (2).

[5] Li-Wei Mai. A Comparative Study Between UK and US: The Student Satisfaction in Higher Education and its Influential Factors. Journal of Marketing Management,2005,21(7–8).

[6] Wu Liu, Xue Yang. Construction of customer satisfaction index model for higher education in China. Journal of public administration, 2007 (01): 84–88.

[7] Guo-Dong Zhao, Shuai Yuan. Study on student satisfaction and influencing factors of blended learning: a case study of Peking University teaching network. China distance education, 2010 (06): 32–38.

[8] Ya-Zheng Li. Research on factors influencing users' intention to continue using online education platform and willingness to pay for courses. University of Science and Technology of China, 2016.

Computational Social Science – Luo, Ciurea & Kumar (eds)
© 2021 Taylor & Francis Group, London, ISBN 978-0-367-70193-2

Reform and practice of innovation and entrepreneurship practice education system for electronic information college students under the perspective of new engineering

H.M. Song, Y. Liu* & X.H. Sun
Hebei North University, Zhangjiakou, China

ABSTRACT: With the continuous deepening of Chinese innovationdriven development strategy, the demand for innovative talents is increasing rapidly. As the main force of innovation and entrepreneurship, college students are essential to cultivating their scientific and technological innovation capabilities. First, the innovation and entrepreneurship education system is elaborated. Then, starting from "training mode, curriculum reform, teacher education, resource platform, quality evaluation," we set up a multidimensional innovative talent training mode, construct a two-way integration curriculum system of professional education and innovative education, and create "three-dimensional" high-quality school-based curriculum resources. We should set up a high-quality education practice platform, then achieve organic integration between professional education and innovation and entrepreneurship education. After continuous reform, we have achieved remarkable results, and put forward future reform strategies.

1 INTRODUCTION

At present, the Western developed countries have cultivated and developed their practical ability of innovation and entrepreneurship to a relatively mature level. The American university students' innovation and entrepreneurship education focuses on improving the comprehensive ability of university students' entrepreneurship, and the goal of education has risen from "employment for the purpose" to "improving the quality." Germany has attached great importance to the smooth transformation of scientific and technological achievements into productive forces. The innovation and entrepreneurship of German university students focus more on areas with high technology content. Japanese universities gradually transformed traditional and utilitarian entrepreneurship education mode into cultivating students' lifelong entrepreneurial spirit and innovative quality. It has formed a more mature innovation and entrepreneurship education system that closely cooperates with enterprises, universities, and the government [1].

With the needs of Chinese economic development, the state has accelerated the implementation of the strategy of innovation-driven development. From the central government to the education sector, the reform of the innovation and entrepreneurship education system has become a national consensus. The Ministry of Education issued a New Engineering Construction Guide in 2017. Engineering majors need to establish a much improved teaching and training system for college students' innovation and entrepreneurship. Following the Fu Dan Consensus and Tian Da Action, "New Engineering" has upgraded traditional engineering majors. As a new concept of engineering education, it has attracted wide attention in the field of education [2]. How to effectively link innovation and entrepreneurship education with the training of new engineering talents is a new development demand [3].

---

*Corresponding author

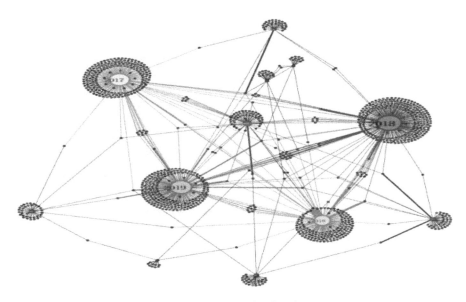

Figure 1.   Visualization of innovation and entrepreneurship education.

2   BASIC PRINCIPLES OF THE INNOVATION AND ENTREPRENEURSHIP EDUCATION SYSTEM

The system of innovative thinking and entrepreneurship education for college students is usually referred to as the *double creation* education system. In recent years, research into the cultivation of innovation and entrepreneurship in China has gradually increased, as show in Figure 1, where colleges and universities attach great importance to the cultivation of students' innovative and entrepreneurial ability, and gradually established a suitable education system [4]. When carrying out related innovation and entrepreneurship education for university students, the sustainable development ability of college students is enhanced as the ultimate goal of educational development. Generally, the basic teaching principles are followed, as outlined next.

2.1   *Resource-sharing principles*

We should make full use of the existing teaching resources, especially classifying similar disciplines into a large group. We should carry out phased construction in a planned way and step by step, through scientific optimization and reorganization, and form a platform for innovation and entrepreneurship in practical education, so as to share educational resources as much as possible and avoid duplication and cross setting. It is important to improve the utilization rate of teaching resources [5].

2.2   *Respect the principles of the national standard*

Closely around the *national standard for teaching quality of undergraduate majors in Colleges and universities* and the concept of *new engineering*, the curriculum system is optimized and restructured within the framework of the established professional talent training system. The reorganization of the curriculum system must adhere to the principles of *advanced, useful, and effective* according to the professional requirements and future development needs, and increase the new curriculum of the *new engineering* concept. Through the construction of school-based textbooks, we should cut down or merge some of the course contents.

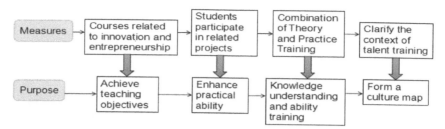

Figure 2. Training mode of innovation and entrepreneurship education system.

2.3 *Standardization operation principle*

In order to ensure the smooth operation of the innovation and entrepreneurship platform, we must establish a scientific and standardized operation mechanism and perfect system. We should set up a leading group, establish a standardized management system, and actively organize beneficial cooperation and exchange activities, so as to effectively ensure the smooth development of innovation and entrepreneurship, and promote this in an orderly way.

3 INNOVATION AND ENTREPRENEURSHIP SYSTEM REFORM CONTENT

The current situation of innovation and entrepreneurship education in colleges and universities is investigated, and the following problems were found in the practice of innovation and entrepreneurship education: first, the separation of innovation and entrepreneurship education; second, the lack of an innovation and entrepreneurship curriculum system; then, the narrow benefits of innovation and entrepreneurship education, and the weak sense of innovation and entrepreneurship of students [6]. Taking the *electronic information* major of application-oriented undergraduate colleges as an example, focusing on the cultivation of innovation and entrepreneurship, the research is carried out over five dimensions: *training mode—curriculum reform—teacher education—resource platform—quality evaluation*.

3.1 *Reform the training plan and curriculum system*

Reconstructing the teaching content and curriculum system suitable for the improvement of students' innovation and entrepreneurship quality in electronic information class is show in Figure 2. In the classroom teaching, we will design a strong experimental course which is conducive to cultivating students' innovative thinking and practical ability, and set up a "frontier technology topic" course leading students to focus on the development of new technologies such as modern electronic instruments, intelligent products, and 5G, so that they can connect professional knowledge with specific products to achieve the purpose of cultivating innovative thinking ability. The innovation and entrepreneurship project of college students will be included in the talent training program to improve students' participation enthusiasm and optimism. Students learn by completing projects, and will not only be able to get professional ability training, but also get a certain amount of credits through the method of completing project learning, and explore the mode of "Teaching—Practice—Teaching—Practice" to enhance the spiral of students' theory and practice of circular training, knowledge understanding, and ability training. The training context of innovation and entrepreneurship is clarified, and a training map formed [7].

3.2 *Upgrading teaching materials, highlighting the innovation, integration, foresight, and adaptability of the courses*

Under the New Engineering concept, the foundation of innovation and entrepreneurship reform is the reconstruction of the curriculum system, and its key lies in the development of a school-based curriculum. The goal of school-based curriculum development is to focus on the standards of the industry and the needs of the enterprises, so as to reflect the frontier development trend and advanced technology of the discipline, and cultivate innovative and entrepreneurial thinking and comprehensive ability. We must focus on the existing knowledge structure of students, through multiple efforts of teachers, industry experts, etc. and strive to achieve the development of the school-based curriculum. Simultaneously, in order to meet the needs of the times, we need to make school-based courses networked and three-dimensional, facilitating students' self-learning.

3.3 *Strengthen the building of teaching staff*

At present, colleges and universities are in the stage of educational transformation, and the development of the country needs a large number of applied talents. Colleges and universities should cultivate students who understand both production technology and management. Therefore, we must create high-quality teachers with *double faculty and dual ability*. The basic characteristics of teachers are: (1) having rich teaching experience in scientific research and practice, capable of innovating teaching methods; (2) having rich theoretical knowledge and professional skills, enabling students to integrate theory and practice perfectly and improving their practical and innovative abilities; (3) while teaching knowledge and skills to students, professors should also cultivate students' inner spirit and quality; and (4) pay attention to the development trend and dynamics of the industry in time, we should grasp the latest trends and apply it to teaching practice [8].

3.4 *Construction of a strong innovation and entrepreneurship practice sharing platform*

The aims are to transform and integrate the existing laboratory of "electronic information" specialty, to set up a laboratory teaching platform, to make the laboratory an important base for teaching, training, technological innovation and discipline competition; to speed up the construction and sharing of experimental teaching platform, to organize a competition for undergraduate innovation and entrepreneurship and some professional competitions; and to make full use of practical innovation platform, experimental projects, analysis data, etc., to stimulate students' innovative potential; actively play the role of students' associations and other organizations such as innovation and entrepreneurship alliance and entrepreneurial associations, hold some lectures, carry out some practice activities; use off-campus social resources to create an innovation and entrepreneurship practice platform outside the school to provide students with development plans. For example, we should create some business incubators and incubator bases for university students; meet the needs of enterprises, rely on projects of school enterprise cooperation, and rationally apply comprehensive resources to carry out social services such as product design, technical training, and so on; relying on laboratories to carry out applied research, we must provide related consulting services and staff training for enterprises; employ some mentors, such as enterprise internal management or technical personnel, to guide the practice of projects, and then promote the incubation of innovation and entrepreneurship projects and the transformation of results [9].

3.5 *Establish a quality inspection and tracking system*

The school will assess the innovation and entrepreneurship of the students and graduates in the school periodically, establish a student information tracking system, update the data, and conduct dynamic analysis in a timely manner. The number of students in entrepreneurship should be counted, the success rate of future innovation and entrepreneurship and their quality are important indicators to evaluate innovation and entrepreneurship education. It is of great help to improve the teaching quality and effectiveness of innovation and entrepreneurship education by formulating an implementation plan for teaching quality monitoring and strengthening education management.

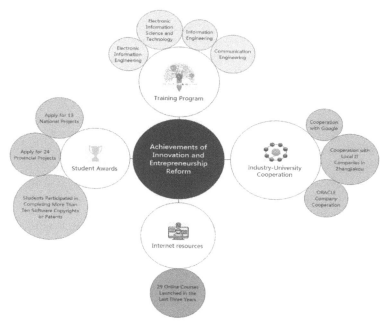

Figure 3. Achievements of innovation and entrepreneurship reform.

4 INNOVATION AND ENTREPRENEURSHIP EDUCATION REFORM RESULTS

Our school has built six innovative practice laboratories, namely, medical Internet of Things, medical informatics, medical data analysis, mobile Internet, apple development laboratory, and Yi Chuang space, which has helped teachers and students in teaching and scientific research. Based on this, some notable achievements have been made in the reform of innovation and entrepreneurship education, as show in Figure 3. (1) Revision of electronic information training programs, including four undergraduate disciplines, such as electronic information engineering, electronic information science and technology, information engineering, and communication engineering, and increasing the proportion of innovation and entrepreneurship practice courses, such as "career planning for college students," "innovation and entrepreneurship foundation," "college students employment guidance and entrepreneurship guidance" courses [5]. (2) Production and cooperation, collaborative education has achieved certain results. The joint Google Corporation and Zhangjiakou local IT company set up the Zhangjiakou Google developer community. The college and ORACLE company jointly established the ORACLE WAP club through school enterprise cooperation, providing a strong support for the cultivation of college students' innovative and entrepreneurial practical ability. (3) The network curriculum resources construction has achieved remarkable results. Our school has launched 29 online courses in the past three years. (4) Teachers actively organize students to take part in innovation and entrepreneurship projects, the results of which are 13 national projects, 24 provincial projects, and students' participation in more than 10 software copyrights or patents, and some products have been applied to related companies.

5 SUMMARY AND FUTURE PROSPECTS

In summary, first, the training mode of innovative talents for electronic information specialty based on the "EIP-CDIO" mode has been established. The cultivation of innovative talents focuses on professional ethics, integrity, and professionalism, and is organically combined with CDIO, to enable students to actively combine theory and practice. Second, according to different training stages of students, targeted courses have been set up, such as: setting up an foundation for freshmen, such as innovation and entrepreneurship general education courses, so that they can easily get started and receive enlightenment education to promote the formation of their innovative and entrepreneurial personality. For sophomores, a professional course on innovation and entrepreneurship in the fields related to this major has been set up, highlighting the specific application and practice of innovative ideas and entrepreneurial skills in professional courses. Practical training courses for students in the third year have been created, setting the requirements for credit practice for innovation and entrepreneurship, setting up courses, like professional application and science and technology innovation projects and so as to cultivate students' comprehensive practical innovation ability. At the same time, a map of electronic information innovation and entrepreneurship education talent training curriculum system has been built. The key points of teacher teaching and student learning are clear, and promote professional integration and innovative thinking stimulation.

ACKNOWLEDGMENT

This research was financially supported by the Research Project of Education and Teaching reform of Hebei North University (GJ2019013).

REFERENCES

[1] Li Zhang, Yueqiu Jiang. Construction path of Applied Universities in the context of "new engineering" background. Fujian tea, 2020, 42 (03): 400–401. (In Chinese).

[2] Jianxin Shen, Yong Lu. The development of school-based curriculum in Local Undergraduate Universities under the concept of "new engineering". Journal of Yancheng Teachers University (Humanities and Social Sciences), 2017, 37 (04): 116–119. (In Chinese).

[3] Qiuxiang Tao, Jiliang Tu, Jianwen Shu, Jie Jia. The path of improving the innovative quality of engineering and technical students in Local Undergraduate Universities in the context of new engineering background Journal of Nanchang Aeronautical University (Social Science Edition), 2018, 20 (01): 97–106+112. (In Chinese).

[4] Yiying Wu, Dong Yang. Knowledge map of innovation and entrepreneurship education in Chinese universities in recent 20 years. Modern university education, 2019 (04): 53–63. (In Chinese).

[5] Peng Li, Lei Zhang. Exploration and Research on the construction of practical education base for college students in innovation and entrepreneurship – Taking Hebei University Of Science and Technology as an example. Journal of Agricultural University of Hebei (Agriculture and Forestry Education Edition), 2017, 19 (06): 127–130. (In Chinese).

[6] Xiaojun Sheng. Countermeasures and suggestions for the construction of college students' innovative education system. Experimental technology and management, 2018, 35 (12): 202–205. (In Chinese).

[7] Wenxiu Li, Ying Bi, in 33, Li Manhong. The practice of innovation and Entrepreneurship Education under the background of new engineering. Chemical higher education, 2018, 35 (02): 1–5. (In Chinese).

[8] Li Zhong. Path analysis of the construction of "double teacher and dual energy" teaching staff in applied universities. Mass literature and art, 2019 (08): 198–199. (In Chinese).

[9] Liang Zou. Ideas and ideas for the construction of innovative and entrepreneurial bases in Applied Universities: take Liaoning University of International Business as an example. Modern business and industry, 2017 (35): 92–93. (In Chinese).

Computational Social Science – Luo, Ciurea & Kumar (eds)
© 2021 Taylor & Francis Group, London, ISBN 978-0-367-70193-2

A discussion of the definition of quality in higher education

Y.X. Ye & Y.J. Chen*
Jinan University, Guangzhou, China

ABSTRACT: This article attempts to illustrate the definition of quality in higher education by using a literature research in four different aspects, including quality learner, quality environment, quality content, and academic quality. Several questions are then raised to consider further study in this area.

1 INTRODUCTION

The quality in education is a focused-on problem all around the world, both in developed and developing countries. To explore the definition of "quality" in education, especially in higher education, the definition and aims of higher education have to be described first. Only by understanding the initial questions, can the questions of quality education be answered profoundly. This article starts by discussing the purposes of higher education. Based on that, we explore what quality of higher education is.

2 THE DEFINITION AND AIMS OF HIGHER EDUCATION

"Quality" is always an area of debate in the field of education, and many studies have already discussed it from various angles. This article focuses on higher education, and seeks to reveal the definition of "quality" in higher education.

The *Oxford Dictionary of Education*, describes higher education as a "Program of study which leads to advanced qualifications. These are usually offered in higher education institutions such as universities, but may also form part of the provision of further education colleges, as in the case of foundation degrees. And it is distinct from the further education sector in terms of funding and purpose." The significance of higher education, as one of the essential links in the educational system to cultivate talents before sending them to various parts of society, cannot be underestimated. As quoted from Thorstein Veblen who published a book called *The Higher Learning in America: A Memorandum on the Conduct of Universities by Businessmen*: "Ideally, and in the popular apprehension, the university is, as it has always been, a corporation for the cultivation and care of the community's highest aspirations and ideals" [1].

Also, there are various and profound definitions of the aims of higher education, such as to cultivate students' positive attitude to life and academic difficulties, with self-confidence, and see things objectively. Moreover, higher education inspires students to be creative and curious in research and life. It also cultivates students' perseverance to overcome difficulties and achieve their goals. In terms of their future careers, it can help students to prepare for and be professional in skilled areas [1–3]. When it comes to the social dimension, experience has proven that universities can provide the best services to the community if they have ambitions for continuous improvement in the quality of their services [4] and, as David Willetts, British Conservative Party politician described, universities are "one of our great national assets," and "They push forward the frontiers of knowledge. They transform people's lives. And they contribute to the health and wealth of our

---

*Corresponding author

nation through their deep involvement in wider society and the economy. That is the key to their continuing success and their world-class status" [5]. Based on the definition and aims of higher education mentioned above, the definition of the quality in higher education is divided into several parts, as follows.

3 "QUALITY" IN HIGHER EDUCATION IN THE CONTEXT OF CHINA

3.1 Overview of quality in higher education

There is no consensus on the definition of quality in higher education [4], although a definition was provided in a UNESCO-CEPES report [6]: "Quality in higher education is a multi-dimensional, multi-level, and dynamic concept that relates to the contextual settings of an educational model, to the institutional mission and objectives, as well as to the specific standards within a given system, institution, programme, or discipline" [7]. This definition only reflects how complex the environment of higher education is and how difficult it is to give a definition to the quality of higher education under such circumstances. Furthermore, quality is also not a static concept that can be measured which means that it's meaning changes under different contexts [8]. As mentioned above, the purposes of higher education include helping students academically and psychologically and nurturing and cultivating students to get ready for their career and make contributions to the society. Therefore, it is important to cover the essential aspects mentioned when defining quality. Also, some basic facilities need to be guaranteed which support the basic functions of higher education.

3.2 Quality learner

Like other schools, higher education also works with the students. Thus, the quality of students plays a role in the whole quality definition. Since only qualified candidates can enter certain universities to continue further study, I am not going to discuss the relevant knowledge level before entering the university or anything related to intelligence quotient. Regardless of the learning levels, the most basic issue is the physical and mental health. Without a relatively healthy condition, students will not be able to withstand the intense study in the university which is why students are asked to complete a physical examination annually. If students are examined and it is verified that their health conditions cannot support their study in higher education, they will need special and extra help to continue their studies. Every qualified learner should have equal access. Discrimination because of gender, age, language, geography, or any other conditions should be eliminated. All learners have the opportunity to enroll in university courses as long as they have reached the relevant academic level [6].

Besides a willingness to study, students in higher education also need support from their family despite most of them being adults. There has been a study into the influence of maternal education for children [9], showing that the mother's education level influences the child's literacy score. Although the object of the study is children and it focuses on their literacy score, it is still believed that it is strong evidence that shows how large the impact from parents is, especially for young adults who have not entered society, living in an "ivory tower." For students in universities, parental support from their family is a psychological crutch, as well as providing funding support.

3.3 Quality environment

The quality environment in higher education includes the quality of the student's learning and living environment, and the teacher's teaching and working environment. Environments should be healthy, protective, and gender-sensitive, and provide adequate resources and facilities [10]. There is research showing that a good-condition and decorated building has an association with the quality and effectiveness of learning for both teachers and students [10].

3.4 *Quality of learning and living environment*

In a paper presented by UNICEF at the meeting of The International Working Group on Education Florence, Italy, in June 2000 entitled "Defining Quality in Education," it was demonstrated that the "Learning environments are made up of physical, psychosocial and service delivery elements." In fact, the learning and the living environment overlaps to some extent as good-quality learning and living are in a mutual-promotional relation. Based on that, the physical environment is essential as the foundation of all the other activities. The quality of the building (e.g., classrooms, air-conditioners, and water resources), library, internal air quality, IT service, sports center, cafeteria, lavatories, and access to specialized facilities and so on are all involved in the physical environment. The quality of the environment can affect the learning outcome indirectly. As the study of Fuller et al. (1999) stated, people who live in more crowded households do less well in exams. What is more, for university itself, the sophisticated and historical buildings and well-equipped campuses will attract top scholars and/or excellent students, "boosting the university's standing in the competitive international league" [11]. As for library, IT services, and other facilities like the experimental lab, these are essential supports for students to do academic research, as mentioned in the purposes of universities that push forward the frontiers of knowledge.

Apart from the physical environments mentioned above, the dormitory also needs to be paid attention to. Because of the mobility of students, universities draw various students together with diverse backgrounds, learning experiences, and habits. Thus, it is important to build a cozy and friendly dormitory environment and promote a harmonious and inclusive atmosphere for all dormitory members.

3.5 *Quality in the teaching and working environment*

Teacher's teaching and working environments are often neglected when talking about the quality of higher education. Research shows that "staff are generally less satisfied with their facilities than students are, which could mean they negatively impact on their research and teaching" [11]. Infrastructure (physical environment), teaching materials, even the size of the class and the relationships among the teachers all influence the teaching experience. The quality of the teacher's teaching and working environment can affect the attitude toward teaching and research. A caring environment can generate teachers' sense of belonging and enthusiasm toward the university.

Furthermore, not only students need academic support, teachers also need academic support and the room to make progress both in the perspective of school and as an individual. The school should offer opportunities to teachers for advanced study. And one more important area is the remuneration of teachers should at least maintain the standards found at the international level [8]. These are the fundamental guarantees that might help to enhance the professional quality of teachers.

3.6 *Quality content*

3.6.1 *The depth of knowledge*

If having a qualitative environment is the foundation of qualitative learning, the content then starts the process of learning. Quality content refers to the intended and taught curriculum of schools [10]. The "quality" of the curriculum should be such that it can offer both conceptual and practical understanding to what students have learned. The curriculum needs to be well designed scientifically in order to facilitate students' learning interest so that they are motivated to access in-depth knowledge. As Biggs (1999) said, one of the essential aims of promoting student engagement and interest is to promote the acquisition of deep rather than shallow knowledge and skills [8,9]. The curriculum lays a solid foundation for students to continue further academic practices. A good curriculum cultivates the spirit of exploration and creation, scientific attitude, and the capacity for scientific research. It is essential in higher education to go deeper and nurture a researcher of the future.

3.6.2 *Internationalization*

Globalization is another important issue in higher education. Since universities have always been affected by international trends, it is necessary to have a more international curriculum or program [6]. The international content can facilitate the students to possess the ability of comprehensive outlook, instead of being limited nationally. An outdated curriculum will not only hinder the development of students, but also hinder the development of schools and countries.

3.6.3 *Career skills*

One of the main purposes of higher education is to help students to get ready for a career, and to fit in with the many and various roles that they will undertake in future life [12]. Thus, it is very important to add career skills to the curriculum. Most universities put this is an extra workshop, and due to the overlap in timing, students may miss some of them. Marilyn Andrews said it should be embedded in academic programs, rather than an add-on so that students have the best opportunity to plan their future and not miss any one of them. She said that as higher education providers, providing the opportunities to students to achieve good academic was not enough, a wider curriculum that can actively expand the benefits should be conducted [12].

3.7 *Academic quality and assessment*

It is said that "The growing tension between enrolment demand, constrained budgets, and greater accountability has resulted in a discouraging environment for the academic profession" [6], but the academic quality needs to be guaranteed since this is the core of the university. There is an Academic Quality Improvement Program to help to assess the academic quality in universities. "AQIP is a modern form of accreditation designed to assist higher education with the quality assessment of programs and services to ensure student success" [4]. Thanks to Yarmohammadian, Mozaffary, and Esfahani (2011), there are nine scales of this program including helping students to learn, understanding the needs of students, continuous improvement, and so on. These assessments can intuitively view the quality of higher education and a good assessment can actually help the university examine its weakness and start to improve them. However, all this assessment is nothing without qualified and committed teachers. The well-equipped facilities in the university cannot achieve greatness and a wonderfully designed curriculum is nothing without qualified academic staff [6].

In the perspective of students and with the experience of having been a student, a qualified teacher in the higher education should not only be knowledgeable, but also have the ability to teach and get along well with students, such as managing the class, communicating with the students, activating the classroom atmosphere, guiding the students to think dialectically and objectively, inspiring students to explore the unknown, and so on. Some professors who have brilliant academic achievements, but knowing nothing about the art of teaching, can result in a drop in the quality of teaching. Teaching should not be the display of knowledge, but needs to be regarded as an art of interaction. As Darling-Hammond (1997) described quality teachers were those who are not only able to help the students to break through academically, but also have insightful views on the subject and the art of pedagogy [10]. Students need more varied assessments to examine the learning progress and outcome rather than just exams using a pen and paper, and therefore teacher's methods of evaluation need to be more ingenious. According to Altbach, Reisberg, and Rumbley (2009, p113): "Until fairly recently, teaching meant 'covering' a body of declarative knowledge— that is, knowledge that could be 'declared' in books or in lectures—while assessment measured how well students received that knowledge based on their ability to regurgitate it in examinations" [6]. The center is slowly changing to being student-centered rather just focusing on knowledge.

The size of the classroom is also essentially important in higher education. "Momentum around increased participation and student mobility in higher education in an increasingly globalizing world is gathering pace and, in the process, changing both the demography and size of university classrooms" [8]. In large classes, the mental distance between the teacher and students is quite far. Some students do not get enough attention from their teacher and it is quite possible to end it up with no interaction with teachers after an entire module, so the students will not be able to get academic feedback.

4 CONCLUSION

To sum up, quality in higher education can be explained from the quality learner, quality environment, quality content, and academic quality. The quality leaner not only means the healthy conditions, both mentally and physically, but also indicates the inclusion of all qualified learners. In terms of the quality environment that needs the support of finance or funds from government and society, this is the foundation of all. All teaching and learning activities are all based on the facilities and the environment. The curriculum is the beginning of the quality academic activity since it provides the outline for the academic learning process. Then it comes to the academic quality and the importance of academic and committed staff. The academic level of staff (teachers and professors) to some extent represents the academic level of the school.

For further research, researchers could focus on the specific features of higher education rather than general education. For example, a focus on features like mobility and diversity in higher education, to explore the deeper meaning of quality. In this case, the definition of quality would be more accurate and specific.

REFERENCES

[1] K. Catcheside, The benefits of a university education, the Guardian. (2012).
[2] A. Fortino, The Purpose of Higher Education: To Create Prepared Minds, The Evollution. (2012).
[3] R. Lambert, S. Smith, Higher education's importance goes well beyond teaching, the Guardian. (2009).
[4] M. Yarmohammadian, M. Mozaffary, S. Esfahani, Evaluation of quality of education in higher education based on Academic Quality Improvement Program (AQIP) Model, Procedia - Social and Behavioral Sciences. (2011), 15: 2917–2922.
[5] H. Swain, What are universities for?, the Guardian. (2011).
[6] P. Altbach, L. Reisberg, L. Rumbley, Trends in Global Higher Education: Tracking an Academic Revolution A Report Prepared for the UNESCO 2009 World Conference on Higher Education. UNESCO 2009 World Conference on Higher Education., 2009.
[7] A.-P. Pavel, The importance of quality in higher education in an increasingly knowledge-driven society, International Journal of Academic Research in Accounting, Finance and Management Sciences. (2012), 2(1): 120–127.
[8] F. Maringe, N. Sing, Teaching large classes in an increasingly internationalising higher education environment: pedagogical, quality and equity issues, High Educ. (2014), 67(6): 761–782.
[9] B. Fuller, L. Dellagnelo, A. Strath, E. S. B. Bastos, M. H. Maia, K. S. L. De Matos, A. L. Portela, S. L. Vieira, How to raise children's early literacy? The influence of family, teacher, and classroom in northeast Brazil, Comparative education review. (1999), 43(1): 1–35.
[10] J. Colby, M. Witt, Defining Quality in Education. A paper presented by UNICEF at the meeting of The International Working Group on Education, Florence, Italy (June 2000). Document no, Retrieved August. (2000), 30: 2005.
[11] A. Marmot, Flashy university buildings: do they live up to the hype? The Guardian. (2015), 21.
[12] M. Andrews, Why our students need co-curricular, not extra-curricular, activities, The Guardian. (2013), 1.

Computational Social Science – Luo, Ciurea & Kumar (eds)
© *2021 Taylor & Francis Group, London, ISBN 978-0-367-70193-2*

Research on student policing management based on political construction of police

Z.J. Sheng & Q. Du
Jinan, China

ABSTRACT: In the student policing management in public security colleges, the most important thing is to adhere to the political construction of police, adhere to the leadership of the party, take Xi's new era of socialism with Chinese characteristics as the command, take moral education as the central link, and always put ideological and political education in the first place to carry out ideological and political education throughout the entire process of teaching management. Besides, we must adhere to strict management of the police, adhere to strict thinking, strict discipline, and train students to be equipped with self-cultivated and self-disciplined learning and working style, and establish a strict policing management system. In accordance with the characteristics and growth of students in the new era, we must innovate education concepts, reform management methods, make full use of modern science and technology, and actively explore a new policing management model to achieve the goal of reforming the strong police and boosting the police with science and technology.

1   INTRODUCTION

At the National Education Conference, Xi comprehensively summarized the new ideas formed in the practice of educational reform and development in China since the 18th National Congress of the Party has been held, and put forward educational reform and development based on the fundamental task of education, and then, made important strategic deployments. As the high educational institution that aims to train professionals for maintaining social security and stability for the country and society, the police college's mission is to deliver qualified professionals in political and legal work. At the 2019 National Public Security Work Conference, Xi put forward new requirements for public security organs and police teams, pointing out that in the new historical period, the situation at home and abroad is complex, and the public security reform must be deepened in an all-round way. The credibility of law enforcement in China's public security organs is centered on the people, insisting on building the police, reforming the police, strengthening the police with science and technology, and rigorously managing the police to achieve a safe and stable political and social environment. General Secretary Xi's speech pointed to the direction for students in public security colleges, put forward requirements, and clarified goals. In the new era, public security colleges must profoundly study and implement the spirit of Xi's speech, insist on comprehensively improving students' police quality and skills, and deepen the reform of the teaching management system. Besides, it is necessary to cultivate a team of reserve police officers with strong combat effectiveness, excellent ideological quality, and strong political stance. The establishment of political police is not only the goal of cultivating students in public security colleges, but also the requirements of students' policing management. In order to adhere to the political construction of police, for one thing, we should fully understand the current status of the student team in the new era, for another, we need fully recognize the problems in policing management of public security colleges, and the third is to always put ideological and political education in policing teaching. Last but not least, we should insist on strict police management, and actively explore the new model of policing teaching and management for public security college students.

2 THE STATUS OF THE STUDENT TEAM IN PUBLIC SECURITY COLLEGES

The current mainstream of students in police colleges is positive and upward. They love the party, the country, and socialism. Besides, they have ideals, pursuits, a sense of responsibility, and a clear responsibility. Also, they are loyal, devoted, contributed and dreamy. What is reflected from this generation of dream catchers is a change, a breakthrough, a new trend, which is in line with the changes of the times. Whether in terms of quantity or quality, the students of contemporary public security colleges have changed a lot compared with before, and the reserve police officers in the new era have made great strides with the experience of their predecessors and their own ideas. In recent years, with the rapid development of network and digital information technology, and the publicity of the public accounts of public security colleges and universities, more and more people have applied for public security colleges.

Current Political and Ideological Situation of Public Security College Students. The nature and tasks of public security organs determine that the people's police of public security organs must have excellent disciplinary style. Public security college students are preparatory policemen, and they will become officers after graduation. Policing management is an inevitable way to cultivate the discipline style of public security college students. Based on the school motto of "loyalty, rigor, unity and dedication", Shandong Police College strictly requires students in daily life management, and at the same time adds liveliness to serious discipline elements, such as occasionally holding activities related to the characteristic culture of public security colleges, adding some relaxing training activities, etc. In a serious and lively environment, the vast majority of our students are loyal to the cause of the party and the people, keeping in mind the mission, through organizing events or learning the deeds of public security heroes on the Internet, to deepen their police awareness, ideological and political awareness, which can constantly maintain the core position of General Party Secretary Xi and the core position of the whole party. Since students spend a lot of time outside of class, club life is very rich and colorful, and various interest groups are in wide range. In reality, students in public security colleges in the new era have broken the excessively serious situation in the past, added colorful modern elements, and at the same time used the developed Internet to deepen their learning, which has laid a solid foundation for the road to engage in policing.

3 PROBLEMS EXISTING IN POLICE MANAGEMENT IN PUBLIC SECURITY COLLEGES

The authority of the coordinator's right to speak is weakened. Compared with ordinary colleges and universities, due to the special nature of public security colleges, management must be strict and personnel requirements must be fully developed. However, under the rapid development of modern Internet, students are more inclined to seek help from the network, classmates and strangers if they have problems rather than consulting with teachers, coordinators, and classmates. Moreover, many students understand the Internet one-sidedly, believing that the information on the Internet and public comment are the mainstream of the times, which leads to their blindness. Sometimes, they can be exposed to the content and management methods earlier than the teacher, and then do it according to their own understanding. Therefore, it weakened the authority of the police school teacher's right to speak, and broke away from the traditional one-way method of imparting knowledge to teachers who have authority and resources in ideological and political education. Over time, some students turned a blind eye or indifference to the school's requirements and notices, and even more directly opposed the school's policing management model.

Some students deviated from the course in police management. Policing management is the main feature of public security colleges, and belongs to a paramilitary management model. Its main contents include the one-day life system, police discipline and police etiquette norms. It is precisely the policing management mode that distinguishes from the nature of ordinary colleges and universities. It is also due to this disciplinary management method that some students deviate from the track in daily life. What they consider is how to make the bed, how to clean the housework, how to organize their own image, etc., but lose interest in learning. In their mind, the most important thing for public security colleges students is to learn policing, not theoretical knowledge, so before the final exam of the semester, it is common to see that many students started to prepare at the last moment. Very few students were able to listen carefully in class, taking notes consistently, and reviewing regularly.

Lack of police awareness in large-scale security activities. Taking Shandong Police College as an example, it often organizes students to participate in security tasks for major events, such as the G20 Hangzhou Summit, the Beijing "Belt and Road' Summit Forum, and the Qingdao SCO Summit. When participating in security activities and walk out of the school, the words and deeds of teachers and students represent the image of the college. Police students in the new era are more curious about all the new things and like to explore on their own. In large-scale security activities, individual students sometimes act alone and do not obey commands. The unauthorized things sometime happened. Some students have not really integrated into the formal police force due to their lack of police awareness, and they have relatively low requirements for themselves in security task. Thus, they cannot really understand the tasks assigned by their superiors, nor can they successfully complete them alone. In this case, participants are sometimes puzzled and questioned by the masses due to lack of relevant handling experience in the work, which is difficult for the relationship between the police and the masses to reach the masses' satisfaction.

It is difficult for students to coordinate between work and study time. Some students did not make adequate psychological preparations before the internship and security. Faced with the conflict between the existing work and their future work, there was a conflicting mentality, so that the students were distracted at work, thinking about how to study; anxious while studying, thinking about how to do a good job. Such a contradictory mental status leads to unclear coordination between students' work and study time, thus it leads to the negative work and study state, and increased psychological pressure as well as unstable mood, which in turn, affect the working state of internship, or security , and the learning effect is not desirable.

4 PERSIST IN BUILDING THE POLICE AND ALWAYS PUT IDEOLOGICAL AND POLITICAL EDUCATION FIRST

Public security colleges are ones under the leadership of the Party. Ideological and political work is basically work of people. It should be guided by Xi's new era socialist thinking with Chinese characteristics, focusing on students, caring for students, serving students, and constantly improving student thinking level, political consciousness, moral quality and cultural literacy. We must speak clearly about politics, firmly establish the "four consciousnesses", firm "four self-confidences", and adhere to the direction of running a socialist school; we must implement the fundamental task of establishing a virtuous person, put Xi's socialist thought with Chinese characteristics in the new era, and practice integrating the core value of socialism into the whole process of teaching and educating, cultivate the socialist cause builders and successors of the comprehensive development of morality, intelligence, physical beauty and labor; follow the laws of ideological and political work, and improve the ability to do ideological and political work, change with the times and advance with the times , put the correct political direction, value orientation throughout the entire process of establishing a school and educating people.

Lead the ideological work of colleges and universities with Xi's new era socialism with Chinese characteristics. Xi's new era of socialism with Chinese characteristics must be taken as the core axis, and Xi's new era of socialism with Chinese characteristics must be taken as the core content of students' ideological and political education. Xi's new era of socialism with Chinese characteristics is rich in content, profound in thought, which is the latest achievement of the theoretical system of socialism with Chinese characteristics. It has made rich practical exploration in some fields, such as deepening reform, innovative development, coordinated development, green development, safe construction, rule of law construction, and cultural confidence, improvement of people's livelihood, party-managed armed forces, especially strict party governance. As the public security institution, it shoulders the important responsibility of cultivating the builders and successors of the cause of socialism with Chinese characteristics. Arming our minds by Xi's new era socialism with Chinese characteristics is an important position, and students of public security institutions are important targets. Therefore, we always take Xi's new-era socialism with Chinese characteristics as a squadron's weekly compulsory theoretical study, to receive party spirit education, and learn current politics. It is necessary to take it as the core content of ideological and political education for public security college students, and we will persist in it for a long time.

Adhere to standardization and innovation, actively explore the institutionalization, standardization, and process of organizational life of public security college students, enhancing the political, epochal, principled and combative nature of public security college students. One is the standardization of organizational life. Implement a supervision mechanism for pre-declaration, supervision during the event, and post-assessment of organizational life. Firmly establish a clear orientation of all party work to the branch, and base on the branch standardization, print and publish the "Party Construction System Compilation", establish the "Party Branch, Party Member Work Manual", determine the concentrated learning day, organization life day, party fee payment day, "Themed Party Day" such as party member political birthdays. The second is to innovate the organizational life mode. Actively explore the establishment and improvement of the "Online Party Branch", implement the "three meetings and one lesson", make the party oath video, "do not forget the original intention" commitment, party building activity day and other full factors upload, full process records, panoramic publicity, in order to achieve party building clearing, working on a web browsing, information on a web search.

A major feature of education in public security colleges is that the professional characteristics of students focus mainly on law enforcement. In the construction of campus culture, it must be led by Xi's new era socialist thinking with Chinese characteristics, and the process of governing the school by law should be promoted; In order to fulfill the goal of ruling the country by law, we need to promote the law and promote the rule of law in a way that teachers and students like to hear, plant the law-abiding soil of law, maintain the authority of the rule of law, improve the legal awareness and legal quality of teachers and students, and promote the conscious study of law, law-abiding and use to set up a harmonious campus atmosphere. In the process of educational reform, public security colleges must also consider the characteristics and growth laws of students, strengthen the cultivation of disciplinary awareness and individual development of public security college students, and ensure the harmony between policing management and student personality development.

5 ADHERE TO STRICT POLICE MANAGEMENT AND ACTIVELY EXPLORE NEW MODES OF STUDENT MANAGEMENT IN PUBLIC SECURITY COLLEGE

In the student education and management of public security colleges, the new realm of managing and governing the party in the new thinking is carefully implemented. It is determined to guide students of public security colleges to strictly cultivate themselves, strictly use power, and be strict with themselves, consciously pursue noble sentiments, consciously stay away from low-level tastes, consciously resist unhealthy trends, promote the values of loyalty, honesty, fairness, truth-seeking, integrity, etc Students of public security colleges should be loyal, clean, and pragmatic in their work. Moreover, they should adhere to the school motto of "loyalty, rigor, unity, dedication", adhere to law enforcement, strict discipline, overall promotion, comprehensive construction, pay close attention to implementation, carry forward the tradition, and establish a strict policing management system. Under the premise of building police and strictly managing police, actively explore new models of student management in public security colleges, realize students as the main body, drive innovation through ideological and political management, strengthen management with reform, strengthen ability with practice, and then achieve the result of reforming and strengthening the police, invigorating the police through technology.

5.1 *Take students as the main body and publicize students' personality*

The life of police college is completely different from that of ordinary colleges and universities. Individuals with strong political positions and distinct personality are needed in their future careers in police service. Therefore, coordinators should timely know the basic situation of the students, such as personality characteristics, family environment, communicative ability, etc., They should communicate with the students regularly, recording one by one, communicating with the frustrated students in a timely manner, and have a deeper understanding of the students in the communication. After understanding the overall characteristics of a team, the coordinators can organize more activities related to personality development in normal times, enhance the relationship with the students in the activities, and further enhance the subjective initiative of the students. To put it another way, the relationship with the students is enhanced, then, the authority of the counselor can be established, and the students with tasks can be united as one command can make the entire group like a wolf group, which can make the management naturally more convenient

5.2 *Strengthen the team building of coordinators in public security colleges*

Coordinators of public security colleges in the new era should change their concepts, adding some elements of the new era to their work, making full use of the convenience brought by the Internet era, actively learning to use the new functions of mobile phones and the Internet, to follow the trend of the era, and understand the psychological state of police college students and enrich their ideological and political education concepts. As to coordinators, they should summarize in innovation, think through experience, integrate their own circle with the circle of public security college students through the Internet, and make friends with students, so that students can feel the warmth of public security colleges. Moreover, coordinators can create their own public account on WeChat, encourage students and parents of this brigade to pay attention to the public account, and mobilize the students to update the public account from time to time, so that the public and the parents of the students can understand the life status of the students and the characteristics of the team. Establish a home-school friendship model. At the same time, what are the requirements of parents to leave a message on the WeChat public account to form a new situation of mutual feedback, break the deadlock that was completely closed in the past, finally, let communication be more active and positive, and make the barriers less in life. Coordinators can weave a good relationship network between students, and parents, and fill in and rectify missing areas in time.

5.3 *Innovation of inner driving force of ideological and political education*

The cultural education in police can help to cast iron army, as well as to build a positive spirit. Vigorously implement cultural education and build an elite student police force. The "police camp culture" is an important part of the construction of the public security team. The profound connotation is the police spirit. Among the requirements of the Shandong Police College for public security college students, the public security college students must have a strong "police awareness". To cultivate the police spirit of the students, the coordinators must do it themselves. They must strengthen the ideological and political education in the context of socialism, enhance the cultural quality and moral level of the students, strengthen the political awareness of the students, resist the impact of negative thoughts, and firmly establish the concept that "grasping ideological politics education is grasping combat effectiveness." In daily meetings, we will explain to the students what will happen when they go to the police, in order to cultivate the students' socialist core values, distinguish right from wrong, strengthen the students' justice, and give some easy-to-understand examples to promote positive social energy. In a specific humanistic atmosphere, students are subtly influenced, internalized into a connotation, focusing on improving the overall quality, and delivering a batch of excellent fresh blood and political backbone to the public security team.

5.4 *Real knowledge through practice, the grassroots is fertile ground*

Coordinators should make full use of the large-scale security activities and internship activities organized by the college. These activities are most suitable for public security combat and can best train the personal qualities of students. Besides, the coordinators should assist the students to develop their own security or internship plan, letting the students understand some of the security or internship situations that may occur, and make all psychological preparations. Next, the coordinators can use the Internet, laws and regulations and combine their work practice to summarize a series of emergency response rules. Firstly, they can train the head of each dormitory, and then the head will explain to each dormitory member, discuss with each other, and form a learning mechanism for mutual learning. In large-scale security activities or internships, the students must be strictly managed, and the day-to-day living system should be performed smoothly, which is also a consideration for the students' own safety. Students who need difficulties in coordination should be helped in a timely manner, and various psychological problems occurred during the activities of the trainees should be guided and adjusted to care for and serve the students. Reasonably guide students to properly handle the relationship between work and study, without delay, to achieve the result of mutual expansion, mutual benefit, and improve efficiency and solve the pressure of students. At the same time, the coordinators can also lead the students to visit the local police history museum with the cooperation of security or internship local police officers to learn the exemplary deeds of the heroes of public security, and inspire students to be a bloody, dare-to-be, promising new era qualified reserve police officer.

ACKNOWLEDGEMENT

This research results from Shandong Police College humanities and social science project "Xi Jinping's New Era Security Concept Guiding Risk Prevention and Control Terrorism in Political Core Area" (Project number: 19CFZJ40).

REFERENCES

[1] Xi Jinping. Held a seminar on important discussions on education at Renmin University of China. China Daily (Overseas Version), 2018-10-10 (01).
[2] Sun Haijie, Zhu Peilin, Gong Tailei. Police school students politicized management scientific and rule of law construction. Reform and Opening, 2014 (21): 95–96.
[3] Yin Yan. The status quo and countermeasures of hidden curriculum construction of loyal education in public security colleges and universities. Journal of Guangxi Police College, 2016, 29(4): 111–115.
[4] Zhang Meizi. On policing management and personalized development of college students. Management Observation, 2009, (1).

Study of the learning mode based on a smart classroom

Y.M. Wu
Xi'an University of Technology, Xi'an, China

ABSTRACT: This article analyzes the characteristics of the smart classroom and designs a learning model on this basis. Before class, during teaching, and after class, the teacher can take advantage of effective methods to optimize the students' learning process, and improve the suitability of the resources and students' learning interest.

1 INTRODUCTION

A traditional learning environment is physically and pedagogically restrictive. In addition, traditional learning environments are limited in their ability to provide immediate knowledge and information.

With the development of information technology, great changes have taken place in people's lives, mainly reflected in the speed, breadth, etc In education activities, it is also inevitable to use information technology to improve the teaching method

The smart learning model is based on the Internet, under the guidance of research-based learning theory, with the support of technology and science, creating an efficient education study method, allowing students to have active learning as participants, and enriching their knowledge.

2 RESEARCH ON THE SMART LEARNING MODEL

For the study of a smart learning model, the existing domestic and foreign research can be roughly divided into research in technical and teaching method applications. In terms of technical design, the development of a smart learning system can start with mobile learning systems as early as possible.

In 2000, the U.S. Department of Education planned and launched research into mobile education, which was established at the University of California, Berkeley's Human–Computer Interaction Laboratory, which has led academics around the world to focus on mobile learning (as cited in Zhang, Zhao, & Li, 2012:204). WEBCT is an online learning system developed by the British Columbia Department of Computer Science, which has an online chat system, learning tracking system, group project organization system, student self-evaluation system, etc.

Based on the above, research into the technical design of smart learning abroad has been presented, and the domestic research of the system in technical design has been carried out with smart learning at the core. Wu Hongyan (2015:127–131), from the perspective of smart learning, and relying on learning analysis technology, constructed a personalized online learning system in the cloud environment, so as to realize a combination of teaching and technology. This system can analyze learners' learning behavior, understand the learning situation, and give appropriate feedback interventions to realize an enhancement of intelligent and adaptive learning resources.

In summary, the learning model based on smart learning mainly refers to intelligent, personalized, and diverse uses for learners in system environment learning services, using removable terminal equipment, to achieve an efficient, autonomous, and open classroom with teaching objectives learning mode. Learner-centered classroom learning focuses on the whole process of intelligent learning and integrates various learning methods. The research is based on learning community theory and constructs a learning model based on a smart class.

3 ANALYZING THE CHARACTERISTICS OF A SMART CLASSROOM

A smart classroom is shown in Figure 1 with the applications of resource sharing, real-time push, intelligent learning analysis, and interactive communication promoting the change to the learning model.

3.1 Resource sharing

Teachers can classify resources according to teaching purposes and objects, facilitating teachers and students to upload and obtain various resources, and are not limited by time and space; they can more effectively integrate all kinds of media resources to bring experience to learners; and the system automatically captures and stores associated generative resources.

3.2 Real-time content push

In response to learning needs, in groups or individual forms, real-time, personalized learning pushes content to students, enabling screen sharing, which greatly improves the participation of students in classroom teaching.

3.3 Intelligent learning analysis

Relying on clustering and other technologies, all data of students are managed in the learning process, with records of all their learning activities. Calculations, comparisons, and diagnostics are combined for implementation analysis to assist the teacher in decision-making.

3.4 Collaborative interaction

Communication is done using intelligent terminal device diversity, such as different size groups collaborative learning. Students can easily use cell phones to take part in the classroom practice in which more than one person or even the whole class takes part. They can think actively and enhance the learning interaction.

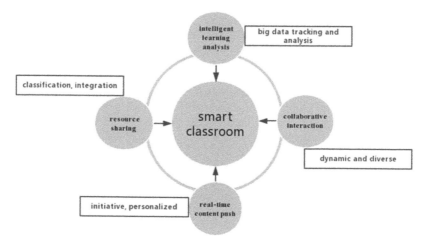

Figure 1. Characteristics of a smart classroom.

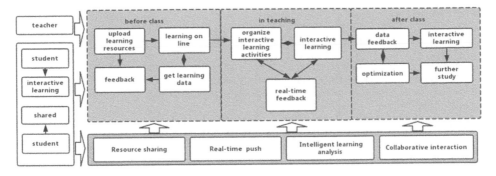

Figure 2. Learning mode based on a smart classroom.

4 DESIGN OF THE LEARNING MODE BASED ON A SMART CLASSROOM

Based on research and theoretical analysis of relevant learning models at home and abroad, it integrates the technical characteristics and changes in teaching and learning forms. This article attempts to break down the various elements in the learning process based on a smart classroom, designing a learning model as shown in Figure 2 based on efficient interactive learning.

Before class, the teacher can take advantage of the layered resource-sharing technology by uploading a preview of the teaching material and setting the time of study. Students, after receiving the notice, preview through the app. After the preview is complete, the application automatically saves the status record. Then, the teacher can supervise and urge the students to preview according to the students' completion status and other information.

In teaching, teachers can also combine students with shared learning results, application collaboration, interactive communication, and instant feedback evaluation technology to organize interactive learning activities, such as group discussions, game-like exercises, and so on. Students apply mobile communication technology to create learning partners and have discussions with each other, share information, and solve problems through collaborative learning. For problems that cannot be solved within the class, students can use the real-time recording and interaction system supported by collaborative interactive communication technology to break the boundary of classroom space and time and seek help from the teacher and other students.

After class, students can access after-school learning interaction through the app: complete the reading, preview, and classroom synchronization exercises issued by the teacher within the specified time, and the system automatically generates statistics of class information for teachers to illustrate the students' learning process and judge student acceptance and teaching effect. Students can also freely choose learning resources and methods to consolidate internalized knowledge according to their actual needs.

5 CONCLUSION

The smart learning model constructed with the help of an intelligent classroom can be used in a variety of courses. Through the powerful function of the smart classroom, it is possible to change the traditional teaching model, facilitate teachers to carry out teaching design, motivate students to learn, and ensure the participation of each student in study practice. Students are no longer just recipients of knowledge. They can also be active contributors in the process of teaching guidance and evaluation.

REFERENCES

[1] Rania Albalawi. Evaluating Tangible User Interface-based Mobile learning, D. Ontario: University of Ottawa, 2013.

[2] Agourram, H., Robson, B., & Nehari-Talet, A. Succeeding the implementation of e-learning systems in a multicultural environment. International Journal of Learning, 2006.

[3] Feistritzer, C.E. Profile of teachers in the U.S.2011. Retrieved from http://www.edweek.org/media/pot2011final-blog.pdf

[4] Agourram, H., Robson, B., & Nehari-Talet, A. Succeeding the implementation of e-learning systems in a multicultural environment. International Journal of Learning, 2006.

[5] Garrison, D. R. E-learning in the 21st century: A framework for research and practice. New York, 2011.

[6] Li Zhifei. Research on visualization teaching design and application in smart classroom. Central China Normal University, 2018 (In Chinese)

Computational Social Science – Luo, Ciurea & Kumar (eds)
© 2021 Taylor & Francis Group, London, ISBN 978-0-367-70193-2

Development of a teaching standard for railway traffic operation management from the perspective of secondary and higher vocational linkage

Q.P. Ye & W.Y. Wu
Wuhan Railway Vocational College of Technology, Wuhan, China

ABSTRACT: This paper analyzes the teaching standards of Railway Transportation Management and Railway Traffic Operation Management. It points out the necessity of setting a national teaching standard under the background of secondary and higher vocational linkage, and puts forward some feasible suggestions for developing the teaching standards of Railway Traffic Operation Management in vocational colleges.

1 INTRODUCTION TO NATIONAL TEACHING STANDARDS FOR RAILWAY TRANSPORTATION MANAGEMENT AND RAILWAY OPERATION MANAGEMENT IN VOCATIONAL SCHOOLS

In order to strengthen the basic construction of vocational education, promote the teaching of vocational education to be scientific and standardized, and establish the high-quality assurance system of vocational education, in April 2014, the Ministry of Education published 230 teaching standards for secondary vocational education (for trial implementation), including teaching standards of Railway Transportation Management in secondary vocational schools. In July 2019, it released another 347 teaching standards for higher vocational colleges, including Railway Traffic Operation Management. These teaching standards have clarified the cultivation objectives, specifications, curriculum, class hours, basic teaching conditions, and requirement of quality assurance.

2 A BRIEF ANALYSIS OF THE NATIONAL TEACHING STANDARDS FOR RAILWAY TRANSPORT MANAGEMENT AND RAILWAY TRAFFIC OPERATION MANAGEMENT IN VOCATIONAL SCHOOLS

The national professional teaching standards have improved the quality of railway transport vocational education. According to these standards, vocational schools can independently formulate a talent cultivation plan. From the teaching standards of vocational schools issued by the Ministry of Education, there are the following characteristics.

2.1 Difference in professional orientation

In the teaching standard of Railway Transportation Management in secondary vocational schools (trial implementation), the corresponding professional posts are distinguished from the main posts in the teaching standard of Railway Transportation Operation Management in higher vocational colleges. Some posts such as station assistant duty attendant, passenger staff, and freight staff are listed in the teaching standard of Railway Transportation Management in secondary vocational schools. The corresponding professional qualification certifications are basically intermediate.

However, it added some posts such as station attendant, passenger attendant, and freight attendant in the teaching standard of Railway Traffic Operation Management for higher vocational colleges, the requirements of corresponding vocational qualification certifications are higher. This shows the difference between the two levels of vocational education in terms of post group.

2.2 Difference in cultivation objectives

No matter the Railway Transportation Management in secondary vocational school or the Railway Traffic Operation Management in higher vocational college, the cultivation objectives are oriented to the railway transportation industry. All graduates can be engaged in railway traffic command and dispatch, passenger transport organization, and freight transport organization. However, the cultivation objectives of Railway Transportation Management in secondary vocational school emphasize service. Besides focusing on service, the cultivation objectives of Railway Traffic Operation Management in higher vocational colleges further emphasize that the students should have organization and management ability. And, in addition to having the same skill requirements as the Railway Transportation Management in secondary vocational education, it is also required to have a more certain technique.

2.3 Differences in the curriculum

Railway Transportation Management in secondary vocational schools includes three specialized basic courses of Railway Transportation Operation Management in higher vocational colleges into the core courses. The three basic courses include railway lines and stations, railway rolling stock, railway signals, and communication equipment. In addition, there is only one course of railway train organization for Railway Transportation Management of secondary vocational schools, but this course is divided into four courses of receiving and delivering train operation, train dispatching and directing railway shunting, station operation plan, and statistics for Railway Transportation Operation Management of higher vocational colleges.

2.4 The main teaching contents and requirements of core professional courses are similar

The teaching standards of Railway Transportation Management in secondary vocational schools and Railway Transportation Operation Management in higher vocational colleges have some deficiencies in professional courses from the perspective of linkage between different levels of vocational education. For example, there are many class hours for the professional core courses for Railway Transportation Management in the secondary vocational school, especially 192 class hours are set aside for passenger transportation and freight transportation. In addition, the main teaching contents and requirements of the core professional courses are almost the same as those of Railway Traffic Operation Management in higher vocational colleges, which does not reflect the difference in professional knowledge.

2.5 The skill training of core courses does not reflect the difference in vocational skill levels

The class hours of professional skills of Railway Transportation Management in secondary vocational schools account for about two-thirds of the total class hours. In addition to 192 class hours for freight and passenger transport, a large number of class hours are provided for skill training. However, the practical integration teaching of Railway Traffic Operation Management in higher vocational colleges does not put forward specific class hours of skill training. From the perspective of teaching standards, the content of skill training cannot reflect the differences between different grades of the same type of occupation.

3 THE NECESSITY FOR ESTABLISHING NATIONAL TEACHING STANDARDS FOR RAILWAY TRAFFIC OPERATION MANAGEMENT UNDER THE BACKGROUND OF THE LINKAGE BETWEEN SECONDARY AND HIGHER VOCATIONAL SCHOOLS

The teaching standards of Railway Transportation Management in secondary vocational schools and Railway Traffic Operation Management in higher vocational colleges issued by the Ministry of Education are based on the premise that secondary vocational schools and higher vocational college set up independently. In 2011, the Ministry of Education issued guidelines on promoting the coordinated development of secondary and higher vocational education, which pointed out that the teaching standards linking secondary and higher vocational education should be gradually formulated.

In recent years, with the different enrollment types of higher vocational colleges, the student sources show diversified, and some higher vocational colleges stratify to cultivate differentiated students. In particular, 3+2 section students account for a proportion of higher vocational education, and for how to develop teaching standards for this part students, the Ministry of Education has not issued the corresponding instruction. The Ministry of Education issued the implementation plan that higher vocational college enrollment had expand in 2019. The plan pointed out that in the modern service industry and other fields, the enrollment scale should be expanded under the linkage between vocational education schools. This means that the Railway Traffic Operation Management in higher vocational colleges would have more secondary vocational students or 3+2 section students in the future. In order to ensure the vocational education quality, it is necessary to jointly establish teaching standards of Railway Transportation Management and Railway Traffic Operation Management in vocational schools.

The National Railway Group Corporation is no longer hiring graduates majoring in Railway Transportation Management in secondary vocational schools. From the perspective of employment and career planning of graduates, some secondary vocational students are bound to enter higher vocational colleges to further improve their professional knowledge. In order to make the cultivation plan link effectively, it is required to establish national professional teaching standards from the perspective of secondary and higher vocational linkage.

4 SOME SUGGESTIONS

Secondary and higher vocational education are of the same type but at different levels. The vocational service orientation of technical and skilled cultivation is the same, but there are fundamental differences in educational content and orientation. In order to ensure the quality of vocational education, the Ministry of Education should introduce national teaching standards from the perspective of the linkage of secondary and higher vocational education as soon as possible. The following factors should be taken into consideration when the teaching standard of Railway Traffic Management is formulated.

4.1 *The development of professional teaching standards shall be coordinated by various parties*

The development of teaching standards for Railway Transportation Management and Railway Traffic Operation Management in vocational education should break the current situation that the two have separately developed the teaching standards for different vocational education stages, and they should be coordinated. Although enterprise experts are involved in developing professional teaching standards, this is not enough, enterprise experts only gave programmatic advice on study tasks. The specific teaching content is not divided according to difficulty and professional level for different grades of vocational education.

The teaching standard should be drawn up with secondary vocational school, higher vocational college, and enterprise experts all participating. The three parties should distinguish the professional grade, content difficulty, and study task for different vocational education stages. On the one hand, it should distribute the learning tasks at different vocational education stages. On the other hand, it can ensure that the learning content corresponds to the examined content of the professional qualification certificate, avoiding duplication or omission of essential knowledge.

4.2 The curriculum system should be hierarchical from the perspective of connection in different vocational education stages

The professional curriculum system should not only cultivate professional skills, but also have appropriate knowledge scope and reflect adequate breadth and depth of knowledge. At the same time it should be step-by-step, reflecting the level of the student. According to the teaching standards of Railway Traffic Operation Management and Railway Transportation Management issued by the Ministry of Education, the course names are roughly the same, although the railway train operation is different. From the perspective of the linkage of secondary and higher vocational education, national teaching standards should be able to distinguish the different levels of vocational education from the course names. The curriculum should be rearranged and, especially for professional core courses, there should be differentiation. For example, the freight organization in secondary vocational schools should set up as railway general cargo transport, railway container transport, and multimodal transport, with the higher vocational college setting up the railway freight technology. The railway passenger service etiquette is set up in the secondary vocational school, and the passenger transport organization and emergency treatment are set up in the higher vocational college.

4.3 The teaching contents should be closely related to the vocational skill standards

The national occupational standard, also known as the national occupational skill standard, is a comprehensive stipulation on the theoretical knowledge and skill requirements of employees based on the occupational classification, according to occupational activities. The national occupational standards are presented in national vocational qualification certificates to prove that workers have mastered the knowledge and skills required for a certain level of occupational posts, and met the requirements of occupational standards, and are qualified for the occupational posts. On December 30, 2005, on the basis of the railway occupational skill standard and the railway occupational skill appraisal standard, the national occupational standard for the railway industry was promulgated. The vocational skill standard of five levels has been established, including junior, intermediate, senior, technician, and senior technician. The knowledge and skill requirements of the five different grades are formulated, respectively, for railway transportation, passenger transportation, and freight transportation. In addition, it is clearly stipulated that different grades of knowledge and skills must be mastered by junior workers, intermediate workers, and senior workers. As can be seen from the railway vocational skill standard, the higher the vocational skill grade is, the greater the required knowledge and skill will be. From the perspective of the linkage between secondary and higher vocational schools, the knowledge and skills required for the occupations with lower grades should be completed in the secondary vocational schools, so that students can register for corresponding vocational qualification certificates.

4.4 The development of specialized course materials from the perspective of the linkage of secondary and higher vocational education

As a part of teaching resources, teaching materials play a very important role in teachers' teaching implementation. At present, most teaching materials of the relevant majors in vocational schools are developed. However, in Zhejiang and Guangdong provinces, where the economy is developed, teaching materials of relevant majors in different vocational schools have been developed from the perspective of the linkage between secondary and higher vocational schools. For example, Zhejiang University Press published the construction techniques used in the secondary and higher vocational schools, but with the background of the linkage between secondary and higher vocational education. From the perspective of the linkage between secondary and higher vocational schools, the development of teaching materials should follow school–enterprise and interschool cooperation. The linkage between Railway Transportation Management in secondary vocational schools and Railway Traffic Operation Management in higher vocational colleges must have teaching materials that are effectively implemented. Some majors have the characteristics of the industry, are not practical, and not widely object-oriented, therefore it is necessary to introduce relevant policy to ensure the development of the secondary and higher vocational textbook system.

Guangdong Province started to develop professional teaching standards and curriculum standards from the perspective of secondary and higher vocational linkage from 2013. The teaching standards and curriculum standards of some general majors, such as tourism management, marketing, e-commerce, and logistics management, have been introduced. At present, the relevant majors with industrial characteristics have not developed the teaching standards for vocational education schools from the perspective of secondary and higher vocational linkage. To improve the quality of vocational education, it is necessary to develop national teaching standards for the secondary and higher vocational relevant majors, especially for the railway transportation industry.

REFERENCES

[1] Minghui Yu, Mihui Li. Analysis on the Elements and Paths of Joint Development of Teaching Standards and Vocational Standards in Vocational Education. Vocational and Technical Education, No. 11, 2019:24–29.
[2] Chunyu Hou, Siwei Li. Research on Development Path of National Professional Teaching Standard in Higher Vocational College—Taking Civil Aviation Communication Technology as an Example. Education Modernization, No. 33, 2019–90.
[3] Chengqing Gong. Construction of Teaching Standard for Integrated Animation and Animation Production Technology. Journal of Guangdong Light Industry Vocational and Technical College, March 1, 2018:51–53.
[4] Nana Du. Development of Teaching Standard for High-Level Integrated Leather Art Design. Modernization Education, 29 July 2018:321–324.
[5] Xuhong Wang, Xiuduan Gong, Linlin Zhang. A Study on Teaching Standards of Printing Media Technology Based on Middle and Higher Vocational Linkages. Higher Education Journal, No. 18, 2018:189–192.

Computational Social Science – Luo, Ciurea & Kumar (eds)
© 2021 Taylor & Francis Group, London, ISBN 978-0-367-70193-2

Research and practice on the construction methods of online and offline one-stop service systems in colleges and universities

S.X. Wang
Network and Informatization Office, Huazhong University of Science and Technology, Wuhan, China

L.B. Zhu
The President's Office, Huazhong University of Science and Technology, Wuhan, China

Y. Xiong*
Network and Computing Center, Huazhong University of Science and Technology, Wuhan, China

ABSTRACT: Facing great pressure of "double first-class" competition, universities in China must enhance the service experience and sense of happiness and belonging of teachers and students. The main reason for this is the lack of integration of online and offline services. The main methods are as follows: building an information platform of a one-stop online service hall, synchronous construction of mobile applications, full sharing with relevant data of university information systems, vigorously promoting the settlement of various business processes, building and making good use of service evaluation function, carrying out process reengineering or optimization, building a one-stop offline service hall, and promoting an online and offline combination. Huazhong University of Science and Technology has established an online one-stop service information platform—The Online Service Hall—and offline one-stop service place—Teachers and Students Service Center—and organically combined the two. The total service items have reached more than 400, and the service experiences of teachers and students have been significantly improved.

1 INTRODUCTION

The modernization of educational governance system and educational governance capacity is one of the main aims of educational modernization, and promoting the reform of "streamline administration, delegate powers, and improve regulation and services" is an important way to promote the modernization of educational governance. Based on the construction of a "one-stop" service combining online and offline, it has become the main form of "streamline administration, delegate powers, and improve regulation and services" reform in Chinese colleges and universities. A study has shown that integrated one-stop student support services are critical for student success (Emma, 2020). This paper mainly studies the methods of "one-stop" service combining online and offline.

First, in recent years, China has attached great importance to the cause of network security and informatization, and proposes that the development of network security and informatization must implement the development idea of "taking the people as the center," so that the people can have a greater sense of gain, happiness, and security in the development of information technology. The State Council has issued a number of documents on the reform of "streamline administration, delegate powers, and improve regulation and services." Second, with fierce competition in China's higher education sector, Chinese universities are facing the huge pressure of "double first-class" construction. We must further improve the work style, strengthen the service concept, improve the service quality, create a management team that is suitable for the requirements of world-class

---

*Corresponding author

universities and world-class discipline universities, effectively improve the management and service level of colleges and universities, and improve the efficiency and transparency of work. The level of understanding makes teachers and students content with teaching and learning, so that teachers and students can be free from complicated issues. Third, with the development of Internet technology, the government and enterprises make great use of mobile Internet and other technologies to provide convenient online and offline services for citizens. Teachers and students on campus easily use the "information gap" between the society services and campus services and hope to obtain better services. Finally, in recent years, the rapid development of university informatization has laid an important foundation for the construction of an online and offline one-stop service, and this makes it possible.

2 MAIN PROBLEMS AND ANALYSIS

2.1 *Main problems*

There are many problems in the management and service in colleges and universities, mainly including: first, the online service is "scattered and disorderly," lacking a centralized platform. Most of the information services in colleges and universities are scattered in various business information systems. Many cross-department and cross-platform services are fragmented, and cannot form a complete online process. There is a lack of unified functions, such as to-do reminder and service evaluation, so that teachers and students have poor online experiences. A centralized platform granting access to all relevant information and services is particularly important (Rudolph et al., 2017). Online activities will be the main venue for universities in the future (Makhalina et al., 2020). Second, there is a lack of centralized service places offline. Teachers and students need to go to various departments to handle affairs for the public, and the service attitude is not good, creating a situation like a door that is hard to enter, with an ugly atmosphere and the matter becomes difficult to handle, resulting in teachers and students complaining. Third, the combination of online and offline is not sufficient. There can be online and offline disconnection or poor docking, or repeated disconnection situations, which frustrate teachers and students. Not only is the burden of work not reduces, but it also increases the burden of constantly filling in online forms. On the whole, the problems of online and offline services and the combination of the two remain serious and need to be improved.

2.2 *Problem analysis*

The main reasons for the above problems are as follows: first, the departments do not pay sufficient attention to them. The awareness of serving teachers and students is not strong, and the management informatization, service informatization, service standardization, and service process are not paid enough attention. Therefore, the investment in management and service personnel, information system construction, and service measures is insufficient. Second, there is a lack of overall planning. Each department only pays attention to its own service content, and lacks unified research and command on the connection and optimization of the service process; and it also lacks a unified information platform to centralize, integrate, and optimize the process. Third, there is a lack of unified offline office space. In order to provide a unified offline service experience for teachers and students, it is necessary to have a centralized location, centralizing as many processing windows as possible, providing more services, and uniformly deploying service facilities, service standards, and service specifications.

3 METHODS DISCUSSION

3.1 *Building of a one-stop online service hall information platform*

A design of Web-based educational services is one of the approaches to university sustainability (Tikhomirov et al., 2015). The construction of a one-stop online service hall information platform created a unified information platform for all kinds of online services. As a process platform, the core of a one-stop online service hall is the process. The process can be divided into the original process and integrated process. The original process refers to all kinds of processes directly developed by using an online service hall platform. The business system is not closely linked, there is less data exchange, involving more audit nodes, and a cross-department process is more suitable for the construction of the original process, giving full play to the flexible and rapid characteristics of the process platform, and quickly completing the development and online process. The integration process refers to the business information system that has been an online operation and more mature business process. For the process which is closely related to a certain business system, it needs to read and update a large amount of data, and has little relationship with other business systems, and will be implemented in the original business system. Only the specific entrance of service items will be integrated into the online service hall. After clicking on the online service hall, you can directly enter the service page in the business system, and at the same time, do it online. The to-do center or task center of the business hall centralizes the to-do items of the business system to the online service hall.

At the same time, the in-depth integration of a one-stop online service hall, information portal, business system, etc. will be built into a unified service entrance, so that teachers and students can find and handle affairs online. First, the one-stop online service hall is integrated into the home page of the information portal. Teachers and students can not only obtain school notices, announcements, official documents, and other public information, but also view and handle the to-do matters of the online service hall through the to-do center integrated into the home page of the information portal. Teachers and students only need to go through the entrance of the online service hall to find out what they want to handle, which is convenient for teachers and students to handle affairs, process approval business, and view the processing results, avoiding manual switching between multiple systems.

3.2 *Construction of a mobile version of a one-stop online service hall synchronously*

Mobile applications are the current development trend of the Internet. Teachers and students are used to getting information and handling various affairs on cell phones. When building an online service hall, the mobile version must be built at the same time. This can be realized by establishing an independent mobile App or WeChat enterprise platform. The core of mobility is to make all the processes of the online service hall mobile or page adaptive processing, so that teachers and students can submit online for processing at the mobile terminal, and managers can approve and process on the mobile terminal. At the same time, the mobile phone App message or WeChat message can be used to remind staff and students to deal with to-do matters, to ensure the timeliness of processing matters. In the process of mobile construction, we should focus on several parts. First, we should deal with the forms adaptively to ensure that they are suitable for mobile operation; second, we should ensure that all the approval functions are implemented on the mobile terminal, and the submission function of teachers and students should also be implemented in the mobile terminal as far as possible, especially complex background management cannot be implemented on the mobile terminal. Third, it is necessary to integrate with the existing mobile applications in schools to prevent the establishment of independent mobile application information islands. Fourth, the style of mobile interface should be specially designed, not a simple copy of PC software versions, but designed according to the characteristics of the mobile terminal and attention paid to user experience.

3.3 *Share the relevant data with the other information system fully*

It should be deeply integrated with the school basic database and relevant business information system database to realize full sharing of data. When teachers and students submit forms, personal basic information is automatically filled in to avoid repetition of form filling. Users only need to fill in a very small amount of information related to business processing, so as to save users'

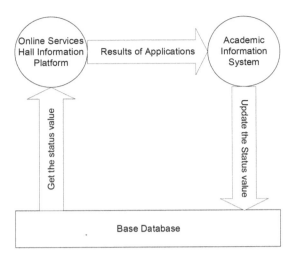

Figure 1.   Data sharing with academic information system.

filling time and solve users' frustration at "filling in forms repeatedly." In order to ensure data sharing, the university should formulate relevant rules and regulations for data sharing, specify the authoritative source of data generation and responsible departments, and according to the principle of "who produces is who maintains," each business information system provides data for the basic database to ensure the maximum sharing of data.

The online service hall should also be deeply integrated with the business information system related to each process. Many processes are not isolated, and relevant data must be obtained from the business information system as the basis for judging the flow node or automatically filling in the form. For example, in the process of undergraduate students' application for overseas exchange, suspension, resumption, or drop-out applications, the system will automatically determine the changes that can be handled according to the current status of students (provided by the academic information system through the interface) before filling in the form. After the process is completed, the online service hall will feed back the results of student status change to the academic information system and academic information system updating the contents of the academic information system database (as shown in Figure 1).

The online service hall information platform is mainly used to complete the process of application, approval, processing, and evaluation. When the process is finished, the data generated by the process should be stored in a standardized way and returned to the corresponding business information system. The business information system will complete further queries, analysis, statistics, and other functions (as shown in Figure 2).

3.4   *Promote the settlement of various business processes vigorously*

The key and difficult point in the construction of the information platform of online service hall lies in how to promote the process settlement of each department. The methods to promote the process settlement mainly include: first, build the process of network and information department first. The network and information departments should take the initiative to change their own management and service methods and work habits, and move all the original offline processes of the department to the online service hall, so as to set a benchmark for other departments and play an exemplary role. The second sector is highly active. Select the departments with a good informatization foundation, strong service awareness, wide service range, willing to try new things, and strong execution as the first batch of departments to actively serve them and jointly play the demonstration and driving effect. Third, establish relevant rules and regulations. Establish

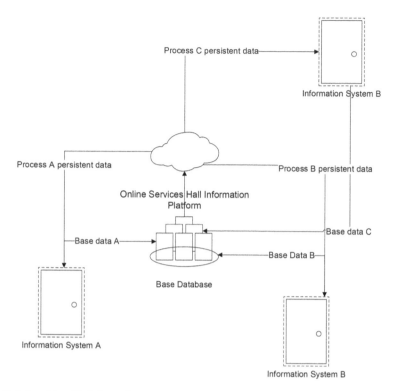

Figure 2.   Relationship with information systems.

management measures for the construction of the information platform of online service hall, clarify the responsibilities of each department, platform construction principles, process construction procedures, performance evaluation, etc. through the system requirements, each department should try its best to put its business into the online service hall. Fourth, make good use of the opportunities. Make good use of the university activities or opportunities such as improving work efficiency, process reengineering, strengthening the construction of work style, and cooperate with relevant departments to carry out process optimization and Internet access. The fifth is to stimulate the enthusiasm of the business department construction process. Help business departments to sort out their business needs, optimize business logic, and pay attention to details. For the process with good operation conditions and high satisfaction of teachers and students, various means are used to vigorously publicize the process, so as to enhance the sense of achievement of the business department in the process construction, and form an incentive effect. The sixth is to improve the process quality, standardize the description and interface of process forms, improve user experience, and optimize process nodes. The principles of process quality first should be adhered to. Immature, low-quality, or defective processes cannot be launched in a hurry. Problems found in online processes should be solved or optimized quickly.

3.5   *Build and make good use of the service evaluation function*

In order to improve the operational quality of the process, we must establish a corresponding mechanism to ensure the standardization, openness, and transparency of the process. The evaluation function is an important function to ensure the quality of the process service. First, each process should be customized according to the standard template, through which users can fully understand the process and matters needing attention. Second, set milestones in key steps of each process to help users track the progress of work in real time, so that all process participants can monitor the process progress, so as to improve the efficiency of work. Finally, after all processes are completed, the applicant should be able to grade and evaluate the service according to the actual experience and feelings in the whole process, combined with the handling efficiency and service attitude of the staff, and the evaluation results should be displayed in real time during the process.

3.6 Carry out process reengineering or optimization activities

Informatization serves as management and is the mapping of management on the network. It is necessary to reengineer and optimize the management and service process, and then realize it in the online service hall by means of information technology. The specific methods are as follows. First, process reengineering is one of the university's main tasks. The president's office, staff union, network and information office, and other departments jointly carry out the work. The second is to determine the leading department of each process. Sort out the departments related to the process, and determine the leading department. The leading department is responsible for sorting out the process and clarifying the tasks and dividing the work for the relevant departments. Third, formulate an incentive or incentive measures. The departments with excellent process optimization and obvious improvement of teachers' and students' experience should be rewarded to make the departments have a greater sense of gain and value.

3.7 Build a one-stop offline service hall

With the development of Internet technology, most services can be realized through the Internet, but there remain many businesses that need to be handled offline, such as stamping the official seal, printing supporting documents, etc. In order to facilitate teachers and students to handle affairs and provide a unified experience for teachers and students, the departments providing services should be centralized. However, the establishment of an offline service hall is not simply to gather the office locations of various departments together, but to handle according to the original process and mode, and to reorganize the approval process and service process through the physical set, so as to realize a transformation from centralization to integration. To realize the continuous platform construction, online and offline coordination, process reengineering normalization, assessment and evaluation standardization, event handling institutionalization, equipment function integration, promote multidepartment joint audit, shorten the process, shorten the processing time, and improve the one-time completion rate. The system includes a window evaluation system, quantitative assessment system, service evaluation system, public service system, classified management system, window partition system, first inquiry responsibility system, and centralized management system.

3.8 Promote the combination of online and offline

Generally speaking, there are three types of service realization methods of business process: one is the purely online process. All operations of this kind of process are completed online, and teachers and students do not need to do any offline processing, such as campus VPN application in an online office hall, work flow in OA system, password retrieval/password modification in business system, etc. The second is a purely offline process. This kind of service only provides offline manual handling, most of which is consulting and solving difficult problems and confidential matters. Third, an online and offline integration process. This kind of process is to submit the application online, the manager will review or approve the application in the background, and then the applicant will submit, pick up, or print the application offline. For example, application for an on-the-job certificate, income certificate, examination and approval of scientific research contract, going abroad for business or private purposes, and student card replacement. In order to improve the service experience of teachers and students, some processes not only provide online services, but also provide offline services. For example, some services are provided for retired staff and overseas students, and online and offline services are also provided to make the service more people-friendly.

4 PRACTICE AND APPLICATION EFFECT

By using the above ideas and methods, Huazhong University of Science and Technology has built an online one-stop service information platform—The Online Service Hall—and an offline one-stop service place—Teachers and Students Service Center—and promoted the combination of online and offline to build a smart service system. The Online Service Hall practices the concept of "Let information run more, let teachers and students run less" and the Teachers and Students Service Center basically realized "Just come once at most and enter one door, you can handle all affairs". The services can be online as much as possible, and the offline part can be self-help as much as possible, and the number of window services should be as small as possible. The Online Service Hall was launched at the end of 2016. Over the past two years, more than 240 online processes have been launched. A total of 21 departments have been involved, of which the leading department is the network and information office, with a total of 25 online services. Among the 149 original processes, there are 115 cross-department processes, involving nine departments. The Online Service Hall had a total of over 860,000 person services till the middle of 2020. Since its operation in December 2017, the Teachers and Students Service Center has been able to handle more than 400 items online and offline. At the same time, it has introduced more than 30 social services in six categories, handling more than 350,000 items for teachers and students, with a service satisfaction rate of almost 100%.

5 CONCLUSIONS

In order to realize the modernization of educational governance and the modernization of educational governance ability, colleges and universities must carry out the reform of "streamline administration, delegate powers, and improve regulation and services." Through the establishment of an online one-stop information platform and offline one-stop service hall, as well as effective online and offline docking and integration, an online and offline intelligent service system is constructed to improve the management and service level of the university and enhance teachers and students' sense of satisfaction.

ACKNOWLEDGMENT

This study has been undertaken with the ?nancial support of the Provincial Teaching Research Project of Hubei Province (2018045) and Huazhong University of Science and Technology Independent Innovation Research Fund Project (2016YXMS107).

REFERENCES

Makhalina O.M., Makhalin V.N., Yaroshchuk A.B. (2020) Overview of Perspective Educational Services of the "Green" Digital Future: Online, Lifelong and Remote Learning. In: Kolmykova T., Kharchenko E. (eds) Digital Future Economic Growth, Social Adaptation, and Technological Perspectives. Lecture Notes in Networks and Systems, vol 111. Springer, Cham.

Power, Emma; Partridge, Helen; O"Sullivan, Carmel (2020). Integrated one-stop support for student success: recommendations from a regional university case study. HIGHER EDUCATION RESEARCH & DEVELOPMENT 39(3):561—576.

Rudolph D., Thoring A., Remfert C., Vogl R. (2017) A Requirements Engineering Process for User Centered IT Services – Gathering Service Requirements for the University of the Future. In: Marcus A., Wang W. (eds) Design, User Experience, and Usability: Understanding Users and Contexts. DUXU 2017. Lecture Notes in Computer Science, vol 10290. Springer, Cham.

Tikhomirov V., Dneprovskaya N., Yankovskaya E. (2015) Development of University"s Web-Services. In: L. Uskov V., Howlett R., Jain L. (eds) Smart Education and Smart e-Learning. Smart Innovation, Systems and Technologies, vol 41. Springer, Cham.

Session 5. Management and economics

Computational Social Science – Luo, Ciurea & Kumar (eds)
© 2021 Taylor & Francis Group, London, ISBN 978-0-367-70193-2

Exploration of teaching reform and innovation on the course of advanced bio-reaction engineering

L.Q. Sun & H.J. Gao
School of Life Science, Beijing Institute of Technology, Beijing, China

ABSTRACT: The implementation of innovation education and engineering quality education in bioengineering postgraduates is an important way to cultivate innovative bioengineering talents. In this paper, combined with the problems encountered in the teaching work, the reform and innovation of teaching content, teaching methods and means, and teaching evaluation system are proposed. The innovative teaching work of talent training in the teaching of bio-reaction engineering is explored, and the experience of reform and innovation summarized.

1   INTRODUCTION

Advanced bio-reaction engineering is a required course for master's degree bioengineering. The main task of this course is to enable students to master the theory of complex biological reaction kinetics, design and analyze bioreactors, and master the detection and optimization of biological reaction parameters, so as to lay a solid foundation for future industrial production and scientific research of the biological industry. In recent years, faced with the problem that the master degree students of bioengineering major come from different undergraduate majors, especially those who have not studied the elementary course of biological reaction engineering, how to effectively carry out the theoretical teaching of a more complex biological reaction dynamics has become a direct obstacle for the biological reaction engineering teaching team. Therefore, it is necessary to continuously carry out the following measures for teaching reform and innovation. Through the opening of relevant experimental links, using hands-on practice to reduce the difficulty of theoretical teaching, the learning effect of students is improved (Meng & Wu, 2006).

2   PROBLEMS IN CURRENT TEACHING

Advanced biological reaction engineering is mainly carried out by theoretical teaching in class, in which students are required to preview before class and finish homework after class. In order for students to master the key points and difficulties of the content learned in the classroom, classroom discussion and extracurricular reading are often carried out; in the class, the students' understanding and mastery of problems are tested by PowerPoint teaching reports, and a group discussion is held. The problems still existing in this teaching method can be summarized as the following three points.

2.1   *Too much course content and too few class hours*

After several rounds of adjustment of the teaching syllabus, the teaching hours of higher biological reaction engineering remain at only 32. In order to complete the same teaching task, teachers need to explain a lot of content in the first half of the teaching time. Students are required to have a certain foundation in order to passively learn the class content. As a result, most of the non-bioengineering students have a poor listening effect and absorbing difficulties.

2.2 There are many formulas, no experiments, abstract, and difficult to solve

Many chapters in the course of biological reaction engineering involve mathematical formulas and the derivation process in reaction engineering. For students from biology and other sciences, the derivation of mathematical formulas in the course content is complicated, the calculation formulas are numerous and difficult to remember, the formulas are abstract, and there is more teaching theory than the practical operation. Teachers cannot fully know each student's mastery of knowledge. The process of formula derivation in multimedia courseware is too fast, so that students are not impressed and cannot fully understand the knowledge. After learning formulas, they are faced with difficulties in how to apply them, which leads to students' lack of interest in theoretical learning and formula derivation, and a lack of learning motivation.

2.3 Unscientific assessment method

In performance accounting, classroom attendance, classroom interaction, homework, and other ways of reflecting the students' usual efforts are used, but it is difficult to accurately evaluate the students' achievements through homework, which leads to the general performance not being fully reflect.

3 REFORM AND INNOVATION OF CURRICULUM CONTENT

Advanced bio-reaction engineering is a required course for a master of bioengineering, and it is also a comprehensive interdisciplinary course. This course involves a wide range of knowledge, reflecting the knowledge community of basic contents such as biology, mathematics, and engineering design.

3.1 Optimize theory course content

Based on the study and analysis of a variety of related textbooks, the theoretical content system of the course was reconstructed. Combined with the scientific research situation of the biology major and the talent training objectives of students, the theoretical content system of the course was optimized and screened. The main aspects of the course content were grasped, and different content modules were collected, such as sterilization of culture medium. In this paper, the process of biological reaction, such as sterilization of air, scale-up of reactor, aeration and agitation, kinetic model of biological reaction process, etc. were combined with the latest subject frontier knowledge to enrich, adjust, and update the content of theoretical teaching, so as to achieve "constant, often new, talk."

3.2 Introduction of practice teaching

On the basis of theoretical teaching, in addition to finding opportunities to guide students to visit the practice and practice base, the school laboratory resources are used to design experiments for engineering practice operation according to the difficulties and key points of theoretical learning. Under the guidance of teachers, while verifying the theory, they fully understand the biological reaction research in laboratory, and realize the difference between a laboratory reaction and industrial-scale reaction. The practice teaching of engineering calculation runs through the whole process of practical teaching, providing opportunities for students to combine theoretical knowledge with production practice organically, and improve the ability of solving practical problems with professional knowledge.

3.3 Using Internet resources to expand teaching

As MOOC courses and other open network course resources are increasingly complicated, in addition to theoretical teaching content, they are also accompanied by a large number of virtual simulation resources online. Combined with the difficulties and other issues of theoretical teaching, full use is made of the advantages of Internet resources, so that students can repeatedly learn and carry out in-depth discussion. This paper analyzes the shortage of resources in our college, and guides students to choosing high-quality online courses as supplementary material. At the same time, teaching resources are divided into different levels to provide multilevel learning content for students with different foundations, which not only meet the basic objectives of professional training, but also achieve high-level needs (He & Luo, 2020)

4 REFORM AND INNOVATION OF THE TEACHING EVALUATION SYSTEM

The assessment of postgraduate courses adopts a variety of assessment methods, which not only introduces the open teaching evaluation system, but also ensures the participation of experts in the field of professional evaluation.

4.1 *Active learning and self-evaluation of the seminar*

At the end of a chapter, students are organized to have a regular discussion. The students are divided into groups according to their own free groups, and each group is assigned a subject closely related to this chapter in the field of biological reaction engineering and the latest frontier field of bioengineering. Students consult and summarize 3–5 pieces of Chinese and foreign literature, make PPT, and select a representative from each group to speak on the stage. According to the scoring standard set by the teacher in advance, the students and teachers grade and comment on the students on the spot as part of their usual performance. This not only exercises the students' ability to consult literature and summarize knowledge, but also trains the students' oral expression ability, and broadens everyone's knowledge. Because the students in the preparation process will have a full understanding of this content, the knowledge learned is better than the effect of teachers directly inculcating knowledge in class. This form of evaluation is welcomed by students. After the students on the stage were finished, there were often heated discussions among the students. Some questions were raised at a great level and insight. The interaction between students and teachers was strong, and the effect was very significant. The expected effect of the teaching evaluation system reform was received.

4.2 *Defense and expert evaluation in practice*

The assessment of practical teaching is not limited to the simple basis of the experimental report, which requires students to seriously think about the experimental practice, summarize the achievements and existing problems of the experimental practice link, and analyze the reasons for the achievements and deficiencies, so as to play their strengths and make up for the deficiencies in the future study. The examination of the experimental practice link is evaluated in the form of addressing the expert's reply. Teachers of relevant courses are invited to serve as experts and as members of the defense committee to give their scores together. This can avoid the situation that the impression score and fairness of a teacher are questioned as happened in the past, and can deepen the understanding of relevant professional issues and strengthen the application in engineering practice through replies and expert opinions. It strengthens the opportunities for teachers and students to contact each other, helps students understand the scientific research achievements of teachers, stimulates students' enthusiasm for further professional learning, helps in continuously understanding the direction of scientific research in the field, broadens students' vision, and is beneficial to cultivating high-quality applied innovative talents (Wu, 2015)

4.3 *Weaken knowledge assessment in written examination and strengthen knowledge application ability assessment*

In order to improve the difficulty of writing the final examination paper, we should change the previous inherent question types, such as filling in the blank, selecting and answering questions, and drawing up questions that focus on the use of knowledge. We should replace simple recitation with analysis, application and knowledge reorganization. Students are allowed to choose the questions with different difficulty coefficients for assessment, which can better coordinate the contradiction between students with different learning abilities and unified test papers. As the course is also an elective course for chemical engineering and biology, it also adopts the method of "examination question selection" which can take into account multiple majors.

The practice in the past five years has shown that the above methods can solve the contradiction that the comprehensive ability of students cannot be fully reflected in the course assessment. This kind of evaluation system has changed the traditional concept of "one exam determines one's life" in the final examination, and obtains the final score by the students' ability improvement performance; it has changed the "cramming teaching mode" of "cramming the whole room" to improve the students' learning initiative; it has changed the traditional examination mode of "one volume examination for a whole year" providing more assessment options and is more in line with the requirements of quality education and innovation education; fully exercising the spirit of team cooperation and collaborative innovation of students, and laying a solid foundation for students to enter production and scientific research in the future.

5 SUMMARY

Advanced bio-reaction engineering is a practical engineering course for postgraduates. This kind of course combined with practice teaching is a basic requirement of the engineering teaching system and the basic guarantee of cultivating new-type technical talents. We should make full use of Internet resources, strengthen the optimization of theoretical teaching content, reform the traditional theoretical teaching mode, explore multiple channels to change the teaching content, teaching methods, practical teaching links and performance evaluation system, pay attention to the cultivation of students' innovative thinking and engineering views, and cultivate contemporary senior bioengineering professionals who can undertake the major tasks and needs of the country in the new era. We must be in a state of constant innovation practice and teaching reform with the traditional education ideas and concepts of collision, and use some literature of the most cutting-edge professional knowledge to reform our teaching styles.

ACKNOWLEDGMENT

This research was financially supported by Laboratory Research Projects in Beijing Institute of Technology (No. 2019BITSYC12) and Comprehensive Reform Project of Graduate Education and Training of BIT.

REFERENCES

He, L. & Luo, X. 2020. Teaching reform of Bioreactor Engineering under the background of "Internet plus". *Modern agricultural science and technology* 6 (1): 256–258.

Meng, T. & Wu, J. 2006. Application of innovative thinking and engineering viewpoint in teaching of biological reaction engineering. *Higher education of chemical engineering* 92 (6): 91–93.

Wu, X. 2015. Teaching reform and innovation of biochemical engineering course based on the cultivation of applied talents. *China Light Industry Education* 5 (6): 76–78

Computational Social Science – Luo, Ciurea & Kumar (eds)
© 2021 Taylor & Francis Group, London, ISBN 978-0-367-70193-2

Research on the operation mechanism of training quality of economics and management graduate students based on system dynamics

L.C. Li, M.R. Fu & J.J. Xiao
Guilin University of Electronic Technology, Guilin, China

Y.Q. Li
Institute of Information Technology of GUET, Guilin, China

S. Ou
Guilin University of Electronic Technology, Guilin, China

ABSTRACT: This paper aims to improve the quality of graduate education in schools of economics and management, and to build an efficient mechanism for graduate training. By using the method of system dynamics, through the investigation and analysis of the operation mechanism of various universities, this paper reveals the factors influencing the quality of postgraduate training and their causal relationship. The results show that the quality of postgraduate training is mainly affected by the quality of tutor, enrollment and selection, classroom teaching quality, thesis quality, graduate innovation ability training, and graduate "double I" education. At the same time, the study provides countermeasures and suggestions for each university to improve the quality of postgraduate training in economics and management.

1 INTRODUCTION

After the Second World War, all countries began pursuing the progress of science and technology and economic development, with international education competition becoming increasingly fierce. China has also kept pace with the developments of the era and continuously deepened its higher education system. Especially in recent years, the establishment of "985" and "211" colleges and universities and the construction of "double first-class" universities demonstrate the importance China attaches to higher education. As the highest level of higher education, this determines the overall core competitiveness of China in the future. Therefore, improving the comprehensive ability of graduate students has become an important goal of domestic education, which is of great significance to the research into the quality of graduate education.

At present, domestic scholars' research into the quality of graduate education is mostly carried out in the form of regression analysis using questionnaires. They analyze individual factors that play a key role in the quality of education in this form, without systematic analysis of the influencing factors. Bin Wu et al. (2009) and Hexia Bi (2014) divided the factors affecting the quality of graduate education in universities into external constraints and internal control factors. Meanwhile, Hexia Bi further put forward reference evaluation factors. Yan Liu (2012) pointed out that the current studies into this aspect still focus on the analysis of the influencing factors, and the most important influencing factors are tutors, enrollment, curriculum arrangement, and experimental conditions. In June of the same year, Yan Liu and Xu Fang confirmed that the academic atmosphere had an important impact on the quality of postgraduate education by analyzing the questionnaire contents. Wenwu Liao et al. (2012) considered the quality of graduate education from five aspects, and put forward countermeasures. Wei Song et al. (2013) took MBA education as an example to discuss the factors influencing the quality of professional degree postgraduate education. Xu Changqing (2017) evaluated the service quality of postgraduate education in Guangdong universities.

In this paper, the system dynamics method is adopted to establish subsystem models from six aspects, namely, the quality of the tutor team, the quality of enrollment and selection, the quality of classroom teaching, the quality of dissertation, the innovation ability training of graduate students, and the "double I" education of graduate students. On this basis, the comprehensive system dynamics model of the factors affecting the quality of graduate education in economics and management was built, and all the influencing factors were studied systematically.

2 THE SYSTEM DYNAMICS MODEL OF THE FACTORS AFFECTING THE CULTIVATION QUALITY OF ECONOMIC AND MANAGEMENT GRADUATE STUDENTS

System dynamics (SD) is a science that closely combines system science theory with computer simulation and studies feedback structure. Feedback is the core of system dynamics, of which the causal loop diagram (CLD) is an important tool to express the internal feedback structure of the system.

In this paper, through the design of questionnaires, field visits, and other methods, the operational measures adopted by colleges and universities to ensure the quality of the training of economic management graduate students were investigated. It was found that the graduate student training quality system is a complex system, which is affected by many factors, and these factors involve many levels. Therefore, by using system dynamics to study the system, the main influencing factors within the system will be uncovered more clearly and intuitively, and targeted opinions and suggestions will be put forward according to these influencing factors. At first, this paper reveals the main factors influencing the quality of graduate education and the causal relationship between each, then the quality of graduate education, the operation mechanism of a large system is divided into subsystems "4 + 2" (that is, the quality of the mentor team, the enrollment of students to select quality, classroom teaching quality, quality of degree thesis, postgraduate student innovation ability training, postgraduate student "double I" education), and by using a Vensim causal loop diagram of the six subsystems build, as shown in Table 1.

2.1 Causal diagram of the subsystem of the mentor team quality

The quality of the tutor team is the primary factor affecting the quality of postgraduate training, especially in the less innovative fields of humanities and social sciences such as economics and management, the development of postgraduate students is often directly related to the ability of tutors. As one of the subsystems that influence the quality of graduate student cultivation, the study on this system is helpful to further grasp the factors that influence the quality of graduate student cultivation. Based on the previous studies, this paper determines the causal relationship of the subsystem of tutor team, as shown in Figure 1.

A causal loop diagram is composed of multiple variables, and the arrows between the variables represent the causal relationship between the two. Each causal chain has polarity, either positive $(+)$ or negative $(-)$. The positive causal chain indicates that the cause increases and the result also increases; a negative causal chain means that if the cause increases, the result must decrease. Figure 1

Table 1. Operation mechanism of training quality of economic management postgraduate students.

Content	"4 + 2" training mode for postgraduate students			
Process training	SD model of mentor team quality	SD model of enrollment and selection quality	SD model of classroom teaching quality	SD model of quality of degree thesis
Competency enhancing	SD model of postgraduate student innovation ability training		"Double I education" of postgraduate	
			SD model of postgraduate information education	SD model of postgraduate internationalization education

670

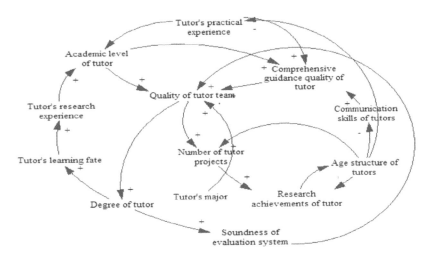

Figure 1.   Causal loop diagram of the subsystem of the mentor team quality.

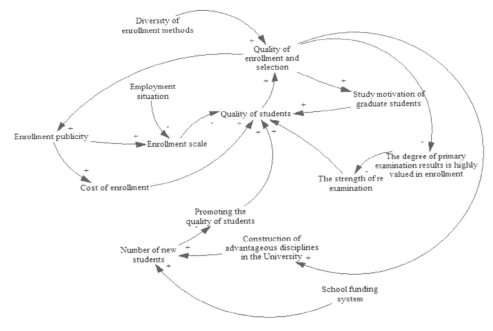

Figure 2.   Causal loop diagram of the subsystem of enrollment and selection quality.

mainly reveals the seven major feedback loops that can be found in the system of the main factors influencing the quality of graduate education, teacher education, teacher's edge, tutor's research experience, academic level, tutor's topic number, scientific research, teacher's age structure, tutor's practical experience, communication skills, and tutor's comprehensive guidance quality.

2.2   *Causal diagram of the subsystem of enrollment and selection quality*

The cultivation quality of graduate students in economics and management is closely related to their own qualities. Only with certain skills and avant-garde thoughts in line with the era can the external cultivation of graduate students, such as supervisors and schools, play a greater role. The causal relationship of the quality subsystem of enrollment and selection is shown in Figure 2.

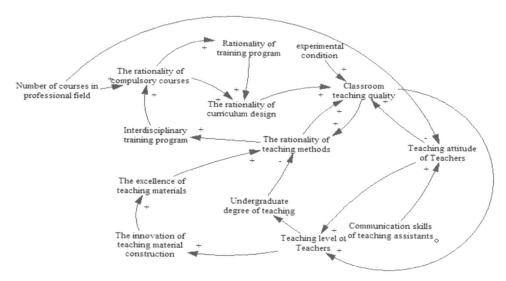

Figure 3. Causal loop diagram of the subsystem of classroom teaching quality.

There are five feedback loops for the causality of the subsystem of enrollment and selection quality. The circuit shows the system of the main factors influencing the quality of graduate education, the major has its own motivation, quality of students, enrollment of postgraduates on first attempt at the achievement level, the level of second interview investigation, the enrollment of students' propaganda, enrollment costs and scale of the enrollment, school advantage discipline construction, push from raw numbers, and push from living quality of students. At the same time, the quality of enrollment is also affected by the diversity of enrollment methods.

2.3 Causal diagram of the subsystem of classroom teaching quality

Apart from the tutor's guidance and the postgraduate's own quality, the main course guidance for the academic achievements during the postgraduate period also plays a very key role. Therefore, the analysis of the classroom teaching system will also be conducive to the analysis of graduate student training quality. The causal relationship of the classroom teaching quality subsystem is shown in Figure 3.

Figure 3 reveals nine positive-feedback loops. It shows that the subsystem main factors affecting the quality of graduate education have the rationality of the teaching methods, teaching the teachers teaching level and teaching of "excellent structuring degree," the construction of teaching material innovation, the excellence of the teaching material content, interdisciplinary training plan, the rationality of the compulsory courses, the rationality of the curriculum, and the rationality of the solution. At the same time, the quality of classroom teaching is also affected by experimental conditions.

2.4 Causality diagram of the subsystem of the dissertation quality

The quality of the dissertation is the most intuitive reflection of the training quality of postgraduates in economics and management. At present, most universities still judge the training quality of graduates based on the results of the dissertation. Analyzing the subsystems of the results of graduate training quality will more intuitively discover the various factors that affect the quality of graduate training. The causal relationship of the dissertation quality subsystem is shown in Figure 4.

There are five positive-feedback loops in the causal relationship of the dissertation quality subsystem. It is not hard to see that the main factors affecting the quality of graduate education are a tutor for grant funding, student, school textbooks and journals, school library information service resources, teacher having the opportunity to participate in scientific research, graduate academic atmosphere, the rationality of the tutors providing advice on the academic, and mentor concerned

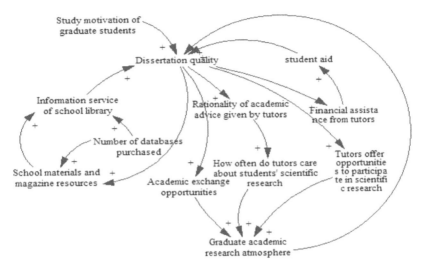

Figure 4.   Causal loop diagram of the subsystem of dissertation quality.

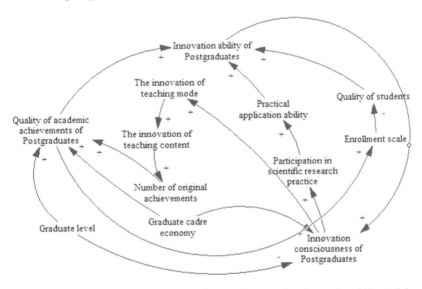

Figure 5.   Causal loop diagram of the subsystem of postgraduate student innovation ability training.

about the frequency of students' scientific research. At the same time, the quality of a graduate's dissertation is also affected by the motivation of the graduate student.

2.5 *Causality diagram of the subsystem of postgraduate student innovation ability training*

The biggest difference between graduate students and undergraduate students lies in the independent analysis and innovation ability of graduate students. Therefore, the cultivation of graduate students' innovation ability is an important level to reflect the quality of graduate students' training. The causal relationship of the innovation ability cultivation subsystem of graduate students is shown in Figure 5.

There are three positive-feedback loops in the cause-and-effect relationship of the training of graduate students' innovation ability subsystem. The three positive-feedback loops show that the main factors affecting the quality of graduate student training in this system include graduate students' consciousness of innovation, participation in scientific research practice, practical

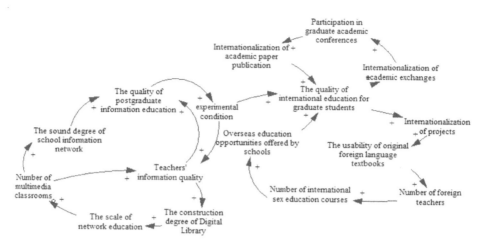

Figure 6.   Causal loop diagram of the subsystem of postgraduate students "double I education".

application ability, innovation in teaching mode, innovation in teaching content, quantity of original achievements, quality of graduate academic achievements, enrollment scale, and quality of student sources.

2.6   Causality diagram of the subsystem of graduate students "double I education"

Nowadays, globalization and informatization have increasingly become the mainstream trend. As the main force of modern scientific and technological innovation, the development of graduate students should conform to the trend of the time, which requires that the training mechanism of graduate education should also carry out the education of relevant contents. "Double I education" refers to postgraduate information education and postgraduate internationalization education, whose purpose is to train the country with informatization and new talents that can be in line with the international internationalization. The causal loop diagram of the "double I education" subsystem of graduate students is shown in Figure 6.

There are two main positive-feedback loops about the quality of information education for graduate students in the causality diagram of the "double I education" subsystem of graduate students, and there are also two positive-feedback loops about the quality of international education for graduate students. These four positive-feedback loop shows that graduate students "double I education" subsystem of the main factors affect the quality of graduate education, including the experimental conditions, the degree of teachers' information quality, the construction of digital library, networked education scale, the number of multimedia classrooms, school information network, and improve the degree and the internationalization degree of academic exchange, the graduate student academic conference participation, academic papers published in the texts of the original internationalization and the internationalization of the project, usability, the number of foreign teachers, the number of international education courses, the school provides opportunities for overseas education.

2.7   Construction of the comprehensive system dynamics model

Due to the quality of the graduate student training mechanism of administration by the mentor, recruit students, teaching, and many other factors mutual influence, the "4 + 2" is integrating the system into effect in the large system of the operation mechanism of graduate student training quality, using system dynamics modeling tools to build "4 + 2" is the system of graduate student training quality of administration system dynamics model of operating mechanism. Figure 7 not only describes the internal causal feedback loop relationship between the seven subsystems of tutor team, enrollment and selection, classroom teaching, dissertation, graduate student innovation

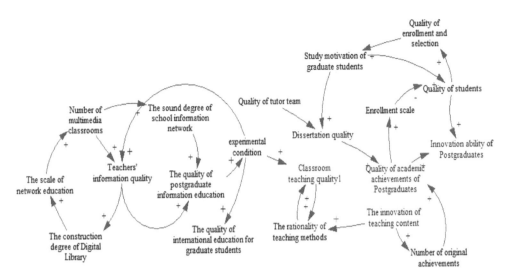

Figure 7. Establishment of a comprehensive system dynamics model for factors influencing the quality of postgraduate student cultivation.

ability training, and graduate student "double I education" training, but also reflects the correlation between each subsystem. The factors influencing the quality of graduate student cultivation are a relatively complex system, with direct or indirect relations among the factors. Therefore, the causal loop relationship among the factors can be more intuitively demonstrated through the establishment of the comprehensive system dynamics model, and targeted suggestions are put forward to improve the operation mechanism of the quality of graduate student cultivation of economics and management.

3 SUGGESTIONS FOR IMPROVING THE TRAINING QUALITY OF ECONOMIC AND MANAGEMENT GRADUATE STUDENTS

3.1 *Strengthen the tutor responsibility system and establish a tutor team in the mode of scientific research collaboration*

Each university should build a dual tutor mechanism of leading teacher and deputy tutor, internal tutor and external tutor, and strengthen the qualification examination mechanism of tutors For example, under the major of MPAcc, there are three mentor teams according to the research directions, namely, the mentor team of accounting informatization, the mentor team of financial and financial big data, and the mentor team of management accounting.

3.2 *Improve the quality of classroom teaching and ensure the quality of academic theses*

Each university should focus on the quality of classroom teaching, from the teaching outline, teaching plan, case teaching method, guiding teaching mode, and other aspects, in order to improve the quality of postgraduate classroom teaching. At the same time, they should adopt the mode of interdisciplinary training and introduce the research method of engineering into the field of liberal arts.

3.3 *Train graduate students' innovation and internationalization ability*

The arrival of the era of big data has caused a shortage of new intelligent talents, and so colleges and universities should seize this opportunity to enhance the innovation and informatization capacity of graduate students. Therefore, colleges and universities should hold a series of activities for postgraduates' academic salons at the second level. At the same time, corresponding supporting facilities should be established based on the Internet +, big data, cloud computing, and information technology environment, information teaching should be introduced, and postgraduate information technology mastery and application ability should be improved.

4 CONCLUSIONS

4.1 *The quality of postgraduate education is affected by many factors*

In the process of postgraduate training, there are many factors that determine the quality of training. The key factors are the tutor team, the quality of students, the quality of teaching, the quality of academic paper training, and the quality of students' informatization. In the future, colleges and universities should consider from multiple perspectives when formulating strategies to optimize the quality of postgraduate training.

4.2 *The influencing factors of postgraduate training quality are interrelated*

By analyzing the feedback results of the above subsystems, it can be found that the tutor team, the quality of students, the quality of teaching, the quality of academic paper training, and the quality of students' informatization are interrelated. A change to one of these factors will lead to a change in the feedback loop of the remaining subsystems, which will have a comprehensive impact on the quality of graduate education.

4.3 *The tutor team is still the main factor affecting the quality of postgraduate training*

In the process of postgraduate training, tutor's guidance plays an important role. The quality of the tutor team is related to the quality of the dissertations and the development of students' comprehensive ability. The construction of a tutor team remains the key to improving the quality of postgraduate training in colleges and universities.

ACKNOWLEDGMENT

This research was supported by Innovation Project of Guang Xi Graduate Education (Nn. JGY2018057, JGY2019076, 2016XWYJ05).

REFERENCES

Bi, H. (2013). Analysis of influencing factors of graduate education quality assessment – Based on empirical research on teachers and students of Universities and colleges in X Province. China Association of Higher Education, 9.

Liao, W., Chen, W. & Guo, D. (2012). Analysis and countermeasure research on the influencing Factors of graduate education quality. Graduate Education Research, (02), 11–14.

Liu, Y. (2012a). An empirical study on the influencing factors of graduate education quality – a case study of five universities in Hebei province. Degree and Graduate Education, (06), 55–57.

Liu, Y. (2012b). Research on factors influencing the quality of postgraduate education in China based on content analysis. Journal of Hebei University (Philosophy and Social Sciences edition), 37(04), 40–44.

Song, W., Zhou, H. & Chen, C. (2013). A study on the factors influencing the quality of professional degree postgraduate education – a case study of MBA education. China Higher Education Research, 3(02), 46–50.

Wu, B. (2009). An Analysis of the Influencing Factors of graduate Education quality. Higher Education Studies in China, 10, 29–31.

Xu, C. (2017). Evaluation of graduate education service quality and analysis of influencing factors. Higher Education Development and Evaluation, 33(05), 30–49, 114–115.

Zhong, Y. (2013). System Dynamics (Second Edition). Science Press. Beijing.

Computational Social Science – Luo, Ciurea & Kumar (eds)
© *2021 Taylor & Francis Group, London, ISBN 978-0-367-70193-2*

Research on the problems and countermeasures of innovation ability training mode of economic and management majors

B.B. Yu
Capital University of Economics and Business, China

ABSTRACT: Innovation education is an important of modern higher education, particularly the specialties of economics and management. The implementation of entrepreneurship education for students can give students the ability to integrate business resources before entering the society, which is beneficial for students looking to find new business opportunities. Based on the necessity of implementing innovation education, this paper explores the practical mode of innovative ability training for college students majoring in economics and management.

1 GENERAL INSTRUCTIONS

Innovation and entrepreneurship have become an inevitable choice for modern economic development. In September 2014, Premier Li Keqiang issued the slogan of "mass entrepreneurship, mass innovation," which set off a new period of "innovation by all and in-novation by all." In the development of higher education, the cultivation of innovative spirit, innovative ability, and entrepreneurial ideas has long been put forward. The outline of the national 12th Five Year Plan for education development clearly puts forward "adhere to people-oriented, comprehensively implement quality education, strive to improve students' sense of social responsibility, innovative spirit and practical ability, and constantly meet the needs of economic and social development for talents and comprehensively improve the national quality." In the outline of the national medium and long-term education reform and development plan (2010–2020), it is mentioned that "top-notch innovative talents" are cultivated in higher education. Although it is impossible for ordinary colleges and universities to cultivate such talents, it provides ideas for the talent cultivation of ordinary colleges and universities, in the training of applied talents. We should highlight the cultivation of students' innovative spirit and ability, that is to say, to cultivate applied innovative talents. This kind of talent not only has good professional knowledge, skills, and quality, but also has good innovation ability training, especially the cultivation of innovative ideas, innovative spirit, and innovative ability. Therefore, for the students majoring in economics and management, schools should pay more attention to training them to have solid theoretical knowledge of economic management, skilled professional skills and good professional quality, be able to creatively solve the practical problems of economic management with the learned theories, and have the spirit of innovation and innovation ability.

2 THE NECESSITY OF INNOVATION ABILITY TRAINING FOR COLLEGE STUDENTS MAJORING IN ECONOMICS AND MANAGEMENT

Colleges and universities have the characteristics of outstanding school running and employment positioning in the social market. As far as the employment rate of students is concerned, it is not lower than that of undergraduate colleges. However, in recent years, with the downsizing of relevant national departments and the layoffs of various enterprises, colleges and universities, especially the employment methods and channels of colleges and universities, have been greatly impacted. The school-based colleges and universities are in a weak position, especially in the employment of the graduates of economics and management, who will face greater pressure. The cultivation of students' independent entrepreneurial ability is an important way for college students majoring in economics and management to get out of difficulties and find jobs. This kind of colleges and universities should take professional knowledge as the standard, and implant the awareness and content of entrepreneurship education into students' teaching and learning, so as to combine education theory with teaching practice, and explore the cultivation mode of innovation ability for students majoring in economics and management. At the same time, we should strengthen the research of entrepreneurship theory and practice, and successfully apply the good research to the promotion of entrepreneurial ability. Only in this way can we truly realize the transformation from employment education to entrepreneurship education and completely change the situation of employment difficulties of economics and management majors in colleges and universities.

First, innovative practice teaching cultivates students' innovative quality. For students majoring in economics and management, theoretical teaching can only cultivate students with broad basic knowledge and reasonable knowledge system structure, but these are only limited to books. Innovative practice teaching can help students apply theory to practice, solve practical problems, deal with a series of difficulties in the real business process, enhance students' innovation consciousness, and stimulate their curiosity and exploration spirit. At the same time, in the process of solving problems, there will be many complex problems. Students need to use interdisciplinary knowledge, experience, judgment, and many cutting-edge theories, which can help students to actively think, develop ideas, and find innovative new insights and methods. In particular, compared with the students of science and engineering, the practice link of economics and management majors is relatively lacking. Innovative practice teaching can make up for the gap between theory and reality and help these students make a good transition before entering the workplace.

The second is the systematic and perfect practical teaching system to improve students' innovative ability to solve problems. Students' innovative ability is reflected in their creativity in solving problems, especially for students majoring in economics and management. The core of management work is to solve the problems of human, material, and financial resources. For example: a human resources major needs to coordinate the relationship between various positions, reasonable performance appraisal, recruitment of excellent employees, training of employees, and so forth, in these jobs. The most important thing is to solve the problems encountered by employees, which requires students to be able to use the theory to solve these practical problems. Through the systematic and perfect practice teaching system, students need to experience all aspects of their professional posts in practice, and they will have the opportunity to encounter many practical problems. This will urge them to learn to actively seek solutions to problems, solve these problems creatively, and effectively improve students' innovative thinking and innovative ability.

Third, the practical teaching system integrated with scientific research thought can enlighten students' academic innovation thought. The goal of undergraduate students majoring in economics and management is to cultivate applied talents, but there are many links in the process of practical teaching that need to use experimental research, questionnaire surveys, and other research methods. These practical links play an enlightening role in students' academic innovation ideas. In the practical link of marketing major, students need to use experimental research methods to compare consumers' perception of new product development ideas in different ways. These practice links enable students to have a preliminary understanding of experimental research methods in marketing research, and not only train students' innovative and practical ability. At the same time, it also inspires the students' academic innovation ideas. These practical contents play a very important role in the cultivation of students' academic innovation ability.

3 PROBLEMS IN THE CULTIVATION OF INNOVATION ABILITY OF STUDENTS MAJORING IN ECONOMICS AND MANAGEMENT

According to the requirements of the national education development plan and the training objectives of application-oriented talents, colleges and universities regard practice teaching as an important way to cultivate innovative ability. In recent years, the proportion of practical teaching in the training plan is increasing, and the investment in practical teaching is also increasing. However, the cultivation of innovation ability through practice is far from enough, especially in the current education system of our country, as there are many problems in the cultivation of students' innovative spirit and innovation ability, which needs to be changed urgently. Therefore, it is necessary to analyze the main problems existing in the cultivation of innovation ability of students majoring in economics and management.

One is that the education system restrains students' innovative thinking mode. Modern education has long adopted the method of finding a "correct" answer to a problem to impart knowledge, which has long restrained students' innovative thinking. College students, especially liberal arts students majoring in economics and management, are under the pressure of entrance examination in the process of receiving primary, junior, and senior high school education. In terms of teaching content, the examination content is the goal, the book knowledge is highlighted, and the test paper is the core. The students mainly accept the knowledge information related to the teaching materials and the examination syllabus. In terms of teaching methods, it mainly cultivates students' examination skills, and trains students' examination ability through exercises. This way of education has restrained the all-around development of students and lacks the cultivation of students' main body and initiative innovation spirit. After entering the university, due to the influence of previous learning methods, students still take examination as the main learning method, and only pay attention to textbook knowledge, have lack of attention to practical and innovative courses, and their innovative thinking and entrepreneurial consciousness are very weak.

Second, the school lacks a systematic innovation ability training system. Innovation comes from practice and is higher than practice, so the cultivation of students' innovative thinking and innovative ability must be based on practice systems, but only relying on the existing practice teaching is not enough to achieve this goal. The current higher education pays increasing attention to practical teaching, and the proportion of practical teaching in the teaching process is also increasing. The practice teaching in many colleges and universities exceeds 30% of the total credits. However, the lack of systematic, comprehensive, and scientific consideration in the process of setting up practical teaching links is often accompanied by the establishment of corresponding practical teaching courses for a certain course to train students' practical ability. These practical courses are independent and cannot connect professional knowledge points into a complete system. Therefore, although students have learned professional theoretical knowledge and carried out corresponding practice, after entering the job, there is still lack of overall understanding of the business, and it is difficult to integrate into the work position quickly.

Third, teachers' lack of innovative thinking mode. For the cultivation of students' innovative thinking and innovative ability, teachers' innovative thinking ability directly affects innovative education. If the teacher lacks the spirit of innovation in the teaching process, he can only lead to the students' rote memorization, test preparation, and lack of interest. First of all, the existing institutions of higher learning, especially ordinary institutions of higher learning, assess teachers' instruction level by level, including expert evaluation, peer evaluation, and student evaluation of teaching, and most of these evaluations rely on the evaluation of teachers' classroom teaching links. These complicated evaluation systems make teachers mainly impart book knowledge and they are unwilling or afraid to innovate teaching contents and teaching methods. Second, teachers themselves receive less innovative thinking education ideas. Many young teachers have gone through the process of undergraduate, masters, and doctoral degrees and pay more attention to the study of theoretical knowledge, especially the teachers of economics and management majors who lack practical teaching experience and cultivation of innovative ability. Therefore, the lack of innovative thinking of teachers themselves hinders the cultivation of students' innovative ability.

4 SUGGESTIONS AND COUNTERMEASURES TO IMPROVE THE INNOVATION ABILITY OF STUDENTS MAJORING IN ECONOMICS AND MANAGEMENT

The cultivation of students' innovative ability has become the core competitiveness of colleges and universities, and it is also an important task of modern higher education. Especially, students majoring in economics and management are in a constantly changing and complicated industry environment. This environment provides the soil for the cultivation of students' innovative ability, and puts forward higher requirements for the training mode and training mechanism of the university. Therefore, in the existing social environment, the following suggestions and countermeasures are put forward for the cultivation of innovation ability of economic and management majors.

One is to build a systematic, perfect, and comprehensive practical teaching system for economic management majors. For economics and management majors, first of all, the construction of practice teaching systems of each specialty must be a complete, orderly, and progressive whole, which includes classroom practice of a certain knowledge point, curriculum practice of a course, comprehensive training of professional comprehensive ability cultivation, and social practice of the combination of specialty and society. These professional practice teaching links realize the combination of theory and practice education, the combination of professional technology and innovation and entrepreneurship, the combination of in-class teaching and extracurricular training, and the combination of single skill and comprehensive ability. Second, the practice teaching system of various majors in the school of economic management needs to be integrated according to the training orientation and training objectives of the whole college, and the comprehensive practice system of management specialty can be established, which can not only let students understand the docking and coordination of various departments in the process of enterprise management, but also enable students to have a more comprehensive understanding of management. For example: human resources, accounting, marketing, information management—these majors can set up a comprehensive management platform for simulated enterprises. Human resources are mainly responsible for personnel arrangement, recruitment, and training of new employees, and performance evaluation of various posts. Accounting majors account for production costs and operating costs of simulated enterprises and assist marketing departments in pricing products. Marketing major is mainly responsible for the operation of simulation enterprises, product pricing, product sales, channel design, and product promotion. Information management majors are responsible for the design and operation of the management system of simulated enterprises. The organic combination of these majors can make students fully feel the content and main problems of real enterprise management in practice. The establishment of this systematic and comprehensive practical teaching system provides an important platform for cultivating students' innovative ability.

The second is to strengthen the cultivation of teachers' innovative spirit and ability. The lack of teachers' innovative ability has a great impact on the cultivation of students' innovative ability. It is very important to strengthen the cultivation and exercise of teachers' innovative ability and spirit, as many college teachers enter university teaching as soon as they graduate. Although they have mastered a lot of theoretical knowledge, they seriously lack practical experience. Teaching students can only talk on paper. Therefore, colleges and universities need to establish a teacher growth mechanism. We should strive for opportunities for teachers to go deep into the enterprise, understand the enterprise, and help the enterprise solve practical problems. Whether we take the way of training in the enterprise or doing management consultation for the enterprise, we can realize the cultivation of teachers' innovation ability. At the same time, the cultivation of teachers' innovative ability also needs to constantly broaden their horizons, understand the advanced teaching methods and research methods at home and abroad, and encourage and subsidize teachers to visit and exchange learning in colleges and universities at home and abroad, which can effectively improve the level of teachers in teaching methods and research fields, and then enhance teachers' innovative thinking and innovative ability.

The third is to establish a multi-dimensional innovation ability training mode formed by innovation projects, competitions, and associations. First of all, through the innovation projects of college students, the students of economics and management have experienced such links as experimental design, questionnaire design, project application writing, opening defense, mid-term defense and final acceptance. These innovative projects guide undergraduates to participate in scientific research in the early stage, conduct preliminary training on students' scientific research ability, stimulate their innovative thinking, and improve their innovation ability. Second, students can understand the entrepreneurial process by participating in the college students' entrepreneurship plan project. The school provides entrepreneurship courses, introduces enterprise mentors, and provides entrepreneurship guidance. Students can learn more about entrepreneurship knowledge and stimulate their entrepreneurial enthusiasm by participating in entrepreneurship plan competition and writing business plans. Third, the school carries out a variety of competition activities of economics and management majors, such as simulated recruitment, marketing planning competition, advertising design competition, marketing competition, enterprise management competition, enterprise resource planning competition, and other competitions to create a good atmosphere for students' innovation practice, encourage students' independent innovation, and enhance their practical interest and enthusiasm. Finally, through participating in various community activities inside and outside the school, students' team cooperation ability, communication ability, and ability to overcome difficulties have been cultivated and trained, which are conducive to the improvement of students' innovation ability.

Economic development has entered an era of globalization, and innovative economy has become an important part of economic development. In this era, college students should have the ability of innovation, in order to continuously develop themselves in the demand of economic value, which requires colleges and universities to consciously cultivate college students' innovation ability, especially the students of economics and management colleges, who need more business innovation minds.

REFERENCES

[1] Zhang Dongliang. Research on collaborative management mechanism of scientific research projects in Zhejiang University: a new paradigm based on Process Reengineering. China basic science, 2018, (02).
[2] Wang Peng. Management mode of National Key Laboratory of University: the conflict between ideal and reality. Modern education management, 2010 (12): 55–57.
[3] Yan Zhonghua. Research on technology management of science and technology projects. Science and technology management research, 2003 (6): 17–19.
[4] Journal of Northwest University of Technology (SOCIAL SCIENCE EDITION), 2014, (1): 105–108.
[5] Wang Xiaofeng. Establishing a big science view and innovating interdisciplinary scientific research organization mode. China higher education, 2011 (2): 24–26.
[6] Li Peng, Liu Yan. The development of German scientific research system and Its Enlightenment on the construction of innovation base in China. Scientific management research, 2011, 29 (2): 52–57.

Computational Social Science – Luo, Ciurea & Kumar (eds)
© *2021 Taylor & Francis Group, London, ISBN 978-0-367-70193-2*

Exploration on the construction of the gold course of mechanics of materials in applied undergraduate civil engineering major

H.S. Guo, Y.S. Luo*, Y.X. Cao, J. Liu & Y. Tu
Hunan Institute of Traffic Engineering, Hengyang City, China

ABSTRACT: At present, most of the mechanics of materials teaching has not got rid of the traditional teaching methods, which exists despite the disadvantages of indoctrination. Therefore, it is necessary to speed up the process of the "gold course" of material mechanics. To this, we can start from two aspects: One is the reform of teaching mode, and the another is reform of teaching content. This article puts forward the teaching mode of four-level thinking to cultivate students' four abilities. Meanwhile, we expound on the reform methods of teaching contents from three aspects: using artificial intelligence to individualize teaching; introducing case base, and integration of ideological and political education. Finally, the gold course system of material mechanics for application-oriented undergraduate civil engineering major has been established.

1   INTRODUCTION

Mechanics of materials is an important basic course for civil engineering majors, which includes theoretical mechanics and structural mechanics (Li, 2015). However, most of the teaching of mechanics of materials includes the problem of indoctrination and mass production. The knowledge receiving efficiency of students is lower and is more difficult to put into practice, so that it can't be effectively connected with their major: leading to the learning objectives is not clear; the learning interest is lower, the teaching resources are wasted, and so forth. At the New Era National Undergraduate Education Congress held in June 2018, the concept of Gold Course was proposed for the first time by the Minister of Education Chen Baosheng, and then "gold course" was written into the documents of the Ministry of Education. Later, at the 11th China University Teaching Forum in November 2018, Wu Yan, director of the Higher Education Department of the Ministry of Education, made a report, "Building China's Gold Course" (Lu, 2018; Song and Liu, 2020; Xu et al., 2020; Chen, 2018), so that "gold course" and "water course" became two hot terms in the field of higher education. Therefore, it is urgent to build a gold course in mechanics of materials.

2   PROBLEM IN THE TEACHING OF MECHANICS OF MATERIALS

Mechanics of materials is a course with strong theory and close connection with engineering practice, which contains a lot of contents, concepts, and formulas. Most of the traditional teaching methods aim to tell students how to understand and memorize concepts and formulas, and to train them to be proficient in problem-solving. However, such a teaching method may lead to students' feeling of boring study and weak interest in in-depth studying, being unable to effectively connect the content which they have learned with their major, and easily confused about learning this course, mainly in the following aspects:

(1) Teaching form is simple, reform of teaching methods is slower, credits are reduced and class hours are insufficient. Mechanical credits and class hours are so reduced that the teaching

---

*Corresponding author

task can't be completed well and by only relying on the traditional classroom teaching method, students' interest in learning becomes lower, and the interaction between teachers and students is less.

(2) Students are in different levels, with different interests and hobbies, different learning objectives, and lack of understanding between teachers and students. The situation of undergraduate students in our university is uneven, and the source of students is diverse. There are science students, liberal arts students, vocational college students and college upgraded students. The knowledge base is quite different, and the teaching method cannot be generalized. It is necessary to assign different tasks and impart different knowledge according to the students' background and learning needs, which cannot be fully reflected in the classroom Teachers can neither fully grasp the knowledge level of each student in the classroom nor teach students according to their aptitude.

(3) Students lack interest in knowledge and learning and are confused about the future. Mechanics of materials belongs to the basic course of specialty, and the knowledge points are abstract, which seems far from the professional knowledge. "Why do we study mechanics? Is it related to our major?" "I think mechanics is very difficult. I want to sleep when I pick up a book. What can I do if I'm not interested?" "What are we going to do and what can we do?" Those questions are often asked by students. Some of them have no goal, no direction of learning.

(4) Weak ideological and political education leads to the students' weak sense of social responsibility. As the main body of college students' training, colleges and universities undertake the task of providing qualified talent for socialist construction (Shi and Liu, 2020). How to cultivate contemporary college students to form scientific socialist core values under the new situation has become the main problem faced by ideological and political education in colleges and universities. Nowadays, there are many criminals with high educational background. They use their professional knowledge to break the law, and their values are distorted. With the development of the Internet, there are always many engineering accidents that occupy the headlines, and many human factors involved in these accidents, such as cutting down on materials for their own interests, regardless of other people's lives, harming others and themselves. Alternatively, some students are addicted to chasing stars all day long. Their dream job is to become a star, but they neglect their social responsibilities and pay little attention to the talents in all aspects needed for the construction of the motherland.

Our university is an application-oriented undergraduate college, aiming to cultivate application-oriented high-skilled talents. Compared with the traditional research-oriented undergraduate, we pay more attention to the cultivation of students' practical ability and the ability to solve practical engineering problems. However, the traditional teaching method is still used in most of the classes teaching mechanics of materials. The teaching method is monotonous, and too much time is spent on formula derivation, complex theory explanation, and so on. The traditional talent training mode has not been able to adapt to the actual needs of society, industry, and demand of students. Therefore, it is necessary to practice the college running concept of "training applied talents" in our university and explore new modes, approaches, and methods of talent training to adapt to the teaching reform.

3 REFORM PLAN AND SUGGESTIONS

3.1 *Innovation of teaching mode*

Since 2020, under the influence of the new virus epidemic situation, network teaching has suddenly appeared, showing its unique advantages, gradually being accepted by the public, and applied in university teaching. With the increase in the proportion of application-oriented undergraduate colleges and universities, society has a higher requirement for contemporary college students. The traditional teaching methods and teaching models can no longer meet social needs. It is necessary to reform the traditional teaching methods according to the changes of the times and market needs. After years of teaching practice and exploration of material mechanics, the author has incorporated the WeChat official account, material mechanics simulation software, computer simulation

Table 1. "Four-level thinking" teaching mode.

Four-dimensional	Four-step			
Theory course	PPT	Blackboard	Video animation	Teaching props
Seminar	Typical exercises	Observation cases	Method debate	Experience exchange
Practice course	Simulation test	Verification test	Demonstration experiment	Design test
Extracurricular development	Online learning	Engineering practice	Participation in scientific research	Thesis

software, and massive open online courses into teaching material mechanics. Which is to say, class-room teaching and case demonstration are combined; online and offline learning are combined; textbooks and practice are combined; and reality with virtual simulation are also combined with animation. It increases the teaching effect and students' interest in learning. At the same time, through hands-on operation and hands-on experience, students' interest in exploring unknown things has been stimulated, imagination has been expanded, and their abilities of understanding and innovation have been also improved by the experiment class. Finally, a "four-level thinking" teaching mode focusing on cultivating students' four abilities have been gradually formed, so that good teaching results are achieved.

Cultivate the "Four Abilities":

(1) Development and Innovation Ability: By the ways of the design experiment and the writing of the thesis, giving full play to the advantages of students' flexible thinking, students' development and innovation abilities are cultivated.
(2) Independent Thinking Ability: By teaching methods such as problem-solving, experiment demonstration, method debate, giving full play to the advantages of different students in many aspects, and so on, students' independent thinking ability is cultivated.
(3) Scientific Research Ability: By the ways of taking part in teachers' scientific research projects, students' learning interest, and scientific research ability can be cultivated, and scientific research level can be improved.
(4) Hands-on Ability: By the tests of simulation, verification, and design, students' hands-on ability is trained.

3.2 Innovation of teaching content

At present, the textbooks of mechanics of materials are still compiled for research-oriented learning needs, with various and complex contents. Most of the cases in the books are related to mechanical, aviation, or other aspects, with relatively little contact with the civil engineering specialty, leading to many students being unable to connect mechanics of material with civil engineering and losing their enthusiasm for learning. Due to the teaching reform in colleges and universities in China, and the continuous increase of new knowledge, most of the course hours have been reduced, and all knowledge points can't be explained in class. In view of these situations, the following measures can be taken.

3.2.1 Integrate artificial intelligence education and customize "personalized" teaching

Students usually have different advantages and interests, and education cannot violate the development of human nature. Under the background of the continuous development of artificial intelligence in the new era, undergraduate teaching should not be mass production and indoc-trination teaching. An effective combination of artificial intelligence and the online and offline combination would teach students following their aptitude to achieve accurate teaching. First, investigate and understand students' development plan for future study and work, and then make different online learning according to students' different goals. For this, we can apply the teaching structure of (offline + online) * (compulsory + elective) and stipulate the minimum number of elective hours each student needs to learn.

For example, the contents closely related to the engineering practice, such as internal force, damage, deformation, and instability of components, can be taken as offline compulsory courses. This can be explained in detail through flipped classes and other forms, so that students can understand the relationship among material mechanics, life, and civil engineering; stimulate learning interest; and change from passive acceptance of knowledge to active learning. However, due to the limitation of teaching hours, some knowledge points cannot be explained in detail. For example, some complex contents such as stress state, strength theory, fatigue load, and some formula derivation can be roughly described in the classroom. At the same time, this content will be added to online elective class hours. Students who plan to continue taking the postgraduate entrance examination later can further learn by electing this part of the lessons.

Meanwhile, the simulation test can be added as an elective lesson. Due to the limitation of hardware, site, time, and other conditions, students cannot carry out the test freely, and the knowledge points of stretching, twisting, and bending are relatively abstract. Therefore, through virtual simulation tests, students can consolidate and deepen the theoretical knowledge of material mechanics and cultivate the ability of students to observe phenomena, put forward problems, and solve problems independently.

Therefore, through the teaching structure of (offline + online) * (compulsory + elective), students can master knowledge points and select the knowledge they need according to their interests or needs. Furthermore, through two years of teaching practice, it has been proved that this teaching method is popular with students, stimulates students' initiative in learning, and improves learning efficiency dramatically.

3.2.2 *Establish case base of mechanics of materials*

Materials mechanics is closely related to engineering practice. While on the one hand, most of the cases in the current teaching books are mechanical components, on the other hand, there are too few cases, which is difficult to understand, and easily leads to the disconnection among stress, deformation, other formulas, and engineering practice. Therefore, it is necessary to establish a civil engineering case base (Fan et al., 2018; Deng et al., 2016; Li and Zhou, 2019; Yang et al., 2020; Chen et al., 2018).

However, the ability of the teachers is limited; the establishment of the database needs the joint efforts of teachers and students. Students are not only the builder of the case database but also the users.

First of all, the teacher explains the relevant knowledge points and leads out the related cases. For example, when explaining the axial tension and compression bar's content, the middle link of the suspension bridge will be taken as an example. On the one hand, the theory is combined with the practice to make the abstract knowledge points easier to understand. On the other hand, through the analysis of the actual engineering cases, the students can understand the utility and importance of mechanics of materials, from passive acceptance to active learning.

In order to stimulate students to observe life actively, discover mechanical models, and get interested in learning mechanics of materials, a cases collection can be included in the usual assessment, and their scores can be calculated according to the clicking rate and feedback of the cases by students, to promote students to carefully collect mechanical models and cases while continuously optimizing the case base, to avoid the situation of copying with homework.

3.2.3 *Integrate ideological and political education and improve social responsibility*

General Secretary Xi Jinping pointed out, "we must persist in taking the moral education as the central link, and integrate the ideological and political work throughout the whole process of education and teaching, to achieve full education and all-round education, strive to create a new prospect for the development of China's higher education." This important discussion clarifies the

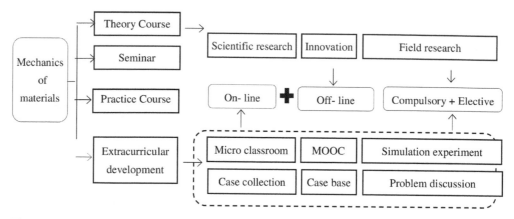

Figure 1. "Golden course" teaching system of mechanics of materials.

task of moral education and human cultivation in colleges and universities, which points out the direction of teaching reform in colleges and universities.

In the teaching process of material mechanics, if the ideological and political education is put into the daily teaching, it is easy to cause students' antipathy, and the teaching effect is just the opposite, so it needs teachers to design ingeniously to connect the ideological and political education to the material mechanics teaching seamlessly (Luo et al., 2015; Ding et al., 2015).

For example, when explaining the basic tasks of material mechanics, the famous buildings in our country can be enumerated, such as Yingxian Wooden Tower, Zhaozhou Bridge, and Hong Kong Zhuhai Macao Bridge. On the one hand, we can strengthen the connection between material mechanics and civil engineering, and find course positioning. On the other hand, we can lead students to realize the glorious history of our country's architecture and enhance national pride. Besides, there are endless engineering accidents. Every year, life is buried in engineering accidents, resulting in a considerable loss of life and financial resources. In the teaching of material mechanics, some related engineering accidents can be interspersed. For example, when the chapter of instability is narrated, students can analyze the collapse accident of Xinjia Hotel in Quanzhou, Fujian Province. By observing the accident characteristics, they can understand the consequences, analyze the causes of the accident, and think about how to avoid the accident in their future work, taking history as a reference, keep the mission in mind, and enhance social responsibility.

Through three years of teaching practice and continuous reform and optimization, a "golden course" teaching system of material mechanics suitable for students of our university has been gradually formed (Figure 1).

4 CONCLUSIONS

The reforming of a gold course of applied undergraduate material mechanics teaching can start from two aspects: the reform of the teaching system and teaching content. The teaching mode of four-level thinking has been established by the authors, aiming to cultivate four abilities of students. Furthermore, in order to meet the learning needs, keep the "freshness" of material mechanics, and increase the social responsibility of the students, the teaching content is simple and profound, and the artificial intelligence education platform is used to establish the (offline + online) * (required + optional) teaching structure. At the same time, cases and ideological and political teaching are adopted. The practice has proved that the establishment of the new teaching system can effectively improve the teaching quality of application-oriented undergraduates and can carry out precise teaching according to the interests and needs of students, which stimulates students' enthusiasm for learning and improves students' practical ability of problem-solving.

ACKNOWLEDGMENT

This research was financially supported by the Education Department of Hunan Province named xjt [2020] No.176; xjt [2020] No. 9; xjt [2019] No. 333; xjt [2020] No. 90.

REFERENCES

Chen, B.S. 2018. Speech at the national conference on undergraduate education in the new era. *China Higher Education,* 3, 4–10.

Chen, Y. & Chen, H.Q. & Pan, J.W. 2018. Exploration and practice of case teaching in mechanics course teaching. *Journal of Machine Design,* 35, 360–362.

Deng, H. & Wang, C. et al. 2016. Applied research of study-discussion case teaching in mechanics course of military academies. *Mechanics in Engineering,* 38, 588–590.

Fan, Q.S. & Yin, Y.J. & Tang, J.J. et al. 2018. Reform and innovation, a decade practice of improvement of the course of the strength of materials. *Mechanics in Engineering,* 40, 543–549.

Ding, K, & Luo, Y.S. et al. Exploration and Practice of New Talent Training Model on Military and Local Cooperation. 2015 International Conference on Education, Management, Information and Medicine, 2015. Atlantis Press.

Li, W.M. 2015. Theory system and logical relationship of mechanics of materials course teaching in civil engineering. *Journal of Architectural Education in Institutions of Higher Learning,* 024, 67–69.

Li, X.L. & Zhou, L.M. 2019. Teaching Reform and Practice of "Mechanics of Materials " Course Based on Flipped Classroom Aiming at " Gold Lessons ". *Science & Technology Vision.*

Lu, G.D. 2018. Harnessing "Water Courses" creating "Golden Courses". *China University Teaching,* 337, 25–27.

Luo, Y.S. & Ding, K. & Jing, S.L. et al. 2015. Thoughts on inter school collaborative innovation and co construction and sharing of excellent resources sharing course of mechanics of materials. *Education Teaching Forum,* 123–125.

Shi, Y.G. & Liu, L. 2020. Exploration of teaching practice of material mechanics based on ideological and political interest teaching. *Guangdong Chemical Industry,* 7.

Song, Y. & Liu, Q.D. 2020. Water Removal and Gold Increase: the Construction Standard and Development Path of "Gold Course" in Colleges and Universities. *Journal of Henan Institute of Education (Philosophy and Social Sciences Edition),* 39–43.

Xu, F. & Fan, J. & Xu, C.G. 2020. Construction of gold course of materials mechanics in provincial universities under the background of emerging engineering education. *Mechanics in Engineering,* 42, 226–231.

Yang, J.N. & Ma, L.S. & Wang, P. 2020. Interactive teaching design and implementation for materials mechanics based on case library construction. *Mechanics in Engineering,* 42, 237–241.

Computational Social Science – Luo, Ciurea & Kumar (eds)
© 2021 Taylor & Francis Group, London, ISBN 978-0-367-70193-2

An analysis of the employment contradiction of tourism management graduates and some solving countermeasures

J.X. Zhang
Business School of Jianghan University, Wuhan, China

ABSTRACT: In recent years, with the rapid development of China's tourism industry, a large number of tourism professionals are in urgent need, but there is a serious dislocation of the supply and demand of undergraduates majoring in tourism management. Tourism management graduates are facing employment difficulties, especially the sudden outbreak of COVID-19 in 2020, which makes the employment of tourism management graduates more severe. Based on a clear understanding of the current situation of employment faced by college students in China, this paper analyzes the contradiction of the employment difficulty of tourism management graduates, and points out that the solution to the employment path of Tourism Management.

1 INTRODUCTION

In recent years, the party and the government are very concerned about the employment of college graduates. At the same time, all fields of society are paying more and more attention to the employment problem of college students. Millions of graduates flood into the society every year, which brings more pressure to the already severe employment situation. It is an indisputable fact that it is difficult for college graduates to find jobs, and the employment of college students in different regions, schools and majors also shows different characteristics. This paper intends to provide some new perspectives and paths to solve the employment problems of college students from the perspective of tourism management graduates.

2 EMPLOYMENT PROBLEMS OF COLLEGE STUDENTS UNDER THE BACKGROUND OF COVID-19

With the further development of China's economic and social system reform, especially since the expansion of college students, the employment problems of college students are increasingly prominent. Through the comparison of the data in the past five years, it is found that since 2015, the number of college graduates in China is still increasing year by year, while the employment rate of college graduates has been declining slowly (Table 1).

2020 is a particular year. The number of college graduates in this year reached 8,740,000, a record high [1]. With the influence of COVID-19, the employment situation of college graduates is more severe. One, 8,740,000 college graduates means that more college students are looking for jobs than ever before, plus the stock of graduates in the past years, which has created more competitive pressures for employment. Two, due to the impact of COVID-19, many units have postponed the recruitment plan, and the recruitment opportunities for campus have also decreased. Third, some enterprises, especially service enterprises, are facing a survival crisis, so it is impossible to attract employment in the near future.

Obviously, the prospect of the employment market of college graduates has been cast a shadow. How to improve the employment rate of college students is a major practical problem facing our country.

Table 1. Statistics of the Number of College Graduates and the Employment Rate of Undergraduates in Recent Years*.

Year*	Number of Graduates (Ten thousand persons)	Employment Rate of Undergraduate Graduates (%)
2015	749	92.20
2016	765	91.80
2017	795	91.60
2018	820	91.00
2019	834	/

*Source: http://www.ncss.org.cn/

Table 2. Main development indicators of China's tourism industry in 2015–2018*.

Year	domestic tourist (100 million persons)	domestic tourism income (100 million CNY)	inbound tourist (Ten thousand persons)	inbound tourism income (100 million USD)	outbound tourist (Ten thousand persons)	total tourism income (trillion CNY)
2014	36.11	30312	12850	1053.80	10728	3.73
2015	39.90	34195	13382	1136050	11689	4.13
2016	44.35	39390	13844	1200.00	12203	4.69
2017	50.01	45661	13948	1234.17	13051	5.40
2018	55.39	51278	14120	1271.03	14972	5.97

*Source: https://www.mct.gov.cn/

3 THE MAIN CONTRADICTIONS RESTRIDTING THE EMPLOYMENT OF TOURISM MANAGEMENT GRADUATES

Since the reform and opening up, China's tourism industry has made great achievements. China has become a big tourist country in the world. Table 2 shows the main development indicators of China's tourism industry in the past five years.

The above data show that China's tourism industry is developing rapidly, and the main development indicators continue to rise, which implies a huge demand for tourism market, and also provides a huge employment space for graduates majoring in tourism management.

But what is the reality? Compared with the severe employment situation in China, the employment situation of tourism management undergraduates is not optimistic. The graduates of tourism management major are faced with the common problems and particularity of other graduates.

3.1 The contradiction between the huge demand of tourism market and the business difficulties of related tourism enterprises

According to the data in Table 2, China's tourism market contains huge potential market demand, but COVID-19 has given us a huge blow.

Since COVID-19 outbreak, China's tourism cconomy has suffered huge losses. Affected by the epidemic, many related tourism enterprises can not carry out business, tourism performance can not be reflected, capital flow has "broken chain", and tourism employees are forced to "lose their jobs". In particular, tourism enterprises such as scenic spots, hotels and OTA have no consumers, and their labor costs, including employee wages, early operation and marketing costs, and other

689

operation and maintenance costs, cannot be recovered. These tourism enterprises are facing huge financial pressure under the state of closure.

It can be said that the whole tourism industry is in difficulty, and some small tourism enterprises may go bankrupt or even face survival crisis. Despite the huge demand of tourism in China, graduates are faced with " difficult to find jobs " or "no job".

3.2 *The contradiction between the huge demand of tourism market and the shortage of tourism talents*

With the huge demand of tourism market, there must be corresponding training system and reserve of tourism talents. After 40 years of development, China has formed a relatively perfect tourism talent training system with doctoral, master, undergraduate, higher vocational and secondary vocational education. According to the national tourism management training statistics in 2017, there are 608 colleges and universities offering tourism management undergraduate majors (mainly including tourism management, hotel management and Convention and exhibition economy and management), and a total of 59,000 undergraduate tourism management majors have been enrolled. In addition, 336 doctoral candidates, 2,832 master candidates 113,000 higher vocational major and 102,000 secondary vocational major and the National Tourism Talent Reserve in 2017 was only 277,168 [2].

However, the actual demand of tourism professionals in China is 8 million, and the talent gap of tourism industry is at least 2 million [3]. This shows that the number of graduates majoring in tourism management can not meet the needs of the tourism market, and there is a large gap of tourism talents. On the one hand, there is a lack of high-end professionals or international talents with high quality, especially the lack of professional and technical talents, service operation and management personnel. On the other hand, due to the lack of low-level jobs and talents in remote areas, many tourism management graduates prefer to work in the city, even at the expense of cross industry employment.

3.3 *The contradiction between the huge demand of tourism market and the lack of specialization of tourism management graduates*

The huge demand of China's tourism market is accompanied by the increasing personalized and diversified needs of tourists, which shows that the tourism industry needs more specialized talents.

Taking research-learning tour as an example, it is a new business in China's tourism market. With the deepening of the concept of quality education and the cross-border integration of tourism industry, the research-learning tour market has great potential. However, there is a lack of specialized talents in research travel. At present, there is a lack of tutors.

According to the relevant data released by the Ministry of Education in 2016, the total number of students in school who meet the age of research-learning tour is about 103 million. At present, there are at least 50 million the talent gap of tutors in China.

In 2018, the report on the development of China's research-learning tour jointly released by China Tourism Research Institute and others shows that the overall scale of China's research-learning tour market will exceed 100 billion CNY in the next three to five years [4].

3.4 *The contradiction between the huge demand of tourism market and the expectation of tourism employment market*

The huge demand of tourism market also indicates the vitality of tourism enterprises and the desire for talents, but the reality is that the requirements of tourism enterprises deviate from the expectations of graduates majoring in tourism management.

On the one hand, tourism enterprises think that graduates are lack of skills and can not find the talents they need. More and more graduates choose as teachers, postgraduate entrance examination or other industries in secondary tourism vocational colleges, and even choose independent employment.

On the other hand, graduates think that the salary given by tourism enterprises is too low, and the salary of tourism enterprises is generally between 3,000 and 5,000 CNY. Because of the expectation deviation between tourism management graduates and tourism enterprises, it is easy to cause their employment difficulties.

4 COUNTERMEASURES FOR THE EMPLOYMENT OF GRADUATES MAJORING IN TOURISM MANAGEMENT

To solve the employment dilemma of tourism management graduates, it needs the joint efforts of the government, schools, enterprises and graduates. As the main part of cultivation, tourism colleges should actively seek the employment path of tourism management graduates, so as to realize the effective docking with tourism market and tourism enterprises. In the near future, we should give full play to the role of counselors, professional teachers and alumni, expand employment channels, organize online training, do the work of graduates, and encourage graduates to cross-border employment. In the long run, we should start from the following aspects:

4.1 *Strengthen the reform of education system and establish a new tourism talent training mechanism*

4.1.1 *Establish a new tourism talent training alliance mechanism*
In order to meet the needs of the tourism industry, under the guidance of the Education Department and the Culture-Tourism Department, tourism colleges, tourism secondary vocational colleges, enterprises related tourism, tourism industry associations, etc. can jointly participate in the establishment of a multi-party deep cooperation talent training alliance mechanism.

4.1.2 *Keep up with the market demand, respond quickly, and cultivate marketable applied talents*
In recent years, new business forms such as parent-child tour, rural tour, ice-snow tour, cruise tour, customized tour, research-learning tour have been emerging in the tourism market, tourism colleges should closely follow the market demand, respond quickly, set up the corresponding professional direction, even the corresponding specialty in Tourism Undergraduate Colleges, and cultivate practical tourism talents suitable for marketing.

4.2 *Let employment education enter the classroom and make employment education run through the whole process of college education*

Tourism Undergraduate Colleges should not only attach importance to the last link of employment education, but should put employment education into the first year of College, and run through the whole process of university education.

We should strengthen the career planning of college students, can invite the elites of tourism industry, human resources experts of tourism enterprises, leaders of human resources and social security departments to give lectures, so as to "check the pulse" of the employment problems in advance, and "consult" throughout the whole process, so as to suit the remedy to students .

At the same time, professional teachers will integrate employment education into classroom teaching to make professional teaching closer to the future market demand.

4.3 *Build a platform to improve the practical ability of college students*

At present, graduates are generally lack of work experience and career experience. Therefore, the colleges must actively adapt to the needs of employers and take the improvement of College Students' practical ability and vocational skills as an important factor to measure the success or failure of education.

The college should take the promotion of tourism management students' professional ability as an important part of the second classroom activities, and strive to build a campus social practice platform, for example "tourism creative design studio".

To further improve the establishment of external "training base", relying on the leading enterprises in the tourism industry, through the cooperation of college and enterprise, college students are selected to regularly participate "practical training" in enterprises, rather than just "taking post practice".

4.4 *Invite tourism elites into the classroom to improve the quality of practical teaching*

We can employ the elites of tourism industry as teachers, and can teach and practice in the college, so as to improve the talent level and professional skills of tourism colleges a, which will help to alleviate the talent demand of enterprises in the future.

5 CONCLUSION

Of course, the employment of the graduates of tourism management major is a systematic project, they also need the policy support of tourism management departments, the revitalization of tourism enterprises' confidence and business recovery, and the correct employment concept of the graduates. Tourism Colleges have a major responsibility.

This paper believes that there are some contradictions in the tourism market, which lead to the employment problems of tourism management graduates, including the contradiction between the huge demand of tourism market and the business difficulties of related tourism enterprises, the contradiction between the huge demand of tourism market and the shortage of tourism talents, the contradiction between the huge demand of tourism market and the lack of specialization of tourism management graduates, the contradiction between the huge demand of tourism market and the expectation of tourism employment market. This paper puts forward the relevant countermeasures to solve the employment of tourism management graduates. In the long run, we should strengthen the reform of education system, establish a new training mechanism for tourism talents, let employment education enter the classroom, make employment education run through the whole process of college education, build a platform to improve the practical ability of undergraduates, employ tourism elites to enter the classroom, and improve the quality of practical teaching.

REFERENCES

[1] Prospective Industry Research Institute. Analysis on the current situation and competition pattern of Chinese college students' employment market in 2020 https://www.sohu.com/a/410193572_473133.2020-07-28
[2] Ministry of Culture and Tourism. Tourism Statistics Bulletin in 2017 https://www.mct.gov.cn/.
[3] Zhang Nan. Analysis of the demand and supply of tourism talent market in China. Tourism industry edition. 2017, 12:223–224
[4] Cloud Growth Research-Learning. In 2020, colleges add tourism category "research travel management and service" major. https://baijiahao.baidu.com/s?id=1647810948356500290&wfr=spider&for=pc.2019-10-19

Computational Social Science – Luo, Ciurea & Kumar (eds)
© 2021 Taylor & Francis Group, London, ISBN 978-0-367-70193-2

Exploration and construction of comprehensive training course based on project oriented

Y. Li
College of Robotics, Beijing Union University, Beijing, China

S. Cong*
College of Urban Rail Transit and Logistics, Beijing Union University, Beijing, China

ABSTRACT: Automation is an engineering specialty, which requires strong practical and application ability. Practical teaching has great influence on the comprehensive quality of students. This paper introduces Automation comprehensive training method based on project oriented. The subject competition projects, the enterprise practical innovation projects and the teachers' scientific research projects are developed as a comprehensive training program, so that it has more practicability and advanced nature. The practical teaching reform greatly improves the width and thickness of the professional practice teaching. Through comprehensive professional training, students can effectively improve the ability to analyze and solve problems as well as practical operation ability, exercise innovative thinking.

1 INTRODUCTION

At present, in the era of information explosion and rapid development of science and technology, the demand for engineering talents is increasing day by day, and the training of engineering talents is also put forward new requirements. As a traditional engineering major, automation needs to transform and upgrade the classic knowledge content and integrate the construction content of artificial intelligence, intelligent control and networking, etc. It is more necessary to strengthen the practice link to cultivate high-quality professional talents with strong engineering practice ability, strong innovation ability and competitive ability.

2 THE BACKGROUND OF PRACTICAL TEACHING

Modeling, simulation, control algorithm design and dynamic characteristics of system analysis are inseparable from mathematics [1]. Data acquisition, processing and control algorithms are inseparable from programming and hardware. Therefore, the automation major requires students to have strong mathematical application ability and software and hardware application ability. From this point of view, automation is a typical applied major combining "hard and soft", involving a wide range of applications, so the teaching of experiment and practice links is particularly important. In recent years we have done a lot of work on the construction of practical environment There are two practical teaching bases on campus and off campus

2.1 *Campus practice environment*

There have advanced Siemens fieldbus control system laboratory and ABB distributed control System Laboratory for major field of control network and control engineering and the modern

---

*Corresponding author

sensor lab, labview experimental system and smart home lab based on wireless sensor network for major field of Information processing and Internet of Things engineering. There are also have a lab for innovation and entrepreneurship Shared by students.

2.2 *Off-campus practice base group*

Relying on high-tech enterprises outside the campus, such as Iron Man, etc., a practice base group outside the campus has been formed. They provide internship sites, graduation design topics for students and form a good industry-school interaction.

2.3 *Existing problem*

A good practical environment in and out of the school provides a foundation for the practice teaching. A whole set of experiments and training project combined with courses have been formed. basic experiments, course design and graduation project of relevant courses have been developed.

However, with the rapid development of control, information, hardware and software technologies, some early practical teaching can no longer meet the actual needs of the current society for professional talents. The practical teaching is in urgent need of updating. More importantly, there are many experiments for a certain course, and few comprehensive course designs with multiple knowledge points integrated at present. Moreover, there are few comprehensive training projects with innovative applications and practical application background.

3 REFORM OF PRACTICAL TEACHING

3.1 *Reform of system*

In order to solve the problems of practice teaching, an integrated and progressive practice teaching system has been developed, as shown in Figure 1. By four-step guide training: cognitive experience for the College freshmen, basic practice for the sophomore, comprehensive training of junior and creative design for senior, it can enable students to realize the transformation of knowledge to ability. The new system of practice teaching embodies the continuity, emphasis and operability.

3.2 *Reform of content*

On the basis of the original curriculum experiment, the content of the comprehensive training has been reformed. The existing training content is revised, and the professional comprehensive training content based on actual projects and cases is developed. The project-guided practical teaching makes students to engage in the technical work and develop the necessary professional abilities in control network, smart home and information management and improve students' engineering awareness, ability to analyze and solve practical problems comprehensively.

Based on the engineering projects of enterprises, the scientific research achievements of teachers, the scientific and technological competition projects of students [2] and the hardware resources of laboratories, the comprehensive practical training projects suitable for cultivating students 'innovative ability are selected and developed. The basic and professional courses are combined to form a comprehensive design and application. In-depth cooperation with enterprises, the research and development of the comprehensive training project of "information collection and processing system based on wireless network" is made. With the research subject as the main body, a comprehensive training program for "industrial ethernet based inspection and control system" is developed. Taking the content of student contest as the main body, the comprehensive training project of "robot control system" is formulated, as shown in Figure 2. The three specialized comprehensive training programs include knowledge points of information collection, communication (wired and wireless), data processing and intelligent control. Each project is implemented by different controllers and communication modes. The comprehensive training programs can be extended

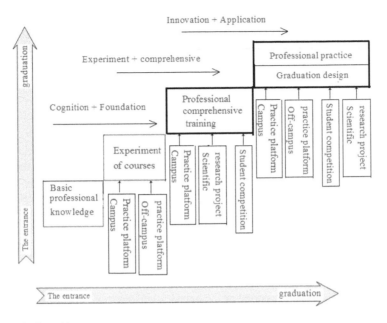

Figure 1. Practical teaching system.

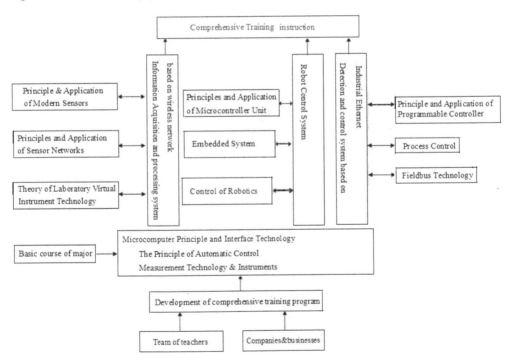

Figure 2. Comprehensive training structure and content.

to process control fields such as temperature, humidity, pressure, flow rate and so on, the field of intelligent detection [3], intelligent home and robot control.

Comprehensive professional training is carried out after the completion of theoretical courses and basic engineering training. It is completed on week 12 of the seventh semester, which reflects the integrity of engineering project learning. By connecting the knowledge points of different into the project, the problem of dividing the knowledge points of different courses can be solved. Moreover, students can choose different professional comprehensive training contents according to their own interests. Students' learning purpose is clear, which improves their comprehensive application ability.

4 CONCLUSION

The reform and implementation of practical teaching provide new ideas and practical project scenes for comprehensive professional training, and realize the integration and transformation of scientific research and teaching. It has the characteristics of strong systematicness, strong engineering, strong expandability and advanced technology, which forms a good atmosphere for students to innovate and apply. In recent years, the students of Automation have achieved very good results in Siemens Industrial Automation Challenge and National Robot Competition.

ACKNOWLEDGEMENT

This research was financially supported by the Industry-University Cooperative Education Project of the Ministry of Education (201801121007)

REFERENCES

[1] Wang Peijin Zhang Wen, Research on automation Specialty Innovation and Engineering Talent Training System under the background of new era. Talent Training & Mechanism Innovation, 2020, 5:5–10
[2] Liu Wen –Sheng Xiong Wei, Practice of Promoting Teaching Reform of Automation Specialty Group through Professional Skills Competition – A case study of Changjiang Institute of Technology. Journal of Changjiang Institute of Technology, 2020.6(37):59–61
[3] Chen Yang, Yang Min, LI Cheng-Fei, Reform and practice of sensor and detection technology course in the background of new engineering. Education Modernization, 2020.5(41):82–84

Computational Social Science – Luo, Ciurea & Kumar (eds)
© *2021 Taylor & Francis Group, London, ISBN 978-0-367-70193-2*

Discussion on content design of safety training of public research platform for biological and medical engineering

L.Q. Sun, M.G. Luo & Y. Cong
School of Life Science, Beijing Institute of Technology, Beijing, China

ABSTRACT: Laboratory safety training has always been an important part of laboratory safety construction. The public platform of biomedical engineering involves many specialties and has the characteristics of interdisciplinary. Its safety training is more complex and difficult than the general laboratory since the different types of lab personnel have different safety training requirements. According to the responsibilities diversity of different categories of personnel, disparate training contents are designed and prepared. This individualized safety training system improves the efficiency and function of safety training, strengthens the professional safety consciousness of laboratory personnel, and reinforces the level of safety construction of public experimental platform.

1 INTRODUCTION

With the increase of national investment in education and scientific research, the construction of laboratories in Colleges and universities is also in full swing. Nowadays, scientific research is moving towards to a new era from single discipline to multi-disciplinary, from single specialty to interdisciplinary. In the laboratory, more and more advanced, large-scale and expensive instruments and equipment are used for different disciplines and research fields. The requirements for the knowledge level, operation skills and management ability of laboratory personnel are also higher and higher. Due to the increase in the number of instruments and equipment and the expansion of laboratory area, the number of laboratory personnel increases, and the requirements of laboratory safety management are also changing. Reasonable allocation of facilities hardware and equipment is the foundation to ensure the safety, however, the safety awareness, professional knowledge and operation skills of laboratory personnel are the key to determine whether the scientific and technological research is safe and effective. Therefore, the level of laboratory safety management is an important part of laboratory construction and laboratory benefit measurement (Tan & Du, 2018). One of the important contents of laboratory safety management is to carry out safety training for laboratory personnel to improve their laboratory safety knowledge, safety awareness and safety operation skills, and laboratory safety access assessment is an important management means (Yuan, 2016). The public research platform of biology and medical engineering is large-scale and interdisciplinary which constructed by our university for the development of integration disciplines on medical industry (Liu & Feng, 2017). The platform supports interdisciplinary scientific research work which focuses on the subjects of electronics, information, optics, biology, medicine, chemistry and materials. There are many kinds of subjects in the platform, involving a variety of scientific research and experimental contents. The traditional laboratory safety training methods are only focused on a small or specific field, and the content of safety training is also restricted around a professional core, which forms a big gap to fulfill the current multi-disciplinary public experimental platform laboratory safety needs. Therefore, we first studied the professional fields and disciplines which involved in the public scientific research platform, and we investigated and analyzed the laboratory safety problems. Then, we conducted the standard basic laboratory safety training

(Yin & Xiao, 2016), following with unique safety training for specific personnel. Last, we evaluated the professional content and learning results (Zhu & Lin, 2019).

There are many personnel whose types are diversity in the laboratory, and each type of personnel has different requirements for laboratory safety. According to the nature of the personnel, we classified the laboratory personnel into six categories, including laboratory principal, laboratory teachers, laboratory full-time personnel, laboratory students, laboratory safety officers and the floating personnel. We established different safety management requirements for different types of personnel, and then designed specific safety training contents (Hong & Zhang, 2017).

2 DIFFERENT PERSONNEL TYPES AND REQUIREMENTS IN THE PLATFORM LABORATORY

2.1 *Persons in charge of the laboratory*

The person in charge of the laboratory mainly refers to the person who is responsible for the large research group. He is very clear about the scientific research content of his research group, understands the details of each experiment in the laboratory, knows the potential danger in his research field, and follows the progress of each teacher and student. The person in charge of the laboratory has sufficient laboratory safety knowledge. As an expert in this field, he can predict the risk type and its possibility of accidents in their research group, he could formulate a perfect safety plan for the research topic, carry out safety training regularly, and more importantly, he should help solving the scientific and experimental problems of the research group, and he could understand the professional laboratory safety management system with details in principle (Li & Zhang, 2019). He are fully responsible for the safety of the research group.

2.2 *Laboratory teachers*

As the main long-term member of the laboratory, the laboratory teacher has sufficient understanding of the experiment content on his laboratory scientific research, and has a full learning of the potential safety hazards and accidents of the scientific research experiment which undertaken by himself, and he can receive safety training regularly and provide laboratory safety guidance to the students. He is clearly about the potential dangerous points of the laboratory, and he can formulate relevant solutions for safety problems. Also, he is matchable to carry out the laboratory safety education for his laboratory students.

2.3 *Full time laboratory staff*

Laboratory full-time personnel refer to the personnel who are responsible for the management and use of laboratory instruments, responsible for the safe running of the laboratory, dutied on the experimental time arrangement and various affairs of his colleagues, and sometimes he is not directly engaged in scientific research and experiment work (Sun & Li, 2018). The full-time laboratory staff should have a full understanding of the instruments and equipment they are maintained, such as how to use, protect and repair; Also he should has a deep learning of all the applications for the equipment, more importantly, he can sense the potential dangers and is capable to give solutions according the specific question. At the same time, he is responsible for the safety guidance and training for students and outsiders to guarantee a basic understanding of safety knowledge, and he can supervise other personnel to carry out experiments according to safety rules and regulations.

2.4 *Students in the laboratory*

In the university laboratory, the students are the main body of the laboratory user and staff, and the students have the characteristics of high mobility, weak laboratory safety awareness. It must first carry out basic safety training for the students before they entering the laboratory, and then carry out targeted safety training for the specific scientific experiment content. Students shoule have a clear understanding of the possible safety accidents and existing safety problems in the scientific experiments, they should receive relevant safety education, participate in laboratory safety access training and examination, and participate in relevant safety exercises to master the methods to solve laboratory safety problems. Every year, they should receive the relevant safety training, and the basic and professional safety training is compulsory to ensure they can pass the examination.

2.5 *Laboratory safety officer*

The laboratory safety officer is the teachers or senior graduate students of a research group usually. In addition to mastering the knowledge which mentioned above, the laboratory safety officer is also encouraged to have a full understanding of the possible safety accidents and dangerous points in the laboratory, they should regularly check the safety situation of the laboratory, organize safety training, and report the laboratory safety inspection results regularly to the laboratory principal and other personnel. He is directly responsible for the platform safety, so he should regularly receive training on laboratory safety of the platform to master the progress of scientific experiments and to evaluate the safety issues of their research group.

The personnel in the laboratory can be basically classified into the above five categories, corresponding to the safety responsibilities of different types. All personnel in the laboratory are required to take the responsibility of laboratory safety learning, training, management and maintance. Once the laboratory safety hazard is found, it can be eliminated in time at the initial stage.

3 DESIGN OF PLATFORM LABORATORY SAFETY TRAINING CONTENT

3.1 *Training contents of full-time laboratory personnel*

The full-time laboratory staff, including the laboratory principal, laboratory teachers, laboratory technical staff, etc., the safety training content should not only include basic laboratory safety knowledge, laboratory safety management theory, laboratory equipment using precautions and safety operation procedures, but also should incorporate the safety department management system to clarify the importance of the safety guidance and education for students and migrant workers (Sun & Li, 2018). In addition to regular safety training carried out by the school, it is also necessary to discuss the safety problems of the scientific research content for the whole research group, and participate in the safety drill and safety plan preview for the full-time laboratory personnel. Also, it is vital to build the safety communication between full-time laboratory personnel and external laboratories, to enhance the safety awareness of full-time personnel in each laboratory (Chen & Zhang, 2016).

3.2 *Training contents for students in laboratory*

According to the characteristics of short time and high mobility of students in the laboratory, the basic laboratory safety training is conducted for the new students in the laboratory. Combined with the characteristics of scientific research, the safety problems of instruments and equipment, reagent materials and experimental methods used in the experiment are discussed together, so as to fully understand the possibility of accidents and the dangerous points in the laboratory. Students should regularly receive safety training from schools, colleges and experimental platforms, finish independent knowledge learning and case analysis through network and lectures, and they can only be allowed to carry out experiments after they passed the laboratory safety access assessment.

3.3 *Training contents of laboratory floating personnel*

Before entering the laboratory, the floating personnel in the laboratory must receive the simplified safety notice of the experimental safety training, which mainly includes the dangerous points in the laboratory and the safety rules when entering the laboratory. In the laboratory, the floating personnel is not supposed to affect the normal lab activity and they should be accompanied and guided by the full-time personnel of the laboratory. They are not allowed to enter the laboratory of the experimental platform without permission, and they are not permitted to operate the laboratory instruments independently. The full-time personnel is responsible for the activity of the floating personnel to avoid safety accidents.

3.4 *Training contents of laboratory safety officer*

Laboratory safety officer is a special group in the above classification, which has the characteristics of great safety awareness, comprehensive safety knowledge and strong sense of responsibility. In addition to the safety training contents mentioned above, it is necessary to strengthen the sense of responsibility of such personnel, and they are encouraged to actively communicate with the laboratory leader or platform safety responsible person in a timely manner, more importantly, they should conduct regular laboratory safety discussion, analysis, and laboratory safety inspection and exchange. Also they need to pay attention to the staff health management of the research group (Yuan & Lu, 2016).

For different types of personnel, it is very necessary to design different contents for training of public experimental platform. Benefited from the design of specific training contents, different types of personnel in the laboratory can have multi-angle and multi-level complementary understanding of laboratory safety, and thereby build a three-dimensional cross network of laboratory safety knowledge in a research group.

4 PLATFORM LABORATORY SAFETY TRAINING MODE

After designing different laboratory safety training contents, how to carry out safety training in various aspects is the main task we are facing. Multi channel training should be carried out from different angles, levels and by means of new media and network. In particular, it is necessary to deal with the relationship between training and learning during normal working hours and rest time, and the relationship between training and learning during normal semester and holiday working hours. The differences in safety training needs in different periods should be met.

4.1 *Training methods should keep pace with the science progress*

Laboratory safety training can not only stay in the old form of classroom teaching. Various media contribute to the effective laboratory safety training, we can carry out the education of safety knowledge through various training software, micro video and micro classroom, thus give full enthusiasm to all types of personnel, facilitates the integrated construction of laboratory safety education and building the laboratory safety culture. For example, we can use WeChat official account and other new means to carry out targeted training in a timely and effective manner, and to overcome the shortages of the timely laboratory safety training. According to the different contents and scientific research fields of different research groups, the specific laboratory safety training modules with different contents were designed. The laboratory safety knowledge was learned and mastered by means of games and competitions, so as to enhance the laboratory safety awareness and improve the laboratory safety skills.

4.2 *Strengthen communication among personnel*

It's important to strengthen the communication between laboratory internal personnel. We can construct the laboratory safety culture by various forms of communication, such as safety knowledge, speech, video, photography and other competitions and exhibitions between platform laboratories are carried out regularly. We should also value the propaganda work of laboratory safety, such as formulate laboratory safety theme day every month, and discuss the safety issues of the platform through website and other media. At the same time, we need to give full rights to the leading and supervising role of laboratory safety officers, and improve the level of laboratory safety management.

4.3 External experience

It's helpful to actively invite laboratory safety professionals and experts from other institutes to the platform for lectures and academic communications. It's also valuable to provide excellent laboratory safety personnel in the platform with opportunities to exchange, learning from other outstanding units. Use the stone of other mountains to improve the laboratory safety construction of the platform. At the same time, we should actively absorb the valuable safety management experience from foreign advanced laboratory and visiting experts. The person in charge of platform laboratory safety should regularly attends relevant academic meetings to improve the management level, they should learn excellent safety training and management experience, and actively improve their own safety management construction. It's a better way to enhance the safety training and managing level by establishing a long-term and good relationship with the public scientific research of other institutions, conducting regular exchanges, learning from each other's strong points.

5 CONCLUSION

The construction of public scientific research service platform in university is the inevitable requirement of scientific research in the new times and it facilitates the development of "big science". The safety training and construction of the platform laboratory is an eternal theme of the platform work. Based on our own characteristics, combined with the practical experience of this platform, we deeply thought about the laboratory safety training, and summarized some preliminary achievements. With the long-term running of the platform, we will further deepen the reform of laboratory safety training and strengthen the construction of laboratory safety culture. We believe that with the joint efforts of all parties, the platform will be able to run smoothly and safely without accidents.

ACKNOWLEDGEMENT

This research was financially supported by Laboratory Research Projects in Beijing Institute of Technology (No. 2019BITSYC12).

REFERENCES

Chen, Y. & Zhang, W. 2016. Design and implementation of laboratory safety management system in Colleges and universities. *Experimental technology and management* 33 (11): 274–278.
Hong, F. & Zhang, T. 2017. Construction and practice of safety access system for teaching laboratories. *Higher education research* 25 (17): 30–33.
Li, Y. & Zhang, W. 2019. Construction of safety education and training system in University Chemical Laboratory. *Experimental technology and management* 36 (7): 232–235.
Liu, T. & Feng, X. 2017. On the construction and operation of public experimental platform of modern agricultural research institutes. *Agricultural science and technology management* 36 (1): 27–30.
Sun S. & Li, H. 2018. Safety management status and Countermeasures of University biological laboratory under the background of "double first class" construction. *Laboratory research and exploration* 37 (11): 298–302.

Tan, D. & Du, C. 2018. Application of safety management in laboratory safety work. *Laboratory science* 21(2): 213–217.

Yin, W. & Xiao, G. 2016. The key and Countermeasures for building a first-class public instrument platform. *Laboratory research and exploration* 34 (6): 273–276.

Yuan, B. & Lu, Z. 2016. Enlightenment of biosafety supervision and management in American university laboratories. *Shanghai preventive medicine* 28 (4): 226–230

Yuan, Z. 2016. Exploration on the safety access mode of University Laboratory. *Higher education research and practice* 35(2): 71–76.

Zhu, J. & Lin, L. 2019. Discussion on strengthening laboratory safety training and management in Institutes of higher learning. *Basic medical education* 21 (5): 367–369.

Computational Social Science – Luo, Ciurea & Kumar (eds)
© 2021 Taylor & Francis Group, London, ISBN 978-0-367-70193-2

Predict the adaptability of medical college students returning to school after COVID-19 using machine learning

Q.T. Xiao, X.R. Liu & J. Jiang
College of Psychology, Army Medical University (Third Military Medical University), Chongqing, China

ABSTRACT: College students across the Chinese mainland have begun returning to school, but they are under 15-day medical observation. It is worth paying attention to whether college students can adjust their mind and rationally cope with the influence of COVID-19 in a collective atmosphere to start their normal study again. This paper aims to establish a predictive diagnostic model of automatic adaptability of 947 medical college students after returning to school with the personal characteristics, socio-demographic, self-evaluation, and collective atmosphere. A data set containing 757 medical students was evaluated with 10-fold cross-validation using seven classifiers, and a data set from another 190 medical students were tested for effectiveness. The CatBoost method had the highest prediction accuracy of 96.32% and the AUC of 80.0%. Machine learning technology can establish a generalized prediction model and discover the latent disease risks caused by the impact of social behavior events, to help managers and psychologists take reasonable countermeasures in advance.

1 INSTRUCTIONS

In December 2019, COVID-19 first appeared in Wuhan, Hubei Province, and rapidly spread to 24 countries nationwide and globally (Hui et al. 2020; WHO 2020a). On 30 January 2020, WHO declared COVID-19 as a public health emergency of international concern (WHO 2020b). To stop the spread of COVID-19 to schools, China's Ministry of Education requested that the spring semester of 2020 be postponed. Extended holidays, long-term staying at home, being unable to go to school to study and participate in social activities, and so on have a great impact on college students' mental health. Studies have found that college students suffer from anxiety and depression under COVID-19, especially those from a young age, economically backward areas, and non-medical backgrounds (Jinghui et al. 2020; Zhaohui et al. 2020). This reflects that COVID-19 has led to the generation of negative emotions among college students (Lei et al. 2020; Ya 2020; Yujie & Mengjie 2020). At the same time, due to the sudden, and widespread, lack of specific drugs, the impact of COVID-19 may affect future health behavior patterns and the lifestyle of college students who are not fully mature mentally. Studies have found that female college students were more affected by COVID-19 than male students, and were more likely to have negative coping styles (Jinghui et al. 2020). College students with a history of drinking alcohol were more likely to have mild and severe depression than those who never drank alcohol. More than half of the negative information received is more likely to induce negative emotions, which needs more social support, such as with the help of family members, friends, and teachers to get rid of the negative emotions (Jinghui et al. 2020).

With the improvement of the epidemic prevention and control situation, college students have returned to school and resumed classes, but they still need medical observation for 15 days. However, there are few studies on whether college students can adjust to the impact of COVID-19 in a collective atmosphere after returning to school and restarting normal learning. In this study, a cross-sectional survey was conducted to develop an automatic prediction model, which was used to find out the potential disease risks of returning college students affected by COVID-19 and to provide references for university administrators and psychologists to make accurate decisions.

2 MATERIALS AND METHODS

After obtaining permission from relevant ethics committees and management departments, 955 medical college students who returned to school and resumed classes were selected as research objects from April to May 2020. Data collection was completed in the last two days of the medical observation period.

2.1 Data collection

Data of age, gender, major, grade, family structure, only child or not, and native place were collected by the investigator from 955 medical college students who have returned to school. Information was also collected on management styles, rules and regulations, interpersonal relationships, trust support, and self-expectations. Meanwhile, the life Event Impact Scale (IES-R) (Sundin & Horowitz 2002) was used to measure the impact of COVID-19 on them. Finally, the CCSAS was used to evaluate their adaptability (Table 1).

Table 1. Methodology in a nutshell.

Step involved	Description
Data collection/ features	Age, sex, professional, grade, family structure, only child or not, native place, management style, rules and regulations of the school, the students' interpersonal relations, trust, self-expectation with the Marlowe-Crowne Social Desirability Scale (MC-SDS), the impact of COVID-19 outbreak on them with IES-R, and adaptability of 955 medical college students during isolation period with CCSAS as decision-making category
Classification with selected	Seven classifiers (KNN, SVM, DT, RF, NB, NN, CB) have been chosen to predict adaptability in medical college students, and statics for each classifier are mentioned in Tables 4 and 5
Evaluation of the classifiers	Confusion matrix and accuracy were evaluated for those machine learning classifiers, and were chosen as the best classifier

2.2 Organizational atmosphere

The organizational atmosphere is a comprehensive reflection of individuals on some organizational characteristics, such as organizational values, interpersonal relationships, management status, development prospects, and individual development in the organization. We used the LS (Litwin & Stringer 1968) to evaluate the organizational atmosphere of college students from four characteristic dimensions: management style, rules and regulations, interpersonal relationship, and trust and support.

2.3 College students' self-expectation

Social expectation refers to the group's expectation of individuals based on their social roles and identities (Congde et al. 2003). College students will form a relatively stable goal according to their specialties, majors, and future career development plans. This goal will be affected by the expectation of the group, resulting in the dependence of college students on the group. To this end, we used the Marlowe-Crowne Social Desirability Scale (MC-SDS) to measure the subjects' dependence on social approval. MC-SDS was initially compiled by Crowne and Marlowe in 1960. In 1987, Edward et al. updated it and eliminated the items of pathological psychology, making it more universal (Xiangdong et al. 1999).

2.4 *College student adaptation*

CCSAS was used to measure the adaptability of the research subjects after returning to school and was used as a decision indicator in the subsequent classification prediction model. CCSAS is developed by the Ministry of Education of China based on drawing on the existing foreign adaptation scale, specifically for Chinese college students, undergraduates, and postgraduates. It has a good conception validity, and its reliability and validity meet the indicators of psychometrics (Xiaoyi et al. 2005).

2.5 *Classification algorithms*

Classification is a machine learning method which classifies data by characteristics into a specific class or group (Aggarwal & Zhai 2012). Different classification algorithms can deal with health-related classification problems very effectively (Jovic et al. 2015). Different classification algorithms such as K-nearest neighbor classifier (KNN), support vector machines (SVM), decision trees (DT), random forest (RF), naïve Bayes classifier (NB), and neural network model (NN) are used to classify the target group.

KNN is an instance-based implicit learning classifier. The classifier searches for k points closest to the new data point in the data vector space and takes the class with the most categories as the class of the data point. This classifier requires a lot of memory and usually works well with fewer dimensions. Some applications of KNN include intrusion detection (Liao & Vemuri 2002), handwritten digits classification (Lee 1991), and so on.

SVM is a classifier that use a decision boundary called a hyperplane to separate different categories of samples. SVM can classify linear and nonlinear data (Burges 1998). SVM have been widely used to classify tumor tissue samples using microarray to express data (Statnikov et al. 2008) and text categorization (Joachims 1998). Therefore, SVM is a technique that can be able to handle classified data well.

DT is a hierarchical classifier that divides data according to rules. There are two types of DT nodes, leaf nodes and nodes. The node represents a classification of attributes, and the leaf represents a category. The application of decision tree classification includes the classification of land remote sensing data (Friedl & Brodley 1997), the diagnosis of ovarian cancer mass spectrometry data (Vlahou et al. 2003), and the classification of Alzheimer's disease by magnetic resonance imaging scanning (Zhang et al. 2014).

RF is an integrated approach that outputs a classified class in a way similar to collective decision making by constructing some decision trees during training (Breiman 2001). Some applications of RF include image classification and microarray-based cancer classification (Statnikov et al. 2003).

NB is a classifier based on probability calculations per attribute for a particular class (Karanasiou et al. 2016). It is based on Bayes' theorem. It assumes that the values of attributes are independent of each other so that the class tag of tuples can be obtained by Bayesian probability calculation.

NN consists of a set of neurons, an input and an output. The output depends on the weight of neurons. Multiple layer perceptron (MLP) is a simple artificial neural network structure (Ahmad et al. 2012). The weighted sum of the current layer must be entered to the next layer. When the input meets or exceeds a certain threshold, the final neuron is activated. When the final output is obtained, all of these transmissions will stop. The number of hidden layers and neurons in NN can be dynamically determined by the application. This paper will choose different NN construction of a different number of hidden layers, including one layer, two layers and three layers.

Ensemble achieves better prediction performance and improves the generalization ability of classifier by training multiple classifiers, which are both common frameworks, Bagging and Boosting. The Bagging methods usually consider homogeneous weak learners, learn these weak learners

Table 2. Confusion matrix.

Actual value			
Predicted value		Positive	Negative
	Positive	TP	FP
	Negative	FN	TN

Table 3. Performance measures of a classifier.

Performance measure	Formula
Accuracy	$(TP + TN)/(P + N)$
Precision	$TP/(TP + FP)$
Recall	$TP/(TP + FN)$
F − Score	$2 \times precision \times recall/$ $(precision + recall)$

independently in parallel, and combine them according to some deterministic averaging process. Boosting method is a highly adaptive method that combines the weak learner sequentially in accordance with a certain deterministic strategy, which emphasizes more on the training samples of the weak learners (Kuncheva 2004). The CatBoost algorithm is adopted in this paper. CatBoost is a classification model implemented based on oblivious trees as a learning device, which can efficiently and reasonably deal with the category characteristics, Gradient Bias and Prediction shift (Dorogush et al. 2018).

2.6 Classifier performance measures

Select the appropriate classifier based on performance metrics. In this study, the obfuscation matrix, accuracy, precision, recall rate, and F-Score were used to evaluate the performance of classifiers. The obfuscation matrix is the basis of performance measurement. In the process of classification training, the evaluation of the optimal solution is defined based on the confusion matrix (shown in Table 2). TP and TN represent the number of positive and negative instances that are properly classified, while FP and FN represent the number of negative and positive instances of misclassified, respectively.

Accuracy, precision, recall, and F-Score were used to evaluate the performance of various classifiers. The values of TP, TN, FP, FN, P, and N were used for the calculation of accuracy, precision, and recall rate. Table 3 shows the calculation formulas of them.

3 RESULTS

Among the 955 college students, 4 students were excluded due to incomplete personal information, and 2 students were excluded due to underperforming. The final number of participants was 947, including 642 males (67.7%) and 305 females (32.3%). Mean age (±standard deviation) was 20.0 (±1.81) years. Among them, 47.5% of college students had no obvious effect on the impact of COVID-19, 36.9% had a mild effect, 14.7% had a moderate effect, and 0.9% had a severe effect.

From 947 medical college students, 757 were randomly selected for classification prediction assessment, and the remaining 190 were used as test data sets to verify the above classifiers. Each instance was predicted, and the predicted results were recorded and compared to the initial decision class. The confusion matrix generated by the seven classifiers after 10 cross-validation was compared (shown in Table 4).

Table 4. Confusion matrix of different classifiers.

Classifier	Classified as	a*	b*
KNN	a	0	7
	b	0	183
SVM	a	0	7
	b	5	178
DT	a	0	7
	b	4	179
RF	a	1	6
	b	0	183
NB	a	2	5
	b	24	159
NN-1*	a	0	7
	b	0	183
NN-2	a	0	7
	b	3	180
NN-3	a	0	7
	b	4	179
CB	a	0	7
	b	0	183

*a: Adaptability needs adjustment and b: Good adaptability; NN-i: i layer in NN.

In addition, the differences of these classifiers in accuracy, precision, recall rate, F-score, and area under the receiver operating characteristic (ROC) curve (AUC) were also compared (as shown in Table 5 and Figure 1).

As can be seen from Table 5 and Figure 1, CatBoost was the most suitable classifier for this data set, with its accuracy (96.32%), Precision (0.93), Recall (0.96), F-score (0.95), and AUC (80.0%).

4 DISCUSSION

Machine learning provides a new prospect for the development of automated disease diagnosis systems, which has become a concern for medical and engineering researchers. In this paper, the organizational atmosphere and its influence on the mental changes of members are the important contents of efforts to predict the impact of organizational atmosphere on medical college students to eliminate the impact of COVID-19, from socio-demographic to medical factors, using machine learning. A significant number of medical undergraduates (148) were impacted by COVID-19 and showed moderate or more adverse symptoms. Therefore, on the one hand, attention should be drawn to the adaptation of these college students after they returned to school, to eliminate the potential risk of disease. On the other hand, measures should also be taken to predict the adaptability of other college students after they return to school.

Arkaprabha Sau et al. (Sau & Bhakta 2017), Sandhya (Sandhya & Kantesa 2019), and Bhakta (Bhakta & Sau 2016) have used machine learning techniques to predict mental health problems of different populations, such as anxiety and depression in the elderly, psychological abnormalities of IT personnel, and psychological and behavioral patterns of college students. However, there are few studies on college students' adaptability to returning to school after the epidemic. Srividyal ct al. found that the ensemble and random forest predicted mental health disorders with an accuracy of nearly 90% for high school, college, and professional people (Srividya et al. 2018). Those fully demonstrate that machine learning technology can be used to detect the psychological conditions of specific groups of people.

Table 5. Evaluation of different classifiers.

Classifier	Evaluation metrics				
	Accuracy (%)	Precision	Recall	F-Score	AUC (%)
KNN	96.32	0.93	0.96	0.95	48.6
SVM	93.68	0.93	0.94	0.93	73.5
DT	94.21	0.96	0.94	0.95	48.6
RF	96.32	1	0.97	0.98	54.8
NB	84.74	0.94	0.85	0.89	72.8
NN-1	96.32	0.93	0.96	0.95	76.3
NN-2	94.74	0.93	0.95	0.94	47.5
NN-3	94.21	0.93	0.95	0.94	53.0
CB	96.32	0.93	0.96	0.95	80.0

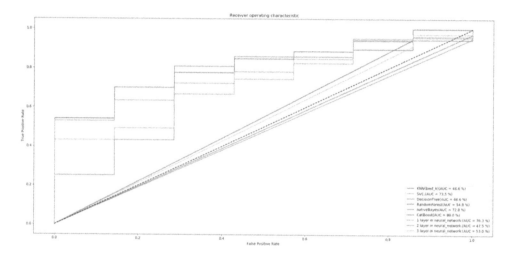

Figure 1. ROC and AUC (%) of seven different classifiers.

In this paper, we evaluated seven different types and strategies of machine learning classifiers, and CatBoos's 10-fold cross-validation test had the highest predictive accuracy. CatBoost model was tested on another 189 medical college students for its validity. Its predictive accuracy was found to be 96.32%, compared with the initial decision class. The research of this paper is to explore and verify a classifier which is more suitable for predicting organizational atmosphere of college students in the future. Moreover, this can also be extended to all college students to discover the COVID-19 outbreak causing potentially high-risk post-traumatic stress disorder symptoms in college students.

5 CONCLUSION

This paper describes the prospect of machine learning technology in predicting the impact of social behavioral events on potential diseases. The performance of the prediction model based on small sample is limited. Multi-dimensional research and large data sets, as well as proper optimization of different classification algorithms, will contribute to the establishment of generalized prediction models. This will contribute to the construction of an efficient automatic prediction system for predicting the possibility of members of an organization or team resisting mental abnormalities with the force of organizational atmosphere in the event of major social events. This can not only help managers to know the health status of the team in a timely way but also help psychologists to take reasonable measures to eliminate the mental health problems caused by adverse events.

FUNDING AND DECLARATION OF INTERESTS

Funding: This study was supported by the AMU project (2019B12). Conflict of interest: None.

REFERENCES

Aggarwal, C. C., & Zhai, C. X. (2012). A Survey of Text Classification Algorithms.
    Ahmad, J., Shafi, I., Ansari, S., & Ismail Shah, S. (2012). Neural network-based approach for the noninvasive diagnosis and classification of hepatotropic viral disease. IET communications, 6(18), 3265–3273.
Bhakta, I., & Sau, A. (2016). Prediction of Depression among Senior Citizens using Machine Learning Classifiers. International Journal of Computer Applications, 144(7), 11–16.
Breiman, L. (2001). Random Forests. Machine Learning, 45(1), 5–32.
Burges, C. J. C. (1998). A Tutorial on Support Vector Machines for Pattern Recognition. Data Mining and Knowledge Discovery, 2(2), 121–167.
Congde, L., Zhiliang, Y., & Xiting, H. (2003). Dictionary of Psychology: Shanghai Education Press.
Dorogush, A. V., Ershov, V., & Gulin, A. (2018). CatBoost: gradient boosting with categorical features support.
Friedl, M. A., & Brodley, C. E. (1997). Decision tree classification of land cover from remotely sensed data. Remote Sensing of Environment, 61(3), 399–409.
Hui, D. S., I, A. E., Madani, T. A., Ntoumi, F., Kock, R., & Dar, O., et al. (2020). The continuing 2019-nCoV epidemic threat of novel coronaviruses to global health - The latest 2019 novel coronavirus outbreak in Wuhan, China. Int J Infect Dis, 91, 264–266.
Jinghui, C., Yuxin, Y., & Dong, W. (2020). Mental health status and its influencing factors among college students during the epidemic of COVID-19. Journal of South Med Univ, 40(02), 171–176.
Joachims, T. (1998). Text categorization with Support Vector Machines: Learning with many relevant features. Paper presented at the Proc. Conference on Machine Learning.
Jovic, A., De Luca, N., Pecchia, L., & Melillo, P. (2015). Automatic classifier based on heart rate variability to identify fallers among hypertensive subjects. Healthcare technology letters, 2(4), 89–94.
Karanasiou, G. S., Tripoliti, E. E., Papadopoulos, T. G., Kalatzis, F. G., Goletsis, Y., & Naka, K. K., et al. (2016). Predicting adherence of patients with HF through machine learning techniques. Healthcare technology letters, 3(3), 165–170.
Kuncheva, L. I. (2004). Combining Pattern Classifiers: Methods and Algorithms: Wiley-Interscience.
Lee, Y. (1991). Handwritten Digit Recognition Using K Nearest-Neighbor, Radial-Basis Function, and Backpropagation Neural Networks. Neural Comput, 3(3), 440–449.
Lei, H., Weidong, M., Hong, G., Yanchao, H., Yan, Z., & Chunyan, Z., et al. (2020). Analysis of negative emotions and their influencing factors during the isolation period under COVID-19. Journal of Xi'an Jiaotong University (Medical Sciences), 1–11.
Liao, Y., & Vemuri, V. R. (2002). Use of K-Nearest Neighbor classifier for intrusion detection. Computers & Security, 21(5), 439–448.
Litwin, G. H., & Stringer, R. A. (1968). Motivation and organizational climate. American Journal of Sociology, 82(4), 1220–1235.
Sandhya, P., & Kantesa, M. (2019). Prediction of mental disorder for employees in IT industry. International Journal of Innovative Technology and Exploring Engineering(No.6), 374–376.
Sau, A., & Bhakta, I. (2017). Predicting anxiety and depression in elderly patients using machine learning technology. Healthcare technology letters, 4(6), 238–243.
Srividya, M., Mohanavalli, S., & Bhalaji, N. (2018). Behavioral Modeling for Mental Health using Machine Learning Algorithms. J Med Syst, 42(5), 88.
Statnikov, A., Wang, L., & Aliferis, C. F. (2008). A comprehensive comparison of random forests and support vector machines for microarray-based cancer classification. BMC Bioinformatics, 9, 319.
Sundin, E. C., & Horowitz, M. J. (2002). Impact of Event Scale: psychometric properties. Br J Psychiatry, 180, 205–209.

Vlahou, A., Schorge, J. O., Gregory, B. W., & Coleman, R. L. (2003). Diagnosis of Ovarian Cancer Using Decision Tree Classification of Mass Spectral Data. Journal of biomedicine & biotechnology, 2003(5), 308–314.

WHO. (2020a). World Health Organization. Novel coronavirus(2019-nCoV)situation report-17.

WHO. (2020b). Statement on the second meeting of the International Health Regulations(2005)Emergency Committee regarding the outbreak of novel coronavirus(2019-nCoV).

Xiangdong, W., Xilin, W., & Hong, M. (1999). Handbook of mental health assessment scale: Chinese Journal of mental health.

Xiaoyi, F., Jianzhong, W., & Xiuyun, L. (2005). Development of Chinese College Student Adjustment Scale. Studies of Psychology and Behavior(02), 95–101.

Ya, W. (2020). Anxiety level and its influencing factors of college students during epidemic prevention and control. Journal of Teacher Education, 7(03), 76–83.

Yujie, Z., & Mengjie, L. (2020). Covid-2019 in Wuhan college students' mental health. Modern business trade industry, 41(14), 52–53.

Zhang, Y., Wang, S., & Dong, Z. (2014). Classification of Alzheimer Disease Based on Structural Magnetic Resonance Imaging by Kernel Support Vector Machine Decision Tree. Progress in Electromagnetics Research, 144, 185–191.

Zhaohui, W., Lili, C., Li, L., Xiaomei, Z., Yimeng, W., & Xiaohong, X. (2020). Effect of COVID-19 on mental health of medical students. Journal of changchun university of traditional chinese medicine, 36(5), 1–4.

Computational Social Science – Luo, Ciurea & Kumar (eds)
© 2021 Taylor & Francis Group, London, ISBN 978-0-367-70193-2

Research on tort liability of self-driving cars

Z.B. Pei, S.W. Cai & X.L. Guo
Dalian Ocean University, Dalian, Liaoning, China

ABSTRACT: Since the 21st century, artificial intelligence technology has developed rapidly and is gradually being applied to people's lives. For example, the emergence of autonomous vehicles in the automobile industry has brought many conveniences to people's travel. But at the same time, there are also some problems in the application of current laws to the civil liability for accidents in autonomous vehicles. This article mainly studies the tort liability of self-driving cars, taking the world's first self-driving car death case as the starting point, and analyzing the behavior of self-driving cars from the perspective of the concept, characteristics, and specificity of torts. It then points out that in the field of autonomous driving, there are legal issues such as complicated identification of the subject of motor vehicle accident liability, difficulty in identifying defects in product quality liability, and producers' claims to develop risk defense to avoid liability. Finally, in response to the above-mentioned problems, it proposes the use of diversified liability subjects, the determination of differentiated liability principles, and the establishment of a social shared risk mechanism to fully protect the legitimate interests of the infringed.

1 OVERVIEW OF TORT LIABILITY FOR AUTONOMOUS VEHICLES

1.1 *The concept of autonomous vehicles*

Self-driving cars are also known as "intelligent connected cars," "unmanned cars," and "wheeled mobile robots." They refer to cars that realize automated driving through the cooperation of sensor devices and artificial intelligence algorithm systems. Among them, the sensor device collects information about the surrounding environment of the car while driving, and the artificial intelligence algorithm system analyzes and processes the information and issues instructions. Due to the different levels of automation, autonomous vehicles are divided into different levels. Now, the six-level classification standard issued by the Society of Automotive Engineers (SAE) in 2014 is widely adopted internationally. *Made in China 2025* divides autonomous vehicles into four levels: DA, PA, HA, and FA. The higher the level, the lower the requirement for driver intervention [1].

1.2 *The particularity of self-driving car infringement*

First, driving behavior is autonomous and unmanned. Unlike traditional cars that require manual operation, the characteristics of self-driving cars are prominently reflected in the unmanned nature of their driving behavior. Self-driving cars are equipped with an intelligent driving control system. In the automatic driving mode, real unmanned driving is realized. The system controls all driving behaviors of the car without user intervention. When a traffic accident occurs, the traditional primary accountability subject, the natural person driver, is therefore absent and cannot be held accountable. Second, the data relevance for driving safety of autonomous vehicles is very special. The safety assurance of autonomous vehicles mainly lies in safe driving technology, intelligent control system, fast network connection, and accurate positioning information. It is not necessary to use human past driving experience as a driving safety standard, but to use relevant data to ensure the safe operation of autonomous vehicles. At present, autonomous vehicles have not established a unified and complete relevant safety standard in various countries, and the level of this standard will affect the determination of liability for motor vehicle traffic accidents [2].

2 MY COUNTRY'S LEGAL ISSUES IN THE FIELD OF AUTONOMOUS DRIVING

2.1 *Issues concerning the determination of liability for motor vehicle traffic accidents*

2.1.1 *The identification of the responsible party is complicated*
Article 76 of *the Road Traffic Safety Law* stipulates that the "vehicle party" shall bear the responsibility for a motor vehicle traffic accident. In theory, "motor vehicle party" usually refers to the owner of the motor vehicle, and adopts the dualism of operation control and operation benefit to determine whether it actually controls the operation and whether it obtains benefits from the operation. In general, the owner of the motor vehicle and the user who leases or borrows the motor vehicle are the guarantors. *The Tort Liability Law* stipulates that when the owner and the user are different, the specific user shall bear the responsibility [3]. The tort liability section of the *Civil Code of the People's Republic of China* that came into effect on January 1, 2021, stipulates that when the owner, manager, and user of a motor vehicle are not the same person due to leases, borrowing, and so on, the damage caused by a traffic accident is if the motor vehicle is responsible, the user of the motor vehicle shall bear the liability for compensation; if the owner or manager of the motor vehicle is at fault for the occurrence of the damage, he shall bear the corresponding liability for compensation. Therefore, some scholars believe that dualism cannot only be used as the criterion for keeping the owner, and the definition of "motor vehicle side" can be discussed based on whether there is the possibility of manual intervention.

In autonomous vehicles, in addition to the motor vehicle, there are also automakers, autonomous driving system operators, Internet service providers, and other producers and sellers. Different entities have different divisions of labor. For example, the obligation of auto manufacturers is to guarantee the quality of self-driving car products; the obligation of autopilot system operators is to collect and process information about the autopilot system; the obligation of network service providers is to ensure network security; for sellers the obligation is to take proper care of it properly. Therefore, when a self-driving car has a traffic accident, responsibility should be allocated among multiple participants based on different causes of the accident.

2.1.2 *Fault imputation is not applicable*
The Tort Liability Law invokes the provisions of Article 76 of *the Road Traffic Safety Law*. The principle of fault shall be applied to traffic accidents between motor vehicles; the principle of presumption of fault shall be applied to traffic accidents between motor vehicles and non-motor vehicles and pedestrians. The motor vehicle party is liable, but the fault of the other party can reduce the motor vehicle party's liability for compensation. The determination of fault is generally based on the subjective "knowledge" or "knowledge duty of care." In the auto-driving mode of the car, the human being as the driver has no substantial behavior to participate in the decision-making of car driving, and its role has been transformed into a "passenger," and the autopilot system has no subjective fault at all. Assuming that the driver is at fault, this means that he has a duty of care.

However, liberating the driver's hands and reducing the driver's burden are a major shining point of autonomous vehicles, and it is obviously inconsistent with this concept to impose strict obligations on them. And when an emergency situation requires manual emergency intervention, the driver may not be able to respond effectively in time. If the driver engages in other activities before taking over and does not focus on monitoring driving, the situation will be even worse [4]. Therefore, it is unreasonable and unrealistic to require drivers to maintain a high degree of attention at all times in an autonomous driving state or in a takeover situation. The fault tort liability rule based on the premise of "drivers breaching the duty of care" should not be directly applied to drivers in autonomous vehicles.

2.1.3 Compensation for damages is not comprehensive enough

Article 2 of the *Regulations on Compulsory Motor Vehicle Traffic Accident Liability Insurance* stipulates that "owners or managers of motor vehicles driving on roads within the territory of China shall take out compulsory motor vehicle traffic accident liability insurance in accordance with the provisions of *the Road Traffic Safety Law*." Article 76, Paragraph 1 of *the Road Traffic Safety Law* stipulates that "If a motor vehicle traffic accident causes personal injury, death, or property loss, the insurance company shall compensate within the limit of liability for compulsory third party liability insurance for motor vehicles; the deficiencies, ..." That is, compulsory traffic accident liability insurance is the first layer of protection to safeguard the interests of all parties and compensate for economic losses.

However, the application of compulsory traffic accident liability insurance is also subject to some restrictions. For example, the insurance is limited to the vehicle personnel and victims other than the insured; for those who are drunk, robbed, or deliberately causing traffic accidents, they only bear the salvage expenses. Liability: According to whether the insured is at fault and the type of compensation expenses, a sub-liability limit is set; the claimant for insurance money is limited to the insured, and so on. The latter two problems will become more severe after the introduction of autonomous driving technology: due to the increase in automation, the possibility of driver's fault will be reduced, which means that the amount of compulsory insurance compensation will be greatly reduced; because the producers of autonomous vehicles are neither, the insurer can hardly be said to be the victim, and it is impossible to recover the compensation from the insurer after assuming the liability for compensation. Due to problems such as unclear subjects of accidents in autonomous driving, difficult to prove causality, and complicated liability determination, the existing insurance system is facing new challenges and cannot guarantee the legal rights and interests of the responsible person and the victim in a timely and comprehensive manner.

2.2 The issue of product quality responsibility recognition

According to the relevant judicial interpretation of the Supreme People's Court, "If a motor vehicle has a product defect that causes damage in a traffic accident, and the party requests the manufacturer or seller to bear compensation in accordance with the provisions of Chapter 5 of *the Tort Liability Law*, the people's court shall support it." Interpreting motor vehicles as "products" has laws to follow, and product quality responsibility can be applied to motor vehicle traffic accidents. With the transfer of autonomous vehicle control functions, responsibilities have also shifted from the driver to the producer. However, judging from the current law, there are also some difficulties in making producers responsible for product quality.

2.2.1 The identification criteria for defects are vague

In Article 46 of the *Product Quality Law*, a defect refers to "the product has an unreasonable hazard that endangers the safety of the person and the property of others; if the product has a national standard or industry standard that protects human health, personal and property safety, it does not meet the standard of." Specifically, divided into design defects, manufacturing defects, and indicator defects. Since various countries and industries have not yet established mature self-driving car-related safety standards, it is impossible to use "national standards and industry standards" as the criteria for judging defects. The identification of "there is an unreasonable danger endangering the safety of persons and other people's property" involves the consumer expectation test (Consumer Expectation Test) and risk-utility test (Risk-Utility Test) [5], The complexity of autonomous driving technology will hinder the reasonable application of these two standards. In judicial practice, both the plaintiff and the judge have to explain whether there are defects, and whether there are alternative, feasible, and safer products based on the appraisal opinions of professional institutions or the testimony of professionals, which brings higher costs.

The development of risk defense may damage the legitimate rights and interests of the infringed

Article 41, paragraph 2, of China's *Product Quality Law* sets the exemptions for producers: 1. The product has not been put into circulation; 2. When the product is put into circulation, the defect causing damage does not exist; 3. The level of science and technology at the time the product is put into circulation cannot yet detect the existence of defects. At the same time, if producers want to be exempt, they need to bear the burden of proof. When self-driving cars cause accidents based on deep learning functions, producers are likely to take advantage of technology to prove that when the car is put into circulation, the defect that caused the damage does not yet exist, or the level of science and technology when the car is put into circulation cannot find the defect. Therefore, it is difficult for the *Product Quality Law* to effectively protect the legitimate rights and interests of the infringed in the field of autonomous driving [6].

3 LEGAL ADVICE ON THE INFRINGMENT OF AUTONOMOUS VEHICLES

3.1 *Promote diversification of responsible entities*

3.1.1 *Owner and user*

Although self-driving cars have reduced the responsibilities of traditional drivers to a certain extent, according to the theory of "who uses them, who benefits and who bears the responsibility" in the civil law, owners and users should be liable for compensation for infringement accidents of self-driving cars.

The principle of corresponding rights and obligations [7]. The owner has property rights to the self-driving car and the system, has all relationships, and has the duty of care for supervision. Therefore, when an accident occurs, regardless of whether the owner is an actual user, there is a legal basis for responsibility and an indirect replacement responsibility. The actual user is the person who gives instructions to the self-driving car and should be responsible for the accidents that occur during the driving of the car. Owners and users should carefully learn how to use self-driving cars, fully read relevant instructions, and make correct start and route settings. In addition, the owner and user should update or regularly maintain the relevant data of the automatic driving system according to the prompts and shall not modify it without authorization. Owners and users are responsible for traffic accidents caused by failure to perform their obligations.

3.1.2 *Producers and sellers*

Self-driving cars are different from traditional cars. Many producers are involved in the research and development, including car manufacturers, autonomous driving system operators, network service providers, and so on. [8]. Self-driving cars have high requirements for safety and accuracy, and automakers play a vital role in the early stages of production and R&D. If the manufacturer's fault causes product defects or other defects, such as brake failure and accidents, compensation shall be made according to law. In short, producers should fully guarantee the safety and stability of autonomous vehicles; otherwise, they will be liable for damages due to infringement.

In the sale of self-driving cars, sellers must fulfill the obligation of inspection and acceptance, proper storage, and the obligation of not substandard or selling fake and inferior products, and the obligation of detailed explanation. The seller's negligence in the circulation and storage links causes damage to the product, or sells counterfeit and inferior products in order to seek illegitimate benefits, resulting in accidents during the operation of autonomous vehicles, and the seller should become the subject of infringement [9].

According to Article 43 of China's *Product Quality Law*, "If a product has a defect that causes damage to the person or the property of others, the victim can claim compensation from the producer of the product or the seller of the product" and Article 15 of *the Consumer Protection Law*, "Consumers whose legitimate rights and interests are harmed when purchasing or using commodities, they can claim compensation from the seller," when a traffic accident occurs in an autonomous vehicle, the manufacturer or seller of the vehicle shall bear the responsibility. Responsibility is based on the law and reasonable on the grounds.

3.1.3 *Third party at fault*

The third party at fault shall be liable for tort due to accidents of self-driving cars due to cyber attacks or illegal modifications. If a cyber attacker invades the autopilot system, changes relevant data, and causes damage to others, he is obviously responsible for his illegal actions; if he illegally modifies the car's hardware system or software system without authorization, causing damage to the car and causing a traffic accident, the modification party bears tort liability [10].

In addition, according to the law, "road managers who cannot prove that they have fulfilled their duties of cleaning, protection, and warning in accordance with laws, regulations, rules, national standards, industry standards, or local standards, shall bear the corresponding liability for compensation." During driving, if the road manager fails to perform the legal obligations and causes the consequences of damage, the victim can claim liability for compensation from the road manager.

3.2 *Responsibility upholds the principle of differentiation*

3.2.1 *No-fault liability*

(1) Owner, user

As mentioned above, in the infringement accidents of self-driving cars, the fault liability rule based on the premise of "driver's breach of duty of care" should not be directly applied. Therefore, it is recommended that the responsibilities of owners and users refer to Germany's strict liability system; that is, owners and users bear responsibility based on the ownership and use rights of autonomous vehicles [11].

The reasons are, first, self-driving cars are also a high-risk object, and the owner and user should take the responsibility first if they open the risk; second, the victim does not have to worry about whether the driver is at fault and how to prove it, which is conducive to finding as soon as possible the subject of the liability for compensation; third, the owner and user have the right of recourse while assuming no-fault liability. When a traffic accident is caused by other subjects, they can recover compensation after assuming the responsibility.

(2) Producer

If a self-driving car has a traffic accident due to its own defects, causing damage to others, refer to the no-fault liability in the *Product Quality Law*. That is, regardless of whether there is a fault, as long as the self-driving car is defective, there is a result of personal and property damage, and there is a causal relationship between the defect and the result of the damage, the producer should be liable for compensation. From the development, design, production to the market, and the entire industrial operation of autonomous vehicles, producers have always occupied a relatively advantageous position, and have strong predictability and overall control of product defects [12].

For the producer's exemption, considering the technological frontier and safety relevance of autonomous driving, it is recommended to exclude the development of risk defense [13]. The development of risk defense in comparative law is not absolute. Although the European Community has relevant provisions in the *Product Liability Directive*, many countries still cannot apply [14].

3.2.2 *Liability for fault*

(1) Seller

Article 42 of *The Tort Liability Law*, "If the seller's fault causes a product to be defective and causes damage to others, the seller shall bear the tort liability. The seller cannot specify the manufacturer of the defective product, nor the supplier of the defective product, the seller shall bear the tort liability," that is, the seller of the self-driving car shall bear the corresponding fault liability in accordance with the general fault liability principle.

If the seller violates the obligation of inspection and acceptance and proper custody, or uses shoddy products as genuine ones, resulting in product quality that does not conform to the agreement, or an accident that causes damage to others, the owner can, after assuming responsibility to the victim, follow the sales contract relationship. The seller is requested to be liable for damages, repair, replacement, and return.

(2) Third party

Self-driving cars need to be connected to the Internet, and the openness of the network provides opportunities for malicious third parties. Furthermore, due to the complexity of the traffic accident itself, the negligence or deliberateness of a third party may also cause the accident. Therefore, the principle of fault liability is applied to it.

According to the third party's fault on the cause of the damage, it can be divided into two situations: the third party's fault is the only cause of the accident, and the third party bears all civil liabilities; the third party's fault is not the only cause of the accident, that is, both parties are meaningless. For joint infringements of liaison, the autonomous vehicle and the third party determine the distribution of liability according to the magnitude of the cause, and if it is difficult to determine, both parties shall bear the liability for compensation equally.

3.2.3 Compensation method

(1) Three-tier insurance structure

In view of the idea of "distribution of damage" in the tort law, self-driving car insurance can also be upgraded on the basis of the current compulsory traffic accident insurance system, and at the same time develop auto commercial insurance and product liability insurance, which is embodied as "traffic compulsory insurance-commercial insurance-products liability insurance" three-tier structure [15].

The *Implementation Rules for the Management of Road Testing of Autonomous Vehicles (Trial)* requires the test subject to purchase traffic accident liability insurance of no less than RMB 5 million per vehicle or provide an accident compensation guarantee letter of no less than RMB 5 million. Compulsory traffic insurance can still have the owner or manager of the car as the insured, but it is recommended that the insurance limit be increased substantially. Because in the automatic driving mode, the incidence of traffic accidents decreases, it is entirely possible to introduce "low premiums and high insurance amounts" insurance to improve everyone's ability to pay. In addition, in the order of insurance compensation, the first order belongs to the compulsory traffic accident insurance; the shortfall after the compulsory traffic insurance compensation is paid by the automobile commercial insurance; when the responsibility or recovery of product quality is involved, the product liability insurance is paid.

(2) Compensation Fund System

In order to fully protect the rights and interests of victims and encourage technological innovation, a compensation fund system can be established for infringement accidents of autonomous vehicles [16]. The compensation fund is mainly funded by the producers, sellers, owners, enthusiasts, or government departments of autonomous vehicles, which has many benefits. First, the compensation fund system helps protect the legitimate rights and interests of the victims in a timely manner. When the responsible party is not determined after the accident, the fund can be used to compensate the victims first. Second, the compensation fund can provide certain relief to the victims when there are exemptions in the tort incident and cannot be held accountable. Third, the compensation fund can also be used to find out the cause of an accident, repair system failures, and improve related technologies.

4 CONCLUSIONS

In summary, the author believes that autonomous vehicles are still products and do not have independent legal personality. Based on the particularity of its infringement, the responsible subjects can be divided into owners and users, producers and sellers, and third parties at fault. Among them, owners, users, and producers apply the principle of no-fault liability, and sellers and third parties apply the principle of fault liability. At the same time, the producer's defense of the development risk is no longer supported and cannot shirk the responsibility for product quality. Finally, the establishment of a three-tier insurance system and a compensation fund system of "Compulsory Traffic Insurance-Commercial Insurance-Product Liability Insurance" can more timely and comprehensively protect the legal rights of victims. In addition to the above legal recommendations, China should also speed up the establishment of relevant safety standards for autonomous vehicles to determine whether products are defective. In the update and iteration of autonomous driving technology, we should adjust the rules and systems in time to make them meet the needs of technological development and ensure that they serve the society.

FUND PROJECT

The 2019 research project of the Agricultural and Forestry Working Committee of the Chinese Association for Degree and Graduate Education (2019-NLZX-YB52)

REFERENCES

[1] *Beijing/Guiding Opinions and Implementation Rules for Road Testing of Autonomous Vehicles*, available at https://mp.weixin.qq.com/s/qIWfkOgzNd_iy-9umyZaBw, accessed May 10, 2019
[2] Li Zongxiang. Research on the Liability for Tort of Unmanned Vehicles: [Master's Thesis]. Dalian Maritime University, 2018.
[3] Wang Liming. Research on Tort Liability Law: Volume One, Renmin University of China Press, 2016.
[4] Hu Yajun. Analysis on the legal liability of driverless car traffic accidents. Journal of Guangxi Institute of Political Science and Law, 2018 (11)
[5] David G. Owen. Design Defects. Mo. L. Rev, 2018(07)
[6] Tian Shaohua, Chen Jie. On the civil liability for damage caused by self-driving car traffic accident. Journal of Hebei University of Economics and Trade (Comprehensive Edition), 2020, 20(01): 38–43+73.
[7] Xiao Ying. Research on the Legal Issues of Artificial Intelligence Autonomous Driving. Hebei University, 2020.
[8] Wang Yinli. On the identification of the infringement subject of driverless cars in traffic accidents. Legal System and Society, 2018 (3)
[9] Si Xiao, Cao Jianfeng. On the Civil Liability of Artificial Intelligence-Taking autonomous vehicles and intelligent robots as the starting point. Law Science, 2017(05)
[10] Li Yongjun. Doubts about the tort law relief of "product self-damage". Journal of China University of Political Science and Law, 2015(03)
[11] Xie Wei. The liability for damages of self-driving cars in traffic accidents. Journal of Chang'an University, 2018 (7)
[12] Niu Feifei. Research on tort liability for auto-driving car traffic accidents. Jilin University, 2020.
[13] Han Xuzhi. The structure of tort liability for autopilot accidents – Also on the three-tier insurance structure of autopilot. Journal of Shanghai University, 2019 (3)
[14] Sun Limin. Research on the Tort Liability of Autonomous Vehicle Traffic Accidents. Journal of Yibin University, 2020, 20(02): 8–15.
[15] Shen Yucen. The design and realization of the rules for the determination of tort liability for autonomous driving accidents in the AI era. Journal of Guangxi Institute of Political Science and Law, 2020, 35(03): 116–122.
[16] Wang Chenfeng. Improving the legal system for insurance of damage caused by self-driving cars on the road. China Information Technology Weekly, 2020-07-13 (016).

Computational Social Science – Luo, Ciurea & Kumar (eds)
© 2021 Taylor & Francis Group, London, ISBN 978-0-367-70193-2

Learning modern technology and developing innovative capabilities—teaching content innovation and teaching practice of modern design course

W. Chen & J. Fan
Jiangsu Open University, Nanjing, China

ABSTRACT: Modern design is a new course in the teaching reform. Learning modern technology and developing innovative capabilities is the teaching goal of this course. In the arrangement of teaching content, the latest scientific and technological achievements are highlighted, which are professional, adaptable, flexible, and extensible, and the application of theory and practice is emphasized. In the teaching method, the change mainly takes the knowledge imparted, and pays attention to the cultivation of the students' innovative thinking, innovative methods, and innovative ability.

1 INTRODUCTION

Learning modern technology and developing innovative capabilities is the goal of teaching the Modern Design course, but also a summary of the course teaching and practical experience. In teaching, how should teachers cultivate students' innovative consciousness and ability? What teaching methods are used to achieve this teaching goal?

Through several years of teaching practice, some methods and experiences have been discovered.

2 TEACHING CONTENT OF THE MODERN DESIGN COURSE

Modern design includes a wide range of content, such as green design, virtual design, innovative design, parallel design, fuzzy design, reverse design, and so on, belonging to the category of modern design [1]. Some of these modern designs involve some deeper basic knowledge, which has exceeded the knowledge of college students, so in the selection, students can accept, and reflect on the professional characteristics. The most hot are green design, virtual design, and innovative design in the modern design textbook.

Green design is a new design concept and method, which is put forward around the theme of how to save resources, use energy effectively, and protect the environment at the same time of the developing economy at the end of the last century. It is considered as one of the effective ways for the design community to realize sustainable development, and has become the hotspot and main content of modern design technology research.

Virtual design is a new multi-disciplinary cross-technology. It involves subjects such as virtual reality technology, computer graphics, and product design. The application of virtual design technology in product design can make the computer simulate the whole product development process, ensure the one-time success of product development, speed up the process of product development, improve the design quality and design efficiency, reduce the cost, make the designers and users merge into one, and design the products with market demand.

Innovative design is one of the important factors to measure comprehensive national strength [2]. For a country, people with a lot of innovative design ability have great potential to develop economy. Therefore, now countries around the world are adjusting economic policies, science and technology policies, and development strategies and using innovative design as an effective weapon to improve product quality and seize market share.

3 TEACHING METHODS OF THE MODERN DESIGN COURSE

Teaching method is an important means to cultivate students' interest and improve students' innovative ability [3]. The teaching content is carried out according to the content of the compiled textbook, while the teaching method has greater flexibility. Different teaching methods can be used in every part of the teaching activities to improve students' innovative ability.

During the classroom teaching of modern design, the courseware based on PowerPoint software is used, which covers all the teaching contents stipulated in the outline, and adds a large number of visual pictures, video materials, scientific research results and computer experiments, which are vivid and greatly increase the interest in learning, and at the same time, the grasp of the teaching content is more profound.

Different teaching methods are adopted for different teaching contents: for the content of green design, taking a large number of pictures as an example to illustrate the importance of sustainable development, the teachers have added some excellent cases and achievements of green design development at home and abroad in the teaching content, and have raised students' awareness of environmental protection and increased their sense of responsibility for environmental protection. On the basis of introducing the excellent virtual software at home and abroad, the content of virtual design focuses on how to use the method of virtual design to design new products, in the form of special lectures, which broaden the students' horizons and enables students to come into contact with advanced software technology through computer experiments and combining theory with practice. On the content of innovative design, on the basis of introducing innovation theory, we focus on introducing examples of innovation, deepening students' understanding of some advanced technologies and their development direction of this major, and focusing on some new innovative methods in the form of special lectures. Students are very interested in it and have mastered some new innovative methods from it.

The content with strong theory and high outline requirements is mainly deductive reasoning and analysis, while the content with narrative is not logical and students are not interested in it. In addition, for the content of relatively independent content, through careful arrangement, the classroom is handed over to the students to organize, let the students design the classroom, and then the teacher comments according to the specific situation.

In the classroom the main line is of teaching activities, but also pays great attention to the continuation of the after-school teaching process. The first is to improve the after-class learning, review and testing mechanism, according to the content of the course appropriate amount of homework, the amount of correction is 100%, and timely review, checking the learning effect. Second, it is to design a group of research problems, let the students complete them independently, and improve their innovative ability. Moreover, it is to open up information feedback channels: teaching teachers to students to announce one's own QQ number, E-mail mailbox, telephone number, to timely answer questions raised after class, and receive feedback information. In addition, a questionnaire evaluation survey was carried out, which focused on teaching content and teaching methods. Make a questionnaire according to the teaching content of each article, test the quality and effect of the teaching process according to the students' completion, and improve continuously. This method has been carried out since the course setting and has received very good results.

Through flexible teaching methods, the students' innovative consciousness is cultivated, and the innovative methods are mastered, and then the innovation ability is improved [4].

4 CORRECT HANDLING OF THE RELATIONSHIPING IN TEACHING

In the course teaching of modern design, attention should be paid to the relationship between modern design and traditional design and the relationship between modern design and computer technology.

4.1 *The relationship between modern design and traditional design*

In the course teaching of modern design, attention should be paid to guiding students to correctly handle the relationship between modern design and traditional design. To this end, first of all show students that modern design is based on the traditional design to a higher level of development. This design is characterized by considering product design as a system engineering, emphasizing creative development design, taking the overall function of the product as the goal, taking the computer as the auxiliary design means, attaching importance to the scheme design, and pursuing the comparison, evaluation, and optimal selection of various schemes. Therefore, compared with traditional design, modern design is a dynamic, scientific, and computerized design. Then, it is further explained to the students that design is a cross-engineering involving multiple disciplines and techniques. It not only needs the guidance of methodology but also depends on various professional theories and professional techniques and cannot be separated from the experience and practice of technical personnel. Modern design is formed by combining new scientific theories and new scientific and technological achievements on the basis of inheriting and developing traditional design. Therefore, learning and using modern design methods is not to completely abandon traditional methods and experiences but to let the vast number of students master new design methods on the basis of traditional methods and practical experience to improve the innovative ability of design. Therefore, modern design and traditional design cannot be completely separated; traditional design in some suitable mechanical and electrical product design is still in use. Of course, modern design is not a panacea, because all kinds of design methods have their specific role and occasion.

4.2 *The relationship between modern design and computer technology*

In the course teaching of modern design, students are also required to correctly understand the relationship between modern design and computer technology. Therefore, emphasize to the students that modern design is not simply referring to computer technology, nor simply referring to the design with the general rules and general approaches of design as the research object. It should include all advanced design theory, design technology, and design method, and it is the integration and unification of all advanced and effective design ideas. Using modern design method for product design can only be through the means of computer-aided design. Therefore, the most important thing to learn the course is to grasp the creative thinking of various designs and master their design skills.

5 INNOVATIONS IN THE CONTENT OF MODERN DISIGN

Setting up the course of modern design is the necessity of the whole course reform of mechanical engineering, the need of social development, the need of science and technology development, the need of market competition, and the need to improve students' ability of innovation.

The content of modern design has the following characteristics:

1) Highlight the latest scientific and technological achievements
   The three parts of green design, virtual design, and innovative design in modern design are the latest research contents and achievements in the field of science and technology.

   The concept of green design was put forward by Danes in the early 1990s; virtual design was a new research field developed in the 1990s; and innovative design was often used in China in the 1990s.

2) Adaptive

The contents of green design, virtual design, and innovative design in modern design not only meet the needs of national economy development and construction but also meet the requirements of students to improve their innovation ability and adapt to market competition. This is one of the important criteria to measure the quality of teaching, therefore, in the introduction of new theories, new technologies, new methods, and other aspects of special attention to practicality, and is included in the reference for students to consult.

3) Flexibility

The content of green design, virtual design, and innovative design in modern design is self-contained: which content is first, which content is later, and which content can be selected according to the actual situation.

4) Extensible

The system and structure of the three parts of green design, virtual design, and innovative design in modern design can adapt to the development of science and technology, and can add new contents and achievements at any time.

5) Emphasis on the application of theory and practice

In the three parts of green design, virtual design, and innovative design in modern design, there are some theoretical expositions and a large number of application examples, which can further enhance students' engineering consciousness.

6 CONCLUDING REMARKS

In the teaching reform, it is one of the effective ways to cultivate students' innovative ability to choose the latest achievements of modern science and technology and adopt flexible teaching methods [5].

For strengthening independent innovation, adapting to the competition of the market, cultivating high-quality talents with innovative spirit, innovative ability and practical ability is the key. Institutions of higher learning, especially vocational colleges with skills, as a high ground for training talents, have an unshirkable responsibility [6]. We should pay attention to the cultivation and improvement of students' innovative consciousness and ability in every link of cultivating students.

ACKNOWLEDGMENTS

This research is supported by The Research on Education Carrier and Platform of Engineering Students Innovative and Entrepreneurship (18-ZD-05) (2018–2020).

REFERENCES

[1] Wei Chen. Modern design. Hefei: Anhui people's publishing house, 2012
[2] Chuan-lin Jiang. Cultivating Innovation Ability: The Inevitable Requirement of Higher Vocational
[3] Education Reform. Teaching and Educating (Higher Vocational Forum), 2015 (18): 40–41
[4] Mao-gang Su. Cultivating Innovation Ability and Improving Students' Quality. Literature Education (2), 2016 (10): 83
[5] Heng Luo. Practice and Thinking of Communication Principle Course Construction [J]. Education and Teaching Forum 2019 (29): 180–181
[6] Hui Shen. Practice and Reflection on the Course Construction of Semiconductor Materials. Education Modernization, 2019 (78): 86–87 + 141
[7] Jian-feng Guan. Exploration on Improving Comprehensive Innovation Ability Based on Curriculum Group Construction and Process Teaching, Contemporary Education Practice and Teaching Research, 2020 (12): 181–182

Computational Social Science – Luo, Ciurea & Kumar (eds)
© 2021 Taylor & Francis Group, London, ISBN 978-0-367-70193-2

Research on the protection of citizens' rights and interests in the law of administrative penalties for public security

C. Xu & Z.B. Pei
Dalian Ocean University, Liaoning, China

ABSTRACT: This paper mainly puts forward suggestions on the legal provisions and law enforcement process in order to achieve the purpose of strengthening the protection of citizens. Law of the People's Republic of China on Administrative Penalties for Public Security stipulates the basis of punishment for acts violating public security administration by public security organs, but the main purpose of this law is not to grant certain powers to public security organs. The primary objective of the law is to protect the rights and interests of citizens, and the process of amending the law proves this. However, after communicating with law enforcement officials and some citizens, the author finds that there are some problems about the protection of citizens' rights and interests in the application of the law. These problems are the excessive administrative discretion space, the hearing system is not perfect, the operability of some clauses is not strong, and the law enforcement personnel do not pay enough attention to the procedural provisions. In order to better protect the rights and interests of citizens, the author adopts the method of combining theory with practice and puts forward countermeasures.

1 THE NEED FOR PROTECTING THE CITIZENS' RIGHTS AND INTERESTS

1.1 *The value objectives of the law of administrative penalties for public security*

In fact, legislation is a process of realizing the legislative value, balancing the needs of interests, and resolving conflicts. The purpose of legislation is to reflect the value goal pursued by legislation. The current Law on Public Security Administration and Punishment is slightly different from the Regulations of the People's Republic of China on Administrative Penalties for Public Security promulgated in 1994 in terms of legislative purpose, but its value target is greatly different. Previous legislation purpose is to "strengthen the administration of public security, maintain social order and public safety, protect the lawful rights of citizens and guarantee the smooth progress of the socialist modernization," and the current legislative purpose is to "maintain the order of public security, safeguard public safety, protect the lawful rights and interests of citizens, legal persons and other organizations, and regulate and guarantee performance of the duties for administration of public security by public security organs and people's police according to law." The increase of words and the change of word order indicate that the Law of Administrative Penalties for Public Security gradually takes "protecting the lawful rights and interests of citizens" as the important core value.

1.2 *Legitimate and reasonable administration can set up a good image of state agencies*

At the present stage, the key problems in China's law enforcement are that there are laws that are not followed, that are not strictly enforced, and that violations of the law are not prosecuted. The law enforcement system is separate in power and responsibility, with multiple enforcement agencies and selective enforcement. Law enforcement is not standardized, strict, transparent, or civilized, and the public has a strong response to unfair and corrupt enforcement. So, in order to show a good image of a state agencies and ease the contradiction between the people and state agencies, law enforcement officials must make sure that every administrative act is legal and rational and make every effort to let the people have every perception of fairness and justice.

1.3 Legitimate and reasonable administration is an inevitable requirement of building a socialist country governed according to law

The life of law lies in its execution, and the authority of law also lies in its execution. Administrative penalties for public security is an important part of the public security management and is an important method of the public security organs to implement administrative management. It is to investigate and punish all kinds who disobey action of public security management, prevent and reduce illegal crime, maintain social public order, and protect the lawful rights and interests of citizens who play an important role. Law enforcement personnel must be around the administration according to law, and to carry out the administrative responsibility, adhere to the rule of law, ruling according to law, and promote administration according to law.

2 LEGAL PROBLEMS IN ADMINISTRATIVE PENALTY FOR PUBLIC SECURITY

2.1 Administrative discretion is too wide

Compared with industrial and commercial administration, public health, urban construction, and other administrative punishments, the power of administration penalty for public security has the characteristics of special coercion, severe sanction, and authoritative guarantee. Its operation is directly related to the rights and interests of citizens such as personal freedom, reputation, and property right. In view of the above characteristics, legitimate and reasonable law enforcement is one of the most important ways to protect citizens' rights and interests, the public security organs must follow the principle of legitimacy in the process of administrative law enforcement, and they cannot make serious unreasonable specific administrative acts. However, the power of administration penalty for public security is active, and it is inevitable that there will be omissions when it is standardized only by the external supervision mechanism, which is not enough to achieve the ideal effect. Therefore, it is necessary to standardize the discretion from the inside, and individualize the discretion within a reasonable range. According to the current situation, law enforcement personnel have independent decision-making power. The determination of the severity of circumstances and the punishment are unified law enforcement agencies, but most of the direct basis of the determination of the severity of circumstances is not publicized. The lack of external supervision results in the phenomenon of excessive administrative discretion.

2.2 The hearing system is imperfect

The hearing system is a legal system formed by the administrative organ informing the reasons for the decision and the right to a hearing before the administrative organ makes a decision affecting the citizens' legal rights and interests of the counterpart, to make sure that the citizens expressing opinions, providing evidence, and the administrative organ hearing opinions and accepting evidence. The fundamental purpose of establishing the hearing system is to ensure citizens' exercise of the right to know and the right to participate, thus limiting the arbitrariness of administrative organs in exercising the power. The provisions mentioned in article 42 of the Law of the People's Republic of China on Administrative Penalties for Public Security, an administrative organ, before making a decision on administrative penalty that involves ordering for suspension of production or business, rescission of business permit or license or imposition of a comparatively large amount of fine, shall notify the party that he has the right to request a hearing. Law of the People's Republic of

China on Administrative Penalties for Public Security has further developed the hearing scope on the basis of the Law of the People's Republic of China on Administrative Penalty. The provisions mentioned in article 98 of the Law of the People's Republic of China on Administrative Penalties for Public Security, states that before deciding to revoke the license or to impose a fine of not less than 2,000 yuan as a penalty for the administration of public security, the public security organ shall notify the person who commits an act against the administration of public security that he has the right to demand the holding of a hearing; and where the said person demands a hearing, the public security organ shall, in a timely manner, hold the hearing according to law. This regulation enhances the operability of the scope of application of the hearing procedure. However, for one of the punishments in the Law of the People's Republic of China on Administrative Penalties for Public Security, administrative detention has not yet provided for a hearing. Administrative detention belongs to the punishment of personal freedom. In view of the personal right, it absolutely takes precedence over the property right; administrative detention is the most severe punishment in administrative penalty for public security. The influence of restricting the personal freedom to the administrative relative person is clearly more important than revoking a license or a large amount of fine. However, the law has not stipulated about that. It only says in article 107, where a person penalized who refuses to accept the decision on a penalty of administrative detention applies for administrative reconsideration or bring an administrative suit, he may apply to the public security organ for a deferred enforcement of the administrative detention. The most severe punishment does not bring it into the scope of application of the hearing system, so it hinders the protection and relief of citizens' rights and interests.

2.3 The operability of some laws is not perfect

The provisions mentioned in article 97 of Law of the People's Republic of China on Administrative Penalties for Public security, where a decision on penalty of administrative detention is made, the family members of the person penalized shall be notified without delay. In addition, article 83 and article 84 also stipulate the duty of law enforcement officers to inform their family members. The family notification system was originally intended to inform the family members about the whereabouts of the parties to prevent trouble, but also to prevent law enforcement agencies from illegally controlling the personal freedom of citizens, and its ultimate purpose is to protect human rights. However, in the process of law enforcement, some people who were sentenced to administrative detention refused to provide their family members' contact information in order to conceal their illegal acts from their families, or because of the resistance, so that law enforcement officials could not fulfill their obligation to inform their families.

In addition, the provisions mentioned in article 10 of Law of the People's Republic of China on Administrative Penalties for Public Security, penalties for acts against the administration of public security are divided into the following types: warning, fine, administrative detention, and revocation of licenses issued by public security organs. But the frequency difference of these four kinds of punishment is big, among which the fine punishment is more prominent. Law of the People's Republic of China on Administrative Penalties for Public Security in involving specific illegal acts and punishment have 54 punishments, including 31 can be fined, 29 kinds of detention that at the same time can be concurrently fined, and 17 kinds of detention at the same time that must be concurrently fined in the situation. It can be seen that the law enforcement in the imposition of a fine is universal. However, if the administrative counterpart refuses to perform after being fined a small amount, only article 51 of the Law of the People's Republic of China on Administrative Penalty can be applied. If the parties fail to carry out the decision on administrative penalty within the time limit, the administrative organ that made the decision on administrative penalty may adopt the following measures: (1) to impose an additional fine at the rate of 3% of the amount of the fine per day; (2) in accordance with law, to sell by auction the sealed up or seized property or things of value or to transfer the frozen deposits to offset the fine; and (3) to apply to a People's Court for compulsory enforcement. In order to enforce the law, law enforcement personnel not only need to use a large number of judicial resources to deal with complicated legal procedures but also make the punishment and deterrence of administrative punishment greatly reduced, if unable to effectively maintain public security and order, and citizens' rights and interests are in a vulnerable state of danger.

2.4 *Law enforcement officials did not pay enough attention to procedural provisions*

For the protection of citizens' private rights, although judicial review can play a certain role, but after all, it is post-relief, and with a certain cost of time, money, and so on. In order to protect citizens' rights and interests in a timely and effective manner, law enforcement agencies should regulate law enforcement, avoid the illegal process and results, and reasonably protect citizens' rights and interests within the legal limits. At present, the more common administrative illegal acts in practice are mostly procedural illegal. Due to the influence of historical development and traditional ideas, some law enforcement officials in China have not completely put procedures and entities on an equal footing. In the actual process of public security administration punishment, there are problems in the three links of making the punishment decision, explaining the reasons and informing the right, making and serving the form of decision for administrative penalty. For example, the provisions mentioned in article 100 and article 101 of the Law of the People's Republic of China on Administrative Penalties for Public security, where the facts about the act committed against the administration of public security are clear and the evidence is irrefutable, for which a warning is to be given or a fine of not more than 200 yuan is to be imposed, the decision on such an administrative penalty for public security may be made on the spot. Where a decision on an administrative penalty for public security is made on the spot, the people's policeman shall show his work certificate to the person who commits an act against the administration of public security and fill out a written decision on the penalty. The penalties for fines of between 200 yuan and 500 yuan can only be dealt with in accordance with normal procedures. In practice, some of the written punishment decisions issued by police stations are sealed in advance and kept or carried by the police handling the case. As a result, some of the police also make punishment decisions on the spot for those who can be fined between 200 yuan and 500 yuan, in violation of the procedural requirements.

3 CAUSE ANALYSIS OF RELATED PROBLEMS

3.1 *The scope of authority and responsibility is unclear*

One of the reasons why law enforcement personnel show too much discretion is that the scope of power and responsibility caused by vague legal language is not clear. Taking the provisions of Article 66 of the Law of the People's Republic of China on Administrative Penalties for Public Security as an example, a prostitute or a person who goes whoring shall be detained for not less than 10 days but not more than 15 days and may, in addition, be fined not more than 5,000 yuan; and if the circumstances are relatively minor, she or he shall be detained for not more than five days or be fined not more than 500 Yuan. In the sentence "be fined not more than 5,000 yuan," the "may" premise is lacking; There is no relevant regulation on the use of "or" in the case of being detained for a maximum of five days or fined for a maximum of 500 yuan; "the circumstances are relatively minor" how to identify, and the lack of relevant provisions. In jurisprudence, it is generally believed that the value of personal freedom is greater than the value of property, so the choice and use of the two should be cautious. However, in the absence of relevant provisions, the discretion space of law enforcement officials appears too extensive. In the case of imperfect laws, the law enforcement process of law enforcement personnel is relatively complex, which includes the interpretation of legal provisions. The acts of law enforcement departments to interpret and supplement laws, in a sense, violate the principle of legality, so it is necessary to make clear the scope of rights and responsibilities of law enforcement officials.

3.2 There are no relevant provisions in the law

Our country's administrative hearing includes three kinds: administrative legislation hearing, administrative penalty hearing, and price decision hearing. At present, only a few laws have introduced the hearing system, and the content is simple, the setting is rough, the procedure is not perfect, and it is difficult to achieve the operation standard in practice. Due to the lack of a unified and feasible hearing system, the administrative organ's hearing behavior has no rules to follow. Other relevant systems can fill the vacancy of the hearing system to some extent. The provisions mentioned in article 107 of the Law of the People's Republic of China on Administrative Penalties for Public Security, where a person penalized who refuses to accept the decision on a penalty of administrative detention applies for administrative reconsideration or bring an administrative suit, he may apply to the public security organ for a deferred enforcement of the administrative detention. If the public security organ believes that such deferment will not create danger to the society, and after the person penalized or his close relative proposes a guarantor who conforms to the conditions provided for in Article 108 of this Law, or after the bail is paid at the rate of 200 yuan per day for the administrative detention, the decision on the penalty of administrative detention shall be deferred from enforcement. Although the guarantor and bail system can postpone the execution of administrative detention, which seems more convenient and effective than the hearing system, it is not the fundamental remedy way after all, contrary to the legal principle. Moreover, there are also many problems in the operation of the guarantor and bail system itself. Therefore, the relief of the rights and interests of the administrative counterpart who is sentenced to administrative detention should still be stipulated from the direction of improving the hearing system.

3.3 The actual effect of law enforcement was not taken into account

The family notification system is set up as the duty of law enforcement officials, to protect human rights, so as to avoid derivative events such as "drug-abusing mothers are caught and young girls starve to death at home" through mandatory regulations. However, people have different perspectives on the system of informing family members. For example, article 66 of the Law of the People's Republic of China on Administrative Penalties for Public Security provides that, a prostitute or a person who goes whoring shall be detained. Contacting article 97 "after the detention to inform the family" of the provisions, some people mistakenly caught whoring will notify the wife. Some people refuse to provide their family members' contact information out of concern for preventing family conflicts or damaging their reputation. In addition, some administrative counterparts summoned, questioned, or detained did not cooperate with law enforcement officials due to their conflicting emotions, which also made it impossible for law enforcement officials to fulfill their obligation to inform their families.

As for the problem of malicious refusal to pay fines, law enforcement officials can only impose a daily late fee of 3%, but most of them are of no help. However, such practices as seizing property auctions and freezing deposit transfers will greatly increase the cost of law enforcement, which is also difficult to operate in practice. It is even more unrealistic to apply to the people's court for compulsory execution, and sometimes the enforcement of the court needs the assistance of the public security organs.

3.4 Program awareness and external supervision need to be improved

Procedural justice is also called the "visible justice," which means that in the process of the administrative legislation, administrative decisions, administrative law enforcement, and administrative judicial process, administrative subject and the administrative relative person shall comply with the statutory administrative procedures to ensure the exercise of administrative functions and powers in accordance with the standardization of the program requirements, realize the impartiality of the administrative process is reasonable, meet the public's predictability and acceptability. Under the influence of China's traditional feudal autocracy, many law enforcement officials have a sense of superiority, and the concept of "attaching importance to substance over procedure" is relatively common. Many law enforcement officials focus on cracking down on illegal acts and fail to fully fulfill the procedural requirements of the law. In the environment where procedural violation is widespread among law enforcement personnel, few administrative law enforcement personnel are punished for this, which leads to the increasingly weak procedural awareness of administrative law enforcement personnel.

In addition, the external supervision system for law enforcement is not perfect and needs to be improved. Article 114 of the Law of the People's Republic of China on Administrative Penalties for Public Security stipulates that, in handling cases of public security, public security organs and the people's police shall voluntarily accept supervision by the society and citizens. However, as there is no clear supervision subject, supervision channel, acceptance subject, and so on in the articles, and there is no unified regulation on the supervision feedback system, the supervision mechanism of society and citizens still has great room for improvement.

4 COUNTERMEASURES AND SUGGESTIONS

4.1 *Show the power list in public*

The most fundamental reason why the legislature authorizes the discretion of the administrative organ is that the discretion can meet the requirement of individual case justice in the application of law to a greater extent under the state of formal rule of law. Therefore, to regulate the discretion is not to object to the discretion equal to the work undertaken by the government, but to object to the discretion beyond the needs of these work and eliminate unnecessary discretion. In order to effectively standardize, in addition to internal constraints, it is suggested to show the power list in public, so that the public, especially administrative counterparts, can know the law enforcement basis and scope of power of law enforcement personnel, so as to achieve the effect of communication and external supervision to some extent. Power list refers to the functions and powers exercised by the government and government departments, which are enumerated in the form of a list. Administrative organs shall perform their functions and exercise their powers in accordance with the lists established by laws and regulations. Administrative organs shall not exercise powers that do not fall within the scope of the lists. Showing the list of powers is to give the masses a legal weapon, encourage them to supervise the actions of administrative law enforcement departments, and enable them to have the consciousness of safeguarding their rights when administrative law enforcement personnel exercise their functions and powers beyond their business scope, and to make a reasonable statement, plea, or reconsideration suit in accordance with the law. This is conducive to the elimination of the phenomenon of transcending authority.

4.2 *Broaden the scope of the hearing*

The personal freedom penalty will give great effect to the administrative relative person's rights and interests, and procedural justice itself is very important, so never give up the idea of a sound system of administrative hearing just because of setbacks in reality. And never make the simple judgement that our country has enough to provide sufficient protection of human rights because of the existence of a reprieve system and administrative litigation. A reasonable hearing system should be designed for the reasons that administrative detention is not easy to be heard. One of the reasons why administrative detention is not easy to be heard is that administrative detention usually needs to be executed immediately, and the holding of a hearing will inevitably bring adverse effects to the execution effect. The second reason why administrative detention is not easy to be heard is that the punishment of restricting personal freedom is usually aimed at administrative counterparts who have illegal acts, serious circumstances, and serious consequences. Such counterparts' personal freedom will not be controlled and they may escape after they have a hearing opportunity. In view of the above problems, the system of hearing agent of administrative detention can be established to ensure the involvement of professional legal personnel such as lawyers. The provisions mentioned in article 42 of Law of the People's Republic of China on Administrative Penalty, the party may participate in the hearing in person, or he may entrust one or two persons to act on his behalf. Administrative detention may also refer to other hearing procedures for administrative penalties such as ordering to suspend production or business, revoking a license, or imposing a relatively large fine, stipulated that the party may participate in the hearing procedure and has the right to hire a lawyer or other agent to accompany or represent him to participate in the hearing. In view of given administrative detention and order to suspend production or business operation, revoke the permit or license, a large amount of fines and other administrative punishment has great difference, we can refer to criminal procedural law, the relevant legal aid practice, set in the administrative detention hearing procedure for lacking a hearing agent, the parties to give legal services of an attorney, for they will be able to more comprehensive knowledge in administrative detention hearing procedure to provide legal aid, enhance the consciousness of them to participate in legal proceedings and motivation.

4.3 *Amend part of the law*

The system of notifying family members is conducive to the protection of citizens' rights and interests. It cannot be abolished only because of various problems in practice. It should be adjusted according to the above problems. At present, it is the duty of law enforcement personnel to notify family members, namely the right of administrative counterparts. It is suggested that appropriate detailed provisions should be made in legal provisions, such as clarifying the meaning of family members. In addition, it is suggested that the duty of law enforcement officials to inform family members should be changed to the duty of interpretation. In addition to the duty of informing family members of minors, the law enforcement officials should only inform adults of their right to inform family members, and explain that they have the right to choose family members to protect privacy. If the adult administrative counterpart expressly waives the right to inform the family member, it shall sign or stamp on the declaration of waiver.

Fine problem in practice did not achieve ideal effect of law enforcement, in addition to the compulsory enforcement issues, Law of the People's Republic of China on Administrative Penalties for Public Security is to impose administrative detention, and impose a fine regulation occupied a certain proportion, a part of public acceptance is not high, the practice often has malicious refuse to pay the fine, so it is suggested to reset the proportion of fines to administrative detention.

4.4 *Improve external supervision*

In practice, many clues of illegal cases come from procuratorial organs. There are few clues for a citizen, legal person, or other organization to lodge a complaint or report. In addition to filing an administrative lawsuit, the administrative counterpart can know the relevant materials at the pretrial exchange stage, but the other approaches are not feasible. Relying solely on existing sources of leads would be a drop in the ocean to monitor the vast administrative enforcement activities. Therefore, it is suggested to extend and protect the administrative counterpart's right to know effectively by means of legislation, and give them the right to read papers under certain conditions. In addition, it should perfect the acceptance system that citizen supervises. The admissibility that supervises to the citizen is the foundation that realizes citizen effectively and safeguards, involve a lot of interest concern inside, a need to have a system to protect certainly thereby. In order to avoid the phenomenon of pushing each other down, not being serious and responsible, pretending to accept a false offer, and deliberately delaying, the relevant staff involved in this phenomenon should be severely punished and actively publicized.

5 EPILOGUE

Today, with the rapid development of economy, China should not only become prosperous and powerful, but also realize the rule of law and equity. The rule of law requires that government power be standardized and citizens' rights be guaranteed. For an administrative organ, its staff members shall, within their statutory functions and powers, conduct lawful and reasonable administrative acts through legal procedures. For citizens, they should strengthen their legal concept, treat the law as a weapon as well as a shield, and know how to protect their rights and interests through legal means to prevent illegal infringement.

REFERENCES

[1] Zheng Liyuan. Application of Discretionary Power in Police Law Enforcement. Academic Frontiers, 2019(07).
[2] PI Zhongxu. Introduction and Omission of The Law on Punishment of Public Security Administration (Revised Draft for Public Opinion Solicitation). Journal of Hubei Police College, 2017(03).
[3] Shi Huadong. Research on the Regulatory Path of Punishment Power in China's Public Security Administration. Youth Law Court, 2017(05).
[4] Zhang Ximing. On the Amendment and Improvement of The Law on Public Security Administration and Punishment. Journal of Liaoning Police College, 2016(06).
[5] Wei Jihua. Discussion on the Protection of Private Rights in The Law on Public Security Administration and Punishment. Journal of Zhongzhou University, 2014(08).

Computational Social Science – Luo, Ciurea & Kumar (eds)
© 2021 Taylor & Francis Group, London, ISBN 978-0-367-70193-2

Research on the cognitive needs of mental health in primary and middle school students and the construction of mental service system

J.M. Fan, X.L. Ni, Y.W. Geng & Y.P. Wang
School of Humanities and Social Science, Xi'an Jiaotong University, Xi'an, China

ABSTRACT: Primary and middle school students have a high demand for tmental health services. However, the primary and secondary schools still do not have a relatively complete mental health service system in China. We can improve the mental health of primary and secondary school students from four aspects of formulating effective mental health education plans, strengthening relevant training and professional supervision of faculties, establishing the parent and school assisted and coordinated social mental services, and give a full play of school club activities.

1 INTRODUCTION

On October 18, 2017, the General Secretary Xi Jinping clearly proposed in the report of the 19th National Congress of the Communist Party of China of "Implementing the Healthy China Strategy," and emphasized "strengthening the construction of social mental service system, and cultivating a self-esteem and self-confidence, rational and peaceful, and positive social mentality." In November 2018, the National Health Commission and other 10 departments have jointly issued the "National Mental Service System Construction Pilot Work Plan." After that, in July 2019, it had further clarified the implementation plan and pilot cities for the construction of the social mental service system pilot work. Currently, the social mental services in China mainly center on the social mental health services. However, the primary and secondary schools in China still do not have a relatively complete mental health service system, lacking relevant professional practitioners, resulting in seriously affecting the provision and access of the mental health services.

On this basis, this research group has carried out a survey and analysis on the current situation of the social mental service system for primary and secondary school students in Shaanxi Province from September to December 2019, in order to explore and discuss accessible methods and ways to promote the construction of the mental health service system in primary and secondary schools and provide government with the theoretical and empirical evidence for the policy formulation and implementation of mental health services.

2 RESEARCH OBJECTS AND METHODS

This study adopts the stratified sampling method to investigate the mental health status and mental health service demands of 2,350 students of different school ages from 12 schools in the six districts and counties in three cities in Shaanxi Province. All data collected has been recorded through Epidata, and it adopts the SPSS 20.0 for statistical analysis.

Table 1. Mental health status of primary and secondary school students.

	Normal Detection Rate	Edge Level Detection Rate	Abnormal Detection Rate
Emotional symptoms	81.20%	7.00%	11.80%
Conduct	75.70%	12.60%	11.70%
Hyperactivity	80.20%	8.80%	11.10%
Peer interaction	54.70%	33.90%	11.40%
Prosocial behavior	83.20%	10.70%	6.20%

Table 2. Gender differences of mental health status of primary and secondary school students.

	Normal Detection Rate		Edge Level Detection Rate		Abnormal Detection Rate	
	Boys	Girls	Boys	Girls	Boys	Girls
Emotional symptoms	82.40%	80.20%	6.80%	7.10%	10.70%	12.70%
Conduct	73.70%	78.10%	12.40%	12.50%	13.90%	9.40%
Hyperactivity	79.20%	81.60%	9.90%	7.70%	10.90%	10.70%
Peer interaction	52.40%	57.00%	34.20%	33.40%	13.40%	9.60%
Prosocial behavior	80.60%	85.80%	11.10%	10.30%	8.30%	3.90%

Table 3. Learning period differences of mental health status of primary and secondary school students.

	Normal Detection Rate		Edge Level Detection Rate		Abnormal Detection Rate	
	Primary School Students	Secondary School Students	Primary School Students	Secondary School Students	Primary School Students	Secondary School Students
Emotion Symptoms	81.90%	80.60%	7.00%	7.10%	11.10%	12.30%
Conduct	74.50%	76.70%	12.40%	12.80%	13.10%	10.40%
Hyperactivity	86.00%	75.10%	7.20%	10.10%	6.80%	14.80%
Peer interaction	46.50%	61.80%	39.50%	29.20%	14.00%	9.00%
Prosocial behavior	84.10%	82.40%	9.30%	11.80%	6.60%	5.80%

3 RESEARCH RESULTS

3.1 Basic situation of the mental health

This study adopts the Strengths and Difficulties Questionnaire (SDQ) compiled by the American psychologist Robert Goodman to evaluate the behavior and emotional problems of primary and secondary school students (Goodman et al., 2000). It also adopts the student self-evaluation version (Jianhua Kou et al., 2007) to evaluate the mental health in the past six months, adopting a three-level rating method of 0–2 from "inconformity" to "complete conformity" to survey the mental health status of primary and secondary school students from aspects such as emotional symptoms, conduct problem, peer interaction, and so forth.

3.2 Mental service demands

3.2.1 Cognition of mental health services

The study shows that 43% of primary and secondary school students believe that mental health services can effectively help solve their troubles in life, while only 5% of primary and secondary school students express that they do not believe that mental health services can effectively help solve the troubles in their life.

3.2.2 Motivation of mental health service

Referring the motivation of asking for mental health services, 38% of primary and secondary school students express that they are likely to ask for mental health services, 20% of primary and secondary school students express that they are unsure whether they will ask for mental health services or not, and 7% of primary and secondary school students express that they will definitely not ask for mental health services. And the reasons for reluctance to ask for mental health services are diversified. At the same time, 33% of elementary and middle school students chose to solve the problem by themselves, and 20% of elementary and middle school students express that they have no idea from where can they get help.

3.2.3 Willingness for accepting the mental health service

The study shows that 85% of students think that schools should establish mental health services organizations; 31% of students express that they need mental health services, while 41% of students are unsure whether they need mental health services or not, and 28% of students think that they do not need mental health services.

3.2.4 Mental health service demand content

Referring to the content of the school mental health services, 15% of students hope that schools can provide mental health knowledge, 21% of students hope that schools can provide services for adjusting academic stress, and 18% of students hope that schools can provide services for emotional adjustment, 13% of students hope that schools can provide services for peer relationships, 11% of students hope that schools can provide services for the relations between teachers and student, 7% of students hope that schools can provide services for behavior modification, and another 8% of students hope that schools can provide memory-related services.

3.3 Analysis on the differences in mental health services

The statistical results show that there exist significant gender differences in the stigma of mental health services of primary and middle school students. Compared with girls, boys show a higher score on the stigma ($t = 3.196, p < 0.01$). And the specific results are as shown in Table 4.

Meanwhile, the statistical results show that there exist significant grade differences in the cognition and demand of mental health services of elementary and middle school students. Specifically speaking, the elementary school students believe that mental health services can effectively help us solve the problems in life ($t = 7.787, p < 0.001$), have a higher demand for mental health services ($t = 2.435, p < 0.05$), and have a stronger willingness to actively ask for mental health services while encountering mental distress or problems ($t = 5.618, p < 0.001$).

Table 4. Gender differences in the cognition and demand for mental health services of primary and secondary school students.

	Boy (n = 1096)	Girl (n = 1254)		
	$\bar{x} \pm s$	$\bar{x} \pm s$	t	p
Credibility in the effectiveness of mental health services	3.59 ± 0.904	3.63 ± 0.781	-1.133	0.257
Demand for mental health services	3.01 ± 1.019	3.06 ± 0.941	-1.21	0.226
Willingness to ask for mental health services	3.17 ± 1.305	3.22 ± 1.331	-0.854	0.393
Stigma of mental health services	2.07 ± 0.912	1.96 ± 0.784	3.196	0.001

Table 5. Grade differences of the cognition and demand for mental health services of primary and secondary school students.

| | Primary School | Secondary School | | |
	Student (n = 1136) $\bar{x} \pm s$	Student (n = 1241) $\bar{x} \pm s$	t	p
Credibility in the effectiveness of mental health services	3.75 ± 0.884	3.48 ± 0.776	7.787	0.000
Demand for mental health services	3.09 ± 1.039	2.99 ± 0.916	2.435	0.015
Willingness to ask for mental health services	3.35 ± 1.394	3.05 ± 1.226	5.618	0.000
Stigma of mental health services	2.05 ± 0.930	1.98 ± 0.762	1.913	0.056

4 COUNTERMEASURES AND SUGGESTIONS

4.1 Set up effective mental health education plan

The middle school shall guarantee a mental health education class plan of at least two hours per week, so that students can understand and recognize the relevant knowledge of mental health services; and the primary schools can appropriately formulate the mental health education plans centering on mental practice activities.

4.2 Strengthen the training and professional supervision of faculties

Since most of the mental health teachers in school are part-time workers, it shall organize regular professional training and supervision for the faculties in primary and secondary schools and carry out professional exchanges and practice sharing activities.

4.3 Establish the parent and school-assisted and coordinated social mental services

It shall adopt the voluntary participation form in elementary and middle schools, regularly inviting psychologists to explain and publicize the mental health education to parents, so that parents can discover their children's mental problems in the early stage, and solve the mental problems together.

4.4 Take advantage of the school club activities to improve the mental health literacy

In schools, it shall improve students' mental health literacy and positive personality shaping through club activities, especially the mental health education practice carried out centering on the club practice activities for primary school students.

ACKNOWLEDGMENTS

The Soft Science Fund of Shaanxi, China (2019KRM193).

REFERENCES

Goodman R, Ford T, Simmons H, et al. Using the strengths and difficulties questionnaire (SDQ) to screen for child psychiatric disorders in a community sample. British Journal of Psychiatry, 2000, 177: 534–539.
Jianhua Kou, Yasong Du, Liming Xia. Formulation of the Shanghai Norm of the Strengths and Difficulties Questionnaire (Student Edition). Chinese Journal of Health Psychology, 2007, 15(1): 3–5.

Computational Social Science – Luo, Ciurea & Kumar (eds)
© 2021 Taylor & Francis Group, London, ISBN 978-0-367-70193-2

On the approaches to the rural migrant workers' children enjoying equal access to compulsory education in China

S.S. Yan & J. Cao
College of Marxism, Zaozhuang University, Zaozhuang, China

ABSTRACT: In the past over seventy years since the founding of the New China, great achievements have been made in the compulsory education. The acceleration of industrialization and urbanization gives rise to a normal of peasants going to cities as rural migrant workers. They also take to the cities their children whose need for education has posed a severe challenge for the equal access to education in China. Currently in the crucial period for the great rejuvenation of the Chinese nation, meeting the rural migrant workers' children need for equal access to compulsory education is of great significance for the rapid development of the compulsory education, the promotion of modernization and eventually the realization of the Chinese dream.

1   INTRODUCTION

Against the background of the increasingly fierce competition in the comprehensive national strength, science and technology, as the first productive forces, are playing a more and more important role in the social advancement and national development. The advanced science and technology is based on the high-quality education, the premise of which is the equal access to education. Since the introduction of the policy of the reform and opening up, great achievements have been made in education in China. However, with more and more rural migrants going to the cities, their children's equal access to education has gradually become a social focus. Broadly speaking, the rural migrant workers' children can be roughly divided into two groups, i.e. "floating" children who follow their parents to live in the cities and "stay-at-home" children who live in their household registration location. The rural migrant workers' children discussed in this paper refer to the floating ones.

2   STATUS QUO OF THE COMPULSORY EDUCATION OF THE RURAL MIGRANT WORKERS' CHILDREN

Rural migrant workers account for the majority of the industrial workers in China, to whom great emphasis has always been attached by CPC and the whole country. According to the statistics released by the Ministry of Human Resources and Social Security, by the end of 2019 the number of the rural migrant workers reached 290 million of which 170 million went to cities as rural migrant workers (http://house.china.com.cn/home/view/1630490.htm). The survey done by some scholars shows that in the phase of the compulsory education in China the percentage of the rural migrant workers' children increased from 6.43% in 2009 to 9.76% in 2015 (Based on the data in Statistic Introduction to the Development of the Chinese Education Cause 2015, compiled by

Development Planning Division of the Ministry of Education), which has posed a great challenge to the equality of compulsory education. In recent years, with the constant adjustment and evolution of the educational policy, for example, the two "oriented" policy made by the State Council in 2003, that is, the government(in charge of the place where the floating children live)-oriented and the public schools-oriented, the rural migrant workers' children have become the subject enjoying equal access to education. In the Report of the 17th National Congress of CPC, it's suggested that we should run education to the satisfaction of the people and that we should optimize the educational structure, promote balanced development of compulsory education, regulate the collection of education-related fees, and ensure that children of rural migrant workers in cities enjoy equal access to compulsory education as other children. In the Report of the 18th National Congress of CPC, it's noted that we should work hard to run education to the satisfaction of the people, ensure that children of rural migrant workers in cities have equal access to education, and help all children gain required knowledge and skills. In the Report of the 19th National Congress of CPC, it's emphasized that we should strive to see that each and every child has fair access to good education. In *Suggestion on Enhancing the Care and Protection of the Stay-at-home Children in Rural Areas* printed and distributed by the State Council in Feb. 2016, it's made clear that the public compulsory schools should be open to the rural migrant workers' children in cities. Though high attention has been paid to the compulsory education of the rural migrant workers' children, in reality they are faced with much less fair access to the compulsory education.

In terms of financial system, the compulsory education expenditure is in the charge of the local governments and scattered in most cases, which makes it difficult to ensure that the rural migrant workers' children can fulfill their compulsory education successfully in the cities after they follow their parents and migrate there. Although we once drew lessons from the "education voucher" system, conducted in America, the system put forward by the famous American economist Milton Friedman favoring the government, especially the local government should offer a sum of specific payment, that is, the education voucher(Milton Friedman. The Role of Government in Education in the collection Economics and the Public Interest, Rutgers University Press, 955), only used for their children's common education fees for which the students' parents have self-determined options of letting the school paying or paying by themselves. In Changxing County, Zhejiang Province, the system education voucher was first explored by distributing vouchers with denominations ranging from 100 yuan to 500 yuan first to students in vocational schools and the private compulsory education ones and then to those in private general high schools and weak high schools [9]. Although to some extent the unfairness in education has been solved, some schools gained comparatively less sum of vouchers. Moreover, the limited vouchers were confined to the funding of part of the students who couldn't cash them. The result is that the chances of the students of poor families accepting education are only increased without giving them the real options for the schools they want to go. These inadequacies make it hard to expand vouchers broadly.

In terms of social situations the rural migrant workers' children face in going to school, in some cities if the they want to attend the public schools they must submit various related evidentiary materials like temporary residence permit, migrant worker verification and the certificate of having no custody in the place they come from, etc. Pus, there exists a phenomenon of some schools collecting fees in disguised forms. All these in fact have raised the threshold of the rural migrant workers' children going to school. Even if they are admitted to the public schools, many schools are not really willing. Actually they are forced to do that under the pressure of policies made by the government. As a consequence, the rural migrant workers' children are very easily subjected to being ignored in practice. Thus, they have no choice but to attend the private schools where the quality of education cannot be guaranteed. What's worse, the series of favorable polices made by the government to protect the relevant rights and interests of the rural migrant workers' children often fail to be implemented due to the inadequate publicity and the less attention paid by the rural migrant workers.

Given the compulsory education status quo of the rural migrant workers' children, it's of necessity to make clear the root of the existing unfairness.

3 THE ROOT OF THE RURAL MIGRANT WORKERS' CHILDREN HAVING UNEQUAL ACCESS TO THE COMPULSORY EDUCATION

3.1 *Subsequent influence of the household registration system*

The household registration system, characterized by the dual urban-rural structure and derived from the planned economic system, has restricted quite many rural school-age children moving to the cities. According to Opinions of the State Council on further promoting the reform of household registration system, released officially in July 2014, it's required that the differentiation between agricultural and non-agricultural residence registration, and the derivative blue stamped residence registration should be abolished, and that all should be uniformly registered as household registration [10]. The goal of 60%, the urbanization rate of the permanent resident population, originally scheduled to be reached by the end of 2020, was basically achieved in 2018. Meanwhile, the goal of 45%, the urbanization rate of the registered permanent urban resident population, predicted to be reached by the end of 2020 has been achieved 43.37% by the end of 2018. However, currently the gap between the urbanization rate of the permanent resident population and the urbanization rate of the registered permanent urban resident population remains quite large (http://news.sina.com.cn/c/2019-03-01/doc-ihsxncvf8747115.shtml), which shows that the implementation of the policy in practice is still far from the prediction of the ideal effect. The remaining of the subsequent influence of the dual household registration system makes it still difficult for the rural migrant workers' children to enjoy comparatively equal access to the compulsory education.

3.2 *Inadequate implementation of supervision mechanism*

With the joint efforts of the authorities at all levels, the percentage of the rural migrant workers' children attending the public schools has remained about 80% in recent years (http://www.china.com.cn/education/2017-10/23/content_41777067.htm). However, till now, an effective supervision system has not yet been established against the background of a growing number of the rural migrant workers' children, which leads to the lack of unclear subject of liability concerning the compulsory education of the rural migrant workers' children. Although a majority of the rural migrant workers' children have already attended the public schools, there exists a phenomenon that some schools deliberately separate them from the local students, which forces quite many of them to choose the migrants' schools designed form them. The migrants' schools are characterized by weak faculty and poor conditions, and once there appear some problems, instead of taking effective measures to solve them, the local government will choose to close these schools. Due to the lack of effective supervision mechanism, quite many rural migrant workers' children have to choose to drop out in the phase of the compulsory education.

3.3 *Exclusiveness in the mind of the urban residents*

With the growing number of the rural migrant workers whose family members also follow in most cases, it will inevitably lead to their children sharing the resources of the public education with the urban children, which is something that the urban residents are unwilling to see, assuming that the rural children's inferiority in many respects such as level of education, life style, behavior and habits, cultural accomplishment and so forth will exert negative impact on their children in the same school. The rural migrant workers' children having unequal access to the compulsory education has something with the discrimination deeply rooted in the mind of the urban residents.

The reasons for the rural migrant workers' children having unequal access to the compulsory education are manifold. It should be noted that the migrant workers' children are also the future of our country. Based on the reality of our country, effective measures should be explored to ensure their equal access to the compulsory education.

4 APPROACHES TO THE PROMOTION OF THE RURAL MIGRANT WORKERS' CHILDREN ENJOYING EQUAL ACCESS TO THE COMPULSORY EDUCATION

4.1 *Enhance the supervision and improve the laws and regulations*

Whether a policy is well-implemented or not depends largely on the constraint of laws and the supervision in the process of implementation. To ensure the rural migrant workers' children really enjoy their rights and interests endowed by the relevant policies, the government should further improve the laws and regulations concerning their equal access to the compulsory education. Based on the current mode of multiple liability subjects at the central, provincial, county and local levels responsible for the fiscal distribution of the compulsory education, the mutual supervision between the subjects should be strengthened to ensure the educational expenditure is really used for the education of the rural migrant workers' children. Meanwhile, the public supervision on the subjects responsible for the implementation of the policies should also be strengthened. Once there appears inadequate implementation, they can reflect it to the authorities at the upper level.

Furthermore, parents are usually obliged to coordinate the school rather than are endowed with rights to actively participate in the education of their children for a long time. On the whole the parents' rights to their children's education, esp. their rights to participate in the school education have not been recognized yet in our country [11]. It should also be legitimately recognized and supported that the students' parents, especially when their parents are the rural umigrant workers, can further participate in their children's education in the school. To sum up, the laws need improving and the supervision strengthening. Only in this way can it be effectively ensured that the rural migrant workers' children enjoy equal access to the compulsory education.

4.2 *Construct the education for all and promote the school-home union*

In the respect of educational subsidy, in recent years the state financial expenditure on education always accounts for over 4% of GDP. However, the solution of the rural migrant workers' children having unequal access to the compulsory education requires not only the material support but also the change of the backward views held by some of the citizens. "Education for all" can contribute to the improvement in the citizens' cultivation and mental outlook to some extent. The goal of education for all, already put forward in *World Declaration on Education for All* and *Frame-work for Action to Meet Basic Learning* passed in World Conference on Education for All held in Thailand in 1990, is to enable people to have autonomous learning with dignity to continuously improve their living and at the same time make contributions to their country and mankind by providing them with knowledge, technique and outlooks on world, values and life, etc. *Frame-work for Action to Education for All,* passed by UNESCO in 2000, involves many specific goals for the realization of the education for all. According to Article 2 all of the children, esp. the female children, those in difficult situations, and those of ethnic groups, can be ensured to receive and finish the free high-quality compulsory primary education.

The concept of education for all has something in common with the concept of lifelong learning advocated in China. For the rural migrant workers, receiving education helps them come to realize the power of knowledge in changing their life. In this way not only can they increase their knowledge and thereafter tutor their children, but also become more active to encourage their children in learning. For the urban residents, receiving education helps them, on one hand, improve their cultivation, and, on the other hand, eliminate their discrimination and prejudice against the rural migrant workers and their children, and thereafter live in harmony with them and their children, which will contribute to the building of a harmonious and civilized society. In addition, the solution of the existing problems in the phase of the compulsory education of the rural migrant workers requires the further improvement of the union between the school education, home education and social education. The rural migrant workers, while increasing their own knowledge, need to get to know they should not only rely on the school for their children's education, and that they, as parents, should also be responsible for that in addition to bringing them up. Only by enhancing the communications between the school and the rural migrant workers can their children's growing healthily be more favored.

4.3 Explore fair measures based on the reality of our country

4.3.1 Lower the threshold of the admission age for the compulsory education
According to the Law on Compulsory Education, the mandatory starting age for the compulsory education is 6 years old. But in reality, the children in the developed areas are funded by their parents to learn relevant knowledge in advance before the age of six, whereas those in less-developed areas are unable to do that because their parents don't have enough money. Moreover, even if the children in developed areas and those in less developed areas go to school at the same age, the latter is usually predicted to lag behind due to the lower level of local education. Therefore, it can be concluded that the children in the less developed areas, compared with those in the developed areas, have already lost in the starting line. Given that, the government can consider the option of letting them receive the compulsory education before the age of six. Their earlier reception of compulsory education, funded officially, can help narrow the gap with those in the developed areas.

4.3.2 Extend the age limit of compulsory education
According to the Law on Compulsory Education, the nine-year compulsory education, conducted in China, means that the children can only receive for free two phases of education, that is, the primary and secondary education. However, in some countries like America the age limit of compulsory education has been extended to 12 years. In China it can be considered to extend the age limit gradually according to the reality in different places so as to let more children of poor families acquire more knowledge at the public expense instead of dropping out at an earlier age due to the lack of money.

4.3.3 Conduct the extra tutoring for the migrant children
When many rural children come to the cities following their parents, they are inferior to the urban children in many respects, such as learning progress, comprehensiveness of knowledge, etc. Extra tutoring can be conducted for those who lag behind largely to bridge the gap. The government can recruit some sympathetic scholars or students with good academic performance to volunteer the unscheduled tutoring of the rural migrant workers' children during the weekends or the summer and winter holidays, to help them review or acquire new knowledge so as to improve their academic performance as much as possible. The local authorities of their household registration can also organize volunteers to tutor them when they go back home during the winter and summer holidays.

4.3.4 Improve the evaluation of the students' performance
The current evaluation of the students' performance in the phase of compulsory education is based on the sheer hundred-mark system. In terms of marks, generally the rural migrant workers' children cannot match up to the urban children. We can draw lessons from the mode of the college students' comprehensive evaluation system, for example, academic scores accounting for 80% and other scores 20%. The teacher's evaluation, peer response, activeness in activities and other aspects concerning a student's virtue, intelligence, physical conditions and mental status are all taken into consideration for the total score, which helps to boost the confidence of the rural migrant workers' students to some extent and promote effectively the educational equality.

Whether the rural migrant workers' children can enjoy equal access to compulsory education is crucial to the future of the cause of education of our country. As President Xin Jinping points out, we must put the people's interests above all else, keep up with people's ever-growing needs for a better life and use our hearts and feelings to exert ourselves to solve the practical problems like employment and education that the masses care about (http://www.cssn.cn/jjx_yyjjx/yyjjx_gd/202005/t20200531_5136921.html). Currently, in the new era characterized by the continuous improvement of the teaching quality and the acceleration of educational modernization, based on the further improvement of educational system and management, and the safeguarding of educational equality, the problem of the rural migrant workers' children having unequal access to the compulsory education must be tackled effectively, and then we can run the education to the satisfaction of the people.

REFERENCES

[1] National People's Congress. *Compulsory Education Law of the Peoples' Republic of China* [EB/OL]. http://www.gov.cn.

[2] Gao Yan, Long Baoxin. *An Analysis of Status Quo and Development of Research on Education of Migrant Children Since the 21st Century. Theory and Practice of Education*, 2018, 38(13), 19–23.

[3] Wang Zelong. *On the Education of the Peasants' Children from the Perspective of Educational Equality. Wuhan University Journal* (Humanity Sciences), 2016, (02).

[4] Zhou Jia. *On the Policies of Integration of the Rural Migrant Workers' Children into Cities. Academic Exchange*, 2015, (12):153–157.

[5] Huang Qingchang, Zhang Yang. *Abolish the differentiation between agricultural and non-agricultural residence registration all throughout the country. China Daily*, 2014-07-31.

[6] Shen Suping. *Parents, State and Education of Children. Comparative Education Review*, 2009, (3): 15.

[7] Liu Xaioman. *Research on and Analysis of the Education Voucher in Changxing County, Zhejiang Province and Relevant Thinking.* Beijing University, 2004: 6, 28, 37, 73.

[8] Milton Friedman. The Role of Government in Education in the collection Economics and the Public Interest. Rutgers University Press, 1955.

[9] Liu Xaioman. Research on and Analysis of the Education Voucher in Changxing County, Zhejiang Province and Relevant Thinking. Beijing University, 2004: 6, 28, 37, 73.

[10] Huang Qingxiang, Zhang Yang. Abolish the differentiation between agricultural and non-agricultural residence registration all throughout the country. China Daily, 2014-07-31.

[11] Shen Suping. Parents, State and Education of Children. Comparative Education Review, 2009, (3):15.

Computational Social Science – Luo, Ciurea & Kumar (eds)
© 2021 Taylor & Francis Group, London, ISBN 978-0-367-70193-2

Research and practice on the reform of electrical basic courses based on the cultivation of applied technology ability

L.L. Tang, J. Liu & H.Y. Jiang
*College of Computer and Information Engineering, Zhixing College of Hubei University,
Wuhan, P.R. China*

ABSTRACT: In order to cultivate students' application technology ability as the ultimate goal, this paper puts forward a comprehensive exploration and practice on the system, teaching content, teaching mode and method, examination mode, and examination method of electrical basic courses in application-oriented colleges and universities, so as to improve the teaching quality of electrical basic courses and students' application ability of circuit and electronic technology. This plays an exemplary role, and forms a comprehensive reform plan of similar courses in similar universities, which are advanced in China, first-class in the province, and can be popularized and applied.

1 INTRODUCTION

The series of basic electrical courses, including circuit analysis, analog electronic technology, digital electronic technology, electronic technology experiment, electronic technology curriculum design, and so forth, play an important role in the cultivation of the electronic information specialty in colleges and universities.

However, most of the applied technology-based universities in our country have a vague orientation and blindly seek to keep up with the comprehensive universities. The curriculum system, teaching content, teaching mode and method, experimental practice system and content, and examination and evaluation mode and method are similar, but they still follow the traditional curriculum system, teaching method, teaching mode and experimental mode, examination and evaluation mode and method formed by comprehensive university for decades, but they are not really followed. They apply the connotation and law of technical talents training objectives to teaching. The main problems are as follows: the logical sequence of curriculum is reversed, the orientation of teaching content is vague, the teaching mode and method are too traditional, the students' learning is seriously divorced from the practical application, and the experimental practice teaching system is not conducive to the cultivation of students' application technology ability.

Therefore, in view of a series of practical problems existing in domestic applied technology-based colleges and universities, this study carries out comprehensive reform and practice from the aspects of curriculum system, teaching content, teaching mode and method, assessment method, and experimental practice teaching system and mode, so as to highlight the cultivation of students' application technology ability and strive to solve the application technology of college graduates. There is a contradiction between the technical ability and the demand target of the employer. The research results are of great significance and promotion value to improve the teaching quality of the basic courses in the applied technology-oriented colleges and universities in our province, and to highlight the cultivation of students' application technology ability.

2 CURRICULUM REFORM SCHEME DESIGN

2.1 In order to highlight the application technology ability training as the guidance, clear the order, and build a new series of electrical basic curriculum system

In view of the problems of unreasonable setting and reverse logical sequence in some of the required prerequisite courses, through in-depth analysis of the knowledge connection level relationship between them and the basic courses of electricity, the pre-course system (engineering mathematics, college physics, etc.) is set in a reasonable logical order, so that the prerequisite knowledge is first, and the teaching of basic electrical courses is emphasized The required content is matched and consistent.

In view of the problems existing in the curriculum system of electrical basic series, such as the confusion of curriculum system setting, the fuzzy knowledge transfer, and the intermittent cultivation of application ability, this paper constructs the classroom theoretical teaching (basic knowledge) → independent experimental teaching (basic experimental training) → curriculum design (design practice training) → electronic innovative application design practice training (comprehensive practice of electronic system design). The hierarchical curriculum system of discipline specialty/electronic design competition and innovation project research (professional technology application improvement) can improve the rapid adaptability of applied technology talents training to social needs.

2.2 Taking the technology application as the main line, the teaching content system of basic content → typical application → improvement content → comprehensive design is constructed, and the teaching content of the series of electrical basic courses is integrated and optimized

In view of the problem of "more teaching content, faster knowledge updating and less class hours," this paper selects basic content, cuts out outdated content, highlights typical application, updates new content, and strengthens comprehensive design, so that the teaching content is optimized, integrated, and updated in time; through the reasonable distribution of the content proportion of in-class teaching, extracurricular learning and practice training, the focus of lectures is highlighted and the proportion of class hours is less.

According to the practical principle of "useful, available and effective," the teaching (teaching material) content should be carefully selected to reduce the theoretical content and eliminate the outdated and redundant knowledge points. The teaching (teaching material) content must be combined with practice and highlight the technical applicability.

2.3 We should make full use of the new teaching resources to explore and practice the teaching mode and method of "taking students as the main body"

In recent years, there are more and more massive open online courses, video open classes, resource sharing classes, rain classes, and other new forms of curriculum resources in colleges and universities at home and abroad. In analyzing how to guide students to effectively use these resources and give full play to the main role of students, this paper discusses the diversified teaching methods of flipped classroom teaching, project-based learning, and network interactive communication. Practice the teaching mode of "taking students as the main body" and realize the teaching goal of "paying attention to guidance, advocating autonomous learning and cultivating application ability."

According to the "flipped classroom" teaching concept of student-centered and teacher led, the new teaching mode as shown in Figure 1 is constructed.

Explore and practice new teaching methods and learning methods, such as education data architecture (EDA) simulation demonstration teaching method, communication learning method based on network interaction, analogy teaching method of drawing inferences from one instance, group teaching method of project research, and so forth.

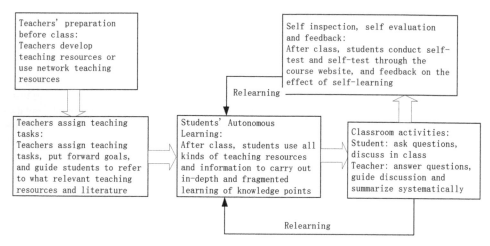

Figure 1. New teaching mode.

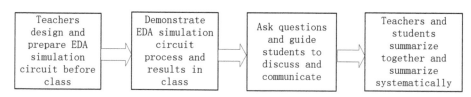

Figure 2. Implementation process of EDA simulation demonstration teaching method.

EDA simulation demonstration teaching method as shown in Figure 2 is the implementation process of EDA simulation demonstration teaching method in the teaching of electrical basic courses.

Communication learning method based on network interaction: in view of the high popularity rate of intelligent terminal equipment among students and the characteristics of students who like chatting, the QQ group or WeChat group of basic electrical courses is established to provide a network interactive communication platform for the communication between teachers and students and between students. At the same time, research and release the corresponding mobile app of the course, so that teachers can publish and discuss in the corresponding discussion group. On the topic, we can also release the learning content and materials for students to learn the course content independently after class. We can also study and establish micro-learning systems. Teachers can arrange homework through the system, conduct regular tests and evaluation feedback, and understand students' mastery of knowledge points of the course.

Analogy teaching method of "drawing inferences from one instance": in circuit, analog electronic technology, digital electronic technology, and experimental teaching, there are many contents with similarity, such as duality of circuit theorem, similarity of amplifier structure, symmetry of digital logic components, and so on, which are typical contents listed in the following table.

Project research group teaching method: Under the teacher's inspiration and guidance, students can actively find practical application problems (topic selection), propose design scheme and demonstration, and take specific technology (design and implementation circuit) to carry out technology application project research activities, and acquire and comprehensively use knowledge and skills. As shown in Figure 3, it is the implementation process of project research group teaching method.

Table 1. Analogy teaching method.

Basic courses of electricity	Teacher's take one	Students' anti-three
Circuit analysis	Kirchhoff's law of current Davining's theorem Loop current method	Kirchhoff voltage law Norton theorem Node voltage method
Analog electronic technology	Triode amplifier Emitter output device Square wave generator	FET amplifier Source output device Triangle wave generator
Digital electronic technology	Encoder Synchronous counter	Decoder Asynchronous counter
Circuit and electronic technology experiment	Verification experiment of Davining's theorem Verification experiment of inverse proportional amplification Verification experiment of synchronous counter	Design Norton theorem verification experiment Design in-phase proportional amplification experiment Design asynchronous counter experiment

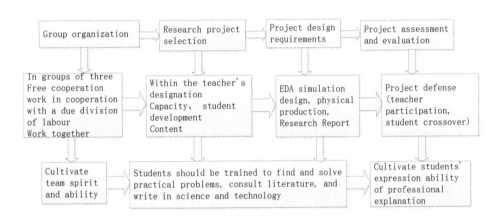

Figure 3. Project research group teaching method.

2.4 With the goal of "learning for application," we should construct a hierarchical practical teaching system and reform the content and mode of practical teaching

In order to solve the problems of "high vision and low hand, can't do it" and "the experimental practice teaching system is not perfect" in the current study of electrical basic series courses, a ladder (basic level, design level, comprehensive level, application and improvement level) experimental and practical teaching system is constructed to set up independent experimental courses, reasonably set up verification experiments, virtual experiments, and comprehensive teaching. According to the proportion of designed experiment and innovative applied experiment, the content of experiment and practice should be reformed. In order to effectively improve the engineering practice ability and innovation ability, we should adopt the experimental mode of "combining virtual with practice," "combining laboratory (centralized) with external experiment (open)," "combining conventional experiment with comprehensive application experiment," and "combining overall arrangement with independent design" power.

The curriculum system of electrical basic practice is divided into levels and steps according to the order of circuit experiment and electronic technology experiment (basic level experiment, separate experiment compulsory course), electronic technology course design (comprehensive design level

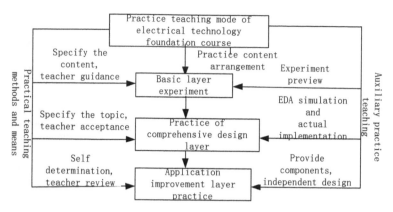

Figure 4.   The establishment of practice curriculum system.

practice, separate course, compulsory course), and electronic design comprehensive application (application improvement level practice, recommended elective), as shown in Figure 4.

Electrical basic series of courses are very practical. The goal of learning circuit and electronic technology courses is to use these technologies to solve practical problems in industry and life. Therefore, the experimental mode should be designed around this goal.

2.5   With the goal of "learning for application," we should construct a hierarchical practical teaching system and reform the content and mode of practical teaching

In view of the disadvantages of the traditional "closed book examination results-based" assessment mode, this paper explores and practices diversified assessment modes and examination methods such as "chapter test + midterm/final examination," "closed book + open book," "classroom discussion + teacher's question," "big assignment design report + acceptance reply," "teacher's evaluation + student's mutual evaluation." We should change the examination of rote knowledge points at the level of knowledge input and storage, highlight the weight and score of knowledge output and application ability, and guide students to change from rote learning and surprise review to learning methods that strengthen their usual learning and understanding and pay attention to the training of application ability.

3   CONCLUSION

The research and implementation object of this paper is electronic information specialty and related science and engineering specialty in our college. The implementation plan highlights application technology ability training, learning for application, and student-centered, which has obvious effects on stimulating students' learning enthusiasm, improving teaching quality of electrical basic courses, and students' technology application ability. It has strong operability and pertinence, and has the universality and practical significance of popularization. At the same time, it also actively responds to the national call for undergraduate education in applied transformation, which is conducive to the cultivation of students' application technology ability in independent colleges, and has a forward-looking, exemplary role and leading promotion value.

ACKNOWLEDGMENTS

The research in this paper was sponsored by the Provincial Teaching Reform Research Program of 2018 Hubei Provincial Education Department (Project No: 2018509).

REFERENCES

[1] Zhang Xueqin, song Jirong, Lu Yongjun, et al. Electrical and electronic experiment curriculum reform and practice oriented to engineering literacy. Laboratory research and exploration, 2019, 038 (005): 216–220

[2] Liang Hongwei, Kai Lingling, Zhang Xiuyan, et al. EDA technology curriculum reform based on practical ability training. Science and education guide, 2018, 000 (020): 28–29

[3] Tang Lili, Huang Qijun. Reform and practice of electronic technology foundation course for Applied Technology Ability Cultivation. Education modernization, 2019 (46): 124–125

[4] Wu Lianghong, Li Mu, Zhao Yanming, et al. Research and Practice on teaching reform of electronic technology course for training outstanding engineers. Contemporary education theory and practice, 2017 (10): 40–43

[5] Zheng Guifang, Wang Xuejun. Research on analog electronic technology curriculum reform based on innovation and practical ability training. Industry and Technology Forum, 2019, 018 (019): 171–173.

Computational Social Science – Luo, Ciurea & Kumar (eds)
© 2021 Taylor & Francis Group, London, ISBN 978-0-367-70193-2

Research on the long-term employment mechanism of college students

D.D. Li
Tianjin University Renai College, China

ABSTRACT: With the increase of college students and graduates year by year, the employment situation is grim. Starting from the internal and external factors of students, this paper analyzes the problems existing in the employment of college students, and puts forward the main points of promoting the employment of college students. This paper examines this issue through guiding students to formulate career planning, enhancing students' psychological quality, integrating emotional education and fine management into the cultivation and education of undergraduate students, developing alumni resources, building a good platform for students' employment, and constructing a long-term mechanism for college students' employment.

1 INTRODUCTION

Since the expansion of college enrollment, the number of college graduates has increased year by year. The number of graduates in 2020 is expected to reach 8.74 million, an increase of 400,000 over the same period of last year [1]. The employment report of Chinese college students in 2019 released by Max Institute shows that the employment rate of undergraduate graduates continues to slow down [2]. How to improve the employment rate of undergraduate graduates and cultivate excellent students for the country has become the focus of the work of colleges and universities. The growth of students is closely related to the quality of their employment. It is of great significance to combine the employment work with the whole staff education, the whole process education, and all-around education, and to establish a long-term employment mechanism for college students.

2 THE RESULTS OF THE INVESTIGATION ON THE EMPLOYMENT BASIS OF COLLEGE STUDENTS

The survey was conducted by issuing questionnaires. The samples involved four grades of undergraduate students, and 1,000 questionnaires were distributed and 1,000 questionnaires were collected.

2.1 *Personal situation of employment*

Students think that college life is free and flexible, and they need a period of adjustment time to better arrange their spare time of university life. Students' understanding of themselves and society is not enough, they do not objectively analyze their own strengths and weaknesses, do not deeply explore their areas of expertise, are not clear about their personal career planning, and do not pay enough attention to career planning.

Students' understanding of the employment prospects of their major is not enough, and their employment confidence is not strong. In addition, students' employment concept needs to be adjusted. Nearly half of the students think that their career goal is to obtain a sense of security and stability. Students expect to earn enough wealth, satisfy their personal interests, acquire enough power, and realize themselves through employment. Students think that the employment situation of college students in the current society is very serious, and they expect to find a suitable job through their own efforts. In the field of employment, most students think that the work they are engaged in is not necessarily consistent with their major.

In the study, 72.68% of the students had short-term social work and part-time experience. Students expect to gain management experience, professional skills, competitiveness, extensive interpersonal relationship from practice, and obtain more stable jobs and better salary and welfare after graduation.

2.2 *The choice of employment*

Students' graduation direction is diversified: 50.66% of the students chose direct employment after graduation, 24.22% of the students chose to take the postgraduate entrance examination during the undergraduate period, 9.69% of the students chose to take the civil service examination during the undergraduate period, 10.13% of the students chose to start a business after graduation, and other students chose to enlist in the army and plan for the western region. The proportion of students choosing direct employment is the highest, followed by postgraduate entrance examination, and finally entrepreneurship and enlistment.

Students believe that family has an important influence on their choice of work place and work field. Most of the students choose large and medium-sized cities because of their good life, great job opportunities, and good local policies. Some students don't think about the future clearly. They don't know what they want to do and what kind of life they want. Students lack comprehensive analysis of the strengths and weaknesses of each city, as well as the awareness of employment policies.

In terms of independent entrepreneurship, 52.42% of the students said that they had the idea of entrepreneurship, because entrepreneurship also has risks. Students expect to know more about entrepreneurship knowledge and simulation practice opportunities, so as to determine whether they are suitable for entrepreneurship through rigorous analysis; 13.66% of the students said that they are preparing to start a business. Some students have registered with the company during their school years. However, there are still many problems in the material resources and various guarantees needed for starting a business. The students expect that the government can provide more favorable entrepreneurial environment, good policies and subsidies, and the school will provide more support to entrepreneurs, so as to reduce the entrepreneurial pressure of entrepreneurs.

3 CONSTRUCTION OF LONG-TERM EMPLOYMENT MECHANISM FOR COLLEGE STUDENTS

3.1 *Establish personal career planning files*

College students' career planning runs through the whole university and lasts for a lifetime. The school should combine the work of students' classified guidance; establish a career planning file with students' personal characteristics from the first year of university, combine the career planning of students with the education of students; run the employment work through the whole staff, whole process, and all-around student training process; and build a long-term personal development plan.

With the help of career planning, students can scientifically and comprehensively analyze their personality, interest, and career orientation. Career planning can not only help students to understand themselves, understand the environment, make up for deficiencies, give full play to their strengths, formulate life goals, and give clear direction of life development, but it can also help individuals to find a suitable job according to their own conditions, and finally achieve the consistency of career planning and personal goals. Career planning is an important process of one's life-long growth, which is of great significance to one's growth. Students have their own career planning and can achieve a better life.

The basic contents of career planning education for college students usually include the following: First, self-assessment. Self-assessment is a comprehensive analysis of self. Through objective and accurate self-assessment, students can make the right choice, choose the suitable career path, and finally make the best career choice. Second, career environment analysis. Career environment analysis is to analyze the internal and external environmental factors that affect their career development, so that students can better plan their career goals and choose their career routes. Third, career planning decisions. Through self-assessment and analysis of professional environment, students can make planning decisions on their own occupation and career goals.

The counselor guides the students to formulate the college students' career planning and supervises the students to implement the plan separately during the university years. The role of counselors is to guide and assist students to implement and adjust their own plans, to guide students to achieve the direction of the goal, to promote the long-term implementation of the plan, so that students can avoid "three minutes of heat, forget later" and give up halfway. Counselors should guide students in different grades according to different tasks in different grades and stages, and make career planning throughout university life, so as to form a complete student growth system and unique student growth files.

3.2 Enhance students' psychological quality

3.2.1 Guide students to know themselves correctly

There is a maxim in the Delphi Temple of Greece: know yourself. Knowing yourself includes knowing what kind of person you are, your personality, advantages and disadvantages, what kind of person you want to be, and so on. If you know yourself too much, you are prone to conceit. If you know yourself too little, you will feel inferior. Therefore, an objective and correct understanding of oneself is the prerequisite for self-perfection.

Counselors should guide students to understand themselves correctly and evaluate themselves objectively, so as to improve themselves. Correct understanding of themselves can firm their own direction of progress and forward goals, so that they will not be shaken by the outside world. Psychologist Sherry Taylor believes that hope is an important part of human nature. A person who constantly improves himself can grow better. In solving problems, students should know how to summarize and reflect and rely on the internal driving force to promote their own progress.

3.2.2 Enhance students' positive experience

Positive experience includes positive emotional experience and pleasure. Plato said that the main purpose of education is to teach young people to find pleasure in the right things. Michael de Montague says that if I study, I only learn what can guide me to live better and make my life more valuable.

Positive experience can make students form a virtuous circle. Students with positive experience are more willing to use positive ways to think about problems and form the way of internal attribution. They will not shift the generation of problems to the outside world, but consider problems from their own internal, and then reset goals to solve problems. Positive emotional experience and pleasure can stimulate people's creativity and imagination. This plays a great role in promoting the employment of students.

3.2.3 Change students' cognitive process

According to Martin Seligman, an American psychologist, optimism refers to that when people explain what has happened, they make persistent, universal, and internal attribution to good events, while they do short-term, specific, and external attribution to bad things. Attribution is acquired. People can change negative attribution into positive attribution through learning, so as to enhance their confidence.

Change the cognitive process of students, so that students can accept the reality, face calmly, form a positive internal attribution, face difficulties from a positive point of view, and then form a positive problem-solving thinking. When training student cadres, counselors can help students to establish positive thinking mode, correct their attribution mode, and strengthen positive attribution mode through a reward and incentive system. Students get a good life from their own practice and realize that failure and success can benefit people a lot.

3.2.4 *Cultivate students' sound personality*

Ingels, a famous American sociologist, thinks that the essence and key to realize modernization in developing countries is personality modernization. College students are in a period when world outlook, outlook on life, and values tend to be stable, and their plasticity is strong. The cultivation of college students' sound personality is not only conducive to the development of students themselves, but also contributes to the modernization of the country. Personality is an important part of personality. The power of character cannot be ignored. Cultivating students' positive quality, shaping good character, and promoting students' life development is conducive to their harvest of happiness.

3.3 *Fine management of employment*

3.3.1 *Clarify the objectives of fine management*

In the process of student employment, counselors provide classified employment guidance for students, refine each category, and determine the ultimate goal. From the beginning of student enrollment, counselors should focus on the goal of training students, management level and efficiency, make practical plans, and take employment goals as a means to promote students' progress.

3.3.2 *Establish a smooth and effective communication platform*

Counselors will carry out data management of student information, build a data management platform, classify students, carry out work through new media and network, establish an effective way to communicate with students, and timely release, collect, and feedback information. Counselors of the same grade should establish a platform for counselors, strengthen communication, and timely exchange employment information and share work experience. Counselors and the employment guidance center should establish a linkage platform to ensure that employment information is received by students in a timely manner.

3.4 *The application of emotional education in employment*

Emotional education refers to taking emotion as an important aspect of human development in the process of education and teaching, following the law of emotion, giving full play to the positive role of emotional factors, enhancing students' positive emotional experience through emotional communication, cultivating and developing students' rich emotions, stimulating their thirst for knowledge and exploring spirit, and realizing the training goal of education.

Emotional education is very important for gratitude education and responsibility education. Emotional education can help students learn to be grateful, make students recognize their responsibilities, and further improve and enhance their employment potential and awareness. Emotional education can enable students to immerse themselves in deep feelings for the country, parents, teachers and friends, and others; think about and examine their own behavior; and decide and choose their current employment behavior with a high sense of responsibility and gratitude. There will be no negative and irresponsible phenomena such as "no employment, first graduation travel," "no employment, staying at home," and so on. Emotional education is conducive to students to develop a positive attitude toward employment and establish a sense of responsibility. Therefore, students will have a sustained and strong desire to develop their employment potential.

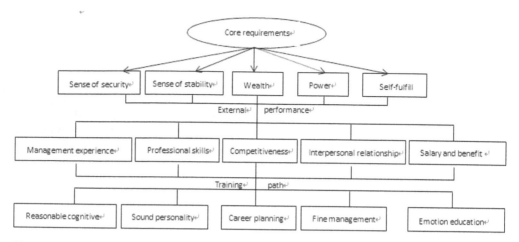

Figure 1. Analysis path of long-term mechanism.

3.5 Utilization of alumni resources and establishment of employment platform

3.5.1 Utilization of alumni resources

Alumni can provide students with practical job experience. Peer education is more acceptable to students than traditional education. Alumni have rich experience in job search and post work. Alumni experience exchange meetings can help students learn from the practical perspective of job search experience and adjust their employment outlook. Outstanding alumni can play a more exemplary role, are more persuasive, and promote students to actively improve their own quality in school. At the same time, alumni can bring more employment resources and more jobs. The recognition of alumni in the enterprise not only publicizes the image of the school but also makes the enterprise willing to accept more students from the school.

3.5.2 Establish more cooperation platforms between schools and enterprises

School enterprise cooperation can cultivate more suitable talents for enterprises. Practice is particularly important for schools. Students in the enterprise internship can apply their rich theoretical knowledge in school to work, which can make up for the lack of professional experience of students. Excellent students will have more opportunities to work in the internship enterprises after graduation.

3.5.3 Establish professional employment guidance team

Professional employment guidance teams can make up for the lack of guiding students' employment work. Professional employment guidance teachers can provide professional guidance according to students' own characteristics, help students narrow down the field of employment, provide more support for students, provide students with whole process and comprehensive assistance, and solve the problems existing in the process of students' employment in time.

4 CONCLUSION

Analysis path is as follows: "Figure 1. Analysis." Through the analysis of the survey results of college students' employment, we fully understand the students' understanding of self and external environment, students' outlook on career choice, and the main factors affecting students' employment. The long-term mechanism of college students' employment is constructed from the following aspects: guiding students to formulate career planning, enhancing their psychological quality, emotional education, fine management, and developing alumni resources to build a good platform for students' employment.

REFERENCES

[1] Dong, Luwanlong. National college graduates employment and entrepreneurship network video conference held to ensure the overall stability of college graduates employment next year. China Education News, 2019-11-01 (3).
[2] MyCOS Research Institute. Employment report of Undergraduates in China. Beijing: Social Sciences Literature Press, 2019.

Computational Social Science – Luo, Ciurea & Kumar (eds)
© 2021 Taylor & Francis Group, London, ISBN 978-0-367-70193-2

On the mode of school-enterprise cooperation in the reform of applied talents training in teaching-oriented universities

S.Y. Du
College of Computer and Information Engineering, Hubei Normal University, Huangshi, China
College of Educational Science, Hubei Normal University, Huangshi, China

X.F. Peng*
College of Computer and Information Engineering, Hubei Normal University, Huangshi, China
College of Arts and Science, Hubei Normal University, Huangshi, China

ABSTRACT: School-enterprise cooperation is a necessary teaching link to the cultivation of applied innovative talents. The research on teaching mode in school-enterprise cooperation is an important research topic in the reform of talent cultivation mode in colleges and universities. CiteSpace visual analysis software is used to analyze the scientific atlas of relevant literature from 2009 to 2018. This paper analyzes the time distribution, key words, research hot spots, and frontiers in the research field of school-enterprise cooperation in application-oriented undergraduate colleges; it uses SPSS software to conduct independent sample t-test on the course scores of C language programming of two classes before and after the cooperation of Hubei Normal University, and finds that there is no significant change in the class scores after the cooperation. This makes us want to study the reform direction of teaching content, teaching mode, and evaluation system of applied innovative talents in school-enterprise cooperation. It provides experience and reference for the cultivation of applied innovative talents in colleges and universities.

Keywords: School-enterprise cooperation, Visual analysis, SPSS t-test, Teaching reform.

1 INTRODUCTION

In recent years, China has built an innovative country, improved the ability of independent innovation, and cultivated applied innovative talents. It is clearly putting forward that it is a major task to improve the ability of independent innovation and build an innovative country. School-enterprise cooperation education is a modern education mode different from traditional education, which plays an important role in the cultivation of applied talents in China. It is urgent to explore the training mode of applied innovative talents, and the test of teaching effect is also very necessary. SPSS software is used to analyze the results of the two classes of Hubei Normal University before and after the school-enterprise cooperation training, and to compare whether there are significant differences between the two classes.

2 VISUAL ANALYSIS OF SCHOOL-ENTERPRISE COOPERATION TEACHING MODE

2.1 *Study time distribution*

The research data came from the literature in China National Knowledge Infrastructure. The retrieval time of the research data was 2009–2018. The retrieval subject was application-oriented undergraduate course, including school-enterprise cooperation, and a total of 1,221 valid literatures were

---

*Corresponding author

Table 1. Keyword frequency of school-enterprise cooperative education literature in China from 2009 to 2018.

Sort	High-frequency keywords	Frequency
1	School-enterprise cooperation	444
2	Applied undergraduate	340
3	Application-oriented universities	152
4	Personnel training	139
5	Application type	129
6	Talent training mode	117
7	Practical personnel	100
8	Practical teaching	85
9	Training mode	64
10	Integration of production and education	61

retrieved. From the period of time, it can be seen that from 2009 to 2018, the number of articles on school-enterprise cooperation in engineering increased year by year, as shown in Figure 1. By 2018, the number had climbed to 400. In general, in the past decade, scholars have paid more attention to the research on school-enterprise cooperation in engineering colleges, and the number of core publications has been on the rise.

2.2 *Keyword frequency statistics*

After merging similar keywords, the analysis is conducted according to the keyword frequency. The top 10 high-frequency keywords are shown in Table 1. Among these 10 high-frequency keywords, those such as "school-enterprise cooperation," "applied undergraduate," and "applied undergraduate colleges" do not indicate the hot issues of school-enterprise cooperative research in applied undergraduate colleges.

2.3 *Research hot spots and frontiers*

The cross nodes in the atlas represent the keywords in the literature, and the larger the nodes, the more times they are cited. Table 2 [1] is obtained by sorting the key words with centrality and higher cited frequency. It can be seen from Table 2 that in the period 2009–2018, relevant scholars' keywords for undergraduate school-enterprise cooperation research include "talent training model," "applied undergraduate," "talent training," "school-enterprise cooperation," "applied undergraduate colleges," "application-oriented," "application-oriented talents," and "practical teaching." Through the co-occurrence cluster analysis of keywords, we can clearly see the focus of relevant scholars on the research of application-oriented university enterprise joint cooperation. The node in Figure 2 represents the keywords and terms extracted from the literature keywords. The larger the node, the more times the phrase appears, which is the research hot spot in this field [2]. It can be seen from the figure that there are five categories in this cluster: major construction problems, application-oriented undergraduate colleges, new engineering courses, reform, and practical teaching system.

The following conclusions can be drawn from the above analysis. First, China's application-oriented undergraduate school enterprise cooperative education research has strong characteristics of the times, its emergence, and development with the support of the government and society, and has a trend of development to other different levels of colleges and universities. Second, through the analysis of hot topics in the field of school enterprise cooperative education research, it can be easily concluded that the research experts have accumulated a lot of rich experience in the innovation of school-enterprise cooperative education system and specialty setting. Third, the evaluation system and theoretical research of school-enterprise cooperative training mode are relatively blank.

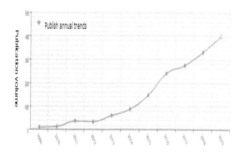

Figure 1. Time distribution of research on school.   Figure 2. Cluster graph of research hot spots.

Table 2. High-frequency keywords of school-enterprise cooperation research.

Year	Frequency	Centrality	Keyword
2010	117	0.32	Talent training mode
2009	340	0.27	Applied undergraduate
2010	139	0.24	Personnel training
2009	444	0.21	School-enterprise cooperation
2009	152	0.21	Application-oriented universities
2010	129	0.21	Application type
2009	100	0.19	Practical personnel
2013	85	0.12	Practical teaching

3 CURRENT SITUATION OF SCHOOL-ENTERPRISE COOPERATION TEACHING MODE IN HUBEI NORMAL UNIVERSITY

In 2015, Hubei Normal University and Shanghai Zhixiang Information Technology Development Co., Ltd., jointly established Dingli College, and the school-enterprise parties jointly carried out the transformation and development of application-oriented innovative talents training. Students in all four majors from The School of Computer Science of Hubei Normal University were transferred to Dingli College for school-enterprise joint training from 2016. In accordance with the requirements of ICT industry chain for high-level technical and skilled talents and national vocational qualification requirements, the reform of school-enterprise cooperative teaching is carried out by organically combining school resources and the practical experience of enterprises. Teaching is carried out in accordance with the mode of school-enterprise cooperation. However, the initial results of school-enterprise cooperation are not good, and there are still many areas to be improved.

Here by SPSS 19 software the school of Hubei Normal University communication engineering (by traditional teaching method) and 1605 class, 1803 class communication engineering (by university-enterprise cooperation joint training) "language C program design" course grades, through the SPSS statistical analysis, the two classes of course grade for independent samples t-test, and it carries on the comparison to two class teaching effects.

3.1 Frequency analysis

Frequency analysis was conducted on the scores of class courses after cooperation, and the analysis results are shown in Table 3. There are 38 people in the class, 8 of whom fall between scores 50 and 60 most frequently. The lowest frequency score was between 20 and 30, with 1 person. The whole interval is close to the normal distribution.

Frequency analysis was conducted on the scores of class courses before cooperation, and the analysis results are shown in Table 4. There are 34 students in the class, and the highest frequency

Table 3. Results of frequency analysis of grade 1803.

		Frequency	Percent	Valid Percent	Cumulative Percent
Valid	.00	1	2.6	2.6	2.6
	1.00	5	13.2	13.2	15.8
	2.00	4	10.5	10.5	23.6
	3.00	8	21.1	21.1	47.4
	4.00	4	10.5	10.5	57.9
	5.00	6	15.8	15.8	73.7
	6.00	6	15.8	15.8	89.5
	7.00	4	10.5	10.5	100.0
	Total	38	100.0	100.0	

Table 4. 1605 Results of frequency analysis.

		Frequency	Percent	Valid Percent	Cumulative Percent
Valid	.00	6	17.6	17.6	17.6
	1.00	7	20.6	20.6	38.2
	2.00	8	23.5	23.5	61.8
	3.00	9	26.5	26.5	88.2
	4.00	3	8.8	8.8	97.1
	5.00	1	2.9	2.9	100.0
	Total	34	100.0	100.0	

scores falling between 70 and 80 is 9. The lowest frequency of scores falling between 90 and 100 is 1, and the whole interval is basically close to normal distribution.

3.2 Independent sample t-test

Different data types require different methods and statistics to complete the specific inspection problem. Since the data-based teaching scores generally belong to continuous variables, the t-test in SPSS can be used to analyze the data [3]. The independent sample t-test method is used for analysis. According to the principle of small probability reduction to proof, the following assumption is made: H0: the results of the two classes are caused by sampling error, and there is no difference between the results of the two classes, that is, the school-enterprise cooperative teaching reform is invalid. H1: In addition to the sampling error, there is indeed a difference between the grades of the two classes [4, 5], that is, the school-enterprise cooperative teaching reform is effective. SPSS was used to conduct independent sample t-test on the data, and the output results are shown in Table 5, Table 6, and Table 7.

3.3 Data analysis

In Table 5, the mean term is the average score of the corresponding two classes, which is 63.55 for class 1803 and 73.548 for class 1605. Standard deviation reflects the dispersion degree of the score. Small standard deviation and high average score indicate that the overall level of students is good [6]. A standard error is an estimate of the reliability of a set of measurements. The standard error is small, and the reliability of measurement is larger; otherwise, the measurement is less reliable [7]. Table 5 shows the results of the comparison of the data of the two classes. Standard deviation term is the standard error of each class, and STD. Error mean is the standard error of the mean of each class. The smaller the standard error of the mean is, the smaller the individual level difference of

Table 5. Grouping data description.

	N	N	Mean	Std. Deviation	Std. Error Mean
G	0	38	63.55	20.363	3.303
	1	34	56.24	13.267	2.275

Table 6. Results of equal variance test.

		Levene's Test for Equality of Variances	
		F	Sig.
G	Equal variances assumed	9.999	.002
	Equal variances not assumed		

Table 7. Test results.

		t-test for Equality of Means			
		T	DF	Sig. (2-tailed)	Mean Difference
G	Equal variances assumed	1.783	70	.079	7.317
	Equal variances not assumed	1.824	64.227	.073	7.317

students is, and the overall level is stable. What we are most concerned about is that Sig.(2-tailed)'s value is $0.079 > 0.05$, indicating that we should accept H0 hypothesis and reject H1, that is, the difference is not significant, fully indicating that there is no significant difference between the two classes' scores. As can be seen from the mean term in Table 5, although the data of class 1803 was greater than that of class 1605, it was not the teaching reform plan that led to the improvement of performance, indicating that the teaching reform plan had no significant effect. If the value of Sig.(2-tailed) is less than 0.05, the situation is reversed.

4 CONCLUSION

This paper uses SPSS software to conduct independent sample t-test on the course scores between class 1803 and class 1605, which are jointly trained by school-enterprise cooperation. It is found that there is no significant difference between the two classes. The school-enterprise cooperation education mode has been carried out in Hubei Normal University for a short time. In the future teaching work, both the university and the enterprise should strive to shoulder the responsibility and jointly complete the task of cultivating applied innovative talents. At the same time, the teaching mode and evaluation system of school-enterprise cooperation training applied innovative talents are comprehensively reformed.

ACKNOWLEDGMENT

This research was supported by a project of study on the Improvement of Learning Quality of Engineering Students in School-Enterprise Cooperation—A Case Study of Dingli College of

Hubei Normal University (NO: 20190157), and Innovative Team of Excellent Young-Middle-Aged Universities in Hubei Province (NO: T201430).

REFERENCES

[1] Shen Xiping, Ding Jiansheng, Li Juansheng. (2007). Using mean and standard deviation to do two independent sample t test in SPSS. Modern preventive medicine, 34 (21): 4066–4067.

[2] Cai Minjun, Liu Renyun. (2010). Computer aided education measurement and evaluation. China water resources and Hydropower Press.

[3] Xu Jihong. (2011). Reform of postgraduate training mechanism in Local Universities Based on the improvement of innovation ability. Journal of Zhejiang University of Technology (SOCIAL SCIENCE EDITION).

[4] Cui Yan, Zhu Dandan. (2013). Discussion on the application of t-test method in teaching. Journal of Science Education (Early issue).

[5] Chen Yue et al. (2015). Methodological function of cite space knowledge map. Scientific research.

[6] Wu Yuntana. (2015). Analysis of the differences in the results of quantitative geography based on t-test. Yinshan Journal: Natural Science Edition.

[7] Guo Fengli. (2017). Research on Influencing Factors of University Graduate Students' willingness to use MOOC.

Computational Social Science – Luo, Ciurea & Kumar (eds)
© 2021 Taylor & Francis Group, London, ISBN 978-0-367-70193-2

Exploration on practical teaching reform of logistics engineering based on new engineering

Y. Fang, Y.S. Luo*, W.X. Jiang & C. Gong
Hunan Institute of Traffic Engineering, Hengyang City, China

ABSTRACT: Based on the construction of new engineering disciplines, this paper takes Hunan Institute of Traffic Engineering as an example and focuses on strengthening students' engineering practice ability and innovation and entrepreneurship ability, and researches the practical teaching reform of logistics engineering. It is constructed from four modules: single skill, comprehensive skills, comprehensive applications, and innovation and entrepreneurship, and it updates the experimental content following the pace of the times, including issues of introducing enterprises to the school, creating a comprehensive training platform, cooperation between schools and enterprises, strengthening the construction of internship bases, proceeding in an orderly way, training students' ability of innovation and entrepreneurship, in order to provide a reference for the training of logistics engineering professionals in local universities in the country.

1   INTRODUCTION

For coping with the new round of scientific and technological revolution and industrial reform, according to the Chinese Educational Modernization 2035, "Several Opinions of the General Office of the State Council on Deepening the Integration of Production and Education," and on Accelerating the Construction and Development of New Engineers Education and "Training Program 2.0," a new engineering construction is raised characterized by "new ideas, new structures, new models, new quality and new systems," and pointed out is the direction for the reform of higher engineering (Lin, 2017; Zhong, 2017).

Colleges need to deepen the integration of production and education and school-enterprise cooperation and grasp the new engineering construction connotation about "new engineering major, engineering science new requirements." It is established that a new requirement of engineering construction is in line with the engineering personnel training model, in order to be training forward-looking, competent industry development needs of high-quality applied engineering personnel (Liu, 2018; Tang & Wang, 2019).

A logistics engineering major has the characteristics of strong practice. It is a very important position which practical ability training in higher education and innovative talent training occupies (Li & Chen, 2020). In the context of the construction of new engineering, through the large practice of applied technical personnel training, it is considerable to turn out some students equipped with practical ability and innovation ability. The professional development program should be based upon the school's orientation and focus on training the student's ability of practical and innovation. It is practicable to seek a suitable practical teaching system for their own development and the local economic needs (Tong et al., 2020).

---

*Corresponding author

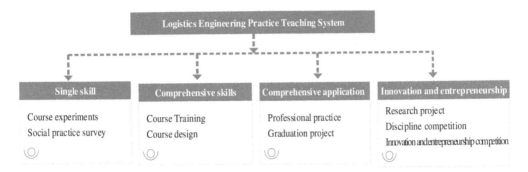

Figure 1. The practical teaching system of logistics engineering.

2 BUILD A PRACTICAL TEACHING SYSTEM FOR LOGISTICS ENGINEERING IN NEW ENGINEERING

In exploring practical teaching in logistics engineering, to fully grasp the development requirements of "Based on southern Hunan, facing Hunan and radiating the whole country" in the Hunan Institute of Traffic Engineering is to implement the goal of training the talents of applied technical talents and to integrate the concept of "new engineering" with the practical teaching system of logistics engineering from the curriculum. Starting with experiments, course training, professional practice, and discipline competition, a practical teaching system is established of single skills, comprehensive skills, comprehensive application, and innovation and entrepreneurship, which lays a practical foundation for cultivating engineering and scientific talents with strong professional background knowledge, innovative and entrepreneurial ability, strong engineering practice ability, and high scientific literacy (Sun & Guo, 2018; Gui, 2019; Qin, 2014; Wang et al., 2020).

3 EXPLORATION ON PRACTICAL TEACHING REFORM OF LOGISTICS ENGINEERING BASED ON NEW ENGINEERING

3.1 *Strengthen experimental teaching and consolidate individual skills*

The experimental teaching of logistics engineering has strong practical characteristics and advantages, in order to cultivate students' strong practical ability and professional knowledge application ability, to cultivate students' strong engineering quality. In order to form an organic whole of the experimental teaching system of logistics engineering, an experimental teaching system was set up to combine theoretical knowledge with experimental teaching and to adapt to the development of disciplines.

Our professional training is conducted to service local economy and to focus on the capacity of application, and then following market demand and professional development, the training plan is adjusted to involve planning and management, machinery and equipment, and information technology in order to train high-quality technical personnel. The updating of experimental content should be combined with the actual enterprise and to keep pace with the times.

The experimental teaching of logistics engineering focuses on training students' ability of logistics planning and implementation, logistics equipment design and application, logistics enterprises, and enterprise logistics information. The module of planning and management mainly involves the design and implementation of enterprise logistics program, logistics center planning and design, transportation, and distribution network design and dispatch. The module of machinery and equipment mainly involves logistics machinery technology, logistics automation technology, logistics equipment design, and application. The module of information technology mainly involves logistics network and intelligent logistics, enterprise resource plan (ERP), and logistics information system.

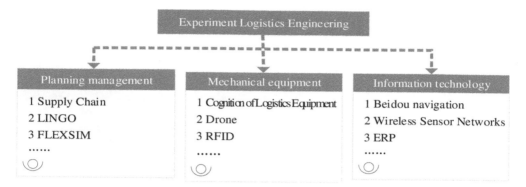

Figure 2.   A single skill module for logistics engineering.

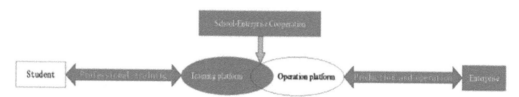

Figure 3.   The structure diagram of the on-campus training organization.

The three experimental modules herein cooperate with each other to consolidate the individual skills of logistics engineering.

3.2   *Build a training platform to improve comprehensive skills*

The practical training platform tends to introduce enterprises into the school. The school and enterprise build a practical training platform together. The resources of college and enterprise should make use of their advantages to build and share productive training bases in schools, and then it is up to not only the training of school talents but also enterprise production and operation objectives. There is complete training equipment, but the utilization rate is low. If the enterprises cooperate on the practical training and combine with their business, it will enhance the construction of intelligent logistics platform and standardize campus management. At the same time, the professional awareness of students will be improved.

In the integrated express service center, it is intended to cooperate with third-party logistics service enterprises, with the aid of training center and logistics points. The teaching activities combine with business activities closely. The function of the intelligent logistics platform is professional training, production and operation, and social services and others, so that under an intelligent campus matching mode, the integrated express service center is formed.

Integrated with the third-party logistics company technology and commercial elements, the express service center achieves the customer's independent pickup process by matching the shelf and goods numbers accurately, and achieves accurate and efficient pickup through the high-shot instrument with the facial recognition and manual assistance. As a result, the intelligent logistics services are convenient and efficient. At the same time, the service center not only can undertake school canteens and supermarket distribution business, but also can accomplish student situated distribution and teaching supplies. In addition, it can undertake third-party international business and propose a whole solution of industry chain logistics for the school and others, through the simulation of center back-end.

The practical teaching work is completed by the school teachers and the enterprise experts, and the main work of operation is the student-related logistics and other relevant majors. The service

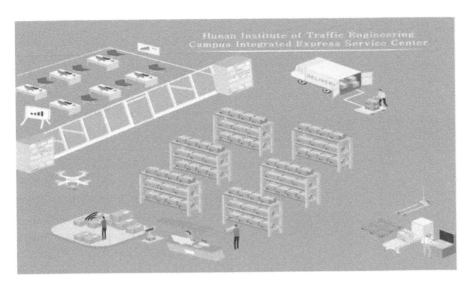

Figure 4.   Campus integrated express service center layout.

Figure 5.   The integrated application module of logistics engineering.

center makes use of the advantages of both sides to integrate all the express and logistics brands on campus, which can change the image of the campus and improve the utilization rate of the logistics training center, so that it not only helps the poor students and enriches their social experience but also improves the competitiveness and the employment rate of university.

3.3   *Build an internship base and enhance the comprehensive application capacity*

Universities are the main suppliers of talent, and enterprises are the main demand side of talent. It is concluded that the education of colleges and universities must be oriented toward enterprise, and vice versa. The promising enterprises should be selected as an off-campus internship base, which is suitably professional and cooperatively collaborative. The base will be much accounted of the continued construction and effective maintenance. Professional teachers as school-enterprise liaison should strengthen communication and docking with enterprises in a timely way, and complete the guidance and management of students in accordance with the "double division system" ideas with engineers.

There are three parts of centralized practice in our professional training, such as professional comprehensive internship, graduation internship, and graduation design. The professional cognitive practice is an independent centralized practice link. The students will visit the enterprise and their logistics parts such as enterprise workflow and logistics operation under the leadership of teachers and enterprise workers. It can enhance their technology and expand their professional vision and enrich the practical knowledge. For the students, there are some requirements of professional comprehensive internship, such as the characteristics and application of various types of logistics equipment, responsibilities and procedures in their work, the knowledge of on-site production

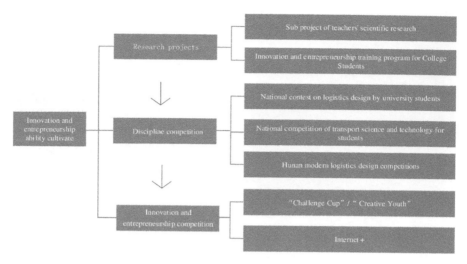

Figure 6. The innovation and entrepreneurship training module for logistics engineering specialty.

management and basic knowledge of safety precautions. By the means of participating in logistics operations and enterprise operations, the students learn about production practices and production equipment and prepare for their next learning work. The graduate internship, which combines theoretical knowledge with production practice, cultivates the student who has the independent ability of induction and has a rigorous scientific attitude for seeking truth from facts. As the students make the graduation design, they will make comprehensive use of the basic theories, basic knowledge, and basic skills, and it will improve their self-study ability and design consciousness and innovative ideas, so that they can analyze and solve practical problems independently. About the theme of the graduation design, 70% of the theme is required to select in production practice. It encourages students to discover the problems in actual production in order to develop their ability to analyze and solve complex engineering problems.

3.4 *Build a scientific research platform to train innovation and entrepreneurial capabilities*

Under the requirements of the new engineering talent training direction, in order to improve students' innovation and entrepreneurship ability and cultivate applied technology talents, our school conducts diversified and three-dimensional exploration of innovation and entrepreneurship ability training. First of all is to build a scientific research platform in order to change the reality of "emphasizing teaching, ignoring scientific research, and lack of innovation" in newly built local applied undergraduate colleges. The school supports teachers in decomposing scientific research projects into sub-topics, selecting some suitable sub-topics to provide students with practical exercises. Throughout teacher guidance and student participation, teachers conduct research on scientific research topics, and students also get training and training opportunities. Second, by participating in the training of teachers' scientific research projects, the outstanding student teams are selected to apply for college students' innovation and entrepreneurship training program projects that can train students to explore and learn. Third, the training plan focuses on students' innovative ability to discover and solve problems, combining professional awareness. It is proposed that students participate in discipline competitions, such as the national contest on logistics design by university students, the national competition of transport science and technology for students, Hunan modern logistics design competitions, and so on. Finally, to strengthen students' entrepreneurial ability and entrepreneurial awareness, it is proposed to encourage students to participate in innovation and entrepreneurship competitions such as the Challenge Cup, Creative Youth, Internet +, and so on.

4 CONCLUSIONS

The logistics engineering major of Hunan Institute of Traffic Engineering divides the training of practical teaching ability into four modules: single skill, comprehensive skill, comprehensive application, and innovation and entrepreneurship. First, according to the training requirements of new engineering talents and the actual needs of logistics engineering talents, the individual skills of students should be improved from three aspects: planning and management, machinery and equipment, and information technology. Second, through introducing enterprises into the school and building a practical training platform, the students can complete the production tasks of enterprises and build a practical training platform while carrying out professional training. Third, through deepening school-enterprise cooperation and strengthening timely communication and docking with enterprises, and it is then completed that the guidance for students' internship work in collaboration with enterprise engineers is according to the idea of "two-teacher system," so as to cultivate students' ability to combine theory with practice and solve practical problems. Finally, under the background of new engineering, talent training puts forward higher requirements for innovation and entrepreneurship ability. Through participating in teachers' scientific research projects, student projects, discipline competitions, and innovation and entrepreneurship competitions, students in our school cultivate their innovation and entrepreneurship ability step by step according to the rules of understanding.

Because the practice teaching of logistics engineering specialty in our school contains objective regularity, in order to consolidate students' engineering practice ability and highlight the goal of innovation and entrepreneurship ability, we will continue to explore the cultivation path of high-quality applied technology talents under the background of new engineering.

ACKNOWLEDGMENT

This research was financially supported by the Education Department of Hunan Province named xjt [2019] No. 333; xjt [2020] No. 90.

REFERENCES

Gui, D.H. 2019. Research on Supply and Demand of Intelligent Logistics Talents Serving Intelligent Manufacturing. *Chinese Vocational and Technical Education*, 19, 89–93.

Li, Y. & Chen, X.T., et al. 2020. Analysis on Matching between Demand and Cultivation of Logistics Skilled Personnel from Perspective of Supply Side. *Vocational and Technical Education*, 41, 26–30.

Lin, J. 2017. The Construction of China's New Engineering Disciplines for the Future. *Tsinghua Journal of Education*, 38, 26–35.

Liu, X.Q. 2018. Emerging Engineering Education, Integration of Production and Education, and Industrial Transformati and Upgrading. *Higher Vocational Education Exploration*, 17, 1–4+15.

Qin, Y.R. 2014. Establishment of Competence Structural Demand and Cultivation System of Applied Logistics Engineering Talents. *Logistics Technology*, 000, 474–476.

Sun, K.X. & Guo, Y.F. 2018. Construction and exploration of engineering practical teaching system for new engineering. *Experimental Technology and Management*, 35, 233–235.

Tang, L.W. & Wang, M.Y. 2019. On Integration of Industry and Education and Curricular Construction of Technology Specialty in Local Colleges. *Journal of Hebei Normal University (Educational Science)*, 21, 101–105.

Tong, Y.R. & Chen, J.P., Li, C.C. 2020. The Construction of Practical Education System for Emerging Engineering Education. *Research in Higher Education of Engineering*, 01, 56–61+122.

Wang, H.J. & Song, H.W. & Zang, A.J. 2020. Construction of Practical Teaching System of Logistics Engineering Major in Applied Undergraduate Universities. *Journal of Shijiazhuang University*, 22, 35–39.

Zhong, D.H. 2017. Connotations and Actions for Establishing the Emerging Engineering Education. *Research in Higher Education of Engineering*, 03, 1–6.

Computational Social Science – Luo, Ciurea & Kumar (eds)
© *2021 Taylor & Francis Group, London, ISBN 978-0-367-70193-2*

Institutional isomorphism and professional accounting education development in Chinese higher education system

G.H. Zhang
Center for Accounting Studies, Harbin University of Commerce, Harbin, China
Institute for Financial and Accounting Studies, Xiamen University, Xiamen, China

X.H. Qu*
School of Economics and Management, Harbin Institute of Technology at Shenzhen, Shenzhen, China
Center for Accounting Studies, Xiamen University, Xiamen, China

ABSTRACT: Employing New Institutional Sociology theory, this paper investigates how and why China's professional accounting education (PAE) was shaped in its own way from emerging, disappearing, re-started, and fast development stages over a century. We find that the development of PAE in China is actually a complex coercive, normative, and mimetic isomorphic process, which provides a comprehensive understanding of contemporary China's PAE from institutional perspectives, benefiting the exchange and collaboration of accounting education among countries. The paper updates and enriches the literatures of PAE development in China.

Keywords: CPA, Professional accounting education, New institutional sociology.

1 INTRODUCTION

Different from the pattern of Western accounting education changes, which are mostly driven by professional bodies (May et al. 1995), accounting education change in China is primarily driven by the government and its political, social, and economic development. The early accounting education in China could be traced back to the Zhou Dynasty (about 11th century BC to 256 BC), when writing and calculation was taken as a special knowledge and taught in class, which was a combination of arithmetic and bookkeeping (Guo 1997). However, in the long history of the Chinese Civil Service Examination System from the years 606 to 1905, Confucianism was deemed the legitimacy, business was repelled, and the subject of accounting was regarded as something unpresentable and unacceptable. Business discipline did not emerge until the early 1900s, when business discipline was formally included in the Articles of Higher Learning in 1902. Thereafter, bookkeeping became a fundamental course of business discipline, but there was no accounting major at that time (Guo 1997). The first CPA in China emerged in 1918, and the accounting discipline was initiated in 1921 in China's universities. The development of CPAs and PAE in China can be summarized into four periods: the emerging period (1918–1948), suspended period (1949–1979), re-emerging period (1980–1990), and fast development period (1991–present).

Emerging period refers to the time span when the first Chinese CPA was approved by the Chinese government in 1918 until the new China was established by the Chinese Communist Party (CCP) in 1949. Suspended period refers to the period when CPAs and accounting firms gradually disappeared across the country after the establishment of the People's Republic (P.R.) of China in 1949 until the reform and open-door policy was implemented in 1979. Re-emerging period refers to the period from 1980 to 1990 when the foreign capital was reintroduced into China and CPAs re-emerged to meet the demand of the introduction of foreign capital. Fast development period refers to the period from 1991 to the present when the stock markets were established and qualification examination

---

*Corresponding author

for CPAs was initiated by the Chinese Institute of Certified Public Accountant (CICPA). Since then, Chinese CPAs and PAE have stepped into a fast development path. In the process of more than a century's development, why was Chinese CPA profession and PAE fluctuated, and why was it not very well developed? What were the underlying reasons for the re-emerging and fast development of Chinese CPA profession and PAE in the fourth period?

So far, the literature has documented the trajectory of accounting education change in China since late 1978 (Tang 1997); institutional pressures are recognized as the key factors that shape the contemporary accounting education change from heavy politically orientated since 1949 to a more internationalized accounting education after the implementation of reform and open door policy in 1979 (Zhang et al. 2014). More detailed aspects of accounting education in China, such as accounting education at university level (Woodbine 2007), the reestablishment of the accounting profession and professional certification (Chan & Rotenberg 1999), and the challenges facing China's accounting education (Wu & Tong 2004) are also discussed. These literatures help us to understand the status quo and the changes of China's accounting education. However, the above-mentioned questions are still not clearly answered. This study aims to answer these questions.

Building upon the theoretical insights from new institutional sociology (NIS), this study investigates how China's PAE was developed and why China's PAE re-emerged and developed at a fast speed in the past 30 years, and it provides updated information about China's PAE changes in the last 30 years. The study may contribute to the existing literatures in both information contents and advancing the country study in accounting education.

2 ISOMORPHISM AND INSTITUTIONAL PRESSURES FOR CHANGE

New institutional sociology (NIS) conceives the idea that organizations may seek to legitimize their existence by conforming to institutional rules and norms in order to gain social legitimacy, facilitate access to resources, and avoid risk (DiMaggio & Powell 1983; Meyer & Rowan 1977; Zucker 1987). DiMaggio and Powell (1983) identified three mechanisms of institutional isomorphic change: coercive isomorphism, mimetic isomorphism, and normative isomorphism.

Coercive isomorphism refers to both formal and informal pressures from other organizations in which they depend upon and from cultural expectations of the society. Referring to the PAE within high education institutes (HEIs), if the institutes are funded by the state, and PAE has more interaction with the state at field level, the more isomorphism will be observed across the country. At organizational level, the more the accounting departments within HEIs are dependent on another organization, the more alike it is to that organization.

Normative isomorphism stems primarily from professionalization and broader cultural expectations of the society within which organizations function. One impetus to isomorphism is the legitimization inherent in the licensing and crediting of educational achievement. The other is the inter-organizational networks that span organizations (DiMaggio & Powell 1983). Referring to PAE, the more accounting professionals involved in the teaching process or more accounting faculty involved in professional organizations, the more isomorphic the PAE will become. At an organizational level, if HEIs rely more on academic credentials to choose accounting faculty members and more faculty members participate in accounting professional organizations, the more alike the PAE in HEIs will be.

Mimetic isomorphic change occurs under conditions of uncertainty or ambiguity. This uncertainty encourages organizations to mimic organizational forms or structures that already exist elsewhere and are, or appear to be, successful adaptations to the environment (DiMaggio & Powell 1983). Referring to the PAE, when the conditions are uncertain or ambiguous, it is likely to mimic the successful models both at organizational level and field level.

3 INSTITUTIONAL ENVIRONMENT OF HIGHER EDUCATION IN CHINA

As mentioned earlier, the long history of Confucian culture in China has deeply impacted its politics, economy, and individual beliefs. The government's requirements and expectations shape organizational behavior. In general, all the universities are under the leadership of the central government. Therefore, whether at organizational level or at field level, coercive pressures for PAE change are significant.

Government control over higher education is highly centralized in the organizational field. The coercive pressures from the central government affect HEIs' structure, source of funds, educational objectives, program setting, student recruitment, as well as faculty and staff employment. Almost every aspect of Chinese HEIs depends on the Ministry of Education (MOE), its provincial branches and higher education bureaus in related ministries. This implies that the homogenization process among HEIs is very soon and PAE in different HEIs is mostly alike. We suppose that PAE in higher education reacts to the government's regulations and plans quickly and HEIs became isomorphic with each other very soon.

4 REVIEW OF CHINESE CPA AND PAE CHANGE FROM INSTITUTIONAL PERSPECTIVES

4.1 The emerging of CPA and PAE (1918–1948)

Although China enjoys a long history of civilization, accountancy was not emphasized due to the significant impact of Confucianism. The dominated Confucianism repelled business and there was no accounting subject before the abolition of the Civil Service Examination System. The First Opium War in the early 1840s broke the balance of China's society and its economic development trajectory. The Treaty of Nanjing in 1842 and the Peace Protocol of 1901 opened China's door, and changed China into a semi-colonial and semi-feudal society.

The inflow of foreign capital and the Westernization Movement changed Chinese cultural expectations and gradually formed normative and mimetic pressures for change. Under the direction of modem-thinking Han officials, Western science and languages were studied; special schools were opened in the larger cities; and arsenals, factories, and shipyards were established according to Western models. In 1902, business discipline was formally included in the Articles of Higher Learning. Thereafter, bookkeeping became a fundamental course of business discipline, but there was no accounting major at that time (Guo 1997).

The institutional change in 1911 ended the Qing Dynasty, and the Bourgeois Democratic Republic was established by Sun Yat-sen. The national capital and bureaucratic capital businesses became prosperous. Due to the lack of Chinese accounting firms, any economic dispute had to turn to foreign accounting firms for judgment. In order to protect the benefit of Chinese business, the government of the Republic of China approved and issued the first Chinese public accountants' regulation in 1918. Promoted by accounting professionals who were mostly graduated from European or Japanese universities in the early 1900s, the accounting department and accounting discipline was firstly set up in Fudan University in 1921, and then in other colleges and universities. The professional body of the National Association of Accountants was also formed in 1925 in Shanghai. Accounting and accounting education then were mostly developed by imitating Western accounting patterns.

In the process of the initiating PAE, clear mimetic and normative isomorphism can be observed. Due to the limited demand and lack of government's strong support, this isomorphic process was very slow, and even was interrupted by the establishment of the P.R. China in 1949.

4.2 Suspended period (1949–1979)

The Chinese Communist Party (CCP) and its government played a leading role in China's political and social life and exerted strong coercive pressures for change since the establishment of the P.R. China. The cultural expectations were totally changed and the capitalist economy was eliminated. Only communist centrally planned economy and socialist ideology were allowed during this period. Socialistic reconstruction from 1950 to 1956 thoroughly eliminated capitalist industry and commerce. As the result, accounting firms and the CPA profession disappeared gradually (Yang 1995). As there was no need for public accountants, PAE halted during this period.

During this period, the government mandates and impact from the outside world exerted strong pressures for change. The in-flow of foreign capital and diversified forms of business enterprises since 1978 could be the direct inducement of the reintroduction of the CPA system. Chinese CPA system was restarted in 1980. Shanghai Accounting Firms was firstly established in 1981. Foreign accounting professional bodies also began to set up their resident representative offices in China. The professional accountants' body, CICPA, was reestablished in 1988.

Mimetic isomorphism was significant in the process of CPA system restarting and change. Western CPA systems and experiences were widely introduced in Chinese academic journals. Foreign experts were also invited to China to share their experiences with Chinese academics and professionals. Chinese accounting firms were established. PAE also restarted. The formal PAE was not introduced in HEIs during this period. But the HEIs also provided night school courses, correspondence courses to the adult on-the-job students. Professional certification, short-term training, and junior college became popular and attracted a lot of on-the-job students (Liu 1988). Two tracks of accounting education, normal university education and adult education, were formed and became popular.

4.4 Fast development period (1991–present)

The fast development of the CPA profession and PAE started after 1991 when Chinese capital markets were set up subsequently in Shanghai and Shenzhen in 1990 and 1991, and qualified candidates could acquire CPA qualification via examination. The deepening of economic reform and the inflow of foreign capitals and foreign professional accounting bodies also changed the cultural expectation to accounting education. The re-emerging of PAE in HEIs was actually pushed by the combined coercive, mimetic, and normative pressures. We observe that PAE grew up quickly once initiated and homogenized in the HEIs very soon.

The coercive pressure for PAE came from the central government first. In 1994, the State council instructed to set up three internationalized CPA training base in Beijing, Shanghai, and Xiamen. As the result, Beijing, Shanghai, and Xiamen National Accounting Institute were set up in 1998, 2001, and 2002, respectively.

Mimetic process accompanied all along with the PAE development in this period as there was too many uncertainty and ambiguity during the developing professional accounting curriculum, teaching contents, and textbooks. Initiated by the CICPA and MOE in 1994, 23 universities became the tentative universities in setting up a CPA specialty in their accounting program aiming at cultivating professional CPAs at the undergraduate level (Ping 1999). The CICPA also examined and approved textbooks for CPA specialty via its leading group of training, which established a committee in charge of editing and approving CPA education textbooks that gave priority to edit and translate the newest international textbooks and were in line with the international accounting tradition (Zhong 1995).

Normative pressures came from the national and international accounting professional bodies and accounting academics, which pushed PAE formally launched in HEIs at undergraduate level in 1994 and then at postgraduate level in 2004. Accounting education became to be both a site of isomorphism and a driver of isomorphism. ACCA launched its first training class for professional accountants in China in 1990 and established cooperative relationship with CICPA (Ni 1998). CGA entered China in 1993 via a Canadian government loan. It cooperated with Chinese government, academics, and HEIs to develop training courses and academic conferences. These two professional bodies also cooperated with Chinese HEIs after 2000 and integrated their credential courses into their partner university's undergraduate curriculum. Many HEIs have established cooperation relationships with ACCA and/or CGA and set up international accounting direction under their accounting major. These HEIs provided similar courses for their ACCA or CGA students.

A Master of Professional Accountancy (MPAcc) program was proposed by Chinese academics who witnessed the successful operation of a professional master program of accounting in Western countries, especially in the United States. Meanwhile, foreign experts are often invited to China to deliver seminars, change academic ideas, and conduct joint research with Chinese scholars. And many Chinese scholars are sent abroad for visiting study and to conduct joint research. This combined normative and mimetic pressure, from organizational field, affects universities' educational objectives, educational programs, curriculum, and textbooks. Conceiving academics' proposal of the solution to the current issues of PAE, a tentative MPAcc program was formally launched in 21 universities in 2004. In 2007, the second cohort universities (4 universities) received the ADCSC's approval. With the joint involvement of ADCSC, MOE, and MOF, the MPAcc program was soon diffused in HEIs. Meanwhile, two additional professional education programs, asset valuation (Master of Valuation) and auditing (Master of Professional Auditing), were also approved by the ADCSC in this institutional isomorphic process.

In addition to the PAE programs at higher education level, in-service-training was also required by the MOE and CICPA for the on-the-job accounting personnel. The programs can be centrally planned programs or demand-driven programs, which well manifest as a complex coercive, normative, and mimetic isomorphic process in shaping the PAE in China.

5 CONCLUSIONS

This study reveals that the development of PAE in China is actually a complex coercive, normative and mimetic isomorphic process from new institutional perspectives. The fast development of PAE in the last 30 years is attributed to the CCP and its government's stable policies in developing globalized economy, the ever-increasing cultural expectations to cultivate internationalized accounting talents, and interaction of national and international professional bodies within HEIs. Coercive pressure from the government, though weaker than before, is still the key for the success of the PAE in China. However, due to the unique institutional settings in China, Chinese HEIs and accounting professional bodies, the government's leadership will still be the main force of the education change in the short run. Therefore, it is reasonable to predict that in the foreseeable future, Chinese PAE will still change in its own particular way, that is, change under the complex process of coercive, normative, and mimetic isomorphism. As for which pressure will dominate in the near future depends on the further reform of Chinese political and economic systems, as well as Chinese economic and capital market development. It is still important to pay attention to the changes of Chinese government policies when involved in Chinese PAE.

ACKNOWLEDGMENT

This study was supported by Shenzhen Key Research Base of Humanities and Social Sciences.

REFERENCES

Chan, M.W.L. & Rotenberg, W. 1999. Accounting, accounting education and economic reform in the People's Republic of China. International Studies of Management & Organization 29: 37–53.
DiMaggio, P.J. & Powell, W.W. 1983. The iron cage revisited: Institutional isomorphism and collective rationality in organizational fields. American Sociological Review 48:147–160.
Guo, D.Y. 1997. The historical starting point and the initial evolution of China's accounting education. Finance and Accounting Monthly 10: 3–6.
Liu, F. 1988. The symposium summary to the accounting education reform organized by Accounting Society of China. Accounting Research 5: 13–18.
May, G.S. & Windal, F.W. & Sylvestre, J. 1995. The need for change in accounting education: An educator survey. Journal of Accounting Education 13:21–43.
Meyer, J.W. & Rowan, B. 1977. Institutionalized organizations: Formal structure as myth and ceremony. American Journal of Sociology 83: 340–363.

Ni, WH. 1998. The development of ACCA in China. Chinese Auditing 12: 52.

Ping, H. 1999. Memorabilia for Chinese Certified Public Accountants industry. Chinese Certified Public Accountants 9:32–34.

Tang, Y.W. 1997.The recent accounting development and internationalization of accounting education in China. Issues in Accounting Education 12:219–227.

Woodbine, G.F. 2007. Accounting education in modern China: An analysis of conditions and observations. Asian Review of Accounting 15: 62–71.

Yang, S.Z. 1995. The Evolution and Development of Chinese Certified Accountant System. Communication of Finance and Accounting 1:11–13.

Zhang, G.H. & Ahmed, K.& Boyce, G. 2014. Institutional changes in university accounting education in post-revolutionary China: from political orientation to internationalization. Critical Perspectives on Accounting 25: 819–843.

Zhong, P. 1995. Tentatively set up CPA specialization in 23 higher educational institutes. Finance and Accounting 10: 5.

Zucker, L.G. 1987. Institutional theories of organization. Annual Review of Sociology 13: 443–464.

Computational Social Science – Luo, Ciurea & Kumar (eds)
© 2021 Taylor & Francis Group, London, ISBN 978-0-367-70193-2

Teaching practice and reform of "process equipment design" course design based on cultivating team cooperation ability

Z.Y. Duan*, H.D. Zhang, H.Y. Zhai, X. Cao, Z. Liu, X.L. Luo & D.Y. Luan
Qingdao University of Science and Technology, Qingdao, China

ABSTRACT: How to effectively support the team consciousness and the cooperation spirit is a very important problem when the engineering education professional certification is carried out This paper is based on the practice of engineering education professional certification and the authors' experience for the "process equipment design curriculum design" reform and teaching practice in recent years. In particular, the work of cultivating students' teamwork ability is briefly introduced to provide some useful reference and help for the curriculum construction under the new situation.

1 INTRODUCTION

In recent years, China has paid increasing attention to the cultivation and the education of undergraduate students. With the advancement of construction projects such as "new engineering," "national first-class specialty," and "engineering education professional certification," the teaching concept of "based on the foundation" and outcome-based education (OBE) has gradually penetrated the hearts of colleges and teachers in the whole body And the professional construction and the talent training are also carried out in a new mode.

The engineering education professional certification is an internationally accepted quality assurance system for engineering education. The OBE concept has developed to follow the result-soriented, student-centered and continuous improvement [1]. The core is to confirm that engineering graduates meet the quality standards established by the industry, and it is a qualification evaluation oriented by training objectives and graduation export requirements [2]. After more than 10 years of efforts, China has made a series of major achievements in professional certification. By the end of 2019, 1,353 specialties of 214 universities have passed the professional certification of the engineering education. For the specialty of Process Equipment and Control Engineering, 26 universities have passed the professional certification. At present, whether they can pass the "professional certification" also has gradually become the standard to measure the quality of the specialty construction.

China's general standards for the professional accreditation formulated by China Engineering Education Accreditation Association are closely connected with international standards in seven aspects: students, training objectives, graduation requirements, continuous improvement, curriculum system, teaching staff and support conditions [3]. Among them, there are special requirements for the graduates' team consciousness and the cooperative spirit in the graduation requirements of engineering majors. On the other hand, the society's development urges the employers to have the team cooperation ability for the personnel training of higher education. However, as far as the current training program and the curriculum system are concerned, the effective support of the team consciousness and the cooperative spirit in the graduation requirements of most majors are relatively weak, and many of them are only supported by the physical education and comprehensive experimental courses, which are insufficient. At present, the professional engineering education

---

*Corresponding author

certification in China has also found that how to effectively improve the team cooperation ability is an urgent problem for most colleges and universities.

The specialty of Process Equipment and Control Engineering in Qingdao University of Science and Technology, founded in 1958 and formerly known as a Chemical Equipment and Machinery specialty, is one of the earliest undergraduate majors established in Qingdao University of Science and Technology. At present, it is a national characteristic specialty, a first-class specialty in Shandong Province and a core specialty of the high-level application-oriented construction in Shandong Province. In 2018, it passed the professional certification of the engineering education with the result of "conditional six years." In November 2017, the joint certification team gave a high evaluation of the professional construction in the school examination. However, in the feedback of "curriculum system," it is pointed out that the teaching syllabus contents and assessment methods of "production practice," "process equipment comprehensive experiment," and "mechanical engineering training" are not fully reflected in the content of cultivating students' team cooperation ability.

Therefore, under the support of the Shandong education reform project "curriculum system construction of cultivating undergraduate innovation and team cooperation ability based on professional certification," the research group in charge of the author carried out relevant reform and research mainly combined with the core problems exposed in the professional certification. Based on the reform of "process equipment design curriculum design," this paper constructs an uninterrupted curriculum system from freshman to senior aiming at highlighting team cooperation ability and innovation ability. Practice shows that it can effectively support the cultivation of innovation and team cooperation ability. This paper focuses on the reform of "process equipment design curriculum design" and the practice of a brief introduction to provide useful help for related majors in effectively supporting the "team cooperation ability" training.

2 INVESTIGATION ON THE CONSTRUCTION OF THE COURSE DESIGN OF PROCESS EQUIPMENT DESIGN

The "process equipment design curriculum design" is not only the core professional course of the Process Equipment and Control Engineering but is also an important practical teaching link [4]. Through curriculum design, students can apply the single and the independent theoretical knowledge in the specific design work comprehensively It not only achieves the purpose of combining the theoretical knowledge with the production practice closely but also cultivates engineering design ability and practical ability. Therefore, the teaching reform and the practice around the "process equipment design curriculum design" have always been the focuses of attention.

Through the research and analysis of the published literature, it is found that the teaching of process equipment design has made a lot of practice and effort in effectively cultivating students' engineering practice ability. For example, the setting of topics in the curriculum design of Inner Mongolia University of Science and Technology strives to improve students' engineering practice ability to the greatest extent [5]; Nanjing Institute of Engineering combines the practical engineering with the curriculum design, and the ability to solve practical problems has been effectively improved [6]. However, there are few teaching reform practices in the aspect of the team cooperation ability training. Tianjin University takes the lead in bringing forth new ideas and takes "three kinds of consciousness" (team consciousness, management consciousness, and economic consciousness) as the core connotation to carry out corresponding teaching practice reform on the course design of process equipment design.

Based on the literature collection and the investigation of team cooperation ability in the curriculum system, the research group found that the current curriculum design teaching system is mainly divided into two categories: one is to adopt the team cooperation system to enhance the cultivation of the team cooperation ability based on improving the individual ability; the other is to adopt the "one person, one question" system to improve the personal engineering practice ability, but the support strength of team cooperation ability is not enough. The detailed curriculum design features of each school are shown in Table 1.

Table 1.   Current situation of course design system construction of process equipment design.

University	Characteristics of Curriculum Design System
Inner Mongolia Institute of Science and Technology	The division of labor of the curriculum design is one person to undertake one task, and the support strength of the team cooperation ability is not enough. In the later stage, students are organized to carry out the summary exchange meeting of curriculum design to summarize the problems and solutions encountered in the design [5].
Nanjing Institute of Engineering	The curriculum design refers to the relevant practical engineering cases Each link is completed independently by individuals, and it is required to master the corresponding pressure vessel specifications, but the team cooperation is relatively weak [6].
Qingdao University of Science and Technology	Before 2016, the curriculum design adopted the "one person, one question" system to cultivate the individual comprehensive ability, but it is insufficient support for the team cooperation ability. After 2016, the practical reform of the curriculum system was gradually carried out to improve the team cooperation ability of graduates.
Tianjin University	The curriculum design is reasonable in setting up links, adopting the form of the team cooperation, and paying attention to the engineering thought and the modern engineer consciousness. Chemical process strength is far ahead in the country, and the process link is introduced based on the equipment design.
North University for Nationalities	The COID [7] engineering education concepts are introduced into the process equipment design and the curriculum design, and the curriculum design is completed in groups, which can well mobilize the students' learning initiative and improve the team cooperation ability

The comprehensive analysis shows that most of the current "process equipment design curriculum design" system focuses on cultivating students' engineering practice ability, but the cultivation of the team cooperation ability and the innovation consciousness is not fully reflected. We need to learn from the advanced experience of Tianjin University and explore a suitable curriculum teaching mode combined with its characteristics.

3   TEACHING PRACTICE REFORM OF CURRICULUM DESIGN

During the professional certification, the certification experts put forward suggestions and requirements for the effective support of the team cooperation ability. Based on the feedback of professional certification experts and the characteristics of Process Equipment and Control Engineering specialty of Qingdao University of Science and Technology, the research group has conducted a full investigation on how to improve the team ability of graduates. Based on the cultivating students' practical ability and learning ability, the teaching mode of "process equipment design curriculum design" is established, which follows the OBE teaching concept and focuses on cultivating students' team cooperation ability and engineering practice ability. The method is as follows:

(1) The system of "one person, one question" was optimized And the teamwork was introduced to complete the curriculum design task. Based on the specialty being set in the school of mechanical and electrical engineering and the number of class hours is three weeks, the curriculum design task book is determined to be a specific process equipment combined with engineering practice, such as heat exchanger, reactor or tower equipment. A team of three to four people will work together to complete the overall tasks of the process calculation, the structural design, the PowerPoint report and the drawing (the combination of hand drawing and computer-aided design). The difficulty and comprehensiveness of curriculum design are weaker than that of Tianjin University. Tianjin University takes a complete set of process equipment as the course design topic, involving the design of various process equipment, including the process calculation, design and drawing, and so on.

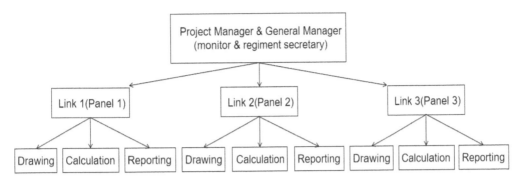

Figure 1. Design division of curriculum system.

(2) Making full use of the advantages of industry professors employed by the province, we employ Xie Yuhui (professorlevel senior engineer) of PetroChina East China Design Institute Co., Ltd., as the consultant and instructor. Xie Yuhui, senior engineer, explained to the students of the organizational structure of the engineering design institute and the division and cooperation of various departments, so that the students had a whole understanding of the engineering design institute in reality. Focus was on the responsibilities of each department of the specific project and the division and cooperation of different departments. On this basis, inspired by the division of labor system of the Design Institute, a class group is taken as a project department (as shown in Figure 1). The team leader is the project manager, the League branch secretary is the chief engineer, and different subprojects are completed by different groups. The members of each group are arranged according to the student number, and the professional performance is not taken as the measurement standard. The completion of the project is completely coordinated by the class collectively and the teacher gives proper guidance, which fully follows the "student-centered" teaching concept. After the reform of the system, all teams are integrated together, which effectively improves the enthusiasm of students. Based on making students get comprehensive and systematic engineering design training and skills training, the team cooperative spirit among students is further enhanced.

(3) The evaluation system has been innovated. The first is the innovation of the performance evaluation method. The total score is divided into individual score and group score (see Table 2 for details). The team cooperation score is regarded as a separate assessment item, fully reflecting the "student-centered" and OBE teaching concept. The results of the group account for 40% of the total score. The evaluation team composed of two to three teachers performs the defense on the spot by reviewing the completion of the overall task of the group, the overall quality of the submitted calculation instructions, and engineering drawings. The individual achievement accounts for 60% of the total score, mainly for the completion and quality of the tasks that each person is responsible for, among which the team cooperation ability and attendance account for 10%. In order to ensure the reasonableness of the evaluation, the evaluation teacher gives a total score of the item according to the number of people in the team For example, the four-person group is 33 points, and the group itself gives the individual score according to each person's contribution to the team. In addition, for the good performance of the team, we will give the team leader a reward score. By reviewing the completion of the group's overall task, the overall quality of the submitted calculation instructions, and engineering drawings, the team performed the defense on the spot

(4) The transformation from "one to many" mode to "many to one" mode (as shown in Figure 2). One person taking on multiple tasks changes to multiple people undertaking one task, and students complete the tasks through mutual communication. Based on maximizing students' practical ability, the team cooperation ability of students is further improved.

Table 2. Evaluation of professional curriculum design results.

Evaluation items		Assessment content	Score
Group achievement (40%)	Data integrity and task completion (15%)	Whether the contents of calculation instructions and drawings meet the requirements of curriculum design	
	Calculation sheet and general assembly drawing (15%)	Whether they can skillfully consult the relevant literature and technical standards; consider the economic efficiency the impact on society and the environment of the design scheme; and inspect the integrity, rationality and correctness of the relevant calculation results. The specification and content integrity of the assessment instructions are reasonable, and the references are reasonable	
	Results of defense (10%)	Be able to accurately describe the instructions and drawings, and answer the specific situation of the questions in the defense process.	
Personal achievement (60%)	Drawing quality (25%)	The workload, content, and format of engineering drawings shall meet the engineering drawing standards stipulated by the state and relevant industries.	
	Technical requirements and specifications (25%)	Whether the dimensioning is standard, the tolerance standard is reasonable, and the technical requirements are comprehensive and fully meet the needs of engineering manufacturing.	
	Teamwork and attendance* (10%)	Understand the meaning of team members or leaders to the whole team, and be able to do their own roles in the team. Being able to share information, communicate and cooperate with members. Have certain organizational management ability and team cooperation ability. High attendance and initiative	
Bonus points (5%)	Outstanding organizational work or other work (5%)	Effectively organize and coordinate the team cooperation and promote the work, or outstanding work in the design process, and make a special contribution to the team.	
Total			

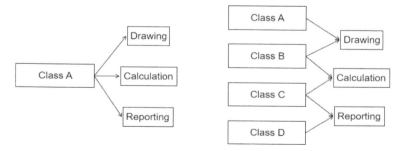

Figure 2. Transition from "one-to-many" to "many-to-one".

4 CONCLUSION

In the reality of scientific research or other work, it is difficult to achieve high quality according to one person's strength. Team cooperation ability is a very important ability of excellent talents. How to effectively cultivate college students' team cooperation ability has become an urgent issue to be solved. Based on the teaching reform practice in the "process equipment design curriculum design" the author briefly introduces how to cultivate team cooperation ability, comprehensively analyzes the current trend of higher education reform and development, and thinks that it is still necessary to carry out beneficial discussion and practice in the following aspects.

(1) The combination of curriculum design and college students' innovation competition enriches the topics of curriculum design. The topic of curriculum design is no longer confined to a single equipment such as a heat exchanger and reactor. The innovative topics related to process equipment are selected as the course design topics, and students are finally allowed to present their design results in the form of physical objects.

(2) Give full play to the advantages of the integration of production, teaching and research, and organically combine the curriculum design with the actual engineering. Based on the established "collaborative education base" and under the premise of obtaining the consent of the enterprise, we can take the complex engineering problems that the enterprise wants to solve as the topic of curriculum design So students can try to solve them in the form of the team, but not take the complex engineering problems as the teaching content [8], only as a means of training, to cultivate students' engineering ideas and team cooperation ability, and truly experience the team's importance in an enterprise.

(3) The teaching time of course design is flexible. The period should be extended appropriately so that students can refine the task, have enough time for communication, and further improve the quality of curriculum design based on enhancing team cooperation ability.

REFERENCES

[1] Zhiyi Li. China engineering education professional certification "last kilometer". Higher Education Development and Evaluation, 2020, 36(03): 1–13+109.

[2] Dexin Xu, Yuxin Zhao, Xuemei Zhou, Ning Li. Research on Innovative Personnel Training System of Measurement and Control Specialty under the Background of Engineering Education Professional Certification. Open Journal of Social Sciences, 2020, 08(03).

[3] Yang Zhao. Construction of Engineering Education Professional Certification System. Proceedings of 2018 4th International Conference on Education & Training, Management and Humanities Science(ETMHS 2018).Ed.. Clausius Scientific Press, Canada, 2018, 282–285.

[4] Jia Li, Qiaoan Tu, Jie Hua Teaching Practice and Exploration of Course Design of Process Equipment Science, Education and Culture, 2008(06): 36–37.

[5] Huiling Jia Exploration on Teaching Reform of Course Design in Process Equipment Design. Education Teaching Forum, 2013(49): 45–46.

[6] Youying Liu, Dasheng Zhu, Chong Li. Practice and Experience of Course Teaching of Process Equipment Comprehensive Curriculum Design. Science and Technology Information, 2012(34): 166.

[7] Tianxia Liu, Zhihao Bai, Chun Du, Guoping Jiang. Teaching Reform and Practice on Process Equipment Design and Course Design Based on COID Concept. Guangzhou Chemical Industry, 2018, 46(02): 141–142+144.

[8] Zhiyi Li. Review and Reflection on the Ten Years of Engineering Education Professional Certification in China. China University Teaching, 2017(01): 8–14.

Computational Social Science – Luo, Ciurea & Kumar (eds)
© *2021 Taylor & Francis Group, London, ISBN 978-0-367-70193-2*

Research on the development strategy of Wuhan urban rail transit industry

H.P. Wang & F. Gao*
School of Business, Jianghan University

Q.H. Liu
School of Business, Jianghan University
Manufacturing Industry Development Research Center on Wuhan City Circle

ABSTRACT: Urban rail transit is the development direction of modern metropolitan transportation, which can solve the problem of traffic jam. This paper analyzes the development status of China's urban rail transit industry. Taking Wuhan as an example, the development status of Wuhan urban rail transit industry is analyzed. Finally, countermeasures such as promoting the development of the whole industrial chain, vigorously developing the operation and maintenance market, and building smart urban rail transit are proposed in this study.

Keywords: Urban rail transit, Industrial chain, Wuhan, Operation and maintenance.

1   INTRODUCTION

The acceleration of urbanization has led to increasingly serious urban road traffic congestion. In order to alleviate the pressure of urban road traffic and facilitate the public travel, urban rail transit emerges as the times require. China's rail transit is in a high-speed development stage, and the number of rail transit industry chain is increasing. Taking Wuhan as an example. Since the first rail transit line was built in 2000, by April 2020, nine operation routes have been built with a total operation mileage of 339 kilometers and 228 stations. In total, more than 700 local enterprises have participated in the construction of Wuhan rail transit.

In September 2019, President Xi Jinping inspected Daxing Airport to take the rail transit, and pointed out that the development direction of modern city traffic is urban rail transit. However, novel coronavirus pneumonia has affected the whole industrial chain enterprises of Wuhan urban rail transit. But in fact, even in the boom of comprehensive resumption of production, the economic income of Wuhan urban rail transit industry is still not optimistic. Therefore, how to solve the problem is extremely urgent. This paper analyzes the current situation of China's urban rail transit development and the urban rail transit development of Wuhan in an all-round way. Subsequently, a development strategy suitable for the actual situation of Wuhan is presented.

2   DEVELOPMENT STATUS OF URBAN RAIL TRANSIT IN CHINA

In the past decade, with the support of high and new technology, the improvement of road heavy load capacity, the breakthrough of core technology and the promotion of high-level talents, China's urban rail transit industry has made remarkable achievements, but also ushered in new opportunities and challenges.

---

*Corresponding author

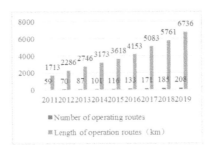

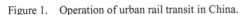

Figure 1.　Operation of urban rail transit in China.　　Figure 2.　Construction planning.

2.1 *Continuous growth of operation scale*

The process of urbanization is speeding up, and the urban permanent population and floating population are constantly increasing. In order to alleviate the pressure of public transport, the scale of rail transit operation in major cities in China has continued to grow. According to the China Urban Rail Transit Association, as of the end of 2019, Chinese mainland 40 cities opened 208 urban rail transit routes, the total length of operation routes 6736.2 km, the new operation line 974.8 km. Figure 1 is the basic operation of China's urban rail transit in 2011–2019. As shown in Figure 2, the approved new routes and investment also continue to increase.

2.2 *Economic income still cannot make ends meet*

Due to the characteristics of high construction cost, large investment in the early stage, long time for capital recovery, and the nature of public welfare, China's urban rail transit has been in a loss state for a long time, mostly relying on government subsidies. For example, the national average operating revenue and expenditure ratio is 72.7% in 2019, among which only Hangzhou, Shenzhen, Beijing, and Qingdao are the cities with more than 100% operating revenue and expenditure ratio, and other cities are in the state of loss. Therefore, on the whole, the operation of urban rail transit is still a common situation, and the economic income of urban rail transit is still not ideal, mainly relying on government subsidies.

2.3 *The downstream industry enters the golden period of development*

Urban rail transit industry can be divided into three stages: construction, vehicle manufacturing and operation and maintenance (Wan et al. 2020). With the continuous development of each city, cities with economic strength and actual demand have successively opened urban rail transit. In a short period of time, the demand for the construction of urban underground space and the assembly and manufacturing of rail transit vehicles will gradually decline, that is, the upstream industry in the industrial chain tends to be saturated. From putting into operation to ending operation, rail transit needs operation and maintenance without exception. Thus, the downstream industry ushered in a golden period of development.

3　DEVELOPMENT STATUS OF WUHAN URBAN RAIL TRANSIT INDUSTRY

Since the first rail transit line was built in 2000, by April 2020, Wuhan has built 9 operation routes, with a total operation mileage of 339 kilometers and 228 stations. More than 700 local enterprises have participated in the construction of Wuhan rail transit.

3.1 *The industrial chain is relatively complete*

In the past 20 years, Wuhan urban rail transit has made remarkable achievements in construction, operation and project development, and its industry has taken shape. Wuhan not only has many universities and institutions engaged in rail transit scientific research, but also a group of enterprises focusing on urban rail transit led by Wuhan Metro Group Co., Ltd. (Wuhan Metro), China Railway SIYUAN Survey and Design Group CO., Ltd., Wuhan CRRC Rail Transit Equipment Co., Ltd. (Wuhan CRRC). More than 700 local enterprises have participated in the construction of rail transit, which integrates scientific research and development, survey and design consultation, engineering construction, operation and maintenance management, equipment manufacturing and other plates and systems to form a complete industrial chain.

3.2 Poor in vehicle manufacturing sector

Although Wuhan is also known as "the city of cars", it mainly focuses on the manufacturing and R&D of small and medium-sized cars, and seldom involves in the manufacturing of urban rail transit vehicles. For the whole Wuhan urban rail transit industry chain, the construction capacity is fair, the operation and maintenance development have great prospects. But the vehicle manufacturing sector is relatively weak. Wuhan has Wuhan CRRC Rail Transit Equipment Co., Ltd. and other enterprises dedicated to the construction of Wuhan localized rail transit equipment manufacturing and maintenance base, which can roughly meet the needs of Wuhan and surrounding areas of rail transit equipment and key projects of industrial development. However, the manufacturing, R&D and development layout of urban rail transit vehicles in China is not in Wuhan, and there is a lack of independent R&D products and brands. Therefore, a large number of vehicle manufacturing parts and raw materials need to be purchased from other provinces, resulting in a large number of raw material costs and logistics transportation costs.

3.3 Large demand for rail transit construction in surrounding cities

With the rapid development of the first-line urban rail transit construction, the second- and third-tier cities have also begun to develop rapidly, and the demand for rail transit construction is gradually increasing.

Taking Yichang City and Xiaogan City around Wuhan as examples, the permanent population of Yichang is 4.1379 million, and the urbanization rate of permanent population is 59.86%. Xiaogan has a permanent population of 4.921 million, including 2.8689 million urban residents, and the urbanization rate is 58.3%. Therefore, urban traffic construction should continue to develop to meet the travel needs of the growing urban population. Due to the clustering effect of the central cities, many commodities and human resources of these neighboring cities have flowed to the central city, increasing the communication between the central city and the neighboring cities. Therefore, the demand for convenient travel has increased correspondingly, and the demand for intercity rail transit has increased.

The development of these non-first-tier peripheral cities cannot only transport resources to the central cities, but also promote the coordinated development of the industrial economy of the central cities, and even force the central cities to develop continuously. The increasing demand for rail transit construction in surrounding cities not only directly increase the economic income for Wuhan urban rail transit, but also form a larger and more complete industrial chain within the scope of the urban circle, and finally form a scale economy.

3.4 The downstream industry has a good development prospect

Savills, a real estate service provider, once released a special report on the subway effect of commercial real estate. The report shows that after the opening of the subway, the property value around Beijing, Shanghai and Guangzhou Metro has been greatly improved, specifically: the property value in the suburbs within 500 m from the subway station has been increased by 20%–25%, while the property value in the suburbs has been increased by 10%–15%. In Wuhan, for example, the price of the first and last subway stations rose the most, with a maximum of 30%. The fastest appreciation rate of real estate along the line is one year before and two years after the opening of subway. House prices will be affected by urban rail transit, which makes the business

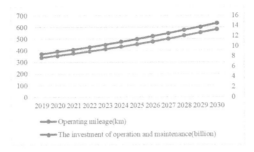

Figure 3.   Forecast of operating mileage of Wuhan urban rail transit in the next 10 years.

value of the surrounding shops increase in multiple. This shows that the investment and business value of rail transit is very considerable.

The operation and maintenance market of urban rail transit is positively related to the operating mileage, and has a certain proportion relationship with the replacement cost of urban rail transit (Wan et al. 2020). By the end of 2019, the total number of vehicles allocated to urban rail transit in China was 6966, with a total of 4.096 billion vehicle kilometers completed in the whole year, with an operating mileage of 6730.27 kilometers, including 338.9 kilometers in Wuhan. China Urban Rail Transit Association predicted that the scale of urban rail transit industry will continue to expand in the next 10 years. The average growth rate of new routes in the past 10 years is 5%. Based on this, it is estimated that by 2030, the operating mileage of Wuhan will reach 579.63 km. According to industry experience, the average construction investment per kilometer is about 1 billion yuan, and 2%–3% will be used for operation and maintenance investment, taking the median of 2.5%. As shown in Figure 3, it is predicted that by 2030, Wuhan will have 14.49 billion yuan invested in operation and maintenance of urban rail transit.

According to the above figure, the operation and maintenance in the middle and lower reaches of Wuhan urban rail transit industry chain in the next 10 years will usher in a golden period of development, which is a major opportunity for the development of Wuhan urban rail transit industry.

3.5   *The whole industrial chain falls into operation crisis under the epidemic situation*

2020 novel coronavirus pneumonia epidemic has brought great influence on the development of the middle and lower reaches of Wuhan urban rail transit industry. This paper analyzes the objective and subjective factors in the background of the high epidemic period and stable period. Due to a series of restrictions such as limited mobility of people, forced outage of public transport and so on in Wuhan during the high epidemic period, the operation of Wuhan Urban Rail Transit has become fragile in the first half of 2020. This is the impact of various regulations and restrictions and other objective factors in the high epidemic period. On the other hand, it is the impact of the public's subjective fear after the epidemic. After the epidemic was effectively controlled, the public's travel freedom was restored. However, the public who experienced the epidemic and did not have to plan their trip are afraid to take high-density public transportation due to fear of being infected, and this fear could not be completely eliminated in a short time. This adds a lot of resistance to the "restart" of Wuhan urban rail transit industry.

4   COUNTERMEASURES FOR THE DEVELOPMENT OF WUHAN URBAN RAIL TRANSIT INDUSTRY

4.1   *Promote the balanced development of the whole industrial chain*

The construction and development, survey and construction, underground space planning and other upstream markets of Wuhan urban rail transit have achieved remarkable achievement.

The downstream markets, such as operation service, maintenance, and additional operation have great development space. However, the midstream market of vehicle manufacturing and material R&D, which involves heavy industry manufacturing, is relatively weak. Although it is difficult to develop, it is necessary to actively supplement the shortcomings.

Wuhan has many institutions of higher learning and enterprise units related to the rail transit industry. They integrate their scientific research strength, technological innovation, enterprise experience, industrial scale, and other resources, jointly overcome technical difficulties, and strive to make up for the short board of the midstream market.

4.2 Vigorously develop operation and maintenance market

It is estimated that in the next 10 years, the scale of Wuhan urban rail transit industry operation and maintenance market will continue to expand, and the amount of investment will increase year by year. Wuhan urban rail transit operation and maintenance market will usher in a golden development period.

On the one hand, we should pay attention to the needs of passengers, improve the service quality of urban rail transit industry, and optimize the travel experience of passengers. A sound feedback and complaint system should be established to guide passengers to communicate actively.

On the other hand, we should pay attention to the service staff management. High quality service is inseparable from high-quality service staff. A group of high-quality, high-level operation and maintenance talents need to be cultivated and used in the operation and maintenance service market of Wuhan urban rail transit to promote the development of the downstream operation and maintenance market.

In addition, we should explore the added value of urban rail transit and make full use of it. Stations of different sizes and sections need to be properly planned. Personalized positioning and different business development models should be considered. The successful positioning of the subway station will create many additional sources of economic income, such as commercial advertising, convenient store rental and so on.

4.3 Construction of smart urban rail transit

In the process of resisting novel coronavirus pneumonia, the new information technology such as big data, cloud computing, Internet of things, artificial intelligence, 5G technology and satellite communication has shown great advantages, which has certain enlightenment for the future development of urban rail transit industry. For example, the use of big data to real-time monitor the passenger transport situation of each train station, the use of the Internet of things to real-time monitor the temperature of passengers, the popularity of self-service ticket machines, etc., are all the embodiment of smart urban rail transit.

We should deeply understand the connotation of smart urban rail transit. By combining traditional services with emerging information technology, the smart systems that provide smart city rail transit services can be constructed to enable passengers to travel smoothly. The development experience of domestic and foreign advanced smart urban rail transit can be combined with the actual situation of Wuhan, to develop and innovate an independent and controllable brand suitable for the actual situation of Wuhan.

ACKNOWLEDGMENT

This paper is supported by the Open Fund Project of Manufacturing Industry Development Research Center on Wuhan City Circle of Jianghan University under grant WZ2016Z01.

REFERENCES

Li, X. 2019. Thoughts and suggestions on the development of Qingdao rail transit Industrial Park. Urban Rail Transit (10): 48–49.

Liu, L. & Dong, Q. & Cheng, M. 2019. The construction path of rail transit professional group connecting with the industrial chain – Taking the urban rail transit professional group as an example. Vocational Education Communication (22): 24–28.

Tian, F. 2014.Current situation and countermeasures of China's rail transit industry development. Chinese and Foreign Entrepreneurs (35): 42.

Wan, C. & Yang, Y. & Qin, X. 2020. Market development analysis of China's urban rail transit after operation and maintenance. Modern Urban Rail Transit (3): 7–11.

Xie, J. & Zhao, F. & Li, X., et al. 2018. Research on the development status of advanced rail transit industry. Mechanical Design 35 (S1): 119–121.

Yang, P. & Huang, M. 2011. Research on the development of rail transit industry in Beijing – Taking Fengtai agglomeration as an example. Beijing Social Sciences (3): 14–19.

Computational Social Science – Luo, Ciurea & Kumar (eds)
© *2021 Taylor & Francis Group, London, ISBN 978-0-367-70193-2*

An empirical study on factors affecting economic benefit of large-scale pig breeding

W.T. Liu & L. Zhou
College of Management of Sichuan Agricultural University, Chengdu, China

G. Fu*
Business School of Sichuan Agricultural University, Dujiangyan, China

C.H. Liu
School of Accountancy of Chengdu College of Arts and Sciences, Chengdu, China

ABSTRACT: Pig breeding in China has shifted from rural households' backyard breeding to large-scale breeding. To better improve the benefit and development of pig breeding industry, this paper selected 238 pig farms in Chengdu as a sample and empirically analyzed the factors influencing the economic benefit of large-scale pig breeding. The results showed the breeding scale, the ratio of technical personnel, the change rate of pork market price, the change rate of piglet price, the change rate of concentrated feed (soybean meal), the scale of investment, and whether the government supports have significant impact on the economic benefit of pig breeding. Therefore, we put forward corresponding suggestions from the farm level and the government level.

Keywords: Large-scale pig breeding, Economic benefit, Factors, Chengdu city.

1 INTRODUCTION

As a large agricultural country, the pig industry plays a non-negligible role in the development of China's agricultural economy. First of all, pigs occupy an important position in China's meat consumer goods, and pigs are also an important part in the Vegetable Basket Project for urban and rural residents. Second, according to the *2018 Statistical Yearbook* published by the National Bureau of Statistics, the total annual meat production was 86.544 million tons, of which 54.518 million tons were pork. As an important part of livestock husbandry, the pig industry has further enhanced the utilization of agricultural by-products and has become an important source of farmers' economic income, accounting for about 13% of the per capita net income of rural residents, which plays a huge role in promoting farmers' income increase. The healthy and stable development of the pig breeding industry is related to farmers' income, employment, and other aspects. Therefore, the healthy and stable development of the pig industry plays an irreplaceable role in the national economy and people's livelihood.

Based on the important position of the pig industry, China's government at all levels has issued various policies and invested a large amount of funds to make the pig industry develop on a large scale. With the support of the government, the development of China's modern pig industry has been promoted. The biggest characteristic of the modern pig industry is that the pig industry is further developing toward a large-scale breeding mode, while the number of backyard farming is gradually decreasing. However, large-scale pig breeding in China has been developing for a relatively short time and is still in its infancy. Problems such as small scale, low technical level, and low production efficiency affect the economic benefit of large-scale pig farms. Therefore, this

---

*Corresponding author

paper selects the large-scale pig farms in Chengdu, Sichuan Province as the research object to study the influencing factors on the economic benefit of large-scale farms, thus putting forward effective measures of promotion.

2 LITERATURE REVIEW

2.1 Foreign research

2.1.1 Environmental management of large -scale pig farms

Foreign scholars have conducted in-depth studies on the environmental governance of large-scale pig farms. For example, Rotz et al. (2002) and Wright et al. (2004) mainly analyzed the environmental pollution problems of large-scale pig farms and proposed their respective solutions on this basis. Rotz et al. (2002) are of the opinion that "improving the conditions of livestock and poultry sheds and grass sowing in farmland can effectively reduce nitrogen consumption in the farms." Economically dissecting two farms, Wright et al. (2004) analyzed the impact of feeding, planting, and other production methods on phosphorus content, and it was concluded that planting herbage on the farm could reduce the loss of phosphorus, thus maintaining the long-term stability of phosphorus content and improving the benefit of the farm.

2.1.2 Cost control of large-scale pig breeding

Martin et al. (1999) analyzed the differences between the mid-eastern United States and China in terms of feed and pig breeding, and put forward the following: "The production efficiency of pig breeding in the mid-eastern United States is more than 8 times that of backyard pig farmers in China. It is about 3.7 times more productive than professional pig farmers in China, and about 1.7 times more productive than large-scale pig farms in southern China." Judging from the number of days of pig rearing, Chinese backyard pig farmers have more days of that than large-scale commercial farmers and professional farmers. The main reason for this is their different feed ratio as well as feed quality, so there is a huge difference in production efficiency. By using the joint farm model, in the study of pig breeding benefit and its influencing factors, Kliebenstein et al. (2013) compared the cost of pig breeding in different regions and found that the feed cost of pig breeding varies from region to region due to the different strictness of agricultural policies. Among all the pig feeding competitors, the feed cost in western Canada is the most advantageous.

2.2 Domestic research

2.2.1 Large-scale development of pig farms

Domestic scholars have conducted in-depth studies on the large-scale development of pig breeding, among which Wang (2010) argues that the natural environment, resource supply, air quality, and professional ability of breeders would all affect the large-scale development of pig breeding. Further studying the specific impact of these factors, he proposed that "only by appropriately developing the breeding scale and improving the breeding methods can the economic and social benefit of large-scale breeding be fully brought into play." Zhang (2010) analyzed the problems existing in the current construction of large-scale livestock and poultry farms and pointed out that imperfect construction and management of that would affect its development to varying degrees, which is not conducive to the rapid and sustainable development of large-scale farms. Zhao (2010) evaluated the suitability of the development scale of pig farming community in Shehong County and came to the conclusion that it's above average and close to good. However, the pig breeding community is an emerging model, with problems such as incomplete facilities and poor breeding management, which have an impact on the large-scale development of breeding communities. Zeng (2013) analyzed the influencing factors in the evolution of scale pig breeding in China, and concluded that farmers would tend to choose larger-scale breeding based on the "rational person" theory, but in practice, there are a variety of factors that would influence farmers in determining the scale of breeding, such as capital, breeding technology, traffic conditions, and government policies; thus, the larger-scale breeding would usually not be chosen. In studying the impact of environmental factors on the productivity of large-scale pig breeding in China, Zuo (2017) found that the growth of pig breeding productivity in farms and professional households would be promoted by infrastructure, rural human capital level, breeding technology, and environmental governance. Leng and Fu (2017) conducted a study on the technical efficiency of pig breeding at different scales based on statistical data from various regions in China and proposed that the scale of pig breeding and the level of regional economic development are not significantly related to the technical efficiency of pig breeding, while whether government policies provide support or not and the level of human resources have a greater impact.

2.2.2 *Cost control of large-scale pig breeding*
Cui (2005), from the perspective of cost control, believes that only by starting from internal control, strictly controlling production costs and expenses in the breeding process, and formulating production supervision and plans can the purpose of cost control be achieved. In combination with the requirements of the Accounting Standards for Enterprises and the farm cost accounting system, Liu (2009) and Xu (2011), respectively, analyzed the accounting methods and items of pig breeding cost. In terms of how to reduce production costs and resist operational risks of pig farms, Yuan (2011) put forward the following opinions in the perspective of farms: first, farms should be built according to local conditions so as to reduce fixed cost expenditure; second, labor and feed costs can be reduced by improving management level and feeding level; third, disinfection work should be done to strictly control the spread and transmission of animal epidemics to reduce the disease rate of livestock. Feed conversion rate is of great significance for pig farms, and the conversion efficiency is crucial to the economic benefit of farms management. In view of the low feed conversion rate in current large-scale farms, Xie and Zhao (2012) conducted in-depth research on the reasons for fluctuations in feed conversion rate and proposed corresponding solutions for different causes. Liu (2015) found the defects of modern pig cost accounting, and standardized large-scale pig cost accounting from three aspects: applicable cost calculation methods, standardized cost calculation process, and basic work of cost calculation. Based on the particularities of pig farm cost accounting, Zeng (2016) argues that the cost accounting of large-scale pig farms not only requires managers to do basic work well, to clarify the accounting objects of farm costs and the cost classification of farms, but also managers need to do a good job in the accounting voucher of large-scale farm costs and the principle and process of cost accounting, so as to better give full play to the production advantages of large-scale farms and do a good job in the financial management of large-scale farms in terms of cost accounting. Lin (2017) proposed that to maximize corporate profits and shareholders' wealth, for pig enterprises, the first is to understand the raising process of pigs, and to do a good job in the distribution of pig feed and other materials as well as the handover and registration of pig herd transfer; the second is to strictly regulate cost accounting according to accounting standards; and the third is to prepare a reasonable cost budget according to true and accurate historical cost data.

2.2.3 *Factors affecting the economic benefit of large-scale farms*
Aiming at the factors affecting the economic benefit of pig breeding in China, some domestic scholars have also conducted relevant research. From the perspective of industrial development, He (2009) believes that the main factors affecting the benefits of local pig industry include industrial policies, market factors, technological level, consumption habits, and epidemic prevention. Gong and Wei (2011) compared the cost and economic benefit data of large-scale farms in the Qianjiang District, and found that the factors affecting the economic benefit of large-scale farms were the technical level and the feeding scale. Based on the technical measures, Xiao and He (2013) proposed six methods to improve the economic benefit of pig breeding, including selecting excellent hybrid breeds, strengthening the management of pig herds breeding, strengthening feed management, strengthening epidemic prevention to eliminate the occurrence of malignant infectious diseases, eliminating invalid breeding pigs in time and putting fat pigs into the market at an appropriate time, improving the service system, and accelerating the promotion of new technologies. And Tai (2015) proposed that in addition to reducing feed costs, increasing sow yield per unit area and

improving economic management level can also improve the economic benefit of pig breeding. Liu (2016) believes that it is an important strategy to obtain ideal economic benefit to comprehensively establish the scientific pig raising concept of "appropriate scale pig raising is the premise, excellent varieties are the foundation, strengthening management is the key, good epidemic prevention is the guarantee, market law is an important link, so is human resources." Through the analysis of the economic characteristics of farmers' pig raising behaviors, Huang et al. (2011) consider from the qualitative perspective that location and traffic environment, policy support, and scientific breeding technology are factors affecting the economic benefit of farmers' pig raising.

2.3 *Literature review*

A summary of the existing literature reveals that although there is a wealth of research literature on pig cost-effectiveness and large-scale farms at home and abroad; however, they have certain limitations. As the largest livestock and poultry breeding industry in China, pig breeding is of great economic significance. Against the background of the reality that China's future pig breeding industry is gradually moving toward large-scale and industrialization, we need to pay more attention to the development of large-scale pig farms, the main group under this development trend. At present, China's large-scale pig breeding is still in the primary stage; the level of industrialization, scientific, and technological are relatively low; and the establishment of the industrialization service system is not perfect enough. Therefore, it is necessary to follow up on it to improve the lag and deficiency of the existing research. There are different situations in different regions in the process of development, and many problems need to be further discovered and solved.

3 THEORETICAL ANALYSIS AND HYPOTHESIS

The most basic goal of pig farmers' operation is to make profits, that is, to have a surplus after deducting the unit cost from the unit pig income, and hopefully to make it as large as possible, with higher unit pig profit and greater unit input-output return. For farms, in addition to the costs that have a direct effect and impact, there are also many factors that influence the economic benefit of large-scale breeding. As can be seen from the previous literature review, different scholars have different views on the research on the influencing factors of the economic benefit of large-scale pig breeding. According to Li (2007), the factors that influence the scale and benefit of pig breeding include the variety of pigs, the scale of resource input, changes in feed market demand, and changes in piglet market demand. Yang (2009) believes that feed ratio, local economic development level, pig price and the quality of the main production body would all affect the economic benefit and efficiency of large-scale breeding to a certain extent. Zhang (2009), from the micro perspective of large-scale farm, believes that factors such as management, feed, variety, and epidemic prevention can affect the economic benefit of large-scale pig farms. He (2009) believes that the main factors affecting local scale operation are capital investment, market changes, and technical level from the perspective of large-scale pig breeding. Huang et al. (2011) believe that geographical location and traffic conditions, government support, degree of specialization, and science and technology have significant influences on the economic benefit of farmers' pig raising.

Drawing on the literature research of existing scholars, and combining with the actual situation of the investigation as well as the availability of data, this paper, from the four directions of production characteristics, social factors, economic factors, and environmental factors, select specific indicators including breeding scale, concentrated feed ratio, technical personnel ratio, piglet price change rate, pig price change rate, concentrated feed (soybean meal) price change rate, capital investment, and whether the government supports as the influencing factors that may cause differences in the economic benefit of pig breeding to study.

(1) Breeding scale. Large-scale breeding, enables farmers to purchase a large amount of feed at one time as well as to trade at a lower transaction price in order to achieve the purpose of

reducing costs. At the same time, it is also conducive to large-scale sales, which enable the sellers to have more bargaining space and obtain higher profits. On the whole, the breeding scale of pig farms will have an impact on the cost and profit of pigs, and thus on the overall breeding benefit. In this paper, the scale of the pig farms investigated is divided into four types: small scale (40 to 100), small-medium scale (100 to 500), medium-large scale (500 to 1,000) and large scale (more than 1000). According to the above analysis, the hypothesis is proposed:

H1: The scale of pig breeding has a significant positive effect on economic benefit.

(2) The ratio of concentrated feed. Feed includes concentrated feed and other feed. The feed ratio varies among different stages and sizes of pig business entities, and there are differences in feed usage and feeding methods, which have different effects on the economic benefit of pigs. In general, the high nutritional value and high feed conversion rate of concentrated feed will increase the growth speed of pigs and shorten the feeding cycle, but its unit cost is also relatively high. When the proportion of concentrated feed is too much, the total feed cost will increase, thus affecting the economic benefit of pig feeding. In this paper, the proportion of the input cost of concentrate in the total feed will be used to reflect the proportion of concentrated feed for pig breeding. Combined with the previous analysis, the assumption is proposed:

H2: The ratio of concentrated feed has a significant positive effect on the benefit of pig scale breeding.

(3) The ratio of technical personnel. The production benefit of a farm will be affected by the production and management technology. The technical level of personnel will directly affect the daily production management of pigs, epidemic prevention, and other aspects, and thus directly and indirectly affecting the management benefit of scale pig breeding. It is generally believed that the higher the technical level of personnel, the better the economic benefit of large-scale pig farms, otherwise, the worse the economic benefit. Based on the above analysis, the hypothesis is proposed:

H3: The ratio of technicians has a significant positive effect on the economic benefit of pig breeding.

(4) Pig price. In terms of economic benefit, when the slaughter price of pigs rises, meaning that the sales income of the pig farm increases, then the profit correspondingly increases. Therefore, the assumption is proposed:

H4: The pig price has a significant positive impact on the economic benefit of pig breeding.

(5) Piglet price and soybean meal price. Piglet cost refers to the cost of purchased or self-breeding piglets. Soybean meal is the main component of concentrated feed, so the price change of feed is measured by the price change of soybean meal. Both piglet price and soybean meal price are based on cost, in which when the market price of both rises, even if the quantity of purchased piglets and the proportion of concentrated feed input remain unchanged, it will inevitably mean an increase in cost. Based on this, the hypothesis is proposed:

H5: Piglet price has a significant negative effect on the economic benefit of pig breeding.

H6: Soybean meal price has a significant negative effect on the economic benefit of pig breeding.

(6) The scale of capital investment. For farmers, the early construction of pig farms and the purchase of fixed assets, and so forth require a large amount of capital investment, as well as a large amount of working capital for the purchase of feed and piglets. The scale of capital available to pig producers is a prerequisite for determining which scale of breeding to choose. Due to the imperfect development of rural finance, pig farmers can obtain very little loans from formal financial organizations, which leads to most of the capital needs being met through informal channels. The capital situation of pig breeding operators determines many aspects of their pig breeding scale and management strategies, which profoundly affects the level of pig output benefit. It is generally believed that, with a large scale of capital investment, the farmers have strong economic strength and can choose more management strategies, which

is more conducive to making better decisions that improve economic benefit, thus obtaining better economic benefits. Therefore, the assumption is proposed:

H7: The scale of capital investment has a positive effect on the economic benefit of pig farms.

(7) Government support. Government support is a favorable condition to promote the development of the pig market and ensure the benefit of pig farmers, which plays an important role in the development of the pig breeding industry. On the one hand, it will be easier to stimulate the enthusiasm of farmers if the government strongly supports the pig breeding industry, carries out certain macro-control, production, and marketing support and (7) government support. Government support is a favorable condition to promote the development of the pig market and ensure the benefit of pig farmers. It plays an important role in the development of the pig breeding industry. On the one hand, it is easier to stimulate the enthusiasm of pig farmers if the government strongly supports the pig breeding industry, carries out certain macro-control, supports production and marketing, and sets up certain protection system. On the other hand, sufficient financial investment can protect small and medium-sized farms in the case of huge market fluctuations. At present, the government's financial and technical support can help pig breeding operators to solve some problems and improve the benefit of pig breeding. Based on the above analysis, a hypothesis was proposed:

H8: Obtaining government support has a positive effect on the economic benefit of farms.

4 EMPIRICAL RESEARCH DESIGN

4.1 Sample selection and data sources

In this paper, the research data come from the following aspects: First, data about pig farms with more than 40 pigs in Chengdu from 2013 to 2017 mainly was obtained from the statistical yearbook and statistical bulletin, as well as by visiting Chengdu Animal Husbandry Bureau and Statistics Bureau. Besides, we obtained data through information published on the official websites of the whole province, cities, and counties and previous related literature. The data studied in this paper are mainly the annual average data of 238 pig farms.

4.2 Variable design

Based on the previous discussion of the hypothesis of influencing factors, the indicators to measure the economic benefit of pig scale breeding are selected as follows: profit per unit pig and input-output rate. The scale of breeding, the ratio of concentrated feed, the ratio of technicians, the change rate of piglet price, the change rate of pork market price, the change rate of concentrated feed (soybean meal) price, capital investment, and whether the government supports, which are taken as the influencing factors that may cause the difference in economic benefit of large-scale pig breeding, as shown in Table 1.

4.3 Model construction

Specific indexes to measure the economic benefit of large-scale pig breeding are profit per unit pig (Y_1) and input-output rate (Y_2), which are taken as dependent variables. The regression model is established with eight influencing factors as independent variables (X_1-X_8), and the ordinary least square (OLS) method is used for linear estimation to analyze the relationship between each influencing factor and the pig economic benefit index. The estimation expression of the specific models is as follows:

$$Y_1 = \alpha + \alpha_1 X_1 + \alpha_2 X_2 + \alpha_3 X_3 + \alpha_4 X_4 + \alpha_5 X_5 + \alpha_6 X_6 + \alpha_7 X_7 + \alpha_8 X_8 + \mu \qquad (1)$$

Table 1. Indicator selection of variables.

Indicator types		Variables	Code	Unit	Definition
Dependent variable		Profit	Y1	yuan	Annual income – annual cost (one farm)
		Input-output rate	Y_2	%	Unit output/unit income or unit income/unit cost (one farm)
Independent variables	Production characteristics	Breeding scale	X_1	/	1 small scale, 2 small-medium scale, 3 medium-large scale, 4 large scale
		Concentrated feed ratio	X_2	%	Annual concentrated feed Cost/Annual total feed cost
	Social factors	Technical personnel ratio	X_3	%	Total number of technicians/personnel
	Economic factors	Pork market price change rate	X_4	%	Pork price in year n/Pork price in year (n – 1)
		Piglet price change rate	X_5	%	Piglet fee in year n/Piglet fee in year (n – 1)
		Concentrated feed (soybean meal) price change rate	X_6	%	Concentrated feed (soybean meal) price in year n/Concentrated feed (soybean meal) price in year (n – 1)
		Capital investment	X_7	10,000 yuan	
	Environmental factor	Government support	X_8	/	0 = no, 1 = yes

Table 2. Variable description statistics.

Variable	Minimum value	Maximum value	Mean value	Standard deviation
Y_1	89.538	958.1415	398.1038	197.30518
Y_2	1.0116	1.5853	1.26407	0.126731
X_1	1	4	1.8757	0.64374
X_2	73.511	97.6421	89.0215	5.348929
X_3	17.5501	74.5764	42.4879	15.226108
X_4	14.95	17.30	1.4765	9.60473
X_5	−15.04	53.93	7.573	23.2934
X_6	−12.81	9.29	−2.414	6.7602
X_7	22.50	568.27	124.7606	89.519174
X_8	0	1	0.4376	0.47891

$$Y_1 = \beta + \beta_1 X_1 + \beta_2 X_2 + \beta_3 X_3 + \beta_4 X_4 + \beta_5 X_5 + \beta_6 X_6 + \beta_7 X_7 + \beta_8 X_8 + \mu \qquad (2)$$

Where α, β refer to constant term; α_i, β_i are parameters need to be estimated; μ refers to random disturbance term.

5 EMPIRICAL ANALYSIS

5.1 Descriptive analysis

Using software to process the data, we obtained the results shown in Table 2. As can be seen from the table, the average profit per pig is 398.10 yuan, the minimum value is 89.54 yuan, the maximum value is 958.14 yuan, and the input-output rate per unit pig is between 1.0116 and 1.5853, which shows that the overall level of economic benefit of pig scale breeding in Chengdu is not high, and the difference between individuals is large. The mean value of breeding scale is 1.8757, which indicates that the whole investigation is mainly based on small-medium-sized breeding scale. As is known in the table, the total concentrated feed ratio in the survey accounts for 89.02% of the total feed costs on average. The ratio of technical personnel is 42.49%, indicating that with the development of pig breeding on a large scale, the corresponding science and technology are also gradually strengthened, and the ratio of technical personnel is continuously increasing. In terms of social factors, the change rate of pork market price fluctuates between −14.95% and 17.30%, with an average value of 1.48%, indicating an overall trend of rising from 2013 to 2017. The average price change rate of piglets is 7.57%, and the maximum value is as high as 53.93%, indicating that the price of piglets fluctuated greatly. The average price change rate of concentrated feed (soybean meal) is −2.41%, resulting in a decrease in cost. The average investment is 1,247,600 yuan, and the investment fluctuates between 225,000 yuan and 5,682,700 yuan, which shows that the capital required by different scales of pig breeding varies greatly, and most of them belong to small-medium-sized ones. With the continuous development of pig industry, 43.76% of farmers have received government support.

5.2 *Analysis of multiple regression results*

As seen from Table 3, the t-value of parameter estimation of each variable in model 1 is 5.3576, 1.7563, 5.4862, 4.4883, −3.7044, and 4.8666, which are all valid significantly at the 1% level of significance. The t-value of parameter estimation of $X8$ is 2.4813, which passed the significance test at 5% level. The F-value of model 1 is 24.1224, which is significant at the 1% level, proving that the linear relationship between dependent variables and independent variables is valid at the 1% level of significance.

In addition, Table 3 shows that the breeding scale, the ratio of technicians, the change rate of pork market price, the change rate of piglet price, the investment scale, and government support are the factors that significantly affect the economic benefit of pig breeding. The F-test statistic of regression analysis of model 1 is valid at the 1% level of significance, which means that the model 1 is correct in form and significant in estimation. The effect degree and significance of the above main influencing factors on the profit per unit pig are summarized as follows.

The regression coefficient of the breeding scale variable is 0.387, indicating that there is a positive correlation between pig breeding scale and economic benefit; at the same time, the parameter estimation result is significantly valid at the 1% level. Therefore, the above correlation is a

Table 3. The results of multiple regression of Model 1.

Variable	Regression coefficient	Standard deviation	T-value
X_1	0.387***	0.013	5.3576
X_2	0.043	0.673	1.7563
X_3	0.323***	0.012	5.4862
X_4	0.107***	0.023	4.4883
X_5	−0.092***	0.054	−3.7044
X_6	−0.015	0.872	−1.4359
X_7	0.312***	0.016	4.8666
X_8	0.134**	0.004	2.4813
Constant	1.362***	0.287	1.9455
$R2$	0.2144		
Adj-R2	0.2103		
F value	24.1224***		

***, **, *indicate significant levels at 1%, 5%, and 10% respectively.

significant positive correlation; that is, the scale of pig breeding has a significant positive effect on the economic benefit, which is the same as Hypothesis 1.

The regression coefficient of the technical personnel ratio variable is a positive number of 0.323, which shows that there is a positive correlation between the ratio of technicians and the economic benefit, while the parameter estimation result is significantly valid at the 1% level of significance. Therefore, this positive correlation is a significant positive correlation, and the ratio of technicians has a significant positive effect on the economic benefit of pig breeding. Thus Hypothesis 3 holds.

The regression coefficient of the variable of the change rate of pork market price is 0.107, that is, there is a positive correlation between the change of pork market price and the economic benefit. Moreover, the parameter estimation result is significantly valid at the 1% level, so the pig price has a significant positive effect on the economic benefit of pig breeding. Therefore, Hypothesis 4 holds.

The regression coefficient of the piglet price change rate variable is −0.092, which is negative, that is, there is a negative correlation between piglet price change and economic benefit. And because the parameter estimation result is significantly valid at the 1% level, it is concluded that piglet price has a significant reverse effect on the economic benefit of pig breeding, which is consistent with Hypothesis 5.

The regression coefficient of the investment scale variable is 0.312, which shows that there is a positive correlation between investment scale and economic benefit; at the same time, the parameter estimation result is significantly valid at the 1% level, so this correlation is a significant positive correlation. The investment scale is a significant positive influencing factor for the economic benefit of pig breeding, which is consistent with Hypothesis 7.

The regression coefficient of the variable of whether the government supports is 0.134, which indicates that there is a positive correlation between government support and economic benefit. Moreover, because the parameter estimation result is significantly valid at the 5% level, this correlation is a significant positive correlation. Government support has a significant positive impact, so Hypothesis 8 is true.

As the two independent variables of concentrated feed ratio and price are not significant, as well as the linear relationship between the dependent variable and the independent variables is not obvious, they are not consistent with Hypotheses 2 and 6.

As seen from Table 4, the t-value of parameter estimation of each variable in model 2 is 3.5456, 4.1261, 5.2725, −4.2093, −3.6127, and 3.8908, respectively, which are all significantly valid at the 1% level. The t-value of parameter estimation of X8 is 4.8573, and it is significantly valid at the 5% level. The F-value of model 2 is 20.5723, which is significantly valid at the 1% level, proving that the linear relationship between dependent variables and independent variables is valid at the 1% level of significance.

Table 4.　The results of multiple regression of Model 2.

Variable	Regression coefficient	Standard deviation	T-value
X_1	0.335***	0.024	3.5456
X_2	0.034	0.715	1.0323
X_3	0.243***	0.16	4.1261
X_4	0.125***	0.005	5.2725
X_5	−0.095***	0.013	−4.2093
X_6	−0.081***	0.031	−3.6127
X_7	0.358***	0.022	3.8908
X_8	0.113**	0.009	4.8573
Constant	0.972***	0.314	1.2964
$R2$	0.2919		
Adj-R2	0.2610		
F value	20.5723***		

***, **, *indicate significant levels at 1%, 5%, and 10% respectively.

In addition, Table 4 shows that the breeding scale, the ratio of technicians, the change rate of pork market price, the change rate of piglet price, the change rate of concentrated feed (soybean meal) price, the investment scale, and whether the government supports are the factors that significantly affect the economic benefit of pig breeding. The F-test statistic of regression analysis of model 2 is valid at the 1% level of significance, which shows that model 2 is correct in form and the estimation significance is significant. The effect degree and significance of the above main influencing factors on the input-output rate are summarized as follows.

The regression coefficient of the breeding scale variable is 0.335, indicating that there is a positive correlation between pig breeding scale and economic benefit, and the parameter estimation result is significantly valid at the 1% level. Therefore, this positive correlation is a significant positive correlation. The scale of pig breeding has a significant positive effect on its economic benefit; thus, Hypothesis 1 holds.

The regression coefficient of the technical personnel ratio variable is 0.243, which is positive, meaning that there is a positive correlation between the ratio of technicians and the economic benefit, and the parameter estimation result is significantly valid at the 1% level. Therefore, this correlation is a significant positive correlation. The ratio of technicians has a significant positive effect on the economic benefit of pig breeding, which is the same as Hypothesis 3.

The regression coefficient of the variable of the change rate of pork market price is 0.125, that is, there is a positive correlation between the change of pork market price and the economic benefit, and because the parameter estimation result is significantly valid at the 1% level, the pig price has a significant positive effect on the economic benefit of pig breeding. Therefore, Hypothesis 4 holds.

The regression coefficients of the piglet price change rate variable and the soybean meal price change rate variable are -0.095 and -0.081, respectively, which are negative. That is, there is a negative correlation between both the piglet price change and the soybean meal price change and the economic benefit, and because the parameter estimation result is significantly valid at the 1% level, it is concluded that changes in piglet price and soybean meal price have significant reverse effects on the economic benefit of pig breeding, which is consistent with Hypotheses 5 and 6.

The regression coefficient of the investment scale variable is 0.358, which indicates that there is a positive correlation between the investment scale and the economic benefit, and because the parameter estimation result is significantly valid at the 1% level, this correlation is a significant positive correlation. The investment scale is a significant positive influencing factor for the economic benefit of pig breeding; thus, Hypothesis 7 holds.

The regression coefficient of the variable of whether the government supports is 0.113. This shows that there is a positive correlation between the government support and the economic benefit, and because the parameter estimation result is significantly valid at the 5% level, this correlation is a significant positive correlation. Whether the government supports or not is a significant influencing factor for economic benefit of pig breeding, which is consistent with Hypothesis 8.

Since the independent variable of concentrated feed ratio is not significant and the linear relationship between the dependent variable and the independent variable is not obvious, it cannot be explored whether there is an influence. Therefore, Hypothesis 2 is not valid.

6 CONCLUSIONS AND RECOMMENDATIONS

6.1 *Conclusion*

Separately from the four directions (production characteristics, social factors, economic factors, and environmental factors), this paper selected indicators including breeding scale, concentrated feed ratio, technical personnel ratio, piglet price change rate, pig price change rate, concentrated feed (soybean meal) price change rate, capital investment, and government support to study the factors influencing the economic benefit of large-scale pig breeding. Based on the analysis of the sample data selected in this paper, the following conclusions can be drawn: (1) The more capital investment and the larger the breeding scale, the more prominent the economic benefit of pig farms, so the breeding scale and capital investment have a positive effect on the economic benefit of large-scale pig breeding. (2) The higher the proportion of technicians in the pig farm, the more significant the economic benefit of it, that is, the ratio of technicians has a positive effect on the economic benefit of large-scale pig breeding. (3) Piglet price and soybean meal price have an important effect on the breeding cost. When the price decreases, the breeding cost decreases and more economic benefit can be obtained under the same income condition. Therefore, changes in piglet price and soybean meal price have a reverse effect on the economic benefit of large-scale pig breeding. (4) The pork market price affects sales income. When the market demand increases, and the price increases, the farmers can obtain an increase in sales income, thus improving the economic benefit. Therefore, the pork market price has a positive effect on the economic benefit of large-scale pig breeding. (5) Government support can reduce the risk of bankruptcy of small and medium-sized farms and promote the development of pig industry to a certain extent, so government support has a positive effect on the economic benefit of large-scale pig breeding.

6.2 *Recommendations*

As the agricultural economy occupies an important position in China, and the pig industry has an important influence on the agricultural economy, how to improve the economic benefit of China's pig industry is of great significance.

First of all, based on the level of farmers, costs should be strictly controlled to improve production efficiency. The main suggestions are as follows: (1) Strictly control the costs in the breeding process, formulate reasonable budget expenses, strengthen personnel management, and reduce labor costs. (2) Train more employees to learn science and technology, study more advanced technologies, and improve the efficiency of pig breeding.

Second, based on the government level, we should do a good job in macro-control of the pig market and formulate relevant policies to support the development of the pig industry. The main suggestions are as follows: (1) Formulate relevant macro-control policies to avoid sharp fluctuations in the market and stabilize pig prices. (2) For small-scale farmers, certain encouraging policies should be given, such as providing financial support or technical support, and giving small-scale farmers protection to develop pig industry. (3) Mobilize senior talents to study advanced scientific breeding technology and improve output efficiency.

ACKNOWLEDGEMENTS

This paper is one of the preliminary achievements of "Research on Scale Difference and Cost-Benefit of Pig Breeding in Sichuan Province," the Project of Sichuan Center for Rural Development Research (No. CR1608).

REFERENCES

Cui, X. 2005. Cost and expense control measures for large-scale pig farms. Henan Animal Husbandry and Veterinary Medicine 26(8): 12–13.

Gong, L. & Wei, C. 2011. Investigation report on economic benefits of large-scale pig farms. Contemporary Animal Husbandry (5): 7–8.

He, L. 2009. Benefit Analysis and large-scale management research of Linhe pig industry. Doctoral dissertation, Inner Mongolia Agricultural University.

Huang, D., Xu E., Zheng S., Chen Y. 2011. Features and economic benefit analysis of peasant households raising pigs in China–investigation in Henan, Hunan and Sichuan Provinces. Shaanxi Journal of Agricultural Sciences (Rural Economy Edition) (8): 13–17.

Kliebenstein, J., Larson, B., Honeyman, M., & Penner, A. 2013. A comparison of production costs, returns and profitability of swine finishing systems. Teen Ink (5): 213–254.

Leng, B. & Fu, R. 2017. Study on technical efficiency of pig breeding in different scales. Rural Economy (11): 51–56.

Li, H. 2007. Pig raising scale and its cost-benefit analysis. Doctoral dissertation, Northwest Agriculture and Forestry University.

Lin, X. 2017. Problems and countermeasures in cost accounting of pig raising enterprises. Times Finance 36(12): 73–75.

Liu, Q. 2009. Discussion on cost accounting method of large-scale pig farm. Commercial Accounting (17): 52–53.

Liu, D. 2015. New exploration on cost accounting of large-scale pig-raising enterprises. China Agricultural Accounting (7): 41–43.

Liu, W. 2016. Technical measures to improve economic benefits of large-scale pig farms. Contemporary Animal Husbandry (12): 1–2.

Martin, L.J., Kruja, Z., & AIexiou, J. 1999. Prospects for hog production and processing in Canada. Miscellaneous Publications (3): 463–468.

Rotz, C.A., Sharpley, A.N., Satter, L.D. Gburek, W.J. 2002. Production and feeding strategies for phosphorus management on daily farms. Journal of Dairy Science 85(11): 3142–3153.

Tai, F. 2015. Measures to improve economic benefits of large-scale pig raising. China Animal Husbandry and Veterinary Abstracts 31(3): 30.

Wang, M. 2010. On the restrictive factors of large-scale breeding–taking large-scale pig farms as an example. Journal of Shangluo University 24(4): 28–30.

Wright, P., Inglis, S., Ma, J., Gooch, C., Aldrich, B., Meister, A., & Scott, N. 2004. Comparison of five anaerobic digestion systems on dairy farms. ASAE Annual International Meeting Papers (4): 213–220.

Xiao, K. & He, J. 2013. Technical measures to improve economic benefits of pig breeding. Heilongjiang Journal of Animal Reproduction 21(3): 48–49.

Xie, X. & Zhao, F. 2012. Causes and countermeasures of low feed conversion rate in pig farms. Chinese Journal of Animal Husbandry 48(6): 65–66.

Xu, Z. 2011. Discussion on cost accounting of pig breeding based on the thinking of cost accounting of animal husbandry on Shanghai farms. China Agricultural Accounting (11): 42–44.

Yang, X. 2009. Analysis of production efficiency and its influencing factors in China's pig industry. Doctoral dissertation, Nanjing Agricultural University.

Yuan, L. 2011. How to reduce production and operation costs in farms. China Pig Industry (4): 13–14.

Zeng, X. 2013. Analysis of influencing factors on the evolution of pig breeding scale in China–taking Santai County in Sichuan Province as an example. Doctoral dissertation, Zhejiang University.

Zeng, L. 2016. Discussion on cost accounting of large-scale pig farms. Livestock and Poultry Industry (10): 34.

Zhang, H. 2009. Factors affecting economic benefits of large-scale pig raising and improving measures. Swine Industry Science 26(4): 52–54.

Zhang, H. 2010. Problems and countermeasures in the construction of livestock and poultry farms (communities). Science of Animal Husbandry and Feed (9): 119–121.

Zhao, X. 2010. Study on the suitability Evaluation of pig breeding community development scale in Shehong County. Doctoral dissertation, Sichuan Agricultural University.

Zuo, Y. 2017. Study on productivity of large-scale pig breeding in China which considering environmental factors. Doctoral dissertation, Southwest University.

Computational Social Science – Luo, Ciurea & Kumar (eds)
© 2021 Taylor & Francis Group, London, ISBN 978-0-367-70193-2

Analysis on the status quo of monitoring terminal subsidies for DSM projects

W. Tang*, P. Wu & Y. Zhang
State Grid Energy Research Institute Co., Ltd.

ABSTRACT: From the perspective of implementation, this paper focuses on the analysis of the necessary tool for demand response monitoring terminal subsidies. This paper analyzes the source of funds, the management of fund distribution and the significance of subsidies, and puts forward relevant suggestions based on the monitoring terminal subsidies in Jiangsu and other provinces and the implementation of demand side management.

1   INTRODUCTION

In recent years, the power grid load has reached new highs, the air conditioning load has increased rapidly, and the peak valley difference has gradually increased. The traditional coping mode is basically to build new units at the power plant side, start peak load regulating units at the grid side, start orderly power consumption in terms of policies, and realize energy-saving transformation on the user side. According to the investment of 9.5 billion yuan per 1 million kilowatts, the construction of power plants and power grids has a huge investment, resulting in a great waste of social resources.

A large number of renewable energy with random and fluctuating characteristics on the power supply side, and the dynamic access of electric heating, air conditioning, electric vehicles, energy storage and other loads on the customer side affect the security and economy of power grid operation. It is urgent to fully awaken and mobilize demand response resources to participate in the flexible interaction of power grid. In order to ensure the safe and economic operation of power grid, it is urgent to implement DSM project to improve the power grid instantaneous balance ability.

There are many literatures on DSM in academic circles. For example, references 1–6 focus on the significance of DSM. From the perspective of implementation, this paper focuses on the analysis of the monitoring terminal, which is a necessary tool for demand response.

2   THE SOURCE OF ONLINE MONITORING TERMINAL SUBSIDY POLICY

2.1   *Use of installing monitoring terminal*

"Enterprise online measurement and control" is an important function of power demand side management, which can monitor the energy consumption data of users, analyze the energy consumption of key enterprises, and timely carry out demand response work. It is very important to promote the enterprise online monitoring function and install the online monitoring terminal for the implementation of demand side management project.

2.2   *Source of subsidy policy*

Since 2010, documents on demand side management have been published intensively at the national level, mainly including the following documents with significant impact:

---

*Corresponding author

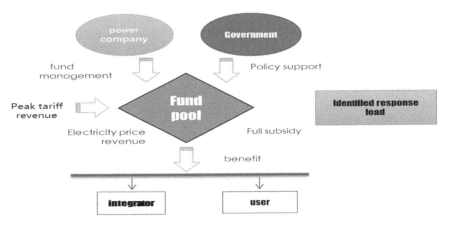

Figure 1.   Capital channels for implementing demand response in Jiangsu Province.

Notice on printing and Distributing Measures for power demand side management issued by national development and Reform Commission, Ministry of industry and information technology, Ministry of finance, state owned assets supervision and Administration Commission of the State Council, State Power Regulatory Commission and state energy administration. The document defines the annual electricity and energy saving index of power grid enterprises (in principle, it shall not be less than 0.3% of the electricity sold in the previous year and 0.3% of the maximum power load) in the electricity sales business area of the relevant grid enterprises, and encourages all regions, relevant departments and units to actively promote the development of DSM.

Subsequently, the Ministry of Finance and the national development and Reform Commission issued the Interim Measures for the central financial incentive fund management of the comprehensive pilot project of power demand side management cities, the notice of the Ministry of Finance and the national development and Reform Commission on Approving the implementation plan of the comprehensive pilot work of DSM cities and allocating the central financial incentive funds, and the general office of the Ministry of industry and information technology on printing and distributing the industry Notice of special action plan for DSM (2016–2020) and other documents.

At the local level, in accordance with the national level document policy spirit, the supporting documents of their respective provinces have been released, and the pilot cities have taken the lead in the implementation, and all provinces and cities have gradually carried out relevant work.

At the provincial level, the management measures for the use of special funds for DSM in each province have been issued. In the management measures, the main sources of subsidy funds are the funds extracted from the surcharges of urban public utilities collected by the sales price surcharges, the electricity revenue from the implementation of differential electricity prices and other funds. The fund uses include: Construction and operation and maintenance of power demand side management platform, public building energy consumption monitoring platform, demonstration project subsidy and demonstration enterprise award, demand response subsidy, publicity, training, evaluation, etc.

It is worth mentioning that the implementation of demand response in Jiangsu Province in the early stage is good, and the income from peak electricity price is the main income of special funds. Due to the different fund income of demand response in each year, the subsidy standard of the next year will be recalculated according to the principle of balance of revenue and expenditure (Figure 1).

2.3   Distribution process

The special funds shall be arranged and used by the provincial finance department and the provincial economic and Information Commission (hereinafter referred to as "the provincial economic and

Information Commission"). First of all, the Department of Finance and the Commission of economy and information technology will issue the documents of DSM projects, which will be applied by energy consuming enterprises and parks. The national development and Reform Commission, the Commission for economy and information technology and the Department of Finance jointly participate in the work of project declaration and review, project database management, project acceptance and evaluation, budget budget preparation and reporting, incentive project supervision and management, and budget performance evaluation.

In general, the provincial economic and Information Commission conducts preliminary examination and project screening on the project materials recommended and reported by various regions according to the relevant requirements of project application, and the projects that meet the requirements are submitted to the expert group for review. According to the principles of openness, fairness and fairness, the provincial economic and Information Commission organizes or entrusts relevant experts (Institutions) to review the project, and the discipline inspection and supervision personnel of the provincial economic and Information Commission participate in the supervision of the evaluation process. According to the principle of selecting the best, the provincial Commission of economy and information technology formulates the support projects and the amount of support, which are reported to the provincial government for approval, and then publicized through the portal website. After the publicity, the final supporting projects and the amount of fund support shall be determined after re verification according to the publicity situation, and the publisher shall ask the Provincial Department of finance to allocate the funds. According to the provincial government's opinions and the provincial economic and Information Commission's fund allocation application, the Provincial Department of Finance timely issues funds according to the provisions. The Provincial Department of finance is responsible for the budget arrangement and allocation of special funds, and carries out performance management, supervision and inspection on the use of funds.

3 MONITORING THE IMPLEMENTATION OF TERMINAL SUBSIDIES

The means of encouraging enterprises to install monitoring terminals in various provinces are basically similar. In addition to the demand response fund subsidies, the other means of encouraging enterprises to participate in DSM projects are basically similar, that is, enterprises participating in DSM projects and actively installing monitoring terminals can have the right to participate in power trading and purchase electricity directly. The general office of industry and information technology issued the notice on organizing and recommending national demonstration enterprises of demand side management in industrial field, which mentioned that enterprises in the national demonstration base of new industrial industries should be recommended first, and those enterprises that have reached "a" in DSM evaluation of industrial field, and those above "AA level" will be directly regarded as demonstration enterprises.

The situation of subsidy for the construction of user terminal monitoring system (monitoring point) varies from place to place. From the perspective of the amount of subsidy funds, it is generally clear that the subsidies in provinces with subsidies come from special funds, and the subsidy standard is 30–80%, and the subsidy shows a downward trend year by year. For example, Ningxia decreases annually, and the subsidy proportion of terminal monitoring system decreases from 80% to 60% and 40% in turn. From the perspective of subsidy situation, some provinces do not subsidize the monitoring terminal investment separately, but the overall project investment subsidy includes terminal subsidy. Subsidies are distributed directly to enterprises and service providers. From the perspective of subsidy publicity, Ningxia economic and Information Commission announced that in 2014, the terminal investment subsidy was 5.9 million yuan, and the subsidy standard was 80% of the amount of each terminal monitoring system (5594 yuan); in 2015, the investment subsidy of terminal monitoring system was 11.92 million yuan, and the subsidy standard was 60% of the equipment investment; in 2017, the investment of 5.97 million yuan, the installation of 2666 sets of monitoring terminals, the investment amount of each set decreased to The average cost of each terminal monitoring system is 2239 yuan. According to the subsidy fund schedule of power demand side terminal monitoring system announced by Ningxia provincial finance department in 2018, the overall subsidy amount is about 6.4 million yuan, the gateway subsidy standard is 100%, and the subsidy standard for other equipment is 40% (taking RMB 5594 per set of equipment as the standard). The subsidy for Inner Mongolia in 2015 and 2016 was 2500 yuan / terminal, while that in 2017 was 2000 yuan / terminal, showing a downward trend year by year.

The progress of monitoring terminal deployment in different provinces is quite different. Although all provinces are actively promoting the implementation of DSM projects, there are increasing differences in the actual process, which is closely related to the enthusiasm of enterprises. In terms of the number of access enterprises, Jiangsu, Ningxia and Guangdong are among the top three provinces. Jiangsu, Wuxi and Zhenjiang are the cities with the most access enterprises. The number of access enterprises in some provinces is only single digit. Jiangsu enterprises attach great importance to the fine management of energy consumption, actively respond to the call of the government, and work closely with service providers such as load integrators to complete the demand response work. At the same time, due to the scale effect, the terminal price is lower than other provinces. The Commission of economy and information technology has no subsidy for the installation of terminals, and the service providers install monitoring terminals for enterprises free of charge. From the current development situation, users and service providers are in a win-win situation. Demand side management has entered a virtuous cycle, and the quality of demand response is constantly improving. In the summer of 2017, Jiangsu province realized the second level automatic demand response of industrial electrical equipment.

4 SIGNIFICANCE OF SUBSIDY POLICY

From the perspective of terminal installation, monitoring is of great significance. Firstly, the installation of monitoring terminal is the basis of DSM. The installation of on-line monitoring device for electricity consumption can not only have a perfect power statistics and management system, but also make the saved and transferred power load measurable, monitored and verifiable. Secondly, the installation of monitoring terminal is an infrastructure for enterprises to enter the spot trading market in the future. It is the general trend for the power reform to enter the deep water area and establish the electric power spot market. The day ahead and within day electric energy trading, electricity prices are different, and the supply and demand situation is different. Accurate measurement of electricity and accurate load curve forecasting are the key points. At the same time, due to the measurement and control function of the monitoring terminal, both sides of the power supply and the user can choose the optimal mode intelligently. According to the power consumption situation, the interactive space between the power supply and the user can be enhanced to achieve the multi-objective optimization and the best economic benefits can be achieved. The installation of monitoring terminal is beneficial to the development of enterprises. By using energy efficiency monitoring, energy consumption diagnosis can be realized, production process can be improved accurately, energy efficiency of terminal can be improved, and enterprise competitiveness can be enhanced. At the same time, enterprises will realize the comprehensive energy utilization in the future. By using the data analysis results collected by the terminal, enterprises can get efficient comprehensive energy services.

5 SUGGESTIONS

From the perspective of terminal subsidies year by year, with the economic development and social progress, the proportion of subsidies has gradually decreased. In some economically underdeveloped areas, the production enterprises that rely on high energy consumption of resources are still in a relatively conservative state, and need to gradually guide their participation enthusiasm. In some economically developed regions, through the transformation and upgrading of production, the owners' ideas are updated, and their understanding of the essence of terminal installation is more profound. They embrace the future with an open mind. They hope that through technological progress, enterprises can really lead in management and other aspects, drive the overall progress of external environment, and form a virtuous circle inside and outside. Ningxia gives enterprises support through clear installation terminal subsidies. After waiting for the atmosphere environment to gradually form, the experience of support policy retrogression is worth learning from the central and western provinces. With the support of the Ningxia government, Ningxia power demand side management public service platform has 501 enterprises, 16708 on-line monitoring systems, and the monitoring capacity is 17 million 70 thousand KVA. It provides strong support for the development of "Internet plus intelligent energy", the establishment of industrial big data system, the service enterprise's energy efficiency diagnosis, and the integration of industrialization and information technology. Due to the rapid development of internal and external environment in Jiangsu Province, a conscious installation atmosphere has been formed. When the opportunity is ripe, Jiangsu can enter the spot market first, and its experience can be adapted to local conditions. It is suggested that the government and power grid enterprises should cultivate the awareness of installation terminal and promote excellent experience to realize the comprehensive development of DSM.

REFERENCES

[1] Exploration and application of power system demand side management. Cao Chensheng. Modern industrial economy and informatization. 2018 (18).
[2] Discussion on the role of power load control system in demand side management. Sun Xuan. Scientific and technological innovation and application. 2016 (33).
[3] Demand side management: a tool for sustainable development. Hans Nelson. Power demand side management. 2007 (05).
[4] Market calls for long-term DSM mechanism. Hu Hongsheng. DSM. 2006 (04).
[5] How should we understand demand side management?. Xu Shidong. China Science and technology information. 2005 (17).
[6] The government should play a leading role in DSM. Zhai xueshu, Wang Bing. Public power. 2005 (11).

Computational Social Science – Luo, Ciurea & Kumar (eds)
© 2021 Taylor & Francis Group, London, ISBN 978-0-367-70193-2

How does strategic orientation impact corporate performance: The path mechanism concerning marketing capability

Q.H. Liu, X. Zhao & H.P. Wang
School of Business, Jianghan University, China

ABSTRACT: This paper divides the strategic orientation into three dimensions: market orientation, entrepreneurial orientation, and learning orientation, which construct a structural model of the relationship among strategic orientation, marketing capability, and corporate performance. Subsequently, the direct mechanism of strategic orientation at corporate performance and the indirect mechanism of marketing capability as intermediary variable are discussed. The results reveal that the direct impact of market orientation and entrepreneurial orientation on corporate performance is less than the indirect effect. Learning orientation has only an indirect impact on corporate performance, and marketing capability is a key factor bridging the relationship between corporate performance and strategic orientation.

Keywords: Strategic orientation, Learning orientation, Entrepreneurial orientation, Market orientation, Marketing capability, Corporate Performance.

1 INTRODUCTION

Strategic orientation is considered as the external performance of company value orientation, which points out the direction and provides development motivation for companies, and is the key factor for companies to obtain excellent performance.

The existing research results have laid a rich theoretical basis for this paper, but also left the following research space: first, the transmission mechanism of the impact of strategic orientation on corporate performance needs to be further clarified. Gao and Liu (2017) believe that culture or values themselves cannot directly produce corporate performance, but its role is to derive certain capabilities for companies and generate performance through such capabilities; secondly, the environment is one of the important influencing factors of organizational behavior and performance, and it regulates the characteristics of the organization (Liu & Wang 2015). Therefore, the economic effect of company strategic orientation is bound to be affected by external environment. Thus, it is of practical significance to explore the influence of strategic orientation at corporate performance based on China's company environment.

In view of this, this paper classifies strategic orientation into three dimensions: market orientation, entrepreneurial orientation, and learning orientation. We construct a structural equation model of the relationship among strategic orientation, marketing capability, and corporate performance and explore the direct mechanism of strategic orientation at corporate performance and the indirect mechanism of marketing capability as intermediary variable.

2 THEORETICAL FOUNDATIONS AND HYPOTHESES

2.1 *The impact of strategic orientation on corporate performance*

Many studies believe that market orientation plays a guiding role in the business operation process, helping companies understand market information, better meet customer needs, and improve corporate performance. Therefore, the following hypothesis is proposed:

H1: Market orientation has a significant impact on corporate performance.

Entrepreneurial orientation shows that companies pay close attention to the changes of customer needs, and companies that pay attention to entrepreneurial orientation can give feedback to the changes of customer needs more quickly and meet the needs of customers in time. Li et al. (2009) conducted a study on 165 new-type companies in Taiwan, which also confirmed that entrepreneurial orientation helps to improve corporate performance. Therefore, the following hypothesis is proposed:

H2: Entrepreneurial orientation has a significant impact on corporate performance.

Learning orientation helps companies to understand the internal situation and external market environment of companies, to formulate a scientific and reasonable development strategy, and ultimately improve the performance of companies, which plays an important role in the development of companies in many aspects. Therefore, the following hypothesis is proposed:

H3: Learning orientation has a significant impact on corporate performance.

2.2 *The impact of strategic orientation on marketing capability*

Market orientation plays a guiding role in the company operation process. The understanding of market information can promote the promotion of company marketing capability. Therefore, the following hypothesis is proposed:

H4: Market orientation has a significant impact on marketing capability.

Knight (1997) found that globalization has caused great turbulence in the international market, brought more marketing opportunities to countries, and made the competition among multinational companies more intense. Thus, innovation and entrepreneurship are very important for companies. Therefore, the following hypothesis is proposed:

H5: Entrepreneurial orientation has a significant impact on marketing capability.

Organizational learning is helpful to the growth of companies' marketing capability. The stronger the learning ability is, the better the company can adapt to the fierce competition environment (Eisenhardt & Martin 2000). Therefore, the following hypothesis is proposed:

H6: Learning orientation has a significant impact on marketing capability.

2.3 *The impact of marketing capability on corporate performance*

A variety of studies have shown that marketing capability is reflected in paying attention to the changes of external market environment, reasonably allocating the knowledge technology and resources within the company, and finally meeting the needs of customers. In this process, marketing capability promotes corporate performance. Therefore, the following hypothesis is proposed:

H7: Marketing capability has a significant impact on corporate performance.

3 DESIGN OF RESEARCH

3.1 *Model design*

Figure 1 shows the research framework.

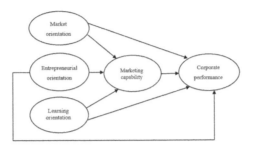

Figure 1.   Research framework.

3.2 *Measurement tools and data collection*

In this paper, we use the mature scale widely used in foreign countries and make adaptive changes to make it both faithful to the original intention and suitable for domestic measurement.

The respondents are senior managers to ensure the quality of the questionnaire. A total of 350 questionnaires were distributed and 297 were recovered, of which 265 are valid, with an effective recovery rate of 75.7%.

4 TESTS OF RELIABILITY AND VALIDITY

4.1 *Descriptive statistical analysis*

The descriptive statistical analysis of the five variables is shown in Table 1. The results show that the absolute values of skewness of each item and kurtosis are less than 3 and less than 10, respectively, indicating that the statistical values of survey data basically follow normal distribution, and subsequent data analysis can be carried out.

4.2 *Test of reliability*

Table 2 shows that the Cronbach's α values of the five variables are all greater than the standard of 0.7; λ^2/df is less than 3, RMSEA coefficient is less than 0.08, and GFI, NFI, CFI are all greater than 0.9, indicating that the scale has good fitting degree.

4.3 *Test of validity*

The correlation coefficients of each dimension in the five variables were all less than 0.7 (Table 3), indicating that the discrimination validity of each variable scale is good.

5 STRUCTURAL EQUATION MODEL ANALYSIS

This paper uses the structural equation model to model the relationship between strategic orientation (including market orientation, entrepreneurial orientation, and learning orientation), marketing capability, and corporate performance.

5.1 *Overall fitting degree of the model*

The indicators of overall fit degree are usually divided into three parts: absolute fitting degree, reduced fitting degree, and value-added fitting degree. The results are shown in Table 4: λ 2 / DF

Table 1. Descriptive statistical analysis of each item.

Variable	Item	Mean value	Standard deviation	Skew-ness	Kurto-sis	Variable	Item	Mean value	Standard deviation	Skew-ness	Kurto-sis
Market orienta-tion	MO1	4.12	1.102	−1.202	0.711	Learning orienta-tion	LO3	2.8	1.28	0.072	−1.02
	MO2	2.86	1.236	0.097	−0.744		LO4	2.35	1.243	0.461	−0.817
	MO3	3.01	1.272	0.007	−0.875		LO5	3.18	1.399	−0.238	−1.175
	MO4	3	1.247	−0.01	−0.825		LO6	2.25	1.225	0.659	−0.57
	MO5	3.86	1.169	−0.807	−0.144		LO7	2.43	1.319	0.493	−0.935
	MO6	2.93	1.357	−0.005	−1.118		LO8	3.49	1.394	−0.506	−1.006
	MO7	3.49	1.276	−0.496	−0.723		LO9	3.68	1.189	−0.742	−0.249
	MO8	3.75	1.232	−0.739	−0.409		LO10	3.56	1.196	−0.632	−0.312
	MO9	3.14	1.253	−0.12	−0.928		LO11	3.51	1.171	−0.553	−0.43
	MO10	3.26	1.259	−0.207	−0.911		LO12	2.74	1.111	0.168	−0.567
	MO11	3.19	1.289	−0.193	−0.968	Marketing capability	MC1	2.68	1.107	0.187	−0.606
	MO12	3.42	1.237	−0.437	−0.669		MC2	2.28	1.211	0.587	−0.61
	MO13	3.14	1.31	−0.155	−1.024		MC3	2.21	1.21	0.641	−0.573
	MO14	3.27	1.239	−0.274	−0.78		MC4	3.55	1.164	−0.456	−0.525
Entrepre-neurial orienta-tion	EO1	4.14	1.06	−1.21	0.896		MC5	3.9	1.076	−0.96	0.399
	EO2	2.51	1.321	0.397	−0.919		MC6	3.8	1.075	−0.696	−0.164
	EO3	2.63	1.362	0.308	−1.056		MC7	4.13	0.99	−1.105	0.798
	EO4	2.49	1.272	0.372	−0.884		MC8	3.95	1.082	−0.966	0.361
	EO5	2.96	1.329	0.003	−1.051		MC9	3.63	1.102	−0.517	−0.423
	EO6	2.93	1.328	−0.008	−1.047		MC10	4.08	0.999	−1.124	0.932
	EO7	3.02	1.31	−0.011	−1.002		MC11	3.9	1.036	−0.816	0.158
	EO8	2.8	1.26	0.169	−0.869	Corporate perfor-mance	EP1	3.65	1.165	−0.658	−0.333
	EO9	2.56	1.238	0.291	−0.836		EP2	3.31	1.249	−0.268	−0.857
	LO1	2.98	1.283	−0.093	−0.98		EP3	3.58	1.219	−0.524	−0.606
	LO2	3.04	1.256	−0.125	−0.902		EP4	3.58	1.195	−0.531	−0.54

Table 2. Reliability test of scale.

	Cronbach's Alpha	λ^2/df	GFI	RMSEA	NFI	CFI
Market orientation	0.855	2.35	0.93	0.073	0.96	0.98
Entrepreneurial orientation	0.791	2.97	0.90	0.077	0.95	0.95
Learning orientation	0.856	2.36	0.93	0.071	0.94	0.92
Marketing capability	0.904	2.82	0.91	0.071	0.94	0.98
Corporate performance	0.845	2.27	0.95	0.075	0.95	0.96

is less than 3, GFI is greater than 0.9, RMSEA is less than 0.08, PNFI and PGFI are both greater than 0.5, NFI and CFI are greater than 0.95, indicating that the structural equation fitting is ideal.

5.2 Test of hypotheses

The hypotheses test data are shown in Table 5. Except for the hypothesis that the path coefficient of H3 is 0.25, less than 0.3, which fails to pass the test, other hypotheses are verified. The specific model path diagram is shown in Figure 2.

5.3 Effect analysis

The direct impact coefficient of market orientation on corporate performance is 0.36, while the indirect effect coefficient is 0.37. Therefore, the direct impact of market orientation on corporate

Table 3. Correlation coefficient of internal dimensions of each scale.

	M1	M2	M3	M4	M5	M6	M7	M8	M9	P1	P2	P3	P4
M1	1												
M2	0.53	1											
M3	0.51	0.58	1										
M4				1									
M5				0.52	1								
M6				0.57	0.56	1							
M7							1						
M8							0.52	1					
M9							0.51	0.54	1				
P1										1			
P2										0.55	1		
P3												1	
P4												0.56	1

Table 4. Overall fitting degree of the model.

Fitting index	Absolute fitting degree			Brief fitting degree		Incremental fitting degree	
Fitting results	λ^2/df	GFI	RMSEA	PNFI	PGFI	NFI	CFI
	2.35	0.90	0.073	0.64	0.58	0.98	0.99

Table 5. Data of hypotheses verification.

Hypotheses	Variable relation	Standardized estimates	T-value	Conclusion
H1	MO→CP	0.36	2.43	Support
H2	EO→CP	0.32	2.17	Support
H3	LO→CP	0.25	0.87	Not support
H4	MO→MC	0.51	2.29	Support
H5	EO→MC	0.45	2.11	Support
H6	LO→MC	0.38	2.41	Support
H7	MC→CP	0.73	3.37	Support

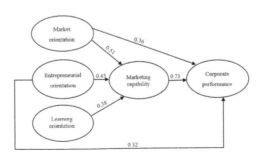

Figure 2. Model path diagram of main effect.

performance is less than the indirect effect. According to this comparison method, the direct impact of entrepreneurial orientation on corporate performance is less than the indirect effect between them. Learning orientation has only indirect impact on corporate performance.

And market orientation and entrepreneurial orientation are more important than learning orientation in influencing corporate performance.

6 CONCLUSION AND ENLIGHTENMENT

6.1 *Conclusion*

Each dimension of strategic orientation has a different impact mechanism on corporate performance. Market orientation and entrepreneurial orientation have a direct and significant impact on corporate performance, while learning orientation does not play a significant role in corporate performance. The indirect impact of market orientation and entrepreneurial orientation on corporate performance is greater than that of direct effect. Marketing capability plays an intermediary role in this indirect effect, and marketing capability plays a complete intermediary role in learning orientation and corporate performance. All these show that marketing capability plays a key role in the strategic orientation process that impacts corporate performance.

6.2 *Management enlightenment*

Due to the significant impact of different strategic orientation on corporate performance, facing the cruel market competition, companies need to implement compound strategic orientation, which can appropriately focus on the proportion of market orientation. Marketing capability plays a key role in the strategic orientation process indirectly affecting corporate performance, and companies need to pay attention to the cultivation of marketing capability.

ACKNOWLEDGMENT

This paper is supported by (1) the Humanities and Social Sciences Research Project of Hubei Provincial Department of Education of China under grant 18G036; (2) the Open Fund Project of Hubei Provincial Key Discipline (Management Science and Engineering) of Jianghan University.

REFERENCES

Eisenhardt, K. M. & Martin, J. A. 2000. Dynamic capabilities: what are they? Strategic Management Journal 21(10/11): 1105.

Gao, F. & Liu, Q. 2017. How market orientation affects corporate performance: A Study on the mediating effect of marketing capability. Luo Jia Management Review (1): 163–171.

Knight, G. A. 1997. Emerging paradigm for international marketing: the born global firm. Michigan: Michigan State University.

Kohli, A. K. & Jaworski, B. J. 1990. Market orientation: the construct, research propositions, and managerial implications. Journal of Marketing 54 (2): 1–18.

Li, Y., Huang, J. & Tsai, M. T. 2009. Entrepreneurial orientation and company performance: the role of knowledge creation process. Industrial Marketing Management 38(4): 440–449.

Liu, Q. 2015. Construction of marketing capability of Chinese companies: from the perspective of organizational learning and knowledge management. Wuhan University Press.

Liu, Q. & Wang, T. 2015. How market orientation affects corporate performance: an integrated study based on marketing capability and environmental uncertainty. Journal of South China University of Technology 17 (03): 13–20.

Zhang, Y. & Li, Q. 2009. Entrepreneurial orientation, dual competence, and organizational performance. Journal of Management Science 12 (1): 137–152.

Computational Social Science – Luo, Ciurea & Kumar (eds)
© 2021 Taylor & Francis Group, London, ISBN 978-0-367-70193-2

Analysis of Sino-US relations: From the perspective of building a new model of major-country relations

X.T. Xiao
School of Foreign Languages, Hubei Polytechnic University, Huangshi, China

ABSTRACT: China's rise has incrementally changed the international order into a bipolar structure, which is bound to undermine the liberal democratic order dominated by the United States
since World War II. In response to this, the United States has launched numerous actions in an
attempt to counter China's rising ambition, thus setting the tone for the Sino-US relations as a great
power rivalry in the decades to come. In this context, Chinese President Xi Jinping first proposed
a "new model of China-US major country relations" in 2012, so as to avoid another cold war
and most importantly to ensure an international order for world peace and common development.
However, the United States did not react positively toward this concept. The cognitive divergence is
unavoidable on both sides, thus leading to an on-going strategic rivalry. Therefore, it is necessary to
conduct in-depth research on the essence for the cognitive divergence and each strategic intention
behind it to avoid direct confrontation.

1 INTRODUCTION

The Sino-US relations are among the most important and complex bilateral relationship in today's
world. As a rising great power, China has been constantly expanding its global influence under the
framework of BRI (Belt and Road Initiative), especially in the Asia-Pacific region, which is bound
to undermine the liberal democratic order dominated by the United States. In response to this,
US President Obama's administration launched the rebalancing to the Asia-Pacific region in 2011
and trans-pacific partnership (TPP) in 2016 with the intention to counter China's rising ambition
in Asia, typically in the South China Sea, while intensifying a "hub and spoke" alliance system
(an asymmetric alliance designed to exert maximum control over a smaller ally's actions) with
South Korea, Japan, and Taiwan in the Asia-Pacific, thus setting the tone for the Sino-US relations
as a great power rivalry in the decades to come. Moreover, in 2017, according to the National
Security Strategy, the United States has identified three main challengers: revisionist forces in
China and Russia, rogue states in North Korea and Iran, and transnational threat organizations,
especially jihadist terrorist groups, which actively compete against the United States and its allies
and partners (Trump 2017). Among all the challengers, great power competition with revisionist
countries, notable with China, has become the top priority for the US political elites.

Along with the mounting tensions created by the United States, will China and the United States,
as the rising power and established power, respectively, necessarily fall into the "Thucydides trap,"
which is coined by Harvard scholar Graham T. Allison in *Destine for War*? Offensive realists reveal
a pessimistic view. According to Offensive realism's viewpoint, state as a rational actor seeks power
and influence to achieve security through domination and hegemony (Steven 2010). On the other
hand, the anarchist nature of the international system forces countries to maximize their share of
world power, seeking superiority rather than equality in order to make themselves safer, and thereby
increasing their chances of survival (Mearsheimer 2001). Therefore, with the rapid rise of China's
economy and military, China and the United States will fall into the "Thucydides trap" due to the
structural paradox, thus catering on antagonistic mindset of many Western scholars and making

"Thucydides trap" doctrine popular in Western academia (Hu & Yu 2017). In this context, Chinese President Xi Jinping first proposed a "new model of China-US major country relations" so as to avoid the tragedy of great power politics and, most importantly, to ensure an international order for world peace and common development. However, the United States did not react positively toward the concept of "a new model of China-US major country relations." In terms of this new concept, the cognitive divergence is unavoidable on both sides, thus leading to an on-going strategic rivalry. Therefore, it is necessary to conduct an in-depth research on the essence for the cognitive divergence and each strategic intention behind it. In addition, we should put more emphasis on promoting concrete cooperation, putting this new type of relation upon more common interests rather than mere conceptual descriptions, and seek common ground while reserving diversities. Sino-US relations should develop in exploration and evolve in practice.

2 THE INITIATIVE OF CHINA-US MAJOR COUNTRY RELATION

2.1 China's rise posing threat to the US global interests

Since the establishment of diplomatic relations between China and the United States in 1979, despite small-scale frictions, the two countries had been in a state of coexistence both politically and economically. When entering the 21st century, China made remarkable achievement in the field of economy, science and technology, military, and diplomacy. This paper mainly focuses on the economic and technological indicators concerning China's rise.

China has become the fastest growing economy in the world, with an eighty-fold gross domestic product (GDP) increase from 1978 to 2017 (Wayne 2019). Figure 1 indicates that China, with continued rapid economic growth, has already surpassed the United States when measuring the GDP (Wayne 2019).

The exponential growth of GDP leads to the increase of industrial capability. As shown in Figure 2, due to its large spending on research and development (R&D), improvement on industrial productivity, relatively cheap labor forces, and the preferential policies and assertive movement toward higher-valued manufacturing chains from the central government, China has become the world's largest manufacturing country in 2016 according to data from the World Bank, while the United States ranked second place. With the increase of its industrial capability, China launched a plan, "Make in China 2025," which made the United States extremely vigilant.

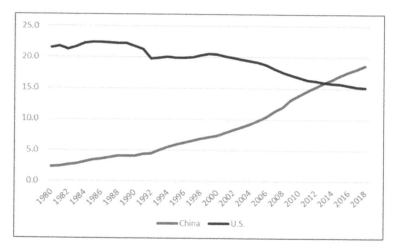

Figure 1. The United States and China GDP as a share of Global Total: 1980–2018. Data source: IMF, World Economic Outlook, April 2019.

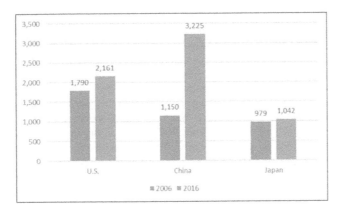

Figure 2.   Gross value-added manufacturing in China, the United States, and Japan. Data source: The World Bank.

In the high-tech industry, China is also approaching the United States in full speed. For instance, the Cloud infrastructure services are regarded as the infrastructure of the next-generation of information society and represent a country's scientific and technological strength. According to the market share of Cloud infrastructure services in the fourth quarter of 2017, China's Alibaba Cloud ranks fifth in market share and is the only company in the top five besides American companies, which shows the competitive situation between China and the United States in the high-tech field (David 2013).

Telecommunication, especially fifth generation mobile communication, is the "neural network" of the information society. At present, Huawei, Ericsson, Nokia, and ZTE are the world's four major telecom equipment giants, two of which are occupied by China. Huawei's sales in 2018 were $105.2 billion and R&D investment was $14.8 billion, significantly surpassing the traditional communications equipment giants Ericsson and Nokia. Compared with US wireless communications giant Qualcomm, Huawei's revenue and R&D investment volume are also leading (Ren 2019). Obviously, China is showing a momentum of catching up with the United States in the field of fifth generation mobile communication, which is likely to be associated with the fourth industrial revolution.

Although China has never tried to challenge or replace the US hegemony in the world, the rapid rise and continuous expansion of China's global influence has necessarily undermined the liberal democratic order led by the West and undermined the US leadership in the world, and the United States perceives it as a real threat to its national security. In 2010, the "pivot to the Asia-Pacific" strategy was launched by the Obama administration in preparation for containing China's rise. In addition, tensions between the two countries have been incrementally escalated on hotspot areas like the South China Sea and Taiwan (White 2012). In the ideological sphere, the United States has embarked on a warfare on public opinion by tirelessly accusing China of violations of human rights, stealing intellectual properties, and illegal monitoring. The rising tensions in these hotspots not only alienated two countries and affected the Sino-US relations in the long run, but they also created more instabilities to world peace and development. How to prevent two sides from falling into the "Thucydides trap" is becoming the top priority both for China and the United States at present.

2.2   The release of "A new model of China-US major country relations"

In this regard, the Chinese leaders have taken the initiative, laying out a framework entitled "a new model of China-US major country relations" for the future development of relations between China and the United States. Since then, China has adhered to the principles of the model with the intention to stop the falling trajectory of the Sino-US relations.

In February 2012, during his visit to the United States, Xi Jinpin first called for efforts to develop the Sino-US relations into "a new model of major-country relations." In May 2012, at the opening ceremony of the China-US Strategic and Economic Dialogue, President Hu Jintao reiterated the desire to build "a new model of China-US major-country relations." In the "Annenberg Estate meeting" between Chinese President Xi Jinping and then US President Obama in June 2013, President Xi Jinping put forward a clear vision for the new model. The fundamental characteristics of the "new model of China-US major-country relations," in his words, can be summarized as "no conflict or confrontation," "mutual respect," and "win-win cooperation" (Wang 2013).

In November 2014, during President Obama's visit to Beijing, President Xi reiterated "a new model of major-country relations" with the intention to extend the Sino-US cooperation in spheres concerning economy, trade, and infrastructure, as well as strengthening the Sino-US cooperation in the multilateral international issues, such as Iran and North Korea nuclear issues, the promotion of stability in Middle East, as well as non-traditional threats concerning climate change.

China has made numerous interpretations on the connotation of the "new model of major-country relations." In contrast, the United States has not responded positively to this concept. The reasons are as follows:

(i) The United States has not explicitly defined the model, nor has it accepted the equal status in this relationship.
(ii) Both China and the United States have deep misgivings and are unwilling to easily accept the concepts and views put forward by the other side.
(iii) China and the United States have different cultural mindsets in terms of strategy. For example, the Chinese tend to think from the top down and reflect on the big concepts first, while Americans are more pragmatic, which means that they are accustomed to think of the details and tend to think from the bottom up.
(iv) China and the United States have differences on the issue of responsibility sharing in the process of building a new type of major power relationship.
(v) The two countries disagree on the prospects and approaches for the development of the "new Sino-US relationship between major powers."

It can be seen that there is a huge cognitive divergence between China and the United States regarding the concept of "a new type of great power relationship." Thus, the strategic rivalry between the two countries was caused. After Donald Trump took office as US president in 2017, a series of trade protection policies were implemented, which marked the starting point of the Sino-US trade friction. The trade war is just a tip of the iceberg in the strategic game between China and the United States. It soon developed into an all-around competition in areas of technology, economy and finance, military, resources, and human exchange, which made the situation a stalemate. Today, the great power rivalries occur between China and the United States, and Russia and United States, but the United States with its allies, namely the European Union, Japan, and South Korea, have also showed signs of discord. The world pattern is transforming into a bipolar structure, where competition and cooperation coexist under the great power politics (Pang 2020). Therefore, it is necessary to understand this new world structure, great power politics, more importantly, to build and expand the "new model of major-country relations" theoretically, so as to cope with the realistic issues.

3 FUTURE POSSIBILIITY AND REALISTIC DILEMMAS

3.1 *Possibilities from deep economic interdependence*

Structurally, China-US relationships have already fallen into the "Thucydides trap," which has not only led to the fluctuation of China-US relationships but also negatively affected the stability and prosperity of the region and the world. However, the "Thucydides trap" between China and the United States is quite different from the historical "Thucydides trap" because there is a close economic interdependence between two sides, which means common interests can be achieved. Therefore, China and the United States must make joint efforts to stick to the bottom line and provide further protection against confrontation. At the same time, we should control differences in our interests, expand the scope of our cooperation, and consolidate the foundation for a new type of relationship between major powers (Jin 2015).

To be specific, the approaches we adhere to are associated with the position of non-conflict, non-confrontation, mutual respect, and win-win cooperation. First, non-conflict and non-confrontation are the prerequisite for building a new type of Sino-US relationship between major powers. Unlike the great power rivalries in history, the world is now in a new historical period. The US, China, and all other countries are a community of economic interests. The ties among countries are getting closer. Confrontation between China and the United States is not good for both sides. Non-conflict and non-confrontation mean following the trend of globalization and reversing the negative expectations of Sino-US relations. The core of Sino-US relations is to resolve mutual strategic mistrust and have long-term confidence in the future of Sino-US relationships. On the other hand, in the course of building the new model of major-country relations, conflicts and divergence are unavoidable, but we can take measures and enhance mechanism to manage and control them effectively (John 2001). Especially in the field of the military, to ensure channels to conduct dialogue and communication is of vital importance to bilateral relationships.

Mutual respect is a fundamental characteristic for this new model and an important guidance. Mutual respect is deeply rooted in ancient Chinese culture and is the embodiment of the great wisdom of the ancients. For instance, more than 2,000 years ago, Confucianism put emphasis on peace and harmony. As two major countries with different social and political systems, differentiated histories and cultures, and converging interests, mutual respect between China and the United States is particularly important in today's globalized world. Only by respecting each other's political systems and development paths chosen by each other's people, as well as each other's key benefits and concerns, can we seek common points while reserving differences. On this basis, China and the United States should expand common ground and resolve differences, so that the differences between China and the United States can coexist harmoniously (Wang 2013).

Win-win cooperation is an important step in realizing this vision. There is huge demand for cooperation between China and the United States in various fields, and the potential is huge. In addition, cooperation is vital in the global governance. The non-traditional threats, from counter-terrorism to cyberspace security, from international nuclear non-proliferation to global climate change, will only be addressed under the cooperation between China and the United States. In short, this win-win result is not only beneficial to China and the United States but also to the survival of mankind.

3.2 *Realistic dilemmas*

The new model of major-country relations mainly focuses on a bilateral relationship, which requires joint-efforts made by China and the United State. However, it is difficult for both sides to control the speed and pattern of the changing dynamics. In a bipolar structure where China and the United States are in head-to-head competition, the balance of regional tensions, trade friction, psychological anxiety, and cognitive divergence between two countries is by no means an easy task. Therefore, we believe that the key issues and further works regarding the new model of major-country relations should focus on the following two aspects:

(1) We should thoroughly study the four major dilemmas in the construction of the "new Sino-US relationship between major powers": theoretical dilemmas, ideological conflicts, cognitive differences, and realistic dilemmas, and explore strategies and ways to resolve these dilemmas.

(2) We need to expand the connotation and extension of the "new model of major-country relations," understand the long-term and asymmetric nature of major power confrontations, make competition moderate, maintain the tolerance of international order, and advocate fairness in major power relations.

4 CONCLUSIONS

Western structural realists argue that China's rise will necessarily bring about the "Thucydides Trap." China and the United States, as the rising power and established power, respectively, will fall into the "tragedy of great power politics." In the context of all-around competition between China and the United States, President Xi put forward the initiative of "a new model of China-US major-country relations." However, the United States has not reacted positively to the concept of "a new model of China-US major country relations" due to the huge cogitative divergence with China. How to view the friction and collision of two countries' grand strategies, and how to extend the "new model of major-country relations" is a major theoretical task needed to be addressed urgently. This paper explores the strategies and approaches based on the new model and points out the current dilemmas based on the principles of "new model of major-country relations." Therefore, it is of important theoretical and realistic significance.

In sum, the past four decades have witnessed numerous economic and trade cooperation and human exchanges between the two countries since 1971, and it is the fruits of generations of efforts in both countries. In a hyper-globalized world with deep economic interdependence, the decoupling between China and the United States cannot easily be achieved by the US political elites' unilateral actions. As long as both sides can control the divergence rationally and focus on the common interests and cultural exchanges between ordinary people to enhance trust, the future of Sino-US relations can be full of potential and possibilities. On the other hand, we should also recognize the long-term nature of great power rivalry and the asymmetric essence in this competition. Making the competition moderate, maintaining the inclusiveness of the international order, and advocating for the fairness of major-country relations are also of great importance.

REFERENCES

David M. L. 2013. A New Type of Major-Power Relationship: Seeking a Durable Foundation for U. S – China Ties. Asia Policy 2013(16): 1–6.

Grove, A.T. 1980. Geomorphic evolution of the Sahara and the Nile. In M.A.J. Williams & H. Faure (eds.), *The Sahara and the Nile*: 21–35. Rotterdam: A. A. Balkema. http://www.state.gov/secretary/remarks/2014/11/233705.htm.

Hu Z., Yu Z. 2017. "Thucydides Trap" and the Sino-U.S. Relations. SOCIALISM STUDIES 2017(6): 143–150.

Jin Canrong 2015 Should Superpowers Slip into the Thucydides' Trap? Analysis on the Competitive and Cooperative Features of Sino-American Relationship. Journal of Hubei University (Philosophy and Social Science): Vol.42 No.3 May 2015.

John J. M. 2001. The tragedy of Great Power Politics. New York: Norton.

John Kerry. Remarks on U.S.-China Relations.

Pang J. 2020. An Analysis of the Political Pattern and Trend of Great Powers in the Era of Great Changes. Frontiers. DOI: 10.16619/j.cnki.rmltxsqy.2020.07.001

Ren Z. 2019. Sino-US technological strength comparison: a decisive battle for a new generation of information technology. http://finance.sina.com.cn/zl/china/2019-08-27/zl-ihytcern3861515.shtml.

Steven E. Lobell. 2010 Structural Realism/Offensive and Defensive Realism. International Studies, Oxford Research Encyclopedia.

Donald J. Trump 2017 National Security strategy of the United States of America. https://www.whitehouse.gov/wp-content/uploads/2017/12/NSS-Final-12-18-2017-0905.pdf

Victor Cha 2009/10 Powerplay: Origins of the US Alliance system in Asia. Quarterly Journal: International Security: 158–196.

Wang Yi. 2013 Toward a New Model of Major-country Relations Between China and the United States. https://www.brookings.edu/on-the-record/wang-yi-toward-a-new-model-of-major-countryrelations-between-china-and-the-united-states/

Wayne M. Morrison 2019 China's Economic Rise: History, Trends, Challenges, and Implications for the United States. Congressional Research Service: 11.

White H. 2012. The China Choice. Why America Should Share Power. Oxford University Press.

Computational Social Science – Luo, Ciurea & Kumar (eds)
© *2021 Taylor & Francis Group, London, ISBN 978-0-367-70193-2*

Research on vehicles route problem of rail transit equipment based on improved ant colony optimization

H. Yu, L. Sun, X.Y. Tong & X.J. Zheng
School of Mechanical Engineering, Dalian Jiaotong University, Dalian, China

ABSTRACT: In order to improve the traditional mode of special car delivery for a certain rail transit equipment material distribution, this paper adopts the milk run cycle material distribution mode. First, a vehicles route problem (VRP) mathematical model based on the milk run circular distribution model is established; then, the ant colony optimization (ACO) and the genetic algorithm (GA) representation, crossover, and mutation are combined into an improved genetic ant colony optimization (GACO) to solve the problem. Then, the proposed algorithm was run five times and contrasted with the standard ACO. Finally, the improved GACO was verified through examples: milk run circular distribution saves the vehicle and distance of material distribution, saves the cost of material distribution of enterprises, and provides reference models and algorithms for related units.

Keywords: Milk-run pick problem, Genetic ant colony optimization, VRP.

1 INTRODUCTION

Intra-factory logistics is an important part of a company's logistics links. It is the link between the various workshops of the company and the logistics lifeline of the entire factory. It is of great significance to study the logistics path of the enterprise, explore the third-party profit of the enterprise, and improve the production efficiency of the enterprise (Brabazon & Maccarthy 2017).

Eilon et al. (1974) proposed vehicles route problem (VRP) for the first time in order to optimize the delivery of goods and make a reasonable plan for the route of the vehicle. Ram Gaur et al. (2018) studied the VRP problem with a fixed number of vehicles and solved it by using dynamic programming. Pisinger and Ropke (2007) proposed a unified heuristic algorithm that can simultaneously solve five different VRP problems: VRP with time constraints (VRPTW), VRP with capacity constraints (CVRP), and VRP with multiple distribution centers (MDVRP), batch distribution VRP (SDVRP), and open VRP (;uner et al. 2017; Utkarsh et al. 2020; Yu et al. 2015). However, there is little research on rail transportation equipment, and there is almost no literature on the logistics and distribution of rail transportation equipment.

The logistics of rail transit vehicles have the characteristics of large volume and long cycle times, so it has a long material preparation time. At present, the delivery model of special car delivery is still used, which is a great waste of cost. This article chooses to adopt the MILK circulation delivery model to solve the problem of material delivery in the enterprise.

2 PROBLEM FORMULATION

The VRP problem mainly solves the problem of vehicle distribution route. There is a logistics center O, which carries out distribution tasks to each workshop N. The distribution vehicles have load restrictions, and the shortest distribution route is required to save their distribution costs. For the mathematical model of logistics path planning, the following assumptions are made on the problem:

1. The sum of transportation cost and inventory cost is the smallest; 2. Limit the total number of transportation vehicles; 3. Each workshop station group can only be served by one delivery vehicle once; 4. The total amount of delivery for each vehicle does not exceed its capacity; 5. Ensure the transportation vehicles depart from the logistics center and finally return with empty containers; 6. Ensure the delivery time does not exceed the workshop demand limit; 7. Ensure the workshops with material requirements have vehicle services.

The objective function Z of the problem is:

$$Z = \min \sum_{i=0}^{N} \sum_{j=0}^{N} \sum_{k=1}^{K} d_{ij} x_{ijk} \tag{1}$$

The decision variables and constraints are:

$$X_{ijk} = \begin{cases} 1, & \text{Vehicle k travels from workshop i to workshop j} \\ 0, & \text{otherwise} \end{cases} \tag{2}$$

$$Y_{ik} = \begin{cases} 1, & \text{Vehicle k is assigned to workshop i} \\ 0, & \text{otherwise} \end{cases} \tag{3}$$

$$Z_b = \begin{cases} 1, & \text{Delivery volume is b} \\ 0, & \text{otherwise} \end{cases} \tag{4}$$

$$\sum_{i=1}^{N+1} X_{oi}^{(k)} = 1 (k = 1, 2, \ldots, k) \tag{5}$$

$$\sum_{i=1}^{N+1} X_{jo}^{(k)} = 1 (k = 1, 2, \ldots, k) \tag{6}$$

$$\sum_{k \in K} y_{ik} \le 1, \forall i \in N_s \tag{7}$$

$$\sum_{i \in N} \sum_{m \in s_i} y_{ik} q_{im} z_b b \le Q \tag{8}$$

$$\sum_{i \in N} x_{ijk} = y_{jk}, \forall j \in N_s, k \in K \tag{9}$$

$$\sum_{j \in N} x_{ijk} = y_{ik}, \forall i \in N_s, k \in K \tag{10}$$

$$\sum_{b \in B} Z_b \le 1 \tag{11}$$

In the above formulation: V represents the number of transport vehicles; Q is the capacity of transport vehicles; K represents a collection of vehicles, $K = \{1,2,\ldots,V\}$; N represents the workshop assembly, $N = \{1,2,\ldots,N\}$; P is a collection of parts, $P = \{1,2,\ldots p_n\}$; B represents delivery batch collection, $B = \{1,2,\ldots B_n\}$; Assemble the parts of the workshop $i \in N_S$; A represents a collection of parts and components, $a_m \in A, m \in M$; D_{ij} represents the distance from workshop i to j; q_{im} represents the volume of part m at workshop i; u_c represents the cost coefficient per unit distance. Eq. (1) denotes the shortest path; Eqs. (2) to (4) are three variables; Eq. (5) and Eq. (6) represent each vehicle's departure from and return to the logistics center; Eq. (7) indicates that each workshop is visited and can only be visited once by one vehicle; Eq. (8) specifies the capacity of each vehicle; Eq. (9) and Eq. (10) express the relationship between two variables; Eq. (11) lays down the demand of each assembly plant which must be completed by the delivery vehicle at one time.

3 AN IMPROVED HYBRID ANT COLONY OPTIMIZATION (GACO)

Ant colony optimization is an optimized algorithm based on long-term research and analysis of ants' foraging behavior. By setting a taboo table to guide the ant's route, it has this unique advantage in solving such problems. But ant colony optimization is prone to "premature" phenomenon in the process of searching for the optimal solution, which is limited to the local optimal solution. Aiming at the shortcomings of the ant colony optimization, it is proposed to introduce the genetic algorithm crossover and mutation mechanism, and finally to solve the VRP problem with the improved hybrid ant colony optimization (GACO) (Hosseini et al. 2014).

The improved hybrid ant colony optimization (GACO) is obtained by combining the duplication, crossover, and mutation mechanisms of the genetic algorithm with the ant colony optimization. The basic idea is this: to add part of the operation of genetic algorithm into each iteration of the ant colony algorithm, so that the optimal solution generated by the ant colony algorithm each time will be the initial population of the genetic algorithm, and then iteratively calculate after the duplication, crossover, and mutation operation to better conduct the global search and expand the search range, thereby avoiding the phenomenon of "precocious maturation."

3.1 Improvement of selection probability of ant colony optimization

The selection probability of common ant colony optimization can be found in the paper by Zheng et al. (2020). The selection of the next moving direction of the ant is mainly divided into exploration and utilization. Exploration is defined as being when the ant is less affected by the pheromone concentration, thereby expanding the search space; utilization is defined as the ant being mainly affected by the pheromone concentration, and the path with high pheromone concentration is selected with a high probability. Between exploration and utilization, there is a balance point q_0. Set a given parameter and generate a random number q. Comparing q with q_0, the ants rule for choosing the next path changes to

$$j = \begin{cases} \arg \max_{k \in allowed_k} ([\tau_{ik}]^\alpha [\eta_{ik}]^\beta), q \leq q_0 \\ J, \text{otherwise} \end{cases} \tag{12}$$

In the above formulation: j represents the node selected according to the transition probability of the standard ant colony algorithm.

When $q \leq q_0$, the ant will use the known maximum transition probability to select the next node, utilization; when $q > q_0$, the ant will transition according to the probability of the standard ant colony algorithm, exploration. Find the balance point by adjusting the value of q_0. In the early stage of the algorithm, it is necessary to quickly obtain a better solution, and the value of q_0 should be as large as possible so that $q \leq q_0$, the ants select nodes through "utilization to speed up the search; in the middle of the algorithm, for purpose of avoiding the algorithm from falling into premature, the search range needs to be expanded. Adjust the value of q_0 to a smaller value, so that exploration is used to select nodes at this time. It is necessary to expand the search range and adjust the value of q_0 to a smaller value so that $q > q_0$. At this time, exploration is used to select nodes; in the later stage of the algorithm, it is necessary to obtain the global optimal solution as soon as possible, restore the value of q_0, and make $q \leq q_0$, ants select nodes by utilization. On the basis of the above analysis, set the value of q_0 to

$$q_0 = \begin{cases} 0.9, Nc < \dfrac{1}{3}Nc\_\max \\ 0.1, \dfrac{1}{3}Nc\_\max \leq Nc < \dfrac{2}{3}Nc\_\max \\ 0.9, Nc < \dfrac{2}{3}Nc\_\max \end{cases} \tag{13}$$

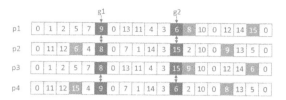

Figure 1.   Improved two-point crossover operation scheme.

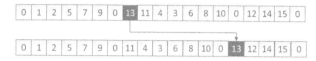

Figure 2.   An example of the mutation operation.

3.2   Combination of genetic algorithm and ant colony optimization

3.2.1   Representation of solutions (coding)

The representation, crossover, and mutation operations of the genetic algorithm are all based on genetic coding, so, first, the route optimization model of the logistics in the factory is coded. Assume that workshops N = {1, 2..., N}, K delivery vehicles, O represents the logistics center, so the length of the chromosome is N + K + 1, where the number of 0 is K + 1, and the rest represents workshop section. For specific encoding methods and copy operations, see the paper by Ho et al. (2008).

3.2.2   Improved two-point crossover

According to the characteristics of the VRP problem, an improved two-point crossover operation is adopted. The steps are as follows.

Select two chromosomes according to the crossover probability. Select two random gene positions g1 and g2 from chromosome P1. Switch the places of genes g1 and g2 with the genes g3 and g4 in the same position of chromosome P2. Switch the places of other genes in P1 in a similar way. Switch the places of other genes in P2 in a similar way.

Next, the adaptive crossover probability was used to select individuals:

$$p_c = \begin{cases} p_{c1} - \dfrac{(p_{c1} - p_{c2})(f' - f_{acg})}{f_{max} - f_{avg}}, f' \geq f_{avg} \\ p_{c2}, f' < f_{avg} \end{cases} \qquad (14)$$

where f_{avg} is the mean fitness of the current population; f_{max} is the maximum fitness of the current population. Usually, $p_{c1} = 0.9$ and $p_{c2} = 0.6$. (Figure 1).

3.2.3   Mutation operation

The author designed a simple mutation operation as follows.

Select a chromosome according to mutation probability. Remove a random gene from the chromosome. Insert the gene into another position of any other route randomly.

The adaptive mutation probability was adopted (Figures 2 and 3).

4   CASE STUDY

To validate the logistics route optimization model, the logistics of the Lushun Base of the Dalian Locomotive Factory is used as an example for simulation analysis. The layout of the plant is shown

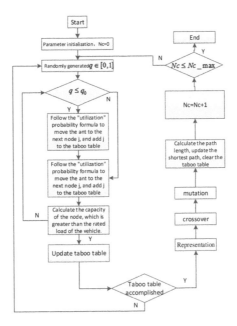

Figure 3.   GACO flow chart.

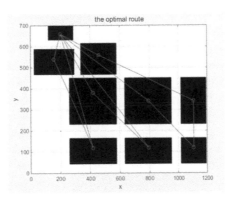

Figure 4.   The optimal route (the black area indicates the location of the workshop).

in Figure 4, and the coordinates of each workshop are obtained by establishing a coordinate system to plan the relative positions between two workshops. This article takes the distribution of materials from the logistics center to the workshops as an example to verify the model and algorithm. Known conditions are the number of delivery vehicles $K = 3$ and the vehicle capacity $Q = 45m^2$. (Table 1).

The relevant data of the test algorithm are the maximum number of iterations NCmax = 100; the number of ants m = 20; the information inspiration factor $\alpha = 1$; Expected heuristic factor $\beta = 2$; pheromone volatilization coefficient $\rho = 0.05$. (Table 2).

After calculation, the average result obtained by running the ordinary ACO for five times is 6,883.8 m, and the average result obtained by the GACO is 6,778.4 m; the optimal result of running the ordinary ACO is 6,791 m, and the optimal result of the GACO is 6,756 m. This results in the best comparison of the improved GACO, which is better and more stable. The optimal solution that satisfies the conditions is 0-1-4-0-2-5-7-8-0-3-6-0, and the optimal vehicle delivery path is:

Route 1: 0-1-4-0, logistics center—urban rail electric—urban rail assembly—logistics center;

Table 1. Date above Logistics Center and workshops.

ID	Coordinate_x	Coordinate_y	Vehicle capacity
0	200	660	0
1	157.5	540	11
2	456.5	560	8
3	418	380	13
4	418	120	29.5
5	792	341	10
6	792	120	27.5
7	1105.5	341	12
8	1105.5	120	11

Table 2. Algorithm 5 runs results.

ID	ACO	GACO
1	6791	6756
2	6981	6791
3	6833	6756
4	6981	6833
5	6833	6756
Average path	6883.8	6778.4

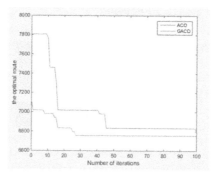

Figure 5. Comparison of convergence curves before and after algorithm improvement.

Table 3. The results of GACO.

ID	Route	Distance	Loading rate
1	0-1-4-0	1601	90%
2	0-2-5-7-8-0	2891	91.1%
3	0-3-6-0	2264	90%

Route 2: 0-2-5-7-8-0, logistics center—urban rail truck—locomotive body—locomotive electrical—locomotive truck—logistics center;
Route 3: 0-3-6-0, logistics center—urban rail car body—locomotive assembly—logistics center (Figure 5) (Table 3).

In contrast, the improved hybrid ant colony algorithm gets better results and runs more stable, while the ordinary ant colony algorithm tends to fall into the local optimal solution. This difference is more obvious in the case with larger data.

In the factory's originally planned distribution scheme for different workshops and different vehicles, eight distribution vehicles were used in eight workshops with a total distance of 12976 m. By comparison, the improved algorithm saves 6,220 m compared to the scheme, five vehicles are saved, and the optimized loading rate is more balanced.

5 CONCLUSION

On account of the VRP problem of rail transit equipment, this article takes advantage of the milk run material circulation distribution method to improve the traditional material distribution method. It saves 6,220 m on the distribution path, shrinks the number of distribution vehicles from the original eight to three, and more than doubles the cost of distribution. The improved three distribution vehicles have a high full load rate during the distribution process, which improves the previous phenomenon of empty vehicles. This article establishes a mathematical model of the VRP problem for rail transit enterprises and utilizes the improved ant colony algorithm to solve the problem through examples, and it achieves a certain optimization effect.

REFERENCES

Brabazon P G, Maccarthy B L. 2017 The automotive order to delivery process: how should it be configured for different markets?. European Journal of Operational Research, S0377221717303508.
Eilon S, Watson-Gandy C D T, Christofides N, 1974, Distribution Management-Mathematical Modelling and Practical Analysis. IEEE Transactions on Systems Man and Cybernetics, 21(6):589–589.
Guner A R, Murat A, Chinnam R B. 2017, Dynamic routing for milk-run tours with time windows in stochastic time-dependent networks. Transportation Research Part E Logs & Transportation Review, 97(JAN.): 251–267.
Ho W, Ho G T S, Ji P, et al. 2008. A hybrid genetic algorithm for the multi-depot vehicle routing problem. Engineering Applications of Artificial Intelligence, 21(4):548–557.
Hosseini S D, Akbarpour Shirazi M, Karimi B. 2014. Cross-docking and milk run logistics in a consolidation network: A hybrid of harmony search and simulated annealing approach. Journal of Manufacturing Systems, 33(4):567–577.
Pisinger D, Ropke S.2007. A general heuristic for vehicle routing problems. Computers & Operations Research, 34(8):2403–2435.
Ram Gaur, D, Mudgal, A, Ranjan Singh, R. 2018. Improved approximation algorithms for cumulative VRP with stochastic demands. Discrete Applied Mathematics, 280:133–143.
Utkarsh A, Behera B K, Panigrahi P K. 2020. Solving Vehicle Routing Problem Using Quantum Approximate Optimization Algorithm.
Yu L, Tianyi X, Zheyong B. 2015. A Two-Phase Heuristic Algorithm for the Common Frequency Routing Problem with Vehicle Type Choice in the Milk Run. Mathematical Problems in Engineering, (2015-10-18), 2015, 2015:1–13.
Zheng L, He Z, Liang W. 2020. VRP Problem Solving Based on Adaptive Dynamic Search Ant Colony Algorithm. Journal of Physics Conference Series, 1487:012030.

Computational Social Science – Luo, Ciurea & Kumar (eds)
© 2021 Taylor & Francis Group, London, ISBN 978-0-367-70193-2

Branding China: What roles can enterprises play in it?

Z.J. Yan & J. Wen
Nanjing University of Science & Technology, Nanjing, P.R. China

ABSTRACT: The ongoing COVID pandemic and China-US rivalry have posed new challenges to China's nation branding efforts. This paper aims to explore how to brand China at the moment. We argue that enterprises generally assume a key role in nation branding. Different types of enterprises in China, including foreign-invested, state-owned, and private enterprises, can play their respective major roles in branding China. Policy implications of this research are also discussed accordingly.

Keywords: China, Nation branding, Enterprises.

1   INTRODUCTION

More or less, China appears to be a unique case in the domain of nation branding. On one hand, the past few decades have witnessed the country's phenomenal transformation and economic growth which is unprecedented in modern times, with its GDP at present being the second in the world only after that of the United States. During this process, in order to be better understood and appreciated by the world, China has vigorously launched a variety of nation branding campaigns, such as the 2008 Olympics, the 2010 Shanghai Expo, and the Confucius Institutes in many other countries, to name just a few. On the other hand, as a quick rising world-class economy as well as the largest exporter in the world, China does not lack international visibility and awareness, but the country does need a good reputation that is conducive to and commensurate with its economic growth. However, due to various reasons, a quick rising China has engendered a certain suspicion and unease among a number of countries, some leading developed countries in particular, which are supposed to be the major audience of China's nation branding. This makes branding China a daunting task.

The outbreak of COVID-19 pandemic and the escalating China-US rivalry seem only to be complicating China's nation branding endeavors. As the COVID-19 pandemic still lingers on and face-to-face human communications, both at governmental and non-governmental levels, are becoming less frequent, mutual mistrust is setting in between China and some major countries. Traditional nation branding techniques, such as public diplomacy, tourism, and mega-events (the Olympic Games, for example), are rendered less effective by COVID-19 pandemic at present and probably in the upcoming couple of years. Under the circumstances, it is necessary for China to locate some new impetus and bring it into full play in order to keep the nation branding momentum. In this paper, we argue that enterprises in China, including foreign-invested enterprises, can play a pivotal role in branding China, not only in difficult times like the COVID-19 pandemic, but also, and more broadly, in conventional nation branding practices. In China's nation branding, the potentials of enterprises, private enterprises in particular, are to be tapped and unleashed, and appropriate policy adjustments should be made to support enterprises' roles in branding China.

The fact that Simon Anholt (1998), a British scholar, coined the term "nation brand" in the late 1990s was by no means accidental [1]. In hindsight, the 1990s was a decade full of epoch-making events, such as the ending of the Cold War, the founding of the World Trade Organization, the emergence of regional economic integration represented by the European Union, the NAFTA, and the APEC, and the rapid spread of the Internet. Additionally, large developing countries like China and India were eagerly embracing the world economy, and multinational firms, with their brands as a major competition tool, were arranging their business operations in a unified global market. Never before had the world experienced such a situation that scholars as well as international organizations were quick to use an umbrella term – globalization – to describe this brand new phenomenon [2]. Globalization has thus become a catchword ever since.

Globalization has brought about to the world many changes – among them "place parity", which is mainly manifested by the convergence of nation states. As the number of countries increases and "peace and development" becomes the central theme of the contemporary world, competition for external resources, such as foreign direct investment, tourism, and markets, is getting intense. Countries need to stand out from the crowd and send out a "me here" signal to those resources owners. As Simon Anholt (2002) rightfully points out that "It does sometimes seem as if globalization is turning the world into a gigantic supermarket where nations are nothing more than products on the shelf, frantically trying to attract the attention of each passing consumer." [3] It was against this backdrop that the concept of "nation brand" was created and quickly embraced by a growing number of countries. Indeed, in a globalized world, in order to cut a larger slice from the globalization pie and maximize their economic welfare, it is necessary for countries to engage in nation branding, which aims to measure, build and manage the reputation of countries.

As one of the pioneers in the study of nation branding, Simon Anholt proposed in 2005 the Nation Brand Hexagon Model as a way to measure the image and reputation of a particular country (see Figure 1). According to this model, nation brand is the sum of people's perceptions of the country's national competence across six dimensions, i.e., exports, governance, culture, people, tourism, and immigration/investment. In practical applications, based on Anholt's Nation Brand Hexagon, the Anholt-GfK Roper Nation Brands Index is released annually since 2008, which can provide national governments and their agencies with a one-of-a-kind resource for actionable insights needed to more effectively manage a country's reputation.

Although Anholt's hexagon framework remains the most broadly used approach to analyzing the nation brand, it fails to elaborate on the drivers behind nation branding movement. If the aim of nation branding is to build a strong nation brand, then the "builders" must be present to make this aim eventuated, but who are the builders? Regretfully, no sufficient clues we can acquire from this hexagon framework. It is our belief that an in-depth identification of nation brand builders, particularly those major ones, is of salient practical value in implementing nation branding strategies.

Conceivably, nation brand builders are stakeholders within the country. This is easy to understand because it is certain that nation branding is solely a country's own job – you just cannot entrust the work of promoting your country's image to some other countries or international institutions. But of course the country can enlist the help from professional teams from abroad in its nation branding campaigns, which is a different story. In a sense, the country is the initiator as well as the ultimate beneficiary of its nation branding efforts.

Within the country, literally speaking, every individual, every organization, and needless to say, governments at all levels, are all stakeholders and in varying degrees involved in the process of nation branding. With regard to the major players, in addition to government, enterprises come to the fore. To our best knowledge, the role of enterprises in nation branding has not been fully explored to date. In Anholt's Nation Brand Hexagon, the role of government is relatively straightforward (see "governance"), but it is less so for enterprises. We argue that enterprises, as a matter of fact, are closely related to all the six elements of the Nation Brand Hexagon (see Table 1). Without an adequate enterprise base, nation branding can only achieve very limited results (for example, as tourist destinations).

NATION BRAND HEXAGON

Figure 1. Simon Anholt's Nation Brand Hexagon.Source: Simon Anholt (2005).

Table 1. Enterprises' relation to Nation Brand Hexagon elements.

No.	Nation Brand Hexagon elements	Enterprises' relation to the element
1	Exports	Enterprises export and create COO effects
2	Governance	Enterprises act both as a governance tool and a major indicator of governance effectiveness
3	Culture	Enterprises are instrumental in forging contemporary culture
4	People	Enterprises help to raise population quality by providing employment and career training
5	Tourism	Enterprises provide tourism facilities and operation contents
6	Immigration/Investment	Enterprises act as a magnet for top-tier immigrants and investment

Source: compiled by the authors.

Based on Table 1, enterprises' role in nation branding deserves to be purposefully singled out for in-depth analysis. In a modern market economy, the importance of enterprises in national economic development cannot be overestimated. The same is true with regard to the role of enterprises in nation branding in the context of globalization. Indeed, in addition to the general contribution to the country in terms of production, employment, tax revenue, R&D, etc., enterprises have a key role in promoting and maintaining the country's international image and reputation.

Firstly, the COO image that enterprises create through exports is one of the most salient indications of the nation brand. Ever since the end of the 19th century, spearheaded by "Made in Germany", COO label has been regarded as a reflection of a country's image. In the present globalized world, economic interconnectedness between countries is getting ever increasingly complicated. Nonetheless, merchandise trade still remains the most basic form of economic exchange among countries. Consumers' perception of other countries are to a large extent shaped by their personal experiences with the products originated from those countries. Naturally, foreign products of high quality and affordable prices are welcomed by local consumers and the attached "Made in" labels make it possible for consumers to transfer their good impression about the products favorably to the place of origin. In today's international marketing field and even daily life, there are "Made in" sagas galore, particularly about "Made in Swiss", "Made in Germany", and "Made in Japan". These anecdotes add to the reputation of the countries involved. As a matter of fact, "exports" is treated by Simon Anholt himself as the number one element in his Nation Brand Hexagon. But remember – export transactions are performed by enterprises.

Secondly, enterprises' products and brands can provide a secure firewall and increase the resilience of the nation brand. Closely related to a country's culture, history, economy, geopolitics, and diplomatic policies, nation brand is much more complicated in nature than a corporate

Table 2. Different enterprises' roles in branding China.

No.	Enterprise types	Enterprises' major roles
1	Foreign-invested	Boosting the image of "Made in China"
2	State-owned	Demonstrating CSR in overseas operations
3	Private	Creating "inverse COO effect" through branded products

Source: compiled by the authors.

brand or product brand. From time to time, the intimacy between countries might turn sour, or polices might be variable and misinterpreted out of various reasons. Under the circumstances, the nation brand equity will face a danger of erosion. However, a strong performance of enterprises' products and brands in the foreign markets can somehow offset the negative impact on the country's reputation. At the very least, it should be known that the origin country of the products and the brands ought to be reckoned with. This was best demonstrated by American products and brands during the 2003 Iraq War [4]. On the other hand, if the public opinion in foreign countries about the country turns negative, while the products and brands from the country are also poorly evaluated by consumers in those countries because of product quality or other social issues, this will probably worsen the image of the nation brand.

Thirdly, enterprises can help boost nation brand image through "inverse COO effect". The world economic landscape is uneven. There are front-runners as well as late comers in globalization. Generally speaking, developed countries have already taken a headstart advantage in industrialization and trade, and they tend to have a favorable COO image in the international market. On top of this, leading corporate and product brands mainly originate from developed countries. All these, plus other factors, contribute to the higher rankings of developed countries in the aforementioned Anholt-GfK Roper Nation Brands Index. However, as Candace White (2012) points out, well-regarded enterprises from developing countries, through "inverse COO effect", can contribute to a favorable national reputation [5]. Different from COO effect, in which the name of the country specified on the "Made in" label of a product affects the product's international acceptance and success, "inverse COO effect" works through a "brand attitude → beliefs → country image" logic. For instance, high quality consumer electronics with Samsung or LG brands lead to people perceiving the reputation of South Korea as "high technology" and "advance economy". Skype, the well-known telecommunications application that is of Estonian provenance, is also a case in point.

3 DIFFERENT TYPES OF ENTERPRISES' ROLES IN BRANDING CHINA

Thanks to the reform and opening-up policy, foreign-invested enterprises and private enterprises have thrived in China along with state-owned enterprises (SOEs) in recent years. Together they make the mainstays of the country's economy today. In China's nation branding efforts, the roles of these different types of enterprises are indispensable.

Generally, there is a "built-in" mechanism of branding China on the part of these enterprises' international business operations. This means that, normally the effective cross-border business activities of the enterprises can help lift China's international image from different aspects (see Table 2). However, in the present situation, necessary measures should be proactively taken to bring into full play these enterprises' roles in branding China.

3.1 *Foreign-invested enterprises: boosting the image of "Made in China"*

Foreign-invested enterprises have long been a major contributor to China's exports. Albeit on a gradual downward trend in recent years, foreign-invested enterprises still accounted for nearly 40 percent of China's total merchandise exports in 2019. By taking advantage of China's cheap skilled labor, multinational companies invest and make technology intensive products in China and sell them abroad. Needless to say, this has raised considerably the international profile of "Made in China". Being a large developing country, China is still in dire need of foreign investment, particular that from world-class multinational companies. At present China has gradually recovered from the impact of the COVID pandemic ahead of other countries. It is urgent to seize the opportunity to attract more top-tier multinational corporations to invest in China by forging a more friendly investment climate. The World Bank ranked China 31st out of 190 economies in its "Doing Business 2020" report. China has good reasons to celebrate her big leap from 46th in 2018 to 31st in 2019. But it should be noted that this position still falls behind China's Hong Kong (3rd) and China's Taiwan (15th), even behind Kazakhstan (25th) and Russia (28th). So there is still much room for improvement in accommodating high-level foreign investment.

3.2 *State-owned enterprises: demonstrating CSR in overseas operations*

State-owned enterprises, especially those owned directly by the central government, play a unique role in branding China, because usually the words "Sino" or "China" appear in their corporate names. 2019 Fortune Global 500 list reveals that more than 50 companies with those titles engage in international business activities. The fact that those SOEs are directly related to China property implies both honor and responsibility. Although their share in China's total merchandise exports has considerably dwindled (less than 10 percent in 2019), the SOEs are rather aggressive in outbound investment in addition to overseas infrastructure construction. In the wake of the COVID-19 pandemic and China-US rivalry, SOEs' overseas operations are going to meet ever-increasing challenges and obstacles. While laying stress on economic results, SOEs should pay more attention to the host countries' politics, economy, laws, and social customs, and take good care of the needs of the local community, supply chain partners, as well as foreign employees affected by the COVID-19 crisis. By obeying the law and sticking to the bottom line of compliance management, SOEs can offer actual support to branding China in their overseas investment and infrastructure construction.

3.3 *Private enterprises: creating "inverse COO effect" through branded products*

The past four decades have witnessed the rise of China's private enterprises. In 2019, private enterprises became China's largest exporter for the first time and contributed 51.9 percent to China's total merchandise exports. A number of leading private brands, Huawei, Lenovo, and Geely for example, have become household names across the world. In spite of the negative impact of the COVID-19 pandemic and China-US rivalry, China is foreseen to keep its title as one of the world major suppliers of manufactured products in the future, which means that a myriad of opportunities are in store for China's private enterprises in terms of export and outbound investment. The aforementioned "inverse COO effect" is most likely to be created by China's private enterprise brands, as partly has been demonstrated by TECNO in Africa and MI in India and Russia. The COVID-19 is making a far-reaching impact on world consumption, production, as well as technical innovation. Private enterprises should focus further on providing healthier, greener, and more sustainable branded products and services to international consumers. A large quantity of internationally well-known corporate and product brands which are organically grown in China will substantially reinforce the positive image of "Made in China'. Meanwhile, domestic policy reforms should be carried on to provide a fairer and stabler environment for the development of private enterprises.

4 CONCLUSION

Well over one decade ago, Theresa Loo (2006) posited that branding China would be "the ultimate challenge in reputation management" [6]. In this dynamic and volatile world, there is no silver bullet for branding China, an emerging economic power that is beginning to have global impact and induce both admiration and suspicion at the same time. China sincerely conveys herself and her goodwill to other countries through nation branding, benefiting both herself and other stakeholders in economic interests. Bringing into full play the roles of all types of enterprises in branding China is definitely a step on the right track and is of strategic importance. Of course, branding China is an all-round endeavor and other nation branding initiatives should also be in place to attain the branding objectives.

ACKNOWLEDGEMENTS

The paper is sponsored by School of Economic Management, University of Science and Technology (Grant No. JGQN1704).

REFERENCES

[1] Anholt, S. 1998. Nation-brands of the twenty-first century. Journal of Brand Management, 5(6): 395–406.
[2] Peter D. 2003. Global shift: Reshaping the global economic map in the 21st century (Fourth Edition). London: SAGE Publications, pp. 7.
[3] Anholt, S. 2002. "Foreword" to the special issue on place branding, Journal of Brand Management, 9(4–5): 229–239.
[4] Khermouch, G., Brady, D., Holmes, S., Ihlwan, M., Kripalani, M., & Picard, J. 2003. Brands in an age of anti-Americanism. BusinessWeek Asian Edition, August 4: http://www.businessweek.com.
[5] Candace L. W. 2012. Brands and national image: An exploration of inverse country-of-origin effect. Place Branding and Public Diplomacy, 8(2): 110–118.
[6] Theresa L. & Gary D. 2006. Branding China: the ultimate challenge in reputation management? Corporation Reputation Review, 9(3):198–210.

Computational Social Science – Luo, Ciurea & Kumar (eds)
© 2021 Taylor & Francis Group, London, ISBN 978-0-367-70193-2

Research on China-Singapore trade relation under the strategy of "the Belt and Road Initiatives"

Z.Y. Zhang & Q. Luo
College of International Trade and Economics, Jilin University of Finance and Economics, Changchun, P.R. China

ABSTRACT: With the acceleration of the ASEAN integration process, trade exchanges be-tween China and ASEAN countries have become increasingly close. Cointegration and Granger causality were used to discuss the relationship between Singapore's FDI to China in this paper and Singapore's exports and imports to China. Data from 1984 to 2018 was taken to test whether Singapore's FDI to China is the Granger reason of Singapore's imports to China, and the long-term and short-term effects are inconsistent: short-term has negative effects, with long-term being positive effects. Singapore's imports to China have a short-term Granger positive effect on Singapore's exports to China, and Singapore's exports to China have a long-term Granger negative effect on Singapore's imports to China.

Keywords: FDI, Export and import, Granger causality test.

1 INTRODUCTION

Singapore is located at the southernmost-tip of Peninsular Malaysia, with an area of only 716.1 square kilometers, making it a veritable "small country". Taking advantage of its abundant port resources and superior geographical location, Singapore has actively developed an export-oriented economy and has been becoming one of the most developed countries in Asia. Because of the rapid development of the economy, Singapore has also expanded the scale of its foreign direct investment. Since 1982, Singapore has begun her direct investment to China. Singapore is also China's important economic and trade partner along "the Belt and Road Initiatives". During the construction of the Maritime Silk Road, the trade cooperation between China and Singapore continued to deepen.

Although facing with the current complex and changeable international situation, China and Singapore are located in the same Asian area, Singapore's direct investment in China still has unstable risk factors. Therefore, in this article, we use Singapore's direct investment in China to conduct an empirical study on China-Singaporean bilateral trade, aiming to create a better business environment for China and improve and perfect policy recommendations.

2 DATA SOURCE AND PROCESSING

The data in this article came from the "Foreign Economic and Trade Yearbook of China" and the United Nations Conference on Trade and Development database. For the selection of data, a time series is adopted, with the time length of 1984–2018, which lasted for 34 years. The objects of quantitative analysis are specifically the following three time series:

Singapore's relative exports to China:

$$x = \frac{(X_{sc}/GDP_s)}{(X_{sw}/GDP_w)} \tag{1}$$

Table 1. ADF unit root test.

variable	Inspection form	ADF	Prob.	Inspection variable	form	ADF	Prob.	Conclusion
$\ln fdi_t$	(cn0)	−5.18	0.00	$\Delta \ln fdi_t$	(nn0)	−6.35	0.00	Stationary
$\ln x_t$	(nn0)	0.44	0.80	$\Delta \ln x_t$	(nn0)	−8.08	0.00	Stationary
$\ln m_t$	(nn0)	−0.74	0.38	$\Delta \ln m_t$	(nn0)	−10.18	0.00	Stationary

(1) The test forms (C, t, K) represent intercept term, trend term and lag order respectively;
(2) Δ represents the first order difference;
(3) Prob. is the p value of MacKinnon single tail test.

Singapore's relative imports to China:

$$m = \frac{(M_{sc}/GDP_s)}{(M_{sw}/GDP_w)} \tag{2}$$

Singapore's relative fdi to China:

$$fdi = \frac{FDI_{sc}}{FDI_{sw}} \tag{3}$$

Where SC means Singapore to China; SW means Singapore to the whole world; S means Singapore; W means the whole world.

The above FDI, GDP, and import and export quotas are all priced in millions of U.S. dollars.

3 TIME SERIES STATIONARITY AND COINTEGRATION TEST

3.1 *ADF unit root test*

Because using the ordinary least squares (OLS) method directly on the selected data may cause the problem of pseudo regression, we use the ADF unit root test to test the stationary and single integral order of the variables. According to the method of selecting the optimal test form, Table 1 shows the results of ADF unit root test.

It can be seen that $\ln fdi_t$, $\ln x_t$, and $\ln m_t$ are all first-order single integral sequences, so the long-term balanced cointegration relationship between variables can be tested through the cointegration relationship.

3.2 *Cointegration relation test and selection of optimal error correction model*

This paper uses the Johansen cointegration relationship test method to investigate whether there is a cointegration relationship among $\ln fdi_t$, $\ln x_t$ and $\ln m_t$ The first step is to determine the optimal lag interval. The test showed that the cointegration optimal lag interval of the overall sample from 1984 to 2018 is 1–6, and the corresponding optimal lag order of VECM was 6. Second is to determine the optimal form of cointegration relationship test. Table 2 reported the results of the cointegration relationship test during the sample period. According to the statistical results, we chose the error correction model and the cointegration equation with a lag order of 6, "the quadratic trend in the sequence space, and the cointegration space including the intercept and have a linear trend" form of VECM. So, there is a unique long-term equilibrium co-integration relationship.

Table 2. Johansen cointegration test and identification of optimal error correction model.

Inspection Form	Sequence	no additional items	no additional items	linear trend	linear trend	quadratic trend
	Cointegration	no intercept no linear trend	intercept no linear trend	intercept no linear trend	intercept linear trend	intercept linear trend
Lags interval: 1 to 6	Trace	2	3	3	3	2
Series: lnfdi,lnx,lnm	Max-Eig	2	3	3	3	2
	AIC	−1.71	−1.67	−1.58	−3.48	−3.83*
	SC	1.14	1.22	1.40	−0.44	−0.69*

*Represents the optimal cointegration test form for final identification.

Table 3. Short term Granger causality test.

dependent variable independent variable	$\Delta \, \text{Lnfdi}_t$		$\Delta \, \text{Lnx}_t$		$\Delta \, \text{Lnm}_t$	
	Chi-sq	SE	Chi-sq	SE	Chi-sq	SE
midrule $\Delta \, \text{Lnfdi}_{t-1} \ldots \Delta \text{Lnfdi}_{t-6}$	–	–	1.22(0.97)	–	21.84(0.00)	−12.44
$\Delta \, \text{Lnx}_{t-1} \ldots \Delta \text{Lnx}_{t-6}$	5.90(0.43)	–	–	–	3.56(0.73)	–
$\Delta \, \text{Lnm}_{t-1} \ldots \Delta \text{Lnm}_{t-6}$	2.60(0.85)	–	12.61(0.04)	0.88	–	–

(1) Chi-sq is a square statistic and prob. is the corresponding adjoint probability;
(2) SE represents short term effect;
(3) If the short-term effect cannot pass the test at 0.1 level, the parameters of the lag term are omitted in the table.

4 GRANGER CAUSALITY TESTS

4.1 Short term Granger causality test

As can be seen from the above, lnfdi_t is the short-term Granger reason of lnm_t. Because its coefficient is negative, Singapore's FDI to China has a short-term negative effect on Singapore's imports from China. The inflow of FDI can help improve the overall technological level and production capacity of Chinese enterprises, thus expanding the scale and proportion of China's exports in high-tech products. However, Singapore's direct investment in China is affected by some factors such as labor proficiency, technological level differences etc, and there is a time lag, which cannot be directly reflected in manufactured products quickly. Therefore, Singapore's FDI to China has a short-term negative effect on Singapore's imports from China.

Second, lnm_t is the short-term Granger reason of lnx_t, and because its coefficient is positive, Singapore's imports to China have a short-term positive effect on Singapore's exports to China. To a certain extent, this shows that the trade exchanges between China and Singapore is getting closer. Under the strategy of "the Belt and Road Initiative", China and Singapore have continuously upgraded their trading partnership. The import and export exchanges between them have continued to develop. The continuous expansion of trade has caused a virtuous circle to continuous economic growth.

4.2 Long term Granger causality test

Through the analysis above, the following conclusions can be considered:

Table 4. Long term Granger causality test.

dependent variable / independent variable	Δ Lnfdi$_t$		Δ Lnx$_t$		Δ Lnm$_t$	
	F-stats	LE	F-stats	LE	F-stats	LE
ε_{t-1}	1.24(0.25)	–	0.1(0.93)	–	21.07(0.00)	–
$\varepsilon_{t-1}, \Delta$Lnfdi$_{t-1} \ldots, \Delta$Lnfdi$_{t-6}$	–	–	0.32(0.91)	–	3.45(0.06)	0.15
$\varepsilon_{t-1}, \Delta$Lnx$_{t-1} \ldots, \Delta$Lnx$_{t-6}$	0.95(0.53)	–	–	–	3.94(0.04)	−0.1
$\varepsilon_{t-1}, \Delta$Lnm$_{t-1} \ldots, \Delta$Lnm$_{t-6}$	0.43(0.85)	–	1.88(0.21)	–	–	–

(1) F is the F statistic and prob. is the corresponding adjoint probability;
(2) LE represents long term effect:
(3) If the long-term effect cannot pass the test at 0.1 level, the parameters of the lag term are omitted in the table.

Singapore's direct investment in China is the long-term Granger reason for Singapore's imports from China, and the long-term effect is positive. In the short term, due to differences in labor proficiency and technical level, Singapore's direct investment in China will to some extent inhibit Singapore's imports from China. In the long run, after reducing or even eliminating the difference between the labor proficiency and technological development level between China and Singapore, Singapore's direct investment in China will promote Singapore's imports from China.

The technology spillover and technology transfer effect of Singaporean companies can reduce or even eliminate the technological gap between China and international high-tech products, which brings the possibility for domestic companies' products to enter the Singapore market. Most Singaporean multinational companies are industries with certain comparative advantages, advanced management experience and production technology. In the long run, they will serve as a model for Chinese companies, improve the export competitiveness of domestic companies, and increase Singapore's influence on China. What's more, Singapore's direct investment in China will also produce a forward and backward industrial linkage effect, which will increase the level of the industrial chains, thereby increasing its external export capacity and increasing Singapore's import trade volume from China.

Singapore's exports to China is the long-term Granger reason for Singapore's imports to China, and the long-term effect is negative. This may be due to Singapore's domestic policy guidance. Pursuing a trade surplus is one of Singapore's important goals of foreign trade and economic relations. Therefore, maintaining a certain degree of trade surplus, that is, exports are greater than imports, which is more beneficial to Singapore's economic development.

5 RECOMMENDATION

In attracting Singapore's direct investment to China, and in order to improve the quality of the investment, first of all, attention should be paid to the selection of Singapore's investment to China. China should encourage Singaporean companies to invest in high-tech industries, environmental protection industries, and tourism industries areas and other ones. This will not only optimize the allocation of resources, but also promote the overall economic development of our country.

Secondly, it is that China should increase its attractive policies for Singapore's financial, communications, and service industries. These three industries are Singapore's advantageous ones. Direct investment to our country can promote the standardized development of our financial industry. We should vigorously develop Singapore's direct investment in China. This will not only promote China-Singapore bilateral economic and trade, but also accelerate our country's economic development.

The third is that, in the future, China's foreign direct investment can learn from Singapore's. Due to its historical development and geographical location, Singapore has a long history of developing foreign direct investment trade models. Foreign direct investment has become the "second wing" that promotes Singapore's economic growth. China should vigorously develop foreign direct investment in a targeted manner, especially in the context of the current "the Belt and Road Initiatives". China should be aware of avoiding blind investment, and focus on ASEAN-related countries. In these ways China can develop its unique ways of foreign direct investment.

ACKNOWLEDGEMENTS

This work was supported by Jilin Provincial Social Science Fund (2019N26, 2020J58).

REFERENCES

[1] Bo Zou, Jianhui Wang, Fushuan Wen.2017.Optimal investment strategies for distributed generation in distribution networks with real option analysis. IET Generation, Transmission & Distribution 11 (3).

[2] Busiswa Nxazonke, Roscoe Bertrum van Wyk.2020.The role of foreign direct investment (FDI) on domestic entrepreneurship in South Africa. Development Southern Africa 37 (4).

[3] Fertõ, Sass. 2020. FDI according to ultimate versus immediate investor countries: which dataset performs better? Applied Economics Letters 27 (13).

[4] Haiyun Liu, Mollah Aminul Islam, Muhammad Asif Khan, Md Ismail Hossain, Khansa Pervaiz.2020.Does financial deepening attract foreign direct investment? Fresh evidence from panel threshold analysis. Research in International Business and Finance 53.

[5] Kazuo Sato.1997.Economic development and financial deepening: The case of Japan. Journal of the Asia Pacific Economy 2 (1).

[6] Peter.A.Petri.2012.The determinants of bilateral FDI: Is Asia different? Journal of Asian Economics 23 (3).

[7] Ved Pal Sheera, Ashwani Bishnoi.2012.Financial deepening in newly industrialised economies of Asia. Int. J. of Economic Policy in Emerging Economies 5 (2).

[8] Yudan Zheng.2013.The effects of option incentives on research and development investment. Int. J. of Applied Management Science 5 (1).

Computational Social Science – Luo, Ciurea & Kumar (eds)
© 2021 Taylor & Francis Group, London, ISBN 978-0-367-70193-2

A critical analysis of China's investments, redundant resources and economic growth in Nigeria

S.A. Imanche
College of Business Administration -Business School of Hohai University, Nanjing, P. R. China
Department of Economics, Faculty of Humanities Management and Social Sciences, Federal University Wukari, Nigeria

Z. Tian* & O.T. Tasinda
College of Business Administration -Business School of Hohai University, Nanjing, P. R. China

ABSTRACT: Recent years have brought astronomical growth of China-Nigeria economic relationships. The relationship increased monumental developmental economic changes in Nigeria, with China growing investments in the country. This paper explored the in-vestment opportunities of China in Nigeria, the volume and nature of Chinese investment in Nigeria, untapped resources in Nigeria, and the effect of the Chinese investments on the Nigerian economy. We adopted secondary data from the Bureau of Statistics, China and the Nigerian Central Bank, through UN COMTRADE. A quantitative analysis method was used for statistical inferences and conclusions. The study findings indicated that the industry which has highly dominated the Nigerian economy as a primary source of income is crude oil. Despite the considerable investment opportunities in Nigeria, Chinese FDI in the Nigerian economy is lower than that of other African nations. This paper concluded and recommended that Nigeria engage more with China to invest more in the country to improve the nation's economic growth and development.

Keywords: China investment, Redundant resources, Economic growth, China, Nigeria.

1 INTRODUCTION

Nigeria, through foreign direct investment (FDI), has been the primary beneficiary of Chinas' investments in Africa. These investments have increased the required capital for businesses, technology, and managerial expertise and skills to enhance business productivity. There have been huge questions surrounding foreign countries' relationships with African countries claiming that it is a mercantile association of business moguls and politicians. However, china's business expansion into Nigeria is not weighing against the nation but has caused significant economic development through considerable business investment and help in the country. According to the Chinese commerce ministry, the primary aim of the Chinese government policy on Nigeria is increasing Chinese multinational organizations in Nigeria's market share. Also, the Chinese government strategy of commerce aimed to enlarge Nigeria's market for produced commodities in China. China sought to enhance existence in the oil and gas sectors of Nigeria by increasing its investment in Nigeria as a strategy of venturing into the ECOWAS market. This concern was raised by the developed economic complementarities between Nigeria and China (Asongu & Ssozi, 2016; Dong & Fan, 2017; Xia, 2020).

Nigeria is considered to be a nation that is richly endowed by natural and human resources. Nigeria, with a population of over 200 million, has an internal market that does not have competitors with the African continent. Nigeria is a member of various global agencies such as the U.N., WTO, A.U.,

---

*Corresponding author

and ECOWAS. Nigeria also has about 2000 industrial developments such as iron complexes, large oil industries, steel rolling industries, up-coming Export Processing Zone (EPZ), food processing, car assembling, and pharmaceutical sectors (Alabi et al., 2018). From the opportunities mentioned above makes Nigeria a potential country for foreign investment from developed countries and have resources redundancy in Coal mining (Olade, 2019), Oil, Gas, and Energy sector (Ike, 2019), Telecommunication (Arawomo & Apanisile, 2018), Agriculture, forestry, and fisheries (Adeleye et al., 2020; Iruo et al., 2018).

In 1971, bilateral interaction was developed between Nigeria and China. During this time, regardless of being opposed by America, Nigeria, together with other developing nations in Asia, Latin America, and Africa, supported the Beijing 21-year campaign on the election of world recognition as a real Chinese government (Pay & Nwosu, 2020). China positioning towards Nigeria reignited again in 2000 after the election of Olusegun Obasanjo as a civilian president in 1999. In October 2000, Nigerian representatives attended the first ministerial Conference on China-African partnership in Beijing. During that period, CCECC was granted a contract to construct 5,000 housing units for athletes who participated in the eighth yearly All-African games hosted in Abuja (Ighodaro, 2018b). In 2001, China and Nigeria signed a protocol to set up the Nigeria Trade Office in China, Trade Promotion Center, and also China Investment Development in Nigeria (Ighodaro, 2018a; Omotoso et al., 2020). Details of binding agreements between Nigeria and China can be extracted from the Nigerian trade hub (Nigeria Trade Hub, 2015); (Egbula & Zheng, 2011) and (Lola & Devadason, 2018).

In the last decade, African nations contributed one-third of the entire oil imports to China. Despite Nigeria being the 6th biggest crude oil supplier in Africa, it contributed 2% of Chinese imports from Africa. In 2010 export to China accounted for 1.6% of oil exports of Nigeria, with the U.S. providing 58.7% and Brazil 11% (Obi, 2019). There were traditional trade partners who acquired blocks in negotiations with the Nigerian government. Various oil organizations from Taiwan, China, and India were provided with RFR on pre-assigned blocks, CNPC Chinese Company was given four blocks with a commitment of investing USD 2 billion for rehabilitating the aging Kaduna oil refinery. Chinese actors obtained bidding grounds and other additional assets in the Obasanjo's tenure. Nigeria incurred a loss of USD 10 billion due to the failure to deal with Asian nations. The crash was because no follow-up strategies were implemented to strengthen the deals, and many of the infrastructure projects were not or only partially developed. However, the imports from China significantly increased by 2.7% yearly to USD 167.15 billion in June 2020, outcompeting the agreement of a 10% decline in the market after the 16.7% decline in May 2020. This indicated the first increment in the inbound movement in 2020 amid the strategies to fight the COVID-19 pandemic. The copper imports in China increased by 101.4% each year, hitting the amount of 656,483 tonnes in June 2020 with the rising of production activities and the decrease of scrap. Iron ore sales increased by 35.3% to the highest from October 2017 to 101.68 million tonnes. Crude oil imports rose to 53.18 million tonnes, equivalent to 12.9 million barrels daily.

China has established one of the most significant and ambitious construction sectors centered on unique civil works skills across the globe. The construction industries are essential for the development of infrastructure in Nigeria and can provide crucial financial aid for African countries like Nigeria. The construction activities have helped address the main development challenge of infrastructural deficiency in Nigeria as the country's investment has expanded. The drive of industrialization of China and the intense FDI inflow into China resulted in the fast development of the manufacturing economy, requiring minerals and oil inputs (Obi, 2019). These inputs outstrip China's domestic resources; thus, the country must acquire these inputs from abroad like Nigeria, which has abundant natural resources. For example, the impact of cessation on importation from China is evident in the Nigerian economy during the pandemic (Adiyoh et al., 2020). Therefore, foreign investment constitutes a significant channel that can lead to the achievement of economic growth. Given these various questions raised, such as:

– Why is China interested in investing in Nigeria?
– Does the total output acquired by China target external or domestic market?
– Does China's FDI target Nigerian business sectors?

Table 1. Autocorrelation Function Test Results.

	GDP	China FDI	Exchange Rate	Trade Volume
GDP	1.00			
China FDI	0.91	1.00		
Exchange rate	0..45	0.21	1.00	
Trade Volume	0.55	0.71	0.53	1.00

2 METHODOLOGY

The study used quantitative analysis utilizing three-step processes, including Unit Root Test, Auto Correlation Function, and Granger Causality test in determining the China FDI inflow into Nigeria and its impact on the Nigerian economy. The study used information analysis of relevant literature reviews from different researchers in selecting macroeconomic valuables for the study findings. The study used annual data of 2003–2019, and the exchange rate figures and GDP were obtained from the Nigerian Central Bank. The Trade Volume between Nigeria and China data and China's outward FDI inflow to Nigeria was obtained from China's National Bureau of Statistics.

$$GDPt = a1 + a2FDIt + a3TV + a4ER + \varepsilon \qquad (1)$$

Where:
GDP = Gross Domestic Product
T.V. = Trade Volume between China and Nigeria
FDI = China outward FDI flow
ε = is the stochastic random term
ER = Exchange Rate

The Autocorrelation analysis was carried out to evaluate the level of correlation in the variable used with delayed data in similar categories. It is known as the cross-correlation outcome of the mark. The autocorrelation function was utilized in the signal processing method for analyzing a series of accurate information like time-domain data. Autocorrelation refers to the condition taking place when the following products in a series correlate. When the mean value is not zero, it shows there was no independent.

Augmented Dickey-Fuller test and Philips-Perron test was employed for checking if the data values are integrated and have a unit root. The study used stationarity testing for the four times series. Variables that have unit root are considered to be non-static in level form, but they change to static level after they are distinguished. These variables are known as integrated of order one and are denoted by 1 (Adu et al., 2018). Granger Causality test was carried out in the study to help in identifying a causal association between the variables used and in determining if the current lagged values of the variables affect one another.

3 RESULTS AND DISCUSSION

3.1 Autocorrelation function test results

First, the selected variables used in the study were tested for serial correlation of the current values of variables and their lagged values for all the time series. An additional correlation test was also estimated to assess the magnitude of correlation among the selected endogenous variables. Table 1 shows the ACT results.

The table shows that there are trends in the series of variables overtime that are correlated but offset overtime time. The results indicate that there is a robust association with the variables, and the results can be used in generalizing the findings.

Table 2. Test Result showing Granger Causality.

Tested Hypothesis	Obs	F-Stats	Prob.
Exchange Rate has no cause effect on China FDI Inflow	17	1.37664	0.4674
China FDI Inflow does not granger-cause Exchange Rate	17	1.07854	0.23675
GDP does not granger-cause China FDI Inflow	17	5.89547	0.0078
China FDI Inflow does not granger-cause trade volume	17	17.6997	0.0058
Trade Volume does not granger-cause China FDI Inflow	17	7.34796	0.0476
China FDI Inflow does not granger-cause Trade Volume	17	5.89646	0.0045
GDP does not granger-cause Exchange Rates	17	1.3585	0.7578
Exchange Rates does not granger-cause GDP	17	6.57636	0.3678
Trade Volume does not granger-cause Exchange Rates	17	2.78547	0.6794
Trade Volume does not granger-cause Exchange Rates	17	3.57585	0.9685
Trade Volume does not granger-cause GDP	17	9.47474	0.0022
GDP does not granger-cause Trade Volume	17	3.78855	0.6587

3.2 Granger causality test result

Table 2 provides estimation results for the granger causality test for China FDI, Nigeria GDP Exchange rate, and bilateral trade between the two countries. The result shows two-way causation between China FDI and bilateral trade between the two partners. The coefficients of China FDI and trade volume between the two countries are positive and statistically significant at 1% and 5% significance levels. The results imply a significant amount of trade intensity results inflow of Chinese FDI into Nigeria. In the same vein, Chinese FDI into Nigeria promotes increased bilateral trade. Also, the study found unidirectional short-run causation running from GDP to China FDI into Nigeria. The coefficient of GDP granger, causing China FDI, was positive and significant at the 1% level. The result implies that Chinese FDI inflows into the Nigerian market are determined by growth in Nigeria's GDP. GDP Is measures economic power and the market size of a country. The one-way causality may be attributed to Nigeria's market size as China seeks an expanded market for its export. The relationship between trade volume between China and Nigeria and Nigeria's GDP was found to be unidirectional, with a short run association running from trade volume to GDP. The result reveals the role of bilateral trade between the countries' impact on the Nigerian economy; as trade between the two countries increases in particular export from Nigeria to China, Nigeria's GDP increases.

The study found no causality between China's FDI inflows and exchange rate in Nigeria. The no causality effect may be attributed to trade imbalances between the two trading partners. China exports more to Nigeria and, in return, receives payment in dollar or yuan; it is in a favorable position trading with Nigeria. Further, massive Chinese investment in Nigeria is in the oil sector, where global market forces determine oil prices. Therefore, the Chinese presence in Nigeria is not influenced by exchange rate volatility due to China's nature of involvement in the country. GDP and exchange rate show no short-run causative effect in the study.

3.3 Unit root test results

The study conducted a stationarity test on the selected variables to determine if the variables are stationary at the level and integrated at order one using unit root testing. This was done after the autocorrelation test. The Augmented Dickey-Fuller (ADF) and Phillip-Peron tests were employed to test for unit roots in the selected variables. The results from both Phillip-Peron and ADF indicate the results were not stationary under each level. The first Difference of each selected variable was taken and checked for stationarity. The variables became stationary after first differencing and integrated at order one, as shown in Tables 3 and 4. The research results acquired suggest that the variables used significantly depict stationarity.

Table 3.　Augmented Dickey-Fuller (ADF) Stationarity Test at Level.

Variables	Level	1st difference	Integration order
China FDI	−2.840 * **	−6.542 * **	I(1)
GDP	−1.023	−4.104 * *	I(1)
Exchange rate	−1.309	−3.516 * **	I(1)
Trade volume	−1.745	−3.086 * *	I(1)

Note: ***0.01, **0.05, *0.1, Critical value at 1%: −3.750; critical value at 5%: −3.000; critical value at 10%: −2.630.

Table 4.　Philips Perron (P.P.) Stationarity Test at Level.

Variables	Level	1st difference	Integration order
China FDI	−4.168 * **	−7.595 * **	I(1)
GDP	−1.734	−3.267 * *	I(1)
Exchange rate	−1.239	−3.517 * **	I(1)
Trade volume	−1.8564	−3.895 * *	I(1)

Note: ***0.01, **0.05, *0.1, Critical value at 1%: −3.750; critical value at 5%: −3.000; critical value at 10%: −2.630

The study findings suggest that GDP contributing effect on the Exchange rate. Similarly, the exchange rate has no contributing effect to GDP. Thus, the two variables have no causal association because of structural imbalance in Nigeria's economy, instability in income from oil, which the nation highly depends on, lack of diversification, and the concept of high demand in foreign exchange related to supply. The study findings also show that unidirectional association exists in trade Amount between Nigeria and China and GDP. Hence, the trade volume between China and Nigeria is significantly beneficial. Lopsidedness exists in the trade amount of the commodities Nigeria imports from China than in the items China imports from Nigeria. Thus, the gap existing between the two countries should be bridged to enhance trade volume between Nigeria and China. Nigeria is highly benefitting from the trade association as the Chinese products contribute to the increased level of consumption in Nigeria as China supplies quality and affordable commodities in the country.

There are various benefits accrued from Chinese FDI into Nigeria, constituting augmentation of domestic capital, technology transfer, skills and expertise, innovation and competition promotion, improved output, employment, revenue performance, and export. The benefits of FDI in a country need to be evaluated against their costs like restrictive business practices, anti-competitive and restrictive trade, abusive transfer pricing, tax evasion, and strained investment flow. Also, excessive impact on economic incidents with adverse effects on the development of industries and country security, transfer of pollution activities, and techs and tax avoidance should be put into consideration. The findings indicate that countries hosting FDI should implement instituted policies that aim at optimizing direct and indirect benefits and optimizing the potential adverse effects. A litmus test should be done to gauge FDI motive for classifying the investments into the market sourcing, resource sourcing, and efficiency-seeking. Efficiency seeking FDI should be considered than the other forms of FDI from the host nation's perspective. For countries to entice efficiency-seeking Foreign Direct Investment, they should ensure macroeconomic stability and implement a predictable, unique, and easy-to-access program environment, and other motives should be developed

However, considering the private FDI and industrial concentration, efficiency incentives should be used as the enforcing element of Chinese FDI inflow into the Nigerian economy. Nigeria should consider using resource-seeking motives, but some FDI categories cannot fit in establishing infrastructure and manufacturing as they are categorized as market seeking. A joint venture in FDI should be employed because it has tremendous potential benefits to the host country's economy. Despite Chinese companies contributing to the growth of the Nigerian economy, these companies have been criticized with claims that the companies were being closed and did not hire local workers. The employment conditions of Nigerians in Chinese companies did not comply with the Nigerian labor laws and the global labor agency. Some of the Chinese companies like Wahum Nigeria Limited have the most inhuman working conditions like 12 hours shifts and many casual employees. Also, the technology transfer from China's FDI was insignificant in Nigeria because most of the Chinese companies brought finished commodities in Nigeria and complete equipment and machinery with Chinese technicians (Casadella & Liu, 2019). This insignificance indicates that host countries for FDI need to implement appropriate regulations and policies to ensure they fully benefit from the FDI.

4 CONCLUSIONS

The study indicates that China's FDI inflows have beneficial impacts on the economic growth of Nigeria. However, the Nigerian government needs to optimize the complementary effects between Nigeria and China. This optimization can be achieved through strengthening high Cooperation and offer an excellent aggregate economic condition in the nation for assurance on higher sustainability and productivity of foreign investment. Also, the Nigerian central administration should invest in inflow inputs from the region with boom in enhancing the financing environment, implement societal contributions essential for supporting investment. Nigerian government should create development banks needed for providing economic and financial support to private investors in the country.–

Despite the high possibilities of Nigeria to acquire higher value from the Chinese FDI and growing influence, the citizens of Nigeria have not entirely focused on these possible advantages. This indicates that more approaches need to be employed to increase the implementation of policies, the establishment of the institution, entrepreneurship, human capital, leadership, and cultural abilities to optimize benefits from the FDI. Nigeria should implement a broad approach to allow active balancing of the involvement of China in the country to support the country's strength and establish a policy for credible development which supports the motives of the Nigerians.

Ethical Approval

No ethical approval was required for this study. The study explored public data.

Funding

Major Social Science Research Projects of Universities in Jiangsu Province'F'sResearch on the Mechanism and Implementation Path of Jiangsu to Create an Independent and Controllable Advanced Manufacturing System't'iNumber2019SJZDA055 and Supported by The fundamental research funds for the central universities'inumber2019B35814.'j

Disclosure of Interest

The authors declare no competing interest.

REFERENCES

Adeleye, N., Adeogun, S., Osabuohien, E., & Fashola, S. (2020). Access to Land and Food Security: Analysis of "Priority Crops" Production in Ogun State, Nigeria (pp. 291–311). https://doi.org/10.1007/978-3-030-41513-6_14

Adiyoh, S., Ze, T., Tougem, T. O., & Dalibi, S. G. (2020). Effect of COVID-19 Pandemic on Small and Medium Scale Businesses in Nigeria. 10. https://doi.org/10.47119/IJRP100561720201305

Adu, O., Edosomwan, O., Abiola, B., & Olokoyo, F. (2018). Industrial development and unemployment in Nigeria: An ARDL bounds testing approach. International Journal of Social Economics. https://doi.org/10.1108/IJSE-10-2017-0448

Alabi, M., Tella, S., Odusanya, I., & Yinusa, O. (2018). Financial Deepening, Foreign Direct Investment and Output Performance in Nigeria. Scientific Annals of Economics and Business, 65. https://doi.org/10.2478/saeb-2018-0007

Arawomo, O., & Apanisile, J. (2018). Determinants of Foreign Direct Investment in the Nigerian Telecommunication Sector. Modern Economy, 09, 907–923. https://doi.org/10.4236/me.2018.95058

Asongu, S., & Ssozi, J. (2016). Sino-African Relations: Some Solutions and Strategies to the Policy Syndromes. Journal of African Business, 17, 35–51. https://doi.org/10.1080/15228916.2015.1089614

Casadella, V., & Liu, Z. (2019). Chinese Foreign Direct Investment (FDI) and Barriers to Technology Transfer in Sub-Saharan Africa: Innovation Capacity and Knowledge Absorption in Senegal. In Globalization and Development (pp. 219–240). Springer.

Dong, Y., & Fan, C. (2017). The effects of China's aid and trade on its ODI in African countries. Emerging Markets Review, 33, 1–18. https://doi.org/10.1016/j.ememar.2017.09.003

Egbula, M., & Zheng, Q. (2011). China and nigeria: A powerful south-south alliance. West African Challenges, 5, 1–20.

Ighodaro, C. A. (2018a). FOREIGN DIRECT INVESTMENT - LED INDUSTRIALISATION IN NIGERIA. Uongozi Journal of Management and Development Dynamics, 28(2), Article 2. https://doi.org/10.2234/uongozi.v28i2.31

Ighodaro, C. A. (2018b). Foreign Direct Investment-Led Industrialisation in Nigeria: Lessons from Chinese Industrialisation. Journal of Management and Development Dynamics, 28(2), 69.

Ike, A. N. (2019). Green Growth and Sustainable Economic Development in Nigeria: Benefits and Challenges. International Journal, 6(1).

Iruo, F. A., Onyeneke, R. U., Eze, C. C., Uwadoka, C., & Igberi, C. O. (2018). Economics of smallholder fish farming to poverty alleviation in the Niger Delta Region of Nigeria. Turkish Journal of Fisheries and Aquatic Sciences, 19(4), 313–329.

Lola, G. K., & Devadason, E. S. (2018). The Engagement of China in Nigeria's Oil Sector: Is the Transformation Positive? Contemporary Chinese Political Economy and Strategic Relations: An International Journal, 4(3), 1025–1060.

Nigeria Trade Hub. (2015). <http://www.nigeriatradehub.gov.ng/>. http://www.nigeriatradehub.gov.ng/

Obi, C. (2019). The Changing Dynamics of Chinese Oil and Gas Engagements in Africa. China-Africa and an Economic Transformation, 173.

Olade, M. A. (2019). Solid Mineral Deposits and Mining in Nigeria: A Sector in Transitional Change. Achievers Journal of Scientific Research, 2, 1–16.

Omotoso, F., Kuti, G., & Oladeji, I. O. O. (2020). Assessment of the (Inter) Dependency of Economic Relations between Nigeria and China: 1999–2019. Assessment, 7(2).

Pay, V. N., & Nwosu, E. (2020). China's Engagement with Africa: Promoting Rentierism? A caSe Study of Sino-Nigerian Economic Relations. Asian Affairs, 1–18.

Xia, Y. (2020). Chinese Investment in Africa: An Empirical Investigation of Trends, Dynamics, and Regulatory Challenges. Handbook of International Investment Law and Policy, 1–31.

Computational Social Science – Luo, Ciurea & Kumar (eds)
© *2021 Taylor & Francis Group, London, ISBN 978-0-367-70193-2*

Application of fuzzy comprehensive evaluation method in university food safety management evaluation

S. Wang, J.Q. Zhang & X.B. Chen
School of Electronics and Information Engineering, University of Science and Technology Liaoning, Anshan, China

H. Li*
School of Business Administration, University of Science and Technology Liaoning, Anshan, China

ABSTRACT: In order to prevent diet-related diseases, canteens should be evaluated. The purpose of this study is to analyze the food safety of university canteens. Our research interviewed 600 college students. According to the four characteristics of canteen staff: personal hygiene, restaurant hygiene, customer satisfaction, and service quality. In this paper, a fuzzy comprehensive evaluation model is used to establish a university canteen evaluation system based on the level of sub-elements, and a food safety score is determined. The simulation result shows that the evaluation result is 85.7730 points, which is very satisfactory. The evaluation result is consistent with the actual situation. Verify the feasibility and effectiveness of the model.

Keywords: Fuzzy comprehensive evaluation, Food safety, Analytic hierarchy processes.

1   INTRODUCTION

Food safety has always attracted the attention of governments and consumers. Therefore, university food safety is considered to be very important (Andaleeb & Caskey 2007). However, little research has been done on universities. Food safety is emphasized in complex situations (Pingali et al. 2005).

Campus food services are dynamic and complex. Food service attributes have become an important part of the quality of life on campus (El-Said & Fathy 2015). Price et al. make customer satisfaction an important factor in campus catering services (Price et al. 2016). Serhan studied the overall satisfaction of canteen customers with campus catering services (Serhan 2019).

Previous research has focused on canteen food influence on student health (Lugosi 2019). There have been many researches conducted on canteens in elementary schools, but university canteens were not involved (Jones et al. 2014). Studies have shown that quality of service factors are equally important (Inkumsah 2011). This research aims to solve this problem through the evaluation method of this article: food safety.

2   MATERIALS AND METHODS

2.1   *Modeling fuzzy comprehensive evaluation algorithm*

The food safety evaluation system is shown in Figure 1.

The target layer contains food safety. The function layer contains staff personal hygiene ($B1$), restaurant hygiene ($B2$), customer satisfaction ($B3$), and service quality ($B4$). The Index layer contains employees wear work caps and masks ($C1$), employees wear or promote health certificates

---

*Corresponding author

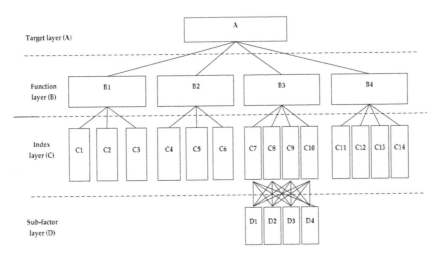

Figure 1.   Food safety evaluation system.

Table 1.   C1–C14 membership.

Membership	Very satisfied	Satisfied	Basic satisfaction	Dissatisfied
C1	0.4500	0.3962	0.1423	0.0115
C2	0.3923	0.4077	0.1731	0.0269
C3	0.3692	0.4385	0.1731	0.0192
C4	0.3462	0.3654	0.2692	0.0192
C5	0.4192	0.3269	0.2308	0.0231
C6	0.3923	0.3538	0.2269	0.0269
C7	0.3462	0.3692	0.2654	0.0192
C8	0.3923	0.3923	0.2000	0.0154
C9	0.3462	0.3231	0.3000	0.0308
C10	0.3462	0.3385	0.2885	0.0269
C11	0.3692	0.3385	0.2731	0.0192
C12	0.4154	0.2692	0.2923	0.0231
C13	0.3923	0.3115	0.2846	0.0115
C14	0.3692	0.3115	0.2923	0.0269

($C2$), and employees often wash work clothes ($C3$). Canteen floor walls and doors are free from dust ($C4$), canteens are cleaned immediately after meals ($C5$), and canteens are well lit and well ventilated ($C6$). The rice noodle oil used in the food is qualified by the national food safety test ($C7$), the food is fresh ($C8$), the food has no expired ingredients ($C9$), and the meat, poultry, and eggs used in the food are quarantine-qualified product ($C10$). Canteen catering staff ($C11$). The canteen handles student feedback in a timely manner ($C12$). The canteen seeks students' opinions ($C13$). Information on the purchase price and operation of main raw materials ($C14$). The data are shown in Table 1. The sub-factor layer contains the price of food is stable ($D1$), the main food and non-staple food price list ($D2$), the main and non-staple food categories are complete ($D3$), and the distinctive flavor stalls are distinctive ($D4$). It is necessary to use the fuzzy comprehensive evaluation algorithm (Zhang et al. 2019). We use analytic hierarchy process to determine the weight, and use fuzzy evaluation analysis to determine the target in the evaluation index. The final evaluation results between 85–95 grades are very satisfactory, 75–85 are satisfied, 55–75 are basically satisfied, and those below 55 are not satisfied. The data are shown in Table 2.

Table 2. D1–D4 membership.

Membership	Very satisfied	Satisfied	Basic satisfaction	Dissatisfied
D1	0.4154	0.3692	0.1769	0.0385
D2	0.3538	0.3769	0.2115	0.0577
D3	0.3692	0.3692	0.2462	0.0154
D4	0.3462	0.3692	0.2615	0.0231

Table 3. Weight calculation and consistency check results.

Factors	weight	λ_{max}	CI	CR
B1	0.1320	4.0328	0.0109	0.0123
B2	0.1320			
B3	0.5848			
B4	0.1513			
C1	0.6000	3,0000	2.22E-16	3.83E-16
C2	0.2000			
C3	0.2000			
C4	0.1047			
C5	0.2583	3.0385	0.0193	0.0332
C6	0.6370			
C7	0.1037			
C8	0.1912	4.1475	0.0492	0.0552
C9	0.5528			
C10	0.1522			
C11	0.0802	4.1155	0.0385	0.0432
C12	0.4246			
C13	0.2820			
C14	0.2132			

2.2 Application of fuzzy comprehensive analytic hierarchy process in food safety management evaluation

This article constructs a reasonable comparison matrix A according to the evaluation scoring criteria obtained by the scoring system. We have obtained the judgment matrix and then performed a consistency check of the data (Lin et al. 2018). Finally, the result we get is shown in the Table 3

3 RESULTS AND DISCUSSION

The comprehensive evaluation model is established. The calculation formula is established as in Equation (1)

$$S = A \circ R = \begin{pmatrix} a1 & a2 & ... & am \end{pmatrix} \circ \begin{bmatrix} r11 & r12 & ... & r1n \\ r21 & r22 & ... & r2n \\ ... & ... & ... & ... \\ rm1 & rm2 & ... & rmn \end{bmatrix} \tag{1}$$

In this paper, a sub-factor layer is added to the index layer, and the evaluation results of the sub-factor layer are calculated. *RW3* is a weight for food service.

$RW3 = A1 \circ D1$

$$= \begin{bmatrix} 0.4154 & 0.3692 & 0.1769 & 0.0385 \\ 0.3538 & 0.3769 & 0.2115 & 0.0577 \\ 0.3692 & 0.3692 & 0.2462 & 0.0154 \\ 0.3462 & 0.3692 & 0.2615 & 0.0231 \end{bmatrix} \circ \begin{bmatrix} 0.3462 & 0.3692 & 0.2654 & 0.0192 \\ 0.3923 & 0.3923 & 0.2000 & 0.0154 \\ 0.3462 & 0.3231 & 0.3000 & 0.0308 \\ 0.3462 & 0.3385 & 0.2885 & 0.0269 \end{bmatrix}$$

$$= \begin{bmatrix} 0.3432 & 0.3684 & 0.2483 & 0.0201 \\ 0.3635 & 0.3663 & 0.2494 & 0.0207 \\ 0.3632 & 0.3659 & 0.2501 & 0.0208 \\ 0.3632 & 0.3650 & 0.2508 & 0.0210 \end{bmatrix}$$

Calculate index layer evaluation results.

$$S_1 = \begin{bmatrix} 0.6000 & 0.2000 & 0.2000 \end{bmatrix} \circ \begin{bmatrix} 0.4500 & 0.0362 & 0.1423 & 0.0115 \\ 0.3923 & 0.4077 & 0.1731 & 0.0269 \\ 0.3692 & 0.4385 & 0.1731 & 0.0192 \end{bmatrix}$$

$$= \begin{bmatrix} 0.4223 & 0.4070 & 0.1546 & 0.0161 \end{bmatrix}$$

$$S_2 = \begin{bmatrix} 0.3944 & 0.3481 & 0.2323 & 0.0251 \end{bmatrix}$$

$$S_3 = \begin{bmatrix} 0.3632 & 0.3650 & 0.2508 & 0.0210 \end{bmatrix}$$

$$S_4 = \begin{bmatrix} 0.3953 & 0.2957 & 0.2886 & 0.0203 \end{bmatrix}$$

Form a matrix called a function layer, remember R:

$$R = \begin{bmatrix} 0.4223 & 0.4070 & 0.1546 & 0.0161 \\ 0.3944 & 0.3481 & 0.2323 & 0.0251 \\ 0.3632 & 0.3650 & 0.2508 & 0.0210 \\ 0.3953 & 0.2957 & 0.2886 & 0.0203 \end{bmatrix}$$

Calculate the result of the final target layer:

$$S = \begin{bmatrix} 0.1320 & 0.1320 & 0.5848 & 0.1513 \end{bmatrix} \circ \begin{bmatrix} 0.4223 & 0.4070 & 0.1546 & 0.0161 \\ 0.3944 & 0.3481 & 0.2323 & 0.0251 \\ 0.3632 & 0.3650 & 0.2508 & 0.0210 \\ 0.3953 & 0.2957 & 0.2886 & 0.0203 \end{bmatrix}$$

$$= \begin{bmatrix} 0.3800 & 0.3585 & 0.2409 & 0.0206 \end{bmatrix}$$

According to the score of the evaluation set, the final value score can be obtained as V

$$V = \begin{bmatrix} 0.3800 & 0.3585 & 0.2409 & 0.0206 \end{bmatrix} \circ \begin{bmatrix} 95 \\ 85 \\ 75 \\ 55 \end{bmatrix} = 85.7730$$

According to the principle of maximum membership, the final result of the evaluation is very satisfactory. The evaluation result of membership degree is 0.3800. According to the evaluation result, the evaluation is 85.7730 points. This shows that the food safety of the cafeteria is very satisfactory. Explain that the respondent has a sufficient level of knowledge in the fields of consumer services, hygiene, and food safety. Our research shows that the respondents' ratings are consistent with the canteen reviews received during the inspection. In the field of human health, the maximum degree of membership is 0.4223, indicating that the evaluation results are very satisfactory. The maximum hygiene of the restaurant is 0.3944, and the evaluation result is very satisfactory. The results of the service quality assessment are very satisfactory.

4 CONCLUSIONS

Based on fuzzy analytic hierarchy process algorithm, a four-layer food safety evaluation model is proposed. This model is used to obtain the results of food safety assessments in a university canteen. The model helps decision makers make accurate decisions and determine the quality of indicators. First, this article proposes to introduce a sub-element layer on the basis of the three-level indicator system. Based on the index layer, this article introduces the sub-element layer of the catering service, thereby introducing the affiliation between $B3$ and $D1 - -D4$. Second, the evaluation results of this article are displayed in the form of a percentage system, and the results are simple. The evaluation result was 85.7730 points, and the evaluation result was very satisfactory. Third, staff hygiene, canteen hygiene, and university canteen service were satisfactory in the evaluation results. However, the canteen service factor is at a satisfactory level, which indicates that canteen service has received more attention in the canteen assessment factor.

ACKNOWLEDGMENTS

The research reported herein was supported by the NSFC of China under Grants Nos. 71571091, 71771112, and 71371092. University of Science and Technology Liaoning Talent Project Grants No. 601011507-03.

REFERENCES

Andaleeb, S.S. & Caskey, A. 2007. Satisfaction with food services: Insight from a college Canteen. Journal of Foodservice Business Research. 10(2), 51–65.

El-Said, O.A., Fathy, E.A. 2015. Assessing university students' satisfaction with on-campus canteen services. Tour. Manag. Perspect. 16: 318–324.

Inkumsah, W. A. 2011. Measuring customer satisfaction in the local Ghanaian restaurant industry. European Journal of Business and Management. 2(3): 153–166.

Jones, B. A., Madden, G. J., Wengreen, H. J., Aguilar, S. S., & Desjardins, E. A. 2014. Gamification of dietary decision-making in an elementary-school canteen. PloS One. 9(4).

Lugosi, P. 2019. Campus foodservice experiences and student wellbeing: an integrative review for design and service interventions. International Journal of Hospitality Management. 83:229–235.

Pingali, P., Alinovi, L., et a. 2005. Food Safety in Complex Emergencies: Enhancing Food System Resilience.

Price S, Viglia G., Hartwell H. et al.. 2016. What are we eating? Consumer information requirement within a workplace canteen. Food Quality and Preference. 53: 39–46.

Serhan, M. 2019. The Impact of Food Service Attributes on Customer Satisfaction in a Rural University Campus Environment. International Journal of Food Science. 1–12.

Zhang, J., Chen, X., & Sun, Q. 2019. An Assessment Model of Safety Production Management Based on Fuzzy Comprehensive Evaluation Method and Behavior-Based Safety. Mathematical Problems in Engineering. 1–11.

Computational Social Science – Luo, Ciurea & Kumar (eds)
© 2021 Taylor & Francis Group, London, ISBN 978-0-367-70193-2

Research on the problems and countermeasures of scientific research base management mode in colleges and universities

L. Li
Capital University of Economics and Business, China

ABSTRACT: Scientific research management in Colleges and universities can better adapt to the development of modern society and actively respond to the challenges brought by the development of the times. At the same time, it can also effectively seize the development opportunities and realize the innovative development of scientific research management in Colleges and universities. Based on this, this paper systematically studies the cur-rent situation of scientific research management mode in Colleges and universities, and puts forward the Optimization Countermeasures of scientific research management mode in Colleges and universities, so as to construct a new mode of scientific research management in Colleges and universities.

1 GENERAL INSTRUCTIONS

The rapid development of modern information technology has brought great challenges and opportunities for the management of scientific research in Colleges and universities. The scientific research management departments of colleges and universities should change their working thinking and concepts in time, conform to the times and the development of science and technology, and pay attention to the optimization and innovation of the management mode of scientific research bases in Colleges and universities, so as to establish a new management mode of scientific research bases in Colleges and universities, lay a solid foundation for the development of scientific research in Colleges and universities, and promote the innovation and development of science and technology in China, and realize the comprehensive national strength and international influence of our country Face lift. To build a talent training platform relying on scientific research base, give full play to the advantages of scientific research base in terms of software and hard-ware conditions, and combine scientific research with personnel training organically, which plays a very important role in enriching the connotation of teaching work in Colleges and universities, guiding and stimulating students' innovation enthusiasm and enriching practical teaching content. Through the research on the positive role of scientific research base in talent training in Colleges and universities, it is conducive to better promote the re-form of talent training mode in Colleges and universities, improve the talent training sys-tem in Colleges and universities, and has a positive significance for the country to do a good job in talent work and implement the strategy of "talent power".

2 TYPES OF SCIENTIFIC RESEARCH BASES IN COLLEGES AND UNIVERSITIES

As an important part of the scientific research system in Colleges and universities, the main purpose of scientific research base is to obtain original innovation achievements and independent intellectual property rights, and to cultivate new industrial technologies. It is an important base to organize and carry out basic research and application research, realize independent innovation, cultivate excellent scientific research talents and carry out academic cooperation, The quality of project implementation and the efficiency of scientific research activities all affect the final research results of the project, thus affecting the level of scientific research achievements of the laboratory and its influence in the industry. Therefore, the realization of the quality management of scientific research projects is conducive to play the application value of interdisciplinary, improve the high-quality scientific research output of the base, promote the continuous effectiveness of process knowledge output, and give full play to the social and economic value of scientific re-search results.

Due to the outstanding scientific research strength and status of colleges and universities, they have all types of scientific research bases in China. Scientific research bases can be divided into two categories according to their functions: knowledge innovation bases and technological innovation bases. Knowledge innovation bases focus on basic research, while technological innovation bases focus on applied research, focusing on technological innovation. In addition, there are also types of international science and technology cooperation bases in China, including the national international science and technology cooperation base under the supervision of the Ministry of science and technology, the international cooperation joint laboratory under the supervision of the Ministry of education, and the discipline innovation and talent introduction base of colleges and Universities under the supervision of the Ministry of education and the State Administration of foreign experts, And the science and technology cooperation base in charge of provincial authorities. Since most of these bases are based on the existing scientific research bases and approved by the competent authorities, their management system is basically consistent with the management system of scientific research bases, so this paper does not list them separately for research.

3   CHARACTERISTICS OF SCIENTIFIC RESEARCH BASES IN COLLEGES AND UNIVERSITIES

One is to improve the rationality of scientific research resources utilization. In view of the uneven allocation of scientific research resources among projects, proper mobilization of scientific research resources can give full play to the role of idle scientific research re-sources, reduce the conflict or overload degree of scarce resources, ensure the reasonable and balanced allocation of resources among projects, and ensure that all projects can be completed on time and with quality.

The second is to strengthen the importance of the project members to the overall scientific research development of the scientific research base. Quality management requires the participation of all staff, and promotes the knowledge sharing among multiple projects, the flexible integration of management process and function, which is conducive to promoting the project members to pay attention to the individual project, better understand the scientific research layout and development objectives of the scientific research base, attach importance to the overall scientific research development of the scientific research base, and enhance the quality awareness of the project.

The third is to improve the monitoring timeliness of multi project parallel execution pro-cess of scientific research base. The construction of the management platform provides a carrier for the realization of various data connectivity, liquidity and sharing, which is conducive to the preparation of project budget during project approval. It can refer to the previous similar project expenditure, and combine with the centralized management of ac-counting subjects of corresponding project categories to formulate reasonable budget; it can automatically remind the project schedule and expected objectives, It can ensure that the project is carried out in strict accordance with the established schedule and target, reduce the delay phenomenon, and automatically compare the expenditure with the budget in real time, avoid the deviation between the expenditure and budget requirements, and successfully conclude the project.

Fourth, the one-stop management platform ensures the linkage and sharing of real-time, accurate and consistent data. Based on the construction of cross system data integration platform, the director of scientific research base, project leaders and project members can realize one-stop query of project data within their own authority, which eliminates the need for multiple systems to query and calculate one by one. It also facilitates the timely filling in, scanning and uploading of scientific research project data, and the platform automatically calculates the corresponding caliber data, it avoids the difference of statistical objects of different statisticians, avoids duplication or omission of data, updates the data in real time, and greatly improves the accuracy and consistency of data.

4 SIGNIFICANCE OF THE CONSTRUCTION OF SCIENTIFIC RESEARCH BASE IN COLLEGES AND UNIVERSITIES

One is to improve the rationality of scientific research resources utilization. In view of the uneven allocation of scientific research resources among projects, proper mobilization of scientific research resources can give full play to the role of idle scientific research re-sources, reduce the conflict or overload degree of scarce resources, ensure the reasonable and balanced allocation of resources among projects, and ensure that all projects can be completed on time and with quality.

The second is to strengthen the importance of the project members to the overall scientific research development of the scientific research base. Quality management requires the partic-ipation of all staff, and promotes the knowledge sharing among multiple projects, the flexible integration of management process and function, which is conducive to promoting the project members to pay attention to the individual project, better understand the scientific research layout and development objectives of the scientific research base, attach importance to the overall scien-tific research development of the scientific research base, and enhance the quality awareness of the project.

The third is to improve the monitoring timeliness of multi project parallel execution pro-cess of scientific research base. The construction of the management platform provides a carrier for the realization of various data connectivity, liquidity and sharing, which is conducive to the preparation of project budget during project approval. It can refer to the previous similar project expenditure, and combine with the centralized management of ac-counting subjects of corresponding project categories to formulate reasonable budget; it can automatically remind the project schedule and expected objectives, It can ensure that the project is carried out in strict accordance with the established schedule and target, reduce the delay phenomenon, and automatically compare the expenditure with the budget in real time, avoid the deviation between the expenditure and budget requirements, and successfully conclude the project.

Fourth, the one-stop management platform ensures the linkage and sharing of real-time, accurate and consistent data. Based on the construction of cross system data integration platform, the director of scientific research base, project leaders and project members can realize one-stop query of project data within their own authority, which eliminates the need for multiple systems to query and calculate one by one. It also facilitates the timely filling in, scanning and uploading of scientific research project data, and the platform automatically calculates the corresponding caliber data, it avoids the difference of statistical objects of different statisticians, avoids duplication or omission of data, updates the data in real time, and greatly improves the accuracy and consistency of data.

5 PROBLEMS IN SCIENTIFIC RESEARCH BASES OF UNIVERSITIES

First, the division of management functions is not clear. The management of scientific research base includes the management of field equipment, scientific research team, scientific research activities and internal operation. Different management contents correspond to different functional departments of the school. The site management corresponds to the asset management department, the equipment management department to the equipment management department, the manage-ment of the scientific research team involves the personnel department, the person in charge of the scientific research base, such as the laboratory director, may also involve the personnel depart-ment and the Organization Department, and the scientific research activities involve the science and technology management department. The science and technology management department is responsible for the management of the scientific research base, and other functional departments cooper-ate to complete the corresponding work. Because the work involves many departments, the division of management functions is not clear in management, which seems that "everyone is in charge but no one cares", which leads to management vacuum and is not con-ducive to the healthy development of scientific research base.

In addition, there are also scientific research laboratories approved by the university it-self, and scientific research bases approved by the corresponding superior departments. There are both connections and differences between scientific research base and scientific research laboratory. In the actual work, the lack of understanding of the relationship and difference between the two will lead to unclear management interface.

Second, resource input is decentralized. Because colleges and universities usually implement secondary management, in addition to the scientific research base at the same level as the college, which can directly obtain the resources support of the University, other types of scientific research base must rely on the college to obtain the investment of the University, This situation will lead to two situations: one is that a certain scientific research base may obtain repeated support from relevant functional departments of the University from multiple channels, including conditional support and financial support, thus resulting in repeated funding; the other is that although the university provides various support to the scientific research base, due to the support obtained by the scientific research base, it needs to rely on the college, It is easy to be affected by the internal management of the college, and the financial and material resources may not be able to reach the scientific re-search base directly, which leads to the decentralization of resource input, and reduces the use efficiency, so that the construction and management of the scientific research base get twice the result with half the effort.

Third, the operation mechanism of scientific research base is not perfect. At present, most of the scientific research bases in Colleges and universities do not have full-time management personnel, and the relationship between scientific research bases and colleges is more difficult to deal with. At the same time, the internal operation mechanism of some scientific research bases is not perfect, and they seldom carry out academic activities in the name of scientific research bases. Teachers have no sense of belonging to the scientific research bases. Only when they fill in the materials or face the assessment of the competent authorities can they feel the existence of the scientific research bases.

6   COUNTERMEASURES FOR OPTIMIZING THE MANAGEMENT MODE OF SCIENTIFIC RESEARCH BASE IN COLLEGES AND UNIVERSITIES

The first is to raise awareness and improve system construction. Scientific research base is a sci-entific research laboratory built by colleges and universities according to their own development needs, and has certain scientific research conditions (research room and scientific research equip-ment), has a strong team, has achieved good scientific research results, reached a certain level, has a good internal operation, after the approval of the procedures of the higher authorities. The scientific research laboratory inside the university is the basis of applying for the construction of scientific research base, but not all the scientific research laboratories in the university can become the scientific research base. Only when the scientific research laboratory meets certain require-ments and is approved by the competent department, can it become a scientific research base. Therefore, only when we understand what a scientific research base is, and what is the connection and difference between a scientific research base and a scientific research laboratory, can we more clearly divide the management functions, formulate corresponding management systems, and clar-ify the management responsibilities of each department, so as to establish rules and regulations for the management of scientific research bases.

Second, overall planning of resource allocation. In order to make the investment in scientific research base of colleges and universities form a resultant force and avoid the waste of resources, it is necessary to strengthen the overall arrangement at the school management level, establish the corresponding joint working mechanism, connect the main functional departments related to the management of scientific research base, discuss and solve the major issues in the construction and development of scientific research base, improve the efficiency of resource utilization, and achieve twice the result with half the effort.

The third is to improve the operation mechanism and supervise and standardize the management. On the one hand, we should strengthen the role of the colleges on which the scientific research bases are based. Without the support of the college, the management of the scientific research base is bound to be unsustainable. The college's support for the scientific research base can be various, including allocating full-time management personnel for the scientific research base through internal deployment, and managing the site and equipment of the scientific research base; requiring the college to support the scientific re-search base to carry out various academic activities; for example, the leaders of the college should attend the annual academic committee meeting of the scientific research base; All kinds of domestic and foreign academic exchange activities organized by the college should be combined with the scientific research base, and it is encouraged to jointly host domestic and foreign academic conferences in the name of scientific research base and the college. On the other hand, strengthen the construction of scientific research base management team. Scientific research personnel are the main force of the scientific re-search base, but the operation and management of the scientific research base will become a problem if there is no management personnel and experimental technicians. The university should support the construction of scientific research base management team from multiple levels, and provide full-time management personnel for scientific research base. At the same time, we should increase the investment of funds and supervise the standardized operation of scientific research bases. The construction and development of scientific research base is inseparable from resource investment. In addition to competing project support from outside the University, the scientific research base also needs stable financial support from the competent department or the school to ensure the normal operation of the scientific research base. More importantly, stable support can enable the scientific re-search base to carry out independent research according to the major needs of the national economy and the scientific research development needs of the school, and improve the innovation ability of the scientific research base.

REFERENCES

[1] Zhang Dongliang. Research on collaborative management mechanism of scientific research projects in Zhejiang University: a new paradigm based on Process Reengineering. China basic science, 2018, (02).

[2] Wang Peng. Management mode of National Key Laboratory of University: the conflict between ideal and reality. Modern education management, 2010 (12): 55–57.

[3] Yan Zhonghua. Research on technology management of science and technology projects. Science and technology management research, 2003 (6): 17–19.

[4] Journal of Northwest University of Technology (SOCIAL SCIENCE EDITION), 2014, (1): 105–108.

[5] Wang Xiaofeng. Establishing a big science view and innovating interdisciplinary scientific research organization mode. China higher education, 2011 (2): 24–26.

[6] Li Peng, Liu Yan. The development of German scientific research system and Its Enlightenment on the construction of innovation base in China. Scientific management research, 2011, 29 (2): 52–57.

Computational Social Science – Luo, Ciurea & Kumar (eds)
© *2021 Taylor & Francis Group, London, ISBN 978-0-367-70193-2*

Thoughts on building an enterprise human resource management information system

S.S. Xu & Y. Zhang
School of Economics & Management, Shenyang Aerospace University, Shenyang , China

ABSTRACT: It is an age of big data now. With the continuous development of science and technology in China, the applications of the information system in enterprise human resource management have become increasingly extensive. Especially, with the development of Internet, more and more enterprises are following the trend of network information in their routine work in the development. The most important part of enterprise development is enterprise human resource management, which is directly relevant to the abilities of enterprises to manage and retain talents efficiently. For this reason, how to better build an enterprise human resource management information system is an issue to be seriously considered by enterprise managers at present. It is necessary to build the human resource management information system in combination with the human resource management requirements of enterprises, integrate system applications and ensure the scientificity of system design, so as to give full play to comprehensive benefits of the enterprise human resource management information system.

1 INTRODUCTION

With the continuous development of science and technology in China, the applications of the information system in enterprise human resource management have become increasingly extensive at present. It is necessary to build the human resource management information system in combination with the human resource management requirements of enterprises, integrate system applications and ensure the scientificity of system design so as to give full play to comprehensive benefits of the human resource management information system.

2 POSITION AND ROLE OF THE HUMAN RESOURCE MANAGEMENT INFORMATION SYSTEM IN ENTERPRISE MANAGEMENT

With the rapid development of market economy in China, the role of talents in enterprise competition and market development has been gradually highlighted. If enterprises want to speed up their development and take a firm stand in the fierce market competition, they must establish appropriate internal talent strategies. Talents are the scarcest resources in the development of enterprises. How to make talents exert their abilities in enterprises for a long time is an issue that enterprise managers must seriously consider in the development of enterprises. For this reason, the establishment of the human resource management information system in enterprises has become an important link which must be well completed in the development of all enterprises.

The human resource management information system ("HRMIS") occupies an important position in modern enterprise management. Since the human resource management information system was started relatively late in China, it lags behind financial management, production management and other information systems. Therefore, it is imperative to strengthen the building of the human resource management information system. We need to adapt to the internal needs of the development of the times and the rapid development of the society, make constant innovation and improvement, change the human resource management philosophy, adopt the advanced information system and highlight the role of the information system in human resource management. The development and design of the human resource management information system depend on the strategic human resource management philosophy, with the ultimate objective to enhance the strategic executive force and the organization management ability of enterprises, increase benefits of human resource management by means of information technology, standardize the current management mode, improve the human resource management system and integrate the human resource management processes, thus strengthening the management efficiency and level.

3 PROBLEMS EXISTING IN THE CURRENT ENTERPRISE HUMAN RESOURCE MANAGEMENT INFORMATION SYSTEM

First, enterprises lack adequate awareness of and do not pay enough attention to building the human resource management information system. Although some enterprises apply the human resource management information system in the development, the degree to which these enterprises are aware of and pay attention to the human resource management information system is not enough, and these enterprises get used to traditional management methods. The application of the human resource management information system in enterprises may bring long-term benefits to enterprises, but the negative attitude and the lack of support and understanding of this system from enterprise managers may make the human resource management information system fail to play its due role. This is the main problem causing the poor application effect of the human resource management information system in enterprises.

Second, enterprises have non-standard operations when applying the human resource management information system. Enterprise employees fail to operate in strict accordance with the instructions for use when using the system, causing the poor effect in use and failure to provide correct data information and data analysis. As a result, enterprise managers have problems when making enterprise decisions, which leads to the loss of profits in enterprise operation. In this case, it may be mistaken that the system is not helpful to enterprise operation and management, which directly leads to the loss of confidence of enterprise managers in use of the human resource management information system.

Third, there are insufficient technical talents in the field of the human resource management information system. This is an important problem of the current enterprise human resource management information system. The lack of technical talents in the field of the enterprise human resource management information system may make it difficult to solve the problems occurring in the operation of the human resource management information system, which has an adverse impact on enterprises. The reason for this is that some enterprises often pay attention to early-stage investment of the human resource management information system but ignore the introduction and training of professional technical talents when building the human resource management information system. This is another serious problem that enterprises currently face when applying the human resource management information system.

4 COUNTERMEASURES AND IDEAS OF BUILDING THE ENTERPRISE HUMAN RESOURCE MANAGEMENT INFORMATION SYSTEM

(i) Raise the attention of enterprise managers to the human resource management information system. In the development process of enterprises, all the long-term development strategies are formulated by enterprise managers who have the right to determine the personnel allocation of enterprises. If enterprises do not pay enough attention to the building of the human resource management information system, it is difficult for the human resource management information system to really play its role in enterprises. To enable the human resource management information system to ensure that accurate and complete enterprise talent data is available for enterprise managers, all employees of different departments in an enterprise are required to actively cooperate with the information input of the human resource management information system when the human resource management information system is applied. After adopting the human resource management information system, enterprises need to serve the established development strategies. At this time, more attention should be paid to the human resource management information system. Changes in the business process and other problems in the human resource management department need to be properly handled to avoid damage to the interests of some employees due to the reference to the human resource management information system, which directly restricts the development of enterprises. Therefore, it is required to ensure that enterprise managers can support the use of the human resource management information system in their enterprises when enterprises are building the human resource management information system.

(ii) Standardize the business system and the guarantee system for the enterprise human resource management information system. In use, the human resource management information system manages information efficiently depending on computers of enterprises. For this reason, enterprise managers must be aware of the fact that in order to get effective information from the human resource management information system, the computer management system is required to be capable of processing information regularly. In addition, the requirements for data receiving, processing, storage and output in the computer management system should be strictly managed. The human resource management information system can process information from various different business modules and give the most reasonable and standardized answers. Using information given by the human resource management information system in combination with the actual situation of the market, enterprises are able to make decisions helpful to their development. The human resource management information system is the best auxiliary tool of enterprise management for enterprise managers. Therefore, we must strictly follow the operating steps to improve the management strength of the human resource management information system in use of the human resource management information system, and should strictly ensure the authenticity of data in data collection, data update or other aspects involved. We should establish relevant systematic guarantee systems, standardize business systems in enterprises and implement them by preparing procedural documents, have rules to follow, and reduce and eliminate various exceptional cases. Only in this way can we ensure that the human resource management information system works normally in enterprises and makes contributions to the development of enterprises.

(iii) Establish a high-quality professional technical team for the enterprise human resource management information system. High-quality and excellent professional technical management talents are the important guarantee to the normal and effective operation of the enterprise human resource management information system. We should improve the professional quality of maintenance management technicians for the enterprise human resource management information system, and establish a highquality team with technical stability. For this reason, a maintenance organization or information center for the human resource management information system can be specially established in enterprises to implement the technology and ensure that maintenance management technicians can solve problems immediately once problems occur when enterprises are applying the human resource management information system. In addition, maintenance management technicians can update and improve the human resource management information system according to the actual development of enterprises. Enterprise managers who need to systematically analyze information and persons in charge of the computer management system should be responsible for the coordination of the whole team, maintenance management progress control, data analysis and acquired data inspection, providing scientific advices when enterprise managers make decisions on the long-term development of enterprises. Enterprises are required to regularly organize technical training for maintenance management technicians for the human resource management information system to ensure their professionalism is ahead of other enterprises, and maintenance management technicians need to regularly check the operation effect in each department. When recruiting professional technicians for the human resource management information system, enterprises need to make careful screening to ensure the professional quality of technicians, and invest a lot of money in the application of the system, for example, in the early stage of building the human resource management information system, and the fund for later maintenance is also the main factor to ensure that the human resource management information system can increasingly bring great benefits to enterprises and bring values to the development of enterprises.

5 SUMMARY

This paper mainly discusses how to establish the human resource management information system in enterprises currently. Enterprises should develop according to the actual situation and ensure that talents for system management work actively to improve the working efficiency of enterprises and bring economic benefits to enterprises.

6 TOPIC SOURCE

Liaoning Provincial Science and Technology Foundation Natural Fund Guidance Project, "Research on the Development and Application of Enterprise Human Resource Management Information System", Project Number: 2019-ZD0242

REFERENCES

[1] Jianzhi Liu Construction of Human Group Information System in Enterprise Group Business management 2018.05. pp 33–37.
[2] Bo Wang Development of Human Resource Management Information System in Large State-owned Enterprises Education Forum 2016.11. pp 14–18.
[3] Qian Zhao Discussion on the Application of Enterprise Human Resource Management Informationization The operation and management 2016.01. pp70–74.
[4] Mingli Zhang Management of modern enterprise human resources information management model construction and implementation exploration 2015.12, pp 28–32.

Computational Social Science – Luo, Ciurea & Kumar (eds)
© 2021 Taylor & Francis Group, London, ISBN 978-0-367-70193-2

Complementary resources and application transformation

P. Wang

Fuzhou University of International Studies and Trade, Fuzhou, China

ABSTRACT: This paper analyzes the historical characteristics of the development of Fujian Province and the great opportunity of SAR + Experimental Zone + Free Trade Zone + New District, analyzes the characteristics of educational resources and teachers in Taiwan from the other side of the Taiwan Straits, and combines the status quo of Taiwan's high-quality teachers. This paper expounds the cooperation of cross-strait teachers in teaching and focuses on the information technology courses in university education under the new situation of Internet +, taking computer application foundation, literature retrieval, and computer network technology, and the characteristics and positioning of their respective courses. Guided by the project-driven teaching method, using a series of related auxiliary means, we pay attention to the combination of theory and practice of teaching characteristics of the achievements of our college characteristics.

1 INTRODUCTION

1.1 *Fujian's characteristics and opportunities*

Fujian Province is located in the southeast coast, with an area of 124,000 square kilometers and a permanent population of 38.39 million. The mountainous and hilly areas of Fujian Province account for about 90% of the total land area of the province; most of these mountains are covered by forests, making Fujian's forest coverage rate reach 65.95%, ranking first in China. Fujian has the second longest coastline in China, with a land coastline of 3,751.5 km.

With the realization of the three links between the two sides of the Taiwan Strait, the geographical position of neighboring Taiwan has become an advantage of our province. The State supports the acceleration of Fujian's development. In addition to the central government's strong investment, it also enjoys many preferential policies, showing a situation of "special zone + Experimental Zone + Free Trade Zone + New Area? and frequent non-governmental exchanges.

1.2 *Educational resources and characteristics in Taiwan*

Taiwan is the largest island in China. It is located on the continental shelf of the southeast coast of the motherland. It faces Fujian across the Taiwan Strait to the West. Its territory includes Taiwan's own island and 21 affiliated islands such as Lanyu, Green island and Diaoyu Island, and 64 islands in Penghu Archipelago. The area of Taiwan's island is 35,873 square kilometers.

Since the development of higher education in Taiwan, the popularization rate of colleges and universities is much higher than that in mainland China. Taiwan, with a population of 23 million, has good national universities such as Taiwan University, Tsinghua University, Jiaotong University, Chenggong University, Central University, Chengchi University, Sun Yat sen University, Zhongzheng University, Taiwan Normal University, and Taipei University, as well as good private universities such as Soochow University, Fu Jen University, and Shixin University. Fujian has a population of 40 million, only Xiamen University, and other good schools.

1.3 Introduction of high-quality teachers from Taiwan

As early as 2009, Fuzhou University of foreign languages and foreign trade became the first batch of Pilot Universities in Fujian Province to cooperate with Taiwan universities. It has established in-depth teaching and research cooperation with 15 universities in Taiwan, including Zhongyuan University, Wenzao Foreign Studies University, Asian University, Yishou University, Shude University of Science and Technology, and Kaohsiung First University of Science and Technology.

In 2014, the school began to vigorously promote the plan of "top 100 teachers from Taiwan," formulated the measures for introducing high-level talents from Taiwan, and started the project of recruiting teachers from Taiwan. In this academic year, there are 23 Taiwan teachers, including 20 with doctorate degree, 20 with assistant professor title or above, 7 over 40 years old, and 16 over 50 years old.

2 SCHOOL NATURE AND ORIENTATION

Fuzhou University of foreign languages and foreign trade is a private full-time regular undergraduate college approved by the Ministry of Education and has independently issued state recognized academic degrees. It is a university with foreign language economy and trade as its characteristics, covering major categories of foreign languages, economics and management, commerce, electronic information technology, and animation. The university is positioned as an application-oriented university and participates in the evaluation of talent training level in higher vocational colleges. In the process of transformation, how can we run the applied vocational education well? It is a problem that we educators should think about!

3 TEACHING REFORM OF INFORMATION TECHNOLOGY COURSES

3.1 Information technology courses

A comprehensive view of social development: from the initial primitive → agriculture → industry → to today's information society. And information development: from the original comparison → language → text → printing → to the present, computer + communication. Information is one of the three major elements of social development in parallel with materials and energy.

Today, the wide application of information technology in the fields of office, industry and agriculture, scientific research and education, finance and commerce, medical care, and military innovation has had a huge and profound impact on economic and social development and has fundamentally changed people's lifestyle, code of conduct, and values. For the contemporary college students who shoulder the mission of building socialism with Chinese characteristics and realizing the great rejuvenation of the Chinese nation, information technology is the knowledge that must be learned and mastered.

3.2 Fundamentals of computer application

The main teaching objectives of this course are to understand information and other related concepts; to master the working principles of a computer; and to focus on the routine applications of a computer. In today's society, it is hard to imagine a young college student who is not familiar with computers. When entering the American university campus, in the classroom, in the library, on the lawn, almost all of the students who come into sight carry portable computers and are busy checking materials, doing homework, and sending and receiving e-mails.

This course is arranged for freshmen. Due to the importance of the provincial computer level one unified examination, the teaching arrangement was divided into two parts: the classroom theory and the laboratory experiment. The teaching hours were as large as eight class hours a week, and there were additional classes. However, the final passing rate was not directly proportional to the time and energy invested. The reason is that the emphasis of conventional application is not grasped. Some students who have not been exposed to the computer have been confused by the theory of system conversion, which leads to fear of difficulties.

Now, according to the law of step by step and grasping the key point of conventional application, we have changed the mode of first theory to practice in the past, and arranged the whole teaching process in the computer room, explaining and demonstrating while practicing, and explaining the problems encountered from the theory again, so that students can understand and accept quickly and have a solid grasp of knowledge. Teachers can adjust the progress of the teaching unit according to the actual situation, especially the help of students in the operation, which makes the whole classroom lively. This kind of student-oriented learning method, only four class hours per week, has achieved good results. In the provincial computer unified examination, our school's passing rate has been maintained at more than 90%, ranking first in the provincial colleges and universities.

3.3 Literature retrieval course

The teaching objectives of this course are to master the basic knowledge of information retrieval; to lay the foundation for the future study materials, graduation thesis, scientific research project needs, and searches; and the focus is to find the required information accurately and quickly.

This course is arranged in the third semester, and all teaching is arranged in the computer room. Because it is after the computer application basis course, and there are only two class hours per week, students find it easy to despise the course, plus the open network environment, so if they don't pay attention to it, they will be in the state of playing online games and other activities.

Therefore, it is necessary to use the project driven method throughout the whole teaching process. The preparation of information materials for the graduation project in the following semester is the main line. Combined with the major, students are required to operate from the topic of interest to the download of a series of related journal papers. The theory guides practice, and the problems encountered in practice are solved in theory. The level of students is clear at a glance. In teaching, we carefully organize all kinds of retrieval from easy to difficult to find the homework questions, and each class should arrange the homework and return it in the form of electronic documents before class. These homework and final examinations account for 50% of the total score. Practice has proved that only in this way can the teaching quality of the whole class be guaranteed.

3.4 Computer network technology course

The purpose of this course is to enable students to systematically master the basic concepts, core principles, and common networking technologies of computer networking. This course is a compulsory course for information majors and an optional course for many related majors. In today's rapid development of information age, electronic information technology plays a very important role in the development of science and technology at home and abroad [1]. The electronic information courses in colleges and universities also have some unique characteristics and properties [2].

Einstein once said, "interest is the best teacher." Electronic information–related professional knowledge updates quickly, and if students do not have a strong interest in learning this major, it is difficult to imagine that these students can meet the requirements of talent needed by society after graduation [3]. In the teaching process, the most important thing is to improve students' interest in the major.

Originated in the United States, the Internet is a communication network that has the greatest impact on the world today, has the closest relationship with ordinary people, and has changed people's daily communication and communication mode. The network adopts TCP/IP protocol. There are many courses related to computer network and protocol, such as communication principle, OSI protocol, generic cabling technology, and TCP/IP protocol principle and network programming. Although they have different emphases, they have a lot of intersection. Among them, computer network technology is more important because it tells the core principle of protocol. The specific measures are as follows.

To solve the problem of repetition of some related courses, the curriculum system was revised. On the basis of formulating the curriculum standard and standardizing the experimental outline, the class hours of the course were expanded. Some course contents of communication principle, OSI protocol, generic cabling technology, and network programming technology were included in the teaching of this course, which had a good effect of connection and connection.

In view of the problem that there is no suitable teaching material at present, only through teachers reading more books and hands-on, and accumulating at least three rounds of teaching experience, can we clearly understand the teaching sequence and key points and make an ideal teaching courseware. The theoretical part focuses on OSI protocol and TCP/IP underlying protocol, and the application layer is simply explained and then turned to practical programming.

Aiming at the problems of theory and practice, the project driven teaching method, which is very suitable for the course, is adopted to improve students' understanding and interest. The whole teaching takes various network programming as the project, with two to three people as a group. Each person has different angles, but it can form a network communication system. First, it starts from simple, and with the deepening of the course, it constantly modifies and accumulates perfectly. Finally, the score of network programming is 30% collecting the total test scores.

According to the different problems of different professional curriculum systems, different emphasis and requirements are adopted. For example, in the public elective courses for non-information majors, the theoretical part of the protocol is emphasized while the programming of the practical part is ignored. The course content also involves network technology. The purpose of the course is to master the working principle and achieve the ability of networking and troubleshooting. However, in the professional courses of information specialty, both protocol and programming should be paid equal attention to, and communication principle, network technology, generic cabling, mobile communication, and other contents should be added according to the professional characteristics. The knowledge added is not necessarily deep, but can be extensive and extensive. The connection between knowledge can improve students' interest and expand their vision.

4 PROMOTION AND DEVELOPMENT

Practice has proved that through the teaching reform, students will have a strong interest in the course; clearly understand the development direction of the discipline; master the knowledge, theory, and method of relevant protocols in digital communication system; be familiar with the methods and means of network programming; have the ability to independently engage in professional technical work related to this course; and have strong ability to analyze and solve practical problems in production ability and good professional quality and innovation consciousness. Many students choose the subject of embedded network terminal application software development in the subsequent graduation project and use the knowledge they have learned to complete the design well.

Taiwan's teachers have a rigorous style and profound academic knowledge. In the day and night, teachers from both sides of the Taiwan Strait, who are one of the milk compatriots of the Chinese nation, work together harmoniously and complement each other with an open mind, which has rapidly improved the overall level of the school and made remarkable achievements in various aspects.

In terms of personnel training, we have carried out the "3 + 1 double campus joint training" with Taiwan's colleges and universities, and have sent more than 900 students to Taiwan's universities to study their counterparts, and nearly 60 professional teachers have been selected to visit and study. In addition, teachers from both sides of the Taiwan Straits have jointly worked out a four-year training program for students, organically combining the advanced practices of Taiwan's vocational education with the actual needs of the mainland.

In terms of academic research, first, with Zheng Zhenduo and Minhai Culture Research Center as the carrier, we employed Taiwan's senior academic talents to join in, and promote the intellectual integration between the two sides. Second, we encouraged Taiwan teachers to actively participate in mainland scientific research. We have been approved to set up more than 10 key projects at provincial level and above and published more than 20 high-level papers in EI and CSSCI.

In recent years, the university has also hosted the fourth (2015) cross strait college students' vocational skills competition and innovative scientific and technological works exhibition sponsored by Fujian Provincial Department of education, which includes 15 vocational skills project competitions, Fujian Taiwan joint training talent Summit Forum, cross-strait college students' innovative scientific and technological works exhibition product presentation meeting, attracting 110 teachers and students from 19 universities in Taiwan. In addition to the universities outside the province, there are 36 colleges and universities in our province, and 75 enterprises have participated in the meeting. It has hosted the cross-strait youth education salon and a series of activities of the fourth cross-strait youth festival and invited 29 experts from both sides of the Strait, as well as more than 200 guests from universities across the Straits, leaders and teachers' representatives of Fuzhou Municipal Colleges and universities. During the same period, more than 400 teachers and students from both sides of the Taiwan Straits stayed in our school for a week's exchange activities. Since its establishment in 2014, it has been held two times, with more than 600 teachers and students from more than 20 universities across the Taiwan Straits.

Feature is the style and form of a thing or a thing that is significantly different from other things. It is determined by the specific environmental factors on which things are generated and developed and is unique to the things to which it belongs.

Some common features of the teaching reform of this course can also be extended to other courses. Its characteristics are taking quality education as the leading factor, cultivating students' professional quality and innovation ability as the goal, stimulating students' interest in learning as the center, improving teaching methods and enriching teaching means; taking theory with practice as the guidance, taking network programming as the operation platform, in order to cultivate students' professional quality and innovation ability The key is the ability to understand, analyze, and solve problems, adjust class hours, and implement practical teaching. Its innovation points are as follows: sorting out teaching courseware suitable for domestic students; forming "three-dimensional" teaching content with a variety of teaching aids; using project driven teaching methods throughout the whole process; and network programming application in the embedded system of characteristic specialty. Its most common key application value is that students can learn and master the knowledge of the course autonomously and interestingly.

5 CONCLUSION

At present, electronic information technology, one of the most vital technical fields in the future, is a complex multi-level and multi-disciplinary technical system. The development of colleges and universities is inseparable from the cultivation of talents [4]. To cultivate high-quality graduates with strong social viability and competitiveness, it is necessary for colleges and universities to scientifically and reasonably construct the training system and establish and improve the teaching mode [5].

Teaching reform is a long-term process. Under the current environment, there will be many difficulties, which require the joint efforts and selfless dedication of all teachers. Only by following the law of scientific school running, closely connecting with social needs and condensing characteristic professional direction, can we highlight the characteristics of running a school and realize the training goal of cultivating high-quality talents with solid discipline foundation and strong practical ability [6].

REFERENCES

[1] Qin Li, Zhang Wendong, Xiong Jijun, Liu Jun, Ren Yongfeng, Liu Wenyi. Analysis on the cultivation mode of innovation and practical ability of electronic majors. Journal of North China University, 2008, 24 (6): 28–29.

[2] Jiang Shuhua, Li Mingqiu, Zhang chenjie. Exploration and practice of examination system reform of electronic courses. Science and technology information, 2009, 32: 252–253.

[3] Xiong Jiefeng, thinking of teaching students according to their aptitude in electronic Undergraduate Teaching. China Science and technology information, 2008, (12): 252–253.

[4] Wang Hongzhi, Chen Geheng, LV Hongwu. Research on the structure and curriculum system of electronic information engineering undergraduate. Journal of Changchun University of technology, 2008, 29 (4): 44–45.

[5] Liu Xin, Chen Lixia, Wang Hong. Research and exploration of practical teaching mode for electronic majors. China modern education equipment, 2008, 66 (8): 86–87.

[6] Wu Lingxi, Zhan Jie, Zhou Renlong. Research on characteristic curriculum system of electronic information science and technology specialty. Contemporary education theory and practice, 2010,2 (1): 105–107.

Computational Social Science – Luo, Ciurea & Kumar (eds)
© *2021 Taylor & Francis Group, London, ISBN 978-0-367-70193-2*

Building standard domestic service brands in China

H.J. Chen*, Z.C. Wang, X.W. Wu & W.D. Wang
School of Humanities and Foreign Languages, China Jiliang University, China

ABSTRACT: This research analyzes the cultural background and state-of-the-art development of domestic service industry in China. On this basis and learning from typical domestic service modes in Japan, the UK, and the Philippines, we propose corresponding countermeasures and suggestions for building domestic service brands conforming to national standards taking Santi Group in Hangzhou as an example.

1   INTRODUCTION

With the implementation of the universal two-child policy and the increasingly serious social aging in China, there is a growing demand for domestic services among the public. At the same time, negative events relating domestic services also emerge one after another. All these force researchers to think about how to set standards to eliminate market chaos and how to build trustworthy and reliable domestic service brands with Chinese characteristics.

2   CURRENT SITUATION OF DOMESTIC SERVICE INDUSTRY IN CHINA AND EXISTING PROBLEMS

2.1   *Lack of uniform service norms and standards*

The domestic service industry is still in its initial development stage in China. Although it involves a huge population and includes 20 classes, which cover more than 200 types of services, it still lacks uniform industry standards. Although enterprises in the industry often provide a dozen days of pre-job training for employees, the training content and relevant requirements are decided by the enterprises themselves and generally less scientific. Furthermore, the industry also faces another problem, that is, it lacks uniform evaluation and identification standards for employees. Some training agencies or domestic service companies issue various kinds of certificates unauthorized by official issuing authorities. This leaves the domestic service market in chaos, such that the public gradually lose their trust on the industry. It is evident that the lack of standards and norms is the biggest problem of the domestic service industry.

2.2   *Random fee collection of domestic services*

It is a matter of fact that there are no definite charging standards for many services. Qilu Evening News once reported an event relating to high laundering fee for curtains. Mr. Wang entrusted a local domestic company to clean a curtain worth about 1200 yuan, while he was charged for 800 yuan of laundering fee afterwards. Then, journalists conducted random research on the local domestic service market and five companies interviewed stated that they applied the agency-dominated business mode. That is, the companies collect information about numerous domestic workers and

---

*Corresponding author

then recommend them to customers. In the mode, these companies take agency fees. The presence of such a mode is fundamentally because that there are no charging standards for many services, so it is unable to guarantee the full transparency of charges.

2.3 *Low quality of domestic workers*

It is a common phenomenon that domestic workers directly take a job without receiving specialized training. According to the survey, domestic workers are commonly less educated migrant workers. Many of them are forced by necessity to be engaged in the industry with the most immediate goal of making money. Characterized by weak sense of identity, they do not take homemaking seriously as a career. Sanxia Evening News on August 14, 2019 reported that Wang, a 52-years-old female nanny, snatched the payment password of her employer after gaining the employer's trust. She stealthily transferred money to her own account. There is another example that a lady accidentally found from a monitor in her house that the infant's nurse recommended by a friend covered her baby with a sweat absorbing towel on the face and left the baby crying on the couch. The specificity of domestic service industry lies in that domestic workers work in the living place of the customers and the service objects are generally special, vulnerable groups such as the young, the old, and the sick. Once the employees have any malicious intention, it is beyond question that the served family is in a potential danger.

3 TYPICAL DOMESTIC SERVICE MODES IN OTHER COUNTRIES

The domestic service industry abroad is relatively developed compared with that in China which has numerous shortcomings.

3.1 *The world-class domestic service brand—Filipino domestic workers*

As we all know, many people first think of the Philippines when it comes to domestic services abroad. Filipino domestic workers have become the world-class brand in domestic service industry and are praised as the most professional nannies. In fact, domestic services are regarded as an extremely respectable work in the Philippines and it is accepted and fond of by many people. In the Philippines, domestic workers are dignified and respectable, which lays a sound foundation for the brand of Filipino domestic workers. On the contrary, although the public do not shied away from jobs like baby-sitters and nannies in China, there are few young people engaged in these occupations. It is less-educated aged females that mainly contribute silently in the industry. Filipino domestic workers had arrived in Beijing, China quietly in the mid-to-late March, 2006. However, due to various causes such as those related to price, policies, and living habits, the market of Filipino domestic workers was not as prosperous as expected. The high price is the primary reason why Beijing residents did not employ Filipino maids. It is reported that the monthly salary of Filipino domestic workers is between 2800 and 3200 yuan, and the actual expenses are far higher.

3.2 *Highly educated nannies in the UK*

When it comes to domestic services in the UK, Norland College has to be mentioned. The nursery training college built in 1892 has a history of nearly 130 years. Taking "Love never disappears" as its motto for family-friendly services, the college has cultivated about 7000 top-ranking nannies across the world. The graduates of Norland College work in rich families and symbolize the most professional domestic service industry. Their most fundamental rule is "Respecting and protecting employers' privacy is the only constant bottom line". As we all know, nannies are equivalent to half of the master of a family. After all, the daily routine of the family is taken care of by these workers in most of the time. In the process, the protection of privacy is a matter of deep concern of employers. What employers care about is whether a nanny can protect rather than gathering or spreading their privacy. While in China, it is often reported that nannies disclose employers' privacy to seek benefits, which is also a major dilemma faced by domestic service industry in China. Therefore, it is quite necessary to learn from the principle of domestic services in the UK.

3.3 Considerateness-based domestic services in Japan

The domestic service industry in Japan introduces the considerateness in the traditional Japanese culture in domestic services, making Japanese influenced deeply by their traditional culture easily accept such services. The key point of considerateness is to "think from other persons' point of view". The culture is embodied in every perspective in the daily life of Japanese. For examples, when holding funeral for a family member, in order not to cause distress to others, the family should smile at guests instead of crying. When incorporating such culture in the domestic service industry, workers smile all the time and pay close attention to and explore the meticulous demands of employers by adhering to the principle that "employers never make mistakes". They take care of their employers' life in a humanized manner which is taken as a service standard. If employers intend to dismiss domestic workers, the workers should be informed 30 days in advance. If employers fail to do so, they need to pay certain allowance for failing to inform the severance in advance. For instance, if the domestic workers are informed of the dismissal 20 days in advance, the employers should pay allowance for 10 days; if the workers are informed 10 days in advance, the employers need to pay allowance for 20 days. If employers do not employ domestic workers any more due to their own reasons and fail to actively help domestic workers to find another job, they should pay allowance higher than 60% of the average wage to retirees. Therefore, domestic service industry is a high-paying industry in Japan. The development of domestic service industry in China calls for innovation, while also needs to inherit traditional Chinese cultures.

4 COUNTERMEASURES AND SUGGESTIONS FOR BUILDING STANDARD DOMESTIC SERVICE BRANDS IN CHINA

4.1 Government-led development and improving the supervision system

Related government departments in China should issue relevant policies and cooperate with domestic service companies to more accurately arrange peer-to-peer services for the domestic workers. The services of domestic workers can be real-timely monitored by requiring workers to punch in and out and using the network and remote technologies. Reputation is paid much attention in the social ecosystem of domestic service industry in the United States. Once workers are complained for three times, they will be dismissed. However in China, customers always have low satisfaction with domestic services. To change such situation, it is necessary to construct a sound supervision system for domestic services and encourage domestic workers to improve their service attitude by using the elimination system. Santi Group, as the president unit of the association of domestic service industry in Hangzhou, assists to launch an integrity evaluation system. These measures are suggested to improve the chaotic and irregular market of domestic services.

4.2 Strict examination and rewarding models

At present, all kinds of certificates prevail in the domestic service industry in China and even some can be bought with tens of yuan. This reflects the non-standard management of the industry. Therefore, it is a pressing need to build a standardized mechanism and set uniform skill training certificate for certain posts and star-rating certificate, so as to resolve chaos in domestic service industry through management.

In addition, the government is also suggested to work with the enterprises to build demonstration domestic service enterprises, issue demonstration certificate to and provide financial support for outstanding demonstration enterprises. This attempts to build an industry benchmark, to let all domestic service enterprises follow these examples and finally facilitate the standardized development of the whole industry. The government departments should be aware of the importance of establishing the industry benchmark, select the best enterprises and channel resources to the demonstration enterprises. In this way, it is expected to help these enterprises to build domestic service brands with Chinese characteristics and contribute to the standardization of domestic service industry in China.

4.3 Strengthening the cultivation of domestic service talents

In the current knowledge age, the competition in all industries comes down to the competition of knowledge. Therefore, to get a firm foothold, everyone should improve their soft power. It is a necessity to cultivate high-quality, specialized personnel, which is the top priority for improving the level of domestic services in China. We should develop training of high-end domestic services to cultivate every domestic worker as a family assistant or housekeeper that can make overall plans, take all factors into consideration, and be competent in all kinds of logistic affairs. It is suggested to carry out systematic and scientific training based on local leading domestic service enterprises, colleges and moderate specialty schools, and professional training agencies. Not only skill courses but also theoretical knowledge and other courses such as sociology, psychology, child nursing, history, literature, education, laws, and financial management should be taught. Besides, a batch of industry–education integrated practice bases characterized by domestic service industry is supposed to be built on this basis. Santi Group in Hangzhou has its own college for domestic service industry, that is, Santi Home Economics College. It sets an advanced typical example in the national modern service industry, changes social prejudice on domestic services, and has trained a large quantity of domestic service talents.

4.4 Providing high-quality after-sale services and building and improving credit system

Similar to the strict regulations set by Santi Group for the management and complaint of service personnel, we can also take measures such as issuing "integrity card". The card allows employers to check previous employers' evaluations and basic professional training certificates of domestic workers before employment. Employers can give certain feedback before the end of each service. Domestic service enterprises reward or punish workers according to these feedbacks. For domestic workers with vicious misconducts, they can be put in the permanent blacklist of domestic services.

4.5 Integrating with traditional cultures and building chinese brands

Although Chinese economy develops constantly and the legal system is constantly improved at present, the domestic service industry still develops at low level and quality and faces numerous problems, for examples: There are few well-known domestic service enterprises in China; The employees generally feature low quality and service level; A complete set of service system has not been formed for the industry. Of course, these problems are implicated to the short development time of the industry; however, we can acquire a lot of beneficial experience by learning from excellent domestic service enterprises in other countries. The considerateness integrated in Japanese domestic services is also worth learning. From the very beginning, Santi Group has adhered to the outstanding traditional culture of "Expending the respect of the aged in one's family to that of other families; expending the love of the young ones in one's family to that of other families." and advocated and insisted on that "As long as everyone gives a little love, our world will be a better place". The company builds its own corporate culture by insisting on the service tenet of "Solving problems and working hard for you" and adhering to the business philosophy of "Quality, value, smile, and gratitude" and the value of "Integrity, compassion, dedication, and innovation".

5 CONCLUSIONS

China, as the largest developing country in the world, needs to actively learn from development experience of other countries, so as to catch up and surpass western developed countries that have developed for hundreds of years. In terms of the domestic service industry, it still develops at a low level in China and can learn from other countries from many aspects. Transformation and upgrading is a pressing need for the domestic service industry in China, which needs to closely follow the pace of the high-quality and rapid development of economy in nowadays, even though there is still a long way to go. In addition, the industry is suggested to pay more attention to build its standard systems and high-standard brands by improving weak links and strengthening advantages, so as to lead the domestic service industry in China to high-speed development.

ACKNOWLEDGEMENTS

This research was supported by Zhejiang Provincial Natural Science Foundation of China under Grant No. LH19G030001, the National College Student Innovation Training Program in 2019 under Grant No. 201910356026, and the College Students' Scientific and Technological Innovation Activities Plan and Xinmiao Talent Project of Zhejiang Province in 2019 under Grant No. 2019R409025.

REFERENCES

[1] Special commentator Zhixin. It is the right time to build a brand of domestic service. Xijiang daily, December 13, 2019 (F02).
[2] Yang Xingdong. Domestic service should work hard on brand. Hunan Daily, August 23, 2019 (006).
[3] Lin Yimin, Dong Tingting. Creating a new benchmark of "model" standardization and brand building. Fujian quality and technology supervision, 2018 (11): 15.
[4] He Ji'an. Combination of standardization and quality management to improve domestic service level. China Standardization Association. Proceedings of the 14th China Standardization Forum. China Standardization Association: China Standardization Association, 2017:1129–1133.
[5] Li Aiping. Particularity and scientific construction of domestic service brand construction. Journal of Luliang University, 2017, 7 (01): 71–74.
[6] Zuo Xiaojuan, Bai Xuemei. Research on the development of high-end domestic service strategy. Economic Research Guide, 2016 (16): 33–34.
[7] Wang Yaqi, Li Lei. Feasibility analysis and strategy of brand building of domestic service enterprises. Chinese and foreign entrepreneurs, 2013 (24): 18–20.

Computational Social Science – Luo, Ciurea & Kumar (eds)
© 2021 Taylor & Francis Group, London, ISBN 978-0-367-70193-2

How sea power serves China's maritime interests

H. Zeng

Institute of Public Policy, South China University of Technology, Guangzhou, P. R. China

ABSTRACT: Early sea power concepts highlight the use of naval power to project influence abroad, whereas modern sea power concepts highlight the use of naval power to safeguard maritime interests. The former refers to gunboat diplomacy which reached its zenith in the 19th century with the ascendency of Western powers. Throughout its history, China has long been a continental power with little maritime ambitions. For lack of a maritime security policy, China was highly vulnerable to foreign invasions from the sea. The Century of Humiliation has always been the driving force behind China's naval modernization. The study aims to contribute to the growing body of literature on sea power by exploring how maritime power serves China's vital interests in an increasingly interconnected and uncertain world. It will identify the potential challenges China would face in its current process of naval modernization. This study will then propose a set of policy considerations for shaping China as a truly global maritime power.

1 INTRODUCTION

The protection of maritime interests has become a major concern for the Chinese government in recent years. While China has an enormous coastline, its military doctrines are primarily land-based. National security concerns have arisen as the nation shifts from an inward-leaning economy to an outward-leaning one. Chinese scholars call for stepped-up measures to protect China's maritime communication lines (i.e., the Malacca Strait) in the event of a naval blockade. This article argues that China needs to: (1) accelerate its maritime transformation from a land power to a sea power; (2) enhance its naval capabilities; and (3) formulate a modern naval strategy for the sake of its critical trade and energy needs and national security.

2 LACK OF NAVAL TRADITIONS

The Beiyang Fleet was China's first attempt to build a modern naval force. The purpose of the fleet was not to secure command of the sea other than to fend off threats from the sea. It was regarded as the strongest in the Far East, but it was short-lived, ending up with a humiliating defeat in the First Sino-Japanese war. Some historians attributed the failure to insufficient budget and bureaucratic corruption (Ye & Mu 2005). It is logical to assume that China's defeat involves a mix of complicated factors, but the fundamental cause could be lack of a maritime security policy, which hindered the Chinese from acquiring sea power.

As opposed to Western powers, which relied on naval power to back up their diplomatic efforts, China engaged with the world through trade and cultural interactions. Ming China's tributary system was founded on this notion, embodied in Admiral Zheng He's voyages to the Indian Ocean. His merits gave birth to a regional cooperation mechanism centered around China. The differences between China and Western powers in handling diplomacy are clear-cut. Running a closed feudal system, the emperors of China saw no need to turn China into a sea power. Ming China was one of the world's most powerful empires, but it never sought to colonize or exercise dominion over any part of the world. On the contrary, it established a Sinocentric order based on Confucianism, which rejected seafaring activities, and passed laws to prohibit the use of military force as a diplomatic tool.

Nevertheless, as a self-sufficient agrarian society, China would inevitably be confronted with obstacles to sea power. If a nation has no ambition to exploit the sea, it will by no means feel the necessity to build a formidable navy, let alone secure command of the sea. However, a nation's prosperity is closely linked to its military potential. The US is a prime example that a nation may rise to great power status if it has command of sea. In theory, command of the sea allows a nation's military and merchant ships to move unimpeded through international waters. But China was historically a nation that rarely depended on foreign trade for national survival. Without maritime security policy on hand, China was left open to foreign interference.

3 MODERN SEA POWER CONCEPTS

Sea power refers to the power to dominate the sea or the ability to conduct seafaring activities, which is an extension of the concept of national sovereignty. US naval historian Alfred Thayer Mahan contended that sea power entails serious military and economic implications (Mahan 1890). Militarily, a nation can control the sea with a good navy. Economically, command of the sea grants a nation access to resources that contribute to its prosperity. If a nation is to dominate world trade, it should own a sizable fleet while on good terms with most other nations.

Whether in peacetime or wartime, a nation must have the military capability to protect its global shipping. As Mahan pointed out, naval power is associated with the rise of great powers. Countries devoid of such resources would gradually be reduced to "consumers of the world order." It is worth noting that the growth in China's naval power, by his logic, will likely transform China into a revisionist power, at the risk of fueling anti-Chinese sentiment worldwide. For this reason, since the beginning of its naval modernization, China has realized that a peaceful international environment is a guarantee of its successful domestic development, and naval power is the direct means to this end.

In this century of economic globalization, sea power is more about cooperation than competition. Sea power in the post-Cold-War era could be defined as managing and exploiting the sea through a series of political, economic, military, and legal means. By this definition, the main task of the navy is to carry out operations involving humanitarian rescue, crisis management, and peace keeping instead of securing command of the sea (Yang & Zhou 2012).

4 CHINA'S RISE AS A GLOBAL MARITIME POWER

Substantial changes in China's energy consumption have created enlarged demands for foreign trade. In this context, China should secure its energy supplies by developing corresponding military power. Ensuring a stable periphery is critical for China's peaceful emergence as a superpower, but to do so, it will be indispensable to possess a navy able to deter foreign military interventions.

For China, the Malacca Strait is a lifeline that represents about 80% of its oil shipment. The over-reliance on maritime transportation via the strait has resulted in what is known as the "Malacca Dilemma," a term coined by former Chinese President Hu Jintao in 2003. The Malacca Strait has been a headache for China, as it is beyond the reach of the Chinese Navy. Naval power is undoubtedly a powerful tool in defense of maritime interests. It is no exaggeration to say that, whoever controls it can asphyxiate China's energy supplies, and can thereby threaten China's energy security (Storey 2006).

A top priority for the Chinese Navy should be to accelerate the transformation from offshore defense to distant sea offense. This will require China to possess a highly informatized navy capable of handling diverse traditional and non-traditional security threats. Though peace and development are the overarching theme of the 21st century, non-traditional threats, such as terrorism, piracy, and regional conflicts, are rattling the nerves of the international community. A navy with strong offensive capabilities, in essence, will enable China to fulfill the responsibilities as a great power.

Mahan's theory of sea power remains relevant in this era of highly interdependent international system, yet nations today seek to address international disputes without armed conflict. The globalization era marks an end to the use of naval power as a source of deterrence. The use of naval power today is rather focused on the protection of commerce. As the global value chain continues to expand, which allows the world to share the dividends of globalization, the perception of national sovereignty begins to erode. Individual sovereign states have their maritime interests interconnected with the international system. Sea power is essential for creating a peaceful international environment which China badly needs. Hegemonism nonetheless exists, but international institutions and norms are important impediments that restrain great powers – particularly the US – from dominating the international system, even if not to a great extent. And consequently, China's quest for sea power is under far less pressure today than it was in the past.

The navy has a crucial role to play in China's national security. The key goal of China's naval modernization is to safeguard China's sovereignty and territorial integrity. Even though China has never used naval forces for aggressive purposes, both Chinese and Western strategists endorse non-peaceful means against nations who interfere with one's maritime affairs. If one's own interests are at stake, force of arms is justifiable but necessary (Lu 2017).

5   MARITIME CHALLENGES FOR CHINA'S NATIONAL SECURITY

Recent US Navy incidents in the Pacific have raised serious questions about US naval power. The US Navy is showing signs of decline as the Chinese Navy is growing by leaps and bounds. In the face of US military pressure, China is rapidly building a blue water navy. More notably, in 2015, China opened its first overseas base in Djibouti, extending its military reach across the Indian Ocean and beyond. And yet, admittedly, the Chinese Navy remains incapable of challenging the US Navy in conventional warfare. China can commission around 14 ships per year. Still, it lagged US military spending by about $400 billion in 2019 (Defense one 2020). Chinese ships are less heavily armed and smaller in displacement than their US counterparts, although both nations operate roughly the same number of warships. China has 2 aircraft carriers in active service, while the US has 11. The US has nearly 800 military facilities in more than 160 countries on all 7 continents. Taking all these factors into consideration, the Chinese Navy will hardly grow to an extent comparable with the US anytime soon – at least not in the near future.

Command of the sea provides the military basis for US hegemony. There can be no doubt that a decline in naval power can endanger the very existence of US hegemony. In order to deter China's naval expansion, the US has more than half of its navy deployed in Asia. Further, the US is providing vital assistance to build up the military capabilities of allies and partners in the region as part of its Indo-Pacific strategy. This could undermine China's security environment but also would have a detrimental impact on China's economic transition.

Compared with traditional sea powers, China is in a position of insecurity with vulnerabilities from the land and sea. The security threats posed by hostile neighbors will take up a considerable part of China's defense resources. Leaving aside territorial disputes, there are complicated ethnic problems in China's border areas, much of which constrain China's investment in sea power. China's sea territory is blocked within the so-called first island chain that extends from western Japan to the northern portion of the Philippines (Hu 2014). Due to the fear that Chinese fleets will be encircled by US forces, the US Island Chain Strategy has arguably helped shape China's naval options to this day. Such geographical constraints greatly limit the Chinese Navy's freedom of navigation in the Pacific. If China is to develop naval power similar to that of the US, it will have to go beyond its geographical constraints, except that the costs it will have to incur will exceed the costs paid by other countries for the same purpose (Robert et al. 2012).

From a geopolitical perspective, Japan, India, and Australia endure as a mighty challenge to China's maritime ambitions. At the core of the first island chain, Japan enjoys a decisive geographical advantage in controlling the sea routes by which the Chinese Navy enters the Pacific. Though not part of the Indo-Pacific strategy, India dominates the sea routes in the Indian Ocean – a major threat to China's energy supply lines. Australia is strengthening ties with its neighboring countries so as to keep China's influence in check in the South Pacific. These countries form the strongest link in the US containment policy.

China's soft power campaign appears ineffective in dispelling the worries about China's growing military might. On the one hand, the Djibouti base heralds the beginning of China's military transformation, putting Western powers under pressure. On the other hand, most Asian countries welcome US military presence in the Pacific as they boost trade with China, further complicating the geopolitical situation in Asia.

6 POLICY CONSIDERATIONS

For any nation to develop into a sea power, apart from naval buildups, it is required to secure command of the sea and, especially, eschew the use of military force in its relations with another (Gao 2019). Building on extensive research, this study will shed light on China's sea power development as a guide for further policy actions:

6.1 *Develop contemporary maritime strategies*

Evidently, the Oceans and the Law of the Sea by the United Nations serves as a stabilizer for the current chaotic world order. China should achieve its goal of sea power within the UN institutional structure. Under the premise of ensuring respect for international law and US leadership, China's ultimate objective should be to adapt to the status quo rather than to alter the status quo, that is to develop maritime strategies that would lead to win-win outcomes and a regional balance of power.

6.2 *Acquire global naval capabilities*

Naval power is the key to sea power. A blue water navy will allow China to project power across the globe. Command of the sea, naval power, and military deterrence are the essential elements of maritime strategies. A navy capable of sustained operations far from home is urgently needed to assure China's national security and maritime interests. China should, in the short term, safeguard the maritime interests in waters under its jurisdiction, and, in the long term, enhance its naval projection capabilities until enough to break out of the first island chain and match those of traditional sea powers.

6.3 *Build warships with advanced combat capabilities*

Aircraft carriers are the ideal manifestation of a nation's military and economic strength. They are expensive to acquire, so only a handful of countries have aircraft carriers in their arsenals. These countries are all part of an exclusive club in which they exchange views on global maritime security issues. Aircraft carrier development is pivotal to China's military transformation. One of the most remarkable advances in China's naval buildup was the commissioning of Liaoning, the first-ever aircraft carrier in the Chinese Navy. This signaled China's entry into the club of aircraft carriers. But some security analysts sought to downplay the event. Retired Chinese admiral Yin Zhuo proclaimed that, whatever sense the aircraft carrier makes, it turns out that China is a rookie at naval warfare while typical sea powers such as the US and Great Britain have more than a century of naval traditions. Robust evidence shows that the Chinese Navy is lagging dramatically behind the US Navy in terms of naval technology and long-range combat experience. The military trends in Asia are more or less characterized by aircraft carrier expansion (Andrew et al. 2015). India and Japan are energetically expanding their navies. Despite the fact that naval modernization programs require a massive commitment of resources and time, China has the shipyards, engineering expertise, and associated infrastructure needed for massive aircraft carrier construction. Developing a strategic mobile fleet with ocean-going combat capabilities will ensure China's naval superiority in the ongoing naval arms race.

Regardless of how fast they grow, the deployment of China's naval forces is demonstrably flawed: Chinese fleets are stationed across the Yellow Sea, the East China Sea, and the South China Sea. Dispersed forces under different command structures are hard to deliver a powerful blow in situations of military conflict. Thus, Chinese naval forces should be redeployed in a more strategically efficient manner to meet future warfare requirements.

6.4 Resolve the "Malacca Dilemma"

The solution to mitigate China's strategic vulnerabilities in energy transport could be to diversify the sources of energy imports and to finance transit routes that would bypass the Malacca Strait. With this in mind, the idea of an "Asian Panama Canal" may be an option worth digging into. Proponents of this idea envisaged constructing a canal across the Kra Isthmus in southern Thailand. Ships passing through the canal could save 3 to 4 sailing days than through the Malacca Strait. The China-Pakistan energy corridor may be an alternative option to consider. Gwadar is rather close to the Persian Gulf and Pakistan is a close ally, notwithstanding that it would be an expensive proposition given the rugged terrain and separatist violence involved.

6.5 Optimize the mechanisms for maritime security

China should optimize the mechanisms for maritime security so that maritime infringements can be reported more promptly to the highest level and maritime law enforcement vessels can be in place where they are needed. Crucial to this undertaking is deeper military-civilian integration. The first step would be to place maritime militias under the command of the China Coast Guard and to mobilize domestic maritime enterprises to join in China's maritime law enforcement operations. In view of the diversity of organizations involved, a coordinating body should be established and given the necessary authority to accomplish these goals. On the whole, the security of China's maritime passages will largely depend on improved coordination between coast guard authorities, the navy, and the maritime safety administration.

6.6 Educate the public on sea power

By learning from the practices and experience of traditional sea powers, we could instill the notion of sea power among the younger generation, and the preferred option for this purpose is via social media channels such as newspapers, magazines, radio, or television. Other options recommended to raise the national awareness of sea power include encouraging publications on maritime sovereignty and adding maritime history as a compulsory course in high schools.

REFERENCES

Andrew, S. et al. 2015. China's aircraft carrier program: drivers, developments, implications. See https://go. gale.com/ps/anonymous?id=GALE%7CA433587942&sid=googleScholar&v= 2.1&it=r& linkaccess= abs&issn=00281484&p=AONE&sw=w.
Defense One 2020. China's defense spending is larger than it looks. See https://www.defenseone.com/ideas/2020/03/chinas-defense-spending-larger-it-looks/164060.
Gao, L. 2019. A study of sea power models and the construction of china's sea power theory. *Asia-Pacific Security and Maritime Affairs* (5): 29–48. (In Chinese).
Hu, B. 2014. Three power objectives of China's maritime power. *Pacific Journal* 22 (3): 77–90. (In Chinese).

Lu, S. J. 2017. Beyond hegemony: china's deterrence-based sea power. *Asia-Pacific Security and Maritime Affairs* (5): 112–127. (In Chinese).

Mahan, A. 1890. *The influence of sea power upon history*. Boston: Little, Brown and Company.

Robert, S. et al. 2012. *Twenty-first century sea power: cooperation and conflict at sea*. New York: Routledge.

Storey, I. 2006. China's "Malacca Dilemma". See https://jamestown.org/program/chinas-malacca-dilemma.

Yang, Z. & Zhou, Y. H. 2012. On the new military revolution and the sea power in the post-Cold-War era. *Pacific Journal* 20 (7): 53–62. (In Chinese).

Ye, Z. C. & Mu, X. H. 2005. Perspectives on China's sea power development strategy. *Studies of International Politics* (3): 5–17. (In Chinese).

Computational Social Science – Luo, Ciurea & Kumar (eds)
© 2021 Taylor & Francis Group, London, ISBN 978-0-367-70193-2

Research and practice on the talent training mode of modern apprenticeship system: Integration of industry and education, progressive training of ability and segmental training

J. Shan & N.N. Yang
Binzhou Polytechnic, Binzhou City, China

ABSTRACT: Modern apprenticeship and the integration of industry and education, school-enterprise cooperation, are the new issues currently facing higher vocational education. The mechanism of vocational education is based on the fusion education system reform as the breakthrough point, which is elaborated in the modern apprenticeship education mechanism of talent training mode of innovative practice, the construction of the curriculum system, the implementation of the "double teachers" teaching pattern, and evaluation in participating mechanism building for the present stage of modern apprenticeship. The integration of production and education ground implementation has important significance.

1 INTRODUCTION

1.1 *Research background*

In August 2014, the Ministry of Education of the People's Republic of China issued their opinions on carrying out the pilot work of modern apprenticeship, and so the pilot work of modern apprenticeship was officially launched. As a kind of new vocational education mechanism, the modern apprenticeship for promoting industry enterprises to participate in the whole process of higher vocational education personnel training, major setting and industry demand docking, course content and professional standards, teaching process and the production process and docking, all improve the quality of talent cultivation and have an important guiding role [1].

On December 5, 2017, the opinions of the general office of the state council on deepening the integration of industry and education (hereinafter referred to as the opinions) were issued and implemented by the general office of the state council. The introduction of the opinions provides policy guarantee for solving the long-standing pain point of disconnection between vocational education teaching and the actual needs of enterprises and provides a policy basis for innovating a diversified school-running system, realizing the organic connection of education chain and industrial chain, and comprehensively implementing school-enterprise collaborative education.

1.2 *Research significance*

Modern apprenticeship and fusion, university-enterprise cooperation education are intended to promote higher vocational education under a new situation of better and faster development of strategic initiatives, which means that vocational education is a new development strategy, the cooperative education mechanism between colleges' construction and talents training mode. For innovative research and exploration to be put forward, we need to note the following aspects.

(1) Deepening the institutional reform of school-enterprise collaborative education. We will explore the mode of integrating industry and education in multiple forms; give full play to

the important role of enterprises in running schools; build a community of interests with in-depth integration of industry and education; promote the organic integration of vocational education and the industrial system; and improve the quality of personnel training.

(2) Innovate the talent training mode to cultivate high-quality technical talents in line with industrial demand.

We will innovate the talent training mode of integration of industry and education, school-enterprise cooperation, give full play to the respective advantages of schools and enterprises, and cultivate high-quality technical talents of innovation who truly meet the job demand and industrial demand of enterprises.

(3) Improve the internal quality assurance system.

Establish and improve the internal quality assurance system suitable for the modern apprenticeship system, explore the construction of the modern apprenticeship system, and improve the quality of education and teaching process and output.

2 RESEARCH AND PRACTICE OF THE TALENT TRAINING MODE OF MODERN APPRENTICESHIP SYSTEM: INTEGRATION OF INDUSTRY AND EDUCATION, PROGRESSIVE AND SEGMENTAL TRAINING OF ABILITY

According to the modern vocational colleges opinions issued by the Shandong Province about apprenticeships pilot work plan notice (faculty, word [2015] no.22), on the notice of the pilot to carry out the modern apprenticeship (faculty into company letter [2015] no.2) [2] documentation requirements, in April 2017, approved by the Shandong Province Education Department, computer network technology, network security) professional training headquarters and Zte's Asia-Pacific cooperation started in Shandong Province, the third batch of modern apprenticeship pilot projects.

2.1 *Build a collaborative education platform based on the integration of industry and education, and solve the problem of the modern apprenticeship system*

Binzhou Vocational College and Zte's training headquarters in the Asia-Pacific region jointly established Binzhou Vocational College Zhongxing College, a secondary school of mixed ownership, to build a platform for school-enterprise integration and collaborative education. The establishment of the collaborative education platform has expanded the role of the university and enterprise in educating talents, and achieved the comprehensive participation of the cooperative enterprise in the professional construction matters such as the revision of the personnel training objectives, the formulation of the personnel training program, and the personnel training process. This provided institutional support for the smooth implementation of modern apprenticeship.

2.2 *Innovation in the talent training mode of modern apprenticeship system: Integration of industry and education, progressive and segmental training of ability*

Through professional research and post-ability analysis, the talent training objectives for this major are determined as follows: For computer network construction, network security, and security operational management of the industry enterprise, culture supports the party's basic line, master the basic knowledge of computer network security, network interconnection, network system integration, network security management, and other professional knowledge; has the development, design, management, maintenance, network security, information security consulting and technical support, such as ability to have good professional ethics and career development. In the first line of production and service, engaged in network security scientific research, teaching, product design and production and management, pre-sales and after-sales technical consultation, network security management and maintenance, and other positions, the moral, intellectual, physical, US, labor comprehensive development of high-quality technical skills talents.

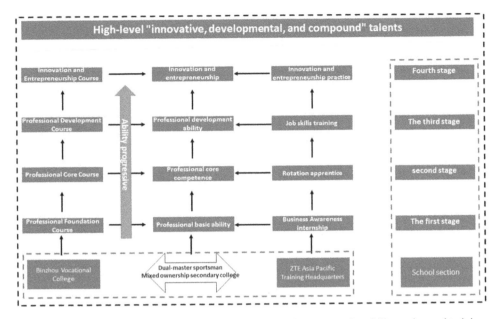

Figure 1. Talent training model of industry-education integration, progressive ability, and staged training

Centering on the goal of talent training, and relying on the school-enterprise integration and collaborative education platform, the talent training mode of industry-teaching integration, capability progressive and segmental training is innovated (Figure 1).

Integration of industry and education: Under the framework of the secondary school with mixed ownership, both the university and enterprise undertake the corresponding personnel training tasks and jointly complete the personnel training work.

Progressive ability: The ability to phase in accordance with the professional basic abilities (ability to implement network security)—professional core abilities (ability to network security operations management), professional development ability (network safety engineering [design] or development), innovation ability, cultivation of innovation and entrepreneurship interest path, level further, gradually increasing.

Section training: The whole process of talent training is divided into four stages.

First stage: Semesters 1 and 2 focus on school education, learning general courses and professional basic courses, and completing corporate culture education through on-campus training base. The content of the courses learned in this stage is mainly the knowledge and skills necessary for the relevant professions, focusing on the cultivation of students' professional quality and providing basic support for their future development.

The second stage: From the first three months of the third semester to the first three months of the fifth semester, the school and enterprise should pay attention to both the study and the core technology courses of network security, such as firewall technology, IPS/IDS, operation and maintenance security, and so forth. This stage focuses on network security, operation, maintenance and management, and other core competence training to enhance students' learning ability for sustainable development.

The third stage: Three months after the fifth semester, mainly enterprise apprenticeship training, in-depth network security engineering design and development in the laboratory, project drill, combat and other forms let students in the real project learn the skills necessary for future posts. This stage focuses on the cultivation of students' abilities in project design and development, innovative thinking, and professional comprehensive ability.

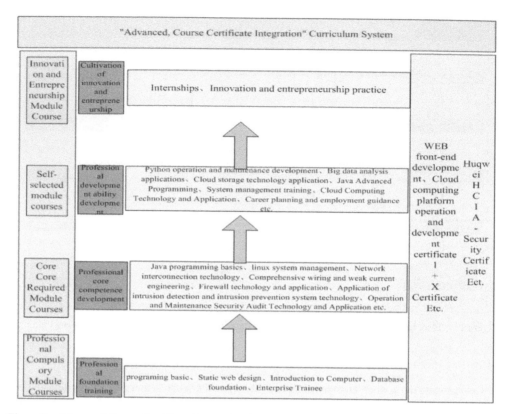

Figure 2. Advanced Course Certificate Integration Course System.

The fourth stage: In the sixth semester, according to the development of students' vocational ability, students will be arranged to find jobs or start their own businesses in relevant positions in Zte's cooperative units such as Beijing Tianrongxin Technology Co., Ltd., Qihoo 360, and other network security companies. This stage focuses on practical practice, mainly to cultivate students' employment competitiveness and innovation and entrepreneurship ability.

2.3 Innovation of curriculum system

According to the needs of the talent training mode and the law of career growth, students' professional ability should be cultivated in accordance with the training process of basic professional ability, core professional ability, professional expansion ability, innovation and entrepreneurship ability, and the content of relevant enterprise certification certificates and 1+X certificates should be integrated to construct the curriculum system of advanced and integrated course and certificate (Figure 2).

2.3.1 Theoretical curriculum system
The first stage of professional compulsory platform courses. The basic supporting courses are set for the basic professional ability and basic professional accomplishment required by the vocational post. The courses at this stage mainly include basic programming, introduction to computer science, and enterprise internship courses.

The second stage is the core compulsory course platform. The main purpose is to improve students' theory and practice and to cultivate their core vocational ability. The courses mainly include

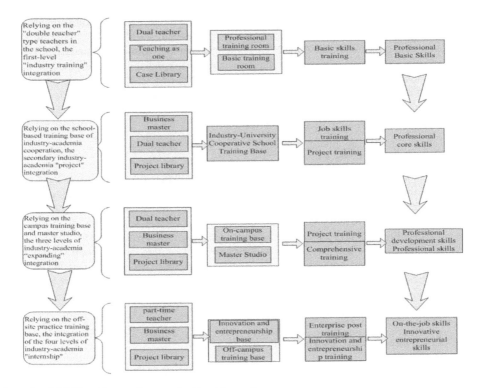

Figure 3.   Multi-level integration practical teaching system.

firewall technology and application, intrusion detection and intrusion prevention technology application.

The third stage is the independent choice of module courses. It mainly trains students' professional development ability and innovation consciousness in new technologies such as cloud computing and big data. Different students can take different courses to broaden and deepen their knowledge and improve their professional ability. It mainly includes Python operation and maintenance development, cloud storage technology and application, and other courses.

The fourth stage is the innovation and entrepreneurship module course. It mainly focuses on students' in-post internship and entrepreneurial practice. Students can work in enterprises or master studios to engage in entrepreneurial practice, so as to accumulate useful experience for future innovation and entrepreneurship. At the same time, students need to complete the graduation project (graduation thesis) related to the practical position or entrepreneurship under the joint guidance of the enterprise master and school teachers.

2.3.2 *Practical teaching system*

The multi-level integration practical teaching system (Figure 3) is constructed, which consists of four levels: practical training of professional basic competence in the first and second semesters, practical training of professional core competence in the third and fifth semesters (the first three months), comprehensive (expansion) competence in the fifth semester (the last three months), and practical training of innovation and entrepreneurship in the sixth semester.

With the core vocational competence training as the main line, the first and second semesters are mainly completed in the professional training room and the basic training room to cultivate students' professional skills. The 3–5 (first 3 months) semester is mainly completed in the training base of the School of Industry and Education Cooperation, through the practical training of post-skills and project training to cultivate students' core skills for future posts. In the fifth semester (the last three months) is completion of the training room and master studio on campus, and cultivating students' professional comprehensive skills and professional development skills through practical training of real enterprise projects and comprehensive practical training. In the sixth semester, practical training on innovation and entrepreneurship was mainly completed in the off-campus training base and innovation and entrepreneurship base of in-depth cooperation. Through in-post practice or innovation and entrepreneurship training, practical skills and innovation and entrepreneurship skills were cultivated.

2.4 *"Double tutor" teaching mode*

In accordance with the formulated computer network technology (network security) basic requirements and skills standards for teachers, guidance teachers' job duties and teaching teachers' job duties and other document standards for students to select on-campus mentors and enterprise mentors, the implementation of students' on-campus mentors and enterprise mentors "double mentors" mode [3]. School teacher professional basic course teaching task, responsible for campus master managers, guide students' theoretical knowledge experiment equipment and research guidance; at the same time, enterprises' double mentor do good communication, help students to the enterprise staff identity transformation, double training school students by enterprises, realize engineering alternation, and cultivate a high-quality network security personnel.

2.5 *Multi-participation assessment and evaluation system*

2.5.1 *Establish an integrated assessment and evaluation system with multiple participants*
Deepening and refining enterprise evaluation, social evaluation and teacher evaluation, university-enterprise cooperation to develop the modern apprenticeship third-party evaluation method of the apprenticeship appraisal system, the prospective employees' practice examination system, the system of prospective employees to staff (graduate), students recall system, and other relevant documents systems, through the system of students assessment, and sets up schools, enterprises, and social integration of the modern apprenticeship appraisal system.

2.6 *Establish the four-stage assessment mode*

There are four stages according to the talent training mode, integrated into the relevant inspection requirements and examination standards. Students' professional skills and quality evaluation can be divided into "learn at school," "in the enterprise professional core courses and apprentice," "the apprentice rotational training," and "field work" four stages to complete implementation, each stage according to different learning contents, training content, set up different evaluation indexes, give prominence to the comprehensive evaluation, and according to the evaluation result they can decide to advance to the next stage of study and eventually be employed by chosen enterprises.

2.7 *Assessment content "four fusion"*

First, integrate the attitude and performance of the apprentice; second, the degree of theoretical knowledge of the specialty (post) of the integrated apprentice; third, the degree of mastery of the professional skills of the integrated apprenticeship practice; fourth, integrated vocational qualification certificate. Among them, the working attitude of apprentices and their internship performance account for 30%, the scores of theoretical examinations account for 30%, the assessment of professional skills accounts for 30%, and the vocational certificates account for 10%. After the end of the in-post internship, the prospective employees can only become regular employees through the enterprise assessment.

Table 1. List of teaching management systems (revisions) that are compatible with modern apprenticeships

The serial number	Name of management system
1	College modern apprenticeship pilot class teaching management implementation measures
2	College computer network technology (network security) professional master basic requirements and skills standards
3	College computer network technology (network security) requirements for apprenticeships and skills standards
4	The college supervises the duties of teachers
5	Job responsibilities of teachers in college
6	College modern apprenticeship practice management system
7	College modern apprenticeship safety measures and disciplinary measures
8	College modern apprenticeship management supervision
9	Thirdparty evaluation of college modern apprenticeship system
10	College modern apprenticeship routine teaching inspection work regulations
11	College modern apprenticeship regular school-enterprise consultation system
12	College modern apprenticeship pilot class student status management rules
13	College apprenticeship assessment system
14	College practice assessment system for quasi-staff
15	The college's associate-employee conversion (graduation) system
16	College student internship recall system

3 ACHIEVEMENTS IN THE CONSTRUCTION OF THE TALENT TRAINING MODE OF THE MODERN APPRENTICESHIP SYSTEM: INTEGRATION OF INDUSTRY AND EDUCATION, PROGRESSIVE AND SEGMENTAL TRAINING OF ABILITY

3.1 The school-enterprise double-principal education mechanism has been formed

By constructing the education platform of mixed ownership secondary colleges, the cooperation between schools and enterprises is further strengthened and the integration of industry and education is promoted. Through the platform, the college and the cooperative enterprise have reached a community of interests in professional co-construction and personnel training, and the enthusiasm and initiative of enterprises to participate in personnel training have been enhanced, forming a long-term mechanism for the school-enterprise joint development of modern apprenticeship.

3.2 The apprenticeship teaching management system has been gradually improved

In the pilot process, through school-enterprise cooperation and multiple participation, a set of systems suitable for the characteristics of the modern apprenticeship system have been gradually established and improved.

3.3 Enriched educational and teaching resources

Taking the pilot of modern apprenticeship system as an opportunity, professional and cooperative enterprises should strengthen the construction of pilot professional education and teaching resources. The HTML web design technology and computer application foundation jointly developed by the university and enterprise have been rated as a provincial teaching resource sharing course, and the teaching materials, courses, practice training base, and information teaching resources have been further optimized and enriched.

3.4 The construction level of teachers has been improved

Through the implementation of the modern apprenticeship pilot double tutor system, with the system construction as the starting point, the college further standardized and improved the construction and management of the double teacher team, and led to the establishment of a professional and part-time, school-enterprise mutual employment, education, and teaching ability, enterprise practice experienced teachers.

3.5 Improved social service capacity

Zhang Chaolun, an engineer from Zte's Asia-Pacific training headquarters, played a leading role in the field of network security and established the Zhang Chaolun master studio. With the advantages of master studios in the areas of apprentice transmission, skill tackling, skill inheritance, skill promotion and modern high-skilled personnel training, teachers' scientific research ability was improved, and the integration of school, enterprise, industry, learning, and research was realized. Up to now, I have applied for three national invention patents, completed eight technical services, completed 10,000 hours of non-academic training, and realized 2.39 million yuan in social training.

3.6 The quality of talent training has been improved, and the satisfaction degree of enterprise personnel has been improved

In the process of talent training, we adhere to the win-win concept of school enterprise and student, take enterprise demand as the guide, and comprehensively cultivate students' ability and quality. The quality and ability of the apprentices are highly compatible with the enterprise and highly recognized by the enterprise. Through the investigation, it is found that the satisfaction of enterprises to students reaches 100%.

4 CONCLUSION

"The capacity of teaching fusion, progressive, training" personnel training mode in the process of the pilot, in systems and mechanisms as the breakthrough point, from the curriculum system the "double tutors" teaching mode and evaluation mechanisms involved in the various propulsion, after three years of exploration practice, the project has achieved good results and has a certain promotion value.

It should be noted that there is still a long way to go in the theory and practice of production-teaching integration and the modern apprenticeship system, which requires us to constantly summarize and explore the implementation process and make contributions to the development of vocational education.

ACKNOWLEDGMENT

Fund project: Binzhou Vocational College college-level educational reform project 2019, number xyjg1910.

REFERENCES

[1] Opinions of the ministry of education on the pilot work of modern apprenticeship. Central people's government of the People's Republic of China.
[2] Guoguangjun.Enlightenment of British National Education and Training System on Modern Apprenticeship Education in ChinaJournal of Hunan Institute of Humanities Science and Technology 2016, (6): 6–6.
[3] Lihuan, Gaohuixia, Rrenchunxiao. Analysis on the Problems Existing in the Practice of Modern Apprenticeship in Higher Vocational CollegesModernization of education 2018, (25): 333–334.

Computational Social Science – Luo, Ciurea & Kumar (eds)
© 2021 Taylor & Francis Group, London, ISBN 978-0-367-70193-2

Analysis of high-tech industry capability of the Belt and Road countries - Based on the revealed comparative advantage index

H.Y. Yan & H.N. Qu*
School of Management, Shanghai University of International Business and Economics, Shanghai, China

X.Y. Ding
School of Economics, Zhejiang University of Technology, Hangzhou, Zhejiang, China

L.Y. Zhan
School of Management, Shanghai University of International Business and Economics, Shanghai, China

ABSTRACT: By calculating the Revealed Comparative Advantage Index (RCA) at different levels like regional area level, country level, industry segment level and industry segment link level, this paper seeks for the countries with outstanding competitive advantages from the areas with extraordinary industry capabilities. Based on that, we focus on advantageous industries in the target countries and further subdivide the industries into different links to precisely locate the trade complementary points for high-tech industry cooperation with countries along the Belt and Road. On the basis of the analysis results, policy implications for facilitating the trade cooperation with strong industry capability countries along the route are addressed.

1 INTRODUCTION

With the continuous development of the Belt and Road (B&R) initiatives, the cooperation among countries along the route has deepened from low-tech industries to high-tech industries. At the same time, with the recovery of trade protectionism in developed countries in the world, the resistance to economic and trade cooperation in high-tech industries has gradually increased. As for the B&R developing countries represented by China, it is becoming increasingly important to find new high-tech industry partners in addition to developed countries

Research based on Revealed Comparative Advantage Index (referred to as RCA) in China roughly started in the early 21st century. Yue (2000) used the United Nations World Trade Data to analyze the comparative advantage of China's foreign trade export product structure and found that China is gradually changing from traditionally exporting resources and labor-intensive products to labor and capital-intensive products. From 2007 to 2012, Shen (2007, 2012a, 2012b) used the RCA to systematically analyze the bilateral trade balance between China and the United States, and focused on the anti-dumping bilateral trade activities. In terms of research between China and the United States, Zhang and Sang (2015) focused on the bilateral trade of technology-intensive products and also pays attention to the issue of trade balance. The study of trade competitiveness using RCA began in the near years. Li (2019) conducted a research on the trade competitiveness of China and South Korea from the perspective of RCA. In terms of B&R, based on the RCA, Cao et al. (2020) studied the structural upgrading of China's exports to the five Central Asian countries under B&R background, focusing on technology added value and quality level analysis. Xu et al. (2015) adopted the network analysis method in constructing a new model for the high-end manufacturing trade in the B&R, and found that China, India and Singapore play a bridge role in the high-end manufacturing trade network of the B&R.

---

*Corresponding author

We can find that there are relatively few studies on the comparative analysis of the high-tech industry capability of countries along the B&R from the perspective of comparative advantage, although very few examined the differences in technological capabilities of countries based on RCA. Therefore, it is of practical significance to use RCA to measure the high-tech capabilities of countries along the B&R, screen out partners with strong high-tech capabilities, and accurately find the complementary trade points of countries along the route in the process of high-tech industry cooperation.

2 RESEARCH DESIGN

Because the technical capabilities required for the production of industrial manufactured goods are high, this study applies the export share of industrial manufactured goods to reflect the high-tech industry capabilities of various countries. Specifically, according to HS (1996) category and chapter catalogue, all products in categories 6 to 20, chapters 28 to 96 of HS code are classified as industrial manufactured products, and then corresponding to the relationship between the six-digit code of HS (1996) and the UN BEC classification, industrial manufactured products is divided into intermediate products and final products, which are used for area analysis, country and industry analysis. When analyzing segment industry, according to the correspondence between HS (1996) six-digit trade data and the International Standard Industrial Classification (ISIC/Rev.3), three sub-industries are divided: office accounting and computing machinery (C30), Broadcast and television communication equipment and instruments (C32), medical precision and optical instruments and watch production (C33).

This study follows the "regional area division" commonly used in the study of B&R to divide the B&R countries into six major areas according to geographical areas. By calculating the corresponding RCA at regional areas level, national level, industry level, and industry link level respectively, we find countries with outstanding performance in a certain segment industry link to precisely locate the trade complementary points for high-tech industry cooperation with countries along the Belt and Road.

3 RESEARCH METHOD AND DATA SOURCE

3.1 *Research method*

This paper uses a comparative analysis method based on RCA. RCA refers to the size of a product's share in a country's total product exports relative to the global product's share in global product exports. The calculation formula is:

$$RCA_{cj,t} = \frac{x_{c,j,t} \Big/ \sum_{j} x_{c,j,t}}{\sum_{c} x_{c,j,t} \Big/ \sum_{c} \sum_{j} x_{c,j,t}} \tag{1}$$

where $RCA_{c,j,t} = $ RCA of country c, product j in period t; $Xc,j,t = $ the export share of country c, product j in period t. The larger the calculated RCA value is, the better the product competitive advantage lies in the global product, which means that the country has strong high-tech industry capability in the certain industry where this product belongs to. It is worth noting that the denominator calculates the export share of the B&R countries, not the whole world.

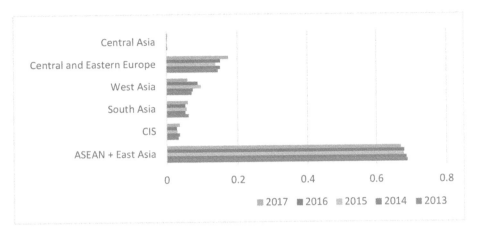

Figure 1.　Export Share of Industrial Products by Each Area from 2013 to 2017.
(Data source: UN Comtrade database, the same below)

3.2　*Data source*

The comparative export share data of the B&R of this study comes from UN Comtrade database for the period 2013–2017, and the RCA calculation of countries, industries, and links focus on the export share data of the B&R in 2017 from the UN Comtrade database.

4　RESEARCH RESULTS

4.1　*Area analysis*

As shown in Figure 1, based on the export share of various trade products of countries along the B&R from 2013 to 2017, this paper divides the countries according to six regional areas. By comparing and analyzing the export share of industrial products that can reflect the capabilities of high-tech industries, it is found that within 5 years , ASEAN + East Asia, Central and Eastern Europe, whose export share of high-tech industries accounted for more than 90% of the total share of the B&R, and the data remained stable, with a large-scale advantage, while the technology industry does not have obvious advantages in other area. So, we chose ASEAN + East Asia and Central and Eastern Europe to further study the specific characteristics of high-tech industry in the countries involved in each area.

4.2　*Country and industry analysis*

Focusing on the export share data of the two major areas of ASEAN + East Asia and Central and Eastern Europe in 2017, we respectively calculate the RCA of the entire area, the RCA of countries included in the area and the RCA of the three sub-industries with the purpose of exploring the high-tech industry capabilities in various countries, areas and high-tech sub-industries. The RCA calculation results in the three sub-industries of each country in the two areas are shown in Figures 2 and 3.

On the basis of the calculation results, we select countries with technical competitive advantages from two areas in three sub-industries as shown in Table 1:

4.3　*Industry link analysis*

Continuously focusing on the export share of 2017, we further segment the product link from the perspective of the industry link: divide the industrial products into intermediate product and

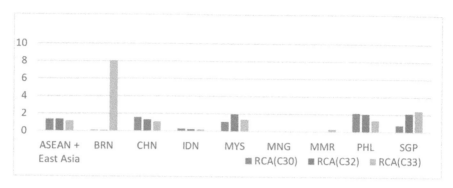

Figure 2. RCA of C30, C32, C33 in ASEAN + East Asia in 2017..

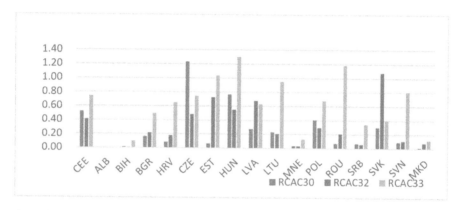

Figure 3. RCA of C30, C32, C33 in central and eastern Europe in 2017.

Table 1. Advantage countries in the two areas in C30, C32, C33.

	ASEAN + East Asia	Central and Eastern Europe
C30	CHN MYS PHL SGP	CZE HUN POL SVK
C32	CHN MYS PHL SGP	CZE EST HUN LVA SVK
C33	CHN MYS PHL SGP BRN	EST HUN CZE LTU ROU SUN

final product according to the production link, and then divide them into intermediate product category 1–5 and final product category 1–3, from high to low according to technical complexity. Combining the three sub-industries with 8 categories of products is to cut out each link of each high-tech industry. By analyzing the countries with outstanding industry level and their outstanding industries, the purpose is to find the countries with trade competitive advantages in each link and the links with trade competitive advantages within each country. RCA is still used for comparative analysis in this process. The top five advantageous countries selected in each link are shown in Table 2.

In the industry C30, compared with China, in intermediate product category 3&4, SGP, PHL, MYS from the East Asia + ASEAN area are more competitive, while countries from Central and Eastern Europe area do not perform as well as East Asia + ASEAN, but HUM has an advantage over China in intermediate product category 3. In the final product category 4, SGP, PHL, MYS

Table 2. Advantage countries in each link from C30, C32, and C33.

	I-1*	I-2	I-3	I-4	I-5	F-1	F-2	F-3
c30			SGP	SGP	CHN	CHN	MYS	CHN
			MYS	PHL	PHL	CZE	SGP	CZE
			HUN	CHN	SGP	HUN	PHL	PHL
			SVK	MYS	MYS	SGP	CZE	MYS
			PHL	HUN	HUN	MYS	CHN	HUN
c32	SVK	SGP	SGP	CHN	CHN	SGP	CHN	MYS
	CZE	MYS	MYS	SGP	MYS	MYS	MYS	CZE
	CHN	CHN	PHL	MYS	PHL	CHN	CZE	CHN
	HUN	CZE	CHN	PHL	SGP	CZE	SVK	SGP
	SGP	HUN	CZE	SVK	HUN	HUN	SGP	SVK
c33	SGP	SGP	MYS	PHL	PHL	SGP	CHN	SGP
	MYS	CHN	CHN	CHN	CHN	CHN	SGP	PHL
	CHN	PHL	HUN	SGP	SGP	LTU	PHL	CHN
	CZE	LTU	EST	CZE	MYS	MYS	EST	CZE
	LTU	MYS	SGP	HUN	LTU	EST	SVN	EST

*"I-1"refers to "intermediate product category 1"; "F-1" refers to "final product category 1"

from the ASEAN + East Asia area, and CZE from the Central and Eastern Europe area all have stronger high-tech industrial capabilities than China.

In the industry C32, the intermediate product category 4&5, China has a significant advantage among B&R countries. However, in the intermediate product category 1, it is found that the Central and Eastern Europe area performs stronger. Among them, SVK and CZA can be used as Chinese technology learning objects in this link. China performs moderately in the intermediate product category 3, SGP, PHL and MYS are stronger than China in this category. Similarly, in final product category 1&3, the high-tech industry capabilities of MYS are superior to China.

In the industry C33, SGP is in the first position in the intermediate product category 1&2 and the final product category 1&3. MYS is also superior to China in the intermediate product category 1, PHL has a competitive advantage over China in the final product category 3. In this industry, the capability of the ASEAN + East Asia area is generally strong.

5 IMPLICATIONS AND LIMITATIONS

5.1 Implications

On the basis of the analysis results, to facilitate China's seeking for new high-tech partners in B&R countries and strengthening cooperation in high-tech industries along B&R, we propose the following three policy implications:

1) Steadily develop the high-tech industry with advantages, complement the industry's shortcomings through trade complementation, to enhance the technology trade complementarities of the countries in the regional production.
2) Adopt a more open cooperation policy to get involved in the regional production cooperation network with an open mind, and enhance the level of mutual benefit among countries.
3) Focus on technological innovation and accelerate industrial transformation and upgrading. Strengthen technological innovation and cooperation in the areas of cultural exchanges, platform construction, major project construction, characteristic park construction, and common technical cooperation research, to promote industrial transformation and upgrading, in the areas of Central and Eastern Europe and ASEAN + East Asia.

5.2 Research limitations

Although the 2013–2017 data shows that the two major areas of the ASEAN + East Asia and Central and Eastern Europe have stable and outstanding capabilities in the high-tech industry, it is still constrained by lacking data representation by only focusing on the 2017 trade data for follow-up research.

REFERENCES

Cao, C., Chen, J., Xia, Y. 2020. A Study on the Upgrading of China's Export Commodities Structure to the Five Countries in Central Asia under the Background of the Belt and Road Initiative—Based on the Analysis of Dominant Comparative Advantages, Technology Added Value and Quality Level. *Journal of Xinjiang University (Philosophy, Humanities and Social Sciences Edition)* 48(01): 48–56.

Hausmann, R. & Rodrik, D. 2003. Economic Development as Self Discovery. *Journal of Development Economics* 72(2):603–633.

He, M. & Zhang, N. N, & Huang, Z.Q. 2016. Analysis on the Competitiveness and Complementarity of Agricultural Products Trade between China and "One Belt One Road" Countries. *Agricultural Economic Problems*37 (11): 51–60 + 111.

Lall, S. 2000. The Technological Structure and Performance of Developing Country Manufactured Exports, 1985–1998. *QEH Working Papers* 28(3):337–369.

Li, B.Y. 2019. Research on Sino-Korea Trade Competitiveness from the Perspective of Dominant Comparative Advantage. *Statistics and Information Forum* 34(07): 44–53.

Shen, G.B. 2007. Explicit comparative advantage, intra-industry trade and Sino-US bilateral trade balance. *Management World* (02): 5–16+171.

Shen, G.B. 2012a. Explicit comparative advantage and US trade effect on anti-dumping of Chinese products. *World Economy* 35(12): 62–82.

Shen, G.B. 2012b. Explicit comparative advantage: Are Chinese products subject to the curse of American anti-dumping? *Financial Research* 38(08): 122–134.

Xu, H.L. & Sun, T.Y. & Cheng, L.H. 2015. Research on the "Belt and Road" high-end manufacturing trade pattern and influencing factors—Analysis of index random graph based on complex network. *Finance and Trade Economics* (12): 74–88.

Yue, C.J. 2000. An empirical analysis of changes in my country's foreign trade export structure and comparative advantage. *International Economic and Trade Exploration* (03): 2–4 + 82–86.

Zhang, B. & Sang, B.C. 2015. Research on Sino-US Bilateral Trade of Technology-Intensive Products—Explicit Comparative Advantage, Intra-Industry Trade and Trade Balance Research. *Asia-Pacific Economy* (01):43–49.

Zhao, D.Q. & Sang, B.C. 2016. International capacity cooperation under the "Belt and Road" initiative: an empirical analysis based on the international competitiveness of industries. *International Trade Issues* (10): 3–14.

Computational Social Science – Luo, Ciurea & Kumar (eds)
© 2021 Taylor & Francis Group, London, ISBN 978-0-367-70193-2

Research on construction of working mechanism of enterprise knowledge management

G.L. Gao, H.Y. Gao & W. Tan
State Grid of China Technology College, Jinan, China

X.H. Pan
State Grid Shandong Electric Power Company, Jinan, China

X.H. Li, S.C. Jin & Z. Zhang
State Grid of China Technology College, Jinan, China

ABSTRACT: In the era of knowledge economy, an important issue facing enterprise universities is how to integrate and utilize all kinds of knowledge resources of enterprises. This paper expounds the function of knowledge management, analyzes the present situation and existing problems of knowledge management in the college, and demonstrates the necessity and feasibility of implementing knowledge management, and puts forward countermeasures and suggestions for the implementation of knowledge management, so as to comprehensively improve the level of knowledge service.

1 INTRODUCTION

With the arrival of the era of knowledge economy, the status and role of knowledge has become increasingly prominent. More and more enterprises apply the concepts, methods and tools of knowledge management to implement enterprise knowledge management and improve the competitiveness of enterprises. How to effectively acquire, accumulate, share, innovate and transform knowledge into benefits has become the focus of enterprise knowledge management research.

As a highly knowledge-intensive enterprise, the college is still in the growth stage of knowledge management. The knowledge management work of the college regards knowledge as an important resource of the college, and studies and constructs the long-term working mechanism of knowledge management through the identification, acquisition, storage, sharing, application and creation of knowledge resources, improves the management efficiency and innovation ability of knowledge resources, so as to enhance the core competitiveness and comprehensive strength of the college.

2 CHARACTERISTICS OF KNOWLEDGE MANAGEMENT

In the knowledge management model, knowledge resources are the core of knowledge management. The knowledge activities carried out around knowledge resources include knowledge identification, creation, acquisition, storage, sharing and application. The outer layer of the model is the necessary elements to ensure the smooth and sustainable implementation of knowledge activities, including enterprise organizational culture, organizational structure and system, technical facilities and so on.

The college is a typical knowledge and technology intensive enterprise and the staff are typical knowledge workers, so the knowledge management has the following characteristics. First, the employees have professional knowledge background, reflect a high degree of autonomy in their work, and can achieve self-guidance and self-management of their work. Secondly, the employees are mostly engaged in creative work. Employees are proficient in specialties, have outstanding innovative ability, and constantly form new knowledge achievements in their work. Finally, the scope of knowledge management includes student training and teaching resources, relevant laws and regulations, various standards and norms, scientific research project materials, laboratory materials, knowledge experts and digital libraries, etc.

3 THE KEY LINKS IN THE CONSTRUCTION OF KNOWLEDGE MANAGEMENT MECHANISM

3.1 *Establishment of knowledge management regulation system*

Knowledge management system needs the support of three elements: technology, management and culture. Knowledge management should create conditions for effective knowledge acquisition, sharing and application. Therefore, in order to construct the long-term working mechanism of knowledge management, it is necessary to establish and practice the knowledge management regulation system.

(1) Clarify the organizational structure and job responsibilities of knowledge resource management. By standardizing the job responsibility requirements involved in the process of knowledge management organization, and the requirements for knowledge contribution, etc., to ensure that knowledge has a continual source. The knowledge resource system of college management and full staff application should be constructed in an all round way.
(2) According to the types of knowledge resources, sort out the work flow of knowledge resource database management, and standardize the management of knowledge resources in and out of the knowledge base. By standardizing the technical standard requirements of the stored knowledge resources, the quality of the stored knowledge resources is ensured.
(3) Standardize the renewal system of knowledge resources, and clarify the requirements of the scope, quantity and frequency of knowledge resources renewal. Ensure the timely updating and updating efficiency of the knowledge base of all relevant departments and posts.
(4) Strictly examine the property rights of knowledge resources, the arrangement and modification of knowledge resources, the disclosure, restriction and secrecy of knowledge resources, and the provisions of the ownership of knowledge resources to ensure the safe dissemination of knowledge in the process of application.
(5) Stipulate positive and negative incentive measures for knowledge resource management, bring knowledge management into the daily work assessment system, combine incentive and assessment, and encourage departments and employees to actively participate in the construction of knowledge resource base, knowledge integration and management.

3.2 *Fully develop and make use of knowledge resources*

The key to knowledge acquisition is to determine the source of knowledge. The knowledge source of the college includes internal and external explicit knowledge and internal and external tacit knowledge. For example, the training and teaching course resources of college staff training and academic education belong to internal explicit knowledge; the on-site experience of part-time trainers in college belongs to external tacit knowledge and so on.

Explicit knowledge is highly structured, and knowledge experts can mine the stock knowledge resources database of various departments of the college by using knowledge system construction methods such as post ability and quality model. Scientific, effective and useful knowledge resources are stored in the knowledge base according to the technical standards and requirements of the knowledge base.

Tacit knowledge can be acquired through various types of knowledge activities, such as organizing expert lectures, professional seminars, online and offline training and teaching courses, knowledge sharing and comments, and other interactive knowledge activities.

3.3 Promote knowledge achievements transformation

Knowledge innovation is the goal of knowledge management, and the sustainable competitive advantage comes from continuous knowledge innovation. Only continuous knowledge innovation can expand the scale of knowledge and improve the quality of knowledge. Knowledge innovation is often the result of the sharing, mining and application of existing knowledge. Through knowledge sharing, mining and application, employees recombine knowledge and create new knowledge value.

By regularly combing the main technical and professional innovation achievements of the enterprise, the knowledge integration achievements will be authoritative and the work will be normalized. According to a series of new materials, new equipments, new processes, new technologies developed and used in the enterprise, and the achievements and technical specifications issued by the enterprise, the training departments are organized to analyze the innovation achievements. The knowledge lists are refined according to the technical skills professional categories, so as to make the innovation achievements knowledgeable, integrated and systematic.

Guide the promotion and application of knowledge integration of professional innovation achievements, and promote the transformation and application of knowledge achievements. The training departments are organized to study the organic docking of new and old knowledge, and to complete the effective integration of new knowledge. The new abilities and knowledge achievements are sorted out into the training course organically. The training materials and courseware of new knowledge are gradually developed, and the new knowledge achievements are applied to technical skills training and centralized training courses of new employees.

4 LEAN MANAGEMENT OF KNOWLEDGE BASE

4.1 The meaning of knowledge resources element

Knowledge element is the key or core knowledge and skills that are refined to the minimum indivisible and independent according to the inherent logic of the knowledge system. Knowledge element, which is the basic unit of knowledge organization, is a knowledge specification representation method proposed and deeply studied by many scholars. They studied knowledge element from the perspective of literature information service, and proposes to realize knowledge resource integration service based on knowledge element.

In this paper, knowledge element is regarded as the minimum knowledge resource unit which is independent and complete and can have an effect on staff training and teaching. Knowledge element has three characteristics, one is integrity, knowledge element can completely represent a certain principle, method or skill. The second is regeneration, because knowledge element is the most basic training and learning unit, through the recombination of knowledge elements, new knowledge elements can be formed. The third is openness. In the process of training, knowledge elements can be combined freely, and the choice and combination of knowledge elements depends on the problems to be solved. On the one hand, employees can choose and combine their own training and learning content according to their own basic and ability needs, on the other hand, trainers can adjust the knowledge element combination of training at any time according to the changes of enterprise development, so that training can keep up with the changing needs of enterprise development.

4.2 The structure of knowledge resource base

Knowledge resource base is a collection of training resources that are stored, managed, indexed and used according to certain rules.

There are certain advantages in constructing knowledge resource base on the basis of knowledge element. First, the regeneration of knowledge elements makes the knowledge base more intelligent. There is a link relationship between knowledge elements, and knowledge elements themselves also have strong vitality, which makes the knowledge resource base based on knowledge elements more intelligent. Second, the knowledge resource base based on knowledge element is an important foundation for the construction of automatic push mechanism of training resources and personalized training. During the development of knowledge element, according to the unified development technical standards, it is associated with the post system to form a knowledge map to provide support for personalized learning of the employees.

4.3 Construct the knowledge resource system of post ability

On the basis of the job training norms, it forms a post ability knowledge system, comprehensively sorts out and organizes the development of training standards, teaching materials, question bank courseware and other types of knowledge resources, so as to realize intensive management of knowledge resources, unified planning, collect stock, and implement them step by step, promote the co-construction and sharing of knowledge resources.

4.3.1 Knowledge resource construction mechanism based on project management

The construction of knowledge resources is a complex and systematic work. In order to improve the management efficiency of knowledge resources construction projects, the construction mechanism of knowledge resources is constructed by using the method of project management.

By combing the elements of knowledge resources construction, comprehensively analyzing the relationship between the demand elements of knowledge resources construction and the expected elements of staff training, the project management system is introduced into the process of knowledge resources construction, and the knowledge resources construction mechanism based on the division of responsibility, project classification and project process assessment of project management is constructed.

The organizational methods of project management are applied to the construction of knowledge resources. Through the construction of resource construction and management mechanism based on project management, the rapid sharing and utilization of knowledge resources is realized, and the construction quality of knowledge resources is improved.

4.3.2 Knowledge resources navigation mechanism based on learning map

By connecting knowledge resources with the classification of post ability and learning path, developing students' learning map covering core posts, the knowledge resources navigation mechanism based on learning maps can be established, the presentation mode of knowledge resources can be optimized to promote the rapid, orderly and continuous training of talents.

(1) Take the work task as the guidance, establish the competency standard of the typical task, identify the main ability items and the difficulty level of the ability elements under the typical task, analyze the post knowledge and skills, and determine the learning content, so that the students' learning content is related to the work task.
(2) Construct the learning development path of students, draw a learning map, and organically combine learning tasks, learning contents, learning methods and career development paths.
(3) According to the knowledge and skills involved in the task, optimize the teaching content and strategy, design and develop an integrated curriculum system such as courses, cases or operation manuals.

5 KNOWLEDGE RESOURCE FULL-CYCLE MANAGEMENT INFORMATION PLATFORM

Knowledge management information platform is a platform that supports the collection, processing, storage, transmission, application and innovation of enterprise knowledge through technical tools such as file management system, search engine, expert system and knowledge resource base, so that enterprise explicit knowledge and tacit knowledge can be transformed into each other.

Based on the knowledge management information platform, an information support platform for the full life cycle management of all kinds of knowledge resources will be built to improve the level of lean management of knowledge resources, and support the integration of online and offline training business. The control of the results and process of knowledge resources construction and management is achieved through information technologies.

6   CONCLUSION

Combined with the current situation of knowledge management in a college, this paper carries out the research on knowledge integration and management with long-term working mechanism as the core. Focusing on the characteristics of knowledge management and the key links of constructing the long-term working mechanism of knowledge management, this paper studies and explores the lean management of knowledge resource base and the construction of knowledge management information platform. The realization of enterprise knowledge management needs to constantly discover and solve the problems in knowledge management in practice, and improve the quality and efficiency through sharing, applying and innovating the knowledge resources of enterprises.

REFERENCES

[1] Gao Guangling, Zhang Peiyun. The Research and Practice on the 220kV Practical Training Course Construction of Intelligent Substation. Journal of State Grid Technology College. 2015, 18(2):48–51.
[2] Chun-ming Leung .From connectivity to next-generation learning. Coiffait & Hill. Blue Skies: New thinking about the future of higher education. London: Pearson Press, 2012.45–48.
[3] GAO Guangling, PAN Xianghua etc. Study on cluster measurement and control device of intelligent substation. The IEEE Conference on Energy Internet and Energy System Integration, 2018.2938–2942.

Computational Social Science – Luo, Ciurea & Kumar (eds)
© 2021 Taylor & Francis Group, London, ISBN 978-0-367-70193-2

Research on the positive role of social responsibility in the development of the e-sports industry

X.Y. Zhang

School of Communication and Arts, Shanghai University of Sport, Shanghai, China

ABSTRACT: E-sports have been developing rapidly in recent years as a new sports industry. The development of e-sports is closely related to the rapid development of Internet information technology and the popularization of live broadcasts. With the increasing scale of the industry, the influence of e-sports is constantly increased. E-sports are different from traditional online games. But being addicted to e-sports will seriously affect the study, work, and life in the traditional concept of people. The e-sports industry must play a positive role and assume social responsibility if it wants healthy development. In this paper, the positive role produced by the social responsibility is analyzed deeply in the development of the e-sports industry. It is suggested that the e-sports industry should actively undertake social responsibility, accelerating industrial upgrading and integration. The overall social influence of the e-sports industry is promoted in a correct and positive way. The industry norms of the e-sports are established, forming a sustainable benign development.

1 INTRODUCTION

In recent years, e-sports has been widely concerned and favored. The e-sports industry has entered gradually into the public. E-sports and online games belong to a broad category of electronic games. E-sports is derived from online games but it is different from online games. E-sports has the basic characteristics of physical exercise, such as the strict time and round limit. It is beneficial to exercise and improves coordination, thinking and responsiveness of participants. The will, quality, competition, cooperation, crisis awareness and team spirit of participants also are improved by e-sports.

For some time, it has been believed that e-sports can make people addicted to it, especially affecting the healthy growth of young people. In fact, there are the game elements whether for online games or the e-sports industry based on online games. E-sports will cause a bad impact if it is not used properly, especially young students. The e-sports and online games both have reason for their development and existence. The key problem is to make sure that they are developed healthily. In such a situation, it is extremely important to establish a proper concept of e-sports. The e-sports industry not only should undertake the social responsibility, but also it can fulfill the social obligation. E-sports industry must be developed healthily, producing positive roles and energy.

The meaning of e-sports and its social and exercise value as well as the existing problems are analyzed in the development of e-sports in China. Some feasible strategies are put forward for the benign development of e-sports [1]. Through detailed data analysis, it is pointed out that China has become a big country of e-sports, and the e-sports industry is developing rapidly. The benign development of the e-sports industry not only can provide the new economic growth for our country but also can add a new highlight to the development of cultural industry [2]. The development status of the e-sports industry at home and abroad is studied in detail, and many factors affecting it in China are analyzed, providing a certain theoretical basis for the sustainable and healthy development of the e-sports industry [3].

The e-sports industry is a new thing, so there is little literature about it. Furthermore, the research on the role and significance of social responsibility is rarely reported in the development of the e-sports industry.

In view of the latest developments of the e-sports industry, this paper describes that the role and significance of social responsibility are analyzed deeply in the development of the e-sports industry, so as to provide some beneficial references for the healthy and sustainable development of the e-sports industry.

2 E-SPORTS AND ITS RELEVANCE TO SOCIAL RESPONSIBILITY

As a sports event, e-sports are based on e-sports games. It is carried out in the virtual environment through hardware and software equipment with information technology. E-sports are such that electronic game competition has reached the level of "competition." It has the characteristics of uniform rules, of fair and competitive. To be specific, e-sports use electronic equipment to carry out the intellectual confrontation between people. Electronics and competition both are the basic characteristics of electronic sports, which highlights its sports property. (1) Electronic is a means. In e-sports, virtual environments are created to carry out, by a variety of hardware and software facilities using science and information technology, competition similar to the traditional sports, such as venues and equipment. (2) Competition is the essential nature of sports, which is different from other electronic games, especially the online games. Although e-sports has many projects, its core must be competition. E-sports have basic characteristics, including ornament, competition, strategy, organization, and recreation similar to traditional sports [4].

Social responsibility refers to the responsibility, mission, and obligation that an organization should bear for country and society. Therefore, an organization should be managed and operated by the conducive way to social development. Social responsibility is that an obligation is not required by economy or law, but it belongs to the voluntary action and social morality. Everyone should have the sense of social responsibility.

E-sports are inseparable from social responsibility. With the increasing development of the e-sports industry, e-sports will form a social organization. The social organization should assume the certain social responsibility and obligation. Therefore, the healthy development of the e-sports industry is inseparable from its commitment to social responsibility, which plays an important role and significance in the healthy development of the e-sports industry.

3 THE POSITIVE ROLE OF SOCIAL RESPONSIBILITY IN THE E-SPORTS INDUSTRY

3.1 Social responsibility promoting the commercialization and industrialization of e-sports

Social responsibility plays a key role in the development of the e-sports industry. Nowadays, e-sports can attract many young, middle-aged, and aged people to participate in them, which has an increasing impact on the development of social economy [5, 6]. The e-sports industry truly integrates into the mainstream culture and public life. People from different regions fully understand the e-sports industry in various ways and feel the infinite charm of e-sports. E-sport is conducive to the integration of local economic and cultural development with the e-sports industry. It further expands the scale of the e-sports industry. E-sports enterprises should assume more social responsibilities. The e-sports fans are guided through the reasonable and correct way. The industrial structure is promoted to improve continuously. The e-sports industry and market have a good prospect and huge potential.

3.2 Social responsibility guiding teenagers to face e-sports correctly

Under the requirement of assuming social responsibility, some schools will open scientific and systematic training courses for e-sports game fans including different ages, talents, and dreams.

They are reasonably guided to establish a correct view of concepts. With the rapid development of network information technology and electronic entertainment game industry, the outlook of life, values, world, history, and other aspects about teenagers will be subtly influenced by various electronic games, including e-sports. There are two advantages for undertaking social responsibility: (1) The e-sports game developers will be avoided to pursue profit blindly. (2) The guidance and correct protection to teenagers should be strengthened to avoid addiction to e-sports games. It is helpful to prevent teenagers from being misled.

3.3 *Social responsibility conducive to build a healthy e-sports ecology*

The e-sports industry has a certain responsibility to build a harmonious society and a good social environment. E-sports enterprises improve the awareness of social responsibility. More and more people are stimulated to assume social responsibilities because of enthusiasm. It is conducive to improving and establishing a standardized development mode of the e-sports industry. The good e-sports ecology is built, which is conducive to the sustainable benign development of the e-sports industry. E-sports can promote the sustainable, healthy, and high-speed development of social economy. The marketing of e-sports advertisement and events, the investment in e-sports, and the consumers are increasing, thus promoting the rapid development of related industries and the employment.

The sustainable healthy development of e-sports provides people with more opportunities and platforms. The talents and those who are interested in e-sports are attracted to participate in it to realize their own life value. At the same time, it is conducive to encourage more e-sports fans and practitioners to act together and undertake more social responsibilities actively. Game players can participate in social services through the platforms. More young people will be brought to participate in social welfare activities, which will be conducive to the construction of social civilization, such as joining hands with various teams and players.

3.4 *Social responsibility is conducive to e-sports research of more profession, standard, system, and science*

The academic activities between academic communities and universities will be increased by the sense of social responsibility in the e-sports industry. More relevant experts and scholars are encouraged to focus their research direction on the emerging e-sports, which can strengthen the scientific research on e-sports.

The e-sports can be combined with education by social responsibility. This combination is promoted by social responsibility, and it is a new attempt [7, 8]. Some universities have opened courses or lectures about the e-sports industry. The direction of e-sports commentary is opened in Shanghai Institute of Sport, majoring the broadcast and the host. On this basis, the first e-sports major will be established in China as a pilot project. Under the requirement of social responsibility, we need to learn e-sports in a formal and systematic way. The e-sports college will be set up, including various professions related to e-sports. It will promote the vigorous development of the e-sports industry

3.5 *Social responsibility for the professionalization of e-sports industry providing new ideal and direction*

As the competent department of e-sports, Sports Information Center of General Administration of Nation has been committed to guiding the healthy development of e-sports in recent years. Social responsibility makes e-sports develop in accordance with the rules of sports, and we should learn from professional sports. At present, the Sports Information Center has started to set up professional e-sports leagues and the relevant management systems are formulated for coaches, referees, and athletes. The relevant organizations are established such as e-sports publicity and promotion committee, and e-sports education and training committee. Social responsibility can help the relevant industries support the development of e-sports industry. So, the professional e-sports will usher in a great development and prosperity of the era. The spirit of transcending oneself and pursuing excellence in e-sports is consistent with the traditional sport projects. Social responsibility is of great significance to the development, extension, inheritance, and spirit of e-sports.

4   CONCLUSIONS

The e-sports industry is at an initial stage, and there are still many problems to be solved. Both opportunities and challenges co-exist in the e-sports industry with great development space. Under the situation of rapid development in the e-sports industry, we will constantly learn and adapt to environment. The content and characteristics in e-sports industry are understood. Their disciplines are analyzed and summarized to have a whole, comprehensive, and systematic understanding and cognition.

We should think exactly of the social responsibility and the development of the e-sports industry. We should consciously take on more social responsibility. The social responsibility can play a positive role of social responsibility in the development of e-sports, making the e-sports industry obtain benign, sustainable, and healthy development.

REFERENCES

[1] Li, F. 2013. The healthy development of e-sports industry. *Science Technology information* (2):334.
[2] Tian, Y. 2017. How to develop benign e-sports industry. *Culture monthly* (12):34–38.
[3] Fan, G. 2007. Research on the influential factors of industrial development of e-sports in China and its system mode. *Market Modernization* (13):353–354.
[4] Li, Y.P. & Huang, H.J. 2016. Recent problems in the development of e-sports industry in China. *Shan dong Sports Science and Technology* 38(6):35–39.
[5] Gao, Y. & Zhao, R.X. & Du, M. 2015. Research on the e-sports industry development in China. *Journal of Harbin Sport University* 33(6):54–58.
[6] Tie, Y. & Zhao, C.F. 2017. Research on E-sports Industry in China. *Sports Culture Guide* (7):100–104.
[7] Mei, Y.F. & He, Z.P. & Li, Y.M. 2017. The influence of e-sports as an official sports specialty on the education of contemporary college students. *The Science Education Article collects*: 399.
[8] Liang, D.Z. 2016. Discussion on the development of e-sports industry in universities. *Cooperative Economy and Technology* (14):12–13.

Computational Social Science – Luo, Ciurea & Kumar (eds)
© 2021 Taylor & Francis Group, London, ISBN 978-0-367-70193-2

Study on resilience of urban planning from the perspective of urban epidemic prevention

C. Peng
Department of Civil Engineering, Guangxi Vocational College of Water Resource and Electric Power, City of Nanning, China

X. Li
School of Management, Xi'an University of Architecture and Technology, City of Xi'an, China

ABSTRACT: The outbreak of 2019-ncov pneumonia in early 2020 had a huge impact on the country and posed a challenge to the level of management of public health emergencies in the country. In the reflection of this epidemic situation, this paper analyzes the necessity of building resilient cities in urban epidemic prevention from the angle of urban planning, and puts forward the planning key points and measures pertinently, with a view to improve the professional scope of urban planning and similar events to predict and handle the ability to ensure urban safety.

1 URBAN EPIDEMIC PREVENTION FEATURES AND RESILIENCE PLANNING CONCEPTS

1.1 *Features of urban public health emergencies and epidemic prevention*

Public health emergencies in the city signify unforeseen sudden occurrences in the urban area, which can cause or have the possibility of causing major public infectious diseases diseases that affect the public, and diseases that have unknown causes, a safety and health event. The main characteristics of urban public health emergencies are multiple causes, such as severe infectious diseases or plagues, and some epidemics occur along with natural disasters. In addition, drugs, food, and other issues will trigger public health emergencies. Second, from the spread of the plague in Europe to this new type of coronavirus pneumonia, the large-scale transmission to a population is due to the mobility of the population, which has a major impact on public health and safety Finally, there is the complexity of the harm. After the incident, not only in the field of health, but also in economic, political and other aspects, it will cause huge impacts. For example, this epidemic poses a severe challenge to our economy.

1.2 *Urban resilience planning concept*

Since the beginning of the 2008 Chicago Climate Action Plan, the concept of "resilient cities" has been put forward globally, and the concept of "resilience planning" for cities has been put forward (Yan Shuiyu et al. 2020). By 2018, China was the first to propose "building a more sustainable and resilient ecological city" in Shanghai, which clearly defined the determination and direction of China's development of resilient cities. In the concept of a resilient city, the vulnerability of the city comes from the level of economic growth, emergency management capability, and so on, and the destruction of the city is caused by various natural events. Therefore, in the construction of a resilient city, the concept of overall planning should be emphasized, and the sustainable development of the city can be realized by combining the analysis of the resource carrying level within the city and the planning scheme according to local conditions.

Previous studies on urban resilience planning usually concentrated on the natural resource conditions of the city and the damage caused to the city by various unpredictable natural disasters (Li Ye et al. 2020). However, in this Corona virus of 2019 epidemic, new problems have been raised on the spatial layout and resilience development of urban planning.

2 KEY POINTS OF RESILIENT CITY PLANNING FROM THE PERSPECTIVE OF URBAN EPIDEMIC PREVENTION

2.1 Create a unitized pattern to form a fully functional community

In accordance with the analysis of the transmission route and isolation method of Wuhan 2019-nCoV, the most critical measure is to cut off the mobility from provinces to cities to communities and to other communities to avoid the spread of infection sources. Since 31 provinces and cities across the country have started the first-level response of emergency plans one after another, the critical measure is to carry out traffic control and to reduce the spread of infection to the greatest extent by cutting off traffic links with key epidemic areas. In terms of urban planning, in response to the same kind of public health and safety events with certain infectious efficiency, in order to meet the needs of traffic control and not affect the normal life of residents as much as possible, the unitary pattern will be the future trend of urban planning (Ma Chao et al. 2020)

In ancient China, there were relevant urban planning schemes for the unitary pattern. For example, Chang'an City in Tang Dynasty, as the "first city in the world" with a population of more than 1 million at that time, its overall urban planning and layout comprehensively considered factors such as natural terrain and military defense. In order to facilitate the management, the city system was established to make the boundaries of each residential area clear, although to some extent, its hindered population mobility has affected the development of social progress, but it can be found that in the face of the outbreak of infectious diseases and other epidemics in modern cities, the urban pattern has its applicability.

When public health and safety incidents of infectious nature occur, unit management should be carried out quickly, which is similar to the community form in the current urban planning. However, the full functionality of the community should be paid attention to, and all the infrastructure necessary for people's lives should be built within the community to provide a basic living guarantee in case of the epidemic. To ensure various energy supplies in community units, such as electricity and gas, in addition to paying attention to the stable supply of water in the water supply and drainage facilities, we should also consider the diversion plan of sewage. In this transmission route of Wuhan COVID-19, fecal transmission is included. Therefore, in future urban pipe network design, we should also consider the diversion emergency plan of sewage pollutants to deal with the same type of public health safety emergencies. In addition to the above guarantees, at the beginning of the epidemic development, there was a phenomenon of patients seeking treatment across regions due to insufficient accommodation level, which further expanded the scope of infection. Therefore, in the design of the unitized pattern of the community, the proportion of relevant functional facilities should be optimized to enable the community under the unitized pattern to meet the basic living guarantee within its jurisdiction and reduce the expansion of adverse conditions.

2.2 Form joint defense and joint control, and set up an emergency response system

For the sake of improving the management efficiency, besides unitized management, we should also pay attention to the linkage between units and form an effective joint defense and control mechanism. For urban planning, it is necessary to create an organic channel for the connection between units, which can become a public open place at ordinary times and a rigid node for the connection of unit patterns during the special period of urban epidemic prevention. In particular, underground space can be fully utilized to make its functions more diversified. In the daily underground space design, in addition to meeting the daily needs, it will also combine the functions of air defense and storage. Through this epidemic situation, new ideas can be developed in the underground space design in urban planning, an epidemic prevention space linking the ground and underground can be created, and a corresponding emergency linkage system can be established.

Just as Gundersen and Huolin put forward their adaptive ability for the change of resilient cities, they can keep the social ecosystem continuously adjusted through reasonable urban planning and have dynamic adaptability. In the joint defense and joint control of the unitized pattern, a dynamic connection can be maintained to realize the organic series connection of each unit, thus improving the city's emergency response capability, absorbing the impact of emergencies on the city and realizing the sustainable development of the city. The characteristics of the urban epidemic are different from the urban damage caused by natural disasters. In this epidemic, it is highlighted that there is a huge contradiction between the ability of urban resources to respond to the epidemic, on the one hand, the lack of resource reserves, and on the other hand, the limited ability to allocate resources, Therefore, after the unit pattern of urban planning is established, all kinds of resource reserve standards can be set to balance the city's needs, to meet the regional use and benign flow under daily conditions, to ensure the dynamic equilibrium of each unit, and in the epidemic situation. It can quickly reach the resource concentration to meet the emergency needs of a certain unit, realize the optimal management of resource reserves and deployment, and improve the city's ability to combat sudden epidemics.

2.3 *The Internet enables smart communities*

Since the disclosure of the "human-to-human" transmission of 2019-nCoV in January 2020, various industries and industries have resumed work in March, and the closed management of provinces, cities, and communities has undergone at least two 14-day closed cycles. In the event of an emergency, the essential lives of the people are guaranteed. On the one hand, it depends on the implementation of emergency plans and measures, and on the other hand, the management level is improved through the "Internet +" technology. The realization of smart communities within the city through Internet technology is the future trend. For example, in this epidemic, the closure of crucial infected areas, or the closure of buildings with confirmed patients, has affected the purchase of residents' daily necessities. The current response method is to utilize WeChat (a mobile application) groups to register and purchase, and property or community personnel to distribute, and requires each owner to submit procurement requirements within a certain time to facilitate collective procurement. After the realization of smart communities, 24 orders are placed every hour to improve convenience and meet diversified needs and to minimize the impact of the closure on residents' lives.

Therefore, from the perspective of urban planning, it can be considered to set a center point in the community network to facilitate the full coverage of the smart community, provide storage, turnover and allocation of various materials, and link property and information technology to improve community management and service level.

In addition to providing convenience in life, the smart community can also register and share information through the Internet. Through this epidemic situation, it can realize the collection of tracked information of people's moving about places, personal sense information, and so forth, so as to effectively lock the whereabouts of relevant personnel when conducting close contact investigation and timely contact and isolation, effectively blocking further disease transmission. From the realization of the above smart community, we can also see the future urban layout of light volume, high turnover and modular urban pattern.

2.4 *Risk prediction of resilient cities*

In response to the urban epidemic situation, timely disposal is an essential content, but risk prediction is also important. Therefore, in the future of urban planning research, it is necessary to strengthen the safety risk assessment of the city, carry out risk prediction in advance in combination with different urban characteristics, make flexible management of the overall urban layout, and plan the places and land for emergency medical treatment, safety disposal, and personnel distribution in advance; for example, the site selection of Huoshenshan Hospital and Leishenshan Hospital in this epidemic is still imperfect In future urban planning, especially in megacities, attention should be paid to the reservation of relevant infrastructure location and land reserve.

3 CONCLUSIONS

Through the reflection on the epidemic situation in COVID-19, there is still much research and analysis space for urban planning. The epidemic situation has tested China's national conditions. Wuhan has accepted various tests in this epidemic situation and has gradually entered the stage of repair and adaptation and grown into a more mature and resilient city. As a significant link in the construction of a resilient city, urban planning can give full play to the city's planning advantages in emergency response and disposal of public safety and health incidents and effectively promote the implementation of various emergency measures. Furthermore, from the perspective of sustainable development, we should learn from the lessons of this epidemic and comprehensively improve the handling capacity of our cities in the face of various public emergencies to ensure urban safety and social stability.

ACKNOWLEDGMENTS

2017 Shaanxi Social Science Funds Program, Research on the Development Mode and Compensation Mechanism of the Circular Economy Industrial Cluster in Southern Shaanxi (Grant No. 2017S014)/Shaanxi Soft Science Research Program (Joint Project) Efforts to Promote the Construction of the Whole Industrial Chain and Acceleration of the Construction of Green Circular Industry System (Grant No. 2018KRLY20)/2018 Guangxi's Middle and Young Teachers Basic Ability Improvement Project: Application and Practice of BIM in Information Treatment of Quality Control in Assembly Construction (Grant No. 2018KY1010).

REFERENCES

[1] Ma Chao, Yun Yingxia, MA Xiaosong (2020). Study on the methods to promote community resilience in urban comprehensive disaster prevention and reduction planning. City Planning Review: 44(6). doi:10.11819/cpr20200610a
[2] Li Ye, Liu Xinghua, He Qing (2020). Urban Transportation System Resilience During the COVID-19 Pandemic. Urban Transport of China:18(3). doi:10.13813/j.cn11-5141/u.2020.0025
[3] Yan Shuiyu, Tang Jun (2020). Progress on the Theory and Practice of Resilient City. Journal of Human Settlements in West China:35(2). doi:10.13791/j.cnki.hsfwest.20200215

Computational Social Science – Luo, Ciurea & Kumar (eds)
© 2021 Taylor & Francis Group, London, ISBN 978-0-367-70193-2

Pogge's five proposals to reshape the global order and to eradicate global poverty

J.F. Sun
Shantou University, Shantou, China

ABSTRACT: In order to redress the problem of global poverty, Thomas Pogge proposes five measures to reshape the global order, namely, to reduce the expected rewards of coups d'état, to eradicate the international borrowing privilege, to eradicate the global resources privilege, to establish the global resources dividend mechanism, and to establish the global health impact fund system. The effective implementation of these proposed measures heavily relies on the strong support from the developed states. However, these measures, if effectively implemented, would cause great loss to the developed states, which renders it implausible for the developed states to support the implementation of these proposed measures.

1 INTRODUCTION

Thomas Pogge, as the leading advocate for the promotion of global justice, has argued relentlessly to reshape the global order to eradicate the global poverty. He cites lots of data to show how severe the global poverty is and how inequal the global wealth distribution is. He says, "It is estimated that 13 percent of all human beings (830 million) are chronically undernourished, 17 percent (1,100 million) lack access to safe water, and 41 percent (2,600 million) lack access to basic sanitation. Some 16 percent (1,000 million) lack adequate shelter. About 31 percent (2,000 million) lack access to essential drugs and 25 percent (1,600 million) lack electricity. Some 17 percent of adults (774 million) are illiterate, and 14 percent of children aged between five and 17 (218 million) are child laborers—often under harsh or cruel conditions: as soldiers, prostitutes, domestic servants, or in agriculture, construction, or textile or carpet production." (Pogge 2008, p. 103) He also points out that "One-third of all human deaths are due to poverty-related causes, such as starvation, diarrhea, pneumonia, tuberculosis, malaria, measles and perinatal conditions, all of which could be prevented or cured cheaply through food, safe drinking water, vaccinations ,rehydration packs, or medicines." (Ibid., p. 104) Pogge then sketches how inequal the global wealth distribution is, saying that "The 'high-income countries,' with 15.7 percent of world population and 79 percent of aggregate global income, enjoy a standard of living of $35,131 annually *per capita*. For the world as a whole, the corresponding figure is $6,987. With annual *per capita* consumption expenditure of about $100, the collective consumption expenditure of the bottom 15 percent is about $95 billion annually, or roughly 0.2 percent of the global product." (Ibid., pp. 104–5) In recent years, despite the great technological and economic progress, the global poverty is rising and the global inequality is escalating (Ibid., pp. 105–6).

After arguing convincingly that the global order plays a causal role in the global poverty (Ibid., pp. 118–22), Pogge proposes five measures to reshape the global order and to eradicate the global poverty, which include measures to reduce the expected rewards of coups d'état, to eradicate the international borrowing privilege, to eradicate the global resources privilege, to establish the global resources dividend mechanism, and to establish the global health impact fund system. Although these measures are all concerned with the global order, the first three measures are more closely related to poor countries themselves which the latter two are more concerned with the global order as a whole. Notwithstanding this difference, for each of these proposed measures to be effectively implemented, the support and cooperation from developed states are vital because the current global order has been shaped and maintained by the developed states.

This article tries to sketch the five measures Pogge proposes to reshape the global order, examine how they are to be implemented and argues how plausible they are.

2 POGGE'S PROPOSAL TO REDUCE THE EXPECTED REWARDS OF COUPS D'ÉTAT

This proposed measure to reduce the expected rewards of coups d'état, together with the other two measures to eradicate the international borrowing privilege and to eradicate the global resource privilege, is concerned with how to entrench democracy in poor countries where democracy has been newly established. In fledgling democracies which used to be haunted by junta or torn apart by military groups, Pogge thinks that, the biggest threat is the potential coups d'état by military groups, which would weaken or overthrow the newly established democracy. To prevent the potential coups d'état, Pogge proposes to reduce the expected rewards of coups d'état even if it is successful, and maintains that, "Measures to deter takeovers must then involve some mechanism that survives a successful takeover and continues to reduce the payoff that the predators reap from their success." (Pogge 2008, p. 158)

Pogge gives an example of Microsoft company. If gangsters take over Microsoft headquarters by force, Bill Gates can simply call the police and the police can arrest the gangsters and return the headquarters to Microsoft (Pogge 2001, p. 9). In this case, the police are preauthorized by the social system to intervene in the case of illegal takeover, which discourages any illegal attempts at the takeover of any personal or collective property under the protection of certain jurisdiction. Inspired by this case, Pogge proposes that taking this Microsoft example as the model, to deter potential coups d'état takeovers "some international analogue to the police" (Pogge 2008, p. 159) so that it can be preauthorized by the democratic regime to intervene in the case that the regime suffers the illegal takeover by coups d'état.

So Pogge proposes to establish an international organization analogous to the international police which has been preauthorized by the fledgling democracies to intervene when illegal takeover by military groups happens.

For this proposed measure to be plausible, three conditions have to be met: First, this kind of international organization has to be sufficiently funded, staffed and equipped with necessary military equipment so as to fulfill its mission to intervene in illegal takeover; Second, it has to be supported by the developed states; Third, there needs clear criteria for the proposed international organization to judge whether the takeover is illegal or not. However, it is very difficult to fulfill each of the conditions. Let's start from the third condition. According to Pogge's proposal, what the international organization can be preauthorized by fledgling democracies to intervene with is only illegal takeover of power in these countries so for the intervention to be legitimate, how to judge whether the takeover is illegal or not becomes vital. If the junta claims that the fledgling democracy is not democratic at all and claims to overthrow it to establish democracy, it would be impossible for the international organization to judge whether to intervene in this takeover or not. This is the criterial problem. The other two conditions are also very difficult to meet. Since poor countries, which possess only tiny portion of the global GDP, cannot afford to support this kind of international organization, without the support from the developed states, this kind of international organization would not be sufficiently funded, staffed and equipped. However, there is no incentive for the developed states to help sustain this kind of international organization because the developed states would not benefit from lending support to it. So it can be seen that Pogge's proposal to establish a preauthorized international organization to intervene in the case of illegal takeover to reduce the expected rewards of coups d'état in fledgling democracies is implausible.

3 POGGE'S PROPOSAL TO ERADICATE THE INTERNATIONAL BORROWING PRIVILEGE

The international borrowing privilege is an aspect of the unjust global order, which means that "any group holding governmental power in a national territory—no matter how it acquired or exercises this power—is entitled to borrow funds in the name of the whole society, thereby imposing internationally valid legal obligations upon the country at large." (Pogge 2008, p. 120) This privilege plays a role in causing and maintaining global poverty in that any successor government whether democratic or undemocratic has to honor and repay debts incurred by any corrupt, brutal, undemocratic, unconstitutional, repressive, unpopular predecessor or else would be punished by the international society (Ibid.). This privilege also provides incentives for military groups to take power in a country and therefore keeps the country in deep poverty.

Pogge proposes to eradicate this privilege by fledgling democracies' adopting a constitutional amendment which requires that money borrowed by unconstitutional governments that acquire or wield power in violation of the democratic constitution would not be paid back by the public finance, (Ibid., 159–60) which means that debts caused by unconstitutional governments in poor countries that adopt this constitutional amendment would neither be honored nor repaid by the countries.

How plausible is this proposal? Pogge assumes the following scenario with Brazil as example. Phase I, Brazil, a fledgling democracy, after establishing democratic constitution, warns foreign banks not to lend money to their unconstitutional rulers in the future and constitutionally requires any future Brazilian government not to honor and pay back these debts foreign banks may lend to their unconstitutional governments. Phase II: A military junta overthrows the democratic government in Brazil by force, disregards the constitution, and borrows lots of money from Citibank. Phase III: Democracy is reestablished in Brazil and the new government dishonors the debts from Citibank and refuses to pay back the debts. Citibank turns to the U.S. government in the hope that the U.S. government would pressure the democratic government in Brazil to pay back the money. (Ibid., p. 160) In this case, Pogge asks, Would the United States put pressure on the Brazilian democratic government to pay back the money borrowed by the former junta by, for example, excluding Brazil from the international financial system? (Ibid.)

To maintain the national interest and to win voter' support, the US government is very likely to intervene to exert pressure on the Brazilian government and urge it to repay Citibank's loans. Due to large national interest involved in this privilege, developed countries are not likely to obey the constitutional amendment adopted by poor countries to eradicate the borrowing privilege, which therefore renders this measure implausible.

Another problem facing the measure is the criterial problem. This measure specifies that international debts incurred by constitutionally undemocratic governments would not be honored by democratic governments so for this measure to be plausible it is compulsory to judge whether a government is constitutionally democratic or not. In some cases, this issue is easy to judge while in many other cases there are lots of controversies. To solve this problem, Pogge proposes to establish the Democracy Panel composed of "reputable, independent jurists living abroad" who understand the constitution in the fledgling democratic country well enough to judge to what extent any group violates the constitution in its exercise of power. (Ibid., p. 162) However, to establish and maintain this Panel is implausible due to lack of support from the international society. Pogge proposes that this Panel should have sufficient human resources to check and judge whether any government in participating countries is constitutionally democratic or not. However, without the financial support from developed countries, this Panel would not be fully financed to be able to fulfill the role Pogge expects. And, it is clear that developed countries are not willing to finance such a Panel whose role violates their interest.

4 POGGE'S PROPOSAL TO ERADICATE THE INTERNATIONAL RESOURCE PRIVILEGE

The international resource privilege, another aspect of the unjust global order, means that any group that takes governmental power in a country, regardless of how it acquired or exercises this power, is recognized as the legitimate owner of the natural resources in the country and can legally transfer the ownership rights of the resources. (Ibid., pp. 118–9) This privilege provides strong incentives for military groups to fight for power in resource-rich countries. Due to the privilege in place in the global order, rich resources in poor countries do not promote their economic progress but lead to their poverty and political chaos.

Pogge proposes to eradicate this privilege by fledgling democracies' adopting a constitutional amendment which rules that ownership rights in its public resources can be legally transferred only by its constitutionally democratic governments, and requires its governments not to recognize ownership rights in its public resources transferred by its unconstitutional governments, (Ibid., 169) which means that any ownership transfer of resources in the country by an unconstitutional government would be contested and rendered illegitimate.

This proposal, if adopted by poor countries and effectively followed by the international society, would significantly reduce coups d'état attempts, maintain the political stability and promote the economic development in the poor countries. However, due to lots of interest implicated in the international resource privilege, developed countries would not honor this constitutional amendment to eradicate the resource privilege. Pogge thinks that if the rich countries can no longer acquire ownership rights in natural resources sold by authoritarian governments at low prices, the result is catastrophic for them since their economies heavily depends upon importing such resources at low prices and would suffer a lot if they can no longer import these resources. (Pogge 2001, p. 21) In sum, if the developed countries would follow this constitutional amendment, the prosperity and economic growth would be largely reduced.

Since so much interest of the developed countries is tied up with this privilege, the constitutional amendment to eradicate it is unlikely to get support from the developed countries, thus making it implausible.

5 POGGE'S PROPOSAL OF GLOBAL RESOURCES DIVIDEND

Reflecting upon the severe global poverty and inequality (Pogge 2008, pp. 103–6), Pogge considers the current global order as unjust from the approach that the poor people are unreasonably and arbitrarily excluded from the use of natural resources without any compensation (Pogge 2001b, 61).

Pogge holds that the natural resources as human beings' common asset should be shared by all human beings in the world whereas in reality the developed countries benefit most from the global natural resources with the global poor excluded from the benefits of the global natural resources. In order to rectify this injustice, Pogge proposes to establish the global resources dividend to reshape the global order.

The global resources dividend (GRD) proposal requires that, "[T]hose who make more extensive use of our planet's resources should compensate those who, involuntarily, use very little." (Pogge 2001b, 66) By compensating, GRD does not require the global resources to be shared equally but requires states and their governments which still control resources in their territory to "share a small part of the value of any resources they decide to use or sell." (Ibid.) Pogge calls the payment states and governments are required to make a dividend "because it is based on the idea that the global poor own an inalienable stake in all limited natural resources." (Ibid.)

Pogge thinks that the GRD, if implemented effectively, would improve global poverty significantly in a few years. He calculates that "a 1% GRD would currently raise about $300 billion annually" (Ibid., p. 67), which equals to "$250 per year for each person below the international poverty line, over three times their present average annual income." (Ibid.) Pogge is very optimistic that when GRD is properly collected and distributed we can eradicate global poverty within a few years with very little cost to the developed countries. He cites the dividend on crude oil products as an example to show the effects of GRD. He calculates that with $300 billion per year as the target of GRD, 18% of it can be achieved through GRD on crude oil "raising the price of petroleum products by about a nickel per gallon." (Ibid., p. 68)

While this proposal gives the global poor what they deserve in the share of global resources and is mainly focused on "resource uses whose discouragement is especially important for conservation and environmental protection" (Ibid., p. 67), it is implausible due to lack of incentives for the developed countries to follow. The GRD proposal is posed toward the developed countries and requires them to distribute certain amount of revenues from natural resources they control, purchase and use to the global poor, for which the full cooperation of the developed countries is vital for this proposal to be effectively implemented. However, what can the developed countries get from the GRD? No interest. Without lucrative interest to gain from this GRD, the developed countries that control the global order would not take the GRD seriously, not to mention to implement it.

6 POGGE'S PROPOSAL OF THE GLOBAL HEALTH IMPACT FUND

This proposal of the global health impact fund is targeted at the medicine crisis faced by people in the poor countries who do not have access to medicines they urgently need for their survival. Pogge analyzes how this medicine crisis is caused by the current medicine patent system. Under the current medicine patent system, pharmaceutical innovators are offered patents for at least 20 years which ensure that pharmaceutical companies with medicine patents enjoy market monopoly for about 10 years (Pogge & Krishnamurthy 2010, p. 7). Poor people cannot afford to buy patented medicines at monopoly prices and the pharmaceutical companies do not have incentives to develop medicines for diseases suffered by people in poor countries. Other than the high medicine price, obstacles to poor people's access to urgently needed medicines include: "lack of local availability of a medicine, lack of available knowledge and information about diseases and their remedies, lack of refrigerators or electricity, and even gross negligence, incompetence and corruption in the health bureaucracy." (Ibid., p. 8) The poor people's lack of access to essential medicines constitutes grave violation of human rights especially when considering that the poor people account for three quarters of the world's population.

Pogge proposes to overcome the drawbacks of the current medicine patent system through the establishment of the health impact fund (HIF), which is "a pay—for—performance mechanism offering to reward the introduction of a product in proportion to its impact on global health." (Ibid., p. 5) Under this system, pharmaceutical companies can voluntarily choose to register their products with this system, offer their registered products at the lowest feasible price of production and distribution for 10 years and receive rewards according to how much health impact their products help deliver.

For the HIF to be plausible, two important problems have to be tackled, namely, how the health impact is calculated, and how the HIF is financed. Concerning the first problem, Pogge holds that "the health impact can be assessed in terms of the number of quality-adjusted life years (QALYs) saved." (Ibid., p. 6) The QALYs is calculated according to the estimates about the length and quality of human lives the registered new medicine has added in comparison to the length and quality of human lives before it is registered and introduced.

As for how the HIF is financed, Pogge hopes that the developed countries would contribute to get the HIF started, saying, "If governments representing one third of global income contribute just 0.03 percent of their gross national incomes (3 of every 10, 000), the HIF could get started with USD 6 billion annually." (Ibid., p. 6) With this budget, "the HIF could support the development of about two new drugs per year, sustaining a stock of about twenty medicines at any given time," (Ibid.) apart from the health impact assessment costs.

Pogge considers the HIF very effective and plausible in providing urgently needed medicine to poor people and improving their health. He holds that the HIF is only complement to the current medicine patent system and that companies are free to register with the HIF. He expects that since the HIF provides revenue for pharmaceutical companies without causing any structural change to the current medicine patent system the developed countries would welcome and support it. However, is it really so plausible?

The crucial problem for the HIF is how it is financed. Although Pogge thinks that if the developed countries "contribute just 0.03 percent of their gross national incomes (3 of every 10, 000), the HIF could get started with USD 6 billion annually," (Ibid.) how could the developed countries be motivated to contribute this amount of GDP to the HIF despite it being very tiny part of their GDPs? What the HIF is essentially concerned about is the health problems in the poor countries. Without proper gain from the HIF, the developed countries would not contribute to the HIF to get it started, which renders the HIF implausible.

7 CONCLUSION

Thomas Pogge has made admirable efforts and contribution to reveal and remedy the injustice in the current global order. Having convincingly argued that the current global order has avoidably caused severe global poverty, Pogge proposes five measures as specified above to reshape the global order and to eradicate global poverty. These measures, if implemented effectively, would significantly reduce global poverty. However, each of the five measures depends on the support and cooperation from the developed countries to be put into practice while there is no incentives for the developed countries to do so, which renders these measures implausible. These measures, supposed to reshape the global order, cannot be implemented without support from the developed countries which shape and maintain and benefit greatly from the current global order.

That these measures are implausible reveals a moral crisis for the developed countries. On the one hand, Pogge argues that these five measures are morally sound which claim what the poor countries deserve. On the other hand, the developed countries would not support or cooperate with these morally sound measures, leaving the poor people still poverty. This moral crisis demonstrates that there is great moral gap between what developed countries and their people claim and what they have done, and they do shoulder responsibility for the global poverty for not filling this gap.

REFERENCES

[1] Pogge, Thomas. 2001a. Achieving democracy. *Ethics & International Affairs* 15(1): 3–23.
[2] Pogge, Thomas. 2001b. Eradicating systemic poverty. *Journal of Human Development* 2(1): 59–77.
[3] Pogge, Thomas. 2008. *World poverty and human rights: cosmopolitan responsibilities and reforms*. New York: Polity.
[4] Pogge, Thomas & Krishnamurthy, Meena. 2010. How not to exclude the poor from advanced medicines: a plea for the health impact fund. *Rights and Development Bulletin* 1(18): 5–12.

Computational Social Science – Luo, Ciurea & Kumar (eds)
© 2021 Taylor & Francis Group, London, ISBN 978-0-367-70193-2

Research on personalized tourism route based on crowdsourcing model

H.F. Li* & W.J. Chen
Dalian Jiaotong University, Dalian, China

ABSTRACT: Crowdsourcing is based on personal choice; taking network technology as the platform; gathering the knowledge, wisdom, strategies, and information of the public; and playing an obvious advantage in solving the complex problems. Nowadays, in the era of pursuing "individuation," more and more tourists not only want to get rid of the time limit brought by group travel but also don't want to spend a lot of time to study and develop their own satisfactory tourism strategies. This paper applies crowdsourcing, a new Internet model, to the tourism industry for an innovative attempt. A personalized travel route recommendation system based on crowdsourcing mode is constructed. At the same time, according to the application field of crowdsourcing, a task of tourism crowdsourcing is designed to achieve the goal of crowdsourcing quality control, and an effective use of crowdsourcing results based on the implicit behavior of tourists is proposed.

Keywords: Crowd-sourcing, Personalized recommendation, Crowdsourcing task verification.

1 RESEARCH BACKGROUND AND PURPOSE

With the rapid and vigorous development of the Internet, people's lives, work, and learning have changed greatly. Tourism has also made great strides toward the development of networks. Team or tour guide has been replaced gradually, as self-help tourism has become the hotspot of today's tourism. At the same time, many online tourism service platforms have emerged and have quickly and accurately occupied the market. These platforms mainly adopt the service mode of booking and group buying, and a large number of background customization work is in the unified charge of tourism professionals.

However, these platforms also have some disadvantages, such as the high cost of time and human resources, and the vigorous development of the information industry makes for all kinds of information flooding, which leads to the convenience of users from the beginning and gradually getting lost. In the huge information base, users often need to spend a lot of time and energy to search for the information in line with their own.

The emergence of crowdsourcing mode and its application in many industries are very successful, which provides the possibility for the realization of truly personalized tourism services and opens an innovative development era for traditional tourism. In this paper, the crowdsourcing model is applied to the tourism industry, aiming to maximize the power of groups in the Internet through effective crowdsourcing quality control and personalized recommendation, get rid of the bad factors existing in the traditional tourism industry, meet the personalized needs of tourism lovers, break through the development bottleneck of the tourism industry, and improve the market competitiveness.

Based on the above background, this paper proposes a personalized travel route recommendation mode using crowdsourcing mode.

2 OVERVIEW OF RELEVANT THEORETICAL KNOWLEDGE

The concept of crowdsourcing was first proposed by Jeff Howe, a journalist of *Wired* magazine in 2006. It is a typical distributed problem-solving method, which is used to describe a method based

---

*Corresponding author

on Internet information platform to assign tasks, obtain ideas, or solve problems. Crowdsourcing refers to when an enterprise or company outsources its internal tasks or problems to other non-fixed social masses through the network technology platform, relies on the external network masses to solve problems, and uses the wisdom and strength of the masses to improve the internal innovation of the company. The crowdsourcing model is simply to involve more people. Thus, crowdsourcing can also be called network crowdsourcing. As more and more organizations get the latest ideas through the Internet, volunteers in the network are willing to spend their spare time solving the problems raised by others, and collect lower pay or no pay compared with the traditional salary standard just to obtain potential reward opportunities, or to realize the value embodiment and self-satisfaction of helping others. The crowdsourcing mode, as a new type of cooperation and innovation mode based on the network information platform between the company and the public, has aroused widespread concern from all walks of life.

In the application of crowdsourcing, crowdsourcing is closely related to different fields of various industries once it appears. From marketing, business model innovation to new product development model, crowdsourcing has its place. In the field of marketing, Pan Haidong discussed the promotion of crowdsourcing mode on value co-creation, community-based customer relationship, and overall marketing value; in the aspect of enterprise business model innovation, he studied the contents of consumer participation innovation, learning organization innovation, science and technology communication innovation, human resource system innovation, enterprise innovation democratization, and so on.

Crowdsourcing model is a new business model, that is, individuals or companies use the Internet as a platform to send out difficult problems or tasks that need to be solved, give full play to and gather the knowledge and skills of the public, solve complex and diverse problems or find creative ideas, reflecting a participatory culture of the public, with the main feature of giving full play to different groups Our wisdom and strength create value together.

Crowdsourcing mode is open, anyone can participate, whether you are a professional or amateur, to realize the sharing of resources, with subject diversity.

Crowdsourcing emphasizes participation. Crowdsourcing comes from the unorganized behavior of tens of thousands of people. The participants come from different corners of the world and engage in different occupations. They may be professionals, amateurs, and so on. They have strong innovation enthusiasm and ability for things, and they often know the needs of similar people better than enterprises.

Crowdsourcing advocates boundless organization. Based on the crowdsourcing model, enterprises can expand the organizational boundary infinitely, let crowdsourcing users from all over the world participate in it, and obtain global talents and resources by establishing a network platform across the cultural boundaries of various countries or companies, so as to serve enterprises.

3 ANALYSIS OF PERSONALIZED TOURISM ROUTE SYSTEM BASED ON CROWDSOURCING MODE

Nowadays, browsing online travel websites to find travel strategies has become the first choice for users to make travel plans. In the past, the search engine is usually a relatively convenient and fast way to obtain tourism information. However, nowadays, there are many tourism websites with different prices, which are more or less linked to the income of the website. Therefore, most of them are popular and common. It is difficult to make a tourism route that meets the needs of a person, that is, relatively personalized. At the same time, the tourism routes launched by the website are all developed by the internal staff of the website, the cost of the website is also large, and the routes are difficult to have characteristics and creativity. Due to the large number of participants, a wide range of knowledge, and fast response of the masses, the crowdsourcing model uses its biggest characteristics to find the right talents at a very low cost, which provides the possibility for the development of personalized tourism routes and gives pride and sense of belonging to those who can play a role in their favorite fields.

According to the above analysis, the main functional requirements of the following systems can be summarized.

Users are used to obtaining. This function is mainly responsible for obtaining the contents and behaviors of web users, making the web platform more aware of users, and providing personalized services for users.

Tourism demand management is mainly responsible for releasing tourists' tourism route demand, that is, the functional demand designed for the employer in crowdsourcing operation mode.

Tourism demand crowdsourcing is mainly responsible for transforming the tourism demand released on the platform into multiple crowdsourcing tasks and then releasing.

In verification of tourist attractions, this function is mainly to improve the quality of crowdsourcing results, apply crowdsourcing mode to the system, maximize the wisdom of the masses in the network platform, and improve the experience of tourists. The so-called scenic spot verification is to make the content of the completed tourism route plan; make use of the characteristics of crowdsourcing; and crowdsourcing workers use their own professional knowledge to evaluate, supplement, or vote the content and put forward their own ideas and views, so as to facilitate the later tourists to learn more quickly according to the content, so as to choose their own satisfactory tourism route plan.

The combination of tourism route schemes is partially divided in the design of crowdsourcing task for tourism route demand, which is completed jointly by crowdsourcing workers without mutual interference, so the collected scheme is not a complete one, and it needs to be recombined according to certain rules.

Travel route recommendation. Facing many travel schemes, when travelers find it hard to choose, the system can use the result recommendation model to actively recommend travel routes that meet users' needs and preferences according to the user characteristics and preferences obtained from behavior analysis, so as to provide tourists with personalized and efficient high-quality services.

Crowdsourcing model is changing the traditional business model in a convenient and fast way. More and more companies are using the information platform network to connect their production and operation with the network public. P & G, IBM, Starbucks, LG, Lego, and other multinational companies have set up such crowdsourcing platforms to use the wisdom and strength of the public to solve the problems faced by the company and update their products, so as to make them more in line with the needs of consumers.

The emergence and development of crowdsourcing model benefit from the improvement of technology and the popularization of network. Nowadays, the Internet has become an indispensable part of people's lives. People can participate in a crowdsourcing platform only by turning on the computer and moving their fingers.

In the real environment, most of the tourists we meet are young people, because young people have the time and energy to try new things. However, in recent years, the survey results show that more and more elderly self-help tourism enthusiasts have joined in the ranks. They maintain an optimistic attitude toward life and are full of passion for life. It can be seen that personalized tourism is no longer the patent of young people, and tourists have expanded to all levels of age.

4 DESIGN OF PERSONALIZED TOURISM ROUTE SYSTEM BASED ON CROWDSOURCING MODE

This paper studies how to design crowdsourcing tasks for tourism demand, so that crowdsourcing tasks have high crowd acceptance rate, and then reduce the delay. Combined with experience and practical investigation, we know that, first, when the complete tourism demand is automatically released to the network as a crowdsourcing task, waiting for the crowdsourcing workers to provide the corresponding tourism scheme, we find that the crowdsourcing task can- not get the effective response of the workers for a long time. Therefore, the local segmentation of tasks is expected to reduce the workload of a single task to attract workers to give effective answers. In order to ensure that tourists can finally get a high degree of satisfaction with the tourism program, we can recruit multiple workers for multiple subtasks, answer them separately without interference, and then generate simple verification tasks automatically according to the answers and recruit additional workers to complete the verification of scenic spots. The personalized travel route recommendation system based on crowdsourcing mode sets a certain monetary reward for each crowdsourcing task.

Based on the verification method of evaluation content, according to the evaluation content submitted by the crowdsourcing workers who analyze and participate in the verification, it is considered as one of the reference factors to judge whether a city scenic spot is an effective scenic spot. The effective scenic spot refers to the scenic spot that really belongs to a city.

Based on the verification method of voting results, the workers participating in the verification can vote according to the scenic spot recommendation tasks completed by the former workers, and express their support or opposition to whether the scenic spot submitted by the scenic spot recommendation workers belongs to a city.

The personalized travel route recommendation system based on crowdsourcing mode will collect a large number of travel plans. If these travel plans are directly returned to travelers, it is not to say how satisfied travelers are with the travel plans, but only these large number of travel plans that have caused problems to them. Therefore, after the personalized self-service travel system combines tourism data to form a large number of tourism schemes, it needs a personalized scheme recommendation method to automatically recommend at least one tourism scheme with high user satisfaction in combination with the needs and preferences of tourists.

At present, there are two ways to obtain user information: explicit and implicit. Under the guidance of the system platform, users actively provide their own information to the system or select or evaluate some services or contents in the system. At present, the most commonly used method of user information explicit acquisition is to ask the user to answer the system preset questions or to evaluate and select the project content specially given by the system. The system locks the user's interest preference by collecting the user's answers.

The sexualization factors discussed in this paper are different from the characteristic factors of travelers, but there is a connection between them. The characteristic factors of travelers classify and combine a group of different travelers for the first time, while the personalized factors endow the same kind or group of travelers with individual differences.

Travel costs. There are great differences among different tourists in the range of variability of tourism spending budget. When tourists are faced with many tourism schemes, the factor of tourism cost budget will undoubtedly become one of the focuses of their route selection.

Number of attractions. Tourists don't know what are the tourist attractions and which ones are worth visiting. They just give the number of tourist attractions to plan based on their own ability and experience. Then, it is obvious that those tourism schemes that include too many or too few scenic spots will not be the wise choice for travelers.

Days of travel. Travelers have certain arrangements for the length of travel time. Therefore, tourism days are also an important factor in tourism programs.

Departure time. In some cases, the date of departure reflects the seasonality of tourists' travel. Some travelers want to visit in the best season of their destination, while others don't care. Therefore, the departure date in the tourism plan will also become one of the concerns of tourists.

Work reputation. Workers in the personalized tourism platform based on crowdsourcing are responsible for providing tourists with tourism solutions. However, providing the credibility of tourism program workers is undoubtedly an important factor for travelers. Without considering the tourism scheme, the tourism scheme provided by workers with high reputation will be favored by tourists.

To sum up, this paper analyzes the main differences of tourism programs and determines personalized factors, namely, tourism cost, number of scenic spots, tourism days, departure date, and workers' reputation, to build a crowdsourcing results recommendation model, aiming to achieve efficient crowdsourcing results of personalized recommendation.

In order to solve the problem of too many crowdsourcing results, which brings users choice puzzlement, a crowdsourcing result recommendation method is designed to help users choose a scheme that the platform thinks best meets their needs, for users to reference and choose. The general idea of the recommendation method: according to a certain tourism demand, all the tourism schemes (crowdsourcing results) provided by all the workers are calculated with the demand similarity, and the tourism scheme with the largest similarity to the demand is recommended to the user as the optimal scheme. After quantifying the five factors of travel expense, number of scenic spots, travel days, departure date, and reputation of workers, it is added to the function calculation of the similarity of travel route scheme and travel demand. The more the final function value is, the more similar the travel scheme provided by workers is to the travel demand of tourists, and the more satisfied the service is, so the travel scheme is recommended to tourists.

5 SUMMARY AND PROSPECT

In order to solve the development bottleneck of tourism industry, this paper attempts to apply crowdsourcing model to personalized tourism services and make an innovation. In recent years, the successful application cases of crowdsourcing mode are numerous, and the correlation research in various aspects has also received widespread attention. In the personalized travel route recommendation system based on crowdsourcing mode, travelers can get at least one crowdsourcing worker's personalized travel plan. At the same time, in order to give full play to the power of group intelligence in the Internet, get rid of all kinds of negative factors existing in traditional tourism services, and improve user experience, this paper designs a tourism crowdsourcing task in terms of weighing the delay of crowdsourcing task and capital expenditure, so as to reduce the possibility of submitting low-quality crowdsourcing results.

Crowdsourcing mode is a kind of innovative thinking and bold practice in the new era of networking. It will be a new direction for the development of enterprise innovation mode in the future to make accurate use of this participatory culture and innovation mode of giving full play to collective wisdom. Based on the network information platform, according to the needs of self-service tourists, this paper designs and constructs a customized personalized tourism route system, which brings convenience to personalized tourism lovers.

ACKNOWLEDGMENT

Natural Science Fund of Education Department of Liaoning Province, JDL2019027
Liaoning Natural Science Foundation, 201800177
The subject of educational science planning in Liaoning Province, JG16DB054

REFERENCES

[1] Jiang Tingting. 2013, On the new development trend of self-service tourism in China. Tourism overview monthly (4).
[2] Wei Shuan Cheng. 2010, The concept of crowdsourcing and the design of crowdsourcing business model of Chinese enterprises. Technology economy and management, (1): 36–39.
[3] Zhang Yufeng. 2010, Innovation of "crowdsourcing mode" on science and technology communication. Science and technology communication (2) (2): 42–44.
[4] Liu Wenhua. 2009, Economic analysis of crowdsourcing. New economy guide, (6): 91–95.
[5] Wang Laixi, Ding Rijia, Wang Yuanchang. 2007, Crowdsourcing: the method of enterprise innovation democratization. Enterprise vitality, (4): 70–71.

Computational Social Science – Luo, Ciurea & Kumar (eds)
© *2021 Taylor & Francis Group, London, ISBN 978-0-367-70193-2*

Research on the evaluation system of employee satisfaction based on AHP

Y.J. Wu*, Y.H. Xiong, X.X. Tang & Y.H. Fu
College of Management, Wuhan University of Science and Technology, Wuhan, China

ABSTRACT: With the development of the economy, low employee satisfaction leads to high employee turnover rate and brings great pressure to enterprise management. In order to better retain and attract talent, it is important to improve employee satisfaction. At present, most enterprises lack a scientific satisfaction evaluation system, so the construction of an employee satisfaction evaluation system is a topic worth exploring. This paper analyzes the factors of employee satisfaction and determines the evaluation index of employee satisfaction, determines the weight of each index through the hierarchical analysis method, and constructs the evaluation index system of employee satisfaction, hoping to help enterprises understand the employee satisfaction situation.

Keywords: Employee satisfaction, AHP (hierarchical analysis method), Evaluation system.

1 INTRODUCTION

In modern enterprise management, treating employees as "customers" is a new value concept. It is very important to explore the potential value of employees and make them contribute knowledge to the enterprise dutifully. In the 21st century, more and more managers have realized the importance of employee satisfaction, and increasing attention has been paid to the survey of employee satisfaction and it has been used as a tool for enterprise management. Through the investigation of employee satisfaction and the systematic analysis of the survey results, enterprises can find out the dissatisfaction of employees and other problems existing in enterprises and put forward targeted countermeasures to the problems, which can effectively solve the management problems of enterprises but also improve employee satisfaction and loyalty, which is conducive to the sustainable development of enterprises. Therefore, it is of great significance to improve the effectiveness of enterprise management through reasonable employee satisfaction survey. At present, most enterprises lack a scientific employee satisfaction evaluation system. This paper mainly uses AHP method to establish a set of scientific employee satisfaction evaluation systems, which is helpful to carry out the survey and results in evaluation of employee satisfaction according to the specific situation of the enterprise, so as to serve the enterprise to retain outstanding talents.

2 INFLUENCE FACTORS OF EMPLOYEE SATISFACTION

2.1 *Literature research*

Khawaja Fawad Latif studies the relationship between training and employee satisfaction and argues that leadership, working environment, and work content are the three main factors that affect employee satisfaction [1]. Monica Izvercian uses grounded theory to classify the factors that affect employee satisfaction into six categories: motivation, social interaction, employee characteristics, organizational environmental characteristics, organizational cognition, and interference factors [2]

---

*Corresponding author

Tang Liqin believes that the influencing factors of employee satisfaction are as follows: first, the influence of salary mechanism; second, the influence of working time and work content; third, the influence of corporate environment and culture [3]. Wang Xiaohong used SPSS software to analyze the collected questionnaire data and concluded that net income was the primary influencing factor of employee satisfaction [4].

From the perspective of domestic scholars' research, scholars pay more attention to the factors influencing employees' satisfaction and analyze employees' satisfaction from multiple aspects of psychology and physiology.

2.2 *Comprehensive analysis results*

By analyzing the influencing factors of employee satisfaction, it can improve employee satisfaction to the enterprise. By collecting, reading, and learning a large number of relevant literature, reviewing the actual investigation and analysis of enterprises by domestic and foreign scholars, and combining the hierarchy of needs theory and incentive-health theory, this paper summarizes and analyzes the different factors affecting employee satisfaction. The main factors are as follows:

(1) Work itself: diversity of work skills, integrity of work content, meaning of work results, autonomy, feedback;
(2) Compensation and benefits: work reward, wages, benefits, and other incentive policies;
(3) Individual development: work promotion, work growth, work development, potential cultivation and promotion, career prospects;
(4) Interpersonal relationship: communication within the department, communication between departments, relationship with direct supervisors, colleagues, and departments;
(5) Managerial leadership: management style, management level, planning and organizing ability, pioneering courage, innovation ability, care for subordinates.

3 CONSTRUCTION OF EMPLOYEE SATISFACTION EVALUATION INDEX SYSTEM

3.1 *Employee satisfaction evaluation index*

Through the summary and analysis of the influencing factors of employee satisfaction, the evaluation indicators are obtained by adopting the three-layer structure of target layer A, criterion layer B, and scheme layer C, as shown in Table 1.

Table 1. Employee satisfaction evaluation index.

Target layer	Criterion layer	Scheme layer
Employee satisfaction A	Work itself B_1	Diversity of work skills C_{11}
		Integrity of work content C_{12}
		Feedback on work results C_{13}
	Compensation and benefits B_2	Basic wages C_{21}
		Merit pay C_{22}
		Social benefits C_{23}
	Individual development B_3	Work promotion space C_{31}
		Professional and technical training C_{32}
	Interpersonal relationship B_4	Coworker relationship C_{41}
		Superior and subordinate communication C_{42}
		Teamwork C_{43}
	Managerial leadership B_5	Management level C_{51}
		Leadership C_{52}

Table 2. Relative importance scale.

Saaty Relative Importance Scale					
Relative importance	1	3	5	7	9
Definition	Equality	A little important	Important	Very important	Most important

Note: 2, 4, 6, 8 is the intermediate value of the above two adjacent degrees.

Table 3. The relative importance of the related indicators of employee satisfaction.

A	B_1	B_2	B_3	B_4	B_5
B_1	1	1/4	2	3	5
B_2	4	1	3	4	6
B_3	1/2	1/3	1	2	3
B_4	1/3	1/4	1/2	1	2
B_5	1/5	1/6	1/3	1/2	1

3.2 Use AHP method to determine the index weight

3.2.1 Construct judgment matrix
(1) Establish comparison criteria, as shown in Table 2.
(2) Establishing a pairwise comparison matrix.

Comparing the evaluation indexes in pairs, and determining the relative importance degree according to the evaluation scale, so as to establish the judgment matrix.

The formula for calculating the weight of evaluation index is as follows:

$$W_i = 1/n \sum_{j=1}^{n} (a_{ij} / \sum_{j=1}^{a} a_{ij}) \tag{1}$$

(1) The judgment matrix of the target layer A and the criteria layer B.
Table 3 shows the judgment matrix of employee satisfaction index.

$$A = \begin{bmatrix} 1 & 1/4 & 2 & 3 & 5 \\ 4 & 1 & 3 & 4 & 6 \\ 1/2 & 1/3 & 1 & 2 & 3 \\ 1/3 & 1/4 & 1/2 & 1 & 2 \\ 1/5 & 1/6 & 1/3 & 1/2 & 1 \end{bmatrix}$$

From the formula 3.1: $W_{B1} = 0.23$, $W_{B2} = 0.47$, $W_{B3} = 0.15$, $W_{B4} = 0.09$, $W_{B5} = 0.06$ and so on, the total ranking weight of indicator layer C is shown in Table 4.

3.2.2 Calculate maximum eigenvalue and consistency check
(1) Calculate the maximum eigenvalue of the judgment matrix:

$$\lambda_{max} = 1/n \sum_{i=1}^{n} (AW)_i / W_i \tag{2}$$

Among them, vector $W = (0.23, 0.47, 0.15, 0.09, 0.06)^T$.
Formula (3.2) gives $\lambda_{max} = 5.2567$.
(2) Consistency check

According to the principle of hierarchy method, the theoretical maximum eigenvalue of A, lambda Max, is applied to test the consistency.

Table 4. The total ranking weight of the layer.

A	B$_1$ 0.23	B$_2$ 0.47	B$_3$ 0.15	B$_4$ 0.09	B$_5$ 0.06	The total ranking weight of the layer	The weight sequence
C$_{11}$	0.57					0.13	3
C$_{12}$	0.14					0.03	11
C$_{13}$	0.29					0.07	5
C$_{21}$		0.11				0.05	6
C$_{22}$		0.31				0.15	2
C$_{23}$		0.58				0.27	1
C$_{31}$			0.75			0.11	4
C$_{32}$			0.25			0.04	7
C$_{41}$				0.20		0.02	13
C$_{42}$				0.40		0.04	8
C$_{43}$				0.40		0.04	9
C$_{51}$					0.33	0.02	12
C$_{52}$					0.67	0.03	10

Table 5. Average random consistency index value table.

Order number	2	3	4	5	6	7	8	9	10	11
RI	0	0.58	0.90	1.12	1.26	1.36	1.41	1.46	1.19	1.52

Table 6. Maximum characteristic value of index and consistency test result table.

	λ_{max}	CI	RI	CR
Work itself B$_1$	3.0000	0	0.58	0
Compensation and benefits B$_2$	3.0037	0.00185	0.58	0.00319
Individual development B$_3$	2.0000	0	0	0
Interpersonal relationship B$_4$	3.0000	0	0.58	0
Managerial leadershipB$_5$	2.0000	0	0	0

Consistency index and consistency ratio:

$$CI = (\lambda_{max} - n)/(n - 1), CR = CI/RI \qquad (3)$$

RI is the mean random consistency index, as shown in Table 5.

When the CI value is larger, the consistency of judgment matrix is worse. In general, when CI <0.1, the consistency of the judgment matrix is considered acceptable; otherwise, pairwise comparison is required.

According to formula (3.3), CI = 0.0642 < 0.1 RI = 1.12 CR = 0.0572 < 0.1, then the consistency of matrix A is acceptable.

To sum up, the consistency test of relevant indicators of employee satisfaction is passed, and the weight distribution is reasonable.

Similarly, the maximum eigenvalue of judgment matrix Bi and the consistency test results can be obtained, as shown in Table 6.

It can be concluded from Table 6 that the consistency test of all indicators in the indicator layer has been passed, and the weight distribution is reasonable.

Table 7. Employee Satisfaction Evaluation Index System.

Objective layer	Accurate layer	Weight	Index level	Weight
Employee satisfaction A	Work itself B_1	0.23	Diversity of work skills C_{11}	0.13
			Integrity of the work content C_{12}	0.03
			Feedback on work results C_{13}	0.07
	Compensation and benefit B_2	0.47	Basic wages C_{21}	0.05
			Merit pay C_{22}	0.15
			Social benefits C_{23}	0.27
	Individual development B_3	0.15	Work promotion space C_{31}	0.11
			Professional and technical training C_{32}	0.04
	Interpersonal relationship B_4	0.09	Coworker relationship C_{41}	0.02
			Supervisor-subordinate communication C_{42}	0.04
			Teamwork C_{43}	0.04
	Managerial leadership B_5	0.06	Management level C_{51}	0.02
			Leadership C_{52}	0.03

3.3 Establish ingress evaluation system for employee satisfaction

Based on the AHP calculation results, establish an employee satisfaction evaluation system, such as Table 7.

4 CONCLUSION AND PROSPECT

For modern businesses, the increase of customer satisfaction is premised on increasing employee satisfaction, which can reach 90% when the latter reaches 80%. The reason is that high-satisfaction employees are able to generate strong enthusiasm and vitality in customer service, and the quality of service provided is better, while low-satisfaction employees lack motivation, and negative attitude at work can also affect customers, resulting in a decline in customer satisfaction, affecting the company's revenue capacity. Through the research, we can find the importance of employee satisfaction, and employee satisfaction will affect the development of enterprises. By improving the hygiene factors and motivational factors that affect employee satisfaction, it can improve employee satisfaction and then improve employee enthusiasm, stimulate their potential, enhance the competitiveness of enterprises, and contribute to the sustainable development of enterprises. By studying the influence factors of employee satisfaction, determining the index of employee satisfaction evaluation, and distributing the weight of each index through AHP, we form an employee satisfaction evaluation system, hoping to provide meaningful guidance for employee satisfaction evaluation!

REFERENCES

[1] Khawaja Fawad Latif. 2012, An integrated model of training effectiveness and satisfaction with employee development interventions. Industrial and Commercial Training, 86–94
[2] Monica Izvercian, Sabina Potra, Larisa Ivascu. 2016, Work Satisfaction Variables: A Grounded Theory Approach. Procedia – Social and Behavioral Sciences, 221
[3] Tang Liqin. 2019, Factors and improvement strategies affecting employee satisfaction in private enterprises. Management Technology of SME (later issue), (01): 102
[4] Wang Xiaohong, Wang Mengxuan, Zhang Pengfei. 2019, Study on the Evaluation of Employee Satisfaction Impact Factors in Military Enterprises. China Management Information, 22(03): 95–98

Computational Social Science – Luo, Ciurea & Kumar (eds)
© *2021 Taylor & Francis Group, London, ISBN 978-0-367-70193-2*

Research on out-going employment security of out-going migrant workers under the COVID-19 outbreak

J. Chen* & R.H. Liu
Chengdu University of Information Technology, Chengdu, China

ABSTRACT: To ensure the stable employment of hundreds of millions of out-going migrant workers during the COVID-19 outbreak, this paper analyzes the various impacts of the outbreak on the industries and small and micro enterprises engaged in the migrant workers, and captures the employment security problems of this group in labor relations, social security, social assistance and the implementation of the government's "stable employment" policy. We put forward counter-measures and suggestions, such as focusing on the construction of the rights and responsibilities of migrant workers, reforming the social security system, relying on community governance for effective employment protection, and vigorously cultivating and developing grass-roots trade union organizations.

Keywords: COVID-19 outbreak, Migrant workers, Employment protection, Social insurance, Social assistance.

1 INTRODUCTION

After SARS in 2003 and H1N1 virus in 2009, China is once again faced with the outbreak and extension of COVID-19 around the Spring Festival in 2020. The epidemic is highly contagious, involving a wide range of people, and the situation is more severe, which has a great impact on informal employment in China. Late outbreak, national economic policy focus to promote employment, All-China Federation of Trade Union also specially issued "Notice on Doing a Good Job in Safely and Orderly Resuming and Resuming Production and Labor Relations during the Prevention and Control of COVID-19 Outbreak": Trade unions at all levels actively assist in job stability, supervise and assist enterprises to provide workers with the necessary labor protection conditions, effectively safeguard the legitimate rights and interests of employees, coordinate labor relations, and mobilize hundreds of millions of employees to seize epidemic prevention and control and achieve economic and social development in 2020 development goals, double victory and contribution [1]. Employment stabilization has become the focus of the work of the CPC Central Committee and local governments at various levels after the epidemic has eased. A large number of employment stabilization policies and measures have been issued, including job stabilization, corporate burden reduction, social insurance relief, unemployment insurance stabilization return, employment subsidies and employment Assistance etc. [2]. According to the "Monitoring Report of Migrant Workers in 2019" released by the National Bureau of Statistics, the migrant workers in 2019 reached 29.07 million, an increase of 2.41 million or 0.8% over the previous year. Among them, 17.425 million were out-going migrant workers, an increase of 1.59 million or 0.9% over the previous year [3]; these large-scale out-going migrant workers are the main force in China's urbanization construction, and they have inherently poor employment stability and low quality employment Under the impact

---

*Corresponding author

of the COVID-19, they are mainly engaged in the service industry and labor-intensive manufacturing, especially some service-oriented small and medium-sized enterprises and small and micro enterprises, individual industrial and commercial households and other informal employment sectors are facing bankruptcy risk. The current question is whether these policies and measures can truly target migrant workers who are in urgent need of assistance, and effectively respond to the impact of the current epidemic on employment. Therefore, it is necessary to design the employment security system and analyze the governance structure for migrant workers in China.

2 ANALYSIS OF THE EMPLOYMENT CHARACTERISTICS OF OUT-GOING MIGRANT WORKERS UNDER THE COVID-19 OUTBREAK

Since the reform and opening up, the rural surplus labor force that has been suppressed for a long time under the urban-rural split policy has formed a huge scale of migrant workers into the city with the reform of the urban-rural dual economic system. In a certain sense, migrant workers' economic mobile employment itself is a typical manifestation of fully competitive market-based employment. The supply and demand mechanism plays an important role in the employment of out-going migrant workers. According to the market-oriented employment mechanism, migrant workers gather large-scale cities to seek non-agricultural jobs. At present, its employment characteristics show the following characteristics:

3 THE SERVICE INDUSTRY HAS BECOME THE MAIN EMPLOYMENT CHANNEL FOR MIGRANT WORKERS, BUT IT HAS BEEN BASICALLY CLOSED DUE TO POPULATION AGGREGATION AND MOBILITY UNDER THE COVID-19 OUTBREAK

According to the data of the "Monitoring Report of Migrant Workers" released by the National Bureau of Statistics in 2019, migrant workers engaged in the tertiary industry accounted for 51.0% of the total migrant workers, an increase of 0.5% from 2018, of which 12.0% were engaged in wholesale and retail, transportation, warehousing and postal services 6.9%, accommodation and catering industry 6.9%, residential service repair and other services accounted for 12.3%, the cumulative proportion of 38.1% is shown in Table 1. It can be seen that the tertiary industry is the main industry that absorbs employment of migrant workers. Compared with agriculture and manufacturing, the service industry is the industry most affected by this outbreak, especially the production and consumption processes such as accommodation, catering, tourism, cultural and sports entertainment. At the same time, due to the typical characteristics of aggregation and mobility, these industries are almost out of business under the impact of COVID-19 outbreak.

Table 1. Industry distribution of migrant workers Units: %, Percentage points.

Industry		In 2018	In 2019	Increase or decrease
The third industry		50.5	51.0	0.5
Among them	wholesale and retail	12.1	12.0	−0.1
	transportation, warehousing and postal services	6.6	6.9	0.3
	accommodation and catering industry	6.7	6.9	0.2
	residential service repair and other services	12.2	12.3	0.1
	other	12.9	12.9	0.0

Source: National bureau of statistics: migrant worker monitoring report, 2019.

Table 2. Industry distribution of migrant workers Units: %, Percentage points.

Industry		In 2018	In 2019	Increase or decrease
The second industry		49.1	48.6	−0.5
Among them	the construction industry	27.9	27.4	−0.5
	manufacturing	18.6	18.7	0.1

Source: National bureau of statistics: migrant worker monitoring report, 2019.

4 THE IMPACT OF THE OUTBREAK HAS BEEN MORE COMPLICATED FOR SMALL AND MICRO BUSINESSES THAT ARE MAINLY EMPLOYED BY OUT-GOING MIGRANT WORKERS

In the case of traditional manufacturing employment, the mainstream employment mode is formal employment, there is a clear relationship between employers, employees, units, and employees, and employees have a very stable labor relationship with the employment sector. With the transformation and upgrading of the social and economic structure, small and micro enterprises have begun to flourish. This kind of employment in the informal sector usually has no clear employment relationship or no formal contract, and the labor relationship is extremely unstable. According to the data of the third national economic census, there are 147 million employees in small and micro enterprises, accounting for 50.4% of all employees in enterprises with legal persons [4]. According to the monitoring report of migrant workers released by the national bureau of statistics in 2019, the proportion of migrant workers in the secondary industry in 2019 is 48.6%, among which the construction industry accounts for 27.4% and the manufacturing industry 18.7%, as shown in Table 2. At present, The secondary and tertiary industries in China's cities have become the main channels to accept the employment of migrant workers, especially the construction industry, manufacturing industry and commercial service industry, which have become the three mainstream directions to absorb this group. Migrant workers are employed in small and micro enterprises in these three industries. Compared with the traditional formal employment model, small and micro businesses are more complicated under the impact of the epidemic, and the difficulties they face are more prominent. The employees are at greater risk of losing their jobs.

5 ANALYSIS OF PROBLEMS EXISTING IN THE EMPLOYMENT SECURITY OF MIGRANT WORKERS UNDER THE UNDER THE COVID-19 OUTBREAK

Since the reform and opening up, China has established an employment security system framework that adapts to the socialist market economic system. The first is the employment protection system based on the labor contract law. Second, we have established a social insurance system covering urban employees, including old-age insurance, medical insurance, maternity insurance, work-related injury insurance and unemployment insurance. Third, the social assistance system covering urban workers. The employment security system plays a very important role in promoting employment, protecting workers' life and resisting employment risks. However, the impact of the sudden COVID-19 outbreak on employment has exposed the current challenges and problems in China's employment security for out-going migrant workers.

5.1 *The labor contract signing rate of out-going migrant workers is low, and some migrant workers are not protected by the "Stable Employment" policy*

The labor contract establishes the labor relationship between the laborer and the employing unit in the form of law, clarifies the rights and obligations of the employing unit and the laborer, and enjoys the protection of labor rights and interests according to law. Whether it is the employment protection system based on the labor contract law or the social security system based on the social insurance, the objects of protection are employees with a clear employment relationship. Such a clear employment relationship is usually based on the signing of a labor contract. According to the "Statistical Bulletin of Human Resources and Social Security Development in 2018" [5]: In 2018, the national employment was 77586 million, of which 43419 million were urban employees, the total number of migrant workers was 28836 million, and the out-going migrant workers were 1726.6 million. The labor contract signing rate of enterprises in China has reached more than 90%, while the National Floating Population Tracking Data (CDMS) shows that in 2017, the labor contract signing rate of out-going migrant workers in China was 64%, the fixed-term labor contract signing rate was 48%, And the long-term labor contract signing rate of non-fixed term was 12% [6]. It can be seen that the rate of out-going migrant workers signing non-fixed term long-term labor contracts is far lower than the rate of signing short-term labor contracts with fixed term, and has shown a downward trend in recent years. From the perspective of labor economics, the signing of long-term labor contracts with no fixed term is conducive to the stability of labor relations, and the fixed term contracts have a negative impact on workers' wages. The labor contract signing rate is lower than the national level and the long-term labor contract signing rate is lower than that of the national level, which shows that the employment protection of migrant workers is still not optimistic. At present, a large number of policies and measures adopted by the country in response to the impact of the epidemic are targeted at those who sign labor contracts, and some migrant workers are "leaking out".

5.2 The low coverage of the social insurance system for urban workers exposes them to various social risks during the COVID-19 outbreak

Due to the poor working environment, high labor intensity and unstable labor relations of migrant workers, the social insurance system is particularly important under the impact of covid-19 outbreak. In terms of pension insurance, most migrant workers still have the traditional concept of family pension. In 2017, the participation rate of pension insurance for out-going migrant workers was 21.65% of the total. China's medical insurance is divided into urban employee medical insurance and urban and rural residents' medical insurance, and the medical insurance for urban and rural residents is a combination of the new type of rural cooperative medical insurance and basic medical insurance for urban residents, since most of the migrant worker took part in the new rural cooperative medical insurance in census register seat, since most migrant workers have participated in the new type of rural cooperative medical insurance in their registered residence, the participation rate of out-going migrant workers in the medical insurance for urban workers in 2017 was 27.25% of the total migrant workers, and the participation rate of the medical insurance for urban workers was still low, making it difficult to enjoy the medical insurance treatment in the places where migrant workers work. In 2017, the participation rate of out-going migrant workers in urban workers' unemployment insurance was 21.73%, and that of urban workers' work-related injury insurance was 17.09% [7]. On the whole, the social insurance coverage and security level of out-going migrant workers are far worse than that of urban workers. In particular, pension insurance, medical insurance and unemployment insurance are especially important for migrant workers, and only about one third of them are covered. The problem of insufficient protection and imbalance makes it exposed to various social risks during the special period of the epidemic.

5.3 The social assistance mainly implements the household registration dependency management, and out-going migrant workers have no chance

China's social security system mainly includes social insurance, social assistance and social welfare, while social assistance is mainly aimed at low-income groups and social welfare is mainly aimed at specific groups. Moreover, social assistance and social welfare are more regionalized and implemented with the management of household registration and dependency. The out-going migrant workers have special status, they do not have rural land security, and they do not belong to urban residents. Once they are unemployed or lose their source of income, most migrant workers have neither social insurance nor local social assistance. This kind of risk is already great in the epidemic period, which directly affects the survival needs of some out-going migrant workers.

5.4 During the outbreak period, the government's "Stable Employment" policies and measures are difficult to play an effective role in targeting the majority of out-going migrant workers

In order to quickly restore various social and economic functions in the late period of the outbreak, local governments at all levels have introduced various policies and measures to "stabilize employment", but they are difficult to play a role in targeting the majority of migrant workers. For example, standardizing the policy of laying off employees can only cover those who have signed labor contracts, but the signing rate of labor contracts for migrant workers is still low. The return of unemployment insurance benefits only applies to those who participate in the unemployment insurance for urban workers. During the epidemic period, the unemployed can receive unemployment insurance benefits if they participate in the unemployment insurance, while the participation rate of the unemployment insurance for migrant workers is only about one third. For the migrant workers who do not participate in the unemployment insurance, when they lose their job or their source of income, they would have been able to support themselves through social assistance. However, due to the special status of migrant workers, they cannot obtain any social assistance at work sites.

5.5 The degree of organization is low, and most out-going migrant workers are unable to realize their interests

To realize employment protection, workers should be able to successfully express their own interests and demands through trade union organizations and realize their right to participate in the dialogue between organizations and social management. According to the 2019 migrant workers monitoring survey report, among migrant workers in cities, 27.6% have participated in activities organized by their communities, an increase of 1.1 percentage points over the previous year. Among them, 3.9% have participated regularly and 23.7% occasionally. The proportion of migrant workers who joined trade unions in the cities was 13.4%, an increase of 3.6 percentage points over the previous year. When it comes to expressing their interests, only 7.8% of migrant workers go to trade unions, women's federations and government departments, and 2.6% go to communities. At present, a large number of out-going migrant workers are widely employed in various social labor fields, and the degree of informal employment is high. Most of the small and micro enterprises they are in have no conditions to establish trade union organizations at all. The labor relations are loose, the degree of organization is very low, and there is a serious lack of organizations and mechanisms for the expression of the right to speak and the protection of rights and interests. In the special period of COVID-19 outbreak, once the rights and interests of labor and employment are damaged, it is difficult to realize the interest demands.

5.6 Countermeasures and suggestions for the employment security of out-going migrant workers during the COVID-19 outbreak

It is foreseeable that with the gradual alleviation of the COVID-19 outbreak, various "stable employment" policies and measures will be effectively implemented, but the hidden employment issues may gradually be exposed, fully paying attention to and protecting hundreds of millions of migrant workers survival and employment, capturing the problems and development dilemmas in the labor and employment security system of this group are of great significance for promoting the long-term reform of employment protection and social security for migrant workers in China.

5.7 During the COVID-19 outbreak, the implementation of various "Stable Employment" policies should focus on and protect the vulnerable group of out-going migrant workers

From the perspective of employment characteristics, migrant workers' employment outside the city is basically subsistence employment. Most out-going migrant workers" educational level is not high, there is no special labor skills and their respective in all sorts of private small and medium-sized enterprises or small micro enterprise engaged in simple labor or services, and these enterprises or industries are most affected by the outbreak, the outbreak period once unemployment or lose their source of income, they will soon be facing predicament survival. For this part of the group, and the outbreak period, in the process of implementing the "stable employment" policy, in the process of return to work and production, the implementation of the policy of steady employment should not only focus on large and medium-sized enterprises, more should attach great importance to the small micro enterprise of weak impact resistant ability and involves residents daily life to return to work and production services, and promoted the effective and full implementation of these enterprises and industries such as burden reduction, job stabilization, social insurance premium reduction, unemployment insurance stabilization return, employment subsidies, and employment assistance. This is not only related to the employment of hundreds of millions of migrant workers out of the country, but also related to social stability in a special period.

5.8 Special treatment should be given to some out-going migrant workers, who have not signed labor contracts in the outbreak, and the social security system should be fully reformed

First of all, local governments should appropriately expand the scope, scale and intensity of employment protection and social assistance. Special treatment will be given to some migrant workers who have not signed labor contracts, and they will also be included in the local social security system, so that they will be treated as urban workers in terms of employment security, and those who meet the conditions during the epidemic should be given urban subsistence allowances. Second, in the long run, it is necessary to reform the social security system for migrant workers. We should vigorously promote the participation of migrant workers in the social security system, focusing on work-related injuries, medical care, unemployment and old-age insurance. In terms of system design, we should reform the employment protection and social security system based on clear employment relationship, break the previous restrictions on the identity, units and regions of the social security system, and move to a residential-based employment promotion and social security system, so as to establish a unified social security system for residents. The government can offer social security subsidies to the employing units that assume social security responsibilities, so as to truly replace the framework of employee insurance with the construction of national insurance. As migrant workers have the characteristics of mobility, it is convenient for them to transfer and continue their social insurance rights and interests in the system design, so as to realize the effective implementation of the policy of "stable employment".

5.9 Local governments should urge enterprises to sign and perform labor contracts with out-going migrant workers in accordance with the law, and establish stable power-responsibility labor relations

In the short term, for out-going migrant workers, the reason why the "stable employment" policy under the COVID-19 outbreak is difficult to effectively implement is that most employers have not signed labor contracts with them in accordance with the law. In the long run, it is necessary to guide and supervise the employing units to sign and perform labor contracts with out-going migrant workers. Labor security supervision and law enforcement departments at all levels should actively urge enterprises to sign labor contracts with migrant workers in accordance with the provisions of the Labor Contract Law and the Implementation Regulations of the Labor Contract Law. The labor contract must be clear in the contract period, the labor remuneration and social insurance premium collects pay and the amount of payment and the relevant responsibility of breach of contract and other specific content, improve propaganda on the significance of migrant workers to sign labor contract, establish a clear and stable labor relations of power and responsibility to protect the employment rights and interests of migrant workers.

5.10 Rely on community governance to achieve effective employment protection for out-going migrant workers

Community governance as a practical application in the field of community governance theory refers to: in some area of government, community organizations, residents and district units, profit-seeking organizations and non-profit organizations, based on market principle, public interests and community identity collaboration, effective supply of public goods, to meet community needs and optimize the process and mechanism of community order [8]. In the COVID-19 outbreak, we can rely on community governance to achieve effective protection of out-going migrant workers out of the country, especially in the urban areas, such as urban living, employment and other immediate needs, the community should give full play to the outbreak period of the grassroots government service consciousness and bear responsibility, for the epidemic period of unemployment or lose their source of income of migrant workers to provide daily necessities and medical care, and set up the mechanism of communication and mutual assistance platform. For out-going migrant workers who have difficulty in finding jobs, communities can provide support through public welfare positions. To supervise and urge enterprises in the area under their jurisdiction to provide migrant workers with comprehensive care and support from work to life. For example, policies and measures such as stable employment, social insurance and epidemic shutdown benefits should be truly and effectively implemented in enterprises in the area under their jurisdiction, and the utility of grassroots government organizations where migrant workers live should be fully exerted.

5.11 Cultivate and develop grass-roots trade union organizations vigorously, and protect the legitimate labor rights and interests of out-going migrant workers effectively

Trade unions are mass organizations that safeguard the interests of the working class. Out-going migrant workers who are important members of the working class have the right to organize and join trade unions, and trade unions have the obligation to protect their labor and employment rights and interests. Due to the large number of out-going migrant workers, high mobility and scattered jobs,and the small and medium-sized enterprises or small and micro enterprises in which they are employed do not have the ability to form trade unions. Therefore, the National Federation of Trade Unions should vigorously cultivate and develop grass-roots trade union organizations in light of local and enterprise conditions. In view of the characteristics of out-going migrant workers living in communities, community trade unions should be established and given the functions of grass-roots trade union organizations so as to attract migrant workers employed in community enterprises to join trade union organizations. For out-going migrant workers with strong mobility and instability, they should establish professional trade unions, industrial trade unions or regional trade unions according to their occupations, and shift from the former single membership model to the multiple membership model, and expand from a single enterprise to the market, industry and community trade unions. During the outbreak, trade unions should make full use of their own resources to help them solve various practical problems, including job stability, unemployment assistance, wage payment during work stoppage, job introduction, employment assistance and other protection of employment rights and interests, so as to help more migrant workers to achieve stable employment and protect their legal labor rights.

ACKNOWLEDGEMENT

This research was financially supported by National Social Science Foundation Project (15XJY005).

REFERENCES

[1] https://www.sohu.com/a/377918516_100010474.

[2] A combination of policies to stabilize employment in various regions, N. China securities journal .2020.

[3] http://www.stats.gov.cn/tjsj/zxfb/202004/t20200430_1742724.html.

[4] http://www.stats.gov.cn/tjsj/zxfb/201412/t20141216_653709.html.

[5] http://www.mohrss.gov.cn/SYrlzyhshbzb/zwgk/szrs/tjgb/201906/t20190611_320429.html.

[6] M.F. Zhao, S.N. Wang, 2020, Analysis on the change trend and characteristics of migrant workers' employment quality, J, Shandong trade union forum, 03, 1–11.

[7] Z. Wang, Employment protection and social security under the impact of covid-19 outbreak, J, The economic aspect, 03(2020) 7–15.

[8] X. Wu, Y. Zhang, 2020, Floating population control in combination with community governance in the covid-19 outbreak, J, Nanjing social sciences, 03, 21–27.

Computational Social Science – Luo, Ciurea & Kumar (eds)
© 2021 Taylor & Francis Group, London, ISBN 978-0-367-70193-2

An integrated research on the benefits, problems, and countermeasures concerning Wuhan rail transit operation

F. Gao*, Y. Peng & H.P. Wang
School of Business, Jianghan University, Wuhan, P.R. China

Q.H. Liu
School of Business, Jianghan University, Wuhan, P. R. China
Manufacturing Industry Development Research Center on Wuhan City Circle, Jianghan University,
Wuhan, P.R. China

ABSTRACT: The continuous development of China's urban rail transit industry plays a key role in easing the pressure of urban transportation. However, due to its high investment and cost, a long time to make profits, and its public welfare characteristics of the government, urban rail transit has been a long-term loss. We analyzed the development status and operation benefits of Wuhan urban rail transit, found out the existing problems of the operation and management, and put forward feasible strategies and suggestions to promote the sustainable development of Wuhan urban rail transit.

Keywords: Rail transit, Operation benefit, Wuhan City.

1 INTRODUCTION

With the continuous and rapid development of China's national economy and the acceleration of the urbanization process, the contradiction of urban infrastructure, especially urban transportation facilities and urbanization development, is gradually emerging. The construction of the urban rail transit system is an important embodiment of the implementation of the "public transport priority" strategy, and it will become the development direction of public transport in large cities in China. In developed cities, the proportion of public transport is generally 60% to 80%, of which the proportion of rail transit is 30% to 45%. No matter how private transportation develops, the status of public transportation as the main body has not changed, and the rail transit often occupies a greater advantage and dominant position.

Since Metro Line 1 was put into operation in 2004, Wuhan rail transit has developed rapidly. Its metro mode, speed, quality, culture, and spirit have also become a vivid portrayal of the Wuhan spirit of "dare to be the first and pursue excellence." However, many problems in Wuhan rail transit system still restrict the development of its rail transit network, such as the huge construction funds of rail transit, the large amount of subsidy funds invested in the later operation and maintenance, and the repayment of huge bank loan interest, which makes the government and enterprises overwhelmed. There are also deficiencies in the strength of absorbing private capital, and the effective management system and business model need to be improved.

Based on the analysis of the development status and benefits of urban rail management in Wuhan, this paper finds that the investment and financing mode of Wuhan rail transit is not suitable for the capital operation of enterprises; the effect of rail transit network is not obvious; and the operation and management technology lags behind; and this paper puts forward corresponding countermeasures and suggestions. Wuhan Metro Group can carry out the work from the following aspects, adopt the mixed-income source mode, self-financing profit and loss mode, improve the operation mode, and

---

*Corresponding author

speed up the operation and management technology update, so as to better promote the sustainable development of Wuhan urban rail transit operation and management mode.

2 ANALYSIS OF WUHAN RAIL TRANSIT OPERATION MODE

By the end of 2019, Wuhan had opened nine rail transit lines with an operating mileage of 339 kilometers, 228 stations, and an average daily passenger flow of 3.35 million people, ranking sixth nationwide (Niu 2019). The construction of Wuhan rail transit is based on the premise of planning. The comprehensive planning of land use along the rail transit line is formulated to optimize the urban functions of the station area, and the requirements of design control are put forward for the rail station project. According to Wuhan Urban Rail Transit Construction Planning (2019–2024), the main capital source of Wuhan rail transit project construction is jointly undertaken by local government and Wuhan municipal government, and the rest is mainly from bank loans. We can find that the financing mode of Wuhan rail transit is mainly government special funds, not in the form of cooperation between social capital and the government. Once the government's financial tightening occurs, the impact on the construction project of Wuhan rail transit is relatively significant.

According to Qin (2014) scholars' research, China has three types of urban agglomerations: national, regional, and regional involving 20 major urban agglomerations in China, among which the largest are the Yangtze River Delta Urban Agglomerations, Pearl River Delta Urban Agglomerations, and Beijing Tianjin Hebei Urban Agglomerations. In order to promote the development of urban agglomerations, the government plans and builds urban rail transit based on urban agglomerations. Wuhan belongs to Wuhan City Circle, which has a population of 31 million and a population density of 5.535 million people per 10,000 square meters. With the rapid economic development of the city group, the construction of domestic intercity rail transit must be adapted. Domestic scholars also learn from the excellent mode of foreign rail transit construction, combined with the actual development of our country, to integrate the research suitable for the development of urban rail transit in China.

3 BENEFIT ANALYSIS OF WUHAN RAIL TRANSIT OPERATION MODE

On the issue of benefits brought by urban rail transit, based on the review of the literature, according to the different levels, we elaborate on the benefit analysis brought by urban rail transit from the social level, economic level, business operator level and individual level of residents.

3.1 Social benefits brought by Wuhan urban rail transit

The construction and operation of Wuhan rail transit project has different benefits to the subjects involved. Because the subjects are connected and cooperated based on the explicit and implicit contracts, the interests of all parties have a common trend. For quasi-public facilities such as metro rail transit projects with obvious external benefits, in addition to the direct and external benefits of the project on social, environmental, and economic aspects, the welfare benefits of users of public facilities should also be considered (Chen et al. 2019).

The social efficiency brought about by urban subway projects can be divided into four categories: social welfare, social economy, social environment, and natural environment. For social welfare, the subway brings residents comfort, accessibility, and safety. In terms of social economy, it drives regional employment, brings value-added utility to land along the line, urban economic development, and induced passenger flow brought by convenient transportation. As for the social environment, the subway construction is carried out underground, which saves the urban pavement land. The last aspect is about the natural environment. Due to the non-carbon emission of the subway, the benefit of pollutant emission reduction is increasing, and the use of energy such as fuel oil is reduced.

3.2 *Economic benefits brought by Wuhan rail transit operation mode*

3.2.1 *Regional economic and social employment impact*

When Tang et al. (2019) and other scholars took Changsha urban rail transit as the research object, they quantitatively studied the impact of urban rail transit construction on regional economy and social employment by building an input-output analysis model of project construction investment. Their research shows that urban rail transit construction has a positive impact on regional economy and social employment, and then they concluded that urban rail transit construction and operation have external benefits, which cannot be reflected in the financial statements of operating enterprises. Therefore, they proposed that when planning the urban rail transit investment project, the government should give full consideration to the contribution of the project investment to the regional economy and social employment. We find that when planning new rail transit projects, the Wuhan municipal government will select areas with great potential for regional economic development to build a rail transit network, including major business districts of three towns in Wuhan, to make regional economic development more collaborative and drive areas with slow economic development with areas with good economic development, to activate the economic activity of the whole city of Wuhan, and to promote the linkage of various regions by convenient means of transportation, so as to promote the employment of regions to a certain extent.

3.2.2 *Impact on housing value along the line*

According to the research of Tang et al. (2013) and other scholars in the early years, they have constructed the spatial lag and spatial error model of the impact of urban rail transit on the value of housing along the line. The research shows that the farther away from the subway station, the smaller the value-added range of housing, and presents the rule of gradual decline. This model can explain why the price of the housing near the rail transit in Wuhan is much higher than that of the housing far from the rail transit. In particular, the houses in some business districts of Wuhan are very close to the metro rail transit of each line. For these kinds of houses located in the transportation hub, the value of the houses will rise. On the one hand, it also indirectly shows that the land price in the area along the rail transit will rise. Specifically, the construction scope of urban rail transit shows the concentration of real estate. The demand for land within the scope of construction is high, and the price will rise accordingly.

3.2.3 *Commercial impact on the space in the rail transit station*

At present, most of the subways and light rails in our country put advertisements in the stations. The main way is to use the advertisement resources scientifically and effectively through outdoor advertisements and in train broadcasting to effectively improve the economic benefits brought by the space in the rail transit station. Urban rail transit stations can realize commercial development and application to the greatest extent. For example, the passenger flow of rail transit stations is large, and the commercial street stores can be set up to improve the income of the metro operation company.

3.3 *Business benefits brought by Wuhan rail transit operation mode*

As for the Wuhan Metro Group, which operates urban rail transit, it has improved the popularity of the enterprise and made it possible to develop new line competition in other cities. For the real estate development enterprises along the line, the value-added of houses and the convenience of transportation improves the sales of houses. For the surrounding commercial enterprises near the subway station, the convenience of transportation has brought more passenger flow, increased the consumer users of commercial enterprises, and indirectly increased sales.

3.4 *Individual benefits brought by Wuhan rail transit operation mode*

For individuals, the construction and operation of metro projects can greatly improve their welfare effect. Individuals not only enjoy convenient subway transportation services, but they also save the cost of alternative fuel vehicles. Near the business circle of Wuhan Metro cluster, shopping and entertainment are more convenient for residents. The combination of metro lines and intercity high-speed rail stations also facilitates the convenience of residents to travel outside.

4 ANALYSIS ON THE PROBLEMS OF WUHAN RAIL TRANSIT OPERATION

4.1 *The mode of investment and financing is not suitable for the capital operation of enterprises*

Wuhan Metro Group, a state-owned enterprise entrusted by the government, is responsible for the operation and management of Wuhan rail transit. When a new line is opened, the funds to be approved mainly include government subsidies and bank loans. BT financing mode is widely used in the construction of large-scale quasi-public facilities projects such as metro projects. On the one hand, the scale of capital required is very large, the amount of investment is high, and the depreciation and interest costs put great pressure on Metro Group. On the other hand, the operation of urban rail transit shows obvious characteristics of high investment, high cost, long recovery, and public welfare of the government. Therefore, the ticket revenue is difficult to cover the daily operating cost expenditure, resulting in the overall loss of Wuhan Metro Group. The profit mainly depends on the government subsidies, which will become increasingly difficult with the decline of government subsidies.

4.2 *Wuhan urban rail transit line-network effect is not obvious*

According to the research of Guo Lei (2019), the urban rail transit industry is characterized by scale economy. The line-network effect can effectively increase passenger flow and transportation efficiency, but the overall line-network scale level is required to be high. At present, only the first-tier cities of Beijing, Shanghai, Guangzhou, and Shenzhen have obvious advantages of line-network effect. The scale of the second-tier cities is growing rapidly, but it still needs a long period of training. Wuhan urban rail transit is the obvious scale of growth as very fast, but the rail transit line-network effect is not enough to drive the income growth. There is no effective way to reduce the operating cost, and the pricing method and ticket price level will not change significantly in a short time, and so the operation of urban rail transit will become the increasing debt burden of Wuhan municipal government.

4.3 *Lag of operation management technology*

In recent years, Wuhan Metro has carried out in-depth operation and management information work and established a centralized, unified, and efficient modern operation and management system. However, with the continuous integration of various professional systems, the data of each professional system has not been shared in real time. This cross-system, multi-process business monitoring technology lags behind, which makes Wuhan Metro Group lack real-time and multi-dimensional operation management analysis ability.

This traditional business data analysis and processing technology is limited by the current processing platform and processing technology, so it cannot comprehensively mine and utilize the data in real time, in a complete and objective way, so that the valuable data cannot be fully obtained and the company cannot make scientific and correct decisions.

5 COUNTERMEASURES AND SUGGESTIONS

5.1 *Adopt the mode of mixed-income source*

Referring to the advanced experience of Hong Kong Metro, Wuhan Metro Group must focus on the transformation of its business model and adopt a mixed revenue source model to increase its other income. The mode of station income and property development along the track will drive the supplement of income sources, expand the benign development of property management plate, and carry out the comprehensive development of the mode of "track+property." Through the development of real estate projects, holding and operating large shopping malls, residential properties, and other properties will ensure the Wuhan Metro enterprises have stable income.

5.2 *Improve the business model with the mode of self-financing*

The operation mode of Wuhan Metro Group needs to be further improved. For example, Hong Kong Metro adopts the mode of separation of powers and self-financing of profits and losses. Although the municipal government has the right to control, it does not have the right to manage and operate, and does not need extra funding from the government, which is worth learning.

5.3 *Speed up the update of operation management technology*

Learning from the research results of Jiang (2020), Wuhan Metro Group should make full use of new technologies such as big data, cloud computing, Internet of things, artificial intelligence, and so on, to build "urban rail transit management information platform based on big data technology." Form a big data platform, conduct in-depth exploration of multi-lines and multi-disciplines, and provide intelligent decision support through analysis to support the integrated application of multi-line and cross-discipline information fusion more effectively. Because Wuhan information technology is at the forefront of the country, the strengthening of Wuhan Metro Group's operation and management technology will reduce the cost of enterprise decision making, and the decision making will become more scientific and multidimensional.

ACKNOWLEDGEMENT

This paper is supported by the Open Fund Project of Manufacturing Industry Development Research Center on Wuhan City Circle of Jianghan University under grant WZ2016Z01.

REFERENCES

J. An. 2020, Study on the economic benefits of comprehensive utilization of urban rail transit spatial resources. Science and Technology Wind, (10): 223.

G. Chen, C. Lu, L. Huang, J. Li. 2019, Social benefit evaluation of urban rail transit projects. Construction Technology, (21): 13–16+21.

China Urban Rail Transit Association. Overview of urban rail transit lines in mainland China in 2019 [EB / OL]. [2020-01-01]. http://www.camera.org.cn/index.php? M = Content & C = index & A = show & catid = 43 & id = 19329.

L. Guo. 2019, Discussion on the current situation and profit model of China's rail transit industry. Science and Technology Economy Guide, 27 (26): 215+210.

Hong Kong Railway Co., Ltd. 2018 annual report. Http://www.mtr.com.hk/ch, 2019-3-7.

Z. Jiang. 2020, Application of big data technology in urban rail transit operation and management. Science and Technology Innovation and Application, (05): 174–175.

Y. Miao, Z. Zhang. 2015, Development and characteristics of rail transit in Tokyo Metropolitan Area. Urban Rapid Rail Transit, 28 (2): 126–130.

W. Niu. 2019, Coordinated planning and management of urban rail transit and land use – Taking Wuhan practice as an example. Transportation Construction and Management, (06): 96.

Y. Qin. 2014, Early planning and construction of suburban railway and intercity railway in China. Journal of Railway Engineering, (1): 3–6.

W. Tang, F. Zhang, Q. Xiao. 2019, The impact of urban rail transit project construction investment on regional economy and social employment. Journal of Railway Science and Engineering, 16 (12): 3160–3166.

W. Tang, F. Zhang, H. Yan, Z. Zhou. 2013, Spatial effect of urban rail transit on housing value along the line. Journal of Hunan University of Science and Technology (SOCIAL SCIENCE EDITION), v.16; No.82 (06): 96–100.

W. Zhan, Y. Xu, Y. Wang. 2018, Application Research on intelligent operation and maintenance system of urban rail transit vehicles. Urban Public Transport, (12): 28–31, 36.

Computational Social Science – Luo, Ciurea & Kumar (eds)
© 2021 Taylor & Francis Group, London, ISBN 978-0-367-70193-2

Study on the mechanism of inclusive leadership inspires employees' job crafting—a moderated double-mediation model

K. Du & C.S. Wu
School of Management, Hefei University of Technology, Hefei, China

L.G. Zhang
State Grid Anhui Electric Power CO.LTO, Hefei, China

ABSTRACT: The employees' job crafting behavior has good benefits for both individuals and organizations. Therefore, how to improve the level of employees' job crafting has become a problem worthy of attention. Current studies pay more attention to the personal factors of employees while ignoring the impact of organizational factors on employees' job crafting. Based on an empirical study of 274 samples, this paper examines the impact of inclusive leadership on employees' job crafting. The results show that inclusive leadership effectively promotes the level of employees' job crafting; the supportive organizational climate and psychological empowerment have a dual-mediation role in this process; power distance not only negatively moderated the relationship between inclusive leadership and psychological empowerment, but also buffered the whole mediating mechanism.

Keywords: Job crafting, Inclusive leadership, Psychological empowerment, Supportive organizational climate, Power distance.

1 INTRODUCTION

The uncertainty of today's world environment, the rapidity and complexity of changes, makes the development of enterprises face new challenges. Therefore, how to stimulate the enthusiasm, initiative, and creativity of employees to ensure the sustainable development of enterprises has become an issue worthy of attention, while as a tool to realize the intrinsic motivation of employees, job design has always been valued by management. Traditional job design theory refers to a top-down process in which managers design jobs for employees according to the principles of specialization, simplification, and scientization of work contents. In other words, the process is led and implemented by managers while employees are in a passive position. Obviously, in such a rigid mode, organizations' flexibility cannot adapt to the rapid changes of the times and environment. Meanwhile, with the rapid development of social economy, employees' demand for autonomy and diversity has become more and more intense. Work is no longer just a means to support the family, but also a way for employees to realize their self-worth and pursue self-transcendence. They hope to combine work with their personal interests, specialties, and motives so as to better exert their potential to actively change their jobs and achieve professional success. Faced with the dilemma that traditional job design theories cannot meet the needs of reality, American scholars Wrzesniewski and Dutton (2001) put forward the concept of job-crafting, which is defined as "the physical or cognitive changes made by individuals in job tasks and work relationships," where employees have become participants, not just recipients. This bottom-up job design method undoubtedly helps employees to show more enthusiasm, initiative, and creativity, since as human vitality is enhanced, the benefits of enterprises increases naturally.

At present, the research on the outcome variables of job crafting has achieved fruitful results, which confirms the positive impact of job crafting on individuals and organizations. Given that job crafting was of critical importance to an organization, it was vital to identify factors that could stimulate employees' job crafting behavior. The research on its predictive factors is rare, and most of them pay attention to employees' personal factors, such as personal motivation (Lyons 2008), self-efficacy (Vough & Paker 2008), active personality (Tims & Bakker 2010; Tims et al. 2012), and career orientation (Jacobs 2011; Leana et al. 2009), and so on. So, is it true that job crafting such an employee's individual initiative behavior has nothing to do with other factors in the organization? Apparently, this is not true–organizational climate, colleagues, leaders, and other organizational contextual factors will also affect employees' perception of job crafting opportunities and the process of job crafting (Zhao Xiaoyun & Guo Cheng 2014). Studies on leadership style show that leaders in an organization would influence employees by shaping the atmosphere of the team or organization. Berg et al. (2013) also emphasized the crucial role of managers in the process of employees' job crafting. As a matter of fact, leadership is not only a vital climate factor in itself, but it can also influence interpersonal relationships or other organizational climate factors within an organization by formulating organizational strategies (Zohar et al. 2008), thus affecting employees' attitudes and behaviors (Carmeli et al. 2010). The effect of the different styles of leadership differ in thousands of ways. As an emerging relational leadership, inclusive leadership is one that is employee-centric, who pays attention to employees' needs, recognizes and encourages employees, tolerates team members' failures, and gives guidance, which is conducive to creating a supportive organizational climate, endowing employees with positive psychological experience, and thus positively affecting the process of employees' job crafting. Therefore, this study attempts to verify the promoting effect of inclusive leadership on job crafting and explore the mediating effect of supportive organizational climate and psychological empowerment. The two variables of supportive organizational climate and psychological empowerment are chosen as mediating variables because, firstly, from an organizational perspective, organizational climate, as an important organizational contextual variable, will have a significant impact on individual attitudes and behaviors. Secondly, from an individual perspective, as one of the important psychological mechanisms, the mediating role of psychological empowerment among different leadership styles and their outcome variables has been proven (Liu Jingjiang & Zou Huimin 2013; Zheng Xiaoming & Liu Xin 2016). Finally, the dual intermediary mechanism between inclusive leadership and its outcome variables has been extensively discussed by scholars (Choi et al. 2015; Xin Xun & Miao Rentao 2018). Then, we further discuss the boundary conditions that affect this intermediary mechanism and study the moderating effect of power distance, a characteristic variable that reflects personal value orientation.

To sum up, this paper will investigate the side-by-side dual mediating role of supportive organizational climate and psychological empowerment between inclusive leadership and employees' job crafting, as well as the moderating role of power distance, which is helpful to analyze the mechanism more clearly and completely.

2 THEORETICAL FOUNDATION AND RESEARCH HYPOTHESES

2.1 Inclusive leadership and employees' job crafting

In the era of the knowledge economy, the internal and external environment of enterprises has undergone tremendous changes. On the one hand, the continuous advancement of economic globalization has intensified the complexity, uncertainty, and competitiveness of the environment; on the other hand, employees' demand for autonomy is getting higher and higher. Meanwhile, they have realized that traditional job design can neither meet their needs for their job nor match their abilities with their work. Obviously, the traditional, relatively unified, and unchangeable job designed by the organization managers is easy to cause the organization to be too rigid to keep pace with the times, which also are not conducive to encouraging employees whose individual initiative and enthusiasm cannot be fully brought into play. In view of this situation, Wrzesniewski and Dutton (2001) improved and developed the theory of job design and put forward the concept of job crafting (which means employees redefine the cognitive, task, and relational boundaries by changing their behavior and cognition of work so that their interests, motivations, and passions are consistent with the requirements of jobs and tasks), then divided them into three dimensions: (1) changing task boundaries, which means employees changing the scope, number, and types of job tasks, including increasing or decreasing the amount of tasks, changing the nature of tasks, changing the distribution of time and energy on multiple tasks; (2) changing relational boundaries, which refers to employees adjusting the quantity or quality of working partnerships; (3) changing cognitive task boundaries, which refers to employees' re-understanding of the meaning and value of work. In terms of motivation for job crafting, Wrzesniewski and Dutton proposed that the need for control, the need for contact with others, and the need for positive self-image are the three major motivations.

Subsequently, Tims et al. (2011), from the perspective of job requirements-resources, through exploratory factor analysis, determined these four dimensions of job crafting: increasing structural job resources (such as resource diversity, development opportunities, and autonomy), increasing social job resources (such as social support, superior assistance, and feedback), increasing challenging job demands (such as high-level work requirements and challenges, increasing work tasks, expanding work scope), and decreasing hindering job demands (such as reducing interpersonal interaction, tasks and decisions that consume excessive energy on one's own emotions).

Domestic scholars have begun to research the relationship between leadership style and job crafting, confirming that the leader, as one of the organizational contextual factors, does affect job crafting. For example, Tian Qitao (2018) and Wang Hongyu and Cui Zhisong (2018) have, respectively, studied the influencing mechanism of servant leadership and coaching leadership on job crafting. Compared with the other forms of leadership that may be conceptually related, inclusive leadership is characterized by unique acceptance, sense of belonging, uniqueness, and inclusiveness (Randel et al. 2017). As a new, different leadership style, the effectiveness of inclusive leadership has been verified (Zhang Ruiying et al. 2018). Based on social information processing theory and organizational support theory, this paper analyzes the influencing mechanism of inclusive leadership on job crafting from the perspective of both organizational and employee individual levels. The concept of inclusion was initially applied in the field of education, and schools were considered to be inclusive of students of different cultures, races, and religions. Nembhard and Edmondson (2006) were the first to introduce "inclusiveness" into leadership research and were the founders of "inclusive leadership." Subsequently, Carmeli et al. (2010) enriched its connotation, describing inclusive leadership as showing openness, availability, and accessibility in interactions with followers, and being able to listen to and care about the needs of followers, which is the special form and the core of "relationship leadership." By shaping a two-way relationship of inclusiveness, democracy, and recognition with employees, inclusive leadership stimulates the three major motivations of employees' job crafting, hence encouraging employees to carry out job crafting behaviors. First, inclusive leaders adhere to people-oriented and balanced authorization (Li Yanping et al. 2012), which enables employees to have greater autonomy and sense of participation in their work, meets their control needs, and increases their structural job resources. Second, inclusive leaders are adept in listening to subordinates' opinions, who have better affinity, tolerance, and supportiveness. They tolerate employees' trial-and-error behaviors at work, thus making them easier to approach and helping to establish a harmonious two-way relationship with employees (Hollander 2009), which not only meets the needs of employees to contact others but also increases social job resources. Third, inclusive leaders tend to praise and encourage subordinates' contributions in words and deeds (Nembhard & Edmondson 2006), and they tolerate subordinates' failures and provide guidance, which makes employees more confident and satisfies the needs of employees to establish positive self-impression. In addition, the openness and inclusiveness of inclusive leadership are also conducive to employees' innovative behaviors (Carmeli et al. 2010), increasing their challenging job demands. The theory of situational strength (Meyer et al. 2010) proposes that organizational contextual factors such as leadership style can give explicit or implicit external feedback on the appropriateness of an individual's specific behaviors, so as to strengthen or avoid such specific behaviors; strong (or weak) organizational context promotes (or hinders) the transformation of individual's cognition, emotion, interest, values, and other psychological factors into specific behaviors. Consequently, inclusive leaders, as an organizational situational force who would give feedback to employees' job crafting behaviors, provide more support and encouragement and tolerate others' mistakes, and they further enhance employees' enthusiasm for job crafting.

Based on this, the following hypothesis is proposed:

H1: Inclusive leadership has a significant positive impact on employees' job crafting.

2.2 *Mediating role of supportive organizational climate*

Organizational climate refers to the recognition of certain events, activities, and procedures of employees and behaviors that may be rewarded, supported, and expected in a certain environment (Schneider 1990) which has stability and persistence and which can distinguish different organizations. However, the early research on organizational climate was too generalized to define the impact of specific climate on specific outcomes. Ehrhart et al. (2014) found that studies on specific organizational climate are differentiated (distinguished from other concepts) and focused (for specific results), which have a tight logical relationship with variables. This means that different organizational climates correspond to specific outcome variables and cannot be generalized. Thus, different organizational climates in specific situations emerged as the times required. Supportive organizational climate refers to "the degree to which, shared by team members, that organizations appreciate their contributions, support them, value their interests, and consider their needs" (Bashshur et al. 2011).

Social information processing theory was put forward by Salancik and Pfeffer (1978), who believed that human beings are adaptive organisms. They often understand and explain the behaviors of themselves and others according to the clues provided by the environment and adjust their attitudes and behaviors according to the obtained information. This theory is often used to explain the mechanism of leadership behavior and individual behavior on the formation of organizational climate or to explain the mechanism of organizational climate on employee behavior (such as voice behavior, helping behavior, etc.) and organizational results (Frazier 2009; Priesemuth et al. 2011). As an important organizational situational force, inclusive leadership can be expected to play a promoting role in the formation of supportive organizational climate. Previous studies have proved that managers' leadership style plays a decisive role in organizational climate: for instance, studies by Zhu Shaoying et al. (2008) and Yang Chunjiang (2011), respectively, show that transformational leadership has a significant positive correlation with team climate and transactional leadership has a positive correlation with supportive organizational climate. Inclusive leadership attaches great importance to the contribution of employees; is good at encouraging, praising, and supporting employees; considers the needs of employees; and establishes democratic, inclusive, and harmonious interpersonal relationships with employees, thereby creating a supportive organizational climate.

Based on this, the following hypothesis is proposed:

H2$_a$: Inclusive leadership has a positive impact on supportive organizational climate.

Organizational support theory was put forward by Eisenberger et al. (1986), who believed that supportive behaviors from organizations are conducive to employees' positive working attitudes and behaviors. The support of the organization provides more opportunities and resources for the work of the employees. At the same time, their psychological security is significantly improved. With the dual assistance of material and spiritual levels, employees gain a continuous stream of motivation, correspondingly generate high-level commitment and dependence on the organization, and work harder to repay the organization with high performance. Undoubtedly, supportive organizational climate will enable employees to obtain more sense of organizational support and help them to develop positive attitudes and behaviors. So far, studies on the outcome variables of supportive organizational climate have been fruitful. Bennett and Bell (2004) found that a high-structure organizational support climate stimulates employees to exert their talents within the scope of rights and the enthusiasm for pursuing self-value realization, which also contributes to improving employee satisfaction. Liao et al. (2009) believed that a supportive organizational climate helps employees to share knowledge and create new knowledge to provide the technological innovation capabilities required by enterprises. Torner et al. (2016) proposed that a supportive organizational climate is related to team effectiveness, innovation, and security. Chinese scholars have also found that a supportive organizational climate helps employees develop positive work attitudes and behaviors: the sense of organizational support creates conditions for employees to break through the original work paradigm and innovate (Zhang Chunyu et al. 2012), and a supportive organizational climate promotes employees' voice behavior (Zhu Yiwen, 2013). This confirms that the supportive organizational climate does have a positive impact on the employee's personal attitude, motivation, and behavior, as well as team effectiveness, team performance, and so forth. Supportive organizational climate attaches importance to the contribution of employees, supports them, and improves employees' sense of self-efficacy. As a key job resource, organizational support helps employees to supplement the resources consumed in the job crafting. Secondly, the supportive organizational climate tolerates risk and embraces uncertainty, gives employees a certain degree of autonomy, encourages them to exert creativity, and meets the control needs of employees. Finally, the support and care from leaders and colleagues help to establish a friendly relationship between employees and managers and satisfies the motivation of employees to contact with others. Besides, social exchange theory points out that when organizations provide more job resources and support to employees, subordinates will show positive attitudes and behaviors at work based on the principle of reciprocity (Hirak et al. 2012). To sum up, supportive organizational climate has a positive effect on job crafting.

Based on this, the following hypothesis is proposed:

$H2_b$: Supportive organizational climate plays an intermediary role between inclusive leadership and employees' job crafting.

2.3 *Mediating role of psychological empowerment*

Psychological empowerment originated from the concern of Coger and Kanungo (1988) on the psychological experience of the authorized person. After that, Thomas and Velthouse (1990) formally proposed the concept of psychological empowerment from the perspective of employee cognition based on the research of Coger and others. This concept believed that this is a stimulated process or psychological state and is the comprehensive cognition of employees on their job roles, including the meaning of work, autonomy, work efficiency, and influence. Among them, the meaning of work refers to the purpose and value of the work perceived by employees; autonomy refers to the degree to which employees can make their own decisions during the work process; work efficiency refers to employees' belief in their ability to complete work tasks; influence refers to the extent to which individuals can affect the outcome of organizational strategy, administration, management, and so on. Research on the antecedents of psychological empowerment has made certain achievements, and leadership style has been proved to be the key factor affecting psychological empowerment. Transactional leadership (Wei Feng et al. 2009), paternalistic leadership (Li Hui et al. 2014), charismatic leadership (Yang et al. 2016), transformational leadership (Yang Chunjiang et al. 2015; Kim & Shin, 2017) and so forth can significantly improve the level of psychological empowerment of employees. From the four dimensions of psychological empowerment, inclusive leadership encourages and appreciates the contribution of employees, which is conducive to helping employees to feel their own value and work significance; adhere to people-oriented style, emphasize empowerment, advocate that leaders and subordinates work together to complete tasks (Li Yanping et al. 2012) rather than leaders decide and subordinates execute, which effectively improve the autonomy of subordinates; they are skilled at listening, allow employees to try and make mistakes, give employees more support and encouragement, thus improving employees' sense of efficacy; they establish a two-way relationship of respect, recognition, response, and responsibility with employees, are open and democratic, and increase the influence of employees in the organization.

Based on this, the following hypothesis is proposed:

H3$_a$: Inclusive leadership positively affects employees' psychological empowerment.

The outcome variables of psychological empowerment include job satisfaction (Li Chaoping et al. 2006; Spreitzer 1995); organizational commitment (Leiqiaoling 2007); organizational citizenship behavior (Ning et al. 2017; Seibert et al. 2011); and job performance (Xiujun Sun 2016). Moreover, studies show that psychological empowerment positively affects proactive behavior (Sun Chunling et al. 2014) and innovative behavior (Li Xiaohong et al. 2017). Apparently, employees are more willing to spend their time and energy studying problems that are of great significance and value to themselves and try different methods to solve these problems (Shalley & Gilson 2004), while job crafting itself is the initiative behavior that is beneficial to employees. As such, it can be inferred that psychological empowerment has a positive effect on job crafting. Specifically, from the four dimensions of psychological empowerment, first of all, the perception of the meaning of work is helpful for the matching of personal and organizational values, which is conducive to the cognitive crafting of employees; secondly, the perception of autonomy and work efficiency enables employees to have more sense of control and full confidence, which promotes the task crafting of employees; thirdly, employees' perception of personal influence at work drives them to strengthen interaction with organizations, leaders, and colleagues in various ways; express their own ideas; and change relational boundaries.

Based on this, the following hypothesis is proposed:

H3$_b$: Psychological empowerment plays an intermediary role between inclusive leadership and employees' job crafting.

2.4 Moderating effect of power distance

The concept of power distance originated from Hofstede's (1980) research on national culture and generally referred to the acceptance of unequal power distribution in a country or society. In the initial research, the power distance mainly focused on the national or social level, while in recent years, more and more scholars have begun to pay attention to the impact of cultural value orientation of individual power distance. At the individual level, power distance is defined as the individual's acceptance of the uneven distribution of power in organizations, institutions, or society (Clugston et al. 2000; Hofstede 1980). The higher the power distance, the higher the degree of individual acceptance of the unbalanced distribution of power; relatively speaking, the lower the power distance, the lower the degree of individual acceptance of the unbalanced distribution of power. The power distance in this study belongs to the individual level and is regarded as a psychological characteristic variable reflecting the differences of individual cultural values.

According to the contingency theory of leadership, there is no universally applicable leadership style, and power distance is often regarded as a pivotal factor to regulate the relationship between leadership behavior and subordinate performance. Such differences in individual cultural values will affect the degree to which employees accept leaders' influence (Cohen et al. 2011). Equity theory requests leaders to treat subordinates equally. However, even if leaders show the same behavior, individuals with different levels of power distance will have different responses, which stem from their different levels of acceptance of leadership behavior and authority (Ensari & Murphy 2003; Gelfand et al. 2007; Tsui 2007). Specifically, for employees with high power distance, they have a higher degree of acceptance of the unbalanced distribution of power and they believe that decision-making power should be concentrated in the hands of leaders, who do not need to consult employees when making decisions; employees only need to implement them. It is necessary for leaders to use authority in their contacts with subordinates. Even leaders and employees should avoid contact outside of work. Obviously, employees with high power distance believe that leaders are superior to others so that the distance between leaders and subordinates should be kept. As such, high power distance employees are not sensitive to the balanced empowerment of inclusive leaders nor their fairly and equally treatment, since their response to leadership behavior does not depend on how leaders treat them (Farh et al. 2007); thus inclusive leadership naturally has less impact on their psychological empowerment. On the contrary, for employees with low power distance, they believe that the relationship between leaders and subordinates is equal, and they have a more acute perception of the leadership's fair treatment, respect, and tolerance. Therefore, the pro-people words and deeds of inclusive leadership will have a strong impact on their psychological empowerment.

Based on this, the following hypothesis is proposed:

H4: Power distance negatively regulates the relationship between inclusive leadership and employees' psychological empowerment. In other words, inclusive leadership has a more significant impact on the psychological empowerment level of employees with low power distance and a relatively weak impact on employees with high power distance.

Hypothesis 3b of this study proposes that psychological empowerment plays an intermediary role between inclusive leadership and employees' job crafting, while hypothesis 4 proposes that because individuals with different levels of power distance have different sensitivities to leaders' pro-people behaviors, the impact of inclusive leadership on psychological empowerment will be different. Based on this, this study further proposes a moderated mediation model, that is, the mediating effect of inclusive leadership-psychological empowerment-employees' job crafting depends on the level of power distance.

Based on this, the following hypothesis is proposed:

H5: Power distance negatively regulates the mediating role of psychological empowerment between inclusive leadership and job crafting, which is shown as a moderated mediation model. Specifically, this mediating effect is relatively strong for employees with low power distance and relatively weak for employees with high power distance.

To sum up, the theoretical model of this study is as follows (Figure 1).

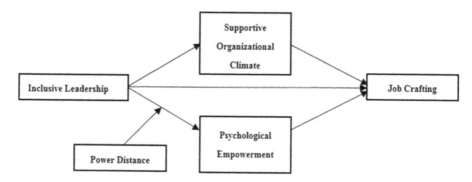

Figure 1. Theoretical model.

3 RESEARCH DESIGN

3.1 Sample

The data of this study come from a number of state-owned and private enterprises in Anhui, Guangdong, Shanghai, and other cities, involving manufacturing, real estate, finance, medical treatment, service, and other industries, which are representative to some extent. The collection of this questionnaire takes the form of a combination of offline paper questionnaires and online electronic questionnaires: the offline questionnaires are aimed at MBA students of the college, while the online questionnaires are aimed at grass-roots employees and managers. In order to minimize the influence of the participants' concerns on the authenticity of the questionnaire, except for complete anonymity, no participants' leaders were present at the questionnaire filling and answering site. Meanwhile, the guidance of the questionnaire emphasizes the anonymity and confidentiality of the survey, and the data are only used for academic research. Additionally, the quality of the questionnaire survey would be improved by beautifying the questionnaire and preparing small gifts for the respondents. Comparatively speaking, the electronic questionnaire can be filled in more freely, with less interference from other factors.

In this study, 400 questionnaires were actually issued, 308 of which were recovered, with a recovery rate of 77%. After eliminating the questionnaires with incomplete information, obvious regularity in answers, or random filling, 274 questionnaires were effective, with an effective recovery rate of 68.5%. Among them, 134 employees were men, accounting for 48.9%, and 140 employees were women, accounting for 51.1%; 170 employees under 35 years old, accounting for 62%; 194 employees were with bachelor degree or above, accounting for 70.8%; there were 52 employees working for 1 to 3 years, accounting for 19%, 32 employees working for 3 to 5 years, accounting for 11.7%, 175 employees working for more than 5 years, accounting for 63.9%; 33 employees were engaged in technical/development category, accounting for 12%, 55 were engaged in financial and marketing, accounting for 20%, 45 were engaged in human resources and administrative, accounting for 15.4%; in terms of post level, there were 108 ordinary employees, accounting for 39.4%, 78 grass-roots managers, accounting for 28.5%, and 65 intermediate managers, accounting for 23.7%; as for the nature of the enterprises, there were 118 state-owned enterprises/institutions, accounting for 43.1%, and 101 private enterprises, accounting for 36.9%.

3.2 *Measures*

The variables involved in this study are maturity scale. In order to test the reliability and validity of the scale, this study randomly selected more than 100 MBA students from the School of Management for the prediction test, adjusted the initial questionnaire according to the situation of the prediction test to ensure that the questionnaire conforms to China's national conditions, and finally formed a formal questionnaire. All variables in the study were measured by Likert's 5-point scoring method, ranging from "1=very inconsistent" to "5=very consistent."

Inclusive leadership: We selected a scale of nine items developed by Carmeli et al. (2010) to measure inclusive leadership, including three dimensions, which has been widely used by domestic scholars, and its reliability and validity have been confirmed. Typical items include 1) openness: "My leadership is willing to listen to my new proposal"; 2) availability: "My leader can always give effective answers to the professional questions I consult"; 3) accessibility: "My leadership encourages me to give him/her feedback on new problems arising from work", and so on. The reliability coefficient (Cronbach's Alpha) of each dimension is 0.842, 0.755, and 0.839, respectively, and the total reliability coefficient of the scale is 0.905.

Psychological empowerment: The 12-item scale by Spreitzer (1995) was employed to measure psychological empowerment, which includes four dimensions: 1) influence: "I have great influence in my department"; 2) the meaning of work: "The work I do is very meaningful to me"; 3) autonomy: "I have a great opportunity to exercise independence and autonomy in how to complete the work"; 4) work efficiency: "I have mastered all the skills required to complete the work", and so on. The reliability coefficient of each dimension is 0.904, 0.873, 0.897, and 0.861, respectively, and the total reliability coefficient of the scale is 0.920.

Supportive organizational climate: We selected a scale of 13 items developed by domestic scholar Ding Yuelan (2018), which was proposed on the basis of organizational support scale developed by Eisenberger et al. (1986) and Ling Wenquan et al. (2006), including three dimensions: 1) working support: "The company can always pay attention to my excellent job performance and give incentives"; 2) attribution support: "The company has great expectations of me and hopes that I will stay in the company all the time"; 3) growth support: "The company has made me learn a lot and made great progress", and so on. The reliability coefficient of each dimension is 0.892, 0.874, and 0.827, respectively, and the total reliability coefficient of the scale is 0.943.

Power distance: The single-dimensional scale of six items developed by Dorfman et al. (1997) was employed to measure power distance, including "leaders should make most decisions without consulting subordinates" and "leaders should not assign important tasks to subordinates." The reliability coefficient of the scale is 0.850.

Job crafting: We selected a scale of 21 items developed by Tims et al. (2012) to measure job crafting, including four dimensions: 1) increasing structural job resources: "I try to improve my learning and working abilities"; 2) increasing social job resources: "I hope the superior can express whether he is satisfied with the work I have done"; 3) decreasing hindering job demands: "I will try my best to avoid being depressed by work"; 4) increasing challenging job demands: "I will actively participate in projects that interest me", and so on. The reliability coefficient of each dimension is 0.899, 0.831, 0.889, and 0.802, respectively, and the total reliability coefficient of the scale is 0.941.

The control variables include four variables: employee's sex, age, education level, and years of service.

3.3 Research methods

Spss 25.0 and Amos 22.0 were used for statistical analysis, and Spss 25.0 is mainly used to analyze the common method deviation, correlation, reliability test of the scale, and multiple linear regression analysis of the data. Meanwhile, Bootstrap analysis is carried out by the software's Process program to further test the mediating effect and regulating effect. Amos 22.0 is mainly used to test the validity of variable measurement and to verify the establishment of dual intermediary model.

4 DATA PROCESSING AND ANALYSIS

4.1 Common method deviation test and discriminant validity analysis

As the questionnaires were completed by employees, the data in this study may have common method deviation. By using Harman's single-factor test method and conducting statistical analysis, it is found that the variance interpretation rate of the factor with the largest characteristic root is 36.745%, which is less than the critical value of 40%, indicating there is no serious common method deviation in the data in this study. In order to test the discrimination validity between variables, Amos 22.0 is used to carry out confirmatory factor analysis on the five variables in the theoretical model. By comparing the measurement model (five-factor model) with the competition model (four-factor, three-factor, double-factor, and single-factor model), it is found that the five-factor model has the best fitting effect (χ^2/df=2.78, RMSEA=0.081, IFI=0.937, TLI=0.921, CFI=0.937, NFI=0.905), indicating that the five-factor model can better represent the structure of measurement factors, and the variables have good discrimination validity, which are indeed five different constructs. Moreover, it further demonstrates that there is no serious common method deviation in the data (Table 1).

Table 1. Confirmatory factor analysis.

Model	χ^2/df	RMSEA	IFI	TLI	CFI	NFI
Five-factor : IL, PE, SOC, PD, JC	2.780	0.081	0.937	0.921	0.937	0.905
Four-factor : IL, PE+SOC, PD, JC	3.034	0.086	0.705	0.692	0.703	0.616
Four-factor : IL+JC, PE, SOC, PD	3.226	0.090	0.677	0.663	0.675	0.591
Three-factor : IL+JC+PE, SOC, PD	3.404	0.094	0.650	0.636	0.649	0.568
Double-factor : IL+JC+PE+SOC, PD	3.721	0.100	0.604	0.588	0.602	0.527
Single-factor : IL+JC+PE+SOC+PD	4.041	0.106	0.557	0.539	0.555	0.486

Note: IL means inclusive leadership, PE means psychological empowerment, SOC means supportive organizational climate, PD means power distance, and JC means job crafting.
Source: This article collates.

Table 2. Descriptive statistical analysis results and correlation coefficient.

	1	2	3	4	5	6	7	8	9
1. Gender	1.000								
2. Age	−.161**	1.000							
3. Education	−0.010	−.471**	1.000						
4. Tenure	−.128*	.852**	−.334**	1.000					
5. IL	−0.108	0.055	−0.049	0.014	1.000				
6. PE	−.157**	.164**	0.029	.152*	.669**	1.000			
7. SOC	−.137*	.172**	−0.083	0.111	.631**	.742**	1.000		
8. JC	−0.045	0.097	0.006	0.058	.533**	.702**	.675**	1.000	
9. PD	0.019	−0.029	0.073	−0.016	0.016	.148*	.170**	0.107	1.000
Mean	1.510	3.070	2.770	4.310	3.674	3.688	3.508	3.787	2.858
SD	0.501	1.510	0.930	1.774	0.661	0.629	0.654	0.533	0.773

Note: n = 274; *P < 0.05, **P < 0.01, ***P < 0.001.
Source: This article collates

4.2 Descriptive statistical analysis

Table 2 shows the mean, standard deviation, and correlation coefficient matrix of each research variable. It can be seen that inclusive leadership is significantly positively correlated with job crafting (r=0.533, P<0.01); inclusive leadership is significantly positively correlated with psychological empowerment (r = 0.669, P < 0.01) and supportive organizational climate (r = 0.631, P < 0.01). Psychological empowerment (r = 0.702, P < 0.01) and supportive organizational climate (r = 0.675, P < 0.01) are also significantly positively correlated with job crafting. These correlations are consistent with theoretical expectations and provide preliminary support for the research hypotheses.

4.3 Hypotheses testing

4.3.1 Double mediation test
In this study, Spss 25.0 was used to conduct hierarchical regression analysis to test the positive impact of inclusive leadership on supportive organizational climate, psychological empowerment, and job crafting. Then the structural equation model of Amos 22.0 was used to verify the dual mediating role of supportive organizational climate and psychological empowerment between inclusive leadership and job crafting. The hierarchical regression results are as follows.

From model 7 in Table 3, it can be seen that after controlling demographic variables such as gender, age, and education level, inclusive leadership has a significant positive impact on employees' job crafting ($\beta = 0.533$, P < 0.001), variance interpretation rate increases by 28%, which means H1 is verified. From model 2, after controlling demographic variables, inclusive leadership has a significant positive impact on supportive organizational climate ($\beta = 0.617$, P < 0.001), variance interpretation rate increases by 37.2%, H2a is verified. From model 4, after controlling demographic variables, inclusive leadership has a significant positive impact on employees' psychological empowerment ($\beta = 0.66$, P < 0.001), variance explanation rate increases by 43.7%, H3a is verified.

According to model 8, when inclusive leadership, supportive organizational climate, and psychological empowerment are simultaneously incorporated into the regression equation for job crafting, the direct effect of inclusive leadership on employees' job crafting becomes no longer significant ($\beta = 0.028$, P > 0.1), while the effect of supportive organizational climate ($\beta = 0.448$, P < 0.001) and psychological empowerment ($\beta = 0.336$, P < 0.001) as intermediary variables is significant. To examine the side-by-side dual mediating role of supportive organizational climate and psychological empowerment between inclusive leadership and job crafting more accurately, this study uses

Table 3. Regression analysis results.

Variable	SOC		PE			JC		
	M1	M2	M3	M4	M5	M6	M7	M8
Control variable								
Gender	−0.112	−0.048	−0.124*	−0.058	−0.058	−0.022	0.033	0.074
Age	0.256*	0.184	0.185	0.108	0.112	0.219	0.157	0.047
Education	−0.004	0.016	0.121	0.132**	0.134**	0.074	0.091	0.022
Tenure	−0.123	−0.055	0.018	0.089	0.085	−0.107	−0.049	−0.071
Independent variable								
IL		0.617***		0.66***	0.631***		0.533***	0.028
Intermediate variable								
SOC								0.448***
PE								0.336***
Regulated variable								
PD				0.133**	0.149**			
Interaction term								
IL*PE					−0.139**			
R^2	0.046	0.420	0.056	0.506	0.524	0.017	0.295	0.554
F	3.241	38.844***	4.001**	45.537***	41.84***	1.134	22.474***	47.19***
ΔR^2		0.374		0.450	0.018		0.282	0.259

Note: n = 274; *P<0.05, **P<0.01, ***P<0.001. Standardized regression coefficient is reported.
Source: This article collates.

Table 4. Confirmatory factor analysis of double intermediary model.

Indicators	χ^2/df	RMSEA	IFI	TLI	CFI	NFI
Numerical	2.952	0.085	0.949	0.934	0.949	0.925

Source: This article collates.

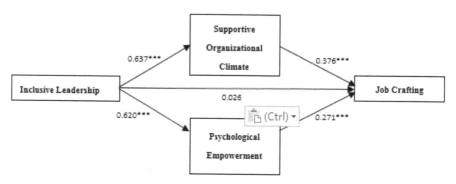

Figure 2. Roadmap of dual mediation model. Note: n = 274; *P < 0.05, **P < 0.01, ***P < 0.001.
Source: This article draws.

structural equation to construct a dual incomplete mediation model. As can be seen from Table 4, all indexes of the model meet the standards and can be accepted, indicating that the model fits well as a whole.

As shown in Figure 2, this paper assumes that the impact of inclusive leadership on employees' job crafting is realized by two indirect paths: (1) Inclusive leadership → Supportive organizational climate → Job crafting; (2) Inclusive leadership → Psychological empowerment → Job crafting.

Table 5. Analysis results of double mediating effect.

| Effect | | Point estimation | 95% confidence interval | |
			Lower limit	Upper limit
Mediating effect of supportive organizational climate	a x b	0.168	0.096	0.257
Mediating effect of psychological empowerment	c x d	0.240	0.137	0.340
Overall mediating effect	a x b+c x d	0.408	0.316	0.505
Direct effect		0.026	−0.070	0.115
Total effect		0.434	0.348	0.512

Note: The effect value is the standardization coefficient.
Source: This article collates.

The coefficients of the four direct paths included in these two indirect paths are significant: Inclusive leadership → Supportive organizational climate (β=0.637, P<0.001), Supportive organizational climate → Job crafting (β=0.376, P<0.001), Inclusive leadership → Psychological empowerment (β=0.620, P<0.001), Psychological empowerment → Job crafting (β=0.271, P<0.001).

In order to ensure the consistency and stability of the analysis results, this paper continues to adopt the Bootstrap method of deviation correction, and takes 5000 Bootstrap samples to verify side-by-side dual mediating role of supportive organizational climate and psychological empowerment between inclusive leadership and job crafting. The analysis results are shown in Table 5: the mediating effect of supportive organizational climate between inclusive leadership and job crafting is 0.168; the mediating effect of psychological empowerment between inclusive leadership and job crafting is 0.240. The mediating effects of the two indirect paths do not contain 0 in 95% confidence interval, which indicates that the mediating role of supportive organizational climate and psychological empowerment between inclusive leadership and job crafting is valid, H2b and H3b are both verified.

In addition, the direct effect is 0.026, P>0.1, and 0 is included in the 95% confidence interval, indicating that the direct effect is not significant. Consequently, supportive organizational climate and psychological empowerment play a full mediating role in inclusive leadership and job crafting.

4.3.2 Moderating effect test

To test the moderating effect of power distance, first, the variables are standardized to reduce multicollinearity, and then the normalized variables are used to construct interaction terms and perform hierarchical regression analysis. The analysis results are shown in Table 3. From model 5, it can be seen that the interaction of inclusive leadership and power distance has a significant negative impact on employees' psychological empowerment (M3, β=−0.139, P<0.01), which indicates that the lower the power distance of employees, the more vulnerable the employees' psychological empowerment is to be affected by inclusive leadership, that is, the stronger the positive effect of inclusive leadership on the psychological authorization of employees with low power distance. Therefore, H4 is supported.

In order to more intuitively demonstrate the moderating effect of power distance on the psychological empowerment relationship between inclusive leaders and employees, the mean value of power distance is added and subtracted by one standard deviation, and a power distance regulatory function chart is made by simple slope analysis, which depicts the relationship between inclusive leadership and employees' psychological empowerment under the conditions of high power distance and low power distance. It can be seen that the slope of low power distance is steeper, which indicates that the same degree of change in inclusive leadership behavior can lead to a greater degree of change in the subordinates' psychological empowerment who have low power distance,

Figure 3. The moderating effect of power distance on the relationship between inclusive leadership and employees' psychological empowerment.

Table 6. Mediating effect under different power distance levels.

	Power distance	Effect	BootSE	BootLLCI	BootULCI
Mediating effect of regulation	eff1 (M-1SD)	0.391	0.058	0.281	0.509
	eff2 (M)	0.328	0.044	0.245	0.418
	eff3 (M+1SD)	0.265	0.048	0.174	0.360
Comparison of mediating effects with regulation	eff2-eff1	−0.063	0.030	−0.123	−0.005
	eff3-eff1	−0.127	0.060	−0.246	−0.011
	eff3-eff2	−0.063	0.030	−0.123	−0.005

Source: This article collates.

that is, the psychological empowerment level of subordinates with low power distance is easier to alter with the changes of inclusive leadership behavior, further supporting H4 (Figure 3).

4.3.3 *Moderated mediation test*

Using model 7 in Spss Process compiled by Hayes (2012) (model 7 assumes that the first half path of the mediation model is moderated, which is consistent with the theoretical model of this study), the moderated mediation model is tested under the condition of controlling gender, age, education, and tenure. Similar to the results of hierarchical regression, the product term of inclusive leadership and power distance has a significant predictive effect on psychological empowerment ($B = -0.1499$, $t = -3.1917$, $P < 0.01$), indicating that power distance can negatively regulate the influence of inclusive leadership on psychological empowerment. Further, as can be seen from Table 6, for employees with lower power distance (M-1SD), the mediating effect value is 0.391; for employees with high power distance (M+1SD), the mediating effect value is 0.265. From the comparison of moderating effects, it can be drawn that there is a significant difference in the influence of high and low power distance on mediation effect (the upper and lower limits do not include 0). Therefore, H5 has been tested.

5 DISCUSSION AND CONCLUSION

5.1 *Research conclusion*

Based on organizational support theory and social information processing theory, this study puts forward a side-by-side dual mediation model to explore the mechanism of inclusive leadership on employees' job crafting and further studies its boundary conditions. Through empirical research, the results show that:

1. Inclusive leadership is positively affecting employees' job crafting. In view of the current situation that the research on job crafting mainly focuses on the outcome variables, this study shifts its attention to the driving factors of job crafting and extends it to organizational contextual factors, which proves that inclusive leadership can stimulate employees' intrinsic motivation, thus promoting employees' job crafting behavior. This provides a brand-new perspective for how to effectively motivate employees to remold their job. At the same time, it enriches the theory of inclusive leadership, as "research into inclusion is still in its infancy" (Mitchell et al. 2015) extends the influence of inclusive leadership on the organization to employees' job crafting behavior and further confirms the effectiveness of inclusive leadership.

2. Supportive organizational climate and psychological empowerment play a side-by-side dual mediating role in the relationship between inclusive leadership and employees' job crafting. In order to reveal the mechanism of inclusive leadership on employees' job crafting more comprehensively, this study explores the intermediary role between the two from multiple paths. From the perspective of organization, social information processing theory points out that leadership behavior has a significant impact on the formation of organizational climate, and organizational climate has an impact on employee behavior. From the perspective of the individual, inclusive leadership has an impact on subordinates' attitudes and behaviors by influencing subordinates' psychological perception (Zhang Lulu 2016), which can significantly enhance the four specific dimensions of employees' psychological empowerment. Meanwhile, employees' psychological empowerment plays a significant role in predicting employees' attitudes and behaviors (Shen Wenzhu 2018), which is also consistent with Tian Qitao's (2018) research. Inclusive leadership, as the core of relational leadership, its openness, availability, and accessibility will undoubtedly promote the formation of supportive organizational climate, improve employees' psychological empowerment, and then positively affect employees' job crafting.

3. Power distance plays a negative moderating role between inclusive leadership and employees' psychological empowerment; further, power distance negatively regulates the intermediary mechanism of Inclusive leadership → Psychological empowerment → Job crafting. The moderating role of power distance in the relationship between leadership behavior and employee performance has been confirmed (Gu Yinhua 2016; Liang Jian 2014). For employees with high power distance, inclusive leadership's pro-people behaviors such as authorization, respect for subordinates, and fair treatment have less influence on their psychological empowerment, and they are not sensitive enough to inclusive leadership's behavior. For employees with low power distance, they are more sensitive to the behavior of inclusive leadership who also believe that leaders should not be over-centralized, the relationship between leaders and subordinates should be equal, and leaders should respect and attach importance to employees, therefore, the pro-people's words and deeds of inclusive leadership will have a significant impact on their psychological empowerment, and the data in Table 3 support this conclusion (M3, $\beta = -0.139$, P<0.01). Power distance also negatively regulates the mediating role of psychological empowerment between inclusive leadership and job crafting—for employees with low power distance, the mediating effect is much higher than that for employees with high power distance. The data in Table 6 support this.

5.2 *Management inspiration*

The conclusion of this study has certain guiding significance for management practice, as follows:

1. Organizations should pay attention to the positive effects of inclusive leadership and attach importance to the cultivation of inclusive leaders. Inclusive leadership would promote the formation

of a harmonious, equal, and supportive organizational climate; appreciate employees' contributions; consider employees' needs; respect employees' interests; tolerate employees' mistakes; give reasonable guidance; encourage employees to put forward their own ideas and opinions; give employees a great sense of security; improve employees' psychological empowerment level; and fully mobilize employees' enthusiasm. Therefore, inclusive leadership is the external driving force that enables employees to be free from worries and dare to reshape their job. On the other hand, leaders with the ability and quality of inclusive leadership are the foundation and prerequisite for creating an inclusive and supportive organizational climate. Enterprises should fully consider the behavioral characteristics of inclusive leadership in the process of leader selection and special training and strive to improve the level of inclusive leadership. In addition, the promotion effect of inclusive leadership behavior on employees should be shown to leaders, timely feedback changes in employee performance improve the incentive mechanism, so that leaders are willing to show more inclusive leadership behavior.

2. Efforts should be made to improve the level of employees' psychological empowerment and stimulate employees' internal driving force to remold their job. Leaders should give appropriate authorization to the employees who are fighting in the front line, recognize their contributions, encourage and appreciate them, listen to and seriously consider their ideas before making decisions, so that employees can feel the significance and autonomy of their work and their influence in the organization. In this way, the level of employees' psychological empowerment will be significantly improved, and the internal driving force of their active work is stimulated, which is conducive to the occurrence of job crafting behavior and forms a win-win situation for employees and enterprises.

3. The effectiveness of leadership behavior not only depends on the leaders themselves; the power distance that reflects the personal value orientation of employees is the boundary condition of this mechanism. Hence, leaders should not give up their inclusive leadership behavior just because they have little effect on individual employees and should realize that for employees with high power distance, leadership behavior has limited influence on their internal psychological empowerment level. However, all employees hope that leaders can treat them in a closer and equal way, and all employees hope to be respected and affirmed by leaders. Therefore, leaders should adhere to inclusive leadership behavior and put people first, believing that employees will repay with better work performance.

5.3 *Research limitations and prospects*

Due to the limitation of some objective conditions, this study has certain limitations, and we hope to solve these problems in future studies.

(1) During the collection of the questionnaire, the leader-employee paired questionnaire was not adopted, and only the employee's personal self-reported questionnaire was obtained, thus leading to possible common method deviation. Although the common method deviation of this data is not serious through software verification, in order to make the research method more scientific, the following research can try to make leaders and employees fill in questionnaires in pairs, respectively. Such non–same sources and non-uniform methods data help to make the research results more objective.

(2) The scales used in this study are developed by Western scholars, except the supportive organizational climate scale. Although these classic scales have been widely used by domestic scholars and have proved their applicability in Chinese situations, and the validity and reliability of each key variable have been verified to be acceptable through data analysis in this study, a more rigorous approach should be to use those scales developed under the background of Chinese organizational culture or make necessary revisions to the scales based on Chinese situations.

(3) The cross-sectional data adopted in this study cannot strictly explain the causal relationship between inclusive leadership and employees' job crafting. In the future, longitudinal data can be used to further test the research conclusion, especially the latent variable development model, to verify the causal relationship of variable incremental changes.

(4) The inclusive leadership and job crafting in this study all appear as overall variables and are not discussed in detail in different dimensions. Future research can consider analyzing the influence of three dimensions of inclusive leadership on reshaping different dimensions of employees' jobs.

ACKNOWLEDGEMENTS

Supported by Institutional Preparation of Management Training Project Panning in 2019 (No: 19JSFW0458).

REFERENCES

Bashshur, M. R., Ana, H., & Vicente, G. R. (2011). When managers and their teams disagree: a longitudinal look at the consequences of differences in perceptions of organizational support. Journal of Applied Psychology, 96(3), 558–73.

Berg, J. M. Dutton, J. E. & Wrzesniewski, A. (2013). Job crafting and meaningful work. In B. J. Dik, Z. S. Byrne, & M.F. Steger (Eds.) Purpose and meaning in the workplace. American Psychological Association.

Carmeli A, Reiter-Palmon R, Ziv E. Inclusive Leadership and Employee involvement in Creative Tasks in the Workplace; The mediating role of psychological safety. Creativity Research Journal, 2010, 22(3):250–260.

Choi, S. B., Tran, T. B. H., & Park, B. I. (2015). Inclusive leadership and work engagement: Mediating roles of affective organizational commitment and creativity. Social Behavior and Personality, 43(6), 931–944.

Christina E Shalley, Lucy L Gilson. (2004). What leaders need to know: A review of social and contextual factors that can foster or hinder creativity. 15(1):0–53.

Conger J A, Kanungo R N. The empowerment process: Integrating theory and practicer. Academy of Management Review, 1988, 13 (3):471–482.

Dorfman, Peter W, Howell, Jon P. Leadership in Western and Asian countries: Commonalities and differences in effective leadership. Leadership Quarterly, 1997, 8(3):233–274.

Eisenberger, R., Huntington R., & Hutchisom, S. (1986). Perceived organizational support. Journal of Applied Psychology, 71(2), 500–507.

Ensari, N., & Murphy, S. E. (2003). Cross-cultural variations in leadership perceptions and attribution of charisma to the leader. Organizational Behavior and Human Decision Processes, 92, 52–66.

Farh, J.-L., Hackett, R. D., & Liang, J. (2007). Individual- level cultural values as moderators of perceived organizational support–employee outcome relationships in China: Comparing the effects of power distance and traditionality. Academy of Management Journal, 50, 715–729.

Gelfand, M. J., Erez, M., & Aycan, Z. (2007). Cross-cultural organizational behavior. Annual Review of Psychology, 58, 479–514.

Hayes, A. F. (2012). PROCESS: A versatile computational tool for observed variable mediation, moderation, and conditional process modeling. Retrieved from http://www.Afhayes.com/public/process2012.Pdf.

Hirak, R., Peng, A. C., Carmeli, A., & Schaubroeck, J. M. (2012).Linking leader inclusiveness to work unit performance: The importance of psychological safety and learning from failures. The Leadership Quarterly, 23(1), 107–117.

Hofstede, G. (1980). Culture's consequences: International differences in work-related values. Beverly Hills, CA: Sage.

Hofstede, G. (2001). Culture's consequences: Comparing values, behaviors, institutions and organizations across nations. Thousand Oaks, CA: Sage.

Hollander, E. (2009). Inclusive leadership: The essential leader-follower relationship. New York: Routledge.

Kim, S., & Shin, M. (2017). The effectiveness of transformational leadership on empowerment. Cross Cultural & Strategic Management.

Leana, C., Appelbaum, E., & Shevchuk, I. (2009). Work process and quality of care in early childhood education: The role of job crafting. Academy of Management Journal, 52(6), 1169–1192.

Leiqiaoling, Zhao Genshen. Research on the relationship between psychological authorization and organizational commitment of knowledge workers. scientific and technological progress and countermeasures, 2007(09):122–125.

Liao, H., Toya, K., Lepak, D. P., & Hong, Y (2009). Do they see eye to eye? Management and employee perspectives of high-performance work systems and influence processes on service quality. Journal of Applied Psychology, 94(2), 371–91.

Li Chaoping, Li Xiaoxuan, Shi Kan, Chen Xuefeng. Measurement of Authorization and Its Relationship with Employees' Work Attitude. Journal of Psychology, 2006(01):99–106.

Li Hui, Ding Gang, Li Xinjian. The Influence of Leadership Style on Employees' Innovative Behavior Based on Patriarchal Leadership Ternary Theor . Journal of Management, 2014, 11(07): 1005–1013.

Ling Wenquan, Yang Haijun, Fang Liluo. Organizational Support of Enterprise Employees. Journal of Psychology, 2006(02):281–287.

Liu Jingjiang, Zou Huimin. Influence of Transformational Leadership and Psychological Empowerment on Staff Creativity. Scientific Research Management, 2013, 34(03):68–74.

Li Xiaohong, Xu Lei, yangweihua. Mechanism of Employee Innovation Behavior: Impact from Leadership. Scientific and Technological Progress and Countermeasures, 2017, 34(21), 154–160.

Li Yanping, Yang Ting, Pan Yajuan, Xu Jia. Construction and Implementation of Inclusive Leadership-Based on New Generation Employee Management Perspective. China Human Resources Development, 2012(03): 31–35.

Lyons, P. (2008). The crafting of jobs and individual differences. Journal of Business and Psychology, 23, 25–36.

Meyer R D, Dalal R S, Hermida R. A Review and Synthesis of Situational Strength in the Organizational Sciences. Journal of Management, 2010, 36(1):121–140.

Mitchell R, Boyle B, Parker V, et al. Managing Inclusiveness and Diversity in Teams: How Leader Inclusiveness Affects Performance through Status and Team Identity. Human Resource Management, 2015, 54(2): 217–239.

Nembhard, I. M., & Edmondson, A. C. (2006).Making it safe: The effects of leader inclusiveness and professional status on psychological safety and improvement efforts in health care teams. Journal of Organizational Behavior, 27(7), 941–966.

Ning, L., Dan, S. C., & Bradley, L. K. (2017). Cross-Level Influences of Empowering Leadership on Citizenship Behavior: Organizational Support Climate as a Double-EdgedSword. Journal of Management.

Priesemuth, M., Schminke, M., Ambrose, M.L., & Folger, M. (2011). Abusive supervision climate: A multiple-mediation model of its impact on group outcomes. Academy of Management Journal, doi: 10.5465/amj.2011.0237.

Randel A E, Galvin B M Shore L M, et al. Inclusive leadership: Realizing positive outcomes through belongingness and being valued for uniqueness. Human Resource Management Review, 2017:S1053482217300517.

Salancik,G.R., & Pfeffer, J.(1978). A social information processing approach to job attitudes and task design. Administrative Science Quarterly, 23, 224–253.

Schneider, B. (Ed.). (1990). Organizational climate and culture. San Francisco: Jossey-Bass.

Seibert, S. E., Wang, G., & Courtright, S. H. (2011). Antecedents and consequences of psychological and team empowerment in organizations: A meta-analytic review. Journal of Applied Psychology.

Spreitzer, G. M. (1995). Psychological Empowerment in the Workplace: Dimensions, Measurement, and Validation. Academy of Management Journal, 38(5), 1442–1465.

Sun Chunling, Zhang Hua, Li He, Song Hong. Research on Mechanism of Influence of Authorization Atmosphere on Initiative Behavior of Project Managers: Mediating Role of Psychological Authorization. Management Review, 2014, 26(07), 196–208.

Thomas K W, Velthouse B A. Cognitive elements of empowerment: An "interpretive" model of intrinsic task motivation. Academy of Management Review, 1990, 15(4):666–681.

Tian Qitao. Research on Mechanism of Service-oriented Leadership Arousing Employees' Work Reshaping Enthusiasm . Soft Science, 2018, 32(06):70–73.

Tims, M., & Bakker, A.B. (2010). Job crafting: Towards a new model of individual job redesign. South African Journal of Industrial Psychology, 36, 1–9.

Tims, M., Bakker, A.B., & Derks, D. (2011). Development and validation of the job crafting scale. Journal of Vocational Behavior, 80, 173–186.

Torner M., Pousette A., & Larsman P, et al. (2016).Coping with paradoxical demands through an organizational climate of perceived organizational support: an empirical study among workers in construction and mining industry.Journal of Applied Behavioral Science, 53(1), 1–25.

Tsui, A. S. (2007). From homogenization to pluralism: International management research in the academy and beyond. Academy of Management Journal, 50, 1353–1364.

Wang Hongyu, Cui Zhisong. How can coach leadership promote the reshaping of employees' work?-A Multilevel Regulated Intermediary Model. Jiangsu Social Sciences, 2018(02):61–71.

Wei Feng, Yuan Xin, Di Yang. Cross-level Study on the Influence of Transaction Leadership, Team Empowerment Atmosphere and Psychological Empowerment on Subordinates' Innovation Performance. Management World, 2009(04):135–142.

Wen Zhonglin, Ye Baojuan. An Adjusted Intermediary Model Test Method: Competition or Substitution?. Journal of Psychology, 2014,46 (05):714–726.

Wrzesniewski, A., & Dutton, J.E.(2001).Crafting a job: Revisioning employees as active crafters of their work. Academy of Management Review, 26, 179–201.

Xin Xun, Miao Rentao. The Effect of Job Remodeling on Employee's Creative Performance-A Moderating Double Intermediary Model. Economic Management, 2018, 40 (05):108–122.

Xiujun, S. (2016). Psychological Empowerment on Job Performance—Mediating Effect of Job Satisfaction. Psychology.

Yang Chunjiang. Research on Leadership Style and Organizational Innovation Atmosphere-Taking Hebei High-tech Enterprises as an Example Scientific and Technological Progress and Countermeasures, 2011, 28 (19):89–93.

Yang Chunjiang, Cai Yingchun, Hou Hongxu. Research on the Influence of Transformational Leadership on Subordinate Organizational Citizenship Behavior from the Perspective of Psychological Empowerment and job embeddedness. Journal of Management, 2015, 12 (02):231–239.

Zhang Chunyu, Wei Jia, Chen Xie Ping, Zhang Jin Fu. A New Perspective of Job Design: Reshaping Employees. advances in psychological science, 2012, 20(08):1305–1313.

Zhang Lulu, Yang Fu, Gu Yinhua. Inclusive leadership: concepts, measurements and relationships with related variables. advances in psychological science, 2016, 24(09):1467–1477.

Zhang Ruiying, Zhang Yongjun, Li Yongxin. New Bottle of Old Wine or Innovation: Is Inclusive Leadership an Independent Leadership Type?. Psychological Science, 2018, 41(05):1158–1163.

Zhao Xiaoyun, Guo Cheng. Work Remodeling: A New Way to Get Meaningful Work and Personal Growth. Psychological Science, 2014, 37 (01):190–196.

Zheng Xiaoming, Liu Xin. The Impact of Interactive Fairness on employee well-being: Mediating Role of Psychological Empowerment and Regulating Role of Power Distance. Acta Psychologica Sinica, 2016, 48 (06):693–709.

Zhu Shaoying, Qi Ershi, Xu Yu. The relationship between transformational leadership, team atmosphere, knowledge sharing and team innovation performance. Soft Science, 2008(11):1–4+9.

Zhu Yu, Qian Di-ting. Research Frontier Analysis and Future Prospect of Inclusive Leadership. Foreign Economy and Management, 2014, 36 (02):55–64+80.

Zohar D, Tenne-Gazit O. Transformational Leadership and Group interaction as Climate Antccedents; A Social Network Analysis .Journal of Applied Psychology, 2008, 93(4):744–757.

Computational Social Science – Luo, Ciurea & Kumar (eds)
© 2021 Taylor & Francis Group, London, ISBN 978-0-367-70193-2

Does job design affects employee knowledge sharing and innovation?—the moderating role of organizational innovation atmosphere

Y.Y. Zhu & C.S. Wu
School of Management, Hefei University of Technology, Hefei, Anhui, China

L.G. Zhang
State Grid Anhui Electric Power Co., Ltd., Hefei, Anhui, China

ABSTRACT: This study developed a moderated mediation model of job design and employee innovation performance. It introduced the organizational innovation atmosphere as the moderating variable at the organizational level and knowledge sharing willingness as the mediating variable at the individual level. The results show that: job design positively affects employee innovation performance; knowledge sharing willingness plays a partial mediating role in the relationship between job design and employee innovation performance; the mediating effect of knowledge sharing willingness is moderated by the organizational innovation atmosphere. These conclusions will enrich the study on the mediating mechanism and boundary conditions of job design on employee innovation performance, and help enterprises better understand the importance of job design and its effect on employee innovation performance through knowledge sharing willingness.

Keywords: Job design, Employee innovation performance, Knowledge sharing willingness, Organizational innovation atmosphere.

1 INTRODUCTION

For a long time, research on job design began to disappear from top journals, and gradually faded out of people's vision, because scholars believe that the basic questions of job design have been answered (Humphrey et al. 2007). Due to the global economic transformation in recent years, the knowledge economy and service economy has become the mainstream of the institutional economy, and the nature of employees' work has changed greatly. The development of the Internet and mobile communications has gradually flattened the existing organizational structure (Hu & Wang 2019). Significant changes in the internal and external work environment make it more necessary to study how to interpret more work results through job design in the new context. Innovation as the main driving force for the development of science and technology, how to stimulate the innovation ability of employees, and make the enterprise invincible in the fierce competition has become the focus of attention from all walks of life (Holzmann & Golan 2016). Because the individual's intrinsic motivation greatly affects the employee's innovation behavior, employees who are motivated by intrinsic motivation are the most innovative (Amabile et al. 1996). And studies have shown that intrinsic motivation is completely dependent on job design (Wang & Zhao 2011). Therefore, it is of theoretical and practical significance to study how to guide employees to take the initiative to innovate through scientific job design, to achieve employee innovation performance and to promote enterprises to achieve innovative results.

Considering the complexity of innovation activities and the high degree of interdependence of innovation performance, and most of the innovation activities occur in rich organizational contexts, it is necessary to consider the impact of job design on employee innovation performance in organizational contexts (Ding & Li 2016). In the context of supporting innovative attempts, innovative behaviors and creative outcomes are more likely to occur (Liu & Shi 2009). Besides, employees will communicate and share knowledge in the organizational context, but knowledge is usually stored in the individual and is an intangible asset (Guo et al. 2020). Whether or not knowledge is shared will involve the willingness of organizational members to share knowledge, it does not happen naturally (Wang et al. 2019). This research introduced the job characteristics model (JCM) to explore the impact of job design on employee innovation performance, to identify main job characteristics of stimulating innovation behavior and improving innovation performance.

2 LITERATURE REVIEW AND RESEARCH HYPOTHESES

2.1 Job design and employee innovation performance

Job design is the construction and adjustment of employees' work content and roles, and determining how these constructions and adjustments affect individuals, teams, and organizations (Grant & Parker 2009). In different periods of social and economic development, the focus of job design is constantly evolving. From engineering guidance that only focuses on the work and on improving work efficiency to the psychological orientation that fully focuses on the psychological needs of employees, job design increasingly considers how to serve the people in the organization (Meng 2016). In the 1970s, Hackman and Oldham (1976) integrated and extended their ideas about job design to JCM, which defined five core dimensions—job autonomy, skill diversity, task integrity, task importance, and feedback. JCM is a highly influential model of work nature research and job design practice, and it is the core method of job design (Oldham & Fried 2016). Therefore, this paper used the five core dimensions of JCM as a job design variable.

Understanding the connotation of innovation and innovative behavior is a premise of understanding innovation performance. The academic community agrees with Amabile's (1998) definition of innovation and believes that innovation is an innovative product, process, service, or strategy produced by an individual or a team, which defines innovation from a product perspective. Scott and Bruce (1994) built a model of individual innovation behavior based on clarifying the antecedent variables of innovation and regarded individual innovation behavior as the embodiment of the joint action of four systems—individual, leadership, organization, and innovation atmosphere. Later, many scholars' research on innovation performance was also based on this model. Innovation performance can be reflected in different subjective levels of individuals, teams, organizations, etc. In the definition of innovation performance, scholars mostly define it from the perspective of the enterprise as a whole, and relatively few define it from an individual perspective. Zhang et al. (2020) believed that innovation performance includes innovative technologies and products that create value for the company, as well as the efforts of R & D personnel to achieve innovative results. Han et al. (2007) believed that employee innovation performance is subordinate to job performance in terms of constructs. It is the behavior that transcends established performance categories and role requirements, which consciously breaks existing practices, produces creative results at the individual level, and promotes the improvement of organizational performance. Wang et al. (2014) also thought that employee innovation performance is an aspect of job performance, and it is a reflection of valuable creative ideas and products at the individual level.

The impact of job design on employee innovation performance can be seen from each dimension of job characteristics: A job with a high degree of autonomy means that individuals can give full play to their abilities when determining how to organize work. Autonomous work is thought to increase core task performance by increasing the likelihood of employees taking on job responsibilities (Saragih 2011), and allowing employees to assume a wider role, acquire new skills and improve self-efficacy (Parker 2007). At the same time, due to stronger self-efficacy, employees tend to be more proactive and more willing to try to solve problems with creative methods (Malik et al. 2015);

jobs requiring skill diversity can make employees more exposed heuristic tasks that can stimulate employees to generate intrinsic work motivation, and the increase of employees' psychological ownership of work makes employees more engaged in work and thus improve work performance (Xie & Gu 2015). Related research results also show that skill diversity is one aspect of job complexity, and job complexity is positively related to the generation of creative results (Chae & Choi 2018). In the follow-up study of new scientific researchers for three consecutive years, researchers found that job integrity has a significant effect on continuous job engagement, and task integrity and job autonomy can also well predict employees' future job engagement (Sun & Teng 2013); Job importance is a relational job characteristic that reflects the relationship needs of employees and connects the impact of employee behavior within the organization (Grant 2007). Grant (2008) proposed that job importance improves job performance by enhancing employees' perception of the impact of their actions on beneficiaries and admiration from others. Research by Liu et al. (2015) also supported this view, arguing that the task importance reflects the degree of behavioral contribution in the role, and affects job performance by affecting the key mental state of employees; Feedback refers to direct and indirect information about productivity and results obtained when completing a task (Hackman & Oldham 1976), it is a key source of task-related information for employees and has an important role in improving employee creativity (Yoo et al. 2019). For employees with high self-efficacy, job feedback can better help them achieve their task goals.

Thus, we make the following assumption:

Hypothesis 1: Job design has a positive impact on employee innovation performance.

2.2 *Knowledge sharing willingness and its mediating effect*

Knowledge sharing is an activity of transferring knowledge and mutual understanding among members within and outside the organization through various communication channels and media (Bavik et al. 2017). Knowledge sharing willingness refers to the willingness of organizational members to share knowledge with others and actively absorb new knowledge (Ye et al. 2018). The higher the employees' willingness to share knowledge, the more likely they are to share knowledge (Wu et al. 2018). There is also a certain relationship between the employees' willingness to share knowledge and the internal employees' emotional and information support. Harmonious interpersonal relationships and emotional support can form a good sharing atmosphere and guide employees to share knowledge.

From a social cognitive perspective, job design affects employees' perceptions of the organization, which in turn affects employee behavior (Grant 2008). Studies have shown that, compared with extrinsic motivation, intrinsic motivation has a stronger and more sustainable effect on the formation of knowledge sharing willingness (Llopis & Foss 2016). Forcing employees to share knowledge through compulsory means such as rewards or punishments will have some effect, but employees usually only provide knowledge that meets the minimum requirements for quantity and quality (Gagne 2009). And job design can promote employees 'internal motivation by providing them with the opportunity to develop their abilities and achieve their growth needs, which will affect the employees' willingness to prosocial behavior to a certain extent (Grant 2007), and promote the permanent change of prosocial behaviors such as knowledge sharing (Pee & Lee 2015). Studies have confirmed that job design stimulates the formation of employees' willingness to share knowledge by promoting intrinsic motivation (Grant 2007). At the same time, due to the complexity of innovation activities and the high degree of interdependence, it is not enough to rely on the resources of individual employees to innovate. It is necessary to interact with other employees to acquire knowledge and exchange information, to improve the innovation performance of individuals and organizations in the process of knowledge diffusion and integration (Liu & An 2016). Employees with a high willingness to share knowledge make common progress in sharing and learning with each other, which can not only improve the overall technical level of the enterprise, but also help solve the difficulties faced by individuals in the work and help to form an innovative atmosphere of continuous pursuit of progress and positive sharing in the organization (Wu et al.

2018). Through knowledge sharing, employees can absorb external knowledge to achieve their innovation performance, and also promote the transcendence development of enterprise innovation performance.

Previous studies have verified the mediating role of knowledge sharing willingness. Researchers studied individual-level factors from the perspective of learning theory—learning goal orientation affects employee innovation performance through the intermediation of willingness to share knowledge (Zhong et al. 2018). Peng et al. (2020) found that dualistic leadership can effectively enhance the internal identity perception of employees, thereby stimulating the willingness to share knowledge and ultimately improve employee innovation performance. Although there is no research on the role of knowledge sharing willingness between job design and employee innovation performance, Foss et al. (2015) explored the mechanism of complementary human resource management practices such as rewards, job design, and work atmosphere that affect the motivation of knowledge sharing, and then affect organizational performance, this provides a basis for the study of the mediating role of knowledge sharing willingness in this paper.

Accordingly, the following hypothesis is proposed in this study:

Hypothesis 2: Knowledge sharing willingness mediates between job design and employee innovation performance.

2.3 *Organizational innovation climate and its moderating effect*

Reviewing previous literature, most scholars tend to study the organizational innovation atmosphere from a subjective perspective. There are differences in individuals' perception of the environment, and individuals' perceptions of the psychological environment is an important factor influencing behavior generation (Amabile 1997). Scholars such as Xie et al. (2018) believed that the organizational innovation atmosphere is a consistent perceived experience of organizational members' attitudes towards innovative behaviors, vision, and goals. Liu and Shi (2009) defined the organizational innovation atmosphere as the subjective cognition of individuals on organizational policies, management behaviors, organizational environment, and innovation support factors, which is essentially the sense of social support experienced by employees during work. This kind of social support comes from a wide range of sources, not only from the organization, but also in the process of getting along with superiors and colleagues, and getting support and encouragement (Ding & Li 2016). These social support factors are essentially extrinsic motivators in the work environment (Liu & Shi 2009).

Existing research has verified that organizational innovation atmosphere affects innovation performance at the organizational level and team level, so does organizational innovation atmosphere affect the relationship between job design and employee innovation performance? Existing literature has studied the effects of cross-level factors on employee innovation performance. For example, Li and Yu (2016) confirmed that organizational innovation atmosphere has a supporting effect on employee innovation ability and innovation behavior, and positively moderates the impact of goal orientation on innovation performance; Oldham and Cummings (1996) studied the impact of employee personality characteristics, job complexity and leadership management style on employee innovation performance, but there is less literature on the interaction between job design and organizational innovation atmosphere. Studies have shown that in workplaces that support innovation, job autonomy can positively influence and play a key role in job-related performance such as process innovation (Giebels et al. 2016) and knowledge creation (Vargas et al. 2016). Besides, Wang et al. (2013) also pointed out that employees can feel the behaviors and results expected by the organization or leaders from the organizational atmosphere, and this perception determines their behaviors. Job design and organizational innovation atmosphere, as internal and external motivating factors in work, inevitably have interactions when influencing employees' innovation performance. If the organizational climate does not support innovation, even if the work can stimulate employees to form innovative ideas and behaviors through effective job design, their willingness to innovate will be weakened by the lack of support from their superiors and colleagues.

All in all, the impact of job design on employee innovation performance will vary depending on the level of organizational innovation atmosphere.

At the same time, this study believes that the relationship between job design and knowledge sharing willingness will also be affected by the organizational innovation atmosphere. According to the theory of planned behavior, an individual's behavioral intention is affected by the environment and the individual's perception of the environment (Ajzen 1991). As can be seen from the above, job design influences employees' knowledge sharing willingness by promoting their sense of belonging, loyalty, and satisfaction to the organization. An organizational climate that supports innovation provides employees with an open exchange of ideas and ideas that helps spread knowledge throughout the organization (Bock et al. 2005). And in organizations with a high degree of innovation, knowledge sharing behaviors are often recognized and encouraged by organizations and leaders, employees tend to interpret knowledge sharing as an act that meets the expectations and interests of the organization and leaders (Zhong et al. 2019); Conversely, if the organizational atmosphere does not support innovation, there is no fair reward, the rights and interests of employees are not valued, and their willingness to share knowledge is more likely to be enhanced or weakened by the work itself. And studies have shown that one dimension of the job characteristics model-job autonomy, its interaction with the cooperative atmosphere of the organization positively affects knowledge sharing behaviors in the organization (Llopis & Foss 2016). Moreover, the organizational climate that supports knowledge sharing moderates the influence of job autonomy and knowledge reward on knowledge sharing motivation.

Accordingly, the following hypotheses are proposed in this study:

Hypothesis 3: Organizational innovation atmosphere positively moderates the relationship between job design and knowledge sharing willingness.

Hypothesis 4: The organizational innovation atmosphere positively regulates the relationship between job design and employee innovation performance

Combining hypothesis 2 and hypothesis 3, this research further proposes that the organizational innovation atmosphere will affect the mediating role of the knowledge sharing willingness in the relationship between job design and employee innovation performance. In other words, with the improvement of the organizational innovation atmosphere, the mediating effect of the knowledge sharing willingness in job design and employee innovation performance has also increased. Based on this, this research makes the following assumption:

Hypothesis 5: The organizational innovation atmosphere is positively regulating the mediating role of knowledge sharing willingness between job design and employee innovation performance (Figure 1).

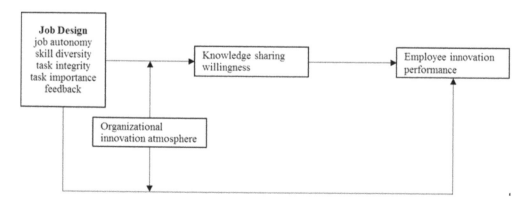

Figure 1.   Research framework.

3   RESEARCH METHODS

3.1   *Research samples*

In this study, data were obtained by questionnaire. The research sample covered enterprises in Anhui, Guangdong, Jiangsu, Zhejiang, and other provinces, mainly in the chemical industry, machinery manufacturing, education, finance, information technology, and other industries. Methods such as anonymous filling can reduce the possibility of homologous bias. Considering the characteristics of variables, most of the respondents who answered the questionnaire were knowledge workers in enterprises and institutions. A total of 394 questionnaires were recovered, of which 278 were valid questionnaires (no missing items in the questionnaire and no contradictions), and the effective recovery rate was 70.56%. The basic situation of the research sample is shown in Table 1.

Table 1.   Sample basic feature distribution.

Statistical variable	Values	Number	Proportion (%)	Statistical variable	Values	Number	Proportion (%)
Gender	Male	160	57.6	Working	<=1	16	5.8
	Female	118	42.4	years	1–3	97	34.9
Age	21–25	54	19.4		3–5	82	29.5
	26–30	95	34.2		5–10	42	15.1
	31–35	65	23.4		>=10	41	14.7
	36–40	27	9.7	Position	Production	50	18.0
	>40	37	13.3	type	Finance	25	9.0
Education level	High school and below	27	9.7		R & D	30	10.8
	Junior college	59	21.2		Administrative	44	15.8
	Undergraduate	163	58.6		Human Resources	26	9.4
	Master	27	9.7		Marketing	47	16.9
	Doctor and postdoc	2	0.7		Other	56	20.1

3.2   *Research tools*

This research questionnaire included four parts: job design, organizational innovation atmosphere, knowledge sharing willingness, and employee innovation performance. To ensure the reliability and validity of this survey, we used a scale that has been empirically tested by scholars at home and abroad many times. The scales used Likert's five-point scoring method, allowing subjects to read the questionnaire carefully and fill it in anonymously. The options are 1–5 points from "completely inconsistent" to "fully consistent".

3.2.1   *Job design*
The scale for measuring job design was job diagnostic questionnaire developed by Hackman and Oldham (1975). It includes five dimensions: job autonomy, skill diversity, task integrity, job importance, and feedback. There are 3 items in each dimension and a total of 15 items.

3.2.2   *Organizational innovation atmosphere*
The organizational innovation atmosphere was measured by a scale compiled by Liu and Shi (2009). After the scale was compiled, many scholars have used the scale to conduct relevant empirical research (Ding & Li 2016, Li & Mei 2018). The scale includes three dimensions of peer support, leadership support, and organizational support. Each dimension has 4 items, for a total of 12 items, such as "I will actively exchange new ideas with colleagues at work"; "Leaders encourage subordinates to provide diverse suggestions and ideas"; "Company rewards employees for coming up with innovative solutions to improve products and services".

3.2.3 Knowledge sharing willingness

The measure of knowledge sharing willingness was based on a scale of knowledge sharing willingness developed by Bock et al. (2005). The scale consists of 5 items such as "I often share professional knowledge learned from education or training courses with colleagues". The reliability of the original scale was 0.930.

3.2.4 Employee innovation performance

Employees' innovation performance was measured using a scale developed by Han et al. (2007). This questionnaire was collected through observations, interviews, and questionnaires, and included three dimensions, including the willingness to innovate, innovative actions, and innovative results, with a total of eight items. Sample questions are "I am willing to acquire new working methods and skills through learning", "I often proactively propose new ideas to improve performance" and "I strive to put innovative ideas into practice".

3.2.5 Control variables

Regarding existing literature, this study used gender, age, education, and years of work as control variables.

4 DATA ANALYSIS AND HYPOTHESIS TESTING

4.1 Reliability and validity test and analysis of homology variance

The reliability of the job design, knowledge sharing willingness, organizational innovation atmosphere, and employee innovation performance scale are respectively 0.806, 0.810, 0.909, and 0.892, which are all higher than the acceptance criteria of 0.7 and has good internal consistency. Confirmatory factor analysis results show that the fitting effect of the four-factor model is significantly better than the other four models, and all the values have reached acceptable standards ($\chi^2/df = 2.373$, TFI = 0.904; CFI = 0.916; IFI = 0.917; RMSEA = 0.07), indicating that the model also has ideal discriminant validity. Also, the average variance extraction amount of all scales is greater than 0.5, which has high convergence validity (See Table 2).

For possible homogeneous variance problems, Harman single factor analysis was used to test. The unexplained variance of the first principal component was 31.89%, which was significantly lower than 50%. Therefore, there is no serious homology error problem.

Table 2. Results of confirmatory factor analysis.

Model	Combination	χ^2/df	RMSEA	CFI	IFI	TLI
Four-factor model	JB; KSW; OIA; EIP	2.373	0.070	0.916	0.917	0.904
Three-factormodel	JB; KSW +OIA; EIP	3.098	0.087	0.870	0.871	0.853
Three-factor model	JB+ KSW; OIA; EIP	4.320	0.109	0.796	0.798	0.771
Two-factor model	JB+ KSW +OIA; EIP	3.294	0.091	0.856	0.857	0.840
Single factor model	JB+ KSW +OIA+EIP	4.867	0.118	0.757	0.758	0.730

JB is job design; KSW is Knowledge sharing willingness; OIA is organizational innovation atmosphere; EIP is employee innovation performance.

4.2 Descriptive statistics and correlation analysis

As can be seen from the Table 3, job design and employee innovation performance are significantly positively correlated (r=0.537, p<0.001). Hypothesis 1 is initially verified. Besides, the four variables are significantly correlated in pairs, and the correlation coefficients between the variables are less than 0.7, which provides the basis for the subsequent testing of the mediating and moderating

Table 3. Descriptive statistics and correlation coefficients.

	Mean	Standard deviation	1	2	3
1 Job Design	3.713	0.452			
2 Knowledge sharing willingness	3.957	0.521	.575**		
3 Organizational innovation atmosphere	3.78	0.586	.571**	.666**	
4 Employee Innovation Performance	3.819	0.599	.537**	.601**	.646**

* means $p < 0.05$; ** means $p < 0.01$; *** means $p < 0.001$.

Table 4. Main effect and mediating effect test.

	Knowledge sharing willingness		Employee Innovation Performance		
Variable	Model 1	Model 2	Model 3	Model 4	Model 5
Gender	−0.084	−0.033	−0.071	−0.026	−0.012
Age	0.097	−0.041	0.238*	0.115	0.133
Education level	−0.028	−0.008	−0.010	0.008	0.011
Working years	0.046	0.084	0.008	0.042	0.006
Job design		0.566***		0.503***	0.261***
Knowledge sharing willingness					0.428***
F	2.436*	123.426***	5.453***	24.685***	34.596***
R^2	0.034	0.336	0.074	0.312	0.434
Ajusted R^2	0.020	0.324	0.060	0.299	0.421

* means $p<0.05$; ** means $p<0.01$; *** means $p<0.001$.

effects. The maximum variance inflation factor (VIF) value in the regression model is 3.35, which is lower than the empirical threshold of 10, and there is no serious multi-collinearity problem in this study.

4.3 Moderated mediation test

4.3.1 Main effect and mediating effect test

Table 4 shows the results of the main and mediating effect tests. Based on controlling statistical variables, Model 4 introduced job design, where the F value is 24.685, R^2 is 0.312, and R^2 is increased by 0.238. The regression result is significant, that is, job design positively affects employee innovation performance. Hypothesis 1 holds. The F value of Model 2 is 123.426, R^2 is 0.336, and R^2 is increased by 0.302. Job design can better predict knowledge sharing willingness ($\beta = 0.566$, $p < 0.001$). Model 5 introduced knowledge sharing willingness based on Model 4, and the result shows that its predictive effect on employee innovation performance is significant ($\beta = 0.428$, $p < 0.001$) and the impact of job design on employee innovation performance is still significant ($\beta = 0.261$, $p < 0.001$), but the regression coefficient of job design variables decreases from 0.428 to 0.261, indicating that the knowledge sharing willingness plays a mediating role between job design and employee innovation performance.

As the traditional hierarchical regression method is used to test the mediation effect, the probability of the first type of error may be higher, so we used the Bootstrap method, which is generally considered to be better, to further test the utility of knowledge sharing willingness. (Zhao et al. 2010). As can be seen from Table 5, the direct effect of job design on employee innovation performance and the mediating effect of knowledge sharing willingness are [0.168, 0.404] and [0.184, 0.344] respectively in the 95% confidence interval, both of which do not include 0, indicating that the mediating effect of knowledge sharing willingness exists.

Table 5. Total effect, direct effect and intermediate effect decomposition.

	Effect size	BootSE	BootCI Lower limit	BootCI upper limit	Relative effect value
Total effect	0.548	0.051	0.446	0.651	
Direct effect	0.284	0.059	0.168	0.404	51.87%
Mediating effect of knowledge sharing willingness	0.264	0.040	0.184	0.344	48.13%

Table 6. Moderating mediation effect test.

Variable	Knowledge sharing willingness Model 6	Employee innovation performance	
		Model 7	Model 8
Gender	−0.032	−0.024	−0.018
Age	−0.100	0.056	0.077
Education level	0.011	0.027	0.025
Working years	0.145	0.103	0.073
Job design	0.246***	0.183***	0.131***
Organizational innovation atomosphere	0.508***	0.505***	0.398***
Job design × Organizational innovation atomosphere	0.114*	0.121***	0.097*
Knowledge sharing willingness			0.211***
F	46.451***	37.763***	35.863***
R^2	0.519	0.495	0.516
Ajusted R^2	0.506	0.482	0.502

* means p < 0.05; ** means p < 0.01; *** means p < 0.001.

4.3.2 Moderating effect test

Model 6 and Model 7 respectively examined the moderating role of organizational innovation atmosphere on job design and knowledge sharing willingness, as well as job design and employee innovation performance. From the test results of Model 6, it can be seen that the interaction item of job design and organizational innovation atmosphere is positively correlated with knowledge sharing willingness ($\beta = 0.114$, p < 0.05), which shows that organizational innovation atmosphere positively moderates the influence of job design on knowledge sharing willingness, Hypothesis 3 holds. From the test results of Model 7, it can be seen that the interaction item of job design and organizational innovation atmosphere has a significant impact on employee innovation performance ($\beta = 0.121$, p < 0.001), indicating that the organizational innovation atmosphere is positively moderating the impact of job design on employee innovation performance. Hypothesis 4 holds. After adding mediation variables, regression coefficients are still significant ($\beta = 0.097$, p < 0.05), indicating that there is a moderating mediation effect (Table 6).

To more intuitively show the moderating effects, we plotted the simple slopes for the relationship between job design and employee innovation performance, as well as job design and knowledge sharing willingness at one standard deviation above and below the mean of organizational innovation atmosphere, as shown in Figure 2 and Figure 3. It can be seen from Figure 2 that when the organizational innovation atmosphere is high (M+1SD), job design has a significant impact on employee innovation performance; While when the organizational innovation atmosphere is low (M-1SD), although job design can also have a positive impact on employee innovation performance, the impact degree is small, indicating that with the enhancement of organizational innovation atmosphere, job design has a gradually increasing positive effect on employee innovation performance. Figure 3 and Figure 2 have the same trend. With the enhancement of the organizational innovation atmosphere, the positive effect of job design on knowledge sharing willingness is gradually

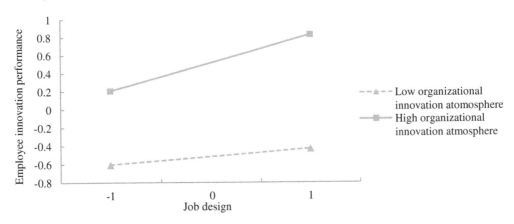

Figure 2. Moderating effect of organizational innovation atmosphere on job design and employee innovation performance.

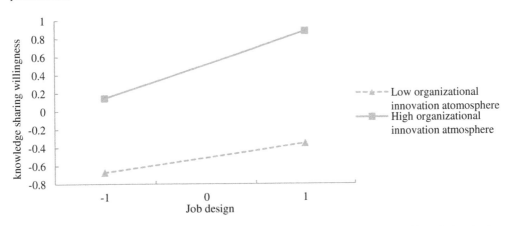

Figure 3. Moderating effect of organizational innovation atmosphere on job design and knowledge sharing willingness.

Table 7. Bootstrap test results with moderated mediation effect.

Organizational innovation atmosphere	Effect size	BootSE	BootCI lower limit	BootCI upper limit
−0.9729	0.035	0.0228	−0.0055	0.0834
0	0.0567	0.021	0.0199	0.1015
0.9729	0.0783	0.025	0.0319	0.1301

The results are based on 5000 repeated sampling samples. All the values reported in the table are standardized coefficients.

increasing. Also, it can be seen from Table 7 that when the value of the organizational innovation atmosphere is one standard deviation below the mean, the indirect effect of job design on the innovation performance of employees through knowledge sharing willingness is not significant. As the level of the organizational innovation atmosphere increases, the mediating effect of knowledge sharing willingness in the relationship between job design and employee innovation performance also shows an increasing trend, that is, with the improvement of the organizational innovation atmosphere, job design is more likely to promote employee innovation performance by enhancing their knowledge sharing willingness.

5 CONCLUSIONS

5.1 *Research conclusions*

First, previous scholars' research on antecedent dependent variables of employee innovation performance mainly focused on individual employee factors and organizational context factors. Among them, job design as an organizational context factor, there is less literature on its relationship with employee innovation performance. This paper validates the impact of job design on employee innovation performance, and the findings further enrich the study of antecedents of employee innovation performance. Secondly, some scholars have studied the mechanism between innovation performance and its antecedent variables, such as influencing employee innovation behavior through psychological ownership (Liu et al. 2016), or improving innovation performance by improving employee innovation self-efficacy (Ding & Li 2016), etc., but the research on the internal mechanism from the perspective of organizational learning is limited. This paper validates the mediating role of the willingness to share knowledge and further enriches the research on the mediating mechanism of job design and employee innovation performance. Finally, the study found that the organizational innovation atmosphere is positively moderating the relationship between job design and employee innovation performance. The degree of innovation atmosphere felt by employees in the organization is also the level of the organization's support for innovation. When employees perceive the organization's strong support for innovation, they will stimulate their willingness to innovate and put more effort into achieving organizational and personal innovation performance. At the same time, the moderating effect of organizational innovation atmosphere on job design and employee innovation performance is achieved in part through the mediating role of knowledge sharing willingness. It can stimulate the employees' willingness to acquire knowledge and exchange information through the role of organizational context characteristics, and promote employees to face innovation challenges better. This study fully considers the moderating role of the organizational innovation atmosphere in improving the willingness to share knowledge and employee innovation performance and improves the logic of the theoretical model. Based on the existing research, it expands the boundary conditions of job design for employee innovation performance.

5.2 *Management implications*

Based on previous scholars' research, this study explored the mechanism of job design on employee innovation performance with knowledge sharing willingness as a mediating variable. Through empirical tests, the research hypotheses have been verified. The conclusion of research is helpful for enterprises to adopt targeted management practices to promote employee innovation.

(1) Stimulate employees' motivation for innovation through humanized job redesign. Specifically, employees can be allowed to exert a variety of knowledge and skills by expanding the scope of work or increasing the type of work; By empowering employees, they can increase the independence and decision-making power of employees in arranging work and determining the procedures for completing tasks; Make sure that employees' work is integrated and identifiable, and enhance their sense of participation; Improve the information feedback mechanism by broadening feedback channels and innovative feedback processes, and give timely responses to employees 'innovative ideas or actions. Humanized job redesign can make employees feel more meaningful and fun during work, stimulate the innovation potential of employees, and help employees actively innovate.

(2) Create an organizational atmosphere that encourages innovation and promotes employees' initiative innovation. From the moderating effect of the organizational innovation atmosphere verified in the foregoing, we can see that while focusing on improving innovation performance through job design, we must also create an organizational atmosphere that encourages innovation. On the one hand, leaders should be able to respect and encourage employees to put forward their opinions and objections in their work, be a good example of innovation, and give full support and assistance to the innovative ideas that employees can implement. Leadership's support and encouragement for innovation can guide employees to continue their innovative behavior. On the other hand,

by establishing and improving a corporate culture that encourages innovation and openness, and timely rewards and commendations at the material level for employees 'innovative achievements, organizational members can form a virtuous circle of consciously conducting innovative activities

(3) Enhance employees' willingness to share knowledge and provide knowledge support for achieving innovative performance. Knowledge sharing willingness conveys the impact of job design on employee innovation performance under the adjustment of organizational context variables. On the one hand, in the process of job design and redesign, employees' perception of the five core dimensions of job characteristics can be improved by employing job enlargement and job rotation, and their autonomy to share and exchange knowledge can be improved by giving them sufficient trust and authorization. On the other hand, the adjustment of a flat organizational structure can make information transmission more convenient and communication smoother. The improvement of knowledge sharing channels has a positive effect on promoting employees' willingness to share knowledge and their conscious behavior of sharing knowledge. By establishing a fair and reasonable reward system and promptly giving material and emotional encouragement to employees who are willing to share knowledge, it is ensured that employees' sharing can receive corresponding returns.

5.3 *Research limitations and future research prospects*

Due to some conditions, this study has certain limitations: first of all, the study analyzes the relationship between variables through cross-section data, and the causal relationship between variables has not been dynamically tested; Secondly, the job design and knowledge sharing willingness are measured using foreign maturity scales. Although their effectiveness has been tested by literature, this paper is based on the Chinese context, and there may still be some misunderstanding in the feedback, which makes this study have certain limitations; Then, 77% of young employees under the age of 35 in the study sample, the applicability of the research conclusions in other age groups needs to be further tested; Finally, there are many factors that affect innovation performance. Some variables such as marital status, job rank, and the nature of the company are not controlled in the study, and the conclusions of the study may be affected.

Due to the limitations of cross-sectional data, future studies can conduct dynamic longitudinal studies to more forcefully verify the causal relationship between variables, to further explore the internal relationship between job design and employee innovation performance, and constantly improve the theoretical model. Future research objects can be extended to employees of all ages, and explore more effective and universal methods to promote employee innovation performance. Besides, this study only considered the relationship between job design and employee innovation performance and the mediation of knowledge sharing willingness in the general sense. Future research can consider covering different industries or different types of enterprises, to improve the practical significance and extensibility of the research from the comparison of multi-industry and multi-type enterprises.

ACKNOWLEDGEMENTS

Supported by Institutional Preparation of Management Training Project Planning in 2019 (NO.: W2019JSFW0458).

REFERENCES

Ajzen, I. 1991. The theory of planned behavior. Organizational Behavior and Human Decision Processes, 50, 179–211.

Amabile, T. M. 1997. Motivating creativity in organizations: On doing what you love and loving what you do. California Management Review, 40, 39–58.

Amabile, T. M. 1998. How to kill creativity. Harvard Business Review, 76, 76–87.

Amabile, T. M., Conti, R., Coon, H., Lazenby, J. & Herron, M. 1996. Assessing the Work Environment for Creativity. Academy of Management Journal, 39, 1154–1184.

Bavik, Y. L., Tang, P. M., Shao, R. & Lam, L. W. 2017. Ethical leadership and employee knowledge sharing: exploring dual-mediation paths. Leadership Quarterly, 29, 322–332.

Bock, G., Zmud, R. W., Kim, Y. & Lee, J. 2005. Behavioral intention formation in knowledge sharing: examining the roles of extrinsic motivators, social-psychological factors, and organizational climate. Management Information Systems Quarterly, 29, 87–111.

Chae, H. & Choi, J. N. 2018. Contextualizing the effects of job complexity on creativity and task performance: Extending job design theory with social and contextual contingencies. Journal of Occupational and Organizational Psychology, 91, 316–339.

Ding, G. & Li, H. 2016. How Job Characteristics Influence Employees' Creative Behavior: A Mediated Moderating Model. Human Resources Development of China, 19–27.

Foss, N. J., Pedersen, T., Fosgaard, M. R. & Stea, D. 2015. Why Complementary HRM Practices Impact Performance: The Case of Rewards, Job Design, and Work Climate in a Knowledge Sharing Context. Human Resource Management, 54, 955–976.

Gagne, M. 2009. A model of knowledge sharing motivation. Human Resource Management, 48, 571–589.

Giebels, E., De Reuver, R., Rispens, S. & Ufkes, E. G. 2016. The Critical Roles of Task Conflict and Job Autonomy in the Relationship between Proactive Personalities and Innovative Employee Behavior. The Journal of Applied Behavioral Science, 52, 320–341.

Grant, A. M. 2007. Relational Job Design and the Motivation to Make a Prosocial Difference. Academy of Management Review, 32, 393–417.

Grant, A. M. 2008. The significance of task significance: Job performance effects, relational mechanisms, and boundary conditions. Journal of Applied Psychology, 93, 108–124.

Grant, A. M. & Parker, S. K. 2009. 7 Redesigning Work Design Theories: The Rise of Relational and Proactive Perspectives. The Academy of Management Annals, 3, 317–375.

Guo, H., Zhang, L., Hong, S. & Huang, S. 2020. The Effect of Prosocial Sensemaking Mechanisms on Knowledge Withholding Intension: A Perspective of Knowledge Leadership. Chinese Journal of Management, 17, 111–120.

Hackman, J. R. & Oldham, G. R. 1975. Development of the Job Diagnostic Survey. Journal of Applied Psychology, 60, 159–170.

Hackman, J. R. & Oldham, G. R. 1976. Motivation through the design of work: test of a theory. Organizational Behavior and Human Performance, 16, 250–279.

Han, Y., Liao, J. & Long, L. 2007. Model of development and empirical study on employee job performance construct. Journal of Management Sciences in China 62–77.

Holzmann, V. & Golan, J. 2016. Leadership to Creativity and Management of Innovation? The Case of the "Innovation Club" in a Production Company. American Journal of Industrial and Business Management, 06, 60–71.

Hu, G. & Wang, X. 2019. Research on the Evolution Logic and Self-organizing Mechanism of Platform Enterprises: A Case Study of Haier Group. China Soft Science, 143–152.

Humphrey, S. E., Nahrgang, J. D. & Morgeson, F. P. 2007. Integrating motivational, social, and contextual work design features: A meta-analytic summary and theoretical extension of the work design literature. Journal of Applied Psychology, 92, 1332–1356.

Li, D. & Yu, B. 2016. Goal Orientation, Climate for Innovation and the Employee Innovation Performance—An Interactive Effect Based on Knowledge Worker. Journal of Industrial Technological Economics, 35, 145–153.

Li, W. & Mei, J. 2018. Research on Influence of Leader Empowerment on Employee Innovation Behavior—A Moderated Mediation Model. Soft Science, 32, 75–79.

Liu, C. & An, L. 2016. Employee Diversity, Knowledge Sharing and Individual Innovation Performance: A Moderated Mediation Model. Science of Science and Management of S.& T., 37, 170–180.

Liu, H., Su, Y. & Wu, N. 2015. Study on the Formation of Organization Responsibility: Effect of Task Significance, Strategic Contribution and Performance Measurability. Human Resources Development in China, 41–50.

Liu, S., Zhang, L., Feng, J. & Wu, K. 2016. I Innovate Because I Am the Owner: A Moderated Mediation Research on the Relationship between Psychological Ownership and Innovative Behavior. Science & Technology Progress and Policy, 33, 128–133.

Liu, Y. & Shi, J. 2009. A Study on the Relationship between the Effects of the Organizational Innovative Climate and those of Motivational Preference, on Employees' Innovative Behavior. Management World, 88–101+114+188.

Llopis, O. & Foss, N. J. 2016. Understanding the climate–knowledge sharing relation: The moderating roles of intrinsic motivation and job autonomy. European Management Journal, 34, 135–144.

Malik, M. A. R., Butt, A. N. & Choi, J. N. 2015. Rewards and employee creative performance: Moderating effects of creative self-efficacy, reward importance, and locus of control. Journal of Organizational Behavior, 36, 59–74.

Meng, L. 2016. Task design incorporating SDT and one's intrinsic motivation: An empirical investigation from acognitive neuroscience perspective. B, Zhejiang University.

Oldham, G. R. & Cummings, A. 1996. Employee Creativity: Personal and Contextual Factors at Work. Academy of Management Journal, 39, 607–634.

Oldham, G. R. & Fried, Y. 2016. Job design research and theory: Past, present and future. Organizational Behavior and Human Decision Processes, 136, 20–35.

Parker, S. K. 2007. 'That is my job' How employees' role orientation affects their job performance. Human Relations, 60, 403–434.

Pee, L. G. & Lee, J. 2015. Intrinsically motivating employees' online knowledge sharing: Understanding the effects of job design. International Journal of Information Management, 35, 679–690.

Peng, C., Lyu, C. & Li, H. 2020. Effects of Ambidextrous Leadership on Employees' Innovation Performance— Serial Mediation Role of Perceived Insider Status and Knowledge Sharing Intention. R&D Management, 32, 72–81.

Saragih, S. 2011. The Effects of Job Autonomy on Work Outcomes: Self Efficacy As An Intervening Variable. International Research Journal of Business Studies, 4, 203–215.

Scott, S. G. & Bruce, R. A. 1994. Determinants of innovative behavior: A path model of individual innovation in the workplace. Academy of Management Journal, 37, 580–607.

Sun, L. & Teng, F. 2013. The Relation between Job Characteristics and Job Engagement of New Science Researchers: A Longitudinal Study. Science and Technology Management Research, 33, 150–154.

Vargas, N., Lloria, M. B. & Roigdobon, S. 2016. Main drivers of human capital, learning and performance. Journal of Technology Transfer, 41, 961–978.

Wang, N., Chen, X. & Jing, C. 2019. Research on the Influence of Material Rewards on Knowledge Sharing: Controversy and Integration. Science & Technology Progress and Policy, 36, 135–143.

Wang, S., Xu, B. & Peng, J. 2013. The impact of organizational climate perception on employees innovation behavior: Based on the medium role of knowledge sharing intention. Science Research Management, 34, 130–135.

Wang, Y. & Zhao, X. 2011. The Analysis of Job Design and Incentives Including Intrinsic Motivation. Journal of Industrial Engineering and Engineering Management, 25, 111–115.

Wang, Z., Xiong, L. & Guo, H. 2014. Impacts of Knowledge Employees' Creativity Personality and Job Characteristics on Individual Innovation Performance Commercial Research, 108–114.

Wu, Z., Zhai, Y., Wang, Z. & Sun, W. 2018. Psychological Contract, Employees Knowledge Sharing Willingness and Innovation Performance: The Moderating Effect Based on the Technology Integration Model. Journal of Shanghai University of International Business and Economics, 25, 59–71.

Xie, Y. & Gu, Q. 2015. Study on the Effect of Job Skill Variety on Employee Creativity and Job Performance: The Perspectives of Psychological Ownership and Work Feedback. Science of Science and Management of S. & T., 36, 162–169.

Xie, Y., Xue, W., Li, L., Wang, A., Chen, Y., Zheng, Q., Wang, Y. & Li, X. 2018. Leadership style and innovation atmosphere in enterprises: An empirical study. Technological Forecasting and Social Change, 135, 257–265.

Ye, L., Liu, Y. & Guo, M. 2018. Influence of Paternalistic Leadership on Skilled Talents' Knowledge Sharing Intention: Based on Self-Concept Perspective. Technology Economics, 37, 55–62+119.

Yoo, S., Jang, S., Ho, Y., Seo, J. & Yoo, M. H. 2019. Fostering workplace creativity: examining the roles of job design and organizational context. Asia Pacific Journal of Human Resources, 57, 127–149.

Zhang, Z., Cheng, H. & Yu, Y. 2020. Relationships among Government Funding, R&D Model and Innovation Performance: A Study on the Chinese Textile Industry. Sustainability, 12, 644.

Zhao, X., Lynch, J. G. & Chen, Q. 2010. Reconsidering Baron and Kenny: Myths and Truths about Mediation Analysis. Journal of Consumer Research, 37, 197–206.

Zhong, J., Deng, J. & Luo, J. 2018. The Effect of Inclusive Leadership on Team Performance and Employee Innovative Performance: A Moderated Mediation Model. Science of Science and Management of S.& T., 39, 137–148.

Zhong, X., Fu, Y. & Wang, T. 2019. Inclusive Leadership, Perceived Insider Status and Employee Knowledge Sharing—Moderating Role of Organizational Innovation Climate. R&D Management, 31, 109–120.

Computational Social Science – Luo, Ciurea & Kumar (eds)
© *2021 Taylor & Francis Group, London, ISBN 978-0-367-70193-2*

Analysis on the coupling effects of strategic emerging industry structure and employment structure in China

L. Liu
School of Management, Hefei University of Technology, Hefei, China
Business School of Fuyang Normal University, Fuyang, China

C.S. Wu & Y.Y. Zhu
School of Management, Hefei University of Technology, Hefei, China

B. Ye & N. Yang
State Grid Anhui Electric Power Co., Ltd, Hefei, China

ABSTRACT: Taking strategic emerging industry data in China in 2010–2017 as the example, this study employs multiple indicators and regression models to explore the coupling effects between strategic emerging industry structure and employment structure in China. According to the research results, strategic emerging industry structure and employment structure in China are in constant fluctuation with an unstable changing rate. Due to the gradual optimization of industry structure, general downtrend of strategic emerging industry structure and employment structure deviation in China demonstrate the gradual rationality of corresponding structural changes and a strong leading role in employment. The variation of the subdivision strategic emerging industry employment proportion plays varying influences on industry output. Therefore, rational adjustment of strategic emerging industry structure has significant meaning in stabilizing the orderly alteration of emerging industry structure and employment structure, exerting the facilitating role of coupling effects in economic growth and increasing the labor employment rate.

Keywords: Strategic emerging industry, Transition of industrial structure, Deviation degree of industrial structure, Employment elasticity.

1 INTRODUCTION

The emerging industry is the strategic forerunner of social economy, and employment serves as the foundation of people's livelihood, both of which prove to be two foremost strategic objectives during the socioeconomic development process. The strategic emerging industry has become the commanding height in future economic competition, and its development has been promoted to the high ground of state strategy. The giant domestic market scale in China provides wide survival and development space for the strategic emerging industry (Huang & Zhang 2019). For industrial enterprises above the designated size in 2019, the added value of China's strategic emerging industry increased by 8.4% in comparison with that of the past year. In this case, this industry gradually becomes the new engine ushering in the economic growth of China and plays a crucial role in stabilizing economic growth, promoting industry upgrade, and pioneering innovative development (Lv 2019). The essence of economic growth is industry development, which in turn determines the development of labor employment. During the process of economic development, both industry structure evolution and employment structural adjustment possess strong correlation. It is proved that the coordinated development of employment structure and industry structure is the inevitable requirement of favorable economic operation (Wang & Tian 2012). As an important parameter determining employment scale, industry structure has the most significant influence on growing job employment. The variation of industry structure necessarily leads to that of employment structure, which means the reliance of employment structure on industry structure to some extent. Additionally, considering the influence of employment structure on industry structure, rational employment structure significantly drives the upgrade of industry structure. Though employment growth in China primarily depends on the pulling efforts of economic growth, the positive role of industry structure adjustment in labor employment has emerged in recent years (Teng et al. 2016). Strategic emerging industry structure refers to the development level of all strategic emerging industries within a specific region as well as the proportional relation among all specific industries. Due to the change of strategic emerging industry structure, there will be a great influence on social employment and economy (Zheng 2012). However, since the strategic emerging industry is mostly subordinate to the high-tech industry, corresponding industry structure variation shows uncertainties during the growth of labor employment. Then, does variation of strategic emerging industry structure promote labor employment, and what about its employment effects? All of these should be answered by further research. With strategic emerging industry data in China in 2010–2017, for example, this study analyzes the variation characteristics of industry structure and employment structure, primarily explores the coupling effects between industry structure and employment structure, and clarifies the interactive mechanism of the two, aiming to offer evidence to the development of China's strategic emerging industry.

2   LITERATURE REVIEW

Originating from physics, the term "coupling" is used to describe the interplay phenomenon between two or among more independent systems (Jiang et al. 2018). Favorable coupling relations support the adjustment and matching of system factors, propel the transition of these factors from the disordered state to the ordered state, and push forward the mutual coordination and promotion of factors among systems (Hu 2015). As shown in considerable studies, there exist mutualistic coupling effects between industry structure and employment structure. Though William Petty never explicitly employs the term "coupling," he is the forerunner of this kind of study. Aware of the income gap among different industries, he realizes that such a gap would give rise to changes in labor employment structure (Petty 2010). Based on the findings of William Petty, Colin Clark further discovers the distribution changes of labor factors among different industries, and proposes the Percy-Clark Theorem, systematically explaining the law concerning industry structure evolution and industry structure variation. Moreover, the theorem also reveals the transition of labor employment structure from the primary industry to the secondary industry and the tertiary industry (i.e., the "upgrading" of economic structure) (Colin 1940). Through combining industrialization and urbanization, Lewis investigates the flow of labor force from agriculture to industry and from countryside to the city in developing countries and proposes the "dual economy model" (Lewis 1954). As pointed out by Kuznets, industry output variation and labor variation among all industrial sectors in national production present consistency over a long period of time (Kuznets 1985). Throughout the study on the influence of industry structure evolution law on employment structure in various countries, Chenery, Elkington, and Syrquin et al. find that the output structure transfer in developing countries mostly runs ahead of pertinent employment structure transfer (Chenery et al. 1970; Syrquin & Chenery 1989). After analyzing related German economic statistics from 1998–2007, Kowalewski discovered the paramount negative influence of industry professionalism on employment structure (Kowalewski 2011). By referring to the competitive industry structure in America's manufacturing industry, Drucker evaluates its role in employment structure (Drucker 2013).

Additionally, Chinese scholars have also performed considerable studies on the coupling relations between industry structure and employment structure. Luo Guoxun explores the relation between industry structure variation and employment structure variation from the perspectives of economic growth and labor productivity (Luo 2000). Zhou Jian'an illustrates the influence of industry structure on labor force input and configuration, and even its decisive role in labor force composition to some degree (Zhou 2006). Combined with the empirical analysis results concerning the VAR model, Liu Pu discovers the action and reaction of industry structure and labor employment, revealing the possibility of propelling fast economic growth and labor employment by optimizing industry structure (Liu 2010). According to the empirical analysis on related data from 1990–2011, Tan Juhua approves of the duality of industry development in labor employment and uncovers the binary causality between labor employment and industry development (Tan 2013). Wang Honghao puts forward the notion of "the emerging industry employment proportion growth rate" or labor force transfer theory, showing the constant transfer of labor force from the traditional industry to the emerging industry, which makes the emerging industry the main subject of employment (Wang 2005). Shi Gaoyan sets forth the variation of the industry output proportion and the labor force employment proportion since the reform and opening up of China, thus discovering the interactivity between industry structure adjustment and labor force transfer (Shi 2014). Sun Qing et al. examine the coordination between industry structure and employment structure as per employment elasticity, structural deviation, and comparative labor productivity in three industries (Sun et al. 2019).

To sum up, research results about the relation between industry structure and employment structure have covered almost all fields at the macroscopic and microscopic layers, involving the integrated application of qualitative, quantitative, and cross-discipline principles and methods. These studies further consummate how industry structure and employment structure affect and act on the knowledge structure of economic growth. However, various conclusions are drawn when failing to grasp the coordinated and symbiotic coupling effects between industry structure and employment structure on the whole. In addition, few studies have been made on the interplay mechanism between two inside the industry, especially those of the coupling relations between strategic emerging industry structure and employment structure. Due to the fast growth of the strategic emerging industry in China, how to conduct a quantitative analysis on the variation characteristics of strategic emerging industry structure and employment structure and how to assess the influence of strategic emerging industry structure variation on labor force employment are of great importance to the clear recognition about the employment effects of the strategic emerging industry, rational adjustment and optimization of emerging industry structure, and driving force of this industry in labor force employment. Therefore, based on related data of China's strategic emerging industry, the variation characteristics of strategic emerging industry structure and employment structure are quantitatively observed in this study. The interaction and coupling effects of the two are also interpreted by theoretical structure deviation, employment elasticity, and the regression model.

3 MODELING AND DATA SOURCE

3.1 *Modeling*

3.1.1 *Structure variation*
Structure variation suggests the variation of industry or employment in a country or a region at two time points. Generally, Moore structure variation value is indicative of the variation of employment structure and industry structure (Li & Zhang,2013; Wang & Dang 2010). In the present study, Moore structure variation value is taken to quantify the variation of strategic emerging industry structure and employment structure.

Here is the formula:

$$\theta_1 = \arccos\left(\frac{\sum_i Y_i(t) * Y_i(t-1)}{\sqrt{\left(\sum_i Y_i(t)^2\right) * \left(\sum_i Y_i(t-1)^2\right)}}\right) \tag{1}$$

$$\theta_2 = \arccos \left(\frac{\sum_i L_i(t) * L_i(t-1)}{\sqrt{\left(\sum_i L_i(t)^2\right) * \left(\sum_i L_i(t-1)^2\right)}} \right) \tag{2}$$

In formulas (1) and (2), included angle θ denotes the angle between two vectors in two periods, when θ_1 is the industry structure variation degree and θ_2 is the employment structure variation degree. The greater the value of θ, the greater the variation degree of industry or employment structure in adjacent years (Li et al. 2006). i is industry; $Y_i(t)$ and $Y_i(t-1)$ are the proportion of the output of the i industry in year t and year $(t-1)$ in gross output of the strategic emerging industry, respectively; $L_i(t)$ and $L_i(t-1)$ are the proportion of the employment in the i industry in years t and $(t-1)$ in gross output of the strategic emerging industry, respectively.

3.1.2 *Structure deviation degree*

The structure deviation degree is the indicator that measures the discrepancy between industry value-added proportion and the corresponding employment proportion (Wu & Feng 2006). Syrquin Chenery illustrates the necessity of keeping a balance between industry structure and employment structure in any country or region, and the part above the rational proportion is the deviation degree of employment structure (Zhang et al. 2012). Based on the principle, the paper introduces the concept of industry structure deviation structure and quantitatively analyzes strategic emerging industry structure deviation.

Here is the formula:

$$S_i = \frac{Y_i}{L_i} - 1 \tag{3}$$

where S_i, the structure deviation degree of the i industry, could be either positive or negative. Y_i is the proportion of strategic emerging industry output, and L_i is the proportion of strategic emerging industry employment. $S_i = 0$ proves the match between industry structure and employment structure in the ideal state. $S_i < 0$ suggests talent surplus in the industry with recessive unemployment in need of labor force transfer. $S_i > 0$ means the lack of industry talents in need of labor force inflow (Dong 2013).

3.1.3 *Employment elasticity*

As the indicator adopted to measure economic growth capacity of absorbing labor employment, employment elasticity refers to the percentage of economic growth corresponding to employment quantity variation (Zhang 2002), which can reflect the employment absorbing ability of the strategic emerging industry.

The formula is shown:

$$\alpha = \frac{M_i}{N_i} \tag{4}$$

In formula (4), α is employment elasticity; M_i is the variation rate of employment quantity in the i industry, and N_i is the gross output variation rate of the i industry. Employment elasticity could be either positive or negative. It is proven to be useful in measuring the absorbing capacity of economic growth for employment. In the condition of $\alpha > 0$, the greater the value, the greater the elasticity of employment, and the greater the pulling effects of economic growth on employment. Similarly, the smaller the value, the lesser the elasticity of employment, and the weaker the pulling effects of economic growth on employment. In the condition of $\alpha < 0$, there exists a negative development trend in employment with economic growth, which is known as the "extrusion" effects of economic growth against employment. Another case is that economic decline contributes to employment growth, which is called the "absorbing" effects (Cai & Li 2010).

3.1.4 *Regression model*

To figure out the functional mechanism of employment structure on industry structure, in this study, the labor force employment proportion of each subdivision industry is viewed as the independent variable, while the proportion of each subdivision industry output is taken as the dependent variable to construct a regression model and explain the influence of employment structure on industry structure.

Here is the regression formula:

$$Y_i = a + bL_i + u \tag{5}$$

Y_i represents the proportion of the strategic emerging industry output; L_i is the proportion of strategic emerging industry employment; a and b are unknown parameters, and u is the disturbing term.

3.2 *Data source*

The definition of the strategic emerging industry has often changed. In January 2020, MIIT published *Strategic Emerging Industry Classification* (2012) (trial), dividing the strategic emerging industry subdivision into seven leading industries, such as the energy-saving industry and the new-generation information technology industry. In November 2018, the National Bureau of Statistics issued *Strategic Emerging Industry Classification* (2018), which gently adjusted the classification of the strategic emerging industry and added the digital creative industry and the related service industry. Considering the accessibility and reliability of data, in the current work, the classification of the strategic emerging industry in the 2020 edition is followed to classify the industry into seven leading industries, and the period 2010–2017 is chosen as the research scope. Primitive data comes from *China Statistical Yearbook*, *Statistical Yearbook of the High-tech Industry in China*, *the Statistical Yearbook of Science in China*, *The Statistical Yearbook of Torch in China*, and *Report for the Strategic Emerging Industry in China*. Some data is from the National Bureau of Statistics, the Strategic Emerging Industry Database, and so on. Non-accessible data, such as the employment quantity in the new energy automobile industry, could be computed by the statistics and data in the approximate industry.

4 VARIATION IN CHINA'S STRATEGIC EMERGING INDUSTRY AND EMPLOYMENT STRUCTURE

During the period 2010–2017, the variation situation of China's strategic emerging industry and industry structure was basically the same. However, the variation degree is different from the poor matching rate. Its employment structure generally falls behind industry structure. Generally, the evolution of the strategic emerging industry and the employment industry basically conforms to the common law of modern economic growth. Following are the specific variation trends and characteristics.

4.1 *Variation in strategic emerging industry structure*

Industry added value means the value increased by production activities within a period, which could accurately reflect the scale and rate of industry production. In the current work, the proportion of China's strategic emerging industry added value in gross domestic product (GDP) is taken to suggest the direction of industry structure variation. Figure 1 shows that during the period 2010–2017, the proportion of strategic emerging industry added value in GDP is on the rise, growing from 4% in 2010 to 10% in 2017. It is predicted that until late 2020, the figure will increase up to 15%, proving the consistency between China's strategic emerging industry and GDP in the variation direction, the favorable development trend of the strategic emerging industry and also the continuous expansion of the industry scale.

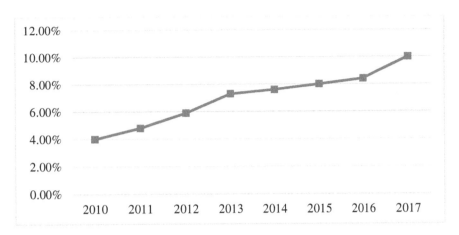

Figure 1. Variation trend of the proportion of China's strategic emerging industry added value in GDP in the period 2010–2017.

Table 1. China's strategic emerging industry structure variation indicator in the period 2010–2017.

Year	$\sqrt{\left(\sum_i Y_i(t)^2\right) * \left(\sum_i Y_i(t-1)^2\right)}$	$\sum_i Y_i(t) * Y_i(t-1)$	$\arccos\left[\dfrac{\sum_i Y_i(t) * Y_i(t-1)}{\sqrt{\left(\sum_i Y_i(t)^2\right)*\left(\sum_i Y_i(t-1)^2\right)}}\right]$
2010–2011	34.966%	34.957%	2.326%
2011–2012	35.371%	35.367%	1.547%
2012–2013	34.904%	34.898%	1.752%
2013–2014	33.668%	33.658%	2.477%
2014–2015	32.008%	31.985%	3.791%
2015–2016	30.079%	30.058%	3.728%
2016–2017	30.350%	30.321%	4.391%

Formula (1) can be adopted to compute the indicator, suggesting the variation of strategic emerging industry structure, as shown in Table 1. In the period 2010–2017, the industry structure variation indicator is maximum in 2016–2017 as high as 4.391%, showing the maximum variation rate of China's strategic emerging industry structure. The range with the second largest value is 3.791% in 2014–2015, showing that the development of the strategic emerging industry is obviously affected by a lot of internal and external factors after the "12th Five-Year Plan," with a fast industry structure variation rate. The range with a small industry structure variation indicator as 1.543% appears in 2011–2012.

4.2 Variation in the strategic emerging industry and employment structure

The variation direction of industry employment quantity could accurately show the scale and rate of industry labor force employment absorbing ability. In the present study, the proportion of strategic emerging industry employment quantity in national employment quantity is employed to show the variation direction of employment structure. Figure 2 shows that in the period 2010–2017, the proportion of strategic emerging industry employment quantity in national employment quantity was increasing stably, when rising from 6.5% in 2010 to 9.5% in 2017 with an annual average growth rate of 6.26%. In this case, it is revealed that the significant pulling effects of the strategic emerging industry on China's employment as a new growth point propel the employment of China.

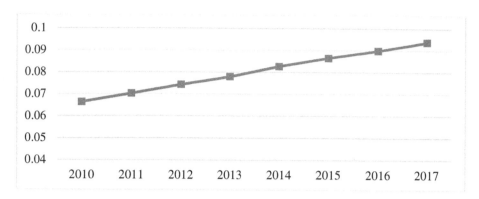

Figure 2. Variation trend of the proportion of China's strategic emerging industry employment quantity in national employment quantity in the period 2010–2017.

Table 2. China's strategic emerging industry structure variation indicator in 2010–2017.

Year	$\sqrt{\left(\sum_i L_i(t)^2\right) * \left(\sum_i L_i(t-1)^2\right)}$	$\sum_i L_i(t) * L_i(t-1)$	$\arccos\left[\dfrac{\sum_i L_i(t) * L_i(t-1)}{\sqrt{\left(\sum_i L_i(t)^2\right) * \left(\sum_i L_i(t-1)^2\right)}}\right]$
2010–2011	37.624%	37.600%	3.531%
2011–2012	35.963%	35.937%	3.802%
2012–2013	34.486%	34.460%	3.865%
2013–2014	33.344%	33.336%	2.084%
2014–2015	32.397%	32.393%	1.504%
2015–2016	31.256%	31.247%	2.327%
2016–2017	30.134%	30.130%	1.664%

Formula (2) can be used to compute the indicator suggesting the variation of strategic emerging employment structure as shown in Table 2. It displays that the employment structure variation indicator is maximum in 2012–2013, which is as high as 3.865%, meaning that in the preliminary development stage of the strategic emerging industry, employment structure variation is featured with the fastest rate. The employment structure variation indicator is the second largest as 2.327% in 2015–2016. In other ways, there is the fast rate of strategic emerging industry employment structure adjustment in the first year of the "13th Five-year Plan." The employment structure variation indicator is insignificant, as it was 1.504% in 2013–2014.

5 ANALYSIS ON THE COUPLING EFFECTS OF STRATEGIC EMERGING INDUSTRY STRUCTURE AND EMPLOYMENT STRUCTURE

Under sufficient market economic conditions, the variation of strategic emerging industry structure and employment structure should be coordinated, and the coupling effects between the two should be mutually promoted, coordinated, and reliant dynamically. With a view of decomposing such dynamic co-movement, this study takes strategic emerging industry structure and employment structure as the specific objective, respectively, to examine the functional mechanism of one on the other and provides support in favor of the high-quality development of China's strategic emerging industry in technical terms.

Table 3. Deviation degree of seven leading strategic emerging industries in China in 2010–2017.

	Industry structure deviation degree S_i							
Year	Energy-saving and environmental protection industry	New-generation information technology industry	Bioindustry	High-end equipment manufacturing industry	New energy industry	New material industry	New energy automobile industry	Industry structure deviation degree (Σ)
2010	−0.745	1.474	1.512	0.668	−0.645	−0.286	4.078	6.056
2011	−0.744	1.357	1.536	0.460	−0.631	−0.279	3.802	5.501
2012	−0.707	1.200	1.380	0.321	−0.617	−0.262	3.067	4.382
2013	−0.700	1.115	1.369	0.305	−0.578	−0.234	2.345	3.623
2014	−0.698	1.095	1.206	0.300	−0.398	−0.245	1.760	3.020
2015	−0.685	1.043	0.869	0.382	−0.361	−0.177	1.529	2.599
2016	−0.671	0.985	0.760	0.250	−0.252	−0.053	1.817	2.837
2017	−0.679	0.993	0.654	0.230	−0.255	−0.083	1.888	2.749
Average	−0.704	1.158	1.161	0.365	−0.467	−0.202	2.536	3.846

5.1 Role of strategic emerging industry structure in employment structure

To avoid the errors caused by the single indicator, in the paper, the indicator structure deviation degree and employment elasticity are cited to observe the "extrusion" and "absorbing" effects of industry structure on employment structure, such as the driving and pulling mechanism. Here are the analytical results.

5.1.1 Analysis on industry structure deviation degree

In formula (3), strategic emerging industry structure deviation degree is computed as shown in Table 3.

From Table 3, it can be seen that in the period 2010–2017, the deviation degree of strategic emerging industry structure decreased from 6.056 to 2.749 with a mean of 3.846, approximate to 0 on a downtrend, revealing that with the progressive optimization of industry structure, corresponding structural changes become more rational, when the energy-saving and environmental protection industry, the new energy industry, the new material industry, and the industry structure deviation degree are negatively close to 0. The decline of the absolute value accounts for the low labor productivity of these industries. The "extrusion" thrust for internal labor force means the outward transition of industry surplus labor force. The positive value of new-generation information technology, bioindustry, the high-end equipment manufacturing industry, and the new energy automobile industry structure deviation degree shows that the high labor productivity of these industries is close to 0. The progressive decrease of value displays "absorbing" pulling effects in external labor force and transfers the labor force inward. Figure 3 presents the variation trend of the deviation degree belonging to each subdivision industry structure and employment structure. Here are the analytical results.

(1) The Pushing Force of the Energy-saving and Environmental Protection Industry, New Energy Industry, and New Material Industry in Strategic Emerging Industry Employment.

The mean of the energy-saving and environmental protection industry, new energy industry, and new material industry structure deviation degree is less than 0 to varying degrees, indicating that the labor force proportion of these industries is above their output proportion. In view of the labor redundancy phenomenon, it is necessary to orderly transfer labor force as per different proportions, thus realizing the coordinated development of industry structure and employment structure. Taking the energy-saving and environmental protection industry, for example, its industry structure deviation degree in 2010–2017 was −0.745, −0.744, −0.707, −0.700, −0.698, −0.685, −0.671, −0.679, and −0.704, respectively. The negative value of the industry structure deviation degree demonstrates that the employment proportion in the energy-saving and environmental protection

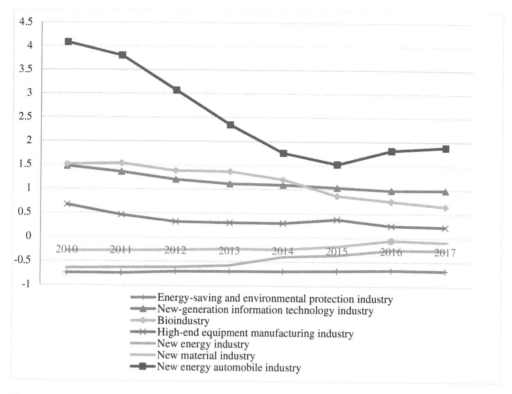

Figure 3. Variation trend of the deviation degree of seven leading strategic emerging industries in China in 2010–2017.

industry gradually exceeds its output proportion, and its deviation degree absolute value generally ranges from 0.67 to 0.75. The imbalance between industry structure and employment structure weakens the employment effects of industry structure and triggers the employment redundancy phenomenon. As capital investment is unchanged, due to the negative correlation between labor force input and marginal productivity, the greater the surplus labor force, the lower the labor productivity, the greater the absolute value of the industry structure deviation coefficient, and the stronger the "extrusion" force in labor force employment. Since 2010, the absolute value of the energy-saving and environmental protection industry structure deviation degree is around 0.7, indicating the necessity of transferring surplus labor force to other industries. Therefore, the industry structure of the energy-saving and environmental protection industry must be adjusted to giving full play to its positive pulling effects on employment.

(2) The Pulling Force of the New-generation Information Technology Industry, Bioindustry, and New Energy Automobile Industry in Strategic Emerging Industry Employment.

The mean of the structure deviation degree of new-generation information technology, bioindustry, the high-end equipment manufacturing industry and the new energy industry is above 0. That is to say, the labor force proportion of these industries is far below its output proportion.

The scarcity of labor force with varying degrees should absorb labor force as per different proportions. Consequently, the coordinated development of industry structure and employment structure is realized. Taking the new energy automobile industry for an example, the deviation degree of industry structure in the period 2010–2017 was 4.078, 3.802, and 3.067, respectively. The maximum of the deviation degree in the period 2010–2013 was 4.078, suggesting that the proportion of new energy automobile industry output was more than that of employment. The lack of employment quantity was consistent with current talent shortage in China's new energy

Table 4. Employment elasticity coefficient of seven leading strategic emerging industries in China in 2010–2017.

	Industry							
Year	Energy-saving and environmental protection industry	New-generation information technology industry	Bioindustry	High-end equipment manufacturing industry	New energy industry	New material industry	New energy automobile industry	Gross employment elasticity
2010	0.321	0.483	0.714	0.647	0.032	0.487	0.741	0.354
2011	0.119	0.616	0.405	1.201	0.291	0.404	0.345	0.312
2012	0.073	0.758	0.693	0.960	0.173	0.400	2.932	0.348
2013	0.179	0.566	0.507	0.529	0.115	0.404	2.848	0.349
2014	0.317	0.415	0.931	0.557	0.116	0.619	2.308	0.420
2015	0.117	0.316	1.805	0.400	0.375	0.280	0.976	0.332
2016	0.113	0.241	0.721	1.014	0.261	0.152	0.378	0.273
2017	0.196	0.334	0.833	0.523	0.451	0.677	0.437	0.338
Mean	0.179	0.466	0.826	0.729	0.227	0.428	1.371	0.341

automobile industry. Therefore, for the new energy automobile industry, more talents should be cultivated. In addition, sufficient talent supply should be provided for China's green low-carbon industry so as to balance industry structure and employment structure and give full play to the positive effects of industry structure variation on employment.

5.1.2 Analysis on industry employment elasticity

Formula (4) is used to compute the employment elasticity coefficient of each subdivision industry in the strategic emerging industry, as shown in Table 4.

From Table 4, it can be found that the gross employment elasticity of the strategic emerging industry fluctuates in 0.420–0.273, with a positive mean of 0.341, which manifests the strong leading role of the strategic emerging industry in employment. Variation trend demonstrates the downward trend of general industry employment elasticity and the fading pulling effects of the strategic emerging industry on employment. The continuous narrowing of employment elasticity indicates that the added value requires less labor increment, which suggests the promotion of labor productivity, while the promotion of labor productivity could be realized by two means, respectively, technical progress and the variation of industry structure and corresponding employment structure. For the strategic emerging industry, the two situations prevail, which is the cause of the reduction in gross employment elasticity. Figure 4 demonstrates the deviation degree concerning the variation of each subdivision industry structure and employment structure. Here are the analytical results.

(1) The New Energy Automobile Industry Has the Maximum Average Employment Elasticity Coefficient and the Strongest Employment Absorbing Ability.

According to Table 4, the employment elasticity coefficient of the new energy automobile industry fluctuated from 2.932 to 0.345 in the period 2010–2017, with a positive mean of 1.461, which reveals the significant pulling effects on employment. Whenever the output of the new energy automobile industry increases by 1 percentage point, corresponding employment increases by 1.461 percentage points. According to its variation trend, the employment elasticity coefficient of the new energy automobile industry first rises and then decreases in a general declining trend, with a maximum in the period 2012–2014. During the period 2010–2017, the average annual growth rate of new energy automobile industry output was 14.9%, and the average growth rate of employment was 12.7%. The results demonstrate that the development of the new energy automobile industry is not completely dependent on labor force and capital input, and technical progress is a key factor in industry development. Meantime, it also benefits from the sustained optimization and update of new energy automobile industry

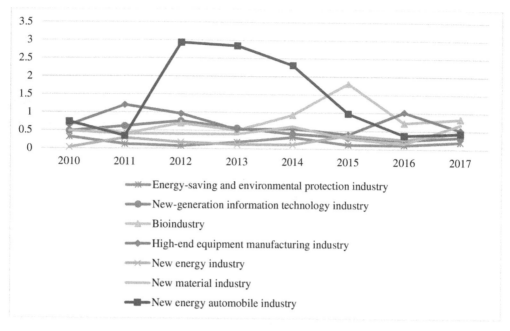

Figure 4. Variation trend of employment elasticity of seven leading strategic emerging industries in China in 2010–2017.

structure. Despite the decline of employment elasticity in the new energy automobile industry, the high value of the employment elasticity coefficient results in the sustained and fast growth of the industry scale and gross employment because the trend of automobile electrification is irresistible in the international society. Together with growing investment of research and development in new energy automobile technology and the continuous progress of new energy automobile-related technology, electronic automobiles are moving in the direction of intelligence and network connection. In China, a large group of non-automobile Internet industries and enterprises bring new thoughts, new creative ideas, new paths, and new production modes to the development of the new energy automobile industry. There is an entire industry system in China's energy automobile industry. Owing to the boom of the industry and its world-leading scale, China has become the country with the fastest new growth, the highest output, and the largest inventory of new energy automobiles (Zhang 2018).

(2) Bioindustry and the High-end Equipment Industry Are Featured with Large Average Employment Elasticity and Strong Employment Absorbing Ability.

As shown in Table 4, the employment elasticity coefficient of bioindustry and the high-end equipment industry fluctuated in 1.805–0.507 and 1.201–0.400 in the period 2010–2017, respectively. The positive figure uncovers the significant pulling effects of the two on employment. In other words, whenever bioindustry output increases by 1 percentage point, corresponding employment will increase by 0.826 percentage points. Whenever high-end equipment industry output increases by 1 percentage point, corresponding employment will increase by 0.729 percentage points. According to its variation trend, with its maximum in 2015 and 2011, respectively, the employment elasticity coefficient of bioindustry and the high-end equipment industry first increases and then decreases in a downtown trend. In the period 2010–2017, the annual average growth rate of bioindustry was 15.9%. The compound growth rate was around 20%, and the average annual employment growth rate was 11.9%. Biotechnology is promoting the green transition of medicine, agriculture, energy, and manufacturing in an all-around way, which boosts the fast growth of biological economy dominated

966

by biological medicine, biological agriculture, and biological manufacturing. The development of bioindustry in China is still prohibited by the insufficiency of high-end talents, weak ability of independent innovation, monopoly of key technologies, and core equipment parts in foreign countries. In 2010–2017, the average annual growth rate of high-end equipment industry output was 10.1%, and the employment average annual growth rate was 14.4%. The high-end equipment industry includes space equipment, intelligent manufacturing equipment, and equipment for people's livelihood. In particular, the level of the intelligent manufacturing equipment industry has become a major sign of measuring the industrialization level of a country. Intelligent manufacturing has acted as a prime trend of high-end equipment industry (Zhou et al. 2018), while the poor internationalization and commercialization, weak intelligent manufacturing equipment technology and innovation ability, poor professionalization and integration ability of equipment for people's livelihood aggravate product homogeneity in the space equipment industry of China.

(3) The New-generation Information Technology Industry and the New Material Industry Have the Poor Average Employment Elasticity Coefficient and Weak Employment Absorbing Ability.

From Table 4, it can be seen that in the period 2010–2017, with a positive mean of 0.466 and 0.428, the employment elasticity coefficient in the new-generation information technology industry and the new material industry fluctuated in 0.758–0.241 and 0.677–0.152, respectively. In this case, it reveals the pulling effects of the two on employment, yet having ordinary employment absorbing ability. Whenever the output of the new-generation information technology industry increases by 1 percentage point, corresponding employment will increase by 0.466 percentage points. Whenever the output of the new material industry increases by 1 percentage point, corresponding employment will increase by 0.428 percentage points. As shown by the variation trend, the employment elasticity coefficient of the new-generation information technology industry first increases and then decreases on a downward trend, with its maximum in 2012. In addition, the employment elasticity coefficient of the new material industry first decreases and then increases on an upward trend, with its maximum in 2017. In the period 2010–2017, the boom of information technology represented by artificial intelligence, big data, cloud computing, the Internet of Things, and the mobile Internet reinforced independent innovation ability and propelled the advancement of China's industry structure. The corresponding industry average annual growth rate was 15.5%, and the employment average annual growth rate was 7.7%. The output growth rate about twice of the employment growth rate showed the significant pulling effects of employment structure variation on employment. In the period 2010–2017, the new material industry had an average annual growth rate of 25% and the employment average annual growth rate of 9.6%. The output growth rate, which was about 2.5 times of the employment growth rate, showed the significant pulling effects of output on employment. As the leading industry in national economy, the new material industry is the foundation of other strategic emerging industries. In the meantime, great progress has been attained in new material research and development and application. Accompanied by the rise of the engineering technical level, the industry system with independent innovation ability takes shape now. At the same time, it provides formidable support for China's high-tech industry to break through technical bottlenecks and realize leapfrog development.

(4) The Energy-saving and Environmental Protection Industry and the New Energy Industry Are Featured with the Minimum Average Employment Elasticity Coefficient and the Weakest Employment Absorbing Ability.

Table 4 shows that in the period 2010–2017, the employment elasticity coefficient in the energy-saving and environmental protection industry fluctuated in the range of 2.932–0.345, with a positive mean of 0.179 and 0.227, respectively. In this case, the poor pulling effects of the two on employment can be seen. Whenever the energy-saving and environmental protection industry increases by 1 percentage point, corresponding employment will increase by 0.179 percentage points. Whenever the new energy industry increases by 1 percentage point, corresponding employment will increase by 0.227 percentage points. As revealed by the variation

Table 5. Seven leading strategic emerging industries and employment regression results in China in 2010–2017.

Industry	Regression formula	R^2	F value
Energy-saving and environmental protection industry	$Y_1 = 0.262 - 0.270\,L_1$	0.814	26.192
New-generation information technology industry	$Y_2 = 0.651 - 0.511L_2$	0.040	0.247
Bioindustry	$Y_3 = 0.119 - 0.030L_3$	0.011	0.067
High-end equipment manufacturing industry	$Y_4 = 0.090 + 0.163\,L_4$	0.043	0.266
New energy industry	$Y_5 = -0.107 + 2.870\,L_5$	0.701	14.060
New material industry	$Y_6 = -0.063 + 1.654L_6$	0.910	60.981
New energy automobile industry	$Y_7 = 0.025 - 0.065L_7$	0.001	0.006

trend, the employment elasticity coefficient of the energy-saving and environmental protection industry first increases and then decreases on a downward trend, with its maximum in 2014. The employment elasticity coefficient of the new energy industry first decreases and then increases on an upward trend, with its maximum in 2017. In the period 2010–2017, the average annual growth rate and the employment average annual growth rate in the energy-saving and environmental protection industry was 16.5% and 2.7% accordingly. The output growth rate was three times more than that of the employment growth rate. Due to its extensive market prospects, especially as of 2017, such as favorable policy environment, softening industry structure and diversified formats of industry develop toward the comprehensive service industry and the energy-saving and environmental protection industry, which are expected to be the pillar industry of the national economy in China. However, considering its heavy reliance on policy, scarcity of primitive innovation, intense market competition, and lack of capital, much effort should be made to improve the current situations of the energy-saving and environmental protection industry. Since the period 2010–2017, the average annual growth rate and the employment average annual growth rate in the new energy industry were 30% and 6.4% separately. The output growth rate is 4.5 times more than that of the employment growth rate. Regarding the inadequate instructions of new energy in policy and technical research and development, technical development bottleneck problems are foreseeable to be broken through.

5.2 Role of strategic emerging employment structure in industry structure

To classify the functional mechanism of strategic emerging industry employment structure in industry structure, in this study, a bivariate regression model is constructed to discuss the influence of strategic emerging industry employment structure on industry output. Taking the proportion of employment quantity in seven leading subdivision industries as the independent variable, and the industry output proportion as the dependent variable, a regression model is built to explain the influence of employment structure on industry structure. The output proportion of seven leading subdivision industries is expressed as $Y_1, Y_2, Y_3, Y_4, Y_5, Y_6$, and Y_7, respectively, and the industry employment proportion is shown as $L_1, L_2, L_3, L_4, L_5, L_6$, and L_7, accordingly. The regression results based on SPSS19.0 are presented in Table 5.

(1) Significant Pulling Effects of the Rising New Energy Industry and New Material Industry Employment Proportion on Industry Output.

According to Table 5, it can be found that there exists a positive linear regression relation between the employment and industry output of the new energy industry and the new material industry, with R^2 above 0.4. Specifically, whenever the new energy industry employment proportion increases by 1 percentage point, the corresponding industry output proportion approximately increases by 2.87 percentage points. The employment proportion could be explained as 70% of output proportion variation. For the new material industry, whenever the employment proportion increases by 1 percentage point, the corresponding output proportion approximately increases by 1.654 percentage points, while the employment proportion could be considered as 91% of output proportion variation. Thus, it can be seen that growing investment in labor force can lead to the growth of new energy industry and new material industry output, indicating that at the increasing return stage, the existing technical level and growing investment in labor force can generate immediate economic benefits, and reveal the low employment threshold and large capacity characteristics of the new energy industry and the new material industry.

(2) Significant "Extrusion" Effects of the Rising Employment Proportion of the Energy-saving and Environmental Protection Industry on Industry Output.

Table 5 suggests a reverse linear regression relation between employment and output in the energy-saving and environmental protection industry. $R^2 > 0.4$ means that whenever the energy-saving and environmental protection industry employment proportion increases by 1 percentage point, corresponding output will decrease by 0.270 percentage points. Affected by the labor force redundancy phenomenon in the industry, surplus labor force spontaneously transfers to other industries. Combined with the scatter diagram, it is obvious that no linear growth appears in the output proportion with the growing employment proportion in the energy-saving and environmental protection industry. However, there is a decrease first and then an increase, showing consistence with the variation trend of the energy-saving and environmental protection industry's employment elasticity and industry structure deviation degree. Consequently, it is necessary to adjust the employment structure of the energy-saving and environmental protection industry, simultaneously raising its labor productivity.

(3) Insignificant "Extrusion" Effects of the Rising Employment Proportion in the New-generation Information Technology Industry, Bioindustry, High-end Equipment Manufacturing Industry, and New Energy Automobile Industry on Industry Output.

By reference to Table 5, there is no linear relation between the employment structure and industry output in four strategic emerging industries, including the new-generation information technology industry, bioindustry, high-end equipment manufacturing industry, and new energy automobile industry. Regression formula $R^2 < 0.4$ has no statistical significance. Especially, for the new-generation information technology industry and the new energy automobile industry, the influence of the labor force employment proportion on industry output is insignificant. Therefore, it is proved that the growth of emerging industry output mainly depends on technical progress, capital investment, operation level promotion, laborer quality, application of advanced technology, and full-factor productivity.

6 CONCLUSION

To conclude, the variation trend of China's strategic emerging industry structure and employment conforms to the general law of industry development, and the coupling liaison effects of the two exert influence on China's social employment and economic growth to a great extent. Based on China's strategic emerging industry development data in the period 2010–2017, this study quantitatively computes the industry structure and employment structure variation indicator, analyzes variation characteristics, applies the industry structure deviation degree and employment elasticity to observe the influence of emerging industry structure variation on employment, and employs the bivariate regression formula to observe the role of employment structure in industry structure. The following conclusions are made.

(1) China's strategic emerging industry structure and employment structure experience dynamic changes. In particular, the industry structure variation direction is the same with the GDP variation direction, while the employment structure variation direction is the same with national employment quantity. Additionally, the industry structure and employment structure variation rate is quite unstable.

(2) China's strategic emerging industry structure and employment structure deviation degree is generally downward and close to 0, which proves that along with the gradual optimization of industry structure, structure changes turn more rational, and demonstrate strong pulling effects on employment. In particular, the mean of the energy-saving and environmental protection industry, new energy industry, and new material industry structure deviation degree is less than 0 with less employment elasticity. In this case, pushing effects on employment exist. For new-generation information technology, bioindustry, and the high-end equipment manufacturing industry, the structure deviation degree mean is above 0. High employment elasticity generates pulling effects on employment, while the new energy automobile industry structure deviation degree is featured with the maximum mean and employment elasticity, generating significant employment pulling effects.

(3) The role of the variation concerning the subdivision strategic emerging industry employment proportion in industry output is different in inspected years. Particularly, the varying employment proportion of the new energy industry and the new material industry has pulling effects on industry output, while the rising employment proportion of the energy-saving and environmental protection industry exerts significant extrusion effects on industry output. The rising employment proportion of the new-generation information technology industry, bioindustry, high-end equipment manufacturing industry, and new energy automobile industry insignificantly influences industry output variation.

Consequently, it is essential to accurately explore problems existing in China's strategic industry structure from the source. Then, it is suggested that the government should formulate strategic emerging industry development plans, enact rational countermeasures, optimize and adjust industry structure, guide the orderly flow of labor force and promote the dynamic equilibrium and coordinated development between strategic emerging industry structure and employment structure to give play to the pulling effects of the strategic emerging industry on employment according to its practical conditions.

ACKNOWLEDGMENTS

This research is supported by the local think tank project, NoW2019JSFW0470. The project title is "Research on Dynamic Analysis Method and Development Strategy of Corporate Strategic Environment."

REFERENCES

Cai, D. & Li, D. 2010. Analysis of the Influence of Economic Growth on Gross Employment Quantity and Structure – Taking Liaoning for Example. Chinese Market (14): 40–42.

Chenery, H. B., Elkington, H. & Sims, C. 1970. A Uniform Analysis of Development Pattern, Cambridge Mass, Harvard University Center for International Affairs.

Colin, C. 1940. The Conditions of Economic Progress, London, Macmillan.

Dong, X. 2013. Study on Vocational Education Hierarchy in China, Tianjin, Doctoral Dissertation of Tianjin University.

Drucker, J. 2013. An Evaluation of Competitive Industrial Structure and Regional Manufacturing Employment Change. 49: 1–16.

Hu, H. 2015. Research on the Coupling Mechanism of Industry- university- research Collaborative Innovation. Science and Technology Management Research 35: 26–29.

Huang, X.H. & Zhang, S.L. 2019. Selection of the Development Path of Chinese Strategic Emerging Industry: Market Incentives in Great Powers. Chinese Industrial Economics (11): 60–78.

Jiang, L., Li, Y.Q. & Dong, W.C. 2018. Study on the Interactivity and Covariance of Chinese Higher Education Structure and Industry Structure – From the Perspective of System Coupling System. Education Science 34: 59–66.

Kowalewski, J. 2011. Specialization and employment development in Germany: An analysis at the regional level 90: 789–811.

Kuznets, S. 1985. Economic Growth in All Countries Beijing, Commercial Press.

Lewis, W.A. 1954. Economic Development with Unlimited Supplies of Labour. 22: 139–191.

Li, B. J., Liu, J. & Yang, X. 2006. Multi-index Analysis on Industry Structure Gap in Henan. Statistics and Decision (9): 67–69.

Li, D. J. & Zhang, H. 2013. Relation and Cause Analysis of Industry Structure and Economic Growth in Beijing. Journal of Northeastern University (social science edition) 15: 148–153.

Liu, P. 2010. Analysis on the Coupling Mechanism of Economic Growth, Industry Development and Labor Employment in China – Based on the Dynamic Empirical Analysis of VAR Model. Economic Issue (4): 24–28.

Luo, G.X. 2000. Variation in Economic Growth, Labor Production, Industry Structure and Employment Struc-ture Quantitative and Technical Economics 26–28.

Lv, Z. 2019. Review about the Follow-up Study on the Cultivation of Strategic Emerging Industry in China. Management World (4): 189.

Petty, W. 2010. Political Algorithm Beijing, China Social Sciences Press.

Shi, G.Y. 2014. Industry Restructuring from the Perspective of Rural Labor Transfer. Modern Economic Information (8): 17–22.

Sun, Q., Han, P. & Ding, Y. Y. 2019. Coordinated Research on Industry Employment Elasticity, Structure Deviation Degree and Comparative Labor Productivity. Statistics and Decision 129–133.

Syrquin, M. & Chenery, H. 1989. Three Decades of Industrialization. The World Bank Economic Review 3: 145–181.

Tan, J.H. 2013. Economic Growth, Industry Structure and Labor Employment: Empirical Economic Problems from China On Economic Problems 55–57.

Teng, Y., Li, T.X. & Yu, Z.Q. 2016. Analysis on Factors and Potentials in Chinese Employment Variation. China Soft Science, 33–42.

Wang, H. H. 2005. Exploration about the Evolutionary Law and Trend of Industry Structure. Gansu Social Sciences (5): 221–225.

Wang, Q.F. & Dang, Y.G. 2010. Moore Value-based Computing of Chinese Employment Structure Retardation Time. Management Review 22: 3–7.

Wang, X. & Tian, X.Z. 2012. Literature Review for the Interactivity between Human Capital and Industry Structure. Journal of Chongqing Technology and Business University (social science edition) 29: 28–34.

Wu, J. & Feng, X. 2006. Analysis on the Dynamic Relation between Industry Structure and Employment Structure in Si-chuan. Finance Sciences (7): 102–109.

Zhang, C.W. 2002. Study on the Variation Trend of Employment Elasticity. Chinese Industrial Economics (5): 22–30.

Zhang, Y. 2018. Report for the Development of Chinese Electric Vehicles. Beijing.

Zhang, Y., Wu, J., Xiao, C., Zhang, K. & Liu, T. 2012. Structural Deviation Degree Based on the Interaction between Emerging Industry Structure and Personnel Structure. Science and Technology Management Study 32: 121–125.

Zheng, X. 2012. Industry Structure and Economic Growth-Exploration of the Problems in the Development of Chinese Strategic Emerging Industry Beijing, Party School of the Central Committee of CPC.

Zhou, J., Li, P., Zhou, Y., Wang, B., Zang, J. & Meng, L. 2018. Toward New-Generation Intelligent Manufacturing. Engineering 4: 11–20.

Zhou, J. A. 2006. Grey Correlation Analysis of Chinese Industry Structure Upgrade and Employment. Finance Theory and Practice 27: 94–98.

Computational Social Science – Luo, Ciurea & Kumar (eds)
© *2021 Taylor & Francis Group, London, ISBN 978-0-367-70193-2*

Research on poverty reduction effect of health insurance schemes on multidimensional poverty of agricultural migrants

X.J. Lu & Y.N. Wang
Faculty of Humanities & Social Science, Dalian University of Technology, Dalian, China

ABSTRACT: The particularity of the agricultural migrants determines that they can participate in various types of health insurance schemes. Based on the data from 2017 wave of China Migrants Dynamic Survey implemented by the National Health Commission, this paper empirically analyzes of the poverty reduction effect of different types of health insurance schemes on the multidimensional poverty of agricultural migrants. The results show that Urban Employees Basic Medical Insurance (UEBMI) has the strongest poverty reduction effect, followed by Urban Residents Basic Medical Insurance (URBMI) and Basic Medical Insurance for Urban and Rural Residents (BMI-URR), and the New Rural Cooperative Medical System (NCMS) has the weakest poverty reduction effect. The four types of health insurance schemes can effectively reduce income poverty, education poverty, and social exclusion, but only UEBMI can significantly reduce health poverty. Based on the analysis above, this paper puts forward relevant suggestions on the reform of health insurance system.

1 INSTRUCTION

The number of agricultural migrants in China has reached 290 million in 2019. Such a large scale of agricultural migrants, while promoting economic and social development of cities, have also suffered from various deprivations, including lower income, poor working conditions, bad health, limited empowerment, and poor education chances. As a new urban poverty group, medical risk is an important cause of poverty for them. One of the fundamental stated goals of health insurance is to reduce the impact of health shocks on individual income and to reduce poverty due to illness for overall population. However, there are obvious urban-rural divisions and occupational differences between different types of health insurance schemes in terms of coverage and standard of treatment. Therefore, analyzing and comparing the effects of different types of health insurance schemes on the multidimensional poverty reduction of agricultural migrants is essential to promote the reform of the health insurance schemes and bring into play the practical effects of medical poverty alleviation.

1.1 *Literature review*

According to Amartya Sen's theory of feasible capacity, poverty is not only income poverty, but also other objective poverty and subjective feeling of welfare (Sen 1976). Scholars believe that multidimensional poverty mainly includes health, education and living standards of individuals or families (Alkire & Santos 2014; Guo & Zhou 2016), which are also the main indicators applied by the United Nations Development Program to evaluate pro-poor policies and measure multidimensional poverty. Agricultural migrants work and live in the city, but their registered permanent residence is in rural areas. Affected by institutional constraints and structural exclusion, it is difficult for agricultural migrants to enjoy all the public services in the cities in which they worked, which also restricts the process of the citizenization of agricultural migrants (Yang 2015; Gao et al. 2018). Therefore, the measurement of multidimensional poverty of agricultural migrants should not ignore the special aspect of social exclusion (Li & Zhang 2019).

The poverty reduction effect of health insurance schemes is mainly through lowering the economic threshold for medical treatment, increasing the application of outpatient and inpatient services to improve health status of agricultural migrants, and reducing personal medical expenses or the burden of family care (Filipski et al. 2015; Wang et al. 2018). The improvement of health status can make up for the lost labor time due to disease, and expands income by increasing labor supply and labor efficiency. Better health status can also help to release more preventive reserves and promote investment in human capital and physical capital, so that agricultural migrants can obtain more income (Kochar, 2004; Song & Song, 2018).

However, few studies have examined the impact of health insurance schemes on the multidimensional poverty of migrant workers. Because of their particularity, agricultural migrants can participate in four types of different public health insurance schemes, namely, New Cooperative Medical Scheme (NCMS), Urban Employees Basic Medical Insurance (UEBMI), Urban Residents Basic Medical Insurance (URBMI), and Basic Medical Insurance for Urban and Rural Residents (BMIURR). Whether distinct types of health insurance have different poverty reduction effects needs to be examined through empirical data. Therefore, this paper intends to analyze and compare the effects of different types of health insurance schemes on the multidimensional poverty reduction of agricultural migrants through empirical analysis.

1.2 *Our contribution*

The contributions of this study are as follows: first, from the perspective of multiple deprivations, a multidimensional poverty measurement system covering the characteristics of agricultural migrants is constructed; second, the poverty reduction effect of different types of health insurance schemes on the overall situation and each dimensions of multidimensional poverty for agricultural migrants are examined; third, based on the conclusion of research, relevant suggestions are put forward for improving the public health insurance system.

1.3 *Paper structure*

The rest of the paper is organized as follows. Section 2 introduces design of the research, including the measuring method of multidimensional poverty, index selection, data sources, and information of samples. Section 3 presents the descriptive statistical results and the logit regression analysis results of health insurance schemes on the multidimensional poverty of agricultural migrants. And the related discussion about the results is also provided at the end of this section. Section 4 summarizes the research conclusions and puts forward some relevant suggestions.

2 RESEARCH DESIGN

2.1 *Selection of multidimensional poverty index and measuring method*

The dependent variable of this research is multidimensional poverty. Combined with previous studies, this paper constructs a multidimensional poverty measurement system of agricultural migrants from four dimensions: income poverty, health poverty, education poverty, and social exclusion. The multidimensional poverty index M_0 is carried out using the A-F double limit analysis method proposed by Alkire and Foster (2011). The formula is expressed as follows:

$$M_0 = \frac{1}{n} \sum_{j=1}^{d} c_i(k) \tag{1}$$

where the weighted score of individual i on all d indicators is the total weighted deprivation score of individual i, equaling to c_i. In poverty identification, two critical values need to be defined. The first is the z critical value used to determine whether an individual is deprived on a

Table 1. Multidimensional poverty indicator system and weights.

Indicators	Variables	Critical Value z	weights
Income poverty	Absolute income poverty	If the per capita annual net income of a family is lower than the national poverty line standard, a value of 1 is assigned; otherwise it is 0.	1/8
	Relative income poverty	Based on 50% of the per capita disposable income of urban residents in each city, if the per capita annual income of the family is lower, a value of 1is assigned; otherwise it is 0	1/8
Health poverty	Self-assessment of health state	Unhealthy, or individuals who cannot take care of themselves, are assigned a value of 1; and healthy, basic health state is assigned a value of 0.	1/4
Education poverty	Education background	Elementary school and below are assigned a value of 1, otherwise 0	1/4
Social exclusion poverty	Feeling discriminated by local people	Basically, agree and completely agree with a value of 1; completely disagree or disagree with a value of 0	1/4

particular index. Then, g_{ij} is equal to the deprived state of the individual i on the index j. If g_{ij} is lower than the critical value z_j, that is, the individual is deprived on a specific index, and $g_{ij} = 1$. Otherwise, it will be regarded as not deprived, with $g_{ij} = 0$. The second is the k critical value used to compare the degree of deprivation of c_i to determine individuals' multidimensional poverty status. If $c_i \geq k$, indicating that the individual suffers more than the tolerance range, the individual i will be regarded as multidimensional poverty; otherwise, it will not be regarded as multidimensional poverty. According to international practice, $k = 1/3$ is generally selected.

The specific indicators of multidimensional poverty are shown in Table 1.

2.2 Independent variable and control variable

The key independent variables are expressed in four dummy variables to indicate whether agricultural migrants participated in NCMS, UEBMI, URBMI, or BMIURR Based on the previous studies on the influencing factors of agricultural migrants' poverty, the control variables of this research are selected from the population and migration characteristics that affect agricultural migrants' poverty, including gender, age, marital status, range of migration, and employment status.

2.3 Dataset

We used the data from the 2017 wave of China Migrants Dynamic Survey implemented by the National Health Commission. Agricultural migrants in this study are defined as the people whose household registration is in the rural areas but are living in the urban areas for more than one month. In order to measure the poverty reduction effect of a single type of health insurance scheme effectively, we ruled out those who had more than one health insurance and included 95,032 individuals in the final sample.

2.4 Characteristics of sample

In order to eliminate the influence of outliers, Winsorize tailing is applied for continuous variables in this study, and the ratio is set to 1%. The age-bracket of the samples is 18 to 61 years old, with an average age of 36.13 years old. Among them, men accounted for 57.5%, while women accounted for 42.5%. Most of them are married, which accounted for 81.57% in the whole. In terms of education status, agricultural migrants are generally not highly educated, with a middle school education and below accounting for 68.33%, and college degree and above accounting for only 10.9%. About 49.11% of agricultural migrants are moving within the province. In terms of employment status, more than half of them are employees, while 37.98% of them are self-employed.

3 RESULTS

3.1 *Multidimensional poverty status of agricultural migrants*

Based on formula (1), it is calculated that 15.69% of agricultural migrants are in a multidimensional poverty, the index of which is 0.0733. Among them, 9,936 people are in absolute income poverty, and the incidence is 10.46%; whereas 26,810 people are in relative income poverty, and the incidence is 28.21%. With the development of the economy, the population of absolute poor has been gradually decreased, and income poverty has changed from absolute poverty to relative poverty. The incidence of health poverty is the lowest among other indicators. There are only 1,449 agricultural migrants bothered by health poverty, accounting for 1.52%. The education poverty and social exclusion poverty of agricultural migrants cannot be ignored. There are 17,506 people in education poverty, accounting for 18.42%; 16,927 people suffer from social exclusion poverty, accounting for 17.81%. Helping agricultural migrants get rid of education poverty and social exclusion has become an important way to alleviate multidimensional poverty.

3.2 *Status of agricultural migrants participating in health insurance*

The results show that 74.89% of agricultural migrants participate in the NCMS, while 2.84% of them participate in the URBMI and 13.29% of them participate in UEBMI. It can be found that the proportion of agricultural migrants participating in urban medical insurance is still low, and the dual division of urban and rural household registration system is still seriously the welfare of agricultural migrants to enjoy urban basic health insurance. The process of promoting the integration of urban and rural residents' health insurance in China is still slow. There are only 3.15% of agricultural migrants participating in BMIURR.

3.3 *The impact of health insurance schemes on multidimensional poverty*

This research applies the Logit regression model to analyze the poverty reduction effects of four types of health insurance schemes on multidimensional poverty. The results are shown in Table 2. The results show that on the basis of controlling individual characteristic variables, the four basic types of health insurance schemes have significant poverty reduction effect. The incidence of multidimensional poverty among NCMS, URBMI, BMIURR, and UEBMI is 0.87, 0.49, 0.69, and 0.37 times higher than that of agricultural migrants without any health insurance. The four types of health insurance schemes can effectively reduce income poverty, education poverty, and social exclusion, but only UEBMI can significantly reduce health poverty.

3.4 *Discussion*

UEBMI has the strongest poverty reduction effect among the four types of health insurance, which can be explained from the level of financing, the scope of insurance coverage, and the level of benefits. On the one hand, UEBMI is paid according to the proportion of individual income and unit benefit, and its total financing level and per capita financing level are far higher than other basic health insurance schemes. On the other hand, compared with other health insurance schemes, which implement policies of low economic threshold and protecting against serious illnesses, UEBMI can be paid for serious illness and hospitalization expenses through social pooling funds. Besides, outpatient and out-of-pocket expenses can be paid through personal accounts. The proportion of

Table 2. Logit regression results of health insurance on poverty reduction of agricultural migrants.

	Multidimensional poverty	Absolute income poverty	Relative income poverty	Health poverty	Education poverty	Social Exclusion Poverty
NCMS (base group: uninsured)	−0.1371*** (0.0380)	−0.2254*** (0.0443)	−0.0733** (0.0317)	−0.0938 (0.1127)	−0.1649*** (0.0372)	−0.0754** (0.0350)
URBMI (base group: uninsured)	−0.7117*** (0.0718)	−0.2824*** (0.0763)	−0.5799*** (0.0568)	0.0757 (0.1731)	−0.6864*** (0.0673)	−0.4551*** (0.0654)
BMIURR (base group: uninsured)	−0.3635*** (0.0649)	−0.1423** (0.0705)	−0.2931*** (0.0528)	−0.1465 (0.1771)	−0.2691*** (0.0613)	−0.2842*** (0.0613)
UEBMI (base group: uninsured)	−1.0025*** (0.0510)	−0.5262*** (0.0553)	−0.5713*** (0.0383)	−0.6257*** (0.1587)	−1.3978*** (0.0533)	−0.4130*** (0.0428)
Gender (base group: female)	−0.3948*** (0.0189)	−0.0183 (0.0218)	−0.0271* (0.0151)	−0.4263*** (0.0543)	−0.7049*** (0.0187)	−0.1018*** (0.0174)
Age	0.0635*** (0.0010)	0.0189*** (0.0012)	0.0239*** (0.0008)	0.1033*** (0.0029)	0.0976*** (0.0010)	0.0109*** (0.0009)
Marital Status (base group: unmarried)	0.5040*** (0.0336)	0.2644*** (0.0341)	0.9970*** (0.0258)	−0.1481 (0.0914)	0.0257 (0.0313)	0.0843*** (0.0254)
Migration Range (base group: inter-provincial migration)	−0.2901*** (0.0188)	0.5429*** (0.0219)	−0.1629*** (0.0149)	0.3194*** (0.0543)	−0.2763*** (0.0185)	−0.3678*** (0.0173)
Employment status (base group: employee) Employer	−0.8211*** (0.0514)	−0.3015*** (0.0545)	−1.0131*** (0.0400)	−0.3632** (0.1486)	−0.5969*** (0.0466)	−0.1498*** (0.0395)
Self-employed	−0.1924*** (0.0199)	0.1756*** (0.0234)	−0.2017*** (0.0162)	0.0230 (0.0569)	−0.1277*** (0.0195)	−0.0728*** (0.0190)
Others	0.2564*** (0.0667)	0.5496*** (0.0691)	0.1912*** (0.0550)	0.4061** (0.1655)	0.1222* (0.0691)	0.0930 (0.0638)
Constant	−3.85605*** (0.0561)	−3.1773*** (0.0612)	−2.3045*** (0.0439)	−8.0769*** (0.1803)	−4.4287*** (0.0550)	−1.6087*** (0.0467)

Note: Standard errors in parentheses; *$p < 0.1$, **$p < 0.05$, ***$p < 0.01$.

UEBMI fund expenditure in total health expenditure is also much higher than that of URBMI and NCMS (Wang et al. 2019). The NCMS has the lowest poverty reduction effect, which is mainly due to the lowest average reimbursement rate. Compared with different inpatient reimbursements, only the town hospitals have a high reimbursement rate, while the ratio of urban hospitals at all levels is lower. In addition, agricultural migrants work in other places all year-round, and the reimbursement procedures are cumbersome, which will affect the poverty reduction effect of NCMS (Table 2).

It is concluded that only UEBMI has a significant impact on health poverty, which is consistent with the previous studies. On the one hand, it may be because the benefit level of UEBMI is the highest among other. On the other hand, the health status of the agricultural migrants who participate in UEBMI is better (Zou & Liu 2016; Chen & Deng 2016). All types of health insurance have significant poverty reduction effect on education poverty, which may be because health insurance can reduce the medical expenses of agricultural migrants, and then more investment can be applied in the reinvestment of their human capital, such as taking correspondence courses to improve their academic qualifications. As for the social exclusion, all types of health insurances have significant poverty reduction effects. As an institutional security for agricultural migrants, health insurance is also a subjective welfare feeling, which helps to reduce social exclusion. Positive subjective welfare feeling is the endogenous driving force for agricultural migrants to get rid of multidimensional poverty, and it is also a higher requirement for poverty alleviation.

4 CONCLUSIONS

The result of this research shows that all four types of health insurance can reduce the multidimensional poverty of agricultural migrants. UEBMI has the strongest poverty reduction effect, followed by URBMI and BMIURR, and NCMS has the weakest poverty reduction effect. The four types of health insurance schemes can effectively alleviate income poverty, education poverty, and social exclusion, but only UEBMI can significantly alleviate health poverty. Therefore, it is recommended that the government should provide more policy incentives to encourage enterprises to include more agricultural migrants in the UEBMI. Secondly, as the main medical insurance for agricultural migrants, it is necessary to improve the portability of the NCMS. Besides, the benefit level of NCMS should be further expanded and the personal co-payment ratio in the payment of medical funds should be further reduced. Finally, it is recommended that the preferential medical subsidies for the poor should be further increased. Furthermore, lower economic threshold of medical insurance will be better for the welfare of the whole society and will encourage low-income groups to make more use of health insurance.

ACKNOWLEDGMENT

This work is supported by the National Social Science Foundation of China (Grant No.17BGL169)

REFERENCES

Alkire, S. & Foster,J. 2011. Counting and multidimensional poverty measurement. Journal of Public Economics. 95(7–8):476–487.

Alkire, S. & Santos, M. E. 2014. Measuring acute poverty in the developing world: Robustness and scope of the multidimensional poverty index. World Development 59(1):251–274.

Chen, H. & Deng P. 2016. Health effect evaluation of the Urban Employee Basic Medical Insurance. Social Security Studies 4:44–52.

Filipski, M. Zhang, Y. & Chen, K.Z. 2015. Making health insurance pro-poor: Evidence from a household panel in rural China. Health Services Research 15:210–223.

Gao, S. Guo, C. & Zhang, Q. 2018. Social exclusion, human ritual expenses and multidimensional poverty alleviation of migrant workers. Finance & Economics 6:110–120.

Guo, X. & Zhou, Q. 2016. Chronic multidimensional poverty, inequality and causes of poverty. Economic Research Journal 6:143–156.

Kochar, A. 2004. Ill-health, savings and portfolio choices in developing economies. Journal of Development Economics 73(1):257–285.

Li, H. & Zhang Z. 2019. Measurement and decomposition of multidimensional poverty of floating population. Inquiry Into Economic Issues 5:182–190.

Sen, A. 1976 Poverty: An ordinal approach to measurement. Econometrica 44(2):219–231.

Song, Y. & Song Z. 2018. The promoting effect and its mechanism of health insurance on the migrants' consumption. Population & Economics 3:115–126.

Wang, W. Zhou, Y. & Hou Q. 2019. The equity of China's basic health insurance system. Journal of Yunnan Agricultural University(Social Science) 13(2):65–70.

Wang, X. Huang, D. & Pu P. 2018. The dual effects of health insurance and the medical expenditure of residents:Mechanistic analysis and empirical research. Modern Economic Science 40(5):1–11.

Yang, J. 2015. Research on the social integration of floating population in China. Social Sciences in China 2:61–79.

Zou, H. & Liu, Y. 2016. Heterogeneous health insurance, self-expense medical and the elderly health. Finance & Economics 6:112–123.

Computational Social Science – Luo, Ciurea & Kumar (eds)
© 2021 Taylor & Francis Group, London, ISBN 978-0-367-70193-2

Brexit: A challenge or an opportunity for China?

H. Zeng
Institute of Public Policy, South China University of Technology, Guangzhou, P. R. China

ABSTRACT: Besides the Covid-19 pandemic, Brexit could be the most far-reaching black swan event in the 21st century, with a tremendous impact on not only the UK, but also on the rest of the world. China is one of the countries affected by Brexit. While academics tend to see Brexit through the lens of European politics, relatively little has been written on the consequences for China. This article employs a holistic vision to explore the impact on the China-UK relationship. The break with the European Union (EU) is reshaping the China-UK relationship in multiple dimensions: China-UK trade, China-EU trade, the Belt and Road Initiative (BRI), RMB internationalization, and studying in the UK. This article examines each of these aspects in turn, before concluding with a menu of options in response to Brexit.

1 INTRODUCTION

The China-UK relationship is of significant relevance to the ongoing discourse of Brexit. In a renewed golden era of political relations, investments between the two nations are rising rapidly and political links expanding. Although Boris Johnson's administration succeeded in leading the UK through Brexit after four years of lengthy negotiations, the future UK-EU relationship remains uncertain. At the heart of this analysis are five distinct impacts. For each, we consider in turn the impact on China and what measures China should take to respond to Brexit. While the uncertainty over Brexit would be bad for business, China should embrace the challenge and search for new opportunities that might arise from it.

2 CHINA-UK TRADE

The vote to leave the EU can have profound effects on both China and UK-based Chinese firms in the short and long run. However, the magnitude of impact remains unclear, depending on the eventual outcome of Brexit. The potential trade impact may affect a range of dimensions such as trade, migration, and regulation.

Following the passing of the Brexit bill through the UK Parliament in January 2020, the center of scholarly attention shifted to the country's future international standing. Though Brexit is being carried out in a smooth and orderly way, foreign businesses take a dim view of the UK's economic prospects. Chinese businesses in the UK are increasingly concerned about whether the UK will be able to secure a deal with Brussels over its continued access to the European Single Market. A statement by Beijing's foreign ministry reads that "China hopes to see a prosperous Europe and a united EU, and hopes the UK, as an important member of the EU, can play an even more positive and constructive role in promoting the deepening development of China-EU ties" (Gov.cn 2015). Foreign direct investment (FDI) into the UK has been on a consistent decline since the EU referendum in 2016. For fear of a no-deal Brexit, foreign businesses have either put their plans on hold or are moving their assets away from the UK. A growing body of evidence suggests that Brexit has damaged business confidence. According to the Sandford Institute for Economic Policy Research, Brexit has trimmed British business investment by 11% and British productivity by 2 to

5% in the three years since the EU referendum (Bloom et al. 2019). The EY Financial Services Brexit Tracker adds that, in 2018, 41% of financial service firms had plans to move operations out of the UK to continental Europe, up from 31% the previous year (EY 2020). Figures released by Baker McKenzie show that Chinese FDI into the UK fell by 76% in 2018 compared to the previous year (Baker McKenzie 2019).

By the same token, the British economy will enter a slower growth trajectory. The most pessimistic scenario is that Brexit will reduce the size of the British economy (London.gov.uk 2018). The UK is expected to sustain a marginal decline in investment in the 10 years following Brexit. Growth in total investment would fall to 1% from 2021 to 2030. This is in line with a study that finds FDI inflows into the UK in 2020 are a quarter lower than the years in the run-up to the EU referendum (PwC 2016).

For years, the role of London as a global financial hub has attracted foreign businesses from across the world to set up their headquarters in the UK. Most Chinese investors conduct business operations in London, mainly engaged in FDI and UK-wide infrastructure projects. For the past decades, the UK has served as a magnifier of Chinese influence in Europe. If the UK no longer enjoys the freedom of movement in goods and people as a consequence of Brexit, trade costs could rise. Any unfavorable changes in investment policy might force Chinese businesses that use the UK as a launchpad from which to plug into the EU to cease investing in the UK or to withdraw their capitals.

However uncertain the post-Brexit China-UK trade relationship may become, the UK government is well aware of China's potential as a trading partner. This is exemplified most clearly by the remarks of former British Prime Minister Theresa May at the 2016 G20 Hangzhou summit. In a statistics report published by the House of Commons, British exports of goods and services to China were worth £30.7 billion, up from £23.4 billion in 2018 (Matthew 2020).

The conventional wisdom in the West has it that Brexit will do more harm than good for the UK in terms of trade and capital flows. According to Wu Zhicheng, director of the Institute of Global Issues, Nankai University: "a complete exit from the European Single Market would result in trade barriers and a decline in investment and trade flows, to the detriment of the UK-EU trade relationship" (Wu 2019). But, amid internal differences among EU members, leaving the EU is not necessarily a bad option. The post-Brexit UK will no longer be subject to EU institutional arrangements and so will regain its power to set its own tariffs and trade agreements without involving EU institutions and other EU members, which might create new possibilities for China to strike deals that it could not with the UK being an EU member.

3 CHINA-EU TRADE

It is clear that the UK needs to find new markets to offset post-Brexit impact. While the UK is still part of the Customs Union during the two-year transition period, where it continues to abide by EU economic rules and enjoy full access to the European Single Market, British companies would face non-tariff barriers due to regulatory divergence. Such barriers would raise trade costs and prompt British investors to divert investment to non-EU markets. There will be no free movement of labor after the transition period. This might generate a sharp decline in capital flows and leave a window of opportunity for China to potentially push for a free trade agreement with the UK. Among emerging economies, China, along with its rising economic clout and international aspirations, is a far more lucrative market to the eyes of the British. A China-UK free trade agreement could be a win-win deal for both camps. It would intensify the UK's trade flows with China and give China more leverage in trade talks (Ren & Bi 2017). But it does not follow from this that the UK will go for a clean break with the EU. Knowing that the corrosive effects of Brexit will take years to manifest, the UK will maintain the closest ties possible with the EU as it negotiates its future relationship with the EU.

A UK outside the EU's corridors of power will definitely have less to offer China. The trade barriers caused by Brexit might elevate the EU's dependence on China for trade (Ren & Bi 2017).

The EU will remain a less attractive partner until a bilateral trade agreement has been agreed upon by the EU and the UK. Negotiating a new relationship of this nature, and ensuring rights and interests of both sides will take five to seven years and require the UK and the EU to minimize differences and seek common grounds. Meanwhile, China may benefit from trade diversion, with a strong possibility of reaching a free trade agreement with the EU down the line.

Germany might be the biggest winner out of Brexit as China's appetite for Germany's high-tech industries surges. China's buying spree of German technology firms took place in the same year as the EU referendum, most notably the takeover of Kuka by Midea Group, a Chinese electrical appliance giant. Acclaimed globally as a pioneer in the field of industrial robotics, Kuka offers technology that allows Chinese manufacturers to reduce their production costs, something which the UK can hardly match. Brexit seems to offer the best window for China to play both sides against the middle.

4 THE BELT AND ROAD INITIATIVE (BRI)

For China, the UK plays a strategically important role in the BRI. The effects of Brexit have been a major area of debate in Chinese academia. A study reflects that exiting the EU could be an opportunity for China to: (1) deepen its trade ties with the UK, the EU, and BRI countries; (2) facilitate the development of labor-intensive industries along the BRI, and (3) integrate BRI economies into its own value chains (Wang et al. 2017).

In addition to the positive impacts, Brexit might weaken the pro-China voices within the EU. In light of the dispute over granting market economy status to China, an EU without the UK as a member is anything but in Beijing's interest. It can be argued that the UK has long been China's chief lobbyist in Brussels, who often provides leadership in fostering consensus among dissenting members on trade issues with China, and should China lose it, the conservative voices within the EU will regain ground and push for measures that would make it difficult for China to materialize the BRI in Europe (Chu 2016). This could also mean new barriers to Chinese companies seeking to invest in Europe. Brexit can reduce future FDI inflows to the UK by about 22% (Dhingra et al. 2017), which may dampen China's FDI coming into the UK. Once the UK leaves, the EU will fall into recession, as both markets will shrink (Liu & Huang 2018). The loss of potential future investment into the EU could significantly impact the ability of Chinese firms to deliver BRI projects in Europe and even amplify the risk of global economic slowdown.

5 RMB INTERNATIONALIZATION

The role that London plays in internationalizing the RMB has remained irreplaceable. In 2014, London became the first foreign city to issue RMB-denominated bonds. This move was seen by the West as China's first attempt to promote the global use of RMB. Ever since then, London has been a place where the RMB is allowed to be directly converted into the Pound Sterling without being converted into the US Dollar and where Chinese-funded banks are qualified for RMB clearing. There has been a steady increase in RMB transactions in the UK over the last four years, with London accounting for 44.46% of global offshore RMB trading in 2019 (Theglobalcity.uk 2019). Given the remarkable and growing role of London as a dominant offshore hub for trading the RMB, the number of Chinese enterprises based on the European continent to conduct investment via London has grown rapidly since 2014. China's efforts to offer RMB-denominated debt reveal an ambition to lead the world out of the existing US-led financial architecture and to shape the international financial system in its favor (Wang 2015).

The Brexit-related exodus of financial firms from London will put London's position as a financial center in question. One of the potential risks to China is the loss of London as an offshore RMB hub. Against the backdrop of Brexit, the People's Bank of China's measures in monetary policy would undergo significant adjustments as a precautionary measure against sharp falls in the Pound Sterling. The expectation that the UK's currency will fall in value is most likely to spur a massive flow of cash out of the UK. The currency volatility in the wake of Brexit could drive Chinese investors to redirect investment to low-risk currencies for hedging purposes. Brexit could also deter the RMB internationalization process in Europe (Zhou 2018).

It is also worthwhile to examine the positive impacts on RMB internationalization. The post-Brexit UK, independent of EU financial regulations, might offer prospects for further cooperation between China and the UK in the financial sector (Zhou 2018). In order for the negative impact to be minimized, and in search of new markets outside the EU, the UK would feel compelled to develop closer ties with China, but only under the premise that London remains a leading RMB offshore center post-Brexit. Under the circumstances in which foreign capitals are flowing out of the UK, the UK government might offer China preferential policies to facilitate Chinese investment and become more receptive to the BRI. China could keep this momentum and further raise the EU's interest in the BRI (Liu 2016). As the starting point of RMB internationalization, the UK's cooperation will be important to the BRI's success in Europe. The benefit from trade diversion could increase the trade flows between China and the EU and the RMB's share in global currency reserves.

6 STUDYING IN THE UK

For some observers, a fall in the Pound Sterling implies an increase in the exchange rate of the RMB against the Pound Sterling, and by that means lowers the cost for Chinese students to study in the UK (Tian 2017). As predicted by Citibank, the Pound Sterling will plunge approximately 10 to 20% against the US Dollar once Brexit comes into full effect. Assuming the tuition fee is £15,000 and the living expense £10,000 per year, one Chinese student can save up to ¥20,000 to ¥50,000 per year (Zhang 2016). Recognizing this important factor, the UK government may introduce policies such as raising prices and taxes. Faced with higher costs in goods and services across the country, British universities might raise their tuition fees—hence, an increase in the cost of studying in the UK.

Under the Withdrawal Agreement, the UK will terminate its EU membership at the end of the Brexit transition period on December 31, 2020. And owing to that, the signed labor agreements between the UK and the EU will become invalid. To put it otherwise, guest workers from the European continent will be required to obtain an employment permit to be eligible to work in the UK. A restricted movement of people may result in limited availability of highly educated talent from EU countries, and, especially in the case of Brexit, some UK-EU education agreements will have to be renegotiated and readjusted. The uncertainty of the situation could discourage EU students from staying in the UK, albeit during the transition period. In the worst-case scenario, British universities can charge higher tuition fees from EU students (Liu 2016).

But evidence suggests that a less-well-off British education system may not affect the willingness of foreign students to study in the UK. The number of student visas issued to Chinese nationals in 2017 exceeded 100,000, 13% higher than the previous year (Gov.uk 2019a).

Thanks to the UK's globally renowned higher education sector, foreign students are a source of billions of pounds that bring major net benefits to British education institutions. Today, they generate more than £10 billion for the British economy per year—a huge boost to local businesses. Chinese students are the UK's biggest contingent of foreign students, worth around £4 billion per year in tuition fees and other spending. Taken together, British universities would be urged by the UK government to be flexible in their admissions to foreign students, whether from the economic or education standpoints (Zhang 2016).

A non-negligible factor that can affect foriegn students is economic recession. Companies leaving the UK tend to downsize and have less job opportunities available. A new finding on the post-Brexit labor market estimates an 80% reduction in inflows of guest workers from the EU due to policy changes in the UK's immigration system (Gov.uk 2019b). Employers may be more willing to hire British nationals than others.

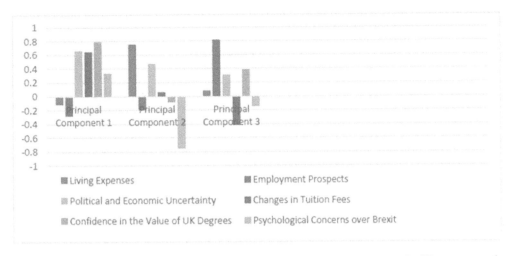

Figure 1. The principal components (primary factors) that can affect foreign students' willingness to study in the UK.

Analyses are made through principal component analysis, based on the results (Yang et al. 2018). This is not a detailed attempt to understand various factors affecting Chinese students' willingness to study in the UK, but rather a quantitative summary of the primary factors that can affect the UK as a destination for studying abroad. As shown in Figure 1, the primary factors are: (1) living expenses; (2) employment prospects; (3) political and economic uncertainty; (4) changes in tuition fees; (5) confidence in the value of UK degrees; and (6) psychological concerns about Brexit.

7 CONCLUSIONS

This study concludes that, though Brexit poses a plethora of challenges, it can be a golden opportunity for China to expand its relationship with both the UK and the EU and advance its geopolitical interests globally. A lack of clarity over the future relationship between the UK and the EU is just one reason why the path to Brexit would be long and uncertain. Considering the complex and uncertain nature of Brexit, China and the rest of the world will feel the impact in both the short and the long term. We suggest policy considerations that: (1) enhance China-UK and China-EU bilateral ties; (2) push for China-UK and China-EU free trade agreements; (3) speed up the RMB internationalization process in Europe; (4) collaborate with Germany in promoting the BRI in Europe; (5) deepen China's participation in multilateral financial institutions; and (6) build support mechanisms for Chinese students studying in the post-Brexit UK.

In the short term, Brexit will allow the UK to take back control of its decision-making and provide the UK with the necessary leverage to establish itself as a pole vis-à-vis the EU, but in the long term, the EU will alienate the UK so as to compete with the UK economically. While the UK will be free to strike trade deals without EU interference, the loss of passporting rights would invite the rest of the world to consider trading with the UK as a lower priority (Gregor 2015). Altering the geopolitical landscape in Europe could leave the UK and the EU in a state of constant rivalry over European affairs. In which case, the EU may take a tougher stance in negotiations with the UK. If the UK adopts a more independent policy in sensitive areas, the possibility of establishing a new stable relationship between the UK and the EU would be highly uncertain.

Brexit may delay or even thwart the EU's integration process, although the UK is outside the Eurozone. The loss of substantial hard and soft power by the EU and the damage that Brexit would do to the EU economy could trigger domino effects that encourage other EU members to leave the

EU. As Professor Zheng Yongnian points out: "With Brexit, the UK sets a precedent for withdrawing from regional alliances, as a catalytic event that heralds the collapse of the EU."

Also, the UK would be affected by the loss of the EU as a counterweight to the United States. The majority of published studies find that Brexit could play into the hands of the United States. The resultant fall in the exchange rate of the Pound Sterling and the Euro could cement the US Dollar's hegemony and, in turn, US dominance over global political and economic processes.

Ironically, though, Brexit could pave the way for the UK to mend the damaged relationship with Russia. Desperate to ride out the post-Brexit economic storm, the UK might call off its sanctions against Russia in exchange for economic benefits. But at a time of rising security pressures from Russia, a diminished EU may be disadvantaged in maintaining Europe's status quo, to the extent of Russia gaining the upper hand in geopolitical terms.

In the final analysis, the exit of a major member would create openings for Beijing to divide and conquer the EU. Although Brexit is a mix of challenges and opportunities, the eventual outcome will depend on the extent to which China and the EU forge their post-Brexit relationships with the UK. China should monitor the developments of Brexit and geopolitical changes during the transition period and thereafter and act in accordance with its national interests. In particular, it should balance the interests of all stakeholders involved with a view to securing a stable international environment.

REFERENCES

Baker Mckenzie 2019. Chinese FDI into North America and Europe in 2018 falls 73% to six-year low of $30 billion. See https://www.bakermckenzie.com/en/newsroom/2019/01/chinese-fdi.

Bloom, N. et al. 2019. The impact of Brexit on UK firms. See https://site.stanford.edu/sites/g/files/sbiybj8706/f/5351-19-019.pdf.

Chu, Y. 2016. A brief discussion on the impact of Brexit on the "One Belt One Road" strategic initiative. *Regional Economic Review* (05): 17–18. (In Chinese).

Dhingra, S. et al. 2017. The local economic impacts of Brexit. Centre for Economic Performance Brexit Analysis (10). See http://cep.lse.ac.uk/pubs/download/brexit10.pdf.

Ey 2020. EY financial services Brexit tracker: Firms go quiet on relocation announcements as focus moves to securing a strong future trading relationship. See: https://www.ey.com/en_uk/news/2020/01/ey-financial-services-brexit-tracker-firms-go-quiet-on-relocation-announcements-as-focus-moves-to-securing-a-strong-future-trading-relationship.

Gov.cn 2015. Chinese President Xi Jinping had met again with British Prime Minister David Cameron. See http://www.gov.cn/xinwen/2015-10/23/content_2952418.htm. (In Chinese)

Gov.uk 2019a. The UK's future skill-based immigration system. See https://assets.publishing.service.gov.uk/government/uploads/system/uploads/attachment_data/file/766465/The-UKs-future-skills-based-immigration-system-print-ready.pdf.

Gov.uk 2019b. 2018 UK visa statistics show 11% growth in China. See https://www.gov.uk/government/news/2018-uk-visa-statistics-show-11-growth-in-china.

Gregor, I. 2015. Brexit: The impact on the UK and the EU. See https://brexit.hypotheses.org/files/2017/01/Global-Counsel_Impact_of_Brexit.pdf.

Liu, X. X. 2016. The impact of Brexit on Chinese students studying in the UK and its counter-measures. *Business* (32): 208–209. (In Chinese)

Liu, X. & Huang, J. 2018. The impact of Brexit and China's response. *Hebei Qiye* (2): 61–62. (In Chinese)

London.gov.uk 2018. Preparing for Brexit. See https://www.london.gov.uk/sites/default/files/preparing_for_brexit_final_report.pdf.

Matthew, M. 2020. Statistics on UK trade with China. See. http://researchbriefings.files.parliament.uk/documents/CBP-7379/CBP-7379.pdf.

PwC 2016. Leaving the EU: Implications for the UK economy. See https://www.pwc.co.uk/economic-services/assets/leaving-the-eu-implications-for-the-uk-economy.pdf.

Ren, Z. & Bi, C. 2017. Research on the effects of Brexit on the "16+1" cooperation. *Economy Shanghai* (06): 69–79. (In Chinese)

Theglobalcity.uk 2019. London RMB business quarterly issue 4. See https://www.theglobalcity.uk/Positive Website/media/research-downloads/London-RMB-Business-Quarterly-Report-Issue-4-WEB.pdf.

Tian, F. 2017. The impact of Brexit on human capital flows between China and the UK. *China National Conditions and Strength* (03): 75–76. (In Chinese).

Wang, Y. Z. 2015. London debt issue represents milestone for China. See http://www.globaltimes.cn/content/948562.shtml.

Wang, Y., Zhang, X. & Zhang, E. 2017. How will Brexit impact China's "One Belt One Road" strategy: An analysis based on GTAP model. *International Economics and Trade Research* 33 (05): 29–39. (In Chinese).

Wu, Z. C. 2019. The impact of Brexit on Europe and the world landscape. See http://news.nankai.edu.cn/nkzs/system/2019/01/17/000429419.shtml. (In Chinese).

Yang, J., Xu, Y. Q. & Su, C. Y., et al. 2018. An economic analysis of the impact of Brexit on Chinese students studying abroad. *Finance Economy* (10): 121–122. (In Chinese).

Zhang, L. 2016. The impact of Brexit on study abroad. See http://www.jyb.cn/zgjyb/201607/t2016 0708_38842.html. (In Chinese).

Zhou, C. H. 2018. The impact of Brexit on China's development. *Modern Business* (22): 59–60. (In Chinese).

Computational Social Science – Luo, Ciurea & Kumar (eds)
© 2021 Taylor & Francis Group, London, ISBN 978-0-367-70193-2

A study inspired by Ukiyo-e

Y.T. Xie
Shanghai University of Medicine & Health Sciences, Centre of Art Education, Shanghai, China

H.Y. Dai
Zhejiang Gongshang University, Faculty of Statistics and Mathematics, Hangzhou, China

ABSTRACT: Japanese elements play very important roles in the fashion field. Fashion designers are addicted to mysterious characteristics from Asian culture. Given the influence of *ukiyo-e*, a Japanese art genre, in the Western art field, it has been considered as the most representative painting art. Ukiyo-e paintings were the genre paintings that depicted the focus of life. The narrative perspective of amorous feelings of the world, the uninhibited but delicate painting style, and the unique color application are all the inspirations for this collection. The purpose of this research is to seek a new breakthrough in ukiyo-e's expression form by extracting, reforming, and combining the traditional patterns with anti-traditional fabric. Then efforts are made to try to apply them in the comfortable and leisure clothing of street style. The key of this collection is the combination of corduroy and color patterns.

1 INTRODUCTION

1.1 *Background of study*

The Japanese culture has long been loved by clothing designers, including such images as the streets of Harajuku, architecture of Japanese style, Buddhist mood, Japanese rock gardens, curled grass, *sakura* (cherry blossoms), and samurai. The ukiyo-e culture and the samurai uniform are very typical cultural elements among designs. Ukiyo-e paintings are mainly genre paintings depicting amorous feelings of the world, which are similar to the painting "On the Riverside Scene of Pure Brightness" of the Song Dynasty, but whose contents are more erotic, indulgent, bold, and uninhabited. The narrative perspective of amorous feelings of the world, the uninhibited but delicate painting style, and the unique color application of ukiyo-e help to establish its unique status in culture.

Modern designers are fascinated by its unique artistry. They trace its tracks in the vicissitudes of history and draw design inspiration from it. From John Galliano to Dries Van Noten, from the haute couture on the T stage to exquisite garments, it is presented on the clothing of different styles in various forms, locations, and ways.

1.2 *Aims and objectives*

The research purpose of this project is to seek ways to make ukiyo-e break away from its traditional depiction, and make innovative fashion usage. Based on traditional ukiyo-e paintings, the traditional patterns are discomposed, extracted, reformed, and combined with anti-traditional fabrics, and then efforts are made to try to apply them in the comfortable and leisure clothing of street style to seek a new breakthrough in the expression form. Among them, different fabrics have different printing effects, and the texture sense and the collocation of the ukiyo-e pattern varies with the color depth of the fabric. Therefore, the combination of fabrics and patterns is the key.

Specifically speaking, the objectives of this project are as follows:

1. Contents and themes, artistic styles, painting techniques, tones, and other characteristics of ukiyo-e are studied.
2. Traditional ukiyo-e paintings are extracted and reformed to create new presentation forms.
3. Efforts are made to apply the digital printing to patterns, and to match up patterns with fabrics.
4. Ukiyo-e patterns are applied in sportswear and street style clothing.

1.3 Scope of study

In the first two sections, the conception, main themes, the development process, painting styles, tones, and other aspects of ukiyo-e are studied and clarified. Section 3 introduces the target consumption group, the collection and development process of design inspiration, the stories of colors and fabrics, and the production process of printings, design sketches, and style pictures. Section 4 summarizes the research and proposes future research directions and ideas.

1.4 Methodology

Research methods include referring to such sources as related books, an atlas, academic journals, articles, databases on the Internet, fashion magazines, and show information. Ukiyo-e paintings are analyzed and reformed to make new clothing patterns. Through printing fabrics, efforts were made to choose suitable fabrics to match up and be combined with the patterns. Then inspiration was drawn from the results to create a new look of clothing.

2 LITERATURE REVIEW

2.1 Origin of Ukiyo-e

In the 17th century, the maximum benefits of the society at that time were vested in merchants, whose lives were greatly improved. Apart from having enough wealth for necessities such as food and clothing, they had the ability to purchase ornamental goods like ukiyo-e paintings to decorate their dwelling places and enrich their spiritual world. They would indulge in entertainment venues of sensual pleasures, like the *Kabukicho*. The hedonistic lifestyle was also portrayed in ukiyo-e paintings, and this became the mainstream at that time.

2.2 Definition of Ukiyo-e

The word *ukiyo-e* originated from the Buddhist language, which dealt with the ethereal world and the reincarnation of the world. Therefore, ukiyo-e paintings were the genre paintings that depicted the focus of life. Ukiyo-e woodcuts were mainly divided into two forms: the "painting books" and the "one painting." The "painting books" were books of illustrations, which first appeared in classical novels in the Edo era in the form of illustrations. Later, with the rise of popular novels, painting books developed at a high speed. The "one painting" referred to a single woodcut printmaking, whose painting technique was more meticulous than the "painting books."

2.3 Categories and painting styles

Japanese author Kafu Nagai summarized the already withered ukiyo-e in his famous work, "The Evil and the Pleasing—On Ukiyo-e": "The most interesting thing may have two kinds of qualities: the evil and the pleasing. Ukiyo-e right has the quality that unities the evil and the pleasing. It makes the world of sensory pleasure have a rich texture, which is heavy, like late-maturing sorghum, whose ears all fall down, plentiful, substantial, and through the sky." At first, ukiyo-e was influenced by the mainstream thought of instant gratification. Most protagonists of the paintings were prostitutes and geishas. Paintings were centered around themes of sexual love, which were straightforward, dissolute, and naked. Ukiyo-e depicted the life of the women living in the red-light districts in detail. The brushwork was meticulous and rigorous and the style was very realistic. Every detail in the scene was finely carved. Grass and flowers in the background, materials and textures of buildings, fabrics and colors on figures' clothes, decorative patterns on hair accessories, and even each piece of human hair were meticulously drawn out. Gorgeous colors matched up with slim lines, and delicate and exquisite brushwork depicted the bold and indulgent world of joy. Contradictory, conflicting, but harmonious, the ukiyo-e style was very unique. Later, with the gradual increase of ukiyo-e painters, themes of ukiyo-e were extended to every field of people's life, including birds and flowers, social affairs, opera allusions, beautiful scenery of resorts, elements of women's life. Painters were from the common folk. Painting themes came from life. Therefore, their paintings could directly reflect the social trends and changes of that time. Painters had various painting techniques and perspectives. All aspects of people's life in the Edo era were depicted meticulously in those paintings, which were full of vitality.

2.4 *Influence*

Ukiyo-e painting developed well during the era of the public's economy and had a profound effect on social life. Painters were from the common folk, and nobody served for the royal elite. The survival soil of ukiyo-e is rich and energetic during that period. Therefore, it has strong vitality. However, in the 1820s, the prevalence of the capitalist economic mode resulted in the loss of this art. People preferred the contents of pornography and more vulgar pursuit. The style of art moved toward death gradually. Although ukiyo-e was replaced by modern typography, it would still play a curtail role in art.

Ukiyo-e was introduced to Europe in the mid-nineteenth century. The paintings of the Asian style caused great repercussion. The upsurge of the Japanese style wave swept through Europe. Painters recommended, purchased, and copied ukiyo-e paintings. They tirelessly learned the Asian painting style and aesthetic composition, were inspired by them, and reflected them in the creation of other paintings, such as Van Gogh's "Portrait of Pere Tanguy" (1887), Manet's "Boy Plays Flute," and so on. Thereby, ukiyo-e was considered as the most representative painting art.

3   DESIGN COLLECTION

3.1   *Design concept and inspiration*

The derivative design of this project is named *ukiyo*, which means "the floating world" in Japanese. Correspondingly, ukiyo-e is a special object for painters in Edo to decorate the way to life. Ukiyo-e is the inspiration of this collection, with its meticulous and rigorous brushwork and realistic style showing the strong power of Japanese culture. The impressions of ukiyo-e are always described with words such as classic, elegance, and traditional. Hence, when it is used to decorate the clothes as the color patterns, designers like to use the original patterns from the famous ukiyo-e style paintings and match them with formal styling dresses. This collection tries to break the normal impressions of ukiyo-e and connect the elements with casual clothing to show the freestyle mode and restore another quality of ukiyo-e, which is simpler and has the possibility to change with historic developments.

3.2   *Target customer*

People who are aged 20 to 30 years old are the target customer of this collection. They accept good education and have a bachelor's degree or above. They may work in fields related to art or fashion, like the artists, fashion bloggers, photographers, and university lecturers in the fashion field. Or they may work in advertising and media fields and have the corresponding positions as those who have enough economic power to purchase luxuries. Therefore, their monthly income should be above

HKD 25,000, and their range of purchasing power is about HKD 7,000 to 10,000. They follow the fashion information and they know who John Galliano, Rick Owens, and Martin Margiela are. They should prefer casual clothing or clothes with abstract patterns. They are energetic, independent, and fashionable. In their spare time, they may like shopping, taking photographs, painting, visiting museums, or are interested in outdoor activities, such as skiing and hiking. They should enjoy and be confident about their lifestyle.

3.3 Coloration and printed patterns development

As mentioned, the color pattern is rooted from the ukiyo-e. Though the themes of ukiyo-e paintings are various and bold, the ensemble of ukiyo-e paintings are still neat, which makes the clothes have a traditional feeling, and they are not suitable for street style. However, when the paintings were distorted, they looked quite different. All the distorted prints in this collection are based on the famous paintings. The paintings look abstract and are closer to graffiti style. The prints are changed through distortion and are similar to Western arts. Some of them are twisty, which makes them look like the famous painting "Skrik" by Edvard Munch (1893). Some of them became fragmentary after segmentation and reassembly. The whole space is unusual, and the stories implied therein disappear. The painting becomes another kind of abstract pattern but with Japanese style (Figures 1 and 2).

The color board is picked from the traditional ukiyo-e paintings. Given to colorful prints, the main color should be simple in order to highlight the color patterns. The main color is a blue tone, ranking the black, gray, and pastel. The indigo blue is similar to denim, which is suitable to connect with color patterns and not as tedious as other dark colors. Blue tones are the basic colors in ukiyo-e arts, especially in wave-themed paintings, so there are also different shades of blue arranged in groups in the collection.

3.4 Fabric development

Corduroy was chosen to be the main fabric, for the following reasons:

1. As the collection is for fall and winter, the corduroy is a thick enough fabric. The unique cord wales of corduroy increase the sense of the fabric texture. There are various cord wales to choose. The different wales have different texture and reflet.

Figure 1.   Color patterns.

Figure 2.   Color board.

2. When printing on corduroy, some part of the fabric can hardly be printed, like the corner of the cord wales. The patterns have more changes, such as third dimensions.
3. The corduroy is lighter than other fabric, though it is as thick as others. This kind of quality of corduroy fits the oversize type and helps to make the design sporty and casual.

Most people choose 100% polyester as the main fabric when the design needs to be applied to a digital print, due to the bright color effect. But this collection used the corduroy with 100% cotton rather than 100% polyester. First, the patterns had faded effect when they were printed on the corduroy with 100% cotton, because the fabric with cotton fibers are harder to dye. The soft nap and cord wales are also obstacles to dye. However, the faded effect is a fit to the characteristic of traditional ukiyo-e arts. They are simple, natural, and historic. Moreover, in the matter of color effect, polyester is better than cotton and the pattern is clearer. However, the feeling and fabric quality of corduroy with 100% polyester are far worse than that with 100% cotton. As a whole, wearable comfort and the quality of the collection are more important.

3.5 Illustrations, production drawings, and styling photos

3.5.1 Outfit

Outfit one includes an oversize coat and a shirt dress. The coat is joined by large area color patterns and black corduroy. There is a huge collar, which is like a cape covering the shoulders and then hangs down naturally. The bulky collar connects with two triangular out pockets in the front of the garment and large accordion pleats in the back. The accordions are three-dimensional and flexible. There are three button tabs on each sides of the front bottom. Ropes are woven together through the button tabs. A full lining is applied. The apricot inner was made by mixed dyed suiting and is soft and quite simple, and the clean color is a good balance to the colorful coat. The collocation between the two fabric textures becomes a contrast to a certain degree. The research summarized the origin, contents, themes, artistic styles, painting techniques, tones, and other characteristics of ukiyo-e and acquired the inspiration through comprehending these elements (Figures 3 and 4).

Outfit two has a knee-length vest, an irregular shirt, and a pair of straight-leg pants. The shirt is the point of this outfit. The left part and right of the shirt are quite anomalistic in that the left part is shorter than the right by about 20 centimeters. Then the front part of the left garment was printed

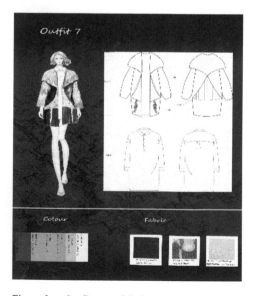

Figure 3.   Outfit one of design.

Figure 4.   Styling photos.

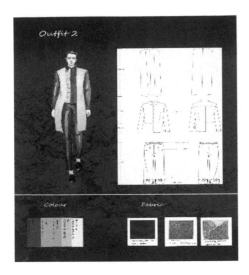

Figure 5.   Outfit two of design.

Figure 6.   Styling photos.

with the colorful pattern. The back part has a large jag along the midline, which corresponds to the front. The straight-leg pants have two big patch pockets in the front. Elastic band is placed for the waistband. The bottom of the trouser leg has a turnover. The leg opening was decorated with the eyelets and ropes to set the revers and respond to the collection details. The pants and the shirts both used corduroy with 16 cord wales, which is repetitive. Hence the long length of the vest is designed to break the repetition. The vest used gray color and cashmere fabric to change the hue of the whole outfit. The vest is very simple, so that the balance can still be kept (Figures 5 and 6).

4   CONCLUSION

This study demonstrated what the ukiyo-e is and detailed the progress to research the new possibilities to combine ukiyo-e with fashion. As Kafu Nagai said, ukiyo-e has many characteristics that can be studied and developed. It is not only a kind of art that can represent Japanese style paintings, but it also shows the aesthetic consciousness of Japan, how they think and the ways to express this. The research summarized the origin, contents, themes, artistic styles, painting techniques, tones, and other characteristics of ukiyo-e and how inspiration was acquired through comprehension.

According to the research and considering the needs of target customers, the collection was designed and named as Ukiyo, the floating world. In terms of literal meaning, the theme is about ukiyo-e, but it is more casual. Collocation in fabric, patterns, and style means a great number of ways can be tried. Due to the time and technology limitations, the experiments of fabric use and patterns met the restriction in a certain degree. So, when doing further research about these issues, the fabric tests and experiments can be increased and fabrics for other seasons can be considered.

REFERENCES

[1] A referenceThis reference has two entries but the second one is not numbered (it uses the 'Reference (no number)' style. Kita, Sandy (September 1984). "An Illustration of the Ise Monogatari: Matabei and the Two Worlds of Ukiyo". The Bulletin of the Cleveland Museum of Art (Cleveland Museum of Art) 71 (7): 252–267.
[2] Hickman, Money L. (1978). "Views of the Floating World". MFA Bulletin (Museum of Fine Arts, Boston) 76: 4–33.

[3] Fleming, Stuart (November–December 1985). "Ukiyo-e Painting: An Art Tradition under Stress". Archaeology (Archaeological Institute of America) 38 (6): 60–61, 75.

[4] Singer, Robert T. (March–April 1986). "Japanese Painting of the Edo Period". Archaeology (Archaeological Institute of America) 39 (2): 64–67.

[5] Thompson, Sarah (Winter–Spring 1986). "The World of Japanese Prints". Philadelphia Museum of Art Bulletin (Philadelphia Museum of Art) 82 (349/350, The World of Japanese Prints): 1, 3–47.

[6] Toishi, Kenzô (1979). "The Scroll Painting". Ars Orientalis (Freer Gallery of Art, The Smithsonian Institution and Department of the History of Art, University of Michigan) 11: 15–25.

[7] Bell, David (2004). Ukiyo-e Explained. Global Oriental.

Computational Social Science – Luo, Ciurea & Kumar (eds)
© *2021 Taylor & Francis Group, London, ISBN 978-0-367-70193-2*

Evaluation of a lawyer's professional ability based on fuzzy analytic hierarchy process

Z.B. Wan, A.Q. Li, Z.H. Huang & Z.H. Zhang
East China Jiaotong University, Nanchang, China

ABSTRACT: Because of the traditional methods of lawyer evaluation, it is difficult to objectively and impartially evaluate the professional ability of lawyers through scientific and comprehensive means, and it is hard to satisfy the needs of rights defenders. Therefore, how to establish a comprehensive and effective evaluation system of a lawyer's professional ability has become a problem to be solved. After exploring the limitations of traditional evaluation algorithms, this paper introduces the fuzzy analytic hierarchy process model to scientifically evaluate a lawyer's professional ability. Simulated experiments also were conducted with Matlab. The results proved that the model has a good practicability for the evaluation of a lawyer's professional ability. It not only avoids the one-sidedness of subjective and objective judgments, but also provides a decision basis for the realization of lawyer portraits and accurate lawyer recommended services, at the same time provides a more objective reference for rights defenders and has a good application prospect.

1   INTRODUCTION

With the deepening of China's legalization system, citizens' awareness of rights protection continues to rise. When people's rights are seriously violated, they usually safeguard their legitimate rights by means of a lawyer. In general, when citizens are seeking legal aid, they always prefer to make the following two choices: looking for a suitable lawyer through relationships and a through searching online.

However, because the profession of law is very skillful, it is difficult for the public to judge lawyers' ability to practice. This means the two methods of lawyer searching have certain limitations. This paper proposes an improved fuzzy analytic hierarchy process (FAHP) to analyze the weights of lawyers' ability factors and to construct a model for assessing lawyers' ability to practice.

2   CONSTRUCTION OF A FUZZY ANALYTIC HIERARCHY PROCESS MODEL

As early as the 1970s, American operations researcher T. L. Satty and others proposed a multi-level and multi-objective decision method combining qualitative and quantitative analysis. It can break down a complex decision problem into several indicators, and it allocates resources and establishes a hierarchical structure according to the contribution of each indicator (Yanmei & Weihua 2008). It is widely used in the fields of safety science and environmental science. Due to the uncertainly and fuzziness of the lawyer's evaluation, this paper combines the analytic hierarchy and fuzzy comprehensive evaluation and introduces the fuzzy consistent matrix in the process of evaluating the lawyer's professional ability, which not only solves the inconsistency of the judgment matrix, but also improves the performance of the algorithm. In order to make the evaluation results more credible, the weighted average model is used to replace the fuzzy operator, and the evaluation set is valuated according to the level interval thought so as to replace the principle of maximum membership (Liu et al. 2018).

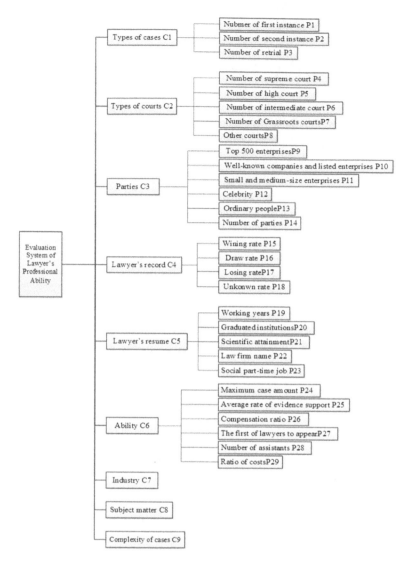

Figure 1.   Evaluation indicators system of a lawyer's professional ability.

2.1   *Establishment of evaluation indicators system*

Through the reading of massive literature material, combining lawyer industry knowledge base and judgment documents data, this study deeply explored the evaluation system of a lawyer's professional ability. With the large amount of lawyers' data and case information in judgment documents, fully considering the analysis of lawyers' data and the actual demand of the Internet lawyers industry, an indicator system of a lawyer's professional ability was initially determined. It consists of the types of cases, types of courts, lawyer's record, parties, lawyer's resume, ability, industry, subject matter, complexity of cases, and so on. Figure 1 shows the evaluation indicators system of a lawyer's professional ability.

2.2   *Construct judgment matrix*

As shown in Table 1, the judgment criteria is of 0.1 to 0.9 scale values. If a_i is compared to a_j, a judgment r_{ij} is obtained; if a_j is compared to a_i, a judgment $r_{ji} = 1 - r_{ij}$ is obtained. The paired

Table 1. 0.1–0.9 Criteria for determining scale value.

Scale	Compare	Description
0.5	Equally important	a_i is equally important as element a_j
0.6	Slightly important	a_i is slightly more important than a_j
0.7	Obviously important	a_i is obviously more important than a_j
0.8	Highly important	a_i is highly more important than a_j
0.9	Extremely important	a_i is extremely more important than a_j
0.1, 0.2, 0.3, 0.4	Inverse comparison	

Table 2. Judgment matrix F1.

O	C1	C2	C3	C4	C5	C6	C7	C8	C9	r_i
C1	0.5	0.8	0.8	0.8	0.7	0.1	0.8	0.7	0.2	4.1
C2	0.2	0.5	0.7	0.8	0.7	0.7	0.8	0.7	0.3	3.9
C3	0.2	0.3	0.5	0.2	0.3	0.4	0.7	0.8	0.6	5.5
C4	0.2	0.2	0.8	0.5	0.7	0.7	0.6	0.5	0.7	3.4
C5	0.3	0.3	0.7	0.3	0.5	0.3	0.7	0.8	0.3	4.8
C6	0.9	0.3	0.6	0.3	0.7	0.5	0.9	0.7	0.8	3.6
C7	0.2	0.2	0.3	0.4	0.3	0.1	0.5	0.9	0.9	5.8
C8	0.3	0.3	0.2	0.5	0.2	0.3	0.1	0.5	0.5	6.5
C9	0.8	0.7	0.4	0.3	0.7	0.2	0.1	0.3	0.3	2.9

Table 3. Judgment matrix F2.

C1	P1	P2	P3	r_i
P1	0.5	0.6	0.8	1.9
P2	0.4	0.5	0.8	1.7
P3	0.2	0.2	0.5	0.9

comparison method was used to compare the indicators of a lawyer's professional ability. The result describes the importance of each indicator through the $A = (a_{ij})n \times n$ symmetric matrix, and finally determines the judgment matrix F1–F7. F1–F2 are shown in Tables 2 and 3; F3–F7 are omitted.

2.3 Determine the Fuzzy Consistent Matrix

Fuzzy matrix $R = (r_{ij})n \times n$, and $i = 1, 2, \ldots, n; j = 1, 2, \ldots, n$, for any i, j satisfied the following formula:

$$r_{ij} = (r_i - r_j)/2s + 0.5 \tag{1}$$

Among that, s represents the level of the fuzzy consistent matrix. The judgment matrix F1–F7 can be transformed into the fuzzy consistent matrix H1–H7. H1–H2 matrix are shown in Tables 4 and 5; H3–H7 are omitted. Finally, if the difference between the corresponding indicators of any two rows in specified in the matrix is constant, then it indicates that the setting of fuzzy consistent matrix is reasonable.

Table 4. Fuzzy consistent matrix H1.

O	C1	C2	C3	C4	C5	C6	C7	C8	C9	$\sum_{j=1}^{s} r_{ij}$
C1	0.500	0.500	0.578	0.528	0.567	0.483	0.633	0.628	0.578	4.995
C2	0.500	0.500	0.578	0.528	0.567	0.483	0.633	0.628	0.578	4.995
C3	0.422	0.422	0.500	0.450	0.489	0.406	0.556	0.550	0.500	4.295
C4	0.472	0.472	0.550	0.500	0.539	0.406	0.556	0.550	0.500	4.545
C5	0.433	0.433	0.511	0.461	0.500	0.417	0.567	0.561	0.511	4.394
C6	0.517	0.517	0.594	0.544	0.583	0.500	0.650	0.644	0.594	5.143
C7	0.367	0.367	0.444	0.394	0.433	0.350	0.500	0.494	0.444	3.793
C8	0.372	0.372	0.450	0.400	0.439	0.356	0.505	0.500	0.450	3.844
C9	0.422	0.422	0.500	0.450	0.489	0.406	0.556	0.550	0.500	4.295

Table 5. Fuzzy consistent matrix H2.

C1	P1	P2	P3	$\sum_{j=1}^{s} r_{ij}$
P1	0.500	0.533	0.667	1.700
P2	0.467	0.500	0.633	1.600
P3	0.333	0.367	0.500	1.200

3 FUZZY COMPREHENSIVE EVALUATION

3.1 Set up the evaluation set

The evaluation set is a qualitative description of virtues or defect degree of the evaluation of the lawyer's professional ability. According to the sorting of fuzzy comprehensive scores from big to small, we can get a reasonable evaluation conclusion to lawyer's professional ability. As can be seen from Figure 1, the indicators are divided into primary indicators $C_i(i = 1, 2, 3, \ldots, 8, 9)$ and secondary indicators $P_i(i = 1, 2, 3, \ldots, 28, 29)$ by consulting relevant data and consulting the leaders of the judicial department and a number of senior lawyers, to finally get the following evaluation set of lawyers' professional ability:

$$Q = \{q_1, q_2, q_3, q_4, q_5\}$$

The elements in the evaluation set Q represent {Strong, Slightly Strong, General, Weak, Very Weak}. If adopting the interval method, then $q_1 - [d_1, d_1], q_2 - [d_2, d_2], \ldots, q_n - [d_n, d_n]$, which means the indicators in the set Q are represented from big to small as $[xh_i, x_{i1}], [x_{i1}, x_{i2}], [x_{i2}, x_{i3}], [x_{i3}, x_{i4}], [x_{i4}, xl_i]$; xh_i and xl_i correspond to scores of upper and lower bounds of the i-th indicator, respectively. Calculated according to the centesimal grade, the full score of each indicator is 100, the lowest score is 0, and the interval is represented as [100, 90], (90, 80], (80, 70], (70, 60], (60, 0].

3.2 Calculate indicator weight

From section 2.3, the fuzzy consistent matrix is determined. Thus, the weights of each type of indicators for each layer of the lawyer can be calculated. The calculation formula is as follows:

$$\omega_i = \frac{1}{n} - \frac{1}{2a} + \frac{\sum_{j=1}^{n} r_{ij}}{na} \tag{2}$$

In formula (2), $i = 1, 2, \ldots, n$, parameter $a \geq \frac{n-1}{2}$, n represent the layer of fuzzy consistent matrix. In general, when $a = \frac{(n-1)}{2}$, the experimental results have the best recognition effect. Combining with the sorting method can also improve the credibility of experimental results. The criterion layer C represents the weight of primary indicators; the weight of each indicator of layer C is ω_{co}. The indicator layer P is expressed as the secondary indicator weight; the sorting weight of each indicator of layer P is $\omega_{PC_i}(i = 1, 2, 3, \ldots, 9)$. According to formula (3), inputting the data of fuzzy consistent matrix into the modeling tool Matlab can get the weights of each indicator of each layer as follows:

The indicator weights of criterion layer C relative to target layer O are:

$$\omega_{co} = (\omega_{c1o}\ \omega_{c2o}\ \omega_{c3o}\ \omega_{c4o}\ \omega_{c5o}\ \omega_{c6o}\ \omega_{c7o}\ \omega_{c8o}\ \omega_{c9o})$$
$$= (0.125\ 0.125\ 0.105\ 0.118\ 0.108\ 0.129\ 0.091\ 0.093\ 0.105)$$

The indicator weights of indicator layer P relative to criterion layer C are:

$$\omega_{pc1} = (\omega_{p1c1}\ \omega_{p2c1}\ \omega_{p3c1}) = (0.400\ 0.366\ 0.233)$$

$$\omega_{pc2} = (\omega_{p4c2}\ \omega_{p5c2}\ \omega_{p6c2}\ \omega_{p7c2}\ \omega_{p8c2}) = (0.200\ 0.230\ 0.205\ 0.170\ 0.195)$$

$$\omega_{pc3} = (\omega_{p9c3}\ \omega_{p10c3}\ \omega_{p11c3}\ \omega_{p12c3}\ \omega_{p13c3}\ \omega_{p14c3}) = (0.167\ 0.146\ 0.157\ 0.157\ 0.183\ 0.190)$$

$$\omega_{pc4} = (\omega_{p15c4}\ \omega_{p16c4}\ \omega_{p17c4}\ \omega_{p18c4}) = (0.350\ 0.275\ 0.217\ 0.158)$$

$$\omega_{pc5} = (\omega_{p19c5}\ \omega_{p20c5}\ \omega_{p21c5}\ \omega_{p22c5}\ \omega_{p23c5}) = (0.229\ 0.174\ 0.174\ 0.249\ 0.174)$$

$$\omega_{pc6} = (\omega_{p24c6}\ \omega_{p25c6}\ \omega_{p26c6}\ \omega_{p27c6}\ \omega_{p28c6}\ \omega_{p29c6}) = (0.147\ 0.177\ 0.197\ 0.197\ 0.150,\ 0.133)$$

In order to more intuitively understanding the importance of each indicator of the lawyer's professional ability, the weights of each indicator at all layers are summarized as shown in Table 6.

3.3 Fuzzy evaluation matrix

The evaluation of the lawyer's professional ability refers to the evaluation of the lawyer's various qualities and performance in related legal work. According to evaluation results, it is possible to have a quite intuitive and clear cognition of the comprehensive ability of lawyers, and it also can provide scientific and reasonable references for lawyer portraits and lawyer recommendation systems. At present, many firms or judicial departments score by simply subjective evaluation or the jury, which consists of leader, colleagues, judges and judicial department, and citizen representation. They usually use the total score method to evaluate. Because there are many artificial factors in the inclusion and a lack of scientific method, it resulted in many potential lawyers being buried, which seriously affected the enthusiasm of the lawyer service industry and brought down the standard of service of the whole industry. Therefore, it is very important and necessary to establish a set of scientific and reasonable and timely evaluation systems of a lawyer's professional ability. However, there are too many factors of lawyer evaluation, besides which most of them are fuzzy, so it is difficult to quantify them with simple values. Therefore, it is more reasonable to use fuzzy principle to evaluate lawyers in a fuzzy way. In the process of the evaluation, it is very important in the choice of the membership function. It directly affects the scientific method and accuracy of the evaluation results, so the role of membership function in the evaluation model is very important.

However, in concrete applications, there are no strict restrictions on membership functions. Membership functions should be built based on application scenarios and methods used in comprehensive experiments. From the set of evaluations in section 3.1, combined with gradient distribution method to get membership (the degree of membership corresponds to the membership function

Table 6. Weight of collection at each layer.

Criteria Layer	$\omega_{cio}(i = 1, \cdots, 9)$	Indicators Layer	Single Sort Weight $\omega_{p_j c_i}$ $(j = 1, \ldots, 5; i = 1, \ldots, 9)$
C1	0.102	P1	0.344
		P2	0.363
		P3	0.293
C2	0.105	P4	0.209
		P5	0.210
		P6	0.202
		P7	0.196
		P8	0.183
C3	0.105	P9	0.171
		P10	0.172
		P11	0.171
		P12	0.174
		P13	0.158
		P14	0.154
C4	0.118	P15	0.289
		P16	0.253
		P17	0.229
		P18	0.228
C5	0.120	P19	0.214
		P20	0.197
		P21	0.209
		P22	0.196
		P23	0.183
C6	0.127	P24	0.162
		P25	0.171
		P26	0.180
		P27	0.176
		P28	0.161
		P29	0.150
C7	0.104		
C8	0.102		
C9	0.117		

below), the membership functions can be constructed as follows:

$$
g_{i1} = \begin{cases} 1 & x_{i1} \leq x_i \leq x_i^h \\ \frac{x_i - x_{i2}}{x_{i1} - x_{i2}} & x_{i2} < x_i < x_{i1} \\ 0 & x_i^l < x_i \leq x_{i2} \end{cases}
\qquad
g_{i2} = \begin{cases} \frac{x_i^h - x_i}{x_i^h - x_{i1}} & x_{i1} < x_i \leq x_i^h \\ 1 & x_{i2} \leq x_i \leq x_{i1} \\ \frac{x_i - x_{i3}}{x_{i2} - x_{i3}} & x_{i3} < x_i < x_{i2} \\ 0 & x_i^l < x_i \leq x_{i3} \end{cases}
$$

$$
g_{i3} = \begin{cases} 0 & x_{i1} \leq x_i \leq x_i^h \\ \frac{x_{i1} - x_i}{x_{i1} - x_{i2}} & x_{i2} < x_i < x_{i1} \\ 1 & x_{i3} \leq x_i \leq x_{i2} \\ \frac{x_i - x_{i4}}{x_{i3} - x_{i4}} & x_{i4} < x_i < x_{i3} \\ 0 & x_i^l < x_i \leq x_{i4} \end{cases}
\qquad
g_{i4} = \begin{cases} 0 & x_{i2} \leq x_i \leq x_i^h \\ \frac{x_{i2} - x_i}{x_{i2} - x_{i3}} & x_{i3} < x_i < x_{i2} \\ 1 & x_{i4} \leq x_i \leq x_{i3} \\ \frac{x_i - x_i^l}{x_{i4} - x_i^l} & x_i^l < x_i < x_{i4} \end{cases}
$$

$$
g_{i5} = \begin{cases} 0 & x_{i3} \leq x_i \leq x_i^h \\ \frac{x_{i3} - x_i}{x_{i3} - x_{i4}} & x_{i4} \leq x_i \leq x_{i3} \\ 1 & x_i^l < x_i \leq x_{i4} \end{cases}
$$

Any element in the evaluation set will be changed due to the data fluctuations of each indicator, The impact factor represented is expressed as g_{ij}; x_i represents the value of the i-th evaluation indicator of the evaluated lawyer. After determining the impact factors, the fuzzy evaluation matrix G can be constructed, and the matrix is shown as follows:

$$G = (g_{ij})_{n \times m} = \begin{cases} g_{11} & g_{12} & \cdots & g_{1m} \\ g_{21} & g_{22} & \cdots & g_{2m} \\ \vdots & \vdots & \vdots & \vdots \\ g_{n1} & g_{n2} & \cdots & g_{nm} \end{cases}$$

The number of evaluation indicators is expressed as n, and the number of elements in the evaluation set is expressed as m.

3.4 Fuzzy comprehensive evaluation

The fuzzy evaluation matrix G reflects the fuzzy relationship between the indicators of each layer and the evaluation set, and g_{ij} describes the membership between them. Assume $B = \omega G =$

$$(\omega_1 \ \omega_2 \ \cdots \ \omega_n) \begin{cases} g_{11} & g_{12} & \cdots & g_{1m} \\ g_{21} & g_{22} & \cdots & g_{2m} \\ \vdots & \vdots & \vdots & \vdots \\ g_{n1} & g_{n2} & \cdots & g_{nm} \end{cases}, \text{ for the weight vector } \omega \text{ and matrix } G, \text{ most people adopt}$$

the max-min fuzzy operators in the fuzzy comprehensive evaluation. However, in practical applications, when the fuzzy operators are processing the fuzzy information, the processing result is inconsistent with the actual results due to the loss of too much data information. If replaced with a weighted average comprehensive evaluation model $b_j (b_j = \sum_{i=1}^{n} \omega_i g_{ij})$, it can obtain a better processing effect and can effectively solve the problem that the processing result is not corresponding with the practical. After transforming:

$$B = \omega G = (b_1 \ b_2 \ \cdots \ b_n)$$

And E can be obtained after normalizing B, each element in E satisfies $e_j = b_j / \sum_{j=1}^{m} b_j$, therefore, the evaluation result can be expressed as:

$$Q = \sum_{j=1}^{m} e_j (d_j + d_j')/2 \tag{3}$$

Take a lawyer in a law firm as an example. The scores of 29 indicators of professional level are shown in Table 7.

In order to verify the correctness and effectiveness of the algorithm, first evaluate the lawyer's professional ability through traditional evaluation methods, and then input evaluation data calculated by gradient distribution method and formula (3) into the software Matlab. The evaluation results of each indicator of criterion layer are as follows:

$$Q_{c1o} \ Q_{c2o} \ Q_{c3o} \ Q_{c4o} \ Q_{c5o} \ Q_{c6o} \ Q_{c7o} \ Q_{c8o} \ Q_{c9o}$$
$$= (85.397 \ 78.735 \ 79.880 \ 81.436 \ 84.110 \ 81.544 \ 81 \ 72 \ 84)$$

This corresponds to the types of cases, types of courts, lawyer's record, parties, lawyer's resume, ability, industry, subject matter, and complexity of cases, respectively. The result of indicator layer $Q_{pc} = 81.017$, the score is between 80 and 90, so it indicates that the comprehensive evaluation of the lawyer's professional ability is slightly strong.

Table 7. Professional level indicators score.

Indicator Layer	Score	q1	q2	q3	q4	q5	Indicator Layer	Score	q1	q2	q3	q4	q5
P1	85	0.5	1	0.5	0	0	P16	88	0.8	1	0.2	0	0
P2	82	0.2	1	0.8	0	0	P17	86	0.6	1	0.4	0	0
P3	94	1	0.6	0	0	0	P18	65	0	0	0.5	1	0.5
P4	80	0	1	1	0	0	P19	82	0.2	1	0.8	0	0
P5	76	0	0.6	1	0.4	0	P20	94	1	0.6	0	0	0
P6	86	0.6	1	0.4	0	0	P21	96	1	0.4	0	0	0
P7	92	1	0.8	0	0	0	P22	72	0	0.2	1	0.8	0
P8	66	0	0	0.6	1	0.6	P23	87	0.7	1	0.3	0	0
P9	84	0.4	1	0.6	0	0	P24	90	1	1	0	0	0
P10	76	0	0.6	1	0.4	0	P25	92	1	1	0	0	0
P11	78	0	0.8	1	0.2	0	P26	75	0	0.5	1	0.5	0
P12	85	0	1	0.5	0	0	P27	68	0	0	0.8	1	0.2
P13	72	0.5	0.2	1	0.8	0	P28	87	0.7	1	0.3	0	0
P14	84	0.4	1	0.6	0	0	P29	84	0.4	1	0.6	0	0
P15	90	1	1	0	0	0							

Table 6 shows that in the evaluation indicator system of a lawyer's professional ability, the track records, case complexity, and the type of courts have a greater impact on the evaluation results, and the complexity of case has the greatest influence on the evaluation results. In addition, according to Table 7, we can know the lawyer has slightly strong ability, rich resumes, and has an advantage in terms of track records. After a great deal of experiment, it was determined that the result is basically consistent with the evaluation results extrapolated by experts. It further proved the correctness and effectiveness of the lawyer's evaluation model.

4 CONCLUSION

In order to solve the non-objective and inaccurate problems in the lawyer's evaluation process, this research combined the advantages of AHP and fuzzy comprehensive evaluation method and proposed an evaluation method based on improved fuzzy hierarchy analysis. This research provides decision-making basis and a theoretical guarantee for improving the service level of lawyers and the satisfaction of rights defenders.

REFERENCES

Liu, Li, Mingyan Wang, and Baocheng Gu. "Software programming ability evaluation model based on FAHP." Journal of Jishou University (Natural Science) 039.002 (2018): 38–43.
Yanmei, Zhou, and L. I. Weihua. "Enhanced FAHP and its application to task scheme evaluation." Computer Engineering & Applications 44.5 (2008): 212–214.

Computational Social Science – Luo, Ciurea & Kumar (eds)
© *2021 Taylor & Francis Group, London, ISBN 978-0-367-70193-2*

Forecasting on China's final consumption in 2020

X.L. Liu
Academy of Mathematics and Systems Science, Chinese Academy of Sciences, Beijing, China
Center for Forecasting Science, Chinese Academy of Sciences, Beijing, China
University of Chinese Academy of Sciences, Beijing, China

ABSTRACT: China has entered an important stage of sustained growth in consumption demand, accelerated upgrading of consumption structure, and obvious enhancement of the role of consumption in stimulating economy. The analysis and prediction of final consumption is of great significance to actively play the leading role of new consumption, achieve stable economic growth with quality and efficiency, and improve people's quality of life. Based on the analysis of the change trend of the final consumption and its structure and the main influencing factors, this paper forecasts the final consumption of China in 2020 by the method of summing sub-items. It is expected that China's final consumption will maintain a sustained growth trend in 2020 without the COVID-19 out broke, but the growth rate will slow down, with a nominal growth rate of about 5.9% year-on-year. If we considered the impacts of COVID-19, the nominal growth rate will be −3.7%.

1 INTRODUCTION

Under the new normal state of the economy, China has entered into an important phase of sustained growth in consumption demand, accelerated upgrading of consumption structure, and obvious enhancement of a consumption-driven economy. According to the National Bureau of Statistics, the contribution rate of final consumption to gross domestic product (GDP) growth reached 59.7%, 64.6%, 58.8%, 76.2%, and 57.8% from 2015 to 2019. In 2019, the total retail sales of consumer goods reached 41,164.9 billion yuan, an accumulative increase of 8.0%, 1.0 percentage points lower than that in 2018. However, the consumption structure still maintains the trend of upgrading, and the sales growth of consumption upgrading goods is obviously faster than that of other commodities, among which, daily necessities, cosmetics, and beverages continue to maintain a double-digit rapid growth. In 2019, China's consumer expectation index, satisfaction index and confidence index were all at historical high levels. From the perspective of consumption preference, consumers' expression of personality, quality, and fashion consumption demand is becoming increasingly prominent. From the perspective of residents' consumption structure, from 2013 to 2019, the proportion of urban residents' consumption in food decreased by 2.6%, while that in health care increased by 2.0%. The proportion of consumption in transportation and communication increased first and then decreased. From 2013 to 2019, the proportion of rural residents in food consumption decreased by 4.1 percentage points, while that in transportation, communication, and health care increased by 2.1 percentage points and 1.7 percentage points, respectively. The consumption proportion of other industries fluctuated slightly, but the overall stability was stable. This shows that the consumption pattern of Chinese residents from material consumption to service consumption is more obvious. The analysis and prediction of final consumption is of great significance to actively play the leading role of new consumption, achieve stable economic growth with quality and efficiency, and improve people's quality of life.

Scholars have explored the influence of different factors on the final consumption (Varlamova & Larionova 2015); used the Organisation for Economic Cooperation and Development data to confirm that population is an important factor affecting consumption (Biljana & Petar 2012; Foster 2015); used micro-survey data to analyze expenditure of different age groups in the United States and Serbia (Gomez & Foot 2005); researched consumption behavior during retirement in the UK (Mao & Xu 2014; Ni et al. 2014; and used China's micro-household survey data to conduct empirical researches about the relationship between demographic transition and household consumption in China. But China's micro-household survey data often lacks the consumption of rural residents. Secondly, based on US household data, Yun et al. (2015) believed that household borrowing and household indebtedness affected current consumption. Fan and Yavas (2018) researched the crowding out effect of mortgage payments on household consumption with China household data. Browning et al. (2013) and Burrows (2018) also explored the impact of household housing debt on household consumption in Denmark and the UK from the perspective of housing prices. Finally, because mobile payment is just emerging and there is less annual data of online payments available, there are few related literatures about the relationship between online payment and household consumption. However, the latest data showed that the popularity of rural e-commerce has promoted rural residents' consumption. Rural online retail sales increased from 180 billion yuan in 2014 to 1,244.9 billion yuan in 2017. In summary, although there are many articles confirming that household consumption was related to income, population, household debt, and other factors, they often focused on one factor and used micro-survey data. There are few articles on multi-factor macro-analysis about final consumption.

2 THE CHANGE TREND OF FINAL CONSUMPTION IN CHINA

In the first three quarters of 2019, the contribution of final consumption expenditures to economic growth reached 60.5%, 17.2 percentage points lower than that of the same period in 2018. From January to October 2019, China's consumer expectation index, satisfaction index and confidence index showed a fluctuating upward trend, at a historical high level. By November 2019, the expectation index, satisfaction index, and confidence index were 128.9, 118.0, and 124.0, respectively, which showed that residents had strong consumption intention.

The proportion of household consumption in final consumption fluctuated between 72.5% and 74.7% from 2001 to 2018, and the proportion of household consumption expenditure dropped to 72.5% in 2018, the lowest since 2000 (Table 1). In the first three quarters of 2019, the general public budget expenditure in China was RMB 17.9 trillion, an increase of 9.4% over the same period in 2018. The progress of financial expenditure had been accelerated, and key expenditure had been effectively guaranteed. Government consumption, as a major component of fiscal expenditure, has also maintained a growing trend.

The residents' income increases steadily. Income is the most direct and dominating factor in determining the consumption. In 2019, the per capita disposable income of China's urban residents was RMB 42359, with the nominal growth of 7.9% and a real increase of 5.0% after deducting the price factor. The per capita disposable income of rural residents was RMB 16021, with the nominal growth of 9.6%. After deducting the price factor, the actual growth was 6.2%, faster than that of urban residents (Figure 1). In 2019, the urban-rural income ratio fell to 2.69 from 2.64 in 2018, and the urban-rural income gap continued to narrow.

From the perspective of favorable consumption factors, information consumption represented by e-commerce has become a new driving force for economic growth. The scale of rural e-commerce in China has steadily increased. The reform of individual income tax has further stimulated consumption potential. Pension service has become a new driving force to expand consumption in China. Family debt has become a main factor affecting the consumption of Chinese residents in the new era. Other factors restricting consumption growth include slow growth of household income and the lag of supply-side transformation, which is difficult to meet the demand of consumption upgrading, resulting in consumption outflow, inadequate policy support system, and imperfect management mechanisms, and so on.

Table 1. Changes of China's final consumption structure from 2001 to 2018.

Year	The proportion of household consumption	The proportion of government consumption	Proportion of rural residents' consumption in residents' consumption	Proportion of urban residents' consumption in residents' consumption
2001	74.0%	26.0%	32.1%	67.9%
2002	74.4%	25.6%	30.9%	69.1%
2003	74.6%	25.4%	30.0%	70.0%
2004	74.7%	25.3%	28.9%	71.1%
2005	74.2%	25.8%	27.8%	72.2%
2006	73.3%	26.7%	26.9%	73.1%
2007	73.3%	26.7%	25.6%	74.4%
2008	73.2%	26.8%	25.0%	75.0%
2009	73.3%	26.7%	24.2%	75.8%
2010	73.4%	26.6%	23.0%	77.0%
2011	73.2%	26.8%	23.3%	76.7%
2012	73.2%	26.8%	22.8%	77.2%
2013	73.2%	26.8%	22.5%	77.5%
2014	73.9%	26.1%	22.4%	77.6%
2015	73.4%	26.6%	22.2%	77.8%
2016	73.4%	26.6%	21.9%	78.1%
2017	72.9%	27.1%	21.5%	78.5%
2018	72.5%	27.5%	21.4%	78.6%

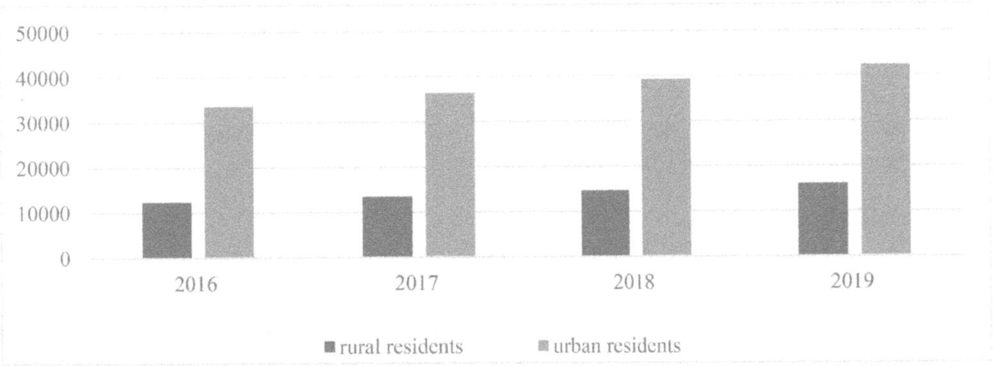

Figure 1. Per capita disposable income of urban residents (yuan/person).
Data Source: National Bureau of statistics of China.

Looking forward to 2020, the main factors conducive to consumption growth are as follows. There have been many favorable policies for consumption in the past two years, such as "the guiding opinions on deeply developing consumption poverty alleviation to help win the battle of poverty alleviation," "opinions on promoting the development of pension services," "opinions on further stimulating cultural and tourism consumption potential," and "opinions on accelerating the development of circulation to promote commercial consumption." The income gap of China's residents is gradually narrowing. In 2019, the per capita disposable income of rural residents grew faster than that of urban residents, and the income ratio of urban and rural residents decreased from 2.69 in 2018 to 2.64; in 2019, the per capita consumption expenditure of rural residents was 13328 yuan, with a nominal growth of 9.9%, which is 2.4 percentage points faster than that of urban residents. The scale of rural e-commerce in China has steadily improved, and the national rural network in 2018 retail sales reached 1.4 trillion yuan, a year-on-year increase of 30.4%; the personal income tax reform further stimulated the consumption potential. In the first three quarters of 2019, China's personal income tax reform reduced 442.6 billion yuan, with an accumulated tax reduction of 1764 yuan per capita. Pension services have become a new driving force to expand consumption, and the proportion of basic pension fund expenditure to pension fund income has increased from 74.2% in 2011 to 86.6% in 2018. People pay more attention to the quality of life and health, and the demand for health and medical services will increase with higher levels, wider coverage, and diversified differentiation. Novel coronavirus pneumonia is the main constraint factor for consumption growth. Some enterprises are operating hard, resulting in unemployment increasing and residents' income growth slowing. The epidemic has had a huge impact on short-term consumption, and the long-term impact also depends on the development of the epidemic. Household debt (mainly housing loans) is the main factor affecting residents' consumption. The crowding out effect of real estate on consumption is obvious; supply side is also a significant factor. It is difficult to meet the demand of consumption upgrading due to the lag of transformation, which leads to consumption outflow; the support of the policy system is not enough, and the management mechanism is not perfect.

3 RESULTS

Based on the analysis of the change trend of the total final consumption, its structure, and the main influencing factors, this paper forecasts the final consumption of China in 2020 by the method of summing sub-items. It is expected that China's final consumption will maintain a sustained growth trend in 2020 without the COVID-19 out broke, but the growth rate will slow down, with a nominal growth rate of about 5.9% year-on-year. If we considered the impacts of COVID-19, the nominal growth rate will be −3.7%.

ACKNOWLEDGEMENTS

This work was supported by grants from the 2019 Chinese Government Scholarship and National Nature Science Foundation of China under Grant No. 71874184.

REFERENCES

Biljana R, Petar V 2012 Household age structure and consumption in Serbia. *Economic Annals*, vol. 57 pp. 79–101.

Browning M, Gørtz M, Leth-Petersen S 2013 Housing wealth and consumption: A micro panel study. *Economic Journal*, vol. 123 pp. 401–428.

Burrows, Vivien 2018 The impact of house prices on consumption in the UK: A new perspective. *Economica*, vol. 85 pp. 92–123.

Fan Y, Yavas A 2018 How does mortgage debt affect household consumption? Micro evidence from China. *Real Estate Economics*

Foster A C 2015 Consumer expenditures vary by age. *Beyond the Numbers*, vol. 4 pp. 1–11.

Gomez R, Foot D K 2005 Age structure, income distribution and economic growth. *Canadian Public Policy*, vol. 29 pp. 141–162.

Mao R, Xu J W 2014 Demographic transition, consumption structure disparities and industrial growth. *Population Research* (In Chinese), vol. 38 pp. 89–103.

Ni H F, Li S T, He J W 2014 Impacts of demographic changes on consumption structure and savings rate. *Population & Development* (In Chinese), vol. 20 pp. 25–34.

Varlamova J, Larionova N 2015 Macroeconomic and demographic determinants of household expenditures in OECD countries. *Procedia Economics & Finance*, vol. 24 pp. 72–733.

Yun KK, Mark S, Yuan M 2015 Aggregate consumption and debt accumulation: An empirical examination of US household behavior. *Cambridge Journal of Economics*, vol. 39 pp. 93–112.

Computational Social Science – Luo, Ciurea & Kumar (eds)
© 2021 Taylor & Francis Group, London, ISBN 978-0-367-70193-2

Text quantitative analysis on administrative punishment for the construction of a case analysis knowledge graph

C.X. Li, J.M. Li* & M.J. Du
School of Information Sciences, 2Business School, Beijing Language and Culture University, Beijing, China

ABSTRACT: The administrative punishment announcements of financial fraud issued by the China Securities Regulatory Commission (CSRC) are unstructured texts, and it is difficult for them to effectively play their role of reference and decision-assisting without pre-processing and resorting. This paper uses quantitative analysis method to analyze the administrative punishment announcements texts of financial fraud issued by the CSRC. It counts the number of the administrative punishment announcements issued from 2011 to 2019, analyzes the trends, and summarizes the categories of financial fraud and punishment methods. This paper analyzes the content of the administrative punishment announcements and designs the conceptual model for the administrative punishment case, which is conducive to the construction of a case analysis knowledge graph of administrative punishment for financial fraud and realizing the text analysis and case reasoning based on artificial intelligence.

1 INTRODUCTION

With the rapid development of China's securities market, financial fraud has frequently appeared in enterprises of different types and sizes, which not only increases the difficulty of auditing and supervision for regulatory authorities but also reduces the confidence of investors in domestic enterprises and causes a trust crisis in the capital market (Bingzhou & Shaojie 2020; Dandan 2019; Yuhe 2018). The administrative punishment announcements for financial fraud issued by the China Securities Regulatory Commission (CSRC) contain specific cases of financial fraud and can provide case information for regulatory authorities and stakeholders to assist them in decision making (Hanlin & Youqing 2019). At the same time, the announcements also have a reference role for the listed companies and internal personnel. However, the administrative punishment announcements for financial fraud currently published on the official website of the CSRC contain a wide range of information such as companies, people, fraud time, and punishment decision, and so on. Moreover, the administrative punishment announcements are unstructured texts and cannot directly display various entities and relationships (Long 2019). At present, the analysis of the financial fraud punishment announcements does not go far enough (Yuan 2020), which could not meet the need for the regulatory authorities and stakeholders. Therefore, how to store and display the text data of those announcements systematically has become an urgent problem to be solved.

As an effective means of complex data analysis and decision support, a knowledge graph is gradually and widely being used in various fields such as justice, finance, and medical care as data resources increase and computing power improves (Feng 2020; Zhen et al. 2019b). A knowledge graph shows rich application value in many aspects, such as assisting intelligent question answering, natural language understanding, and big data analysis (Yiming et al. 2019). It can connect multiple data nodes to discover hidden information (Hengqi et al. 2019), and at the same time it can process a large number of professional documents of enterprises and ensure high processing

---

*Corresponding author

accuracy (Pengfei et al. 2020). By combining a knowledge graph with administrative punishment announcements, the administrative punishment announcements can effectively play their role of reference and decision-assisting (Qingfeng 2019).

This paper makes a quantitative analysis of the administrative punishment announcements texts of financial fraud issued by the CSRC and proposes the need to construct a case analysis knowledge graph based on administrative punishment announcements texts of financial fraud. Then, through the analysis of the content of the administrative punishment announcements texts, this paper designs the conceptual model of case analysis to facilitate the construction of a case analysis knowledge graph and help the regulatory authorities to efficiently understand the information in the administrative punishment announcements and enhance the decision-making ability of stakeholders.

2 RELATED WORKS

This paper will make quantitative analysis and then analyze the content of the announcements texts to lay the foundation for the construction of a case analysis knowledge graph. Related works mainly include the following: (1) Based on knowledge management theory, Feng Xinling and others compared and distinguished mapping knowledge domain and a knowledge graph from the perspective of knowledge acquisition, knowledge organization, knowledge storage, knowledge sharing, and knowledge innovation (Xinling et al. 2017). This research helps this paper figure out the difference between mapping knowledge domain and knowledge graph. (2) From the perspective of "artificial intelligence+", Li Zhen et al. introduced the relationship between knowledge graph and artificial intelligence and analyzed connotation of an educational knowledge graph from the perspective of knowledge model and resource management and then introduced the classification and technical framework of an educational knowledge graph (Zhen et al. 2019a). The ontology construction method of the educational knowledge graph proposed by this research provided ideas for this paper. (3) Ma Can built a knowledge graph for the "smart court." This research extracted important structured knowledge from a large number of case judgment text. It was oriented toward the case and associated important knowledge such as courts, judges, plaintiffs, and defendants, and stored the data in a Neo4j graph database, which broke the independence between cases (Can 2019). When building the structure of a knowledge graph, the research figured out five entity categories, six relationships for the court judgments text, which provided ideas for the construction of the knowledge graph of the punishment announcements for this paper. (4) Qiao Gangzhu and others have carried out an intelligent reasoning research on theft cases based on the knowledge graph. This study deeply explored the implicit relationships involved in theft cases and presented an ontology-based method for constructing knowledge graph theft cases (Gangzhu et al. 2019). However, it was not efficient for them to use manual methods to extract knowledge.

To sum up, there are few related studies on text quantitative analysis of knowledge graph construction in existing research, and there is a lack of research focusing on the construction of a knowledge graph for case analysis of financial fraud administrative punishment. Therefore, this paper will make a quantitative analysis of the administrative punishment announcements texts; analyze the concepts, entities, and relationships therein; and design the conceptual model to facilitate knowledge graph construction.

3 TEXT QUANTITATIVE ANALYSIS

This section mainly includes two subsections: data source and selection, data analysis.

3.1 Data source and selection

There are many regulatory contents involved in the administrative punishment announcements, including financial supervision, environmental pollution supervision, and production quality supervision (Peihua 2020). In this paper, the selected administrative punishment announcements are

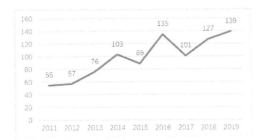 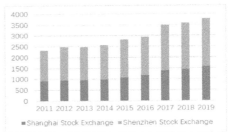

Figure 1. The number of announcements.
Data source: Official website of China Securities
Regulatory Commission.

Figure 2. The number of listed companies.
Data source: Statistical yearbooks of Shanghai
Stock Exchange and Shenzhen Stock Exchange.

issued by the CRSC, and the regulatory content is financial fraud of listed companies in the Shanghai Stock Exchange and Shenzhen Stock Exchange. The relevant regulations are the Securities Law of the People's Republic of China and The Administrative Punishment Law of the People's Republic of China. So far, this paper has collected 882 administrative punishment announcements for financial fraud from 2011 to 2019 by Python crawler. This paper also collects the related regulations texts. The administrative punishment announcements for listed companies and the relevant regulations studied in this paper are unstructured text data.

3.2 *Statistics and analysis on punishment announcements*

This paper adopts quantitative analysis methods (Jiayin & Xiaozhe 2016; Rui 2016). It adds up the number of announcements issued from 2011 to 2019 and studies its quantity change trend, and introduces the listed company's quantity for comparison. This study defines the types of financial fraud based on the Securities Law and the keywords of financial fraud described in the announcements, and finds frequent financial fraud. Then it counts and analyzes punishment methods.

Statistics of punishment announcements quantity. The annual quantity of financial fraud administrative punishment announcements issued by the CSRC is shown in Figure 1, and Figure 2 shows the number of listed companies in China from 2011 to 2019. We can see that: (1) Except for 2015 and 2017, the number of announcements is generally on the rise, and the total number of listed companies in recent years is increasing. It can be inferred that with the increase of the number of listed companies, the number of financial frauds is also increasing; (2) Since 2014, the number of administrative punishment announcements has risen to a new height, and the number has begun to break through 100. This is because with the rapid development of the securities market, the regulatory authorities have also strengthened auditing and supervision; (3) In 2016, the number increased fastest with the growth rate of 41.8%. This was mainly because in 2016, the CSRC carried out first special law enforcement activities to maintain the order of the securities market (Xun & Yan 2017).

Classification of financial fraud. Based on the regulations and the keywords in the administrative punishment announcements, this paper divides financial fraud into two categories: accounting fraud and capital fraud. The subcategories are described next.

Accounting fraud

- Accounting confirmation violation: Intentionally causing the discrepancy between account books' time and accrual confirmation time, and the discrepancy between the definition and characteristics of accounting elements.
- Accounting measurement violation: Intentionally causing the discrepancy between account quantity and the actual quantity.

Figure 3. The proportion of financial fraud.
Data source: Based on the punishment announcements collected in this study.

- Accounting report violation: Including the following four subcategories:
  a) False record: Information disclosure obligors record non-existent facts in the information disclosure document, for instance: false increase accounts receivable, fixed assets
  b) Delayed disclosure: Failing to disclose information at legal time
  c) Major omission: Incomplete disclosure of information
  d) Misleading statements: Statements that make investors misjudge

Capital fraud

- Fraudulent listing: An issuer who does not meet the conditions for issuance but obtains the issuance approval by deception.
- Capital contribution violations: False declaration of registered capital, false capital contribution, and capital withdrawal.
- Unauthorized use of funds: Not using funds according to regulations.
- Illegal operation: Operating securities business without the approval of relevant departments.
- Occupy company assets: The people of the enterprise take advantage of their position to illegally occupy the property as their own.
- Illegal guarantee: Without the approval of any internal decision-making body, the actual controller, chairman of the board, or other authorized person directly instructs the relevant personnel to stamp the company seal on the guarantee contract.
- Manipulating stock price: Stock investors control the future stock price trend by controlling stock investment information that have reference significance for other investors.
- Illegal trading of stocks: The securities staff buy and sell stocks in violation of work regulations.
- Insider trading: Insiders buy or sell securities based on inside information.

Figure 3 shows the proportion of financial fraud categories that occurred between 2011 and 2019.

Figure 3 shows that from 2011 to 2019, report violations occurred more frequently in accounting fraud, where false records, major omissions, delayed disclosure, and misleading statements accounted for 42%, 32%, 15%, and 11% of accounting fraud, respectively. In the category of capital fraud, insider trading, manipulation of stock prices, and illegal trading of stocks occurred more frequently and accounted for 57%, 21%, and 18% of this category, respectively. On the whole, insider trading occurred most frequently in recent years and accounted for 34% of the total, followed by false records that accounted for 17% and major omissions that accounted for about 13%.

Punishment methods. This paper divides punishment methods into seven types in accordance with the Administrative Penalty Law of the People's Republic of China. Figure 4 (where only the punishment methods of financial fraud with high frequency are listed) shows the different punishment methods adopted by the CSRC for different categories of financial fraud. It can be seen from Figure 4 that the most frequent punishment method used by the CSRC is fines, followed by warnings, and the third is confiscation of illegal income, while the rectification is relatively less. We can see that when it comes to financial fraud, the CSRC tends to implement the punishment decision in compliance based on the regulations, which could make listed companies take warning.

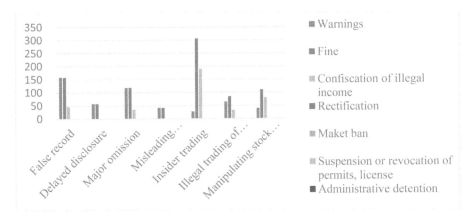

Figure 4. Punishment methods.
Data source: Based on the punishment announcement collected in this study.

Through the above-mentioned analysis, it can be found that with the development of China's securities market, the phenomenon of financial fraud is constantly emerging. The regulatory authorities closely follow the requirements of national laws and continuously strengthen the supervision. However, there is much information involved in the financial fraud cases, which is not effective for the analysis of regulatory authorities and stakeholders. Therefore, it is necessary to construct the knowledge map.

4 CASE ANALYSIS

4.1 Content analysis of administrative punishment announcements text

This section studies the content of punishment announcements and analyzes the concepts, entities, and relationships required to facilitate the construction of the knowledge graph of the punishment announcements. The information contained in the announcement on administrative punishment of financial fraud issued by CSRC is mainly summarized as follows:

- Information of litigant: Basic information of the litigant of financial fraud; litigant includes the person and the company
- Summary of the case: A brief summary of the facts of financial fraud
- Detailed description of illegal facts: Specific acts of financial fraud
- Laws and regulations: The laws and regulations formulated by the state, and also the basis for the CSRC's judgment
- Defense opinions: Defense reasons for financial fraud, and the CSRC's handling process and results of the defense reasons
- Punishment decision: Punishment methods decided by the CSRC for the financial fraud in accordance with laws and regulations
- Announcement: The release time of the punishment announcements and the issuing authority

Figure 5 shows the content of the punishment announcement.

4.2 Conceptual model

Through the preceding analysis of the administrative penalty announcement text, this paper combs the concepts, entities, and relationships in the construction of the knowledge graph and designs a conceptual model diagram, as shown in Figure 6. This paper sorts out nine concepts and 18

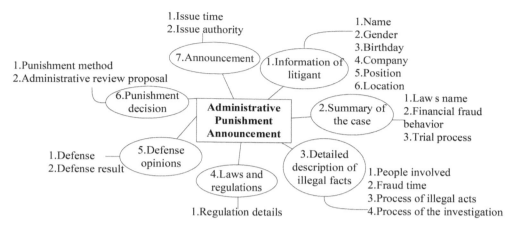

Figure 5. The content of the punishment announcement.

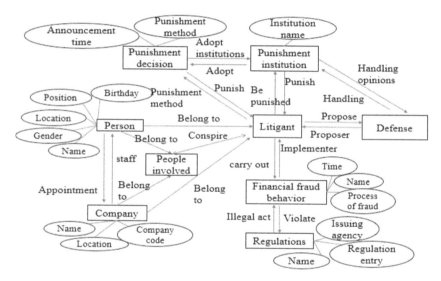

Figure 6. Conceptual model.

relationships. The nine concepts include person, punishment agency, litigant, person involved, company, and financial fraud. The relationships among them include punishment relationship and collusion relationship. This paper will take the two concepts of "litigant" and "punishment agency" as examples to introduce the conceptual model diagram:

- Litigant includes person and company. When the litigant is a natural person, it has attributes such as name, gender, and position. When the litigant is a legal person (i.e., a company), it has attributes such as name, company code, and company location.
- The attribute of punishment agency is the institution name.
- Litigant -> punishment agency: The litigant is punished by the punishment agency.
- Punishment agency -> litigant: The punishment agency punishes the litigant.

This paper sorts out basic information involved in the financial fraud case and designs the conceptual model in the administrative punishment announcements by analyzing the case information in order to provide a research idea for the construction of a case analysis knowledge graph of financial fraud administrative punishment.

5 SUMMARY

This paper uses a quantitative analysis method. Taking the announcement of the administrative punishment for financial fraud issued by the CSRC from 2011 to 2019, and related laws and regulations, this paper analyzes the quantitative changes of financial fraud, the categories of financial fraud, and punishment methods. Based on the quantitative analysis of the texts, this paper puts forward the demand of constructing the case analysis knowledge graph of administrative punishment and has done three jobs. The first is to have sorted out the content of administrative punishment announcements. The second is to have analyzed concept, entity, and relationship, which are contained in the knowledge graph. The last is to have designed the conceptual model. This paper puts forward a research foundation for the construction of the administrative punishment case knowledge graph.

ACKNOWLEDGMENTS

This work was supported by the Science Foundation of Beijing Language and Cultural University (supported by the Fundamental Research Funds for the Central Universities, and the Research Funds of Beijing Language and Culture University) (Approval number: 18PT02, 19YJ040004, 20YCX157), by the Discipline Team Support Program of Beijing Language and Culture University (No. GF201905), by a grant from the National Science Fund of China (No. 61972052), by the Special Fund of Beijing joint construction project, and by the Computer Foundation Education Research Institute of National Colleges and Universities (No. 2020-AFCEC-116).

REFERENCES

Bingzhou W and Shaojie 2020 Interest-driven, intermediary endorsement and financial fraud of listed companies—Based on the 2008–2017 CRSC Penalty Announcement *Accounting and Communications* pp 1–5.

Can M 2019 Construction method and research of knowledge graph for "wisdom court" *Guizhou University*.

Dandan W 2019 Causes and Countermeasures of Audit Failure *Beijing Jiaotong University*.

Feng S 2020 The Application of Knowledge Graph in Network Security of Financial Institutions *Financial Technology Era*, pp 82–85.

Gangzhu Q, Tingting F and Guochen Z 2019 Research on Intelligent Reasoning of Legal Documents in Theft Cases Based on Knowledge Graph *Application of Computer Systems* pp 206–213

Hanlin Z and Youqing W 2019 Research on Information Disclosure of Listed Companies—Based on 2013–2017 CSRC Penalty Announcement *Institute of Management Science and Industrial Engineering. Proceedings of 2019 9th International Conference on Education and Social Science* (ICESS 2019) pp 2146–2153

Hengqi H, Juan Y, Xiao L and Yunjiang X 2019 Summary of Knowledge Graph Research Application of Computer Systems pp 1–12

Jiayin Z and Xiaozhe Y 2016 Quantitative analysis of my country's Internet financial policy text *Zhejiang Finance* pp 3–10

Long W 2019 Identification of corporate financial report fraud in the era of big data *Modern Economic Information* pp 199–200

Peihua Z 2020 Negative Reputation and Corporate Finance—Empirical Evidence from Listed Companies for Penalty for Violation *Finance and Trade Economics* pp 50–65

Pengfei Z, Zhixiang Y, Wei B and Xudong H 2020 Applied Research on Construction Method of Knowledge Graph Oriented to Green Standards *Standard Science* pp 68–73

Qingfeng W 2019 Research and Implementation of Enterprise Portrait Technology Based on Knowledge Graph *Harbin Institute of Technology*

Rui X 2016 Case Study of Huaze Cobalt Nickel Audit Failure *Shandong University*

Xinling F, Sheng H, Taichun X, Qunhui W and Yijun L 2017 A Comparative Analysis of "Science Knowledge Graph" and "Google Knowledge Graph"—Based on the Perspective of Knowledge Management Theory *Journal of Information* pp 149–153

Xun W and Yan W 2017 Securities market audit failure and supervision—An analysis based on the CSRC punishment notice in 2001–2016 *China Certified Public Accountants* pp 86–91

Yiming Z, Jie Z, Xiaolong H, Yijie S and Hao L 2019 Intelligent question answering and medication recommendation system based on knowledge graph of traditional Chinese medicine *Electronic Technology and Software Engineering* pp 134–135

Yuan Y 2020 Analysis of the characteristics of accounting firms subject to administrative punishment by the CSRC—Based on the announcement of the administrative punishment by the CSRC from 2009 to 2019 *Chinese & Foreign Entrepreneurs* p 251

Yuhe Y 2018 Financial Fraud Listed Companies' Internal Control Defects and Countermeasures-Based on the Analysis of the 2016–2017 Penalty Announcement by the China Securities Regulatory Commission *Modern Business* pp 121–123

Zhen L, Dongdai Z and Yong W 2019a Educational Knowledge Graph from the Perspective of "Artificial Intelligence+": Connotation, Technical Framework and Application Research *Distance Education Journal* pp 42–53

Zhen L, Xiaoxiao D, Dongdai Z and Tingting T 2019b Human-computer collaborative construction method and application research of knowledge graph in adaptive learning system *Modern Education Technology* pp 80–86

Computational Social Science – Luo, Ciurea & Kumar (eds)
© *2021 Taylor & Francis Group, London, ISBN 978-0-367-70193-2*

Exploration and practice of curriculum for ideological and political education: Taking signals and systems as an example

C. Ji & C.H. Cao
School of Computer, Northeastern University, Shenyang, China
Northeast University PBL Teaching Innovation Research Center, Shenyang, China

G.Y. Zhang, R. Geng & J. Wang
School of Computer, Northeastern University, Shenyang, China

ABSTRACT: Carrying out ideological and political education in the teaching of professional courses has become an important link for colleges and universities to fulfill the fundamental task of cultivating people with moral standing. This article runs through the ideological and political content in the "Signals and Systems" course, realizes the educational transformation from an ideological and political course to an ideological and political content of curriculum," and proposes a specific curriculum ideological and political teaching plan, so as to achieve effective connection between ideological and political case and curriculum content. Through ideological and political curriculum, education can stimulate students' motive of the power of learning; improve students' political literacy, and firm students' political ideas.

1 INTRODUCTION

The National Conference on Ideological and Political Work in Colleges and Universities in 2016 stressed that ideological and political work should be carried out throughout the whole process of education and teaching; new media and new technologies should be used to make work come alive; and traditional advantages of ideological and political work should be highly integrated with information technology, so as to enhance the sense of times and appeal (National Conference on Ideological and Political Work in Colleges and Universities 2016). In 2019, General Secretary Xi pointed out at the school's ideological and political theory teacher's symposium that it is necessary to adhere to the unity of explicit and implicit education, and to tap the ideological and political resources contained in the curriculum, so that various courses and ideological and political courses are in the same direction (Study and Implement the Spirit of General Secretary Xi Jinping's Important Speech at the Symposium for Teachers of Ideological and Political Theory in schools 2019).

In this context, as a professional teacher, we need to think deeply about how to introduce ideological and political education into the professional classroom. In the teaching of specialized courses, the elements and links of ideological and political of curriculum and professional ideological and political should be added to upgrade the former pure professional knowledge teaching to the level of a higher education system with Chinese characteristics. Let our students know what kind of person they want to be and what kind of contribution they want to make to the motherland in the future.

2 THE NECESSITY OF "SIGNALS AND SYSTEMS" COURSE IDEOLOGICAL AND POLITICAL

"Signals and Systems" is an important professional basic course for communication and electronic information majors. It is also a postgraduate entrance examination course. Through the study of this course, students can not only master the basic theories and methods of signals and systems, but more importantly, it cultivates students' interest in the study of this major.

The course of "Signals and Systems" has many theoretical contents, strong logic, and complicated mathematical derivation. Most of the current teaching situation of this course is dominated by professional knowledge taught by teachers, and students' vision is often limited to the study of theoretical knowledge of professional courses. Students often ignore such problems: Why study this course? What is the impact that learning this course well will have on my future career? How will it establish a correct outlook on life and values? What is the present development situation of China's communication industry? And these are the fundamental problems that stimulate students' motive of the power of learning (Zhang et al. 2020).

To this end, according to the characteristics of the "Signals and Systems" course, we dig deeper into the ideological and political elements contained in the course. In class, in addition to teaching basic knowledge, practical cases are introduced to carry out the ideological and political of curriculum, to guide young students to establish a correct outlook on life, values, and career; give full play to the implicit educational function of professional courses; and promote the educational transformation from ideological and political course to ideological and political content of curriculum (Gong & Lan 2020).

3 THE INTRODUCTION OF THE IDEOLOGICAL AND POLITICAL CONTENT OF THE "SIGNALS AND SYSTEMS" COURSE

The main contents of "Signals and Systems" course include time domain analysis of signal and system, and signal and system transform domain analysis, and so forth. Now, taking the textbook *Signal and Linear System Analysis* compiled by several teachers of the communication major of Northeastern University, where the author works, as an example (Ji et al. 2018), the key knowledge points and ideological and political elements of this course are integrated, the course content is re-planned, the ideological and political connotation is excavated, and the breakthrough point of the ideological and political content of curriculum is proposed. The teaching units, teaching contents, and corresponding ideological and political breakthrough points of the "Signals and Systems" course are shown in Table 1.

Table 1. Teaching unit and teaching content of ideological and political content of curriculum.

Teaching unit	Teaching content	Ideological and political breakthrough point
Time domain analysis of continuous time signals and systems	1.1 Description and classification of signals 1.2 Commonly used continuous time signals and features 1.3 Basic operation and decomposition of signals	The ancients used the beacon and wolf smoke (light signal) and drumming gold (sound signal) to send alarms or commands, so as to understand the wisdom of the ancient Chinese and promote patriotism.
Frequency domain analysis of continuous time signals and systems	3.1 Orthogonal decomposition of signals 3.2 Fourier series analysis of periodic signals	The time domain signal is decomposed into sinusoidal signals, which opens the door for frequency domain analysis, and then produces a series of important research achievements such as sampling theorem and Parseval energy conservation theorem, thus revealing the importance of basic theory. In this way, students are encouraged to lay a good foundation in order to make achievements on the road of future growth and contribute to the country and society.

(Continued)

Table 1. (Continued)

Teaching unit	Teaching content	Ideological and political breakthrough point
Frequency domain analysis of continuous time signals and systems	3.3 Spectrum of periodic signals 3.4 Spectrum of aperiodic signals—Fourier transform	Take the spectrum analysis of ECG signals as an example to make students cherish their health and resonate.
Frequency domain analysis of continuous time signals and systems	3.8 Fourier Transform of sampling signal and sampling theorem	Derived from the sampling theorem to strictly abide by various standards and regulations, and develop good behavior habits to enhance students' awareness of observing discipline and law.
Frequency domain analysis of continuous time signals and systems	3.11 Ideal low pass filter	Constructing a filter using an idealized model makes the problem simple and easy to analyze. In the face of problems, students are encouraged not to be intimidated by difficulties. Complex problems can be disassembled, and problems can be simplified by using limit thinking and equivalent thinking.
Frequency domain analysis of continuous time signals and systems	3.12 Modulation and multiplexing	The frontier knowledge related to communication, such as quantum communication technology, power carrier communication and other new technologies, and the application of these technologies in military and daily life, are interspersed in the content of modulation and demodulation to improve students' learning motivation and enhance their professional confidence.
Complex frequency domain analysis of continuous time signals and systems	4.1 Laplace transform 4.2 Properties of the Laplace transform	Introduce Laplace's contribution to this course and his great achievements, so that students experience the research spirit of scientists in the learning process, so as to set lofty goals and make their own contributions to the development of society.
Complex frequency domain analysis of continuous time signals and systems	4.5 Complex frequency domain analysis of linear systems	Analyze the same system from the perspective of time domain, frequency domain, and complex frequency domain, and solve the relationship between excitation and response, so it can lead to the need to look at the problem from multiple angles. Many things or problems have a certain degree of complexity and require all-around analysis, thinking from multiple angles and sides, in order to fully understand things correctly.
Complex frequency domain analysis of continuous time signals and systems	4.9 System stability	The introduction of negative feedback can make the system more stable, thus leading students to be good at reflection, so that they can find their own shortcomings, which plays a very important role in the process of human growth.
Time domain analysis of discrete time signals and systems	5.6 Use MATLAB to realize time domain analysis of discrete time signals and systems	This leads to practice as the criterion for testing theory, which is an important scientific accomplishment, which in turn cultivates students' materialistic worldview and rigorous style. In the process of completing the experimental project, students are cultivated to keep improving, use team work, and pursue the excellence of the craftsman spirit.
Z-domain analysis of discrete time signals and systems	6.1 Z-transform of discrete time signals 6.2 Basic properties of Z-transform 6.7 Frequency response characteristics of discrete time systems	The application of Z transformation in the military field, especially radar systems, can enhance students' patriotism and realize that the pursuit of science is endless.

4 SUMMARY

The ideological and political construction of the "Signals and Systems" course aims to integrate the ideological and political education of college students with the teaching of professional courses, cultivate the students' belief to keep improving, set up the goal of pursuing excellence, stimulate students' desire for knowledge, cultivate students' engineering consciousness and innovation ability, and improve their ideological and political consciousness and psychological quality.

ACKNOWLEDGMENTS

This paper is supported by 2017 "Cooperation Between Production and Education" Education Content and Curriculum System Reform Project of the Ministry of Education (201702185012); this paper is supported by 2020 Northeastern University PBL Teaching Method Research and Application Project (NEUJX04332, NEUJX04306); this paper is supported by Northeastern University Teacher Development Program (DDJFZ202005).

REFERENCES

Chengying Gong, Conghua Lan. Exploration on Curriculum Construction of "Signals and Systems" based on Ideological and Political of Curriculum, Major and Curriculum Construction, 2020(4):19–23.

Ce Ji, Dingde Jiang, Qingyang Song, Yao Yu. Signal and Linear System Analysis [M], Beijing: Science Press, November 2018.

National Conference on Ideological and Political Work in Colleges and Universities, Ministry of Education Government Portal, December 2016.

Study and Implement the spirit of General Secretary Xi Jinping's important Speech at the Symposium for Teachers of Ideological and Political Theory In schools, Ministry of Education Government Portal, March 2019.

Rufeng Zhang, Jing Xiang, Yajuan Zhang, Jinlong Cui, Xinxin Feng. Discussion on Ideological and Political of Curriculum of "Signals and Systems". Light and Textile Industry and Technology, 2020, 49(05):111–112.

Computational Social Science – Luo, Ciurea & Kumar (eds)
© 2021 Taylor & Francis Group, London, ISBN 978-0-367-70193-2

Author index

Amenduni, F. 227

Cai, S.W. 711
Cao, C.H. 1013
Cao, J. 473, 734
Cao, P.Z. 620
Cao, S.Q. 531
Cao, X. 770
Cao, X.Y. 233
Cao, Y. 525, 566
Cao, Y.J. 255, 394, 468
Cao, Y.X. 682
Chai, J.J. 603
Chen, B.C. 283
Chen, G. 625
Chen, H.J. 856
Chen, H.S. 596
Chen, J. 531, 910
Chen, K. 215
Chen, L. 191
Chen, M.C. 136
Chen, W. 154, 718
Chen, W.H. 293
Chen, W.J. 132, 900
Chen, X.B. 283, 302, 836
Chen, X.F. 508
Chen, X.H. 20
Chen, Y.J. 636
Chen, Z.Z. 202
Cong, S. 693
Cong, Y. 697
Cui, H.Z. 54, 185
Cui, X. 271

Dai, C. 561
Dai, H.Y. 986
Ding, M.Y. 431
Ding, X.Y. 875
Dong, H.F. 376
Du, J.H. 34
Du, J.J. 546
Du, K. 924
Du, M.J. 1005
Du, Q. 641

Du, S.Y. 752
Duan, Z.Y. 770

Fan, J. 718
Fan, J.M. 730
Fan, Y. 376
Fang, Y. 758
Feng, H.W. 436
Feng, J.M. 321
Feng, Q.S. 321
Feng, W.F. 221
Fu, G. 309, 782
Fu, M.R. 669
Fu, Y. 86
Fu, Y.H. 905

Gao, F. 776, 918
Gao, G.L. 881
Gao, H.J. 665
Gao, H.Y. 881
Gao, J.C. 546
Gao, P.B. 504, 551
Gao, W.X. 28
Geng, R. 1013
Geng, Y.W. 730
Gong, C. 758
Gong, J.J. 546
Graciano, F. 451
Gu, X.Y. 144
Guo, H.S. 682
Guo, J.P. 102
Guo, M.J. 28
Guo, W. 486
Guo, X.L. 711

He, X.K. 11, 136
Hsu, H.C. 539
Hsu, H.L. 539
Hu, H.H. 191
Hu, H.Q. 590
Hu, X.L. 566
Hu, Y. 603
Huang, J.W. 576
Huang, P.H. 596
Huang, Z.H. 993

Imanche, S.A. 829

Ji, C. 1013
Jia, S.L. 221
Jiang, B.C. 389
Jiang, H.Y. 740
Jiang, J. 703
Jiang, J.Y. 389
Jiang, N. 66
Jiang, W.X. 758
Jin, C.H. 576, 590
Jin, S.C. 881

Kang, T. 334, 422
Ke, Z.Y. 480
Kuzminykh, V. 102

Lei, Z.C. 341
Li, A.Q. 993
Li, C. 66
Li, C.H. 86
Li, C.X. 1005
Li, D.D. 746
Li, D.Y. 546
Li, H. 836
Li, H.F. 132, 900
Li, J.M. 1005
Li, J.P. 446
Li, L. 356, 841
Li, L.C. 271, 669
Li, L.J. 442
Li, S.S. 102
Li, W. 520
Li, X. 890
Li, X.H. 495, 556, 881
Li, Y. 693
Li, Y.H. 520
Li, Y.L. 442
Li, Y.Q. 603, 669
Li, Z. 504, 551
Li, Z.W. 584
Li, Z.Y. 525
Lian, S.Z. 136
Liang, J. 446
Liao, C.Y. 79

Liao, X.H. 365
Ling, D.X. 191
Liu, C. 20, 546
Liu, C.H. 782
Liu, C.S. 584
Liu, D.Q. 210
Liu, G.Q. 86
Liu, J. 682, 740
Liu, J.J. 546
Liu, L. 956
Liu, M.R. 369
Liu, Q.H. 776, 799, 918
Liu, R.H. 910
Liu, T. 215
Liu, W.B. 525
Liu, W.T. 782
Liu, X. 197
Liu, X.L. 267, 1001
Liu, X.R. 703
Liu, X.Y. 34
Liu, Y. 356, 630
Liu, Y.C. 179
Liu, Y.X. 504, 551
Liu, Z. 770
Lu, Q. 309
Lu, X. 571
Lu, X.J. 972
Lu, Y. 144
Luan, D.Y. 770
Luo, M.G. 697
Luo, Q. 824
Luo, X.L. 770
Luo, Y.S. 682, 758
Lv, Y. 66

Ma, X.L. 500
Ma, Y. 514
Meng, X.Y. 419
Miao, L.Y. 401, 442
Millman, C. 356

Ni, X.L. 730
Nie, H. 136

Ou, S. 669
Ouyang, W.M. 436

Pan, X.H. 881
Pei, Z.B. 3, 250, 711, 722
Peng, C. 890
Peng, X.F. 752
Peng, Y. 918
Poce, A. 227

Qi, Z.G. 11
Qin, X.M. 353
Qu, H.N. 875
Qu, X.H. 764
Qu, Z. 571

Rehman, M.U. 86
Ruan, Y.J. 45

Sha, J.Q. 566
Shan, J. 867
Shao, X.Y. 615
She, J.H. 215
Shen, Y.J. 96
Sheng, C. 309
Sheng, Z.J. 641
Song, H.M. 630
Sun, J. 546
Sun, J.F. 894
Sun, J.L. 59
Sun, L. 811
Sun, L.Q. 665, 697
Sun, X.H. 630
Sun, X.M. 197

Tan, H. 233
Tan, Q.K. 329
Tan, W. 881
Tang, H. 59
Tang, L. 108
Tang, L.L. 740
Tang, P. 459
Tang, W. 329, 356, 794
Tang, X.D. 571
Tang, X.X. 905
Tang, Z.B. 154
Tasinda, O.T. 829
Tian, J.H. 514
Tian, Z. 829
Tong, X.Y. 811
Tu, Y. 682

Wan, Z.B. 993
Wang, C. 54, 185, 486
Wang, F. 451
Wang, G.M. 603
Wang, H. 163
Wang, H.P. 776, 799, 918
Wang, J. 1013
Wang, L.S. 376
Wang, P. 348, 850
Wang, R.M. 576
Wang, S. 836

Wang, S.H. 609
Wang, S.X. 656
Wang, T. 376
Wang, W.D. 856
Wang, W.H. 571
Wang, X. 34
Wang, X.C. 520
Wang, X.Q. 419
Wang, Y. 92, 566
Wang, Y.G. 486
Wang, Y.N. 972
Wang, Y.P. 730
Wang, Z. 11
Wang, Z.C. 856
Wei, H. 302
Wen, J. 818
Wu, C.S. 924, 942, 956
Wu, D.L. 546
Wu, J.X. 154
Wu, P. 329, 794
Wu, S.S. 508
Wu, W.W. 504, 551
Wu, W.Y. 651
Wu, X.W. 856
Wu, Y. 202
Wu, Y.F. 625
Wu, Y.J. 905
Wu, Y.M. 647

Xia, L. 277
Xia, L.X. 79
Xia, N. 293
Xia, Y. 79
Xiahou, J.B. 271
Xian, X. 197
Xian, X.L. 243
Xiang, B.L. 20
Xiang, Y.H. 221
Xiao, J.J. 669
Xiao, Q.T. 703
Xiao, X.T. 805
Xie, B.X. 468
Xie, Y.M. 446
Xie, Y.T. 986
Xiong, Y. 656
Xiong, Y.H. 905
Xu, B.B. 102
Xu, C. 3, 722
Xu, F.H. 609
Xu, M.H. 185
Xu, M.S. 54, 185
Xu, S.S. 846
Xu, X.L. 451

Xue, H.J. 382
Xue, Z. 28

Yan, H.Y. 875
Yan, M.X. 321
Yan, Q.Z. 514
Yan, S.S. 473, 734
Yan, T.M. 250
Yan, W.Y. 596
Yan, Z.J. 818
Yang, D.D. 334, 422
Yang, J.J. 584
Yang, N. 956
Yang, N.N. 867
Yang, X.L. 20
Yao, S. 525
Ye, B. 956
Ye, Q.P. 651
Ye, X.Y. 191
Ye, Y.X. 636
Yi, H. 20
Yu, B.B. 677
Yu, H. 197, 811
Yu, J.B. 277
Yu, J.F. 102
Yu, J.H. 255, 468
Yu, M. 365
Yuan, J.Y. 596

Zeng, F. 408
Zeng, H. 861, 979
Zeng, Y.L. 154
Zha, Y.Y. 54
Zha, Z.X. 508
Zhai, H.Y. 770
Zhan, L.Y. 875
Zhan, Y. 401, 442
Zhang, C. 197
Zhang, D.J. 341
Zhang, G.H. 764
Zhang, G.Y. 1013
Zhang, H.D. 770
Zhang, J. 486, 590
Zhang, J.Q. 283, 836
Zhang, J.X. 688
Zhang, L.B. 86
Zhang, L.G. 924, 942
Zhang, M.J. 102
Zhang, M.M. 54
Zhang, N.H. 136
Zhang, P. 546
Zhang, P.Y. 356
Zhang, Q. 215
Zhang, X.Y. 886
Zhang, Y. 20, 427, 794, 846
Zhang, Y.H. 233
Zhang, Z. 881

Zhang, Z.H. 993
Zhang, Z.Y. 179, 824
Zhao, J. 45
Zhao, W. 495
Zhao, X. 799
Zhao, Y. 45
Zheng, X.J. 811
Zhong, Y.D. 408
Zhou, H.Z. 11
Zhou, L. 782
Zhou, W.J. 463
Zhou, X. 108
Zhou, X.C. 495
Zhou, X.L. 59
Zhou, Y. 576
Zhou, Y.B. 620
Zhou, Z.K. 341
Zhu, L.B. 656
Zhu, L.P. 259
Zhu, N.N. 369
Zhu, S.H. 446
Zhu, S.W. 102
Zhu, W.H. 480
Zhu, X.H. 609
Zhu, X.L. 255
Zhu, Y.Q. 73
Zhu, Y.Y. 942, 956